Lecture Notes in Artificial Intelligence 10462

Subseries of Lecture Notes in Computer Science

YongAn Huang · Hao Wu
Honghai Liu · Zhouping Yin (Eds.)

Intelligent Robotics and Applications

10th International Conference, ICIRA 2017
Wuhan, China, August 16–18, 2017
Proceedings, Part I

Springer

Editors
YongAn Huang
School of Mechanical Science and Engineering
Huazhong University of Science and Technology
Wuhan
China

Hao Wu
School of Mechanical Science and Engineering
Huazhong University of Science and Technology
Wuhan
China

Honghai Liu
Institute of Industrial Research
University of Portsmouth
Portsmouth
UK

Zhouping Yin
School of Mechanical Science and Engineering
Huazhong University of Science and Technology
Wuhan
China

ISSN 0302-9743 ISSN 1611-3349 (electronic)
Lecture Notes in Artificial Intelligence
ISBN 978-3-319-65288-7 ISBN 978-3-319-65289-4 (eBook)
DOI 10.1007/978-3-319-65289-4

Library of Congress Control Number: 2017948191

LNCS Sublibrary: SL7 – Artificial Intelligence

Printed on acid-free paper

This Springer imprint is published by Springer Nature
The registered company is Springer International Publishing AG
The registered company address is: Gewerbestrasse 11, 6330 Cham, Switzerland

Preface

The International Conference on Intelligent Robotics and Applications (ICIRA 2017) was held at Huazhong University of Science and Technology (HUST), Wuhan, China, during August 16–18, 2017. ICIRA 2017 was the 10th event of the conference series, which focuses on: (a) fundamental robotics research, including a wide spectrum ranging from the first industrial manipulator to Mars rovers, and from surgery robotics to cognitive robotics, etc.; and (b) industrial and real-world applications of robotics, which are the force driving the research further.

This volume of *Lecture Notes in Computer Science* contains the papers that were presented at ICIRA 2017. The regular papers in this volume were selected from more than 350 submissions covering various topics on scientific methods and industrial applications for intelligent robotics, such as soft and liquid-metal robotics, rehabilitation robotics, robotic dynamics and control, robot vision and application, robotic structure design and mechanism, robot learning, bio-inspired robotics, human–machine interaction, space robotics, mobile robotics, intelligent manufacturing and metrology, benchmarking and measuring service robots, real-world applications, and so on. Papers describing original works on abstractions, algorithms, theories, methodologies, and case studies are also included in this volume. Each submission was reviewed by at least two Program Committee members, with the assistance of external referees. The authors of the papers and the plenary and invited speakers come from the following countries and areas: Australia, Austria, China, Cyprus, Germany, Hong Kong, Japan, Korea, Singapore, Spain, Switzerland, UK, and USA.

We wish to thank all who made this conference possible: the authors of the submissions, the external referees (listed in the proceedings) for their scrupulous work, the six invited speakers for their excellent talks, the Advisory Committee for their guidance and advice, and the Program Committee and the Organizing Committee members for their rigorous and efficient work. Sincere thanks also go to the editors of the *Lecture Notes in Computer Science* series and Springer for their help in publishing this volume in a timely manner.

In addition, we greatly appreciate the following organizations for their support:

National Natural Science Foundation of China
School of Mechanical Science and Engineering, HUST
State Key Laboratory of Digital Manufacturing Equipment and Technology, China
State Key Laboratory of Robotic Technology and System, China
State Key Laboratory of Mechanical System and Vibration, China
State Key Laboratory of Robotics, China

August 2017

YongAn Huang
Hao Wu
Honghai Liu
Zhouping Yin

Organization

Honorary Chair

Youlun Xiong	HUST, China

General Chair

Han Ding	HUST, China

General Co-chairs

Naoyuki Kubota	Tokyo Metropolitan University, Japan
Kok-Meng Lee	Georgia Institute of Technology, USA
Xiangyang Zhu	Shanghai Jiao Tong University, China

Program Co-chairs

YongAn Huang	HUST, China
Honghai Liu	University of Portsmouth, UK
Jinggang Yi	Rutgers University, USA

Advisory Committee Chairs

Jorge Angeles	McGill University, Canada
Tamio Arai	University of Tokyo, Japan
Hegao Cai	Harbin Institute of Technology, China
Xiang Chen	Windsor University, Canada
Toshio Fukuda	Nagoya University, Japan
Huosheng Hu	University of Essex, UK
Sabina Jesehke	RWTH Aachen University, Germany
Yinan Lai	National Natural Science Foundation of China, China
Jangmyung Lee	Pusan National University, Korea
Ming Li	National Natural Science Foundation of China, China
Peter Luh	University of Connecticut, USA
Zhongqin Lin	Shanghai Jiao Tong University, China
Xinyu Shao	HUST, China
Xiaobo Tan	Michigan State University, USA
Guobiao Wang	National Natural Science Foundation of China, China
Michael Wang	The Hong Kong University of Science and Technology, SAR China
Yang Wang	Georgia Institute of Technology, USA

Huayong Yang	Zhejiang University, China
Haibin Yu	Chinese Academy of Science, China

Organizing Committee Chairs

Feng Gao	Shanghai Jiao Tong University, China
Lei Ren	The University of Manchester, UK
Chunyi Su	Concordia University, Canada
Jeremy L. Wyatt	University of Birmingham, UK
Caihua Xiong	HUST, China
Jie Zhao	Harbin Institute of Technology, China

Organizing Committee Co-chairs

Tian Huang	Tianjin University, China
Youfu Li	City University of Hong Kong, SAR China
Hong Liu	Harbin Institute of Technology, China
Xuesong Mei	Xi'an Jiaotong University, China
Tianmiao Wang	Beihang University, China

Local Chairs

Kun Bai	HUST, China
Bo Tao	HUST, China
Hao Wu	HUST, China
Zhigang Wu	HUST, China
Wenlong Li	HUST, China

Technical Theme Committee

Gary Feng	City University of Hong Kong, SAR China
Ming Xie	Nanyang Technological University, Singapore

Financial Chair

Huan Zeng	HUST, China

Registration Chair

Jingrong Ge	HUST, China

General Secretariat

Hao Wu	HUST, China

Sponsoring Organizations

National Natural Science Foundation of China (NSFC), China
Huazhong University of Science and Technology (HUST), China
School of Mechanical Science and Engineering, HUST, China
University of Portsmouth, UK
State Key Laboratory of Digital Manufacturing Equipment and Technology, HUST, China
State Key Laboratory of Robotic Technology and System, Harbin Institute of Technology, China
State Key Laboratory of Mechanical System and Vibration, Shanghai Jiao Tong University, China
State Key Laboratory of Robotics, Shenyang Institute of Automation, China

Contents – Part I

Soft, Micro-Nano, Bio-inspired Robotics

Ankle Active Rehabilitation Strategies Analysis Based on the Characteristics of Human and Robotic Integrated Biomechanics Simulation 3
Zhiwei Liao, Zongxin Lu, Chen Peng, Yang Li, Jun Zhang, and Ligang Yao

Dynamic Drive Performances of the Bionic Suction Cup Actuator Based on Shape Memory Alloy 14
Yunhao Ge, Jihao Liu, Bin Li, Huihua Miao, Weixin Yan, and Yanzheng Zhao

Piezoelectric Micro-Pump Suction Cup Design and Research on the Optimal Static Driving Characteristics 26
Enguang Guan, Yunhao Ge, Jihao Liu, Weixin Yan, and Yanzheng Zhao

The Research on the Method of Gait Planning for Biped Robot 39
Pan Li, Kaichao Li, and Yifan Wei

Control Strategy and Experiment of a Novel Hydraulic-Driven Upper Extremity Exoskeleton 51
Zirong Luo, Guohen Wu, Xing Li, and Jianzhong Shang

Research on the Locomotion of German Shepherd Dog at Different Speeds and Slopes 63
Weijun Tian, Qi Zhang, Zhen Yang, Jiyue Wang, Ming Li, and Qian Cong

Prehension of an Anthropomorphic Metamorphic Robotic Hand Based on Opposition Space Model 71
Guowu Wei, Lei Ren, and Jian S. Dai

Multi-directional Characterization for Pollen Tubes Based on a Nanorobotic Manipulation System 84
Wenfeng Wan, Yang Liu, Haojian Lu, and Yajing Shen

A Review on Vibration Characteristics of Carbon Nanotubes and Its Application Via Vacuum 94
Dongliang Huang, Zhan Yang, and Lining Sun

A Novel Soft Robot Based on Organic Materials: Finite Element Simulation and Precise Control 103
Fanan Wei, Jianghong Zheng, and Changle Yu

3D Motion Control and Target Manipulation of Small Magnetic Robot 110
Jingyi Wang, Niandong Jiao, Yongliang Yang, Steve Tung, and Lianqing Liu

A Locomotion Robot Driven by Soft Dielectric Elastomer Resonator. 120
Chao Tang, Bo Li, Changsheng Bian, Zhiqiang Li, Lei Liu, and Hualing Chen

Design and Test of a New Spiral Driven Pure Torsional Soft Actuator 127
Jihong Yan, Binbin Xu, Xinbin Zhang, and Jie Zhao

Design of a Soft Pneumatic Actuator Finger with Self-strain Sensing. 140
Yi-Dan Tao and Guo-Ying Gu

A Programmable Mechanical Freedom and Variable Stiffness Soft Actuator with Low Melting Point Alloy 151
Yufei Hao, Tianmiao Wang, and Li Wen

Investigate of Grasping Force for a Soft Robot Hand Under Pulling Force and Varying Stiffness 162
Haibin Yin, Qian Li, Junfeng Li, and Mingchang He

Toward Effective Soft Robot Control via Reinforcement Learning. 173
Haochong Zhang, Rongyun Cao, Shlomo Zilberstein, Feng Wu, and Xiaoping Chen

The Calibration Method of Humanoid Robot Based on Double Support Constraints 185
Fei Liu and Li Tang

Rehabilitation Robotics

Real-Time Collision Avoidance Algorithm for Surgical Robot Based on OBB Intersection Test. 195
Yao Qiu, Zhiyuan Yan, Yu Miao, and Zhijiang Du

Exploration of a Hybrid Design Based on EEG and Eye Movement 206
Junyou Yang, Yuan Hao, Dianchun Bai, Yinlai Jiang, and Hiroshi Yokoi

Optimal Design of Electrical Stimulation Electrode for Electrotactile Feedback of Prosthetic Hand 217
Boya Wang, Qi Huang, Li Jiang, Shaowei Fan, Dapeng Yang, and Hong Liu

A Directional Identification Method Based on Position and Posture of Head for an Omni-directional Mobile Wheelchair Robot 230
Junyou Yang, Chunwei Yu, Rui Wang, and Donghui Zhao

Design of an Wearable MRI-Compatible Hand Exoskeleton Robot 242
Kun LIU, Yasuhisa Hasegawa, Kousaku Saotome, and Yosiyuki Sainkai

A Preliminary Study of Upper-Limb Motion Recognition with Noncontact Capacitive Sensing 251
Enhao Zheng, Qining Wang, and Hong Qiao

Multi-class SVM Based Real-Time Recognition of Sit-to-Stand and Stand-to-Sit Transitions for a Bionic Knee Exoskeleton in Transparent Mode 262
Xiuhua Liu, Zhihao Zhou, Jingeng Mai, and Qining Wang

EMG-Based Control for Three-Dimensional Upper Limb Movement Assistance Using a Cable-Based Upper Limb Rehabilitation Robot 273
Yao Huang, Ying Chen, Jie Niu, and Rong Song

A Real-Time Intent Recognition System Based on SoC-FPGA for Robotic Transtibial Prosthesis 280
Jingeng Mai, Zhendong Zhang, and Qining Wang

Motion Planning and Experimental Validation of a Novel Robotic Device for Assistive Gait Training 290
Tao Qin, Hao Zhang, Peijun Liu, Fanjing Meng, and Yanyang Liu

Impedance Control of a Pneumatic Muscles-Driven Ankle Rehabilitation Robot 301
Chi Zhang, Jiwei Hu, Qingsong Ai, Wei Meng, and Quan Liu

Gait Recognition Using GA-SVM Method Based on Electromyography Signal 313
Ying Li, Farong Gao, Xiao Zheng, and Haitao Gan

Estimating 3D Gaze Point on Object Using Stereo Scene Cameras 323
Zhonghua Wan and Caihua Xiong

Overall Kinematic Coordination Characteristic of Human Lower Limb Movement 330
Bo Huang and Caihua Xiong

Eye Gaze Tracking Based Interaction Method of an Upper-Limb Exoskeletal Rehabilitation Robot 340
Quanlin Li, Caihua Xiong, and Kai Liu

Investigation of Phase Features of Movement Related Cortical Potentials for Upper-Limb Movement Intention Detection 350
Hong Zeng, Baoguo Xu, Huijun Li, Aiguo Song, Pengcheng Wen, and Jia Liu

Human-Machine Interaction

Mobile Terminals Haptic Interface: A Vibro-Tactile Finger Device for 3D Shape Rendering 361
Xingjian Zhong, Juan Wu, Xiao Han, and Wei Liu

Towards Finger Gestures and Force Recognition Based on Wrist Electromyography and Accelerometers 373
Bo Lv, Xinjun Sheng, Weichao Guo, Xiangyang Zhu, and Han Ding

Man-Machine Interaction for an Unmanned Tower Crane Using Wireless Multi-Controller 381
Songbo Ruan, Yeping Peng, Guangzhong Cao, Sudan Huang, and Xiangyong Zhong

Web-Based Human Robot Interaction via Live Video Streaming and Voice 393
Jiahui Shi, Hongbin Ma, Jialiang Zhao, and Yunxuan Liu

Object-Shape Recognition Based on Haptic Image 405
Yi Gong, Juan Wu, Miao Wu, and Xiao Han

Personal Desktop-Level Jet Fighter Simulator for Training or Entertainment . . . 417
Xinye Zhao, Yitao Wang, Wenming Zhang, and Xiaowei Zhang

Full-Pose Magnetic Estimation Based on a Two-Stage Algorithm for Remote Hand Rehabilitation Training 428
Hui-Min Shen

Preprocessing and Transmission for 3D Point Cloud Data 438
Zunran Wang, Chenguang Yang, Zhaojie Ju, Zhijun Li, and Chun-Yi Su

Dexterous Hand Motion Classification and Recognition Based on Multimodal Sensing 450
Yaxu Xue, Zhaojie Ju, Kui Xiang, Chenguang Yang, and Honghai Liu

Static Hand Gesture Recognition with Parallel CNNs for Space Human-Robot Interaction 462
Qing Gao, Jinguo Liu, Zhaojie Ju, Yangmin Li, Tian Zhang, and Lu Zhang

Robust Human Action Recognition Using Dynamic Movement Features 474
Huiwen Zhang, Mingliang Fu, Haitao Luo, and Weijia Zhou

Static Ankle Joint Stiffness Estimation with Relaxed Muscles Through Customized Device 485
Renjie Xiong, Cheng Sun, Muye Pang, Kui Xiang, and Zhaojie Ju

IMU Performance Analysis for a Pedestrian Tracker 494
Jianwei Zheng, Minhui Qi, Kui Xiang, and Muye Pang

A Remote Online Condition Monitoring and Intelligent Diagnostic System for Wind Turbine 505
Detong Kong, Wei Liu, Zhanli Liu, and Hongwei Wang

Rehabilitation Training for Leg Based on EEG-EMG Fusion 517
Heng Tang, Gongfa Li, Ying Sun, Guozhang Jiang, Jianyi Kong, Zhaojie Ju, and Du Jiang

A Review of Gesture Recognition Based on Computer Vision 528
Bei Li, Gongfa Li, Ying Sun, Guozhang Jiang, Jianyi Kong, Zhaojie Ju, and Du Jiang

Hand Gesture Recognition Using Interactive Image Segmentation Method . . . 539
Disi Chen, Gongfa Li, Jianyi Kong, Guozhang Jiang, Ying Sun, Du Jiang, and Zhaojie Ju

Simulation of 2-DOF Articulated Robot Control Based on Adaptive Fuzzy Sliding Mode Control 551
Feng Du, Gongfa Li, Zhe Li, Ying Sun, Jianyi Kong, Guozhang Jiang, and Du Jiang

Dynamical System Algorithm Specification Analysis and Stabilization. 560
Charles C. Phiri, János Botzheim, Cristina Valle, Zhaojie Ju, and Honghai Liu

A Review of Upper and Lower Limb Rehabilitation Training Robot 570
Wenlong Hu, Gongfa Li, Ying Sun, Guozhang Jiang, Jianyi Kong, Zhaojie Ju, and Du Jiang

External Force Detection for Physical Human-Robot Interaction Using Dynamic Model Identification. 581
Dewen Wu, Quan Liu, Wenjun Xu, Aiming Liu, Zude Zhou, and Duc Truong Pham

Mechanical Design and Human-Machine Coupling Dynamic Analysis of a Lower Extremity Exoskeleton 593
Bo Li, Bo Yuan, Jun Chen, Yonggang Zuo, and Yifu Yang

Intention-Based Human Robot Collaboration 605
Guoqiang Liang, Xuguang Lan, Hanbo Zhang, Xingyu Chen, and Nanning Zheng

Swarm Robotics

A Knowledge-Based Intelligent System for Distributed and Collaborative Choreography 617
Xinle Du, Haoqin Ma, and Hongwei Wang

Distributed Consensus Control of Multi-USV Systems 628
Bin Liu, Hai-Tao Zhang, Yue Wu, and Binbin Hu

Distributed Event-Triggered Consensus Control of Neutrally Stable Linear Multi-agent Systems 636
Bin Cheng and Zhongkui Li

Approximate Dynamic Programming for Relay Deployment in Multi-robot System 648
Song Yao, Yunlong Wu, and Bo Zhang

GSDF: A Generic Development Framework for Swarm Robotics 659
Xuefeng Chang, Zhongxuan Cai, Yanzhen Wang, Xiaodong Yi, and Nong Xiao

A Case Study on the Performance of Gazebo with Multi-core CPUs 671
Hai Yang and Xuefei Wang

CNP Based Satellite Constellation Online Coordination Under Communication Constraints 683
Guoliang Li, Lining Xing, and Yingwu Chen

Underwater Robotics

Research on Fault-Tolerant Control Method of UUV Sensor Using Walcott-Zak Observer 699
Zheping Yan, Yingming Bi, and Tao Chen

The Tracking Control of Unmanned Underwater Vehicles Based on QPSO-Model Predictive Control 711
Wenyang Gan, Daqi Zhu, Bing Sun, and Chaomin Luo

Motion Analysis of Wave Glider Based on Multibody Dynamic Theory 721
Xiao-tao Li, Fang Liu, Li Wang, and Hu-qing She

Distributed Formation Control of Autonomous Underwater Vehicles Based on Flocking and Consensus Algorithms 735
Wuwei Pan, Dapeng Jiang, Yongjie Pang, Yuda Qi, and Daichao Luo

Pitch Angle Active Disturbance Rejection Control with Model Compensation for Underwater Glider 745
Dalei Song, Tingting Guo, Hongdu Wang, Zhijian Cui, and Liqin Zhou

Numerical Simulation Research in Flow Fields Recognition Method Based on the Autonomous Underwater Vehicle 757
Xinghua Lin, Jianguo Wu, Dong Liu, and Lili Wang

Trajectory-Keeping Control of AUV Based on RNM-ADRC Method Under Current Disturbances for Terrain Survey Mission 766
Tao Chen, Hang Gao, Da Xu, Chuang Wan, Yuzhu Wang, and Zheping Yan

UUV-Six Degrees of Freedom Positioning Method Based on Optical Vision 779
Wei Zhang, Ximeng Wang, Lifeng Gao, and Shilin Wei

Hydrographic and Meteorological Observation Demonstration with Wave Glider "Black Pearl" 790
Can Li, Hongqiang Sang, Xiujun Sun, and Zhanhui Qi

Simulation for Path Planning of OUC-II Glider with Intelligence Algorithm 801
Yuhai Liu, Xin Luan, Dalei Song, and Zhiqiang Su

RETRACTED CHAPTER: The Summary of Underwater Gliders Control Strategies 813
Yuhai Liu, Xin Luan, Dalei Song, and Zhiqiang Su

System Construction for Distributedly Controlling the Thrusters of X4-AUV 825
Xiongshi Xu, Keigo Watanabe, and Isaku Nagai

A Localization Method Using a Dynamical Model and an Extended Kalman Filtering for X4-AUV 834
Keigo Watanabe, Takanori Yamaguchi, and Isaku Nagai

Experiment Study of Propulsion Property of Marine Mobile Buoy Driven by Wave 846
Zongyu Chang, Guiqiao Lu, Guangchao Du, Zhongqiang Zheng, Yuanguang Tang, Jiliang Wang, and Xin Lu

Design of Thermal Power Generation System Based on Underwater Vehicles. 857
Rui Wang, Hongwei Zhang, Guohui Wang, and Zhesong Ma

Design and Simulation of a Self-adaptive Fuzzy-PID Controller for an Autonomous Underwater Vehicle. 867
Jianhong Zhao, Wei Yi, Yuanxi Peng, and Xuefeng Peng

Retraction Note to: The Summary of Underwater Gliders Control Strategies E1
Yuhai Liu, Xin Luan, Dalei Song, and Zhiqiang Su

Author Index 879

Contents – Part II

Industrial Robot and Robot Manufacturing

An NC Code Based Machining Movement Simulation Method for a Parallel Robotic Machine 3
Xu Shen, Fugui Xie, Xin-Jun Liu, and Rafiq Ahmad

Trajectory and Force Generation with Multi-constraints for Robotic Belt Grinding 14
Yangyang Mao, Huan Zhao, Xin Zhao, and Han Ding

Fractional-Order Integral Sliding Mode Controller for Biaxial Motion Control System 24
Xi Yu, Huan Zhao, Xiangfei Li, and Han Ding

Pose Estimation with Mismatching Region Detection in Robot Bin Picking 36
Zhe Wang, Lei Jia, Lei Zhang, and Chungang Zhuang

A Five-Degree-of-Freedom Hybrid Manipulator for Machining of Complex Curved Surface 48
Yundou Xu, Jianhua Hu, Dongsheng Zhang, Jiantao Yao, and Yongsheng Zhao

A Dynamic Real-Time Motion Planning Method for Multi-robots with Collision Avoidance 59
Yonghong Zhang, Huan Zhao, Congcong Ye, and Han Ding

Reverse and Forward Post Processors for a Robot Machining System 70
Fusaomi Nagata, Yudai Okada, Takamasa Kusano, and Keigo Watanabe

Research on Robot Grinding Technology Considering Removal Rate and Roughness 79
Shaobo Xie, Shan Li, Bing Chen, and Junde Qi

Electromechanical Coupling Dynamic Model and Speed Response Characteristics of the Flexible Robotic Manipulator 91
Yufei Liu, Bin Zi, Xi Zhang, and Dezhang Xu

Correction Algorithm of LIDAR Data for Mobile Robots 101
Wenzhi Bai, Gen Li, and Liya Han

Research and Application on Avoiding Twist Mechanism Based on Relative Rotation Platforms . . . 111
Guobin Yang, Lubin Hang, Jiuru Lu, Zhiyu Fu, Wentao Li, and Liang Yu

Optimal Motion Planning for Mobile Welding Robot. . . . 124
Gen Pan, Enguang Guan, Fan Yang, Anye Ren, and Peng Gao

Off-Line Programmed Error Compensation of an Industrial Robot in Ship Hull Welding . . . 135
Guanglei Wu, Delun Wang, and Huimin Dong

Study and Experiment on Positioning Error of SCARA Robot Caused by Joint Clearance. . . . 147
Changyu Xu, Huimin Dong, Shangkun Xu, Yu Wu, and Chenggang Wang

Real-Time Normal Measurement and Error Compensation of Curved Aircraft Surface Based on On-line Thickness Measurement . . . 157
Yuan Yuan, Qingzhen Bi, Limin Zhu, and Han Ding

Feasibility of the Bi-Directional Scanning Method in Acceleration/deceleration Feedrate Scheduling for CNC Machining . . . 171
Jie Huang, Xu Du, and Li-Min Zhu

A Feed-Direction Stiffness Based Trajectory Optimization Method for a Milling Robot. . . . 184
Gang Xiong, Ye Ding, and LiMin Zhu

Mechanism and Parallel Robotics

Kinematic Analysis and Performance Evaluation of a Redundantly Actuated Hybrid Manipulator . . . 199
Lingmin Xu, Qiaohong Chen, Leiying He, and Qinchuan Li

Topology Optimization of the Active Arms for a High-Speed Parallel Robot Based on Variable Height Method. . . . 212
Qizhi Meng, Fugui Xie, and Xin-Jun Liu

Stiffness Analysis of a Variable Stiffness Joint Using a Leaf Spring . . . 225
Lijin Fang and Yan Wang

Design of a Series Variable Stiffness Joint Based on Antagonistic Principle . . . 238
Shipeng Cui, Yiwei Liu, Yongjun Sun, and Hong Liu

Two-Degree-of-Freedom Mechanisms Design Based on Parasitic Motion Maximization . . . 250
Zhenyang Zhuo, Yunjiang Lou, Bin Liao, and Mingliang Wang

Novel Design of a Family of Legged Mobile Lander 261
Rongfu Lin and Weizhong Guo

Designing of a Passive Knee-Assisting Exoskeleton for Weight-Bearing 273
Bo Yuan, Bo Li, Yong Chen, Bilian Tan, Min Jiang, Shuai Tang, Yi Wei, Zhijie Wang, Bin Ma, and Ju Huang

Development of HIT Humanoid Robot 286
Baoshi Cao, Yikun Gu, Kui Sun, Minghe Jin, and Hong Liu

Design of a Robotic Laparoscopic Tool with Modular Actuation. 298
Kai Xu, Huichao Zhang, Jiangran Zhao, and Zhengchen Dai

Preliminary Development of a Continuum Dual-Arm Surgical Robotic System for Transurethral Procedures 311
Kai Xu, Bo Liang, Zhengchen Dai, Jiangran Zhao, Bin Zhao, Huan Liu, Liang Xiao, and Yinghao Sun

Optimal Design of a Cable-Driven Parallel Mechanism for Lunar Takeoff Simulation 323
Yu Zheng, Wangmin Yi, and Fanwei Meng

Optimal Design of an Orthogonal Generalized Parallel Manipulator Based on Swarm Particle Optimization Algorithm. 334
Lei Peng, Zhizhong Tong, Chongqing Li, Hongzhou Jiang, and Jingfeng He

Research on a 3-DOF Compliant Precision Positioning Stage Based on Piezoelectric Actuators 346
Guang Ren, Quan Zhang, Chaodong Li, and Xu Zhang

Analysis for Rotation Orthogonality of a Dynamically Adjusting Generalized Gough-Stewart Parallel Manipulator. 359
ZhiZhong Tong, Tao Chen, Lei Peng, Hongzhou Jiang, and Fengjing He

A Delta-CU – Kinematic Analysis and Dimension Design 371
Jiayu Li, Huiping Shen, Qinmei Meng, and Jiaming Deng

Research on the Synchronous Control of the Pneumatic Parallel Robot with Two DOF. 383
Shaoning Wang, Tao Wang, Bo Wang, and Wei Fan

Analysis on Rigid-Elastic Coupling Characteristics of Planar 3-RRR Flexible Parallel Mechanisms 394
Qinghua Zhang and Qinghua Lu

Advanced Parallel Robot with Extended RSUR Kinematic for a Circulating Working Principle 405
Stefan Tobias Albrecht, Hailin Huang, and Bing Li

Design and Analysis of a New Remote Center-of-Motion Parallel Robot for Minimally Invasive Surgery 417
Jingyuan Sun, Shuo Wang, Hongjian Yu, and Zhijiang Du

Accuracy Synthesis of a 3-R2H2S Parallel Robot Based on Rigid-Flexible Coupling Mode 429
Caidong Wang, Yihao Li, Yu Ning, Liangwen Wang, and Wenliao Du

A Singularity Analysis Method for Stewart Parallel Mechanism with Planar Platforms 441
Shili Cheng, Guihua Su, Xin Xiong, and Hongtao Wu

Performance Research of Planar 5R Parallel Mechanism with Variable Drive Configurations 453
Weitao Yuan, Zhaokun Zhang, Zhufeng Shao, Liping Wang, and Li Du

Experimental Study on Load Characteristics of Macro-Micro Dual-Drive Precision Positioning Mechanism 464
Jing Yu, Ruizhou Wang, and Xianmin Zhang

Machine and Robot Vision

Solving a New Constrained Minimization Problem for Image Deconvolution 475
Su Xiao, Ying Zhou, and Linghua Wei

An Object Reconstruction Method Based on Binocular Stereo Vision 486
Yu Liu, Chao Li, and Jixiang Gong

A New Method of Determining Relative Orientation in Photogrammetry with a Small Number of Coded Targets 496
Hao Wu, Xu Zhang, and Limin Zhu

A Camera Calibration Method Based on Differential GPS System for Large Field Measurement 508
Haijun Jiang and Xiangyi Sun

A New Pixel-Level Background Subtraction Algorithm in Machine Vision 520
Songsong Zhang, Tian Jiang, Yuanxi Peng, and Xuefeng Peng

A New Chessboard Corner Detection Algorithm with Simple Thresholding 532
Qi Zhang and Caihua Xiong

FPGA-Based Connected Components Analysis Algorithm Without Equivalence-Tables . . . 543
Luxiang Ling, Zhong Chen, Shuai Li, and Xianmin Zhang

A Methodology to Determine the High Performance Area of 6R Industrial Robot . . . 554
Nianfeng Wang, Zhifei Zhang, and Xianmin Zhang

Recognition of Initial Welding Position Based on Structured-Light for Arc Welding Robot . . . 564
Nianfeng Wang, Xiaodong Shi, and Xianmin Zhang

A Dual-Camera Assisted Method of the SCARA Robot for Online Assembly of Cellphone Batteries . . . 576
Kai Feng, Xianmin Zhang, Hai Li, and Yanjiang Huang

A Fast 3D Object Recognition Pipeline in Cluttered and Occluded Scenes . . . 588
Liupo Zheng, Hesheng Wang, and Weidong Chen

Implementation of Multiple View Approach for Pose Estimation with an Eye-In-Hand Robotic System . . . 599
Kai Li, Chungang Zhuang, Jianhua Wu, and Zhenhua Xiong

Research on Extracting Feature Points of Electronic-Component Pins . . . 611
Yongcong Kuang, Jiayu Li, Jinglun Liang, and Gaofei Ouyang

Efficient Combinations of Rejection Strategies for Dense Point Clouds Registration . . . 623
Shaoan Zhao, Lin Zuo, Chang-Hua Zhang, and Yu Liu

Reconstructing Dynamic Objects via LiDAR Odometry Oriented to Depth Fusion . . . 634
Hui Cheng, Yongheng Hu, Haoguang Huang, Chuangrong Chen, and Chongyu Chen

A Robot Teaching Method Based on Motion Tracking with Particle Filter . . . 647
Yanjiang Huang, Jie Xie, Haopeng Zhou, Yanglong Zheng, and Xianmin Zhang

Self Calibration of Binocular Vision Based on Bundle Adjustment Algorithm . . . 659
Duo Xu, Yunfeng Gao, and Zhenghua Hou

Statistical Abnormal Crowd Behavior Detection and Simulation for Real-Time Applications . . . 671
Wilbert G. Aguilar, Marco A. Luna, Hugo Ruiz, Julio F. Moya, Marco P. Luna, Vanessa Abad, and Humberto Parra

Driver Fatigue Detection Based on Real-Time Eye Gaze Pattern Analysis . . . 683
Wilbert G. Aguilar, Jorge I. Estrella, William López, and Vanessa Abad

Onboard Video Stabilization for Rotorcrafts 695
Wilbert G. Aguilar, David Loza, Luis Segura, Alexander Ibarra, Thomas Abaroa, and Ronnie Fuertes

Evolutionary People Tracking for Robot Partner of Information Service in Public Areas 703
Wei Quan and Naoyuki Kubota

Robot Grasping and Control

Study on the Static Gait of a Quadruped Robot Based on the Body Lateral Adjustment 717
Qingsheng Luo, Bo Gao, and Rui Zhao

Analyses of a Novel Under-Actuated Double Fingered Dexterous Hand. 727
Rui Feng and Yifan Wei

LIPSAY Hand: A Linear Parallel and Self-adaptive Hand with Y-Shaped Linkage Mechanisms 739
Jian Hu, Ke Li, Wenzeng Zhang, Xiangrong Xu, and Aleksandar Rodic

A Novel Parallel and Self-adaptive Robot Hand with Triple-Shaft Pulley-Belt Mechanisms. 752
Qingyuan Jiang, Shuang Song, and Wenzeng Zhang

A Novel Robot Finger with a Rotating-Idle Stroke for Parallel Pinching and Self-adaptive Encompassing 764
Jingchen Qi, Linan Dang, and Wenzeng Zhang

Remote Live-Video Security Surveillance via Mobile Robot with Raspberry Pi IP Camera 776
Xiaolong Jing, Changyang Gong, Zhenyu Wang, Xudong Li, and Zhao Ma

Dynamic Identification for Industrial Robot Manipulators Based on Glowworm Optimization Algorithm 789
Li Ding, Wentao Shan, Chuan Zhou, and Wanqiang Xi

A Method of Computed-Torque Deviation Coupling Control Based on Friction Compensation Analysis 800
Yao Yan, Le Liang, Yanyan Chen, Yue Wang, and Yanjie Liu

Fuzzy PD-Type Iterative Learning Control of a Single Pneumatic Muscle Actuator 812
Da Ke, Qingsong Ai, Wei Meng, Congsheng Zhang, and Quan Liu

Cascade Control for SEAs and Its Performance Analysis 823
Yuancan Huang, Yin Ke, Fangxing Li, and Shuai Li

One of the Gait Planning Algorithm for Humanoid Robot Based on CPG Model 835
Liqing Wang, Xun Li, and Yanduo Zhang

Design of an Active Compliance Controller for a Bionic Hydraulic Quadruped Robot 846
Xiaoxing Zhang, Xiaoqiang Jiang, Xin Luo, and Xuedong Chen

Motion Control Strategy of Redundant Manipulators Based on Dynamic Task-Priority 856
Weiyao Bi, Xin-Jun Liu, Fugui Xie, and Wan Ding

A Boundary Control Method for Suppressing Flexible Wings Vibration of the FMAV 869
Yunan Chen, Wei He, Xiuyu He, Yao Yu, and Changyin Sun

Proxy Based Sliding Mode Control for a Class of Second-Order Nonlinear Systems 879
Guangzheng Ding, Jian Huang, and Yu Cao

Numerical Methods for Cooperative Control of Double Mobile Manipulators 889
Víctor H. Andaluz, María F. Molina, Yaritza P. Erazo, and Jessica S. Ortiz

Author Index 899

Contents – Part III

Sensors and Actuators

Modeling of Digital Twin Workshop Based on Perception Data 3
Qi Zhang, Xiaomei Zhang, Wenjun Xu, Aiming Liu, Zude Zhou, and Duc Truong Pham

A Stable Factor Approach of Input-Output-Based Sliding-Mode Control for Piezoelectric Actuators with Non-minimum Phase Property 15
Haifeng Ma, Jianhua Wu, and Zhenhua Xiong

Design of Quadrotor Unmanned Aerial Vehicle 25
Mofei Wu, Zhigang Cheng, Lin Yang, and Lamei Xu

Dust Detection System Based on Capacitive Readout IC MS3110. 35
Xiaoqin Tong

Design and Modeling of a Compact Rotary Series Elastic Actuator for an Elbow Rehabilitation Robot 44
Qiang Zhang, Benyan Xu, Zhao Guo, and Xiaohui Xiao

A Vibro-tactile Stimulation and Vibro-signature Synchronization Device for SSSEP-Based Study 57
Huanpeng Ye, Tao Xie, Lin Yao, Xinjun Sheng, and Xiangyang Zhu

Improved Indoor Positioning System Using BLE Beacons and a Compensated Gyroscope Sensor. 69
Jae Heo and Younggoo Kwon

Stretchable sEMG Electrodes Conformally Laminated on Skin for Continuous Electrophysiological Monitoring 77
Wentao Dong, Chen Zhu, Youhua Wang, Lin Xiao, Dong Ye, and YongAn Huang

Isotropy Analysis of a Stiffness Decoupling 8/4-4 Parallel Force Sensing Mechanism. 87
Jiantao Yao, Danlin Wang, Xueyan Lin, Hong Zhang, Yundou Xu, and Yongsheng Zhao

Preliminary Results of EMG-Based Hand Gestures for Long Term Use 98
Peter Boyd, Yinfeng Fang, and Honghai Liu

Research on Variable Stiffness and Damping Magnetorheological Actuator for Robot Joint 109
Xiaomin Dong, Weiqi Liu, Xuhong Wang, Jianqiang Yu, and Pinggen Chen

Physical Field-Enhanced Intelligent Space with Temperature-Based Human Motion Detection for Visually Impaired Users 120
Jiaoying Jiang, Kok-Meng Lee, and Jingjing Ji

Optimal Design and Experiments of a Wearable Silicone Strain Sensor 130
Tao Mei, Yong Ge, Zhanfeng Zhao, Mingyu Li, and Jianwen Zhao

Mobile Robotics and Path Planning

Research and Implementation of Person Tracking Method Based on Multi-feature Fusion 141
Fang Fang, Kun Qian, Bo Zhou, and Xudong Ma

Method and Experiment of the NAO Humanoid Robot Walking on a Slope Based on CoM Motion Estimation and Control. 154
Qingdan Yuan, Zhigang Xi, Qinghua Lu, and Zhihao Lin

TVSLAM: An Efficient Topological-Vector Based SLAM Algorithm for Home Cleaning Robots. 166
Yongfu Chen, Chunlei Qu, Qifu Wang, Zhiyong Jin, Mengzhu Shen, and Jiaqi Shen

Development of Wall-Climbing Robot Using Vortex Suction Unit and Its Evaluation on Walls with Various Surface Conditions. 179
Jianghong Zhao and Xin Li

Motion Planning and Simulation of Multiple Welding Robots Based on Genetic Algorithm 193
Yongsheng Chao and Wenlei Sun

Leader-Follower Formation Control Based on Artificial Potential Field and Sliding Mode Control 203
Xu Wang, Hong-an Yang, Haojie Chen, Jinguo Wang, Luoyu Bai, and Wenpei Zan

Trajectory Tracking by Terminal Sliding Mode Control for a Three-Wheeled Mobile Robot. 215
Jia-Xin Shao, Yu-Dong Zhao, Dong-Eon Kim, and Jang-Myung Lee

Research and Development of Ball-Picking Robot Technology 226
Hengbin Yu, Shoujun Wang, Haibo Zhou, Lu Yang, and Xu Zhou

Mobile Indoor Localization Mitigating Unstable RSS Variations and Multiple NLOS Interferences 237
Kyuchang Kwon and Younggoo Kwon

The Integrated Indoor Positioning by Considering Spatial Characteristics 246
Dongjun Yang and Younggoo Kwon

Characterization of the Sick LMS511-20100Pro Laser Range Finder for Simultaneous Localization and Mapping 254
Wenpeng Zong, Guangyun Li, Minglei Li, Li Wang, and Yanglin Zhou

Performance Metrics for Coverage of Cleaning Robots with MoCap System 267
Kuisong Zheng, Guangda Chen, Guowei Cui, Yingfeng Chen, Feng Wu, and Xiaoping Chen

A General Batch-Calibration Framework of Service Robots 275
Kuisong Zheng, Yingfeng Chen, Feng Wu, and Xiaoping Chen

Autonomous Navigation Control for Quadrotors in Trajectories Tracking 287
Wilbert G. Aguilar, Cecilio Angulo, and Ramón Costa-Castello

On-Board Visual SLAM on a UGV Using a RGB-D Camera 298
Wilbert G. Aguilar, Guillermo A. Rodríguez, Leandro Álvarez, Sebastián Sandoval, Fernando Quisaguano, and Alex Limaico

Projective Homography Based Uncalibrated Visual Servoing with Path Planning 309
Zeyu Gong, Bo Tao, Hua Yang, Zhouping Yin, and Han Ding

A Fully Cloud-Based Modular Home Service Robot 320
Yili Wang, Naichen Wang, Zhihao Chen, and Wenbo Chen

People Tracking in Unknown Environment Based on Particle Filter and Social Force Model 335
Yang Wang, Wanmi Chen, and Yifan Luo

Time-Jerk Optimal Trajectory Planning for a 7-DOF Redundant Robot Using the Sequential Quadratic Programming Method 343
Li Jiang, Shaotian Lu, Yikun Gu, and Jingdong Zhao

Nonlinear Control of Omnidirectional Mobile Platforms. 354
Víctor H. Andaluz, Oscar Arteaga, Christian P. Carvajal, and Víctor D. Zambrano

Virtual Reality and Artificial Intelligence

Leaf Recognition for Plant Classification Based on Wavelet Entropy and Back Propagation Neural Network 367
Meng-Meng Yang, Preetha Phillips, Shuihua Wang, and Yudong Zhang

A Registration Method for 3D Point Clouds with Convolutional Neural Network 377
Shangyou Ai, Lei Jia, Chungang Zhuang, and Han Ding

Tool Wear Condition Monitoring Based on Wavelet Packet Analysis and RBF Neural Network. 388
Tao Li, Dinghua Zhang, Ming Luo, and Baohai Wu

Research on Modeling and Simulation of Distributed Supply Chain Based on HAS 401
Wang Jian, Huang Yang, and Wang ZiYang

Robust EMG Pattern Recognition with Electrode Donning/Doffing and Multiple Confounding Factors 413
Huajie Zhang, Dapeng Yang, Chunyuan Shi, Li Jiang, and Hong Liu

A Robot Architecture of Hierarchical Finite State Machine for Autonomous Mobile Manipulator. 425
Haotian Zhou, Huasong Min, Yunhan Lin, and Shengnan Zhang

A Diagnostic Knowledge Model of Wind Turbine Fault. 437
Hongwei Wang, Wei Liu, and Zhanli Liu

Robust Object Tracking via Structure Learning and Patch Refinement in Handling Occlusion. 449
Junwei Li, Xiaolong Zhou, Shengyong Chen, Sixian Chan, and Zhaojie Ju

Graspable Object Classification with Multi-loss Hierarchical Representations. 460
Zhichao Wang, Zhiqi Li, Bin Wang, and Hong Liu

A Robotized Data Collection Approach for Convolutional Neural Networks . . . 472
Yiming Liu, Shaohua Zhang, Xiaohui Xiao, and Miao Li

Virtual Simulation of the Artificial Satellites Based on OpenGL 484
Yang Liu, Yikun Gu, Zongwu Xie, Haitao Yang, Zhichao Wang, and Hong Liu

Wearable Rehabilitation Training System for Upper Limbs Based on Virtual Reality 495
Jianhai Han, Shujun Lian, Bingjing Guo, Xiangpan Li, and Aimin You

Active Gait Rehabilitation Training System Based on Virtual Reality 506
Bingjing Guo, Wenxiao Li, Jianhai Han, Xiangpan Li, and Yongfei Mao

A Realtime Object Detection Algorithm Based on Limited
Computing Resource 517
Fei Liu, Yanbin Wang, Yimin Wei, Chunxue Li, and Li Tang

Aerial and Space Robotics

Linearity of the Force Leverage Mechanism Based on Flexure Hinges. 527
Jihao Liu, Enguang Guan, Peixing Li, Weixin Yan, and Yanzheng Zhao

A Kind of Large-Sized Flapping Wing Robotic Bird: Design
and Experiments 538
Erzhen Pan, Lianrui Chen, Bing Zhang, and Wenfu Xu

A Trajectory Planning and Control System for Quadrotor Unmanned
Aerial Vehicle in Field Inspection Missions 551
Gang Chen, Rong Wang, Wei Dong, and Xinjun Sheng

Attitude and Position Control of Quadrotor UAV Using PD-Fuzzy
Sliding Mode Control 563
Jong Ho Han, Yi Min Feng, Fei Peng, Wei Dong, and Xin Jun Sheng

Integrated Design and Analysis of an Amplitude-Variable Flapping
Mechanism for FMAV. 576
Peng Nian, Bifeng Song, Wenqing Yang, and Shaoran Liang

Modeling and Hover Control of a Dual-Rotor Tail-Sitter
Unmanned Aircraft 589
Jingyang Zhong, Bifeng Song, Wenqing Yang, and Peng Nian

Experimental Study on Dynamic Modeling of Flapping Wing Micro
Aerial Vehicle 602
Shaoran Liang, Bifeng Song, Wenqing Yang, and Peng Nian

A Novel Low Velocity Robotic Penetrator Based on Ampere Force 613
Jingkai Feng, Jinguo Liu, and Feiyu Zhang

Research of the Active Vibration Suppression of Flexible Manipulator
with One Degree-of-Freedom 624
Luo Qingsheng and Li Xiang

Space Robotic De-Tumbling of Large Target with Eddy Current Brake
in Hand 637
Jiayu Liu, Baosen Du, and Qiang Huang

Development of Modular Joints of a Space Manipulator with Light Weight and Wireless Communication 650
Liang Han, Can Luo, Xiangliang Cheng, and Wenfu Xu

Accurate Dynamics Modeling and Feedback Control for Maneuverable-Net Space Robot. 662
Yakun Zhao, Panfeng Huang, and Fan Zhang

Research on Space Manipulator System Man Machine Cooperation On-Orbit Operation Mode and Ground Test 673
Dongyu Liu, Hong Liu, Bainan Zhang, Yu He, Chao Luo, and Yiwei Liu

Collaborative Optimization of Robot's Mechanical Structure and Control System Parameter 686
Yuanchao Cheng, Ke Li, Fan Yang, Songbo Deng, Yuanyuan Chang, Yanbo Wang, and Xuman Zhang

Nonlinear MPC Based Coordinated Control of Towed Debris Using Tethered Space Robot 698
Bingheng Wang, Zhongjie Meng, and Panfeng Huang

Design and Experimental Study on Telescopic Boom of the Space Manipulator. 707
Shicai Shi, Qingchao He, and Minghe Jin

Trajectory Planning of Space Robot System for Reorientation After Capturing Target Based on Particle Swarm Optimization Algorithm 717
Songhua Hu, Ping He, Zhurong Dong, Hongwei Cui, and Songfeng Liang

An Iterative Calculation Method for Solve the Inverse Kinematics of a 7-DOF Robot with Link Offset 729
Shaotian Lu, Yikun Gu, Jingdong Zhao, and Li Jiang

Kinematic Nonlinear Control of Aerial Mobile Manipulators 740
Víctor H. Andaluz, Christian P. Carvajal, José A. Pérez, and Luis E. Proaño

Mechatronics and Intelligent Manufacturing

Design, Modeling and Analysis of a Magnetorheological Fluids-Based Soft Actuator for Robotic Joints. 753
Daoming Wang, Lan Yao, Jiawei Pang, and Zixiang Cao

Control of a Magnet-Driven Nano Positioning Stage with Long Stroke Based on Disturbance Observer 765
Letong Ma, Xixian Mo, Bo Zhang, and Han Ding

Experimental Research of Loading Effect on a 3-DOF Macro-Micro Precision Positioning System 777
Lingbo Xie, Zhicheng Qiu, and Xianmin Zhang

Development of Control System for the Assembly Equipments of Spacer Bar Based on PLC 788
Hong He, Congji Li, Xiaoqin Li, Zegang Wang, and Peng Xu

HADAMARD Transform Sample Matrix Used in Compressed Sensing Super-Resolution Imaging 796
Mei Ye, Hunian Ye, and Guangwei Yan

Numerical Simulation of Forming Process Conditions and Wall Thickness for Balloon 808
Xuelei Fu, Hong He, and Wenchang Wang

Tri-Dexel Model Based Geometric Simulation of Multi-axis Additive Manufacturing 819
Shanshan He, Xiongzhi Zeng, Changya Yan, Hu Gong, and Chen-Han Lee

Development of Rubber Aging Life Prediction Software 831
Hong He, Kai Liu, Xuelei Fu, and Kehan Ye

A Reliability Based Maintenance Policy of the Assembly System Considering the Dependence of Fixtures Elements Across the Stations. 843
Shiming Zhang, Feixiang Ji, and Yinhua Liu

Efficient Cutter-Freeform Surfaces Projection Method for Five-Axis Tool Path Computation 855
Xiyan Li, Chen-Han Lee, Pengcheng Hu, Yanyi Yang, and Fangzhao Yang

Optimization of Milling Process Parameters Based on Real Coded Self-adaptive Genetic Algorithm and Grey Relation Analysis 867
Shasha Zeng and Lei Yuan

HybridCAM: Tool Path Generation Software for Hybrid Manufacturing. 877
Xiongzhi Zeng, Changya Yan, Juan Yu, Shanshan He, and Chen-Han Lee

Author Index 891

Soft, Micro-Nano, Bio-inspired Robotics

Ankle Active Rehabilitation Strategies Analysis Based on the Characteristics of Human and Robotic Integrated Biomechanics Simulation

Zhiwei Liao, Zongxin Lu, Chen Peng, Yang Li, Jun Zhang(✉), and Ligang Yao(✉)

School of Mechanical Engineering and Automation, Fuzhou University, No. 2, Xueyuan Road, Fuzhou 350116, Fujian, China
zhang_jun@tju.edu.cn, ylgyao@fzu.edu.cn

Abstract. Ankle rehabilitation exercise therapy mainly includes passive, active, and resistance rehabilitations. Active rehabilitation exercise is highly important during ankle rehabilitation, how to combine the ankle and the rehabilitation robot has elicited considerable research attention. This paper proposed a combination strategy for the ankle and the rehabilitation robot based on a 3D modeling of the ankle and the robot developed individually based on simulation software (AnyBody). By integrating the 3D models of the human body and the robot, setting constraints between the pedal of the robot and the pelma of the human body and the degrees of freedom of the ankle is constrained as 3, a man-machine integration model for ankle active rehabilitation strategy analysis was established. Then, human muscle characteristics were established under the combination of different variables and the range of ankle motion were carried out by the human body active movement with ankle plantar/dorsal flexion motion, this further leads to strategies for ankle active rehabilitation. Finally, this paper designs the driving function for the robot based on Fourier function, and the force condition for the ankle movement during rehabilitation exercise was evaluated from the perspective of biomechanics. This study would provide a fundamental reference for the further formulation of active rehabilitation strategies and the control of rehabilitation robots.

Keywords: Ankle active rehabilitation · Biomechanics · Simulation · Rehabilitation strategies

1 Introduction

Ankle joint consists of many facet joints, bones, and tissues; the joint is one of the most complex joints in the human skeleton [1]. The ankle joint is the maximum load bearing joint of the human body and is easily injured, especially in athletes and the elderly [2]. Exercise therapy is one of the main methods of rehabilitation, and patients under the help of auxiliary equipment are required to complete the rehabilitation objective in a reasonable workspace. The exercise therapy of the ankle is a complex and lengthy process that includes three stages of early, mid-term, and late rehabilitation. Given the development of robot technology, the ankle rehabilitation robot became extensively investigated [3–13].

Y. Huang et al. (Eds.): ICIRA 2017, Part I, LNAI 10462, pp. 3–13, 2017.
DOI: 10.1007/978-3-319-65289-4_1

The development of biomechanics technology enabled research on ankle rehabilitation based on biomechanics to become one of the most attractive research fields. Warnica et al. [14, 15] investigated the effects of muscle activation and ankle stiffness on the amplitude and frequency of the postural sway and pointed out that the active contraction of muscles may play a different role in the balance control of ankle orthosis. Naito et al. [16] proposed a method to estimate muscle length parameters based on measured data. Benjamin et al. [17] investigated the effects of mechanical assistance on the biomechanical properties of the associated muscle system and completed the dynamic analysis of the biologically activated muscle and the skeleton stiffness. Cheung et al. [18] established a finite element model of ankle by using MRI images; the effects of the hardness and pressure of the pelma and the stress distribution of the bones are proposed. Zhang et al. [19–21] proposed a method for the prevention therapy and rehabilitation exercise of the ankle joint. Tang et al. [22] established the system and mechanical model of the 3D musculoskeletal system of the human body; the conversion between standard musculoskeletal system and individual musculoskeletal model is constructed. King et al. [23] analyzed the causation of ankle injury through the anatomy of the ankle joint and proved that ankle stability can be enhanced by increasing the muscle strength or ligament extension. Liu et al. [24] discussed the effects of force changes of the gastrocnemius muscle on the biomechanical mechanism of the pelma, thereby showing that the force changes of the gastrocnemius muscle significantly influence the pressure of the pelma and the stress of the plantar fascia.

This paper established an interaction model between the patients and a novel ankle rehabilitation robot and proposed a combination strategy for the ankle and the rehabilitation robot based on the 3D modeling of the ankle and the robot developed individually. Based on characteristics analysis of the human muscle under the combination of different variables and the range of ankle motion driven by the human body active movement with ankle plantar/dorsal flexion motion, strategies for the ankle active rehabilitation are proposed. Finally, the driving function for the robot based on the Fourier function is further provided, and the ankle movement during rehabilitation exercise is evaluated from the perspective of biomechanics.

2 Modeling of the Novel Robot and the Human Ankle

The 3D modeling of the novel ankle rehabilitation robot and the human ankle is fundamental for the analysis of the active rehabilitation strategies for the integrated robot and human system. To realize the three types of ankle movements for exercise therapy of ankle rehabilitation as well as the design of an ankle rehabilitation robot with the advantages of a simple structure, small size, light weight, easy to carry, and easy to move, a novel robot composed of series and parallel mechanisms is proposed, as shown in Fig. 1. The series part of the robot has one DOF, including moving platform, s-joint, rocker, bearing rod, and rotating platform, this part determines the abduction/adduction rehabilitating motions of the robot. The parallel part of the robot has two DOFs, including moving platform, three push-rods, and base platform, this part controls the plantar/dorsal flexion and the inversion/eversion rehabilitating motions of the robot. Three revolute pairs along the X-, Y-, and Z-axes realized three types of ankle rehabilitation movements. The plantar/dorsal flexion motion is realized along the X-axis, whereas the inversion/eversion motion is realized along the Y-axis, and the abduction/adduction motion is realized along the Z-axis.

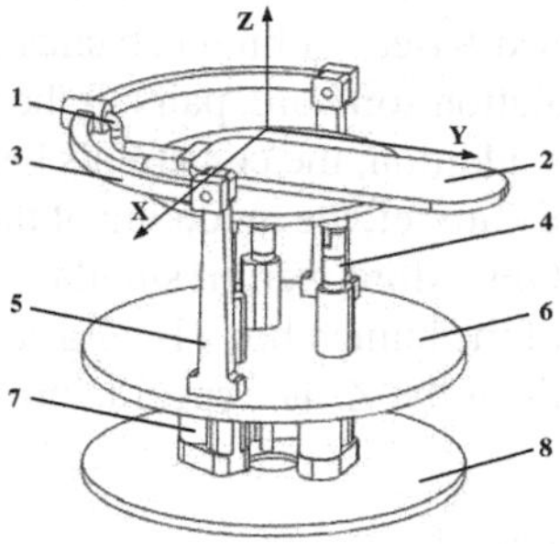

1. S-Joint 2. Moving platform 3. Rocker 4. Pushrod 5. Bearing rod 6. Rotating Platform 7. Stepper motor 8. Base platform

Fig. 1. Diagram of the ankle rehabilitation robot

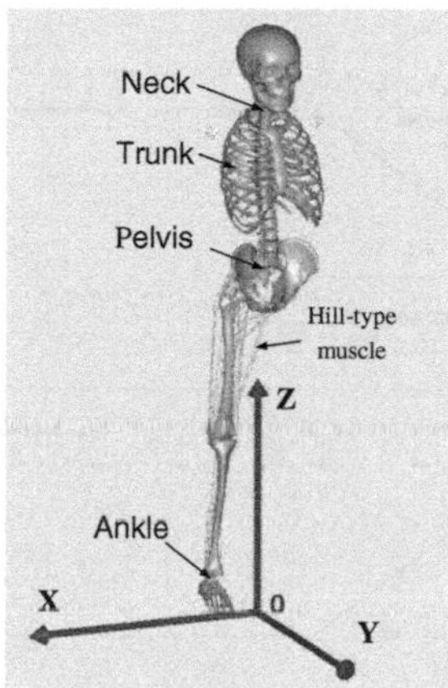

Fig. 2. Human body model

The 3D modeling of the human body is developed based on the AnyBody model library. Furthermore, the head, trunk, and right leg are preserved; the remaining parts of the human body are removed to increase the speed of calculation. In the human body model, the Hill-type muscle model is selected as real muscle because this model is more similar to the characteristics of a real human body. The model is constructed based on European standards, which recommend a default height of 180 cm and weight of 75 kg, as shown in Fig. 2.

3 Combination for the Robot and the Ankle Based on Biomechanics Simulation

Based on the 3D modeling of the robot and the human ankle, a combination of the robot and the ankle is further proposed based on biomechanics simulation software. Before importing the robot to the simulation software, pairs of the robot must be determined to avoid redundant constraints. In addition, the constraints between the pedal of the robot and the pelma of the human body are set; the modeling of the human body with the robot is constrained to only three DOFs. Moreover, to simulate the rehabilitation process of patients in reality, the posture of the human body is adjusted, and the center point of the ankle joint is located on the axis of the Z. The specific process is shown in Fig. 3.

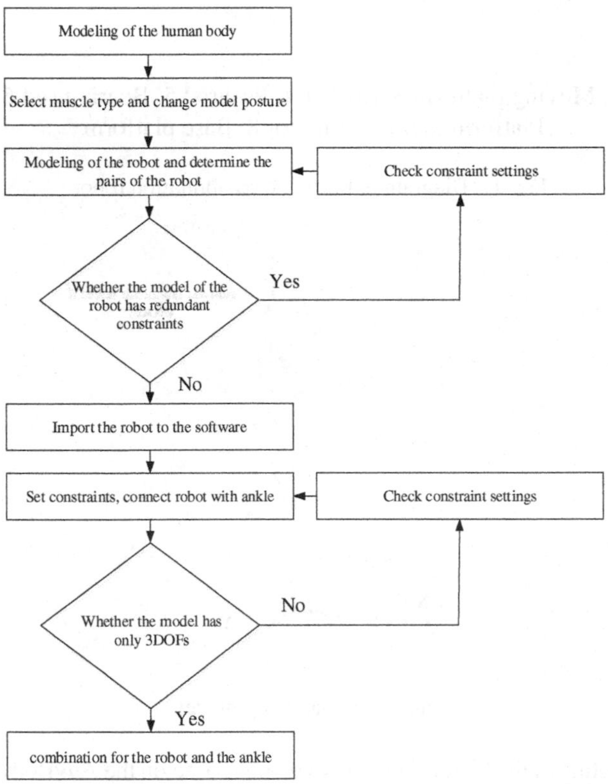

Fig. 3. Modeling and combination processes for the robot and the ankle

As shown in Fig. 3, the specific modeling and combination processes for the robot and the ankle are proposed. After the processes are completed, the pelvis of the human body, the three rotation directions of the trunk (Lateral Bending, Rotation, and Extension), and the rotation of the neck are fixed, and the combination modeling of the robot and the ankle only has three DOFs to achieve three types of movement. This paper studies the plantar flexion and dorsiflexion movement of the ankle joint alone to analyze the active rehabilitation strategies of ankle rehabilitation, as shown in Fig. 4.

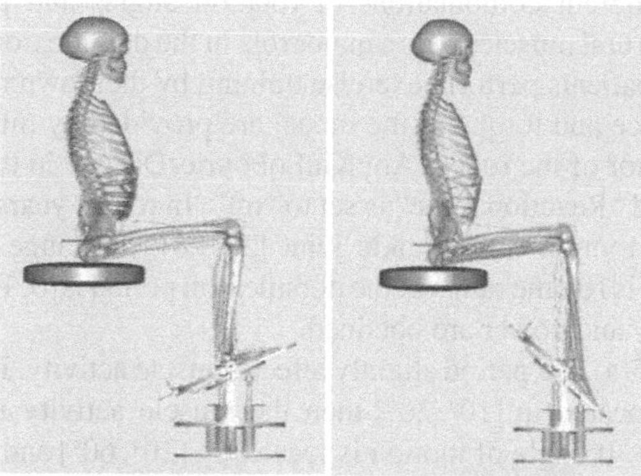

(a) Dorsiflexion (b) Plantar flexion

Fig. 4. Plantar flexion and dorsiflexion

In Fig. 4, the active rehabilitation process is the active contraction of the muscle to drive ankle movement because the moving platform is fixed with the human pelma; thus, the three pushrod movements are driven by the movement of the moving platform. The anterior and posterior crural muscles play major roles in the dorsiflexion and plantar flexion motions, respectively. The work in this section provides the basis for the investigation on the ankle active rehabilitation strategies.

4 Ankle Active Rehabilitation Strategies Analysis

Ankle rehabilitation after a series of ankle joint activity and strength training in early rehabilitation facilitates a certain recovery. In mid-term rehabilitation, active exercise is the main means of rehabilitation; fully active patients complete ankle movement by contracting the muscles around the ankle joint. In this section, the activity and power of related muscles and the passive traction of the muscle from different angles are analyzed. The maximum muscle activity is the percentage of the muscle strength, which performs the highest muscle activity. Relaxed muscles correspond to the highest percentage of 100%; percentages of more than 100% indicated no real meaning. The muscle power (Pmet) refers to a rough estimate of the total metabolic power consumed by all relevant muscles during active contraction in units of watts (W); the greater the power, the greater consumption per unit time. Muscle activity and muscle power in the rehabilitation process of patients cannot be increased sharply to avoid ankle pain. This paper analyzed the active rehabilitation strategies of patients based on the plantar and dorsal flexion motions. The purpose of the ankle active rehabilitation strategies in this section is to determine the appropriate rehabilitation range of patients.

4.1 Characteristics of the Muscles Under the Dorsiflexion of Active Rehabilitation

This section focuses on the maximum muscle activity and muscle power of the anterior crural muscles in different combinations of rotation angles and periods of motion because the anterior crural muscles play a major role in the dorsiflexion motion. In active ankle rehabilitation, patients perform exercise training by their own movement to drive the robot, and the force and torque of the model are provided by muscle contractions. On this basis, the motor of the robot "AnyKinEqFourierDriver" in the simulation software is turned off, and "Reaction.Type" is set to "off". In recent years, researchers have studied the range of motion of the ankle joint [26–28]; the range of dorsiflexion is $[10°, 60°]$, the interval is 10°, the range of the dorsiflexion period is [0, 10], and the interval is 2 s. Muscle activity and power are obtained.

As shown in Fig. 5(a), the period slightly affects muscle activity, if period $T > 4$ and angle of motion is located in $[10°, 20°]$ then the muscle activity increases with the increases of the angle. If angle of motion is located in $[20°, 60°]$ and muscle activity is maintained at 40%, then minimal changes are observed. However, if period $T < 4$ and angle $\theta > 20°$, then muscle activity has significant changes. If period $T < 2$ and angle $\theta > 60°$, then muscle activity is close to 100%. As shown in Fig. 5(b), the period and the angle has minimal effects on muscle activity, if period $T > 6$, but if period $T < 4$ and angle $\theta > 20°$, then the muscle power has a sharp increase with the decrease of the period and the increase of the angle. Based on the above, this paper shortens the range of the angle in to $[20°, 40°]$, and the interval is 1° to analyze the relationship among muscle activity, angle, and period of the movement (Fig. 6).

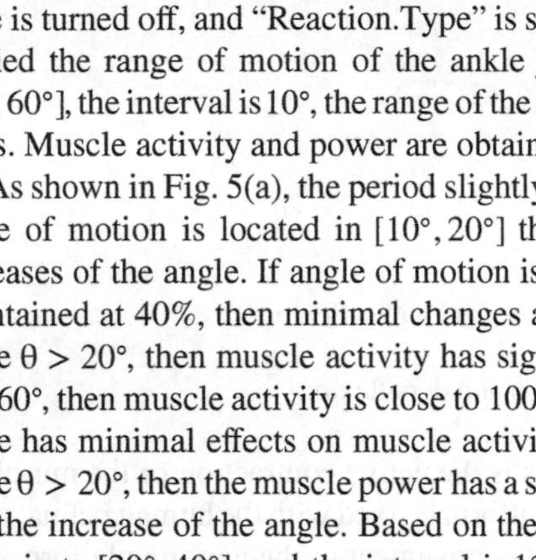

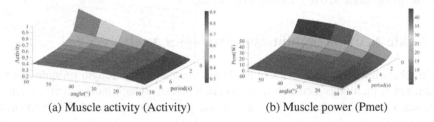

(a) Muscle activity (Activity)

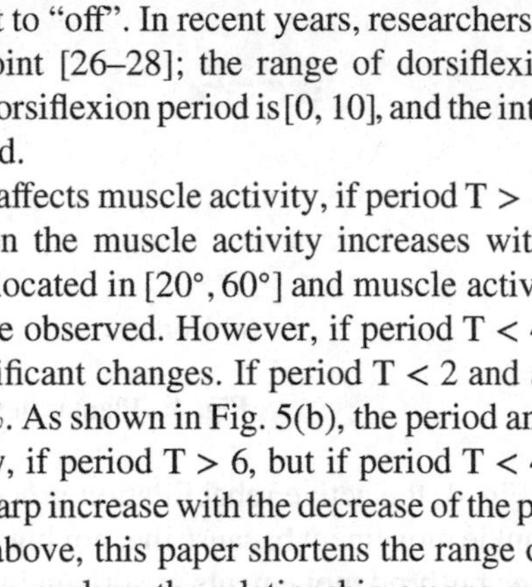

(b) Muscle power (Pmet)

Fig. 5. Active dorsiflexion motion

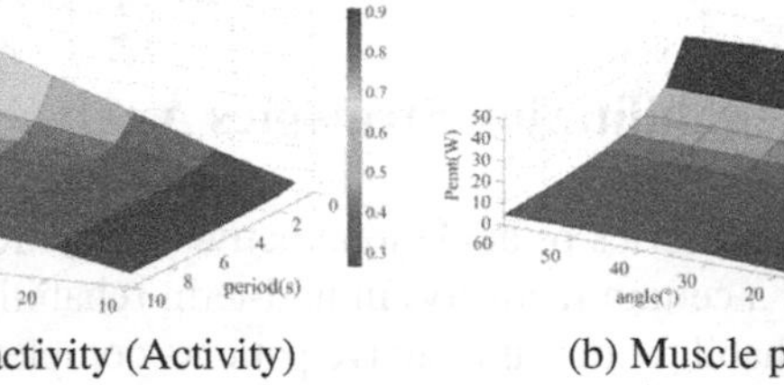

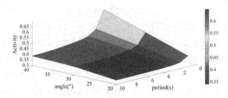

Fig. 6. Active dorsiflexion motion with 20°–40° maximum muscle activity

As shown in Fig. 6, if period $T > 6$ and angle $\theta < 30°$, then muscle activity increases with the increases in the angle; if period $T < 6$, muscle activity showed an increasing trend. The faster the active dorsiflexion, the greater the muscle consumption and the influence of angle change on muscle power is relatively small, if angle $\theta > 30°$ and period $T > 6$, then the period and the angle has minimal effects on muscle activity. Continuing to increase the angle has no real meaning on the effect of the muscle activity. Based on above, the maximum angle of the active dorsiflexion motion can be selected at 30°, and the period of the motion not less than 6 s is appropriated.

4.2 Characteristics of the Muscles Under the Plantar Flexion of Active Rehabilitation

Similar to the previous section, this section focuses on the maximum muscle activity and the muscle power of the posterior crural muscles in different combinations of rotation angles and periods of motion. The range of the plantar flexion is −60°-−10°, the interval is 10°, the range of the plantar flexion period is 2 s–10 s, and the interval is 1 s. Muscle activity and power are obtained (Fig. 7).

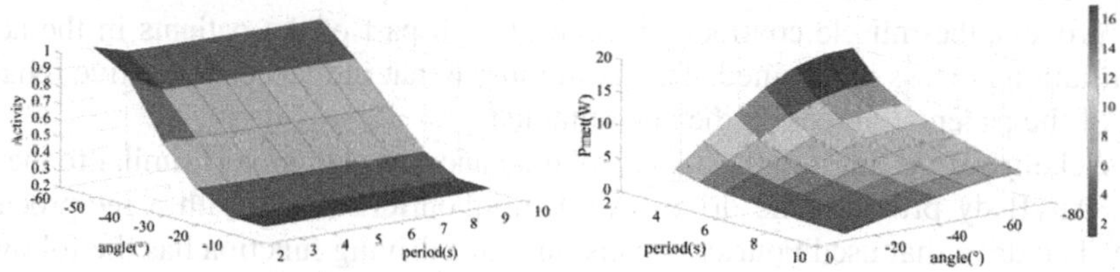

(a) Muscle activity (Activity) (b) Muscle power (Pmet)

Fig. 7. Active plantar flexion motion

As shown in Fig. 7(a), period slightly affects muscle activity, and if period $T > 4$, then period has minimal effect on muscle activity. However, the angle of motion has considerable effect on muscle activity. If the angle of plantar flexion motion is located in −20°-−10°, the muscle activity shows minimal changes and is maintained at 20%. However, if the angle is located in −60°-−20°, the muscle activity has significant changes. Furthermore, if the angle reaches 60° and period $T < 3$, then muscle activity is close to 100%. As shown in Fig. 7(b), the period and angle affect muscle power, especially if the period is located in 3 s–6 s and the angle is located in −60°-−20°. Muscle power increases sharply with the decreases of the period and the increases of the angle.

Based on the abovementioned discussion, this paper shortens the range of the angle by −30°-−20° with an interval of 1 s to analyze the relationship among muscle activity, angle, and period of the movement (Fig. 8).

As shown in Fig. 8, if the angle is located in −20°- −24°, then muscle activity increases slowly with the increases in the angle. If the angle is located in −24°- −30°, then muscle activity increases more rapidly, but muscle activity has slight decreases as period increases. Thus, the period has a minimal effect on muscle activity during the motion of plantar flexion. Based on the above, the maximum angle of the active plantar flexion motion can be selected at 24°, and the period of the motion is selected the same as the dorsiflexion motion, in which not less than 6 s is appropriate.

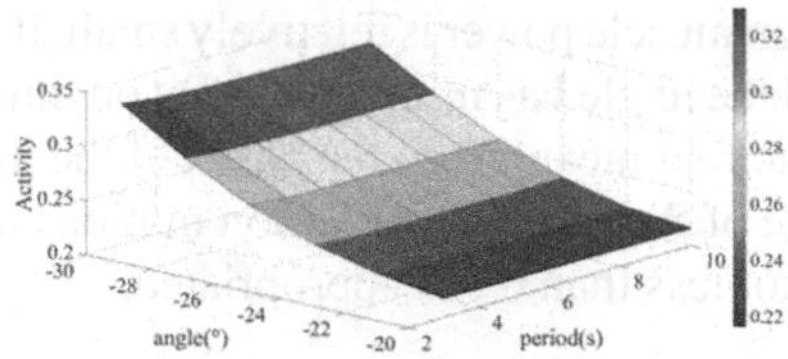

Fig. 8. Active plantar flexion motion at 30°–-20° maximum muscle activity of the posterior crural muscles

5 Driving Function Design and Rehabilitation Exercise Evaluation

In this section, the three types of ankle motion are realized based on the parameters obtained from the simulation of the active motion and the driving function of the rehabilitation mechanism designed by the rehabilitation strategies. Moreover, by measuring the active muscle contraction force of the anterior and posterior crural muscles in the entire process, the muscle contraction force of each part of the patients in the active rehabilitation process is obtained. The main muscle that influences the active rehabilitation of the patients' ankle can then be obtained.

The change of the angle in the ankle's plantar and dorsal flexion is similar to the sine curve. AnyBody provides the drive "AnyKinEqFourierDriver" with a reciprocating stroke. The drive that used Fourier expansion as the driving function has the following form:

$$\theta(t) = \sum_{i=1}^{n} A_i \sin\left(\omega_i t + B_i\right), i = 1, 2, 3, \ldots, n, \tag{1}$$

where A and B are the Fourier coefficient, and ω_i is the angular velocity, which can be expressed as

$$\omega_i = (i-1)2\pi f, \tag{2}$$

where f is the frequency, and the unit is Hz. In Eqs. (1) and (2), by taking n equal to 2, we obtain

$$\theta(t) = A_1 sinB_1 + A_2 \sin(2\pi f + B_2) \tag{3}$$

The sine functions with angles that vary with time can be used to obtain different driving functions with different amplitude, phase angle, and frequency by putting different A, B and f into equations.

Based on the parameters of dorsiflexion and plantar flexion, we can obtain that the maximum angle of dorsiflexion is 30° and the maximum angle of plantar flexion is 24° and that the maximal muscle activity has a significant change in the increased speed. Therefore, the angles of dorsiflexion and plantar flexion are set as 30° and 24°, respectively, and the period is 6 s. Afterwards, simulating the motion of ankle in these settings and obtaining the driving function become simple, as shown below.

$$\theta(t) = \frac{3\pi}{20}\sin(\frac{\pi}{3}t - 0.035\pi) + \frac{\pi}{60}, \tag{4}$$

where $A_1 = \frac{\pi}{60}, A_2 = \frac{3\pi}{20}, B_1 = \frac{\pi}{2}, B_2 = -0.035\pi, T = 6$.

We input A, B, and T to the drive "AnyKinEqFourierDriver" and then add to the hinge Z. Thus, the rehabilitation robot can complete the plantar and dorsal flexion motions as shown in Eq. (4). The active contraction forces of the anterior and posterior crural muscles are obtained in a period (Fig. 7).

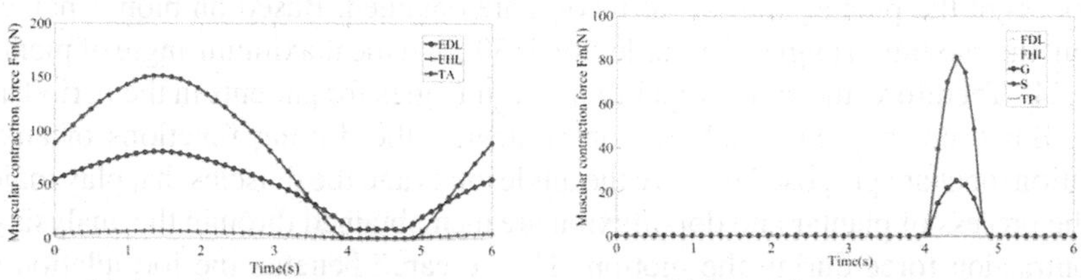

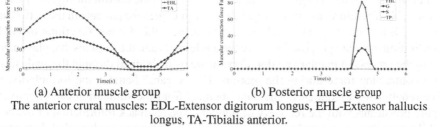

(a) Anterior muscle group (b) Posterior muscle group

The anterior crural muscles: EDL-Extensor digitorum longus, EHL-Extensor hallucis longus, TA-Tibialis anterior.

The posterior crural muscles: FDL-Flexor digitorum longus, FHL-Flexor hallucis longus, G-Gastrocnemius, S-Soleus, TP-Tibialis posterior.

Fig. 9. Active contraction force Fm of the anterior and posterior crural muscles

As shown in Fig. 9, based on the drive function, the first half of the period is the active dorsiflexion motion of the patients, and the second half is the active plantar flexion motion of the patients. As shown in Fig. 9(a), the anterior crural muscles are important in active contraction when the patients are in active dorsiflexion motion. During dorsiflexion motion, given the increasing of the dorsiflexion angle, the active contraction force Fm of each muscle is also increased, and it gradually decreased during the recovery of the dorsiflexion motion. Among them, the TA and EDL play major roles in the anterior crural muscles, and the maximum muscle contraction forces reached 150 and 80 N, respectively. In addition, the TA also has a minimal inference of the movement, but this is not evident. As shown in Fig. 9(b), the posterior crural muscles are important in active contraction when the patients are in active plantar flexion motion. During the plantar flexion motion process, only S and G play major roles in the posterior crural muscles; the remaining muscles have minimal influences. Given the increasing plantar flexion angle, the active contraction force Fm of S and G is also increased and gradually decreased during the recovery of the plantar flexion motion. Among them, S play a significant role in the plantar flexion motion, and the maximum muscle contraction force reached 82 N. Furthermore, G has slight inference of the movement, and the maximum muscle contraction force reached 23 N.

6 Conclusions

In this paper, the muscle activity and power of the related muscles during the active rehabilitation of the ankle are analyzed based on the novel ankle rehabilitation robot and its analysis software AnyBody. By analyzing the parameters such as the periods and angles that occur during sharp movements, the limit positions of patients during active ankle rehabilitation are determined; this determination guides the design of active rehabilitation strategies for patients in this period. This paper examines the plantar and dorsal flexion of the ankle rehabilitation exercise of patients. The limit positions of the ankle in the process of the plantar and dorsal flexion are obtained. Based on biomechanical simulation, the maximum angle of dorsiflexion is 30°, and the maximum angle of plantar flexion is 24°. Therefore, the active rehabilitation strategies for patients in the period are designed. Based on the new rehabilitation strategies, the driving functions of ankle rehabilitation robot are proposed to drive the ankle joints and the muscles that play major roles in the process of plantar and dorsiflexion are then obtained through the analysis of active contraction force during the motion. This research benefits the formulation of active rehabilitation strategies. The data obtained from the simulation can effectively provide physicians with the recovery of patients. The feedback information of the new rehabilitation strategies are obtained in time, thereby making the ankle rehabilitation of patients more scientific and effective. This strategy effectively avoided the possibility of patients' injury during an ankle active rehabilitation exercise.

References

1. Zhang, M., Claire, D.T., Xie, S.: Effectiveness of robot-assisted therapy on ankle rehabilitation – a systematic review. J. Neuroengineering Rehabil. **10**(1), 30 (2013)
2. Wang, J.: Rehabilitation of ankle joint and foot. Chin. Rehabil. Theor. Pract. **14**(12), 1197–1198 (2008)
3. Girone, M.J., Burdea, G.C., Bouzit, M.: Rutgers ankle orthopedic rehabilitation interface. ASME Dyn. Syst. Control Div. DSC **67**, 305–312 (1999)
4. Dai, J.S., Zhao, T., Nester, C.: Sprained ankle physiotherapy based mechanism synthesis and stiffness analysis of a robotic rehabilitation device. Auton. Robots **16**, 207–218 (2004)
5. Saglia, J.A., Dai, J.S., Caldwell, D.G.: Geometry and kinematic analysis of a redundantly actuated parallel mechanism that eliminates singularities and improves dexterity. J. Mech. Des. **130**(12), 1786–1787 (2008)
6. Takemura, H., Onodera, T., Ding, M., et al.: Design and control of a wearable stewart platform-type ankle-foot assistive device. Int. J. Adv. Robot. Syst. **9**, 1–7 (2012)
7. Xie, S.Q., Jamwal, P.K.: An iterative fuzzy controller for pneumatic muscle driven rehabilitation robot. Expert Syst. Appl. Int. J. **38**(7), 8128–8137 (2011)
8. Agrawal, A., Sangwan, V., Banala, S.K., Agrawal, S.K., Stuart, A.: Binder-macleod, de-sign of a novel two degree-of-freedom ankle-foot orthosis. J. Mech. Des. **129**, 1137–1143 (2007)
9. Yoon, J., Ryu, J.: A novel reconfigurable ankle/foot rehabilitation robot. In: Proceedings of the 2005 IEEE International Conference on Robotics and Automation, Barcelona, Spain, pp. 2290–2295 (2005)
10. Yoon, J., Ryu, J., Lim, K.S.: Reconfigurable ankle rehabilitation robot for various exercises. J. Robot. Syst. **22**, S16–S33 (2006)

11. Yu, H.: Design of parallel ankle rehabilitation robot system. YanShan University (2006)
12. Liu, G., Gao, J., Yue, H., et al.: Design and kinematics analysis of parallel robots for ankle rehabilitation intelligent robots and systems. In: 2006 IEEE/RSJ International Conference on IEEE, pp. 253–258 (2006)
13. Zhang, L.X., Sun, H.Y., Qian, Z.M.: Kinematics analysis and simulation of horizontal lower limbs rehabilitative robot. J. Syst. Simul. **22**(8), 2001–2005 (2010)
14. Warnica, M.J., Weaver, T.B., Prentice, S.D., et al.: The influence of ankle muscle activation on postural sway during quiet stance. Gait Posture **39**(4), 1115–1121 (2014)
15. Weaver, T.B., Glinka, M.N., Laing, A.C.: Stooping, crouching, and standing – characterizing balance control strategies across postures. J. Biomech. **53**, 90–96 (2017)
16. Hisashi, N., Yasushi, A., Ayu, M., et al.: Identification of individual muscle length parameters from measurements of passive joint moment around the ankle joint. J. Biomech. Sci. Eng. **7**(2), 168–176 (2012)
17. Robertson, B.D., Sawicki, G.S.: Influence of parallel spring-loaded exoskeleton on ankle muscle-tendon dynamics during simulated human hopping. In: 33rd Annual International Conference of the IEEE EMBS, Boston, Massachusetts, USA, pp. 583–586, 30 August–3 September 2011
18. Cheung, J.T., Zhang, M., Leung, A.K., et al.: Three—dimensional finite element an analysis of the foot during standing-a material sensitivity study. J. Biomech. **38**(5), 1045–1054 (2005)
19. Zhang, Q.X., Zhang, L., Wang, G.X.: Effect of local muscle fatigue on the proprioception of the ankle joint. Chin. Sport Sci. **31**(3), 68–74 (2011)
20. Zhang, Q.X.: The test-retest reliability of ankle muscle force perception with different target moments. In: Conference of the Fourteenth National Conference on Sports Bio-Mechanics, pp. 404–407 (2010)
21. Zhang, Q.X., Zhang, L.: The test-retest reliability of ankle joint position and the perception of muscle force. J. Clin. Rehabil. Tissue Eng. Res. **14**(35), 6520–6524 (2010)
22. Tang, G.: Simulation analysis of human body based on biomechanics. In: Shanghai: School of Mechanical and Power Engineering, Shanghai Jiao Tong University (2011)
23. Jin, Y., Tang, Z.: Analysis of the causes of the injury from the anatomical structure of the ankle joint. Phys. Educ. Rev. **35**(7), 87, 88 (2016)
24. Liu, Y., Zhou, S.Y., Zhen, Y.J., et al.: Effects of gastrocnemius muscle force on foot biomechanics based on finite element analysis. J. Med. Biomech. **31**(5), 37–442 (2016)

Dynamic Drive Performances of the Bionic Suction Cup Actuator Based on Shape Memory Alloy

Yunhao Ge[1], Jihao Liu[2], Bin Li[2], Huihua Miao[1], Weixin Yan[1(✉)], and Yanzheng Zhao[1]

[1] Robotics Institute of Shanghai Jiao Tong University, Shanghai 200240, China
{gyhandy,xiaogu4524,yzh-zhao}@sjtu.edu.cn

[2] State Key Laboratory of Mechanical System and Vibration, Shanghai Jiao Tong University, Shanghai 200240, China

Abstract. This paper proposes a design of SMA-based bionic suction cup actuator, the core of which is that the bias spring cooperates with the SMA spring to produce a displacement at the varying temperature caused by the drive current. To characterize the dynamic drive performance of the actuator, the thermodynamic model is established by the constitutive model of SMA material of Tanaka series. The simulation model for the bias SMA spring actuator in MATLAB/SIMULINK is built. Moreover, the effects of different load and different driving environment on the displacement of SMA actuator as well as the corresponding reaction time are investigated. The steady displacement response enlarge with the drive current increasing, while the load readjusts the initial deformation of the actuator and also enlarges the final steady shearing displacement. Further, based on the actuator driving performance, the adsorption characteristics of the bionic suction cup are studied and verified experimentally.

Keywords: Bionic suction cup · SMA spring actuator · Dynamic drive performance · Adsorption characteristics

1 Introduction

Micro bionic suction cup actuator design mainly considers the bionic, driving characteristics and efficiency. Vacuum pressure suction actuated by the tissue's elastic deformation has relatively excellent stability and adhesion performance [1], and is the easiest to be imitated. Bionic suction cup drives need to have a similar muscle nature, not only have a large output displacement, but also should have the appropriate drive frequency and weight ratio.

Considering the new functional materials applied in bionic suction cup, because the output power of the ion-exchange polymer metal composite(IPMC) driver is very small [2], while the drive voltage of electroactive polymer (EAP) is as high as the kilovolt [3], compared with piezoelectric actuator, the amount of deformation of the SMA drive is much greater. Especially SMA actuators have the advantages of low driving voltage, high driving force, no noise and high weight ratio. This paper adopts the SMA spring actuator driving the miniature bionic suction cup.

Y. Huang et al. (Eds.): ICIRA 2017, Part I, LNAI 10462, pp. 14–25, 2017.
DOI: 10.1007/978-3-319-65289-4_2

The spring-like SMA actuator is capable of generating the larger strain of which the bionic suction cup takes advantage to generate vacuum pressure. The SMA actuator's principles are on the basis of the shape memory effect which generates the force and displacement in the process of martensitic phase transformation as the temperature changes [4].

The relationship between the force and deformation of the SMA actuator is no linear [5]. It is hard to establish an accurate mathematical model, while the study methods of the dynamic characteristics are different under muiti-driving environment. However, the dynamic performance of SMA actuators under different currents and different loads directly affects the reliability and the performance of bionic suction cup. This paper focuses on the driving performance of bidirectional SMA actuators and experiments on the adsorption characteristics of micro-bionic suction cup based on SMA actuator.

2 The Thermo-Dynamical Model of SMA Spring

In order to estimate the adsorption performance of bionic suction cup actuated by SMA, it's essential to set up the theoretical model for the driving performance of SMA actuator. The paper briefly introduces Liang-Rogers constitutive equation of SMA material, then deduces the thermo-dynamical model of SMA spring actuators that reflects the relationship between the output force/displacement and the input current [6].

The martensitic volume fraction (ξ) depends on the temperature (T) and the stress (σ) during the process of the phase transformation [7]. Liang founded the dynamic model using sine basis functions for the phase transformation of the SMA material. When SMA spring cools, the martensitic phase transformation yields:

$$\xi = \frac{1-\xi_A}{2}\cos[a_M(T-M_f)+b_M\sigma]+\frac{1+\xi_A}{2} \quad (\mathrm{M_f}+\frac{\sigma}{C_M} \le \mathrm{T} < \mathrm{M_s}+\frac{\sigma}{C_M}) \tag{1}$$

While SMA spring is being heated, the austenitic phase transformation yields:

$$\xi = \frac{\xi_M}{2}\{\cos[a_A(T-A_s)+b_A\sigma]+1\} \quad (\mathrm{A_s}+\frac{\sigma}{C_A} \le \mathrm{T} < \mathrm{A_f}+\frac{\sigma}{C_A}) \tag{2}$$

ξ_A and ξ_M respectively refer to the initial martensitic volume fraction in the austenitic and martensitic phase transformation; a_A, a_M, b_A and b_M are material constants, $\mathrm{M_s}$ and $\mathrm{M_f}$ are the initial and final temperature in the martensitic phase transformation; $\mathrm{A_s}$ and $\mathrm{A_f}$ are the initial and final temperature in the austenitic phase transformation; C_A and C_M are material constants.

The equations listed above reflect the constitutive model of the SMA wire under the one-dimension tension or compression, but there exist sheering stress τ and sheering strain γ in the spring's deformation process. We can obtain the new one under the shearing condition [8].

$$\tau - \tau_0 = G_{sma}(\gamma - \gamma_0) + \frac{\Theta}{\sqrt{3}}(T - T_0) + \frac{\Omega}{\sqrt{3}}(\xi - \xi_0) \tag{3}$$

The equivalent stress equation in the elastic-plastic mechanics $\sigma = \sqrt{3}\tau$; the equivalent strain equation $\varepsilon = \sqrt{3}\gamma$. Suppose that the actuator's load force is Fsma, the shear displacement is y, the number of active coils is n, spring diameter and wire diameter are D and d respectively, then according to the coil springs theory, we can deduce the shearing stress τ and shearing strain γ of the SMA spring heating:

$$\gamma = \frac{8F_{sma}D}{G_{sma}\pi d^3} \quad \tau = \frac{8F_{sma}D}{\pi d^3} \tag{4}$$

Suppose there is neither stress nor strain in the spring at the initial moment, we can obtain:

$$y = \frac{\pi n D^2}{dG_{sma}}\tau - \frac{\pi n D^2}{dG_{sma}}\frac{\Theta}{\sqrt{3}}(T - T_0) - \frac{\pi n D^2}{dG_{sma}}\frac{\Omega}{\sqrt{3}}(\xi - \xi_0) \tag{5}$$

Uniting Eqs. (3) and (5), we can obtain the simplified relationship among the load force (Fsma), the shear displacement (y), the martensitic volume fraction (ξ) and the temperature (T):

$$F_{sma} = \frac{d^4 G_{sma}}{8nD^3}y + \frac{\pi d^3}{8D}\frac{\Theta}{\sqrt{3}}(T - T_0) + \frac{\pi d^3}{8D}\frac{\Omega}{\sqrt{3}}(\xi - \xi_0) \tag{6}$$

The SMA actuator responding to the temperature change generates the force and displacement, which can be caused by heating and cooling. The electric heating devices is easy to control with the external power supplying, while the air cooling being realized through the heat exchange with the environment. To create the micro bionic suction cup, this paper assembles the SMA actuator with the electric heating and the air cooling. Figure 1 illustrates the heat transfer process, in which the heat actuating the SMA spring is Joule heat that dissipates through conduction, convection and radiation. Almost ninety percent of the heat passes into the air in convection, and the remaining mainly radiates, while the heat by conduction is negligible.

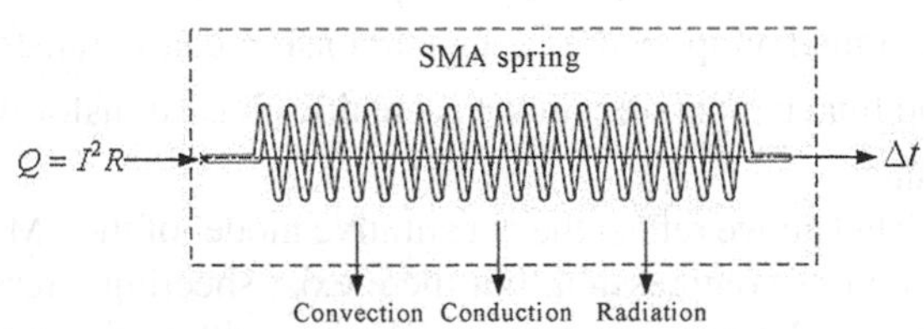

Fig. 1. SMA spring's heat transfer analysis

We can obtain the heat diffusion equation of SMA actuator during the process of the temperature change:

$$\frac{\partial t}{\partial \tau} = \frac{h}{\rho c}\left(\frac{\partial^2 t}{\partial x^2} + \frac{\partial^2 t}{\partial y^2} + \frac{\partial^2 t}{\partial z^2}\right) + \frac{\dot{\Phi}}{\rho c} \tag{7}$$

In the equation above, t denotes the SMA's temperature; τ denotes time; c denotes the specific heat; ρ denotes the density; h denotes the heat conductivity coefficient and Φ denotes the heat intensity respectively. Since SMA model conforms the simplified lumped parameter conditions, the heat conduction has nothing to do with the model's temperature and the coordinate. The Eq. (7) can be simplified [9].The heat intensity in the electric heating process is:

$$\dot{\Phi}_t = \frac{\Delta \Phi_V}{\Delta V} = \frac{I^2 R}{V} \tag{8}$$

Here R and V are SMA material internal resistance and volume respectively. The negative heat-intensity in the heat-dissipation is:

$$\Phi_W = \frac{\Delta \Phi_W}{\Delta V} = \frac{hA(t - t_\infty)}{V} \tag{9}$$

t_∞ is the environment temperature; A is the heat radian surface area of the SMA actuators. Based on the Eqs. (8) and (9), we can obtain the temperature change of the SMA actuator heated by the electric component:

$$\rho c V \frac{dt}{d\tau} = I^2 R - hA(t - t_\infty) \tag{10}$$

In the actuator's natural cooling process, the first stem of the upper equation does not exist any longer. Given the driving current is I, we can obtain the martensitic volume fraction of the phase transformation at the current temperature determined by Eq. (10). Then we can calculate the spring's shear force and displacement based on the thermo-dynamic model of the SMA spring.

3 Dynamic Driving Performance of Two-Way SMA Spring Actuator

The two-way SMA actuators must have the capability to execute double action reciprocating in order to generate or release the vacuum pressure. The actuators equipped with the SMA spring is of the two-way shape memory effect which extends when heating and contracts when cooling. In terms of the types of the bias elements, the two-way SMA actuators can be divided into the bias one using the conventional spring and the differential one that are equipped with additional SMA spring that produces the restoring force [10]. The paper estimates and analyzes the driving performance of the bias SMA spring actuator by theoretical methods and experiment.

Based on the principles above, the paper demotes a novel bias spring actuator as shown in Fig. 2, which is made up of the outer shell, the guide rod, the support plate, the movable part, the bias spring and the SMA spring that together drive the guide rod bidirectionally linear moving subject to the alternative temperature.

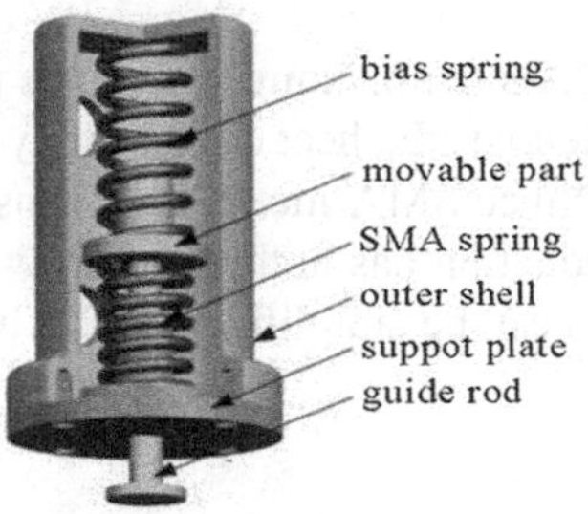

Fig. 2. Design of the bias SMA spring actuator

Suppose the length of SMA spring and bias spring are l_s and l_b in the natural state with no load, then the sum length of two springs assembled in the outer shell is compressed to l; the elastic modulus of the bias spring is K_b. The springs' deformation are $\Delta l_s{}^i$ and $\Delta l_b{}^i$ respectively at the initial moment. When subjected to variable temperature, the SMA spring and the bias spring will produce the shearing displacements that is respectively represented by sign $\Delta l_s{}^h$ and $\Delta l_b{}^h$. And the shearing displacements meet the geometrical relationship all the time:

$$\Delta l_b^{i/h} = (l_b + l_s) - (l + \Delta l_s^{i/h}) \tag{11}$$

At the initial moment, the SMA spring generates shear force $F_s{}^i$ that is equal to bias spring force $F_b{}^i$, hence the SMA spring is only of martensite at the atmospheric temperature, and we can obtain:

$$\frac{d^4 G_{sma}}{8knD^3}\Delta l_s^i = F_s^i = F_b^i = K_b \Delta l_b^i \tag{12}$$

Based on Eqs. (11) and (12), we can obtain the shearing displacement of the SMA spring and the bias spring $\Delta l_s{}^i$ and $\Delta l_b{}^i$ at the initial state. In the process that the two-way actuator is subject to the alternative temperature, the SMA spring's shearing force $F_s{}^h$ and the bias spring's one $F_b{}^h$ yield:

$$\frac{d^4 G_{sma}}{8knD^3}\Delta l_s^h + \frac{\pi d^3}{8kD}\frac{\Theta}{\sqrt{3}}(T - T_0) + \frac{\pi d^3}{8kD}\frac{\Omega}{\sqrt{3}}(\xi - \xi_0) = F_s^h = F_b^h = K_b \Delta l_b^h \tag{13}$$

The spring's temperature T, martensitic volume fraction ξ and the shearing modulus G_{sma} are introduced into Eq. (13), in combination with Eq. (11), we can obtain the shearing displacement from the SMA spring and bias spring are Δl_s^h and Δl_b^i respectively. In the operation process, the shearing displacement Δl generated by the bias spring actuator is:

$$\Delta l = \Delta l_s^i - \Delta l_s^h \tag{14}$$

According to the theoretical model above, the paper founds the simulation model about the bias spring actuator in MATLAB/SIMULINK as shown in Fig. 3. Table 1 lists the feature parameters in the simulation about the SMA spring and the bias spring. We heated the SMA spring driven by the DC regulated power supply in the experiments. The simulation result in the SIMULINK was compared in terms of the records of the temperature change and the shearing displacement under the variable loads and the driving currents, aiming to prove the validity of the theoretical model for the bias SMA spring actuators.

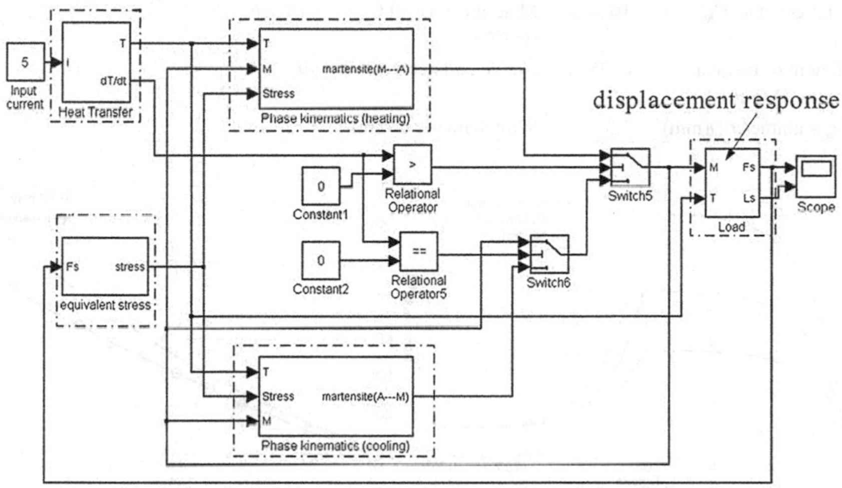

Fig. 3. Simulation model of bias SMA spring actuator

As shown in Fig. 4(a), the dynamic curve reflects the simulation temperature of the SMA Spring heated under various driving currents. At the beginning, the velocity of the spring's temperature change accelerates in accordance with the heating current at the same case of ambient temperature and heat dissipation condition. The SMA spring's temperature tends to reach a steady value though a period of thermal equilibrium. In general, the larger the heating current is, the longer it takes to gradually return to the atmosphere temperature by air cooling.

Table 1. The feature parameters of bias SMA spring actuator

SMA spring				Bias spring	
Density ρ (kg/m3)	6500	Specific heat c_p(J/Kg°C)	320	Elastic modulus k_b (N/m)	850
Resistance R (Ω)	0.15	Heat conduction coefficient h(W/m^2°C)	120	Initial length l_b (mm)	24
Ambient temperature T∞ (°C)	20	Martensitic elastic modulus D_M(Pa)	28e9		
Austenitic elastic modulus D_A (Pa)	75e9	Poisson's ratio μ_{SMA}	0.33		
The initial temperature of the martensitic phase transformation M_s(°C)	40	The finial temperature of the martensitic phase transformation M_f (°C)	20		
The initial temperature of the austenitic phase transformation A_s (°C)	40	The final temperature of the austenitic phase transformation A_f (°C)	60		
Material constant C_M (Pa/°C)	10.3e6	Material constant C_A (Pa/°C)	10.3e6		
Coefficient of thermal expansion Θ (Pa/°C)	0. 55e6	Active coil number(n)	9		
Spring's diameter D(mm)	7.3	Wire diameter d(mm)	1.3		

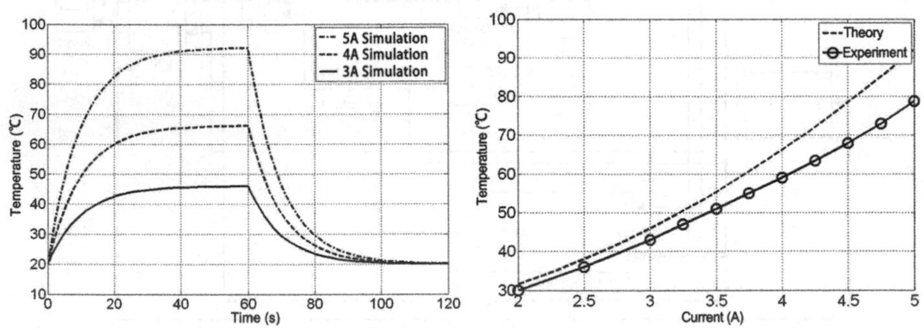

(a) (b)

Fig. 4. (a) Temperature response of SMA spring (b) Relationship between stable temperature and actuating currents

Figure 4(b) illustrates that the SMA spring can be subject to a certain steady temperature to the driving current in both practical experiment and simulation, which conforms the SMA spring's simulation models. Nevertheless, there exists some error between the theoretical simulation and the experimental that mainly results from the constitutive model error and the feature parameter error.

Figure 5 shows the displacement response of the bias SMA spring actuator stimulated by different amplitude current whose duty cycle is 0.5 in 160 s. Subject to 4A in the driving current with no load, the spring generates the shearing displacement in the simulation that is almost close to the experiment result. It indicates the theoretical model of SMA spring and bias actuator are valid.

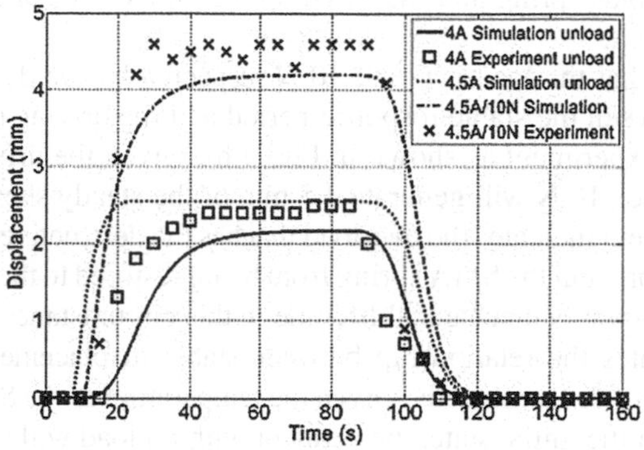

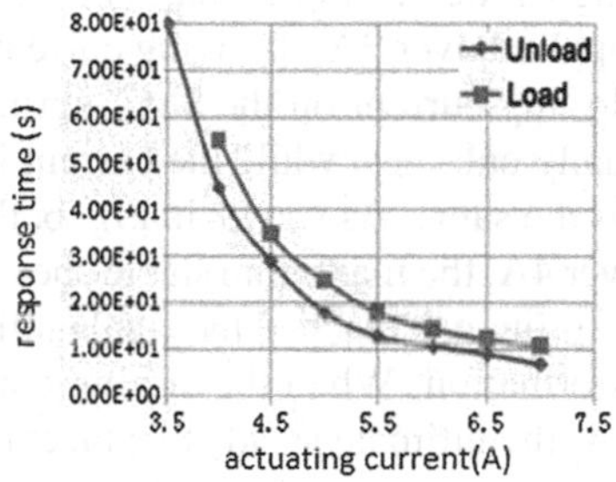

Fig. 5. (a) Displacement response of the bias SMA spring actuator; (b) Relationship between response time and actuating currents

Given that the constant driving current is 4A, the SMA spring starts to generate the shearing displacement at 15 s; when the heating temperature gradually exceeds the starting point of austenitic phase transformation, the SMA spring's equivalent elastic modulus will become stronger, which will extend to push the guide rod. At the end of the austenitic phase transformation, shearing displacement and the equivalent elastic modulus are stable and the guide rod will no longer move outward. When the driving current is larger, SMA spring's temperature will reach a much higher stable point more rapidly, so does the actuator's displacement. When SMA spring is in the air cooling process, the springs' shearing displacement gradually decreases. Based on the simulation and experiment result in the case that two-way actuator is loaded with nothing, we can find that the driving current has no effect on the spring's shearing restoration generated from the shearing deformation of the bias spring mainly.

According to the observation, the dynamic response curve of the shearing displacement is subject to variable loads. Comparing to the cases under various loads in Fig. 5(a), it will take the spring longer heating period to generate the shearing displacement than one with no loads in the same conditions. Due to the fact that the stress has an effect on the martensitic phase transformation of the SMA material, the starting point of austenitic phase transformation is higher under the load. So it will cost much longer time to preheat to generate obvious shearing displacement. Meanwhile, the load readjusts the initial deformation of the bias spring and the SMA spring and still enlarges the final steady shearing displacement.

Under the case that the load is 0 and 10 N respectively, the dynamic relationship curve mapping between the stable response period and the driving current are derived from the practical experiment as shown in Fig. 5(b). Given the driving current is 4A, the actuator that loads 10 N will generate 4.5 mm of the steady shearing displacement in 35 s. The experiment indicates the overload that has the destructive effect on the shape memory effect will prevent the SMA spring from being restored to the initial state, which will shorten the two-way's mechanical lifetime in this circumstance.

Figure 6 illustrates the relationship between stable displacement and the driving current in two cases of different loads. Given the temperature of the SMA spring is close to the atmosphere in the initial state, the actuator with no load will start to generate the shearing displacement as long as the driving current is over 2.4A. The same actuator with a certain load must be applied over 3A of driving current to generate the displacement. With the increasing driving current on the SMA spring, the stable value of the shearing displacement gradually enlarges, while the current is over the ultimate point, the steady displacement stays the same. As shown in Fig. 6, the driving current applied to the actuator unloaded is over 4A, the maximum stable shearing displacement is about 2.5 mm, which means SMA spring has reached the stable ultimate point that is the end of the austenitic phase transformation. When the driving current on the actuator with the load of 10 N reaches 4.5A, the ultimate steady displacement will be 4.5 mm.

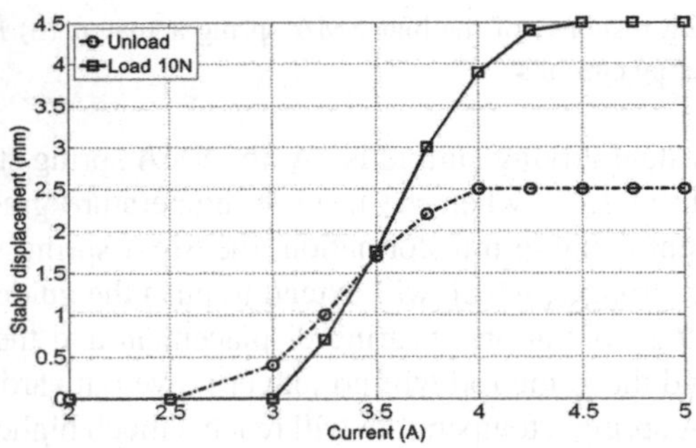

Fig. 6. Relationship between stable displacement and actuating currents

4 Experimental Investigation of the Adsorption Characteristics of Bionic Suction Cup

The dynamic response of the SMA actuator includes the generation and disappearance of vacuum pressure. Vacuum pressure requirements are quickly generated or canceled to match the footwall movement of the climbing robot. As Fig. 7 shows, the lower end of the actuator is connected with the elastomer plate to form a bionic suction cup prototype.

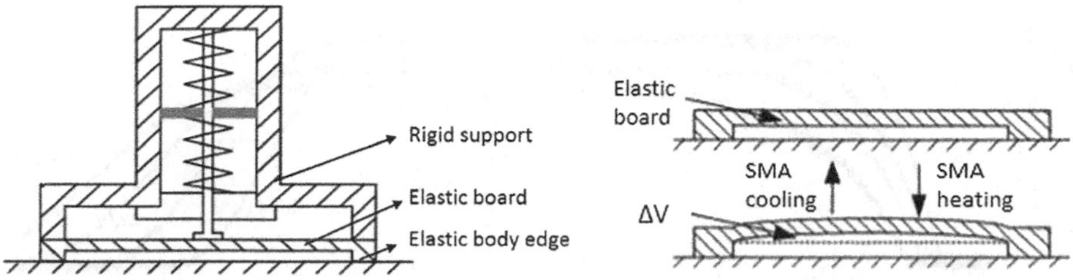

Fig. 7. Application of SAM actuators for the bionic suction cup

To test the vacuum pressure response of the prototype under different drive current, the full 5 V bias pressure sensor MPX5100 with the range of 100 kPa was installed at the lower surface of the adsorption surface, and was drilled on the suction surface, so the inlet of the vacuum pressure sensor communicates with the air cavity of the prototype. A constant current source has been used to heat the SMA actuator, and the oscilloscope was used to record the vacuum pressure response in the bionic suction cup, the test equipment is shown in Fig. 8.

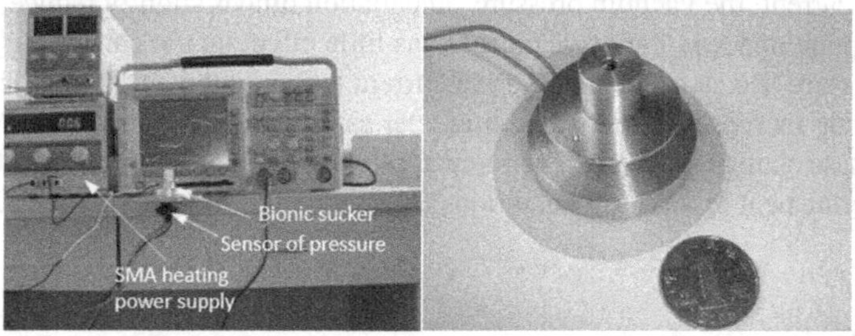

Fig. 8. Experimental equipments of vacuum pressure response testing

Figure 9(a) shows the response curves for the generation of vacuum pressure under different drive currents. When the drive current is constant, the response curve is similar to that of the biased SMA spring actuator output displacement. Under the drive current 4.75A, the bionic suction cup can produce the maximum vacuum pressure of about 14000 Pa, with the peaking time of 25 s. As the drive current increases, the steady-state vacuum pressure will also increases, while the peaking time of the maximum vacuum pressure value decreases Bionic suction cup can almost reach the same steady-state vacuum pressure under the drive current of 4.5A or 4.7A, which illustrates that the SMA spring undergoes a complete austenite transformation under the current over 4.5A, and both the output displacement and the vacuum pressure can not increase with increasing current.

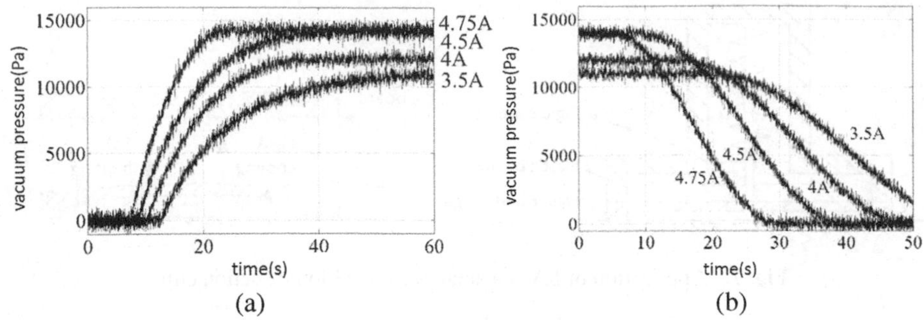

Fig. 9. (a) Waveform of vacuum pressure's generation; (b) Waveform of vacuum pressure's cancelling

To study the response curve during the disappearance process: firstly, the vacuum pressure within the suction cup achieve steady-state value under different drive current. Then the SMA spring heating current should be cancelled, and the process of suction cup vacuum pressure response curve in Fig. 9(b) should be recorded. With the increase of the drive current, the vacuum pressure cancellation time is slightly reduced, but from the overall point of view, the drive current has little effect on the cancelling process of vacuum pressure. Because when the drive current is larger, the amount of deflection of the bias spring increases, resulting in a greater return force. Meanwhile, SMA spring austenite phase change is deeper, the force applied to the bias spring is also larger; and the SMA spring heat dissipation conditions are exactly the same, with about 20 s disappears time.

5 Conclusions

This paper uses the SMA spring actuator to drive the micro bionic suction cup. To investigate the driving performance of the SMA actuator, the paper deduces the thermodynamic model for SMA spring based on the constitutive model, and estimates the dynamic driving performance of the SMA bias actuator by simulation and experiments, which prove the validity of theoretical model. When drive current is applied to the SMA spring actuator, the actuator will generate the increasing displacement linearly with the temperature arising that will reach a steady ultimate point in the case that there exists thermal equilibrium between the SMA and the air. The stronger the driving current is, the higher the temperature of the spring is, and the larger the steady displacement response will be. Meanwhile, the load has profound impact on the steady displacement and respond time of the actuator. Finally, the vacuum pressure response curve of bionic suction cup under different current is similar to that of the biased SMA spring actuator output displacement, which verifies the feasibility of the design.

Acknowledgments. This work is supported by the National Natural Science Foundation of China under Grant No. 51475305 and 61473192.

References

1. Nachtigall, W.: Animal attachments: minute, manifold devices. biological variety-basic physical mechanisms-a challenge for biomimicking technical stickers. In: Bionik. Springer, Berlin, Heidelberg. pp. 110–111 (2005)
2. Kim, B., Ryu, J., Jeong, Y. et al. A ciliary based 8-legged walking micro robot using cast IPMC actuators. In: Proceedings of the IEEE International Conference on Robotics and Automation. Taipei. pp. 2940–2945 (2003)
3. Walker, I.D., Dawson, D.M., Flash, T., et al. Continue robot arms inspired by cephalopods. In: Proceedings of the SPIE Unmanned Ground Vehicle Technology, vol. 5840, Orlando, USA. pp. 303–314 (2005)
4. Huang, W.: Shape memory alloys and their application to actuators for deployable structures. University of Cambridge Department of Engineering, England (1998)
5. Huang, W.: On the selection of shape memory alloys for actuators. In: Materials & Design, vol. 23, Issue 1, pp. 11–19 (2002)
6. Liang, C., Rogers, C.A., et al.: A multi-dimensional constitutive model for shape memory alloys. J. Eng. Math. **26**(3), 429–443 (1992)
7. Mammano, Scirè: G., Dragoni, E.: Effects of loading and constraining conditions on the thermomechanical fatigue life of NiTi shape memory wires. J. Mater. Eng. Perform. **23**(7), 2403–2411 (2014)
8. Hu, B.: Bio-inspired miniature suction cups actuated by shape memory alloy. Int. J. Adv. Rob. Syst. **6**(3), 151–160 (2009)
9. Yu, H.: Rigid/compliant coupling wheeled micro robot based on shape memory alloy. Shanghai Jiao Tong University, Shanghai (2006)
10. Yang, B., Zhang, X., Yan, X.: Relationship between bias spring parameters and output performances of SMA actuators. J. Beijing Univ. Aeronaut. Astronaut. **41**(4), 707–712 (2015)

Piezoelectric Micro-Pump Suction Cup Design and Research on the Optimal Static Driving Characteristics

Enguang Guan[1], Yunhao Ge[2], Jihao Liu[1(✉)], Weixin Yan[2], and Yanzheng Zhao[2]

[1] State Key Laboratory of Mechanical System and Vibration, Shanghai Jiao Tong University, Shanghai 200240, China
yihongyishui@sjtu.edu.cn

[2] Robotics Institute of Shanghai Jiao Tong University, Shanghai 200240, China
yzh-zhao@sjtu.edu.cn

Abstract. This paper first introduces the design and principles of a negative pressure suction cup with integrated piezoelectric valveless micro-pump. Laser engraving technique is used to build the multilayer prototype. In order to improve the adsorption performance of the suction cup, the static driving characteristics of the piezoelectric actuator have been optimized: based on a series of assumptions according to piezoelectric composite oscillators, a theoretical mechanics model of circular single-chip piezoelectric actuator has been constructed to explore the lateral deformation of the piezoelectric actuator in the effect of simple electric load. By the method of finite element simulation, the model is verified. Further, through optimizing its geometry design, the piezoelectric actuator has been enabled of the best static deformation characteristics.

Keywords: Piezoelectric valveless micropump · Static drive characteristics · Suction cup · Piezoelectric actuator

1 Introduction

Among the principles of biologic adhesion mechanisms [1], negative pressure suction actuated by the tissue's elastic deformation is facile to be imitated with excellent stability and adhesion performance. While the traditional apparatuses that pump the gas from the suction cup is no possible to realize miniaturization any more, a gas micropump integrated into the sucker has function of rapidly regulating negative pressure to supply steady negative pressure. The micropump is the common core actuating element and execution unit in the microfluidic handling system, of which the membrane micropump is the hottest theme in the research field at present. The principle of the membrane micropump is that the membrane's vibration actuated by external energy field forces the pump volume change to transmit the fluid in the direction from inlet to outlet, which is determined by the valves of inlet and outlet that have very good performance in terms of directing flow. Hence, we present a scheme to integer a

Y. Huang et al. (Eds.): ICIRA 2017, Part I, LNAI 10462, pp. 26–38, 2017.
DOI: 10.1007/978-3-319-65289-4_3

micropump into the miniature suction cup, which takes advantage of the fluid transmission characteristics to supply stable negative pressure.

In this paper, the theoretical mechanics model of the circle piezoelectric unimorph actuator is founded to be used to solve transverse deformation of the circle chip actuator excited by the constant voltage; the finite element method is adopted to simulate and verify the model. According to the theoretical mechanics model derived in the section, the paper optimizes of the actuator's geometric dimensioning design on purpose to achieve the optimal static deformation characteristics.

2 Miniature Suction Cup Integrating with the Piezoelectric Valveless Micropump

2.1 Gas Micropump Design

Wall climbing robots generally work at the atmosphere, so the micropump integrated into the sucker is required on superior air transport properties, which has been utilized to transport nothing but the liquid so far. The compression ratio of gas is extremely strong comparing with pumped liquid. As a consequence, in order to improve the transmission characteristics it's essential to enlarge the pressure change of the pump chamber [2]. By definition of the fluid volume compression ratio:

$$dP = -dV/V/\kappa \tag{1}$$

Where κ denotes the fluid compressibility, dV denotes the volume change, V denotes the initial volume of the fluid. In the case of the liquid with lower compressibility κ, the pressure difference dP aroused by minute volume change dV will be obviously larger than that of the coercible gas. By Eq. (1), we obtain that enlarging volume change ratio dV/V is in order to amplify dP, so there exists two approaches to improve the performances of the gas micropump: the first solution is enlarging the volume change as much as possible by optimizing the designing of the piezo actuator and adjusting the driving voltage and frequency; the other is to decrease the initial volume V of the micropump chamber. This paper presents the designing of the gas micropump from the aspects of the actuator and structure as follows.

The adsorption performance of the micropump suction cup depends directly on the actuator's property. It means the actuation's deflection should be as large as possible to enlarge the high pressure change, meanwhile the dynamic regulation on the negative pressure in the sucker requires very high drive frequency response.

Figure 1 illustrates the circular piezoelectric unimorph actuator and its driving principle. The piezo actuator is a composite component that consists of transverse isotropy piezoelectric layer, isotropy elastic membrane layer, bonding layer and electrodes layer. The bonding layer glued together the elastic membrane and the piezoelectric chip whose outmost shell coated with silver electrodes facilitates welding wiring. Once a voltage applied on the piezoelectric material, the piezoelectric layer immediately responds to the radial and axial strain deformation. The axis deformation is negligible except the radial on an account of the thickness much less than radius.

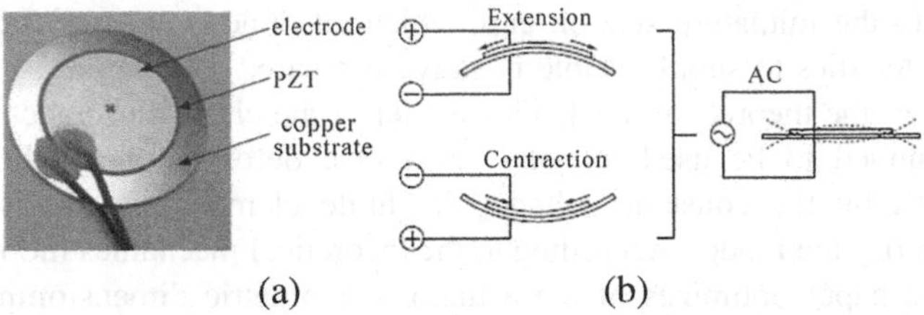

Fig. 1. (a) Piezoelectric unimorph actuator; (b) Driving principle of piezoelectric unimorph actuator

In the case that the piezoelectric layer extends radially, the whole chip actuator upward bends, or downward in contrast as Fig. 1 shown. The actuator will produce continuous bending vibration in the cycle of the alternative source.

According to the different forms of micro-valves, the membrane micropumps can be divided into the valve or the valveless [3–6]. The driving frequency of the valve micropump is lower in the same size, which is because of mechanical fatigue and air pressure loss resulting from the valve start lag, and the machining and fabrication of the moving valves is difficult. However, the valveless micropump overcomes the disadvantages above, which employs the tapered tubes with functions of directing the fluid flow. It will have a profound prospect. In the paper, the sucktion cup integrates with a valveless micropump on the basis of cone-shaped tubes.

As previously mentioned, the most direct measure to decrease the initial volume is to reduce the pump chamber's thickness. Nevertheless, if the pump chamber's thickness is less than the deflection at the center of the chip actuator, the chamber wall will interfere with the piezoelectric actuator and constrain actuator's deformation. And the pump chamber is too thin to be machined. In order to resolve the contradiction, Anders Olsson presents a valveless gas micropump with dual actuators, which speeds up the change ratio of gas volume while the initial volume staying the same. The related researches indicates this type of the gas pump have the ability to transport the maximum net flow 8 ml/min, and load ultimate back pressure 5 kPa [7].

2.2 Principles of the Micropump Suction Cup

As Fig. 2 shown, it is the principles of the mciropump sucker with double piezoelectric actuators referring to Ref. [7, 8]. The piezoelectric layer of the circular unimorph actuator excited by AC source deforms radially, driving the elastic membrane layer to concave/convex, which changes the pump volume along with the cycle of the current. So a pump cycle can be divided into a 'supply mode' and a 'pump mode'. During the 'supply mode', the cavity volume increases and a larger amount of fluid flows into the cavity through the input element, which acts as a diffuser whose cross section enlarges along the transportation direction, than through the output element, which acts as nozzle whose geometry is just the reverse of the diffuser. However, during the 'pump mode', when the cavity volume decreases, a larger amount of fluid flows out of cavity through the output element, which acts as a diffuser, than through the input element,

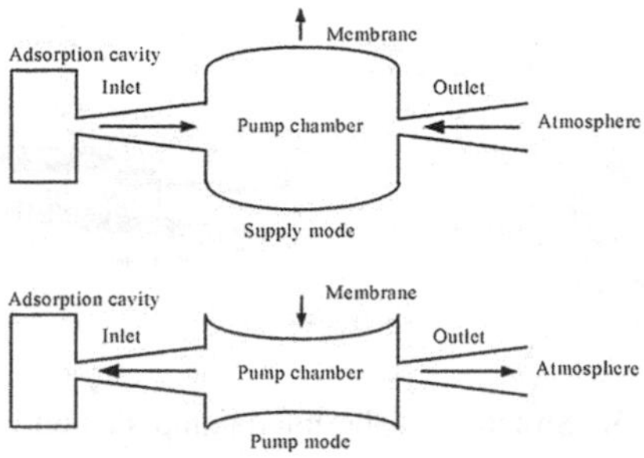

Fig. 2. Principle of the micropump sucton cup

which acts as a nozzle. The result for the complete pump cycle of the diffuser/nozzle pump is that a net volume has been transported from the input to the output side of the pump. When actuated by the piezoelectric actuator, the pump chamber vibrates generating the continuous differential flow, so the gas is pumped from the adsorption cavity to generate stable negative pressure continuously.

2.3 Micropump Suction Cup Prototype

The conventional mechanical manufacturing technologies can't meet the precision of the primal dimensions, such as the minimum cross section width of the tapered tubes, the vibration cavity's thickness and so on. The studying prototype is fabricated using laser carving technic and lamination. The lamination is that the micropump suction is a stack of many thin films that are fabricated respectively, and it is an effective method to set up the prototype of complex constructions, which is available to solve the problems that the micro-channels in the micropump are too precise to machine and assemble.

As Fig. 3 shows, it is the lamination micropump sucker presented in the paper, consisting of top/bottom covers, upper/lower baseplates, a pump chamber layer and a suction cavity layer. The testing pipe joints can connect to the hole through the end cover made of aluminum alloy by traditional machining, and it will be uninstalled in order to release the loads in practice. There exists a groove for the power line on the top cover. Meanwhile, the eccentric holes of 1 mm in diameter in the upper and lower baseplates open to the venthole in top cover and the tapered tube on the bottom cover. The chamber layer with a pump chamber and tapered tubes is fabricated by the laser carving devices. To investigate the relationship between the thickness of the pump chamber with sucker's adsorption performance, the experiment demands various thickness of blackening stainless steel boards. In theory, the thinner the pump chamber is, the better the gas micropump performs. However, the board has a trend to curl with its thickness reducing, which means it is difficult to ensure the machining accuracy. So we choose two boards of the thickness 0.1 mm and 0.2 mm as the pump chamber layers that the upper/lower baseplates are of the same stuff with. The eccentric hole in the bottom cover connects to the adsorption cavity at the center of the adsorption cavity layer made of silicone rubber that is of high resistance of air loss. It's essential to preserve seal between the sucker and the attached substance by either using the circular

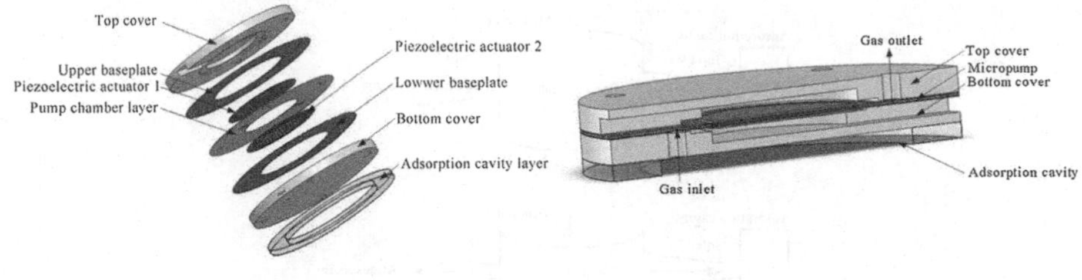

Fig. 3. Structure of the micropump suction cup

elastic edge or ensuring outer radius of the layer larger than the cavity's. As Fig. 4 shown, there are the prime parts as previously mentioned.

The assembly procedures are as follows: The first step was to brush epoxy resin of uniform thickness on the both sides of the pump chamber layer. The next step was to glue upper/lower baseplates on the pump chamber layer with three layers' eccentric holes aligning, which turned out the pump chamber. In addition, it was essential to prevent the tapered tubes and holes from block and to focus attention on the bubbles in the bonding process. The micropump with doubles actuators was the upon pump chamber pasting firmly two circle piezoelectric unimorph actuators with the epoxy resin, and it was fixed between top/bottom covers with the ventholes and holes aligning. In order to ensure the air impermeability, it's valid to fill silicon rubber boards between ends covers and sides of the micropump, and screw down the edge of the whole micropump sucker. The final step was to paste the adsorption cavity layer on the bottom cover using the universal glue. The test experiment for the assembly impermeability was that blowing air through the pipe linked to the joint of the suction cup soaking in the water. If some bubble escaped from the other vent, the micropump sucker cup passed the test successfully.

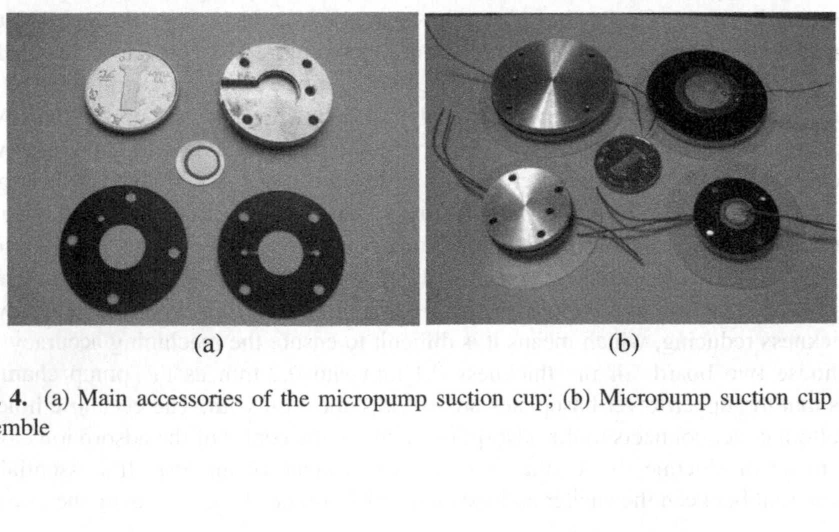

(a) (b)

Fig. 4. (a) Main accessories of the micropump suction cup; (b) Micropump suction cup after assemble

3 Optimizing Static Driving Characteristics of the Piezoelectric Actuator

As mentioned above, the actuator's deflection is critical to the volume change of the gas micropump, which the research indicates is relative to some key factors: the geometric dimensioning, material properties, driving voltage and driving frequency [9]. And the geometric dimensioning and material property have effect on the static driving characteristics of the actuator, while the driving frequency on the dynamic performance. Based on a series of assumption, the theoretical mechanics model of the circle piezoelectric unimorph actuator is founded to be used to solve transverse deformation of the circle chip actuator excited by the constant voltage; through optimizing its geometry design, the piezoelectric actuator has been enabled of the best static deformation characteristics.

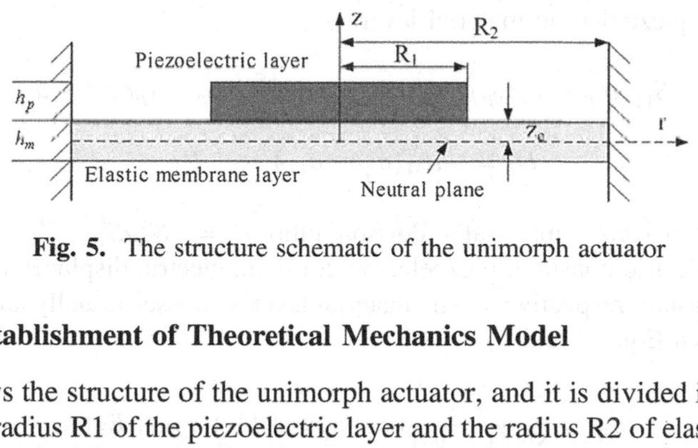

Fig. 5. The structure schematic of the unimorph actuator

3.1 The Establishment of Theoretical Mechanics Model

Figure 5 shows the structure of the unimorph actuator, and it is divided into two parts, including the radius R1 of the piezoelectric layer and the radius R2 of elastic membrane layer. The theoretical mechanics model of the unimorph actuator has been discussed in a large number of literatures [10, 11], but it's too complex to calculate. To simple the analytic analysis, some hypothesis is presented as follows base on the structure of the circular piezoelectric composite [12, 13]:

1. The thickness of the piezoelectric layer and elastic membrane layer is far less than its own radius, and the deformation is weaker comparing to the actuator's dimension, which is the reason why the calculations about the actuator's deformation refer to the thin plate theory.
2. Both the bonding layer and electrode layer are much thinner than any of the material layer or elastic membrane, so their effect on the other is neglected.
3. The piezoelectric material is homogeneous isotropic and perfectly elastic, so is the elastic membrane.
4. The polarization direction of the piezoelectric layer is along the direction of the actuator thickness paralleling to z-axis, so the deformations actuated by electric filed in other direction are neglected also.

Base on the small deflection theory of round thin axisymmetric plate, the force and moment equilibrium equations of the actuator without external load [11] yield:

$$\frac{dN_r}{dr} + \frac{N_r - N_\theta}{r} = 0 \quad \frac{dM_r}{dr} + \frac{M_r - M_\theta}{r} = Q_r = 0 \tag{2}$$

In these equations, N_r, N_θ denote resultant force in radial direction and tangent direction of the composite respectively, M_r, M_θ denote resultant moment in radial direction and tangent direction; Q_r denotes shearing force that is negligible in the case that the actuator is loaded nothing. Base on the hypothesis mentioned above, the actuator's strain-stress relationship is of linear dependence, so the relationships among radial strain (ε_r), shearing strain (ε_θ), reference plane displacement (u) and rotation angle (ω) are derived [11]:

$$\varepsilon_r = \frac{du}{dr} - (z - z_c)\frac{d^2\omega}{dr^2} \quad \varepsilon_\theta = \frac{u}{r} - (z - z_c)\frac{1}{r}\frac{d\omega}{dr} \tag{3}$$

The piezoelectric equations in the polar coordinates are described in terms of the chip actuator's piezoelectric material layer:

$$\varepsilon_r = S_{11}^E(\sigma_{rp} - v_p\sigma_{\theta p}) - d_{31}E_3 \quad \varepsilon_\theta = S_{11}^E(\sigma_{\theta p} - v_p\sigma_{rp}) - d_{31}E_3 \tag{4}$$

$$D_3 = -d_{31}(\sigma_{rp} + \sigma_{\theta p}) + \xi_{33}^T E_3 \tag{5}$$

Where, piezoelectric material's Poisson ratio $V_p = -S_{12}^E/S_{11}^E$, ξ_{33}^T 、d_{31} 、D_3 、E_3 and S_{11}^E are dielectric constant, piezoelectric constant, electric displacement and elastic compliance constant respectively. The material layer's stresses radially and tangentially are derived from Eqs. (4 and 5):

$$\sigma_{rp} = \frac{1}{S_{11}^E(1 - v_p^2)}(\varepsilon_r + v_p\varepsilon_\theta + (1 + v_p)d_{31}E_3) \tag{6}$$

$$\sigma_{\theta p} = \frac{1}{S_{11}^E(1 - v_p^2)}(v_p\varepsilon_r + \varepsilon_\theta + (1 + v_p)d_{31}E_3) \tag{7}$$

And the stresses of the membrane layer yield:

$$\sigma_{rm} = \frac{1}{S_m(1 - v_m^2)}(\varepsilon_r + v_m\varepsilon_\theta) \quad \sigma_{\theta m} = \frac{1}{S_m(1 - v_m^2)}(v_m\varepsilon_r + v_m\varepsilon_\theta) \tag{8}$$

Where, v_m and S_m denote Poisson ratio and elastic compliance constant of elastic membrane material. The radial and tangential force and moment per unite length in the composite of the piezoelectric layer and elastic membrane where $r < R_1$:

$$N_{r1} = \int_0^{h_p} \sigma_{rp}dz + \int_{-h_m}^0 \sigma_{rm}dz \quad N_{\theta 1} = \int_0^{h_p} \sigma_{\theta p}dz + \int_{-h_m}^0 \sigma_{\theta m}dz \tag{9}$$

$$M_{r1} = \int_0^{h_p} \sigma_{rp}(z - z_c)dz + \int_{-h_m}^{0} \sigma_{rm}(z - z_c)dz \tag{10}$$

$$M_{\theta 1} = \int_0^{h_p} \sigma_{\theta p}(z - z_c)dz + \int_{-h_m}^{0} \sigma_{\theta m}(z - z_c)dz \tag{11}$$

In the equations above, z_c denotes the position of neutral plane in the coordinate system shown in Fig. 5, where there is neither radial strain nor shearing strain. The position of the reference plane is calculated based on the equation as follows:

$$\int_0^{h_m} \sigma_{rm}(z - z_c)dz + \int_{h_m}^{h_m + h_p} \sigma_{rp}(z - z_c)dz = 0 \tag{12}$$

With the aid of Eqs. (9–11) and (3), (6–8), the equilibrium equations about the piezoelectric composite board are represented in terms of radial displacement $u_1(r)$ and shearing displacement $\omega_1(r)$ in the neutral plane [11]:

$$\frac{d^2u_1(r)}{dr^2} + \frac{du_1(r)}{rdr} - \frac{u_1(r)}{r^2} = 0 \quad \frac{d^3\omega_1(r)}{dr^3} + \frac{d^2\omega_1(r)}{rdr^2} - \frac{1}{r^2}\frac{d\omega_1(r)}{dr} = 0 \tag{13}$$

As for the radius r between R_1 and R_2 on the elastic membrane layer, the resultant force and resultant moment per unit length yield respectively:

$$M_{r2} = \int_{-h_m}^{0} \sigma_{rm}(z - z_c)dz \quad M_{\theta 2} = \int_{-h_m}^{0} \sigma_{\theta m}(z - z_c)dz \tag{14}$$

Similarly, the equilibrium equations about the elastic membrane layer are modified using the radial displacement $u_2(r)$ and shearing displacement $\omega_2(r)$ in the neutral plane, which are similar to Eq. (13) [11]. As Fig. 5 shown, the deflection at the center of the clamped actuator is limited, so we can obtain the boundary conditions:

$$u_1(0) < \infty; \quad \frac{d\omega_1(r)}{dr}\Big|_{r=0} < \infty; \quad u_1(R_2) = 0; \quad \omega_1(R_2) = 0; \quad \frac{d\omega_1(r)}{dr}\Big|_{r=R_2} = 0 \tag{15}$$

The continuity boundary conditions yield in the case $r = R_1$:

$$\frac{d\omega_1(r)}{dr}\Big|_{r=R_1} = \frac{d\omega_2(r)}{dr}\Big|_{r=R_1}; \quad \omega_1(r)\big|_{r=R_1} = \omega_2(r)\big|_{r=R_1} \tag{16}$$

$$u_1(R_1) = u_2(R_1); \quad N_{r1}(R_1) = N_{r2}(R_1); \quad M_{r1}(R_1) = M_{r2}(R_1) \tag{17}$$

Base on the equilibrium equations and boundary conditions mentioned above, the approximate analytic solution for the deflection of the piezoelectric actuators applied the constant DC source is obtained:

$$\omega(r) = \begin{cases} \omega_1(r) = \dfrac{C_1\{2R_1^2\ln(\frac{R_1}{R_2}) + [1-(\frac{R_1}{R_2})^2]r^2\}U}{C_2 - C_3(\frac{R_1}{R_2})^2 + \frac{1}{2}h_p^4 S_m^2(1+\upsilon_p)(\frac{R_1}{R_2})^4}, & r \le R_1 \\ \omega_2(r) = \dfrac{C_1\{2R_1^2\ln(r) - R_1^2[2\ln(R_2)-1] - (\frac{R_1}{R_2})^2 r^2\}U}{C_2 - C_3(\frac{R_1}{R_2})^2 + \frac{1}{2}h_p^4 S_m^2(1+\upsilon_p)(\frac{R_1}{R_2})^4}, & R_1 \le r \le R_2 \end{cases} \tag{18}$$

Where the constant C_1, C_2, C_3 denote respectively:

$$C_1 = 3d_{31}h_m S_{11}^E S_m(h_m + h_p) \tag{19}$$

$$C_2 = 4S_{11}^E h_p h_m^3 S_m + 6S_{11}^E h_m^2 h_p^2 S_m + 4S_{11}^E h_m h_p^3 S_m + \frac{1}{2}h_p^4 S_m^2(1+\upsilon_p) + \frac{2h_m^4 S_{11}^{E2}}{1+\upsilon_p} \tag{20}$$

$$C_3 = 4S_{11}^E h_p h_m^3 S_m + 6S_{11}^E h_m^2 h_p^2 S_m + 4S_{11}^E h_m h_p^3 S_m + h_p^4 S_m^2(1+\upsilon_p) \tag{21}$$

Due to the equations mentioned above, we obtain the volume change resulting from the deformation of the piezoelectric actuator:

$$\Delta V = \int_0^{R_1} \omega_1(r) 2\pi r dr + \int_{R_1}^{R_2} \omega_2(r) 2\pi r dr \tag{22}$$

3.2 Mechanical Model Validation

The equations about the actuator's deformation and the chamber's volume change have been derived according to the assumptions in the last section. The finite element method is an engineering simulation solution, and it is confirmed to be viable to analyze the characteristics of the miniature piezoelectric actuator. The result from the finite element simulation for the actuator's static characteristics is compared with the analytical solution, which aims to verify the theoretical mechanics model.

The finite element model of the circle piezoelectric unimorph actuator is set up in the simulation software ANSYS. As mentioned above, both the bonding layer and electrode layer are so thinner than the other components' dimension that they have few effects on the actuator's driving characteristics. That is the reason why the model was a composite of the piezoelectric layer and elastic membrane layer while the other layers are neglected. The piezoelectric chip actuator employs Pz26 as piezoelectric layers and coppery board as elastic membrane layer, whose dimensions are shown at Table 1.

Table 1. Dimensions of the piezoelectric material (Pz26) and Cu

Dimensions	Values
Piezoelectric material layer (Pz26) radius R_1 (mm)	5
Piezoelectric material layer (Pz26) thickness h_p (mm)	0.2
Elastic membrane layer (Cu) radius R_2 (mm)	6
Elastic membrane layer (Cu) thickness h_m (mm)	0.15

In order to ensure the simulation accuracy, the model of the piezoelectric layer was reconstructed based on the 3D piezoelectric coupling unit SOLID-5, which was an element of eight-node hexahedron, and the copper membrane layer's model based on SOLID-95. In addition, it was absolutely necessary to define the parts according to the parameters in Table 1. Then ordering vlgue, it was to glue together the piezoelectric layer and the copper membrane. Considering the speed of the simulating calculation, the simulation model was simplified to the semicircle meshed by the hexahedron as Fig. 6 shown. Meanwhile the model was defined by the boundary conditions including circle copper edge clamped, symmetric displacement constraints on the semicircle straight edge and constant current applying on the sides of the piezoelectric layer.

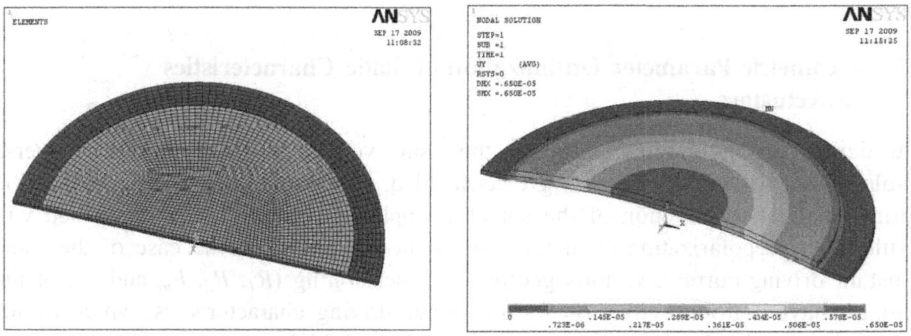

(a) (b)

Fig. 6. (a) Piezoelectric actuator meshed; (b) Deformation of the actuator after adopting voltage

Figure 7 demonstrates the dynamic curves of the deformation (a) and volume change (b) excited by various driving voltages base on the finite element method(FEM) and theoretical mechanics analytical method(TM). As Fig. 7 shown, the analytic result approximately equals to the solution simulated by the finite element method. For examples, the maximum deformation of the actuator is 7.32 μm and 6.5 μm respectively, and volume change value is 0.35 μl and 0.31 μl in the case that driving voltage is set to 200 V. However, the coupling effects between the piezoelectric layer and elastic membrane layer are ignored in the finite element method so that the result is smaller than the analytical one. The simulation results also indicate the actuator's transverse displacement and volume change approximately linearly increase with the growth of the voltage.

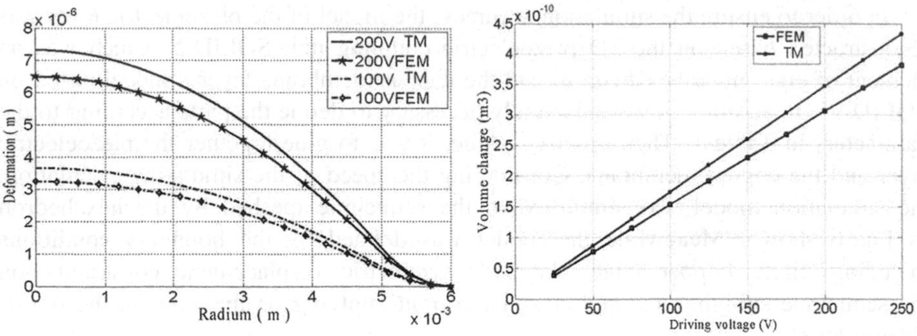

(a) (b)

Fig. 7. (a) Comparion of deflection curves; (b) Comparion of volume changes

3.3 Geometric Parameter Optimization of Static Characteristics of Actuators

The demonstration in Fig. 7 indicates the rising voltage will enlarge the transverse displacement. However, the strong electric filed, whose direction is opposite to the initial polarization direction of the actuator, applies on the piezoelectric, which will result in the depolarization even more the material failure. In the case of the same constant driving current, various geometric dimensioning (R_1, R_2, h_m and h_p) of the actuator have profound effect on its own static driving characteristics, which means there are the optimal values of radium ratio R_1/R_2 and thickness ratio h_m/h_p in terms of volume change of the pump chamber [9]. The finite element method is extremely valid to analyze the static driving characteristics, but it is required to rebuild the simulation model considering the various geometric dimensions. Referring to the literature [7], we optimize the actuator's dimensions according to the derived analytical solution for the deflection of the circle piezoelectric unimorph actuator, which will enhance the static driving characteristics.

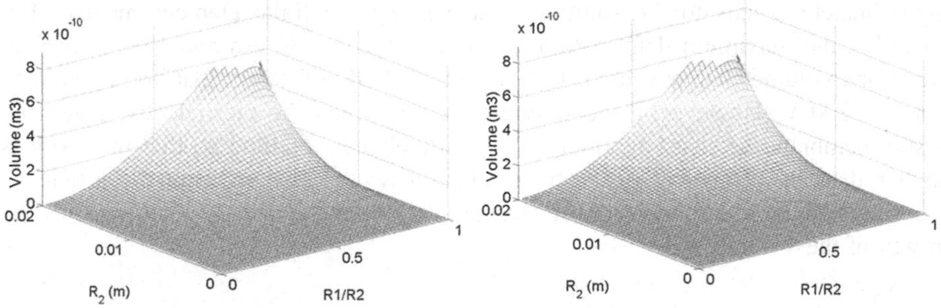

(a) (b)

Fig. 8. (a) The influence of the ratio of the radius between various materials on the volume change; (b) The influence of the ratio of the thickness between various materials on the volume change

The piezoelectric layer's thickness is 0.22 mm, and copper layer's thickness is 0.15 mm, the copper layer's radius is in the range of 0 to 20 mm, the radius ratio of the piezoelectric layer and copper layer varies between 0–1. As Fig. 8 shown, the piezoelectric composite board has a trend to bend with the increase of the radius, which is the reason why the pump volume change rises as the growth of the radius of the copper layer on condition that the radius ratio remained the same. We find the optimal radius ratio 0.94 that results in the maximum volume change from the Fig. 8(a), but the radius ratio have few relationship with the volume change when the copper layer's radius is smaller. In the case of Fig. 8(b), it demonstrated the dynamic curve for the pump volume change in terms of the radius of the piezoelectric layer 5 mm, copper layer's radius 6 mm, the thickness of the copper layer in the range of 0 to 0.5 mm, the thickness ratio of the piezoelectric layer and copper layer between 0–2 and driving voltage 30 V. It indicates there is the optimal thickness ratio 0.17 in the case that the copper layer is thinner, which results in the expected volume change.

4 Conclusion

The article presents a miniature sucker cup integrating a piezoclectric valveless micropump, and describes designing and principle of the prototype on the lamination. We optimize the static driving characteristics of the circle piezoelectric unimorph actuator aiming at enhancing the sucker's adsorption performance. Basc on some assumptions, we take advantage of the finite element method to model the simplified semicircle actuator in the simulation software ANSYS, which confirms the validity of the analytical solutions. The theoretical mechanics analytical results indicates there exist the optimal ratio of the radius and thickness between the piezoelectric layer and elastic membrane layer, which realizes the maximum pump chamber change actuated by the same voltage.

Acknowledgments. This work is supported by the National Natural Science Foundation of China under Grant No. 51475305.

References

1. Nachtigall, W.: Animal attachments: minute, manifold devices. biological variety-basic physical mechanisms-a challenge for biomimicking technical stickers. In: Bionik, pp. 110–111. Springer, Berlin, Heidelberg (2005)
2. Gerlach, T.: Pumping gases by a silicon micro pump with dynamic passive valves, solid state sensors and actuators. In: 1997 International Conference on TRANSDUCERS 1997 Chicago, vol. 1, pp. 357–360. IEEE (1997)
3. Stehr, M., Messner, S., Sandmaier, H., et al.: The VAMP – a new device for handling liquids or gases. Sens. Actuators, A **57**(2), 153–157 (1996)
4. Linnemann, R., Woias, P., Ditterich, J.A.: A self-priming and bubble-tolerant piezoelectric silicon micropump for liquids and gases. In: The Eleventh Annual International Workshop on Micro Electro Mechanical Systems. Piscataway, NJ. pp. 532–537 (1998)

5. Richter, M., Linnemann, R., Woias, P.: Robust design of gas and liquid micropumps. Sens. Actuators, A **68**(1), 480–486 (1998)
6. Schabmueller, C.G.J., Koch, M., Mokhtari, M.E., et al.: Robust design of gas and liquid micropumps. J. Micromech. Microeng. **12**(4), 420–424 (2002)
7. Olsson, A., Stemme, G., Stemme, E.: The first valve-less diffuser gas pump. In: Proceedings of IEEE the Tenth Annual International Workshop on Micro Electro Mechanical Systems, pp. 108–113. Nagoya, Japan (1997)
8. Cheng, G., Liu, G., Yang, Z., et al.: A piezoelectric micro-pump based on the lamination. Piezoelectrics & acoustoopt. **28**(2), 243–245 (2006)
9. Cui, Q.: Research on multi-flied coupling modeling and digital simulation of a piezoelectric valveless micropump. Shanghai Jiao Tong University, Shanghai (2009)
10. Li, S., Chen, S.: Analytical analysis of a circular PZT actuator for valveless micropumps. Sens. Actuators A. **104**(2), 151–161 (2003)
11. Mo, C., Wright, R., Slaughter, W.S., et al.: Behavior of a unimorph circular piezoelectric actuator. Smart Mater. Struct. **15**, 1094–1102 (2006)
12. Liu, G.: The design theory and experimental study of piezoelectric micro-pumps in series based on lamination. Jilin University, Jilin (2006)
13. Sun, X.F.: Designing theory and structural optimizing technology study on double-vibrator piezo-pum. Jilin University, Jilin (2009)

The Research on the Method of Gait Planning for Biped Robot

Pan Li[1], Kaichao Li[1(✉)], and Yifan Wei[2]

[1] Beijing University of Aeronautics and Astronautics, Beijing 100191, People's Republic of China
12071145@buaa.edu.cn
[2] Farragut High School, Koxville, TN 37934, USA

Abstract. Because of its great flexibility and high environmental adaptability, biped robot has been widely applied in many kinds of fields, including scientific exploration, daily service and so on. Biped robot, which has great potential both in social and economic aspects, is a cutting-edge research area in the field of robot study. In this manuscript, a humanoid gait planning method of biped robot based on the hip height and attitude compensation is proposed. Firstly, the sagittal and lateral kinematics models of the robot are established. The poses at critical moments in the process of walking of the robot are analyzed. At the same time, a complete gait planning is accomplished by using the cubic spline interpolation method, and the smooth trajectory obtained of each joint of the robot verifies the feasibility of the proposed method. Finally, the walking experiment is carried out to further verify the feasibility of the planning method.

Keywords: Biped robot · Zero Moment Point · Cubic spline interpolation · Gait planning

1 Introduction

The biped robot is a nonlinear system with multiple degrees of freedom, complex structure and strong coupling. As a result, it is a hot spot to achieve its stable walking control in the field of biped robot research [1]. The gait is the foundation of robot walking. The rationality of gait has a direct impact on the overall performance of the robot. So it is of great value to study the gait planning method and perform gait simulation in practice.

The biped robot has similar appearance and joint configuration to the lower body of human, and it is necessary for biped robot to have an anthropomorphic gait in order to be able to adapt to complex environment. Biomechanics studies show that, in the cycle of human's walking, the height of the hip is cyclical in the vertical plane and the hip has undulating movement in the horizontal plane in order to maintain body balance and reduce the energy consumption during walking. The traditional robot gait planning method assumes that the the height of hip joint remains constant and the robot torso remains perpendicular to the ground [2]. This assumption simplifies the robot pose model during walking to reduce the complex coordination between the degrees of freedom and reduce the amount of calculation in the planning process. Nevertheless, it

Y. Huang et al. (Eds.): ICIRA 2017, Part I, LNAI 10462, pp. 39–50, 2017.
DOI: 10.1007/978-3-319-65289-4_4

does not accord with the human walking mode. This manuscript presents a gait planning method based on hip height and attitude compensation by studying the difference between human walking and traditional walking mode of robot. By establishing the sagittal and lateral kinematics models of the robot, analyzing the poses at critical moment during walking, the gait planning for the biped robot is completed with the help of cubic spline interpolation method. According to the simulation result by MATLAB, it can be found that the trajectory of each joint of the robot is smooth, which verifies the feasibility of the proposed method in theory. Compared with traditional method, this method shows an obvious advantage in lateral joint angle and lateral offset distance. Finally, experiments are performed, from which the feasibility of this method is verified in practice.

2 Mathematical Model of the Robot

The mathematical model of the biped robot is the foundation of the gait planning. A small steering gear biped robot with 10 degrees of freedom is studied in this manuscript (initial parameters shown in Table 1), which is shown in Fig. 1. Each leg has 5 degrees of freedom of rotation. Thereinto, the hip and ankle both have one sagittal and one lateral degree of freedom, while the knee only has a sagittal degree of freedom. For the reason that the axis of the ankle do not intersect with that of the hip, an eleven-bar mechanism model is established according to the structural characteristics of the biped robot, which is shown in Fig. 2.

Table 1. Initial parameters of robot

Number of joint	0	1(10)	2(9)	3(8)	4(7)	5(6)
Length (mm)	60	17	53	52	55	43
Mass (g)	288	95	32	138	32	84

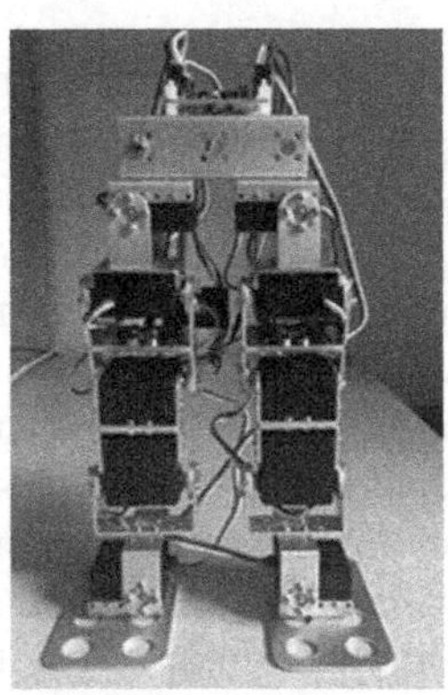

Fig. 1. Biped robot

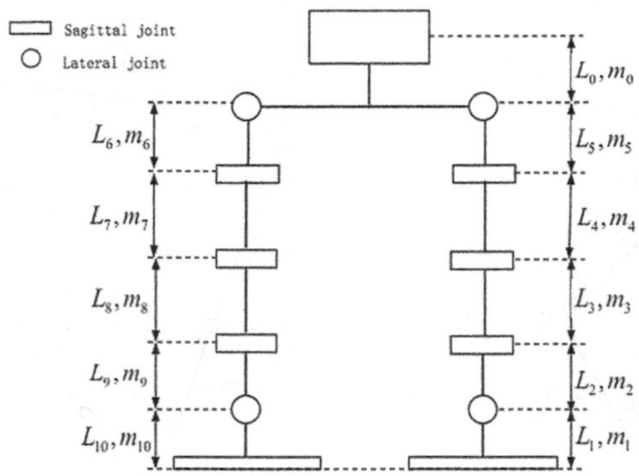

Fig. 2. Link model of Biped Robot

Kinematics analysis of biped robot is the prerequisite for the gait research. The purpose of kinematics modeling is to find out the kinematic relationship between the joints and the rigid body of the robot, which is the basis for gait planning [3]. The kinematics modeling of robotics includes forward kinematics and inverse kinematics modeling. With the geometric parameters of each rod of the robot and the angle of each joint, the forward kinematics can find out the position of each joint of the robot. The inverse kinematics can analyze the movement of the joint angles through the given position and orientation of the key part of the robot.

The biped robot has six sagittal joints and four lateral joints. As the motion coupling of the sagittal and lateral joints is small in the process of walking, the motion can be decoupled into sagittal and lateral movements and the influence of the lateral motion on the sagittal model can be neglected while the influence of leg bending on the lateral model while moving in the sagittal position is considered [4]. In addition, it can be considered that in the course of the movement, the robot's two feet do not finish any movement in the lateral direction.

The origin of the reference coordinate system is located at the junction of the lateral plane of the supporting leg's ankle joint, the horizontal plane and the radial plane of the center of the two hip joints. The direction of the X-axis is the sagittal direction of the robot, and the Y-axis is pointing to the positive direction of the left leg of the robot while the Z-axis is aligned vertically. The kinematic model of the robot is established, which is shown in Fig. 3(a) is the sagittal kinematic model, and (b) is the lateral kinematics model.

According to the kinematic model, the forward kinematics equation of the robot is obtained through geometric constrain. Then, the coordinates of the centroid of each link can be calculated. The first derivation and the second derivation are velocity and acceleration, respectively. Through the cosine theorem and the relationship between the angles of joints, each joint angle of the robot can be solved. The first derivation and the second derivation are angular velocity and angular acceleration, respectively. All of these provide the basis for the calculation of the ZMP (Zero Moment Point) of the robot.

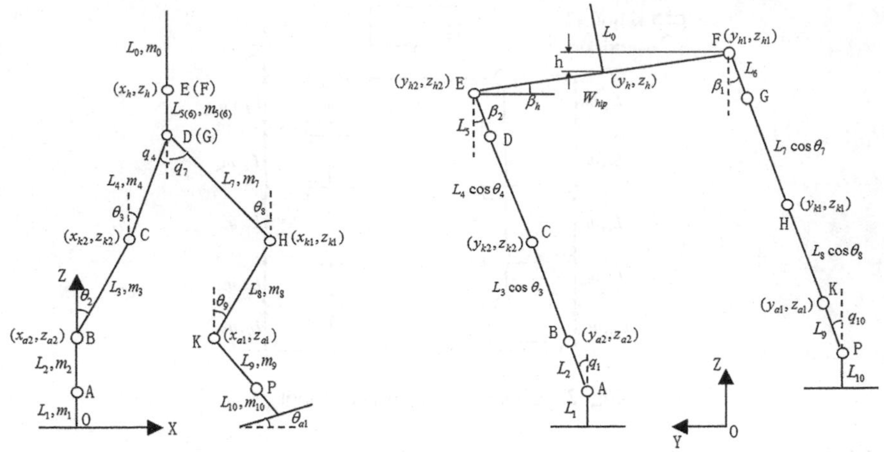

(a) **Sagittal model** (b) **Lateral model**

Fig. 3. Kinematics model of robot

3 Gait Planning Method of Biped Robot

The gait planning includes two aspects: attitude planning and ZMP (Zero Moment Point) trajectory planning [2]. ZMP is currently the most widely used stability criterion in gait planning [5], which refers to the intersection point of the support surface and the extension cord of the resultant force of gravity and inertia force. The computational formula is as follows.

$$x_{ZMP} = \frac{\sum_{i=1}^{n} m_i(\ddot{z}_i + g)x_i - \sum_{i=1}^{n} m_i \ddot{x}_i z_i}{\sum_{i=1}^{n} m_i(\ddot{z}_i + g)} \quad (1)$$

$$y_{ZMP} = \frac{\sum_{i=1}^{n} m_i(\ddot{z}_i + g)y_i - \sum_{i=1}^{n} m_i \ddot{y}_i z_i}{\sum_{i=1}^{n} m_i(\ddot{z}_i + g)} \quad (2)$$

In the formula (1) and formula (2), m_i is the mass of each link, and (x_i, y_i, z_i) is the coordinate of the center of gravity of each link.

A gait generation method based on hip height and attitude compensation is presented in this manuscript. In this method, during the one-leg support period of the cycle, the height of the hip joint from the ground fluctuates within a certain range when the swing leg is swinging forward, and the maximum height value and the minimum height value are set in the planning process. Moreover, the left and right hip joints do not remain parallel, but swing up and down relative to the center of the two hip joints in the vertical direction.

The specific gait planning method is as follows. Firstly, according to the position and orientation of the main nodes of the biped robot at the critical moment, the trajectory of each joint throughout the gait cycle is planned by spline interpolation. Secondly, the ZMP of the robot is calculated and the hip trajectory optimization parameters within the appropriate range is traversed to find the maximum stability margin of the robot. Finally, after obtaining the optimal solution, the motion of the other joints is determined according to the kinematic equation of the biped robot, and the aim of walking stably is achieved.

3.1 Gait Planning

The walking process of the biped robot can be divided into three parts: the starting gait, the cycle walking and the stop gait. Figures 4, 5 and 6 show the radial and lateral poses of the biped robot at critical moments for these three kinds of gait, respectively. In the following figures, h is the maximum height of the swing of the hip in the vertical direction; k is the number of cycles; T_{sd}, T_{sm}, T_{sc} are the parameters for the starting gait; T_d, T_m, T_e are the parameters for the cycle walking. $T_{e0}, T_{ed}, T_{em,} T_{ec}, T_{ee}$ are the parameters for the stop gait; X_1, X_2 are the sagittal optimization parameters; y_1, y_2, y_3 are the lateral optimization parameters. Specific planning process is described as the below part. Starting: The biped robot moves from the upright state, lowering the height of the hip to the supposed height during walking, and the two hip joints are deflected by the lateral roll to the support leg; then the swing leg swings forward, and the two hip joints continue to move to the direction of the support leg. At the same time, the hip joint of the swing leg swings up relative to the center of the two hip joints, and the hip joint of the support leg moves downwards. The swing leg takes a step and then touches the ground. The hip re-parallels, and the starting state completes.

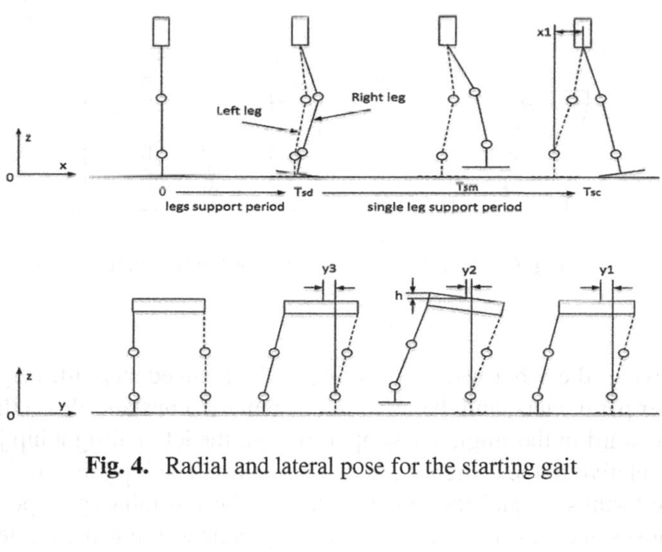

Fig. 4. Radial and lateral pose for the starting gait

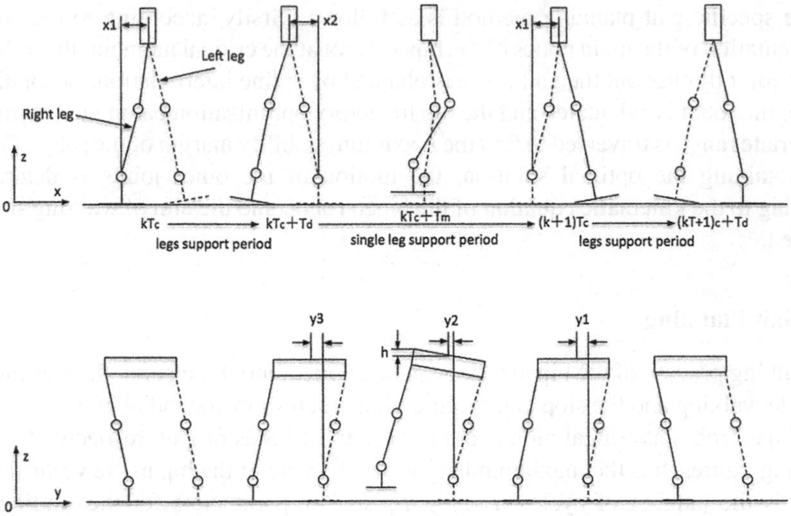

Fig. 5. Radial and lateral pose for the cycle walking

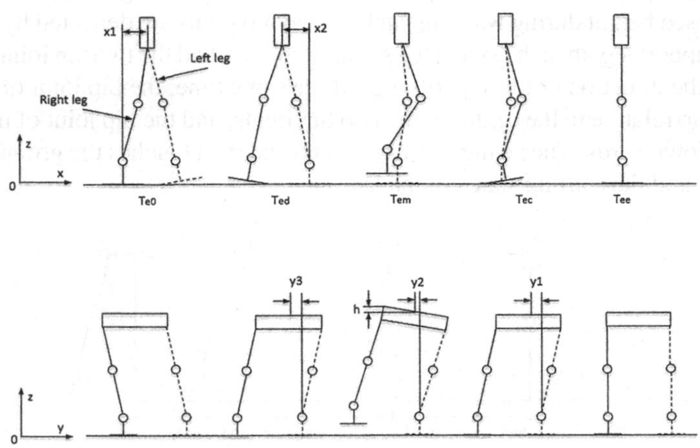

Fig. 6. Radial and lateral pose for the stop gait

Cycle walking: the robot walks according to the planned step, lift height, walking cycle and other parameters, and the travel speed remains constant. When the swing leg is swinging forward in the single leg support period, the left and right hip joints swing up and down relative to the center of the two hip joints. The hip joint corresponding to the swing leg swings up and then swings down. The hip joint corresponding to the support leg moves in the opposite direction. During cycle walking, the single leg support period and the two legs support period alternate periodically. Due to the symmetry of the robot and the periodicity of walking, if gait is planned in one cycle, then the robot can walk continuously.

Stop: the swing leg start leaving the ground, and the center of gravity of the robot is shifted to the support leg; Then the swing leg swings forward, and the two hip joints continue to move to the direction of the support leg; At the same time, the hip joint corresponding to the swing leg swings up relative to the center of the two hip joints, and the hip joint corresponding to the support leg moves downwards; The swing leg swings to the highest point; The swing leg takes a step and the hip re-parallels; The center of gravity of the robot moves to the center of two legs, and the center of gravity of the robot rises to the initial upright position. The stop state completes, and the robot stops moving.

The initial gait parameters are shown in Table 2. Due to the symmetry of the robot and the periodicity of walking, four cycles are taken to study in the complete walking process of the robot. The following x-axis range shows the whole trajectory cycle of robot in 14 s: 0–3 s for the starting gait, 3 s–11 s for the cycle walking, and 11 s–14 s for the stop gait.

Table 2. Initial gait parameters

Parameter	T_c	T_d	T_m	D_c	L_c	H_c	H_{max}	H_{min}
Value	2 s	0.5 s	1.2 s	85 mm	75 mm	20 mm	205 mm	195 mm

The foot movement trajectory obtained is shown in Fig. 7.

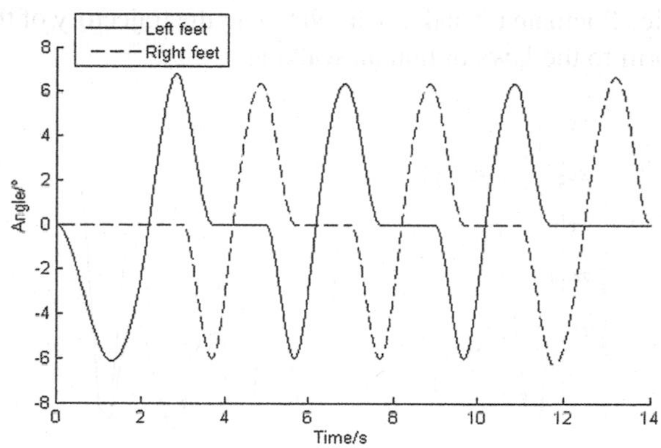

Fig. 7. Foot movement trajectory

The ankle trajectory obtained is shown in Fig. 8.

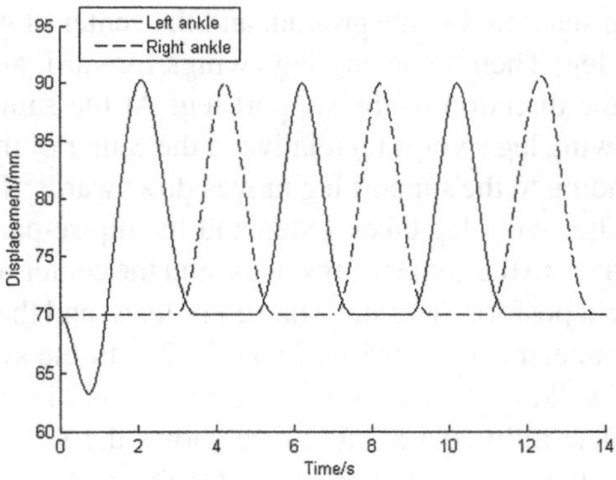

Fig. 8. Ankle trajectory in Z direction

The trajectory planning of the hip is critical to the robot gait planning. After planning the trajectory of the foot and ankle joint, as long as the completion of the hip trajectory planning, the variation curve of each joint angle can be solved through inverse kinematics equation, and finally the walking gait of robot can be obtained.

The hip trajectory obtained is shown in Fig. 9 (CTHJ refers to the center of the two hips). The trajectory of the center of two hips is a regular fluctuation curve. The left and right hip trajectories fluctuate up and down relative to the trajectory of the center of two hips, which conform to the laws of human walking.

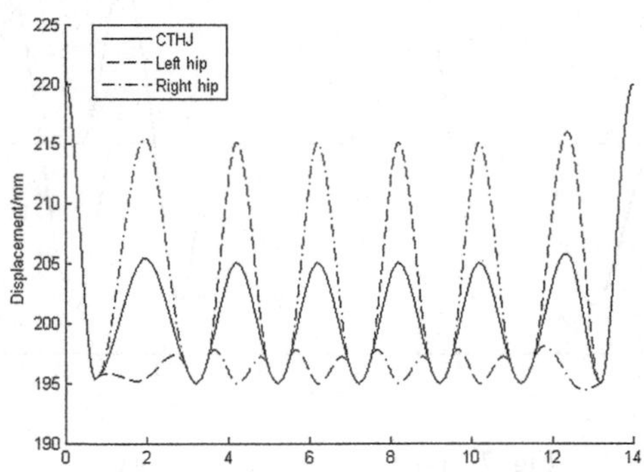

Fig. 9. Hip trajectory in Z direction

3.2 Gait Performance Analysis

The parameter h refers to the maximum distance that the hip joint swings up and down in the vertical direction. When $h = 0$, the two hips remain parallel; when $h > 0$, the two hips are parallel in the legs support period, and two hips swing up and down in the single leg support period. Two sets of data ($h = 0$ mm and $h = 10$ mm) are taken to contrastively analyze the effect of hip height and attitude compensation on lateral ZMP trajectory, lateral centroid trajectory and robot joint angle values. As is explained, when h = 0, the two hips remain parallel and it is exactly what traditional method assumes. When h = 10 mm, two hips swing up and down in the single leg support period, and it is exactly what the method based on hip compensation assumes.

Due to the periodicity of walking, the angle of the joint of the left leg is taken as an example to analyze the impact of the method based on the compensation of the hip joint on the joint angles. The joint angle values are shown in Fig. 10 (h = 0 and h = 10 mm).

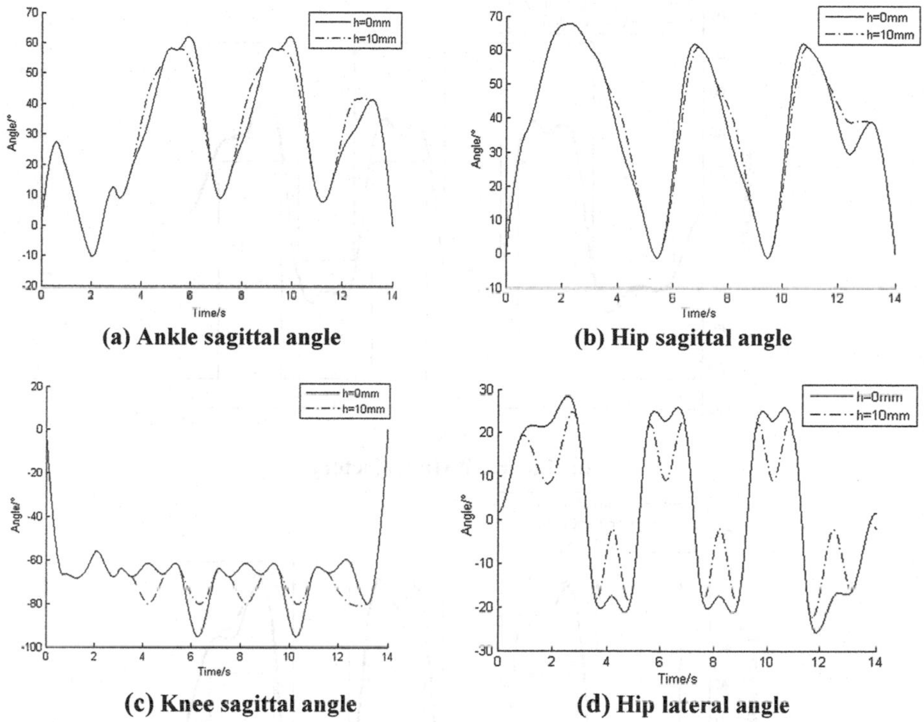

(a) Ankle sagittal angle

(b) Hip sagittal angle

(c) Knee sagittal angle

(d) Hip lateral angle

Fig. 10. Trajectory of each joint angle of robot

The joints of the robot have to move in a certain range to avoid the damage to the robot, so the maximum value of joint angle should be as small as possible. While the other gait parameters are same, the maximum forward angle value corresponding to h = 10 mm is smaller than that corresponding to h = 0 mm. Therefore, the planning method based on the hip compensation can protect the mechanical structure of the robot.

The lateral joint angle of robot is basically the same in the legs support period. But in the single leg support period, the lateral joint angle corresponding to h = 10 mm is smaller than that corresponding to h = 0 mm. That is to say, this planning method can reduce the lateral joint angle in the single leg support period, which is beneficial for the robot to walk steadily.

The ZMP trajectory and the centroid trajectory are shown in Fig. 11 (h = 0 mm and h = 10 mm). With the same gait parameters, the lateral ZMP trajectory and centroid trajectory corresponding to h = 10 mm are located in the outer side of these corresponding to h = 0 mm. This indicates that the swing of the hip can make the lateral ZMP trajectory and centroid trajectory move outward at the same lateral offset of the robot. Therefore, under the same lateral stability margin, the gait planning method based on the hip compensation can effectively reduce the lateral offset of the robot.

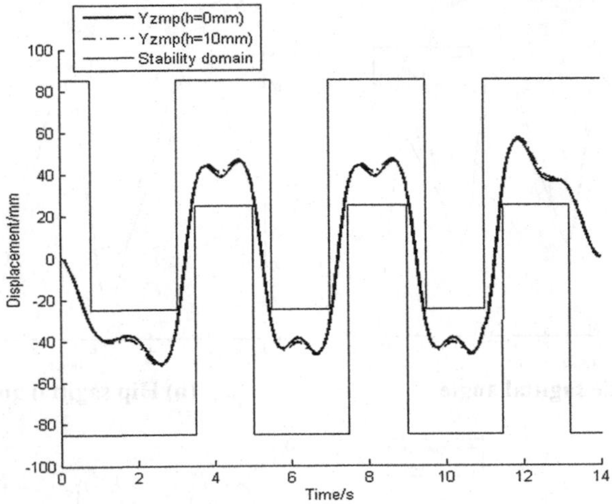

(a) Lateral ZMP trajectory

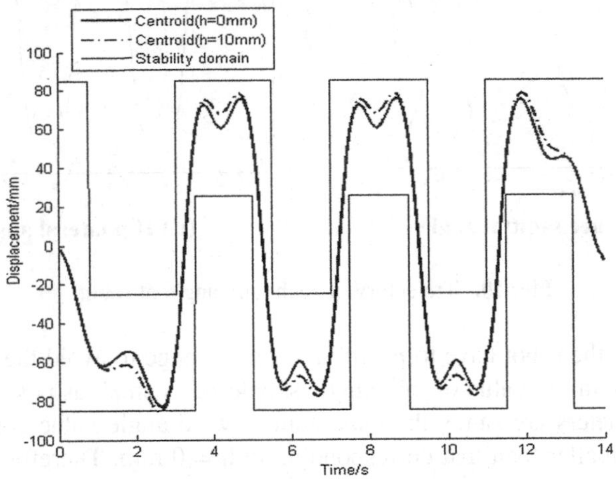

(b) Centroid trajectory

Fig. 11. Comparison of ZMP and centroid trajectories in the Y direction

4 Robot Walking Experiment

The angle values of each joint are transformed into the rotation angle of the steering gear to control the movement of the robot through the control software. Figure 12 shows the biped robot walking process which includes three parts: 0–3 s for the starting gait, 3–7 s for the cycle walking, 7–10 s for the stop gait. The robot is in upright state at the initial and end moment.

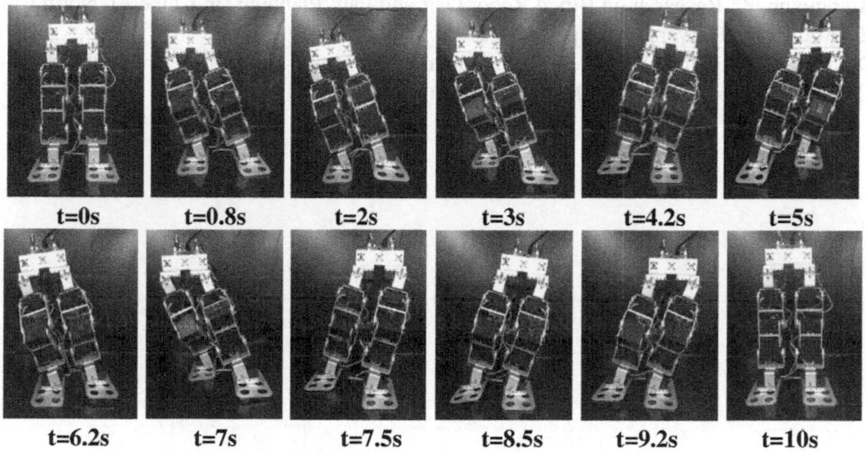

Fig. 12. Biped robot walking process

As can be seen from the figure, in the process of walking, the hip is always parallel to the ground in the legs support period, and the hip swings up and down in the single leg support period. The left and right legs alternately swing forward, and the robot gait meets the design requirements. Meanwhile steering gears run smoothly and the robot walks stably without any dumping or rollover.

5 Conclusion

In this manuscript, the mathematical model of the biped robot is given on the basis of D-H method. The trajectories of feets, ankles and hips are planned by the gait generation method based on hip compensation, and the angle of each joint is solved through inverse kinematics equation. The simulation results of MATLAB show that smooth trajectories can be obtained by this planning method, which proves that it is feasible. At the same time, under the condition of identical parameters, the impact of the hip compensation on the joint angle during walking is analyzed by comparing itself with the conventional planning method($h = 0$ mm), and its regulating effect on the lateral ZMP trajectory and centroid trajectory is demonstrated. Finally, the robot prototype experiment is carried out to further verify the feasibility of the planning method. The experimental results show that the robot gait meets the design requirements and the walk movement is stable.

References

1. Xiaolu, Y., Li, Y., Juntong, C.: Design of ATmega88 and delphi-based biped walking robots (in Chinese). Microcomput. Appl. **23**, 38–40+43 (2014). doi:10.3969/j.issn.1674-7720.2014.23.023
2. Bo, Z., Zhijiang, D., Lining, S., Wei, D.: Research on gait planning of biped walking robot (in Chinese). Mach. Electron. **04**, 52–55 (2008). doi:10.3969/j.issn.1001-2257.2008.04.017
3. Lina, Z.: Design and Study of the Gait Planning and Stability Control of Biped Walking Robot on Focus (in Chinese). Harbin Engineering University, Harbin (2008)
4. Xiaoguang, Z.: Research on Biped Robot Gait and Path Planning (in Chinese). North China Electric Power University, Beijing (2012)
5. Duyong, C., Nanfeng, X.: Research on humanoid robot two-step gait planning algorithm (in Chinese). Comput. Eng. Des. **09**, 1985–1988 (2010)

Control Strategy and Experiment of a Novel Hydraulic-Driven Upper Extremity Exoskeleton

Zirong Luo(✉), Guohen Wu, Xing Li, and Jianzhong Shang

National University of Defense Technology, Changsha 410073, Hunan, China
luozirong@nudt.edu.cn, 15111460906@163.com

Abstract. In this paper, we introduce a novel hydraulic-driven upper extremity exoskeleton and present new type of force feedback control strategy, and the control experiment is also carried out. The experimental results show that the first generation upper extremity exoskeleton successfully follows the wearer's hand movements and achieve load capacity of 20 kg. The experimental results also indicate that the new type force feedback control can effectively reduce the interaction force between the wearer and the upper extremity exoskeleton and the oscillation. Compared with the control method simply using interaction force as the control input, the method with speed compensation controller can effectively reduce the amplitude and oscillation of the interaction force.

Keywords: Upper extremity exoskeleton · Hydraulic driven · Human-robot interaction technology · Force feedback control

1 Introduction

Exoskeleton technologies can bring new capabilities to fighting forces and improve endurance and safety in industrial settings [1, 2, 3]. The first functional load-carrying and energetically autonomous exoskeleton was demonstrated at U.C. Berkeley, walking at the average speed of 0.9 m/s (2 mph) while carrying a 34 kg (75 lb) payload. The original BLEEX sensitivity amplification controller, based on positive feedback, was designed to increase the closed loop system sensitivity to its wearer's forces and torques without any direct measurement from the wearer. And then Lihua Huang [4] from U.C. Berkeley presented an improved control scheme which added robustness to changing BLEEX backpack payload. Lockheed Martin [5] developed an unpowered, lightweight industrial exoskeleton called the FORTIS exoskeleton.

For the control of the exoskeleton, the wearer is directly responsible for most of the tasks in human computer interaction, such as command generation, environment perception, and motion feedback. The sensing and control system of exoskeleton needs to quickly perceive the wearer's intention to move, and to complete the precise following of the wearer's movements.

Kazerooni developed a 6-DOF upper extremity power-assist system which uses hydraulic pressure as the power source, through the interaction between a six component force sensor to measure the wear and the exoskeleton as control input [6].

Y. Huang et al. (Eds.): ICIRA 2017, Part I, LNAI 10462, pp. 51–62, 2017.
DOI: 10.1007/978-3-319-65289-4_5

Then, Caldwell developed a set of 7-DOF upper extremity exoskeleton, which also used the force sensors to measure the human-computer interaction force [7]. While Cavallaro estimated the torque needed by the upper extremity joints through EMG signal, and it was applied to a 7-DOF upper extremity exoskeleton [8]. However, the follow effect and the oscillation problems haven't been fully solved by the above achievements.

In this paper, a 3-DOF power-assist exoskeleton is studied, and the principle of realizing the exoskeleton control with the interaction force of the wearer and the exoskeleton is analyzed, and hence the direct force feedback control model of upper extremity exoskeleton is established. In the fast force feedback control model, the introduction of speed compensation is proposed to reduce the interaction force between the wearer and the exoskeleton, and the introduction of the compensation can greatly reduce the oscillation in the control. Finally, the effectiveness of the control model in the external force feedback control is verified by experiments.

2 The Control Strategy of the Upper Extremity Exoskeleton

2.1 The Role of Force in Control System

The wearer and the 3-DOF exoskeleton can be seen as one integrated control system composed of two sets of complete but interactive control systems, as shown in Fig. 1. Separately, the wearer controls his body movements through the brain, and the exoskeleton controls the rotation of the joint through the outer bone controller. Interactively, the wearer is regarded as the control input of the interaction system, and the controlled object is the upper extremity exoskeleton.

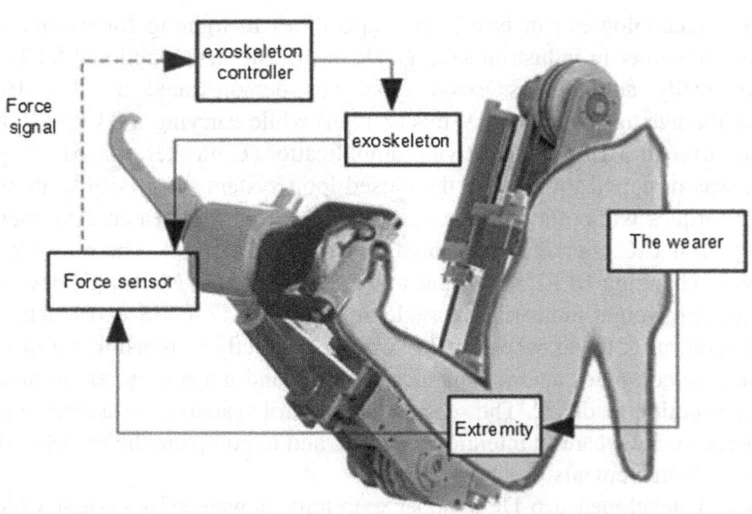

Fig. 1. The interactive interface of the exoskeleton

In this paper, we use one three dimensional force sensor to collect the interaction force between the wearer's hand and the handle of exoskeleton in order to analyze the intentions of wearers. As shown in Fig. 2(a), once the wearer gives his hand a force F, hands will begin to move relative with the exoskeleton's handle; but the exoskeleton has not realized the tracking in action yet, it still maintains the previous state, So the interaction force *f* is generated

$$F = f \tag{1}$$

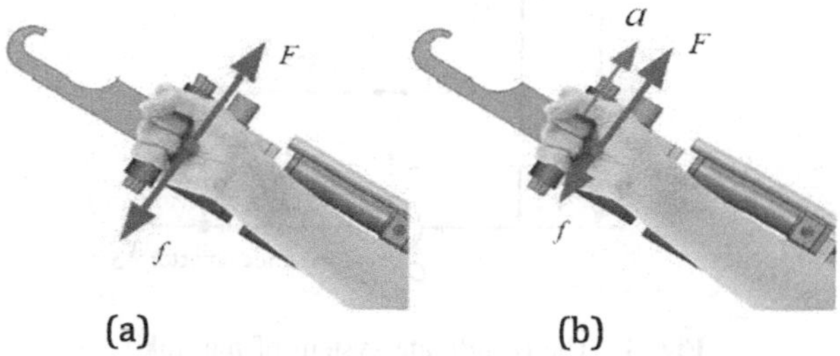

Fig. 2. The force of hands in x direction

At this moment, in order to reduce the force f the end of exoskeleton should move accelerately, as shown in Fig. 2(b). There is a balance equation among the force of the human hand, the force of the exoskeleton and the inertia force of the hand

$$f = F - M.a \tag{2}$$

In order to achieve the compliance following of the exoskeleton and the hands, the force *f* should be as small as possible. Under the constant force, through the Eq. (2), it is found that the effective way is to adjust the acceleration of the hands and the end of the exoskeleton. Also through Eq. (2), the force *f* reflects the size and direction of the acceleration, which can be used to predict the moving direction of the wearer's hand. The ideal value of *f* is 0. Equation (2) changes to $F = M.a$. Combining with the Eq. (1), we can get

$$a = Kf$$

Namely,

$$\begin{pmatrix} a_x \\ a_y \\ a_z \end{pmatrix} = \begin{pmatrix} k_x & 0 & 0 \\ 0 & k_y & 0 \\ 0 & 0 & k_z \end{pmatrix} \begin{pmatrix} f_x \\ f_y \\ f_z \end{pmatrix} \tag{3}$$

Where, K is the corresponding coefficient between the detected force and the acceleration, which is selected according to the debugging situation in real object.

2.2 The Relationship Matrix Between Force and Joint Angular Velocity

The spatial description of the end effector of the exoskeleton with respect to the fixed global coordinate system can be analyzed by D-H description method for kinematic analysis. A three degree of freedom upper extremity exoskeleton is shown in Fig. 3.

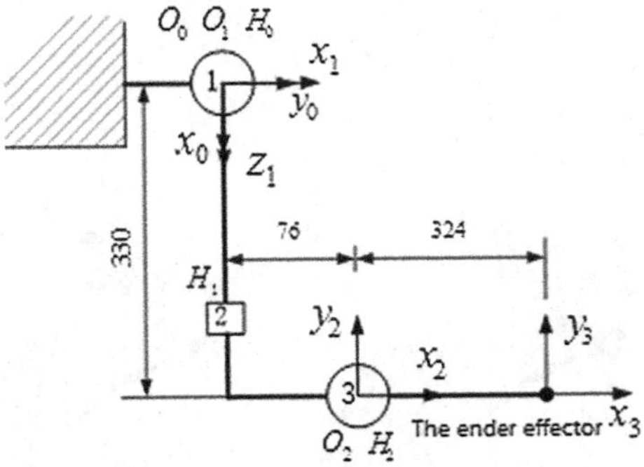

Fig. 3. The coordinate system of the link

In the coordinate system of the link, the transformation matrix can be described as

$$^{i-1}A_i = Rot(x_{i-1}, \alpha_{i-1})Trans(x_{i-1}, a_{i-1})Rot(z_i, \theta_i)Trans(z_i, d_i) \tag{4}$$

Where, a_i is the distance between z_{i-1} and z_i along x_i; x_i is the included angle between z_{i-1} and z_i around x_i; d_i is the distance between x_{i-1} and x_i along z_{i-1}; θ_i is the included angle between x_{i-1} and x_i around z_{i-1}. The transformation formula from the link to the moving coordinate is as follows

$$^0A_4 = {}^0A_1{}^1A_2{}^2A_3 \tag{5}$$

Through coordinate transformation, the relationship between the coordinates of the end effector of exoskeleton and the series of joint angle is established, and after the derivation of the both sides, we can get the Jacobian matrix of the upper extremity exoskeleton

$$\begin{aligned} X &= f(\theta), \\ \dot{X} &= J(\theta)\dot{\theta}, \end{aligned} \tag{6}$$

The inverse kinematics problem can be solved by Jacobi matrix inversion: $\dot{\theta}=J(\theta)^{-1}\dot{X}$

When there is an interaction force in x, y and z direction of the three dimensional force sensor, the corresponding displacement Δx, Δy, Δz are generated, and the corresponding joint angle change analysis is shown as Fig. 4(a)–(c)

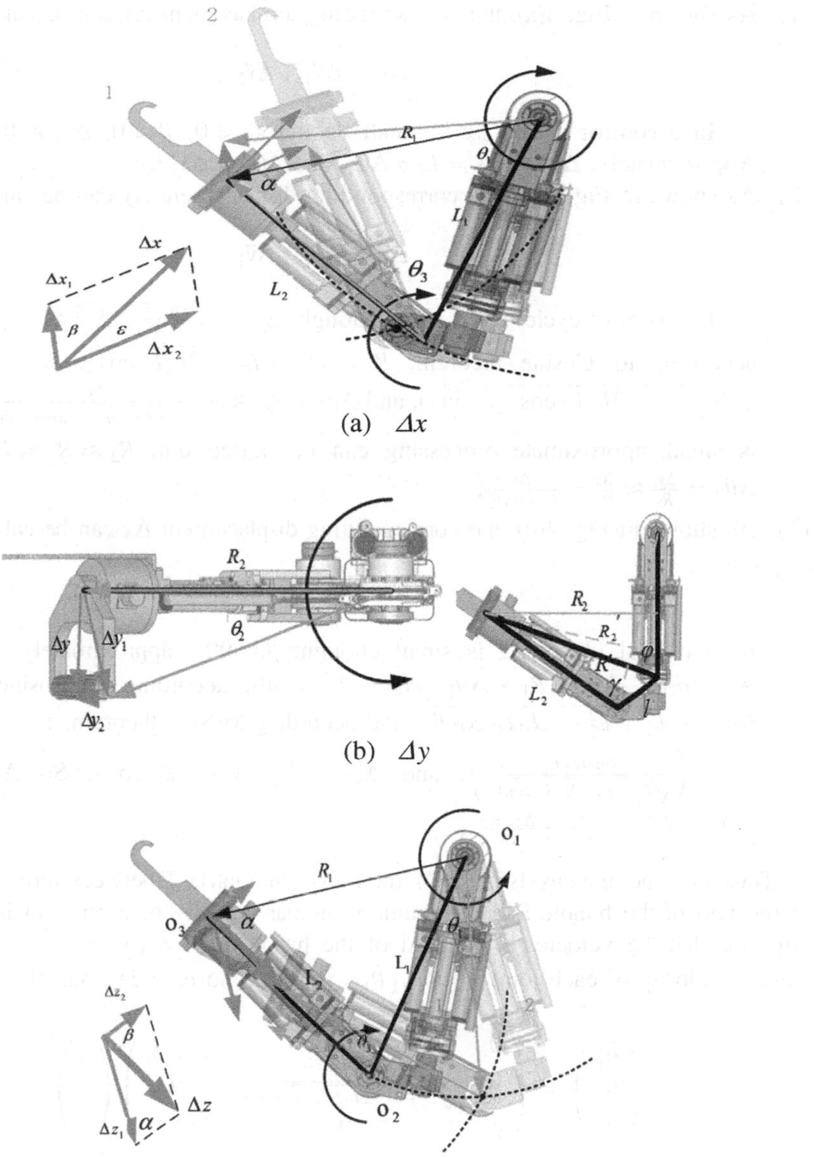

(a) Δx

(b) Δy

(c) Δz

Fig. 4. The corresponding change of angle

(1) As shown in Fig. 4(a), the corresponding displacement Δx can be calculated as

$$\Delta\vec{x} = \Delta\vec{x}_1 + \Delta\vec{x}_2 \tag{7}$$

In a control cycle, Δx is small enough, $\varepsilon \approx 0$, $\beta \approx 0$, $\Delta x_1 \approx 0$, $\Delta x_2 \approx 0$. Approximately, $\Delta x = \Delta x_2 = L_2 \bullet \Delta\theta_3$, and $\Delta\theta_3 = \Delta x / L_2$;

(2) As shown in Fig. 4(b), the corresponding displacement Δy can be calculated as

$$\Delta\vec{y} = \Delta\vec{y}_1 + \Delta\vec{y}_2 \tag{8}$$

In a control cycle, Δy is small enough, $\Delta y_1 = \sqrt{\Delta y^2 + \Delta y_2^2} \approx \Delta y$, $\Delta\theta_2 = \frac{\Delta y}{R_2}$, according to Cosine theorem, $R' = \sqrt{l^2 + L_2^2 - 2L_1L_2\cos\gamma}$, $R_2 = R'\sin\varphi = \sqrt{l^2 + L_2^2 - 2L_1L_2\cos\gamma} \times \sin\varphi$, and $\Delta\theta_2 = \frac{\Delta y_1}{R_2} \approx \frac{\Delta y}{\left(\sqrt{l^2 + L_2^2 - 2L_1L_2\cos\gamma \times \sin\varphi}\right)}$, when l is small, approximate processing can be carried out: $R_2 \approx R_2' = L_2 \times \sin\theta_3$, $\Delta\theta_2 = \frac{\Delta y_1}{R_2} \approx \frac{\Delta y}{R_2'} = \frac{\Delta y}{L_2 \times \sin\theta_3}$;

(3) As shown in Fig. 4(c), the corresponding displacement Δx can be calculated as

$$\Delta\vec{z} = \Delta\vec{z}_1 + \Delta\vec{z}_2 \tag{9}$$

In a control cycle, Δz is small enough, $\beta \approx 90°$, approximately $\Delta\vec{z_1} \perp \overrightarrow{o_1o_3}$, $\Delta\vec{z_2} \perp \overrightarrow{o_2o_3}$, $\Delta z_1 = R_1 \times \Delta\theta_1$, $\Delta z_2 = L_2 \times \Delta\theta_3$, according to Cosine theorem, $R_1 = \sqrt{L_1^2 + L_2^2 - 2L_1L_2\cos\theta_3}$, and according to Sine theorem, $\alpha = \angle o_1o_2o_3 = \arcsin\left(\frac{\sin\theta_3 \times L_1}{\sqrt{L_1^2 + L_2^2 - 2L_1L_2\cos\theta_3}}\right)$, and $\Delta z_1 = \frac{\Delta z}{\sin\alpha}$, $\Delta z_2 = \Delta z\cot\alpha$; So $\Delta\theta_1 = \frac{\Delta z_1}{R_1} = \frac{\Delta z}{L_1\sin\theta_3}$, $\Delta\theta_3 = \frac{\Delta z_2}{L_2} = \frac{\Delta z\cot\alpha}{L_2}$.

Based on above analysis in (1–3), the transform matrix T between moving velocity of the end of the handle and the rotation angular velocity of each joint is obtained. Suppose that the velocity of the end of the handle is $V = \begin{pmatrix}\dot{x} & \dot{y} & \dot{z}\end{pmatrix}$, the rotation angular velocity of each joint is $\omega = \begin{pmatrix}\dot{\theta}_1 & \dot{\theta}_2 & \dot{\theta}_3\end{pmatrix}$, so $\omega = TV$, namely,

$$\begin{pmatrix}\dot{\theta}_1 \\ \dot{\theta}_2 \\ \dot{\theta}_3\end{pmatrix} = \begin{pmatrix} 0 & 0 & \frac{1}{L_1\sin\theta_3} \\ 0 & \frac{1}{\sqrt{l^2 + L_2^2 - 2L_1L_2\cos\gamma \times \sin\phi}} & 0 \\ \frac{1}{L_2} & 0 & \frac{\cot\alpha}{L_2} \end{pmatrix} \begin{pmatrix}\dot{x} \\ \dot{y} \\ \dot{z}\end{pmatrix} \tag{10}$$

Where, $\alpha = \arcsin\left(\frac{\sin\theta_3 \times L_1}{\sqrt{L_1^2 + L_2^2 - 2L_1L_2\cos\theta_3}}\right)$.

After the derivation of $\omega = TV$, the geometrical transformation of acceleration can be obtained, the transform matrix T can be applied equally to detected components of acceleration in three directions.

2.3 The Control Strategy with Speed Compensation

The displacement data of the hydraulic cylinder is collected by the displacement sensor, and the linear relationship between the displacement of the hydraulic cylinder and the rotation angle of the joint is obtained by the geometric relation. It is

$$\begin{pmatrix} s_1 \\ s_2 \\ s_3 \end{pmatrix} = \begin{pmatrix} r_1 & 0 & 0 \\ 0 & r_2 & 0 \\ 0 & 0 & r_3 \end{pmatrix} \begin{pmatrix} \theta_1 \\ \theta_2 \\ \theta_3 \end{pmatrix}.$$

The on-off state of servo valve controls the movement direction and displacement of the hydraulic cylinder, the size of the valve opening size determines the speed of the hydraulic cylinder, the change rate of the opening size determines the acceleration of the hydraulic cylinder. The output of the servo valve is not only related to the angular acceleration $\dot{\omega}$ of the theoretical demand, but also the current angular velocity ω requires the servo valve to provide the corresponding flow rate. Speed compensation is introduced, so the control output of the servo valve is

$$E = f(\dot{\omega}) + f(\omega) \tag{11}$$

Where, $f(\dot{\omega})$ is the corresponding output of $\dot{\omega}$, $f(\omega)$ is the corresponding output of ω, combined with the above geometric relations, the relationship between output and speed signal and force signal can be obtained

$$E = f(\omega) + f(\dot{\omega}) = A_o R\omega + B_o R\dot{\omega} = A_o RTV + B_o R\dot{T}Kf \tag{12}$$

Where, $A_0 = \begin{pmatrix} a_1 & 0 & 0 \\ 0 & a_2 & 0 \\ 0 & 0 & a_3 \end{pmatrix}$, $B_0 = \begin{pmatrix} b_1 & 0 & 0 \\ 0 & b_2 & 0 \\ 0 & 0 & b_3 \end{pmatrix}$ are conversion factors related to hydraulic cylinder and hydraulic source, which can be determined according to the actual situation of the hydraulic system, such as hydraulic cylinder cross section, hydraulic pressure, the size of the valve port and etc.

The speed of the end of handle can be obtained by the integral of the acceleration, but the speed error will be accumulated with the increase of time due to the existence of system error. Therefore, speed compensation can be introduced by two methods:

(1) The angular velocity can be directly calculated through the displacement values collected by the displacement sensor, then, the current output value can be directly provided according to $E = A_0 R\omega + B_0 R\dot{T}Kf$;

(2) Assuming that the exoskeleton tracks well, the output value of the last control cycle can be used as the output value of the speed of last moment, namely, $E' = A_0 R\omega$, the current output is $E = E' + B_0 R\dot{T}Kf$, this method can be understood as a kind of incremental control method based on force signal.

3 Experiment and Discussion

In the experiment, the controlled object is shown in Fig. 5, which has 3 DOFs, and can be regarded as a serial mechanism.

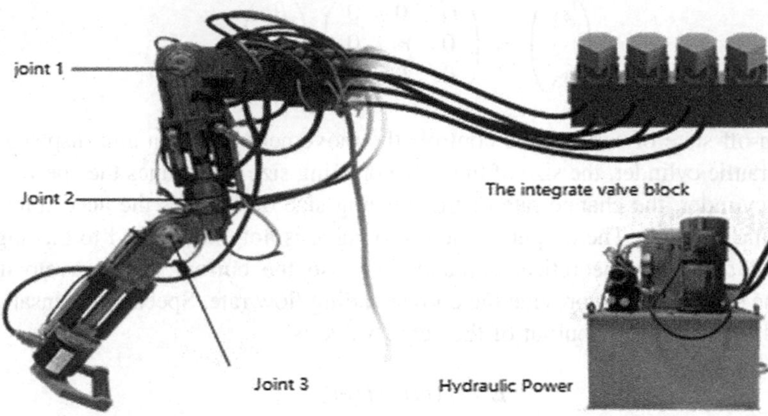

Fig. 5. The physical system

Figure 6 shows the control schematic of upper limb exoskeleton experiment system. The hardware system of the whole control loop is mainly composed of computer, sensors and data acquisition system.

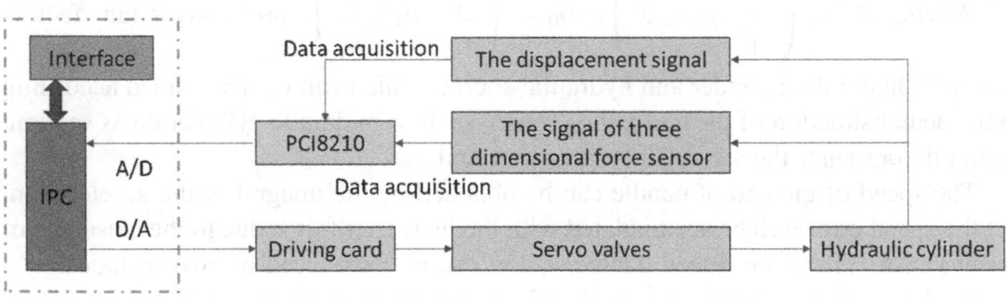

Fig. 6. The experiment system schematic of upper limb exoskeleton

According to Eq. (10),

$$T = \begin{pmatrix} 0 & 0 & \frac{1}{L_1 \sin\theta_3} \\ 0 & \frac{1}{\sqrt{l^2 + L_2^2 - 2L_1 L_2 \cos\gamma} \times \sin\phi} & 0 \\ \frac{1}{L_2} & 0 & \frac{\cot\alpha}{L_2} \end{pmatrix} \quad (1)$$

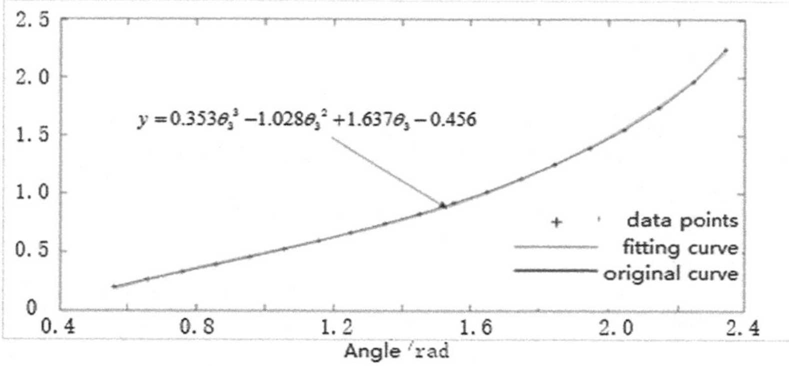

Fig. 7. The fitting results by cubic curves

Because $\frac{\cot\alpha}{L_2}$ is too complicated, in this example, it is fitted by cubic curves, as shown in Fig. 7.

So T is simplified as

$$T = \begin{pmatrix} 0 & 0 & \frac{1}{L_1 \sin\theta_3} \\ 0 & \frac{1}{L_2 \sin\theta_3} & 0 \\ \frac{1}{L_2} & 0 & \frac{\cot\alpha}{L_2} \end{pmatrix} = \frac{1}{L_2}\begin{pmatrix} 0 & 0 & \frac{0.96}{\sin\theta_3} \\ 0 & \frac{1}{\sin\theta_3} & 0 \\ 1 & 0 & 0.353\theta_3^3 - 1.028\theta_3^2 + 1.637\theta_3 - 0.456 \end{pmatrix} \quad (2)$$

Putting the simplified results into the control model, then, $E = A_0RTV + B_0R\dot{T}Kf$ can be obtained, and applying the control model to the experiment subjects. Adjust the parameters R, A_0, B_0, K to the best state. In fact, in the debugging process, not all the unknown parameters are debugged, which will lead the work to increase exponentially, and it's not necessary. Make $k_v = A_0RT$, $k_f = B_0R\dot{T}K$, and then set V and f as the inputs to debug the control model, thus, only two PID parameters are enough.

Fig. 8. The test path

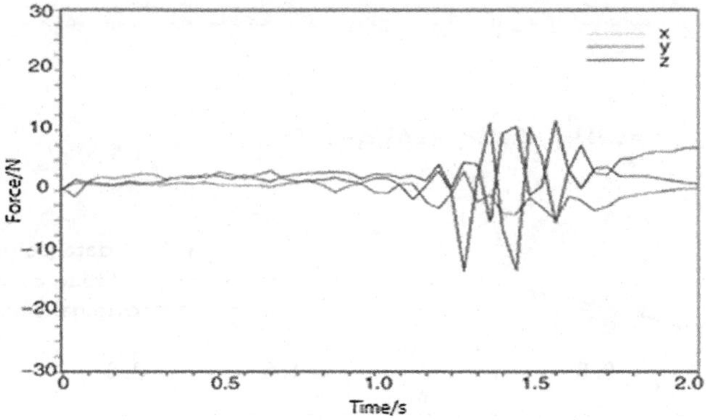

(a) Strategy without velocity compensation

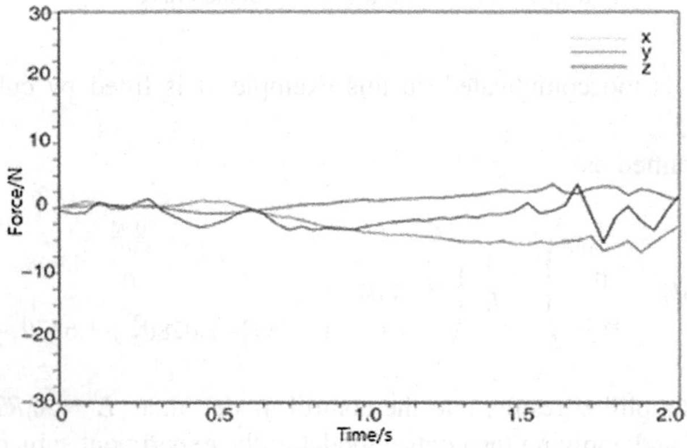

(b) Strategy I with velocity compensation

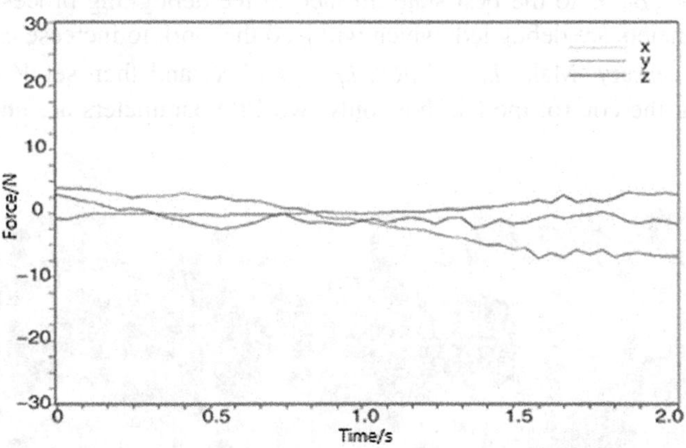

(c) Strategy II with velocity compensation

Fig. 9. The force curve of different control strategies

To validate the proposed control strategy in Sect. 2.3, the controlling tracking experiment is carried out to observe the influence of the angular velocity compensation on the control of exoskeleton; and the criterion is the magnitude of interaction force. In order to reduce the error of operation, each strategy is operated for 10 times.

As shown in Fig. 8, in the experiment, the operator pulls the end of the handle from point A to point B with a 20 kg object, then, the magnitude of the wearer's force can be detected.

Figure 9(a), (b) and (c) show the interaction force curves using different strategies, corresponding with strategy without velocity compensation, with velocity compensation by strategy I and by strategy II. We can see that:

(1) Compared with the simple force feedback control strategy (see Fig. 9(a)), the force feedback control strategy with the added speed compensation (see Fig. 9(b) and (c)) has not only less interaction force input, but also reduces the interaction force between the wearer and the exoskeleton. At the same time, with the addition of the velocity compensation, the oscillation is also obviously weakened, which shows that the direct force feedback control model with velocity compensation is more close to the real model of the upper extremity exoskeleton.

(2) In the two strategies with velocity compensation of direct force feedback control algorithm, the curve of strategy I (see Fig. 9(b)) is gentler than that of strategy II (see Fig. 9(c)). It may be caused by the deviation of the amount of compensation for the opening of the servo valve. Different from strategy I, strategy II is only concerned with the change of acceleration reflected by force.

4 Conclusion

In this paper, the exoskeleton control method focused on force signal with the introduction of angular velocity compensation is analyzed based on the interaction force between the wearer and the exoskeleton. The experimental results show that the hydraulic upper extremity exoskeleton can realize the wearer hand tracking and easily lift 20 kg weight. Compared with the control method simply using interaction force as the control input, the method with speed compensation controller can effectively reduce the amplitude and oscillation of the interaction force. It is suggested that in the future research, we can optimize the control hardware and reduce the micro oscillation.

References

1. Bergamasco, M., Frisoli, A., Avizzano, C.A.: Exoskeletons as man-machine interface systems for teleoperation and interaction in virtual environments. Springer Tracts Adv. Robot. **31**, 61–76 (2007)
2. Carignan, C.R., Naylor, M.P., Roderick, S.N.: Controlling shoulder impedance in a rehabilitation arm exoskeleton. In: IEEE International Conference on Robotics and Automation, IEEE explore, California, USA. pp. 2453–2458 (2008)

3. Romilly, D.P., Anglin, C., Gosine, R.G., Hershler, C., Raschke, S.U.: A functional task analysis and motion simulation for the development of a powered upper-limb orthosis. IEEE Trans. Rehabil. Eng. **2**(3), 119–129 (1994)
4. Kazerooni, H., Steger, R., Huang, L.: Hybrid control of the Berkeley lower extremity exoskeleton (BLEEX). Int. J. Robot. Res. **25**(25), 561–573 (2006)
5. Exoskeleton Technologies. http://www.lockheedmartin.com/us/products/exoskeleton.html. Accessed 03 Sept 2016
6. Kazerooni, H., Guo, J.: Human extenders. J. Dyn. Syst. Meas. Contr. **115**(2B), 281–290 (1993)
7. Caldwell, D., Tsagarakis, N., Kousidou, S., Costa, N., Sarakoglou, I.: Soft exoskeletons for upper and lower body rehabilitation–design, control and testing. Int. J. Humanoid Rob. **4**(3), 549–574 (2007)
8. Cavallaro, E., Rosen, J., Perry, J., Burns, S.: Real-time myoprocessors for a neural controlled powered exoskeleton arm. IEEE Trans. Biomed. Eng. **53**(11), 2387–2396 (2006)

Research on the Locomotion of German Shepherd Dog at Different Speeds and Slopes

Weijun Tian, Qi Zhang, Zhen Yang, Jiyue Wang, Ming Li, and Qian Cong(✉)

Key Laboratory of Bionic Engineering of Ministry of Education, Jilin University, Changchun 130022, Jilin, China
congqian@jlu.edu.cn

Abstract. Quadruped can take the initiative to adjust their gait for adapting to different external environment. Its superior coordination ability provides bionic design inspiration for the quadruped robot. The motion of three German Shepherd Dogs on a treadmill was recorded using a three-dimensional motion capture system VICON MX. The speed of the treadmill was respectively set at 4 km·h^{-1} and 10 km·h^{-1}, and the slope was set at 0° and 20°. Workstation, Polygon and MATLAB were utilized for data processing to obtain the joint angles of the German shepherd dog's forelimbs and hind limbs. The motion frequency of the dog increases, it indicates that the joints move faster to adjust the speed of the treadmill. As the speed of the treadmill increases, the cycle and the stance phase of each limb decrease, the percentages of the stance phase in total cycle are 68.0%, 49.0%, respectively for 4 km·h^{-1}, 10 km·h^{-1} at 0° slope of the treadmill and 65.6%, 47.1% for 20° slope of the treadmill, it indicates that speed affected the time characteristics, while the slope had little effect on the time characteristics. The joint angles of forelimb and hindlimb show that movement of different joints between different speeds and slopes are various.

Keywords: Dog · Kinematics · Joint angle · Swing phase · Stance phase

1 Introduction

The traditional terrain-machine is lack of mobility under complicated road condition [1]. Compared with traditional machine, legged robot especially the bionic legged robot has strong ground adaptability. Researchers focus on quadruped robot in developing the bionics robot because of its better carrying capacity and stability than biped robot, and simpler configuration than hexapod and eight-legged robot. Quadruped such as canine and goat have outstanding locomotion capability in extremely rough unstructured environment. They can adjust their gestures to ensure that they move safely, stably and efficiently when the topography changes. The behavior of each animal is described by the spatio-temporal features [2, 3] of its joint angles [4] and patterns of locomotive cycles. An application of the measurement system is designed to study the motion in the shoulder joint of a dog [5]. Values of stride parameters were compared between Greyhounds and Labrador Retrievers, and apparent differences in the trotting gait between these two breeds are mainly attributable to differences in size, and that

Y. Huang et al. (Eds.): ICIRA 2017, Part I, LNAI 10462, pp. 63–70, 2017.
DOI: 10.1007/978-3-319-65289-4_6

dogs move in a dynamically similar manner at the trot [6]. The kinematics and kinetics of dog in different gait patterns were investigated [7–10]. The foot-ground contact biomechanics of German Shepherd Dog (GSD) in normal walking, trotting and jumping gaits were investigated using a pressure plate system [11]. The quantitative gait analysis showed that there was no significant difference between the movements of the left and right sides of the dogs in a trot gait [12].

In this work, we focused on the kinematic characteristics of German Shepherd Dog (GSD), and the motion properties on treadmill at different speeds and slopes were investigated.

2 Materials and Method

The study was approved by the Institutional Review Board Committee of Jilin University, Changchun, China. Three adults GSDs (body weights 35.3 ± 1.5 kg) were the subjects of this work, and 10 reflective markers were placed on the left side of the GSD, respectively scapula, humerus, elbow, carpus, and metacarpophalangeal joint of the forelimb, coxa, trochanter, knee, ankle, and metatarsophalangeal joint of hindlimb, as shown in the Fig. 1. GSDs were trained at the treadmill, and the speeds were set at 4 $km \cdot h^{-1}$ and 10 $km \cdot h^{-1}$, the slopes of treadmill were set at 0° and 20°. The kinematic data were collected using an 8-camera three-dimensional motion capture system (Vicon MX, Vicon Motion Systems Co., UK), and processed in the software of Workstation. From the motion of the metatarsophalangeal, it was determined when the limbs were in contact with or off the treadmill surface, and the stance phase (contact with the ground of single limb) and swing phase (off the ground of single limb) were divided. The joint angles (less than 180°) of forelimb (hip, knee and ankle) and hindlimb (shoulder, elbow and wrist) were computed using Matlab based on the collected data, and changes in these joint angles of forelimbs and hindlimbs were analyzed.

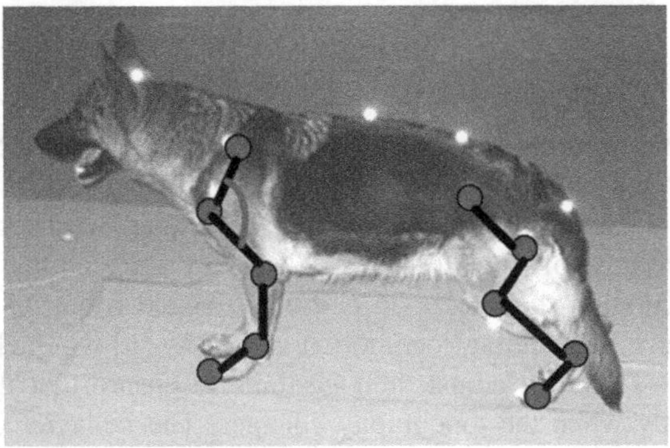

Fig. 1. GSD and markers

3 Results and Discussion

3.1 Joint Angles of Forelimb

The period of motion is normalized, and joint angular curves of forelimb over a gait cycle are shown in Fig. 2. Figure 2(a), (b), (c) and (d) are the angles of forelimb at treadmill speed of 4 km·h^{-1} and 10 km·h^{-1} when the slopes of the treadmill are 0° and 20°.

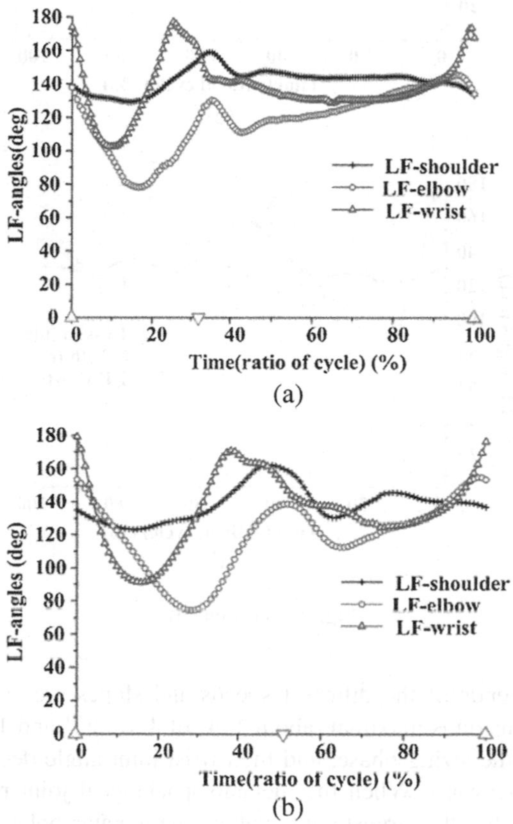

Fig. 2. Joint angles of forelimb at treadmill speed of 4 km·h^{-1} (a) and 10 km·h^{-1} (b) when the slope of the treadmill is 0°, and corresponding angles of 4 km·h^{-1} (c) and 10 km·h^{-1} (d) when the slope of the treadmill is 20°. Open uptriangle is used to mark the moment when limb detaches from the ground, and open downtriangle is the moment when limb hits the ground.

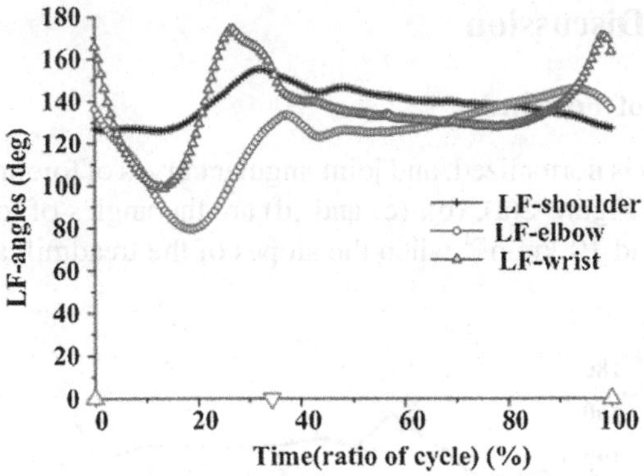

(c)

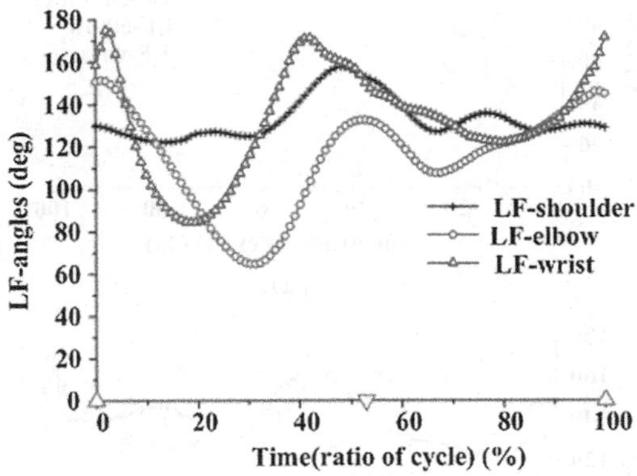

(d)

Fig. 2. (continued)

The rules and trends of the different speeds and slopes are similar. The Figures show that wrist extension is maximal (about 174° of 4 $km{\cdot}h^{-1}$ and 179° of 10 $km{\cdot}h^{-1}$) at the beginning of the swing phase, and then wrist joint angle decreases because the wrist joint flexes backward. When the metatarsophalangeal joint reaches the highest point, the wrist joint has the largest angle value. The metatarsophalangeal joint moves forwards, so that the wrist joint extends forwards, which leads to increase of the wrist joint angle. After the wrist joint extends forwards to the maximum, it flexes forward until the forelimb contact with the treadmill. At the beginning of the stance phase, the wrist joint flexes gently so that it can provide cushion when the impact happens. At the end of the stance phase, wrist joint extends for the next swing. The shoulder and elbow joints extend and flex simultaneously, the maximums decrease with the backward swing of forelimb because the shoulder and elbow joints flex backward. Then these two joints swing forwards freely, and the joint angles increase to the maximums while the stance phase begins. During the stance phase, the shoulder and elbow flex to adapt to the impact.

Figure 2 shows that the slope of the swing angle is greater in the swing phase than that in the stance phase, it indicates that the joints move faster to adjust the speed of the treadmill. When the speed increases, the time of the stance decreases. Specifically, the percentages of the stance phase in total cycle are 68.0%, 49.0%, respectively for 4 km·h^{-1}, 10 km·h^{-1} at 0° slope of the treadmill and 65.6%, 47.1% for 20° slope of the treadmill.

Table 1. Extrema and ranges of forelimb angles (°)

		0° slope		20° slope	
		4 km·h^{-1}	10 km·h^{-1}	4 km·h^{-1}	10 km·h^{-1}
Shoulder	Max.	159.0	162.2	155.1	157.7
	Min.	129.3	123.3	126.0	122.1
	Ra	29.7	38.9	29.1	35.5
Elbow	Max.	145.5	155.0	145.4	151.4
	Min.	78.2	74.5	79.6	64.5
	Ra	67.3	80.5	65.9	86.9
Wrist	Max.	177.0	179.2	174.0	174.8
	Min.	102.4	91.2	98.8	84.3
	Ra	74.6	87.9	75.2	90.5

The extrema and ranges of forelimb angles are listed in Table 1. The differences among angle ranges of forelimb at different speeds on the same inclination are larger than 6.4° (the shoulder joint between 4 km·h^{-1} and 10 km·h^{-1} the at 20° slope), and the largest difference is 21.0° (the elbow joint between 4 km·h^{-1} and 10 km·h^{-1} the at 20°slope). The change of the angle ranges on different inclination at the same speeds are less than 6.6°.

3.2 Joint Angles of Hindlimb

The joint angular curves of hindlimb over a gait cycle are shown in Fig. 3. Figure 3(a), (b), (c) and (d) are the angles of hindlimb at treadmill speed of 4 km·h^{-1} and 10 km·h^{-1} when the slopes of the treadmill are 0° and 20°.

The trends of the different speeds and slopes are also similar. Since the beginning of the stance phase, hip extends so that joint angle increases until to the maximum, and then hindlimb begins to swing, the hip flexes from this moment. At the end of the stance phase, hip extends for the next swing. Knee begins to flex gently, and there is one wave in the swing phase. During the stance phase, ankle and knee flex simultaneity, and there are two waves and one peak troughs in the ankle curves.

When the speed increases, the time of the stance decreases. Specifically, the percentages of the stance phase in total cycle are 64.9%, 44.0%, respectively for 4 km·h^{-1}, 10 km·h^{-1} at 0° slope of the treadmill and 66.3%, 43.8% for 20° slope of the treadmill.

The extrema and ranges of forelimb angles are listed in Table 2. When speed increases, the angle ranges of knee and ankle expand significantly on the same inclination. The difference among angle ranges of hip at different speeds is larger when the slope of treadmill is 0°, while that difference is inconspicuous when the slope of treadmill is 20°. The difference among angle ranges of knee on different inclination at the same speeds are less than 2.7°, and that of hip is 8.3° at 4 km·h^{-1}, 6.4° at 10 km·h^{-1}.

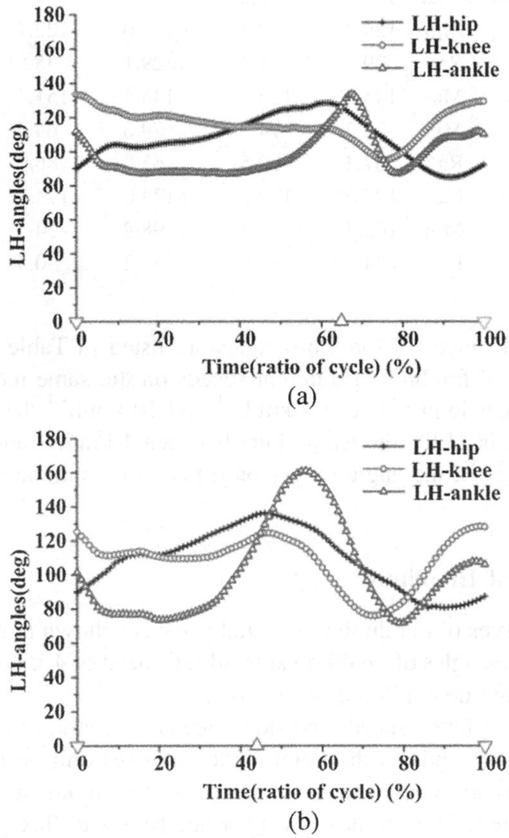

Fig. 3. Joint angles of hindlimb at treadmill speed of 4 km·h^{-1} (a) and 10 km·h^{-1} (b) when the slope of the treadmill is 0°, and corresponding angles of 4 km·h^{-1} (c) and 10 km·h^{-1} (d) when the slope of the treadmill is 20°. Open uptriangle is used to mark the moment when limb detaches from the ground, and open downtriangle is the moment when limb hits the ground.

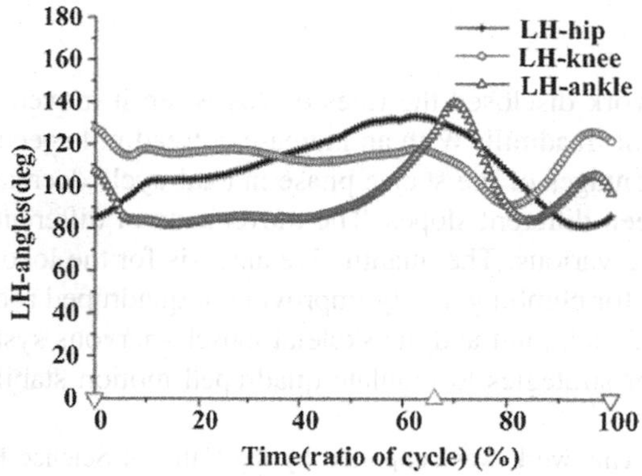

(c)

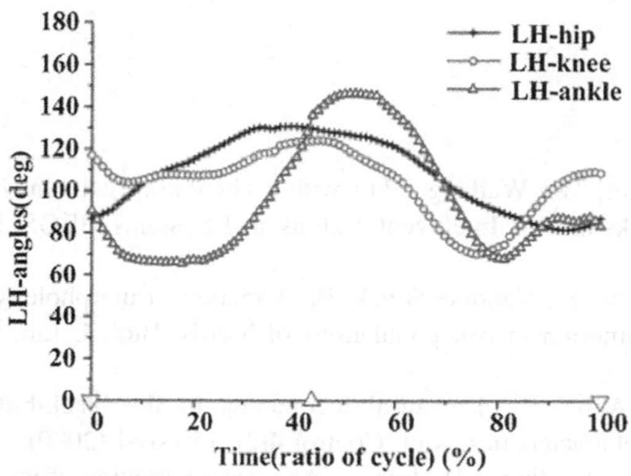

(d)

Fig. 3. (continued)

Table 2. Extrema and ranges of hindlimb angles (°)

		0° slope		20° slope	
		4 km·h^{-1}	10 km·h^{-1}	4 km·h^{-1}	10 km·h^{-1}
Hip	Max.	128.4	136.0	133.3	130.7
	Min.	84.8	80.5	81.4	81.6
	Ra	43.6	55.5	51.9	49.1
Knee	Max.	133.5	128.4	127.2	124.0
	Min.	94.8	75.8	91.2	70.2
	Ra	38.7	52.5	36.0	53.8
Ankle	Max.	133.3	160.9	139.6	146.3
	Min.	86.8	71.6	81.7	65.5
	Ra	46.5	89.2	57.9	80.9

4 Conclusions

The results of this work disclosed the rules of dog when it moved at different speeds and different slopes of treadmill. With an increase in treadmill speed from 4 km·h^{-1} to 10 km·h^{-1}, the percentages of the stance phase in total cycle decreased, and there was little variation between different slopes. The movements of different joints at different speeds and slopes are various. The quantitative analysis for the locomotion at different slopes will be useful for climbing ability improving of quadruped robot. Further studies on the mechanism of each joint and the skeletal-muscle-nervous system will help us to understand the motor strategies to regulate quadruped motion stability.

Acknowledgements. This work was supported by the National Science Foundation of China (Grant No. 51305157) and the Scientific and Technological Developing Program of Jilin Province (20130522105JH).

References

1. Bares, J., Whittaker, W.: Walking robot with a circulating gait. In: Proceedings of IEEE International Workshop on Intelligent Robots and Systems (IROS 1990), pp. 809–816, Ibaraki (1990)
2. Damme, R.V., Aerts, P., Vanhooydonck, B.: Variation in morphology, gait characteristics and speed of locomotion in two populations of lizards. Biol. J. Lin. Soc. **63**(3), 409–427 (1998)
3. Verstappen, M., Aerts, P.: Terrestrial locomotion in the black-billed magpie. I. Spatio-temporal gait characteristics. Mot. Control **4**(2), 150–164 (2000)
4. Li, H., Dai, Z., Shi, A., Zhang, H., Sun, J.: Angular observation of joints of geckos moving on horizontal and vertical surfaces. Chin. Sci. Bull. **54**(4), 592–598 (2009)
5. Kinzel, G.L., Hillberry, B.M., Hall, A.S., Van Sickle, D.C., Harvey, W.M.: Measurement of the total motion between two body segments—II description of application. J. Biomech. **5** (3), 283–293 (1972)
6. Bertram, J.E., Lee, D.V., Case, H.N., Todhunter, R.J.: Comparison of the trotting gaits of labrador retrievers and greyhounds. Am. J. Vet. Res. **61**(7), 832–838 (2000)
7. Tian, W.J., Cong, Q., Menon, C.: Investigation on walking and pacing stability of German shepherd dog for different locomotion speeds. J. Bionic Eng. **8**(1), 18–24 (2011)
8. Tian, W.J., Cong, Q., Jin, J.F.: Joint angles and ground reaction forces of German Shepherd Dog. J. Jilin Univ. (Eng. Technol. Edn.) **39**, S227–S231 (2009)
9. Qian, Z.H., Miao, H.B., Ren, L., Ren, L.Q.: Lower limb joint angles of German Shepherd Dog during foot-ground contact in different gait patterns. Jilin Daxue Xuebao **45**(6), 1857–1862 (2015)
10. Qian, Z.H., Miao, H.B., Shang, Z., et al.: Foot-ground contact analysis of German Shepherd Dog in walking, trotting and jumping gaits. Jilin Daxue Xuebao **44**(6), 1692–1697 (2014)
11. Abdelhadi, J., Wefstaedt, P., Nolte, I., Schilling, N.: Fore-aft ground force adaptations to induced forelimb lameness in walking and trotting dogs. PLoS ONE **7**(12), 5806–5819 (2012)
12. Gillette, R.L., Zebas, C.J.: A two-dimensional analysis of limb symmetry in the trot of labrador retrievers. J. Am. Anim. Hosp. Assoc. **35**(6), 515–520 (1999)

Prehension of an Anthropomorphic Metamorphic Robotic Hand Based on Opposition Space Model

Guowu Wei[1(✉)], Lei Ren[2], and Jian S. Dai[3]

[1] University of Salford, The Crescent, Salford M5 4WT, UK
g.wei@salford.ac.uk
[2] University of Manchester, Manchester M13 9PL, UK
lei.ren@manchester.ac.uk
[3] King's College London, Strand, London WC2R 2LS, UK
jian.dai@kcl.ac.uk

Abstract. This paper presents constrained prehension of an anthropomorphic metamorphic robotic hand based on opposition based model. Structure of the metamorphic hand is briefly introduced and grasp evaluation based on the opposition-space model is presented. Kinematics of the robotic hand is then established leading to the formulation of grasp constraint laying background for investigation of dexterity and manipulability of the metamorphic robotic hand.

Keywords: Metamorphic robotic hand · Anthropomophic hand · Prehension · Kinematics · Grasp constraint

1 Introduction

Since Dr. Engelberger [1] patented the first industrial robot in the early 1960s, there is a tremendous surge of activity in robotics research; and robot hand, as end-effector of robots in operation today has also attracted growing interest since the Berlichingen hand made for a knight in 1509 [2]. However, the modern robotic hand research only started since 1960s when the Belgrade hand was developed. Although most of the industrial applications prefer two or three-fingered grippers with simple design and robust grasp, teleoperation in hazardous or unstructured environments provided much of the impulsion for developing dexterous multifingered robot hands and a number of them have been presented by researchers in the past three decades, including to name but a few, the Stanford/JPL hand [3], the Utah/MIT hand [4], the DLR hand [5] and the UBH3 hand [6].

Most of the early multifingered hands have a common feature of having a rigid palm and base structure. With more functionality requiring for secure grasping and manipulation of various and complex objects, some of the recent anthropomorphic hands such as the Shadow Hand [7] and the DLR/HIT Hand II [8] and the more recently development of two anthropomorphic hands by IIT (Italian Institute of Technology) [9] and the Bionics [10] have used articulated palms of two to three movable sections to increase dexterity and manipulability of the hands.

Y. Huang et al. (Eds.): ICIRA 2017, Part I, LNAI 10462, pp. 71–83, 2017.
DOI: 10.1007/978-3-319-65289-4_7

However, in contrast to a human hand with a foldable and flexible palm, the above robotic hands lack desired versatility, they can neither provide adaptability of hand morphology nor perform the advanced in-hand manipulation.

The introduction of using the metamorphic mechanism as the palm for the metamorphic robotic three-fingered hand [11] marked a turning point and shed light on using the reconfigurable mechanism as a palm of robotic hands to improve their dexterity and manipulability. Metamorphic mechanisms [12] are a class of mechanisms that are capable of altering topological configurations from one to another with a resultant change in the mobility of mechanisms. Originated from artworks, this class of mechanisms can be extracted from origami folds that change topological structure during motion so as to performing multifunction for various tasks. For robot hand involving metamorphic mechanism, it is expected that it can demonstrate more dexterity, versatility and manipulability for grasps of different kind of conditions and environments including the unstructured environment.

Design a robotic hand that is as dexterous as human hand is always a dream of robotic researchers, based on the metamorphic mechanism, several versions metamorphic robotic hands [11,13] have been developed but grasp performance, dexterity and manipulability of these hands demand further investigation. Using the opposition space prehension models proposed by Iberall [15], this paper establishes kinematics of the anthropomorphic metamorphic robotic hand and explicitly formulates the grasp man and grasp constraint laying background for investigation and measurement of dexterity, versatility, adaptability and manipulability of the proposed metamorphic robotic hand.

2 Opposition-Space-Model Based Prehension of the Anthropomorphic Metamorphic Hand

2.1 Mechanical Structure of the Anthropomorphic Metamorphic Robotic Hand

Based on the previous development of the metamorphic three-fingered hand stemming from an origami fold as having been presented in the patent by Dai [11], a dexterous anthropomorphic metamorphic hand is developed by Wei et al. [13] as illustrated in Fig. 1. The hand consists of a reconfigurable palm and five fingers including a four-DOF thumb, and three-DOF index finger, middle finger, ring finger and litter finger. The reconfigurable palm is a spherical five-bar linkage containing five links l_1 to l_5 with the base link l_5 connected to the wrist. Two actuated joints A and E are introduced for adjusting position and orientation of the spherical reconfigurable palm. Joint A is in particular used to change the structure of the reconfigurable palm by rotating the crank link l_1, forming a four-bar linkage at a metamorphic phase. A 3-phalanx thumb of the hand is amounted at link l_2 with joint T providing the fourth degree of freedom,

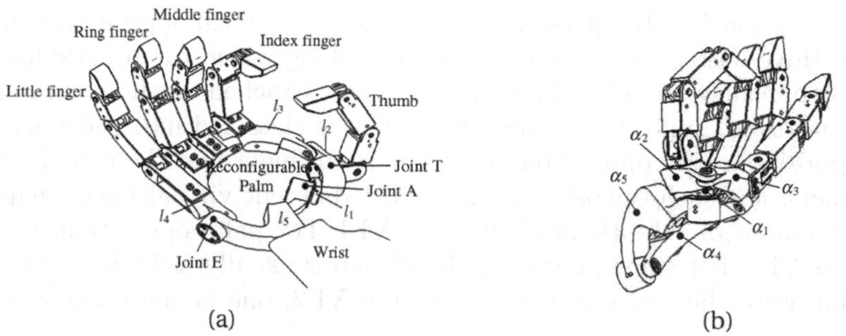

Fig. 1. (a) Mechanical structure of an anthropomorphic metamorphic hand; (b) Metamorphic hand with its palm in a reconfigured position.

a 3-phalanx index finger is mounted at link l_3, and 3-phalanx middle, ring and little fingers are amounted at link l_4. In accordance with the size of an adult's hand, the maximum radius of the spherical linkage palm is assigned as 50 mm. The angles corresponding to links l_1 to l_5 (see Fig. 1b) are α_1 to α_5 complying with $\alpha_1+\alpha_2+\alpha_3+\alpha_4+\alpha_5=2\pi$. In order to increase the dexterity of the palm, both human hand arrangement and rotatability criterion of spherical linkage [14] are considered so that the angles of links are assigned as $\alpha_1 = 25°$, $\alpha_2 = 40°$, $\alpha_3 = 70°$, $\alpha_4 = 112°$ and $\alpha_5 = 113°$. In this case, the fundamental representation of the link angles $\{\bar{\alpha}\}_{i=1}^{5} = \{25°, 40°, 67°, 68°, 70°\}$ satisfies

$$\alpha_1 + \alpha_2 + \alpha_5 = \alpha_3 + \alpha_4. \tag{1}$$

Thus, excluding the indeterminate position where all links fall on a great circle, all joints have full rotatability except for the joint between link 4 and link 5.

2.2 Opposition-Space-Model Based Prehension

Robotic hand is desired to be adaptable to perform specified grasp task and its essential function is prehension. Designed in the above structure, the reconfigurable palm of the robotic hand is foldable, flexible and operated with two degree of freedom. Its operation varies the configuration that subsequently changes the finger base position and finger orientation and subsequently the hand orientation and posture to adapt to different tasks including pinching and twisting. In order to investigate the dexterity of the metamorphic hand, in this section, opposition space model [15] is employed and prehension of the robotic hand is analysed and classified. In the opposition space model, it is proposed that each force applied for grasping is presented by a virtual finger (VF) and each VF is defined as a function unit. Fingers can be grouped into one VF to apply a force or torque opposing to forces or torques applied by other VFs. Once a robotic finger is mapped into a VF, its physical characteristics, i.e., link lengths, joint range of motion and degrees of freedom are mapped into abstract state variables that describe the VF. In opposition space model, the term opposition is used to describe three basic directions, or primitives, along which the robotic hand can apply force, relative to a hand coordinate frame attached on the palm in Fig. 2. According to the opposition, hand prehension is classified into pad opposition, palm opposition, side opposition, a third virtual finger and the combination of the formers. For each opposition, there exist different virtual finger mappings. For pad opposition, the thumb is used as VF1. For palm opposition, the palm is used as VF1. For side opposition, the thumb is usually used, however, in the adduction grip, the index finger is used. For VF2, one or more fingers can be used.

Pad Opposition. Pad opposition refers to the case that the direction of the opposing forces applied by robotic hand on the specified object parallels to the X-axis. In this case, the robotic hand is expected to exert small forces and impart fine motions to perform the prehensile actions such as pinching a pin, operating scissors and chucking chalk. For the metamorphic hand, these prehensile actions can be realized by change the configuration of the reconfigurable palm and orientation of the thumb as illustrated in Fig. 2. Figure 2c indicates that the size of the palm is reduced when the hand is used to pinch small spherical shaped objects like a small ball and this initially reflects the advantage of this metamorphic hand of a reconfigurable palm.

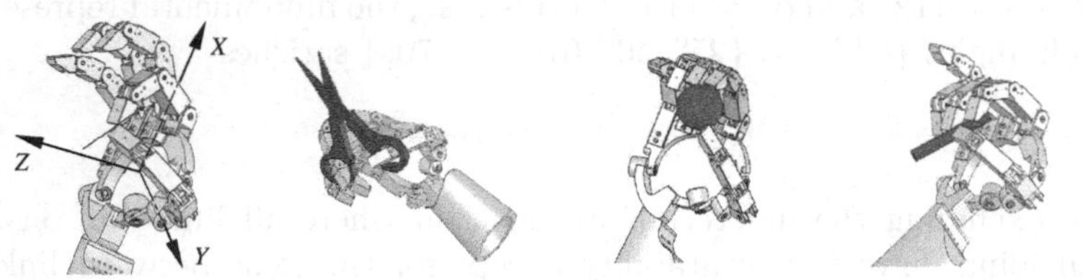

(a) Pinch a pin (b) Operate scissors (c) Pinch a small ball (d) Chuck chalk

Fig. 2. Pad opposition

Palm Opposition. Palm opposition is pertinent to the case that direction of the opposing forces applied by robotic hand on the object is generally perpendicular to the palm (along Z-axis). In this occasion, the robotic hand is expected to match and create larger forces of a stable grasp; this includes grasping big ball and griping sticks. Simulation in Fig. 3 shows that the metamorphic anthropomorphic hand can fulfil these prehensile actions which are realized by driving crank link 1 to adjust the size of the hand, driving link 4 to change orientation of the fingers and driving joint T to further adjust orientation of the thumb. The reconfigurable property of the palm is especially used in Fig. 3d when the hand is used to grasp a stick (for example bar of a hammer) where great force is demanded so that the thumb needs to wrap over the dorsum of the other digits to resist certain forces and couples by providing a powerful buttress on the lateral side.

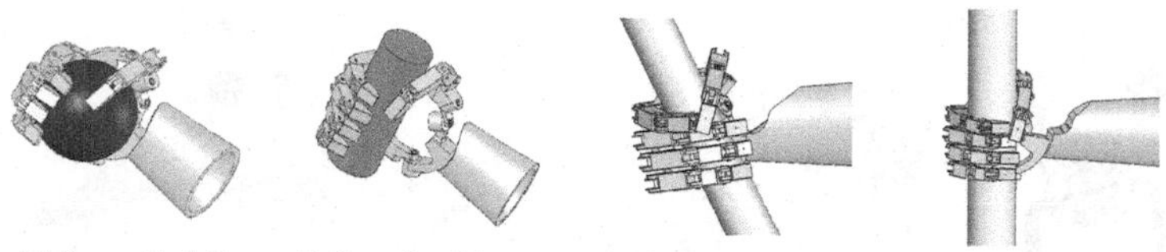

(a) Grasp a big ball (b) Open-fist stick grasp (c) Diagonal volar grasp (d) Closed-fist stick grasp

Fig. 3. Palm opposition

Side Opposition. Side opposition accords with the case that the direction of opposing forces are applied by robotic hand is generally aligned with Y-axis of the hand coordinate frame on the palm. These forces are exerted to the object by putting the thumb pad against the object in opposition to the radial side of a finger in the cases such as pinching a coin or turning a key in Fig. 4b and c, or by putting the radial sides of the index finger and middle finger on both sides of the object to realize the gripping or hold of the small, light object such as a cigarette in Fig. 4a. One can find that although there is no adducted motion between the index finger and the middle finger, this metamorphic hand can fulfil the adduction grip benefiting from the utilizing of the reconfigurable palm which contributes to change the poses of the fingers.

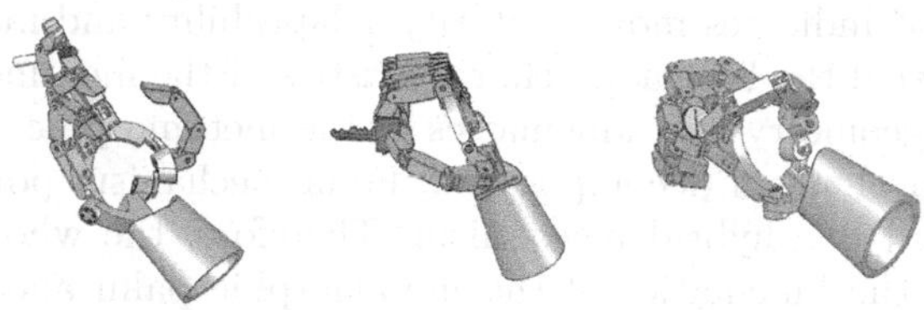

(a) Grip a cigarette (b) Turn a key (c) Pinch a coin

Fig. 4. Side opposition

A Third Virtual Finger and Combination Grasp. A third virtual finger relates to the case of gravity associated grasp in which the hand applies a force against gravity to handle a board or to grasp a hook (see Fig. 5a). This is so-called hook grip where forces from index finger, middle finger, ring finger and little finger need to be exerted on the object continuously for long periods. The metamorphic hand can fulfil this hook grip but the strength of the grip is determined by strength of the tendons which drive the fingers. Further, the versatility of the metamorphic hand, which might not be so much acknowledged, is that it can not only realize any of the aforementioned postures, but also fulfil more complex prehensile actions by combining the above grasping types. In the combined grip, the metamorphic hand can perform the prehensile actions such as holding a cup in Fig. 5b, holding a pencil in Fig. 5c, and simultaneously griping a rod, and pinching a small ball in Fig. 5d.

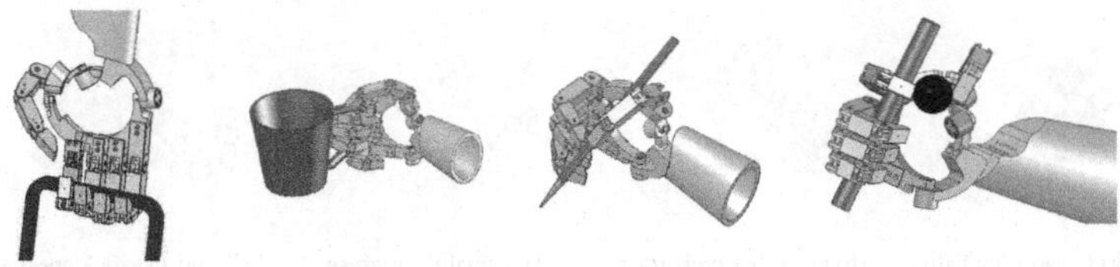

(a) Grasp a hook
(b) Hold a cup
(c) Hold a pencil
(d) A combine grip

Fig. 5. VF3 and combination grasp

3 Geometry and Kinematics of the Metamorphic Hand

The beneficial effect of the reconfigurable palm can be indicated by its ability of changing palm configuration and changing finger postures to suit various tasks. The motion of l_1 and l_4 results in the change of the palm topological structure. When joint E is fixed at a certain value, the palm operates as a spherical four-bar linkage by evolving into a one-DOF phase. This results in an instant metamorphic phase. When l_2 overlaps the base link l_1, and two links are locked, the palm evolves into a four-bar phase and becomes another one-DOF phase in Fig. 1b. This results in an innate metamorphic phase. While an instant metamorphic phase can always be achieved, the innate metamorphic phase needs to be considered in the mechanical design stage. Thus, with the reconfigurable palm, the metamorphic hand indicates more dexterity, adaptability and manipulability.

In order to reveal the kinematic characteristics of the metamorphic hand, in this section, the geometry and kinematics of the metamorphic hand is investigated based on mechanism decomposition. From mechanism point of view, the metamorphic hand is a hybrid mechanism. Therefore, the whole hand can be decomposed and the kinematics of the metamorphic palm and fingers can be separately studied and then integrated leading to the investigation of the hand kinematics. Closed-form solutions are obtained leading to the prehension and grasping study of the proposed anthropomorphic metamorphic robotic hand.

3.1 Geometric Constraint of the Reconfigurable Palm

Figure 6a gives the schematic diagram of the reconfigurable palm. This is a spherical five-bar linkage with the base link l_5 connected to the wrist. The right-hand-side of the base link connects the first input link l_1 at joint A and the left-hand-side connects the second input link l_4 at joint E. The five fingers are mounted at points F_1 to F_5 respectively. The angle between joint B and OF_1 is δ_1 and angles between joint D and OF_2, OF_3, OF_4, and OF_5 are δ_2, δ_3, δ_4 and δ_5. For various configurations of the palm, points F_1, F_2, F_3, F_4, and F_5 form various pentagons as illustrated in Fig. 6a.

In the spherical five-bar linkage, joints A and E are active joints, and joints B, C and D are passive joints. In order to derive the geometric constraints of this reconfigurable palm, coordinate frames are set up in Fig. 6a in such a way that, for all the local coordinate frames of links l_1 to l_5, they are all centred at

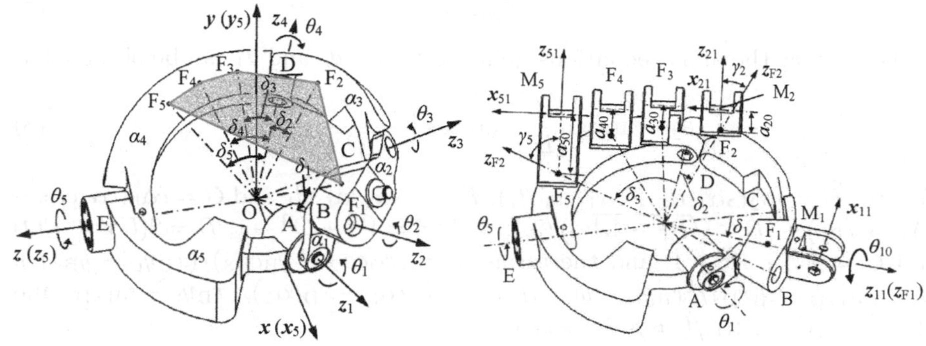

(a) (b)

Fig. 6. (a) Parameters of the reconfigurable palm; (b) Parameters of finger base.

point O with z_i-axis aligned with proximal joint of the link l_i, y_i-axis directed along $\boldsymbol{z}_i \times \boldsymbol{z}_{i+1}$ and x_i-axis determined by y_i and z_i with the right-hand rule. A global coordinate frame is set up at point O and has its z-axis aligned with joint E and its y-axis directed along $\boldsymbol{z}_5 \times \boldsymbol{z}_1$, coinciding with y_5 in Fig. 6. Based on this, given the values of angles θ_1 and θ_5, coordinates of points B, C and D can be obtained in the global coordinate frame as

$$\boldsymbol{p}_{\mathrm{B}} = \begin{bmatrix} x_{\mathrm{B}} \\ y_{\mathrm{B}} \\ z_{\mathrm{B}} \end{bmatrix} = \mathbf{R}(y, \alpha_5)\mathbf{R}(z_1, \theta_1)\mathbf{R}(y_1, \alpha_1)\boldsymbol{k} = \begin{bmatrix} c\alpha_1 s\alpha_5 + s\alpha_1 c\alpha_5 c\theta_1 \\ s\alpha_1 s\theta_1 \\ c\alpha_1 c\alpha_5 - s\alpha_1 s\alpha_5 c\theta_1 \end{bmatrix}, \quad (2)$$

$$\begin{aligned} \boldsymbol{p}_{\mathrm{C}} = \begin{bmatrix} x_{\mathrm{C}} \\ y_{\mathrm{C}} \\ z_{\mathrm{C}} \end{bmatrix} &= \mathbf{R}(z_5, \theta_5)\mathbf{R}(y_4, \alpha_4)\mathbf{R}(z_4, \theta_4)\mathbf{R}(y_3, \alpha_3)\boldsymbol{k} \quad (3) \\ &= \begin{bmatrix} c\alpha_3 s\alpha_4 c\theta_5 - s\alpha_3\,(s\theta_4 s\theta_5 - c\alpha_4 c\theta_4 c\theta_5) \\ c\alpha_3 s\alpha_4 s\theta_5 + s\alpha_3 (s\theta_4 s\theta_5 + c\alpha_4 c\theta_4 c\theta_5) \\ c\alpha_3 c\alpha_4 - s\alpha_3 s\alpha_4 c\theta_4 \end{bmatrix} \end{aligned}$$

and

$$\boldsymbol{p}_{\mathrm{D}} = \begin{bmatrix} x_{\mathrm{D}} \\ y_{\mathrm{D}} \\ z_{\mathrm{D}} \end{bmatrix} = \mathbf{R}(z_5, \theta_5)\mathbf{R}(y_4, \alpha_4)\boldsymbol{k} = \begin{bmatrix} c\alpha_4 c\theta_5 \\ s\alpha_4 s\theta_5 \\ c\alpha_4 \end{bmatrix}. \quad (4)$$

Where s and c denote the sine and cosine functions, and $\boldsymbol{k}$ is a unit vector as $\boldsymbol{k} = [0, 0, 1]^{\mathrm{T}}$.

From Fig. 6a, the geometric constraints of the spherical five-bar linkage yield,

$$\boldsymbol{p}_{\mathrm{C}}^{\mathrm{T}}\boldsymbol{p}_{\mathrm{B}} = \cos\alpha_2, \quad (5)$$

$$\boldsymbol{p}_{\mathrm{C}}^{\mathrm{T}}\boldsymbol{p}_{\mathrm{D}} = \cos\alpha_3, \quad (6)$$

$$\boldsymbol{p}_{\mathrm{C}}^{\mathrm{T}}\boldsymbol{p}_{\mathrm{C}} = 1. \tag{7}$$

By solving the about equations, joint angles θ_2, θ_3 and θ_4 can be obtained as

$$\theta_2 = \arctan\frac{F}{E} \pm \arccos\left(\frac{G}{\sqrt{E^2+F^2}}\right) \tag{8}$$

Where, $E = s\alpha_2(s\alpha_1 c\alpha_5 + c\alpha_1 s\alpha_5 c\theta_1)$, $F = -s\alpha_2 s\alpha_5 s\theta_1$ and $G = c\alpha_5(c\alpha_1 c\alpha_5 - s\alpha_1 s\alpha_5 c\theta_1) + (B \mp \sqrt{B^2 - 4AC})/2A$, with $A = V^2 + Q^2 + 1$, $B = 2(UV + PQ)$ and $C = U^2 + P^2 - 1$, and the terms $U = (y_{\mathrm{D}} c\alpha_2 - y_{\mathrm{B}} c\alpha_3)/(x_{\mathrm{B}} y_{\mathrm{D}} - y_{\mathrm{B}} x_{\mathrm{D}})$, $V = (y_{\mathrm{B}} z_{\mathrm{D}} - z_{\mathrm{B}} y_{\mathrm{D}})/(x_{\mathrm{B}} y_{\mathrm{D}} - y_{\mathrm{B}} x_{\mathrm{D}})$, $P = (x_{\mathrm{B}} c\alpha_3 - x_{\mathrm{D}} c\alpha_2)/(x_{\mathrm{B}} y_{\mathrm{D}} - y_{\mathrm{B}} x_{\mathrm{D}})$ and $Q = (z_{\mathrm{B}} x_{\mathrm{D}} - x_{\mathrm{B}} z_{\mathrm{D}})/(x_{\mathrm{B}} y_{\mathrm{D}} - y_{\mathrm{B}} x_{\mathrm{D}})$.

$$\theta_4 = \arccos(\cot\alpha_3 \cot\alpha_4 - z_{\mathrm{C}}/s\alpha_3 s\alpha_4) \tag{9}$$

with $z_{\mathrm{C}} = \frac{-B \pm \sqrt{B^2-4AC}}{2A}$, and

$$\theta_3 = \arctan\frac{F'}{E'} \pm \arccos\left(\frac{G'}{\sqrt{E'^2+F'^2}}\right). \tag{10}$$

Where $E' = s\alpha_3(c\alpha_1 s\alpha_2 + s\alpha_1 c\alpha_2 c\theta_2)$, $F' = -s\alpha_3 s\alpha_1 s\theta_2$ and $G' = c\alpha_3(c\alpha_1 c\alpha_2 - s\alpha_1 s\alpha_2 c\theta_2) - c\alpha_4 c\alpha_5 + s\alpha_4 s\alpha_5 c\theta_5$.

The above gives the motion characteristics of the articulated palm.

3.2 Palm Integrated Whole Hand Kinematics

Fingers of the robotic hand are connected to the above reconfigurable palm through finger bases attached at points F_1, F_2, F_3, F_4 and F_5 as in Fig. 6b. In order to relate palm motion to finger motion, local coordinate frames F_i-$x_{\mathrm{F}i}y_{\mathrm{F}i}z_{\mathrm{F}i}$ are set up at points F_1 to F_5 with $z_{\mathrm{F}i}$-axis directed along OF_i, $y_{\mathrm{F}1}$ directed along $\boldsymbol{z}_{\mathrm{F}1} \times \boldsymbol{z}_3$, $y_{\mathrm{F}2}$ directed along $\boldsymbol{z}_{\mathrm{F}2} \times \boldsymbol{z}_4$, and $y_{\mathrm{F}i} (i = 3, 4, 5)$ directed along $\boldsymbol{z}_{\mathrm{F}i} \times \boldsymbol{z}_5$. Local coordinate frames M_i-$x_{i1}y_{i1}z_{i1}$ $(i = 1, 2, \ldots, 5)$ of the MCP joints of the fingers are set up with x_{i1}-axis aligned with the ith MCP joint and z_{i1}-axis directed along $\mathrm{F}_i\mathrm{M}_i$. The angle between $z_{\mathrm{F}i}$ and z_{i1} is γ_i and the distance between F_i and M_i is a_{i0} as in Fig. 6b. It should be pointed out herein that γ_1 equals 0.

From the above analysis, the coordinate transformations from the finger base coordinate frames to the global coordinate frames can be obtained as

$$\mathbf{R}_{\mathrm{F}i} = \begin{cases} \mathbf{R}(y, \alpha_5)\mathbf{R}(z_1, \theta_1)\mathbf{R}(y_1, \alpha_1)\mathbf{R}(z_2, \theta_2)\mathbf{R}(y_2, \delta_1) & \text{if } i = 1 \\ \mathbf{R}(z, \theta_5)\mathbf{R}(y_4, \alpha_4)\mathbf{R}(z_4, \theta_4)\mathbf{R}(y_3, \delta_2) & \text{if } i = 2 \\ \mathbf{R}(z, \theta_5)\mathbf{R}(y_4, \alpha_4 - \delta_i) & \text{if } i = 3, 4, 5 \end{cases}. \tag{11}$$

Thus, the homogeneous transformation matrix from the finger base coordinate frame to the global coordinate frame can be derived as

$$\mathbf{D}_{\mathrm{F}i} = \begin{bmatrix} \mathbf{R}_{\mathrm{F}i} & \mathbf{R}_{\mathrm{F}i}\boldsymbol{k}' \\ 0 & 1 \end{bmatrix} (i = 1, 2, \ldots, 5), \tag{12}$$

where $\boldsymbol{k} = [0, 0, \mathrm{R}]^{\mathrm{T}}$ and $\mathbf{R}_{\mathrm{F}i}\boldsymbol{k}'$ gives the position vector of point F_i in the global coordinate frame. R is the radius of the sphere on which all the links move.

Then the homogeneous transformation from coordinate frames of the MCP joints to the global coordinate frame can be given according to Fig. 6b as

$$\mathbf{D}_{\mathrm{M}i} = \begin{cases} \mathbf{D}_{\mathrm{F}i}\mathbf{D}_{\mathrm{FM}i}\mathbf{D}_{10} & (\mathrm{i} = 1) \\ \mathbf{D}_{\mathrm{F}i}\mathbf{D}_{\mathrm{FM}i} & (\mathrm{i} = 2, 3, \ldots, 5) \end{cases}, \tag{13}$$

Where, $\mathbf{D}_{\mathrm{FM}i} = \begin{bmatrix} c\gamma_i & 0 & -s\gamma_i & -a_{i0}s\gamma_i \\ 0 & 1 & 0 & 0 \\ s\gamma_i & 0 & c\gamma_i & a_{i0}c\gamma_i \\ 0 & 0 & 0 & 1 \end{bmatrix}$ denotes the transformation from coordinate frame M_i-$x_{i1}y_{i1}z_{i1}$ to coordinate frame F_i-$x_{\mathrm{F}i}y_{\mathrm{F}i}z_{\mathrm{F}i}$. It should be noted that for the thumb finger base, it has γ_1 equals 0. And $\mathbf{D}_{10}$ presents the additional degree of freedom for the thumb as $\mathbf{D}_{10} = \begin{bmatrix} \mathbf{R}(z_{\mathrm{F}1}, \theta_{10}) & 0 \\ 0\ 0\ 0 & 1 \end{bmatrix}$ with $\mathbf{R}(z_{\mathrm{F}1}, \theta_{10})$ denoting the rotation matrix about $z_{\mathrm{F}1}$ of θ_{10}.

Eventually, given the coordinates of the fingertips in the finger base coordinate frames as $\mathbf{T}_{ft} = e^{[\mathbf{s}_{i1}]\theta_{i1}}e^{[\mathbf{s}_{i2}]\theta_{i2}}e^{[\mathbf{s}_{i3}]\theta_{i3}}\mathbf{M}$, with $\mathbf{M} = (\mathbf{I}, \boldsymbol{p}_{bt})$ and $\boldsymbol{p}_{bt} = [a_{i1} + a_{i2} + a_{i3}, 0, 0]^T$, where a_{i1}, a_{i2}, and a_{i3} are the lengths of the three phalanxes of the ith finger, the coordinate of the fingertips can be expressed in the global coordinate frame as

$$\mathbf{T}_{f_i} = \mathbf{D}_{M_i}\mathbf{T}_{ft_i}, \tag{14}$$

From Eq. (14), kinematics of the metamorphic hand can be obtained and as shown in [13]. With this kinematic analysis, grasp and prehension of the proposed metamorphic hand presented in Sect. 2.2 can be formulated and investigated.

4 Constraint Representation of Grasp and Prehension

4.1 The Grasp Map and Grasp Constraint

In [16], it is stated that to grasp an object with n fingers contacting it, the grasp map with respect to the object coordinate frame can be given as

$$\mathbf{G} = \left[\mathbf{Ad}^T_{g^{-1}_{oc_1}}\mathbf{B}_{c_1} \quad \cdots \quad \mathbf{Ad}^T_{g^{-1}_{oc_n}}\mathbf{B}_{c_n}\right], \tag{15}$$

with $\mathbf{B}_{ci}$ indicating the wrench basis corresponding to different contact type. And grasp constraint of a multifingered hand grasping have the form

$$\mathbf{J}_h(\theta, x_o)\,\dot{\theta} = \mathbf{G}^T\boldsymbol{V}^b_{po}, \tag{16}$$

where matrix $\mathbf{J}_h(\theta, x_o)$ denotes the hand Jacobian as

$$\mathbf{J}_h(\theta, x_o) = \begin{bmatrix} \mathbf{B}^T_{c_1}\mathbf{Ad}^{-1}_{g_{s_1c_1}}\mathbf{J}^s_{s_1f_1}(\theta_{f_1}) & & 0 \\ & \ddots & \\ 0 & & \mathbf{B}^T_{c_n}\mathbf{Ad}^{-1}_{g_{s_nc_n}}\mathbf{J}^s_{s_nf_n}(\theta_{f_n}) \end{bmatrix},$$

with $\dot{\theta} = (\dot{\theta}_{f_1}, \cdots, \dot{\theta}_{f_n})$ giving the joint velocities, and $\boldsymbol{V}^b_{po}$ is the body velocity of the object expressed in the global coordinate frame.

4.2 Grasp and Prehension Constraint

Considering the opposition-space-model based prehension discussed in Sect. 2.2, and assuming that the contacts type between the fingertips and the objects are of point contact with friction, the grasp and prehension constraints of the metamorphic hand are formulated and presented in this section.

Taking the pinch a small ball in pad opposition indicated in Fig. 2c as an example. Assuming that the thumb, the index finger and the middle finger are involved in this grasp, and the tips of the three fingers are in contact with the ball. Without loss of generality, it is assumed that the contact points between the three fingers and the ball lying on the same plane which passes through the centre of the ball. The object coordinate frame $P\text{-}x_o y_o z_o$ is set up at the centre of the ball with its x_o-axis and z_o-axis lie on the same plane which is formed by the three contact points. Contact frames $C_i\text{-}x_{c_i} y_{c_i} z_{c_i}$, $i = 1$, 2 and 3 are located at the contact points with z_{c_i}-axis directing towards P and y_{c_i}-axis parallel to y_o-axis. Then, the position of the contact frame with respect to the object frame can be given as

$$\mathbf{R}_{pc_i} = \begin{bmatrix} c\varphi_i & 0 & -s\varphi_i \\ 0 & 1 & 0 \\ s\varphi_i & 0 & c\varphi_i \end{bmatrix} \text{ and } \boldsymbol{p}_{pc_i} = \begin{bmatrix} r_1 s\varphi_i \\ 0 \\ r_1 c\varphi_i \end{bmatrix}. \tag{17}$$

Where φ_i denotes the angle between x_{ci}-axis and x_o-axis with $i = 1$, 2 and 3, and r_1 is the radius of the ball.

The grasp map for each finger can then be obtained by transforming the standard wrench basis of point contact with friction into the object coordinate frame as

$$\mathbf{G}_i = \begin{bmatrix} \mathbf{R}_{pc_i} & 0 \\ \widehat{\boldsymbol{p}}_{pc_i}\mathbf{R}_{pc_i} & \mathbf{R}_{pc_i} \end{bmatrix} \mathbf{B}_{c_i} = \begin{bmatrix} c\varphi_i & 0 & -s\varphi_i \\ 0 & 1 & 0 \\ s\varphi_i & 0 & c\varphi_i \\ 0 & -r_1 c\varphi_i & 0 \\ r_1 c(2\varphi_i) & 0 & -r_1 s(2\varphi_i) \\ 0 & r_1 s\varphi_i & 0 \end{bmatrix}, \tag{18}$$

where, the wrench basis for each finger is $\mathbf{B}_{c_i} = \begin{bmatrix} \mathbf{I}_{3\times 3} \\ \mathbf{0}_{3\times 3} \end{bmatrix}$, and $i = 1$, 2 and 3.

Substituting Eq. (18) into Eq. (15), the grasp map for the desired prehension can be obtained as $\mathbf{G} = \begin{bmatrix} \mathbf{G}_1 & \mathbf{G}_2 & \mathbf{G}_3 \end{bmatrix}_{6\times 9}$. In the above equations, the parameter φ_i can be determined in the object coordinate frame once the grasp is formed and the contact points are identified. The force-closure property of this grasp can then be detected by convexity conditions.

Further, referring to Fig. 6, each of the fingers is an serial RRR chain, its Jacobian with respect to the finger base coordinate frame has the form

$$\mathbf{J}^s_{f_it_i} = \begin{bmatrix} 0 & a_{i1}s\theta_{i1} & a_{i1}s\theta_{i1} + a_{i2}s(\theta_{i1}+\theta_{i2}) \\ 0 & -a_{i1}c\theta_{i1} & -a_{i1}c\theta_{i1} - a_{i2}c(\theta_{i1}+\theta_{i2}) \\ 0 & 0 & 0 \\ 0 & 0 & 0 \\ 0 & 0 & 0 \\ 1 & 1 & 1 \end{bmatrix}. \tag{19}$$

Where, a_{i1}, a_{i2}, and a_{i3} are the lengths of the three phalanxes of the ith finger, and θ_{i1}, θ_{i2}, and θ_{i3} are the joint angles of the three joints in the ith finger.

According to Eq. (16), in order to formulate the grasp constraint for the grasp of a ball using the metamorphic hand, $\mathbf{Ad}^{-1}_{g_{f_ic_i}}$ needs to be constructed. Based on the geometric relations of coordinate frames established in Fig. 6, there exists

$$\mathbf{Ad}^{-1}_{g_{f_ic_i}} = \mathbf{Ad}^{-1}_{g_{pc_i}} \mathbf{Ad}^{-1}_{g_{op}} \mathbf{Ad}^{-1}_{g_{f_io}}. \tag{20}$$

In the object coordinate frame, the first term on the right-hand side of Eq. (20) can be written as

$$\mathbf{Ad}^{-1}_{g_{pc_i}} = \begin{bmatrix} \mathbf{R}^T_{pc_i} & -\mathbf{R}^T_{pc_i}\widehat{\boldsymbol{p}}_{pc_i} \\ \mathbf{0} & \mathbf{R}^T_{pc_i} \end{bmatrix}. \tag{21}$$

Substituting Eq. (17) into Eq. (21), it yields

$$\mathbf{Ad}^{-1}_{g_{pc_i}} = \begin{bmatrix} c\varphi_i & 0 & s\varphi_i & 0 & r_1c(2\varphi_i) & 0 \\ 0 & 1 & 0 & -r_1c\varphi_i & 0 & r_1s\varphi_i \\ -s\varphi_i & 0 & c\varphi_i & 0 & -r_1s(2\varphi_i) & 0 \\ 0 & 0 & 0 & c\varphi_i & 0 & s\varphi_i \\ 0 & 0 & 0 & 0 & 1 & 0 \\ 0 & 0 & 0 & -s\varphi_i & 0 & c\varphi_i \end{bmatrix}. \tag{22}$$

Assume that with respect to the global coordinate frame $O - xyz$ set in the metamorphic hand, the orientation and position of the object coordinate frame, which can be measured through real-time vision system, are given as $\mathbf{R}_{op}$ and $\boldsymbol{p}_{op}$, the second term on the right-hand side of Eq. (20) can be obtained as

$$\mathbf{Ad}^{-1}_{g_{op}} = \begin{bmatrix} \mathbf{R}^T_{op} & -\mathbf{R}^T_{op}\widehat{\boldsymbol{p}}_{op} \\ \mathbf{0} & \mathbf{R}^T_{op} \end{bmatrix}. \tag{23}$$

Finally, substitute terms $\mathbf{R}_{of_i}$ and $\boldsymbol{p}_{of_i}$ in Eq. (14) into the following equation

$$\mathbf{Ad}^{-1}_{g_{f_io}} = \mathbf{Ad}_{g_{of_i}} = \begin{bmatrix} \mathbf{R}_{of_i} & \widehat{\boldsymbol{p}}_{of_i}\mathbf{R}_{of_i} \\ \mathbf{0} & \mathbf{R}_{of_i} \end{bmatrix}, \tag{24}$$

the third term on the right-hand side of Eq. (20) can be obtained.

Substituting Eqs. (19)–(24) into Eq. (16) gives the hand Jacobian for the pad opposition prehension illustrated in Fig. 2c as

$$\mathbf{J}_h = \begin{bmatrix} \mathbf{J}_{11} & \mathbf{0} & \mathbf{0} \\ \mathbf{0} & \mathbf{J}_{22} & \mathbf{0} \\ \mathbf{0} & \mathbf{0} & \mathbf{J}_{33} \end{bmatrix}_{9\times 9}, \tag{25}$$

where, for $i = 1$, 2 and 3 the 3×3 sub-matrix $\mathbf{J}_{ii}$ is $\mathbf{J}_{ii} = \mathbf{B}_{c_i}^T \mathbf{Ad}_{g_{f_i c_i}}^{-1} \mathbf{J}_{f_i t_i}^s$. This Jacobian matrix can be readily obtained through symbolic computer programme system such as Matlab^{TM} once all the essential variables are given.

The grasp map and the grasp constraint derived above can be used to evaluate the properties of grasps performed by the metamorphic hand. The grasp map helps to identify whether a grasp performed by the metamorphic hand is force-closure grasp and the grasp constraint helps to determine whether a grasp is manipulable. If for any object motion $\boldsymbol{V}_{op}^b$ there exists $\dot{\theta}$ which satisfies Eq. (16), the grasp executed by the metamorphic hand is supposed to be manipulable.

Using the above formulation, all the prehension cases presented in Sect. 2.2 can be formulated and grasping quality can be measured through the use of convex hull.

5 Conclusions

Prehension of an anthropomorphic metamorphic robotic hand was investigated in this paper based on the principle of opposition space model. Structure of the robotic hand was introduced and grasp evaluation and functionality of the hand was presented using opposition space model. Kinematics of the proposed robotic hand was then established, and grasp map and grasp constraint of the hand were formulated providing explicit mathematical model for representing the opposition space model based prehension. The study presented in this paper hence has laid background for dexterity and manipulability investigation of metamorphic robotic hand and provided insights into grasping measurement for robotic hand development.

References

1. Nof, S.Y.: Handbook of Industrial Robot. Wiley, New York (1985)
2. Childress, D.S.: Artificial hand mechanisms. In: Mechanisms Conference and International Symposium on Gearing and Transmissions (1972)
3. Salisbury, J.K., Craig, J.J.: Articulated hands: force control and kinematic issues. Int. J. Robot. Res. **1**(1), 4–17 (1982)
4. Jacobasen, S.C., Iversen, E.K., Knutti, D.F., Johnson, R.T., Biggers, K.B.: Design of the Utah/M.I.T. dexterous hand. In: IEEE International Conference on Robotics and Automation, pp. 1520–1532 (1986)
5. Tomovic, R., Berkey, G.A., Karplus, W.J.: A strategy for grasp synthesis with multifingered robot hand. In: IEEE International Conference on Robotics and Automation, pp. 83–89 (1987)

6. Lotti, F., Tiezzi, P., Vassura, G.: UBH3: investigating alternative design concepts for robotic hands. In: IEEE International Conference on Robotics Automation, pp. 135–140 (2004)
7. Aminzadeh, V., Walker, R., Cupcic, U., Elias, H., Dai, J.S.: Friction Compensation and Control Strategy for the Dexterous Robotic Hands, Chap. 62, pp. 697–706. Springer, Heidelberg (2013)
8. Liu, H., Wu, K., Meusel, P., Hirzinger, G., Jin, M., Liu, Y., Fan, S., Lan, T., Chen, Z.: A dexterous humanoid five-fingered robotic hand. In: IEEE International Symposium on Robot and Human Interactive Communication, Munich, Germany, pp. 371–376 (2008)
9. NewScientist (2010). http://www.newscientist.com/article/mg20627566.800-robots-with-skin-enter-our-touchyfeely-world.html
10. Fischman, J.: Bionics: National Geographic, pp. 35–53, January 2010
11. Dai, J.S.: Robotic hand with palm section comprising several parts able to move relative to each other (2004). (Patent WO/2005/105391, 10 November 2005; International Patent PCT/GB2005/001665, UK Patent GB04 095 48.5)
12. Dai, J.S., Jones, J.R.: Mobility in metamorphic mechanism of foldable/erectable kinds. ASME Trans. J. Mech. Des. **121**, 375–382 (1999)
13. Wei, G., Dai, J.S., Wang, S., Luo, H.: Kinematic analysis and prototype of a metamorphic anthropomorphic hand with a reconfigurable palm. Int. J. Humanoid Robotics **8**(3), 459–479 (2011)
14. Liu, Y., Ting, K.: On the rotatability of spherical N-bar chains. ASME Trans. J. Mech. Des. **116**(5), 920–923 (1994)
15. Iberall, T.: Human prehension and dexterous robot hands. Int. J. Robot. Res. **16**(3), 285–299 (1997)
16. Murray, R.M., Li, Z., Sastry, S.S.: A Mathematical Introduction to Robotic Manipulation. CRC Press, Boca Raton (1994)

Multi-directional Characterization for Pollen Tubes Based on a Nanorobotic Manipulation System

Wenfeng Wan[1], Yang Liu[1], Haojian Lu[1], and Yajing Shen[1,2](✉)

[1] City University of Hong Kong, No. 83, Tat Chee Ave, Kowloon, Hong Kong SAR
yajishen@cityu.edu.hk

[2] City University of Hong Kong Shenzhen Research Institute, Shenzhen 518000, China

Abstract. Pollen tubes' main function is to transport gametes to ovules. Mechanical properties of pollen tubes affect their growth and penetration. Most existing systems for characterizing pollen tubes can only characterize pollen tubes from one direction. However, considering pollen tubes' nonuniform properties, results got from one fixed direction don't necessarily represent pollen tubes' overall properties. In order to characterize pollen tubes from multi-direction instead of one direction, a nanorobotic system is proposed herein. The system contains two robots, robot 1 for sample assembly and robot 2 for sensor assembly. Robot 1's rotation degree enables pollen tubes to be characterized from multi-direction. During experiments, the pollen tube is bent at different angles from 0° to 360°. Bending forces at different angle are quite different. The results demonstrate that pollen tubes are inhomogeneous along circumferential direction and justify the necessity to characterize pollen tubes from multi-direction. Experiment results can be used to measure pollen tubes' stiffness at different direction and analyze how pollen tubes penetrate through pistil.

Keywords: Pollen tubes · Bend · Nanorobotic system · Mechanical characterization · Multi-direction · Nonuniform properties

1 Introduction

Almost all our food ultimately come from plants. During the growth of many plants, such as corns and fruits, pollen tubes are an indispensable part [1]. Pollen tubes are part of male gametophyte and are responsible for transporting gamete cells to ovules of the pistil. Besides, pollen tubes are an ideal model for studying plant cells' properties [2, 3]. They are easy to culture and are able to keep polarity and developmental identity in vitro. They are among the fastest growing cell (up to 1 cm/h) and this property facilitates live imaging. Until now, pollen tubes have been widely studied with respect to their electrical, mechanical, chemical, biological and other properties [4–6]. These studies are exerting positive effect on various fields, including plant biology, biomechanics, intra- and intercellular signaling.

Mechanical stimulus has been shown to largely affect properties of pollen tubes, such as deformation, elongation, penetration, transfer of sperm cell and finally fertilization [2]. Pollen tubes can respond to external mechanical force by adjusting intracellular

Y. Huang et al. (Eds.): ICIRA 2017, Part I, LNAI 10462, pp. 84–93, 2017.
DOI: 10.1007/978-3-319-65289-4_8

signals. Many systems have been designed to measure cells' mechanical properties. Traditionally, pressure probe and cell compression have been applied to measure volumetric Young's modulus and fracture force [7, 8]. However, the cell's weakest part in both methods will be first to rupture. Measuring results only represent properties in the weakest spot, and are not applicable for cells with anisotropic properties. Micro/nano-indentation systems have allowed characterization of pollen tubes with a resolution of micrometer and nanometer [9]. For example, a real-time cellular force microscope is designed to measure stiffness of pollen tubes and analyze the effect of mechanical intervention on calcium fluxes [6]. Nevertheless, this manipulation system doesn't have the rotation degree. This drawback leads to some pollen tubes' areas, for instance, side and below areas, are impossible to measure.

Microfluidic systems mainly made of polydimethylsiloxane (PDMS) have been widely adopted to conduct various research on pollen tubes, including bending, compression and investigation of elongation growth [10–12]. Microfluidics can isolate, locate and provide growth medium for individual pollen tubes. They can also miniaturize functional component, and transparent PDMS facilitate observation and image analysis. For instance, Nezhad AS et al. use a microfluidic system to bend pollen tubes by fluidic flow, and then elastic modulus is estimated according to the bending geometry by finite element analysis [13, 14]. Although this group are the first to bend pollen tubes, there are some drawbacks in their study. First, elastic modulus is not achieved by direct measurement but by finite element modeling. During modeling, some inaccurate assumptions have been made, including isotropic cell wall, cytoplasm being modelled as a solid and pollen tube being modelled as a rigid body. The modelling results are unstable and highly dependent on cell wall thickness. This may be caused by those inaccurate assumptions. Also, the experiment setup can only allow bending of pollen tubes from one fixed direction, instead of multi-direction. However, plant cells' mechanical properties are highly anisotropic because of various types of molecules and linkages. This indicates that measurement results from one fixed direction in the paper don't necessarily represent pollen tubes' overall properties.

Because of nanorobotic systems' high resolution and high controllability, they have been adopted for various characterization and manipulation tasks, such as measuring micro/nano material's properties [15], assembling [16] and cell characterization [17]. A nanorobotic system with rotation degree can rotate samples and therefore enables samples to be imaged from multi-direction [18, 19]. For example, a rotation robot has been used to observe defects on magnetic microwires so that defects on all the surfaces can be studied instead of just on partial surfaces [20].

A nanorobotic manipulation system has been designed in this paper in order to bend pollen tubes from multi-direction. This system consists of two components: robot 1 and robot 2. Pollen tubes are assembled on robot 1. Robot 1 has three degrees of freedom: one rotation degree and two translational degrees. The rotation degree enables pollen tubes to rotate and thus to be bent from different direction. The two translational degrees are to move pollen tubes to robot 1's rotation axis so that during rotation, pollen tubes' locations won't drift two much. Force sensors are assembled on robot 2, which has three degrees of freedom: three mutually perpendicular translational movements. These three degrees are aimed to adjust pollen tubes' bending location and bending amplitude by controlling force sensor's location. The experiment results indicate that bending force of a same amplitude at different direction vary as much as four times. This large difference is in accordance with pollen tubes' anisotropic properties. The results justify the necessity to characterize pollen tubes from multi-direction as well as the proposed system's wide applications in characterizing and manipulating anisotropic bio-specimen and other materials.

2 Nanorobotic Manipulation System

The nanorobotic manipulation system has two robotic components (Fig. 1). Robot 1 is for sample assembly, and robot 2 is for sensor assembly. Robot 1 consists of two linear nanopositioners (ECS3030, Attocube Inc., Germany) and one rotatory nanopositioner (ECR3030, Attocube Inc., Germany). Robot 2 consists of three linear nanopositioners (ECS3030, Attocube Inc., Germany). Figure 2 shows robot 1's horizontally linear movement, vertically linear movement, rotary movement and three movements together. Robot 2's three linear movements are mutually vertical.

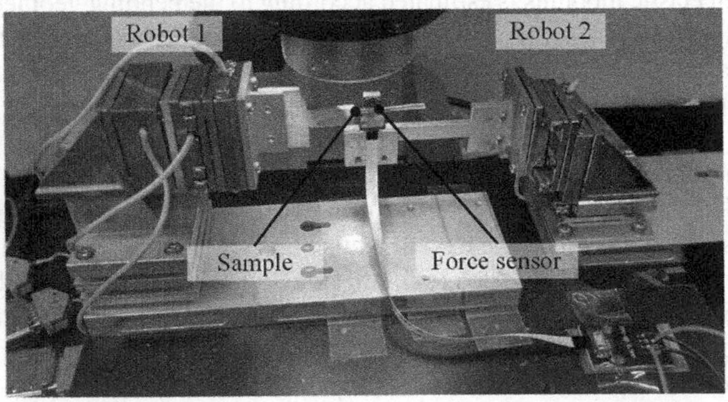

Fig. 1. The nanorobotic manipulation system. It consists of two parts: robot 1 for sample assembly and robot 2 for force sensor assembly

The actuator of linear and rotary nanopositioners is made of piezoelectrical ceramics. Nanopositioners' movement mechanism is slip-stick mechanism. These nanopositioners have built-in optical encoders for closed loop positioning. The resolution and travel range of linear nanopositioners are 1 nm and 20 mm, respectively, and the resolution and travel range of rotary nanopositioners are 1 u° and 360° endlessly.

One important aspect for rotatable robot 1 should be pointed out. When to-be-measured areas of pollen tubes are not on rotary nanopositioner's rotation axis, pollen tubes would have large position shifts during rotation. In this case, it becomes complex and time-consuming to adjust positions of pollen tubes and force sensors after pollen tubes' each rotation. In order to address this issue, an automatic alignment method has been proposed to locate pollen tubes to robot 1's rotation axis. The detailed alignment method and alignment results can be found in [18, 19].

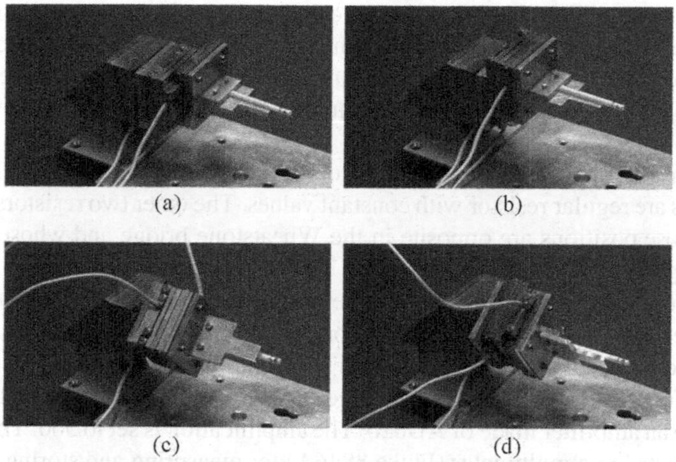

Fig. 2. Demonstration of robot 1's movements: (a) horizontally linear movement, (b) vertically linear movement, (c) rotation movement and (d) three movements together

3 Experimental Setup

The experimental setup is shown in Fig. 3. The six nanopositioners of the nanorobotic manipulation system are driven by two controllers (ECC 100, Attocube Inc., Germany), each controller for three nanopositioners.

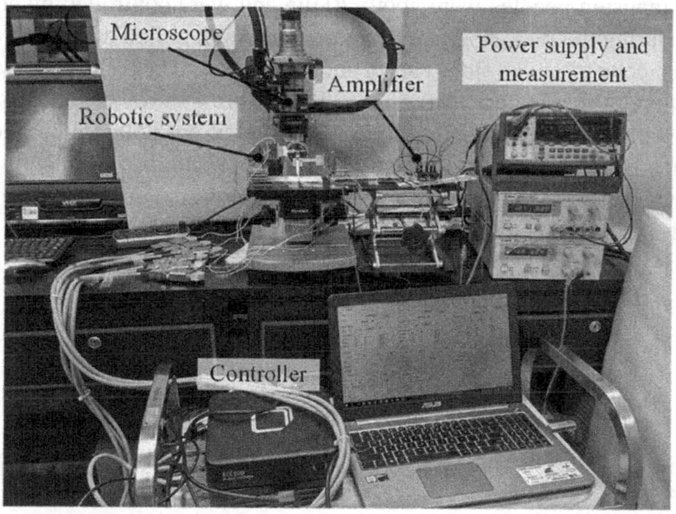

Fig. 3. Experimental setup for multi-direction characterization of pollen tubes

The force sensor mainly contains a silicon cantilever (PRSA-L300-F50-STD, SCL-Sensor.Tech., Austria). A piezo-resistor is integrated at the bottom of the cantilever to fulfill self-sensing, which eliminates the optical measurement part in conventional AFM systems. This makes the measurement system cheap, simple and compact. The cantilever is of rectangular shape with a length of 305 μm and a width of 110 μm. The cantilever's tip has a height of 5 μm and a diameter smaller than 15 nm. The cantilever's force sensitivity is 5 nN/μV.

At the base of the cantilever, four resistors form a Wheatstone bridge. Two of the four resistors are regular resistor with constant values. The other two resistors are piezo-resistors whose positions are opposite in the Wheatstone bridge and whose resistance would change as the cantilever deforms. Two pots of the Wheatstone bridge are connected to a voltage supplier. The voltage of the other two pots is the output. When the cantilever deforms, the two piezo-resistors' resistance would change, and therefore the output voltage would also change. This is the cantilever's mechanism to measure force, transforming force signals to voltage signals. The output of the cantilever is connected to an amplifier made of AD620. The amplification is set to 500. The amplifier is then connected to a multimeter (Fluke 8846A) for measuring and storing data.

Pollen grains were collected from hippeastrum striatums bought from Flower Market Road, Hong Kong. Pollen grains were dehydrated with a fan for 12 h and then were stored in a refrigerator at −20°. Before being cultured, pollen grains were rehydrated in a humid closed container. The culture medium contains 10% sucrose, 5 mM MES, 5 mM KNO_3, 0.13 mM $Ca(NO_3)_2$ and 0.16 mM H_3BO_3. The pH value of the culture medium was adjusted to be 5.5 by KOH. Pollen grains were cultured for 10 h before being picked up by tungsten needles. During culture, pollen tubes would grow out of pollen grains.

Tungsten needles were used to pick pollen tubes out the culture medium and were fabricated by chemical etching. The solution for chemical etching was 6 M NaOH. The diameters of tungsten needles were about 20 μm and were controlled by etching time and current. During the pick, 500 μl culture medium with pollen tubes in it were transported with a pipette to a glass side. Then a pollen tube would be picked out by a tungsten needle from the culture medium on the glass side. Figure 4 shows the process of picking pollen tubes with a tungsten needle. It is easier to pick out pollen tubes from culture medium on the glass slide than to directly pick out pollen tubes from culture medium on petri dishes, because the more water there is, the stronger the adhesion force is between pollen tubes and culture medium.

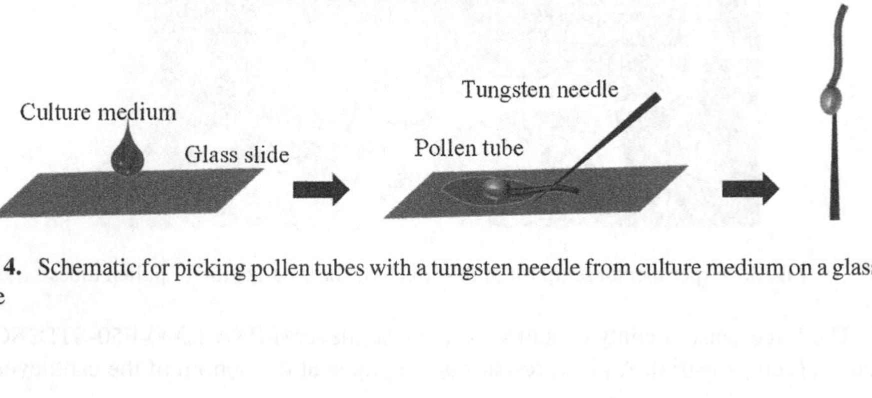

Fig. 4. Schematic for picking pollen tubes with a tungsten needle from culture medium on a glass slide

4 Results

Figure 5 shows pollen grains in culture medium before culture and after 10 h culture. 10 h is chosen as the culture period so that pollen tubes are long enough for picking. Also, the culture time shouldn't be too long, since in that case, pollen tubes would burst because of a lack of nutrition and space. Pollen tubes' average diameter is about 18 μm.

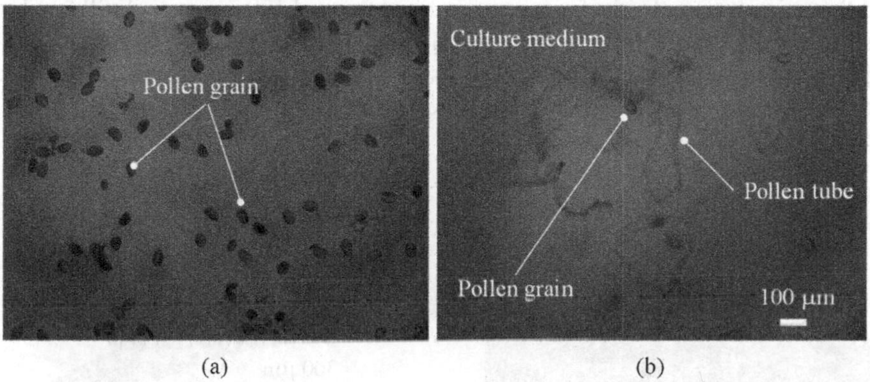

Fig. 5. Pollen grains (a) before culture and (b) after 10 h culture. Pollen tubes would grow out of pollen grains after culture

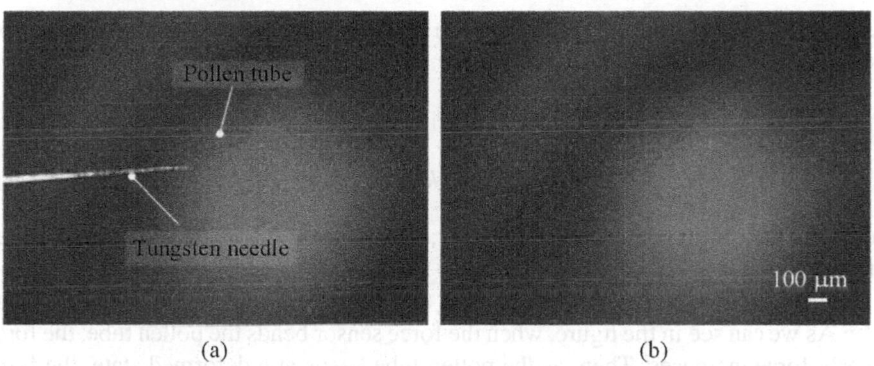

Fig. 6. Pollen tubes after being picked out by tungsten needles

Figure 6 shows pollen tubes after they are picked up by a tungsten needle. During the picking up process, the angle formed by tungsten needles and pollen tubes should be 180° as approximately as possible. This relative position between them would facilitate following characterization experiments. As we can see in Figs. 5(b) and 6, pollen tubes in water environment are relatively plump because of being filled with water. After being pick out and exposed to air environment, pollen tubes become concave because of losing water. It should be pointed out that although pollen tubes in water environment are relatively plump, there contours are still nonuniform along both axis direction and angle direction.

During the bending experiment, the bending location was 350 μm away from the connection between the tungsten needle and the pollen tube. The bending magnitude was 100 μm. Figure 7 shows the schematic of the bending experiment. The bending has been conducted at four different angles: 0°, 90°, 180° and 270°. The pollen tube was first bent at rotation angle 0° and bent, and then the pollen tube was rotated to 90°, 180° and 270° sequentially, and bent at those three angles. At each angle, the pollen tube was bent three times. Namely, the force sensor moved forward 100 μm, stayed still for about 20 s, moved backward 100 μm, waited until the pollen tube changed back to its initial shape. After the pollen tube recovered to its initial shape after de-formation, the force sensor moved forward 100 μm again and then repeated the above procedure for totally three times.

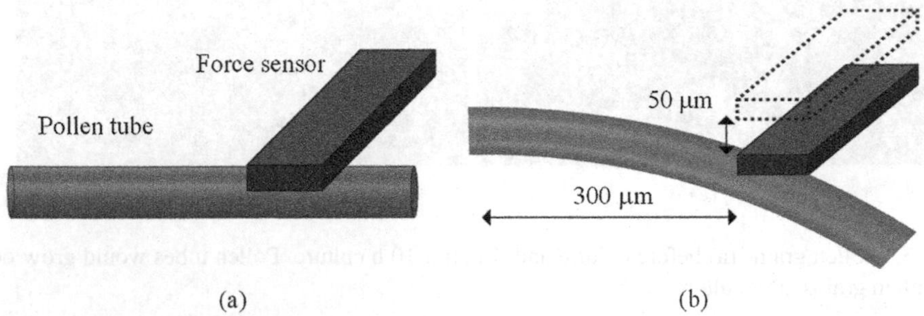

Fig. 7. Schematic of the bending experiment: (a) before the force sensor bending the pollen tube, (b) after the force sensor bending the pollen tube

The bending force is given in the following equation,

$$F = s \cdot v/k \tag{1}$$

where s is the cantilever's force sensitivity; k is the magnification and v is measured voltage.

Figure 8 shows the recorded force over the whole bending procedure for different angle. As we can see in the figure, when the force sensor bends the pollen tube, the force sensor's force increases. Then, as the pollen tube keeps at a deformed state, the force decreased gradually, which means that the pollen tube encounters slight yield. After the force sensor moves backward, the force decreases sharply, and at the same time, the pollen tube is recovering back to its initial shape.

After we got the recorded force for different angle over the whole bending procedure, we calculated the average bending force. With Matlab, we first detected the force's sharp increase, then calculated the average value for the 15 data before the sharp increase and the average value for the 15 data after the sharp increase. The difference between the two average values was the bending force. Because the pollen tube was bent three times at each angle, we would get three such bending forces, the average of which was defined as the bending force.

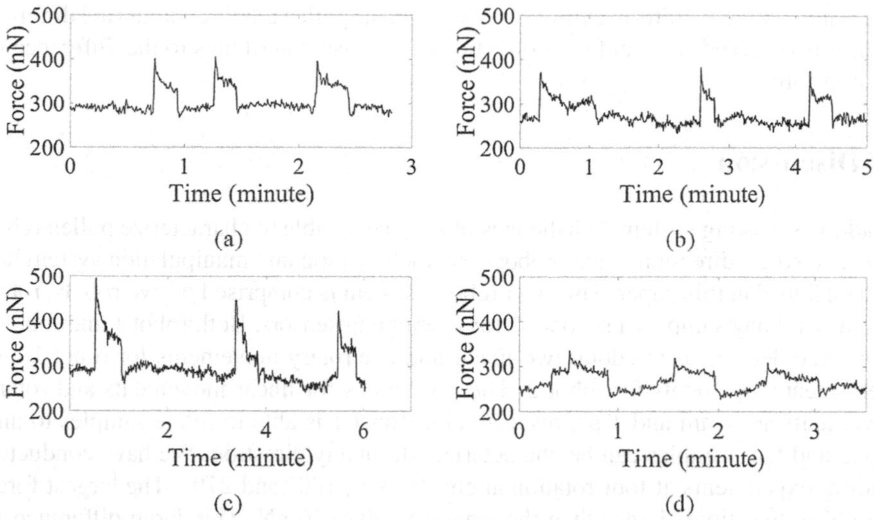

Fig. 8. Bending force over time at different angles: (a) 0°, (b) 90°, (c) 180° and (d) 270°

As shown in Fig. 9, bending forces at angle 0°, 90°, 180° and 270° are 69 nN, 76 nN, 120 nN and 26 nN, respectively. The standard deviations are 4 nN, 5 nN, 30 nN and 6 nN, respectively. The largest value 120 nN is four times larger than the smallest value, 26 nN, and is about two times as larger as the other two values. The large difference of bending force at different angle proves that pollen tubes' properties are nonuniform from different direction. The bending experiment results justify the necessity to characterize pollen tubes from multi-direction.

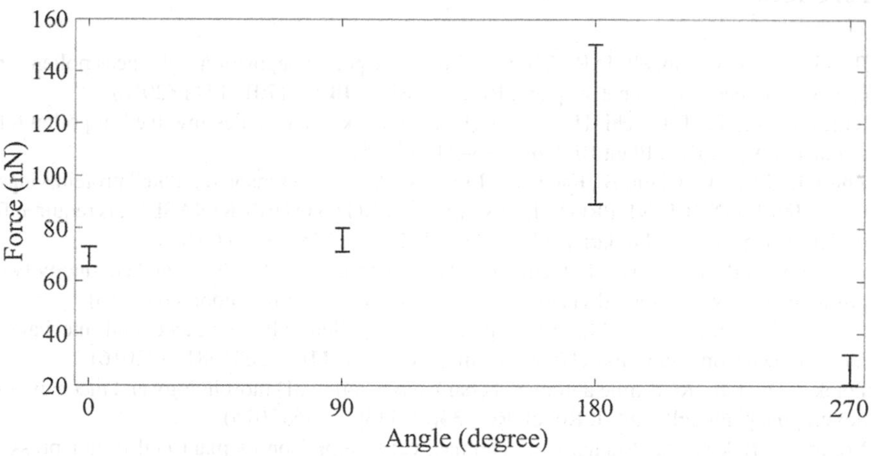

Fig. 9. Bending force for the pollen tube at different direction

This inhomogeneous property is caused by pollen tubes' complex molecular linkages and various types of polysaccharides. Also, as shown in Fig. 6, the picked-up pollen tube is not a straight uniform cylinder. As a result, the pollen tube's contours are different when it is observed from different directions. This also contributes to the difference in bending force.

5 Discussion

To address existing systems' challenges of only being able to characterize pollen tubes from one single direction, a nanorobotic characterization and manipulation system has been designed in this paper. This nanorobotic system is comprised of two robots, robot 1 for assembling samples and robot 2 for assembling sensors. Both robot 1 and robot 2 have three degrees of freedom, two linear and one rotary movements for robot 1 and three linear movements for robot 2. The resolutions for linear movements and rotary movements are 1 nm and 1 μ°, respectively. Robot 1 is able to rotate samples to any angles and thus samples can be characterized from any direction. We have conducted bending experiments at four rotation angles 0°, 90°, 180° and 270°. The largest force 120 nN is four times larger than the smallest value, 26 nN. This large difference in bending force at different direction proves pollen tubes' nonuniformity and the nanorobotic system's effectiveness in characterizing and manipulating samples from multi-direction. In the future, we would compute pollen tubes' elastic modules and flexural rigidity according to the bending force. We would also apply our nanorobotic system to conduct other characterization and manipulation for pollen tubes or other samples.

Acknowledgement. This work is practically supported by Shenzhen Basic Research Project (JCYJ20160329150236426), and GRF of Hong Kong (CityU 21201314).

References

1. Bruckman, D., Campbell, D.R.: Timing of invasive pollen deposition influences pollen tube growth and seed set in a native plant. Biol. Invasions **18**(6), 1701–1711 (2016)
2. Higashiyama, T., Takeuchi, H.: The mechanism and key molecules involved in pollen tube guidance. Annu. Rev. Plant Biol. **66**, 393–413 (2015)
3. Zhou, L., Lan, W., Chen, B., Fang, W., Luan, S.: A calcium sensor-regulated protein kinase, CALCINEURIN B-LIKE PROTEIN-INTERACTING PROTEIN KINASE19, is required for pollen tube growth and polarity. Plant Physiol. **167**(4), 1351–1360 (2015)
4. Agudelo, C., Packirisamy, M., Geitmann, A.: Influence of electric fields and conductivity on pollen tube growth assessed via electrical lab-on-chip. Scientific reports 6 (2016)
5. Shamsudhin, N., et al.: Massively parallelized pollen tube guidance and mechanical measurements on a lab-on-a-chip platform. PLoS ONE **11**(12), e0168138 (2016)
6. Felekis, D., et al.: Real-time automated characterization of 3D morphology and mechanics of developing plant cells. Int. J. Robot. Res. **34**(8), 1136–1146 (2015)
7. Wang, L., Hukin, D., Pritchard, J., Thomas, C.: Comparison of plant cell turgor pressure measurement by pressure probe and micromanipulation. Biotech. Lett. **28**(15), 1147–1150 (2006)
8. Tomos, A.D., Leigh, R.A.: The pressure probe: a versatile tool in plant cell physiology. Annu. Rev. Plant Biol. **50**(1), 447–472 (1999)

9. Felekis, D., Muntwyler, S., Vogler, H., Beyeler, F., Grossniklaus, U., Nelson, B.J.: Quantifying growth mechanics of living, growing plant cells in situ using microbotics. IET Micro Nano Lett. **6**(5), 311–316 (2011)
10. Shamsudhin, N., et al.: Probing the micromechanics of the fastest growing plant cell_the pollen tube. In: 2016 38th Annual International Conference of the IEEE Engineering in Medicine and Biology Society (EMBC), pp. 461–464 (2016)
11. Nezhad, A.S.: Microfluidic platforms for plant cells studies. Lab Chip **14**(17), 3262–3274 (2014)
12. Horade, M., Kanaoka, M.M., Kuzuya, M., Higashiyama, T., Kaji, N.: A microfluidic device for quantitative analysis of chemo attraction in plants. Rsc Adv. **3**(44), 22301–22307 (2013)
13. Nezhad, A.S., Naghavi, M., Packirisamy, M., Bhat, R., Geitmann, A.: Quantification of the Young's modulus of the primary plant cell wall using Bending-Lab-On-Chip (BLOC). Lab Chip **13**(13), 2599–2608 (2013)
14. Agudelo, C.G., Sanati Nezhad, A., Ghanbari, M., Naghavi, M., Packirisamy, M., Geitmann, A.: TipChip: a modular, MEMS-based platform for experimentation and phenotyping of tip-growing cells. Plant J. **73**(6), 1057–1068 (2013)
15. Zhou, C., et al.: A closed-loop controlled nanomanipulation system for probing nanostructures inside scanning electron microscopes. IEEE/ASME Trans. Mechatron. **21**(3), 1233–1241 (2016)
16. Zimmermann, S., Tiemerding, T., Fatikow, S.: Automated robotic manipulation of individual colloidal particles using vision-based control. IEEE/ASME Trans. Mechatron. **20**(5), 2031–2038 (2015)
17. Yajing, S., Masahiro, N., Zhan, Y., Seiji, K., Michio, H., Toshio, F.: Design and characterization of nanoknife with buffering beam for in situ single-cell cutting. Nanotechnology **22**(30), 305701 (2011)
18. Shen, Y., Wan, W., Lu, H., Fukuda, T., Shang, W.: Automatic sample alignment under microscopy for 360° imaging based on the nanorobotic manipulation system. IEEE Trans. Robot. **33**(1), 220–226 (2016)
19. Shen, Y., Wan, W., Zhang, L., Yong, L., Lu, H., Ding, W.: Multidirectional image sensing for microscopy based on a rotatable robot. Sensors **15**(12), 31566–31580 (2015)
20. Wan, W., Lu, H., Zhukova, V., Ipatov, M., Zhukov, A., Shen, Y.: Surface defect detection of magnetic microwires by miniature rotatable robot inside SEM. AIP Adv. **6**(9), 095309 (2016)

A Review on Vibration Characteristics of Carbon Nanotubes and Its Application Via Vacuum

Dongliang Huang, Zhan Yang(✉), and Lining Sun

Jiangsu Provincial Key Laboratory of Advanced Robotics,
Collaborative Innovation Center of Suzhou Nano Science and Technology,
Soochow University, Suzhou 215123, China
yangzhan@suda.edu.cn

Abstract. With the progress of nanotechnology and the development of micro electro mechanical system which makes it possible to manufacture high precision micro and nano level vibration based sensors. The purpose of this paper is to provide a brief overview of study on the study on vibration characteristics of carbon nanotubes such as resonance frequency and quality factors which have a great influence on the application of the CNTs. We also discussed the application of the Vibration of Carbon Nanotubes. We conclude with a look at the future of the Study on Vibration Characteristics of Carbon Nanotubes and more application of the CNT resonators.

Keywords: Vibration characteristics · Carbon nanotubes · Application · Vacuum

1 Introduction

Since their discovery in 1991 by Ijima, more and more scientists were interested in CNTs because of their particular molecular structures, their unique electronic, mechanical properties [1, 2] and the great application prospect in many fields [3, 4]. The young's modulus of carbon nanotubes is about 1 Tpa, and the shear modulus is up to 0.5 TPa [5], and the tensile strength is as high as 30 GPa [6]. In addition, due to the vibration characteristics of the carbon nanotubes is excellent [7, 8], More and more researchers are focusing on the use of carbon nanotubes as mechanical resonators to fabricate nano electromechanical systems [9]. The high frequency vibration characteristics of carbon nanotubes (CNTs) have wide application prospects in the fields of emitter [10], oscillator [11], sensor and charge detector [12]. The physical properties and application characteristics of carbon nanotubes are closely related to the vibration characteristics of carbon nanotubes [13]. The quality factor is very important as one of the vibration characteristics of the carbon nanotubes [14], which has a great influence on the accuracy and resolution of the vibration device. High quality resonance system, which provides high frequency resolution and energy storage time, plays an important role in many fields of Physics. The combination of high Q value and high resonance frequency is an important prerequisite for applications such as Nano-Gyroscope and single atom mass sensing [15] and so on. So, more and more researchers begin to study the vibration characteristics of carbon nanotubes and the methods to improve them and on this basis, the application of

Y. Huang et al. (Eds.): ICIRA 2017, Part I, LNAI 10462, pp. 94–102, 2017.
DOI: 10.1007/978-3-319-65289-4_9

vibration based on carbon nanotubes was studied. In this paper, we illustrated the present study by enumerating several typical studies and concluded with a look at the future of the Study on Vibration Characteristics of Carbon Nanotubes and more application of the CNT resonators.

2 Review on Vibration Characteristics of Carbon Nanotubes

The Philippe Poncharal team of Georgia Institute of Technology designed and implemented the vibration test of the cantilever beam carbon nanotubes under the electric field excitation [16]. The experiment was carried out under transmission electron microscope. Application of a time (t)-dependent voltage to the nanotubes $\left[V(t) = V_d cos(\omega t)\right]$ caused a time-dependent force and dynamic deflections which was used to excite the CNT vibrate resonantly. In this experiment, the first order resonance and the two order resonance of a carbon nanotube with a length of 6.25 μm and a diameter of 14.5 nm were obtained by changing the frequency of the excitation voltage (Fig. 1).

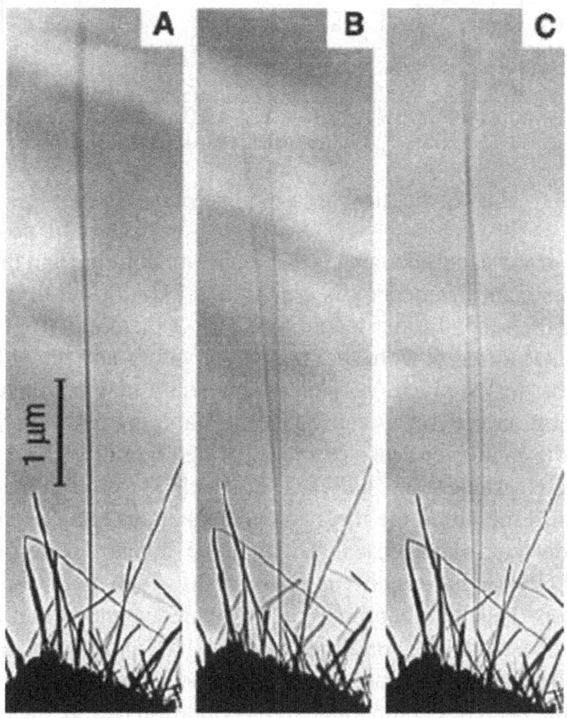

Fig. 1. (A) In the absence of a potential, the nanotube tip (L = 6.25 μm, D514.5 nm) vibrated slightly because of thermal effects. (B) Resonant excitation of the fundamental mode of vibration (v_1 = 530 kHz); (C) Resonant excitation of the second harmonic (v_2 = 3.01 MHz) [16].

In the experiment the authors also found that the quality factor and elastic modulus of carbon nanotubes vibration, when the elastic modulus of carbon nanotubes is 0.098 TPa, quality factor Q = 170 experimental observations obtained; when the elastic modulus increases to 0.73 TPa, the quality factor of carbon nanotubes was up to 500 Q. The elastic modulus is greatly affected by the diameter of the carbon nanotubes, and the elastic modulus is inversely proportional to the diameter, which means that the larger the diameter of the carbon tube is, the smaller the elastic modulus is (Fig. 2).

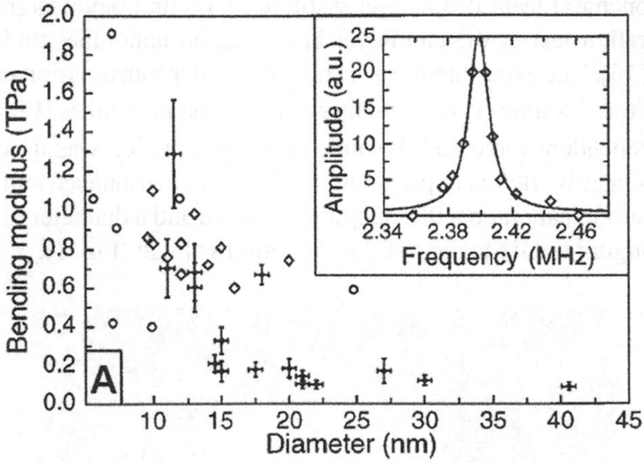

Fig. 2. Elastic properties of nanotubes [16].

In this paper, Carbon nanotubes are cantilever beam vibration and the length is about 6.25 μm and the resonance frequency is around 530 kHz for the first order resonance and 3.01 MHz for the second order resonance. The Q factors is from 170 to 500. All these values are quite different from the theoretical values and the Q factors were too low to meet the practical application. This phenomenon may be related to the energy loss during the vibration which was caused by the Damping in the environment.

In addition to the cantilever beam vibration, bridge vibration fixed at both ends has also been researched by some teams. Edward, A. et al. [17] from Technische Universiteit Delft, Holland, Laird measured the mechanical resonances of an as-grown suspended carbon nanotube, detected via electrical mixing in the device. The resonance frequency of the device is 39 GHz and the quality is up to 35000 which meets the requirements of practical application. This high combination of frequency and Q factor was attributed to a thermal excited state which is probability below 10^{-8} and a relaxation time of 140 s which is associated with a lower mode of microsecond relaxation time. They found that as the tunneling time becomes comparable to the vibration period, the effect of the electron tunneling on the mechanical resonance depends on the frequency (Fig. 3).

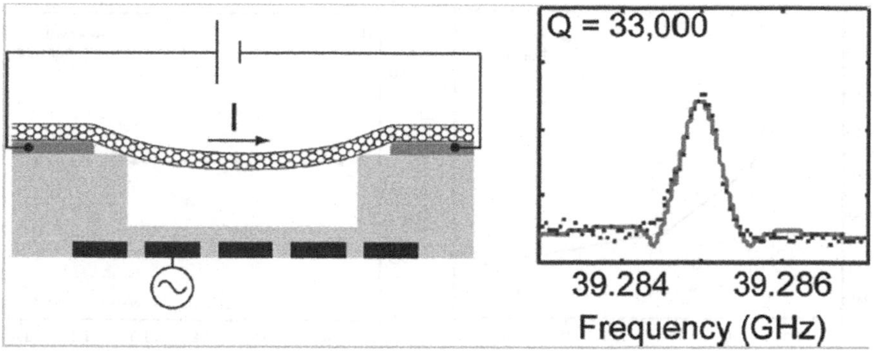

Fig. 3. Schematic of the device and measurement circuit and vibration characteristic curve [17].

The energy loss which was caused by the surface damage can be reconciled in carbon nanotubes where surface damage is eliminated by growing the nanotube in the final fabrication step. Electrical measurements were performed in a dilution refrigerator at a mixing chamber temperature of ~100 mK with a bias $V_{sd}^{DC} = 2$ mV applied across the device.

In order to obtain different vibration characteristics under different applications, Study on vibration characteristics of carbon nanotubes by various factors should be conducted. Ajit K. Vallabhaneni et al.'s works [18] examines the quality factor (Q factor) of the resonance with the axial and transverse vibration of single-walled carbon nanotubes (SWCNT) resonators by using molecular dynamics simulation (MD). The effects of the device on the device length, diameter, and chirality, as well as the temperature on the resonant frequency and quality factor of these devices are investigated. The quality factor (Q) associated with transverse vibration is found to increase with increasing device length ($Q \sim L^h$, where $0.8 < h < 1.4$) and decrease with increasing device diameter ($Q \sim D^{\mu}$ where $1.4 < \mu < 1.6$), while the Q associated with axial vibration is almost independent of length and diameter. For both vibrational modes, Q shows a temperature dependence $Q \sim T^{-\alpha}$. The relationship between the resonance frequency and the length and diameter of carbon nanotubes has been showed in Fig. 4. The Fig. 4 showed that the resonance frequency decreases with the increase of length of the CNTs and the increase of diameter has little effect on the resonance.

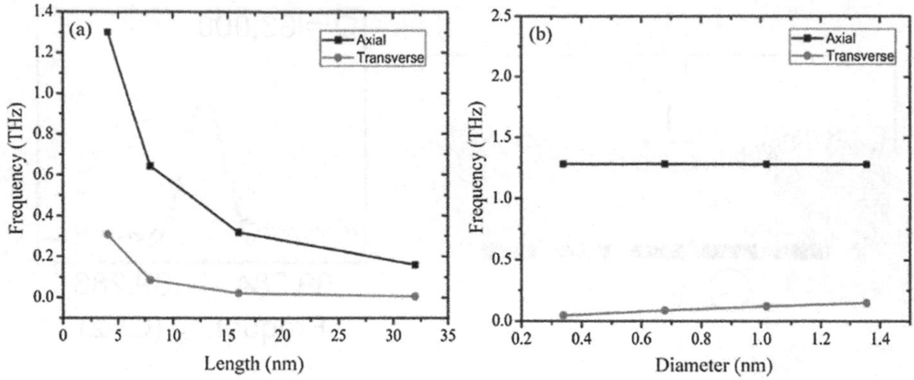

Fig. 4. Variation of resonant frequency with respect to size at 10 K. (a) resonant frequency as a function of length for a CNT and (b) resonant frequency as a function of diameter [18].

Figure 5 showed the effect of temperature, length and diameter on quality factor. From this fig, we can see that when other factors are constant, the quality factor decreased with the increase of temperature and increased with increase of the length of the CNTs and decreased with the increase on the diameter of the CNTs. In this paper, the effects of ambient temperature, the length and diameter of carbon nanotubes on the resonant frequency and quality factor of carbon nanotubes were investigated by molecular dynamics simulation and the feasibility needs to be verified by experiments.

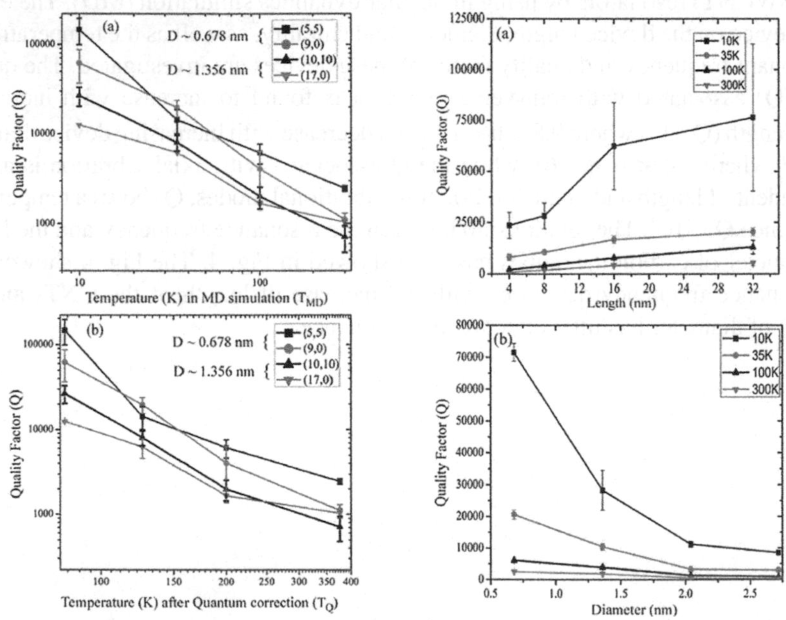

Fig. 5. Quality factors plotted as function of temperature for transverse vibrations and variation of quality factor for transverse vibrations with size [18].

3 Review on the Applications of Carbon Nanotube Vibration

3.1 Nanotube Radio

Zettl team built a full-featured, fully integrated radio receivers, orders of magnitude smaller than any previous radio, from a single carbon nanotube [19]. The operation of carbon nanotubes is quite different from the traditional radio. Although the traditional radio is essentially electrical, but the function of the carbon nanotube radio is at least partly mechanical. Briefly, the electromagnetic wave transmitted from incoming radio shocks to the nanotube forcing its physical vibration through its action on the charged tip. The vibration is significant, when the frequency of the incident wave coincides with the nanotube bending can be tuned in the operation process, therefore, nanotube radio, like any good radio, can be tuned to receive electromagnetic spectrum only preselected frequency band. Already, this explains at least superficially how the nanotube serves at both antenna and tuner of radio (Fig. 6).

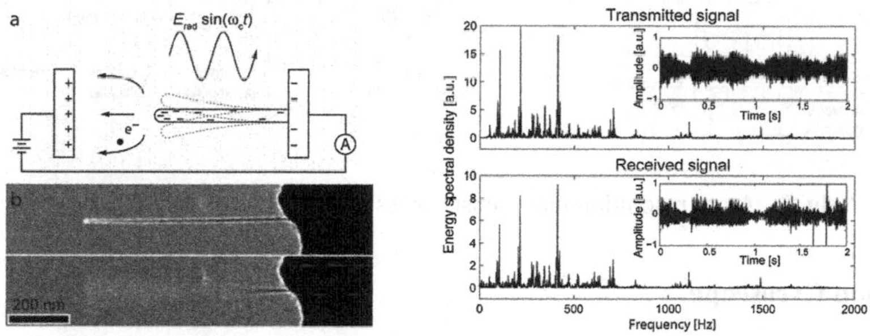

Fig. 6. Nanotube radio and its function [19].

3.2 Mass Sensor

A. Zettl team constructed a 10^{-25} g level mass sensor. Using a carbon nanotube-based nanomechanical resonator, an atomic-resolution mass sensor have been constructed. atoms or molecules fall on the resonator, they cause a change in the mechanical resonance frequency, which can be inferred from the mass of the absorbed particles. We were able to measure the quality of a single gold atom by using this technique [20]. This device is essentially a mass spectrometer with a mass sensitivity of 1.3×10^{-25}kg Hz$^{-1/2}$ or, 0.40 gold atoms Hz$^{-1/2}$. Using this extreme mass sensitivity, we observe atomic mass shot noise, which is analogous to the electronic shot noise measured in many semiconductor experiments. Unlike conventional mass spectrometers, the nano mass spectrometer does not need to test samples for potentially destructive ionization, is more sensitive to macromolecules and may eventually be incorporated into a chip (Fig. 7).

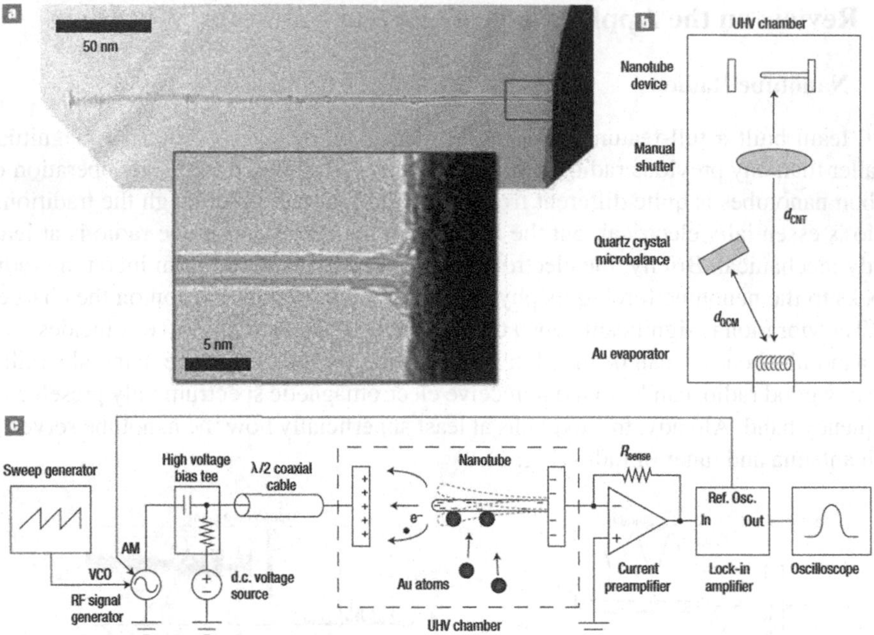

Fig. 7. Atom resolution mass sensor mass sensor and its schematic [20].

3.3 Nano Gyroscope

A gyroscope is a device that measures angular velocity or Angle. MEMS gyroscope is now widely used in various occasion. However, MEMS gyroscope can not meet the requirements of higher accuracy and smaller size. So, nanoscale gyroscope with vibratory carbon nanotubes (CNT) has been presented. This gyroscope has higher productivity, higher resolution, lower energy cost, larger measurement scales and smaller size. Zhan Yang et al. had presented a gyroscope based on Coriolis effect and fabricated by Carbon Nanotube (CNT) [21]. This gyroscope was designed based on Coriolis effects. A CNT is vibration at X direction, when there a rotation around z direction, the acceleration will happen to the y direction and CNT will vibrate in y direction. Field emission was induced to detect the motion in y direction to calculate the rotation speed. A CNT which is 1 μm in length was induced as the resonator of the gyroscope. A rotation speed of 0.1 rad/s, 1 rad/s, 10 rad/s, 20 rad/s, 100 rad/s, 1000 rad/s and 10000 rad/s were simulated. The gyroscope is good sensitive till 100 rad/s. One of the great advantages of the nano gyroscope is that its size is reduced to a level of microns or less, making it more useful and promising (Fig. 8).

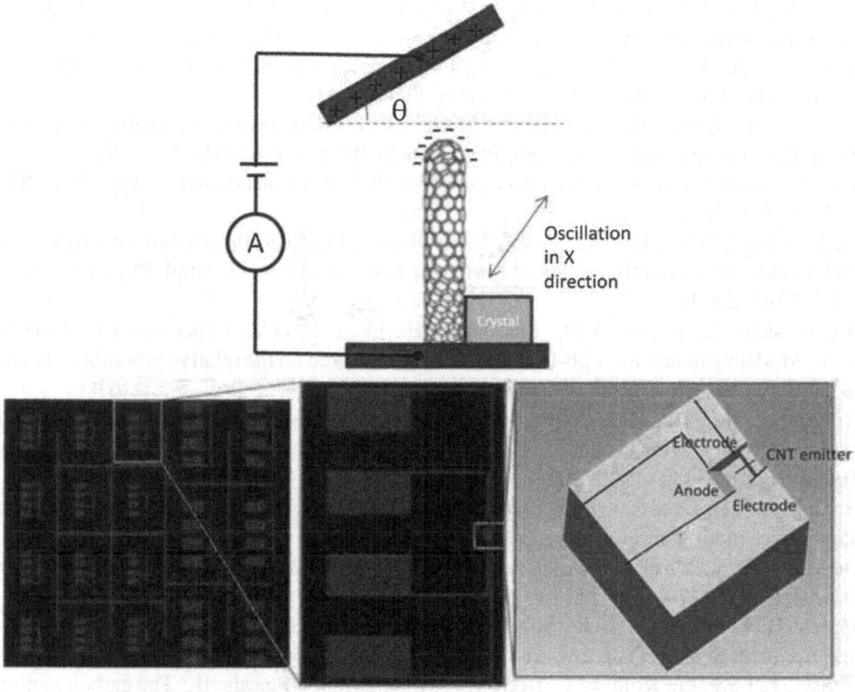

Fig. 8. Structure design of nano gyroscope and its schematic [22].

4 Conclusion and Future Work

This article provided a brief overview of the study on vibration characteristics of carbon nanotubes and several application of carbon nanotube vibration. The first paper used static electricity to drive the vibration of carbon nanotubes and the first order resonance frequency is about 530 kHz and the quality factor is up to 500. The other kind of bridge vibration fixed at both ends whose resonance frequency is up to 39 GHz with 35000 in Q factor. The method of molecular dynamics (MD) simulation was employed to study the relationship between the diameter, length of CNTs and temperature and characteristics of the CNTs. Several application of the CNT resonators have been implemented like nanotube radio, mass sensor, nano gyroscope. In the future, the characteristics of the carbon nanotubes need to be further studied to obtain a more stable and controllable vibration characteristics. The application field of vibration should be further widened and the stability of the application has been achieved as well as commercial value should also be excavated.

References

1. Chen, J., Hamon, M.A., Hu, H., Chen, Y., Rao, A.M., Eklund, P.C.: Solution properties of single-walled carbon nanotubes. Science **282**, 95–98 (1998)

2. Saito, R., Dresselhaus, G., Dresselhaus, M.S.: Physical properties of carbon nanotubes. In: The 12th International Winterschool on Electron, pp. 467–480 (1998)
3. Sun, Y.P., Fu, K., Lin, Y., Huang, W.: Functionalized carbon nanotubes: properties and applications. Acc. Chem. Res. **35**(12), 1096–1104 (2003)
4. Galpaya, D., Wang, M., George, G., Motta, N.: Carbon nanotubes: synthesis, structure, properties, and applications. J. Appl. Phys. **116**(5), 053518-1–053518-10 (2001)
5. Yao, N., Lordi, V.: Young's modulus of single-walled carbon nanotubes. J. Appl. Phys. **84**(4), 1939–1943 (1998)
6. Li, F., Cheng, H.M., Bai, S., Su, G., Dresselhaus, M.S.: Tensile strength of single-walled carbon nanotubes directly measured from their macroscopic ropes. Appl. Phys. Lett. **77**(20), 3161–3163 (2000)
7. Bichoutskaia, E., Popov, A.M., Lozovik, Y.E., Ershova, O.V., Lebedeva, I.V., Knizhnik, A.A.: Modeling of an ultrahigh-frequency resonator based on the relative vibrations of carbon nanotubes. Phys. Rev. B **80**(16), 165427 (2009)
8. Amjadipour, M., Dao, D.V., Motta, N.: Vibration analysis of initially curved single walled carbon nanotube with vacancy defect for ultrahigh frequency nanoresonators. Microsyst. Technol. **22**(5), 1115–1120 (2016)
9. Huang, X.M.H., Zorman, C.A., Mehregany, M., et al.: Nanoelectromechanical systems: nanodevice motion at microwave frequencies. Nature **421**(6922), 496 (2003)
10. De Heer, W.A., Chatelain, A., Ugarte, D.: A Carbon Nanotube Field-Emission Electron Source. Science **270**(5239), 1179–1180 (1995)
11. Mahar, B., Laslau, C., Yip, R., Sun, Y.: Development of carbon nanotube-based sensors—a review. IEEE Sens. J. **7**(2), 266–284 (2007)
12. Chaste, J., Fève, G., Kontos, T., Berroir, J. M., Glattli, C., Plaçais, B.: The carbon nanotube mesoscopic transistor, dynamics and optimisation as nanosecond pulsed charge detector. Journées De La Matière Condensée De La Société Française De Physique Jmc (2008)
13. Natsuki, T., Ni, Q.Q., Endo, M.: Analysis of the vibration characteristics of fluid-conveying double-walled carbon nanotubes. Carbon **46**(12), 1570–1573 (2008)
14. Hüttel, A.K., Steele, G.A., Witkamp, B., Poot, M., Kouwenhoven, L.P., van der Zant, H.S.: Carbon nanotubes as ultrahigh quality factor mechanical resonators. Nano Lett. **9**(7), 2547–2552 (2009)
15. Jensen, K., Kim, K., Zettl, A.: An atomic-resolution nanomechanical mass sensor. Nature Nanotechnol. **3**(9), 533 (2008)
16. Poncharal, P., Wang, Z.L., Ugarte, D., Heer, W.A.D.: Electrostatic deflections and electromechanical resonances of carbon nanotubes. Science **283**(5407), 1513–1516 (1999)
17. Laird, E.A., Pei, F., Tang, W., Steele, G.A., Kouwenhoven, L.P.: A high quality factor carbon nanotube mechanical resonator at 39 GHz. Nano Lett. **12**(1), 193–197 (2011)
18. Vallabhaneni, A.K., Rhoads, J.F., Murthy, J.Y., Ruan, X.: Observation of nonclassical scaling laws in the quality factors of cantilevered carbon nanotube resonators. J. Appl. Phy. **110**(110), 0343121–0343127 (2011)
19. Jensen, K., Weldon, J., Garcia, H., Zettl, A.: Nanotube radio. Nano Lett. **7**(11), 3508–3511 (2007)
20. Jensen, K., Kim, K., Zettl, A.: An atomic-resolution nanomechanical mass sensor. Nature Nanotechnol. **3**(9), 533 (2008)
21. Yang, Z., Nakajima, M., Shen, Y., Fukuda, T.: Nano-gyroscope device using field emission of isolated carbon nanotube. Int. Symp. Micro-Nanomechatronics Hum. Sci. **7268**, 256–261 (2013)

A Novel Soft Robot Based on Organic Materials: Finite Element Simulation and Precise Control

Fanan Wei[1(✉)], Jianghong Zheng[1], and Changle Yu[2]

[1] School of Mechanical Engineering and Automation, Fuzhou University, Fuzhou 350116, China
weifanan@fzu.edu.cn
[2] School of Ocean, Fuzhou University, Fuzhou 350116, China

Abstract. Considering the inherent safety and high flexibility, soft robot is drawing extensive attention from both industry and academics. Soft robots made of organic materials show the merits of considerably low cost and ease of fabrication. But, the precise control of soft robot is still of great challenge. A sophisticated understanding of the motion mechanics of organic soft robot is essential for accurate control of this type of robot. In this paper, PDMS/PEDOT: PSS system, which is developed for soft robot, is modelled and simulated using finite element method. Through parameters tuning, the dependences of PDMS/PEDOT:PSS system deformation on layer thicknesses and electrical current are acquired and compared with the experimental results available in reference. Furthermore, we proposed a three layer structure for soft robot, and investigated the relationship between deformation of this structure and layer thicknesses. Our work proposed an approach for precise control for soft robot through a combination of finite element simulation and experiment data. Additionally, the three layer structure we proposed is demonstrated to be a promising solution for high response speed soft robot.

Keywords: Soft robot · PEDOT:PSS · PDMS · Finite element simulation

1 Introduction

With the promotion of mechanical engineering, automation technology and the introduction of internet, robot is ready to change the whole industry, especially in the developing countries, thoroughly. The safety and robustness of robots have been the challenging issues along the robot history. The massive application of robot requires effective solution to these issues.

Soft robot, due to its inherent characteristic, provides a promising solution to the issues of safety and robustness. The relatively low Young's modulus of the soft materials ensures that human body can be avoided from hurt once in contact with such kind of robot. On the other hand, the high resilience of soft materials employed in soft robot can protect the robot from lots of possible damages. Therefore, because of its outstanding merits, soft robot has witnessed a soaring rise in the past decade.

Y. Huang et al. (Eds.): ICIRA 2017, Part I, LNAI 10462, pp. 103–109, 2017.
DOI: 10.1007/978-3-319-65289-4_10

At the beginning, pneumatic actuation was the most prevalent driving approach for soft robot [1]. Currently, almost all the available commercial soft robots are actuated through tuning of air pressure [2]. But, the control of pneumatic actuation could be considerably rough; so soft robot based on pneumatic driving method is not suitable for tasks requires high precision. Numerous effort has been devoted to search for alternative actuation methods that meet accurate control requirement.

Organic materials, whose deformation can be controlled by electric voltage or current, can provide a route to accurate motion control for soft robot. Piezoelectric ceramics are such a type of materials. But most ceramics are extremely stiff. What's more, the deformation of piezoelectric ceramics is extremely tiny, which is too small to be competent for practical task. Then, researchers all over the world have been attempting to develop organic system that respond to voltage and current changes. So far, there are two feasible routes towards the goal: one is thermal expansion, the other one is liquid absorption and desorption.

Soft robot based on hybrid system with CNT, graphene and polymers are extensively investigated for thermal expansion actuation [3]. The small magnitude of thermal expansion leads to a limited motion range of the soft robot. On the contrary, hybrid system with a hydrophilic layer and a hydrophobic layer showed significant deformation under varied humidity. This humidity sensible hybrid system works with the hydrophilic layer absorb or desorb water under different environment [3]. Silvia Taccola et al. further improved this actuation approach by introducing electric current heating into the system [3]. They used PEDOT:PSS (poly (3, 4-ethylenedioxythiophene): poly (styrenesulfonate))/PDMS (polydimethylsiloxane) as the two layer system. PEDOT:PSS is not only hydrophilic, but also conductive. When electric current go through PEDOT:PSS layer, the hydrophilic layer is heated, water get away from it, and subsequently, it shrinks. Through this system, deformation angle as large as 360° was achieved. This actuation method enables the soft robot to complete tasks requires large motion magnitude. But its accurate control still calls for a full understanding of the fundamental mechanics during its deformation.

In this paper, with the purpose of clarifying the working mechanism of the humidity sensible soft robot, we choose the PEDOT:PSS/PDMS system as an example and investigate it with finite element simulations conducted in COMSOL. The structure parameters and voltage are changed in the simulation so as to elucidate the deformation response to the controllable variables. The temperature distribution in PEDOT: PSS/PDMS double layers through the deformation process is studied as well. In order to enhance the response speed of the system, we propose a three layer (sandwich) structure and demonstrate the feasibility of such a novel structure through finite element simulations.

2 Modelling and Simulation

The PDMS/PEDOT:PSS system is composed of two layers: one hydrophilic layer and another hydrophobic layer. The hydrophilic layer, PEDOT:PSS, can absorb water in moist environment, and desorb water under dry condition. Thence its volume will change according to the environment humidity. Meanwhile, the hydrophobic layer,

PDMS, will keep its volume constant when humidity changes. As a result, the two layer structure will roll towards the PEDOT:PSS side in dry condition and towards PDMS side in wet condition, respectively. PEDOT:PSS, which is conductive as well, can heat when electric current go through it. Subsequently, water will evaporate from PEDOT:PSS layer. Then the structure will roll towards PEDOT:PSS side. Therefore, deformation of this system can be controlled by voltage. Since the water absorption and desorption processes under joule heating is alike to that of thermal expansion. To be exact, it should be thermal shrink. Considering the complexity of water absorption and desorption processes, we simplify the system by replacing the processes with thermal shrink of PEDOT:PSS layer.

The geometry of the investigated system is shown in Fig. 1. The structure parameters are set according to the actual device fabricated in the reference [3]. And initial thickness values for PDMS and PEDOT:PSS are set to be 100 μm and 0.6 μm, which are in line with the experimental data. Voltage is applied between the two gold electrodes, as indicated in Fig. 1. Considering the high coupling between the solid mechanics field, electric field and temperature field in the studied system, we adopted COMSOL Multiphysics to perform the simulation work. So as to save simulation time, the mesh size is set to be 'normal' in COMSOL.

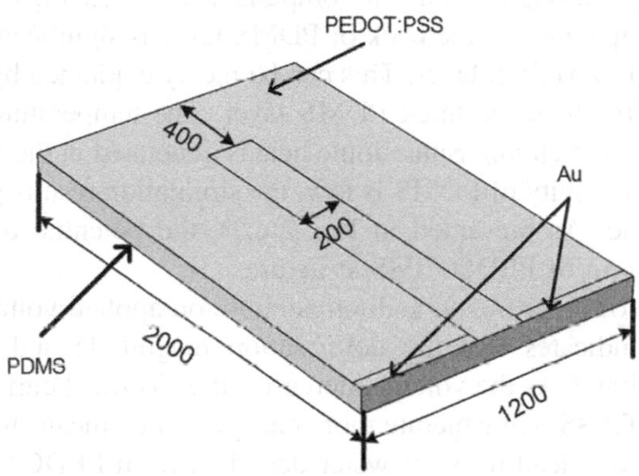

Fig. 1. Geometry structure of the the layer system studied. (Thickness of PEDOT:PSS is relatively small compared with that of PDMS)

Major parameters adopted in the finite element simulation are presented in Table 1. First of all, simulation with all parameters fixed is conducted. Then, we change the applied voltage between 0.2 and 2.4 V, with an interval of 0.2 V. And simulations with the thicknesses of PDMS and PEDOT:PSS layers varied are conducted so as to clarify the dependence of deformation on the device structure. Finally, we build up model for the sandwich structure (the three layers structure), and perform simulation to demonstrate whether it works.

Table 1. Major parameters adopted in the finite element simulation.

Parameters	Value	Units
Coefficient of thermal expansion of PDMS	3e−4 [4]	1/K
Equivalent thermal expansion coefficient of PEDOT:PSS	−1e−3	1/K
Heat capacity of PDMS	1460	J/(kg K)
Heat capacity of PEDOT:PSS	1980 [5]	J/(kg K)
Thermal conductivity of PDMS	0.15 [3]	W/(m K)
Thermal conductivity of PEDOT:PSS	0.2 [6]	W/(m K)
Poisson's ratio of PEDOT:PSS	0.34 [7]	1

3 Results and Discussions

In this paper, we attempted to study the deformation process of PEDOT:PSS/PDMS based soft robot through finite element simulation. Using the hydrophilic and hydrophobic double layer structure, the deformation of the structure upon applied external voltage is mimicked. As shown in Fig. 2, the deformation field, temperature field and the potential field in the system are acquired through the simulations simultaneously. The device is observed to deform largely towards the PEDOT:PSS layer with current go through it. And a comparison between Figs. 2(b.2) and 1(b.3) indicates that the temperature at the back of PDMS layer is significantly lower than the temperature in the PEDOT:PSS layer. This can be nicely explained by the considerably low thermal conductivity of the thick PDMS layer. The temperature is of reasonable value after heating by the circuit. Since Joule heat is generated in the PEDOT:PSS layer and the thermal conductivity of PDMS is low, the simulation results goes well with the fundamental principle. As presented in Fig. 2(c.2), the potential decreasing is symmetrical in the two arm of PEDOT:PSS structure.

The dependence of deformation and temperature on applied voltage is presented in Fig. 3. The result indicates that the deformation magnitude and temperature both increase with the voltage. As the voltage increases, the electrical current increases; then the circuit in PEDOT:PSS will generate more energy, which means higher temperature. Higher temperature will lead to more water desorbed from PEDOT:PSS. Then larger deformation is expected.

Figure 4 gives the dependence of deformation on the thickness of PEDOT:PSS. From the simulation data, we can see that deformation increases almost linear with thickness of PEDOT:PSS layer. This can be explained by the relatively thin PEDOT: PSS layer. And the resistance of the circuit will decrease linearly with the layer thickness; then the generated heat will increase linearly. Therefore, the deformation distribution is expected to increase linearly with thickness of PEDOT:PSS layer.

The three layer structure (PEDOT:PSS/PDMS/PEDOT:PSS) is modeled in COMSOL and simulated. By applying different voltage at both the PEDOT:PSS layers, the deformation field, as well as temperature and potential fields are shown in Fig. 5. The simulation demonstrates that the structure can deform towards both sides by tuning the applied voltage on both sides. Thus, the proposed structure will be a solution to the low response speed of this type of soft robot. Furthermore, it enables the structure to deform towards both sides by controlling the applied voltage in the two PEDOT:PSS layers.

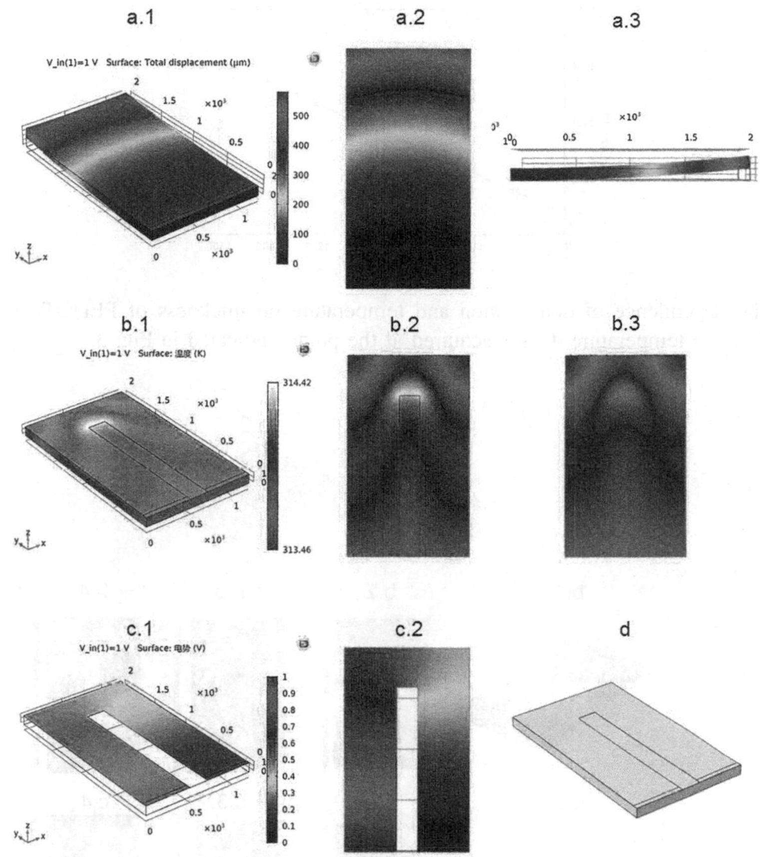

Fig. 2. Finite element simulation results with applied voltage of 1 V under 25 °C. (a.1–a.2) the deformation distribution observed at xyz-view (a.1), xy-view (a.2) and yz-view (a.3). (b.1–b.3) the temperature distribution simulation result. (b.2) and (b.3) present the temperature distribution at both the front and back sides of the device. (c.1) and (c.2) shows the potenitial distribution on the PEDOT:PSS surface.

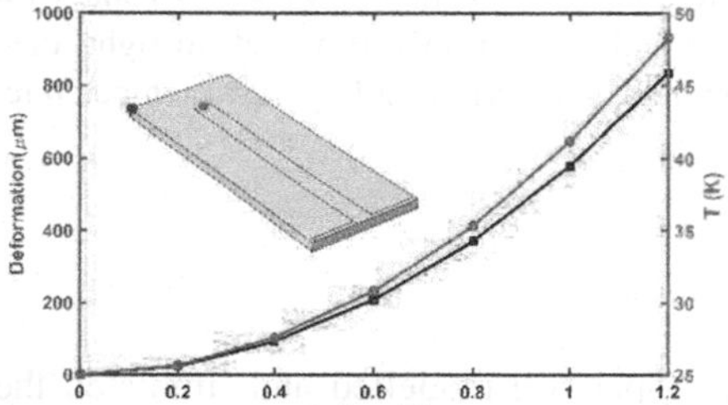

Fig. 3. The dependence of deformation and temperature on applied voltage between the Au electrodes. The deformation data is acquired at the point indicated by the blue circle; while the temperature data is acquired at the red point. The applied voltage changes from 0 V to 1.2 V, with an interval of 0.2 V.

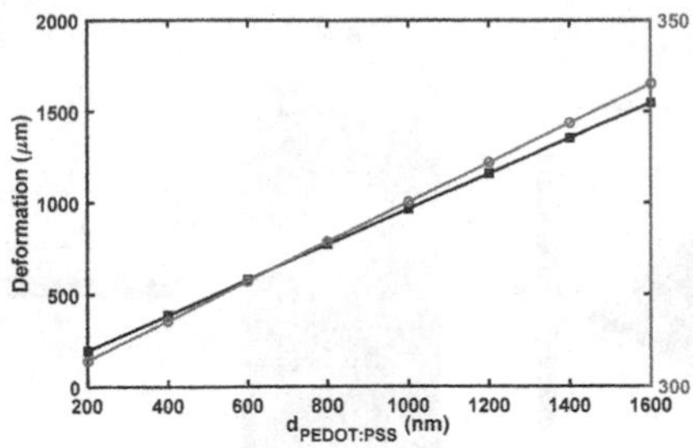

Fig. 4. The dependence of deformation and temperature on thickness of PEDOT:PSS layer. Deformation and temperature data is acquired at the points indicated in Fig. 3.

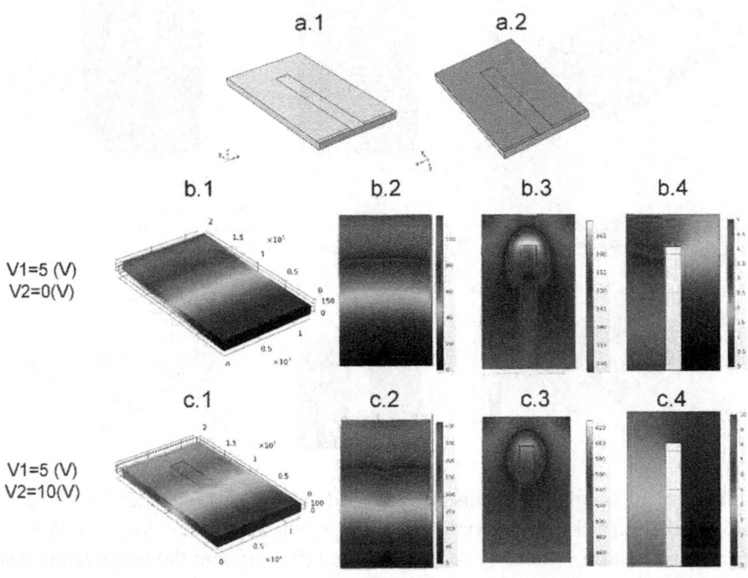

Fig. 5. Finite element simulation results for the sandwich structure (PEDOT: PSS/PDMS/PEDOT:PSS). (a) presents the geometry structure from forward and rear viewpoints. (b) displays the simulation results corresponding to $V_1 = 5$ (V) and $V_2 = 0$. (c) shows the results corresponding to $V_1 = 5$ (V) and $V_2 = 10$ (V). From left to right, the sub figures for the field distribution are deformation field at xyz view and xy view; temperature distribution and surface potential distribution.

4 Conclusions

To be concluded, in this paper, we modelled and simulated the deformation process upon electrical current exerted in the PEDOT:PSS layer in the organic soft robot. Through simulation work, we clarified the dependence of deformation on the thickness of PEDOT:PSS layer and the applied voltage. The simulation results indicate that we can figure out the exact transfer function from voltage and geometry parameters to the soft robot motion. Finally, simulation of the proposed three layer structure verified it as a possible solution to the low response speed of soft robot.

Acknowledgement. The authors need to thank the Science and Technology Department of Fujian Province, China for the funding (No. 2017J01748).

References

1. Bartlett, N.W., Tolley, M.T., Overvelde, J.T.B., Weaver, J.C., Mosadegh, B., Bertoldi, K., Whitesides, G.M., Wood, R.J.: A 3D-printed, functionally graded soft robot powered by combustion. Science **349**(6244), 161–165 (2015)
2. Shepherd, R.F., Ilievski, F., Choi, W., Morin, S.A., Stokes, A.A., Mazzeo, A.D., Chen, X., Wang, M., Whitesides, G.M.: Multigait soft robot. Proc. Natl. Acad. Sci. USA **108**(51), 20400–20403 (2011)
3. Taccola, S., Greco, F., Sinibaldi, E., Mondini, A., Mazzolai, B., Mattoli, V.: Toward a new generation of electrically controllable hygromorphic soft actuators. Adv. Mater. **27**(10), 1668–1675 (2015)
4. Schubert, B.E., Floreano, D.: Variable stiffness material based on rigid low-melting-point-alloy microstructures embedded in soft poly(dimethylsiloxane) (PDMS). RSC Adv. **3**(46), 24671–24679 (2013)
5. Scholdt, M., Do, H., Lang, J., Gall, A., Colsmann, A., Lemmer, U., Koenig, J.D., Winkler, M., Boettner, H.: Organic semiconductors for thermoelectric applications. J. Electron. Mater. **39**(9), 1589–1592 (2010)
6. Liu, J., Wang, X.J., Li, D.Y., Coates, N.E., Segalman, R.A., Cahill, D.G.: Thermal conductivity and elastic constants of PEDOT:PSS with high electrical conductivity. Macromolecules **48**(3), 585–591 (2015)
7. Lang, U., Naujoks, N., Dual, J.: Mechanical characterization of PEDOT:PSS thin films. Synth. Met. **159**(5–6), 473–479 (2009)

3D Motion Control and Target Manipulation of Small Magnetic Robot

Jingyi Wang[1,2], Niandong Jiao[1(✉)], Yongliang Yang[3], Steve Tung[1,4], and Lianqing Liu[1(✉)]

[1] State Key Laboratory of Robotics, Shenyang Institute of Automation, Chinese Academy of Sciences, Shenyang 110016, China
{ndjiao, lqliu}@sia.cn
[2] University of Chinese Academy of Sciences, Beijing 100049, China
[3] Department of Mechanical Engineering, Michigan State University, East Lansing, MI 48824, USA
[4] Department of Mechanical Engineering, University of Arkansas, Fayetteville, AR 72701, USA

Abstract. The Electromagnetic robots have received much attention because of their advantages of control agility and good precision. Most of the electromagnetic robots are controlled in two-dimensional motion. However, the environment *in vivo* is complicated and two-dimensional control is difficult to meet the complicated situation. In this paper, we propose a three-dimensional (3D) control method for the locomotion and manipulation of small magnetic robots. The robot can be controlled to move in 3D direction using visual feedback with an expert control algorithm. The velocity of the robot is nearly proportional to the applied current in the coils, and can reach 1 mm/s. To verify its performance, the robot is used to manipulate microspheres into a 3 × 3 array. The robot is expected to be an agile tools that could play important roles in micromanipulation and biomedical treatment.

Keywords: Magnetic robot · Expert control · Three-dimensional motion · Micromanipulation

1 Introduction

Wireless controlled microrobots have promising prospects in the field of survey and manipulation in blood vessels and inner tissues due to their miniaturization, high controllability [1]. Microrobots with various actuation modes have been developed such as electromagnetic [2–7], thermal [8], chemical bubble [9], swimming tail [10, 11], bacterial [12, 13], and hybrid [14]. The electromagnetic actuation method has many advantages such as high controllability, great accuracy, and a large actuating force. With the method, the microrobot can be controlled to move in a microfluidic chip [15], in 3D space [16], as well as in vivo [17]. However, the environment in vivo is complicated, and there are still many challenges for microrobot such as complex 3D motion and agile micromanipulation.

Y. Huang et al. (Eds.): ICIRA 2017, Part I, LNAI 10462, pp. 110–119, 2017.
DOI: 10.1007/978-3-319-65289-4_11

In this paper, we propose a novel electromagnetic manipulation system which can control the wireless controlled robot to achieve three-dimensional motion. With the expert control method, the driving force is reducing when the robot is nearing the target point. The tracking path is separated to several segments with target points. The moving speed is controlled to reduce to zero when the robot arrives at every target point. With the automatic control method using an expert control algorithm, the robot can be controlled to achieve an accurate three-dimensional motion in liquid environment. Furthermore, as a micromanipulation tool, the robot has realized placing microspheres in an array with high precision.

2 Methods

2.1 Theoretical Analysis of Magnetic Robot

The magnetic torque (**T**) in three direction is generated as:

$$T = V\mathbf{M} \times B = V \begin{bmatrix} \mathrm{M}_x \\ M_y \\ M_z \end{bmatrix} \times \begin{bmatrix} B_x \\ B_y \\ B_z \end{bmatrix} \tag{1}$$

where V, **M**, and **B** indicate the volume and magnetization vector of the robot and the magnetic flux vector of the magnetic field in three direction, respectively.

The magnetic force is given by [18]:

$$\mathbf{F} = V(\mathbf{M} \cdot \nabla)\mathbf{B} \tag{2}$$

the ∇ is the gradient operator:

$$\nabla = \begin{bmatrix} \frac{\partial}{\partial \mathbf{x}} \\ \frac{\partial}{\partial \mathbf{y}} \\ \frac{\partial}{\partial \mathbf{z}} \end{bmatrix} \tag{3}$$

the uniform magnetization of the robot and the magnetic flux are formed:

$$\mathbf{M} = \begin{bmatrix} M_x \\ M_y \\ M_z \end{bmatrix}, \mathbf{B} = \begin{bmatrix} B_x \\ B_y \\ B_z \end{bmatrix} \tag{4}$$

the magnetic field generated by the current flows through a spiral electromagnetic coil, and can be calculated by the Biot-Savart law as follows [18]:

$$\mathbf{B} = \frac{\mu_0 I}{4\pi} \int_L \frac{d\mathbf{l} \times \mathbf{h}}{|\mathbf{h}|^3} \tag{5}$$

where I is the current through the coil, L is the integral path, **h** is the full displacement vector from the wire element to the point where the field needs to be calculated, dl is a vector of the differential element of the current through the wire, and µ0 is the magnetic permeability ($\mu_0 = 4\pi \times 10^{-7}$ H/m).

The magnetic fields at the any position (x, y, z) can be described as the multiplication of the electromagnetic field value per unit current $\hat{B}(x, y, z)$ and the current (**I**) applied can be expressed as:

$$\mathbf{B}(\mathbf{x}, \mathbf{y}) = \hat{\mathbf{B}}(\mathbf{x}, \mathbf{y}, \mathbf{z})\mathbf{I} \tag{6}$$

The magnetic flux can be expressed as the function of the current **I**:

$$\begin{bmatrix} B_x \\ B_y \\ B_z \end{bmatrix} = \begin{bmatrix} \hat{B}_{x,1} & \hat{B}_{x,2} & \hat{B}_{x,3} & \hat{B}_{x,4} & \hat{B}_{x,5} & \hat{B}_{x,6} \\ \hat{B}_{y,1} & \hat{B}_{y,2} & \hat{B}_{y,3} & \hat{B}_{y,4} & \hat{B}_{y,5} & \hat{B}_{y,6} \\ \hat{B}_{z,1} & \hat{B}_{z,2} & \hat{B}_{z,3} & \hat{B}_{z,4} & \hat{B}_{z,5} & \hat{B}_{z,6} \end{bmatrix} \begin{bmatrix} I_1 \\ I_2 \\ I_3 \\ I_4 \\ I_5 \\ I_6 \end{bmatrix} \tag{7}$$

where $B_x(I_n)$, $B_y(I_n)$ and $B_z(I_n)$ (n = 1, 2, 3, 4, 5, 6) denote the three axis directional magnetic fields generated by the n th coils. The magnetic force can be expressed as:

From Eq. (7), the magnetic force, the magnetic flux, and the gradient magnetic fields used to propel the robot in any position, can be calculated using the current in every coil. Therefore, the force that controls the small robot can be changed by changing the current in the electromagnetic coils.

2.2 Close-Loop Control of the Robot

The 3D motion control is achieved by using visual feedback control. The close-loop system using image processing and the expert control algorithm is set up to control the small robot. The structure diagram of the visual feedback expert control is shown in Fig. 1.

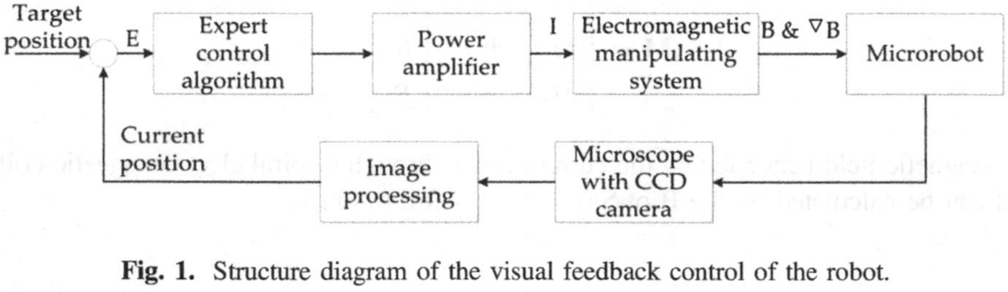

Fig. 1. Structure diagram of the visual feedback control of the robot.

The control of the robot is a point-by-point path-tracking motion through visual feedback control. The driving force is reducing when the robot is nearing the target point. The tracking path is separated to several segments with target points. The moving speed is controlled to reduce to zero when the robot arrives at every target point. This algorithm makes the robot move accurately with higher precision and more robustness.

3 Experiments

3.1 Fabrication Process of Magnetic Robot

The electromagnetic manipulation robot was fabricated with NdFeB magnetic powder and polydimethylsiloxane (PDMS) through mold fabrication. The mold was fabricated using an engraving machine (Benchtop Engravers EGX-600/400, Roland DGA Corporation, Irvine, CA, USA). The mold was carved on an acrylic plate. The PDMS and NdFeB magnetic powder were mixed at a weight ratio of 1:1. The mixture was stirred for 30 min. The mixture was put into the mold and the redundant mixture was removed. The robot was degassed in a vacuum box to remove bubbles, and then placed in the oven to bake at 60 °C for 4 h. The size of the robot is about 5 mm, and we define the robot with size of half centimeter small robot.

3.2 Design of the Electromagnetic Manipulation System

The electromagnetic manipulation system consists of six electromagnetic coils. The electromagnetic manipulation system can be used to manipulate the robot in three dimensions. Two of the pairs of the electromagnetic coils were set to be orthogonal in the horizontal plane, whereas the other pair was set in the vertical plane. The horizontal electromagnetic coils were installed on the height-adjustable pillar. The height-adjustable pillar and the nylon plastic frames are convenient to adjust the vertical and horizontal position of the electromagnetic coils, respectively. The manipulation region of the system is limited as a result of the constant structure. To overcome this disadvantage, we propose a novel electromagnetic manipulation system that has position-adaptable electromagnetic coils. By regulating the six-position adaptable electromagnetic coils, the manipulation area can be adjusted to expand or shrink. A long-focus microscope with a charge-coupled device (CCD) was used to observe and provide visual feedback through the upper electromagnetic coil in the z direction.

The electromagnetic manipulation system is shown in Fig. 2. It consists of six iron-core electromagnetic coils, a motorized long-focus microscope (Navitar 1-62317, Navitar Inc., New York, USA), a side observational microscope, two data acquisition cards (NI PCI-6229 DAQ Card, National Instruments Corporation, Austin, USA), a direct current (DC) power supply, a micron 3D positioning stage, and a power amplifier. The 3D positioning stage is used to adjust the position of the robot in the container. The specific parameter of the electromagnetic manipulation system can be found in our previous work [19].

Fig. 2. The whole electromagnetic manipulation system.

The control algorithm, programmed by LabVIEW (Labview 2012, National Instruments Corporation, Austin, USA), processes this information, generates the required output signal, and then transmits it to the data acquisition card. The electromagnetic manipulation system can be used to accurately manipulate the robot.

4 Results

4.1 Velocity Measurement

In order to measure the motion velocity, various applied current were applied. Initially, it started to accelerate from its stationary state due to the magnetic force. With the rapid increase in robot speed, the fluid viscous force was instantaneously increased. The robot can instantaneously reach a uniform velocity in a fluidic environment [20]. As shown in Fig. 3, the average velocity of the robot increased with an increase in applied current in the coils.

Due to the experimental results the velocity of the robot is approximately proportional to the applied current in the coils. The velocity in the tracking process is not permanent. The robot is first accelerated and then moves with a constant velocity. Finally, it decelerates gradually.

4.2 Vertical Motion of the Small Robot

Based on the expert control algorithm using visual feedback, 3D motion experiments were executed in liquid environment (silicone oil). First, the robot (5 mm) was recognized and the position information was acquired. Then, a rectangular paths on the yz plane was drawn for the robot tracking. Then the robot can accurately propel along the predetermined paths automatically. Figure 4 shows the screen-shots of the vertical motion. The red paths was defined as the tracked paths for the robot. And the blue arrow indicates the direction of movement.

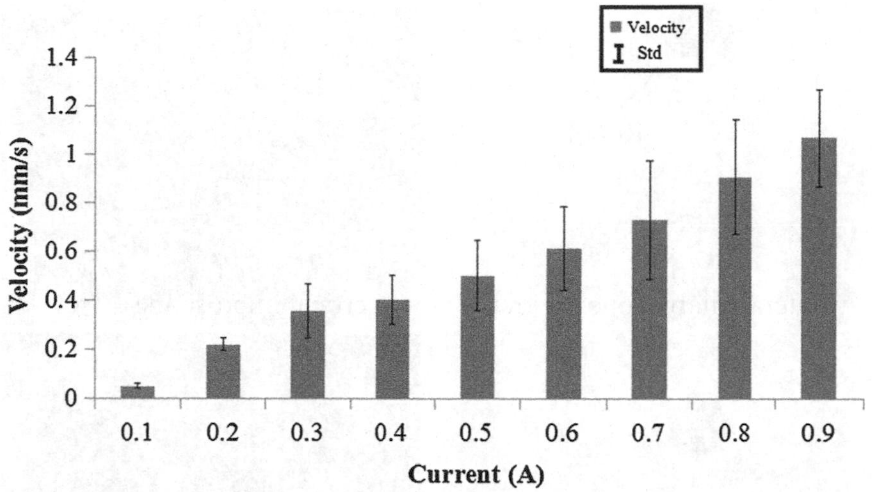

Fig. 3. Velocity measurement of the robot according to the current through the electromagnetic coils.

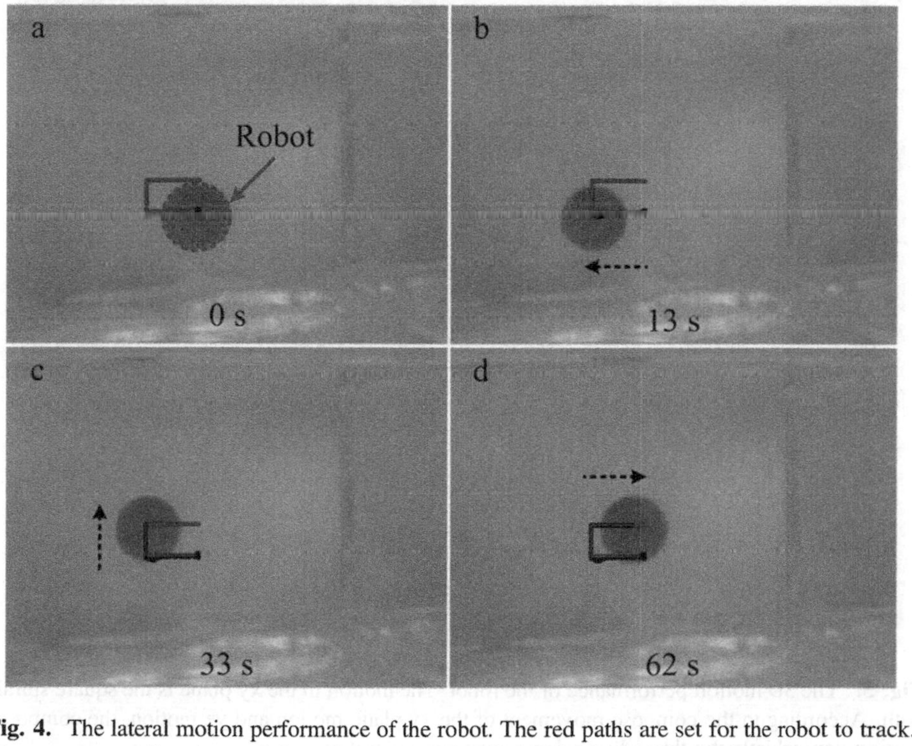

Fig. 4. The lateral motion performance of the robot. The red paths are set for the robot to track. The movement directions of the robot are marked in black dotted lines. The robot tracks the rectangular path. (Color figure online)

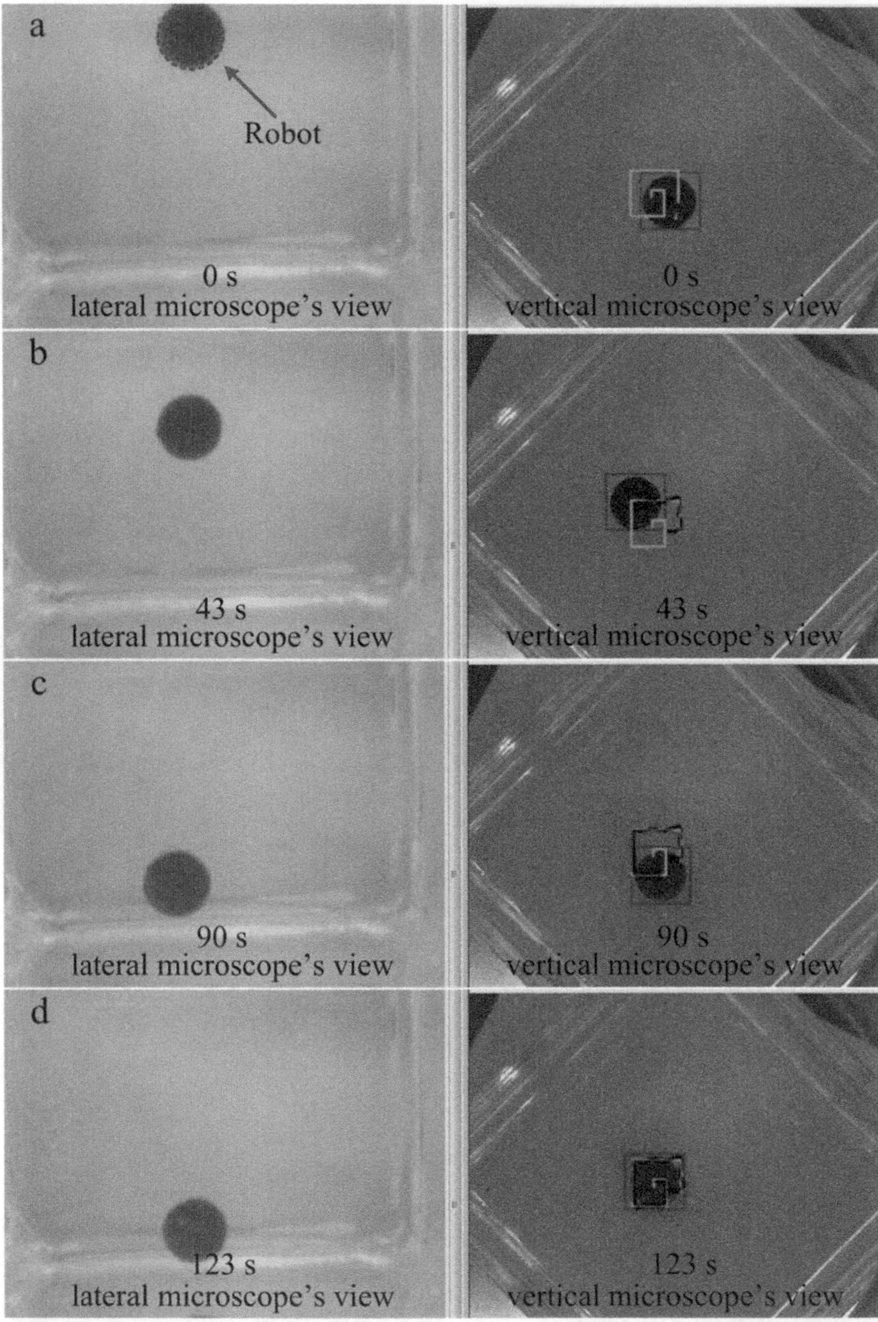

Fig. 5. The 3D motion performance of the robot. The motion in the xy plane is the square spiral path. According to the compose movement of the xy plane motion and yz motion, the compose motion can achieve the three-dimension movement control.

4.3 3D Motion of the Small Robot

The robot can be controlled to move in liquid environment in 3D motion based on the expert control algorithm using visual feedback. The predefined square spiral path in the xy plane was drawn in the vertical microscope's view. Then the robot was controlled to track the square spiral path. Meanwhile, the predefined path in the yz plane was drawn in the lateral microscope's view. The robot was controlled to track the path in the yz plane due to lateral electromagnetic coils. According to the compose movement of the xy plane motion and yz motion, the compose motion can achieve the three-dimension movement control. As shown in Fig. 5, the 3D motion of small robot is achieved in three dimension.

4.4 Manipulating Microspheres

The robot can be controlled to manipulate zirconia microspheres. Nine microspheres whose diameter is approximately 1 mm are placed on the substrate at random. The robot is controlled to propel towards the microspheres, and it pushed every microsphere to desired place.

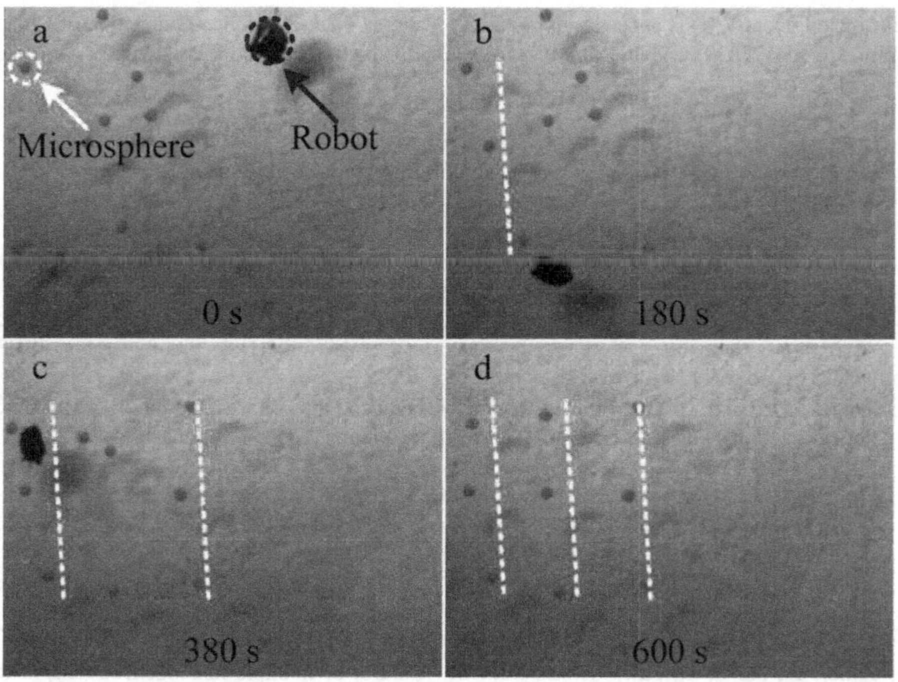

Fig. 6. Manipulation of 3 × 3 array microspheres. The robot and zirconia microsphere are marked in blue and red circles. (a) The zirconia microspheres are placed at random. (b) The robot pushes the microsphere into a row. (c) The robot manipulates the microspheres into the third row. (d) The robot completes the 3 × 3 array manipulation. The manipulation procedure takes approximately 600 s. (Color figure online)

As shown in Fig. 6, the robot was controlled to push the three zirconia microspheres into a row. Then the robot moved to push microspheres to second and third row. The microspheres are manipulated into a 3 × 3 array successively, which verified the mobility and accuracy of the small robot. The procedure of manipulation takes approximately 600 s.

5 Conclusions

This paper proposed an 3D motion control for a small robot by means of an electromagnetic manipulation system. The electromagnetic manipulation system was designed with six position adaptable electromagnetic coils, which are used to adjust the manipulation region. The robot can be controlled to move in three-dimension using visual feedback with an expert control algorithm. The performance of the robot was verified by moving in three-dimensional liquid environment and manipulating microspheres into a 3 × 3 array. In the next work, we will reduce the size of the robot to micrometer scale and integrate microgripper on the robot. With the development of the magnetic robots, they may play a role in intravascular survey, drug delivery, and minimal invasive surgery in the near future.

Acknowledgments. This paper is supported by National Natural Science Foundation of China (Grant No. 61573339) and the CAS/SAFEA International Partnership Program for Creative Research Teams.

References

1. Nelson, B.J., Kaliakatsos, I.K., Abbott, J.J.: Microrobots for minimally invasive medicine. Annu. Rev. Biomed. Eng. **12**, 55–85 (2010)
2. Go, G., Choi, H., Jeong, S., Lee, C., Ko, S.Y., Park, J.-O., Park, S.: Electromagnetic navigation system using simple coil structure (4 coils) for 3-D locomotive robot. IEEE Trans. Magn. **51**, 1–7 (2015)
3. Tottori, S., Zhang, L., Qiu, F., Krawczyk, K.K., Franco-Obregón, A., Nelson, B.J.: Magnetic helical micromachines: fabrication, controlled swimming, and cargo transport. Adv. Mater. **24**, 811–816 (2012)
4. Kim, S., Qiu, F., Kim, S., Ghanbari, A., Moon, C., Zhang, L., Nelson, B.J., Choi, H.: Fabrication and characterization of magnetic robots for three-dimensional cell culture and targeted transportation. Adv. Mater. **25**, 5863–5868 (2013)
5. Zhang, Y., Yue, M., Guo, D., Wang, D., Yu, H., Jiang, S., Zhang, X.: Characteristics of spatial magnetic torque of an intestine capsule micro robot with a variable diameter. Sci. China Ser. E: Technol. Sci. **52**, 2079–2086 (2009)
6. Diller, E., Floyd, S., Pawashe, C., Sitti, M.: Control of multiple heterogeneous magnetic robots in two dimensions on nonspecialized surfaces. IEEE Trans. Robot. **28**, 172–182 (2012)
7. Chowdhury, S., Jing, W., Cappelleri, D.J.: Controlling multiple robots: recent progress and future challenges. J. Micro-Bio Robot. **10**, 1–11 (2015)

8. Shen, X., Viney, C., Johnson, E.R., Wang, C., Lu, J.Q.: Large negative thermal expansion of a polymer driven by a submolecular conformational change. Nat. Chem. **5**, 1035–1041 (2013)
9. Búzás, A., Kelemen, L., Mathesz, A., Oroszi, L., Vizsnyiczai, G., Vicsek, T., Ormos, P.: Light sailboats: laser driven autonomous robots. Appl. Phys. Lett. **101**, 041111 (2012)
10. Hwang, G., Braive, R., Couraud, L., Cavanna, A., Abdelkarim, O., Robert-Philip, I., Beveratos, A., Sagnes, I., Haliyo, S., Régnier, S.: Electro-osmotic propulsion of helical nanobelt swimmers. Int. J. Robot. Res. **30**, 806–819 (2011)
11. Huang, C., Lv, J.-A., Tian, X., Wang, Y., Yu, Y., Liu, J.: Miniaturized swimming soft robot with complex movement actuated and controlled by remote light signals. Sci. Rep. **5**, 17414 (2015)
12. Martel, S., Felfoul, O., Mathieu, J.-B., Chanu, A., Tamaz, S., Mohammadi, M., Mankiewicz, M., Tabatabaei, N.: Mri-based medical nanorobotic platform for the control of magnetic nanoparticles and flagellated bacteria for target interventions in human capillaries. Int. J. Robot. Res. **28**, 1169–1182 (2009)
13. Martel, S., Tremblay, C.C., Ngakeng, S., Langlois, G.: Controlled manipulation and actuation of micro-objects with magnetotactic bacteria. Appl. Phys. Lett. **89**, 233904 (2006)
14. Li, D., Choi, H., Cho, S., Jeong, S., Jin, Z., Lee, C., Ko, S.Y., Park, J.O., Park, S.: A hybrid actuated robot using an electromagnetic field and flagellated bacteria for tumor-targeting therapy. Biotechnol. Bioeng. **112**, 1623–1631 (2015)
15. Wang, J.Y., Jiao, N.D., Tung, S., Liu, L.Q.: Magnetic robot and its application in a microfluidic system. Robot. Biomim. **1**, 18 (2014)
16. Diller, E., Sitti, M.: Three-dimensional programmable assembly by untethered magnetic robotic micro-grippers. Adv. Funct. Mater. **24**, 4397–4404 (2014)
17. Ocegueda, K., Rodriguez, A.: A simple method to calculate the signal-to-noise ratio of a circular-shaped coil for MRI. Concepts Magn. Reson. Part A **28**, 422–429 (2006)
18. Jiles, D.C.: Introduction to Magnetism and Magnetic Materials. CRC Press, Boca Raton (1998)
19. Wang, J.Y., Jiao, N.D., Tung, S., Liu, L.Q.: Automatic path tracking and target manipulation of a magnetic microrobot. Micromachines **7**, 212 (2016)
20. Yesin, K.B., Vollmers, K., Nelson, B.J.: Modeling and control of untethered biorobots in a fluidic environment using electromagnetic fields. Int. J. Robot. Res. **25**, 527–536 (2006)

A Locomotion Robot Driven by Soft Dielectric Elastomer Resonator

Chao Tang[1,2], Bo Li[1,3], Changsheng Bian[1,2], Zhiqiang Li[1,2], Lei Liu[1,2], and Hualing Chen[1,2(✉)]

[1] School of Mechanical Engineering, Xi'an Jiaotong University, Xi'an, 710049, China
hlchen@mail.xjtu.edu.cn

[2] State Key Laboratory for Strength and Vibration of Mechanical Structures, Xi'an Jiaotong University, Xi'an, 710049, China

[3] State Key Laboratory for Mechanical Manufacturing Systems Engineering, Xi'an Jiaotong University, Xi'an, 710049, China

Abstract. Dielectric elastomers are a variety of electroactive polymer that deform due to the electrostatic interaction between two electrodes with opposite electric charge. Dielectric elastomer actuators are recognized as one of the most promising soft actuation technologies that have a widely application in soft robots. In this paper, an ultralight locomotion robot based on dielectric elastomers resonance mechanism has been prepared. During operation, a suitable excitation frequency and voltage level were determined experimentally, which provide guidance to optimize on-demand locomotion of the robot. The structure of our robot is very simple and only weight 9.6 g but have a speed of 65 mm/s benefit by dielectric elastomer resonator. The resonance mechanism increases the energy conversion efficiency of dielectric elastomer actuators which has huge potential application in soft robots.

Keywords: Dielectric elastomer · Soft active material · Locomotion robot · Resonator

1 Introduction

In recent years, new and unconventional (non-mechanical) drive mechanisms have been explored that utilize pneumatics [1–3], magnetic fields [4], electricity [5], light [6, 7] or humidity [8]. They mainly use new actuations composed of smart materials. Among potential candidates, electroactive polymers (EAP) seem to have great potential to be a new means of actuation. In spite of the technical difficulties their application areas are rapidly expanding especially in robotic fields since the actuation mechanism of the polymers is similar to the human muscle. Among the various kinds of EAPs, dielectric elastomers can be considered to be prospective because they are very soft and their deformation is much greater than that of any other existing one. The deformation of dielectric elastomers can be used in various ways to produce actuation [9].

The dielectric elastomer actuators were discovered a decade ago that an applied voltage may cause dielectric elastomers to strain over 100% [10]. Because of this large

Y. Huang et al. (Eds.): ICIRA 2017, Part I, LNAI 10462, pp. 120–126, 2017.
DOI: 10.1007/978-3-319-65289-4_12

strain, dielectric elastomers are often called artificial muscles. In addition to large voltage-induced strains, other desirable attributes of dielectric elastomers include fast response, no noise, light weight, and low cost. The discovery has inspired intense development of dielectric elastomers as actuators for diverse applications [11].

Dielectric elastomer actuators, which provide a method of soft actuation, are used widely in robots. The stretched film-type actuators [12], the roll actuators [5], foldable actuators [13] and minimum-energy structure actuators [14] are the typical configurations of the actuator employing dielectric elastomer. The main drive-mechanism of these actuators relies on static electro-deforming characteristic or forced vibration, and the energy conversion efficiencies are not ideal. In this work, an ultralight locomotion robot, which combines a DER with ultralight foam, was prepared, and a suitable driving condition was obtained through experiments. The structure of our robot is very simple, it is ultralight, and moves quickly.

2 The Vibration Characteristics of dER

The principle of operation of a dielectric elastomer transducer is shown in Fig. 1. A membrane of a dielectric elastomer is sandwiched between two electrodes. For the dielectric elastomer to deform substantially, the electrodes are made of an even softer substance, with mechanical stiffness lower than that of the dielectric elastomer. A commonly used substance for electrodes is carbon grease. When the transducer is subject to a voltage, charge flows through an external conducting wire from one electrode to the other. The charges of the opposite signs on the two electrodes cause the membrane to deform.

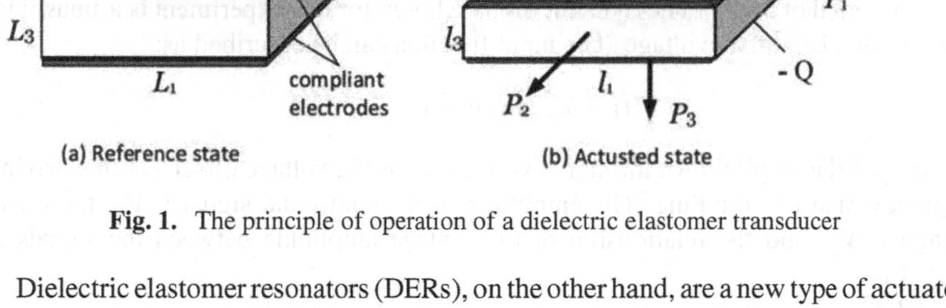

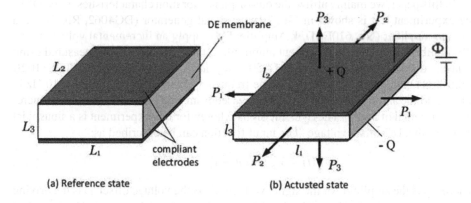

Fig. 1. The principle of operation of a dielectric elastomer transducer

Dielectric elastomer resonators (DERs), on the other hand, are a new type of actuators where a pre-stretched dielectric elastomer film resonates sharply in response to a dynamic loading voltage with a certain frequency. The structure of the DER used in this paper is shown in Fig. 2(a). A polyacrylic elastomer (VHB™ 4910 form 3 M™) was pre-strained bi-axially to 400% × 400%. The membrane was sandwiched between two

1 mm thick polymethyl methacrylate (PMMA) frames (70 mm × 70 mm) with an open hole (60 mm diameter). A carbon grease electrode (NO. 846 from MG Chemicals) was painted on both sides of the membrane, with the diameter of 40 mm, and two pieces of copper foil were connected to the carbon grease electrodes.

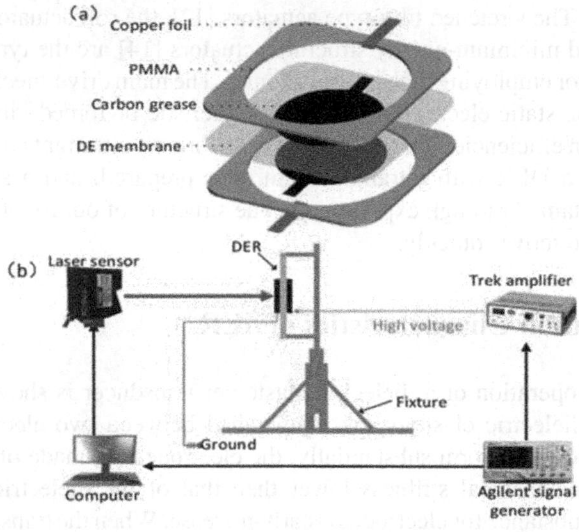

Fig. 2. (a) The structure of the DER, (b) the experiment test of the out-of-plane vibration characteristic of DER

In this paper, we mainly utilize the out-of-plane vibration characteristics of the DER, the experiment test is shown in Fig. 2(b). A signal generator (DG4062, Rigol) and a voltage amplifier (No. 610E, Trek AmplifierTM) supply an incremental voltage to the DE membrane, and the displacement amplitude at the center of DER is measured using a laser displacement sensor (LK-G150, Keyence). An oscilloscope (TBS1102B Tektronix) monitored the waveform. The frequency changed from 11 Hz to 310 Hz in 300 s, and the amplitude of the voltage increased from 5000 V to 7000 V with an incremental interval of 500 V. The dynamic signal chosen for this experiment is a sinusoidal signal with a DC-offset voltage. The input function can be described as:

$$V(t) = V_{ac} \sin \omega t + V_{dc} \tag{1}$$

where V_{ac} is the amplitude of the signal voltage, V_{dc} is the voltage offset, ω is the driving frequency, and t is the time. The amplitude of the sinusoidal signal is V_{ac}, the offset voltage is V_{dc}, and the relationship of the voltage amplitude between the signals is $V_{dc} = \frac{1}{2} V_{ac}$.

The results of the sweep frequency measured for different voltages are plotted in Fig. 3. As shown in Fig. 3(a)–(e), there are two primary resonance regions. Several other resonant frequencies were also observed for the range above 310 Hz. However, the DER is susceptive to breakdown failure in this frequency range. In addition, when the excitation voltage exceeded 7000 V, the DER failed at breakdown either [15]. Therefore, we limit both the excitation voltage and frequency to a safe range to ensure stable operation. There are only two resonance region at Fig. 3(a), and the width of the 2nd resonance region is wider than the 1st resonance region, along with the increasing of the voltage, there are many resonance peaks appear at low frequencies. This phenomenon mainly because of the DE membrane suffers loss of tension at high voltage. The DE membrane will crumpling under a static high voltage according to our previous study [16]. The driving mechanism of our robot is based on the out-of-plane vibration of DE membrane. The amplitude and frequency of the vibration have a great influence on the movement speed of our robot. The maximum displacement amplitude comparison of the 1st and the

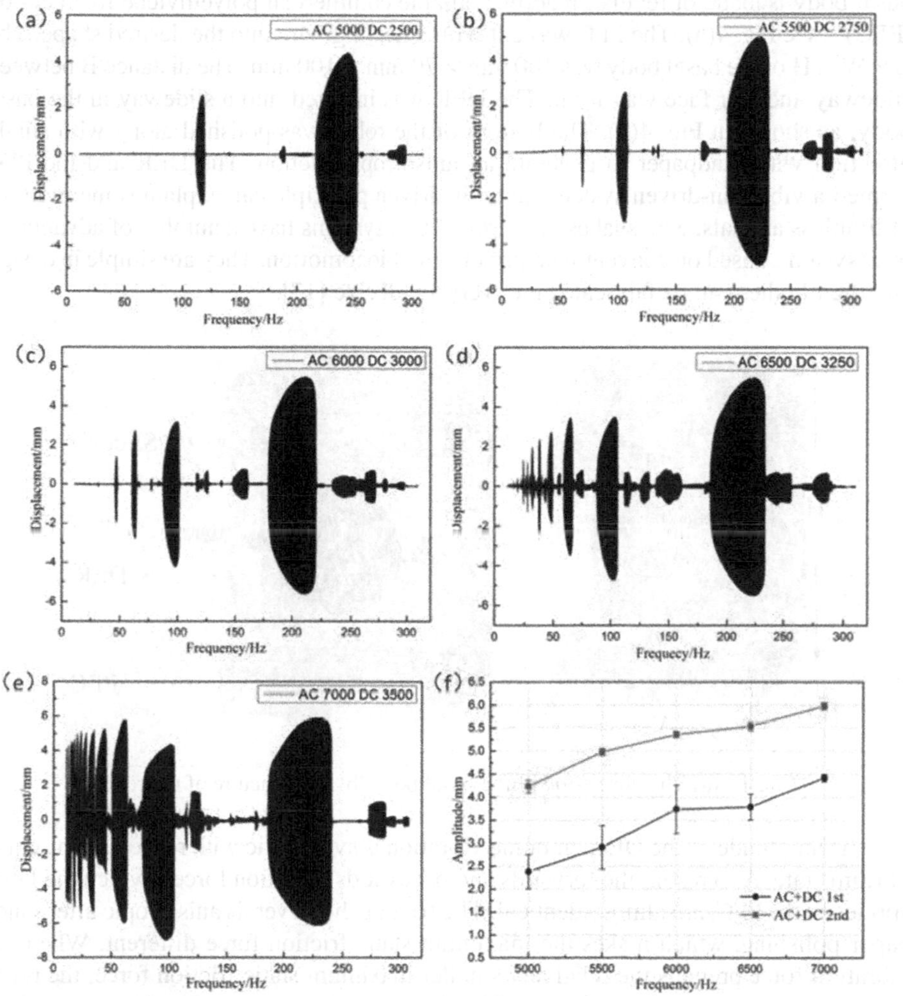

Fig. 3. (a)–(e) The amplitude of dielectric elastomer membrane vibration for different voltages and frequencies, (f) the maximum displacement amplitude comparison of the 1st and the 2nd resonance frequency

2nd resonance frequency is discussed in Fig. 3(f), the displacement amplitude of the 2nd resonance frequency is higher than the 1st resonance frequency, and the amplitude errors of the second resonance frequency are smaller than the first resonance frequency, which is reflected in the error bars. The amplitude errors of second resonance frequency are small which means the robot can move stably with different DERs if we choose this driving frequency.

3 The Structure and Locomote Performance of the Robot

The robot designed in this paper contains two parts: the DER and the basal body. The basal body is made of regular, readily available commercial polyethylene foam cotton (PFC) – see Fig. 4(a). The PFC was cut with a heating wire into the desired shape. The L × W × H of the basal body was 100 mm × 70 mm × 100 mm. The distance B between slideway and rear face was 1 cm. The DER was inserted into a slideway in the basal body, as shown in Fig. 4(b). The bottom of the robot was polished along with single direction with sandpaper to generate an anisotropy friction. The DER and the PFC formed a vibration-driven system, and this driven principle can explain some motions of limbless animals, e.g., snakes or worms. Such systems have a number of advantages over systems based on conventional principles of locomotion. They are simple in design and their bodies can be fabricated into very small size [17].

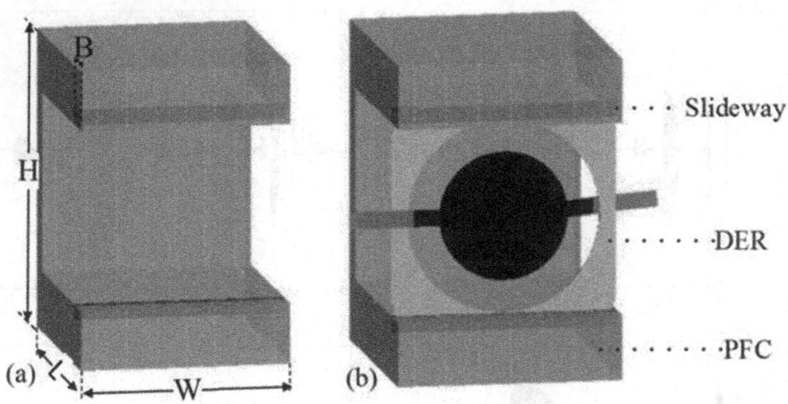

Fig. 4. (a) The dimension of the basal body (b) the structure of the robot

The amplitude of the DE membrane vibration is symmetric with respect to the equilibrium state. As a result, the forwards and backwards vibration forces, which the DER applies to the PFC, are almost identical. The friction, however, is anisotropic after sandpaper polishing, which makes the maximum static friction force different. When the vibration force prevails the relatively smaller maximum static friction force, the robot begins to move along that direction. The direction of movement can be corrected by changing the position of the DER along the slideway. When the DER moves to the left of the slideway, the robot will turn right and vice versa. Straight locomotion is realized by precisely positioning the DER. The weight of the DER is about 4.7 g and the weight of the PFC is 4.9 g. The speed of the robot is about 65 mm/s when driven by a sinusoidal signal at 220 Hz, and the amplitude of the sinusoidal signal is 6500 V with a DC offset of 3250 V. The speed of the robot with a loading weight of 2.5 g is about 56 mm/s. A series of locomotion video snapshots of the robot are shown in Fig. 5. The structural density of our robot is 14 kg/m^3 (mean structural density is defined as the total weight of a robot divided by the whole volume) which is lighter than most robot driven by piezoelectrics (PZTs) or shape memory alloys (SMAs), let alone the robot driven by motor.

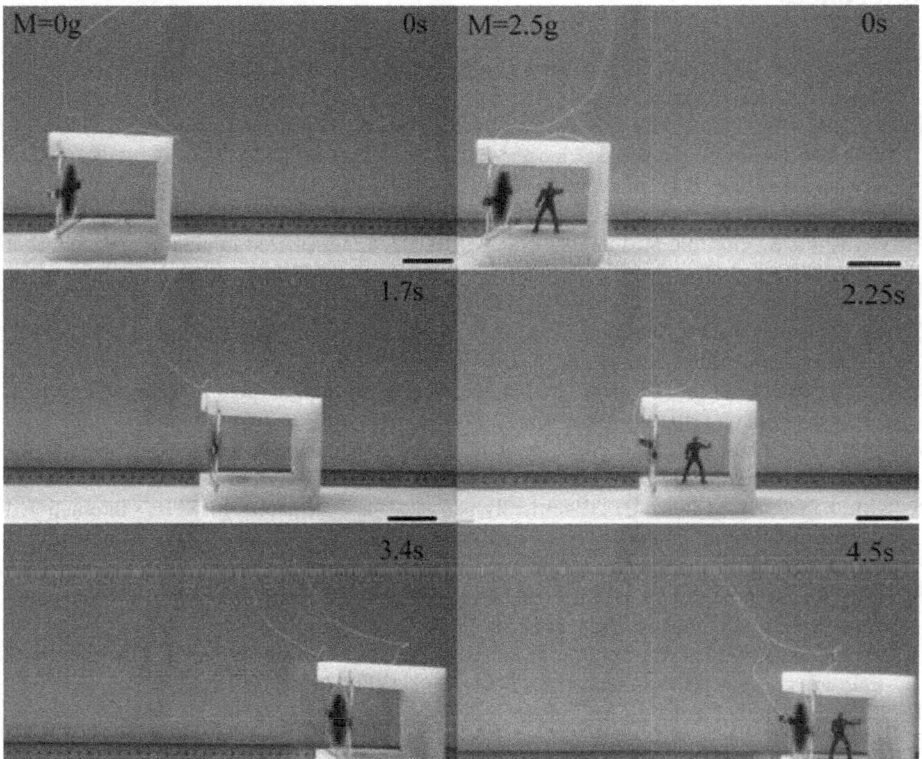

Fig. 5. A locomotion video snapshot of the robot with a 2.5 g toy and a 12 g toy. M is the weight of the toy; the scale bars are 5 cm

4 Conclusion

In summary, an ultralight locomotion robot actuated by a dielectric elastomer resonator was prepared in this study. The robot was made of two components: the DER and the PFC body. The direction of its movement can be adjusted by changing the position of the DER. The displacement amplitude for sweep frequency testing shows that the second resonance region, when the input signal is AC with a DC offset, is better suited to move the robot. This robot has an ultralight simple structure, and the open area in its center is available to carry a battery, a high voltage amplifier module, and a circuit board, which enable the robot to move freely. The resonance mechanism for the dielectric elastomer presented in this paper has an excellent property which can be used for the design of soft robotics.

Acknowledgements. This research was supported by the National Natural Science Foundation of China (Grant No. 91648110, 11402184, and 11321062).

References

1. Tolley, M.T., Shepherd, R.F., Mosadegh, B., et al.: A resilient, untethered soft robot. Soft Robot. **1**(3), 213–223 (2014)
2. Bartlett, N.W., Tolley, M.T., Overvelde, J.T.B., et al.: A 3D-printed, functionally graded soft robot powered by combustion. Science **349**(6244), 161–165 (2015)
3. Wehner, M., Truby, R.L., Fitzgerald, D.J., et al.: An integrated design and fabrication strategy for entirely soft, autonomous robots. Nature **536**(7617), 451–455 (2016)
4. Miyashita, S., Guitron, S., Ludersdorfer, M., et al.: An untethered miniature origami robot that self-folds, walks, swims, and degrades. In: 2015 IEEE International Conference on Robotics and Automation (ICRA), pp. 1490–1496. IEEE (2015)
5. Pei, Q., Rosenthal, M., Stanford, S., et al.: Multiple-degrees-of-freedom electroelastomer roll actuators. Smart Mater. Struct. **13**(5), N86 (2004)
6. Hu, Y., Wu, G., Lan, T., et al.: A graphene-based bimorph structure for design of high performance photoactuators. Adv. Mater. **27**(47), 7867–7873 (2015)
7. Park, S.J., Gazzola, M., Park, K.S., et al.: Phototactic guidance of a tissue-engineered soft-robotic ray. Science **353**(6295), 158–162 (2016)
8. Kim, J., Seo, Y.B., Choi, S.,H.: Electrically-activated paper actuators. In: SPIE's International Symposium on Smart Materials, Nano-, and Micro-Smart Systems, pp. 158–162. International Society for Optics and Photonics (2002)
9. Jung, K., Koo, J.C., Lee, Y.K., et al.: Artificial annelid robot driven by soft actuators. Bioinspiration Biomimetics **2**(2), S42 (2007)
10. Pelrine, R., Kornbluh, R., Pei, Q., et al.: High-speed electrically actuated elastomers with strain greater than 100%. Science **287**(5454), 836–839 (2000)
11. Suo, Z.: Theory of dielectric elastomers. Acta Mech. Solida Sin. **23**(6), 549–578 (2010)
12. Henke, E.F.M., Schlatter, S., Anderson, I.A.: A Soft Electronics-Free robot. arXiv preprint arXiv:1603.05599 (2016)
13. Sun, W., Liu, F., Ma, Z., et al.: Soft mobile robots driven by foldable dielectric elastomer actuators. J. Appl. Phys. **120**(8), 084901 (2016)
14. Zhao, J., Niu, J., McCoul, D., et al.: Phenomena of nonlinear oscillation and special resonance of a dielectric elastomer minimum energy structure rotary joint. Appl. Phys. Lett. **106**(13), 133504 (2015)
15. Huang, J., Shian, S., Diebold, R.M., Suo, Z., Clarke, D.R.: The thickness and stretch dependence of the electrical breakdown strength of an acrylic dielectric elastomer. Appl. Phys. Lett. **101**(12), 122905 (2012)
16. Li, B., Liu, X., Liu, L., Chen, H.: Voltage-induced crumpling of a dielectric membrane. EPL (Europhys. Lett.) **112**(5), 56004 (2015)
17. Fang, H.B., Xu, J.: Controlled motion of a two-module vibration-driven system induced by internal acceleration-controlled masses. Arch. Appl. Mech. **82**(4), 461–477 (2012)

Design and Test of a New Spiral Driven Pure Torsional Soft Actuator

Jihong Yan(✉), Binbin Xu, Xinbin Zhang, and Jie Zhao

Harbin Institute of Technology, Harbin 150001, People's Republic of China
jhyan@hit.edu.cn

Abstract. Owing to the twist degree of freedom (DOF), soft torsional motion can increase flexibility and quickly achieve positional and attitude adjustment in complex and narrow spaces. Compared with bending actuator, there is much less research on soft torsional actuators. Current soft torsional actuators often accom-pany with other motion couplings, so it is difficult to provide pure twist. Based on the princi-ple of spiral chambers with pneumatic driving, a new type of torsional actuator module is designed in this paper. Combined with finite element simulation, the ge-ometric parameters of the module are optimized and then fabrication is carried out by two stages. In order to control the module, a kinematic model, which is the relationship between the air pressure and the twist angle, is established by means of experimental calibration. Finally, a test platform is set up, which is used for measuring the static characteristics of the designed module. The maximum ob-tainable torsion angle and torque are obtained separately through the experiments on torsion angle test and torsion torque test.

Keywords: Soft torsional actuator · Pure twist · Kinematics · Static characteristics

1 Introduction

Due to low morphology, high energy-weight ratio and inherent compliance, soft robots have infinite DOF and can adapt to complex environments, which make them become an active area in robotic research in recent years. Soft bending actuators have been extensively studied, because they have the bending ability to mimic the motion of mollusks like earthworms, snakes, worms, elephant trunks and human hand. However, there are few studies on torsional actuators. Through the observation of soft creatures in nature, most animals have evolved with twisting DOF, such as in the shoulders, wrists, ankles and including the mentioned-above mollusks. Torsional actuators can effectively improve the flexibility of motion and have many advantages. Firstly, increase flexibility. Compared to a single bending actuator, torsional actuators can conveniently implement the same twisting operations and require smaller space. Besides, they are able to achieve complex positon and attitude quickly, which can be used in some twist joints that require compact structure and simple cable transmission, but do not need precise positional and force control. Secondly, reduce resistance. When crossing the narrow space like a hole or tunnel, soft torsional actuator can reduce the resistance by generating torsional

Y. Huang et al. (Eds.): ICIRA 2017, Part I, LNAI 10462, pp. 127–139, 2017.
DOI: 10.1007/978-3-319-65289-4_13

deformation. Thirdly, cooperate movement. By operating with other functional components, such as bending and elongation modules, torsional actuators can realize a lot of complex motion.

As generating direct twisting motion is difficult, it is more common to design a torsional actuator based on other forms of motion. For example, rotary Peano actuators [1, 2], antagonistic shape actuators [2–4], fiber weave actuators [2, 5, 6], and pleated chamber actuators [7, 8], most of them adopt the principle of strain difference and depend on the radial or axial expansion deformation. They generate twist motion by the guide of confined layer like strings or fibers. As it is difficult to control the other direction deformation, the twisting motion is always coupled with radial or axial expansion and bending distortion. At present, research on soft torsional actuator is as follows:

Rotary Peano actuators are designed based on the principle of linear Peano actuators, utilizing the contraction occurring over the width of a long tube when it is inflated. Hideyuki et al. [1] developed a "flat tube" made of urethane, which can change its cross-sectional shape from flat to round when pressurized. When helically stacked into a cylinder, it can produce linear motion under pressurization while two positions with a circumferentially different phase between up and down are connected to each other by a wire at the same time, it can generate rotational motion in conjunction with extension. Siddharth et al. [2] made a rotary Peano actuator using the same contraction principle of the linear Peano actuators, which utilizes the contraction occurring over the width of a long tube when it is inflated. Rotary motion is generated by virtue of their helical arrangement on a cylinder but also with the change of radial, axial dimensions when pressurized.

Antagonistic shape actuators use spiral shaped structures. If the fabric at the central axis of the actuators is assumed to be inextensible, these actuators can produce rotational motion when the helix chambers are inflated. Siddharth et al. [2] introduced the principle of an antagonistic shape actuator and designed a fluidic torsional actuator by combining two oppositely oriented helix like structures into a single structure to generate pure torsional motion with minimal linear motion. Ellen T. et al. [3] utilized helical soft tubes made of fluid elastic silicon as fluid elastic actuating element. By putting them on the surface of the heart, they can produce twist and untwist motion.

Fiber weave actuators are defined by the constraints due to the nearly inextensible fibers on its surface. Just like the principle of McKibben actuators, these actuators generate torsional motion by the guide of the asymmetric arrangement of fibers. Panagiotis et al. [5] designed a mechanically programmable soft pneumatic actuator with silicone rubber and polymer fibers, which can achieve a wide range of motions, including axial extension, radial expansion, and twisting. In order to maximize the torsional motion, they performed finite element simulations of the actuators by simply varying the parameter of fiber angle. These soft actuators can be used in flexible and compliant endoscopes, pipe inspection devices, and assembly line robots. Hong Kai et al. [6] made an improvement on the previous developed actuators. When fabricating the soft module, they used corrugated outer layer to reduce the radial expansion and realized twisting motion through the oblique corrugated outer layer.

Pleated chamber actuators depend on the fold structure like the accordion. When the center is constrained and the chambers are inflated, the pleated chamber actuators can generate true rotary motion different from regular torsional motion. Oleg Ivlev et al. [7] developed a rotary actuator with pleated rotary elastic chambers which used a hard motor shell to generate torque from 11 Nm to 20 Nm. This actuator is mainly used for motion therapy devices for the lower extremities, providing patient friendly rehabilitation. Kargov et al. [8] also developed a likewise actuator module with the structure of pleated chamber. They combined this module and elongation module together to make an arm prototype.

These soft torsional actuators above have many problems, including coupling motion, not pure twist, difficulty in matching pressure in different chambers, etc., this paper uses the drive mode of the fluid elastic actuator (FEA) to directly generate the torsional motion by inflating spiral chambers. First of all, the structure design of the new soft torsional module is introduced. Then the geometric parameters of the soft module are optimized based on the results of finite element analysis and the fabrication is also completed. In order to establish the kinematics model of the module, a test platform consisting of a pneumatic driven system and a measurement platform is developed to calibrate the kinematics. Finally, it has been used in the test of torsion angle and torsion torque to obtain the static characteristics of the soft torsional actuator module.

2 Fabrication of Soft Torsional Module

2.1 Structure Design

As shown in Fig. 1, the soft torsional module has been designed based on the design idea of screw type twisting.

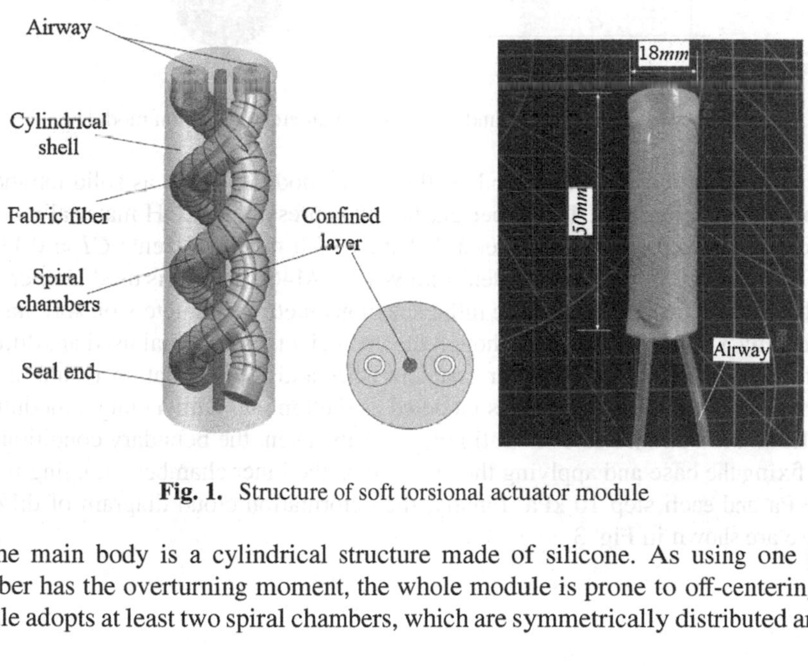

Fig. 1. Structure of soft torsional actuator module

The main body is a cylindrical structure made of silicone. As using one spiral chamber has the overturning moment, the whole module is prone to off-centering, the module adopts at least two spiral chambers, which are symmetrically distributed around the center of the actuator, to improve the twisting stability and avoid twisting offset. The inner spiral chamber is a hollow cylinder with spiral thin fiber line surrounded outside which is used to limit radial deformation and provide more efficient elongation. One end of spiral chamber is sealed with a harder silicone and the other end is connected to the airway for gas actuation. In addition, an inextensible material embedded in the center of the whole cylinder is used to limit axial elongation. When spiral chambers are inflated simultaneously, they generate elongation motion. But the constraint layer limits elongation and turns the driving force into moment, realizing torsional motion.

2.2 Geometric Parameter Optimization

Modeling. Figure 2 is the cross-sectional view of the module, in which the geometrical structure is affected by five variables: center distance of chambers $D = 2R$, diameter of module D_r, spiral angle α, diameter of inner spiral chambers $d = 2r$ and the whole length L. In order to obtain the response of twisting performance to every geometric parameter, Finite Element Method (FEM) software ABAQUS is used and controlling variable method is adopted to optimize the structure design.

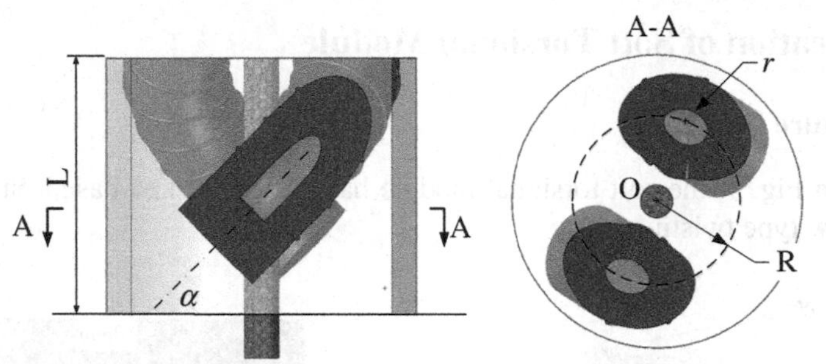

Fig. 2. The configuration and geometric parameters of torsion modules

When establishing the FEM models, the whole module is split as solid tetrahedral quadratic hybrid elements. The hyper-elastic incompressible YEOH material model is used to describe the nonlinear material behavior with the coefficients $C1 = 0.11$ and $C2 = 0.02$. The coefficients are chosen from Wacker M4601 which is used in references [9, 10]. Here, we just care about the influence of geometric parameters on soft material described by YEOH model. Even though the coefficients of material used are different from fabrication, they have similar characteristics and the simulation results is also applicable. The constraint layer was modeled as shell model with Young's modulus of $E = 31.067$ MPa and a Poisson's ratio of $\nu = 0.36$. Then, the boundary conditions are set by fixing the base and applying the pressure on the inner chambers, ranging from 0 to 60 kPa and each step 10 kPa. Finally, the deformation cloud diagram of different pressure are shown in Fig. 3.

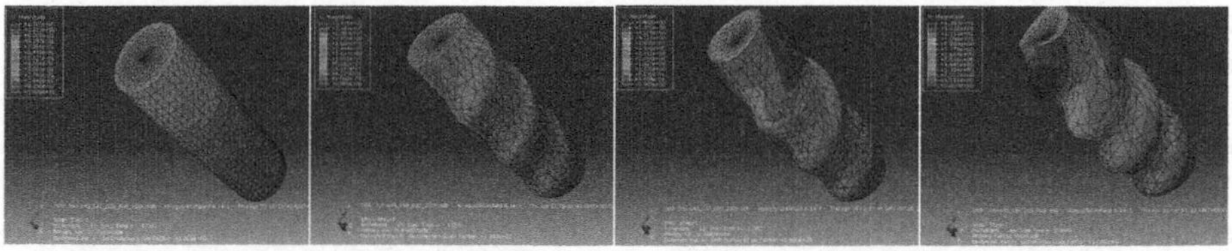

a) P=10KPa b) P=30KPa c) P=40Kpa d) P=60KPa

Fig. 3. FEM simulation results of torsional module

Results. As the designed module is required to adapt to narrow space, its diameter should not be too large. Here, D_r is fixed as 18 mm for simulation. Firstly, the other parameters are kept unchanged and the center distance of chambers is set to 6 mm, 8 mm, 10 mm, and 12 mm respectively. It is shown in Fig. 4(a) that as the center distance of chambers increases, the torsion angle also increases. However, when the pressure is relatively lower, it is not clear whether the elastic deformation or the air pressure of the soft module plays the leading role. As the air pressure reaches a certain value (about 35 kPa) and begins to dominate, the relationship between air pressure and torsion angle becomes linear.

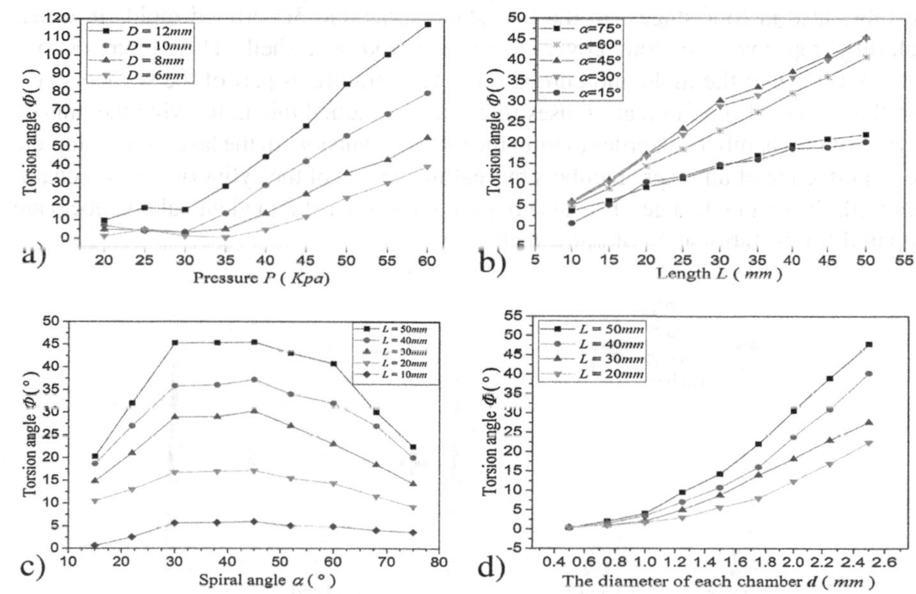

Fig. 4. Influence of different geometric parameters on the torsion angle. The parameters are (a) center distance of spiral chamber, (b) length, (c) spiral angle, (d) diamater of each chamber

In order to obtain the influence of the other geometric parameters on the torsion angle, the pressure in the linear region is chosen as the fixed parameter, i.e., $P = 50$ kPa and the other three parameters are changed respectively. Through analyzing a large amount of simulation of different modules, we can get the following results: Fig. 4(b) shows that when the whole length of the module becomes longer, the torsion angle becomes larger. Figure 4(c) shows that when the spiral angle locates between 30° and 45°, the torsion angle seems to be lager. The reason for that is as spiral angle become lager, the spiral chambers helically stacked into a cylinder are self-interference and difficult to deform, leading to smaller torsion angle. Therefore, it is suitable to choose the spiral angle by 45°. Figure 4(d) shows that as the diameter of inner spiral chambers increases, the torsion angle increases.

Considering the size of designed module should not be too large, the length of the whole module is chosen as 50 mm. As larger center distance of chambers leads to thinner outer wall and bubble deformation, it is also necessary to choose the suitable center distance. By thinking the above influence, the final selected parameters are: $\alpha = 45°$, $d = 2$ mm, $D = 10$ mm, $\mathrm{L} = 50$ mm, $\mathrm{D}_r = 18$ mm.

2.3 Fabrication

Overall, the process of the soft torsional module comprises the fabrication of the inner spiral chambers and the outer cylindrical shell.

Fabrication of the inner spiral chambers. As shown in Fig. 5, the spiral chambers are fabricated in four stages. In the first stage, assemble 3D printed molds in order, including top cover, bottom cover, cylindrical rod and shell. Then pour silicone Ecoflex-0030 into the mold and cure at room temperature as part of the second stage. The third stage of the fabrication uses fine fiber or braided thread to twine the outside surface of the chambers, in order to limit the radial expansion. In the last stage, complete the manufacture of an inner chambers by sealing the end of the cylindrical spiral chambers with the mold silica gel. It is also required that at least two identical chambers are prepared for the fabrication of outer shell.

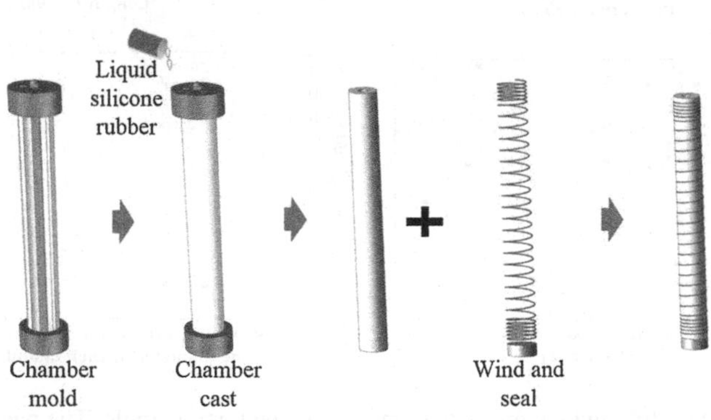

Fig. 5. Fabrication process of inner chamber

Fabrication of outer cylindrical shell. The outer cylindrical shell is also fabricated in four parts (Fig. 6). Firstly, install the finished inner chambers on the 3D printed spiral supporting frame to make a spiral configuration. Secondly, assemble other 3D printed molds in order, consisting of top cover, bottom cover, shell, and place confined layer in the center. In the third stage, pour silicone Ecoflex-0030 into the mold, then put the spiral configuration into the mold and cure at room temperature. Lastly, take out the curing module by dismantling the outer mold and withdrawing the support. Connect a medical silicone tubes to the other end of inner chambers for gas actuation. Thus, the fabrication of a soft torsional actuator module is completed.

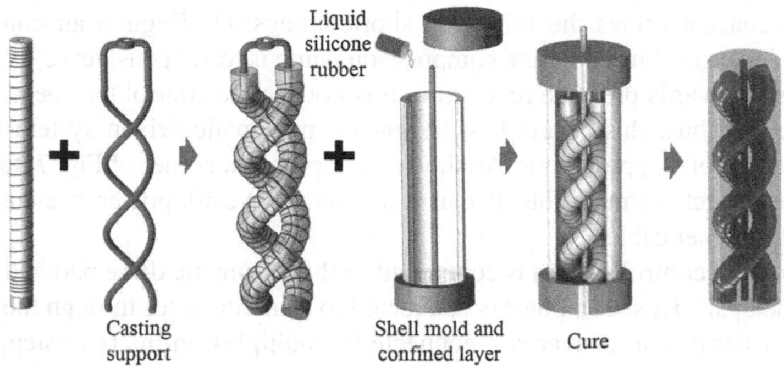

Fig. 6. Fabrication process of outer shell

3 Test Platform of Soft Torsional Module

The test platform consists of a pneumatic driven system and a measurement platform. The pneumatic driven part takes responsible for inflating the module and the experimental measuring part is used for testing the static characteristics of the module. As shown in Fig. 7, considering the collaboration of multi-modules and the scalability of

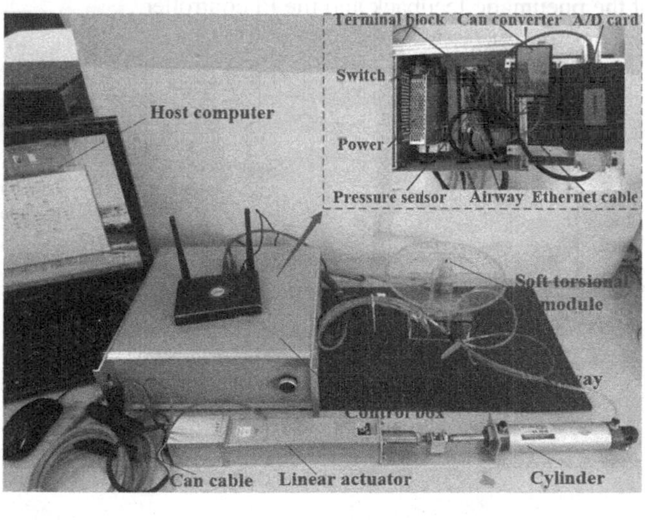

Fig. 7. The composition of test platform

subsequent experiments, the whole system mainly contains five parts: host computer, control box, linear actuator, cylinder and measurement platform.

3.1 Pneumatic Driven System

At present, research institutions and scholars at home and abroad often use air/liquid pump and electromagnetic control valve to drive and control the soft robots. However, this design concept brings the following shortcomings: (1) Regular air compression pump have large size but miniature compression pump is too expensive; (2) As they are mostly biased towards pressure regulation, it is not easy to control the speed and flow in the process. Thus, this article has designed a pneumatic driven system based on adjustable speed of stepper motor. As shown in upper right corner of Fig. 7, the control box includes switch, terminal block, can converter, A/D card, power, pressure sensor, airway and Ethernet cable.

The circuit of control system is comprised of the pneumatic drive part and the pressure feedback part. Host computer is connected to Can-converter through the Ethernet and the other side of Can-converter is connected to multiple Can-interface stepper motor driver by cascade connection. Each stepper motor drives a linear telescopic rod which drives the cylinder to fill and bleed the soft module. The acquisition part uses A/D acquisition card and pressure sensors to measure the pressure of the output of each cylinder, and then transmits the collected data back to the host computer to realize pressure feedback loop control.

Soft robots always have the characteristics of coupling motion and nonlinear deformation. Actually, some special tasks often need several modules cooperating to finish, through analyzing the features of environment and types of tasks, it has been proved effective by using the closed-loop control of pressure and open-loop control of position system. Figure 8 shows how the system works. P_d is expected value of input pressure and P_r is the pressure feedback collected by the pressure sensors. A closed-loop system is composed of the pneumatic feedback and the PI controller.

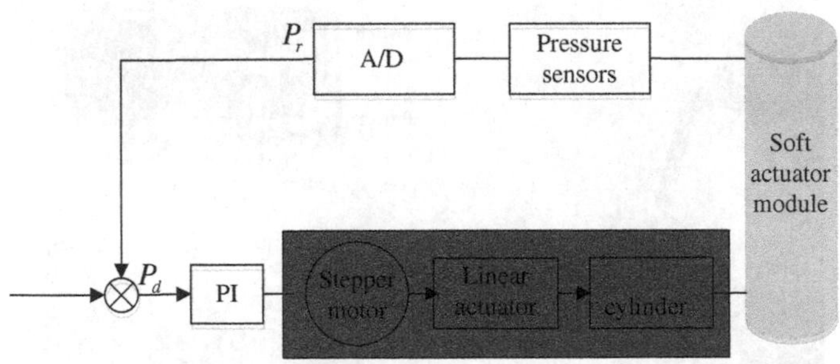

Fig. 8. Working principle of pneumatic driven system

3.2 Measurement Platform

A measurement platform is set up to measure the relationship between the pressure and torsion angle (details in Fig. 11(a)). It consists of the base, whole circle instrument and rotation pointer. The bottom of the torsional module is fixed to the center of the whole circle instrument and a rotation pointer is mounted on top to indicate the torsion angle. When used for test, it needs to cooperate with pneumatic driven system for driving. The airways of the driven system connect to the airway of the soft module, thus, the pressure can be adjusted by the host computer.

4 Experiments

Since the test platform is completed, it can be used for kinematic calibration and characteristics measurement including the torsion angle test and torsion torque test.

4.1 Kinematic Calibration

For the purpose of controlling the soft torsional module, kinematics modeling is needed. Due to the nonlinearity of soft material, it is difficult to establish the kinematics model. However, from the FEM analysis results shown in Fig. 4, when the pressure increases, the torsion angle also increases. Though it shows nonlinear at the initial part, it seems to be linear relationship when the pressure reaches certain value. Thus, the kinematic calibration method can be used to solve the kinematics.

Kinematic fitting. Now, we use the test platform above to calibrate the kinematics. Firstly of all, the angle measuring increment is set to 5°, and the measuring range is 0–10°. Then, the torsional module is inflated to each specific angle and the air pressure is recorded. Finally, the data on air pressure - torsion angle (see in Fig. 9, blue curve) are obtained by averaging the values of five measurements.

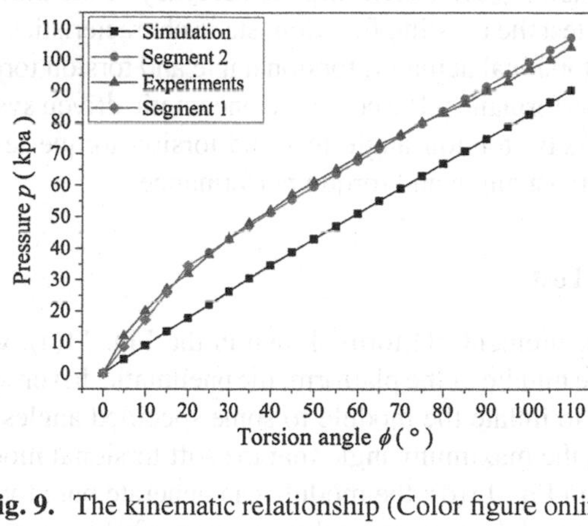

Fig. 9. The kinematic relationship (Color figure online)

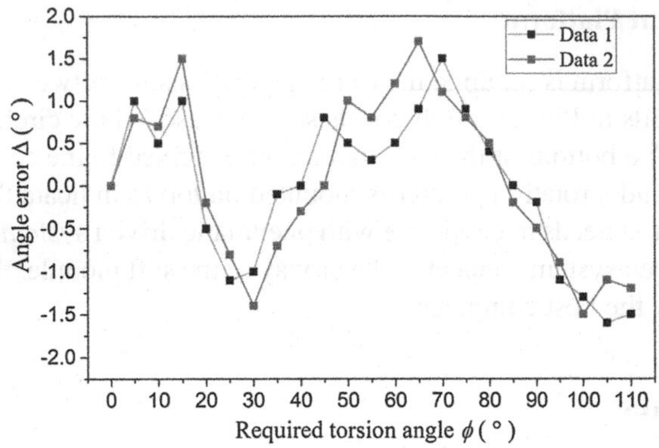

Fig. 10. The angle error between simulation and experiment

The result of kinematic fitting as shown by the green and red line in Fig. 8 is consistent with the measured curve (blue). In the lower part with a torsion angle <20°, as the air pressure is relatively lower, it is not clear whether the elastic deformation or the air pressure of the soft module plays the leading role. So we use a segmentation linear function to describe each corresponding relationship separately when the angle is within the range 0–20° and more than 20°. The final segmentation fitting results are plotted with green and red lines in Fig. 9.

Kinematic Verification. As the segment kinematics model is established, the motion control of the soft torsional module can be carried out. To verify the accuracy of the kinematics, the test experiment has been adopted. Firstly, a set of specified angles are chosen and the segment kinematics is used to calculate the values of the required air pressures. Secondly, calculated air pressures are used to inflate the torsional module to reach actual torsion angles. Through subtracting the actual torsion angles from the specified angles chosen, the angle error is shown in Fig. 10 above, which presents the torsion angle error is about ±1.7°, verifying the accuracy of the kinematics model.

In order to further test the twisting function, static characteristics tests are conducted on soft module. To a torsional actuator, torsion angle and torsion torque are two important indicators of its performance. Based on the pneumatic driven system, the following two experiments, namely, torsion angle test and torsion torque test, are designed to obtain the maximum twist angle and torque performance.

4.2 Torsion Angle Test

According to the measurement platform shown in the Fig. 11(a), when soft torsional module is placed in the middle of the platform, the pneumatic driven system is controlled by the host computer to inflate the module to some specified angles. By increasing the pressure step by step, the maximum angle that the soft torsional module can achieve is measured. As shown in Fig. 11(d), the module can generate pure torsional motion with minimal linear motion and the torsion angle can reach 120°.

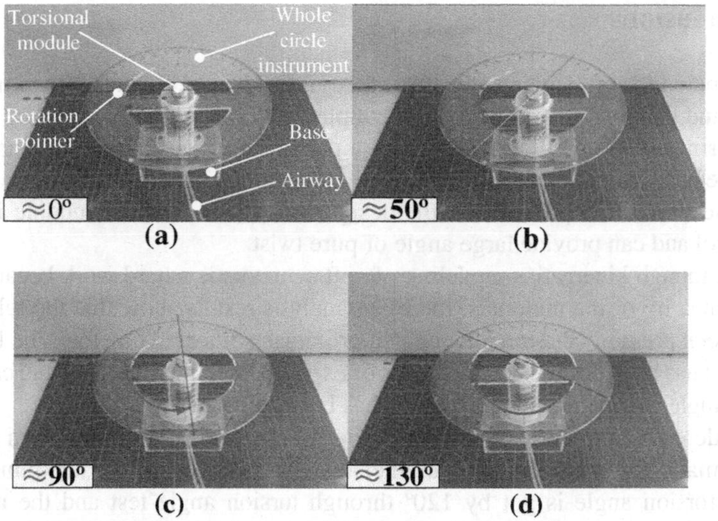

Fig. 11. Experiments of torsion angle test

4.3 Torsion Torque Test

A force sensor is added on the measurement platform for torque test, as shown in Fig. 12(a). The rotation pointer and force sensor equal to the force arm and force. The initial angle is preset at zero displacement and the relationship between air pressure and torque can be measured by increasing the pressure gradually. Through averaging the values of several measurements, the final result is shown in Fig. 12(b). The maximum obtainable torque corresponding to zero displacement is 0.026 Nm.

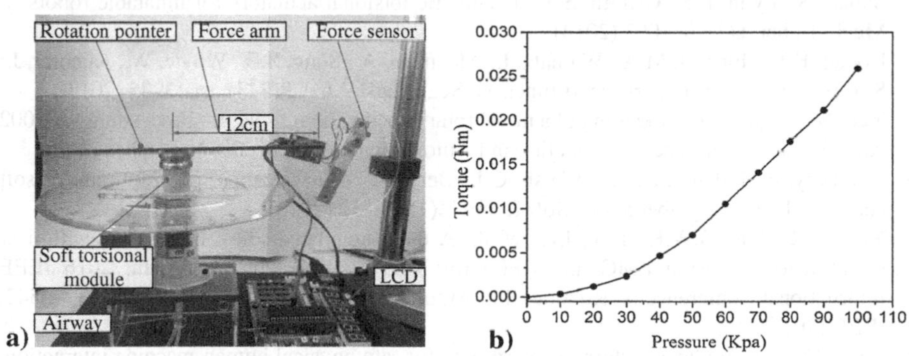

Fig. 12. Experiments of torsion torque test. (a) experimental device; (b) Torque characteristics for the module

5 Conclusions

(1) Compared to soft bending actuators, there is little research on soft torsional actuators and the torsion motion is often coupled with other kinds of motion. Based on the principle of spiral chambers with a pneumatic drive, a new torsional module has been designed and fabricated, whose geometric parameters have been optimized by the FEM. It is proved that the designed torsional actuator module is easy to control and can provide large angle of pure twist.

(2) Even though kinematics modeling of soft actuators is a hard work because of the nonlinearity of the materials, the FEM analysis results show that the relationship between pressure and torsion angle approximates linear. Therefore, the kinematic model has been established by means of experimental measurement and calibration and angle control precision of ±1.7° has been achieved.

(3) In order to obtain the static characteristics of the module, a test platform including pneumatic driven part and measurement part has been set up. The maximum obtainable torsion angle is got by 120° through torsion angle test and the maximum obtainable torque corresponding to zero displacement is got by 0.026 Nm through torsion torque test.

Acknowledgments. Research supported by Self-Planned Task (No. SKLR201501A05) of State Key Laboratory of Robotics and System.

References

1. Tsukagoshi, H., Kitagawa, A., Kamata, Y.: Wearable fluid power composed of transformed flat tube actuators. In: 2002 IEEE/RSJ International Conference on Intelligent Robots and Systems, Hamburg, pp. 1178–1183. IEEE (2002)
2. Sanan, S., Lynn, P.S., Griffith, S.T.: Pneumatic torsional actuators for inflatable robots. J. Mech. Robot. **6**(3), 031003 (2014)
3. Roche, E.T., Horvath, M.A., Wamala, I., Alazmani, A., Song, S.E., Whyte, W., Kuebler, J.: Soft robotic sleeve supports heart function. Sci. Transl. Med. **9**(373), eaaf3925 (2017)
4. Yee, N., Coghill, G.: Modelling of a novel rotary pneumatic muscle. In: Proceedings of 2002 Australiasian Conference on Robotics and Automation, vol. 27, p. 29, November 2002
5. Connolly, F., Polygerinos, P., Walsh, C.J., Bertoldi, K.: Mechanical programming of soft actuators by varying fiber angle. Soft Robot. **2**(1), 26–32 (2015)
6. Yap, H.K., Ang, B.W.K., Lim, J.H., et al.: A fabric-regulated soft robotic glove with user intent detection using EMG and RFID for hand assistive application. In: 2016 IEEE International Conference on Robotics and Automation (ICRA), Stockholm, pp. 3537–3542. IEEE (2016)
7. Ivlev, O.: Soft fluidic actuators of rotary type for safe physical human-machine interaction. In: 2009 IEEE International Conference on Rehabilitation Robotics, Kyoto, pp. 1–5. IEEE (2009)
8. Kargov, A., Breitwieser, H., Klosek, H., Pylatiuk, C., Schulz, S., Bretthauer, G.: Design of a modular arm robot system based on flexible fluidic drive elements. In: 10th International Conference on Rehabilitation Robotics, Noordwijk, pp. 269–273. IEEE (2007)

9. Polygerinos, P., et al.: Towards a soft pneumatic glove for hand rehabilitation. In: 2013 International Conference on Intelligent Robots and Systems (IROS), Tokyo, pp. 1512–1517. IEEE (2013)
10. Soft Robotics Toolkit. http://softroboticstoolkit.com/book/pneunets-modeling. Accessed 3 June 2017

Design of a Soft Pneumatic Actuator Finger with Self-strain Sensing

Yi-Dan Tao and Guo-Ying Gu(✉)

State Key Laboratory of Mechanical System and Vibration, School of Mechanical Engineering, Shanghai Jiao Tong University, Shanghai 200240, China
guguoying@sjtu.edu.cn

Abstract. Benefit from high compliance, lightweight and natural motion, soft pneumatic actuators (SPAs) have shown great potential for using as fingers in robotic hands. Currently designed SPAs lack strain feedback, which limits their further applications. Soft fingers using SPAs require sensors embedded in their bodies for strain feedback, and most of the reported sensors used in SPAs are commercial products which suffer from low compliance, bulky volume and integration complexity. This work reports a SPA finger with self-strain sensing by integrating a SPA and a dielectric elastomer (DE) film capacitive strain sensor. The single-chamber with fiber reinforced SPA is chosen for bending motion. The DE film capacitive strain sensor is designed with a simple sandwiched structure which is easy-to-fabricate and thick enough for high compliance. The materials chosen to fabricate the SPA and the sensor are the same which helps to make the integration natural and closely bonded. The actuation response as well as sensing outputs are experimentally analyzed and results show that the sensing output change linearly with bending angles, which can be used for further position feedback control.

Keywords: SPA finger · Bending · Self-strain sensing

1 Introduction

Hands serve prehensile functions for us to interact with external world with high dexterity [1], and loss of the ability to move the fingers, whether partial or total, can greatly limit daily activities [2]. So it is necessary for patients with hand dyskinesia to do rehabilitative treatment using external actuators. Nowadays there are many actuators developed to help in hand rehabilitation, most of these devices are rigid and powered by motors. Applications requiring higher power density, more compliant, lighter and smaller actuators such as piezoelectric actuators [3], shape memory alloy (SMA) based actuators [4] and SPAs [2, 5–9]. Although piezoelectric actuators can be compliant, most of them are designed to be applied in high micromanipulation due to their limited range of motion. Compared with SMA based actuators, the SPAs show significant potential for both rehabilitation and gripping assistance [10, 11] as they are easier to fabricate, undergo more natural deformation, and response faster.

Y. Huang et al. (Eds.): ICIRA 2017, Part I, LNAI 10462, pp. 140–150, 2017.
DOI: 10.1007/978-3-319-65289-4_14

Recent developments of SPAs can be divided into two categories according to the structural design: the single-chamber with fiber reinforced actuators [5, 12, 13] and multi-chambered actuators [2, 6]. These two kinds of SPAs can both achieve bending motion with compressed air to mimic the motion of human fingers. According to fabrication process, multi-chambered actuator usually requires casting the cavity layer and cover layer separately and then sticking them together [14], so it has worse air tightness than the single-chamber with fiber reinforced actuator which only needs one casting molding. These SPAs are designed to achieve a certain kind of motion and few of them are made with sensory feedback. In order to enable the closed-loop control of SPAs, the integration of sensors is necessary. Traditional pneumatic actuators usually use displacement and force sensors such as linear encoders and load cells for displacement and force feedback [15], but these rigid sensors are not suitable for SPAs because they will reduce compliance. Commercially available flexible sensors usually suffer from low sensitivity and signal drift [16], so there are some researches focused on custom soft deformable sensors such as liquid microchannel sensors [17–19], braided fiber mesh [20] and optical waveguides [21, 22]. Park et al. [18] present a pneumatic artificial muscle (PAM) actuator with liquid-metal-based resistive sensing microchannel embedded, where the air pressure stretches the elastomer layer of the PAM to make a motion as well as change the resistivity of the conductive micro-channel for sensory feedback. This kind of sensor can be sensitive, but it relies on expensive material and requires complex process. To simplify the fabrication process, the conductive braided fiber mesh is adopted. Then, the displacement of the SPA can be calculated from the change of the resistance of the mesh [20]. However, this design is only suitable for a few types of SPAs and it is generally incompatible with large strains. To address these problems, Zhao et al. [21] report a soft pneumatic hand with stretchable optical waveguides integrated for strain sensing. Based on the principle of total reflection of light, the optical waveguide consists of two optical media. The compressed air causes bending of the pneumatic finger which results in the deformation of the optical waveguide and makes the received light intensity change. Although the optical waveguide can undergo large deformation and it is easy to fabricate, it contains rigid parts such as light-emitting diodes (LED) and photodetectors which increases the overall size of the SPA and limits the compliance.

In this work, a SPA finger with DE film capacitive strain sensor integrated is developed. In order to possess high air tightness, the single-chamber with fiber reinforced actuator is adopted. The DE film capacitive strain sensor is designed as a sandwiched structure in which a DE membrane with carbon grease coated on both sides is used as the sensing capacitor and the other two membranes serve as sealing layers. Benefiting from the simple structure and thin thickness, this DE film sensor is compliant, stretchable and easy-to-fabricate. The novelties of this SPA finger lie in the facts that: (i) the use of DE film capacitive strain sensor for self-sensing; (ii) the use of same material to fabricate SPA and sensor, which achieves a robust integration of actuation and sensing.

The remainder of the paper is organized as follows. Section 2 presents the design of the SPA finger and the details about experimental setup. In Sect. 3, the bending angles of the SPA finger under different air pressure levels as well as the sensing outputs are measured, then the response behavior and sensing relations are experimentally analyzed. Finally, conclusions are drawn in Sect. 4.

2 Design and Fabrication

The designed self-strain-sensing SPA finger is shown in Fig. 1, which consists of a single-chamber with fiber reinforced actuator for bending actuation, a DE film capacitive strain sensor for strain sensing and the outmost protective layer. The details of the design and fabrication of the actuator and the sensor are described below.

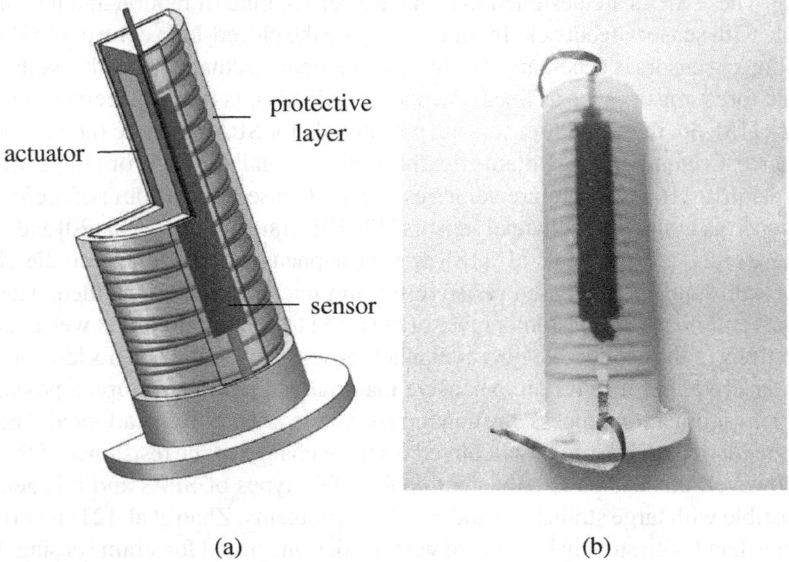

Fig. 1. Description of the designed self-strain-sensing SPA finger: (a) 3D model with a partial sectional view; (b) a fabricated sample

2.1 Actuator

The structure of the single-chamber with fiber reinforced actuator is shown in Fig. 2. The steps to fabricate the actuator are: (1) casting the liquid silicone rubber (Dragon Skin 30, Smooth-On) into the molds to form the actuator body with a single air chamber; (2) winding the thread around the actuator body along the entire length to limit the circumferential extension; (3) gluing the flexible gauze to the flat face of the actuator body to limit the elongation of the flat face and thus make the actuator bend under the pressurized air.

2.2 Sensor

Dielectric elastomers, such as polydimethylsiloxane (PDMS) and VHB membranes (a commercial product from 3 M company) are generally used for fabricating soft stretchable sensors. In this work, in order to make a robust integration of the actuator and the sensor, the silicone rubber used in fabricating the actuator is adopted for making the DE film capacitive strain sensor. The structure of this sensor is shown in Fig. 3.

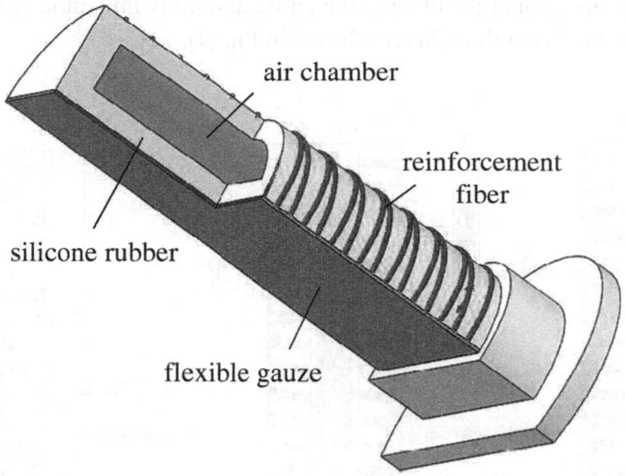

Fig. 2. Description of the single-chamber with fiber reinforced actuator

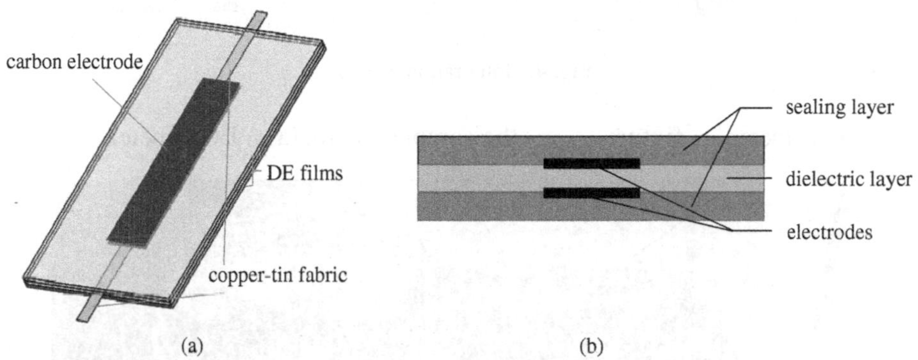

Fig. 3. Description of the DE film capacitive strain sensor: (a) 3D model; (b) cross-sectional view

This sensor is composed of three DE films, in which one film with carbon grease coated on both sides (serve as compliant electrodes) serves as the sensing capacitor, other two films serve as the sealing layer to protect the sensing capacitor from damage and pollution. The sensing principle relies on the phenomenon that the strain causes changes of the capacitor's plate area and distance, which contributes to the capacitance variation. DE films are fabricated by casting the liquid silicone rubber to the coating machine (ZAA2300, Zehntner) and heating at 60 °C for 20 min to form the 200 μm thick films. After coating carbon grease on the dielectric layer film to form the capacitor, the sealing layer films are glued to the capacitor with liquid silicone rubber and then heat the sample at 60 °C for 20 min.

2.3 Integration and Experimental Setup

The self-positional-sensing SPA finger is fabricated by gluing the sensor to the bending surface of the actuator and then placing the entire assembly into another mold to encapsulate the body in a 2 mm thick layer (shown in Fig. 4).

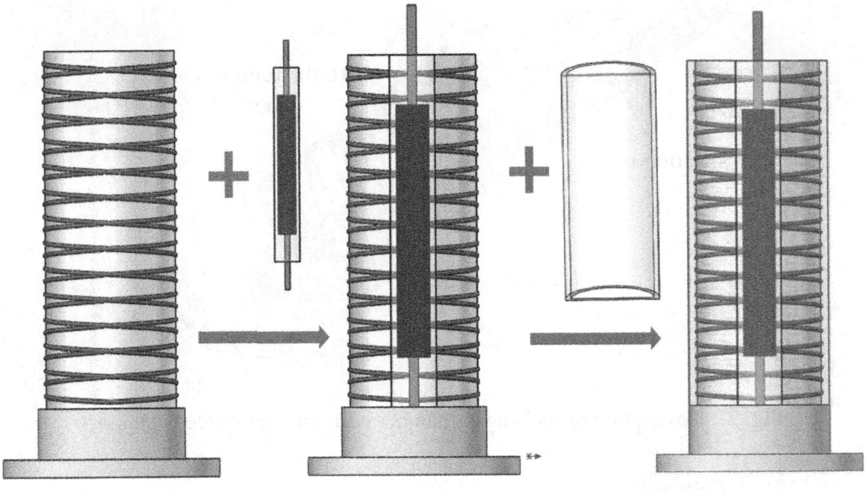

Fig. 4. Integration process

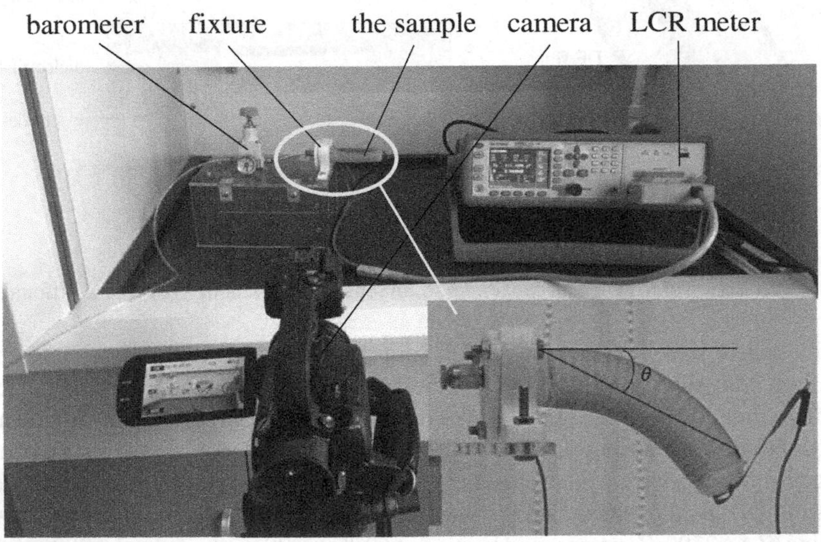

Fig. 5. Experimental setup

An evaluation platform is developed to characterize the finger (shown in Fig. 5): a barometer is used to control the air pressure; a recording camera is fixed for the convenience of measuring the curvature of the finger, the bending angle of the finger is defined as shown in Fig. 5; an Inductance-Capacitance-Resistance (LCR) meter (E4980AL, Keysight) is used to measure the capacitance of the sensor.

3 Results

3.1 Actuation and Sensing Response

The finger sample was mechanically characterized for both actuation and sensing. It was actuated by applying several levels of air pressure (increased from 0 to 100 kPa by a step size of 10 kPa and then decreased to 0), and corresponding bending angles as well as capacitances of the sensors were measured. After several loading cycles, the average results are shown in Fig. 6.

3.2 Characterization of Bending Response

For each bending response loop, the hysteresis appears in the curve (Fig. 6 (a)) can be calculated as follow:

$$H_{\theta p} = \frac{\Delta_{\max}\theta}{\theta_{\max}} \times 100\% \tag{1}$$

$$\Delta_{\max}\theta = \max\left\{\left|\theta_{ip} - \theta_{rp}\right|\right\} \tag{2}$$

where $H_{\theta p}$ is the hysteresis of bending response; θ_{max} is the maximum bending angle of the finger; θ_{ip} is the bending angle when the air pressure is p during the pressure increasing process; θ_{rp} is the bending angle when the air pressure is p during the pressure reducing process.

Considering the results from several bending response loops, the maximum hysteresis of the finger sample is 11.79%, and the corresponding air pressure is 30 kPa. In addition, the linearity between the air pressure and the bending angle of the pressure increasing process ($0.9962 \leq R^2 \leq 0.9982$) is larger than which of the pressure reducing process ($0.9916 \leq R^2 \leq 0.9920$).

3.3 Sensing Analysis

The results of sensing outputs (Fig. 6(b)) show that a hysteresis loop appears in the sensing output loop, which can be calculated as follow:

$$H_{c\theta} = \frac{\Delta_{\max}c}{c_{\max}} \times 100\% \tag{3}$$

$$\Delta_{\max}c = \max\{|c_{i\theta} - c_{r\theta}|\} \tag{4}$$

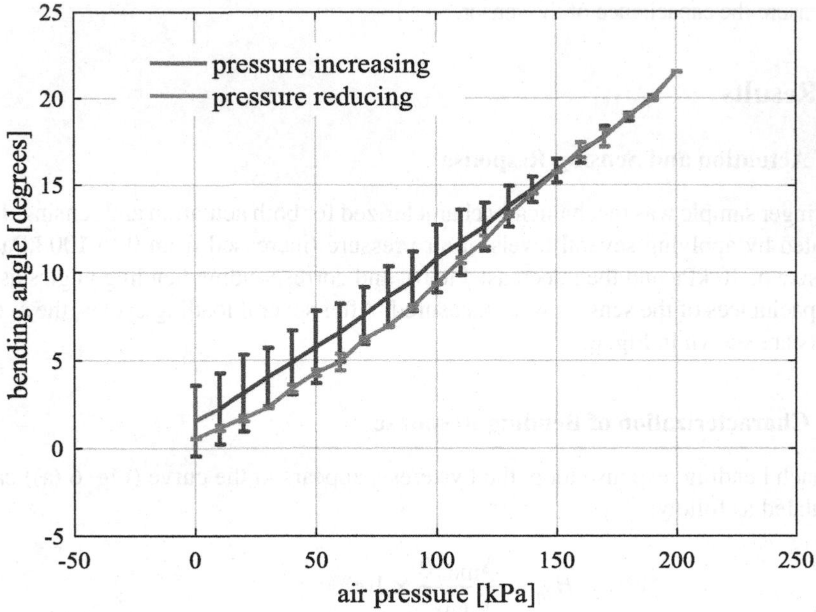

(a)

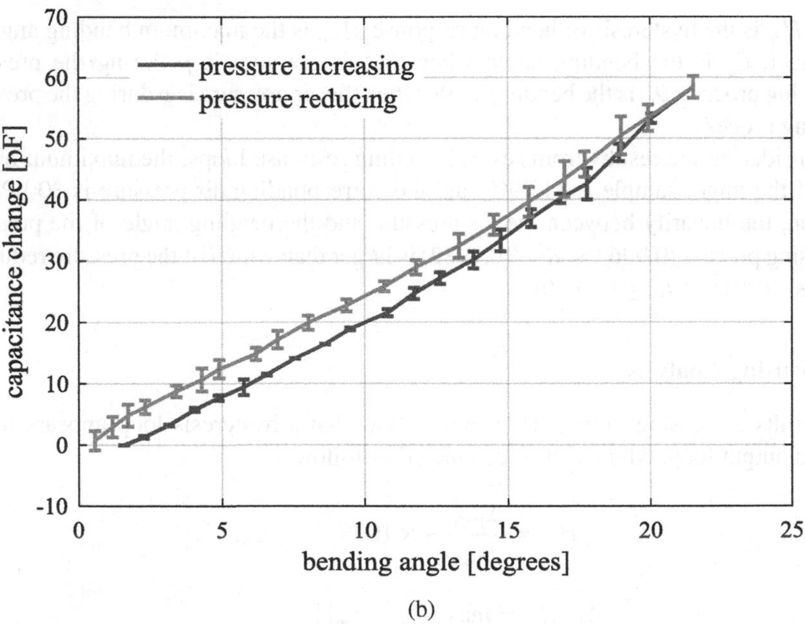

(b)

Fig. 6. Experimental results: (a) bending response; (b) sensing output where $H_{c\theta}$ is the hysteresis of sensing output; c_{max} is the maximum capacitance change of the sensor; $c_{i\theta}$ is the capacitance change when the bending angle is θ during the pressure increasing process; $c_{r\theta}$ is the capacitance change when the bending angle is θ during the pressure reducing process; capacitance change is calculated as the current capacitance value minutes the initial capacitance value. Considering the results from several sensing output loop, the maximum hysteresis of the sensor is 9.34%.

To characterize the sensing performance, the least square method (LSM) is used to fit a linear relationship between the capacitance change and the bending angle, which can be supposed as:

$$c = k\theta + b \tag{5}$$

where k and b are slope and intercept of the fitting line respectively, which can be calculated as follows:

$$k = \frac{n(\sum c_i\theta_i) - (\sum \theta_i)(\sum c_i)}{n(\sum \theta_i^2) - (\sum \theta_i)^2} \tag{6}$$

$$b = \frac{(\sum \theta_i^2)(\sum c_i) - (\sum \theta_i)(\sum c_i\theta_i)}{n(\sum \theta_i^2) - (\sum \theta_i)^2} \tag{7}$$

with c_i, θ_i ($i = 1, 2, \ldots, n$) being the average value of several experimental measured data. So the linear relationship can be expressed as:

$$c_i = k_i\theta + b_i \tag{8}$$

$$c_r = k_r\theta + b_r \tag{9}$$

where c_i, k_i (= 2.8326) and b_i (= −0.6806) represent the pressure increasing process; and c_r, k_r (= 2.6396) and b_r (= 0.9065) for the pressure reducing process.

The slope k is chosen as the indication of the sensitivity of the sensor, which is 2.8326 (pF/kPa) for the pressure increasing process and 2.6396 (pF/kPa) for the pressure reducing process.

The linearity of the sensor are $R^2 = 0.9947$ and $R^2 = 0.9979$ for the pressure increasing process and the pressure reducing process respectively.

Finally, an extra actuation experiment is carried out for evaluate the positional prediction by using sensing output (shown in Fig. 7). It can be seen that the predicted results are in good agreement with the experimental results, and the precision of the positional prediction can be improved by adding experimental results for fitting the relationship between the capacitance change and the bending angle.

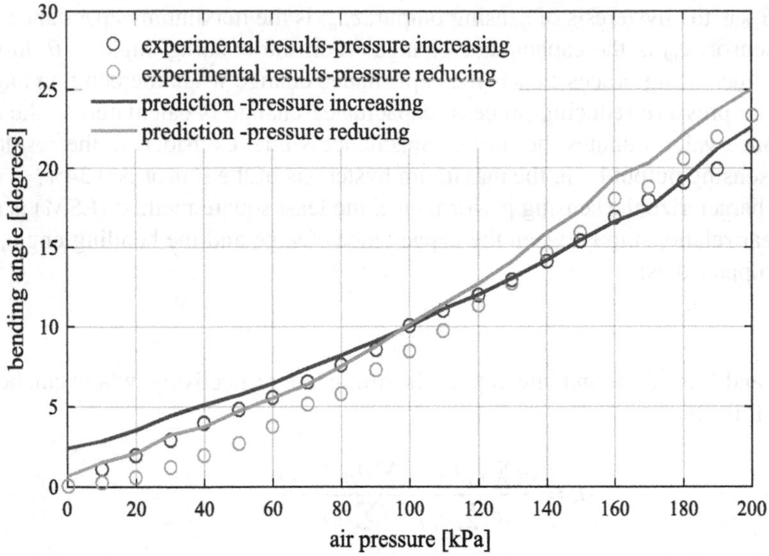

Fig. 7. Bending results

4 Conclusion

In this study, a SPA finger with self-strain sensing is designed by integrating the SPA and the DE film capacitive strain sensor together. In order to possess high air tight-ness, the single-chamber with fiber reinforced actuator is adopted. The DE film capacitive strain sensor is designed as a simple sandwiched structure which makes it compliant, stretchable and easy-to-fabricate. This work proposes the method to fabricate the SPA and the sensor. The response behavior and sensing characteristics (sensitivity, linearity and hysteresis) of the SPA finger are experimentally analyzed. Results show that there is a high linear relationship between the sensing output and the bending angle which can be used for positional feedback to further control the finger.

Future work will focus on two aspects. One is design and integrating several kinds of stretchable sensors to expand the sensation of the SPA finger, such as force sensors and temperature sensors. The other is to develop a feedback controller for precise positional tracking of the self-sensing SPA finger.

Acknowledgments. This work was in part supported by the National Natural Science Foundation of China under grant No. 51622506, and the Science and Technology Commission of Shanghai Municipality under grant 16JC1401000.

References

1. Chossat, J.-B., Yiwei Tao, Duchaine, V., Park, Y.-L.: Wearable soft artificial skin for hand motion detection with embedded microfluidic strain sensing. In: 2015 IEEE International Conference on Robotics and Automation (ICRA), pp. 2568–2573 (2015). doi:10.1109/icra.2015.7139544
2. Polygerinos, P., Lyne, S., Zheng Wang, N.L.F., Mosadegh, B., Whitesides, G.M., Walsh, C. J.: Towards a soft pneumatic glove for hand rehabilitation. In: 2013 IEEE/RSJ International Conference on Intelligent Robots and Systems (2013). doi:10.1109/iros.2013.6696549
3. Liang, C., Wang, F., Tian, Y., Zhao, X., Zhang, H., Cui, L., Zhang, D., Ferreira, P.: A novel monolithic piezoelectric actuated flexure-mechanism based wire clamp for microelectronic device packaging. Rev. Sci. Instrum. **86**(4), 045106 (2015). doi:10.1063/1.4918621
4. Paik, J.K., Wood, R.J.: A bidirectional shape memory alloy folding actuator. Smart Mater. Struct. **21**(6), 065013 (2012). doi:10.1088/0964-1726/21/6/065013
5. Bishop-Moser, J., Kota, S.: Design and modeling of generalized fiber-reinforced pneumatic soft actuators. IEEE Trans. Rob. **31**(3), 536–545 (2015). doi:10.1109/tro.2015.2409452
6. Sun, Y., Song, Y.S., Paik, J.: Characterization of silicone rubber based soft pneumatic actuators. In: 2013 IEEE/RSJ International Conference on Intelligent Robots and Systems, pp. 4446–4453 (2013). doi:10.1109/iros.2013.6696995
7. Wirekoh, J., Park, Y.-L.: Design of flat pneumatic artificial muscles. Smart Mater. Struct. **26**(3), 035009 (2017). doi:10.1088/1361-665x/aa5496
8. Wang, Z., Polygerinos, P., Overvelde, J.T.B., Galloway, K.C., Bertoldi, K., Walsh, C.J.: Interaction forces of soft fiber reinforced bending actuators. IEEE/ASME Trans. Mechatron. **22**(2), 717–727 (2017). doi:10.1109/tmech.2016.2638468
9. Connolly, F., Walsh, C.J., Bertoldi, K.: Automatic design of fiber-reinforced soft actuators for trajectory matching. Proc. Natl. Acad. Sci. **114**(1), 51–56 (2016). doi:10.1073/pnas.1615140114
10. Low, J.-H., Ang, M.H., Yeow, C.-H.: Customizable soft pneumatic finger actuators for hand orthotic and prosthetic applications. In: 2015 IEEE International Conference on Rehabilitation Robotics (ICORR) (2015). doi:10.1109/icorr.2015.7281229
11. Polygerinos, P., Wang, Z., Galloway, K.C., Wood, R.J., Walsh, C.J.: Soft robotic glove for combined assistance and at-home rehabilitation. Robot. Auton. Syst. **73**, 135–143 (2015). doi:10.1016/j.robot.2014.08.014
12. Galloway, K.C., Polygerinos, P., Walsh, C.J., Wood, R.J.: Mechanically programmable bend radius for fiber-reinforced soft actuators. In: 16th International Conference on Advanced Robotics (ICAR) (2013). doi:10.1109/icar.2013.6766586
13. Maeder-York, P., Clites, T., Boggs, E., Neff, R., Polygerinos, P., Holland, D., Stirling, L., Galloway, K., Wee, C., Walsh, C.: Biologically inspired soft robot for thumb rehabilitation. J. Med. Devices **8**(2), 020934 (2014). doi:10.1115/1.4027031
14. Mosadegh, B., Polygerinos, P., Keplinger, C., Wennstedt, S., Shepherd, R.F., Gupta, U., Shim, J., Bertoldi, K., Walsh, C.J., Whitesides, G.M.: Pneumatic networks for soft robotics that actuate rapidly. Adv. Func. Mater. **24**(15), 2163–2170 (2014). doi:10.1002/adfm.201303288
15. Erin, O., Pol, N., Valle, L., Park, Y.-L.: Design of a bio-inspired pneumatic artificial muscle with self-contained sensing. In: 38th Annual International Conference of the IEEE Engineering in Medicine and Biology Society (EMBC) (2016). doi:10.1109/embc.2016.7591146
16. Zhao, H., Jalving, J., Huang, R., Knepper, R., Ruina, A., Shepherd, R.: A helping hand: soft orthosis with integrated optical strain sensors and EMG control. IEEE Robot. Autom. Mag. **23**(3), 55–64 (2016). doi:10.1109/mra.2016.2582216

17. Morrow, J., Shin, H.-S., Phillips-Grafflin, C., Jang, S.-H., Torrey, J., Larkins, R., Dang, S., Park, Y.L., Berenson, D.: Improving soft pneumatic actuator fingers through integration of soft sensors, position and force control, and rigid fingernails. In: 2016 IEEE International Conference on Robotics and Automation (ICRA) (2016). doi:10.1109/icra.2016.7487707
18. Park, Y.-L., Wood, R.J.: Smart pneumatic artificial muscle actuator with embedded microfluidic sensing. In: 2013 IEEE SENSORS (2013). doi:10.1109/icsens.2013.6688298
19. Lu, T., Wissman, J., Ruthika, M.C.: Soft anisotropic conductors as electric vias for ga-based liquid metal circuits. ACS Appl. Mater. Interfaces **7**(48), 26923–26929 (2015). doi:10.1021/acsami.5b07464
20. Felt, W., Remy, C.D.: Smart braid: air muscles that measure force and displacement. In: 2014 IEEE/RSJ International Conference on Intelligent Robots and Systems (2014). doi:10.1109/iros.2014.6942949
21. Zhao, H., O'Brien, K., Li, S., Shepherd, R.F.: Optoelectronically innervated soft prosthetic hand via stretchable optical waveguides. Sci. Robot. **1**(1), eaai7529 (2016). doi:10.1126/scirobotics.aai7529
22. Yun, S., Park, S., Park, B., Kim, Y., Park, S.K., Nam, S., Kyung, K.-U.: Polymer-waveguide-based flexible tactile sensor array for dynamic response. Adv. Mater. **26**(26), 4474–4480 (2014). doi:10.1002/adma.201305850

A Programmable Mechanical Freedom and Variable Stiffness Soft Actuator with Low Melting Point Alloy

Yufei Hao, Tianmiao Wang, and Li Wen(✉)

School of Mechanical Engineering and Automation, Beihang University, Beijing, China
liwen@buaa.edu.cn

Abstract. Soft robotic technologies have been widely used in the fields like bio-robotics, wearable devices, and industrial manipulations. However, existing soft robots usually require multiple pneumatic/fluidic channels for pressurizing soft material segments in series or in parallel to achieve multiple mechanical degrees of freedom. In this study, we demonstrated a soft actuator embedded with Low-Melting-Point Alloy (LMPA), with which the mechanical degrees and stiffness can be selectively controlled. The LMPA was embedded in the bottom of the actuator, with the Ni-Cr wires serpentining under different positions of the LMPA layer. Through a reheating- recrystallizing circle, the actuator can self-heal and recover from the crack state. The melting process of the LMPA under different currents and different sections, the variable stiffness, the self-healing properties, and the programmable mechanical freedom of the actuator was explored through experiments. The results showed that the LMPA could be melted about 10 s under the current of 0.7 A. With the LMPA, the bending force and the elasticity modulus of the actuator could be enhanced up to 16 times and 4,000 times separately. Moreover, up to six motion patterns could be achieved under the same air pressure inflated to a typical single-chamber soft actuator. The combination of Low-Melting-Point Alloy and the soft actuators may open up a diversity of applications for future soft robotics.

Keywords: Variable stiffness · Low-Melting-Point alloy · Soft actuator · Programmable mechanical freedom

1 Introduction

Soft robotics, which has promising features such as lightweight, low-cost, easy fabrication and high compliance [1], is a multi-disciplinary research area involving chemistry [2], material science [3], biology [4–8] and mechanics [9]. Soft robotic technologies have been widely used in actuation [10], sensing [11], and the nonlinear dynamics control [12]. Besides, soft robots have been widely studied in the applications like manipulation [13], locomotion [14], wearable devices [15] and invasive surgery [16]. However, there exist two drawbacks that limit the further development of soft robots. The materials of soft actuators are mostly silicon rubbers which has high compliance, however, restrict the capacity of the robots in many applications that require high mechanical strength. Besides, most of the soft actuators only have one kind of motion

Y. Huang et al. (Eds.): ICIRA 2017, Part I, LNAI 10462, pp. 151–161, 2017.
DOI: 10.1007/978-3-319-65289-4_15

pattern when they were in operation [10]. For the dexterous movement, multiple pneumatic/fluidic channels for pressurizing soft material segments in series or parallel were needed [16, 17].

Recently, to improve the stiffness without diminishing the mobility of the soft robots, several materials with tunable stiffness have been used in soft robotics. For example, the viscosity of the macro-particle of magnetorheological (MR) or electrorheological (ER) fluids could be changed under the stimulus of the electric or magnetic field [18, 19]. However, these materials require external electromagnets or high voltage capacitors and have limited functional times. By packing particles into a membrane, the stiffness could also be changed when a vacuum pressure was exerted, so-called jamming [20, 21]. However, the jamming effect requires much volume to achieve a great stiffness. Shape memory polymer, whose elasticity modulus could be changed if the temperature is above the glass transition temperature, was also used for variable stiffness [22]. However, the response is slow due to the poor thermal conductivity of the SMP material. For the change of the mechanical freedoms, shape memory polymer [23], selectively-placed flexible conformal covering [24] or paper-elastomer composites [25] were used. However, the major mechanical property was prescribed during the design and fabrication stage and cannot be changed on working.

Moreover, previous methods of changing stiffness or freedoms have two major drawbacks: (1) the relative stiffness changing is limited; (2) the structure would not recover if they were broken. In contrast, Low-Melting-Point alloys (LMPAs), which can transform from solid state to liquid state by heating are ideal choice to overcome the two problems. The LMPAs have two major advantages: (1) they can increase their elastic stiffness up to several thousand orders [26]; (2) They can heal themselves from crack or external impact by a reheating-recrystallizing process [27]. Recently, researchers started to apply the LMPA to the soft manipulators [28, 29], the metal-elastomer foams [27] and the fibers [30]. However, no study addressed a robotic device that allows the mechanical freedoms and stiffness to be changed simultaneously.

In this article, we demonstrated an LMPA embedded soft actuator with programmable mechanical freedoms and stiffness. By selectively activating the Ni-Cr wires under the LMPA layer, we can choose which sections of the LMPA layer to be melted. Thermal actuation experiments were conducted to explore the melting process of the LMPA under different currents and the influence of heating one section on the others. To test the stiffness variation and self-healing property of the actuator, we conducted mechanical experiments to measure the bending stiffness and elasticity modulus of the actuator when the LMPA was in the solid state and liquid state separately. Finally, various motion patterns were realized on a soft actuator with a single channel to verify the variable mechanical freedoms.

2 Materials and Methods

2.1 The Design and Fabrication of the Soft Actuator with LMPA

The detailed structure of the LMPA embedded soft actuator is depicted in Fig. 1. Figure 1a shows the prototype and the CAD model of the actuator. The actuator is composed of two parts: the top extensible part and the multilayer bottom part. The top part is a rippled structure which could make the soft actuator bend in two directions. For the variation of the mechanical freedom and stiffness of the soft actuator, a thin layer of LMPA was embedded into the bottom part of the soft actuator. Ni-Cr wires were used to selectively melt the different sections (section I, II and III) of the LMPA because of the high electrical resistivity of the material. The detailed structure of the multilayer bottom part is shown in Fig. 1b. The structure has three kinds of layers: the LMPA layer, the Ni-Cr wire layer, and the silicon rubber (dragon skin 10, smooth-on Inc., USA) seal layer. The LMPA layer was a 1.2 mm thick rectangular plate with many through holes. Because the LMPA layer has no adhesive force on the silicon rubber, by making many though holes in the LMPA layer, the silicon rubbers above and underneath the LMPA could be connected via the holes. Thus making the structure more compact. The LMPA is a type of alloy with the following composition by weight: 32.5% bismuth, 51% indium, and 16.5% tin. We choose this material for its low melting point which is about 62 °C and the high elasticity modulus. The Ni-Cr wire layer was fabricated by pouring the uncured silicon rubber on the pre-twined Ni-Cr wires with a diameter of 0.1 mm. For

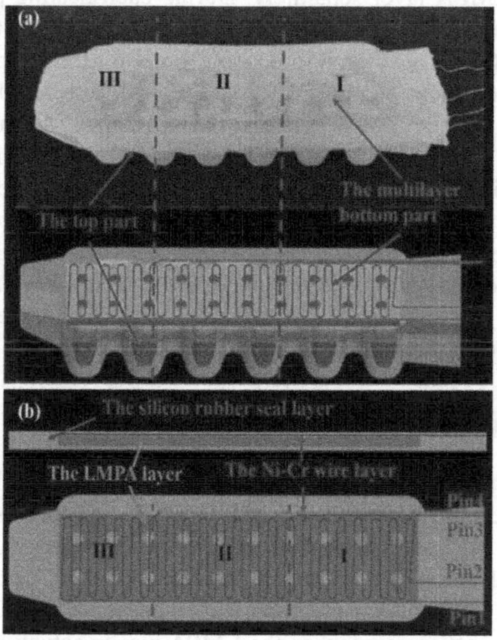

Fig. 1. The design of the soft actuator with variable mechanical freedom and stiffness. (a) The structure composition of the soft actuator. The structure has three mechanical freedoms by heating the different sections (section I, II and III) of LMPA. (b) The detailed structure of the multilayer bottom part of the actuator. The part has three difference layers. By selectively powering any two of the four pins, the corresponding sections of the LMPA layer would be melted: apply currents to pins 1 and 2, the three sections will all be melted; apply currents to pins 1 and 3, the section II and III will be melted; apply currents to pins 1 and 4, the section III will be melted; applying currents to pins 2 and 3, the section I will be melted; apply currents to pins 2 and 4, the section I and II will be melted; apply currents to pins 3 and 4, the section II will be melted.

the aim of selectively melting the three sections of the LMPA layer, two additional wires were separately connected to the middle of the whole Ni-Ci wire. Totally four pins were provided for the power. The total thickness of the layer was 0.7 mm. By applying currents on any two of the four pins, the corresponding LMPA above the surface of the wires should be melted. Thus the mechanical freedom and the stiffness of the actuator could be changed. The silicon rubber layer was used to bond the LMPA layer and the Ni-Cr wire layer together and seal the LMPA layer in an enclosure space.

2.2 The Melting Process Test of the LMPA

Two experiments were conducted to test the thermal conduction behavior of the LMPA. For simplification, only the bottom part was used for the test. The first experiment was to test the temperature ascending process and the melting speed of the LMPA under different currents. For this experiment, the multilayer bottom part was suspended in the air by several wires under the bottom to minimize the heat loss through contact with other media. Four constant electrical currents (0.4 A, 0.5 A, 0.6 A and 0.7 A) were used to melt the whole LMPA layer separately. During each heating process, the surface temperature of the sample was recorded by an infrared imager (Ti400, Fluck Thermography, America). After the heating, it took about 4 min for the sample to return to room temperature before the next heating process. The data was finally addressed by the supporting software (SmartView 4.1, Fluck Thermography, America). When heating only one section of the LMPA, it should be better that the heater will not affect the adjacent segments significantly and the stiffness transform between these segments is smooth and gradient. So the other experiment was designed to test the influence of heating one section of the LMPA on the adjacent sections. For this purpose, only the middle section of the LMPA was heated under the current of 0.5 A. Then the surface temperature of the whole sample was recorded, and the temperature distributions between the adjacent segments were compared for further analysis.

2.3 Bending Stiffness and Elasticity Modulus of the Actuator

When the LMPA is at the solid state, the mechanical property of the soft actuator will be dominated by the rigid LMPA layer, while the silicon rubber structure will be the dominate part when the LMPA is melted. Besides, it can recover its mechanical properties from the crack through a reheating and recrystallizing circle. For the stiffness variation test of the soft actuator, a comparison experiment was conducted to test the bending force of the soft actuator when the LMPA was in actuation and non-actuation state. Besides, we also tested the bending strength of the actuator after several reheating-recrystallizing cycles to verify the healing property of the LMPA. For simplification, only the multilayer bottom part was tested. The experiment setup was illustrated in Fig. 2. The sample was fastened to the fixator. The connector, fixed to the force sensor which was fastened to a robot arm, was used to contact the sample and exert a force on the sample. The lateral distance between the connector and the base of the sample was 25 mm. During the test, the connector first contacted the surface of the sample, then moved downward with a speed of 0.5 mm/s. After moving 2 mm, the connector suspended 5 s before the next moving. The total displacement was 20 mm for the whole test. During the process, the force date was acquired by a LabVIEW program at a sample rate of 50 Hz. Totally twelve samples were tested for each condition of the LMPA (liquid, solid and cycle test) separately.

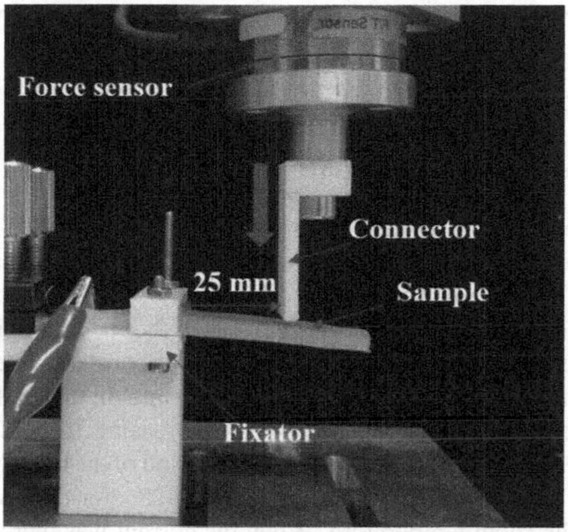

Fig. 2. The experiment setup for the bending stiffness test of the actuator. The sample was fastened to the fixator. The connector was fixed to the robot arm via the force sensor. The lateral distance between the connector and the base of the sample was 25 mm. During the test, the robot arm was moved downward with a speed of 0.5 mm/s and suspended 5 s for each 2 mm displacement.

To verify the elasticity modulus of the LMPA for the self healing property, we fabricated three bottom part samples for the elasticity modulus measurements before and after the samples were broken. For each sample, we first tested the strain and stress of the LMPA layer; then the strain and stress was tested again after a remelting and refreezing process for the sample. The experiment was conducted on the universal material test machine (exceed model E44, MTS, America). During the test, the sample was fastened by the two clamps of the machine. The initial distance between the two clamps was 30 mm. Then the upper clamp was moved upwards at the speed of 0.5 mm/s until the sample was broken. The force and the displacement values were recorded at the sample rate of 20 Hz.

2.4 Variable Mechanical Freedoms of the Soft Actuator

By applying currents to any two of the four pins, the corresponding sections of the LMPA layer would be melted. Thus the mechanical freedom will be changed. To analysis the motion patterns of the soft actuator when different sections of the LMPA layer were melted, we fixed the actuator to a base, then powered the matched pins according to Fig. 1b to melt the corresponding sections of the LMPA. After the metal was melted, we inflated the actuators with the pressure up to 30 kPa and captured the motion process of the actuator. Besides the programmable mechanical freedoms, the actuator can sustain its shape under the deflation state after the LMPA was solidified. To verify this property, we first melted the LMPA and inflated the soft actuator to 40 kPa and captured the motion profile of the actuator. Then waited about 5 min for the LMPA to refreeze and deflate the actuator. The profile of the final state was compared with the previous one.

3 Results and Discussion

3.1 The Thermal Behavior of the LMPA

Figure 3 shows the results of the thermal actuation test of the LMPA. As Fig. 3b shows, for all the four currents, the melting process of the LMPA experienced about three stages. For the first stage, the temperature increased to nearly 56 °C at a linear speed, during which the LMPA only absorbed the energy to reach the melting point but kept in the solid state. For the second stage, the temperature kept constant. The LMPA absorbed energy to transform from the solid state to the liquid state. In the final stage, the LMPA was in the liquid state, and the temperature went on increasing. It could also be concluded from Fig. 3b that with a higher current, the melting period of the LMPA would be sharply cut down. When the current was 0.4 A, it took about 28 s for the LMPA to transfer to liquid state from the room temperature. When the current was increased to 0.5 A, the melting time was about 18 s. Increasing the current to 0.7 A, the melting time would be shorted to about 10 s. A simple model could estimate the melting speed for a given LMPA structure:

$$Ri^2t = \mathrm{mc}\Delta T + \mathrm{mL} \tag{1}$$

Where R is the resistance of the heater, i is the current applied to the heater, m is the mass of the metal, c is the specific heat capacity, ΔT is the temperature difference between the room temperature and the melting point and L is the latent heat of the metal. As the model shows, the power of the heater should be enough to melt the metal structure fast. For this purpose, the Ni-Cr wire was used as the heater for its high electrical resistivity, which is about 62.3 times to that of copper. Besides, the melting time with the relation of current was a quadratic function. That is the reason that the melting speed would increase sharply by increasing the current a little.

Through the pictures in the first row of Fig. 3a, we can see that the surface temperature of the LMPA was equally distributed. This demonstrated that the spaces between the helical Ni-Cr wire would not result in the uneven heating of the LMPA, which may lead to some part of the LMPA cannot be melted. The pictures in the second row of Fig. 3a show the infrared images of heating only section II of the LMPA layer, and Fig. 3c shows the temperature distribution of the middle line (depicted in the last image of Fig. 3a) of the LMPA layer. As these results show, the high temperatures only concentrate on section II (the area between the actuated wires), while the temperatures of the unheated sections decreased drastically with the distance from the heating area increasing. This demonstrated that the heat was mostly absorbed by the section II of the

LMPA, then conducted to the adjacent sections in a sharply decreased speed. The sharp temperature distribution variation demonstrates that the adjacent sections have little interactions when heated separately, which guaranteed that the mechanical freedom would keep constant when heated several times.

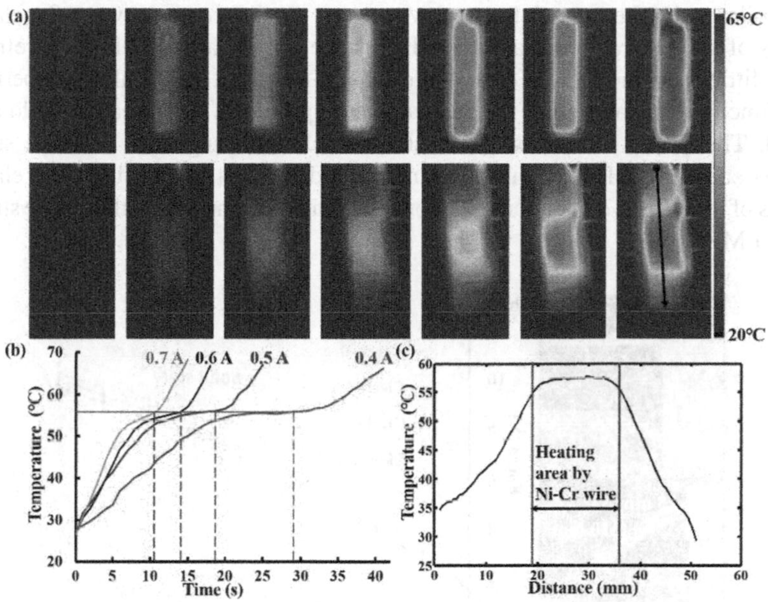

Fig. 3. The thermal behavior of the actuator. (a) The infrared thermal images at different times with the current of 0.4 A. The upper row shows the results of heating the whole LMAP layer, while the lower row shows the results of heating section II of the LMPA layer. (b) The melting process of the LMPA under different currents. As the curves show, the temperature of the LMPA first increased to the melting point, then the LMPA transformed from the solid state to the liquid state with the temperature keeping constant. After that, the temperature went on increasing. (c) The temperature distribution of the middle line of the LMPA when section II was heated. The line and the direction were indicated in the last image of panel b. The red dashed lines show the position of the heating wire. (Color figure online)

3.2 The Variable Stiffness of the Actuator

With the LMPA, the stiffness of the soft actuator could be improved significantly when the metal was at the solid state. However, this rigid structure has little influence on the dexterity of the soft actuator if it was melted. As Fig. 4a shows, the sample even bend downward under the effect of gravity if the LMPA was melted. However, the bending stiffness could be significantly improved when the LMPA was in the solid state. As Fig. 4b shows, the bending force increased linearly with the increasing of the deflection. Besides, the force data is much bigger compared to that when the LMPA was in the liquid state under the same deflection distance. For example, the force of the solid state is about 16 times of that of the liquid state under the deflection of 20 mm, which demonstrated soundly that the stiffness of the actuator could be significantly enhanced when the LMPA was in the solid state. It could also be deduced that the embedded LMPA could improve the mechanical property of the soft actuator. Furthermore, the soft actuator can recover within 4 min after crack, which could be verified by Fig. 4b and Table 1. As the results show, the bending forces under different heating-cooling cycles were similar under the same deflection. Moreover, the difference between the elasticity modulus of the pre-fractured sample and the same sample after a reheating-refreezing circle is little. This means that the actuator has an excellent self-healing property. The shape, functionality, as well as the mechanical properties of the actuator, could also be restored. These kind of effects are insurmountable by using other variable stiffness materials such as SMP [22] and jamming ground coffee [21]. Besides, the elasticity modulus of the sample could increase about 4,000 times compared to that of the structure without LMPA.

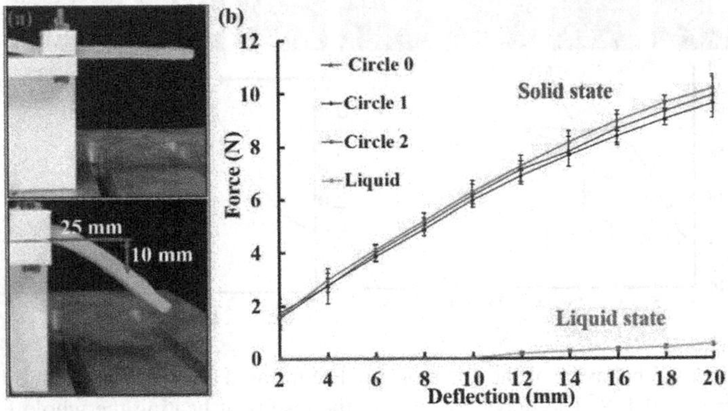

Fig. 4. The results of the bending stiffness test. (a) The real pictures of the sample when the LMPA under the solid state and the liquid state. (b) The force versus deflection of the sample when the LMPA under the solid state and liquid state.

Table 1. The results of the elasticity modulus test of the soft actuator.

Sample state	Elasticity modulus (GPa)	Standard deviation
Pre-fracture	6.12	0.649
Aft-healing	5.92	0.436
Pure rubber	1.5×10^{-4} [31]	–

3.3 The Programmable Mechanical Freedom of the Actuator

By activating the pin matches in Fig. 1b, the corresponding sections of the LMPA layer would be melted, and the mechanical freedom can be changed. This property is well demonstrated in Fig. 5. As the figure shows, the actuator will have various motion patterns under the same pressure if different sections of the LMPA were melted. When pressurized, the un-melted LMPA will restrict the bending so only the melted sections could bend. This method is more applicable and functional than the fiber-reinforced actuators [15]. By varying the angle of the fiber, the fiber-reinforced actuator can achieve different motion patterns, but these patterns cannot be changed if the structure was finally fabricated. By modifying the motion patterns during work, the actuator will have more application values. For example, it can mimic the variable stiffness bionic structures such as the muscle, and it can contribute to much more grasping strategies if it is used for gripping. Apart from the motion dexterity, the actuator can keep its actuated shape after the LMPA was recrystallized. As Fig. 5b shows, section III of the actuator bent under the pressure of 40 kPa, with large bubbles at the upper rippled part. However, the actuator will keep the same state if the LMPA was solidified, with zero energy consumed. This attribute is very energy-efficient for the long term operations. Such as the long-distance transportation after the gripper grasp the objects.

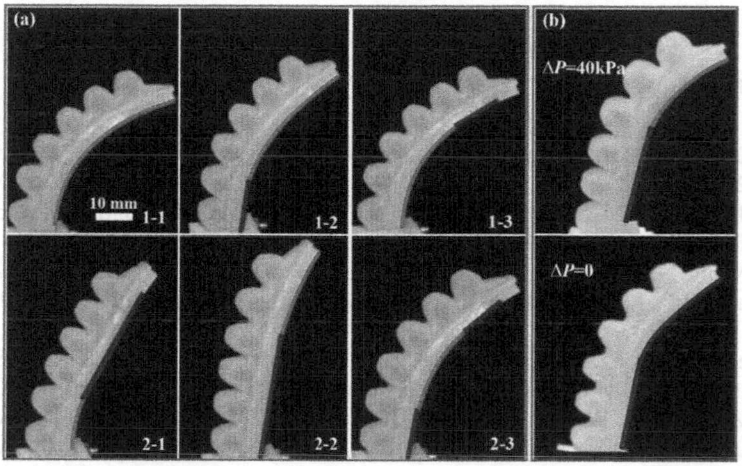

Fig. 5. (a) The various motion patterns of the actuator under the same pressure (30 kPa). "1-1" shows the result by melting the whole LMPA layer; "1-2" shows the result by melting section II and III; "1-3" shows the result by melting section I and II; "2-1" shows the result by melting section I; "2-2" shows the result by melting section III; "2-3" shows the result by melting section II. (b) The shape-retain capability of the actuator. When the LMPA was in solid state (lower picture of panel b), the actuator can still keep the pressurized state (upper picture of panel b) with no additional energy consumption. In all the pictures, the red lines demonstrate that the corresponding sections were melted, while the blue lines show that the relevant sections were solid. (Color figure online)

4 Conclusion

In this paper, a soft actuator with programmable mechanical freedom and variable stiffness was demonstrated. The function of the actuator was achieved by melting different sections of the LMPA layer using the Ni-Cr wires. The thermal experiments showed that the heating method could guarantee the homogeneous temperature distribution, and heating one section of the LMPA layer had little influence on the neighboring sections.

Through mechanical experiments, the self-healing property of the actuator was verified. The bending force of the actuator could increase at least 16 times, and the tensile strength could increase about 13 times owing to the metal. The various motion patterns of the actuator were also studied. By heating different sections of the LMPA layer, the actuator allows for up to 8 motion patterns during working with the same channel under pressurization. Besides, with the LMPA, the actuator can also keep the actuated state with zero power consumptions. The combination of soft robots and LMPA is a promising approach to overcome the drawback of the soft robots, such as the light load capacity, monotonous motions and poor stiffness for some strength needed situations.

Acknowledgments. This work was supported by the National Science Foundation support projects, China under contract number 61633004, 61403012, and 61333016; the Open Research Fund of Key Laboratory Space Utilization, Chinese Academy of Sciences (No. 6050000201607004).

References

1. Rus, D., Tolley, M.T.: Design, fabrication and control of soft robots. Nature **521**(7553), 467–475 (2015)
2. Ilievski, F., Mazzeo, A.D., Shepherd, R.F., Chen, X., Whitesides, G.M.: Soft robotics for chemists. Angew. Chem. **123**(8), 1930–1935 (2011)
3. Mac Murray, B.C., An, X., Robinson, S.S., van Meerbeek, I.M., O'Brien, K.W., Zhao, H., Shepherd, R.F.: Poroelastic foams for simple fabrication of complex soft robots. Adv. Mater. **27**(41), 6334–6340 (2015)
4. Lauder, G.V., Wainwright, D.K., Domel, A.G., Weaver, J.C., Wen, L., Bertoldi, K.: Structure, biomimetics, and fluid dynamics of fish skin surfaces. Phys. Rev. Fluids **1**(6), 060502 (2016)
5. Ren, Z., Yang, X., Wang, T., Wen, L.: Hydrodynamics of a robotic fish tail: effects of the caudal peduncle, fin ray motions and the flow speed. Bioinspir. Biomim. **1**(1), 016008 (2016)
6. Wehner, M., Truby, R.L., Fitzgerald, D.J., Mosadegh, B., Whitesides, G.M., Lewis, J.A., Wood, R.J.: An integrated design and fabrication strategy for entirely soft, autonomous robots. Nature **536**(7617), 451–455 (2016)
7. Wen, L., Weaver, J., Lauder, G.: Biomimetic shark skin: design, fabrication and hydrodynamic testing. J. Exp. Biol. **217**(10), 1637–1638 (2014)
8. Wen, L., Weaver, J., Thornycroft, P.M., Lauder, G.: Hydrodynamic function of biomimetic shark skin: effect of denticle pattern and spacing. Bioinspir. Biomim. **10**(6), 066010 (2015)
9. Polygerinos, P., Wang, Z., Overvelde, J.T., Galloway, K.C., Wood, R.J., Bertoldi, K., Walsh, C.J.: Modeling of soft fiber-reinforced bending actuators. IEEE Trans. Rob. **31**(3), 778–789 (2015)
10. Connolly, F., Polygerinos, P., Walsh, C.J., Bertoldi, K.: Mechanical programming of soft actuators by varying fiber angle. Soft Robot. **2**(1), 26–32 (2015)
11. Arabagi, V., Felfoul, O., Gosline, A.H., Wood, R.J., Dupont, P.E.: Biocompatible pressure sensing skins for minimally invasive surgical instruments. IEEE Sens. J. **16**(5), 1294–1303 (2016)
12. Marchese, A.D., Tedrake, R., Rus, D.: Dynamics and trajectory optimization for a soft spatial fluidic elastomer manipulator. Int. J. Robot. Res. **35**(8), 1000–1019 (2015)
13. Hao, Y., Gong, Z., Xie, Z., Guan, S., Yang, X., Ren, Z., Wang, T., Wen, L.: Universal soft pneumatic robotic gripper with variable effective length. In: 35th Chinese Control Conference (CCC), Cheng Du, China, pp. 6109–6114. IEEE (2016)

14. Bartlett, N.W., Tolley, M.T., Overvelde, J.T., Weaver, J.C., Mosadegh, B., Bertoldi, K., Whitesides, G.M., Wood, R.J.: A 3D-printed, functionally graded soft robot powered by combustion. Science **349**(6244), 161–165 (2015)
15. Park, Y.L., Chen, B.R., Young, D., Stirling, L., Wood, R.J., Goldfield, E.C., Nagpal, R.: Design and control of a bio-inspired soft wearable robotic device for ankle-foot rehabilitation. Bioinspir. Biomim. **9**(1), 016007 (2014)
16. Ranzani, T., Gerboni, G., Cianchetti, M., Menciassi, A.: A bioinspired soft manipulator for minimally invasive surgery. Bioinspir. Biomim. **10**(3), 035008 (2015)
17. Gong, Z., Xie, Z., Yang, X., Wang, T., Wen, L.: Design, fabrication and kinematic modeling of a 3D-motion soft robotic arm. In: 2016 IEEE International Conference on Robotics and Biomimetics, Qing Dao, China, pp. 509–514. IEEE (2016)
18. Pettersson, A., Davis, S., Gray, J.O., Dodd, T.J., Ohlsson, T.: Design of a magnetorheological robot gripper for handling of delicate food products with varying shapes. J. Food Eng. **98**(3), 332–338 (2010)
19. Taniguchi, H., Miyake, M., Suzumori, K.: Development of new soft actuator using magnetic intelligent fluids for flexible walking robot. In: 2010 International Conference on Control Automation and Systems (ICCAS), Gyeonggi-do, Korea, pp. 1797–1801 (2010)
20. Wei, Y., Chen, Y., Ren, T., Chen, Q., Yan, C., Yang, Y., Li, Y.: A novel, variable stiffness robotic gripper based on integrated soft actuating and particle jamming. Soft Robot. **3**(3), 134–143 (2016)
21. Wall, V., Deimel, R., Brock, O.: Selective stiffening of soft actuators based on jamming. In: 2015 IEEE International Conference on Robotics and Automation (ICRA), Seattle, USA, pp. 252–257. IEEE (2015)
22. Yang, Y., Chen, Y.: Novel design and 3D printing of variable stiffness robotic fingers based on shape memory polymer. In: 6th IEEE International Conference on Biomedical Robotics and Biomechatronics (BioRob), Singapore, pp. 195–200, June 2016
23. Firouzeh, A., Salerno, M., Paik, J.: Soft pneumatic actuator with adjustable stiffness layers for Multi-DoF actuation. In: 2015 IEEE/RSJ International Conference on Intelligent Robots and Systems (IROS), Hamburg, Germany, pp. 1117–1124. IEEE (2015)
24. Galloway, K.C., Polygerinos, P., Walsh, C.J., et al.: Mechanically programmable bend radius for fiber-reinforced soft actuators. In: 2013 IEEE International Conference on Robotics and Automation, Karlsruhe, Germany, pp. 1–6. IEEE (2013)
25. Martinez, R.V., Fish, C.R., Chen, X., et al.: Elastomeric Origami: programmable paper-elastomer composites as pneumatic actuators. Adv. Func. Mater. **22**(7), 1376–1384 (2012)
26. Schubert, B.E., Floreano, D.: Variable stiffness material based on rigid low-melting-point-alloy microstructures embedded in soft poly (dimethylsiloxane) (PDMS). RSC Adv. **3**(46), 24671–24679 (2013)
27. Van Meerbeek, I.M., Mac Murray, B.C., Kim, J.W., Robinson, S.S., Zou, P.X., Silberstein, M.N., Shepherd, R.F.: Morphing metal and elastomer bicontinuous foams for reversible stiffness, shape memory, and self-healing soft machines. Adv. Mater. **28**(14), 2801–2806 (2016)
28. Alambeigi, F., Seifabadi, R., Armand, M.: A continuum manipulator with phase changing alloy. In: 2016 IEEE International Conference on Robotics and Automation (ICRA), Stockholm, Sweden, pp. 758–764. IEEE (2016)
29. Zhao, R., Yao, Y., Luo, Y.: Development of a variable stiffness over tube based on low-melting-point-alloy for endoscopic surgery. J. Med. Devices **10**(2), 021002 (2016)
30. Tonazzini, A., Mintchev, S., Schubert, B., Mazzolai, B., Shintake, J., Floreano, D.: Variable stiffness fiber with self-healing capability. Adv. Mater. **28**(46), 10142–10148 (2016)
31. https://www.smooth-on.com/products/dragon-skin-10-medium/. Accessed 21 Apr 2017

Investigate of Grasping Force for a Soft Robot Hand Under Pulling Force and Varying Stiffness

Haibin Yin(✉), Qian Li, Junfeng Li, and Mingchang He

School of Mechanical and Electronic Engineering,
Wuhan University of Technology, Wuhan 430076, China
chinaliuyin@whut.edu.cn

Abstract. The purpose of this research is to analyze and test grasping force of soft robot hand with variable stiffness, as the essential part of the robot hand, each finger is composed of 7 SMA fibers, five of which are used as skeleton of finger with variable stiffness while others provide pulling force. Firstly, the structure and materials of the fingers are introduced. Secondly, the computations of grasping forces and stiffness of fingers were implemented based on Cosserat theory and stiffness model, respectively. Moreover, the variable stiffness of fingers was measured by applying the different heating currents for the SMA simultaneously. Results indicates that the increase of stiffness around 61.7% from the low stiffness to high stiffness. Finally, the pulling forces with the variable stiffness and different grasping forces were measured, which demonstrate that grasping forces can be adjusted by varying stiffness.

Keywords: Soft robot hand · Variable stiffness · Grasping force · SMA

1 Introduction

Compared with the traditional rigid robot hand, the adaption of soft hand to environment is more powerful, and the potential threat to people is also smaller. However, soft robot hands are also faced with the problem of insufficient rigidity, which leads to the shortage of grasping force. Lots of researchers have studied the soft hands with variable stiffness, which can adapt the environment and grasping capacity.

According to published literature, the driving methods of soft robot hands can be divided into three types: rope-driven, fluid-driven and actuated by intelligent material. Yang et al. [1] designed a rope-driven soft robot hand with three joints in the thumb and two joints for other four fingers. Carrozza et al. [2] designed a tendon-driven prosthetic hand by abrasive injection to overcome its some limitations of function and control. Germany Karlsruhe research Center [3] proposed a man-made hydraulic driving control of the robot hand. Brown et al. [4] designed a spherical soft hand which could grasp objects by changing the air pressure of the hand. George et al. [5] developed a new type of aerodynamic robot. These studies need extra motors or pumps to provide the pulling force or pressure, and conduct low power-density for whole system. At the same time, many researchers adopted smart materials such as electro-conjugated fluid [6] and SMA [7] to drive soft hands directly by imputing current.

Y. Huang et al. (Eds.): ICIRA 2017, Part I, LNAI 10462, pp. 162–172, 2017.
DOI: 10.1007/978-3-319-65289-4_16

Besides of the driving force, the soft hands need the capacity of variable stiffness. Some studies focused on the filling methods to adjust the stiffness of soft hand [8], but these filled fluid and gas are required to package in a soft cavity, which enlarge the volume of soft hands. Some researcher applied cable or tendon to control the stiffness of soft robot [9]. however, these methods are also faced with the same problem with the driving method of motors and pumps. Some smart materials, such as SMP [10], low melting-point alloy [11] process high power-density for variable stiffness but lower pulling strength. Chenal et al. [12] embedded SMA in the soft fiber and adjust the SMA tension to achieve different stiffness. Wang et al. [13] developed a variable stiffness component by packaging SMA.

This paper constructs a soft hand with variable stiffness by using two SMA fibers, one shrinkage produce driving force, another type SMA will increase stiffness with the increasing of temperature. Compared with the methods mentioned above, this soft finger is with more concise structure and higher power-density.

2 A Soft Hand

Robot hand achieves the grasping operation in space requires at least three fingers and each finger should meet the conditions of force balance, so we have designed a soft robot hand as shown in Fig. 1. This hand has a total of three fingers, where Finger 2 and Finger 3 are completely identical in size and structure while Finger 1 is longer. Enlarge structure of fingers are shown in Fig. 2, (a) represents the side view, (b) represents the front view. SMA-1 with super elasticity is used as the bone structure to support the finger and makes sure to obtain an initial stiffness of soft finger, and four SMA-3 wires are placed in parallel to the SMA-1 wire to realize the variable stiffness of soft finger. To actuate the soft finger, SMA-2 are used as actuator. All SMA wires are fixed in brackets, which are made from PLA by using a 3D printer. The structure sizes and material parameters are listed in Table 1.

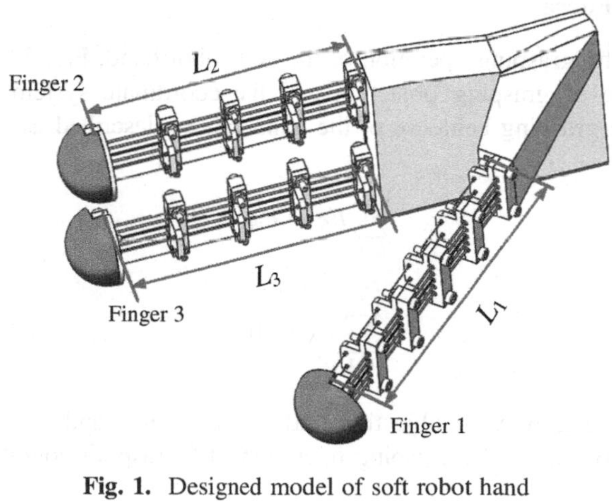

Fig. 1. Designed model of soft robot hand

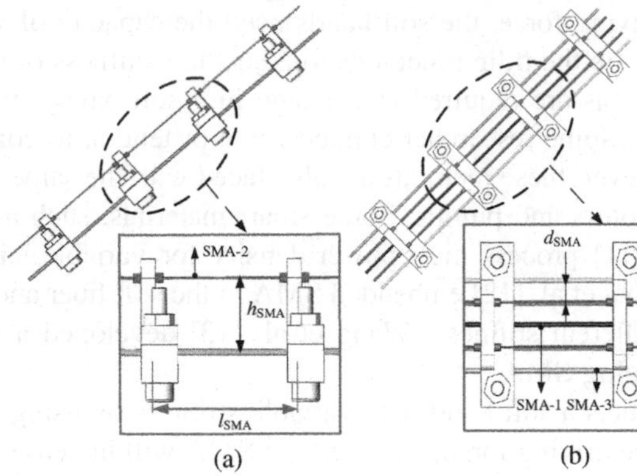

Fig. 2. Enlargement structure of fingers (a) side view; (b) front view

Table 1. Structure sizes and material parameters of fingers

SMA-1	E_1	53 Gpa
	Diameter d_1	0.5 mm
SMA-2	Diameter d_2	0.15 mm
SMA-3	E_M	24 Gpa
	E_A	47 Gpa
	Diameter d_3	0.5 mm
Fingers	Equivalent density ρ	43.56 g/cm^3
	Mass of fingertip	0.0027 kg
	$[L_1, L_2, L_3]$	[100, 80, 80] mm
	$[h_{SMA}, d_{SMA}, l_{SMA}]$	[8, 3, 20] mm

3 Analysis of Grasping Force and Stiffness

3.1 Grasping Force

Figure 3 shows the grasping operation of the soft robot hand, Fig. 3(a) and (b) represents the scheme of grasping objection and the coordinate system for any finger, respectively. The grasping behavior of the hand can be described as

$$\sum_{i=1}^{3} F_{fi} = F_{mg}, \tag{1}$$

$$\sum_{i=1}^{3} F_{Ni} = 0, \tag{2}$$

where F_{mg}, F_{fi} are gravity of objection and friction force, and $F_{fi} = \mu F_{Ni}$, μ is the friction factor. To realize the grasping operation, the grasping condition is given by

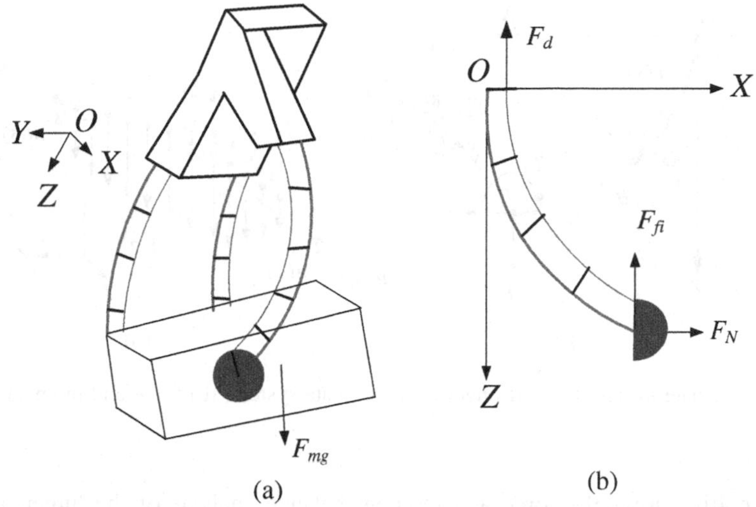

Fig. 3. Grasping operation of the soft robot hand (a) scheme of grasping objection; (b) coordinate system for any finger

Eq. (2), where F_{Ni} determines the friction force and reflects the grasp capability, so F_{Ni} is defined as grasping force.

3.2 Model of Grasping Force Based on Cosserat Theory

Cosserat elastic rod theory [14] regards the rod as a series of continuous material point, and the connection of grasping force, pulling force and deformation can be established by combining the analytical result of each point.

Figure 4(a) shows the coordinate system for soft finger. Point s in the coordinate system O-XYZ and local coordinate system o-xyz can be expressed as $\boldsymbol{r}(s)$ and $\boldsymbol{r}^l(s)$, respectively. The deformation of infinitely small unit in the $\boldsymbol{r}^l(s)$ can be regarded as three movements $\boldsymbol{v}^l = [v_1^l, v_2^l, v_3^l]^T$ and three rotations $\boldsymbol{w}^l = [w_1^l, w_2^l, w_3^l]^T$. In addition, the transformation of the local coordinate system and the global coordinate system can be described as

$$\boldsymbol{r}(s) = \boldsymbol{R}(s)\boldsymbol{r}^l(s), \tag{3}$$

where $\boldsymbol{R}(s)$ is the rotation matrix and the superscript l represents local coordinate system.

The change in the length of the rod caused by three direction movements can be given as

$$\frac{d\boldsymbol{r}^l(s)}{ds} = \dot{\boldsymbol{r}}^l(s) = \boldsymbol{v}^l(s). \tag{4}$$

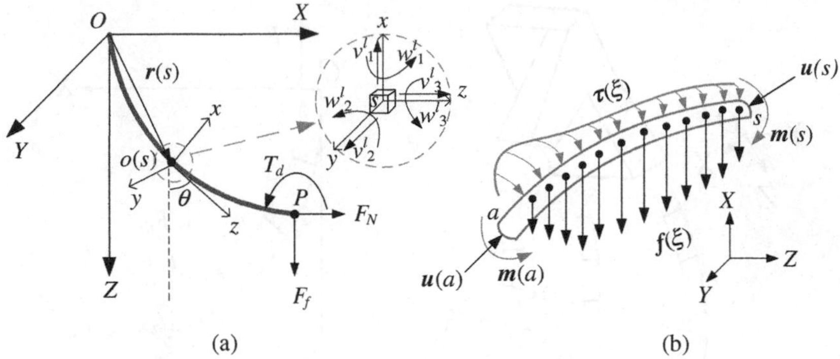

Fig. 4. Coordinate system for soft finger (a) coordinate system; (b) force and moment balance analysis

Figure 4(b) shows the force and moment balance analysis of the finger, and the force balance can be expressed as

$$\boldsymbol{u}(s) - \mathbf{u}(a) + \int_a^s \boldsymbol{f}(\xi)d\xi = 0, \tag{5}$$

for the torque balance

$$\boldsymbol{m}(s) - \boldsymbol{m}(a) + \boldsymbol{r}(s) \times \boldsymbol{u}(s) - \boldsymbol{r}(a)\boldsymbol{u}(a) + \int_a^s [\boldsymbol{t}(\xi) + \boldsymbol{r}(\xi) \times \boldsymbol{f}(\xi)]d\xi = 0, \tag{6}$$

where $\boldsymbol{u}(a)$ and $\boldsymbol{u}(s)$ represent the concentration force, $\boldsymbol{m}(a)$ and $\boldsymbol{m}(s)$ denote the concentration torque, $\boldsymbol{f}(\xi)$ and $\boldsymbol{\tau}(\xi)$ represent distribution force and distribution torque, respectively. $\boldsymbol{r}(s)$ is the distance from point s to origin, $\boldsymbol{r}(a)$ is the distance form point a to origin.

The relationship between the shear strain and the contact force of soft finger in the local coordinate system is given by

$$\boldsymbol{u}(s) = \boldsymbol{R}(s)\boldsymbol{u}^l(s) = \boldsymbol{R}(s)diag(K_1, K_2, K_3)(\boldsymbol{v}^l(s) - {}^*\boldsymbol{v}), \tag{7}$$

where $K_1 = K_2 = GA$, $K_3 = EA$, G represents the shear modulus, E is the elastic modulus, A denotes the cross-sectional area, and the initial state of the rod can be expressed as ${}^*\boldsymbol{v} = [0, 0, 1]^T$.

The relationship between internal torque and the bending torsion of soft finger in the local coordinate system is given by

$$\boldsymbol{m}(s) = \boldsymbol{R}(s)\boldsymbol{m}^l(s) = \boldsymbol{R}(s)diag(W_1, W_2, W_3)\boldsymbol{w}^l(s), \tag{8}$$

where $W_1 = EI_x$, $W_2 = EI_y$, $W_3 = GI_p$, I_x and I_y represent the moment of inertia correspond to the x-axis and y-axis in local coordinate system, respectively, I_p is the polar moment of inertia.

Compress/tension deformation and shear deformation of the finger can be neglected in this paper, so $\boldsymbol{v}^l(s) = [0, 0, 1]^T$. Combining with Eqs. (3) and (4)

$$\dot{\boldsymbol{r}}(s) = \boldsymbol{R}(s)\boldsymbol{v}^l(s) = [\sin\theta, \quad 0, \quad \cos\theta]^T. \tag{9}$$

Since the distribution force of the model is the finger's own weight, and the value of the distribution force is equal to the linear density, the distribution force can be expressed as $\boldsymbol{f}(\xi) = \rho A g e_g$, where g is the gravitational acceleration, $e.g.$ represents the unit vector of gravitational acceleration, ρ denotes the finger density, A is the cross-sectional area of finger. Here $e.g. = [0, 0, 1]^T$, so Eq. (5) can be expressed as

$$\mathbf{u}(s) = [C_1, \quad C_2, \quad -\rho A g s + C_3]^T, \tag{10}$$

the end of finger is acted by $F = \boldsymbol{u}(L) = [F_N, 0, F_f]^T$, so $C_1 = F_N$, $C_2 = 0$, $C_3 = F_f\text{-}\rho AgL$,

$$\boldsymbol{u}(s) = [F_N, \quad 0, \quad F_f + \rho AgL - \rho Ags]^T. \tag{11}$$

As the fingers are not affected by the distribution torque, substituting Eqs. (9) and (11) into Eq. (6)

$$\dot{\boldsymbol{m}}(s) = [0, \quad -\cos\theta F_N + \sin\theta(F_f + \rho Agl - \rho Ags), \quad 0]^T. \tag{12}$$

Since static model of fingers can be simplified in this paper, considering $\boldsymbol{v}^l = {}^*\boldsymbol{v}$ and the deflection in a plane, Eq. (8) can be expressed as

$$\boldsymbol{m} = [0 \quad m_y \quad 0] = [0, \quad EI\dot{\theta}(s), \quad 0]^T. \tag{13}$$

According to Eqs. (9), (12) and (13), the differential is represented by

$$\begin{cases} \dot{x}(s) = \sin\theta, \\ \dot{z}(s) = \cos\theta, \\ \dot{\theta}(s) = \frac{m_y}{EI}, \\ \dot{m}_y(s) = -\cos\theta F_N - \sin\theta(F_f + \rho AgL - \rho Ags), \end{cases} \tag{14}$$

the boundary conditions of the finger are

$$x(0) = 0, \quad z(0) = 0, \quad \theta(0) = 0, \quad m_y(L) = F_d h_{\mathrm{SMA}}. \tag{15}$$

From Eqs. (14) and (15), the relationship between the grasping force F_N of the soft finger and the pulling force F_d can be obtained when the rigidity is constant.

3.3 Stiffness Model

The elastic modulus of SMA-3 is dependent of the content of martensite or austenite and represented by

$$E_3(\xi(T,\sigma)) = \xi(T,\sigma)E_M + (1-\xi(T,\sigma))E_A, \tag{16}$$

the elastic modulus during heating can be expressed as

$$\begin{aligned} E_3(\xi(T,\sigma)) = &\left\{\tfrac{\xi_M}{2}\cos[\alpha_A(T-A_s)+b_A\sigma]+\tfrac{\xi_M}{2}\right\}E_M \\ &+\left\{1-\tfrac{\xi_M}{2}\cos[\alpha_A(T-A_s)+bA\sigma]+\tfrac{\xi_M}{2}\right\}E_A, \end{aligned} \tag{17}$$

the elastic modulus during cooling can be expressed as

$$\begin{aligned} E_3(\xi(T,\sigma)) = &\left\{\tfrac{1-\xi_A}{2}\cos\left[\alpha_M(T-M_f)+b_M\sigma\right]+\tfrac{1+\xi_A}{2}\right\}E_M \\ &+\left\{1-\tfrac{1-\xi_A}{2}\cos\left[\alpha_M(T-M_f)+b_M\sigma\right]+\tfrac{1+\xi_A}{2}\right\}E_A. \end{aligned} \tag{18}$$

Figure 5(a) shows the relationship between elastic modulus of the SMA-3 and heating temperature, elastic modulus of the SMA-3 with complete component of martensite and austenite are 24 GPa and 47 GPa, respectively. Phase change temperatures of Martensite and austenite are 25 °C and 50 °C, respectively. The results of simulation and experiment are agreement during the cooling and heating processes. Figure 5(b) shows the test result of current and temperature, temperature of SMA-3 rises from 25 °C to 140 °C while current increase from 0 A to 2.5 A. Combing these experiment data in Fig. 5, we could obtain the relationship between current SMA-3 and elastic modulus as shown in Fig. 6, which can guide the control of variable stiffness.

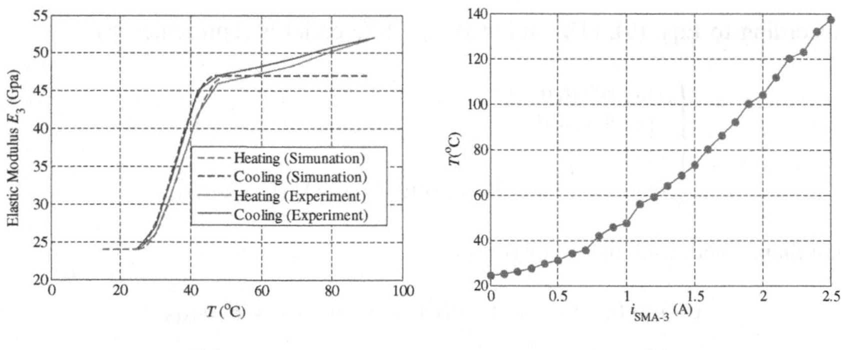

Fig. 5. Results of simulation and experiment (a) relationship between elastic modulus and temperature of SMA-3; (b) relationship between current and temperature of SMA-3

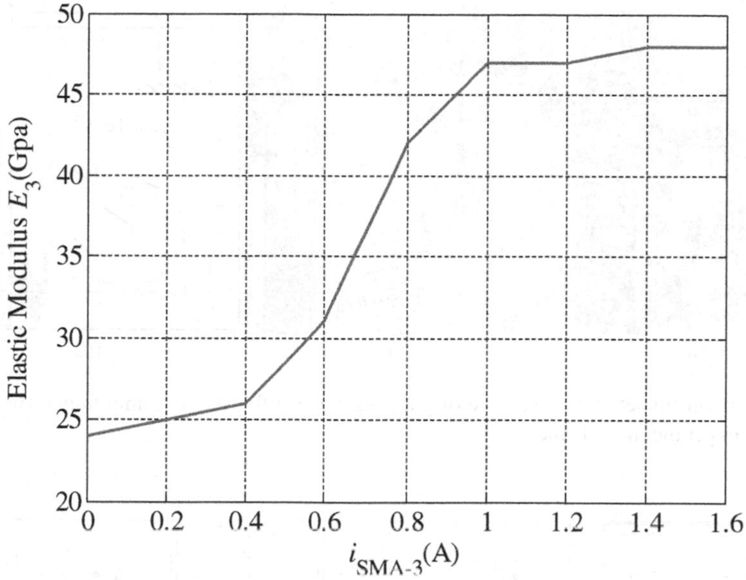

Fig. 6. Relationship between current and elastic modulus of SMA-3

4 Experiment of Grasping Force with Variable Stiffness

4.1 Experiment Scheme

Figure 7(a) and (b) show the experimental setups for measurement of grasping force with 80 mm and 100 mm fingers, respectively. Figure 7(c) represents the experiment scheme. SMA-2 and SMA-3 are heated by current $i_{SMA\text{-}2}$ and $i_{SMA\text{-}3}$ to produce the pulling force and variable stiffness. As shown in Fig. 6, elastic modulus E_3 are 24 Gpa, 37 Gpa, 47 Gpa under input current $i_{SMA\text{-}3}$ 0 A, 0.7 A, 1.2 A, respectively, while the range of input current $i_{SMA\text{-}2}$ is 0 A to 0.36 A. Two force sensors located in the vertical and horizontal directions are used to measure pulling force (F_d) and grasping force (F_N). Through A/D converter, the signal of force sensor are imported into Arduino board finally, and displayed on the PC in real time. In addition, there are many parameters that affect the grasping force of the soft finger, such as the length of SMA-2 (L) and the distance of objection (D), which are fixed in this experiment.

4.2 Results and Discussion of the Grasping Force

Figure 8 shows the relationship between grasping force and pulling force in experiment and simulation for 80 mm and 100 mm fingers, respectively. The short finger will need greater pulling force before the fingertip to touch with force sensor, and then grasping force can be measured. The pulling force before touching sensor of 80 mm and 100 mm fingers range from 1–2 N, 0.5–1 N, respectively. In addition, the grasping force disappears to zero quickly when fingertip is driven to divorce from the objection

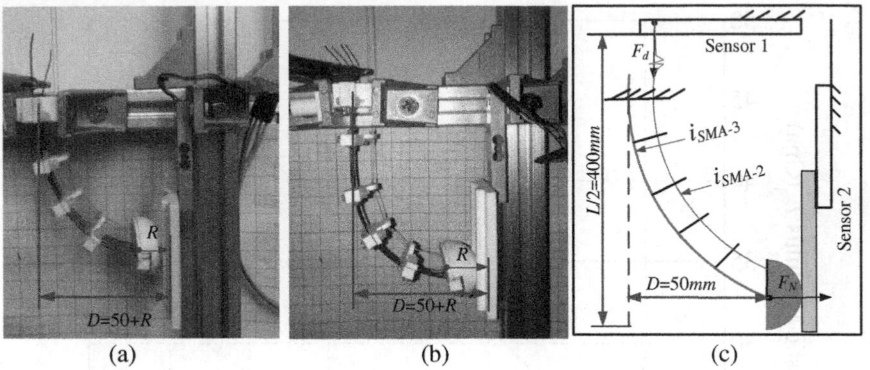

Fig. 7. Experiment setups for measure of grasping force stiffness (a) 80 mm finger; (b) 100 mm finger; (c) experimental scheme

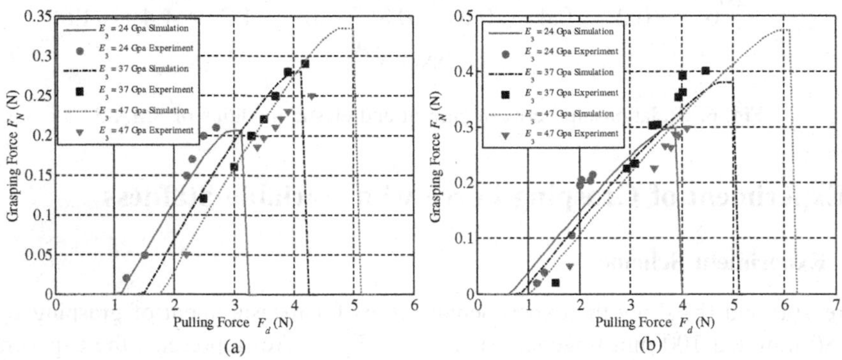

Fig. 8. Relationship between grasping force and pulling force in experiment and simulation (a) 80 mm finger; (b) 100 mm finger

under critical pulling force in simulation. However, the critical pulling force doesn't act on the fingers in experiments. Experimental results indicates that maximum grasping force for 80 mm and 100 mm fingers can increase 41.9% and 86.5%, respectively. Moreover, these experimental results are located in the position for small and larger of pulling force, few of data appears in middle area.

To explain the experimental phenomenon, the relationship between pulling force and current $i_{SMA\text{-}2}$ for 80 mm and 100 mm fingers are measured and shown in Fig. 9(a) and (b). The finger stiffness related to each SMA stiffness can be expressed as $EI = E_1 I_1 + 4E_3 I_3$. When elastic modulus E_3 of SMA-3 increases from 24 Gpa to 47 Gpa, the whole stiffness of fingers increases 61.7%. The pulling forces show same changing trend, when the stiffness increase, the maximum pulling force for 80 mm and 100 mm fingers increase 60% and 76%, respectively. However, the maximum pulling force decline slightly when elastic modulus is E_3 = 47 Gpa. In addition, the pulling force has saturation phenomena when current $i_{SMA\text{-}2}$ rise up to 0.3 A. These phenomenon

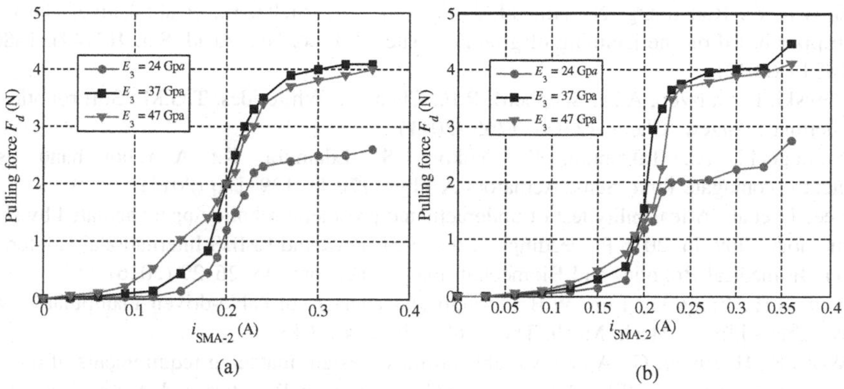

Fig. 9. Relationship between $i_{SMA\ 2}$ and pulling force (a) 80 mm finger; (b) 100 mm finger

produce because the length of SMA-2 is limited. The pulling force results from the shrinkage of SMA-2, so the pulling force is limited when the shrinkage is limited, and the consumption of shrinkage is larger for higher stiffness finger.

5 Conclusions

A soft robot hand with variable stiffness was designed in this paper. Cosserat theory and stiffness model were established to analyze the relationship between pulling force and grasping force in different stiffness.

Theoretical and experimental results show the soft robot hand grasping capacity with variable stiffness. Since the length of SMA-2 impact the grasping force, more detailed influence and analysis should be considered for robot optimization in future work.

Acknowledgment. The research work was supported by the National Natural Science Fundament of China under Grants No. 51575409 and 11202153.

References

1. Yang, J., Pitarch, E.P., Abdel-Malek, K., et al.: A multi-fingered hand prosthesis. Mech. Mach. Theor. **39**(6), 555–581 (2004)
2. Carrozza, M.C., et al.: A cosmetic prosthetic hand with tendon driven under-actuated mechanism and compliant joints: ongoing research and preliminary results. In: Proceedings of the 2005 IEEE International Conference on Robotics and Automation, ICRA 2005, pp. 2661–2666. IEEE (2005)
3. Kargov, A., et al.: Development of a miniaturised hydraulic actuation system for artificial hands. Sens. Actuators A Phys. **141**(2), 548–557 (2008)

4. Brown, E., Rodenberg, N., Amend, J., Mozeika, A., Steltz, E., et al.: Universal robotic gripper based on the jamming of granular material. Proc. Nat. Acad. Sci. **107**(44), 18809–18814 (2010)
5. Lievski, F., Mazzeo, A.D., Shepherd, R.F., Chen, X., Whitesides, T.G.M.: Soft robotics for chemists. Angew. Chem. **123**(8), 1765 (2011)
6. Yamaguchi, A., Takemura, K., Yokota, S., Edamura, K.: A robot hand using electro-conjugate fluid. Sens. Actuators A Phys. **170**(1), 139–146 (2011)
7. Lee, J., et al.: A feasibility test of underactuated prosthetic robotic fingers actuated by shape memory alloy. In: 2016 Proceedings of the IEEE RAS and EMBS International Conference on Biomedical Robotics and Biomechatronics, Singapore, pp. 26–29 (2016)
8. Yeo, S.H., Yang, G., Lim, W.B.: Design and analysis of cable-driven manipulators with variable stiffness. Mech. Mach. Theor. **69**, 230–244 (2013)
9. Wolf, S., Hirzinger, G.: A new variable stiffness design: matching requirements of the next robot generation. In: IEEE International Conference on Robotics and Automation, ICRA 2008, pp. 1741–1746 (2008)
10. Shintake, J., Schubert, B., Rosset, S., Shea, H., Floreano, D.: Variable stiffness actuator for soft robotics using dielectric elastomer and low-melting-point alloy. In: 2015 IEEE/RSJ International Conference on Intelligent Robots and Systems (IROS), pp. 1097–1102 (2015)
11. Kim, Y.J., Cheng, S., Kim, S., Iagnemma, K.: A novel layer jamming mechanism with tunable stiffness capability for minimally invasive surgery. IEEE Trans. Rob. **29**(4), 1031–1042 (2013)
12. Chenal, T.P., Case, J.C., Paik, J., et al.: Variable stiffness fabrics with embedded shape memory materials for wearable applications. In: 2014 IEEE/RSJ International Conference on Intelligent Robots and Systems, IROS 2014, pp. 2827–2831 (2014)
13. Wang, W., Rodrigue, H., Ahn, S.H.: Smart soft composite actuator with shape retention capability using embedded fusible alloy structures. Compos. Part B Eng. **78**, 507–514 (2015)
14. Antman, S.S.: Nonlinear problems of elasticity. Appl. Math. Sci. **1**(2), xviii, 831 (2005)

Toward Effective Soft Robot Control via Reinforcement Learning

Haochong Zhang[1], Rongyun Cao[1], Shlomo Zilberstein[2], Feng Wu[1], and Xiaoping Chen[1(✉)]

[1] University of Science and Technology of China, Hefei 230027, Anhui, China
{solomonz,ryc}@mail.ustc.edu.cn, {wufeng02,xpchen}@ustc.edu.cn
[2] University of Massachusetts Amherst, Amherst, MA 01003-9264, USA
shlomo@cs.umass.edu

Abstract. A soft robot is a kind of robot that is constructed with soft, deformable and elastic materials. Control of soft robots presents complex modeling and planning challenges. We introduce a new approach to accomplish that, making two key contributions: designing an abstract representation of the state of soft robots, and developing a reinforcement learning method to derive effective control policies. The reinforcement learning process can be trained quickly by ignoring the specific materials and structural properties of the soft robot. We apply the approach to the Honeycomb PneuNets Soft Robot and demonstrate the effectiveness of the training method and its ability to produce good control policies under different conditions.

Keywords: Soft robot control · Reinforcement learning · PneuNets

1 Introduction

Recently, soft robotics have attracted growing attention from multiple disciplines including robotics, materials science, bionics, and AI. In contrast to hard-bodied robots [4,15], soft robots are made of soft and/or extensible materials from the surface to the motion mechanism. Soft robots have shown remarkable potential in realizing some performance which conventional rigid robots can hardly perform even after several decades of research. For example, a soft robot has an instinctive advantage on grabbing irregular, deformable or fragile objects, which is needed absolutely in domestic services. A soft robot can also fulfill manipulation friendly and safely when it works closely with humans, which is also beyond the capabilities of conventional service robots [14,19,22]. However, unlike rigid robots, which can be easily and accurately modeled and controlled using mature theories and methodologies, modeling and controlling of soft robots present new challenges to AI and robotics [11,23].

While softness can be an advantage of typical soft robots [14], it also makes them particularly vulnerable to a variety of environmental impacts [25]. In domestic service scenarios, a soft robot may interact with a wide variety of

Y. Huang et al. (Eds.): ICIRA 2017, Part I, LNAI 10462, pp. 173–184, 2017.
DOI: 10.1007/978-3-319-65289-4_17

objects in the environment, where some of the information of these objects and the robot itself cannot be effectively perceived. For example, hardness or weight of an object that the robot tries to pick up can potentially interfere with the interaction. As a result, it will seriously affect the behavior and control of the robot. Therefore, to improve the robustness of soft robot control, especially under the insufficient perception, is critical for the development of soft robots.

Even when all the information about the environment and the soft robot itself is provided, building an effective model for a particular task is still a substantial challenge [21]. It may involve very complex physical mechanics analysis that can only be established for a specific task [26]. With small changes to the task or environment, the model may be significantly different. To date, there is no principled way for modeling soft robots due to the diversity of material and structure (e.g., pneumatic, hydraulic, cables, electro-active polymers) [14]. Moreover, complex mechanics geometric models and simulator for soft robots are often difficult to adequately reflect the real characteristics of the hardware [8]. Various details that are ignored by the model tend to introduce hurdles that compromise performance [6].

To address these challenges, we propose a simple, effective and integrated reinforcement learning (RL) framework into soft robotics for the soft robot control problem. In contrast to other approaches for controlling soft robots [22], our approach has the advantage of ignoring the specific properties of the materials and partial structural characteristics of soft robots, thereby simplifying the modeling task. Besides, since we apply reinforcement learning directly to the soft robot hardware, the various features of the soft robot can also be directly reflected in the results of reinforcement learning. Hence, we can obtain a soft robot control policy that is well aligned with the actual hardware.

Although reinforcement learning has substantial advantages over existing methods, it also presents some challenges.

Firstly, reinforcement learning generally requires that the target problem be abstracted as a Markov decision process (MDP) [1]. Therefore, how to derive a set of representations that can adequately reflect the nature of soft robots, but also facilitate the use of reinforcement learning algorithms is the first problem we face. In this paper, we introduce a class of abstract representations of soft robot control problems in the representation part.

Secondly, in the training part, the performance of reinforcement learning is often dependent on an effective exploration of the problem space. Due to the presence of actions with infinite degrees of freedom and the continuous state space of soft robots [22], to obtain an accurate result, a large-scale search effort is necessary. Nonetheless, the possible loss of the robot hardware and the cost of time limit the scale of training. And since we anticipate the tasks to be handled by soft robots to become progressively more complicated, even though the performance of soft robots continues to improve, the above costs cannot be ignored. And while simulators have been used in the past to address that challenge, developing a realistic simulator of our soft robot is extremely difficult. This prevents us from relying on simulators for training. To address this, we use a

Fig. 1. Platform with soft robot arms, control circuit, air pumps and valves.

combined simulation and physical two-step training to enhance the performance of our resource-bounded physical robot training.

Intuitively, training results must be efficiently performed to ensure the overall system performance. In the experiments, we demonstrate that different execution conditions have significant impacts on the performance of the policy. This has greatly affected the effectiveness of reinforcement learning in our setting. In the execution part, we explore the possibility of policy open-loop and closed-loop execution and carefully evaluate the performance under different conditions. In order to perform open-loop control, we use the simulator to maintain an internal state to implement a "pseudo-closed-loop" control. And we rely on external sensors to achieve closed-loop control.

The remainder of this paper is organized as follows. In Sect. 2, we begin with the reinforcement learning implementation with the state representation, the assumptions and the algorithm in details. In Sect. 3, we implement and test the methods we discussed above on a physical soft robot platform to comprehensively show the performance of the methods. Section 4 provides an overview of the related work. Section 5 concludes the paper and summaries the contribution and future work.

2 Effective Soft Robot Control

As shown in Fig. 1, we use a soft robot constructed with honeycomb pneumatic network (HPN) [24]. It has many advantages, such as structural stability, flexibility, and crush-resistance. The structure of the soft robot makes it possible to carry out large-scale and powerful motion on a light self weight. Here, we try to achieve effective control of the soft robot arm. The difference between this problem with general soft robot control is that we are more concerned with the position of the soft robot arm end coordinates and the policy performance under environmental influences.

We ensure that our method is applicable to other piecewise soft robots. Under this assumption, the entire soft robot consists of relatively independent sections. Each section has the property of infinite degrees of freedom. An advantage of a piecewise soft robot is that it is easy to expand, for example by adding new sections to the end of the soft robot (Figs. 2 and 3).

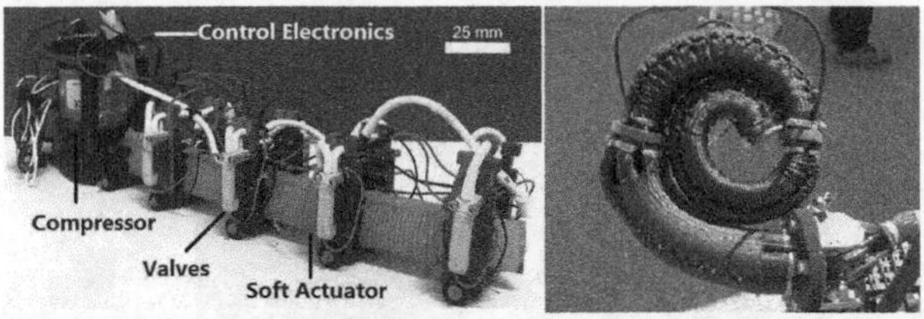

Fig. 2. Some other examples of piecewise soft robots [16,17].

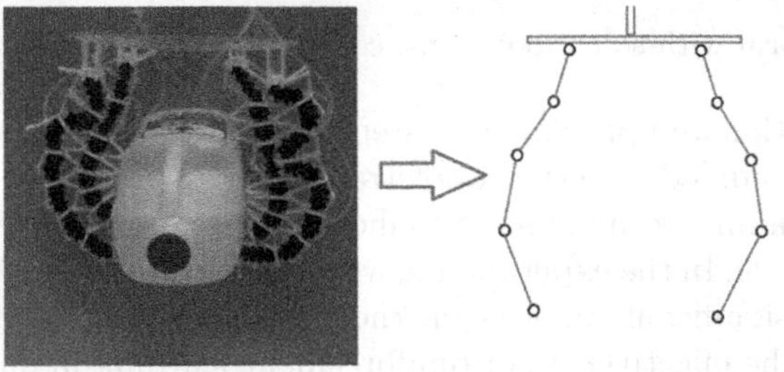

Fig. 3. Illustration of the state representation. The states consist of spatial coordinates of each point in the illustration. Each arm state has five coordinates. Section 3 describes the specific method to obtain states in real time.

2.1 The MDP Representation

To apply reinforcement learning, the first step is to abstract the problem and represent it as a Markov decision process (MDP). We describe below the fundamental MDP elements, including states (S), actions (A), and the reward function ($R(s,a,s')$).

States. A convenient representation of the state space should be able to adequately reflect the configuration of the soft robot, and must also be easy to manipulate. To balance these two factors, we use a set of predetermined points to represent the state of the piecewise soft robot. Based on the piecewise property, the state representation of the entire soft robot is the combination of the states of each section.

We use three coordinates to represent a section state, respectively, the beginning, the center and the end. The start and end points of the two adjacent sections are overlapped. So, for a soft robot connected by n sections, each state contains $n*2+1$ continuous space coordinates p_n.

$$\begin{aligned} State &= \{s_1, s_2, s_3, ...\} \\ \forall s &\in State \\ s &= \{p_1, p_2, p_3, ...\} \\ &= \{(x_1, y_1, z_1), (x_2, y_2, z_2), (x_3, y_3, z_3), ...\} \end{aligned} \quad (1)$$

Actions. Action abstraction is an essential expression of soft robot characteristics. Here we are also faced with a dilemma: a too simple abstraction will severely limit the capacity of soft robots, while an overly complex abstraction could compromise the model's ease of use and versatility.

Therefore, in combination with our observation of a large number of soft robots [14], we define the action as the movement of one of the motion section in discrete directions.

$$Action = \{left, right, long, short, ...\} \times \{all\ sections\} \tag{2}$$

This is an abstraction of action, not the action that can be used directly in the soft robot. For reinforcement learning, this is appropriate. But in order to implement on soft robots, we must define these actions integrated with the physical implementation of the soft robot. For cable-based soft robots, these actions mean different cable adjustments. And for pneumatic soft robot, these actions mean different air pressure adjustments. More specifically, we can increase or decrease each air bag air pressure to define motion section actions.

For a two motion mechanisms (M_a and M_b) motion section there are 4 actions: (1) **left**: M_a increases a unit pressure, M_b remains unchanged. (2) **right**: M_a remains unchanged, M_b increases a unit pressure. (3) **long**: M_a increases a unit pressure, M_b increases a unit pressure. (4) **short**: M_a decreases a unit pressure, M_b decreases a unit pressure.

In practice, "a unit pressure" may represent a unit of air pressure change, may also represent a unit of the inflating volume, but also may represent a period of inflation. The representation of an action is closely related to a particular physical implementation, in the last case "a unit" even means a unit of time.

Different action representations lead to different models. In either case, the problem of uncertainty needs to be dealt with. The uncertainty of the action is taken into account in the design of our algorithm, while the uncertainty of the observation is ignored.

Reward Function. The definition of a reward function depends on the task we are trying to complete. We define the reward function as a linear correlation function of the states in order to simplify the representation of the problem.

For example, we want to move the end of the soft robot to a specified target position p_T. We define the reward function $R(s, a, s')$ as:

$$\begin{aligned} R(s,a,s') &= distance(s, s_T) - distance(s', s_T) \\ distance(s, s_T) &= |p_E - p_T| \\ &= |x_E - x_T| + |y_E - y_T| + |z_E - z_T| \end{aligned} \tag{3}$$

where p_E and p'_E is the end coordinate of the soft robot state s and s', s_T is the target coordinate.

2.2 The RL Algorithm

The designed learning procedure include two steps, simulation step and physical step. The main reason for doing so is to decrease physical learning cost and increase the quality of policy. Simulation trained policy can improve the efficiency of physical training. The mainly difference between these two steps is learning algorithm interaction with the simulator or real world.

Q-Learning. Based on above settings, we use function approximation Q-Learning [3] to train the control policy. Each episode, soft robot executes a sequence of actions from the initial state. On the basis of the above state, action, reward representation, we introduce a "final action" a_F. This action does not affect the shape of the robot, but indicating this episode will be stopped. The entire learning process consists of repeating episodes from the initial state to a_F.

Because of the continuous space and nonlinear setting of the problem, we map the state space into a high-dimensional linear space. This data structure can be seen as a neural network with only one hidden layer. We use the data structure to fit the $Q(s, a)$ function. The advantage of this approach lies in the simplicity of implementation and maintenance. But as the number of soft robot sections increases, the dimensions of the superposition space increase at a faster rate to keep performance.

From the definition of reward, we can see that if we do not have any *discount factor* γ or $\gamma = 1.0$, the intermediate state is not important. For this task, we want to perform as few steps action as possible. Therefore, we modify the Q function to be the mean value function associated with the step numbers.

$$\begin{aligned} Q(s_t, a_t) = {} & Q(s_t, a_t) \\ & + \alpha_t [\frac{1}{t+1} R(s_t, a_t, s_{t+1}) \\ & + \frac{t}{t+1} \max_a Q(s_{t+1}, a) - Q(s_t, a_t)] \end{aligned} \tag{4}$$

With probability ϵ, the algorithm chooses an action at random and with probability $1 - \epsilon$ choose an action $a = \arg\max_a Q(s, a)$.

Simulation Step. We developed the simulator based on the work of [6] to achieve the same effect as possible. Simulator parameters are adjusted empirically and the uncertainty of motion, hypothesis of normal distribution, is introduced. The observation of the states and the execution of the actions is relatively straightforward in the simulator.

We use the simulator to sample in the state space before reinforcement learning. We randomly execute actions and compute the Q value in each sample. By sampling the state space, we use the gradient descent algorithm to compute the weight of this network. We use the $Q(s, a)$ function trained in this way as the initial Q_0 in simulation step to reach better performance.

Physical Step. In addition to the soft robot experiment platform that we also use the OptiTrack[1] motion capture system (MCS) [5,7] as a visual sensor to observe soft robot state. Our method requires that states be fully observable, which

[1] http://optitrack.com/.

means that we ignore sensor errors. Experimental platform has two air pumps for pumping and inflatable. Air pumps with the valve opening and closing to achieve the actions we defined. More detailed information about the experimental platform will be described in the Sect. 3.

We use the simulation trained $Q(s, a)$ function as the initial Q_0 in the physical step. In addition, there are not any structural differences between the two steps in the algorithm. The main difference is the parameter setting and implementation of the algorithm.

It is also very important that we tried the reinforcement learning in physical environment only. But very unfortunately, after a long period of training, we still can not get the policy can be effectively executed. In a few cases, the policy may even fall into a non-stopable situation.

2.3 Execution and Control

How to execute effectively after learned a policy is also our main concern. We will discuss this issue from the perspective of closed-loop and open-loop (Fig. 4).

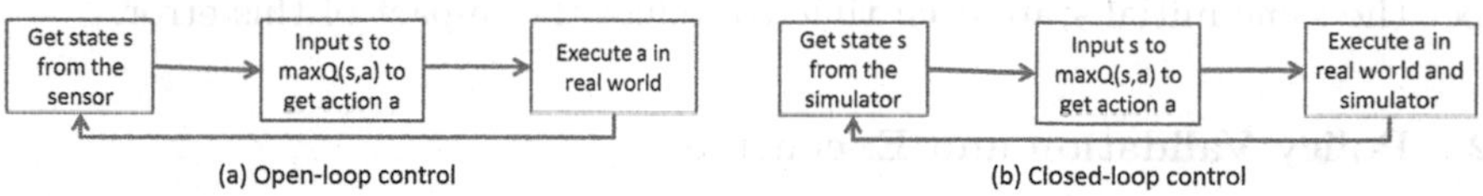

Fig. 4. The primary difference between open-loop and closed-loop control. The closed-loop control is shown above, and the open-loop control is shown below.

Closed-loop control is relatively simple. Each time we obtain the current state of the soft robot from the actual sensor. We then input this state into the Q function to get and execute the optimal action of the policy (Figs. 5 and 6).

For open-loop control, we cannot observe the actual state of the soft robot directly. So, the simulator is used to maintain an internal state of the robot. For each execution cycle, we use the internal state as the state of the entity robot. And in each cycle, we execute the same action in both the simulation and the real world. It is easy to see that this situation lead to unavoidable errors. But for

Fig. 5. Soft Robot 3D prints hard components and EPDM (Ethylene Propylene Diene Monomer) airbags.

Fig. 6. The markers we put on the robot.

some scenarios where closed-loop control cannot be achieved, this compromise is worthwhile.

3 Experiments

3.1 Platform Setup

We use a 3-section soft robot arm to complete the validation of the experiment. In this experimental platform, we use two air pumps to complete the pumping and filling in each section. The control circuit of the valve enables us to achieve almost continuous action. Hence we limited each movement period to 100 ms.

The soft robot arm composed by honeycomb cells consists of soft actuator and variable hexagonal frame. This structure is a combination of soft and hard components. If stay in a relaxed state, the cell is approximately hexagonal. We use 3D printing to produce all the structural and hard components. The soft actuator is made of EPDM (Ethylene Propylene Diene Monomer) with high temperature vulcanization molding.

Another important motion error source is air-tightness of the air bag. We select the same initial state each time to reduce the impact of this error.

3.2 Policy Validation and Execution

We executed the policy we learned several times to verify the final performance. In particular, we tested the performance of different policies under different loads weight. More specifically, we selected 0 g, 10 g, 20 g, 50 g and 100 g. At each cycle, we get the current state of the soft robot arm from the MCS and then execute the action with the highest value in the RDF constructed Q function (Figs. 7 and 8).

Figure 9 shows the execution results of different environment (different load) and different experiment settings in 4 cases: (1) Use the simulator trained policy

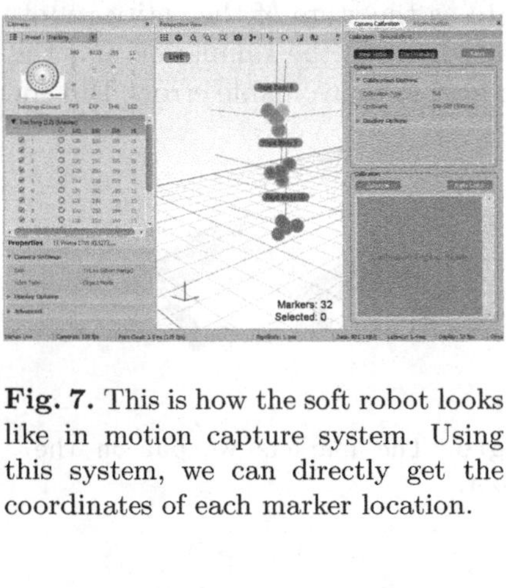

Fig. 7. This is how the soft robot looks like in motion capture system. Using this system, we can directly get the coordinates of each marker location.

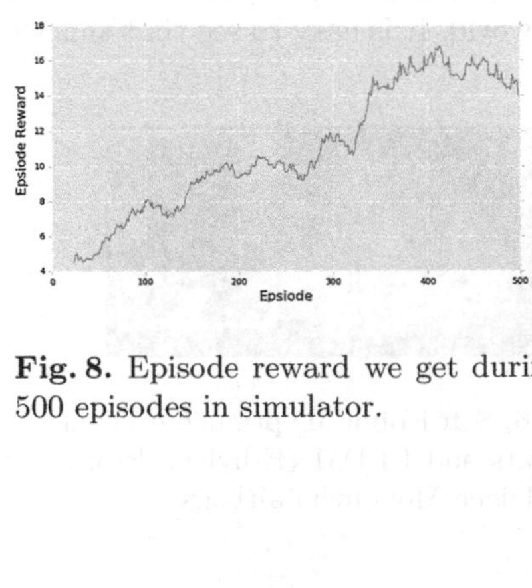

Fig. 8. Episode reward we get during 500 episodes in simulator.

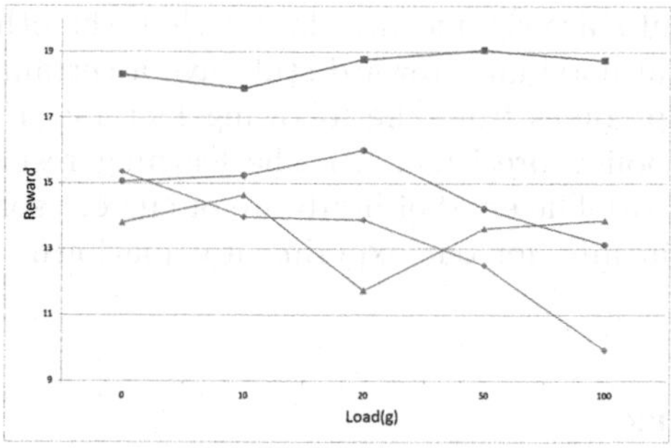

Fig. 9. The average reward of 10 times execution of different policies and different load. Which are using simulator trained policy only (●), simulator trained policy and MCS (◆), reality trained policy and MCS (▲), simulator, reality trained policy and MCS (■). Here, the reward is not Q value but the sum of the reward for each step. According to our manual measurement results, the theoretical maximum reward is about 20.15.

and maintains an internal state in simulator. (2) Use the simulator trained policy and sensors to acquire real states in real time. (3) 50 episodes reality environment trained policy and use sensors to acquire reality states in real time. (4) 50 episodes real environment trained policy based on simulator trained result and use sensors to acquire real states in real time.

In the first case above, we implement the open-loop control of the soft robot. The result is affected by 2 problems: the simulator are not fully accurate and internal state cannot effectively adapt the variety of load. Intuitively, we can not simply perform the same sequence of actions in different scenes.

In the cases 2, 3 and 4, we use the sensors to observe the soft robot's own state in real time to achieve a closed-loop control. In these cases, we improve the performance of policy execution in different loads.

In case 3, We tried to use physical robots directly to train the control policy. However, due to lack of training samples, in many cases the robot will execute a very strange action. In some cases, we observed that the execution process can not converge, so we must manually stop the execution of the policy (Fig. 10).

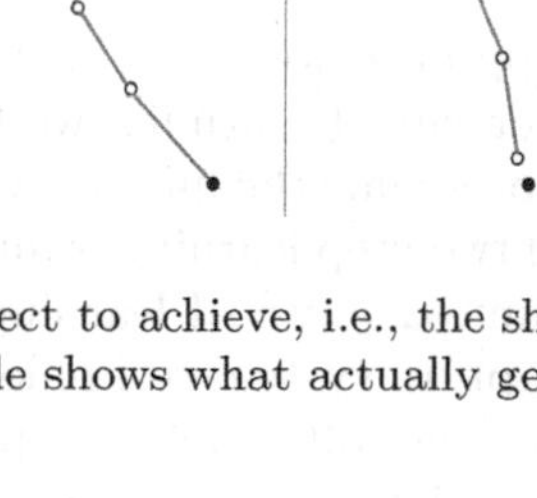

Fig. 10. Left side is what expect to achieve, i.e., the shortest sequence of actions and closest to the target. Right side shows what actually gets from our RL algorithm.

In case 4, the results are significantly better than the other cases. But compare to the theoretical maximum reward still have a certain gap. According to our analysis, this gap comes from the following factors: (1) Discretize actions cause a decrease in policy precision. (2) The learning result does not achieve the optimal solution. (3) The error of hardware of current soft robot experiment platform. More specifically, for our experimental platform, air tightness is the main source of error.

4 Related Work

Piecewise constant curvature kinematics, an approximation of the continuum and soft robot kinematics model, has been widely used in a variety of control tasks. [26] made a review of kinematic modeling for continuum robots. The use of complex curve modeling often leads problems to in sophisticated physical model analysis. But the complexity of the physical model has brought a wide range of adaptation challenges.

Because many biological systems consist of soft materials, biomimetics is often a motivation for the work on soft robots. Researchers have attempted to analysis and utilize a number of biological organs. For example, the robotic elephant trunk manipulator [10] and the OctArm continuum manipulator [18]. Based on an analysis of the morphological features of the octopus arm (connective tissue, arm density, transverse muscles, longitudinal muscles), Laschi [13], Mazzolai [17] and their groups designed the octopus soft robot arm.

To improve the accuracy of the kinematics model, [9] presented a feed-forward neural network learning method to deal with the inverse kinetics elements calculation for a cable-driven non-constant curvature soft manipulator. Because there are small differences between all the individual soft manipulators, they all had to be trained separately in order to represent the relation between the position of the robot tip and the cables' output force.

Kober and Peters [12] summaries the relationship between robotic and reinforcement learning. Several challenges like the curse of dimensionality, real-world, modeling and goal are mentioned. Moreover, MDP-based methods are widely used by other robot systems such as soccer [2,27] and disaster response [20,28].

5 Conclusion and Future Work

This paper presents a new approach for the control of a piecewise soft robot. Through the model-free reinforcement learning, we finally get the action policy for a specific task. In order to overcome the limitations of soft robot hardware, we use the simulator and physical two-step learning framework to improve the policy performance in the case of limited physical learning samples. After the policy is obtained, we use motion capture system to implement the soft robot closed-loop control, thereby enhancing the soft robot's adaptability to the environment. And we using the simulator to maintain an internal state to implement open-loop control. All the above ideas have been Implemented and verified in a physical soft robot platform.

There are many future work for soft robot control via reinforcement learning. Firstly, the value function or policy migration between the simulation and the reality needs a more solid theoretical guidance, analysis and proof. Secondly, more complex and diversified reinforcement learning methods have the potential to improve system performance. Finally, this approach can be extended to more complex forms and tasks such as obstacle avoidance and locomotion.

Acknowledgments. Feng Wu was supported in part by National Natural Science Foundation of China (No. 61603368), the Youth Innovation Promotion Association of CAS (No. 2015373), and Natural Science Foundation of Anhui Province (No. 1608085QF134).

References

1. Bai, A., Wu, F., Chen, X.: Bayesian mixture modelling and inference based thompson sampling in monte-carlo tree search. In: Proceedings of NIPS, pp. 1646–1654 (2013)
2. Bai, A., Wu, F., Chen, X.: Online planning for large markov decision processes with hierarchical decomposition. ACM Trans. Intell. Syst. Technol. **6**(4) (2015). Article No. 45
3. Baird, L., et al.: Residual algorithms: Reinforcement learning with function approximation. In: Proceedings of ICML, pp. 30–37 (1995)
4. Chen, Y., Wu, F., Shuai, W., Wang, N., Chen, R., Chen, X.: Kejia robot - an attractive shopping mall guider. In: Proceedings of ICSR, pp. 145–154 (2015)
5. Chen, Y., Wu, F., Wang, N., Tang, K., Cheng, M., Chen, X.: KeJia-LC: a low-cost mobile robot platform - champion of demo challenge on benchmarking service robots at RoboCup 2015. In: Proceedings of RoboCup, vol. 9513, pp. 60–71 (2015)
6. Cheng, B., Sun, H., Chen, X.: Evolving honeycomb pneumatic finger in bullet physics engine. Robot Intell. Tech. App. **3**, 411–423 (2015)
7. Cheng, M., Chen, X., Tang, K., Wu, F., Kupcsik, A., Iocchi, L., Chen, Y., Hsu, D.: Synthetical benchmarking of service robots: a first effort on domestic mobile platforms. In: Almeida, L., Ji, J., Steinbauer, G., Luke, S. (eds.) RoboCup 2015. LNCS (LNAI), vol. 9513, pp. 377–388. Springer, Cham (2015). doi:10.1007/978-3-319-29339-4_32
8. Duriez, C.: Control of elastic soft robots based on real-time finite element method. In: Proceedings of ICRA, pp. 3982–3987 (2013)
9. Giorelli, M., Renda, F., Ferri, G., Laschi, C.: A feed-forward neural network learning the inverse kinetics of a soft cable-driven manipulator moving in three-dimensional space. In: Proceedings of IROS, pp. 5033–5039 (2013)
10. Hannan, M.W., Walker, I.D.: Kinematics and the implementation of an elephant's trunk manipulator and other continuum style robots. J. Robotic Syst. **20**(2), 45–63 (2003)
11. Inoue, T., Hirai, S.: Modeling of soft fingertip for object manipulation using tactile sensing. In: Proceedings of IROS, pp. 2654–2659 (2003)
12. Kober, J., Peters, J.: Reinforcement learning in robotics: a survey. In: Reinforcement Learning, pp. 579–610 (2012)

13. Laschi, C., Cianchetti, M., Mazzolai, B., Margheri, L., Follador, M., Dario, P.: Soft robot arm inspired by the octopus. Adv. Robot. **26**(7), 709–727 (2012)
14. Laschi, C., Mazzolai, B., Cianchetti, M.: Soft robotics: technologies and systems pushing the boundaries of robot abilities. Sci. Robot. (2016)
15. Lu, D., Zhou, Y., Wu, F., Zhang, Z., Chen, X.: Integrating answer set programming with semantic dictionaries for robot task planning. In: Proceedings of IJCAI (2017)
16. Luo, M., Pan, Y., Skorina, E.H., Tao, W., Chen, F., Ozel, S., Onal, C.D.: Slithering towards autonomy: a self-contained soft robotic snake platform with integrated curvature sensing. Bioinspir. Biomim. **10**(5), 055001 (2015)
17. Mazzolai, B., Margheri, L., Cianchetti, M., Dario, P., Laschi, C.: Soft-robotic arm inspired by the octopus: II. From artificial requirements to innovative technological solutions. Bioinspir. Biomim. **7**(2), 025005 (2012)
18. McMahan, W., Chitrakaran, V., Csencsits, M., Dawson, D., Walker, I.D., Jones, B.A., Pritts, M., Dienno, D., Grissom, M., Rahn, C.D.: Field trials and testing of the OctArm continuum manipulator. In: Proceedings of ICRA, pp. 2336–2341 (2006)
19. Pfeifer, R.: soft robotics - the next generation of intelligent machines. Invited talk on IJCAI (2013)
20. Ramchurn, S.D., Huynh, T.D., Wu, F., Ikuno, Y., Flann, J., Moreau, L., Fischer, J.E., Jiang, W., Rodden, T., Simpson, E., Reece, S., Roberts, S., Jennings, N.R.: A disaster response system based on human-agent collectives. J. Artif. Intell. Res. **57**, 661–708 (2016)
21. Renda, F., Giorelli, M., Calisti, M., Cianchetti, M., Laschi, C.: Dynamic model of a multibending soft robot arm driven by cables. IEEE Trans. Robot. **30**(5), 1109–1122 (2014)
22. Rus, D., Tolley, M.T.: Design, fabrication and control of soft robots. Nature **521**(7553), 467–475 (2015)
23. Shibata, M., Hirai, S.: Soft object manipulation by simultaneous control of motion and deformation. In: Proceedings ICRA, pp. 2460–2465 (2006)
24. Sun, H., Chen, X.-P.: Towards honeycomb pneunets robots. In: Kim, J.-H., Matson, E.T., Myung, H., Xu, P., Karray, F. (eds.) Robot Intelligence Technology and Applications 2. AISC, vol. 274, pp. 331–340. Springer, Cham (2014). doi:10.1007/978-3-319-05582-4_28
25. Tolley, M.T., Shepherd, R.F., Mosadegh, B., Galloway, K.C., Wehner, M., Karpelson, M., Wood, R.J., Whitesides, G.M.: A resilient, untethered soft robot. Soft Robot. **1**(3), 213–223 (2014)
26. Webster, R.J., Jones, B.: Design and kinematic modeling of constant curvature continuum robots: a review. Int. J. Robot. Res. **29**(13), 1661–1683 (2010)
27. Wu, F., Chen, X.: Solving large-scale and sparse-reward DEC-POMDPs with correlation-MDPs. In: Proceedings of RoboCup, pp. 208–219 (2007)
28. Wu, F., Ramchurn, S., Chen, X.: Coordinating human-UAV teams in disaster response. In: Proceedings of IJCAI, pp. 524–530 (2016)

The Calibration Method of Humanoid Robot Based on Double Support Constraints

Fei Liu(✉) and Li Tang(✉)

Ludong University, Yantai, Shandong, China
liufeildu@163.com, lilicon_4@163.com

Abstract. Aiming at the problem of calibration in humanoid robot, this paper presents a calibration method combining online and offline process. According to the constraint condition of the double support phase in the bipedal walking, we define the calibration problem as a nonlinear least squares problem and use the Levenberg-Marquardt method to solve the problem. We use NAO humanoid robot to collect data online and then the calibration process is performed offline on the local computer, so it does not affect the normal movement of the robot and improves the adaptability of the robot calibration.

Keywords: Calibration · Nonlinear least squares problem · Humanoid robot

1 Introduction

Robots often need to be recalibrated because of assembly error, joint wear or link deformation caused by strong collision in the long-term use. For example, NAO humanoid robot [1] in the RoboCup SPL game will continually walk on the field, kick the ball or block the ball and execute other actions. Two robots may have a strong collision with the ball, causing one or even two robots to fall into the ground. Therefore, the robots in such a fierce competition need to be recalibrated even if the robots have been calibrated in advance.

Calibration problem is a classic problem in the field of robotics [2–4]. The problem can be further divided into model-based parametric calibration and model non-parametric calibration. Most work on model-based parametric calibration has concentrated on kinematic-model-based calibration [3]. The general idea is to formalize the kinematic calibration problem as a linear or nonlinear equation, which is a function of the geometric parameters, the joint positions, and the location (position and orientation) of the end-effector frame [4]. There are many kinematic self-calibration methods in the field of humanoid robot calibration [5–8]. NAO robot is a typical humanoid robot, which is widely used in RoboCup SPL competition and research projects. At present, many methods have been proposed to calibrate the NAO humanoid robot. For example, Kastner et al. [9] proposed an automatic calibration method, which firstly observed two checkerboards attached to the robot's feet by a camera on the robot's head under several different kinematic configurations, and then the calibration problem was finally solved utilizing the Levenberg-Marquardt algorithm [10]. Maier et al. [11] presented one approach to calibrate the kinematic model of a humanoid robot based on observations of

Y. Huang et al. (Eds.): ICIRA 2017, Part I, LNAI 10462, pp. 185–192, 2017.
DOI: 10.1007/978-3-319-65289-4_18

its monocular camera and a reduced number of configurations, which consisted of the joint angle offsets of the whole body including the legs, as well as the camera extrinsic and intrinsic parameters. Röfer et al. [12] proposed a calibration policy which firstly got both feet in a planar position by changing the joint offsets of feet and legs by manual, and then used camera automatic calibration to obtain the tilt angle of the body with respect to the feet.

When walking on planar ground, both feet should be on the same plane during the double support phase. We can record all the joint angles during double support phase and compute the positions of both feet by kinematic model. If the computed positions of both feet are not same, it indicates that there may be some errors in the kinematic model. Further, we can formulate the position difference of both feet as a nonlinear least squares problem. Using the Levenberg-Marquardt algorithm and the joint angles measured, the 22 parameters involved (22 joint offsets) are estimated by minimizing the sum of squared residuals of positions during double support.

Our method is divided into two phases: the first phase is the data collection phase and the second phase is the calibration phase. In the first phase, when the robot falls repeatedly, we design a strategy that allows the robot to enter a bipedal stance and then start recording the robot's joint angles' data. In the second phase, the data set is transmitted to the local computer through the wireless network and then the calibration process is performed offline on the local computer. The advantage is that it does not require the robot to stop the movement for specialized calibration.

The remainder of this paper is organized as follows: in Sect. 2, the kinematic model is discussed. Section 3 presents the calibration method. The experimental results are presented in Sect. 4. Finally, we provide a brief summary in Sect. 5.

2 Kinematic Model

NAO robot has 25 DOF (degrees of freedom), which includes 5 DOF on left leg, 5 DOF on right leg, 1 DOF on the hip, 2 DOF on the head and 6 DOF on each arm. The kinematic model is shown in Fig. 1. In Fig. 1, the robot-relative coordinate system is described by its origin lying at the center of both hips, the x-axis pointing forward, the y-axis pointing to the left, and the z-axis pointing upward. The position of left foot sole in the robot-relative coordinate system is computed as follows:

$$\begin{aligned}
haunch &= Trans_Y(YHipOffset) \\
pelvis &= haunch \cdot Rot_X(\tfrac{pi}{4}) \cdot Rot_Z(LHipYawPitch) \cdot Rot_X(-\tfrac{pi}{4}) \\
hip &= pelvis \cdot Rot_X(LHipRoll) \cdot Rot_Y(LHipPitch) \\
tibia &= hip \cdot Trans_Z(UpperLegLength) \cdot Rot_Y(LKneePitch) \\
shrank &= tibia \cdot Trans_Z(LowerLegLength) \\
ankle &= shrank \cdot Rot_Y(LAnklePitch) \cdot Rot_X(LAnkleRoll) \\
P_{FL} &= ankle \cdot P_{fl}
\end{aligned} \tag{1}$$

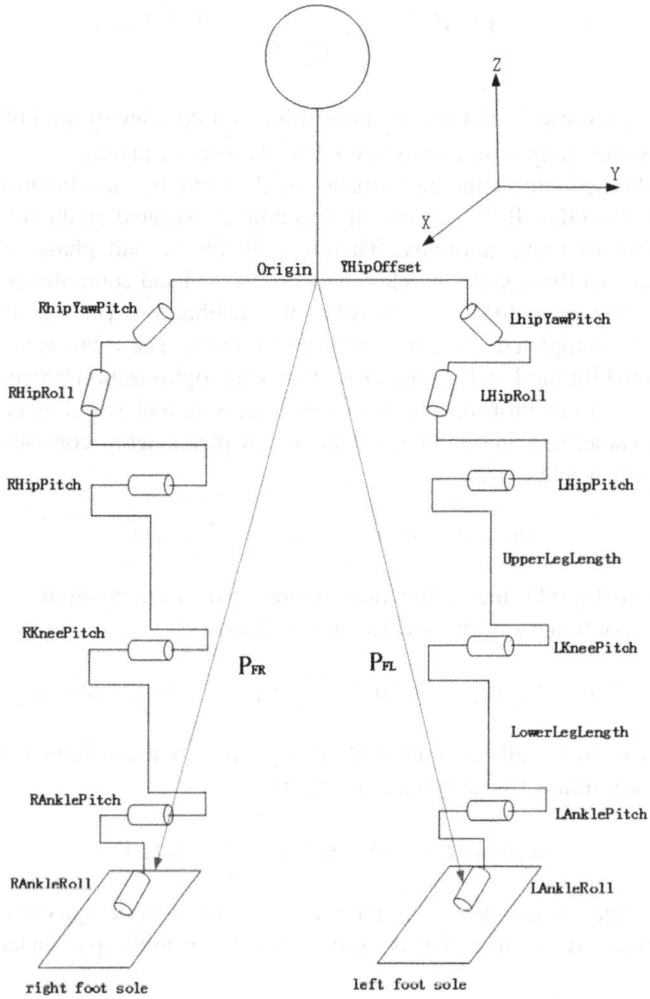

Fig. 1. Kinematics model of NAO robot.

3 Calibration Method

Our calibration method is divided into two phases: the first phase is the data collection phase and the second phase is the calibration phase. In the first phase, during bipedal standing, if the difference between position of right feet P_{FR} and left foot P_{FL} is more than the threshold β,

$$||P_{FR} - P_{FL}|| > \beta \tag{2}$$

And the robot starts to record the joint angles' data d_i of each cycle during double support phase, which is added to the data set D.

$$\begin{aligned} D &= \{d_1, d_2, \ldots, d_i, \ldots, d_n\}, i = 1, 2, 3, \ldots, n \\ d_i &= (J_{fl}^i, J_{fr}^i), \end{aligned} \tag{3}$$

where J_{fl}^i and J_{fr}^i represent joint angles' data of left and right leg of the i-th cycle. After recording data, the calibration process gets into the second phase.

There is a limited computing performance on the robot because the main frequency of CPU is only 1.6 GHz. If the calibration algorithm is executed on the robot, the robot will not be able to move normally. Therefore, in the second phase we choose to transmit the data set through the wireless network to the local computer and the results of the optimization are uploaded to the robot after calibration. The whole process can be automatically completed without human intervention. The robot can continuously locomote meanwhile the local computer performs the optimization process.

Considered that the error may be caused by joint wear and assembly error, the joint angle and link values should be calibrated. Each new parameter q_k consists of measured value φ_k and offset value η_k.

$$q_k = \varphi_k + \eta_k, \varphi_k \in d_i, k = 1, 2, \ldots, m \tag{4}$$

According to Eq. (1), the Z direction difference δ in the position of both feet in robot-relative coordinate system is calculated as follows:

$$R_D = \{\delta | \delta = P_{FL}^Z(q_k) - P_{FR}^Z(q_k) \wedge q_k = \varphi_k + \eta_k \wedge \varphi_k \in d_i \wedge d_i \in D\} \tag{5}$$

Where R_D contains all the collected data points, so the calibration problem is formulated as a nonlinear least squares problem:

$$argmin_\eta \Sigma_{\delta \in R_D}(\delta^2), \eta = \{\eta_1, \eta_2, \ldots, \eta_m\} \tag{6}$$

In order to find a parameter set η that minimizes the sum of squared residuals, the Levenberg-Marquardt method [10] is used in which the initial parameter η_i is set to zero.

4 Experimental Results and Analysis

In our experiment, NAO humanoid robot walks forward at a constant speed on the planar ground for many times. We firstly analyze the residual between the left and right foot on the Z axis in double support phase. As shown in Figs. 2 and 3, there are differences of the left and right foot and this result indicates that the left and right foot is not on a plane during double support phase.

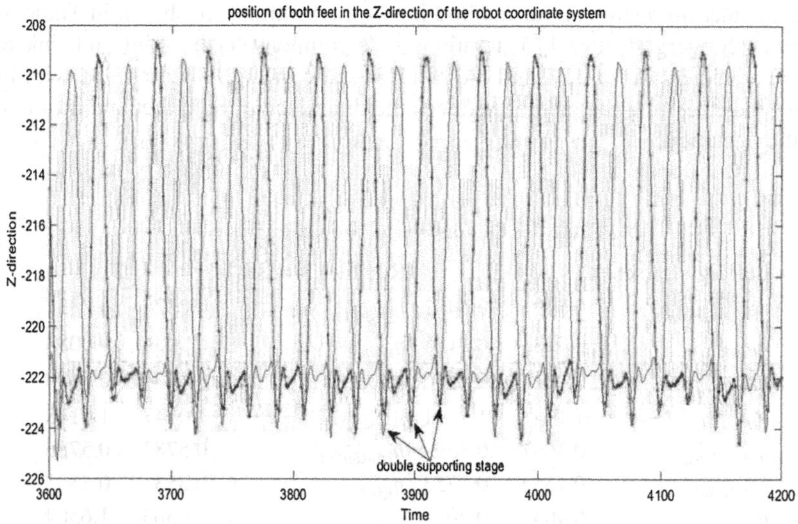

Fig. 2. Position of both feet in the Z-direction of the robot coordinate system.

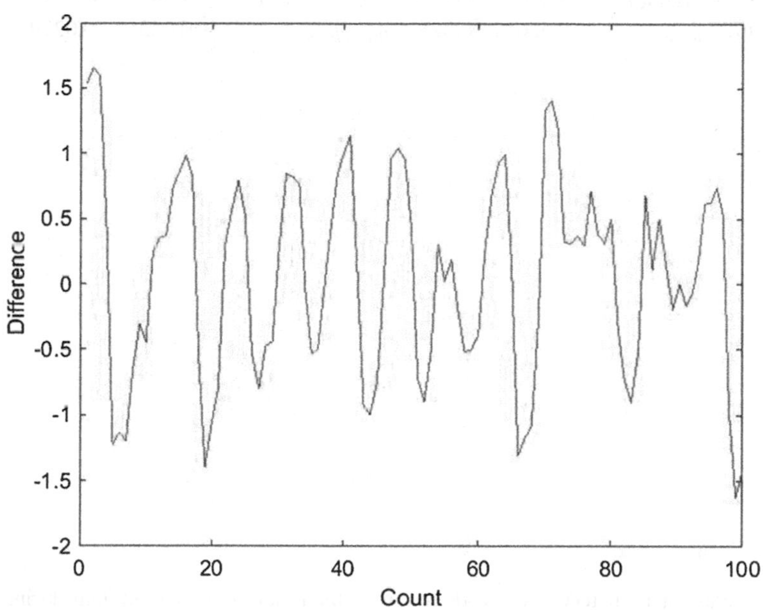

Fig. 3. Residual of both feet in the Z-direction of the robot coordinate system during double support phase.

The calculation results of our method in two calibrations are shown in Table 1. The calibrations converged after 175 iterations. We compensate the joint and link offsets with the parameters in Table 1 and the result is show in Fig. 4. Compared with Fig. 3, the result indicates that the method reduces the difference of both feet and improves the kinematic accuracy.

Table 1. Parameters of two calibrations.

Left leg's offset	1	2	Right leg's offset	1	2
$\eta_{lHipYawPitch}$	0.6491	0.6481	$\eta_{rHipYawPitch}$	0.232	0.2331
$\eta_{lHipRoll}$	−0.0133	−0.0135	$\eta_{rHipRoll}$	−0.0084	−0.0083
$\eta_{lHipPitch}$	0.2714	0.2716	$\eta_{rHipPitch}$	0.6404	0.6403
$\eta_{lKneePitch}$	−1.114	−1.1141	$\eta_{rKneePitch}$	−1.0147	−1.0148
$\eta_{lAnklePitch}$	0.3691	0.369	$\eta_{rAnklePitch}$	0.5787	0.5786
$\eta_{lAnkleRoll}$	0.1355	0.1353	$\eta_{rAnkleRoll}$	0.353	0.3531
$\eta_{lpi/4}$	−0.404	−0.4038	$\eta_{rpi/4}$	−1.6565	−1.6504
$\eta_{lYHipOffset}$	0	0	$\eta_{rYHipOffset}$	0	0
$\eta_{LupperLegLength}$	−0.0934	−0.093	$\eta_{RupperLegLength}$	−0.0383	−0.0381
$\eta_{LlowerLegLength}$	−0.0557	−0.0554	$\eta_{RlowerLegLength}$	−0.02	−0.0199
$\eta_{LfootHeight}$	−0.0261	−0.026	$\eta_{RfootHeight}$	0.023	0.0229

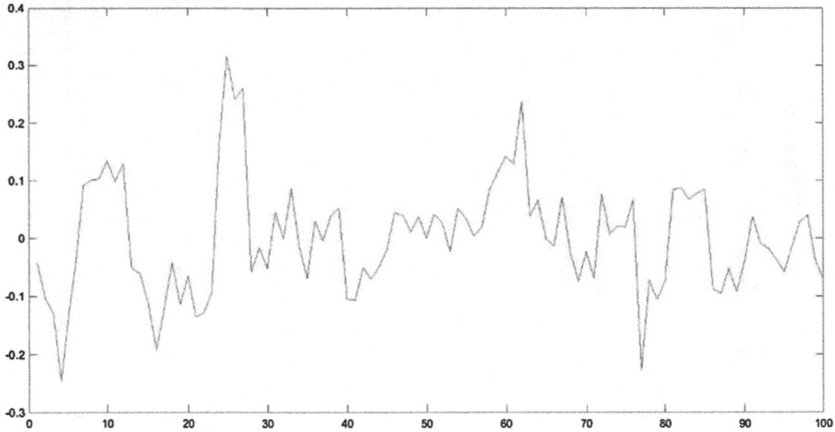

Fig. 4. Residual of both feet in the Z-direction of the robot coordinate system during double support phase after compensation.

As can be seen from the following results of $\eta_{lKneePitch}$ and $\eta_{rKneePitch}$, the left and right leg's knee pitch joint deviations are big during the experiment because the robots always walk in a bent knees posture and it has great damage to the knee joint. Also from the results of $\eta_{lpi/4}$ and $\eta_{rpi/4}$, the robot's left and right hip assembly positions are different which is caused by the assembly error. From the results of $\eta_{lYHipOffset} = \eta_{rYHipOffset} = 0$, it is indicated that this method can't compensate the width variable of the hip because there are not the two parameters in the nonlinear least squares equations.

5 Conclusions

This paper presents a calibration method of humanoid robot based on double support constraints. According to the constraint condition of double support phase in the bipedal walking, we define the calibration problem as a nonlinear least squares problem and use the Levenberg-Marquardt method to solve the problem. The NAO humanoid robot is used to collect data online at constant speeds and then the calibration process is performed offline on the local computer, so it does not affect the normal movement of the robot and improves the adaptability of the robot calibration. The experimental results show that this method has good effect on improving the kinematics model. However, the calibration results at different speeds are quite different, so a more generalized solution will be discussed in the future.

Acknowledgement. This work has been funded by Program for Shandong Science and Technology (2012YD03111), Laboratory of Robotics in Ludong University, Multi-agent Systems Laboratory and Information Center in the University of Science and Technology of China.

References

1. NAO Humanoid Robot. https://www.ald.softbankrobotics.com/en/cool-robots/nao. Accessed 16 June 2017
2. Goswami, A., Quaid, A., Peshkin, M.: Complete parameter identication of a robot from partial pose information. In: 1993 IEEE International Conference on Robotics and Automation, Atlanta, GA, USA, pp. 168–173. IEEE Press (1993)
3. Elatta, A., Gen, L.P., Zhi, F.L., Daoyuan, Y., Fei, L.: An overview of robot calibration. Inf. Technol. J. **3**(1), 74–78 (2004)
4. Khalil, W., Besnard, S., Lemoine, P.: Comparison study of the geometric parameters calibration methods. Int. J. Robot. Autom. **15**(2), 56–67 (2000)
5. Nickels, K., Huber, E.: Hand-Eye calibration of RoboNaut. https://ntrs.nasa.gov/archive/nasa/casi.ntrs.nasa.gov/20100033083.pdf. Accessed 16 June 2017
6. Strobl, K.H., Hirzinger, G.: Optimal hand-eye calibration. In: 2006 IEEE/RSJ International Conference on Intelligent Robots and Systems, Beijing, China, pp. 4647–4653. IEEE Press (2006)
7. Birbach, O., Bäuml, B., Frese, U.: Automatic and self-contained calibration of a multi-sensorial humanoid's upper body. In: 2012 IEEE International Conference on Robotics and Automation, Saint Paul, MN, USA, pp. 3103–3108. IEEE Press (2012)
8. Pradeep, V., Konolige, K., Berger, E.: Calibrating a Multi-arm multi-sensor robot a bundle adjustment approach. In: Khatib, O., Kumar, V., Sukhatme, G. (eds.) Experimental Robotics. Springer Tracts in Advanced Robotics, vol. 79, pp. 211–225. Springer, Heidelberg (2014). doi:10.1007/978-3-642-28572-1_15

9. Kastner, T., Röfer, T., Laue, T.: Automatic robot calibration for the NAO. In: Bianchi, R.A.C., Akin, H.L., Ramamoorthy, S., Sugiura, K. (eds.) RoboCup 2014: Robot World Cup XVIII. LNCS, pp. 233–244. Springer, Heidelberg (2015). doi:10.1007/978-3-319-18615-3_19
10. Madsen, K., Nielsen, H.B., Tingle, O.: Methods for non-linear least squares problems. http://www.imm.dtu.dk/pubdb/views/edoc_download.php/3215/pdf/imm3215.pdf. Accessed 16 June 2017
11. Maier, D., Wrobel, S., Bennewitz, M.: Whole-body self-calibration via graph-optimization and automatic configuration selection. In: 2015 IEEE International Conference on Robotics and Automation, Seattle, WA, USA, pp. 5662–5668. IEEE Press (2015)
12. Röfer, T., Laue, T., Kuball, J., Lübken, A., Maaβ, F., Müller, J., Post, L., Richter-Klug, J., Schulz, P., Stolpmann, A., Stöwing, A., et al.: B-Human team report and code release 2016. https://github.com/bhuman/BHumanCodeRelease/raw/master/CodeRelease2016.pdf. Accessed 16 June 2017

Rehabilitation Robotics

Real-Time Collision Avoidance Algorithm for Surgical Robot Based on OBB Intersection Test

Yao Qiu, Zhiyuan Yan, Yu Miao, and Zhijiang Du(✉)

State Key Laboratory of Robotics and System, Harbin Institute of Technology, Harbin 150080, Heilongjiang, China
duzj01@hit.edu.cn

Abstract. To improve flexibility and adaptability, a surgical robot normally has a number of arms and redundant degrees of freedom, which may cause collision between robotic arms. In this paper, a real-time collision avoidance algorithm based on OBB (Oriented Bounding Box) intersection test was presented to solve this problem. The cube model of each arm was established based on OBB. According to the distance between any two OBBs, probable collision components can be found. Along with the kinematics of the surgical robot, maximum motion range of each joint can be obtained under the condition of no collision. Then the concept of *Collision Index* was proposed to represent the relationship between the angular position and the maximum motion range of each joint. In order to avoid collision, *Collision Indexes* were used to limit the motion of each joint. Experimental results show that the algorithm is effective and could avoid collision between manipulators.

Keywords: Collision avoidance algorithm · OBB · Surgical robot · RALP

1 Introduction

Compared with traditional surgery, robot assisted minimally invasive surgery (RMIS) has many advantages, e.g. less trauma, less blood loss, faster recovery and lower operating fatigue [2]. At present, Da Vinci robot and Zeus surgical robot are widely used in all kinds of thoracic and abdominal surgery. In 2016, 17979 surgery cases were performed by Da Vinci surgical robot in China, about 140 times as many as 134 cases in 2008. The proportion of partial nephrectomy is associated with acquisition of the surgical robot [8], which means RMIS is being increasingly accepted by more and more patients, and Da Vinci system was considered latest surgical technique [5,10], and there are much advanced research about surgical robot is in progress [7]. However, there are still some deficiencies in surgical robot. The structure is complex, with at least three arms, and 11 degrees of freedom in each arm (instrument manipulator). Complex structures may lead to collisions between robotic arms in the operation process. At the

Y. Huang et al. (Eds.): ICIRA 2017, Part I, LNAI 10462, pp. 195–205, 2017.
DOI: 10.1007/978-3-319-65289-4_19

same time, in order to ensure that there is a relatively large public workspace between micro instruments, the trocar positions are generally concentrated. This may result in a relatively close distance between manipulators, which is likely to lead to collision too. In addition, doctor's improper operation can also cause collision, such as frequent reset operation.

Therefore, it is often necessary to consider the probability of collision between manipulators in the preoperative planning. Li and Zhang define the distance between some components as the collision probability [6,9], which was used as one of the optimization objective to reduce the collision rate when optimizing the preoperative position. However, clinical practice of robotic surgery indicates that it is very easy to avoid collision at the beginning of surgery. This is because the arms are far away from each other in the initial setup of surgical robot. Collisions are often caused by the doctor's operating habits such as frequently resetting the robot. Therefore, the real-time collision avoidance algorithm is of great significance for surgical robot. Bosscher presents a collision avoidance algorithm based on speed control. The algorithm needs real-time detection of the obstacles in the external environment by vision system [1], and can not solve the collision problem of the multiple arms of the surgical robot. At present, little attention has been paid to solve the collision problem in surgical robot. This paper proposes a real-time collision avoidance algorithm.

In Sect. 2.1, the geometric model of each component is established, and distances between all components are calculated in Sect. 2.2, then PCG (Possible Collision Group) is also obtained based on these distances. Section 2.2 gives the collision detection principle between two OBBs. Next, the concept of *Collision Index* is proposed in Sect. 2.3, which is calculated to limit motion of all joints. In the end, several experiments are presented to verify the effectiveness of the proposed algorithm in Sect. 3.

2 Method to Avoid Collision

To realize collision avoidance, firstly the geometric model of the component is established, and the position of each component can be solved by the angle of all joints and the kinematics of the robot. Then we should find these component groups most likely to collide. Next, maximum motion ranges of all components could be calculated to ensure there is no collision between any group in PCG. comparing current angle of any joint and its motion range, to judge whether the motion of this joint will immediately lead to collision. In the end, the *Collision Indexes* of all joints are used to restrict joint movement to avoid collision between arms.

2.1 Geometric Model

Robotic arms (except endoscope manipulator) have 11 degrees of freedom, including 4 passive joints and 7 active joints. The 4 passive joints of all arms

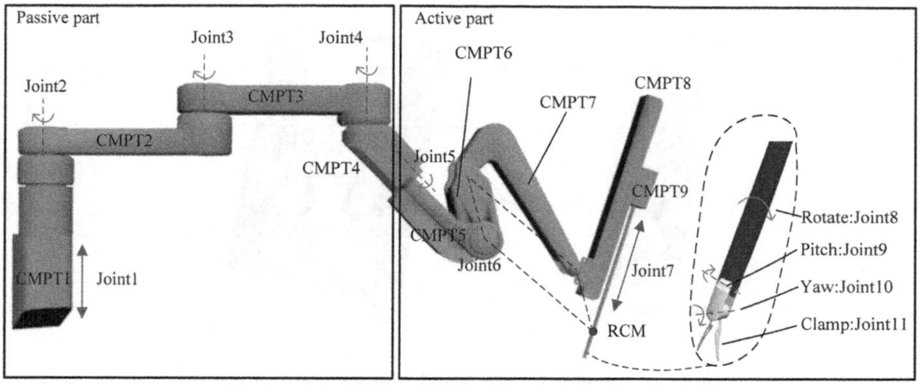

Fig. 1. Manipulator model

keep still during the surgery, the 8–11 joint of the active joints respectively control the rotation, pitch, deflection as well as the clamping of the end instrument, which do not affect the collision, and the joint 7 is a moving joint, which moves along the axis of the instrument and does not affect the collision between the manipulators. In conclusion, the collision between robotic arms is only caused by joint 5 and 6 of each manipulator, the potential collision parts are components 6, 7 and 8. So the settings of joints which could lead to collision are defined as Φ, and corresponding settings of probable collision components are defined as O (suppose the robot has three arms: left arm, middle arm and the right arm. In Eqs. (1) and (2). 'l', 'm', 'n' represent three arms respectively. The number in subscripts represent serial number of joint or component.)

$$\Psi = \{J_{l5}, J_{l6}, J_{m5}, J_{m6}, J_{r5}, J_{r6}\} \tag{1}$$

$$O = \{C_{l6}, C_{l7}, C_{l8}, C_{m6}, C_{m7}, C_{m8}, C_{r6}, C_{r7}, C_{r8}\} \tag{2}$$

To describe the collision between manipulators, it is necessary to model each component $o_i \in O$. Both real-time and accuracy should be taken into consideration. As shown in Fig. 1, components 5, 6, 7, 8 are similar to cube. In order to facilitate the calculation and improve the accuracy, a simple cube model is considered. Take the component 6 as an example, as shown in Fig. 2.

Figure 2 shows the geometric model of component 6, which consists of the following parts:

(1) *Base Line*: the geometric center line of the component. Its position and direction represent the position and attitude of the whole component;

(2) *Detection Area* (green in figure): the outermost region of the component geometry model. This area is used to detect intersection between convex hull component. When there is intersection between two *Detection Areas* of any component, the joint should be immediately slowed down to avoid collision. The thickness of the *Detection Zone* is determined by the velocity of the joint and delay of the system;

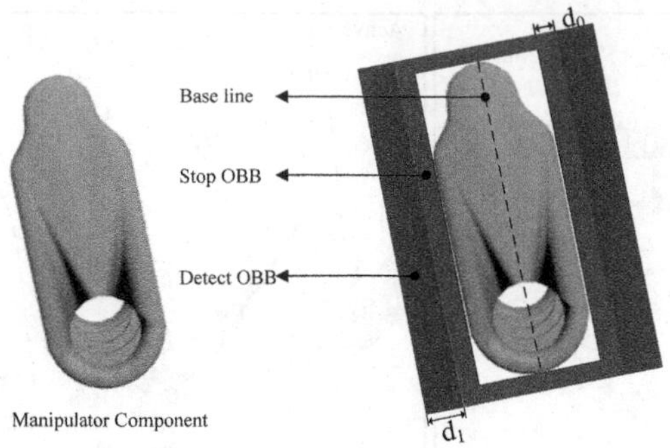

Fig. 2. Geometric model (Color figure online)

(3) *Stop Area* (red in the picture): collision is not allowed between different *Stop Areas.* Once there is intersection between two *Stop Areas* of any component (except adjacent components from the same manipulator), movement of all joints should be stopped immediately to avoid collision. The thickness of the *Stop zone* d_0 is shown in Fig. 2, considering the low speed of the manipulator movement during the operation, and the workspace of the manipulator can be reduced if d_0 and d_1 are too large. In this case set $d_0 = 0.8\,\text{cm}$, $d_1 = 0.7\,\text{cm}$. The *Detection Area* and *Stop Area* are two OBBs with different sizes.

Based on the above analysis, any component can be abstracted into the following mathematical model:

$$\mathrm{o}_i \Leftrightarrow \begin{cases} \text{Center} \\ \text{Base Line} \\ \text{OBB}\{\text{Detection}\} \\ \text{OBB}\{\text{Stop}\} \end{cases} \tag{3}$$

2.2 Distance Between Components

$\mathrm{o}_i, \mathrm{o}_j$ are two different components, their base lines respectively L_i and L_j, the corresponding endpoints are A, B and C, D. Intersection points between common perpendicular of line L_i, L_j and the two lines are P_1 and P_2. Then the shortest distance between L_i and L_j can be expressed as:

$$d(\mathrm{L}_i, \mathrm{L}_j) = \begin{cases} |\mathrm{P}_1 - \mathrm{P}_2| & (\mathrm{P}_1 \in \mathrm{L}_i \cap \mathrm{P}_2 \in \mathrm{L}_j) \\ \min\{\,|\mathrm{A}-\mathrm{P}_2|\,,|\mathrm{B}-\mathrm{P}_2|\,\} & (\mathrm{P}_1 \notin \mathrm{L}_i \cap \mathrm{P}_2 \in \mathrm{L}_j) \\ \min\{\,|\mathrm{C}-\mathrm{P}_1|\,,|\mathrm{D}-\mathrm{P}_1|\,\} & (\mathrm{P}_1 \in \mathrm{L}_i \cap \mathrm{P}_2 \notin \mathrm{L}_j) \\ \min\{\,|\mathrm{A}-\mathrm{C}|\,,|\mathrm{A}-\mathrm{D}|\,,|\mathrm{B}-\mathrm{C}|\,,|\mathrm{B}-\mathrm{D}|\,\} & (\mathrm{P}_1 \notin \mathrm{L}_i \cap \mathrm{P}_2 \notin \mathrm{L}_j) \end{cases} \tag{4}$$

Where $|\bullet|$ indicates distance between two points. The distance between components o_i and o_j could be represented by shortest distance of their base lines, that is $d(\mathrm{o}_i, \mathrm{o}_j) = d(\mathrm{L}_i, \mathrm{L}_j)$. The base distance between components o_i,o_j is defined as $k_{ij} = \frac{k_i + k_j}{2}$ (shown in Fig. 3(a)). k_i, k_j and are the diagonal lengths

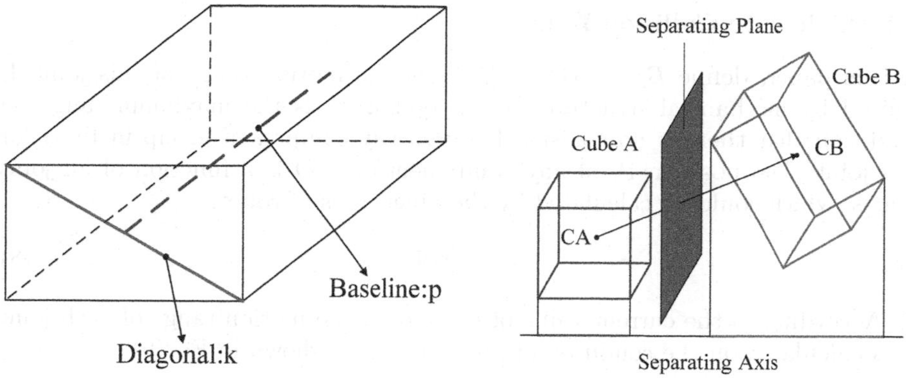

(a) Cube Model (b) Intersection Test

Fig. 3. Cube and intersection test

of the o_i,o_j models, respectively. It is easy to prove that when $d(L_i,L_j) > k_{ij}$, collision between o_i and o_j is impossible. When calculating the distance between the components, we should find out the PCG (Possible Collision Group).

Definition 1. *All component groups satisfy* $d(o_i,o_j) \leq 1.5k_{ij}$ *(except components from the same arm) belong to* PCG*:*

$$\mathrm{PCG} = \bigcup_{o_i,o_j \in O(i \neq j)}^{d(o_i,o_j) \leq 1.5k_{ij}} (o_i,o_j) \tag{5}$$

If the two bounding boxes in the space do not intersect, there must be a separate line SA (Separating Axis), such that the projection of this two convex hulls A and B on the shaft do not overlap with each other [3]. At the same time, there must be a separating plane SP (Separating Plane) perpendicular to this axis, and the two cubes are separated by SP, as shown in Fig. 3(b). If the cube A, B do not intersect, then there is at least one separation axis SA parallel to a line in the set SAG (Separating Axis Group) [4].

$$\mathrm{SAG} = \{\mathbf{Ax}, \mathbf{Ay}, \mathbf{Az}, \mathbf{Bx}, \mathbf{By}, \mathbf{Bz}\} \cup [\,\mathbf{Ax}\ \mathbf{Ay}\ \mathbf{Az}]^T \times [\,\mathbf{Bx}\ \mathbf{By}\ \mathbf{Bz}\,] \tag{6}$$

$$\exists \mathbf{L} \in \mathrm{SAG}, s.t. |\mathrm{proj}(\mathbf{T}, \mathbf{L})| > 0.5|\mathrm{proj}(\mathrm{A}, \mathbf{L})| + 0.5|\mathrm{proj}(\mathrm{B}, \mathbf{L})| \tag{7}$$

In Eq. (6), $\mathbf{Ax}, \mathbf{Ay}, \mathbf{Az}, \mathbf{Bx}, \mathbf{By}, \mathbf{Bz}$ respectively represent three direction vectors of the cube A and B. proj() is the projection operation in Eq. (7), $\mathbf{T}$ is a vector whose start and end points are composed by center points of the two convex hulls, $\mathbf{T}$ = CB − CA, as shown in Fig. 3b. Equation (7) means that if the projection of the vector $\mathbf{T}$ on the separation axis $\mathbf{L}$ is larger than the mean value of projection of cube A and B on the same axis, the two cube A, B do not intersect with each other [4].

2.3 Calculate Collision Index

In this paper, define $B_i = \left[B_i^{\min}, B_i^{\max}\right]$ as the motion range of this joint J_i limited by mechanical structure, corresponding β_i is the maximum range on condition that there is no collision between any component group in PCG for this joint. The position P_i of any component $\mathrm{o}_i \in \mathrm{O}$ is a function of all joint angles, which could be calculated by the kinematics of robot:

$$\mathrm{P}_i = f(\Phi) \tag{8}$$

According to the current value of each joint, the motion range of each joint β_i is calculated on the condition of no collision. As shown in Eq. (9).

$$\begin{aligned} &\beta_i = F(\Phi) \\ &s.t.\ \ \forall \mathrm{o}_i, \mathrm{o}_j \in \mathrm{O}, \mathrm{o}_i \otimes \mathrm{o}_j \end{aligned} \tag{9}$$

where β_i could be obtained through dichotomy shown in Fig. 4. The motion range $\beta_i = (\beta_i^{\min}, \beta_i^{\max})$ could be obtained based on the angle of θ_i, maximum motion range $(B_i^{\min}, B_i^{\max})$ and the solution accuracy in the case of no collision. In Eq. (9), $\mathrm{o}_i \otimes \mathrm{o}_j$ represents that two convex hulls of components o_i and o_j (except components on the same arm) do not overlap with each other. After calculating the range of motion of each joint, the relationship between β_i and θ_i will be compared. If the θ_i is very close to the boundary of β_i, indicating collision is about to happen, this joint should be slowed down.

Definition 2. *The relationship between current angle and maximum motion range of a joint is defined as* Collision Index (Δ_i). *As shown in the following Equation.*

$$\Delta_i = \begin{cases} +\infty & \beta_i^{\min} = B_i^{\min} \cap \beta_i^{\max} = B_i^{\max} \\ |\beta_i^{\min} - \theta_i| & \beta_i^{\min} \neq B_i^{\min} \cap \beta_i^{\max} = B_i^{\max} \\ |\beta_i^{\max} - \theta_i| & \beta_i^{\min} = B_i^{\min} \cap \beta_i^{\max} \neq B_i^{\max} \\ \min\{|\beta_i^{\max} - \theta_i|, |\beta_i^{\min} - \theta_i|\} & \beta_i^{\min} \neq B_i^{\min} \cap \beta_i^{\max} \neq B_i^{\max} \end{cases} \tag{10}$$

In Eq. (10) *Collision Index* represents the relationship between current angle and maximum motion range of the joint. When joint J_i moves to maximum mechanical range B_i, there is no collision between any group in PCG. The *Collision Index* of this joint is set as ∞, otherwise *Collision Index* is equal to the difference between $\beta_i^{\min}(\beta_i^{\max})$ and θ_i (considering the direction of motion). The smaller the absolute value of *Collision Indexs*, the more likely it is to collide. When the absolute value of *Collision Index* is greater than 5°, the robot runs normally. Otherwise, there should be a warning and this joint should be slowed down or stopped, so collision avoidance could be realized by joint movement restriction from *Collision Index* (Δ_i).

$$\mathrm{J}_i \Rightarrow \begin{cases} \text{run} & if(|\Delta_i| > 5) \\ \text{warning and stop} & otherwise \end{cases} \tag{11}$$

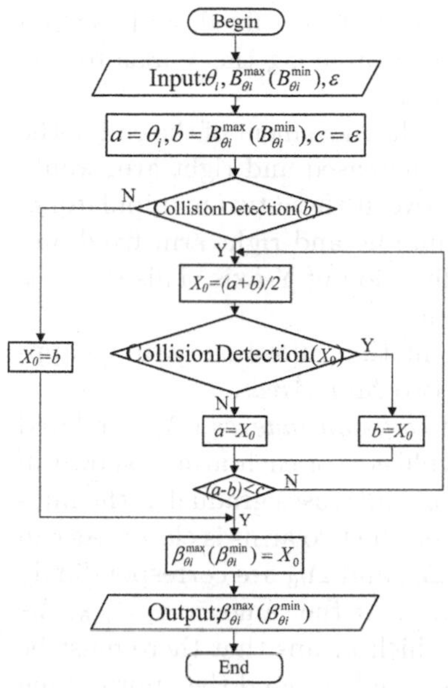

Fig. 4. Dichotomy principle

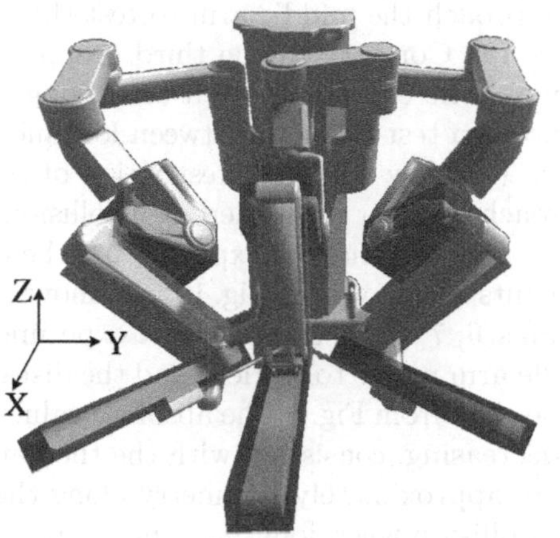

Fig. 5. Robot setup

3 Experiments and Conclusion

3.1 Experiments

In this experiment, robot-assisted laparoscopic prostatectomy (RALP) was taken as an example. The setup of passive joints are shown in Table 1.

Table 1. Setup of passive robot joint

Joint	M1	M2	M3	R1	R2	R3	L1	L2	L3
Angle(°)	70	−140	72.5	73.1	−79.4	−50.9	−69.4	80.7	47.3

The initial setup of surgery robot is shown in Fig. 5, the end points of micro instruments of three arms were located in the target workspace origin $(0, 0, 0)$. Following five cases were used to test the effectiveness of the algorithm:

(1) Keep the left and right arm fixed, move the end of the middle arm along Y-axis,that is $x_m = z_m = 0$. y_m was increased and middle arm would move to left to test the collision between left and middle manipulators.

(2) The same as first group, keep the left and right arm fixed, and the end point of the middle arm moves in a negative direction along the Y-axis to test the collision between right and middle manipulators.

(3) Keep the middle and left arm fixed, control the movement of the end of the right arm along Y-axis, set $x_r = z_r = 0$. y_r was increased and right arm would approach the middle arm to test the collision between these two manipulators.

(4) Contrary to the third groups, keep the middle and right arm fixed and move the end point of left arm along negative direction of Y-axis. This group is used to test collision between left and middle arm.

(5) Now removing restriction of Δ_i on joint motion, move one arm to approach another until there is a collision between two *Stop Areas*.

In five groups of experiments, the changes of *collision indexes* (Δ) of related joints are shown in Fig. 10, the movement and collision of each arm is shown in Figs. 6, 7, 8 and 9. In the first experiment, when y_m increases gradually, the middle arm moves to the left, and the distance between the two arms is closer. As can be seen from Fig. 6, the absolute values of Δ_{m5}, Δ_{l5} and Δ_{l6} are correspondingly decreasing, consistent with the theoretical analysis. At the same time, Δ_{m5}, Δ_{l5} are approximately symmetry along the X-axis, which means that there must be a collision when joint θ_{m5} and θ_{l5} rotate the same angle around the approaching direction of the two arms at any time. During this process, Δ_{m6} has been ∞, indicating that there will not be collision even if joint 6 moves to the maximum range. Similarly, as the y_m increases, the distance between right arm and endoscope manipulator was also increased, so the *Collision Index* of joint 5,6 of right arm maintained ∞. When y_m was increased to $y_m = 8$, the Δ_{m5} was reduced to 4.43°, and the component 8 of middle and left arms are about to collide (red in Fig. 6c). At this point, system makes prompts and prevents the middle arm from moving to left.

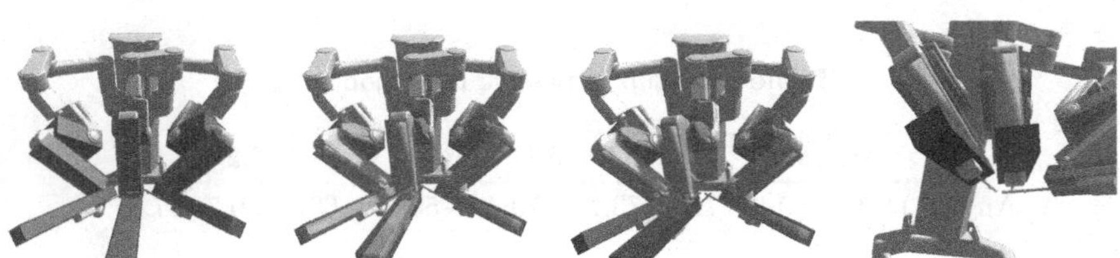

(a) y_m=0 (b) y_m=4 (c) y_m=8 (d) y_m=8

Fig. 6. First group: the middle arm move to left (y_m increase) (Color figure online)

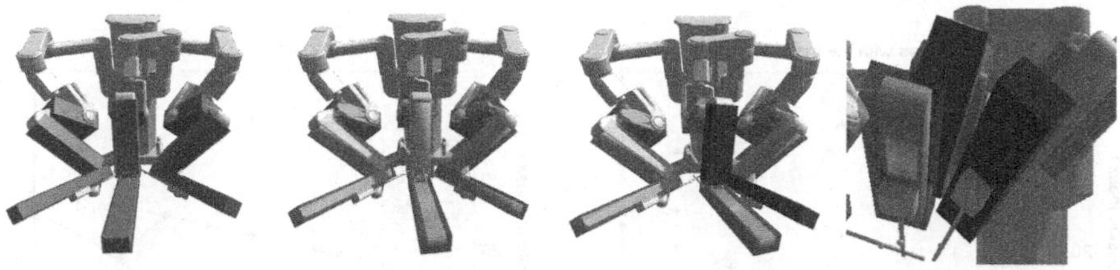

(a) y_m=0 (b) y_m=-4 (c) y_m=-10 (d) y_m=-10

Fig. 7. Second group: the middle arm move to right (y_m decrease)

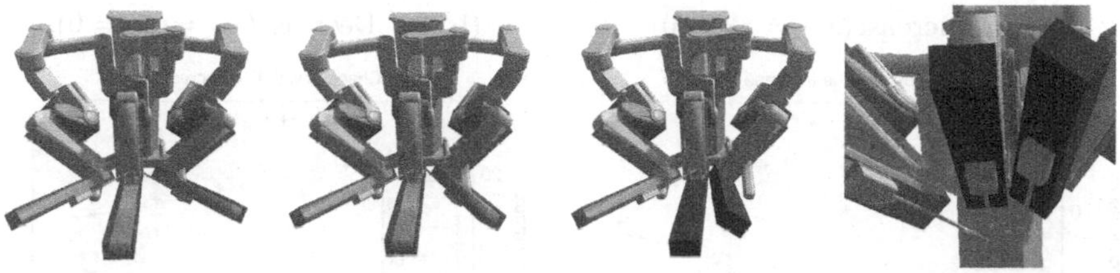

(a) y_r=0 (b) y_r=4 (c) y_r=8 (d) y_r=8

Fig. 8. Third group: the right arm move to left (y_r increase)

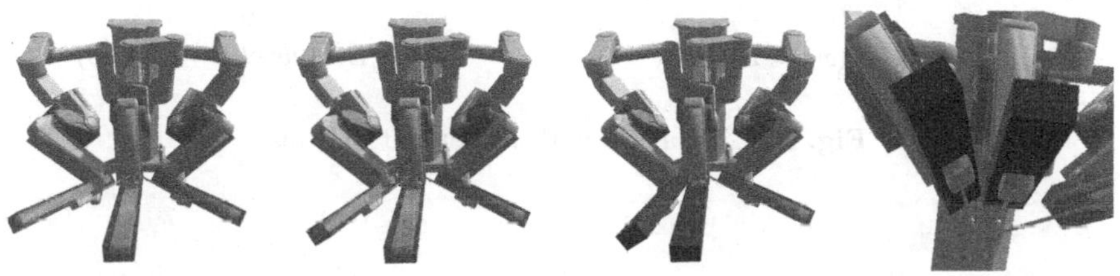

(a) y_l=0 (b) y_l=-4 (c) y_l=-7 (d) y_l=-7

Fig. 9. Fourth group: the left arm move to right (y_l increase)

In the second group of experiments, when the y_m was gradually reduced, the distance between the middle arm and the right arm was decreased, and the process was similar to that of y_m increased in first group, so we did not explore it in detail here. In the third and the fourth groups, the right arm and left arm moved to the middle, the *Collision Index* of each joint was still in accordance with theoretical analysis. In the fifth set of experiments, restrictions of Δ on the movement of each joint was cancelled. When there is a intersection between *Stop Zones*, the system makes a prompt and stop movement of each joint, as shown in Fig. 11.

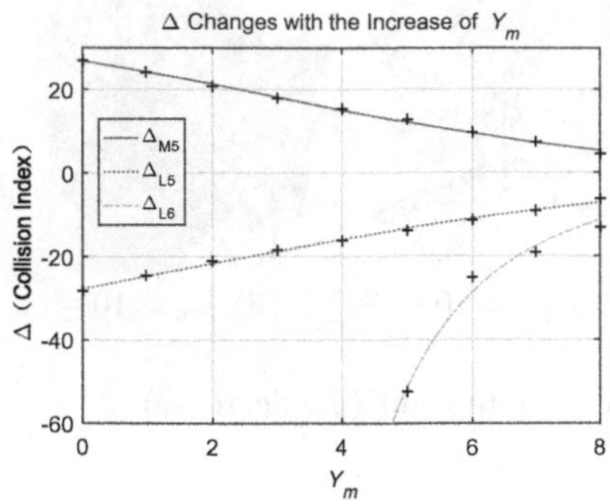

(a) y_m Increase($x_m = z_m = 0$)

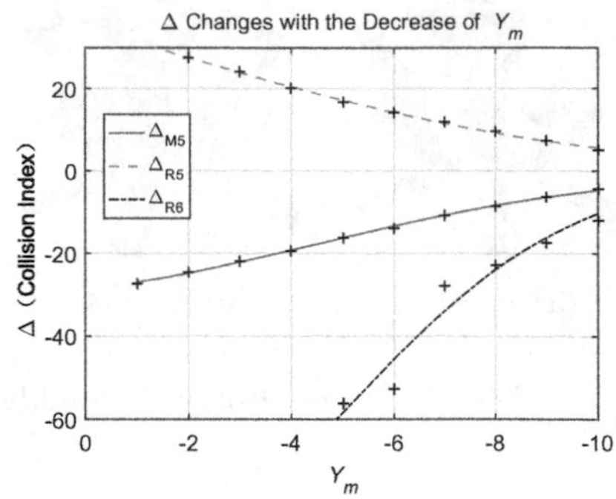

(b) y_m Decrease($x_m = z_m = 0$)

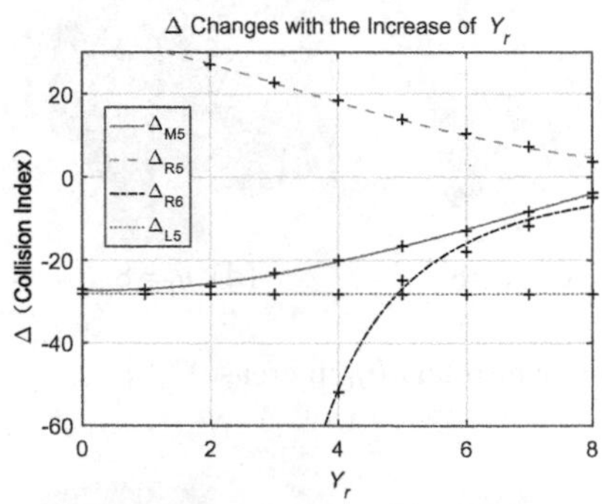

(c) y_r Increase($x_r = z_r = 0$)

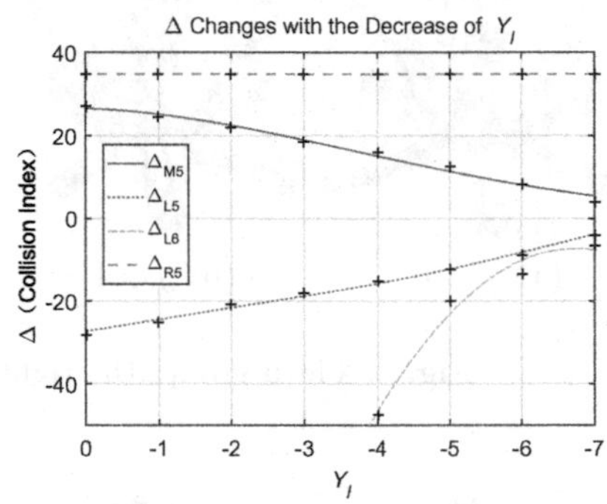

(d) y_l Decrease($x_l = z_l = 0$)

Fig. 10. Δ changes with movement of arms

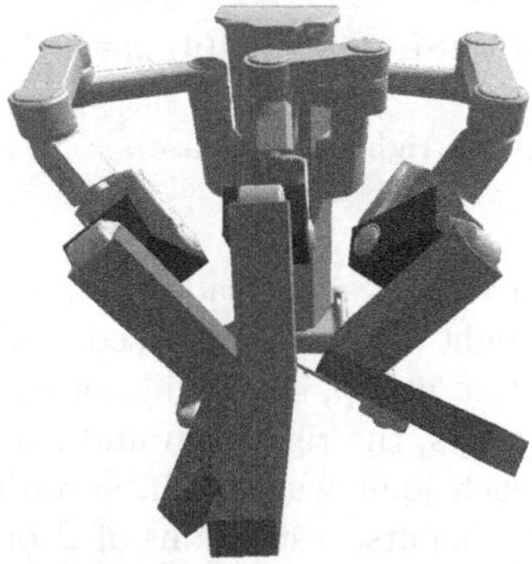

(a) Collision between *Stop Area* of left and middle arm

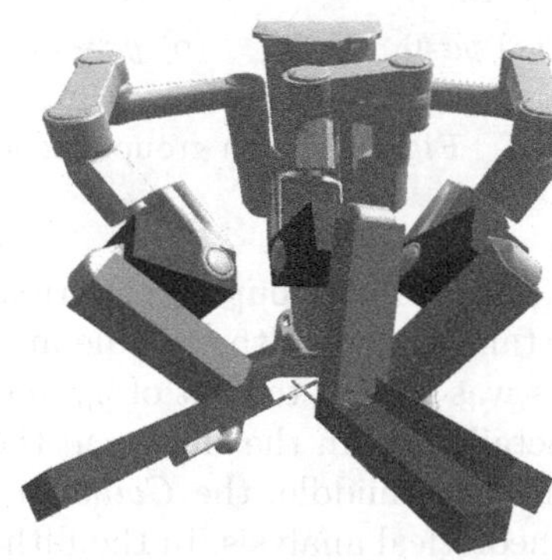

(b) Collision between *Stop Area* of right and middle arm

Fig. 11. Fifth group: collision detection of *stop area*

3.2 Conclusion

Based on the above analysis, the real-time collision avoidance algorithm proposed in this paper can detect the maximum motion range of each joint in real time, and calculate the *Collision Index* and control the motion of each joint to avoid collision. At the same time, in the process of operation, it is necessary to detect the collision between each component in the *Stop Area*. If collision occurs between *Stop Areas*, the motion of all joints must be immediately stopped to protect the robot from collision or damage. The purpose of the establishment of intersection detection between *Stop Zones* is to prevent the failure of collision avoidance based on *Detection Area* due to the fast movement of components, which is equivalent to the second protection of the system. Therefore the real-time collision avoidance algorithm based on the *Detection Area* and the *Stop Area* can protect the robot arm from collision.

Acknowledgement. This research is supported by National Natural Science Foundation of China (61403107). The work is conducted in State Key Laboratory of Robotics and System (Harbin Institute of Technology). Thanks to Dr. Yan for his guidance in theory, and supports from Professor Du and Professor Dong.

References

1. Bosscher, P., Hedman, D.: Real-time collision avoidance algorithm for robotic manipulators. Ind. Robot **38**(2), 113–122 (2009)
2. Estey, E.P.: Robotic prostatectomy: the new standard of care or a marketing success? Can. Urol. Assoc. J. (Journal de l'Association des urologues du Canada) **3**(6), 488–490 (2009)
3. Gottschalk, S.: OBBTree: a hierarchical structure for rapid interference detection. In: Conference on Computer Graphics and Interactive Techniques, pp. 171–180 (1997)
4. Huynh., J.: Separating axis theorem for oriented bounding boxes (2009). http://www.jkh.me
5. Lago, P., Lombardi, C., Vallone, I.: From laparoscopic surgery to 3-d double console robot-assisted surgery. In: IEEE International Conference on Information Technology and Applications in Biomedicine, pp. 1–4 (2010)
6. Li, G., Wu, D., Ma, R., Huang, K.: Pose planning for robotically assisted minimally invasive surgery. In: International Conference on Biomedical Engineering and Informatics, pp. 1769–1774 (2010)
7. Sivarajan, G., Taksler, G.B., Walter, D., Gross, C.P., Sosa, R.E., Makarov, D.V.: The effect of the diffusion of the surgical robot on the hospital-level utilization of partial nephrectomy. Med. Care **53**(1), 71–78 (2015)
8. Wang, W.D., Zhang, P., Shi, Y.K., Jiang, Q.Q., Zou, Y.J.: Design and compatibility evaluation of magnetic resonance imaging-guided needle insertion system. J. Med. Imaging Health Inf. **5**(8), 1963–1967 (2015)
9. Zhang, F., Yan, Z., Du, Z.: Preoperative setup planning for robotic surgery based on a simulation platform and gaussian process. In: IEEE International Conference on Mechatronics and Automation (2016)
10. Zoppi, M., Molfino, R., Cerveri, P.: Modular micro robotic instruments for transluminal endoscopic robotic surgery: New perspectives. In: IEEE/ASME International Conference on Mechatronics and Embedded Systems and Applications, pp. 440–445 (2010)

Exploration of a Hybrid Design Based on EEG and Eye Movement

Junyou Yang[1], Yuan Hao[1(✉)], Dianchun Bai[1], Yinlai Jiang[2], and Hiroshi Yokoi[2]

[1] School of Electrical Engineering,
Shenyang University of Technology, Shenyang, Liaoning Province, China
junyouyang@sut.edu.cn, 282142358@qq.com

[2] School of Intelligent Robotics,
The University of Electro-Communications, Chofu, Tokyo, Japan
yokoi@mce.uec.ac.jp

Abstract. This study presents a novel hybrid interface based on both electroencephalography (EEG) and eye movement. The detection of combination EEG with eye movement provides a new means of communication for patients whose muscular damage are unable to communicate. And this method can translate some brain responses into actions. In this paper, based on the motor imagery, event related synchronization/desychronization (ERS/ERD) were tested by using time-frequency spectrum and brain topographic mapping. A features extraction algorithm is proposed based on common spatial pattern (CSP), then the support vector machine (SVM) were carried out to classificate data. An EEG recording device integrated with an eye tracker can be complementary to attain improved performance and a better efficiency. The eye movement signals (via eye tracker of Tobbi) and EEG signals of ERS/ERD are as the input of hybrid BCI system simultaneously while subjects follow movement of the arrows in each direction. The recognition accuracy of the entire system reaches to 86.1%. The results showed that the proposed method was efficient in the classification accuracy.

Keywords: Brain-computer interface (BCI) · Eye tracker · Electroencephalography (EEG) · Prosthesis · Common spatial pattern · Support vector machine classifier · Hybrid BCI

1 Introduction

People suffering from neuromuscular disorders such as stroke, brain or spinal cord injury, cerebral palsy and muscle atrophy are unable to control motor nerve because they loss control of their voluntary muscles. As a new communication system, brain-computer interface (BCI) can provide a means of communication method and environment control for those patients who have severe motor dysfunction but have normal brain functions [1–3].

Since these patients have normal minds, there are many kinds of noninvasive techniques to measure the brain activities, one of them is electroencephalography (EEG) which has high time resolution, less environmental limits, and requires relatively

Y. Huang et al. (Eds.): ICIRA 2017, Part I, LNAI 10462, pp. 206–216, 2017.
DOI: 10.1007/978-3-319-65289-4_20

inexpensive equipment [4]. Although BCI techniques have been improved remarkably, their usage has not to be implemented totally.

In order to improve the recognition accuracy of motion intention, the hybrid BCI has been paid attention [5–7]. The hybrid BCI integrates the EEG with other physiological signals, which makes up the deficiency of the existing brain computer interface. Hybrid BCI can combine the advantages of two or more kinds of brain computer interface to achieve the desired requirement while avoiding their drawbacks, such as improving speed of the system, accuracy, control complexity, etc. By improving the scientific technology, hybrid-BCI will be greatly extended. Some patients with motor dysfunction still have the cognitive function of the brain and the control ability of eye movement. Thus, the tracking of eye movement can bring more convenience to these patients.

The development of eye movement technique has a long history and has made great progress [8–10]. Eye movement can reflect the brain cognition and have an effect on the brain activity. The application field of eye tracker is very extensive. Eye movement input does not require intensive training and it operates fairly fast [11]. What's more, eye movement unaffectedly feedback the subject's condition of attention concentration. The hybrid BCI system which combines eye tracking with a noninvasive BCI is expected to make up for each other's shortcomings and to build a better one. There is a great chance for progress in this field by combining EEG and eye movement data. This method could make some tasks performable for not only the physically disabled but also older people.

When people do unilateral limb movement or motor imagery task that can lead to cortical rhythm amplitude suppression (ERD) of rhythm on the same side of the brain, and enhancement (ERS) over major sensorimotor areas on the opposite side of the brain. The common spatial pattern (CSP) has gained considerable interests by researcher as an effective relevant discriminative method [12, 13]. Based on that, F. Jamaloo et al. proposed CSP method which could automatically select and discriminate corresponding relationship between frequency bands and CSP features. Furthermore, a deeper research on EEG signals of motor imagery had been conducted [14]. However, consideration of the complexity of the feature vectors in the method and the discrimination accuracy hard to achieve the necessary level in application. Hence, the method of common spatial pattern algorithm and time-frequency analysis for feature extraction on the channels selection and frequency range of EEG signals was employed. In this paper, we select the SVM as the classification method for EEG signals because the SVM can result in high performances for a wide range of classification and pattern recognition applications.

This paper is divided into five parts. Section 2 presents the process of signal acquisition. A brief overview of the signals processing is given in Sect. 3. Some important results are listed in Sect. 4. Finally, Sect. 5 concludes the paper.

2 Signal Acquisition

2.1 Subjects

The test data were collected from 5 student volunteers (4 males and 1 female) with an average age of 24 (standard deviation (SD): 3.13). All of them have the normal or corrected to normal vision. None of the participants had the prior experience with eye tracking. Prior to experiments, each subject was instructed of the procedure and purpose of the study.

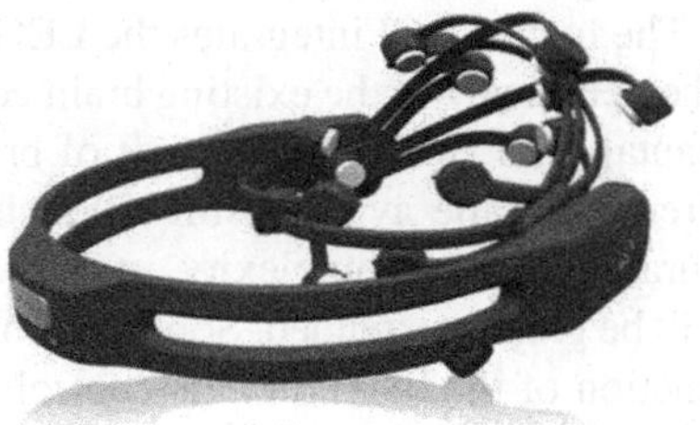

Fig. 1. Emotiv EPOC EEG Headset

2.2 Design and Procedure

The whole experiment is performed in a quiet laboratory environment with constant lighting. An advanced accuracy test procedure was used to monitor the quality of the collected data. The volunteers seated in a chair were instructed to watch a monitor located about 60 cm away at eye level. Correct sitting position was employed to ensure the accuracy and stability of eye positional data. A Tobii X120 eye tracker was positioned below the monitor. The monitor and eye tracker were placed on a table. This is an arrow task mode, that subjects were instructed to move their eyeballs toward the specified directions (left/right). Once all electrodes had good contacts with the scalp, a laptop (Thinkpad T450, Lenovo Inc) started to initiate the OpenViBE software for connecting to the Emotiv headset. Calibrate the eye tracker and check that the eye track was record correctly, and then start the test. At first 3 s, a "+" cursor will appear on the display, which implies that it will begin to perform an imagination task; at 4 s, a red arrow appear on the display (1. image_left; 2. image_right), that the motor imagery task that must be performed by the subject, imagined hand movements according to the instructions. The left arrow pointing indicates that the subject should be expected to perform a left movement of imagery motor. With the same principle, the right arrow indicates that the subject should be expected to move in the right hand. Until a "+" cursor appears again, the imagination stops.

2.3 Data Acquisition

EEG signals were acquired using the Emotiv EPOC EEG headset (Fig. 1). This device captures EEG signals across 14 channels at 128 Hz sampling rate, and 16 bit resolution, according to the international 10-20 system, including (AF3, F3, F7, FC5, T7, P7, O1, O2, P8, T8, FC6, F4, F8, AF4, and references CMS/DRL), which describes the locations of scalp electrodes. The references are used to normalize the channel data between different runs. In addition, the eye movement was recorded by the eye tracker of Tobii. The experiments were conducted with a Tobii X120 eye tracker at 120 Hz sampling frequency which was connected to a 24-inch flat panel screen with a resolution of 1980×1200 pixels.

3 Data Analysis

3.1 Gaze Detection

Kalman Filter algorithm was applied to obtain optimal data by removing blink for the eye tracker data. The performance of the eye tracker was measured by using a single gaze data from the experiment of per subject. Eye tracker signals (gaze x and gaze y coordinates) were separated according to label indexes.

A discrete control system is introduced into the system. "*x*" denote the state quantity and "*z*" denote the measured quantity. "*k*" and "*k*−1" are different point of time. "*A*" and "*B*" are equation of coefficient. "*u*" is defined as controlled amount. "*w*" is noise of process and "*v*" is noise of measuring. State prediction equation is got as

$$x(k) = A * x(k-1) + B * u(k) + w(k). \tag{1}$$

Measurement equation is

$$z(k) = H * x(k) + v(k). \tag{2}$$

$$x(k|k-1) = x(k-1|k-1). \tag{3}$$

$$x(k|k) = x(k|k-1) + Kg(k) * (x(k) - x(k|k-1)). \tag{4}$$

$$p(k|k-1) = p(k-1|k-1). \tag{5}$$

$$p(k|k) = (1 - Kg(k)) * p(k|k-1). \tag{6}$$

$$Kg(k) = p(k|k-1)/(p(k|k-1) + R^2). \tag{7}$$

In experiment, "*A*" is set to be 1. "*u*" is set to be 0. "*R*" is covariance of gaze point. It is acquired by calculating covariance of eye movement data. "*p*" is represented covariance of state quantity. "*Kg*" is as the Kalman gain. The equations are used to optimize the measured and collected data.

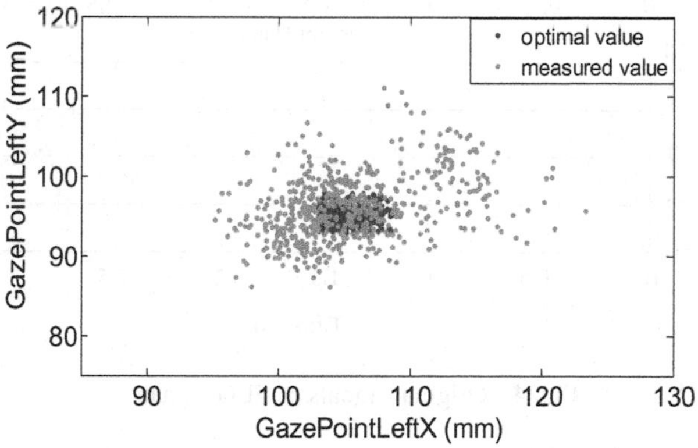

Fig. 2. Scatterplot of the gaze point coordinates (Color figure online)

Figure 2 is a scatterplot of the gaze point coordinates. Blue dots are original position of the gaze point. It is obvious that these datum seem to dispersion. Red dots are the coordinate position of filtered gaze point via Kalman. It is clear that these datum seem to concentration.

3.2 EEG Data Time-Frequency Analysis

The raw EEG signals are collected when a subject is implied with a multi-directional arrow task by the proposed hybrid interface as shown in Fig. 3. The raw EEG signals are shown in Fig. 4, the energy of which tend to disperse. We can get that the EEG signals are very weak and sensitive to the noise.

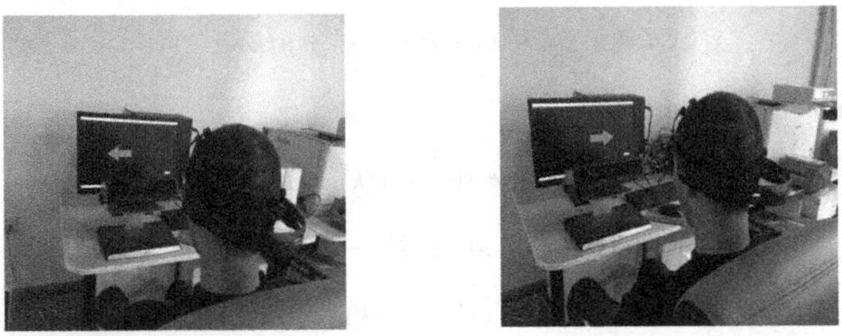

(a) Left motion (b)Right motion

Fig. 3. Scenario during experiment

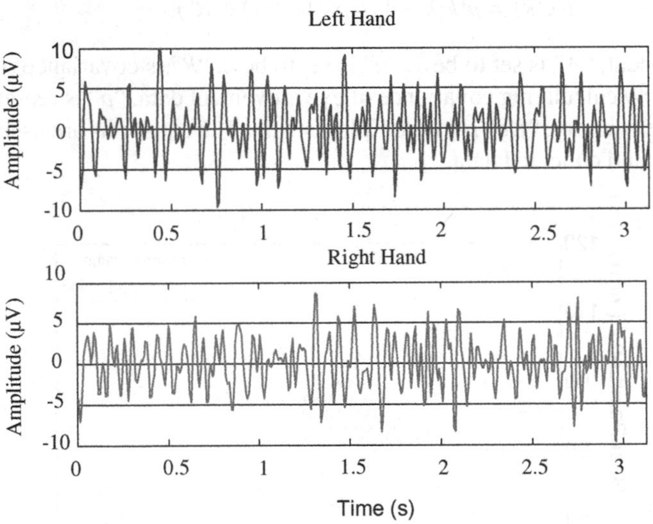

Fig. 4. Original signals of EEG signals

In fact, the EEG signals are non-stationary and vary slowly. The frequency components change over time and the dynamic characteristics can be obtained by analyzing the time-varying frequency spectrum of event. To reveal the change and distribution, the EEG signals are analyzed based on time frequency by the Short-Time Fourier Transform (STFT) [15]. There are three steps to analyze the dimensional time-frequency. Firstly, include the framing, then add windows and finally conduct the Fourier transform. The STFT model is given as,

$$STFT(t, \omega) = \int_{-\infty}^{+\infty} x(u)h^{*}(u - t)e^{-j\omega t}du, \tag{8}$$

where ω is angular frequency, $x(u)$ is the signal of EEG at the u time to be analyzed, the STFT is computed using $h(u - t)$ as the window function. Figures 3(b)–5(a) shows the time-frequency spectrogram of the most significant for imagining left and right hand movements respectively. The horizontal axis represents time, vertical axis gives frequency. The brain electrical mapping with strong time-frequency specificity is shown in Figs. 3(d)–5(c). Using time-frequency distribution diagram to describe the variation of EEG energy characteristics in the process of motor imagery with time and frequency presented an efficient method.

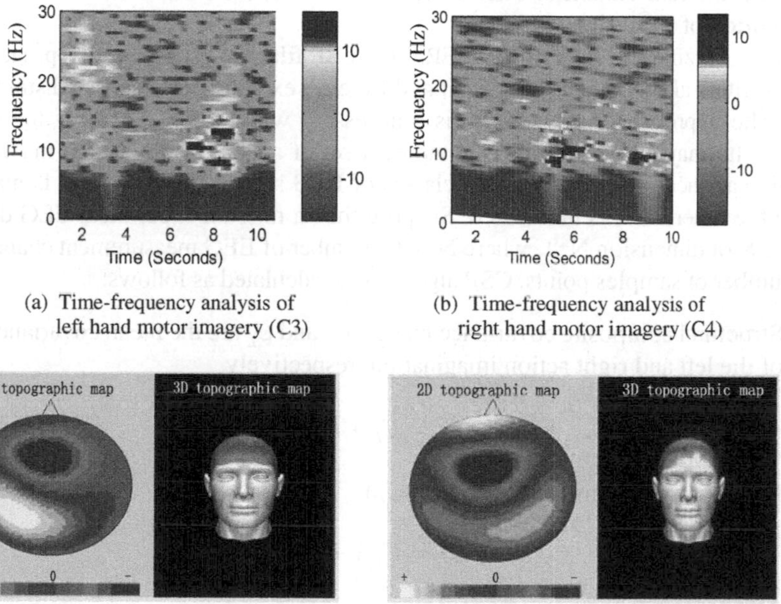

(a) Time-frequency analysis of left hand motor imagery (C3)

(b) Time-frequency analysis of right hand motor imagery (C4)

(c) Spatial distribution map of Left hand motor imagery (C3)

(d) Spatial distribution map of right hand motor imagery (C4)

Fig. 5. Time frequency spectra of EEG topographic map and C3, C4 channels

The time-frequency distribution maps describe the changes of EEG characteristics of the subjects with time and frequency. Because the energy distributions of different imaginary movements are different in time-frequency domain, the right hand motor imagery at C4 guiding has a strong energy between 8–25 Hz, C3 guiding has a certain specific spectrum change between 9–13 Hz. Brain topographic mapping analyzed, the ERD/ERS phenomenon is proved with the power change of the electrode. On the whole, the ERD/ERS phenomenon appeared in different imagined hand movement patterns.

Therefore, characteristic frequency band and the characteristic period are selected, based on time-frequency analysis and brain topographic map, and EEG signals in the characteristic area are extracted as the input signal of CSP algorithm.

3.3 Feature Extraction

Because of the individual differences, the ERD/ERS phenomenon caused by different limb movements is different in spatial distribution. The time during the imagery period are chosen as the feature for classification. Many methods can be used to extract the feature vector of EEG, such as fast-time Fourier transform, wavelet transform, auto regressive (AR) model estimation and so on. However, these approaches just focus on energy characteristics of EEG in each channel which neglect the related relationship between different channels. Therefore, in this study, we extract the feature of spatial distribution of EEG signals.

For analyzing the state of ERD/ERS, the CSP filter is a supervised approach to find spatial filters allowing analyzing spatial difference existing between two distinct classes [16]. The appropriate spatial filter is a necessary way to easily classify the provided signals. Its major objective is to compute a set of spatial filters, which maximize the signals variances gap between two classes of EEG signals that one class is maximized while the other class is minimized. Suppose that, a mission experiment EEG data are a matrix X of dimension N*T, where N is the number of EEG measurement channels, T is the number of samples points. CSP algorithm is calculated as follows:

(1) Structural composite covariance matrix. C_R and C_L are the mean covariance matrix of the left and right action imagination, respectively,

$$C_C = \overline{C_L} + \overline{C_R}, \tag{9}$$

(2) Decompose the covariance matrix (C_C),

$$C_C = U_C \lambda_C U_C^T, \tag{10}$$

where U_C is the eigenvectors matrix, λ_C is the nonzero eigenvalues of a diagonal matrix.

(3) Principal component analysis is used to preserve all eigenvalues and eigenvectors. Structural whitening transformation matrix is achieved,

$$P = \lambda^{-1/2} u U_C^T, \tag{11}$$

(4) Then the covariance matrix C_R and C_L are transformed,

$$S_L = P\overline{C_L}P^T, \tag{12}$$

$$S_R = P\overline{C_R}P^T, \tag{13}$$

(5) CSP spatial filter can be obtained,

$$Z = WX, \tag{14}$$

where W represents a spatial filter corresponding to the projection matrix.

(6) Structural signal characteristics. Select the eigenvector of λ_C which corresponds to the largest eigenvalue, and constructs the spatial filter,

$$f_j = \log(VAR_i / \sum_{i=1}^{n} VAR_i), \tag{15}$$

where VAR_i is the variance in the time series after filter.

Each line of S_R and S_L is a "CSP" component, which is directly used as a criterion for classification. By formula (7), a set of feature vectors are as sample feature vectors.

3.4 Classification

To classificate the brain evoked response of imagination, the SVM is adopted. SVM classifier is trained by the training samples, and then the class labels of the test samples are expected. At the same time, SVM algorithm has the advantages such as complete theory and good adaptability.

By appropriate nonlinear mapping, input vectors are mapped into a high dimensional feature space, which are linear separable [17]. Thus, LIBSVM toolbox is used for implement of SVM algorithm.

Suppose the equation of D-dimensional space classified plane as $x \cdot \omega + b = 0$, where the linear separable sample set is (x_i, y_i), $i = 1, 2, \ldots$, n, $y \in \{-1, +1\}$, and then satisfies the following equation:

$$y[(\omega \cdot x_i) + b] - 1 \geq 0, i = 1, 2, \ldots, n. \tag{16}$$

Based on the *Lagrange* function, the optimal classification function is obtained as,

$$f(x) = sng\{\sum_{i=1}^{n} \alpha_i^* y_i K(x_i \cdot x) + b^*\}. \tag{17}$$

Error penalty factor C is a main parameter that affects the performance of SVM. To some extent, the wrong points for the degree of punishment are controlled by the value of C [18]. Therefore, particle swarm optimization (PSO) is used to obtain optimal parameters, that the particle number is 20.

4 Results

Table 1 presents the accuracy. In this paper, time domain, frequency domain and spatial domain methods are used to classify EEG signals of the imaginary task. Since CSP spatial filter method has an advantage in EEG signal recognition of the imaginary limb movement, we used it to extract the feature of 16 guide signals, and the 6 feature vectors are obtained as shown in Fig. 6.

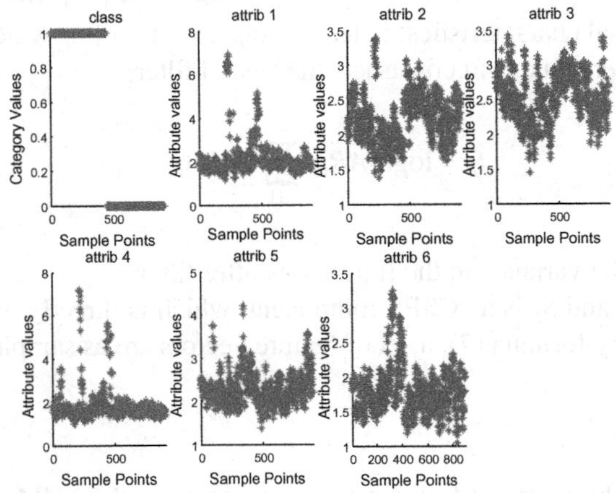

Fig. 6. EEG signals fractal dimension visualization

According to Table 1, EEG data are filtered in 8–30 Hz by FIR bandpass filter. Furthermore, filtered, data can be extracted by CSP algorithm, and then the data set is classified by the trained SVM. In the SVM parameter selection, the parameter of particle swarm optimization algorithm is 20, and the average classification accuracy of EEG signal is 80.1%.

Table 1. Classification results of 5 subjects under different modes

Subject	Accuracy		Classification rate
	EEG	Gaze detection	
S1	83.3%	95%	89.1%
S2	80.8%	93%	86.9%
S3	82.9%	90%	86.4%
S4	75.1%	92%	83.6%
S5	78.3%	91%	84.6%
Mean	80.1%	92.2%	86.1%

Before confirming the effectiveness of hybrid system, the motor imagery of brain was evaluated to their classification rate. According to the classification results, if EEG signals were used as the only signal source, the correct classification rate is low which cannot meet the needs of system. The classification accuracy of hybrid BCI control mode is read out and compared with the data selected from single signal source. The result shows that human motion intention is accepted by hybrid system.

As far as the concerned experiment, the classification accuracy of the proposed system is 86.1%. As a result, the hybrid system can not only solve the problem of the muscle fatigue when using the eye electric signal but also improve the recognition accuracy of the signal. The proposed system can also reduce the serious consequences caused by error classification. Thus, it is more suitable for practical application of artificial limb system can achieve expected design goal in the proposed system.

5 Conclusion

In this paper, we proposed a hybrid BCI system that integrates motor imagery based on EEG with eye tracking. The experiment results demonstrated that the integration system significantly improves performance of motor imagery task. The CSP method was used to analyze the EEG correlation between different leads and the SVM was used for classification. Subjects' brain information and eye movement module were adopted as input information while 2 different commands tests were carried out. Classification accuracy and stability are significantly improved by combination eye tracker signals. The results indicated that the integration method dramatically improves the task performance that the classification accuracy rate can reach to 86.1%. Therefore, adopting CSP can effectively extract the EEG signal features. It is proved that the introduced hybrid interface is a valuable technique and promising for disabled persons.

References

1. Wolpaw, J.R., McFarland, D.J.: An EEG-based brain-computer interface for cursor control. Electroencephalogr. Clin. Neurophysiol. **78**, 252–259 (1991)
2. Schalk, G.D., McFarland, J.T.: BCI2000 a general purpose brain computer interface (BCI) system. IEEE Trans. Biomed. Eng. **51**(6), 1034–1043 (2004)
3. McFarland, D.J., Wolpaw, J.R.: Brain-computer interface for communication and control. Commun. ACM **54**, 60–66 (2002)
4. Wolpaw, J.R.: BCI meeting 2005-workshop on signals and recording methods. IEEE Trans. Neural Syst. Rehabil. Eng. **14**(2), 138–141 (2006)
5. Pfurtscheller, G.: The hybrid BCI. Front. Neurosci. **4**(30), 1–11 (2010)
6. Allision, BZ.: Extending BCIs through hybridization and intelligent control. J. Neural Eng. **9** (2012)
7. Millan, J.R.: Combining brain-computer interfaces and assistive technologies: state-of-the-art and challenges. Front Neurosci. **4**, 161 (2010)
8. Wechsler, Q.H., Duchowski, A.T.: Special issue: eye detection and tracking. Comput. Vis. Image Underst. **98**, 1–3 (2005)

9. Kawato, T.: Detection and tracking of eyes for gaze camera control. Image Vis. Comput. **22**, 1031–1038 (2004)
10. Kim, J.: A simple pupil-independent method for recording eye movements in rodents using video. J. Neurosci. Methods **138**, 165–171 (2004)
11. Vertegaal, R.: A Fitts law comparison of eye tracking and manual input in the selection of visual targets. In: International Conference on Multimodal Interact (ICMI), pp. 241–248 (2008)
12. Blankertz, B., Kawanabe, M.: Invariant common spatial patterns: alleviating nonstationarities in brain-computer interfacing. In: Advances in Neural Information Processing Systems, Cambridge, Canada, vol. 20, pp. 113–120 (2008)
13. Koike, Y.: A real-time BCI with a small number of channels based on CSP. Neural Comput. Appl. **20**, 1187–1192 (2011)
14. Ang, K.K., Chin, Z.Y., Zhang, H.: Filter bank common spatial pattern (FBCSP) in brain-computer interface. In: 2008 IEEE International Joint Conference on Neural Networks, Hong Kong, pp. 2390–2397 (2008)
15. Hayashi, Y., Kiguchi, K.: A study of features of EEG signals during upper-limb motion. In: 2015 IEEE International Conference on Advanced Intelligent Mechatronics (AIM), Busan, pp. 943–946 (2015)
16. Waldert, S., Demandt, E.: Hand movement direction decoded from MEG and EEG. J. Neurosci. **28**(4), 1000–1008 (2008)
17. Acl, B., Mohebbi, M.: Support vector machine-based arrhythmia classification using reduced features of heart rate variability signal. Artif. Intell. Med. **44**(1), 51–64 (2008)
18. Lanez, E.: Mental tasks selection method for a SVM-based BCI system. In: 2013 IEEE International Systems Conference (SysCon), Orlando, FL, pp. 767–771 (2013)

Optimal Design of Electrical Stimulation Electrode for Electrotactile Feedback of Prosthetic Hand

Boya Wang, Qi Huang(✉), Li Jiang, Shaowei Fan, Dapeng Yang, and Hong Liu

State Key Laboratory of Robotics and System, Harbin Institute of Technology, Harbin 150001, China
huangqi856@hit.edu.cn

Abstract. Electrotactile feedback plays an important role in the prosthetic hand system, whereas electrotactile method has the problems such as interference with electromyography (EMG) signals and low spatial resolution. This paper proposes a method to optimize the design of electrotactile electrode array to address above problems. According to the theory of nerve excitation and the electric field model on biological tissues, we proposed two optimizing targets for design: Noise Factor and Focusing Degree. Based on theoretical derivation of the relationship between electrode parameters and optimizing indices, we find that concentric circle electrode (CCE) with biphasic stimulation is superior over single circle electrode (SCE) on both indices. Use finite element analysis, we construct a more accurate electric field model to analysis. The result of FE analysis shows that CCE still had superiority over SCE when choosing optimized electrode parameters. Furthermore, we find out that when designing electrode array, the CCE have a huge advantages against SCE for its capacity of resisting disturbance so that we can design an electrode array with high density.

Keywords: Electrical stimulation electrode · Prosthetic hand · Noise factor · Focusing degree

1 Introduction

Amputees lose their hands because of illnesses or accidents, which loses not only the tool to interact with the external world, but also the original perception of the environment. To some extent, prosthetic hand can improve amputee's life quality and make up for their psychic trauma [1]. Tactile feedback is an essential part for amputees to get sensations. A common method is electrotactile feedback, using current to stimulate human tissue, exciting the subcutaneous nerve and producing many different feelings [2]. Now the usual method of electrotactile feedback on prosthetic hand is TENS (Transcutaneous Electrical Nerve Stimulation) with many advantages, such as precise stimulation site, a strong sense of excitement and low power consumption [3]. Contacting with human tissue directly, the electrode used by TENS is an essential part of HMI (Human-Machine Interface). Compared with electrode array, different single electrodes have their own limitations and can't be placed more on limited areas (usually forearm) to deliver

Y. Huang et al. (Eds.): ICIRA 2017, Part I, LNAI 10462, pp. 217–229, 2017.
DOI: 10.1007/978-3-319-65289-4_21

complex information. But few researches pay attention to the design of array electrodes used for TENS and there are no guidelines on the design [4]. Besides, researches on electrotactile feedback applied to the prosthetic hand mainly focus on the process of electrical stimulation. But in real time control of prosthetic hand, the collecting of EMG signals and electrotactile feedback are acquired simultaneously. As a result, electrical stimulation signals usually saturate the amplifier and reduce the signal-to-noise ratio, affecting EMG signal acquisition. The usual way is using data processing methods [5–7].

In this paper, we propose two evaluating indices: Noise Factor and Focusing Degree. The Noise Factor (NF) describes the interference coefficient of electrical stimulation noise to EMG sensors and the Focusing Degree (FD) describes the intensity of electrical stimulation at a certain depth of the skin. The value of NF and FD decide whether the electrical stimulation feedback can be used to achieve high spatial resolution, low noise interference and low crosstalk in an EMG control system when using array electrical stimulation feedback. So we first establish the mathematical model of the electrical stimulation electrode and the EMG sensor on the skin, putting forward the physical definitions of NF and FD as well as getting mathematical relationship between NF, FD and electrode parameters. In order to verify the accuracy of mathematical description, we build up the skin model of single layer and the physiological model of forearm. Based on these models, we get the change regulations of NF and FD with the stimulation of Single Circle Electrode (SCE). In order to decrease NF and improve FD, we design Concentric Circle Electrode (CCE) and finish verifying the changes of CCE and its array.

2 Theoretical Model

In today's prosthetic hands with a bidirectional interface, EMG collecting electrode and electrical stimulation (ES) electrode are placed on amputee's stump to control and apperceive the states of prosthesis respectively. But the waveform of ES overlaps with EMG signals in both frequency domain and time domain, which results in the difficulty to filter out ES noise and make myoelectric control system work properly. Moreover, the electric field in the tissue under the condition of ES has an important influence on the effect of stimulation, directly affecting perception of nervous system.

2.1 Models and Assumptions

In order to analyze ES and EMG acquisition theoretically, we set up a model, as shown in Fig. 1. We set the distance between the centers of electrical stimulation electrode and EMG sensor as R, the distance between the two pick-up electrodes of EMG sensors as d. The electrodes are placed on the surface of the skin, whose resistivity is set as ρ. The skin is homogeneous and the properties don't change with time. The center of the stimulation electrode is set as the origin and the direction of X axis is from the center of stimulation electrode to EMG sensor along the surface of the skin. Physiologically, there're many kinds of receptors in skin. Meissner's corpuscle, Merkel cell and Pacinian corpuscle are named as "tactile primary color" in the skin of the human body [8]. The depth of Meissner's corpuscle and Merkel cell is about 1 mm and Pacinian corpuscle is located in the depth of over 2 mm. The Merkel cell is sensitive to the tactile perception with sustained pressure and touch, and its corresponding afferent fibers usually parallel to the axial direction of the forearm [9]. Thus we assume receptors is located just below the electrode (exactly underneath the center of stimulation electrode, red point in Fig. 1) and a single nerve fiber excitation at depth of $h = 1$ mm is considered.

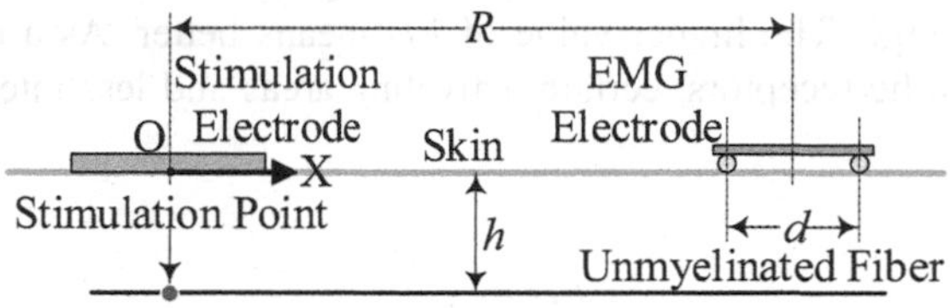

Fig. 1. Placement of stimulation electrode and EMG sensor (Color figure online)

Electrical stimulation set up a space electrical field. The stimulation signal is propagated to the EMG sensor via the potential difference between the pick-up electrodes [10]. The interference is closely related with the current/voltage on the electrode as well as the electrode parameters. For further research, we define the potential difference relevant to electrode parameters as Noise Factor (NF). The smaller noise factor means smaller interference of the ES electrode to the EMG sensor.

Because the goal of electrode design is to give amputees a variety of sensory feedback, a single electrode should be able to stimulate a limited range of nerve precisely. We use the nervous model proposed by Frank Rattay in 1989 [11], researching on the neural excitation under ES. According to his theory, the neural fiber is equivalent to an electrical network composed of many elements, as shown in Fig. 2(a). As for unmyelinated fibers, the influence of extracellular current sources is given by (1), defined as AF (Activating Function), where $V_e(x, t)$ is the potential outside the membrane and x is the length coordinate of the fiber.

$$AF = \frac{\partial^2 V_e(x,t)}{\partial x^2} \tag{1}$$

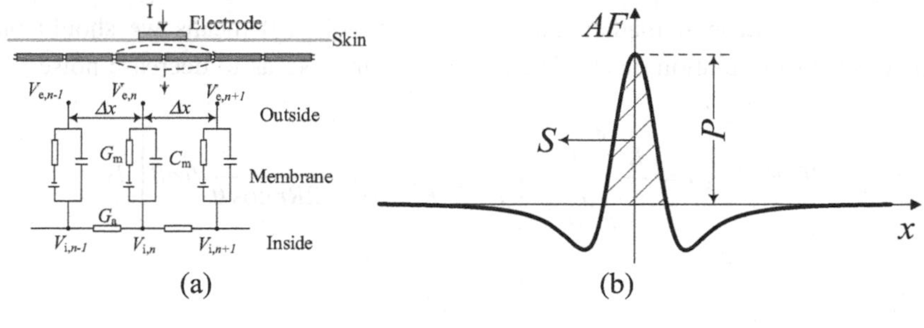

Fig. 2. An electrical network model of unmyelinated fibers and Activating Function

If AF is positive in a tissue area, nerves lying in the tissue area will be activated and generate action potentials. Hyperpolarization is produced where AF is negative. As shown in Fig. 2(b), there is an obvious peak value for AF at a certain depth under electrode. Higher positive AF value means higher electrical stimulation sensitivity [9]. We define a parameter, Focusing Degree (FD), to characterize this property and evaluate electrical stimulation. The definition is given by (2). *P* is the peak value of AF and *S* is the integral of AF's positive part along the direction of the nerve, just as the shaded area shown in Fig. 2(b). Generally speaking, we need AF to have larger peak value in smaller range. The higher value of *FD* means better. As a result, it leads to stimulation of specific receptors, certain activating areas and less interference to other electrodes.

$$FD = \sqrt{\frac{P^2}{S}} \tag{2}$$

2.2 Stimulation Using Single Circle Electrode

To get a quantitative analysis of the influence of the parameters of electrode on NF and FD, considering model's generality, we first set up a mathematical model of a Single Circle Electrode (SCE). The only parameter is the radius r_E of the electrode, as shown in Fig. 3(a).

Any point on stimulation electrode can produce a potential different on the two pick-up electrodes of the EMG sensor as $U_r = \int_{R-d/2}^{R+d/2} \rho J_r dr$, where J_r is the current density. Any conductive object in the electric field can be regarded as a combination of point charges. Thus the Noise Factor (*NF*) is defined as the potential difference on EMG electrodes, as shown in Eq. (3), where $K = \frac{\rho I}{2\pi^2}$, I is the current exerted on electrode.

$$NF(R, d, r_E) = \frac{K}{r_E^2} \int_{R-d/2}^{R+d/2} \left[\int_0^{2\pi} \int_0^{r_E} \frac{1}{R^2 + r^2 - 2Rr\cos\theta} \frac{R - r\cos\theta}{\sqrt{R^2 + r^2 - 2Rr\cos\theta}} dr d\theta \right] dx \tag{3}$$

Because of the placement of electrodes, the distance between two electrodes R is much larger than radius r_E of stimulation electrode ($R \gg r_E$), we get a simplified Eq. (4) for *NF*. Through mathematical derivation, $\frac{\partial NF}{\partial R} < 0$ means we should place EMG sensor to stimulation electrode as far as possible, so as to decrease noise.

$$NF(R, d, r_E) = \frac{K}{r_E^2} \int_{R-d/2}^{R+d/2} \left[\int_0^{2\pi} \int_0^{r_E} \frac{1}{R^2 + r^2 - 2Rr\cos\theta} dr d\theta \right] dx \tag{4}$$

The electrode placed on the surface of the tissue generates an electric field in the tissue, activating nerve to build an information path and letting amputee get the feedback from prosthetic hand. An current element $d\sigma$ on stimulation electrode with current I can produce a potential $\frac{Id\sigma}{2\pi r}$ at the distance r. According to superposition theorem, stimulation factor is defined as Eq. (5), where $K = \frac{\rho I}{2\pi^2}$.

$$G(r_{\mathrm{E}}, x, h) = \frac{K}{r_{\mathrm{E}}^2}\int_0^{2\pi} d\theta \int_0^{r_E} \frac{r}{\sqrt{r^2 + x^2 - 2rx\cos\theta + h^2}} dr \tag{5}$$

We change the order of integration and derivation to get Eq. (6). Because r_{E} has the same order of x and y. So we cannot ignore the influence of $\frac{1}{\sqrt{x^2+h^2+r^2-2xr\cos\theta}}$. We can only use numerical method to calculate and analyze this *Complete Elliptic Integral of the First Kind*. So we find that with the increase of r_{E}, *FD* is becoming smaller, making electrical stimulation of specific parts of human body.

$$AF = \frac{\partial^2 G(r_{\mathrm{E}}, x, h)}{\partial x^2} = \frac{K}{r_{\mathrm{E}}^2}\int_0^{2\pi}\int_0^{r_{\mathrm{E}}} \frac{2x^2 - h^2 - 4xr\cos\theta + 3r^2\cos^2\theta - r^2}{(x^2+h^2+r^2-2xr\cos\theta)^2} \bullet \frac{1}{\sqrt{x^2+h^2+r^2-2xr\cos\theta}} r dr d\theta \tag{6}$$

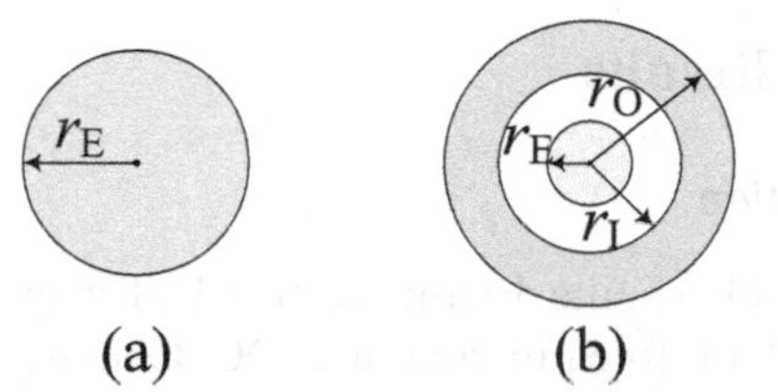

Fig. 3. Single Circle Electrode and Concentric Circles Electrode

2.3 Stimulation Under Concentric Circles Electrode

In Sect. 2.2, we find the increase of electrode's radius has a negative effect on stimulation result as expected, which raises *NF* and reduces *FD*. On the other hand, considering the processing convenience, electrode can't be made too small. In order to ensure the decrease of *NF* and increase of *FD* when electrode's size is acceptable, we design a type of concentric electrodes, Concentric Circles Electrode (CCE). Inner electrode is a SCE with radius r_{E} and outer electrode is a ring electrode with inside radius r_{I} and outer radius r_{O} ($r_{\mathrm{I}} < r_{\mathrm{O}}$), as shown in Fig. 3(b). We exert same currents with different polarity on inner electrode and outer electrode individually. Without the loss of generality, the current polarity on outer electrode is defined as positive (+). Normalizing r_{I}, r_{O} to r_{E}, we also have the definition as shown in Eq. (7).

$$r_{\mathrm{I}} = \alpha r_{\mathrm{E}},\ r_{\mathrm{O}} = \beta r_{\mathrm{E}},\ s = \sqrt{\beta^2 - \alpha^2}\ (\alpha > 1, \beta > 1) \tag{7}$$

Similar derivation as above, *NF* is defined in Eq. (8) and stimulation factor at any point (*x*, h) in the field is defined in Eq. (9).

$$NF(r_{\mathrm{E}},\alpha,\beta,R,d)=\frac{K}{s^2r_{\mathrm{E}}^2}\int\limits_{R-d/2}^{R+d/2}\int\limits_{0}^{2\pi}\int\limits_{\alpha r_{\mathrm{E}}}^{\beta r_{\mathrm{E}}}\frac{rdrd\theta dx}{R^2+r^2-2Rr\cos\theta}-\frac{K}{r_{\mathrm{E}}^2}\int\limits_{R-d/2}^{R+d/2}\int\limits_{0}^{2\pi}\int\limits_{0}^{r_{\mathrm{E}}}\frac{rdrd\theta dx}{R^2+r^2-2Rr\cos\theta} \tag{8}$$

$$G(r_{\mathrm{E}},x,h,\alpha,s)=\frac{K}{r_{\mathrm{E}}^2}\int_0^{2\pi}d\theta\int_{\alpha r_{\mathrm{E}}}^{\beta r_{\mathrm{E}}}\frac{r}{\sqrt{r^2+x^2-2rx\cos\theta+h^2}}dr-\frac{K}{s^2r_{\mathrm{E}}^2}\int_0^{2\pi}d\theta\int_0^{r_{\mathrm{E}}}\frac{r}{\sqrt{r^2+x^2-2rx\cos\theta+h^2}}dr \tag{9}$$

Through mathematical derivation, we find $NF > 0$ which means the noise generated by stimulation electrode is unavoidable theoretically. Besides $\frac{\partial NF}{\partial \alpha} > 0$ and $\frac{\partial NF}{\partial s} > 0$ means the increase of α, s leads to higher noise. Also, we get $\frac{\partial FD}{\partial r_{\mathrm{E}}} < 0$, $\frac{\partial FD}{\partial \alpha} < 0$, $\frac{\partial FD}{\partial s} < 0$. The enlargement of all the three parameters will cause decline of *FD*.

3 Simulation and Results

3.1 Design of Simulation

We build up the skin's FE (Finite Element) model of single uniform layer and the physiological FE model of forearm and use DC Conduction module of ANSYS Maxwell 16.0 to finish our task. Due to the characteristics of human tissue, we should use a certain period of current to operate electrical stimulation [12]. In the design of all the models, only the parameters of the electrode are considered. So we exert DC current/voltage in simulation for reduction of the amount of calculation. In simulation, we study on the comparison of both models and get rules to design CCE and its array.

The skin model of single uniform layer is established to verify the accuracy of theoretical derivation. The model is a hemisphere with radius of 1 m and the location of electrode is its center (Fig. 4(a)). Outer surface of hemisphere is set as sink to ensure the conduction path from electrode to infinity. SCE is exerted DC current of −10 mA and CCE is exerted DC current of ±10 mA (flowing into the surface is set to positive), as shown in Fig. 5.

The physiological model of forearm is based on the hypothesis that all the biological tissues are isotropic homogeneous. There are two ways to model human tissues. One is building up model of human body from MRI or CT images and the other is based on average dimensions of tissues using cylinder or plane [13]. We choose the latter one and the model is based on the research result of Andreas Kuhn [14]. But we use the hierarchical structure of cube instead of concentric cylinder structure to reduce the amount of calculation and the parameter we use are in Table 1. Model (shown in

Fig. 4(b)) has the similar parameters as an able-bodied adult human arm with total length 500 mm, width 70 mm and height 50 mm. The center of the electrode coincides with the surface of the skin layer when the single electrode is stimulated. Line of centers is coincidence with the axis of the model and line of center's midpoint is coincident with the center of the skin layer when using electrode array. SCE is exerted DC voltage −20 V and CCE is exerted DC voltage of ±20 V (limit value of the commercial simulator), as shown in Fig. 5.

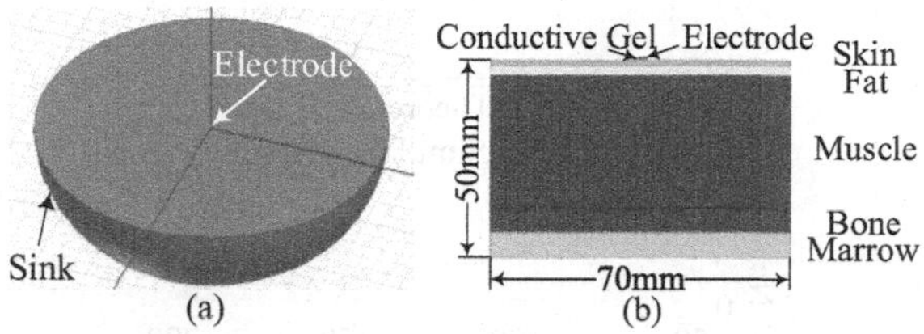

Fig. 4. FE models of single uniform layer and forearm

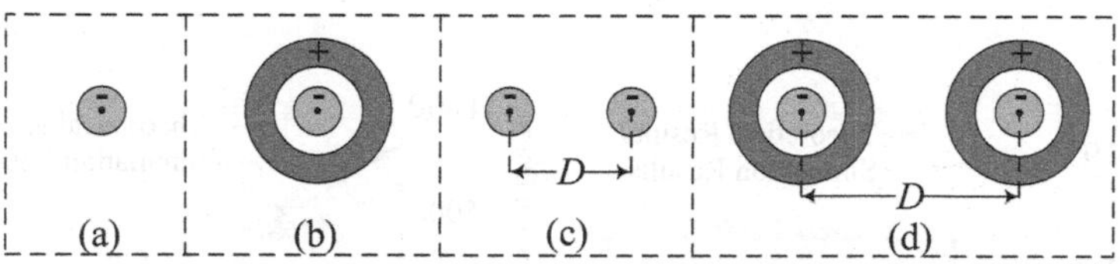

Fig. 5. Configurations of single electrodes and electrode arrays

Table 1. Parameters of forearm's FE model

Tissue	Relative permittivity	Conductivity/S $\cdot$ m^{-1}	Thickness/mm
Marrow	10,000	0.0800	6.5
Bone (cortical)	3,000	0.0200	6
Muscle	40,000	0.1111	33.5
Fat	25,000	0.0300	2.5
Skin	6,000	0.0025	1.5
Conductive Gel	1	0.0033	0.5

3.2 Results

The simulation results are mainly based on the evaluation of Noise Factor and Focusing Degree, divided into three parts: comparison of theoretical model and simulation model, comparison of CCE and SCE and stimulation with electrode arrays.

Comparison of Theoretical Model and Simulation Model
Theoretical and simulation results of *NF* at different values of *R* are shown in Fig. 6. When SCE's radius is 1 mm, the two results are almost the same, with maximum error of 2‰. When r_E changes from 1 mm to 3 mm (step 0.5 mm), *NF* at $R = 200$ mm (common distance in experiment) varies with the radius, just shown in Fig. 7. *NF* decreases with the increase of *R* and *NF* increases with the increase of r_E, the same result as theoretical derivation. There is a certain error between theoretical and simulation results of *FD* (Fig. 7). The reason is probably that we ignore the edge effect in theoretical model. When the size of electrode goes larger, it will have stronger "edge effect", letting the simulation results bigger than theoretical ones.

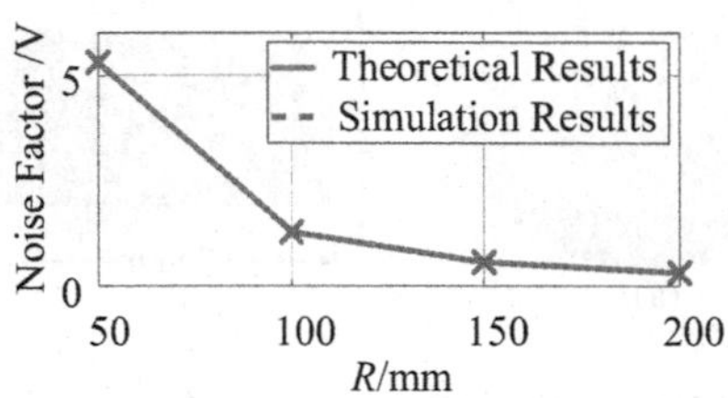

Fig. 6. SCE's theoretical results and simulation results of NF with different *R*

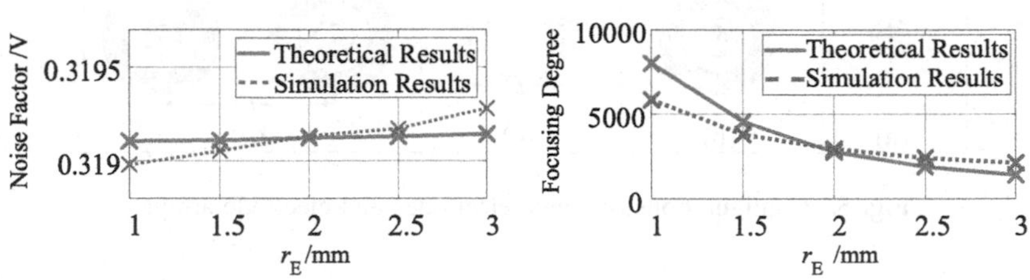

Fig. 7. SCE's NF and FD with change of r_E

Because there are three parameters of CCE, we use Variable-Controlling Approach to evaluate the influence of different parameters. The comparison between theoretical and simulation values of different values of r_E, α, *s* are shown in Figs. 8, 9 and 10 (*NF* at $R = 200$ mm is considered). The results have the same trends that the increase of three parameters will cause decline of *FD* and increase of *NF*. But there're still errors between theoretical and simulation results. It's possible that it is the ignorance of edge that cause the error.

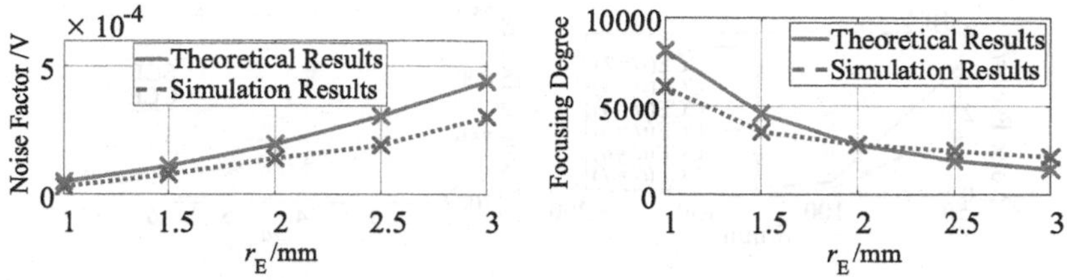

Fig. 8. CCE's NF and FD with change of r_E ($\alpha = 2$, $s = 2.236$)

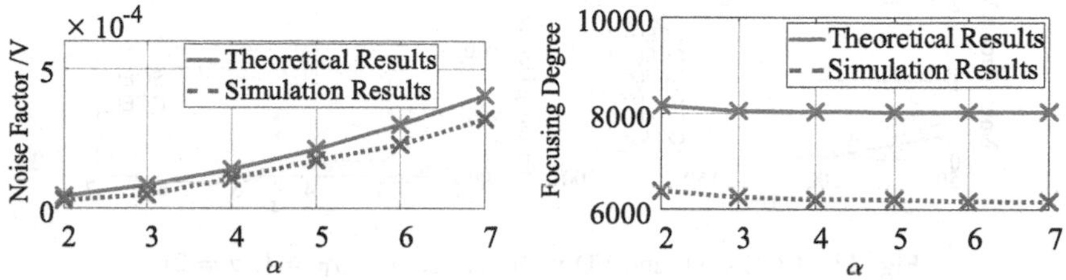

Fig. 9. CCE's NF and FD with change of α ($r_E = 1$, $s = 2$)

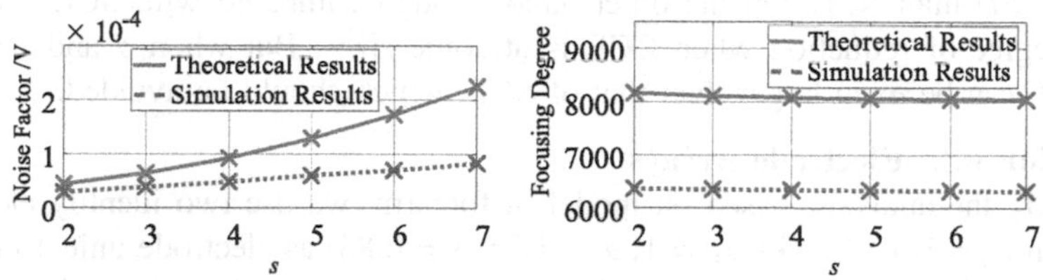

Fig. 10. CCE's NF and FD with change of s ($r_E = 1$, $\alpha = 2$)

Comparison of CCE and SCE

Based on model of forearm, we compare *NF* and *FD* of CCE electrode and SCE electrode with same radius r_E to determine the impact of three parameters. Results are shown in Figs. 11, 12 and 13. The increase of α and r_E will reduce *FD*. But it is different from the theoretical model that the increase of s greatly increases *FD*.

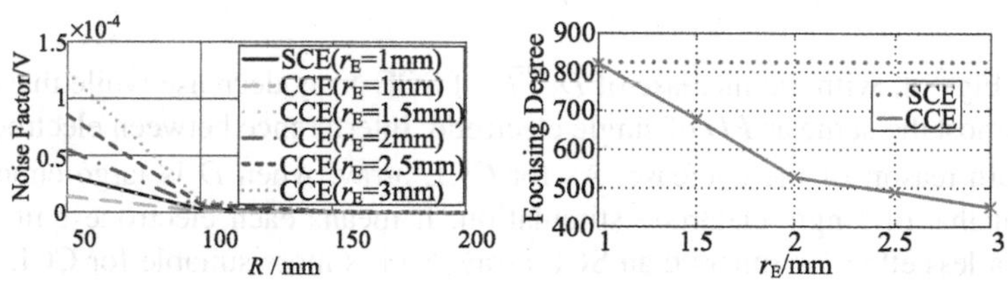

Fig. 11. CCE's NF and FD with change of r_E ($\alpha = 2$, $s = 2.236$)

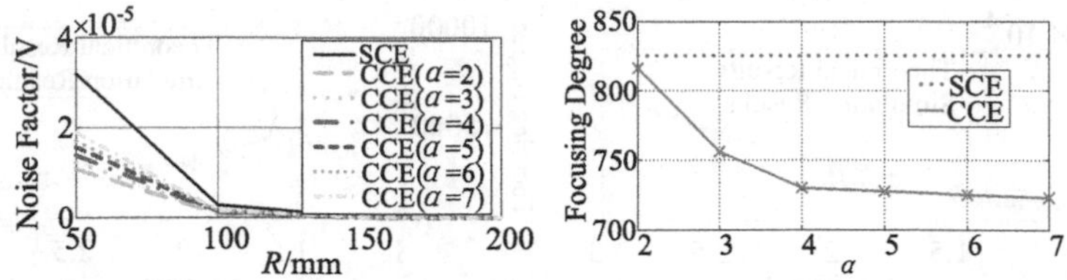

Fig. 12. CCE's NF and FD with change of α ($r_E = 1$, $s = 2$)

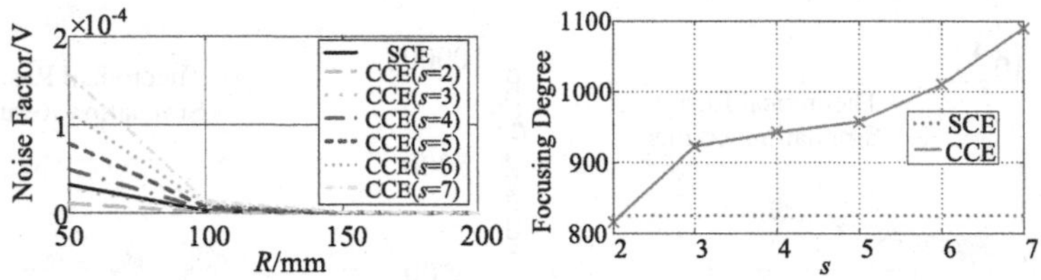

Fig. 13. CCE's NF and FD with change of s ($r_E = 1$, $\alpha = 2$)

As for *FD*, we should set r_E no more than 1 mm, α less than 2 and s more than 2 to get larger *FD* than SCE. On the other hand, though compared with SCE, *NF* has a certain degree of reduction when CCE is at some sizes. But when α and s become larger, *NF* can be even larger than that of SCE, which should be avoided.

Stimulation with Electrode Arrays

To simplify the analysis, based on model of forearm, we use two identity electrodes (SCE with $r_E = 1$, CCE with $r_E = 1$, $\alpha = 1.25$, $s = 0.83$) as electrode units to form an electrode array, because the array with two units is the simplest electrode array and electrode array with more electrode units could be seen as the expansion of the array with two units.

In simulation, we research the influence of the parameter, the center distance *D* (Fig. 6(c) and (d)), on *FD* and *NF*. Results (shown in Fig. 14) show that *NF* of CCE array is obviously much smaller than that of SCE array (less than one third) and the *NF* increases along with the increase of *D*. We defined Eq. (10), where FD_1 and FD_2 are *FD*s of two electrode.

$$\overline{FD} = \frac{1}{2}(FD_1 + FD_2) \tag{10}$$

From Fig. 14, with the increase of *D*, $\overline{FD}$ of SCE array decrease while that of CCE array is almost the same as *FD* of single electrode. Interference between electrodes may be the main reason of the decrease. As for CCE array, when *D* is large enough, $\overline{FD}$ remains at that of single electrode stimulation. It means each electrode unit of CCE array has a less effect on others than SCE array. So it's more suitable for CCE array to realize ES of prosthetic hand. However, the anti-interference ability of CCE doesn't mean that we can place as many units as we want in a limited area on forearm. The comfort and discrimination of amputees will be major constraints.

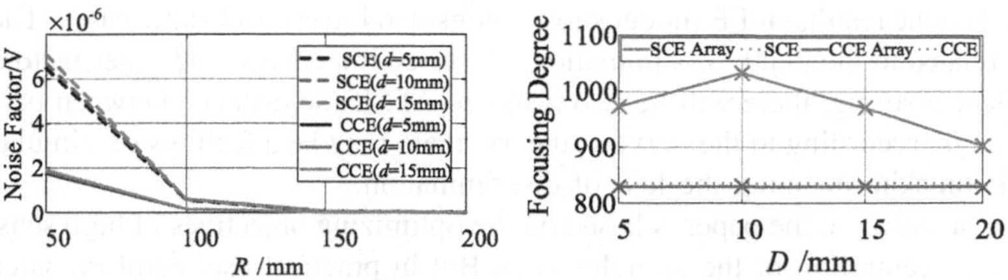

Fig. 14. NF and FD of CCE and SCE's electrode array with change of *D*

4 Discussions

This paper presents a new type of electrical stimulation (ES) electrodes, Concentric Circle Electrode, and a method to optimize the design of the electrode with theoretical model and FE model. Compared with traditional Single Circle Electrode, opposite polarity signals exert on inner and outer electrode of CCE separately to realize high spatial resolution and low interference with EMG signals, which are quantitatively described by the indices of *FD* and *NF* respectively.

By using Variable-Controlling Approach, FD and NF values of CCE with varying parameters (r_E, α, s) are investigated and compared with those of SCE with fixed r_E. The increase of r_E and α will cause the increase of *NF* and the decrease of *FD* at the same time, whereas the increase of s will cause the increase of *NF* and *FD* at the same time. The change trend of s is contrary to the result of theoretical model. It is mainly because the electrical field is established in multilayered medium, causing the stimulus waveform to be completely deformed at the concerned depth, leading to the failure of theoretical model. In the design of CCE, we should choose r_E and α from small values (r_E no more than 1 mm, α less than 2), and s from large values (more than 2) to realize the selectivity of stimulation. Considering the limited space of forearm and the balance between *NF* and *FD*, parameter s of CCE could not be too big, especially for CCE array. Besides, the distance between EMG sensor and stimulation electrode is an important factor that affects *NF*. In the design of bidirectional HMI, we should place the two electrodes as far as possible to decrease *NF*, which can help us get EMG signal with a higher noise-signal ratio.

For stimulation with electrodes array, CCE also have superiority over SCE for resisting disturbance and decreasing noise. Furthermore, stimulation of CCE with different polarities prevents the stimulation of deeper sensory-motor structures (e.g., motor nerves and muscles) [15]. We choose the center distance as 5 mm, 10 mm, 15 mm and 20 mm to simulate. In general, with the increase of the electrode spacing, *NF* increases and *FD* remains as almost the value of single electrode. Thus better distinguish can be achieved when using CCE array. But in the design, the electrode spacing should be large enough to avoid losing discrimination of each unit for amputees. Meanwhile increased spacing will restrict the quantity of electrode units in a limited area. Thus we should make a trade-off between them through experiments.

According to some researches, when the size of electrode goes large, AF will have the "edge effect". In our theoretical model, we ignore edge effect to simplify the formula. But the results of FE model show a considerable error in some cases. The edge effect is related to the depth of stimulation, electrode impedance and other factors [16]. If the effect is strong, there will be two peaks of AF. If the distance between peaks are large enough, according to the nervous theory, there may be a feelings of stimulating at two parts on skin, meaning the loss of discrimination.

The conclusion in the paper is based on the optimizing objectives of high sensibility and high concentration of the stimulus area. But in practical use, comfort, safety and other factors should also be taken into consideration, which can only be verified on the basis of human experiment. Therefore, we will carry out clinical experiments in vitro and vivo to explore the property of different electrodes in the future.

5 Conclusions

We proposed a method to optimize the design of electrode for electrical stimulation. By using theoretical model and finite element model, we find Concentric Circle Electrode can have superiority over Single Circle Electrode, having larger Focusing Degree and smaller Noise Factor. Furthermore, for electrode array, the configuration of CCE can avoid the crosstalk of each unit greatly.

Acknowledgments. This work is supported by National Natural Science Foundation of China (No. 51675123), Foundation for Innovative Research Groups of National Natural Science Foundation of China (No. 51521003), Self-Planned Task of State Key Laboratory of Robotics and System (No. SKLRS201603B), and the Research Fund for the Doctoral Program of Higher Education of China (No. 20132302110034).

References

1. Arieta, A.H., Yokoi, H., Arai, T., Wenwei, Y.: FES as biofeedback for an EMG controlled prosthetic hand. In: 2005 IEEE Region, Tencon 2005, pp. 1–6 (2005)
2. Zhang, Q., Hayashibe, M., Fraisse, P., Guiraud, D.: Fes-induced torque prediction with evoked emg sensing for muscle fatigue tracking. IEEE/ASME Trans. Mechatron. **16**(5), 816–826 (2011)
3. Antfolk, C., D'Alonzo, M., Rosén, B., Lundborg, G., Sebelius, F., Cipriani, C.: Sensory feedback in upper limb prosthetics. Expert Rev. Med. Devices **10**(1), 45 (2013)
4. Thierry, K., Andreas, K.: Electrodes for transcutaneous (surface) electrical stimulation. J. Autom. Contr. **18**(2), 35–45 (2008)
5. Luo, Z., Yang, G.: Study of myoelectric prostheses based on fuzzy control and touch feedback. In: International Conference on Neural Networks and Brain, ICNN&B 2005, vol. 3, pp. 1815–1819 (2005)
6. Dosen, S., Schaeffer, M.C., Farina, D.: Time-division multiplexing for myoelectric closed-loop control using electrotactile feedback. J. Neuroeng. Rehabil. **11**(1), 138 (2014)
7. Frigo, C., Ferrarin, M., Frasson, W., Pavan, E., Thorsen, R.: Emg signals detection and processing for on-line control of functional electrical stimulation. J. Electromyogr. Kinesiol. **10**(5), 351–360 (2000). Official Journal of the International Society of Electrophysiological Kinesiology

8. Kajimoto, H., Kawakami, N., Tachi, S., Inami, M.: Smarttouch: electric skin to touch the untouchable. Comput. Graph. Appl. **24**(1), 36–43 (2004). IEEE
9. Zhu, K.H., Li, P., Chai, G.H., Sui, X.H.: Effects of stratum corneum and conductive gel properties on sensory afferents recruitment by 3D TENS computational modeling. In: International IEEE/EMBS Conference on Neural Engineering, pp. 506–509 (2015)
10. Jiang, L., Huang, Q., Zhao, J., Yang, D., Fan, S., Liu, H.: Noise cancellation for electrotactile sensory feedback of myoelectric forearm prostheses. In: 2014 IEEE International Conference on Information and Automation (ICIA), pp. 1066–1071 (2014)
11. Rattay, F.: Analysis of models for extracellular fiber stimulation. IEEE Trans. Biomed. Eng. **36**(7), 676–682 (1989)
12. Mclachlan, J.C.: Transcutaneous electrical nerve stimulation. Lancet **337**(8743), 742 (1991)
13. Gomez-Tames, J., Gonzalez, J., Yu, W.: Influence of fat thickness and femur location on nerve activity computation during electrical stimulation. In: IEEE International Conference on Intelligent Engineering Systems, pp. 51–54 (2013)
14. Kuhn, A., Keller, T., Lawrence, M., Morari, M.: A model for transcutaneous current simulation: simulations and experiments. Med. Biol. Eng. Comput. **47**(3), 279–289 (2009)
15. D'Alonzo, M., Dosen, S., Cipriani, C., Farina, D.: Hyve: hybrid vibro-electrotactile stimulation for sensory feedback and substitution in rehabilitation. IEEE Trans. Neural Syst. Rehab. Eng. **22**(2), 290 (2014)
16. Kronberg, G., Bikson, M.: Electrode assembly design for transcranial direct current stimulation: a FEM modeling study. Eng. Med. Biol. Soc. **2012**(4), 891–895 (2012)

A Directional Identification Method Based on Position and Posture of Head for an Omni-directional Mobile Wheelchair Robot

Junyou Yang, Chunwei Yu(✉), Rui Wang, and Donghui Zhao

School of Electrical Engineering, Shenyang University of Technology, Shenyang, Liaoning Province, China
junyouyang@sut.edu.cn, yuchunweiemail@163.com

Abstract. To assist the daily life for the heavy disabled, a new intelligent wheelchair robot has been developed which has the Omni-directional movement function in authors' lab. To conduct the lift support task, the robot must firstly know the intention of users. In this paper, we focus on the research on the recognition of the directional intent of users. Therefore, a new directional identification method based on the head posture of users was proposed. A fuzzy reasoning method is proposed to predict the users' behavior intent based on the principle of Kalman Filter and spatial data fusion. The wheelchair robot also can realize Omni-directional mobile control of the intelligent wheelchair. Prototype were conducted using the proposed developed robot and the results demonstrate its feasibility and efficacy.

Keywords: Wheelchair robot · Fuzzy reasoning method · Directional intent identification

1 Introduction

In recent years, social aging has become a serious problem in many countries. As a developing country, china is also facing a serious threat of aging society. In addition, the occurrence of traffic accidents and diseases have also led to a significant increase in the number of disabled year by year [1]. Motor dysfunction leads to much inconvenience to the elderly and the disabled and causes a heavy physical and mental burden to their families and the society. Therefore, it is quite necessary to exploit a robot which can assist the daily life for the elderly to improve the quality of life and alleviate the burden of their families and the society. Furthermore, the desired robot can achieve the object that help the disabled to improve their degrees of freedom of action and to reintegrate into the society.

Therefore, researchers of the other country have focus on the developing of the intelligent device that assist the elderly and disable people. Most of them are developing the intelligent wheelchair robot to help the elderly and disable people to complete the daily action [2–4]. The standard electric wheelchairs are usually installed with computer or some sensors. This standard electronic wheelchair can be controlled by

Y. Huang et al. (Eds.): ICIRA 2017, Part I, LNAI 10462, pp. 230–241, 2017.
DOI: 10.1007/978-3-319-65289-4_22

joystick. From 1986, the study of navigation for the intelligent wheelchair was conducted by the computer vision [5]. Then Connell and Viola who worked in IBM T.J. Watson Research Center place a seat on a mobile robot platform and achieved the navigation functions such as walking and obstacle avoidance [6]. Moreover, the researchers in the world developed many kinds of intelligent wheelchair robot such as Wheelesley Robot by Massachusetts Institute of Technology, NavChair Robot by University of Michigan. It can be seen that the transmission structure of intelligent wheelchair is same as the standard electronic wheelchair, and it have not the omni-directional motion function. To help the elderly to conduct pick up task, some intelligent wheelchairs are installed with a manipulator [7]. Human-computer interaction technology about intelligent wheelchair mainly contain the following: joystick control, key control, touch screen control and computer control. These control methods are only suitable for people who has a relatively slight disability, a high level of control and a better sense of control. For those with a higher degree of disability and lower limb activity, the control methods for the intelligent wheelchair include: voice control, respiratory control, gesture control, biological signal control [8]. However, those human computer interactions of intelligent wheelchair are seriously affected by the environment such as voice control affect by noise, gesture control affect by light intensity, biological signal control affect by time variability of biological signals. In addition, those devices are expensive. Thus, above method cannot truly meet the demand of the elder with lower limb dysfunction movement. The present paper proposed a new directional identification method based on the head posture of users, and control the movement of the wheelchair. It is inconvenient for the elderly to control the wheelchair by head while walking since it is dangerous for patients with high degree of spinal cord injury and motor neuropathy to control the robot and walking simultaneously.

In this paper, a fuzzy reasoning method is proposed to judge the intention of the elderly based on the posture of their head [9–11]. In detailed, a Kalman Filter is firstly used to fuse the attitude sensor data [12–14], so as to obtain the accurate position and attitude information of the elderly head. Then, a fuzzy reasoning method is proposed to identify the directional of users. Finally, the effectiveness of the proposed method is verified by experiment, and the goal of helping the elderly and disabled by improving their movement ability.

2 Kinematics of the Omni-directional Directional Mobile Wheelchair and Head Posture Movement of Users

Wheelchair robot with omnidirectional movement function is shown in Fig. 1.

The movement of the wheelchair is composed by the motion of the four wheels shown in Fig. 2, which can achieve the omnidirectional movement by the different speed of the four wheels.

Fig. 1. Omni-directional wheelchair robot

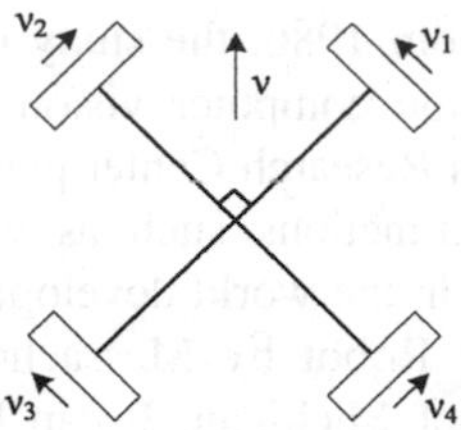

Fig. 2. Wheel distribution structure

2.1 Kinematics of Omni-directional Mobile Wheelchair

According to the wheelchair's structure shown in Fig. 2, a structural model is developed shown in Fig. 3.

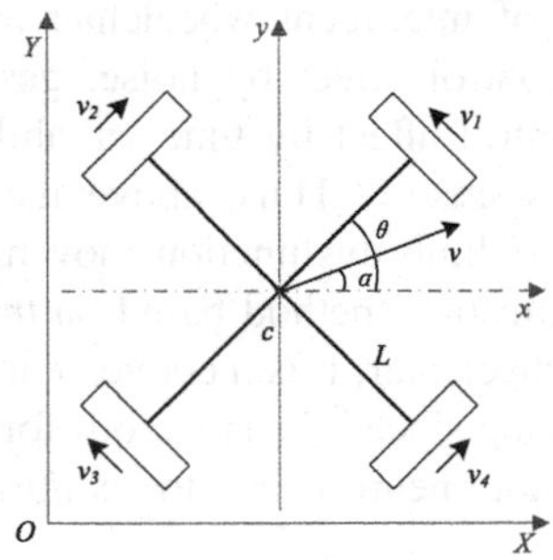

Fig. 3. Structural model of the machine

The parameters and coordinate system are as follow:

$\sum(X, Y, O)$: Absolute coordinate system; $\sum(x, y, c)$: Translation coordinate system determined by local of the wheelchair robot; L: Distance between the geometric center; v: The velocity of the wheelchair robot; α: Angle between x-axis and the movement direction of the wheelchair robot; θ_i: Angle between the i wheel and the x-axis measured from the center of the wheelchair robot (i = 1, 2, 3, 4), the kinematics equation of wheelchair robot can be described as:

$$v_1 = -v_x \sin\theta + v_y \sin(\pi/2 - \theta) + \dot{\theta}L \tag{1}$$

$$v_2 = v_x \cos\theta + v_y \cos(\pi/2 - \theta) - \dot{\theta}L \tag{2}$$

$$v_3 = -v_x \sin\theta + v_y \sin(\pi/2 - \theta) - \dot{\theta}L \tag{3}$$

$$v_4 = v_x \cos\theta + v_y \cos(\pi/2 - \theta) + \dot{\theta}L \tag{4}$$

where v_x, v_y are the x and y component of velocity of wheelchair robot, and the relationship in (1), (2), (3), (4) imply:

$$v_1 + v_2 = v_3 + v_4 \tag{5}$$

Because of that the Omni-directional wheelchair is controlled by the head posture of users, we define the angle along the *X*-axis and *Y*-axis as Ax and Ay, respectively, and the head rotation angular as Adz. Then we can get the following equations:

$$\begin{aligned} v_x &= P_x \cdot Ax \\ v_y &= P_y \cdot Ay \\ \dot{\theta} &= P_Z \cdot Adz \end{aligned} \tag{6}$$

where *Px, Py, Pz* are conversion coefficient between the angular velocity of human head along *X*-axis, *Y*-axis and *Z*-axis respectively. The control of the Omni-directional wheelchair robot by head position information of human will according to the formula (6).

2.2 Kinematics Analysis of the Human Head Position

Since that the Omni-directional wheelchair system is controlled by the human head posture, it is necessary conduct the kinematics analysis of the human head position. Firstly, a rectangular coordinate system $\sum(X, Y, O)$ at the center of the head is established as shown in Fig. 4.

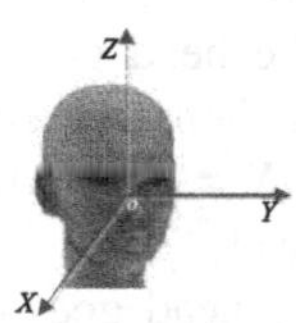

Fig. 4. The head coordinate system

X axis denotes the back direction of head, Y axis denotes the right direction of head, while *Z* axis denotes the above direction of head.

Then, the coordinate system is defined as $\sum(X_i, Y_i, Z_i, O_i)$ (i = 1, 2, 3, 4) considering the head's rotation. In this study, we will identify four directional intents by four head postures, such as: forward, leaning to the left, the left rear forward and the clockwise rotation shown Fig. 5, respectively. The four representative head movements are detailed analyzed as follows:

(1) The head forward
    As shown in Fig. 5(a), when the head forward it rotates around *Y*-axis, then the relative coordinate system turns into $\sum(X_1, Y_1, Z_1, O_1)$, where $\angle Z_1OX = \alpha =$ Ax.

(2) Leaning to the left of the head
As shown in Fig. 5(b). When the head leans to the left, it rotates around X-axis, and then the relative coordinate system turns into $\sum(X_2, Y_2, Z_2, O_2)$, where $\angle Z_2OY = \beta = Ay$.

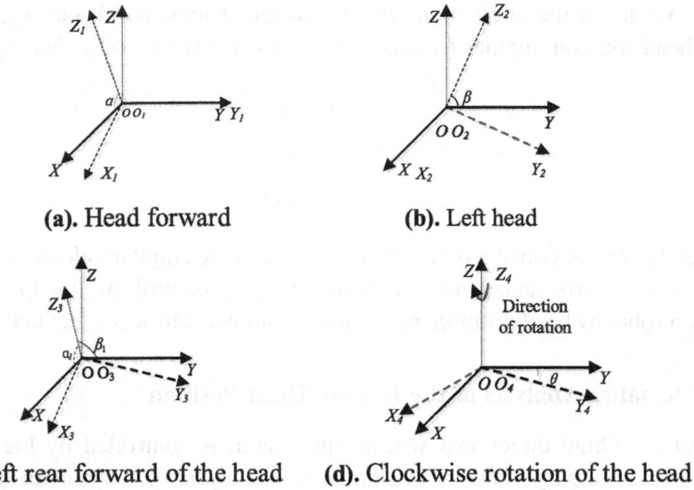

Fig. 5. Head movement model

(3) The left rear forward of the head
As shown in Fig. 5(c), when the head goes to the left rear forward, it rotates around X-axis and Y-axis, and then the relative coordinate system turns into $\sum(X_3, Y_3, Z_3, O_3)$, where $\angle X_3OX = \alpha_1 = Ax$, $\angle Y_3OY = \beta_1 = Ay$.
(4) The clockwise rotation of the head
As shown in Fig. 5(d), when the head goes clockwise, it rotates around Z-axis, and then the relative coordinate system turns into $\sum(X_4, Y_4, Z_4, O_4)$, where $\angle X_4OX = \angle Y_4OY = \theta$. When we do the differential operation in θ, we can conclude that the head rotation angular around the Z-axis is $\dot{\theta} = d\theta/dt$.

Based on the above discussion, we can control the wheelchair to move by using the head pose information.

3 MEMS Sensor Data Fusion and Directional Controller Design

To develop the directional intent identification method based the head position and posture of user, we firstly utilize the Microelectronic Mechanical Systems (MEMS) sensors to measure the angle of the head in space. And then, we extract the patients' position and posture information of the head by data fusion using a Kalman Filter.

Finally, we use a fuzzy reasoning method to process the obtained date control the motion of the Omni-directional wheelchair robot.

3.1 Design of the Directional Recognition Fuzzy Reasoning Method

The rules of fuzzy reasoning can be explained in fuzzy conditional statement. The fuzzy reasoning is mainly composed of four parts: fuzzification, fuzzy inference, deblurring and fuzzy knowledge base. The flow chart of fuzzy reasoning is shown in Fig. 6.

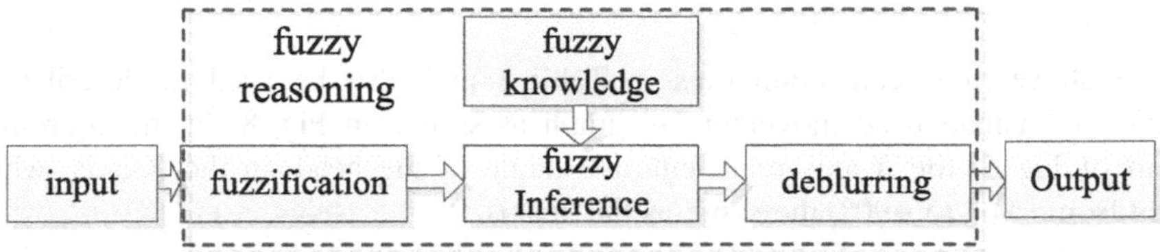

Fig. 6. The block diagram of fuzzy reasoning

Fuzzification is mainly through the membership function of the input data for fuzzy processing. The input and output of the system are X-axis and Y-axis angle information are set to: *Ax* and *Ay*. The following relations as shown in formula (7). Due to the normal range of human head is −90–90° therefore the domain of the system is −90–90°.

$$X = Ax,\ Y = Ay \tag{7}$$

In this system, we define a discrete output field as Z = $\{1, 2, 3, 4, 5, 6, 7, 8, 9\}$. The nine values in the field are represented as "Front (F)", "Behind (B)", "Left (L)", "Right (R)", "Front left (FL)", "Front right (FR)", "Behind left (BL)", "Behind right (BR)", "Stop (S)" by linguistic variables. The three domains X, Y, Z are represented as Angle Axis of X, Angle Axis of Y. Move Direction and the domain X and Y is still a non-fuzzy ordinary variable. According to the difference of rotation angle, we divide the head rotation angle of human into seven fuzzy sets, namely "Negative big (NB)", "Negative medium (NM)", "Negative small (NS)", "Zero stop (ZS)", "Positive big (PB)", "Positive medium (PM)", "Positive small (PS)". The membership functions of each fuzzy set are shown in Fig. 7.

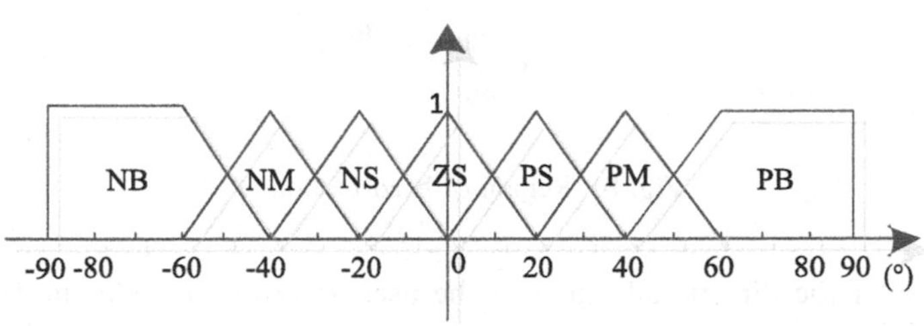

Fig. 7. Membership functions of X and Y

The knowledge based of the fuzzy reasoning system are set up as the if-then rules:

Rule 1: IF Angle Axis of X is NB
AND Angle Axis of Y is NB
THEN Move_Direction is BL
⋮
Rule 49:
IF Angle Axis of X is PB
AND Angle Axis of Y is PB
THEN Move_Direction is FR

The above rules can summaries in Table 1 and also be used to describe the direction of human head movement in graph as shown in Fig. 8. In the coordinate system of Fig. 5, the *X*-axis represents the angle of the head on the X-axis, which ranges from −90° to +90°, the same as the *Y*-axis.

Table 1. Fuzzy inference rule table

X\Y	NB	NM	NS	ZS	PS	PM	PB
NB	BL	BL	B	B	B	BR	BR
NM	BL	BL	B	B	B	BR	BR
NS	L	L	BL	B	BR	R	R
ZS	L	L	L	STOP	R	R	R
PS	L	L	FL	F	FR	R	R
PM	FL	FL	F	F	F	FR	FR
PB	FL	FL	F	F	F	FR	FR

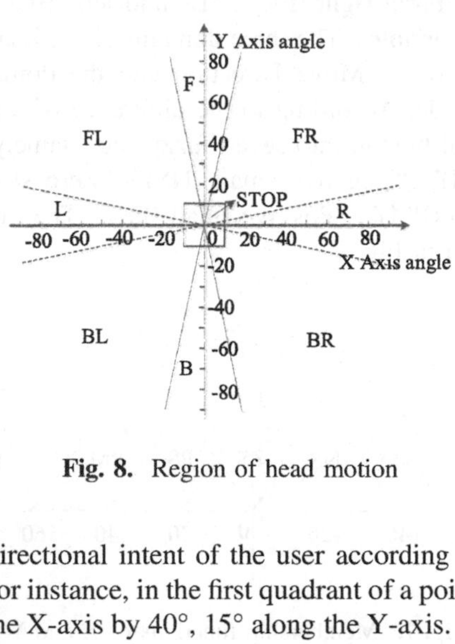

Fig. 8. Region of head motion

We will infer the directional intent of the user according to rules in Table 1 and head region in Fig. 8. For instance, in the first quadrant of a point (40, 15), it shows that the head moves along the X-axis by 40°, 15° along the Y-axis. In this case, through the fuzzy inference rules we can infer that the patient's motion intention is FR of the movement, and then the wheelchair robot move FR.

According to the actual situation, when the head is moving along the axis direction, the sense of direction is relatively strong, so the division of the corresponding motion region is narrow.

Through the above analysis, we can infer the patient's movement intention by the head position and status. However, in order to realize Omni-directional motion of wheelchair, the system collects the data of the *Z*-axis gyroscope and controls the rotation of the wheelchair.

3.2 Attitude Sensor Data Fusion

The accelerometer and gyroscope sensors were used to measure patients' location and status information. For stationary and slow moving objects, the accelerometer is used to measure the acceleration of the object, and it is can accurately measure the angle and the angle changes of the object as shown in Fig. 9(a). We can see the waveform of the angle is quite accurate. However, there is a lot of interference in movement, as shown in Fig. 9(b), therefore the data cannot be used directly.

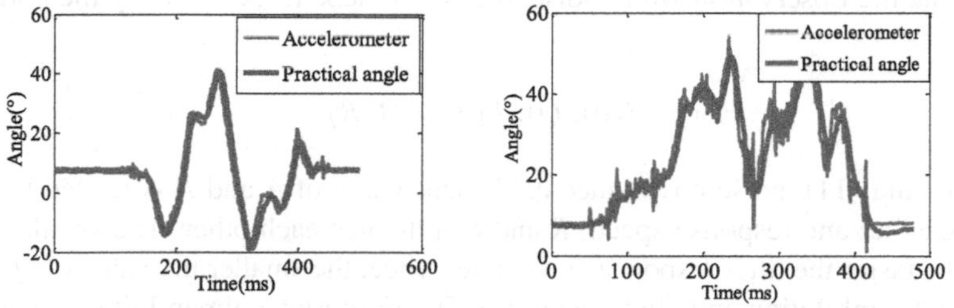

(a) Slow movement (b) Movement under the disturbance condition

Fig. 9. Acceleration waveforms when objects movement

Gyroscope is used to measure the angular of object, assuming that the angular is $w(t)$, the real angle θ can be got through formulate (8).

$$\theta = \int w(t)dt \tag{8}$$

Actually, there is a lot of noise in practice, the interference signal will accumulate as well which will result in inaccurate data. The accuracy of the two kinds of sensors is not enough to measure the patients' location and status information. Therefore, we used a Kalman Filter to do data fusion of the sensors.

Kalman Filter with the minimum mean square error based on the estimated optimum criterion to seek out a set of recursive estimate algorithm, the working flow chart is shown in Fig. 10.

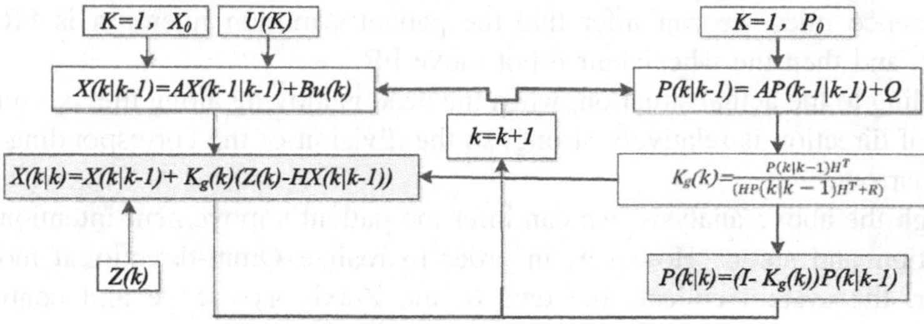

Fig. 10. The flow chart of Kalman Filter

The controller of discrete system is:

$$X(k) = AX(k-1) + BU(k) + W(k) \tag{9}$$

$$Z(k) = HX(k) + V(k) \tag{10}$$

where, A is the system matrix, B is the input matrix of the system, H is the output matrix of the system. $W(k)$ and $V(k)$ are the random signal which represent the process white noise and the observation white noise, they are independent and obey the normal distribution:

$$P(w) \sim N(0, Q), P(v) \sim (0, R) \tag{11}$$

In the formula (11), noise covariance coefficient value of A and B is related to the final filtering effect and response speed, R and Q influence each other. The smaller the value of R, the faster the filter response and convergence, the smaller the value of Q, the stronger the filter inhibition and eliminate noise. The proposed Kalman Filter is mainly containing five equations, including two prediction equations and three update equations which is shown as follows: the two prediction equations are:

$$X(k|k-1) = AX(k-1|k-1) + Bu(k) \tag{12}$$

$$P(k|k-1) = AP(k-1|k-1)A^T + Q \tag{13}$$

And the three update equations are:

$$Kg(k) = P(k|k-1)H^T/(HP(k|k-1)H^T + R) \tag{14}$$

$$X(k|k) = X(k|k-1) + Kg(k)(Z(k) - HX(k|k-1)) \tag{15}$$

$$P(k|k) = (I - Kg(k))P(k|k-1) \tag{16}$$

where $U(k)$ equals θ, $Z(k)$ is an angle measured by the angular velocity meter. With Kalman Filter, we can achieve sensor data fusion and obtain the head' position and posture of patients with accurate information. After parameter adjustment, the parameters of this system are $Q = 0.1$, $R = 0.7$ and have a rapidity and stability from the Fig. 11.

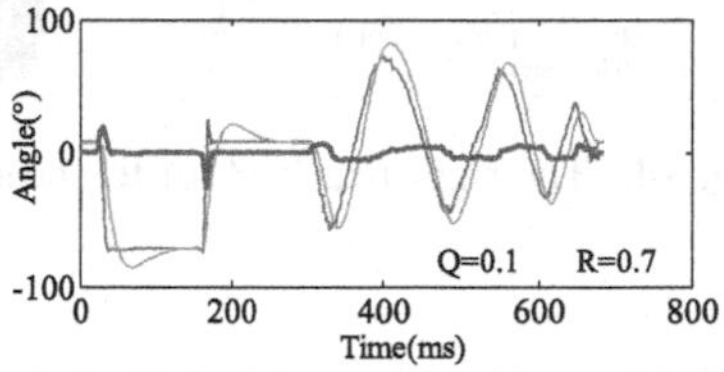

Fig. 11. $Q = 0.7$, $R = 0.1$

4 Experiment

This part is mainly about the actual experiment of the wheelchair movement direction and data analysis. Our experiment to validate the effectiveness of the proposed method recruited three healthy subjects (two males and one females), coded A, B and C. The average age of the subjects was 25, the average height was 169 cm, and the average weight was 66.3 kg. Experimental result is shown as follows (Figs. 13, 14 and 15).

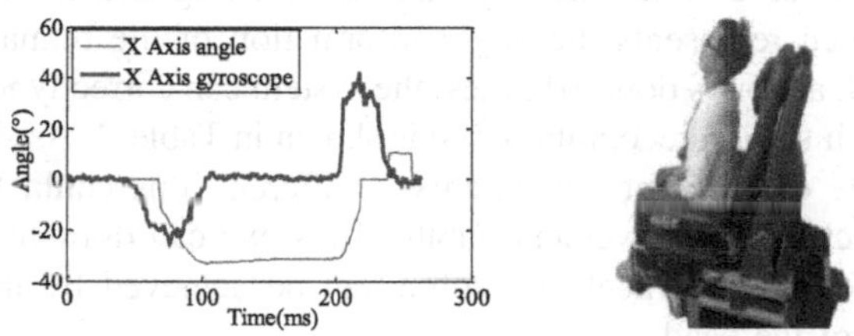

Fig. 12. The head hypsokinesis waveform and demonstration

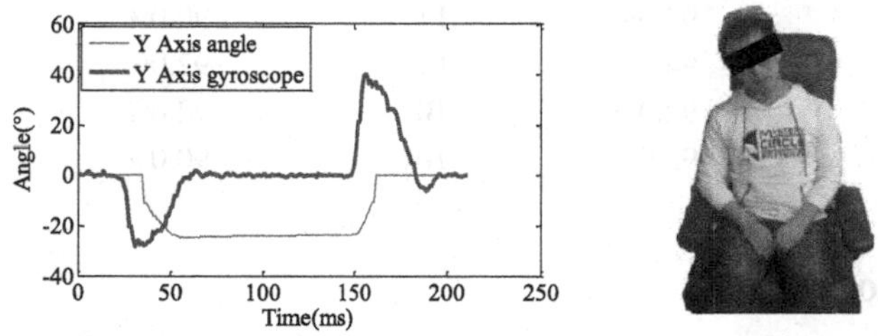

Fig. 13. Leaning to the right of the head and demonstration

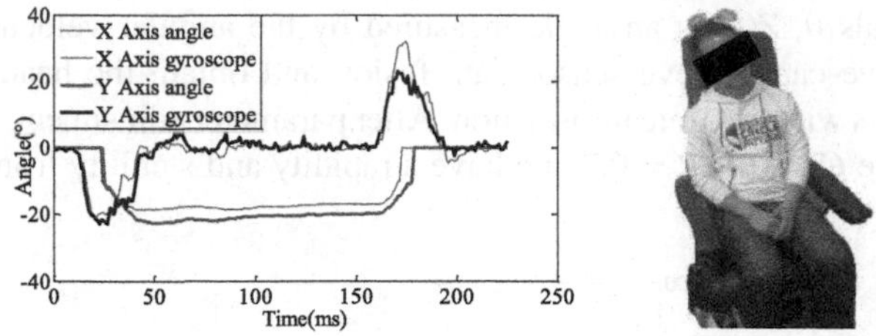

Fig. 14. The right rear back of the head

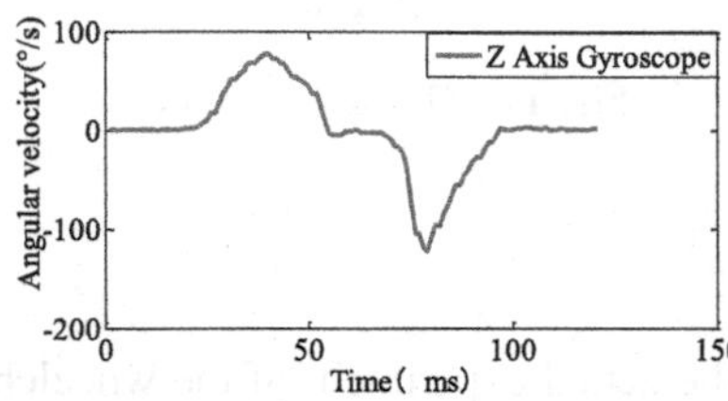

Fig. 15. The counterclockwise rotation of the head

From the analysis of the above experiments: the thick curve represents the angular velocity along the X axis during the head movement, so that the waveform of the curve is peaked. The thin curve is obtained by Kalman Filtering and the waveform is relatively smooth, which represents the angle information of the human head from the Fig. 12. When each action is done 50 times, the system can correctly identify the action and the directional intention recognition rate is shown in Table 2. As shown in Table 2, the recognition rate of different movements is different, this could be due to sensor errors or the subjects' head movement habits. Thus, we can draw the conclusion that the accurate control of the wheelchair robot can be achieved by the change of the position and state of the head.

Table 2. The recognition rate of patients' motor intention

Motor intention	Recognition rate	Motor intention	Recognition rate
F	94.0%	FL	90.0%
B	94.0%	FR	92.0%
F	92.0%	BL	92.0%
R	92.0%	BR	90.0%

5 Conclusion

In order to improve the motion ability of the heavy disabled patients and elderly with their daily life and diet, an Omni-directional wheelchair robots are developed for inferring the movement intention elderly. In this paper, a new directional intention control method of wheelchair robot based on patient head position and state control is proposed. Using fuzzy reasoning and data fusion technology to identify the wheelchair robot. According to the prototype production and the actual test verification, the feasibility and effectiveness of the proposed control method are invalid. In the future, the control method and a new inference method will be considered to improve the performance of the intelligent wheelchair.

References

1. Yamazaki, K., Ueda, R., Nozawa, S., Kojima, M., Okada, K.: Home-assistant robot for an ag ing society. Proc. IEEE **8**(100), 2429–2441 (2012)
2. Wang, H., Kang, C.U., Ishimatsu, T., Ochiai, T.: Auto-navigation of a wheelchair. Artif. Life Robot. **1**(3), 141–146 (1996)
3. Gomi, T., Ide, K.: The development of an intelligent wheelchair. In: Intelligent Vehicles Symposium, pp. 70–75 (1996)
4. Matsumoto, T.Y., Ogsawara, T.: Development of intelligent wheelchair system with face and gaze based interface. In: IEEE International Workshop on Robot and Human Interactive Communication, pp. 262–267 (2001)
5. Tomari, M.R.M., Kobayashi, Y., Kuno, Y.: Development of smart wheelchair system for a user with severe motor impairment. Procedia Eng. **41**, 538–546 (2012)
6. Rockland, R.H., Reisman, S.: Voice activated wheelchair controller. In: Bioengineering Conference, pp. 128–129 (1998)
7. Kuno, Y., Yoshimura, T., Mitani, M., Nakamura, A.: Robotic wheelchair looking at all people with multiple sensors. In: Proceedings of IEEE International conference on Multisensor Fusion and Integration for Intelligent System, MFI, pp. 341–346 (2003)
8. Rofer, T., Mandel, C., Laue, T.: Controlling an automated wheelchair via joy stick/head-joystick supported by smart driving assistance. In: IEEE International Conference on Rehabilitation Robotics, pp. 743–748 (2009)
9. Wang, H.O., Tanaka, K., Griffin, M.F.: An approach to fuzzy control of nonlinear systems: stability and design issues. IEEE Trans. Fuzzy Syst. **4**(1), 14–23 (1996). IEEE Press
10. Ahmad, S., Siddique, NH., Tokhi, MO.: Modular Fuzzy Logic Controller for Motion Con-trol of Two-Wheeled Wheelchair. InTech, Rijeka (2012)
11. Won, S.H.P., Golnaraghi, F., Melek, W.W.: A fastening tool tracking system using an IMU and a position sensor with Kalman filters and a fuzzy expert system. IEEE Trans. Ind. Electron. **56**(5), 1782–1792 (2009)
12. Rehbinder, H., Hu, X.: Drift-free attitude estimation for accelerated rigid bodies. Automatica **40**(4), 653–659 (2004)
13. Neto, P., Mendes, N., Moreira, A.P.: Kalman Filter-based yaw angle estimation by fusing inertial and magnetic sensing. Springer International Publishing, vol. 35, no. 3, pp. 244–260 (2015)
14. Saito, H., Watanabe, T.: Kalman-filtering-based joint angle measurement with wireless wearable sensor system for simplified gait analysis. IEICE Trans. Inf. Syst. **E94-D**(8), 1716–1720 (2011)

Design of an Wearable MRI-Compatible Hand Exoskeleton Robot

Kun LIU[1(✉)], Yasuhisa Hasegawa[1], Kousaku Saotome[2], and Yosiyuki Sainkai[2]

[1] Intelligent Robotics and Bio-Mechatronics Laboratory, Nagoya University, Nagoya, Aichi, Japan
liu2@robo.mein.nagoya-u.ac.jp

[2] Sankai Laboratory, University of Tsukuba, Tsukuba, Ibaraki, Japan
http://www.mein.nagoya-u.ac.jp/index.html,
http://sanlab.kz.tsukuba.ac.jp/

Abstract. This paper proposes an wearable MRI-compatible hand exoskeleton robot that supports a subject moving his fingers voluntarily or involuntarily in high electromagnetic field. The hand robot consists of four exoskeletal fingers excluding a thumb, which is fabricated with nonmagnetic materials through 3D printing. In order to work in an MRI environment, pneumatic actuators are applied to drive the joints of the wearable robot. Potentiometers are installed in the MP and PIP joints of four fingers to measure the angles of finger's motions. Basic performances of the robot are evaluated by flexion rang of fingers, time delay and fingertip force. In the future, the compatibility of robot in MRI environment will be confirmed through measurement experiments of a subject's brain activity.

Keywords: Finger motion assistance · Brian activity · MRI compatibility

1 Introduction

1.1 Background

Recently, robotic technology has been applying to rehabilitation such as robot therapy for patients with stroke and brain injury [1]. Some gait training robots, such as LOKOMAT and HAL are successfully developed, which make it possible to recover from lower limbs motor disorder for stroke patients [2,3].

Moreover, with the development of neuroscience, the reorganization and plasticity of human brain induced by motion skill learning are being revealed gradually [4,5]. Functional Magnetic Resonance imaging (fMRI), which measures brain activity of a human in real time by detecting changes of blood flow [6], plays an important role in observing brain activation while motion learning is executed by a subject.

However, most of current exoskeleton robots could not be used in MRI environment to measure brain activity because magnetic materials are forbidden in

Y. Huang et al. (Eds.): ICIRA 2017, Part I, LNAI 10462, pp. 242–250, 2017.
DOI: 10.1007/978-3-319-65289-4_23

MRI environment. Some researchers developed some gait-like MRI compatible devices to study lower limb movement, however there are also plenty of patients with hand deficit [7]. Hence, a hand robot corresponding to a high magnetic field environment is therefore required.

1.2 Purpose

The purpose of this study is to develop an wearable MRI-compatible hand exoskeleton robot that supports a human moving his fingers in high electromagnetic field voluntarily and involuntarily. This robot executes passive and active movements when a subject wears this robot. The final purpose of this study is to evaluate which kind of motion learning will improve motor function for patients with stroke. We will therefore measure human brain activity in real time by fMRI while this assistive hand robot is used to provide finger motions for a subject.

2 Development of an Wearable MRI-Compatible Hand Exoskeleton Robot

The developed hand robot is shown in Fig. 1. The exoskeleton of the hand robot consists of four exoskeletal fingers excluding the thumb, which is constructed with nonmagnetic materials through 3D printing. This robot is designed for supporting tapping motion while a subject wears the exoskeleton hand. Air pressure provides extension power for each fingers independently by air tubes.

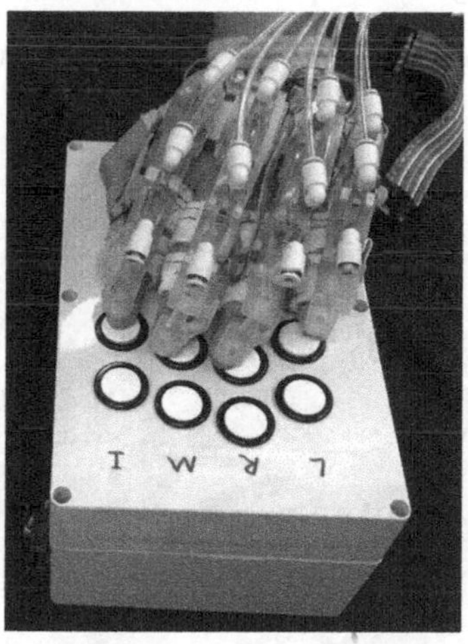

Fig. 1. Exoskeleton hand and button box

2.1 Materials Selection

This hand assistive robot will be applied around an MRI scanner, a high magnetic field. There are plenty of challenges associated with using the device in MRI scanner. Any devices used for performing movements during MRI must be designed particularly for the special environment and fabricated with low magnetic or non-magnetic materials [8]. There are mainly two reasons. Firstly, the robot should not disturb the imaging process and impact the scanner environment. Secondly, the magnetic field should not affect the function of robot [9,10]. Considering these constrains, we choose pneumatic actuators to drive the joints of the wearable robot. And the exoskeleton is fabricated with nonconductive resin materials through 3D printing, shown in Table 1.

In addition, potentiometers are used to measure the angle of joints. These potentiometers contain low magnetic materials, and the influence is ignorable for MRI machine, because they are located around outside of MRI scanner.

Table 1. Materials selection

Part	Material	Magnetism
Exoskeleton	FullCure720	Non
Joint axis	POM resin	Non
Screw	SUS304	Non
Potentiometer	SV01 series	Low

2.2 Exoskeleton Design

The exoskeleton is designed for human left hand that consists of four independent exoskeletal fingers.

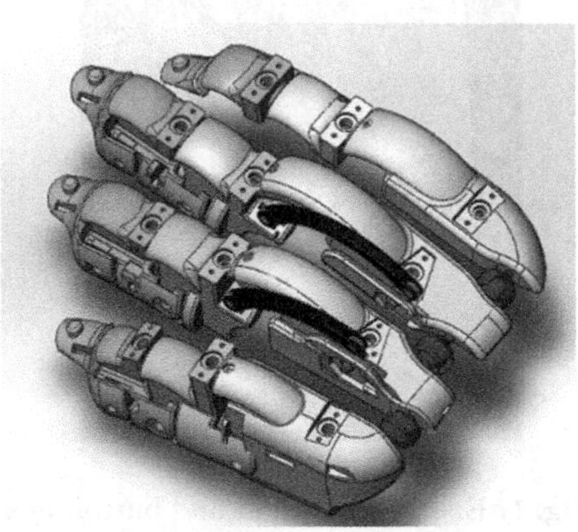

Fig. 2. Assembly drawing of exoskeleton hand

Figure 2 refers to an assembly drawing of the exoskeleton hand. Each finger comprises three joints, but only PIP and MP joints are independently driven by air pressure controlled by solenoid valve. In addition, each finger is assembled with a ball joint, which allows each finger moving in three degrees of freedom. Therefore this exoskeleton gives 17 degrees of freedom for movement in totally. Potentiometers are installed in MP and PIP joints of four fingers to measure the angles of the motions.

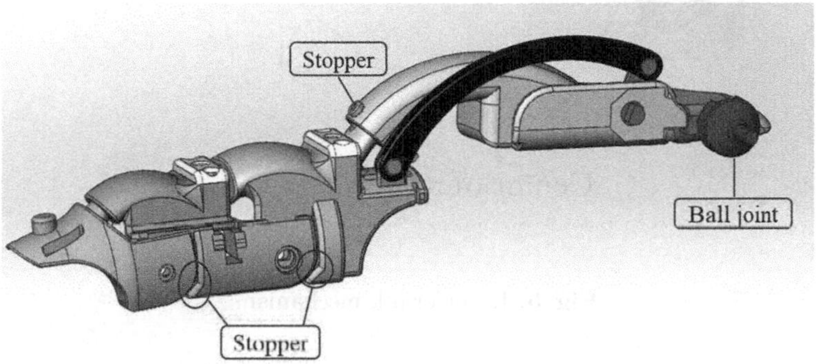

Fig. 3. Structure of middle finger

Figure 3 shows the structure of middle finger. Each finger is connected with a ball joint, and each joint is designed a stopper in case of hyperextension of finger.

Figure 4 is a slider structure to flex PIP joint all of finger. The air gets in from the tube fitting, and then drives slider to move. An O-ring is used to keep a good air tightness with less friction while a joint moves passively.

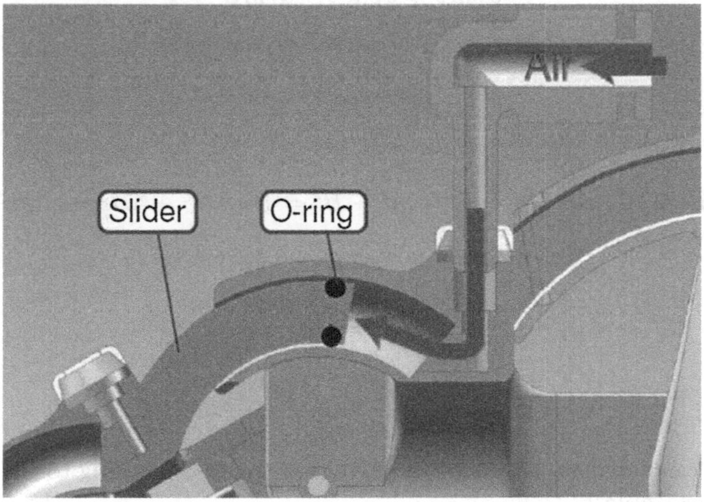

Fig. 4. Slider structure to flex PIP joint

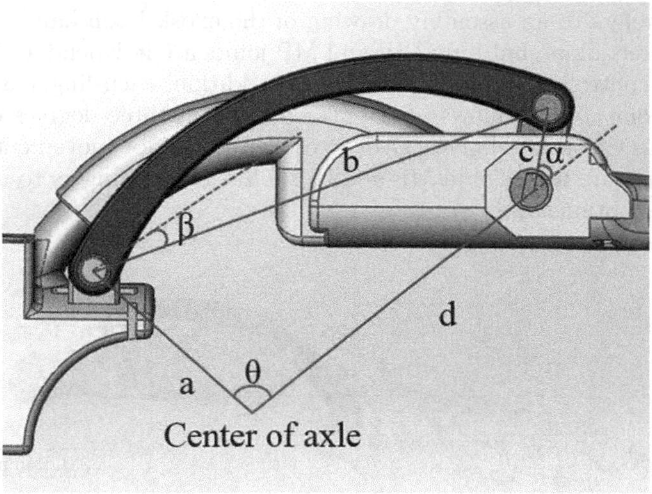

Fig. 5. Lever crank mechanism

Figure 5 refers to a lever crank mechanism of MP joints of middle finger and ring finger. Potentiometers are connected with the cran link, so we can measure joints angles when moving. To describe the calculation process of the angle of potentiometer, we use some characters to instead of angles and length. We assume that α is the angle of potentiometer, β is the angle between b and the parallel line of d, θ is the angle of straight line of a and d.

$$a\cos\theta + b\cos\beta = c\cos\alpha + d \tag{1}$$

$$a\sin\theta = b\sin\beta + d\sin\alpha \tag{2}$$

$$b^2\cos^2\beta = (-a\cos\theta + c\cos\alpha + d)^2 \tag{3}$$

$$b^2\sin^2\beta = (a\sin\theta - c\sin\alpha)^2 \tag{4}$$

If we suppose $\mathrm{A} = d + a\cos\alpha$, $\mathrm{B} = -c\sin\alpha$,

$$2Ba\sin\theta - 2Aa\cos\theta = b^2 - a^2 - A^2 - B^2 = \sqrt{4A^2a^2 + 4B^2a^2}\sin(\theta+\gamma) \tag{5}$$

$$b^2\sin^2\beta = (a\sin\theta - c\sin\alpha)^2 \tag{6}$$

$$\sin(\theta+\gamma) = \frac{b^2 - a^2 - A^2 - B^2}{\sqrt{4A^2a^2 + 4B^2a^2}} \tag{7}$$

$$\theta = \sin^{-1}\frac{b^2 - a^2 - A^2 - B^2}{2a\sqrt{A^2 + B^2}} - \tan^{-1}\frac{A}{B} \tag{8}$$

Hence, we can calculate the angle of movement by $\Delta\theta$ according to the parameters listed in Table 2.

Table 2. Parameters of joints

	a	b	c	d	α_0	β_0
Middle finger	25.6 [mm]	57 [mm]	10 [mm]	37.72 [mm]	17.1°	102°
Ring finger	23.4 [mm]	51 [mm]	10 [mm]	38.87 [mm]	47.4°	105°

2.3 Control Method

Figure 6 refers to the block diagram for control. This robot supports four fingers excluding a thumb of a human hand moving independently by corresponding to pneumatic actuators.

The power for finger flexion is provided by an air compressor, which is able to export stable air pressure for motion. Eight five-meter air tubes are used to transfer power for each joints, because the compressor and the control box are not permitted taking into MRI room. The output of air is controlled by solenoid valves. Rubber bands provide power for finger extension. A microcomputer based on linux is used to control solenoid valves, record angles of joint motion and tapping time.

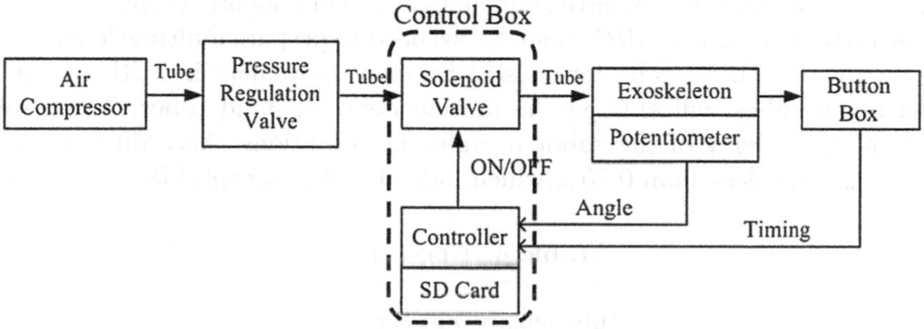

Fig. 6. Block diagram for control

3 Performance Evaluation

3.1 Flexion Range of Fingers

A potentiometer installed in each joint is used to measure the angle of the joint. Although these potentiometers contain low magnetic materials, the influence is ignorable for MRI machine.

According to the result of measurements, the flexion range of each joint is shown in Table 3. The exoskeleton covers 70% of rang of motion of the human fingers. Although the movable angle of fingers are smaller than human hands, it is enough to support passive tapping motions in our experiments.

Table 3. Flexion range of fingers

	MP joint	PIP joint
Human finger	90°	100°
Index finger	68° (76%)	65°(65%)
Middle finger	66° (73%)	68°(68%)
Ring finger	64° (71%)	68° (68%)
Little finger	64° (71%)	74° (74%)

3.2 Time Delay

This wearable robot will be applied in an MRI environment directly. The air compressor and control box have some magnetic components, therefore they are not permitted taking into MRI room. So we need to prepare multiple length of air tubes to provide air pressure for exoskeletal hand movement in MRI. Increasing length of air tubes will enlarge the motion delay of hand robot. The Table 4 shows the time delay of the robot in different conditions. According to result, the time delay is less than 0.15 s, which indicates it is acceptable.

Table 4. Time delay

Tube length	Pressure	
	70 psi	90 psi
5 m	0.1 s	0.07 s
10 m	0.14 s	0.12 s

3.3 Fingertip Force

For clarifying the supporting force for finger motion, we use a high accuracy electronic scale to measure fingertip force of each finger. The air pressure is 90 psi and the air tube is 5 m. According to test results, all fingers generate near 2N fingertip force, which is enough to assist exoskeleton finger moving involuntarily. The result is shown in Fig. 7.

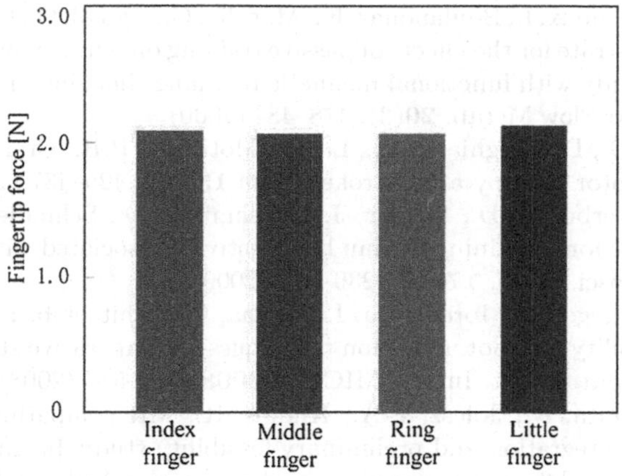

Fig. 7. Fingertip force

4 Conclusion and Future Work

This study described an MRI compatible tapping assistive robot. Pneumatic actuators were therefore used to drive the joints of the wearable robot. We then tested flexion range of each joint and fingertip force to ensure that the exoskeleton was able to support a subject moving his hand involuntarily. In the future, we will assess the performance of the robot in an MRI environment by measurement experiments of a subject's brain activities. Finally, the compatibility of the assistive robot in MRI environment will be confirmed.

Acknowledgement. This work was supported by JSPS KAKENHI Grant Numbers 17H05906.

References

1. Jezernik, S., Colombo, G., Keller, T., Frueh, H., Morari, M.: Robotic orthosis lokomat: a rehabilitation and research tool. Neuromodulation: Technol. Neural Interface **6**(2), 108–115 (2003)
2. Ikeda, T., Matsushita, A., Saotome, K., Hasegawa, Y., Matsumura, A., Sankai, Y.: Muscle activity during gait-like motion provided by MRI compatible lower-extremity motion simulator. Adv. Robot. **30**(7), 459–475 (2016)
3. Hollnagel, C., Brugger, M., Vallery, H., Wolf, P., Dietz, V., Kollias, S., Riener, R.: Brain activity during stepping: a novel MRI-compatible device. J. Neurosci. Meth. **201**(1), 124–130 (2011)
4. Karni, A., Meyer, G., Jezzard, P., Adams, M.M.: Functional MRI evidence for adult motor cortex plasticity during motor skill learning. Nature **377**(6545), 155 (1995)
5. Doyon, J., Benali, H.: Reorganization and plasticity in the adult brain during learning of motor skills. Curr. Opin. Neurobiol. **15**(2), 161–167 (2005)

6. Carel, C., Loubinoux, I., Boulanouar, K., Manelfe, C., Rascol, O., Celsis, P., Chollet, F.: Neural substrate for the effects of passive training on sensorimotor cortical representation: a study with functional magnetic resonance imaging in healthy subjects. J. Cereb. Blood Flow Metab. **20**(3), 478–484 (2000)
7. Takahashi, C.D., Der-Yeghiaian, L., Le, V., Motiwala, R.R., Cramer, S.C.: Robot-based hand motor therapy after stroke. Brain **131**(2), 425–437 (2008)
8. Mehta, J.P., Verber, M.D., Wieser, J.A., Schmit, B.D., Schindler-Ivens, S.M.: A novel technique for examining human brain activity associated with pedaling using fMRI. J. Neurosci. Meth. **179**(2), 230–239 (2009)
9. Fischer, G., Krieger, A., Iordachita, I., Csoma, C., Whitcomb, L., Fichtinger, G.: MRI compatibility of robot actuation techniques-a comparative study. Med. Image Comput. Comput. Assist. Interv. MICCAI **2008**, 509–517 (2008)
10. Chinzei, K., Hata, N., Jolesz, F.A., Kikinis, R.: MR compatible surgical assist robot: system integration and preliminary feasibility study. In: International Conference on Medical Image Computing and Computer-Assisted Intervention, pp. 921–930. Springer, Heidelberg (2000)

A Preliminary Study of Upper-Limb Motion Recognition with Noncontact Capacitive Sensing

Enhao Zheng[1(✉)], Qining Wang[2], and Hong Qiao[1,3,4]

[1] The State Key Laboratory of Management and Control for Complex Systems, Institute of Automation, Chinese Academy of Sciences, Beijing 100190, China
enhao.zheng@ia.ac.cn
[2] The Robotics Research Group, College of Engineering, Peking University, Beijing 100871, China
[3] CAS Center for Excellence in Brain Science and Intelligence Technology, Shanghai 200031, China
[4] University of Chinese Academy of Sciences, Beijing 100049, China
http://www.compsys.ia.ac.cn/EN/index.html

Abstract. This study explores the noncontact capacitive sensing strategy on upper-limb motion recognition. The noncontact capacitive sensing system for upper limb comprises a sensing front-end, a sensing circuit and a graphic user interface. The sensing front-end is designed with the thermoplastic band which is surrounded on the forearm. Different from the capacitive sensing methods in previous works for lower limbs, the system in this study measure the muscle contractions without strong physical interaction to the environment. After development of the system, one healthy subjects participated in the experiment. Eight forearm motions, including six basic wrist joint motions and two gestures were investigated. With the selected feature set and the designed classification method, the capacitive sensing strategy produced over 99% recognition accuracy for the eight-motion recognition. The preliminary results demonstrate that the noncontact capacitive sensing approach is a promising solution to the upper-limb motion recognition.

Keywords: Noncontact capacitive sensing · Human upper limb · Motion recognition

1 Introduction

Human-machine interface of upper-limb is one of the major research interest in wearable robots including intelligent robotic prostheses and exoskeletons. One of the key issues in this area is the accurate recognition of human upper-limb motion intent. It is a crucial step towards the smooth and stable control of upper limb wearable robots [1]. The primary part in realizing the human-machine interface is to extract human intent information with wearable sensing techniques. There are several options in terms of sensing positions on human body, i.e. the supraspinal (brain) areas, the peripheral nerves and the muscle skeletal systems [2]. Among

Y. Huang et al. (Eds.): ICIRA 2017, Part I, LNAI 10462, pp. 251–261, 2017.
DOI: 10.1007/978-3-319-65289-4_24

the sensing positions, the brain areas and the peripheral nerves require invasive recordings to obtain the accurate information, the potential risks in which hinder the applications of the sensing methods currently in wearable robotic control [3]. The muscle contractions contains various motion intent information and it can be measured in noninvasive way. Therefore, the signals of muscle contractions are more frequently used in human-machine interface.

Among the muscle contraction signals, the surface electromyogram (EMG) attracts the most attentions of the researchers. As the signal can be measured in noninvasive way and the it is the direct reflection of human motor neural signals from the spinal cord. There are several methods in translating the sEMG signals to specific human motion intentions. Some researchers designed their algorithms based on machine learning techniques, especially in robotic prosthesis control [4–9]. In the study of [6], the authors introduced a myoelectric control system based on a selective multiclass one-versus-one classification scheme, aiming to reject the unknown data patterns in practical use. sEMG signals were sampled from 10 healthy subjects and 5 long transradial amputees. The performance of one-versus-one classification method was compared with nine popular classifiers and the results demonstrated that the one-versus-one significantly outperformed ($p < 0.01$) all other classifiers. In the study of [8], the authors studied the upper-limb myoelectric motion pattern recognition with a generic electrode grid on targeted muscle reinnervation (TMR) patients. In this study, 4 patients who had undergone TMR surgery participated in the experiments, and a generic electrode grid was placed on the reinnervated muscles to cover the control sites. 5 sets of upper limb motions including 29 classes were investigated and the LDA based classifiers were designed. With different evaluation methods, the results showed that grid-like electrode arrangement yields significantly lower classification errors for classifiers with a large number of movement classes (>9).

Some researchers addressed the problems of continuous motion estimation and motion intent decoding based on sEMG signals [10–15]. A few of them studied machine learning methods to continuously decode the upper-limb motion [10,11]. The method is limited by the intense training procedures and specificity of the tasks [13]. Some other researchers introduced models into the motion decoding algorithms to achieve better performance [13–15]. For instance, the authors of [15] introduced a synergy-based approach to control the robotic arm with multiple degrees of freedom (DOFs) using sEMG signals. The synergy model constructs a function between the low-dimensional neural commands and the high-dimensional muscle contraction patterns. In the study [15], the authors validated the method in 2-DOF robotic control tasks on ten able-bodied subjects over days. The results showed that the synergy-based method could achieve more robust performance to co-contraction between antagonist muscles compared to the conventional muscle-pair method.

Above all, the sEMG-based studies took advantage of the muscle contraction information to achieve the human-machine interface. Although lots of works have been done by previous studies, challenges still exist. For the sEMG sensing methods, the electrodes have to be firmly adhesive to the skin. The condition of the skin, such as the sweats, negatively affect the signal quality which further decrease the recognition performance [16]. Secondly, the sEMG signals are affected by the sensing sites of the electrodes [17]. To insure the recognition performance, the sensing positions have to be configured before the usage each time. Additionally, the sEMG signals are weak signals, and they are prone to external noises and disturbances, which increase the expense of the signal processing procedure. Although post-processing algorithms were studied to reduce the negative influence of the signal source, the problems still exist.

In order to extract the neural information for realizing the human-machine interface, some researchers made attempts in other signal sources. For instances, in the studies of [18–22], the researchers proposed a capacitive sensing methods for recognizing the lower-limb motion modes. The methods have been proved to be a promising alternative to sEMG methods. It can provide a solution to human motion sensing if pasting metal electrodes on human skin is not available. However, the capacitive sensing was only applied on lower-limb motion recognition. The feasibility of the methods on upper-limb motion has never been studied or evaluated. There are distinct differences of the muscle contraction patterns between the lower-limb ambulation and the upper-limb motion. The capacitive sensing methods that have been applied on lower-limb motion recognition cannot be directly used for the upper-limb.

In this study, we explored the feasibility of the upper-limb motion recognition with noncontact capacitive sensing approach. We firstly designed a sensing front-end for forearm motion sensing which was dressed on the forearm. We then carried out experiment on a healthy subject and pre-processed the capacitance signals. Thirdly, the characteristics of the signals were analyzed and the motion classification performance was evaluated. With the preliminary study, the feasibility of the capacitive sensing on upper-limb motion sensing and motion recognition is proved.

2 Methods

2.1 Noncontact Capacitive Sensing System for Upper Limb Motion Sensing

We designed a prototype of the sensing system for upper-limb motion sensing. The system consists of one sensing front-end (sensing band), a signal sensing circuit and the data sampling graphic user interface (GUI). The sensing front-end was dressed on the forearm to measure the forearm motions. The sensing circuit was designed to convert the capacitance signals to digital data. After the capacitance signals were sampled, all data were transmitted to the computer via a serial port cable. The GUI was used to receive the sensor data and to control the experiment procedure. The signals were visualized in real time by the GUI and stored in the computer automatically after each trial. As shown in Fig. 1(a), the sensing front-end of the noncontact capacitive sensing system was made of a thermoplastic band which could be reshaped when heated up to 70 °C. The thermoplastic band was reshaped based on the profile of the forearm of the subject. Six copper meshes were fixed on the inner surface of the band, acting as the electrodes of the capacitive sensing system. There was a gap on the sensing band (see Fig. 1(a)) and a bandage was fixed on the gap to adjust the tightness of the band when dressed on human body. The placement of the sensing front-end on the forearm was shown in Fig. 1(b). It was placed at the prominent part of the muscles, about one third of the total length of the forearm to the elbow joint. The gap was on the medial side of the arm (the direction was shown in Fig. 1(a)). In order to avoid the interference to upper limb motion, the wires of the copper electrodes were routed through the small holes of the thermoplastic band. As shown in Fig. 1(b), each copper mesh, the human body and the cloth between them form an equivalent capacitor, with the copper mesh and the human body being the electrodes of the capacitor, the cloth being the dielectric. When there are forearm motions that bring about muscle contractions, the gap between the human body and the copper mesh will change. The change of the distance between the two equivalent electrodes will accordingly change the capacitance values. By recording the capacitance signals, we extract the forearm motion patterns. The sensing circuit was designed to sample the capacitance signals and convert them into digital signals. There was a time-to-digital module on the circuit, and it was used to measure the change-and-discharge cycle time of the equivalent capacitor. By implementing appropriate discharging resistor, the actual capacitance values were calculated. The sampling rate of the capacitance signals were 100 Hz. All the sensor data were synchronized and transmitted to the computer. A GUI was designed to control the experiment procedure, visualize the signals and store the sensor data. The data of each trial were stored in a file for the subsequent analysis.

2.2 Experiment Protocol

In this study, one male healthy subject was employed to take part in the experiment. The subject was 29 years old, 180 cm in height and 75 kg in weight. Written and informed consent were provided by the subject before the experiment. The subject wore the sensing front-end on his right forearm as shown in Fig. 1(b). One copper mesh was placed inside the cloth arbitrarily to make the human body connect with the signal ground. The sensing band was reshaped based on the subject's right forearm. The bandage on the gap of the sensing band was adjusted based on the subjects oral feedback to make the system comfortably dressed on the arm. During the experiment, the subject sat in front of the computer screen with GUI. During the measurement of each time, the GUI provided the raw capacitance signals in real time to the subject. In this study, eight forearm motions were investigated, i.e. wrist flexion/extension, wrist pronation/supination, wrist ulna/radius deviation, fist and palm. Each motion was performed for five trials. For all the motions, the subject started the motion from the neutral position of the forearm, in which the hand was placed vertical to the level ground in relax. For each motion, the subject was asked to perform the task to his maximum extent. In each trial, after the measurement began

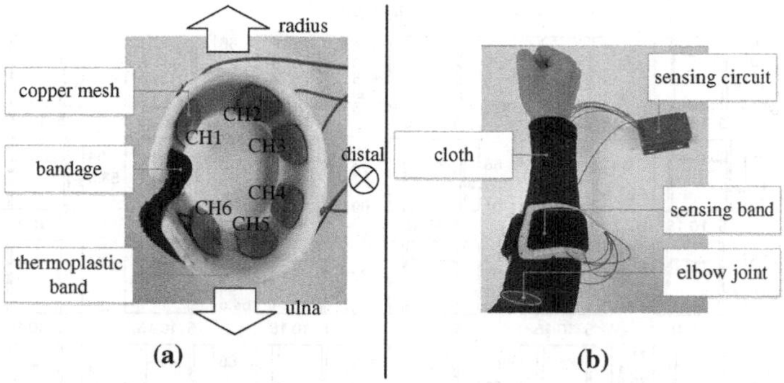

Fig. 1. The sensing front-end of the noncontact capacitive sensing system and the placement on human forearm. (a) The sensing front-end of the system. The white band was the thermoplastic band which was reshaped based on the subject's forearm. The copper meshes were fixed on the inner surface of the band with 0.1 mm thickness. The wires (blue) were routed through the holes of the thermoplastic band. The black part was the bandage to adjust the tightness. CHx (x =1, 2,... ,6) represent the channel number of the electrode. (b) The placement of the sensing band on the right forearm. There was cloth between the band and human skin to insulate the electrodes from the human skin. (Color figure online)

(controlled by the GUI panel), the subjects wait for 2–3 s with his right forearm in neutral position, then the subject performed the indicated task (motion) for about 15 s then came back to the neutral position for another few seconds. Each trial lasted 20 s. There was short-time (2 min) rest in the middle of the experiment.

2.3 Data Processing

After the experiment, there were 40 groups of data collected. We firstly filtered the data with a 4th-order Butterworth low band pass filter to remove the high-frequency noises. The cut-off frequency was 15 Hz. Compared with our previous works [20,21], the notch filter was not used in this study. In the experiment, the upper limb motions were not periodic motions as that of lower limb ambulation. The baselines of the signals contains motion information. We therefore kept the baselines of the capacitance signals. In this study, we used machine learning methods to recognize the upper limb motion patterns. The sliding windows were used to segment the filtered data [21]. The window length was 200 ms and the overlap was 10 ms (one sample). On each sliding window, the features that convey motion information were calculated for all the channels (six signal channels in total in this study). As there were no previous studies that involve upper limb motion recognition with noncontact capacitive sensing, we made initial trials and checked the feature distributions on phase plots (details see below). After the features were determined and calculated, they were combined in series to make up a feature vector. The feature vector served as the input to the subsequent

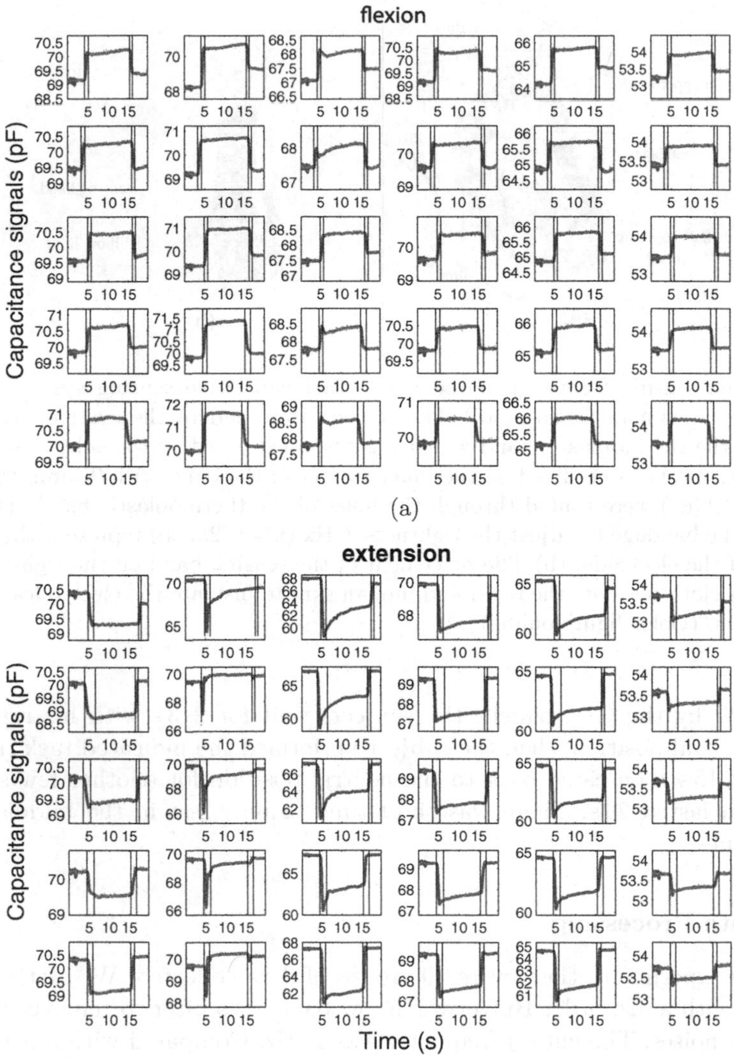

(b)

Fig. 2. The capacitance signals of the wrist flexion (a) and the wrist extension (b). In the figures, each row represents a experiment trial, while each column represents a signal channel. The red dotted lines in the subplots are the raw capacitance signals, and the blue lines are the filtered signals. The vertical black lines are the labeled timing points that the motion occurs. (Color figure online)

classifier for training and testing. In this study, the classifier was quadratic discriminant analysis (QDA) classifier. The QDA classifier is a type of Gaussian discriminant analysis. The contribution of the data for each class (motion) was consumed to be multivariate Gaussian distribution. The mean vector and the covariance matrix of each class were fitted based on the training data. The QDA classifier was efficient in computation while produce accurate recognition results [6,7]. In each trial, the data include both the neutral position and the specific motion pattern. We manually labeled the data based on the signal changes (see the black lines of Fig. 2).

2.4 Evaluation Method

In this study, we used N-fold cross-validation (LOOCV) to train and test the classification method. In LOOCV, the data of one fold served as the testing set, while the rest data were used for training. The procedure was repeated for N times until all the data were used for the testing set. In this study, we measured five trials of data for each upper-limb motion pattern. N was therefore set to be five. Each sliding window could produce a classification judgment. The overall recognition error (RE) was defined as:

$$RE = \frac{N_{mis}}{N_{total}} \times 100\% \quad (1)$$

where N_{mis} is the number of mis-recognized testing data and N_{total} is total number of testing data. We separately evaluated the gait initiation and termination and ambulation modes. For different upper-limb motion pattern evaluation, we used confusion matrix to illustrate the recognition performance of certain motion patterns. Detailed presentation of confusion matrix could be found in [20].

3 Results

3.1 Capacitance Signal Processing

We firstly checked the raw capacitance signals of all the upper-limb motions. As shown in Fig. 2, the blurs in the raw capacitance signals were removed by the filter. There were clear changes between the neutral position and the motion pattern. Taking the two motion patterns that were shown in Fig. 2 as an example, the differences of the signal amplitude were obvious, which was benefit for the subsequent analysis.

3.2 Feature Calculation

As there were no previous experiences on capacitive sensing for upper-limb motion recognition, we used phase plots to evaluate the possible features. We chose $AVE(x)$, $STD(x)$, $TAN(x)$, $MAX(x)$, $MIN(x)$ and $MAX(x)-MIN(x)$ as the candidates, where x represents the data of each sliding window. $AVE(x)$ was the average value of the data in each sliding window. $STD(x)$ was the standard deviation. $TAN(x) = (x_{end} - x_1)/L$, where x_{end} was the latest data that coming into the sliding window and x_1 was the eldest one. L was the length

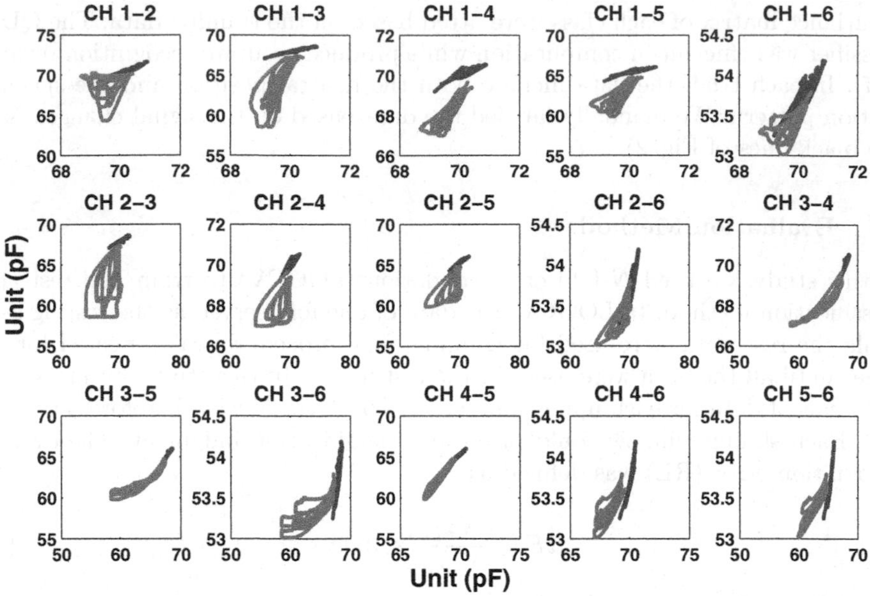

Fig. 3. The phase plot of the capacitance features for wrist flexion and extension. Each sub-figure represents one combination of two signal channels, as shown in the subplot title. In the subplots, the blue dots stand for the data of flexion, while the red dots the extension. In this figure, we took $AVE(x)$ as an example. (Color figure online)

of the sliding window. $MAX(x)$ and $MIN(x)$ were the maximum and minimum of the data of the sliding window, respectively. The phase plots can reveal the relationships of the two variables. Here the variables are the features that are calculated for two capacitance signals. We took all the combinations of the channel-pairs (15 in total) and compared the phase plots of different motion patterns to evaluate their distribution. The phase plots of $AVE(x)$ for wrist flexion/extension were demonstrated in Fig. 3. As shown in Fig. 3, the distribution of $AVE(x)$ for the two motion patterns can be separated easily. We checked all the feature candidates and all the motion patterns. Finally, we chose $AVE(x) + STD(x) + TAN(x) + (MAX(x) - MIN(x))$ as the feature set for the subsequent recognition.

3.3 Recognition Results

With the selected feature set, the QDA classifier and 5-fold LOOCV, the capacitive sensing system produced an accurate recognition results. The average recognition accuracy for all the eight motion patterns were 99.9%. The detailed recognition accuracies for each motion patterns were shown in Table 1. For all the motion patterns except palm, the approach could achieve 100% recognition accuracy. The lowest accuracy was 99.3% for palm and 0.7% of them were misclassified as fist.

Table 1. Confusion matrix (mean) for eight motion patterns with QDA classifier (%)

Targets	Estimation motions							
	flexion	extension	pronation	supination	ulna_d	radius_d	fist	palm
flexion	100.0	0.0	0.0	0.0	0.0	0.0	0.0	0.0
extension	0.0	100.0	0.0	0.0	0.0	0.0	0.0	0.0
pronation	0.0	0.0	100.0	0.0	0.0	0.0	0.0	0.0
supination	0.0	0.0	0.0	100.0	0.0	0.0	0.0	0.0
ulna_d	0.0	0.0	0.0	0.0	100.0	0.0	0.0	0.0
radius_d	0.0	0.0	0.0	0.0	0.0	100.0	0.0	0.0
fist	0.0	0.0	0.0	0.0	0.0	0.0	100.0	0.0
palm	0.0	0.0	0.0	0.0	0.0	0.0	0.7	99.3

4 Discussion and Conclusion

In this study, we addressed the problems of noncontact capacitive sensing strategy on upper-limb motion recognition. The designed sensing front-end could effectively record the forearm muscle contractions in different motion patterns (see Fig. 2). With the selected feature set and a simple classifier, the system produced accurate results in motion recognition for eight forearm motions. Compared with the existing sEMG-based studies [6,7], the noncontact capacitive sensing approach yielded similar results in recognition of the motion tasks of the experiment (see Table 1). The results of this preliminary study proved the feasibility of the capacitive sensing on upper-limb motion recognition and suggested a promising alternative solution to muscle signal measurement.

The study only involved with one able-bodied subject and several simple forearm motion patterns. There are three main limitations in this study. Firstly, the noncontact capacitive sensing method was only evaluated on one able-bodied subject. In the experiment, the sensing band was reshaped purely based on the shape of the subject's forearm. Variations on arm shape and muscle contraction patterns exist among different individuals. The repeatability of the capacitance signals, the feature distributions and the recognition performance have yet to be evaluated. Secondly, only several basic wrist joint motion patterns were investigated in the experiment. More complicated motions such as more gestures and the combinations of the joint motions have not been studied. Thirdly, the tasks only involved with static motions. The influences of motion transitions, the unpredicted motion patterns and the random environmental disturbances on recognition results were not analyzed.

Nevertheless, the capacitive sensing on upper limb is a promising issue that worth being exploited. Future works will be carried out on the following aspects. First of all, experiments on more complicated motions including transitions and unpredicted motions will be carried out, and more subjects will be recruited to take part in the experiments. More recognition algorithms will be designed and evaluated. Secondly, the effects of the environmental disturbances, such as re-wearing the sensing band will be studied. The optimization of the methods and the improvement of the system will be made based on the experiment results. Thirdly, the existing capacitive sensing based motion recognition is designed with the machine learning techniques, both for the lower limb and upper limb. Future studies will be made on the exploration of the physical significance of the capacitance signals in muscle contraction measurement. Intuitive motion recognition with capacitive sensing will be explored.

Above all, the preliminary study broaden the application area of the capacitive sensing on human body sensing. The capacitive sensing strategy proved the feasibility of the method on upper limb human-machine interfaces. The experiment results also demonstrated that the possibility of the approach for wearable robotic control. Future endeavors will be made on more extensive experiments and more complex motion tasks.

Acknowledgements. This work was supported by the National Natural Science Foundation of China (NO. 91648207).

References

1. Pons, J.L.: Wearable Robots: Biomechatronic Exoskeletons. Wiley, Hoboken (2008)
2. Farina, D., Jiang, N., Rehbaum, H., Holobar, A., Graimann, B., Dietl, H., Aszmann, O.C.: The extraction of neural information from the surface EMG for the control of upper-limb prostheses: emerging avenues and challenges. IEEE Trans. Neural Syst. Rehabil. Eng. **22**(4), 797–809 (2014). doi:10.1109/TNSRE.2014.2305111
3. Nicolas-Alonso, L.F., Gomez-Gil, J.: Brain computer interfaces, a review. Sensors **12**(2), 1211–1279 (2012). doi:10.3390/s120201211
4. Huang, Y., Englehart, K.B., Hudgins, B., Chan, A.D.: A Gaussian mixture model based classification scheme for myoelectric control of powered upper limb prostheses. IEEE Trans. Biomed. Eng. **52**(11), 1801–1811 (2005). doi:10.1109/TBME.2005.856295
5. Oskoei, M.A., Hu, H.: Support vector machine-based classification scheme for myoelectric control applied to upper limb. IEEE Trans. Biomed. Eng. **55**(8), 1956–1965 (2008). doi:10.1109/TBME.2008.919734
6. Scheme, E.J., Englehart, K.B., Hudgins, B.S.: Selective classification for improved robustness of myoelectric control under nonideal conditions. IEEE Trans. Biomed. Eng. **58**(6), 1698–1705 (2011). doi:10.1109/TBME.2011.2113182
7. Young, A.J., Hargrove, L.J., Kuiken, T.A.: Improving myoelectric pattern recognition robustness to electrode shift by changing interelectrode distance and electrode configuration. IEEE Trans. Biomed. Eng. **59**(3), 645–652 (2012). doi:10.1109/TBME.2011.2177662
8. Tkach, D.C., Young, A.J., Smith, L.H., Rouse, E.J., Hargrove, L.J.: Real-time and offline performance of pattern recognition myoelectric control using a generic electrode grid with targeted muscle reinnervation patients. IEEE Trans. Neural Syst. Rehabil. Eng. **22**(4), 727–734 (2014). doi:10.1109/TNSRE.2014.2302799
9. Zhu, X., Liu, J., Zhang, D., Sheng, X., Jiang, N.: Cascaded adaptation framework for fast calibration of myoelectric control. IEEE Trans. Neural Syst. Rehabil. Eng. **25**(3), 254–264 (2017). doi:10.1109/TNSRE.2016.2562180

10. Artemiadis, P.K., Kyriakopoulos, K.J.: EMG-based teleoperation of a robot arm in planar catching movements using ARMAX model and trajectory monitoring techniques. In: Proceedings of 2006 IEEE International Conference on Robotics and Automation (ICRA), pp. 3244–3249. IEEE Press, Orlando (2006). doi:10.1109/ROBOT.2006.1642196
11. Artemiadis, P.K., Kyriakopoulos, K.J.: An EMG-based robot control scheme robust to time-varying EMG signal features. IEEE Trans. Inf. Technol. Biomed. **14**(3), 582–588 (2010). doi:10.1109/TITB.2010.2040832
12. Vogel, J., Castellini, C., Van Der Smagt, P.: EMG-based teleoperation and manipulation with the DLR LWR-III. In: 2011 IEEE/RSJ International Conference on Intelligent Robots and Systems (IROS), pp. 672–678. IEEE Press, San Francisco, (2011). doi:10.1109/IROS.2011.6094739
13. Antuvan, C.W., Ison, M., Artemiadis, P.: Embedded human control of robots using myoelectric interfaces. IEEE Trans. Neural Syst. Rehabil. Eng. **22**(4), 820–827 (2014). doi:10.1109/TNSRE.2014.2302212
14. Ikemoto, S., Kimoto, Y., Hosoda, K.: Surface EMG based posture control of shoulder complex linkage mechanism. In: 2015 IEEE/RSJ International Conference on Intelligent Robots and Systems (IROS), pp. 1546–1551. IEEE Press, Hamburg (2015). doi:10.1109/IROS.2015.7353573
15. Lunardini, F., Casellato, C., d'Avella, A., Sanger, T.D., Pedrocchi, A.: Robustness and reliability of synergy-based myocontrol of a multiple degree of freedom robotic arm. IEEE Trans. Neural Syst. Rehabil. Eng. **24**(9), 940–950 (2016). doi:10.1109/TNSRE.2015.2483375
16. Sensinger, J.W., Lock, B.A., Kuiken, T.A.: Adaptive pattern recognition of myoelectric signals: exploration of conceptual framework and practical algorithms. IEEE Trans. Neural Syst. Rehabil. Eng. **17**(3), 270–278 (2009). doi:10.1109/TNSRE.2009.2023282
17. Young, A.J., Hargrove, L.J., Kuiken, T.A.: The effects of electrode size and orientation on the sensitivity of myoelectric pattern recognition systems to electrode shift. IEEE Trans. Biomed. Eng. **58**(9), 2537–2544 (2011). doi:10.1109/TBME.2011.2159216
18. Zheng, E., Chen, B., Wei, K., Wang, Q.: Lower limb wearable capacitive sensing and its applications to recognizing human gaits. Sensors **13**(10), 13334–13355 (2013). doi:10.3390/s131013334
19. Chen, B., Zheng, E., Fan, X., Liang, T., Wang, Q., Wei, K., Wang, L.: Locomotion mode classification using a wearable capacitive sensing system. IEEE Trans. Neural Syst. Rehabil. Eng. **21**(5), 744–755 (2013). doi:10.1109/TNSRE.2013.2262952
20. Zheng, E., Wang, L., Wei, K., Wang, Q.: A noncontact capacitive sensing system for recognizing locomotion modes of transtibial amputees. IEEE Trans. Biomed. Eng. **61**(12), 2911–2920 (2014). doi:10.1109/TBME.2014.2334316
21. Zheng, E., Wang, Q.: Noncontact capacitive sensing-based locomotion transition recognition for amputees with robotic transtibial prostheses. IEEE Trans. Neural Syst. Rehabil. Eng. **25**(2), 161–170 (2017). doi:10.1109/TNSRE.2016.2529581
22. Zheng, E., Manca, S., Yan, T., Parri, A., Vitiello, N., Wang, Q.: Gait phase estimation based on noncontact capacitive sensing and adaptive oscillators. IEEE Trans. Biomed. Eng. (2017). doi:10.1109/TBME.2017.2672720

Multi-class SVM Based Real-Time Recognition of Sit-to-Stand and Stand-to-Sit Transitions for a Bionic Knee Exoskeleton in Transparent Mode

Xiuhua Liu[1], Zhihao Zhou[1], Jingeng Mai[1], and Qining Wang[1,2(✉)]

[1] The Robotics Research Group, College of Engineering, Peking University, Beijing 100871, China
qiningwang@pku.edu.cn

[2] Beijing Innovation Center for Engineering Science and Advanced Technology (BIC-ESAT), Peking University, Beijing 100871, China

Abstract. Real-time locomotion intent recognition is a challenge in lower-limb exoskeletons. In this paper, we present a multi-sensor based locomotion intent prediction system for sit-to-stand and stand-to-sit transition recognition the subject wears a knee exoskeleton in transparent mode. The desired reference torque for movement control is obtained from a direct torque control loop. The feedback torque is estimated by an inner current control loop. Five able-bodied subjects were recruited in experiments. The classifier is based on multi-class Support Vector Machine. Four kinds of modes and four kinds of transitions are tested in this study. Recognition accuracy during steady periods is $99.68\% \pm 0.07\%$ for five able-bodied subjects. And during transition periods, all the transitions are correctly detected and no missed detections was observed for all the trials of the five subjects. Preliminary experimental results show that the proposed method is capable of performing real-time intent recognition and consequently reduces the interaction force between human body and the exoskeleton.

Keywords: Locomotion intent prediction · Multi-class SVM · Gait transitions · Knee exoskeleton

1 Introduction

Since population ageing is taking place in most major areas of the world, it requires more assistive devices to meet the needs of the elderly people in daily activities. In most cases, lower-limb exoskeletons should provide the assistance only when the assistance is needed for the motion and should not cause interference to the wearer. To achieve this goal, transparent mode is widely-used in lower-limb exoskeletons, where the lower-limb exoskeleton has no interaction with the wearer and no corrective force is being applied [1]. However, it is difficult to achieve transparency since the inertia and the mechanical friction the

Y. Huang et al. (Eds.): ICIRA 2017, Part I, LNAI 10462, pp. 262–272, 2017.
DOI: 10.1007/978-3-319-65289-4_25

lower-limb exoskeleton itself which must be compensated by the controller leading to large interaction forces between the wearer and the lower-limb exoskeleton. Zanotto et al. tried to use closed loop controllers which can mask the inertia and friction of the actuator when disturbances are inside the closed-loop rejection bandwidth [1]. Other studies decrease interaction forces by specific mechanical structures [2–4] and certain algorithms [5,6].

As one of the most demanding functional tasks in daily life, sit-to-stand and stand-to-sit (STS) transitions may not be performed smoothly because of aging, stroke or other neurological impairments [7]. Consequently, physicians and physiotherapists emphasize STS training for patients with leg injuries, e.g. cerebrovascular accident. Robotic systems, such as lower-limb exoskeletons, can provide additional assistance to the affected limb to restore functional activities. Then the first challenge is how to recognize human intent. Several studies proposed locomotion pattern recognition for sitting, standing and STS with different sensors [8–14]. However, few results are reported for subjects wearing real exoskeletons. Because of the physical interaction between the exoskeleton and the human body, the recognition method may be different when the subject wears a real exoskeleton.

To solve these problems, our previous studies proposed a recognition method about STS based on the multi-sensor fusion information of interior sensors of a light-weight bionic knee exoskeleton (BioKEX) [15] and analyzed STS in off-line mode [17]. However, real-time recognition of STS with the knee exoskeleton is still a challenge. In this paper, to decrease the burden of the exoskeleton inertia on the wearer, we proposed an online recognition method which could successfully recognize four locomotion modes and four locomotion transitions about STS based on the multi-sensor fusion information of interior sensors of a light-weight BioKEX by using interaction torques at the interfaces between the robot and the user's leg to generate a feedback close loop on the joint actuators. Signals of angle and three inertial measurement units (IMUs) are measured. In addition, we propose a method based on multi-class Support Vector Machine to avoid the time-consuming and sensitive error mentioned above and improve recognition performance. Before online test, we had trained the classifier model in offline mode for each subject. Therefore, we did not need time to train classifier mode when conducting online test which saved much time compared with [16]. We used leave-one-out-cross-validation (LOOCV) when training the classifier model. In experiments, we evaluated recognition performance in steady periods and transition periods individually which were essential for clinical application. Furthermore, we compared the desired reference torque and the feedback torque, and analyzed the angle and angular velocity to evaluate the transparency.

2 Methods

2.1 Bionic Knee Exoskeleton

We used the wearable exoskeleton for knee joint assistance developed by our group [15] (see Fig. 1), to conduct the experiments. It consists of a sync-drive, a symmetric wearable structure, linkage mechanism, a belt driving and a inline rotation encoder for speed control. The BioKEX works as follows. The brusherless DC motor fixed on the bracket which connects with the thigh part drives the pulley to work then the pulley drives the nut upward movement to provide the wearer with power when the wearer stretches out the leg; the slider drives the nut down to give the wearer resistance when the wearer bends the leg.

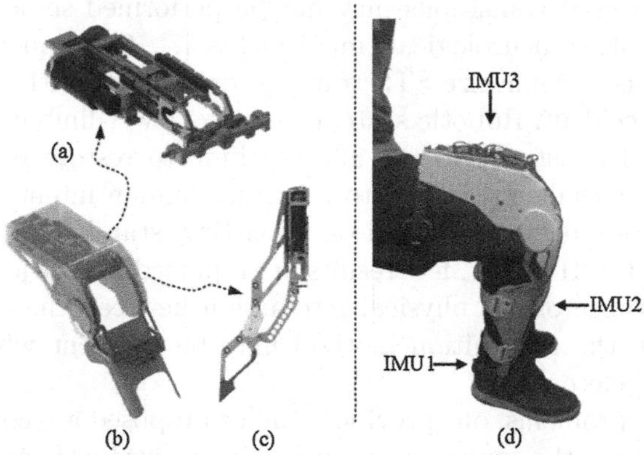

Fig. 1. (a) The internal structure of the knee exoskeleton; (b) the external structure of the knee exoskeleton; (c) the lateral linkage mechanism of the knee exoskeleton; (d) the bionic knee exoskeleton prototype. A subject with the BioKEX worn in initial position with three inertial measurement units (one on the thigh, one on the shank and the other on the foot).

2.2 Measurement System

In order to obtain as much useful locomotion information as possible, the positions of sensors were carefully selected (see Fig. 1(d)). A potentiometer was installed at the knee which can record its displacement and can be used to indirectly calculate the knee joint angle and angular velocity. Three IMUs (SparkFun Electronics Inc., Boulder, CO, USA) were placed on the front of thigh part, shank part of exoskeleton and the back of the shoe to measure accelerations and gyroscopes along two perpendicular axes in the sagittal plane. To measure the interaction force between the wearer and exoskeleton, one bilateral load cell (FUTEK Inc.) is mounted on the push rod. The measurement system is described as Fig. 2. The IMU board is built with an gyroscope for estimating the orientation of the IMU board, an accelerometer and a magnetometer for complementing errors of the orientation. Force and position information are fed into controller with the sampling rate at 50 Hz. STM32F103 is used as the micro control unit (MCU) of the IMU board for data collection. From the measurement system we can see the process of data transmission: Signals collected by MCU were transmitted to the receiver circuit from the control circuit with the serial port. The J-Link can download the program to the MCU (STM323F407) the main frequency of whom was 128 M. Brushless motor driver was utilized with rated voltage 24 V and rated current 30 A. The type of motor we used was Brushless EC with rated output power 200 W.

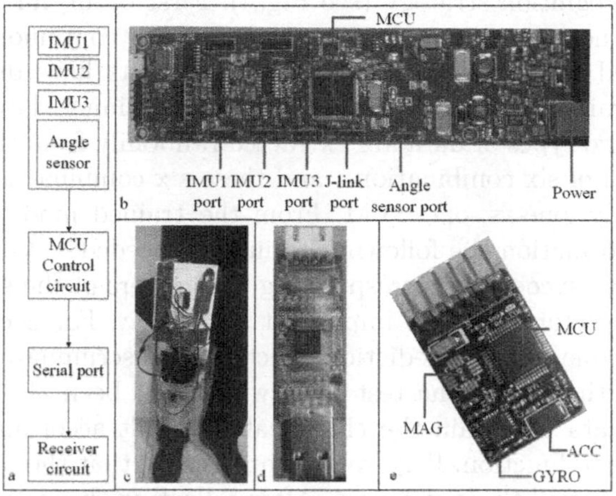

Fig. 2. (a) The structure of data transmission in the designed system; (b) MCU with main frequency 128 MHZ Control circuit used in this study; (c) the BioKEX designed by our group with a subject on; (d) the J-Link; (e) IMU board embedded with an accelerometer (ACC), a gyroscope (GYRO) and a magnetometer (MAG).

2.3 Control System

Figure 3 shows our control strategy: it consists of a direct torque loop and a current loop. The feedback torque is the input to the torque controller and the output is the reference signal for current controller. The torque control loop determines the desired torque of the exoskeleton. Therefore, the desired torque will be force controller and there is no prescribed reference signal. Therefore, U_d, which is the desired voltage for the motor, can be derived according to the torque control algorithm. After obtaining the desired voltage, an inner current controller is employed.

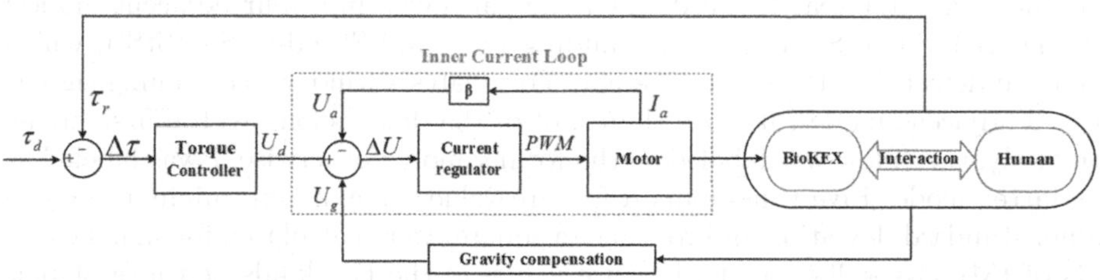

Fig. 3. Overall control strategy

2.4 Recognition Algorithm

Signals of angle and IMUs are quasi-cyclic and varied with modes. To decrease variances of feature values within class and improve recognition performance, we design the recognition strategy (see Fig. 4). First of all, for offline classifier training model, one gait cycle was segmented into four continuous modes which could be detected with angle signal. Different subjects have different models. The first step training model was to classify the training data based on the labeled results: two types of data are extracted randomly from the four types of data, with a total of six combinations, and these six combinations were trained separately, namely, one-vs.-one SVM. From the trained model, we can know the necessary information the following online test needed: totalSV-the number of support vectors, svcoef-the corresponding parameters of the support vectors, SVs-all support vectors, rho-one important parameter. For each combination, we had designed one unique prediction function to discriminate the two different modes. And then enter the test data which had been classified the same as the training data to obtain the classification result according to the corresponding prediction function. It is worth emphasizing that the modes should be detected reliably for all the gait cycles. After collecting the signals of angle and IMUs, angle signal was used to label the gait cycle and feature values were calculated. Classification error usually tended to decrease as the number of feature values increased. However, larger feature set dimension increased computation burden, and recognition performance could even deteriorate when inappropriate feature values were added to the feature set. We used greedy algorithm to sequentially add new feature values to the feature set. Finally, a minimal set of relevant biomechanical features were selected. Since feature sets may be different, each subject had a certain feature set. According to the optimal feature set, we write down the selected signal channels and the corresponding time-domain feature values for the following online test. We modified the SVM to apply to our problems. For online classifier training, first, the optimal feature value of current gait mode was calculated and the corresponding classifier was selected. After training models, we conducted the online experiments. We have mentioned that each subject had a unique feature set. Therefore, after collecting sensory signals, we calculated feature values according to the unique feature set during the online experiments. Then the recognition result was obtained with the classifier. In our research, we divided the gait cycle into four adjacent modes: Sitting (SI), Sit-to-Stand (SiSt), Standing (ST) and Stand-to-Sit (StSi), which could be detected with signals of angle. For analysis windows containing signals of two adjacent modes, if more than half of the data belong to the first mode, the analysis window was labeled as the former mode; otherwise it was labeled as the latter mode. Five time-domain feature values (maximum, minimum, mean value, standard deviation and root mean square) were calculated for signal channels of IMUs as well as angle. Feature values of the two kinds of sensor signals were combined together to generate the feature set (see Fig. 4). Support Vector Machine (SVM) classifier was selected as the classifier for all the four modes considering the computational cost of the processing unit in exoskeleton.

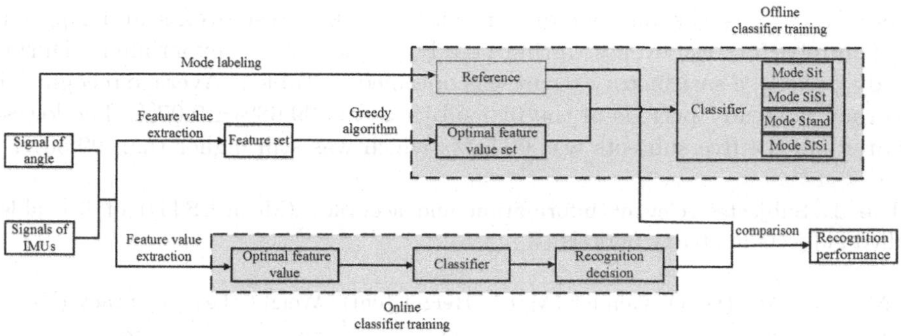

Fig. 4. Block diagram of the locomotion intent prediction system.

2.5 Experiment Protocol

The subjects relevant information in this study was listed in Table 1 and subjects were provided informed written consent before the experiment. During the experiment, the subject wore the exoskeleton BioKEX on the right leg and performed a series of activities. The experiment consisted of 60 trials and there was no interval between every two trials. We considered that the subject was seated, where the initial position of leg was bent a right angle, as a starting point for the experiment. Then the subject sat for 10 s, stood up, stood for 10 s, and sat down during the experiment. In order to make a reasonable assessment of the performance of the classification, a locomotion period was divided into a steady locomotion period and a locomotion transition periods. And we defined four critical moments, namely boundary, for each cycle. The first critical moment was "SiSt_On", the moment at the beginning of SiSt; The second critical moment was "SiSt_Off", the moment at the end of SiSt; The third critical moment was "StSi_On", the moment at the beginning of StSi; The fourth critical moment was "StSi_Off", the moment at the end of StSi. And the boundary between SI and SiSt was "SiSt_On", the boundary between SiSt and ST was "SiSt_Off", the boundary between ST and StSi was "StSi_On", and the boundary between StSi and SI was "StSi_Off". To obtain reliable experimental results, we used leave-one-out-cross-validation (LOOCV) when training the classifier model. We had a statistical calculation with a stopwatch that one transition process took about 2 s.

3 Experimental Results

3.1 Recognition Accuracy in Steady Locomotion Periods

We calculated the average classification accuracy (ACA) during steady locomotion periods by

$$ACA = N_{correct}/N_{total} * 100\% \quad (1)$$

where $N_{correct}$ is the number of correctly classified test events and N_{total} is total number of test events during steady periods of the experiment. During steady periods, a satisfactory result was obtained in Table 1. Average recognition accuracy in steady periods of the five subjects was 99.68% ± 0.07%. The lowest accuracy of the five subjects was 99.59%, which was still higher than 99.50%.

Table 1. Subjects' relevant information and accuracy (Mean ± STD) of five able-bodied subjects in steady periods (%)

No	Age (year)	Gender (M/F)	Height (cm)	Weight (kg)	Accuracy (%)
1	23	F	163	50	99.59
2	27	M	172	60	99.65
3	27	M	173	64	99.68
4	30	M	180	80	99.74
5	30	M	170	60	99.77
Average	27 ± 3	NULL	171 ± 6	62 ± 11	99.68 ± 0.07

3.2 Recognition Performance in Locomotion Transition Periods

To evaluate the recognition performance in transition periods, we calculated prediction time and numbers of missed detections (see Table 2) during transitions. For transitions from one locomotion mode to another, we defined the critical moment above. There are two rules a correct locomotion transition should follow. First, more than 30 consecutive correct recognition decisions were made. Second, after the first condition was satisfied, no false decisions occurred in the remaining transition period [20]. We calculated the prediction time (*PT*) by

$$PT = t_c/t_{pre} \tag{2}$$

where t_c is the critical moment of the transition, t_{pre} is the moment of the first correct decision. All transitions could be detected (see Fig. 5). In addition, no missed detections was observed for all the trials of the five able-bodied subjects (see Table 2).

Table 2. Performance of transition detection (Mean ± SEM) for five able-bodied subjects. Note: a negative value of prediction time indicates the decision is made after the critical moment.

Transition	Prediction time (ms)	Number of missed detections
Stand-StSi	−202 ± 19	0
StSi-Sit	−71 ± 21	0
Sit-SiSt	−266 ± 8	0
SiSt-Stand	343 ± 44	0

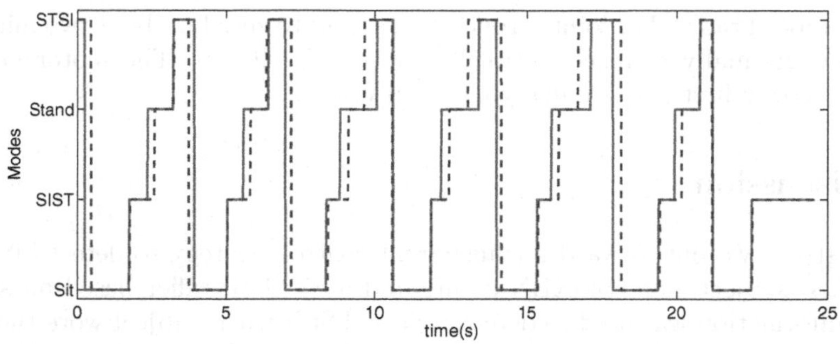

Fig. 5. Representative result of recognition performances of online test experiments. The dash line denotes decision stream made by the locomotion intent recognition system. The solid line denotes the offline reference for the same online test experiments.

3.3 Interaction Torques

In the experiment of transparent mode control, we first set the knee angle of the exoskeleton to zero and use our strategy to achieve its zero-torque function. In the test, after wearing the exoskeleton correctly, the subject began to accomplish the transitions between standing and sitting slowly. After a period of time (about two minutes), the subject began to accomplish the transitions between standing and sitting quickly. The signal of load cell of the exoskeleton, angle and the current of the motor were sampled and evaluated the performance and results of zero torque control are shown in (Fig. 6). The joint angle values were always between 0° and 90° regardless of the speed of the subject. For joint speed, it was close to zero for about two minutes, and then gradually changed between

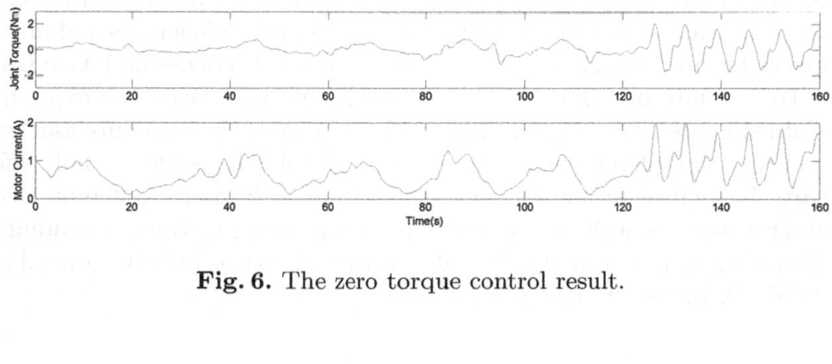

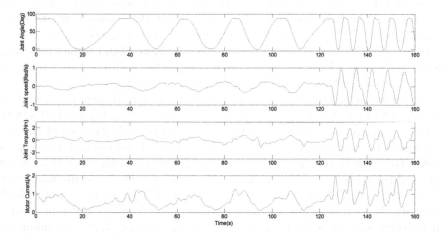

Fig. 6. The zero torque control result.

1 rad/s and −1 rad/s. For joint torque, it was close to zero for about two minutes, and then gradually changed between 2 N.m and −2 N.m. The motor current changed slowly first and gradually changed fast.

4 Discussion

In this study, we put forward a transparent control strategy to detect SiSt and StSi when subject wore BioKEX online and a SVM classifier based on simple sensor information was used to detect SiSt and StSi when subject wore BioKEX and the strategy is able to predict the intent of the wearer in real time. We have solved the problems that the recognition expended much time and the error was sensitive. To this end, we trained the classifier model in offline mode for each subject before online test. Therefore, we did not need time to train classifier mode when conducting online test which saved much time compared with [16]. To reduce the sensitivity of the error and obtain reliable experimental results, we used leave-one-out-cross-validation (LOOCV) when training the classifier model. Recognition accuracy, prediction time and zero control result were used to evaluate the recognition performance. The method could recognize locomotion modes online with high accuracy and all of them can be detected. Our recognition method is promising to improve the performance of the STS for the following reasons. First, in order to make the proposed method meet daily life's needs appropriately, we verified this method with sensors on BioKEX not directly just on subject. Second, we obtained a satisfactory result: Average recognition accuracy was 99.68% ± 0.07%, however most of other studies is about 98.5% [18,19]. Besides, to decrease computation burden and assure the recognition performance of test data, we utilize greedy algorithm to reduce the dimension of feature value. Importantly, our method could detect all transitions though not all of them were in advance. As mentioned above, it took about 2 s when finishing a transition process and the longest delay time occurred in the transition of SI → SiSt which was 309 ms. Therefore, our delay time could meet the requirements to detect the transitions. Finally, our strategy can decrease the interaction force between the wearer and the exoskeleton during Sit-to-Stand and Stand-to-Sit transitions, namely it has a good transparency. From the results of all the five subjects we can say that our method is reliable to detect the transitions.

The proposed method for recognizing SiSt and StSi posture transitions online showed high detection performance using only the embedded IMUs in the exoskeleton in transparent mode. However, there exists deficiencies and improvement spaces for our recognition method. First, we test our method with healthy people. To evaluate our method more reliably, we will carry out experiments with disabled people. Second, our method is used to distinguish only four modes which can not completely meet the needs of disabled people in daily life. In the future, We will continue current study to classify many different locomotion patterns such as walking, ascending and descending stairs, ascending and descending slopes, and even at different speeds with other assistive control strategy of BioKEX to test its recognition performance.

Acknowledgments. This work was supported by the National Natural Science Foundation of China (No. 91648207), the Beijing Municipal Science and Technology Project (No. Z151100000915073), and the Beijing Nova Program (No. Z141101001814001).

References

1. Zanotto, D., Lenzi, T., Stegall, P., Agrawal, S.K.: Improving transparency of powered exoskeletons using force/torque sensors on the supporting cuffs. In: Proceedings of the IEEE International Conference on Rehabilitation Robotics, Seattle, pp. 1–6 (2013)
2. Zanotto, D., Stegall, P., Agrawal, S.K.: ALEX III: a novel robotic platform for gait training-design of the 4-DOF leg. In: Proceeidngs of the IEEE International Conference on Robotics and Automation, Karlsruhe, pp. 3914–3919 (2013)
3. Veneman, J.F., Kruidhof, R., Hekman, E.E.G., Ekkelenkamp, R., Van Asseldonk, E.H.F., van der Kooij, H.: Design and evaluation of the LOPES exoskeleton robot for interactive gait rehabilitation. IEEE Trans. Neural Syst. Rehabil. Eng. **15**(3), 379–386 (2007)
4. Bae, J., Kong, K., Tomizuka, M.: A cable-driven human assistive system and its impedance compensation by sensor fusion. In: Proceedings of the ASME Dynamic Systems and Control Conference, Cambridge, pp. 777–783 (2010)
5. Kazerooni, H., Racine, J.L., Huang, L., Steger, R.: On the control of the Berkeley lower extremity exoskeleton (BLEEX). In: Proceedings of the IEEE International Conference on Robotics and Automation, Barcelona, pp. 4353–4360 (2005)
6. Khosravi, S., Arjmandi, A., Taghirad, H.D.: Sliding impedance control for improving transparency in telesurgery. In: Second RSI/ISM International Conference on Robotics and Mechatronics, Tehran, pp. 209–214 (2014)
7. Kralj, A., Jaeger, R.J., Munih, M.: Analysis of standing up and sitting down in humans: definitions and normative data presentation. J. Biomech. **23**, 1123–1138 (1990)
8. Masse, F., Gonzenbach, R., Paraschiv-Ionescu, A., Luft, A., Aminian, K.: Wearable barometric pressure sensor to improve postural transition recognition of mobility-impaired stroke patients. IEEE Trans. Neural Syst. Rehabil. Eng. **24**(11), 1210–1217 (2016)
9. Van Lummel, R.C., Ainsworth, E., Lindemann, U., Zijlstra, W., Chiari, L., Van Campen, P., Hausdorff, J.M.: Automated approach for quantifying the repeated sit-to-stand using one body fixed sensor in young and older adults. Gait Posture **38**(1), 153–156 (2013)
10. Rodríguez-Martín, D., Samà, A., Pérez-López, C., Cabestany, J., Català, A., Rodrguez-Molinero, A.: Posture transition identification on PD patients through a SVM-based technique and a single waist-worn accelerometer. Neurocomputing **164**, 144–153 (2015)
11. Chen, B., Wang, X., Huang, Y., Wei, K., Wang, Q.: A foot-wearable interface for locomotion mode recognition based on discrete contact force distribution. Mechatronics **32**, 12–21 (2015)
12. Taslim Reza, S.M., Ahmad, N., Choudhury, I.A., Ghazilla, R.A.R.: A fuzzy controller for lower limb exoskeletons during sit-to-stand and stand-to-sit movement using wearable sensors. Sensors **14**(3), 4342–4363 (2014)
13. Chen, B., Zheng, E., Fan, X., Liang, T., Wang, Q., Wei, K., Wang, L.: Locomotion mode classification using a wearable capacitive sensing system. IEEE Trans. Neural Syst. Rehabil. Eng. **21**(5), 744–755 (2013)

14. Zheng, E., Chen, B., Wei, K., Wang, Q.: Lower limb wearable capacitive sensing and its applications to recognizing human gaits. Sensors **13**(10), 13334–13355 (2013)
15. Liao, Y., Zhou, Z., Wang, Q.: BioKEX: a bionic knee exoskeleton with proxy-based sliding mode control. In: Proceedings of the IEEE International Conference on Industrial Technology, Seville, pp. 125–130 (2015)
16. Tsukahara, A., Kawanishi, R., Hasegawa, Y., Sankai, Y.: Sit-to-stand and stand-to-sit transfer sup-port for complete paraplegic patients with robot suit HAL. Adv. Robot. **24**(11), 1625–1638 (2010)
17. Liu, X., Zhou, Z., Wang, Q.: Recognizing sit-stand and stand-sit transitions for a bionic knee exoskeleton. In: ASME Design of Medical Devices Conference, Minneapolis (2017)
18. Geravand, M., Korondi, P.Z., Werner, C., Hauer, K., Peer, A.: Human sit-to-stand transfer modeling towards intuitive and biologically-inspired robot assistance. Auton. Robots **41**(3), 575–592 (2017)
19. Mamun, K.A., Mace, M., Gupta, L., Verschuur, C.A., Lutman, M.E., Stokes, M., Vaidyanathan, R., Wang, S.: Robust real-time identification of tongue movement commands from interferences. Neurocomputing **80**, 83–92 (2012)
20. Chen, B., Zheng, E., Wang, Q.: A locomotion intent prediction system based on multi-sensor fusion. Sensors **14**(7), 12349–12369 (2014)

EMG-Based Control for Three-Dimensional Upper Limb Movement Assistance Using a Cable-Based Upper Limb Rehabilitation Robot

Yao Huang[1,2], Ying Chen[1,2], Jie Niu[1,2], and Rong Song[1,2(✉)]

[1] Key Laboratory of Sensing Technology and Biomedical Instrument of Guang Dong Province, School of Engineering, Sun Yat-sen University, Guangzhou, People's Republic of China
songrong@mail.sysu.edu.cn

[2] Guangdong Provincial Engineering and Technology Center of Advanced and Portable Medical Devices, School of Engineering, Sun Yat-sen University, Guangzhou, People's Republic of China

Abstract. Voluntary residual motor efforts to the affected limb of patients after stroke have not been involved enough in most rehabilitation robot control strategies. In this paper, a natural integration between human and machine is proposed by using the surface electromyography (EMG) signals from six muscles which mainly contribute to the upper limb movement. A linear state space model is trained, which can estimate the real-time movement intention by using EMG signals, to calculate the movement needed forces and then provided by a cable-based upper limb rehabilitation robot. Ten healthy subjects are recruited to complete the tasks with and without robot assistances. The performances of the subjects with the assistances are compared to that of the subjects without assistances. Results show that the forces from the model were real-time continuously estimated and accurate. Furthermore, there is no significant difference in the group mean root mean square error (RMSE) and muscle activations between the task without assistance and with assistance. These results show that the robot using the state space model could provide physiologically appreciate assistance to the subject, and the robot could conduct the rehabilitation training combined with the voluntary residual motor efforts. Clinical test will be carried out to validate the feasibility of the robot-aided rehabilitation using myoelectrical control.

Keywords: Rehabilitation robots · Active control · State space methods · Electromyography

1 Introduction

Stroke is one of the diseases which have the highest cause of permanent disabilities. Nowadays, various rehabilitation robots has been useful methods of motor rehabilitation for stroke patients due to its superiority in efficiency, precision, and controllability [1, 2]. Over past four decades, the main control strategies are based on the kinematic signals

Y. Huang and Y. Chen—Contributed equally to this work.

Y. Huang et al. (Eds.): ICIRA 2017, Part I, LNAI 10462, pp. 273–279, 2017.
DOI: 10.1007/978-3-319-65289-4_26

[3, 4], and some famous control strategies based on these signals like PID control and impedence control has been popularly used.

However, these control strategies take less care of voluntary residual motor efforts to the affected limb of patients after stroke but more control accuracy. Electromyography (EMG) signals as know a easily captured physiological signal, are highly related to the muscular forces and joint torque, and can reflect movement intention [5]. It is useful and popular to take EMG signals in controlling rehabilitation robots. The earliest EMG-based control maps the EMG signals to binary control signals which only discretely control the robot [5]. Some continuous EMG-based control strategies are further proposed to estimate joint torque for single joint [6, 7]. Since few EMG-based control strategies can continuously estimate the position or torque of multi-joints, these paper propose a linear state space model to estimate movement intention of human from upper limb EMG signals and conduct experiments to verified the performance of this EMG-based control strategies by using a cable-based upper limb rehabilitation robot.

2 Method

2.1 Cable-Based Rehabilitation Robot

A cable-based rehabilitation robot for providing assistance is shown in Fig. 1. The robot is composed of a cube frame, an end-effector, a driven system, a data acquisition system and a computer. The driven system contains three motors with drivers, a motion control board and cables which connects the motors, the cube frame and the end-effector. The end-effector was pulled by three cables, resulting in 3 degrees of freedom. A motion capture system (OptiTrack, NaturalPoint, USA), six surface EMG electrode with filter amplifiers and a data acquisition cards (SCB 68, National Instruments, USA) are formed to collect both kinematics and EMG data during robot-aided movements. The actual position of the end-effector and desired position are programmed to be shown on the screen of the computer.

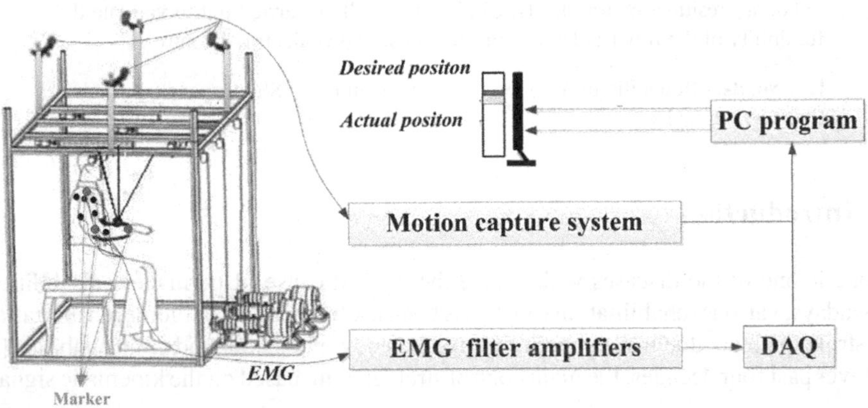

Fig. 1. Architecture of the cable-based rehabilitation robot.

2.2 An EMG-Based Control Strategy

An EMG-based control diagram is proposed to estimate the movement-needed force based on the EMG signal which consists of EMG signal processing, EMG-activation model, and a state space model.

EMG Processing. EMG signals of six muscles which mainly contributes to upper limb movements (brachioradialis (BR), biceps (BIC), triceps (TRI) and anterior (DA), middle (DM), posterior (DP) parts of deltoid) are recorded by data acquisition system of the robot from subjects. After the raw EMG signals are amplified to 5000 times, a band-filtered (10–400 Hz) by a 4th-order Butterworth filter are used. For obtaining the envelope of the EMG signals, a low-pass filtered with a fourth-order Butterworth filter with a cutoff frequency of 4 Hz are used.

EMG-to-Activation Model. In order to obtain the muscle activation in real time, the EMG-to-activation model is built. A second-order discrete linear model proposed by Lloyd and Besier [8] is used for neural activation.

$$u(t) = \alpha e(t-d) - \beta_1 u(t-1) - \beta_2 u(t-2) \tag{1}$$

where $e(t)$ is value of EMG envelope at time t, $u(t)$ is neural activation at time t, d is the electromechanical delay (80 ms), α is the gain coefficient, β_1 and β_2 are the recursive coefficients. According to the relationship between the neural activation and the muscle activation [9], the muscle activation is calculated by:

$$a(t) = \begin{cases} d\ln(cu(t)+1), & 0 \le u(t) < 0.3 \\ mu(t)+b, & 0.3 \le u(t) < 1 \end{cases} \tag{2}$$

where c, m and b are set at 0.3512, 0.8854 and 0.1155, respectively.

A State Space Model. To map muscle activations to the movement-needed force of the upper limb joints in real-time, a state space model is established.

The dynamic model of subject with robot is used to obtain the movement-needed force during robot-human movement:

$$\mathbf{T} = \mathbf{J}^{-1}\mathbf{F} \tag{3}$$

where **T** is the matrix of movement-needed force, **J** is the matrix of the unit-vectors of the three cables, and **F** is the external force matrix from the subject to the end-effector of the robot.

The relationship between the muscle activations and the movement-needed force then is identified offline by a state space method:

$$\begin{aligned} x_{k+1} &= \mathbf{A}x_k + \mathbf{B}a_k + w_k \\ F_k &= \mathbf{C}x_k + v_k \end{aligned} \tag{4}$$

where k = 1, 2… is the integral number of sampling period, a_k is the vector of six muscle activations, F_k is the external force matrix, w_k and v_k are both zero-mean-Gaussian noise. A, B and C are obtained by offline identification using the prediction-error minimization algorithm.

By using the known A, B and C, the state space model can be programmed for online movement-needed force estimation from EMG signals.

2.3 Experiment

Participants. Ten healthy men (mean age 22 ± 1.3 yrs) with no experience with this robot were invited to join this study, after providing written informed consent. All participants were able to lift their upper limbs with right handed and declared that they have no health issue. All subjects provided their informed consent prior to participating in this study. All experimental procedures were approved by the human ethic committee at the Sun Yat-sen University.

Apparatus. Three reflective markers were pasted in the dorsal center of wrist, elbow and shoulder of every participant. The sampling rate of the motion capture system was set at 100 Hz, and raw position data were filtered by a second-order forward and backward Butterworth filter while the cut-off frequency was set at 6 Hz. The EMG signals of the six muscle were recorded at 1000 Hz. The desired and the actual wrist positions were visualized on a computer screen in the front of the participants.

Procedure. Participants seated at a chair and was softly tied to decrease extra compensation from trunk. Each participant went through three rounds of multi-joint upward tracking movements with right arm to the place 0.25 m away from the initial point along a straight line. During the first round, which was the modeling round, the robot offered no assistance to participants. The muscle activations from EMG signals and movement-needed force calculated by dynamic model were then used to identify the parameters of the state space model after this round. During the second round, participants were asked to finish tracking movements freely without robot. For the third round, participants were assisted by the robot with the EMG-based control. Every round required the participant to execute 6 laps consecutively.

Measurements and Data Recording. The root mean square errors (RMSE) of the desired trajectory and the actual trajectory of the wrist were calculated to evaluate the performance the EMG-based control strategy. The mean normalized muscle activation of each muscles was recorded.

Data Analysis. The paired-samples t-tests were used to examine differences in RMSE, and mean normalized muscle activation per muscle during movements without and with robot assistance.

3 Results

Figure 2 shows the performance of the state space model in estimating of the movement-needed forces. The dash grey line is force calculated by the dynamic model, while the solid dark line is force estimated by the EMG-based model during a 5-second trial. The two forces of three cables show similar trends with the dash line, and their values are closed.

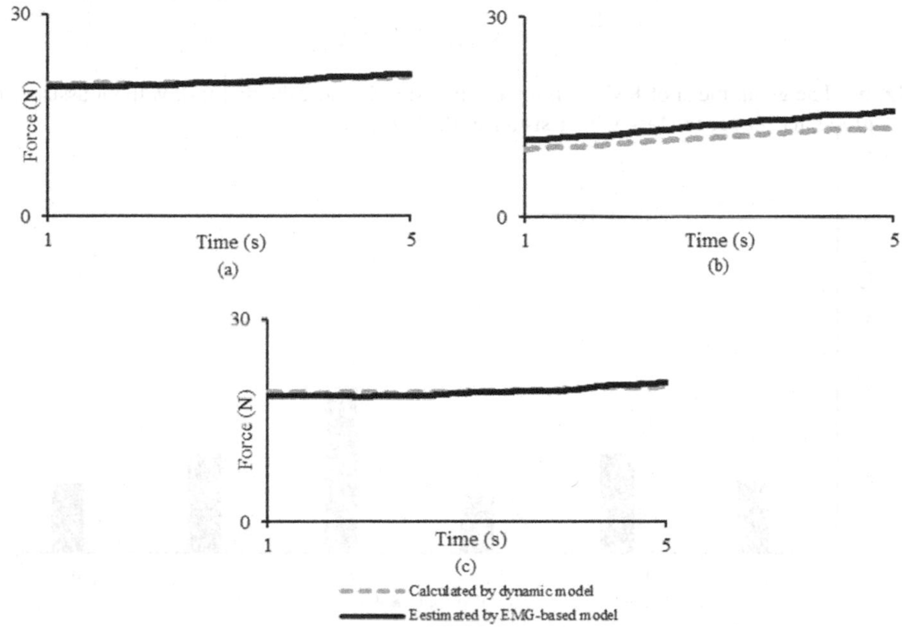

Fig. 2. The movement-needed forces calculated by the dynamic model (the dash gray line) and forces estimated by the EMG-based model (the solid dark line), (a) Cable 1, (b) Cable 2, (c) Cable 3.

Figure 3 presents the group mean of RMSE in three dimensional space. As the results of the paired T–test between without and with robot assistance of each dimension, there are no significant difference all three dimensions.

The muscle activations of the six muscles (BR, BIC, TRI, DA, DM and DP) during without and without robot assistances are demonstrated in Fig. 4. Comparing the normalized activations of the six muscles in the task without assistance and with assistance, both DA and DM are mainly contributed for the upward tracking movement. There are decreases in muscle activation of all muscles except the TRI as the assistances are provided. However, no significant difference are found in all these muscles.

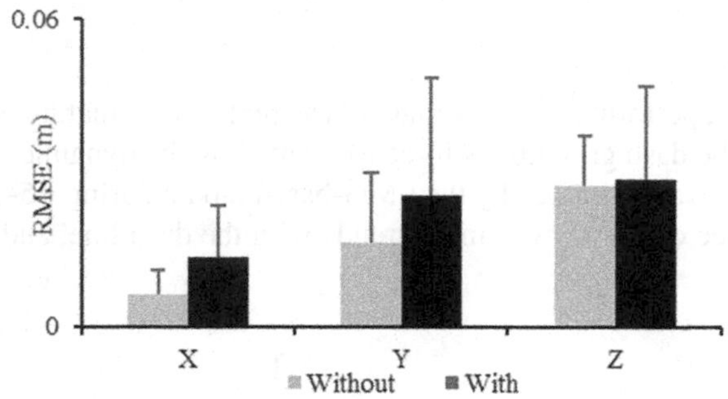

Fig. 3. The group mean of RMSE in three dimensional space, the first task without assistance (the gray bar); the second task with assistance (the black bar).

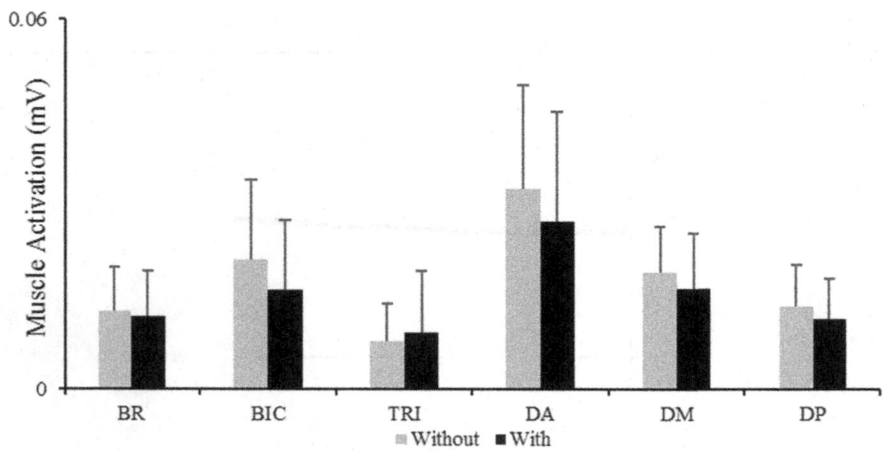

Fig. 4. The group mean muscle activations (brachioradialis (BR), biceps (BIC), triceps (TRI) and anterior (DA), middle (DM), posterior (DP) parts of deltoid) of subjects during performing two tasks; the first task without assistance (the gray bar); the second task with assistance (the black bar).

4 Discussion

The performance of the state space model in estimating of the movement-needed forces prove that the state space model for continuously estimating the movement-needed forces from EMG signals are possible and accurate. However, most of the of human-robot cooperation using EMG-based control either are restricted to continuous control in one-dimensional space, or are represented in a discrete mode [6, 7]. Although a state space model is used by Artemiadis et al. [10], their model is not used for rehabilitation robot and without considering interaction between human and machine.

The statistic results of RMSE and muscle activation prove that subjects can achieve same position control and same muscle usage when performing tasks without and with robot assistance from the EMG-based control strategy. The results of muscle activations show that the assistance from the robot under EMG-based control can only slightly reduce the muscle effort that mainly responsible for the movement but without affecting their contributions. The slightly decrease is similar to previous studies [11], and it is proved that healthy subjects show better ability to control their muscles and forces with external assistance [12].

These results can further prove that the robot using the state space model could provide physiologically appreciate assistance to the subject, and the robot could conduct the rehabilitation training combined with the voluntary residual motor efforts. Clinical test will be carried out to validate the feasibility of the robot-aided rehabilitation using myoelectrical control.

Acknowledgments. The project was supported by the National Natural Science foundation of China (Grant No. 61273359 and 91520201), the Guangdong Science and Technology Planning Project (Grant No. 2014B090901056 and 2015B020214003) and the Guangzhou Research Collaborative Innovation Projects (Grant No. 201604020108).

References

1. Riener, R., Nef, T., Colombo, G.: Robot-aided neurorehabilitation of the upper extremities. Med. Biol. Eng. Comput. **43**(1), 2–10 (2005)
2. Krebs, H., Volpe, B., Aisen, M., Hogan, N.: Increasing productivity and quality of care: robot-aided neuro-rehabilitation. J. Rehabil. Res. Dev. **37**(6), 639–652 (2000)
3. Ferris, D., Sawicki, G., Domingo, A.: Powered lower limb orthoses for gait rehabilitation. Top. Spinal Cord Inj. Rehabil. **11**(2), 34–49 (2005)
4. Kong, K.: Proxy-based impedance control of a cable-driven assistive system. Mechatronics **23**(1), 147–153 (2013)
5. Dipietro, L., Ferraro, M., Palazzolo, J., Krebs, H., Volpe, B., Hogan, N.: Customized interactive robotic treatment for stroke: EMG-triggered therapy. IEEE Trans. Neural Syst. Rehabil. Eng. **13**(3), 325–334 (2005)
6. Song, R., Tong, K.-Y., Hu, X., Li, L.: Assistive control system using continuous myoelectric signal in robot-aided arm training for patients after stroke. IEEE Trans. Neural Syst. Rehabil. Eng. **16**(4), 371–379 (2008)
7. Potvin, J.R.: Mechanically corrected EMG for the continuous estimation of erector spinae muscle loading during repetitive lifting. Eur. J. Appl. Physiol. Occup. Physiol. **74**(1–2), 119–132 (1996)
8. Lloyd, D.G., Besier, T.F.: An EMG-driven musculoskeletal model to estimate muscle forces and knee joint moments in vivo. J. Biomech. **36**(6), 765–776 (2003)
9. Manal, K., Buchanan, T.S.: A one-parameter neural activation to muscle activation model: estimating isometric joint moments from electromyograms. J. Biomech. **36**(8), 1197–1202 (2003)
10. Artemiadis, P.K., Kyriakopoulos, K.J.: EMG-based control of a robot arm using low-dimensional embedding's. IEEE Trans. Robot. **26**(2), 393–398 (2010)
11. Kwon, S., Kim, Y., Kim, J.: Movement stability analysis of surface electromyography-based elbow power assistance. IEEE Trans. Biomed. Eng. **61**(4), 1134–1142 (2014)
12. Suzuki, K., Mito, G., Kawamoto, H., Hasegawa, Y., Sankai, Y.: Intention-based walking support for paraplegia patients with Robot Suit HAL. Adv. Robot. **21**(12), 1441–1469 (2007)

A Real-Time Intent Recognition System Based on SoC-FPGA for Robotic Transtibial Prosthesis

Jingeng Mai[1], Zhendong Zhang[1], and Qining Wang[1,2(✉)]

[1] The Robotics Research Group, College of Engineering, Peking University, Beijing 100871, China
{jingengmai,qiningwang}@pku.edu.cn

[2] Beijing Innovation Center for Engineering Science and Advanced Technology (BIC-ESAT), Peking University, Beijing 100871, China

Abstract. This paper presents the design and implementation of a real-time intent recognition hardware system for robotic transtibial prosthesis, based on system-on-chip and field-programmable gate array (SoC-FPGA). The proposed system integrates the software programmability of an ARM-based processor with the hardware programmability of an FPGA. A hardware prototype was developed and a SVM-based pattern recognition algorithm was implemented with high-level synthesis technology. Experiments on a transtibial amputee subject demonstrated that the proposed system costs shorter decision time in identifying four lower-limb movement phases (sitting, standing, sit-to-stand and stand-to-sit).

Keywords: Real-time intent recognition · SoC-FPGA · SVM · Robotic transtibial prosthesis

1 Introduction

Transtibial prostheses play an important role in amputees' daily lives. Compared with passive lower-limb prostheses, robotic transtibial prostheses can adjust the output torque or damping behaviors automatically with algorithms and strategies executed on embedded controller [1–5]. In order to make the robotic transtibial prostheses performing better interaction and coordination with human movements, it is important to recognize and predict human action intention [6–8].

On-board computing performance is an important factor affecting the real-time intent recognition. On the one hand, high performance computing capacity can reduce the recognition decision time to support the prosthesis running in higher control frequency; on the other hand, with the high performance hardware, the system can process more sensing information and carry out more optional algorithms, to handle complex recognition tasks.

It is an effective way to accelerate the computing performance, by using the hardware with parallel processing ability, such as field-programmable gate array (FPGA) and graphics processing unit (GPU). Compared with GPU, FPGA has an advantage of lower consumption [9]. Only a few studies have realized on-board

Y. Huang et al. (Eds.): ICIRA 2017, Part I, LNAI 10462, pp. 280–289, 2017.
DOI: 10.1007/978-3-319-65289-4_27

system with FPGA in robotic transtibial prostheses. For example, Huang et al. tried to implement real-time linear discriminant analysis (LDA) algorithm based on FPGA for intent recognition in lower limb prosthesis system [10].

However, there are still problems in real-time intent recognition for robotic prostheses. First, how to implement more complicated pattern recognition is a challenge. Compared with the LDA algorithm which can be completely expressed as matrix operations, other algorithms like multi-class support vector machine (SVM) and recurrent neural networks (RNN), are usually expressed as a combination of logic, mathematics and matrix operations [11,12]. For the matrix operations (consisting of a large number of multiplications and additions), FPGA's parallel computing resources can be fully utilized to improve performance. However, for some sequential operations, FPGA normally runs slower than application-specific integrated circuits (ASICs), because the FPGA frequency is lower than the ASICs which constructed with similar fabrication process [13]. In addition, it is important to let hardware designers efficiently build and verify algorithms, by giving them better control over optimization of their design architecture.

In this paper, a real-time intent recognition system based on SoC-FPGA for powered lower limb prosthesis is presented. This system supports multi-class SVM runs on a single chip combined with ARM (a kind of ASICs) and FPGA. And algorithm acceleration optimization method based on high-level synthesis technology is proposed, to support algorithm efficiently builded with C code, and optimized for parallelization. At last, experiments are designed to verify algorithm correctness and evaluate system performance.

2 System Architectures

Based on the robotic transtibial prosthesis developed by our research team [14], we designed and implemented a hardware system for real-time intent recognition, including micro controller unit for real-time data acquisition, computer console

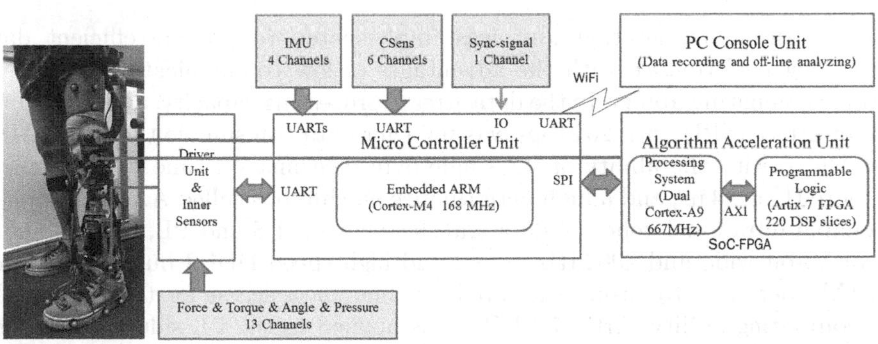

Fig. 1. System architecture of real-time intent recognition system for robotic transtibial prosthesis

unit for PR algorithm off-line analyzing and algorithm acceleration unit for real-time algorithm execution. The system architecture is shown in Fig. 1.

(1) **Robotic Transtibial Prosthesis**
The intelligent powered lower limb prosthesis is PKU-RoboTPro-II, with a rated power of 150 W, the total weight of 1.75 kg. This prosthesis which combined with high performance motor, multi-channel inner sensors and advanced control strategy can improve the amputees' gait symmetry and walking stability on different terrains including level ground, stairs, and ramps [14].

(2) **Micro Controller Unit**
The micro controller unit (MCU) was designed to receive additional multi-channel sensor data which used for intent recognition, including 2 IMU modules (each IMU module contains two channel angle data) and 1 CSens module (6 independent channels), and 1 channel sync-signal which is generated by a motion capture system. The core of the MCU is a cortex-M4 processor with maxim frequency of 168 MHz.

(3) **Computer Console Unit**
The computer console unit was designed for off-line PR algorithm model analyzing. This unit receives multi-channel sensor data sent by MCU with WIFI, and records the raw data from an optical motion capture device, and combines them for PR model training.

(4) **Algorithm Acceleration Unit**
The algorithm acceleration unit (AAU) was developed to accelerate PR algorithm. The AAU consists of a system-on-chip and a field programmable gate array (SoC-FPGA), and is packaged in a single chip (Xilinx Zynq-7020), supporting high performance of concurrent computing. The AAU receives real-time sensor data stream which is sent by MCU, and carries out data processing, and executes intent recognition algorithm.

3 Implementation of Algorithm Acceleration Based on SoC-FPGA for Real-Time Intent Recognition

3.1 Algorithm Acceleration Unit Hardware Capabilities

The algorithm acceleration unit was implemented to provide efficient data processing performance with the advantages of low communication delay and parallel computing ability. In the data stream processing capacity, a serial peripheral interface (SPI) with 25 Mbps was built for data transmission between AUU and MCU, with the ability of the single-byte transmission time between them no more than 0.4 us; and a high performance on-chip bus called AXI was applied to enable ultra-low latency of data transfer between PS and PL, with the ability of 32-bit wide and 200 MHz speed; and high speed DDR3 bus was used for DRAM operation, to ensure the rapid dynamic data access for the PS side. In the computing ability, Artix-7 FPGA was applied in the PL side with the key features of 85k programmable logic unit, 220 DSP Slices processing array, and 560 KB RAM resources, to support a high performance of parallel computing ability (Fig. 2).

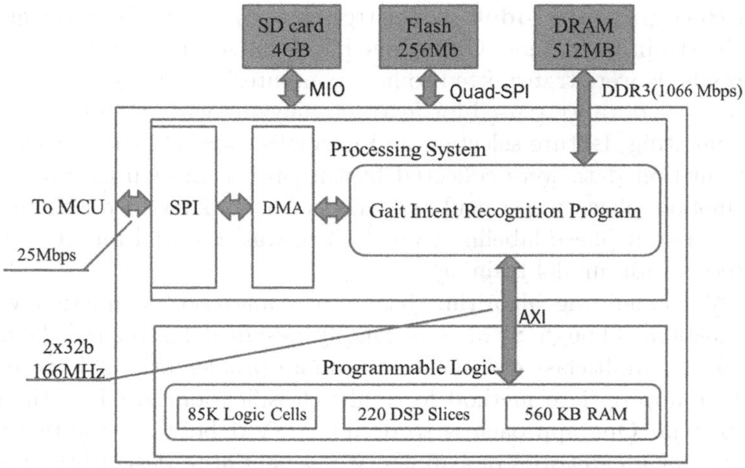

Fig. 2. Sata flow schema of algorithm acceleration unit

3.2 Steps of Real-Time Intent Recognition Implementation on SoC-FPGA

The process of real-time intent recognition implementation usually contains three steps, as shown in Fig. 3. The first step is to build appropriate classifiers for intent recognition. This step is based on the existing training samples to get an optimized PR model. It is usually a time consuming process due to the need for sufficient sample in the training to get a better recognition effect. So, it is usually an off-line analysis process on the PC using the MTALAB tool. The second step is to implement the recognition algorithm with C language which can be running on embedded MCU hardware. It is the purpose to validate and evaluate the operational effect of the on-line recognition algorithm in the embedded hardware, to confirm the correctness of the recognition result, to obtain the running time of the recognition process. The third step is to accelerate the online algorithm. The purpose is to take advantage of the parallel computing resources in SoC-FPGA to reduce the processing time. The key method for this step is to evaluate which parts of the online algorithm are suitable for hardware acceleration and implement them in SoC-FPGA hardware.

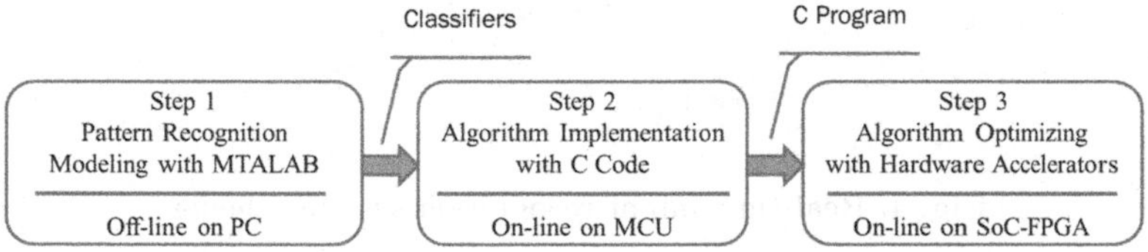

Fig. 3. Steps of real-time intent recognition implementation on SoC-FPGA

Pattern Recognition Model. The pattern recognition model was constructed using off-line training method. On the one hand, the sensing data which collected by the prosthesis were transmitted to host computer as raw dataset for training. And then, these data were used for feature analyzing with MATLAB, including feature computing, feature selection and normalization. On the other hand, the kinematic motion data were collected by the optical motion capturing device, and the motion phase was calculated and segmented based on these data to obtain the motion phase labeling value, which was essential for the supervised pattern recognition model training.

The SVM clustering algorithm based on Gauss kernel function was used for classification. Though SVM is originally designed for binary classification and the use for multiclass classification is more problematic, it is necessary to construct an appropriate method for multi-classification based on the original SVM algorithm. One approach is to design a SVM between any two types of samples. So, for k categories of samples, it is essential to design k(k−1)/2 SVM sub classifiers. Assuming that there are four kinds of A, B, C, D results, sub classifiers (A, B), (A, C), (A, D), (B, C), (B, D), (C, D) should be selected for model training, and then, a voting method is used to get the final results based on these sub classification results.

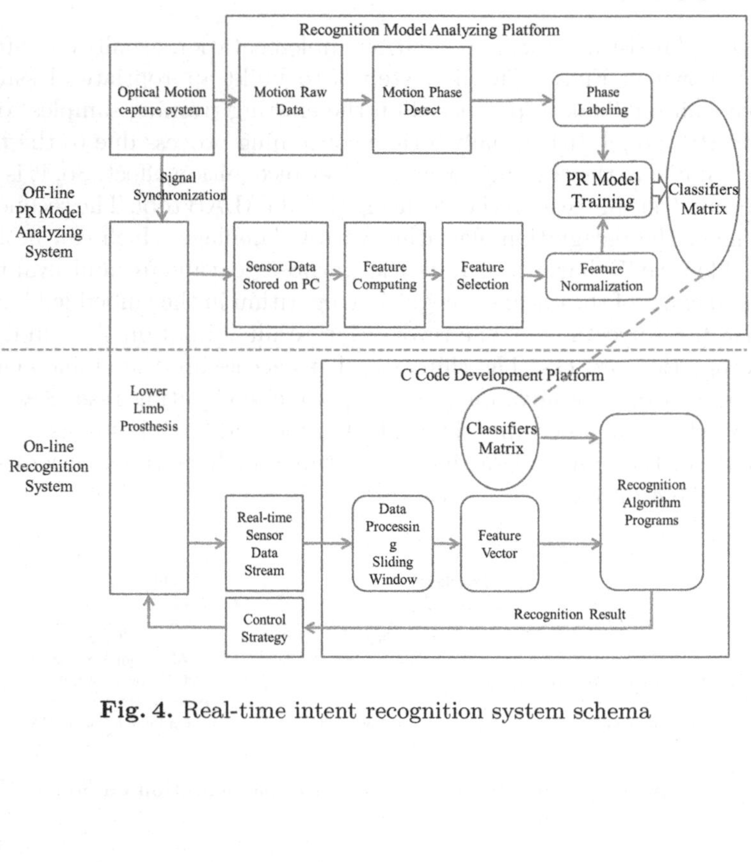

Fig. 4. Real-time intent recognition system schema

On-Board Recognition Algorithm Implementation. As shown in Fig. 4, first, the SoC-FPGA periodically received multi-channel sensing data from the intelligent prosthetic limb, and formed a sliding window with a plurality of groups of data. Then, in each data processing cycle, the characteristic information of each channel was calculated, including mean value, maximum value, minimum value and standard deviation, and the eigenvalues of multiple channels constitute a set of feature vectors. And then, the feature vector and the weight matrix which obtained by off-line training were used to calculate the classification decisions, and, voting strategy was used to select the final results. In order to facilitate the optimization process, the recognition algorithm was expressed as matrix operations, including multiplication, addition, multiplication and norm, and all the recognition algorithm was implemented with C codes.

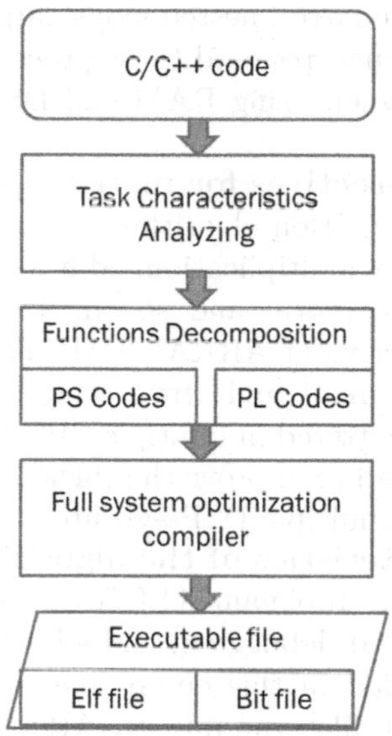

Fig. 5. The work flow of algorithm acceleration optimizing

Algorithm Acceleration Optimizing with High-Level Synthesis Technology. Recognition algorithm was optimized with high-level synthesis (HLS) technology combined with SoC-FPGA hardware to reduce execution time. HLS is an automated design process that interprets an algorithmic description of a desired behavior and creates digital hardware that implements that behavior [15,16]. It works at a higher level of abstraction, starting with an algorithmic description in a high-level language such as SystemC and Ansi C/C++. The default behavior of HLS is to execute functions and loops in PS hardware in a sequential manner such that the programs are executing with an accurate reflection of the C/C++ code. But, if only PS hardware was used, the disadvantage of using sequential execution will be obvious, when a large number of nested loops occur in the algorithm.

It is an appropriate way to address this bottleneck using algorithm acceleration method with PL hardware resources, using some optimization directives to enhance the performance of the hardware function, allowing pipelining which substantially increases the performance of the functions. There are some key strategies for the acceleration implementation:

- **Analyzing the characteristics of the algorithm task**
  C code in the algorithm can be decomposed into two categories, one does not support hardware optimization procedures, the other is to support hardware optimization procedures. For the former, some structural features can help us to easily distinguish, such as system functions, recursive functions, etc. For the latter, nested loops, large array operations, matrix operations are typical structures. For the latter, nested loops, large array operations, matrix operations are typical structures, and these processes can be used to achieve significant acceleration when using RAM and DSP arrays in the FPGA for parallel processing.
- **Using optimization directives for matrix operations**
  The core part of the recognition algorithm can be described as some matrix operations, such as matrix multiplication, addition, point multiplication, vector normalization, matrix norm and so on. Taking matrix multiplication as an example, the directive of ARRAY_PARTITION was used to improve the parallelism of array read and write, and the element values of two-dimensional arrays are scattered in multiple FPGA internal RAM. The directive of PIPELINE was used to improve the efficiency of computing operations, multiple operations into multiple DSP synchronization operations.
- **Analyzing the characteristics of the algorithm task**
  An integrated development tool named SDSoC was used for SoC-FPGA codes programming, building and debugging, and all algorithmic application codes were written and compiled in this environment. The result of compilation is to generate elf format files running on ARM hardware on PS side and bit files running on FPGA hardware on PL side, and then development tool automatically packages the two files into a bin image file. At last, the image file was stored to the SoC-FPGA system SD card for execution.

4 Experiments and Prototype

4.1 Experimental Methods

To evaluate the performance of the real-time intent recognition system, a male unilateral transtibial amputee subject volunteered to participate in the study and provided written consent prior to testing. The target of pattern recognition classification contains four states, sit down (SD), stand up (ST), sit down to stand (SD→ST) and stand up to sit down (ST→SD). The experiment consists of several sessions: offline training, online identification algorithm verification, algorithm acceleration verification and performance evaluation.

i. In off-line amputee subject wore transtibial prosthesis, and repeated 20 times up and down movement; meanwhile, the optical motion capture system recorded motion data, meanwhile, the MCU system sampled and sent sensor data to PC, including IMUs and capacitance sensor data, with the sampling frequency of 100 Hz. And, a trigger signal sent by the optical motion capture system was used to ensure data synchronization with the MCU.
ii. In on-line recognition algorithm validation session, the PR algorithm based on SVM classifiers was implemented on the MCU. The data processing period was 10 ms to ensure that there was enough time for the recognition algorithm performed in one cycle. The sliding window length and the window increment were set to 15 data points and 1 data point, with corresponding time of 150 ms and 10 ms. When the algorithm was executed, the intent recognition result together with original sensor data were transferred to the PC side for verification of recognition correctness.
iii. In algorithm acceleration verification session, the PR algorithm was implemented on the AAU. The AAU received sensor data which were collected by MCU, with data processing frequency up to 500 Hz, and made intent decisions in each period. The sliding window length and the window increment were still set to 15 data points and 1 data point, with corresponding time of 30 ms and 2 ms. In addition, the intent decisions made by the AAU were sent back to MCU, and then, all original sensor data together with intent decisions were transferred to the PC side.
iv. In order to evaluate the execution time of the recognition algorithm on different hardware environment, the start and stop the algorithm were reflected in the high and low level of MCU and AAU IO port, and then the high level duration was measured as the execution time.

4.2 System Performance in Real-Time

As shown in Fig. 6, A hardware prototype of MCU plus SoC-FPGA was developed. The MCU board and SoC-FPGA board were join with plug-in connectors. As shown in Figs. 7 and 8, the recognition decision time was 3.6 ms when SVM-based PR algorithm applied in MCU environment with cortex-M4 processor. And it was reduced to 0.05 ms when the same algorithm executed in SoC-FPGA

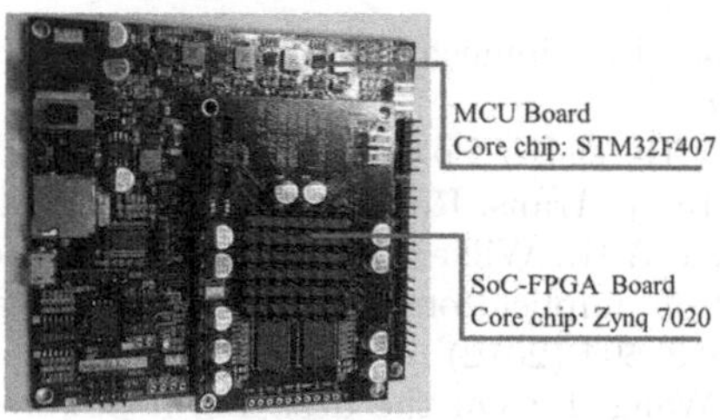

Fig. 6. Hardware prototype of MCU plus SoC-FPGA

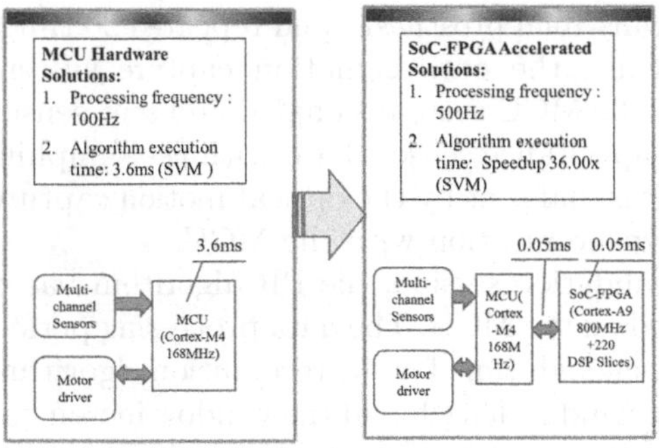

Fig. 7. Framework comparison

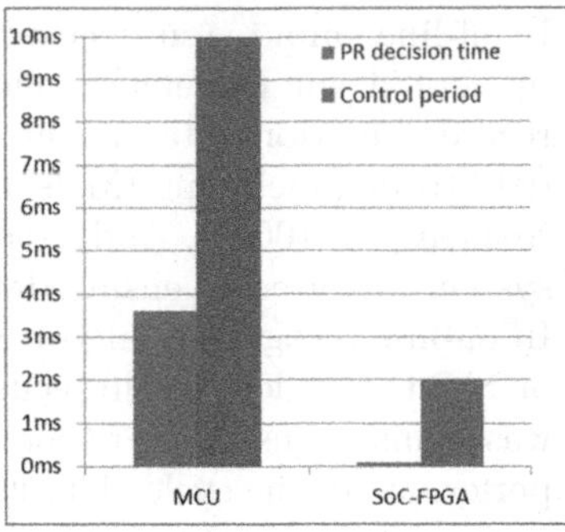

Fig. 8. Performance comparison

hardware. Considering the extra time of data transmission was no more than 0.05 ms (transferring 120 Byte data with 25 Mbps SPI bus), the total execution time no more than.

5 Conclusions

This paper presented a hardware system based on SoC-FPGA for on-board intent recognition, and proposed an efficiently implementation method for algorithm acceleration based on high-level synthesis technology. The designed hardware system constructed with a MCU board and a SoC-FPGA board. Meanwhile, SVM-based intent recognition algorithm was implemented with C codes and executed in SoC-FPGA board. The whole system were tested on an amputated subject. Experiments demonstrated that the SoC-FPGA implementation of the recognition algorithm achieved a speedup 36 over the MCU system, and the control frequency of the accelerated system was up to 500 HZ.

Acknowledgments. This work was supported by the National Natural Science Foundation of China (No. 91648207), the Beijing Municipal Science and Technology Project (No. Z151100000915073), and the Beijing Nova Program (No. Z141101001814001).

References

1. Winter, D.A., Sienko, S.E.: Biomechanics of below-knee amputee gait. J. Biomech. **21**(5), 361–367 (1988)
2. Au, S.K., Weber, J., Herr, H.: Powered ankle-foot prosthesis improves walking metabolic economy. IEEE Trans. Robot. **25**(1), 51–66 (2009)
3. Sinitski, E.H., Hansen, A.H., Wilken, J.M.: Biomechanics of the anklefoot system during stair ambulation: implications for design of advanced anklefoot prostheses. J. Biomech. **45**(3), 588–594 (2012)
4. Zhu, J., Wang, Q., Wang, L.: On the design of a powered transtibial prosthesis with stiffness adaptable ankle and toe joints. IEEE Trans. Ind. Electron. **61**(9), 4797–4807 (2014)

5. Cherelle, P., Grosu, V., Matthys, A., Vanderborght, B.: Design and validation of the ankle mimicking prosthetic (AMP-) foot 2.0. IEEE Trans. Neural Syst. Rehabil. Eng. **22**(1), 138–148 (2014)
6. Varol, H.A., Sup, F., Goldfarb, M.: Multiclass real-time intent recognition of a powered lower limb prosthesis. IEEE Trans. BioMed. Eng. **57**(3), 542–551 (2010)
7. Chen, B., Wang, Q., Wang, L.: Adaptive slope walking with a robotic transtibial prosthesis based on volitional EMG control. IEEE/ASME Trans. Mech. **20**(5), 2146–2157 (2015)
8. Young, A.J., Simon, A.M., Fey, N.P., Hargrove, L.J.: Intent recognition in a powered lower limb prosthesis using time history information. Ann. Biomed. Eng. **42**(3), 631–641 (2014)
9. Tian, X., Benkrid, K.: High-performance quasi-Monte Carlo financial simulation: FPGA vs. GPP vs. GPU. ACM Trans. Reconfigurable Technol. Syst. **3**(7), 1–22 (2010)
10. Zhang, X., Huang, H., Yang, Q.: Implementing an FPGA system for real-time intent recognition for prosthetic legs. In: DAC Design Automation Conference, pp. 169–175 (2012)
11. Groleat, T., Arzel, M., Vaton, S.: Hardware acceleration of SVM-based traffic classification on FPGA. In: International Workshop on Traffic Analysis and Classification, pp. 443–449 (2012)
12. Chang, A.X.M., Martini, B., Culurciello, E.: Recurrent neural networks hardware implementation on FPGA. Comput. Sci. (2015)
13. Kuon, I., Rose, J.: Measuring the gap between FPGAs and ASICs. IEEE Trans. Comput. Aided Des. Integr. Circ. Syst. **26**(2), 203–215 (2007)
14. Feng, Y., Zhu, J., Wang, Q.: Metabolic cost of level-ground walking with a robotic transtibial prosthesis combining push-off power and nonlinear damping behaviors: preliminary results. In: Proceedings of the 38th Annual International Conference of the IEEE Engineering in Medicine and Biology Society, pp. 5063–5066 (2016)
15. Coussy, P., Morawiec, A.: High-Level Synthesis: From Algorithm to Digital Circuit. Springer, Heidelberg (2008)
16. Gajski, D.D., Dutt, N.D., Wu, A.C.-H., Lin, S.Y.-L.: High-Level Synthesis: Introduction to Chip and System Design. Kluwer Academic Publishers, Boston (1992)

Motion Planning and Experimental Validation of a Novel Robotic Device for Assistive Gait Training

Tao Qin[1(✉)], Hao Zhang[2], Peijun Liu[3], Fanjing Meng[1], and Yanyang Liu[3]

[1] School of Mechanical and Automotive Engineering, Hubei University of Arts and Science, Xiangyang 441053, China
heu_qt@163.com

[2] Fine Arts College, Hubei University of Arts and Science, Xiangyang 441053, China

[3] Department of Rehabilitation Medicine, Xiangyang Central Hospital, Hubei University of Arts and Science, Xiangyang 441021, China

Abstract. Walking is an essential part of every day's mobility. To enable non-ambulatory gait-impaired patients the repetitive practice of this task not only in hospital, rehabilitation settings, but also in community, and homes, a novel robotic device for assistive gait training has been designed followed the end-effector principle. Deriving inspiration from the foot motion attitude of floor walking and the existing elliptical trainer, some ingenious constrain mechanisms were introduced to generate a human-like walking gait and the desired foot and metatarsophalangeal (MTP) joint motions including toe and heel rising, dorsiflexion of MTP joint. The motion sequence of bilateral footplates and the rotation speeds of bilateral cranks were planned. A robotic prototype was developed as a platform for evaluating the design concepts and motion planning. The experimental results validated the feasibility and effectiveness of the design and motion planning.

Keywords: Assistive gait training · Rehabilitation robot · Motion planning

1 Introduction

Walking is the most basic and important capability that enables humans to pursue their activities of daily living (ADL) and realize the high-quality lives [1]. However, because of the factors obstructing the walking ability such as aging and central nervous system (CNS) disease (e.g., stroke and spinal cord injury (SCI)), more and more patients are suffering from lower limb motor dysfunction and gait disorder [2]. Particularly stroke is the leading cause of permanent disabilities or one-sided paresis of lower limbs and still influences the patients' life quality several years after they are discharged to home [3]. So improving and regaining walking ability are the desirable outcome for many gait-impaired patients. The studies show that guide rehabilitation training can farthest restore or compensate for the lost functions and prevent the formation of hemiplegics gait and the onset of disuse syndrome [2]. So many robotic devices have been

Y. Huang et al. (Eds.): ICIRA 2017, Part I, LNAI 10462, pp. 290–300, 2017.
DOI: 10.1007/978-3-319-65289-4_28

developed to replace the traditional physical efforts training and provide intensive repetitive motions and store the training data to assess quantitatively rehabilitation level. Representatives of some commercial developments are multi-degrees of freedom (DOFs) system which can facilitate to achieve the mult-DOFs strengthening training of each joint. Hocoma's Lokomat with 8-DOFs is a driven gait orthosis on a treadmill to be an effective intervention of treadmill training for the gait-impaired patients. It improves the therapy outcome by providing highly intensive and individualized training in a motivational environment of constant feedback [4]. The G-EO-System of Reha Technology AG supports and gently leads the patients in the continuous walking practice situations with two of symmetrical 3-DOFs footplates. The footplates control ankle joint and the foot motion to imitate various walking situations such as walking on a level, climbing and descending stairs [5]. Some other robotic devices are still at a research state or under development, such as the 6-DOFs gait rehabilitation robot with upper- and lower-limb connections of Gyeongsang National University [6], the Gait-Master series of University of Tsukuba [7], etc. The above mentioned robotic devices are complex and difficult to operate because they are integrated with multiple drivers, so it is difficult to widespread in community and home settings. Furthermore, these devices cannot generate the relative rotation of the metatarsophalangeal (MTP) joint between forefoot and rearfoot (see in Fig. 1) during human normal walking on level ground. The relative rotation at MTP joint from the moment of heel off to toe off is a critical component of normal walking [8]. Therefore, the adequate motion between forefoot and rearfoot should be available with the robot in order to achieve more natural motions of the MTP joint during gait training.

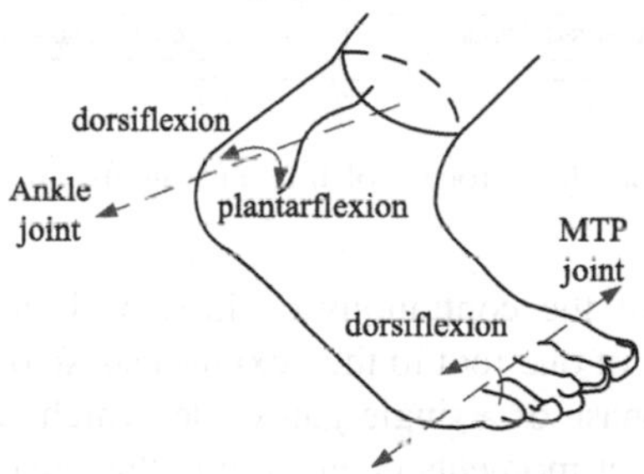

Fig. 1. Motion of ankle joint and MTP joint in the sagittal plane

Moreover, most of the current existing robotic devices for gait training have limited used in community and home settings due to the issues of design, cost, and demands on clinical staff. Taking the above factors and the inherent defects of the existing gait rehabilitation devices into account, a novel robotic device for assistive gait training was developed to enable the gait-impaired patients the repetitive gait training based on the end-effector principle, which is shown in [9]. A two segment footplate and the ingenious constrain mechanisms were introduced to generate relative rotation between the fore- and rear-footplate to achieve the dorsiflexion of MTP joint. And this gait training robot is an affordable and automatic rehabilitation training device that can simulate more authentic gait of floor walking, and be easy to use in various places, such as the hospital, rehabilitation settings, community, and patients' homes.

2 Gait Analysis

Human walking is a complex task for humans to pursue their daily lives. It can be viewed as a problem of foot position and orientation. A complete process of human walking on level ground begins with the feet standing still and upright (i.e., the neutral position). Then one side of the legs swings forward and into the single support phase until the moment of the heel strike. After that, both legs are beginning into the continuous and alternating cycle walking locomotion. Finally, a complete walking process ends with both legs get back to the state of the neutral position. That is a complete walking process can be defined as three steps: initial step, continuous cycle step and terminal step, which is shown as Fig. 2.

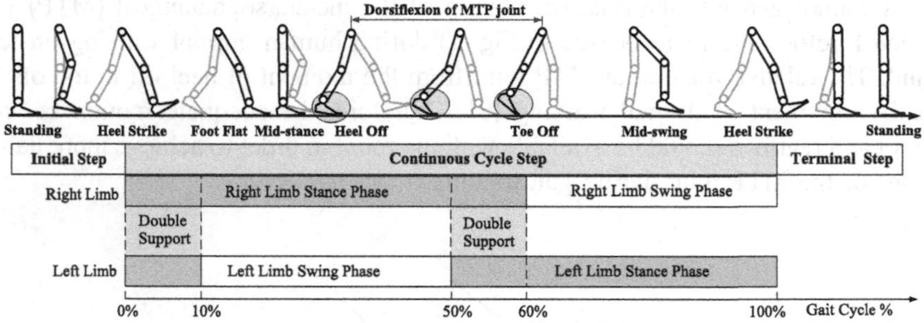

Fig. 2. A complete process of human walking on level ground

A single cycle step of the continuous cycle steps is a time interval of motion occurring from heel strike of one foot to the next heel strike of the same foot. There are stance phase and swing phase of a single gait cycle, which account for approximately 60% and 40%. Five distinct moments occur during the stance phase: heel strike, foot flat, mid-stance, heel off, and toe off [10]. The foot exhibits a rolling-like motion on the ground around two different rotation centers: heel and MTP joint during the stance phase of floor walking. After heel off, the ankle joint begins plantarflexion forcing the MTP joint to dorsiflex until the toe is rolling off the ground so as to elevate the arch and increase foot stability [11] (see in Fig. 2). Thus, the MTP joint motion has an important role during human walking, particularly during the stance phase. A flexible foot could be simply modeled as two segments with one revolute joint at MTP joint to represent more closely natural foot motion of floor walking, which will achieve more natural and stability gait than a single rigid foot [12].

3 Motion Planning of the Robotic Device for Gait Training

3.1 Mechanism Description

Original elliptical trainer, mostly using the symmetrical crank-rocker mechanisms and the footplates are fixed on the linkages, can simulate the walking gait in the sagittal plane and help the patients regain the strength and flexibility for walking [13]. However, the dorsiflexion angle of the foot is inaccurate when using the elliptical trainer because the motion attitude of footplates on the linkages cannot be adjusted. To improve this problem and generate natural walking gait, a novel robotic device for assistive gait training integrated with two sets of symmetrical gait mechanisms has been suggested as shown as Fig. 3. Two gait mechanisms are driven independently to control the motion of their respective footplate. The footplate is designed as fore-footplate and rear-footplate, which are articulated by a hinge. The middle of the rear-footplate is hinged at the linkage, and the rear-footplate tail is hinged at the inclined slide rail. The footplate rotation angle can be adjusted by the inclined slide rail so as to generate more natural foot and ankle motion attitudes (e.g., toe and heel rising, plantarflexion and dorsiflexion of ankle joint) when the robotic device is running. Moreover, when the user's feet are fastened firmly to the footplates, the structure of the suggested footplate can generate relative rotation between the fore-footplate and rear-footplate at the hinge when the guide wheel is sliding on the constraint guide rail to simulate the dorsiflexion of MTP joint of floor walking. The motion attitude of unilateral footplate after the crank rotated clockwise in a circle is shown as right of Fig. 3. Comparing with the foot attitude of the cycle step in Fig. 2, it can be seen that the footplate motion attitude of the robot is in accordance with the foot motion attitude of floor walking.

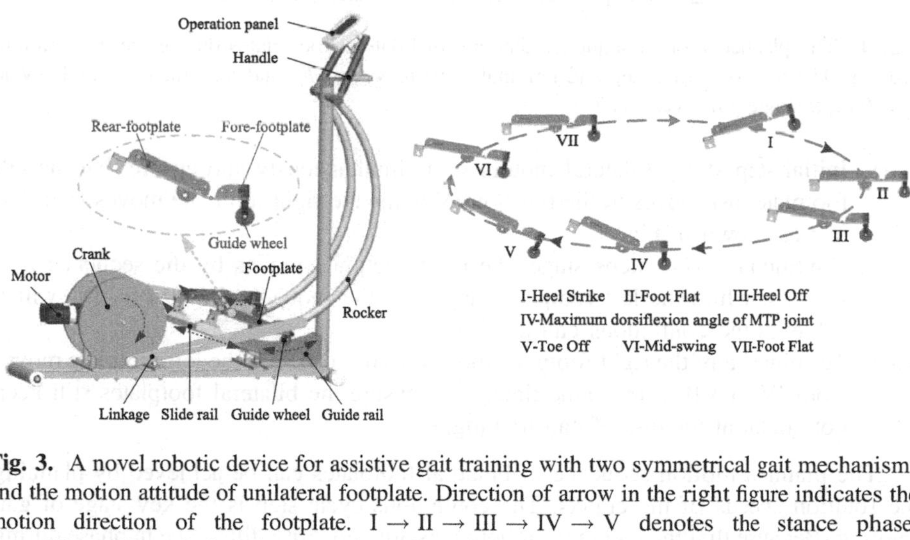

Fig. 3. A novel robotic device for assistive gait training with two symmetrical gait mechanisms and the motion attitude of unilateral footplate. Direction of arrow in the right figure indicates the motion direction of the footplate. I → II → III → IV → V denotes the stance phase, V → VI → VII → I denotes the swing phase, and III → IV → V denotes the dorsiflexion stage of MTP joint.

The suggested gait training robot not only inherits the characteristics and advantages of the existing elliptical trainer, but also provides adequate solution for the foot motion attitude and the MTP joint motions by using the ingenious constrain mechanisms without an additional driver. And two sets of symmetrical gait mechanisms are driven independently by bilateral motors to control the motion laws of the bilateral footplates, respectively. So planning the rotation speeds of bilateral cranks can achieve the coordinated gait laws of the two footplates to simulate the natural walking gait of floor walking.

3.2 Motion Planning

Two sets of gait mechanisms of this robot are driven independently to control the motion laws of the bilateral footplates, respectively. The patients can remain standing on the horizontal footplates at the beginning and end of gait training. Therefore, the gait training process can be divided into the following three stages: initial step, continuous cycle steps and terminal step as shown in Fig. 2. In order to facilitate the gait training process, the initial position of the right and left footplate are set at the position VII and II shown in Fig. 3, respectively. And the motion sequence of bilateral footplates is planned as follows (see in Fig. 4).

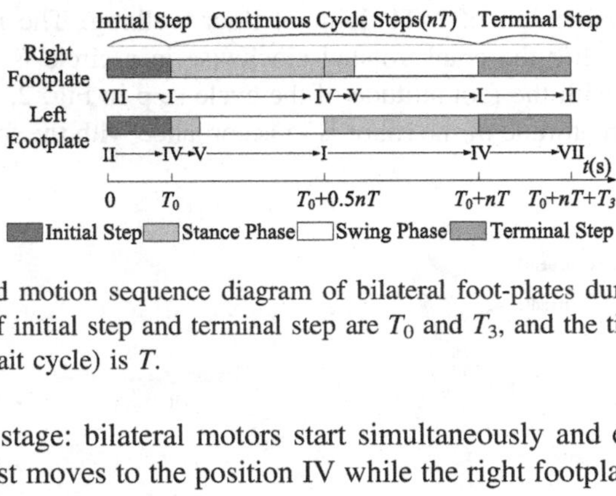

Fig. 4. The planned motion sequence diagram of bilateral foot-plates during the gait training process. The time of initial step and terminal step are T_0 and T_3, and the time of a single cycle step (i.e., a single gait cycle) is T.

(i) Initial step stage: bilateral motors start simultaneously and ensure that the left footplate just moves to the position IV while the right footplate moves from VII to I as shown in Fig. 3.
(ii) Continuous cycle steps stage: the right footplate moves by the sequence I → IV → I, while the left footplate moves by the sequence IV → I → IV, which achieves the continuous gait training.
(iii) Terminal step: the right footplate moves from I to II, and the left footplate moves from IV to VII at the same time. That ensure the bilateral footplates still keep horizontal at the end of gait training.

The planned motion sequence of bilateral footplates can be achieved by planning the rotation speeds of the cranks. The continuous cycle step is the key stage of gait training. Because that the gait laws of both legs are only 50% difference in phase during a single gait cycle, and the stance and swing phase are approximate 60% and 40% of a single gait cycle [10]. In order to ensure the phase distribution relationships and the motion speed of the footplate is continuous and smooth, the rotation speed of the right crank during the cycle step is planned as:

$$\begin{cases} \omega_{1\text{Rstance}} = \frac{1}{T}[\chi_1 \cos(2\pi t/T_1) - \omega_0 - \chi_1], & 0 \le t \le T_1 \\ \omega_{1\text{Rswing}} = \frac{1}{T}[\chi_2 \cos(2\pi t/T_2) - \omega_0 - \chi_2], & T_1 \le t \le T \end{cases} \quad (1)$$

Where

T_1 is the time of stance phase,
T_2 is the time of swing phase,
T is a single gait cycle, $T = T_1 + T_2$,
ω_0 is the rotation speed of the crank at the moment of the phase switch occurs.
χ_1 and χ_2 are coefficients.

A single gait cycle is completed after the crank rotated clockwise in a circle, so

$$\int_0^{T_1} \omega_{1\text{Rstance}} dt + \int_{T_1}^{T} \omega_{1\text{Rswing}} dt = 2\pi \quad (2)$$

The coefficients χ_1 and χ_2 can be deduced by combining Eqs. (1) and (2).

In order to ensure the rotation speeds of the crank and footplate change continuously and smoothly at the transition between the continuous cycle step and both the initial step and terminal step, the rotation speed of the right crank in the initial step and terminal step are planned by using cubic splines.

$$\begin{cases} \omega_{1\text{Rinitial}} = a_1 t^3 + a_2 t^2 + a_3 t + a_4, & 0 \le t \le T_0 \\ \omega_{1\text{Rterminal}} = b_1 t^3 + b_2 t^2 + b_3 t + b_4, & 0 < t < T_3 \end{cases} \quad (3)$$

Where

T_0 is the time of initial step,
T_3 is the time of terminal step,
$a_i, b_i (i = 1, 2, 3, 4)$ are coefficients.

And the constraints are:

$$\begin{cases} \omega_{1\text{Rinitial}}|_{t=0} = 0 \\ \omega_{1\text{Rinitial}}|_{t=T_0} = \omega_{1\text{Rstance}}|_{t=0} \\ \dot{\omega}_{1\text{Rinitial}}|_{t=T_0} = 0 \\ \int_0^{T_0} \omega_{1\text{Rinitial}} dt = \Delta\theta_{1\text{Rinitial}} \end{cases} \begin{cases} \omega_{1\text{Rterminal}}|_{t=0} = \omega_{1\text{Rswing}}|_{t=T} \\ \omega_{1\text{Rterminal}}|_{t=T_3} = 0 \\ \dot{\omega}_{1\text{Rterminal}}|_{t=0} = 0 \\ \int_0^{T_3} \omega_{1\text{Rterminal}} dt = \Delta\theta_{1\text{Rterminal}} \end{cases} \quad (4)$$

Where

$\Delta\theta_{1\text{Rinitial}} = \theta_{1\text{I}} - \theta_{1\text{VII}}$, $\Delta\theta_{1\text{Rterminal}} = \theta_{1\text{II}} - \theta_{1\text{I}}$,

$\theta_{1\text{I}}, \theta_{1\text{II}}, \theta_{1\text{VII}}$ are the rotation angle of the crank when the footplate is at the position I, II and VII as shown in Figs. 3 and 4.

The coefficients $a_i,\ b_i(i = 1, 2, 3, 4)$ can be deduced by combining Eqs. (3) and (4). So the planned rotation speed of the right crank during the process of gait training is:

$$\omega_{1R} = \begin{cases} \omega_{1Rinitial}, & 0 \le t \le T_0 \quad \text{Initial step} \\ \omega_{1Rstance}, & 0 \le t \le T_1 \quad \text{Cycle step} \\ \omega_{1Rswing}, & T_1 \le t \le T \\ \omega_{1Rterminal}, & 0 \le t \le T_3 \quad \text{Terminal step} \end{cases} \tag{5}$$

The planned rotation speed of the left crank during the process of gait training can be obtained by using the same method as above.

According to the above motion planning methods, the rotation speeds of the cranks depend on the time $T,\ T_0,\ T_3$. So the training speed can be adjusted conveniently by choosing a different gait time. Suppose $T = 10$ s, $T_1 = 6$ s, $T_2 = 4$ s, $T_0 = 3$ s, $T_3 = 3$ s. The planned rotation speeds of bilateral cranks are shown in Fig. 5.

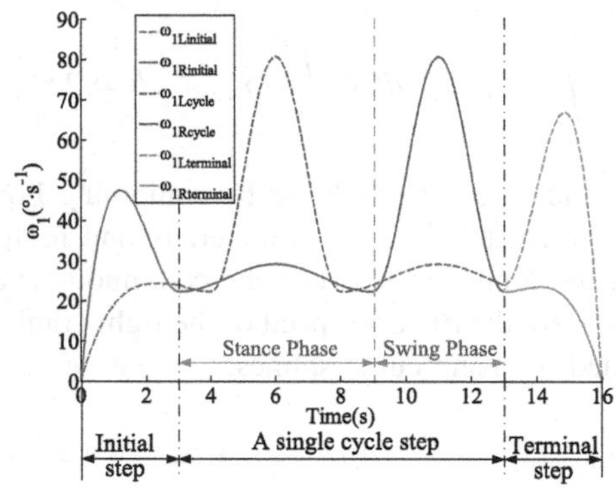

Fig. 5. The planned rotation speeds of bilateral cranks including a single cycle step

4 Experiments and Discussions

Considering the actual situation that the individuals with lower limbs motor dysfunction are manifested as decreased range of the lower limbs joints motion [2], the robotic prototype was developed. And the experiment platform consists of three main hardware parts: the gait training robot, Quanser Hardware-in-the-Loop (HIL) control system (Quanser, Canada), and host PC, as shown in Fig. 6. The real-time control board QPID of Quanser HIL control system is the interface between the gait training robot and the control software QUARC in the host PC. The control command and information of the sensors can be transmitted by the D/A ports and I/O ports on the real-time control board QPID. With the aim to achieve different gait training for the patients, two sets of symmetrical gait mechanisms can be precisely controlled independently by the position-current-controller loop.

In order to evaluate the feasibility of the design and the necessity of motion planning, set the cranks rotate without load at a uniform rotation speed $36^\circ \cdot s^{-1}$ and the

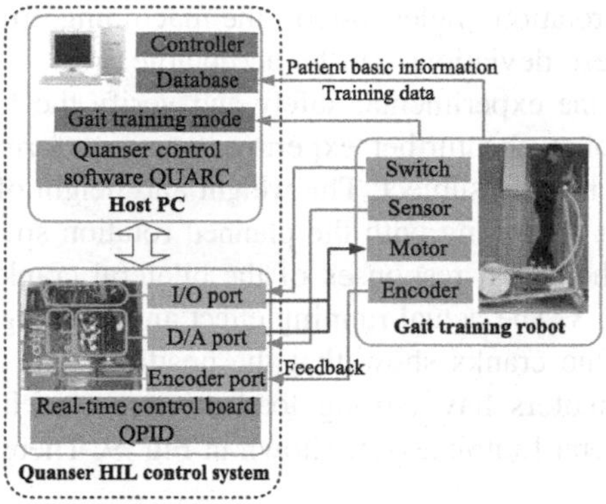

Fig. 6. The experiment platform

planned rotation speed as shown in Fig. 5, respectively. For the convenience of description, the actual responses of the right footplate in two cycle steps are chosen to analysis as shown in Fig. 7. The rotation angles of the right rear-footplate θ_{rpR} and fore-footplate θ_{fpR} are within the maximum plantarflexion angle and dorsiflexion angle of the foot during normal floor walking, which can promote greater comfort and prevent re-injure for the patients during training. And the fore-footplate (green and golden curves in Fig. 7) keeps horizontal with the ground from the terminal stance phase to the pre-swing phase of a single cycle step, which simulates the process that the forefoot is level until the toe is rolling off the ground during normal floor walking. Comparing with the curves when the crank is rotating at a uniform rotation speed, the phase distribution relationship of the stance phase and swing phase is changed when the crank is rotating at the planned rotation speed. The stance and swing phase are about 60% and 40% of a single cycle step, which is more tally with the actual phase relationship during normal floor walking. There are some deviations in the actual

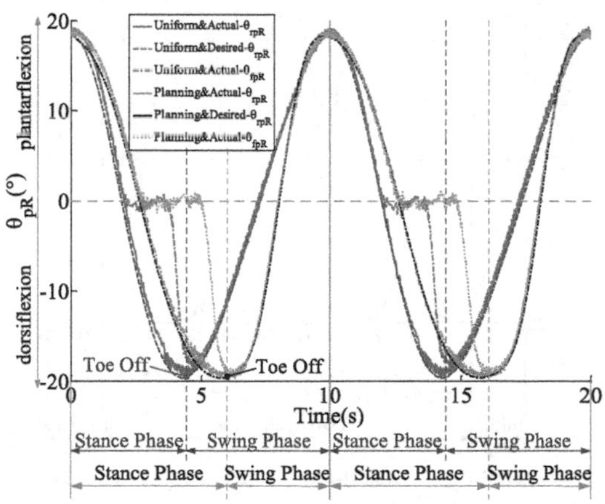

Fig. 7. The actual response of the rotation angle of right foot-plate in two cycle steps when the crank is rotating with a uniform rotation speed and the planned rotation speed, respectively. (Color figure online)

tracking curves of the rotation angles, due to the machining error and the accuracy of the tilt sensors, and these deviations can be acceptable.

In order to ensure the experimental safety and verify the function feasibility and load capacity of the robot, the further experiment was performed based the position control strategy with a healthy subject. The weight and height of the subject are 62 kg and 170 cm. The robot is running with the planned rotation speeds of bilateral cranks as shown in Fig. 5. The actual responses of the bilateral cranks including two cycle steps are shown in Fig. 8. The actual running effect and tracking curves of the rotation angles and speeds of the cranks show that the position controller has good control performance and the motors have strong load driving capacity. The corresponding rotation angles of bilateral footplates are shown in Fig. 9. There is a little lag between

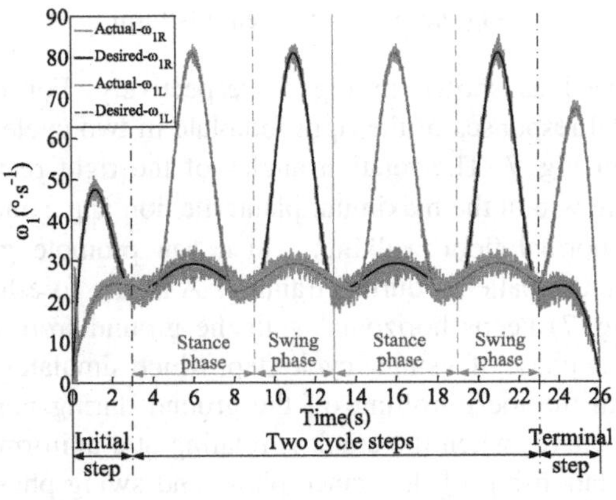

Fig. 8. The actual response of the planned rotation speeds of bilateral cranks including two cycle steps

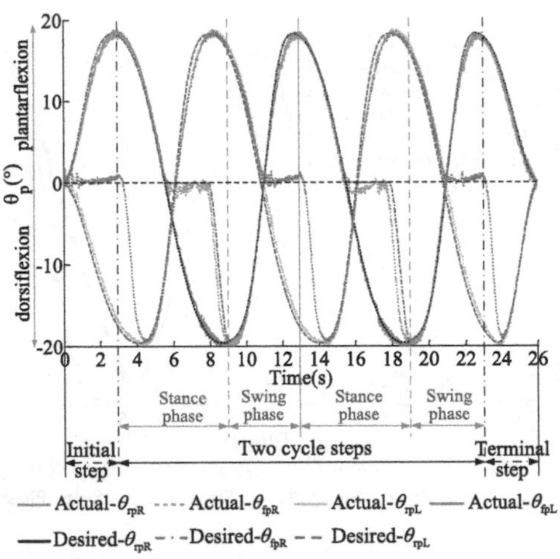

Fig. 9. The actual response of the rotation angles of bilateral footplates including two cycle steps when the bilateral cranks are rotating with the planned rotation speeds.

actual experiment curve and the desired curve due to the rotation speeds of the footplates are changing during gait training. And the fore-footplate rotates relative to the rear-footplate from the rotation angle of the rear-footplate is zero to the moment that the rotation angle of the rear-footplate is negative maximum. That is the dorsiflexion of MTP joint and is consistent with the actual motion characteristic of the foot during human normal walking on level ground.

5 Conclusion

In this paper, a novel robotic device for gait training has been developed to help the nonambulatory gait-impaired patients improve gait training efficiency, and it enables the robot generate a human-like walking gait and provide adequate solution for the dorsiflexion of MTP joint by using the ingenious constrain mechanisms without an additional driver. Based on the gait characteristics and phase distribution relationship during human normal walking on level ground, the motion sequence of bilateral footplates and the rotation speeds of bilateral cranks are planned. The experiments of robotic prototype running without load and with a subject have been completed and the results are encouraging and verify the feasibility and effectiveness of the design and motion planning. The developed robotic device has advantages of simple structure, easy operation, low power and more authentic gait. That is of great significance for gait training. Future work will focus on the research of advanced control strategies and motion planning strategies to make the gait training more personalized and comfortable.

Acknowledgments. This work was supported by an open project of electromechanical-automobile discipline of Hubei Province under grant No. XKQ2017015.

References

1. Hirata, Y., Iwano, T., Tajika, M., Kosuge, K.: Motion control of wearable walking support system with accelerometer based on human model. In: Kulyukin, V.A. (ed.) Advances in Human-Robot Interaction, pp. 322–327. InTech, Rijeka (2009)
2. Kubo, K., Miyoshi, T., Kanai, A., Terashima, K.: Gait rehabilitation device in central nervous system disease: a review. J. Robot. **2011**(5), 1–14 (2011)
3. Lloyd-Jones, D., Adams, R.J., Brown, T.M.: Heart disease and stroke statistics-2010 update: a report from the American heart association. Circulation **121**(7), e46–e215 (2010)
4. Hocoma. https://www.hocoma.com/solutions/lokomat/. Accessed 23 Apr 2017
5. Hesse, S., Waldner, A., Tomelleri, C.: Innovative gait robot for the repetitive practice of floor walking and stair climbing up and down in stroke patients. J. NeuroEng. Rehabil. **7**(1), 30 (2010)
6. Jungwon, Y., Novandy, B., Yoon, C.H., Park, K.J.: A 6-DOF gait rehabilitation robot with upper and lower limb connections that allows walking velocity updates on various terrains. IEEE/ASME Trans. Mechatron. **15**(2), 201–215 (2010)

7. VR Lab: University of Tsukuba. http://intron.kz.tsukuba.ac.jp/gaitmaster/gaitmaster_e.html. Accessed 23 Apr 2017
8. Nawoczenski, D.A., Baumhauer, J.F., Umberger, B.R.: Relationship between clinical measurements and motion of the first metatarsophalangeal joint during gait. J. Bone Joint Surg. **81**(3), 370–376 (1999)
9. Zhang, L.X., Qin, T., Song, C.Y., Cheng, S.: Prototype development and experimental study on a footpad-type walking rehabilitation robot. High Technol. Lett. **23**(9), 939–945 (2013)
10. Winter, D.A.: Biomechanics and Motor Control of Human Movement, 4th edn. University of Waterloo Press, Ontario (1991)
11. Yoon, J., Ryu, J.: A novel locomotion interface with two 6-DOF parallel manipulators that allows human walking on various virtual terrains. Int. J. Robot. Res. **25**(7), 689–708 (2006)
12. Gilchrist, L.A., Winter, D.A.: A two-part, viscoelastic foot model for use in gait simulation. J. Biomech. Eng. **29**(6), 795–798 (1996)
13. Burnfield, J.M., Shu, Y., Buster, T., Taylor, A.: Similarity of joint kinematics and muscle demands between elliptical training and walking: implications for practice. Phys. Ther. **99**(2), 289–305 (2010)

Impedance Control of a Pneumatic Muscles-Driven Ankle Rehabilitation Robot

Chi Zhang[1,2], Jiwei Hu[1,2], Qingsong Ai[1,2(✉)], Wei Meng[1,2], and Quan Liu[1,2]

[1] School of Information Engineering, Wuhan University of Technology, Wuhan 430070, China
qingsongai@whut.edu.cn

[2] Key Laboratory of Fiber Optic Sensing Technology and Information Processing, Wuhan University of Technology, Ministry of Education, Wuhan 430070, China

Abstract. Pneumatic muscle is a new type of flexible actuator with advantages in terms of light weight, large output power/weight ratio, good security, low price and clean. In this paper, an ankle rehabilitation robot with two degrees of freedom driven by pneumatic muscle is studied. The force control method with an impedance controller in outer loop and a position inner loop is proposed. The demand of rehabilitation torque is ensured through tracking forces of three pneumatic muscle actuators. In the simulation, the constant force and variable force are tracked with error less than 10 N. In the experiment, the force control method also achieved satisfactory results, which provides a good support for the application of the robot in the ankle rehabilitation.

Keywords: Pneumatic muscle · Ankle rehabilitation · Impedance control

1 Introduction

With the gradual growth of population over the age of 60 years, China has entered the aging society. The aged people have a significant decline in the degree of physical activity of limbs, which brings a lot of inconvenience to their daily life. In addition, the number of patients with joint and muscle injury is increasing rapidly and their rehabilitation problems are becoming more and more serious. Medical theory has proved that appropriate amount of scientific rehabilitation training can improve the recovery effect for ankle injury patients after the completion of surgery [1]. Traditional ankle rehabilitation needs high physical demanding work from therapists. However, because of the lack of rehabilitation surgeons, patients can not get enough rehabilitation training, which reduces the rehabilitation effect. The assistance of ankle robot can reduce the burden of rehabilitation physicians and help ankle injury patients speed up the recovery. The parameters in ankle rehabilitation training can also be recorded to provide the basis for physicians to develop a rehabilitation plan for next stage.

Pneumatic muscle is a new type of pneumatic actuator with output characteristics to human muscle. It is composed of an inner rubber tube and an almost non elongated diamond braid wrapped around the rubber. The two ends have packaging and fixing devices (see Fig. 1). The internal rubber tube expanding under the gas pressure leads to the increase of pneumatic muscle diameter and the decrease of its length [2]. And the

Y. Huang et al. (Eds.): ICIRA 2017, Part I, LNAI 10462, pp. 301–312, 2017.
DOI: 10.1007/978-3-319-65289-4_29

external braided mesh limits pneumatic muscles to be shortened without limitation. If the end is subjected to the external tension, the pneumatic muscle can produce the tension. Pneumatic muscle has the advantages of light weight, large output power/dead-weight ratio, good safety, cleanness, low price and so on [3]. Based on these characteristics, the application in rehabilitation field has been studied and many rehabilitation robots using pneumatic muscles as actuators have been developed [4]. However, due to the flexible rubber material, friction of woven web [5], nonlinear of compressed air and the uncertainty of robot model, the robot control system is very complex with strong nonlinearity and parts of time-varying parameters [6].

Fig. 1. PESTO pneumatic muscle

TU Diep Cong Thanh et al. used nonlinear PID controller to improve the control performance of the two axis pneumatic muscle mechanical arm. They combined the traditional PID controller and neural network and proposed a neural network nonlinear PID controller, which was suitable for the control object with strong nonlinearity, uncertainty and disturbance. The experimental results showed that the controller had good control performance and anti external interference [7]. Lin Chih-Jer et al. studied the hysteresis characteristics of pneumatic muscles. They established the double pneumatic muscle system PI model used as the feedforward compensation of a sliding mode controller to reduce the tracking error [8]. Shameek Ganguly et al. studied the position control of a single freedom degree manipulator driven by pneumatic muscle and put forward a new method to establish an accurate model for the system [9].

Patients can move along the predetermined trajectory through these position control methods to improve the patients' movement ability and the joints' mobility [10]. However, the simple position control will put the patients in a passive state and the output torque can not be given quantitatively, which may lead to secondary damage. The impedance control can enhance the interaction between patients and robots during the rehabilitation training, so that patients can participate in the rehabilitation training more actively.

2 The Ankle Rehabilitation Robot

2.1 Ankle Model

In this paper, the robot platform driven by pneumatic muscle is mainly used for ankle rehabilitation. The ankle joint is one of the most complex skeletal structures of human (see Fig. 2). The ankle model has three rotational freedom: the varus and valgus motions around the X axis, the plantar flexion and dorsiflexion motions around the Y axis and the adduction and abduction motions around the Z axis. Carl Mattacola, University of Kentucky, deeply studied the rehabilitation courses of ankle injury patients and pointed out that the rotational motions around X and Y axis played a major role in ankle rehabilitation. In this paper, the robot platform just right has these two freedom and the workspace of the robot can reach the ankle motion range in these two directions.

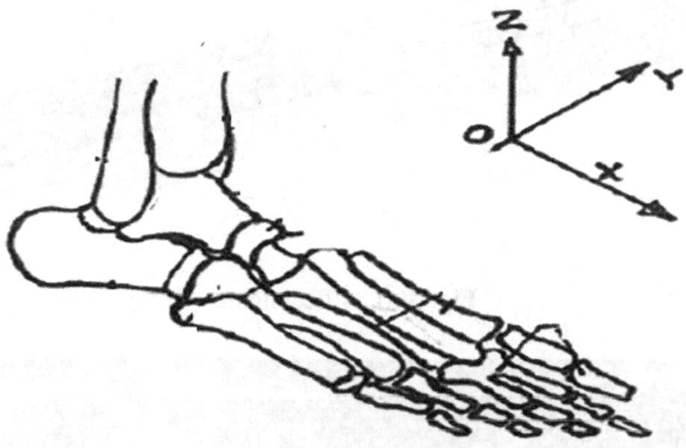

Fig. 2. Ankle joint

2.2 Robot Platform

The effect figure and practicality picture of this ankle rehabilitation robot is shown in Fig. 3a and 3b. The robot is driven by three pneumatic muscles that pull the wire rope. In order to reduce the height of the robot platform and facilitate the rehabilitation training of patients, the pneumatic muscles are placed in horizon. One end of the pneumatic muscle is connected with the force sensor, which is also fixed on the platform frame. The other end is connected with the wire rope. After changing directions through the fixed pulley and then passing through the three holes on the fixed platform, the three wire ropes are connected with the moving platform. Pneumatic muscle drives the wire rope in this way and then drives the platform to complete the corresponding action. The fixed hole on the fixed platform is made of plastic material with smooth surface to reduce the friction between the wire rope and the hole. A rigid supporting rod is vertically arranged between the moving platform and the fixed platform. The lower end is fixed with the fixed platform and the upper end is connected with the moving platform through a Hooke hinge. The Hooke hinge limits the rotational motion of the moving platform in the Z axis direction to ensure that the moving platform has only two degree of freedom. The robot is also equipped with displacement sensors and force sensors to measure the position and force information of three pneumatic muscle actuators during operation.

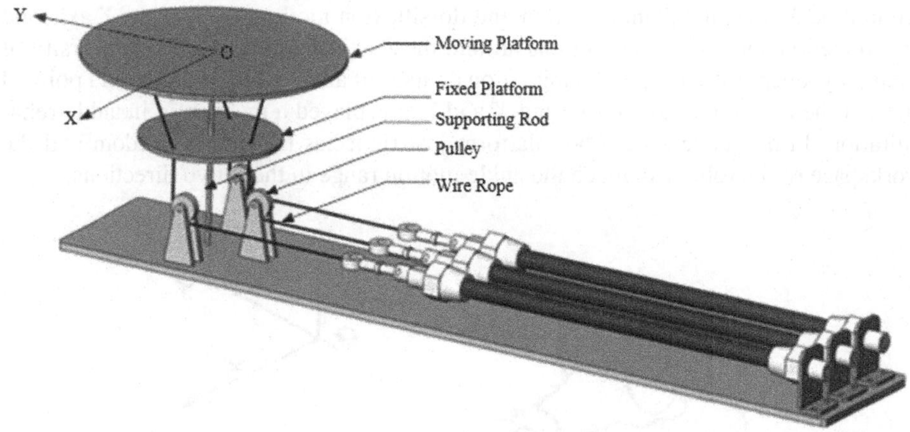

Fig. 3(a). Effect figure

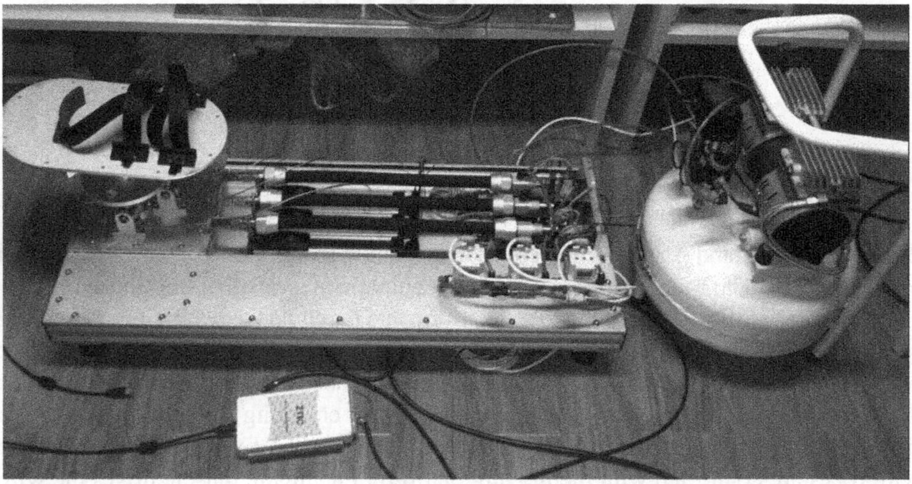

Fig. 3(b). Practicality picture

The system schematic diagram of the robot is shown in Fig. 4. The robot communicates with the host computer written by LabView through the data acquisition card. The pneumatic muscle's gas is supplied by the air source and the quantity is determined by the input voltage of the proportional valve. The data acquisition card is connected with the host computer through USB interface. On the one hand, it can convert the analog signals from displacement sensors and force sensors into digital signals through A/D and then send the signal to the host computer to process. On the other hand, it can convert the digital control signal calculated by the control algorithm into analog voltage input signal through D/A and then send the signal to the proportional valve. The proportional valve can adjust the air input of the pneumatic muscle according to the voltage signal and then achieve corresponding movement.

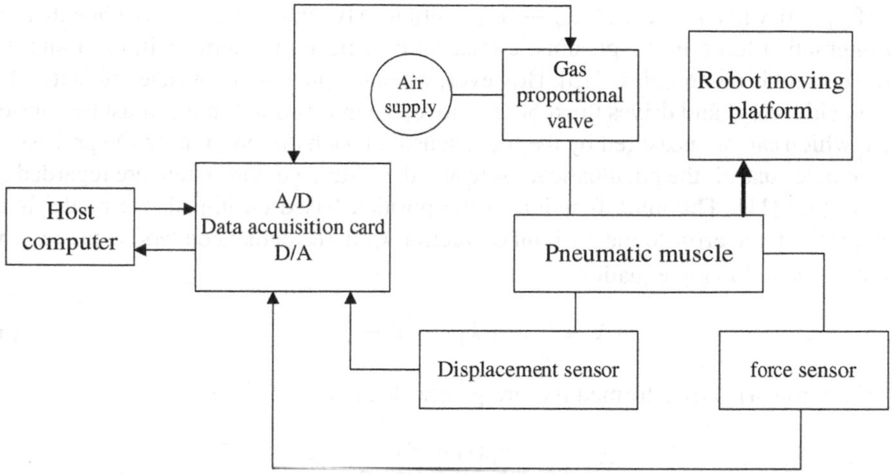

Fig. 4. System working principle diagram

3 Impedance Control of Ankle Rehabilitation Robot

3.1 Impedance Control Model of Pneumatic Muscle

The impedance controller is comprised of an inlayer position control unit and an outlayer impedance control unit [11]. In this paper, the ankle rehabilitation robot only installs the force sensors in each pneumatic muscle, the interaction force between the robot and the external environment can be reflected by the force in joint space. Therefore, we can track the output force of each pneumatic muscle through impedance control of each driven branch to realize the force control [12].

Equation (1) is a commonly used target impedance model, which describes the relationship between the interaction force and the position [13].

$$M_d(\ddot{X} - \ddot{X}_d) + B_d(\dot{X} - \dot{X}_d) + K_d(X - X_d) = F - F_e \tag{1}$$

Because the single pneumatic muscle only moves in one direction, the equation can be simplified as one-dimensional form [14].

$$m_d(\ddot{x} - \ddot{x}_d) + b_d(\dot{x} - \dot{x}_d) + k_d(x - x_d) = f_d - f_e \tag{2}$$

m_d b_d k_d, respectively, are the inertia parameter, damping parameter and stiffness parameter.f_d f_e, respectively, are the desired interaction force and actual interaction force. x is the pneumatic muscle trajectory. When the pneumatic muscle moves in the free space without load, the interaction force f_e is 0.

$$m_d(\ddot{x} - \ddot{x}_d) + b_d(\dot{x} - \dot{x}_d) + k_d(x - x_d) = f_d \tag{3}$$

If $f_d = 0$, when $t \to \infty$, $x - x_d \to 0$, x continuously approaches x_d. So when there is no interaction force in the pneumatic muscle, only position control will exist and the force control is meaningless [15]. However, when the pneumatic muscle interacts with the outside world and drives the robot to move, the interaction force f_e must be considered, which can be measured by the force sensor of each driven joint. In the process of impedance control, the pneumatic muscle and the external environment are regarded as one system [11]. The main function of the position based on impedance model is to convert the force error to the position correction x_f. In the actual conversion process, x_f satisfies the following equation:

$$m_d\ddot{x}_f + b_d\dot{x}_f + k_dx_f = f_d - f_e \tag{4}$$

Equation (4) is transformed into frequency domain:

$$X_f(s) = \frac{F_d(s) - F_e(s)}{m_ds^2 + b_ds + k_d} \tag{5}$$

In the target impedance model, m_d, b_d, k_d, respectively, correspond to the acceleration, velocity and position of the control object. They can be adjusted in the actual control process, which can improve the dynamic response speed and steady-state tracking effect of the control system. The stiffness parameter k_d is mainly related to the input pressure. And the relationship between them can be expressed as:

$$k = \alpha_1 p + n\frac{dp}{dL} + \alpha_0 \tag{6}$$

P is the pneumatic input pressure and it corresponds to the input voltage of the proportional valve. α_1, n, α_0, need to be adjusted by experiment. block diagram of the Pneumatic muscle impedance controller is shown in Fig. 5.

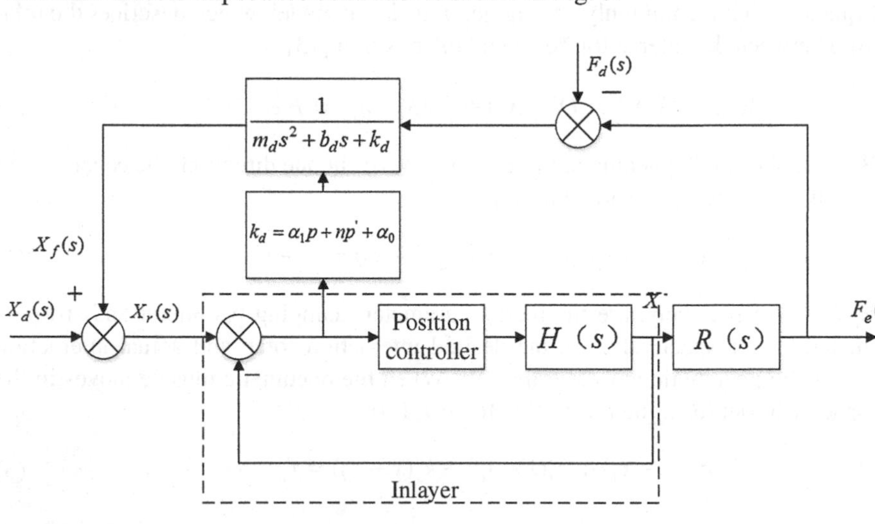

Fig. 5. Block diagram of control principle

Impedance control can provide a suitable impedance force based on the patient's rehabilitation status. Patients need to overcome this force in rehabilitation training so as to achieve the purpose of active rehabilitation. This method is suitable for patients with a certain ability to exercise.

When the actual impedance control is carried out on the robot platform, firstly, the output force should be determined according to the recovery status of the ankle. After the force is determined, the desired output force is tracked by establishing pneumatic muscle impedance control.

4 Experiment

4.1 Simulation

Considering that the actual control object is a single pneumatic muscle, Simulink is used to simulate the impedance control of the model.

In the simulation model, in order to simulate the force model in the real environment, the contact environment between the robot and the patient's lower limb is simplified to the stiffness and damping system. The relationship between the contact force and the motion of the platform is established as follows:

$$f_e = k(x - x_0) - d(\dot{x} - \dot{x}_0) \tag{7}$$

In the equation, k is the simulation stiffness of contact environment, d is the damping coefficient, $x - x_0$ denotes the position variation of the robot platform and $\dot{x} - \dot{x}_0$ denotes the velocity variation.

Generally, the desired acceleration $\ddot{x}_d$ is 0. So the Eq. (1) can be deformed as follows:

$$\ddot{x} = m_d[f_d - f_e + b_d(\dot{x}_d - \dot{x}) + k_d(x_d - x)] \tag{8}$$

In this impedance control model, we can obtain the correction of acceleration $\ddot{x}$ by desired interaction force f_d, actual interaction force f_e, actual displacement x_e and desired displacement x_d. Then displacement correction will be calculated through two integration. The impedance controller used in this model can get better simulation results. Constant force tracking simulation was done and the desired driving force was set to 20 N. The simulation result was shown in Fig. 6.

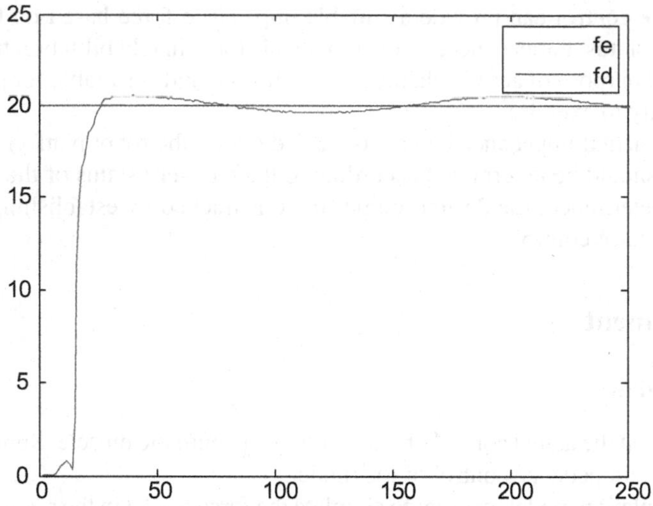

Fig. 6. Constant force tracking

Then variable force tracking simulation was done and the force was set to:

$$y = 20 * \sin t \tag{9}$$

The simulation result was shown in Fig. 7.

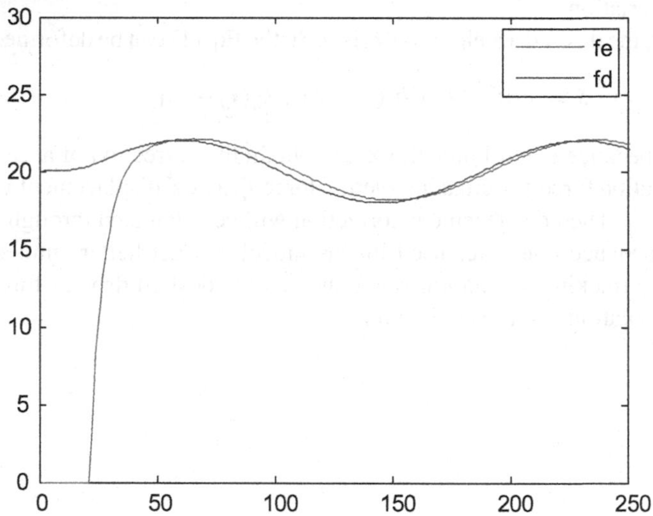

Fig. 7. Variable force tracking

4.2 Impedance Control Experiment

Based on impedance control method of the pneumatic muscle, the experiment is carried out on the physical robot platform. Because of the characteristics of pneumatic muscle and the application background of ankle rehabilitation, in experiment, the inertia parameter m_d and the damping parameter b_d use fixed value. The stiffness parameter k_d is adjusted in real time according to the output voltage of the controller.

In the constant force tracking experiment, the desired forces of three pneumatic muscles are set 60 N, 70 N, 85 N, respectively (Figs. 8, 9 and 10).

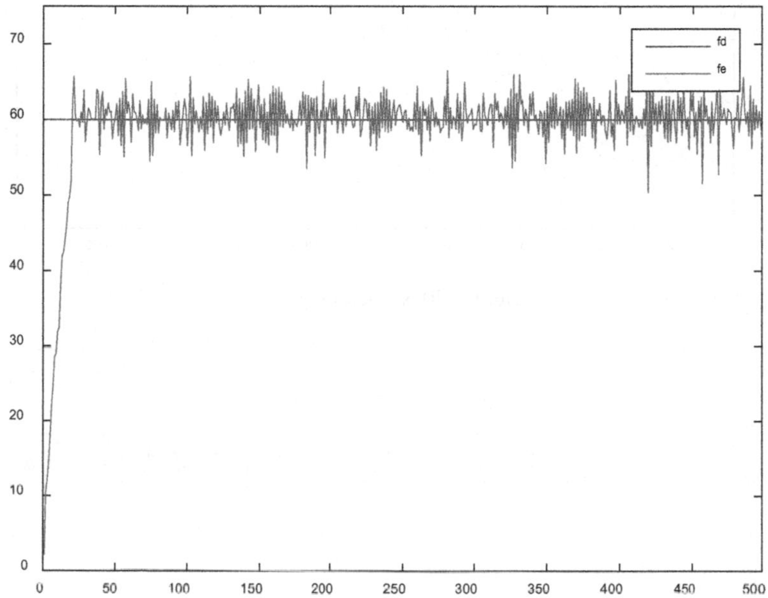

Fig. 8. 60 N tracking graph

The results show that the response time is a little long when the robot tracks the constant force. After that, the tracking force will gradually stabilize. But there are still some fluctuations, and the range of the force error is less than 10 N. This is mainly because in the initial stage of impedance control, actuator output force is almost 0, which leads to the great force deviation. The impedance model adjusts the actuator to achieve desired output force through converting the force error to the displacement correction. This conversion process takes a long time, resulting in a longer response time. It is also an important reason for the force fluctuation.

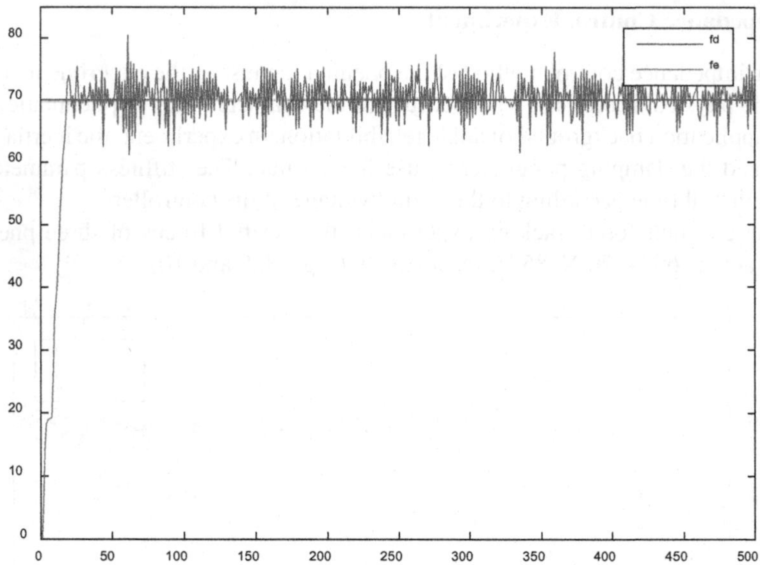

Fig. 9. 70 N tracking graph

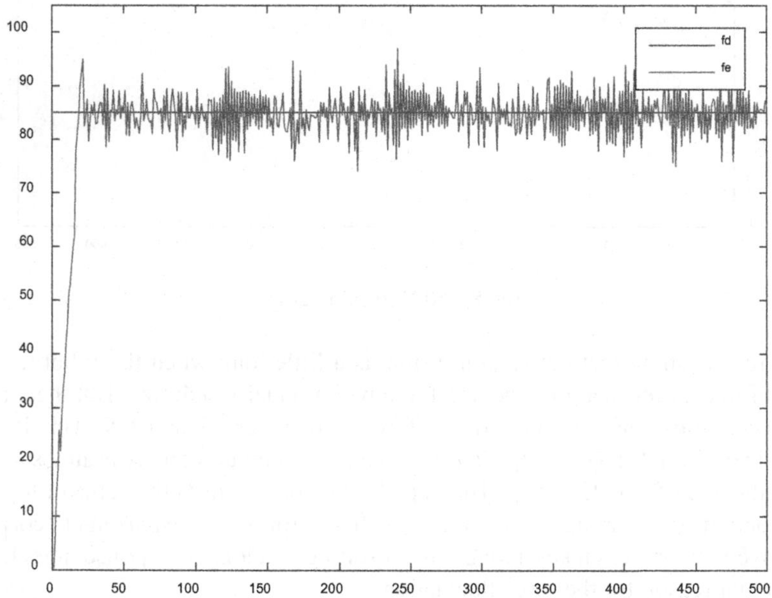

Fig. 10. 85 N tracking graph

5 Conclusion

In this paper, the ankle rehabilitation robot driven by three pneumatic muscles is presented as an experimental platform. An impedance control method based on position inner loop is proposed as the control method. The force control is realized through the force tracking of three pneumatic muscles. This control strategy makes up for the lack of human-computer interaction defect because of only position control. In the actual experiment, the force tracking in the error range of 10 N is realized, which proves the feasibility of this method.

Acknowledgments. This research is supported by National Natural Science Foundation of China under grants No. 51675389, 51475342.

References

1. Bradley, D., et al.: NeXOS-the design, development and evaluation of a rehabilitation system for the lower limbs. Mechatronics **19**(2), 247–257 (2009)
2. Inoue, K.: Rubbertuators and applications for robots. In: Proceedings of the 4th IEEE International Symposium on Robotics Research, Cambridge, pp. 57–63 (1988)
3. Chou, C.P.: Measurement and modeling of McKibben pneumatic artificial muscle. IEEE Trans. Robot. Autom. **12**(1), 90–102 (1996)
4. Gaylord, R.H.: Fluid actuated motor system and stroking device. U.S. Patent 2238058, 22 July (1958)
5. Doumit, M., Fahim, A.: Michael Munro. analytical modeling and experimental validation of the braided pneumatic muscle. IEEE Trans. Robot. **25**(6), 1282–1291 (2009)
6. Wickramatunge, K.C., et al.: Study on mechanical behaviors of pneumatic artificial muscle. Int. J. Eng. Sci. **48**(2), 188–198 (2010)
7. Tu, D.C.T., Ahn, K.K.: Nonlinear PID control to improve the control performance of 2 axes pneumatic artificial muscle manipulator using neural network. Mechatronics **16**(9), 577–587 (2006)
8. Lin, C.J., Lin, C.R.: Hysteresis modeling and tracking control for a dual pneumatic artificial muscle system using Prandtl-Ishlinskii model. Mechatronics **28**, 35–45 (2015)
9. Ganguly, S., Garg, A.: Control of pneumatic artificial muscle system through experimental modeling. Mechatronics **22**(8), 1135–1147 (2012)
10. Perez Ibarra, J.C.: Adaptive impedance control for robot-aided rehabilitation of ankle movements. In: 2014 5th IEEE RAS & EMBS International Conference on Biomedical Robotics and Biomechatronics (BioRob), São Paulo, Brazil (2014)
11. Proietti, T., Crocher, V.: Upper-limb robotic exoskeletons for neurorehabilitation: a review on control strategies. IEEE Rev. Biomed. Eng. **9**, 4–14 (2016)
12. Chen, S.H., Lien, W.M.: Assistive Control System for Upper Limb Rehabilitation Robot. IEEE Transactions on Neural Systems & Rehabilitation Engineering A Publication of the IEEE Engineering in Medicine & Biology Society **24**(11), 1199–1209 (2016)
13. Prashant, K.: Impedance control of an intrinsically compliant parallel ankle rehabilitation robot. IEEE Trans. Industr. Electron. **63**(6), 3638–3647 (2016)

14. Shahid, H., Sheng, Q.: Adaptive impedance control of a robotic orthosis for gait rehabilitation. IEEE Trans. Cybern. **43**(3), 1025–1034 (2013)
15. Meng, W., Liu, Q.: Recent development of mechanisms and control strategies for robot-assisted lower limb rehabilitation. Mechatronics **31**, 132–145 (2015)

Gait Recognition Using GA-SVM Method Based on Electromyography Signal

Ying Li, Farong Gao[(✉)], Xiao Zheng, and Haitao Gan

School of Automation, Institute of Intelligent Control and Robotics, Hangzhou Dianzi University, Hangzhou 310018, China
frgao@hdu.edu.cn

Abstract. To improve the recognition accuracy of the lower limb gait, a classification method based on genetic algorithm (GA) optimizing the support vector machine (SVM) was proposed. Firstly, electromyography (EMG) signals were collected from four thigh muscles related to lower limb movements. Then the values of variance and integral of absolute were extracted as the useful features from de-noised EMG signals. Finally, the penalty parameter and the kernel parameter were optimized by GA. The results show that the GA-SVM classifier can effectively identify five gait phases of the extremity motion, and the average accuracy is increased by 6.56%, higher than the non-parameter-optimized SVM method.

Keywords: Gait recognition · EMG · Feature extraction · Genetic algorithm · Support vector machine

1 Introduction

Gait is a move pattern of the lower limb walking, which has the characters of periodicity, continuity and repeatability. In most cases, a gait cycle is divided into the stand phase and swing phase according to the heel strike and toe off, and the stand phase and swing phase can also be subdivided into several phases due to the diversity and complexity of the gait model [1]. The electromyography (EMG) signal, which reflects the instantaneous muscle activity, is generated with the contraction and relaxation of the skeletal muscle in walking stage. Therefore the characteristics of EMG can be implemented to recognize human motion [2, 3]. The EMG signal was used to identify limb movements in the 1990s. The presented recognition algorithms include neural network (NN), hidden Markov model (HMM), support vector machine (SVM), Bayes classification, etc. [4–6]. Considering that the EMG signal is weak and unstable [7], it is still faced with the challenges to classify walking gait not only from the perspective of classifier but also from the view of recognition accuracy.

With the development of SVM [8], the SVM is widely applied in the human movement recognition due to the advantages of versatility, robustness [9]. Compared to

This work is supported in part by National Natural Science Foundation of China (U1509203, 61372023).

Y. Huang et al. (Eds.): ICIRA 2017, Part I, LNAI 10462, pp. 313–322, 2017.
DOI: 10.1007/978-3-319-65289-4_30

the artificial neural networks (ANN) and locally-weighted projection regression, the SVM performed better to discriminate five finger grasping motions [10]. The finger movements were also classified where the optimizing SVM with updating decision function was proposed in [11]. Furthermore, the forearm and hand movements were identified by SVM combined with wavelet analysis [12]. As reported in [13], continuous phases of a gait cycle were classified by SVM in different walking movements, including walking, stair ascent, and stair descent.

The major problem encountered in the SVM model is how to determine the parameter values of the penalty and the kernel function. Some methods are introduced to optimize parameters, such as the trial and error method, grid-search, cross validation, particle swarm optimization (PSO), genetic algorithm optimization (GA), etc. [14–16]. The PSO-SVM classifier was utilized to classify EMG signals for neuromuscular disorders diagnosis [17], and it was concluded that PSO-SVM classifier can be an effective tool for quantifying the disorder level. Schizas et al. compared the performance of SOMs, K-means, MLP-NN, and genetic-based classifier, it was shown that both ANN and GA model had promising results [18]. A hybrid method combining SVM with GA was proposed for feature weighting in [19]. GA-SVM model was applied in the discrimination of the surrounding rock [20], and it turned out that had a high classification capability for generalization and prediction accuracy. Kanitz et al. exploited GA-SVM to decode individuated finger movements based on EMG [21], the recognition results were similar to other classifiers. Actually, the performance of SVM classifier depends on the parameter selection of the penalty and kernel functions, which have some limitations, including the scale of data size, time consuming, and difficulties in finding the optimal parameters.

In this paper, the GA-SVM method was proposed to classify gait phases based on EMG signals. The paper is organized as follows. In Sect. 2, the EMG features are extracted, and some basic principle and model are described. Section 3 provides an optimized method of GA-SVM to identify five gait phases with the experimental data in walking motion, in which the performances are discussed with classification accuracy. Finally, the conclusions are summarized in Sect. 4.

2 Methods

2.1 EMG Feature Extraction

The EMG signal is usually considered as the random signal that follows the zero-mean Gauss distribution, whose variance changes with the signal intensity. The proposed EMG features contain time domain features, frequency domain features, time-frequency domain features, and nonlinear dynamic features [22]. Owing to the simple and efficient operation, the time domain features are extracted widely. In this paper, two time domain features, integral of absolute value I and variance V, are chosen to represent the level of muscle contraction, for each feature sequence which are calculated as follows,

$$I = \frac{1}{N}\sum_{i=0}^{N-1}|x_i| \quad (1)$$

$$V = \frac{1}{N-1}\sum_{i=0}^{N-1}(x_i)^2 \quad (2)$$

where x_i is the i-th EMG sample sequence, and N is the length of EMG sample segment, respectively. Thus, the feature vectors can constitute a feature matrix $\boldsymbol{Z}=[\boldsymbol{I}, \boldsymbol{V}]$, i.e.,

$$\boldsymbol{Z} = [\boldsymbol{Z}_{mj}] \quad (3)$$

where $m = 1, 2, \ldots, M$ is the sequence of feature vectors, $j = 1, 2, \ldots, K$ with $K{=}kl$, in which k is the number of selected muscles, and l the number of feature vectors, respectively.

2.2 Support Vector Machine

The Support vector machine (SVM) was proposed by Vapnik as a classification technique [8], and was widely used in pattern recognition due to its robustness and versatility. SVM is a popular machine learning method based on statistical learning theory, in which the vectors are mapped into high dimensional feature space, the non-linear relationships between input and output are found in a new high-dimensional space. That is, the best location is found to create linear boundaries through the non-linear transformations. The SVM system structure is illustrated in Fig. 1.

In Fig. 1, $\boldsymbol{Z}_1, \boldsymbol{Z}_2, \ldots, \boldsymbol{Z}_M$ are the input feature vectors, and $K(\boldsymbol{Z}, \boldsymbol{Z}_1)$, $K(\boldsymbol{Z}, \boldsymbol{Z}_2), \ldots, K(\boldsymbol{Z}, \boldsymbol{Z}_M)$ are the inner product functions of the support vector machine. The decision function of the output classification $f(\boldsymbol{Z})$ can be defined as,

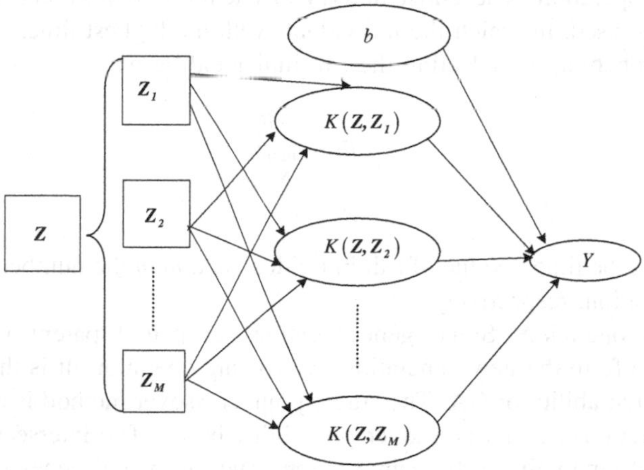

Fig. 1. The support vector machine system

$$f(\mathbf{Z}) = \text{sgn}(w\mathbf{Z} + b) = \text{sgn}\left(\sum_{j=1}^{M} a_j Y_j K(\mathbf{Z}_j, \mathbf{Z}) + b\right), 0 \le a_j \le c \tag{4}$$

where, $b \in R$ represents the bias, a_j is Lagrange multiplier corresponding to each training sample, c the penalty parameter, and w weight coefficient, respectively.

Thus, the samples are transformed from the low-dimensional space to the high-dimensional space by kernel functions, and are classified in the high dimensional space by the decision function. To a certain extent, the problems of the SVM model are how to decide on the parameters to acquire good classification results. Therefore, it is important for classification results to select an appropriate method of the parameter optimization in specific problems.

2.3 GA-SVM Classification

As a parallel optimization algorithm, genetic algorithm (GA) has been proposed to the optimization of SVM parameters. The parameter combinations of optimal fitness are acquired through multi-iteration selection, crossover and mutation operation after coding samples according to the chromosome. The contents of GA-SVM are as the following:

(1) Chromosome code. The initial population is generated by chromosome coding, and the solution of the original problem is transformed into the range that the genetic algorithm can handle. The penalty parameter c and kernel function parameter g are encoded by using the binary method, which are encoded as binary strings.
(2) Fitness calculation. The populations are selected based on the principle of survival of the fittest by computing individual fitness. In the case of cross certification, the recognition rate of training set via SVM classifier is selected as the fitness value. The higher the recognition rate, the greater the fitness value, the individual is kept.
(3) Selection operation. The elitist individual retention strategy and roulette wheel method are used, in which the individuals with the highest fitness are preserved. Let the probability of selecting the i-th individual be p_i,

$$p_i = \frac{F_i}{\sum_{j=1}^{num} F_j} \tag{5}$$

where F_i is the fitness value of i-th individual, and num the number of individuals in a population, respectively.
(4) Crossover operation. Some gene locations of paired parent individuals are replaced to form the new generation by crossing operation. It is the guarantee of global search ability of GA. The single point crossover method is exploited to do cross operation. Let the crossover probability be P_c. The intersection points are determined randomly in the paired parent, and the partial genes of some paired individual are exchanged at the intersection.

(5) Mutation operation. Some gene positions of individuals are changed to form a new individual, which can improve the local search ability of GA, and ensure the diversity of the population to avoid being trapped in the local optimal solution. For the binary encoding, gene loci to be mutated are decided randomly, in which mutation are operated through the method of reversing bit by bit. The mutation probability is P_m.
(6) Termination condition. The termination condition is set as maximum evolutional generation, which is end for the GA iteration.
(7) Parameter initialization. The parameters are needed to set for GA initialization, which are population size, chromosome string length, maximum evolutional generation, crossover probability and mutation probability, respectively.

The proposed GA-SVM algorithm is depicted in Fig. 2.

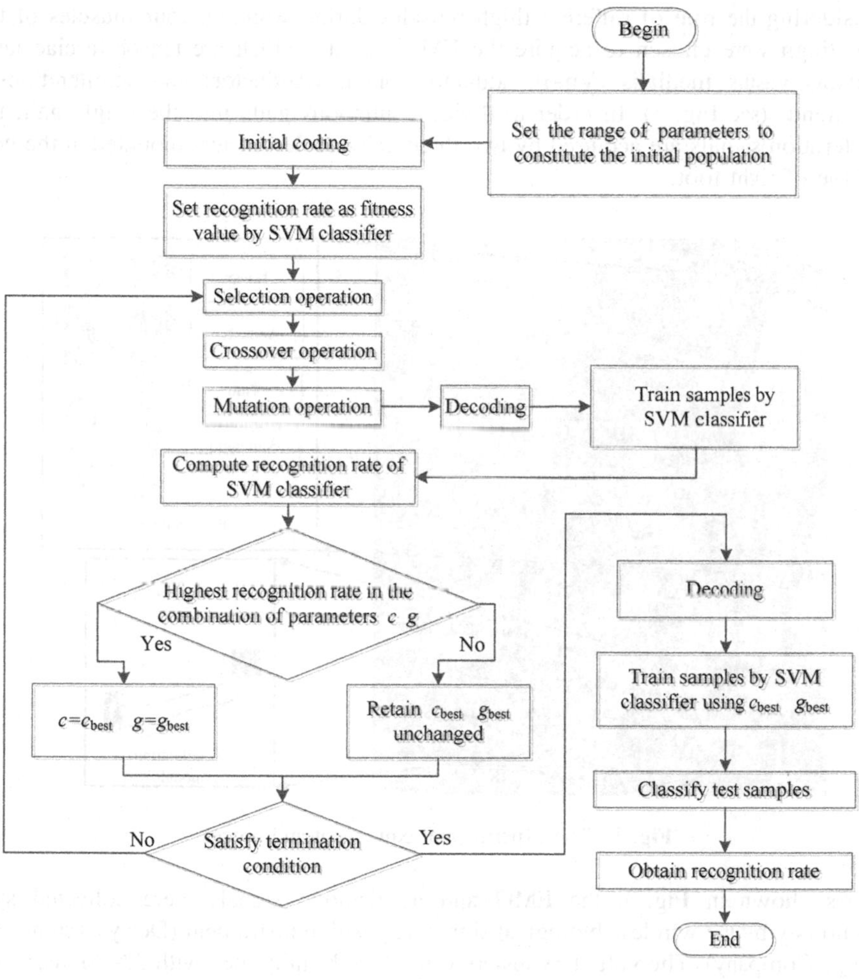

Fig. 2. The flow chart of proposed GA-SVM algorithm

In Fig. 2, the proposed algorithm consists of four steps: (1) set the range of penalty parameter c and kernel function parameter g to constitute the initial population through chromosome coding; (2) set the recognition rate as fitness value; (3) calculate the recognition rate corresponding to each c, g after GA operation, and retain optimal parameters as c_{best}, g_{best} when the recognition rate reaching the maximum; (4) set the termination condition as maximum number of iterations, if the termination condition is not satisfied, the third step is returned to continue the GA operation, when the termination condition is satisfied, the optimal parameters are obtained to train samples by SVM classifier using c_{best}, g_{best} and acquire final classify results.

3 Experiments and Results

3.1 Data Acquisition

Considering the role of different thigh muscles during walking, four muscles of the right thigh were chosen to acquire the EMG signals, which are tensor fasciae latae (Tensor) vastus medialis (Vmo), adductor longus (Adductor) and semitendinosus (Semitend) (see Fig. 3). In order to divide continuous gaits into the single gait, the acceleration signals are acquired by two three-axis accelerometers mounted at the heel and toe of right foot.

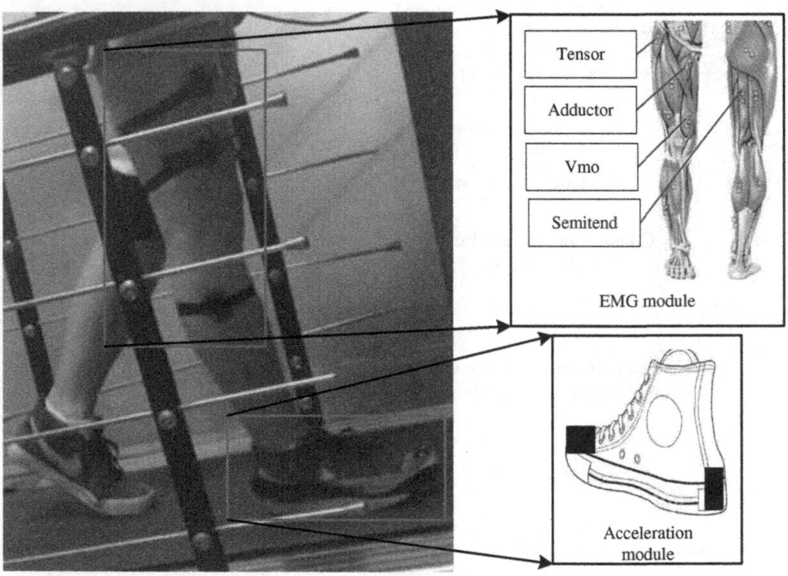

Fig. 3. The experimental equipment and process

As shown in Fig. 3, the EMG and acceleration signals were collected synchronously by the wireless biological signal acquisition instrument (Delsys Trigno, US Delsys Company). The voluntary testers were three healthy men with 23–26 years old, which were arranged to walk at the speed of 1.5 m/s on the treadmill (BROTHER

WL-332). To reduce the interference of random factors, the volunteers were asked to walk 10 steps continuously. Sampling frequencies of EMG signal was 2000 Hz, and the acceleration signal was 150 Hz. The collected data were transmitted to the computer through Bluetooth device.

3.2 Data Analysis

During the experimental process, the EMG and acceleration signals were often influenced by physical noise, the surrounding environmental noise, and device noise, etc. It is necessary to do de-noising processing. In this paper, the modulus maxim method was adopted to eliminate noise, whose principles are different variation performances of modulus maxima between signal and noise under different wavelet scales.

Continuous gaits were divided into single gait based on acceleration signals referring to literature [23]. A single gait was divided into stance phase and swing phase through the calculation at the occurrence time for the heel strike and toe off. Then stance phase and swing phase were subdivided into five phases according to proportion [1], i.e., pre-stance phase, mid-stance phase, terminal stance phase, pre-swing phase and terminal swing phase.

After de-noising processing and gait division, eight-dimension feature vectors were extracted for four-channel EMG signals, which were abstracted using the sliding window method in five gait phases. The length of window was 300 samples and the increment was 30 [4]. Corresponding to Eq. (3), the feature vectors were calculated as $\boldsymbol{Z}_m=[\boldsymbol{I}_{m1},\boldsymbol{V}_{m1},\boldsymbol{I}_{m2},\boldsymbol{V}_{m2},\boldsymbol{I}_{m3},\boldsymbol{V}_{m3},\boldsymbol{I}_{m4},\boldsymbol{V}_{m4}]$.

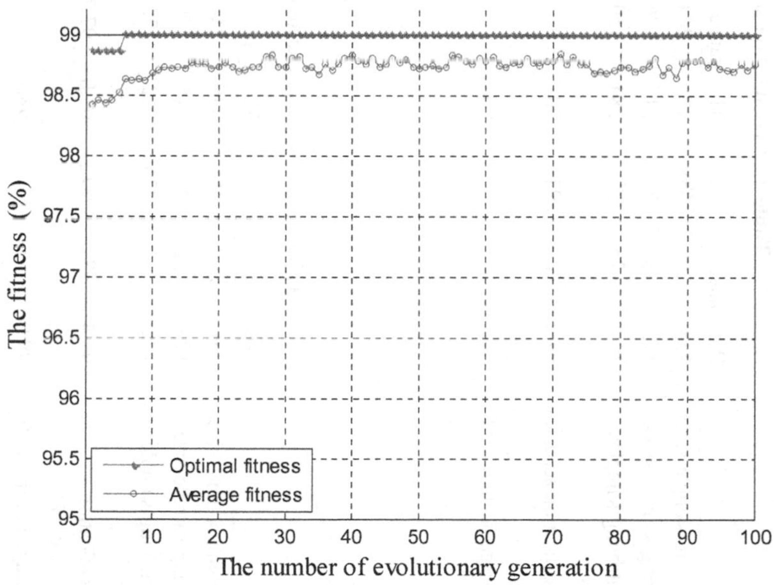

Fig. 4. The fitness curve of GA-SVM

Table 1. Classification accuracy of five gait phases before optimization.

	Pre-stance			Mid-stance			Terminal stance			Pre-swing			Terminal swing		
Training sample sizes	258	200	358	475	264	434	335	275	405	143	140	223	182	150	262
Testing sample sizes	173	133	238	315	176	289	225	183	270	96	93	148	123	100	174
Identifying samples sizes	163	125	224	284	159	260	205	169	255	86	84	137	112	89	156
Classification accuracy (%)	94.2	94.0	94.1	90.2	90.3	90.0	91.1	92.3	94.4	89.6	90.3	92.6	91.1	89.0	89.7
Average accuracy (%)	94.1			90.2			92.6			90.8			89.9		

Table 2. Classification accuracy of five gait phases after GA-SVM optimization.

	Pre-stance			Mid-stance			Terminal stance			Pre-swing			Terminal swing		
Training sample sizes	258	200	358	475	264	434	335	275	405	143	140	223	182	150	262
Testing sample sizes	173	133	238	315	176	289	225	183	270	96	93	148	123	100	174
Identifying samples sizes	172	131	236	312	172	282	223	180	266	94	91	145	120	96	168
Classification accuracy (%)	99.4	98.5	99.2	99.0	97.7	97.6	99.1	98.4	98.5	97.9	97.8	98.0	97.6	96.0	96.6
Average accuracy (%)	99.0			98.1			98.7			97.9			96.7		

3.3 Results and Discussion

The GA-SVM classifier was constructed according to the flow chart of GA-SVM algorithm (see Fig. 2). The iterative process of GA-SVM is depicted in Fig. 4. The best fitness curve represents the best individual in each generation, and the average fitness curve denotes the whole fitness. With the increment of the iteration times, the average fitness gradually reaches the optimal value. When the termination condition is satisfied, the optimization process ends.

The parameters of GA-SVM classifier were set referring to [24], the maximum evolutionary generation as the termination condition is 100, the maximum number of population S_{pop} is 20, the crossover probability P_c is 0.7, the mutation probability P_m is 0.9%, the ranges of penalty parameter is 0–1000 and the kernel function is 0–10. When the maximum evolutionary generation is reached, the SVM is trained with the optimal parameters (c_{best}, g_{best}) to obtain the best recognition result.

The recognition rates of five gait phases are illustrated in Tables 1 and 2 before and after optimization corresponding to three individuals.

In Table 1, the recognition rates of five gait phases reached from 89.9% to 94.1% before optimization. As shown in Table 2, the classification accuracy was improved significantly after GA-SVM optimization, which achieved the range from 96.7% to 99.0%. The average recognition rate was enhanced by 6.56%, and the classification accuracy is stable among three individuals. As a result, it exhibits a higher accuracy in gait recognition for GA-SVM method to select parameter values than the SVM defaults.

4 Conclusion

The SVM has been proved to be an effective method for limb movement recognition based on EMG, and its performance depends on the reasonable parameters. In this paper GA was applied to optimize the penalty and kernel function parameters for gait recognition. The results verified that the classification accuracy of GA-SVM is significantly improved compared to SVM with default parameters. It can be concluded that the GA-SVM classifier can effectively identify lower limb gaits with higher classification accuracy.

References

1. Vaughan, C.L., Davis, B.L., O'Connor, J.C.: Dynamics of Human Gait. Human Kinetics Publishers, Champaign (1999)
2. Young, A.J., Smith, L.H., Rouse, E.J., et al.: Classification of simultaneous movements using surface EMG pattern recognition. IEEE Trans. Biomed. Eng. **60**(5), 1250–1258 (2013)
3. Wang, J.J., Gao, F.R., Sun, Y., et al.: Non-uniform characteristics and its recognition effects for walking gait based on sEMG. Chin. J. Sens. Actuators **29**(3), 384–389 (2016)
4. Huang, H., Kuiken, T.A., Lipschutz, R.D.: A strategy for identifying locomotion modes using surface electromyography. IEEE Trans. Biomed. Eng. **56**(1), 65–73 (2009)

5. Rueterbories, J., Spaich, E.G., Larsen, B., et al.: Methods for gait event detection and analysis in ambulatory systems. Med. Eng. Phys. **32**(6), 545–552 (2010)
6. Taborri, J., Palermo, E., Rossi, S., et al.: Gait partitioning methods: a systematic review. Sensors **16**(1), 66–86 (2016)
7. Li, Y., Gao, F.R., Chen, H.H., et al.: Gait recognition based on EMG with different individuals and sample sizes. In: 35th Chinese Control Conference (CCC) on Proceedings, pp. 4068–4072. IEEE, Chengdu (2016)
8. Vapnik, V.N.: The nature of statistical learning theory. IEEE Trans. Neural Networks **8**(6), 1564 (1997)
9. Quitadamo, L.R., Cavrini, F., Sbernini, L., et al.: Support vector machines to detect physiological patterns for EEG and EMG-based human-computer interaction: a review. J. Neural Eng. **14**(1), 011001 (2017)
10. Castellini, C., Smagt, P.V.D.: Surface EMG in advanced hand prosthetics. Biol. Cybern. **100**(1), 35–47 (2009)
11. Kawano, S., Dai, O., Tamura, H., et al.: Online learning method using support vector machine for surface-electromyogram recognition. Artif. Life Robot. **13**(2), 483–487 (2009)
12. Lucas, M.F., Gaufriau, A., Pascual, S., et al.: Multi-channel surface EMG classification using support vector machines and signal-based wavelet optimization. Biomed. Signal Process. Control **3**(2), 169–174 (2008)
13. Huang, H., Zhang, F., Hargrove, L.J., et al.: Continuous locomotion-mode identification for prosthetic legs based on neuromuscular-mechanical fusion. IEEE Trans. Biomed. Eng. **58**(10), 2867–2875 (2011)
14. Wang, J., Wu, X., Zhang, C.: Support vector machines based on K-means clustering for real-time business intelligence systems. Int. J. Bus. Intell. Data Min. **1**(1), 54–64 (2005)
15. Lin, S.W., Ying, K.C., Chen, S.C., et al.: Particle swarm optimization for parameter determination and feature selection of support vector machines. Expert Syst. Appl. **35**(4), 1817–1824 (2008)
16. Huang, C.L., Wang, C.J.: A GA-based feature selection and parameters optimization for support vector machines. Expert Syst. Appl. **31**(2), 231–240 (2006)
17. Subasi, A.: Classification of EMG signals using PSO optimized SVM for diagnosis of neuromuscular disorders. Comput. Biol. Med. **43**(5), 576–586 (2013)
18. Schizas, C.N., Pattichis, C.S., Middleton, L.T.: Neural networks, genetic algorithms and the K-means algorithm: in search of data classification. In: International Workshop on Combinations of Genetic Algorithms and Neural Networks, pp. 201–222. IEEE (1992)
19. Phan, A.V., Nguyen, M.L., Bui, L.T.: Feature weighting and SVM parameters optimization based on genetic algorithms for classification problems. Appl. Intell. **46**(2), 455–469 (2017)
20. Qiu, D., Shucai, L.I., Zhang, L., et al.: Application of GA-SVM in classification of surrounding rock based on model reliability examination. Int. J. Min. Sci. Technol. **20**(3), 428–433 (2010)
21. Kanitz, G.R., Antfolk, C., Cipriani, C., et al.: Decoding of individuated finger movements using surface EMG and input optimization applying a genetic algorithm. In: 33rd International Conference of the IEEE EMBS, Baston (2011)
22. Zardoshti-Kermani, M., Wheeler, B.C., Badie, K., et al.: EMG feature evaluation for movement control of upper extremity prostheses. IEEE Trans. Rehabil. Eng. **3**(4), 324–333 (1996)
23. Boutaayamou, M., Schwartz, C., Stamatakis, J., et al.: Development and validation of an accelerometer-based method for quantifying gait events. Med. Eng. Phys. **37**(2), 226–232 (2015)
24. Lessmann, S., Stahlbock, R., Crone, S.F.: Genetic algorithms for support vector machine model selection. In: International Joint Conference on Neural Networks, Vancouver, pp. 3063–3069 (2006)

Estimating 3D Gaze Point on Object Using Stereo Scene Cameras

Zhonghua Wan(✉) and Caihua Xiong

State Key Lab of Digital Manufacturing Equipment and Technology, Institute of Rehabilitation and Medical Robotics, Huazhong University of Science and Technology, Wuhan 430074, Hubei, The People's Republic of China
{d201377148,chxiong}@hust.edu.cn

Abstract. 3D eye gaze estimation in real environment is still challenging. A novel method of scene-based 3D gaze estimation is proposed in this paper. As this model combines two models, the 2D nonlinear polynomial mapping model of traditional regression-based gaze estimation and the 3D visual axis linear ray model of traditional geometry-based gaze estimation, it includes two steps. The first step is to estimate the visual axis from the pupil center in an eye camera image. The second one is to estimate the 3D gaze point which is the intersection between the visual axis and the scene object, which can be obtained by stereo scene cameras. As the 3D gaze points are on the object, rather than outside or inside the object like geometry-based 3D gaze estimation, this method is potential for human robot interaction in real environment. Through a simple test, the accuracy of our 3D gaze estimation system is acceptable.

Keywords: Eye tracking · 3D gaze estimation · Stereo scene cameras

1 Introduction

As we all know, human hand grasping requires eye-hand coordination [1]. Thus, eye gaze tracking is a natural and promising approach for human robot interaction. Eye gaze tracking usually detects eyes in the images to estimate and track gaze point where a person is looking at. Gaze may be 2D gaze point in a monitor or 3D gaze in real environment [2].

If users want to interact with robot through eye gaze tracking, the robot must know 3D gaze point coordinates in real world [3]. Although eye gaze tracking has been actively researched for 30 years, most work focuses on 2D eye gaze tracking. 3D eye gaze tracking is still young and remains challenging [4].

In this paper, we present a model of scene-based 3D gaze estimation with an acceptable accuracy. As this method combines the polynomial model of traditional regression-based gaze estimation and the visual axis ray model of traditional geometry-based gaze estimation, it is convenient to estimate 3D gaze and free from environmental constraints. The hardware just needs three cameras and one infrared LED, while geometry-based 3D gaze estimation needs at least two camera and two infrared LED for one eye. More importantly, the 3D gaze points of our method must be on the object surface, while other

Y. Huang et al. (Eds.): ICIRA 2017, Part I, LNAI 10462, pp. 323–329, 2017.
DOI: 10.1007/978-3-319-65289-4_31

methods may be outside or inside the object. With these unique features, this method can be used to human robot interaction in real environment.

2 Related Work

Gaze estimation consists of 2D gaze estimation with gaze point on a monitor, and 3D gaze estimation with gaze point in real environment. Human robot interaction needs 3D gaze point. 3D gaze estimation can be classified into geometry-based 3D gaze estimation and regression-based 3D gaze estimation.

Geometry-based gaze estimation [5–7] computes visual axis from the image features based on eyeball geometric model, optical reflection and refraction principle. The intersection of the visual axis and a monitor is 2D gaze point. The intersection of two visual axis is 3D gaze point. In 2009, Hennessey [8] reported the first system for geometry-based 3D gaze estimation in real environment. However, as a remote eye tracking system with limited operating space, it is unfit for human robot interaction.

Regression-based gaze estimation assumes a mapping function between image features and gaze point. Regression-based 2D gaze estimation [9] assumes a polynomial mapping function between the vector from pupil center to glint position and 2D gaze point in a plane. Regression-based 3D gaze estimation needs to add new features which are related to the observation distance Z. In 2012, JW Lee [10] adopted many image features, including the relative positions of the first and fourth Purkinje images to the pupil center, inter-distance between these two Purkinje images, and pupil size. However, as the fourth Purkinje images is hard to extract and the correlation between pupil size and distance is weak, it is limited by the experimental environment. In 2017, Li Songpo [11] thought that there was a strong correlation between the distance of two pupil centers and the observation distance, while Sigut J [12] employed the length of the major axis of the iris ellipse.

Scene-based 3D gaze estimation derives from regression-based 3D gaze estimation. The main difference to obtain the observation distance is that the former employs the scene information from the frontal cameras, rather than image features from eye cameras. In 2014, Takemura [13] employed a Visual SLAM technique to estimate the 3D gaze point.

3 The Model of Scene-Based 3D Gaze Estimation

In this paper, we present a 3D gaze estimation method, which can guarantee the 3D gaze point on the interest object. This model combines two models, the 2D nonlinear polynomial mapping model of traditional regression-based gaze estimation and the 3D visual axis linear ray model of traditional geometry-based gaze estimation. Our method has two steps. Firstly, we detect the pupil center in the image, then map the pixel coordinates to the visual axis direction. After that, 3D gaze point can be acquired by the intersection of visual axis direction with scene object, which can be acquired by the stereo scene camera. In order to realize this goal, we need two-stage calibration. The first is stereo scene camera calibration, which is user-independent. The second is gaze direction mapping calibration, which determines the parameters of the gaze direction mapping functions. The hardware needs two frontal scene cameras, one eye camera and one infrared LED.

3.1 The Basic Model

Our goal is to compute the 3D gaze point coordinates (X_g, Y_g, Z_g) from pupil center coordinates (u, v).

Assumption 1: It is a one-to-one mapping between the pupil center and visual axis l, namely $(u, v) \overset{f}{\longleftrightarrow} l\colon \begin{cases} X = f_x(u, v, Z) \\ Y = f_y(u, v, Z) \end{cases}$

Assumption 2: The gaze point is the intersection between visual axis and object surface, namely $(X_g, Y_g, Z_g) \in S$, $(X_g, Y_g, Z_g) \in l$.

Based on the above assumptions, the model of scene-based 3D gaze estimation can be divided into two steps. The first one is to estimate the visual axis l from pupil center pixel coordinates (u, v). The second is to compute the gaze intersection (X_g, Y_g, Z_g) between visual axis l and 3D scene point cloud S. These steps can be shown as below,

$$(u, v) \overset{f}{\longleftrightarrow} l, \text{ the intersection between } l \text{ and } S \Rightarrow (X_g, Y_g, Z_g) \tag{1}$$

Figure 1 is an example of 3D gaze estimation, among which the left figure is looking at a cube, while the right is a tetrahedron.

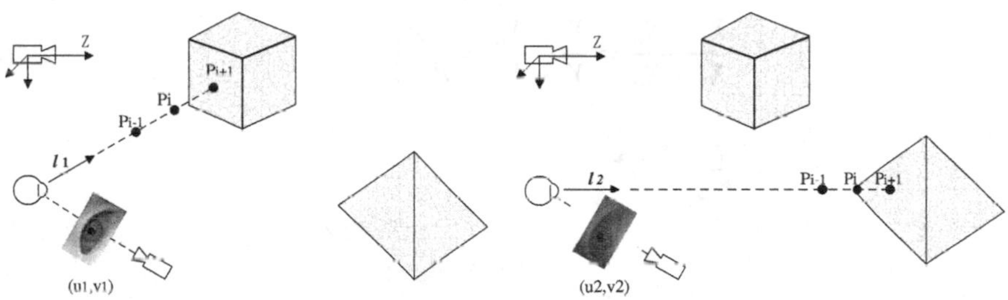

Fig. 1. Scene-based 3D gaze estimation model. Pupil center (u, v) is mapped to visual axis, which intersects with 3D scene object.

In the above second step, as 3D scene is discrete point cloud, there isn't analytic expression of the scene object, and therefore it is impossible to solve the gaze point through analytical method.

Verifying a solution is simpler than solving. Thus the numerical method is used to solve the 3D gaze point. As shown in Fig. 1, we calculate multiple alternative points in the visual axis, then select that one which is on the scene point cloud. This model can be shown as below,

$$(u, v) \overset{f}{\longleftrightarrow} l: \begin{cases} X = f_x(u, v, Z) \\ Y = f_y(u, v, Z) \end{cases}, \text{ select } P_i(X_i, Y_i, Z_i) \in S \quad (2)$$

Now there is a core issue how to set the mapping function *f*. This is a typical regression problem, including many methods. In this paper, the mapping function is derived based on the two-dimensional nonlinear mapping and the three-dimensional ray geometry.

3.2 Determine the Mapping Function

The mapping function *f* consists of two parts. The first part is the two-dimensional nonlinear mapping between the pupil center (*u*, *v*) and one point on the visual axis at a distance of zero. The second part is the three-dimensional linear ray geometry which can calculate other points on the visual axis by collinear relationship.

When Z is Zero. It is a traditional regression-based 2D gaze estimation model, which assumes a two order polynomial mapping function between the vector (u, v) from pupil center to glint position and 2D gaze point (X, Y) in a plane at a distance of zero. Of course, there are other forms, such as reducing the number of terms. But it has been proved that the two order is more reasonable than the higher order.

$$\begin{aligned} X &= f_x(u, v) = a_0 + a_1 u + a_2 v + a_3 uv + a_4 u^2 + a_5 v^2 \\ Y &= f_y(u, v) = b_0 + b_1 u + b_2 v + b_3 uv + b_4 u^2 + b_5 v^2 \end{aligned} \quad (3)$$

When Z isn't Zero. Assumption 3: The range of the human eye is a cone, as shown in Fig. 2. Other points can be obtained by collinear relationship, as below,

$$\frac{X_1 - X_0}{X_2 - X_0} = \frac{Y_1 - Y_0}{Y_2 - Y_0} = \frac{d}{d + Z} \quad (4)$$

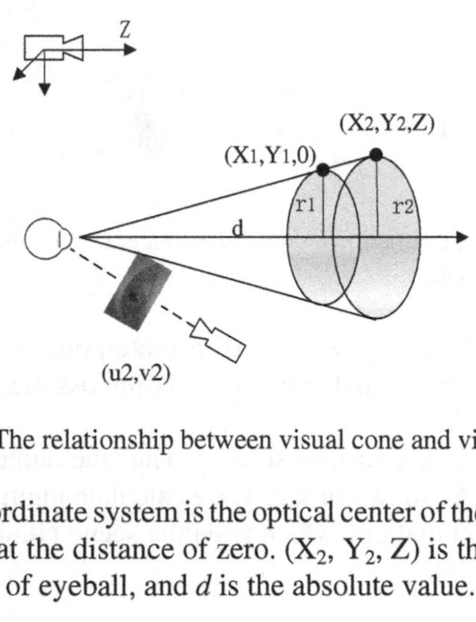

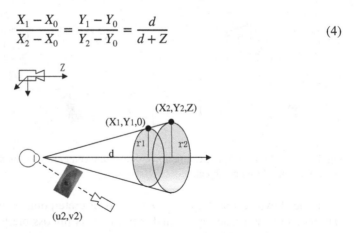

Fig. 2. The relationship between visual cone and visual axis.

The origin of the coordinate system is the optical center of the left front scene camera. (X_1, Y_1, 0) is the point at the distance of zero. (X_2, Y_2, Z) is the point at distance of Z. (X_2, Y_2, *d*) is the center of eyeball, and *d* is the absolute value.

Thus, the solution can be obtained as below,

$$\begin{aligned} X_2 &= X_1 + \frac{Z}{d}(X_1 - X_0) \\ Y_2 &= Y_1 + \frac{Z}{d}(Y_1 - Y_0) \end{aligned} \tag{5}$$

Finally, we can get the mapping function from the pupil center to the point on the visual axis as below,

$$\begin{aligned} X &= f_x(u, v) + \frac{1}{d}\left(f_x(u, v) - X_0\right) \cdot Z \\ Y &= f_y(u, v) + \frac{1}{d}\left(f_y(u, v) - Y_0\right) \cdot Z \end{aligned} \tag{6}$$

4 Experimental Results

We conducted an experiment to test the feasibility and accuracy of our proposed model. The architecture of the scene-based 3D gaze estimation is shown in Fig. 3. The model consists of two parts. The monocular eye detection module detects and locates the pupil and the glint in the eye camera image, as shown in Fig. 4. The 3D gaze estimation module calculates the visual axis direction based on the regression mapping function, and then obtains the 3D gaze point coordinates by combining the scene object information. An example of obtaining scene object 3D coordinates with stereo scene cameras is shown in Fig. 5.

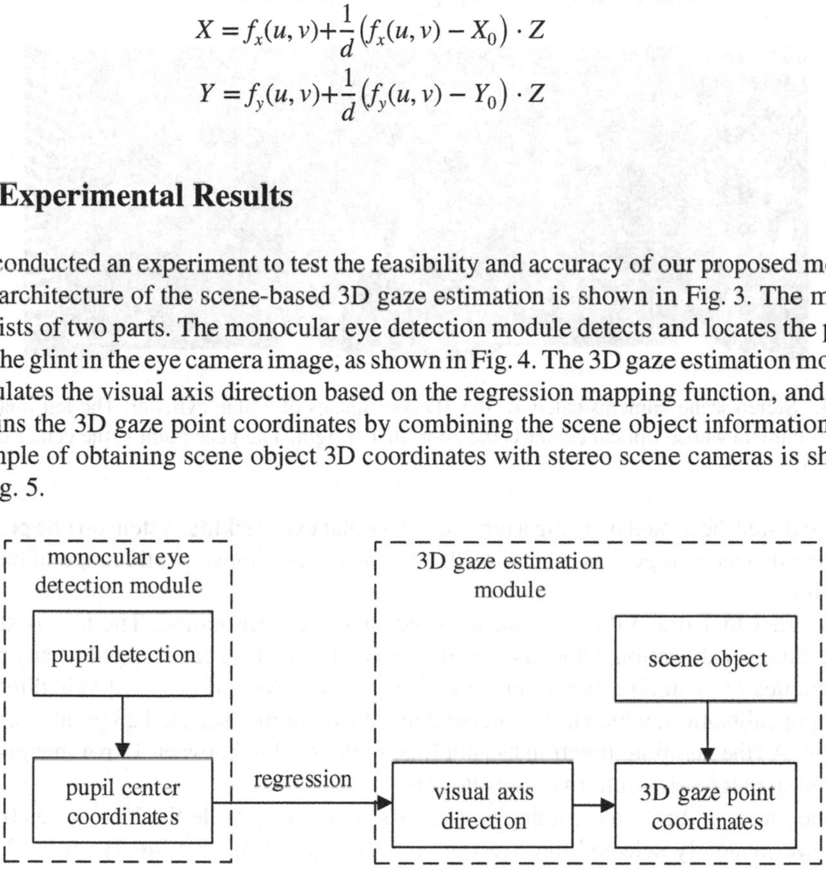

Fig. 3. Architecture of the scene-based 3D gaze estimation

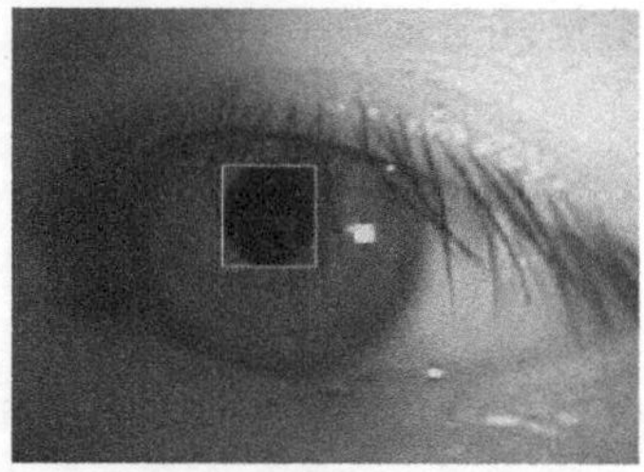

Fig. 4. An example of pupil detection and glint detection

Fig. 5. Stereo scene cameras calculate the 3D coordinates of a blue cylinder. The left image is the left camera whose optical center is the coordinate origin. The gaze point is the center of the cylinder.

We tested the model by using a remote monocular eye tracking system on one person. Our test distance range was 30 cm to 70 cm, which was almost a valid range of human grasping.

In order to achieve this system, we need two-stage calibration. The first is stereo scene camera calibration, which is user-independent. Then we can triangulate any point coordinates by matching points in stereo images. The second is visual axis direction mapping calibration, which is user-dependent. We randomly selected 23 points for calibration. As the mapping function has nonlinear relationship between 15 parameters, we divided the regression into two steps iteratively.

The mean calibration error angle of X axis was 0.65°, while the Y axis was 0.64°. Then we randomly selected three test points, namely [11.5, −3.1, 46.2] cm, [6.9, 2.6, 55] cm, [19.3, 5.1, 58] cm. The mean error angle of X axis was 1.08°, while the Y axis was 1.85°. This was meant that there were 1 cm deviation of X axis at a mean distance of 53 cm, while the Y axis was 1.7 cm. The accuracy of our gaze estimation system is satisfactory for many applications, since the viewfield of the fovea is approximately 1°.

5 Conclusion

In this study, we present a novel model of scene-based 3D gaze estimation, which means that stereo scene cameras are used to obtain the object information. This method combines the advantages of the regression-based gaze estimation and the geometry-based gaze estimation, it is able to estimate 3D gaze in real environment and easy to setup. The hardware just needs three cameras and one infrared LED, while geometry-based 3D gaze estimation needs at least two camera and two infrared LED for one eye. Through a simple test, the accuracy of our 3D gaze estimation system is acceptable. As our head-mounted eye tracking system is in development, we tested our model in a remote eye tracking system, which is off-line. We will improve our eye tracking system and test our model in human robot interaction in the future.

Acknowledgement. This work was supported in part by the National Natural Science Foundation of China under Grants 91648203 and 51335004, the Program of International S&T Cooperation of China under Grant 2016YFE0113600, and the Science Foundation for Distinguished Young Scholars of Hubei Province under Grant 2015CFA004.

References

1. Carrasco, M., Clady, X.: Exploiting eye–hand coordination to detect grasping movements. Image Vis. Comput. **30**(11), 860–874 (2012)
2. Duchowski, A.T.: Eye Tracking Methodology: Theory and Practice. Springer, Heidelberg (2007)
3. Frisoli, A., Loconsole, C., Leonardis, D., Banno, F., Barsotti, M., Chisari, C., et al.: A new Gaze-BCI-Driven control of an upper limb exoskeleton for rehabilitation in real-world tasks. IEEE Trans. Syst. Man Cybern. Part C Appl. Rev. **42**(6), 1169–1179 (2012)
4. Hansen, D.W., Ji, Q.: In the eye of the beholder: a survey of models for eyes and gaze. IEEE Trans. Pattern Anal. Mach. Intell. **32**(3), 478–500 (2010)
5. Shih, S.-W., Liu, J.: A novel approach to 3-D gaze tracking using stereo cameras. IEEE Trans. Syst. Man Cybern. Part B Cybern. **34**(1), 234–245 (2004)
6. Zhu, Z., Ji, Q.: Novel eye gaze tracking techniques under natural head movement. IEEE Trans. Biomed. Eng. **54**(12), 2246–2260 (2007)
7. Villanueva, A., Cabeza, R.: A novel gaze estimation system with one calibration point. IEEE Trans. Syst. Man Cybern. Part B Cybern. **38**(4), 1123–1138 (2008)
8. Craig, H., Peter, L.: Noncontact binocular Eye-Gaze tracking for point-of-gaze estimation in three dimensions. IEEE Trans. Biomed. Eng. **56**(3), 790 (2009)
9. Morimoto, C.H., Koons, D., Amir, A., Flickner, M.: Pupil detection and tracking using multiple light sources. Image Vis. Comput. **18**(4), 331–335 (2000)
10. Lee, J.W., Cho, C.W., Shin, K.Y., Lee, E.C., Kang, R.P.: 3D gaze tracking method using Purkinje images on eye optical model and pupil. Opt. Lasers Eng. **50**(5), 736–751 (2012)
11. Li, S., Zhang, X., Webb, J.: 3D-Gaze-based robotic grasping through mimicking human visuomotor function for people with motion impairments. IEEE Trans. Bio-med. Eng. (2017)
12. Sigut, J., Sidha, S.-A.: Iris center corneal reflection method for gaze tracking using visible light. IEEE Trans. Biomed. Eng. **58**(2), 411–419 (2011)
13. Takemura, K., Takahashi, K., Takamatsu, J., Ogasawara, T.: Estimating 3-D point-of-regard in a real environment using a head-mounted eye-tracking system. IEEE Trans. Human-Mach. Syst. **44**(4), 531–536 (2014)

Overall Kinematic Coordination Characteristic of Human Lower Limb Movement

Bo Huang(✉) and Caihua Xiong

School of Mechanical Science and Engineering,
Huazhong University of Science and Technology, Wuhan, China
{d201577164,chxiong}@hust.edu.cn

Abstract. The human lower limb movement characteristic is still a research focus, and the individual kinematic coordination characteristic of human specific lower limb movement (e.g. walking, running et al.) has been explored. However, the overall coordination characteristic of diverse human lower limb movements is rarely studied and we always ignore the fact that the coordinated movement doesn't always coordinate all joints simultaneously. In order to investigate these issues, we firstly use principal component analysis (PCA) to study the overall kinematic coordination characteristic of human lower limb movements. The four DoF are chosen to analyze: hip flexion/extension, hip adduction/abduction, knee flexion/extension and ankle plantarflexion/dorsiflexion, and the result shows that the first two principal components contain primary human lower limb movement information, so it suggests the possibility that it exists the overall kinematic coordination. But the result of evaluating the effectiveness of the overall kinematic coordination characteristic shows that hip adduction/abduction movement and ankle plantarflexion/dorsiflexion movement can't be reproduced well by the first two principal components. And the correlation coefficient analysis and agglomerative hierarchical cluster analysis results also show that hip adduction/abduction and ankle plantarflexion/dorsiflexion are relatively independent and the other joint movements are more correlated. So we conclude that the independent joint movement should be considered separately when studying the coordination characteristic, and correlation coefficient analysis and agglomerative hierarchical cluster analysis can give us some useful guides.

Keywords: Human lower limb · Kinematic coordination · Correlation analysis · Agglomerative hierarchical cluster analysis

1 Introduction

Human lower limb motor ability is very important in our daily life, it endows human many motor functions, such as walking, running, hopping, and so on. Here two legs consist of at least 14 DOF, and corresponding complex musculoskeletal system, it seems very difficult to achieve these movements. But in fact, human can achieve these functions easily. So many researchers have worked on the study of this human lower limb motor ability. In recent years many researchers have suggested that human can achieve complex movement by rearranging and combining simple motor primitives in different

Y. Huang et al. (Eds.): ICIRA 2017, Part I, LNAI 10462, pp. 330–339, 2017.
DOI: 10.1007/978-3-319-65289-4_32

levels, such as kinematic, kinetic and neural level [1]. The primitives can allow the use of simper controls. Moreover, the kinematic coordination (primitives) in human lower limb has also been explored by different research groups. Nancy St-Onge et al. [2] used PCA dimensionality reduction method to analyze the interjoint coordination of eight types of human lower limb movement, and their result showed that only two synergies even one synergy can explain the main human movement information. Federico L. Moro et al. [3] analyzed the kinematic coordination of human whole body motion and showed that five kinematic Motion Primitives (kMPs) were sufficient to reconstruct periodic motions (i.e. walk, run et al.) and two kMPs were sufficient for discrete motions (reaching the ball position with hand), and they confirmed that the kMPs can be used to transfer the nature of human movement to robot.

To our knowledge, the current works in lower limb kinematic coordination focused on the respective coordination characteristic of specific motion pattern, in other words, they analyzed the respective kinematic coordination characteristic of specific human limb movements separately and compared their similarities and differences. But the overall kinematic coordination characteristic of different human lower limb natural movements was rarely explored. Of course, the prerequisite is that here exists the overall kinematic coordination characteristic of human lower limb movement. But no doubt, it is a more valuable issue for many applications, such as motor control, biomechanics et al. For instance, the lower limb exoskeleton developed so far can't be coordinated with human natural motion [4], so we can endows exoskeleton more natural and richer motor abilities by using the overall kinematic coordination characteristic. So this paper investigates the problem and tries to confirm the existence of the overall kinematic coordination characteristic.

The kinematic coordination analysis's aim is to describe the high dimensionality human limb movement in a lower dimensionality. So the data dimensionality reduction techniques is always used to analysis this issue, and the method is based on the hypothesis that all the selected joint movements exist pairwise coordinated relationship during human movement [5]. But kinematic coordination doesn't always happen in all joint movements, so it ignores the rationality of the hypothesis, as a result, the human lower limb movement may not be reproduced well by the primary kMPs or synergies even though they contain the most information of human movement. So we conclude that it is necessary to figure out the coordinated relationship between different joint movements. Namely, we may need to recognize which joint movements, indeed, are correlated. In fact, similar problem has been found in human lower limb kinematic coordination analysis by Nancy St-Onge et al. [2], their work suggested that the desynchronization between different human lower limb joint movements led to more synergies being needed to describe the relationship between different joint movements. Moreover, similar phenomenon has also been found on the synergistic characteristic analysis of the human hand grasping tasks in our group [5, 6]. To explore the issue, we firstly check the effectiveness of the overall kinematic coordination characteristic by reconstructing the original movements, and evaluate the relative reconstruction error for each joint movement, the result confirms our reasoning. Then this paper further investigates the coordinated characteristic between different joint movements based on correlation analysis and agglomerative hierarchical cluster analysis, and provides a more reasonable method to explore kinematic coordination characteristic. Overall, this paper's main highlights can be summarized as follows:

(1) This paper investigates the overall kinematic coordination characteristic of diverse human lower limb movements, and confirms the possibility of its existence.
(2) Based on correlation analysis and agglomerative hierarchical cluster analysis, we proposal a more reasonable method to study kinematic coordination characteristic.

The remainder of this paper is organized as follows. Section 2 describes the experimental paradigm, data recording and data analysis method. Section 3 gives the kinematic coordination analysis result. The movement reconstruction result and coordinated characteristic analysis between different human lower limb joint movements are presented in Sects. 4 and 5. Finally, Sect. 6 concludes this paper.

2 Materials and Methods

2.1 Subjects

Four healthy male subjects (height: 167–178 cm weight: 53.3–82.9 kg) participated in the experiments.

2.2 Experimental Paradigm

To explore the overall kinematic coordination characteristic of human lower limb natural movement, the experimental paradigm design mainly considered the common human lower limb movements in daily life. The main chosen movement types can be found in the Table 1. In addition, hopping and sitting-standing were also selected.

Table 1. Human lower limb movement tasks in different ground conditions

Level	Cross slope	Longitudinal grade	Stepping over obstacle	Stair
Walking	±15.3°	±8.2°	L:30 cm, W:30 cm H:10 cm/20 cm/30 cm	W:30 cm H:15 cm
Running	±15.3°	±8.2°	L:30 cm, W:30 cm H:10 cm/20 cm/30 cm	No

In the Table 1, the first row lists all the ground conditions during walking or running. In the column of "Cross slope", the "+" represents that left side of ground is higher than right side, vice versa. In the column of "Longitudinal grade", the "+" represents upslope, the "–" represents downslope. In the column of "Stepping over obstacle", L is obstacle length, W is obstacle width, H is obstacle height, and the obstacle height is set to three grades. In the last column, W is stair width, H is stair height, the total step number of the stair is three, and we only asked the subjects to go upstairs. The "No" represents that the corresponding movement type was not be made due to some practical limits of experiment. In walking and running trials, all subjects were asked to move along a 7 m walkway except for the "stair", but the incline surface was only 2 m in the middle of the walkway, and their two ends were extended to two level surfaces. For all aforementioned trials, they were also asked to swing their arms normally and initiate the movement with left and right leg, respectively, and the speed was their preferred chosen. All the movement types were recorded five times, the below movement types were same, too.

The hopping tasks have two types: one-legged hopping and two-legged hopping, and the movement direction also has four types (i.e. forward, backward, leftward, rightward). The one-legged hopping was made by the left or right leg. All the hopping tasks asked subjects to move a step. In addition, the sitting-standing task asked subjects to sit down on the chair and then stand up. The chair height was set to three grades: 305 mm, 435 mm, 555 mm, and the chair was placed in the appropriate location behind the subjects.

2.3 Data Recording

Human lower limb movement data was collected using Vicon Motion Capture System with 10 cameras. 20 reflective markers were pasted to corresponding human lower limb skin surface. The sampling frequency was 100 Hz. The joint kinematic data (i.e. joint movement trajectory) was calculated by the Plug in Gait provided by Vicon software.

2.4 Data Analysis

For walking and running, a gait cycle was defined by two successive heel contact event (the heel contacts with the ground), corresponding to the local minimal value of heel marker height [7]. Each walking or running trial remained only one gait cycle data which was selected according to left heel or right heel but the "stair" didn't remain only one cycle data (see below). For the hopping tasks, the beginning time was defined as when the corresponding joint movements began to change significantly (i.e. one-legged hopping: left or right leg joint movements, two-legged hopping: two legs joint movements), the ending time was the time when toe height was local minimal value. For the "stair" and sitting-standing, the beginning time was same as hopping tasks, but the ending time was defined as when corresponding joint movements stopped nearly. Because the primary human lower limb movement is in the sagittal plane, and hip adduction/abduction is also important in some tasks, such as hopping to the left or right. Thus this paper only selected the four joint movements of left leg: hip flexion/extension, hip adduction/abduction, knee flexion/extension and ankle plantarflexion/dorsiflexion. The right leg movement was not taken into account in this paper.

To explore the overall kinematic coordination characteristic, all of one subject's movement data was pooled together as a "motion pattern", and the lower limb joint movement trajectories can be described by a posture sequence matrix:

$$\mathbf{Q} = \left[\mathbf{q}_1 \ \cdots \ \mathbf{q}_i \ \cdots \ \mathbf{q}_n\right] \in \Re^{4 \times n} \tag{1}$$

Where the i-th column $\mathbf{q}_i = \left[q_{1i}\, q_{2i}\, q_{3i}\, q_{4i}\right]^T \in \Re^{4\times 1}$ represents a specific human lower limb posture at the moment, q_{1i}, q_{2i}, q_{3i} and q_{4i} represent the corresponding angle values of hip flexion/extension, hip adduction/abduction, knee flexion/extension and ankle plantarflexion/dorsiflexion, respectively.

3 Kinematic Coordination Analysis

The kinematic coordination analysis's aim is to describe the high dimensionality human limb movement in a lower dimensionality. So the data dimensionality reduction techniques is always used to analysis this issue, and the most common method to get kinematic coordination characteristic is principal component analysis(PCA) [3], so this paper firstly analyze the overall kinematic coordination characteristic by PCA. By using PCA on the matrix **Q**, the original movement data of human lower limb can be represented as follows:

$$\mathbf{q}_i = \left[u_1\, u_2\, u_3\, u_4\right] \times \left[p_{1i}\, p_{2i}\, p_{3i}\, p_{4i}\right]^T + \overline{\mathbf{q}}_i \tag{2}$$

Where u_i is i-th principal component vector, p_{1i}, p_{2i}, p_{3i}, p_{4i} are the corresponding first, second, third, fourth principal component values. $\overline{\mathbf{q}}_i \in \Re^{4\times 1}$ is the average of $\mathbf{q}_i$. As we know, the i-th principal component vector and variance are equal to the corresponding i-th eigenvalue-eigenvector pair $\left(\lambda_i, e_i\right)$ of the covariance matrix associated with the posture vector $\mathbf{q}_i$, and $\lambda_1 \geq \lambda_2 \geq \lambda_3 \geq \lambda_4 \geq 0$. Thus we can find the complicated human lower limb movement can be decomposed into four simple movements along four orthogonal principal component vectors, namely four eigenmovements along four orthogonal eigendirections. Moreover, the eigenvalues can be used to evaluate the percentage of movement variability along corresponding eigendirections. Thus an index to represent the cumulative proportion of total variance (CPV) explained by first k principal components can be defined as

$$CPV = \frac{\lambda_1 + \lambda_2 + \cdots + \lambda_k}{\lambda_1 + \lambda_2 + \cdots + \lambda_4} \tag{3}$$

The index can be used to determine the appropriate number of principal components (eigenmovements) which should be retained.

The kinematic coordination analysis result (Fig. 1) shows that the first two principal components can explain 91.75% of total variance, which means that the two eigenmovements contain the primary information of human lower limb movement. Thus it confirms the possibility of the existence of the overall kinematic coordination characteristic.

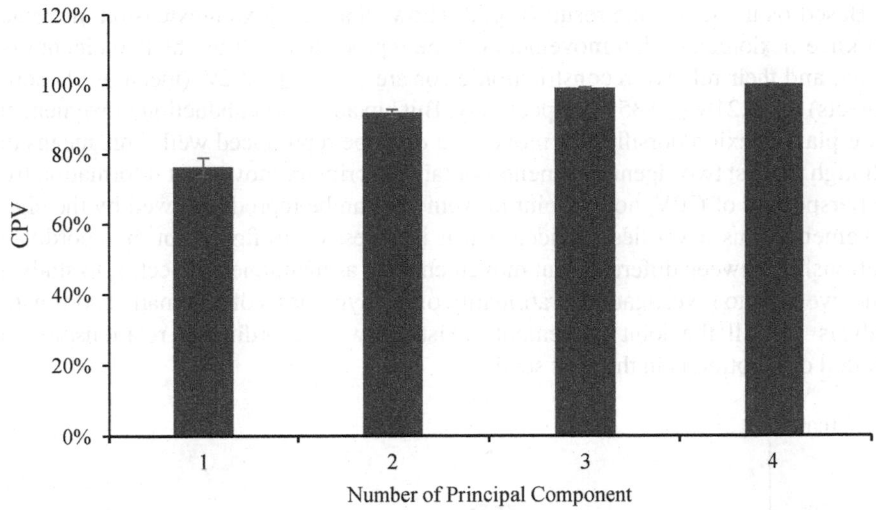

Fig. 1. The cumulative proportion of total variance explained by first k principal components. Error bars indicate the standard deviation across subjects.

4 Movement Reconstruction

As mentioned before, we guess that the recent method to analyze all joint movements directly without considering the detailed coordinated relationship between joints isn't very appropriate, it is necessary to figure out the coordinated relationship between different joint movements. To study the issue, we checked the effectiveness of the overall kinematic coordination characteristic in this section. The method is reconstructing the original movements by the first two eigenmovements, and we evaluated the reconstruction error of each joint movement quantitatively.

The reconstruction method is

$$\tilde{\mathbf{q}}_i = [u_1\ u_2] \times [p_{1i}\ p_{2i}]^T + \bar{\mathbf{q}}_i \tag{4}$$

The relative reconstruction error of the i-th joint movement can be evaluated as follows:

$$rre = \frac{\sum_{j=1}^{n} \left\| \tilde{q}_{ij} - q_{ij} \right\|^2}{\sum_{j=1}^{n} \left\| q_{ij} - \bar{q}_i \right\|^2} \tag{5}$$

Where $\tilde{q}_{ij}$, q_{ij} are reconstructed and original joint angle of i-th joint movement in j-th limb posture, respectively.

Based on the index, the result (Fig. 2) shows that hip flexion/extension movement and knee flexion/extension movement can be reproduced well by the two eigenmovements, and their relative reconstruction error are 3.27% ± 2.92% (mean ± SD across subjects) and 0.21% ± 0.35%, respectively. But hip adduction/abduction movement and ankle plantarflexion/dorsiflexion movement can't be reproduced well. This means that although the first two eigenmovements contain the primary movement information from the perspective of CPV, not all joint movements can be reproduced well by the eigenmovements, thus it verifies our idea that it is necessary to figure out the coordinated relationship between different joint movements. So as mentioned in Sect. 1, to study the issue, we need to investigate the rationality of the hypothesis of kinematic coordination analysis that all the joint movements exist pairwise coordinated relationship. The detailed description is in the next section.

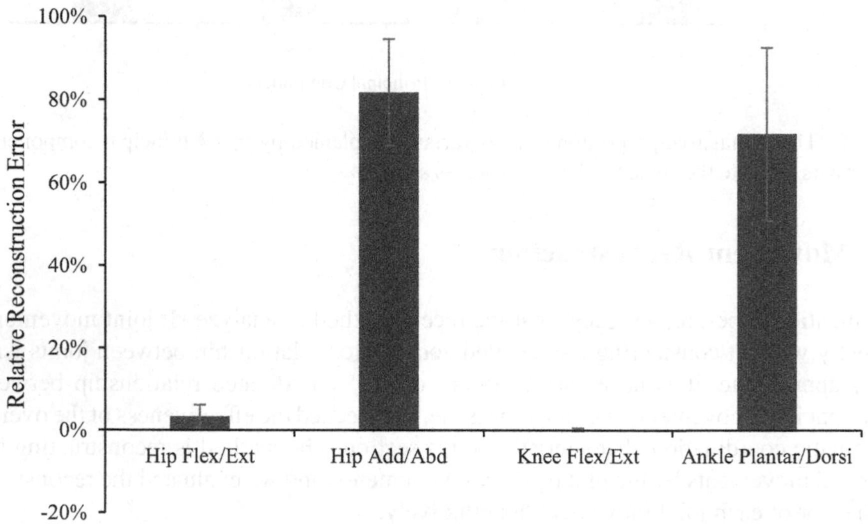

Fig. 2. The diagram illustrates the relative reconstruction error of each human lower limb joint movement. Error bars indicate the standard deviation across subjects.

5 Correlation Analysis

To solve the issue as mentioned before, we investigated the pairwise coordinated relationship between joint movements and recognized which joint movements, indeed, are correlated. The coordinated relationship between joint movements is measured by Pearson correlation coefficient between joint movement sequence vectors, and the larger correlation coefficient value implies stronger dependence between two joint movements [6].

Figure 3 shows the absolute values of correlation coefficients (still called correlation coefficient in the following text for simplicity) between different two joint movements. The result suggests that the value of correlation coefficient between hip flexion/extension movement and knee flexion/extension movement is 0.62 ± 0.04, which means they are significantly correlated. The coordinated relationship between them also verifies the existence of the overall kinematic coordination. But hip adduction/abduction movement and ankle plantarflexion/dorsiflexion movement are only low correlated or not correlated with other movements. Although there are some variation across subjects, we can find that hip adduction/abduction movement and ankle plantarflexion/dorsiflexion movement are more independent compared with hip flexion/extension movement and knee flexion/extension movement. Comparing with the movement reconstruction result, we can also find that the most correlated joints can be reproduced well by the lower dimensionality eigenmovement, namely the hip and knee flexion/extension movements, but the relatively independent joint movements can't be reproduced well as the former, and you can find the corresponding relative reconstruction errors of hip adduction/abduction and ankle plantarflexion/dorsiflexion are very high. The consistency suggests that the method to analyze the kinematic coordination characteristic of all joint movements directly, indeed, is not appropriate, and the independent joint movement should be considered separately.

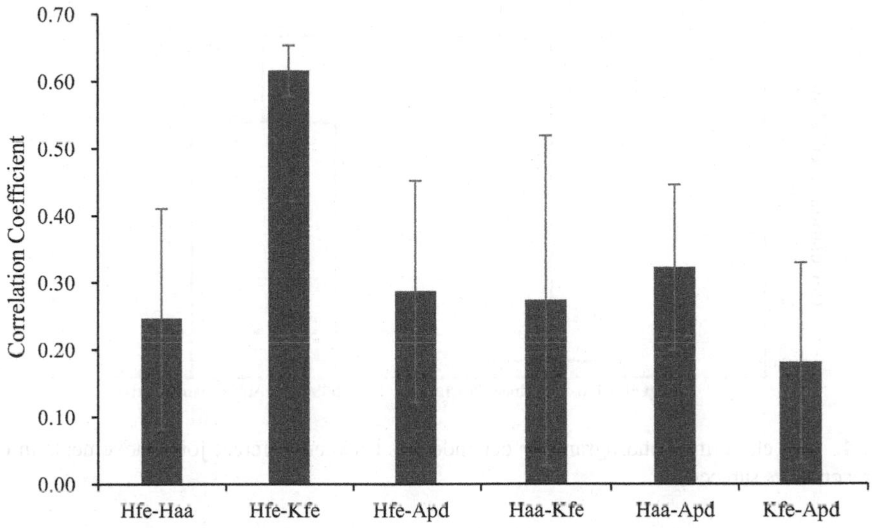

Fig. 3. Correlation coefficients between different joint movements (Hfe: hip flexion/extension, Haa: hip adduction/abduction, Kfe: knee flexion/extension, Apd: ankle plantarflexion/dorsiflexion). Error bars indicate the standard deviation across subjects.

To describe the dependences or correlated relationships between all the joint movements more intuitively, we use agglomerative hierarchical clustering method [8] to analyze the joint coordinated relationship. Agglomerative hierarchical clustering method proceeds by a series of successive mergers. It starts with the individual objects, here each joint movement can be regarded as an object. The most similar objects are first grouped, and these initial groups are merged according to their similarities. Eventually, as the similarity decreases, all subgroups are fused into a single cluster. The similarity measures for pairs of joint movements can be defined as the absolute value the Pearson correlation coefficient between joint movement sequence vectors. This is because we consider that the coordinated relationship is only related to the absolute value of correlation coefficient. The single linkage algorithm is selected to determine the similarities between different subgroups:

$$S_{UV} = \max\left\{ S_{ij} \middle| i \in U, j \in V \right\} \tag{6}$$

Where S_{UV} is the similarity between subgroup (or cluster) U and cluster V, and S_{ij} is the similarity between i-th element of U and j-th element of V.

The clustering dendrogram can be found in Fig. 4. From the Fig. 4, we can find hip flexion/extension and knee flexion/extension belong to a small cluster, and hip adduction/abduction movement and ankle plantarflexion/dorsiflexion movement tend to belong to another one, but the similarity between them is low. The result is similar to Fig. 3, but it can clearly show that the hip adduction/abduction movement and ankle plantarflexion/dorsiflexion movement should be considered separately.

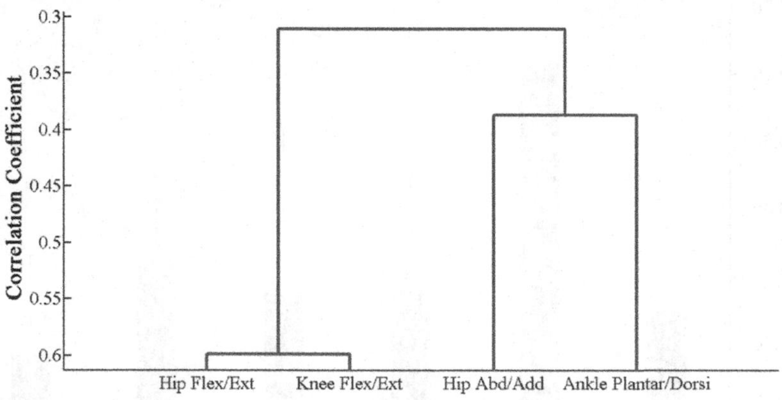

Fig. 4. The clustering dendrogram for dependences between different joint movements in one representative subject.

Moreover, we calculated the overall kinematic coordination characteristic for only hip flexion/extension and knee flexion/extension. The result showed that only the first eigenmovement can explain 83.93% movement information, and the relative reconstruction error of hip flexion/extension and knee flexion/extension are only 38.29% and 5.81%, respectively. It implies that it is useful to consider only the correlated joint movements.

6 Conclusion

This paper explores the overall kinematic coordination characteristic of different human lower limb natural movements, and confirms the possibility of its existence. Moreover, we conclude that the method to analyze the kinematic coordination characteristic of all joint movements directly, indeed, is not appropriate, and the independent joint movement should be considered separately. But if the variance of independent movement is large, it will be retained by the index (CPV) and reproduced well by the primary eigen-movement, too. But this is not taken in account in this paper. Anyway, Because similar conclusion has also been verified in the synergic characteristic study of human hand grasping tasks [5], and we suppose that the idea can also be extended to other similar motor synergy study, such as muscle synergy, even we suppose that the method can be used to divide diverse human lower limb movements into different similar subgroups, it will also facilitate the understanding of motor coordination, but it need further work to verify. Of course, this work also has some other limits, for instance, here is only four subjects, but statistical test may need more data from different subjects.

Acknowledgements. This work was supported in part by the National Natural Science Foundation of China under Grants 91648203 and 51335004, the Program of International S&T Cooperation of China under Grant 2016YFE0113600, and the Science Foundation for Distinguished Young Scholars of Hubei Province under Grant 2015CFA004.

References

1. Giszter, S.F.: Motor primitives - new data and future questions. Curr. Opin. Neurobiol. **33**, 156–165 (2015)
2. St-Onge, N., Feldman, A.G.: Interjoint coordination in lower limbs during different movements in humans. Exp. Brain Res. **148**(2), 139–149 (2003)
3. Moro, F.L., Tsagarakis, N.G., Caldwell, D.G.: On the kinematic motion primitives (kmps) – theory and application. Front. Neurorob. **6**(5), 10 (2012)
4. Cornwall, W.: In pursuit of the perfect power suit. Science **350**(6258), 270–273 (2015)
5. Liu, M., Xiong, C.: Synergistic characteristic of human hand during grasping tasks in daily life. In: Zhang, X., Liu, H., Chen, Z., Wang, N. (eds.) ICIRA 2014. LNCS, vol. 8917, pp. 67–76. Springer, Cham (2014). doi:10.1007/978-3-319-13966-1_7
6. Xiong, C.H., Chen, W.R., Sun, B.Y., Liu, M.J., Yue, S.G., Chen, W.B.: Design and implementation of an anthropomorphic hand for replicating human grasping functions. IEEE Trans. Rob. **32**(3), 652–671 (2016)
7. Ivanenko, Y.P., Cappellini, G., Dominici, N., Poppele, R.E., Lacquaniti, F.: Modular control of limb movements during human locomotion. J. Neurosci. **27**(41), 11149–11161 (2007)
8. Johnson, R.A., Wichern, D.W.: Applied Multivariate Statistical Analysis, 5th edn. Prentice hall, Upper Saddle River (2002)

Eye Gaze Tracking Based Interaction Method of an Upper-Limb Exoskeletal Rehabilitation Robot

Quanlin Li, Caihua Xiong(✉), and Kai Liu

State Key Lab of Digital Manufacturing Equipment and Technology,
Institute of Rehabilitation and Medical Robotics,
Huazhong University of Science and Technology,
Wuhan 430074, Hubei, The People's Republic of China
{liquanlin, chxiong, liukai1986}@hust.edu.cn

Abstract. Stroke is one of the leading causes of long-term disability today. Rehabilitation robot can benefit the patients to perform intensive and repetitive task-specific rehabilitation training to enhance motor recovery with less effort. Engagement in rehabilitation training, namely active rehabilitation is a key factor for patient to get effective recovery. In this paper, we propose an upper-limb exoskeletal underactuated rehabilitation robot with only 2 actuated degrees of freedom (DOFs) for task-specific rehabilitation training. To ensure the patients' engagement in performing training, we used an eye gaze tracking based interaction method to guide the end-point of the robot to move on its workspace, a 2D surface, as a result of performing task-specific rehabilitation training. Finally, some validation experiment is conducted on the interaction efficiency.

Keywords: Eye gaze tracking · Patient engagement in training · Rehabilitation robotics · Upper extremity exoskeleton

1 Introduction

Stroke, one of the leading causes of long-term disability today, is currently affecting 6.8 million people in the United States alone, and this count is expected to rise to over 10 million people by the year 2030 [1]. Also in China, the number of people suffering from stroke, already up to 11 million, is supposed to reach 31 million by 2030 if without proper prevention plan [2]. Patients with stroke usually exhibit a disability of physical function of the upper body and have difficulty lifting or grasping [3]. To regain stroke survivors' independence and decrease the cost of therapy and care, recent research in stroke rehabilitation has accentuated the need for more effective therapy than the current standard of care. Particularly, effective rehabilitation of upper limb function is of greater importance because of the limitations stroke survivors experience in performing activities of daily livings (ADLs) [4]. Fortunately, research has shown that intensive, repetitive motion training can improve therapeutic outcomes both during the acute phase immediately after the stroke and in the longer term [5]. The key factors involved in this successful therapy are task-specific training [6], high intensity and

Y. Huang et al. (Eds.): ICIRA 2017, Part I, LNAI 10462, pp. 340–349, 2017.
DOI: 10.1007/978-3-319-65289-4_33

duration [7], repetitive training [8], the active participation of the patient [9], and maximal challenge [9]. While the increasing the intensity of this highly tailored post-stroke therapy can enhance the motor recovery, quite a number of skilled therapists and significant effort is needed to help the patients to complete motion training. To solve this tough problem, researchers have proposed several robotic rehabilitation systems to aid the therapist in providing consistent, repeatable training with less effort [10–14]. Robot-assisted rehabilitation training benefits the patients with stroke so much, while there is also evidence that passive movements are insufficient to alter motor recovery [15], and that patients must be actively engaged and attempting to move [9, 16] so as to gain the beneficial effects of robotic rehabilitation.

Engagement in robot-assisted rehabilitation training, namely active rehabilitation training, means that it is required to detect the motion intention of the patients, which is then decoded to interact the robot device through a control interface. To facilitate the interaction between human users and assistive robots to ensure engagement in training, researchers have been exploring natural, interpersonal communication signals and biosignals employed inside the human body, and are attempting to enable the user to intuitively convey control commands to assistive robots using these natural signals. For example, surface elcectromyography (EMG) and force-based interfaces are chosen to control an elbow supports robot actively for men with muscular dystrophy [17], while eye gaze tracking and brain-computer interface(BCI) seems to be promising in control an upper limb exoskeleton for rehabilitation in real-world tasks [13]. In addition, Substantial research has been conducted on signals like speech [18, 19], facial expression [20, 21]. Particularly, Gaze, referred to as where a person is looking, taking great advantages since it is natural, effortless, rich in information and easily available for most human beings and especially for those with severe disabilities [22, 23], shows great benefit in commanding a robotic system for certain types of operation such as reaching motion and grasping motion in upper limb rehabilitation training task. Obviously, gaze is encouraging for intuitive human-robot-interaction (HRI), but the development of applications with gaze-based HRI is limited by the gaze tracking technology. Generally, to help the patients to complete task-specific training intuitively with the assistance of a robot in real 3D environment, one needs to explicitly estimate the 3D coordinate of a person's gaze. Although there is a few researches of 3D gaze tracking in real environment in the past decades, the achieved accuracy is too low for practical applications.

As a promising method to achieve engagement in robotic control in real 3D environment, especially in rehabilitation robotic control, 3d-gaze tracking constantly attracts researchers to push this technology and the applications based on it forward. An exciting accuracy of about 2 cm in depth direction in 3D gaze tracking was reported in a newly work of 3D-Gaze-based robotic grasping for people with motion impairments [24]. However, this proposed method based on features extracted from eye image estimates the 3D-gaze point via a trained neural networks, which may need a complex training procedure and may need a calibration due to head mounted device slip from one's head. A 3D gaze cursor is also developed for end-point grasp control based on a model-free approach, with the limitation that the eyes should show repeatability in how they are looking at a target [25] which means that any displacement between the eyes and the gaze tracker will lead to unpredictable errors in coordinates of the gaze point.

Due to poor accuracy in 3D gaze estimation, a 2D gaze cursor was used to select color-coded objects in users' field of view, and then the 3D coordinates of the target selected by the user was estimated by a Kinect device for robotic grasping [13]. The user can reaching and grasping the target selected by gaze precisely, at the expense of an additional 3D construction device standby and a calibration procedure between the device and the robot.

Currently the 3D gaze interaction technology is not so accurate and elegant enough to power the patient affected with stroke to complete active rehabilitation task, however in the early stage of stroke and or suffering from severe disabilities, the patient maybe only can communicate with the environment with gaze and BCI. In comparison with BCI interaction method, gaze communication technology takes more advantages since effortless, rich in information and easily available [22, 23]. To adopt the gaze modality for intuitive HRI for an upper limb rehabilitation robot, some specific settings may be needed, including the gaze information decoding and the control strategy. In this paper we propose an eye gaze tracking based interactive method of an upper-limb exoskeletal rehabilitation robot to ensure the patients engagement in the robot-assisted rehabilitation training. Particularly, the contributions of this papers are:

We propose an underactuated upper-limb exoskeletal rehabilitation robot with only 2 actuated DOFs, while it can help the patients to perform most of the reaching task in activities of daily livings (ADLs). Besides, less actuated DOFs of a rehabilitation robot can contribute to less weight and real-time human-like trajectory generation for reaching task rehabilitation.

We propose a novel interaction method by using the 2D gaze information between the patients and our underactuated upper-limb exoskeletal rehabilitation robot. With only 2 actuated DOFs and 3 coupled DOFs, the workspace of the end-point of the rehabilitation robot is a surface of 2 DOFs. A new interaction strategy which interprets the gaze information for the real-time motion guiding of the robot end-point in its 2D surface workspace.

Due to 2D gaze information interacting with a robot whose end-point workspace is 2D surface workspace, a simple one-point calibration procedure is proposed to avoid complex and strenuous preparation before rehabilitation training for the patients who are in the early stage after stroke or with severe disabilities.

This paper is mainly organized as follows. Section 2 introduces the design of the underactuated upper-limb exoskeletal rehabilitation robot with 2 actuated DOFs, 3 coupled DOFs, and 3 passive DOFs. And then the 2d gaze-based intuitive interaction method between the patients and the rehabilitation robot is presented in details in Sect. 3. Finally, in Sects. 3.1 and 3.2 experimental validation result is shown to demonstrate the effectness of the intuitive interaction method between the patients and the rehabilitation robot.

2 Underactuated Rehabilitation Robot

In this section, we investigate the postural synergy characteristics of human upper limb in some common reaching task in ADLs, and then present an underactuated rehabilitation robot whose end-point workspace is a surface of 2 DOFs.

2.1 Postural Synergy of Upper Limb Reaching Movements

The human upper limbs show significant synergetic characteristics at the joint displacement level in ADLs reaching movements. To analyze such features, we model the arm as a five-joint open kinematic chain, with the definition of the joints being (1) SHR (shoulder humeral rotation), (2) SAA (shoulder abduction/ adduction), (3) SFE (shoulder flexion/ extraction), (4) EFE (elbow flexion/ extension) and (5) RUSP (radio-ulnar supination/ pronation), from shoulder's proximal side to the distal side. Due to limited contribution to the arm's reaching movements [26, 27], the wrist articulation, which allows flexion/extension and radial/ulnar deviation, is excluded in the kinematic model of the arm in this reaching ADLs movement analysis. For the mirror relationship between the right arm and the left arm, the right arm is chosen to perform this synergy characteristics analysis of upper limb ADLs Reaching Movements.

The synergy characteristics of the upper limb is depend on the movements selected to be analyzed. In this work, a robot is needed to be designed to perform task-specific rehabilitation training, so we choose 5 ADLs reaching movements including touching the head, mouth, left ear, left shoulder, right ear and right shoulder based on the existing studies [26, 28–30] and the advices of medical staff. To get the desired synergy characteristics, motion capture experiments was carried out among 16 healthy subjects (mean age 24.4 years old, with height of 170 ± 3 cm and weight of 61 ± 5 kg). All the subjects performed the activities within the same time period, and the joint space trajectories were measured by the motion capture system (VICON F20-6, Oxford Metrics Ltd.) at a frequency of 50 Hz. And then the synergetic analysis was performed based on PCA. Let $x = [x_1 \quad x_2 \quad \cdots \quad x_5]$ be the posture vector representing a measured joint angle of the right arm. Posture vectors of all the subjects during all five movements collected in the experiment can be organized into a posture matrix $X \in \Re^{5\times n}$ denoted by (1),

$$\mathbf{X} = [\mathbf{x}_1 \quad \mathbf{x}_2 \quad \cdots \quad \mathbf{x}_5] \tag{1}$$

where n is the total number of samples.

Direct use of conventional PCA to X will lead to relatively complex mechanism since each joint angle is dependent on multiple input. A joint grouping technology is used based on the analysis of cluster analysis. Let ρ_{ij} be the absolute value of the correlation coefficient between the ***i***-th and the ***j***-th joint trajectories. Trajectory of SHR is highly correlated with that of SAA with $\rho_{1,2} = 0.95$, as well as the SFE and EFE with $\rho_{1,2} = 0.86$ while $\rho_{34,5} = 0.63$ and $\rho_{12,345} = 0.26$. Results above shows that the five joints in the arm chain can be classified to two groups, one group including SHR, SAA and the other group including SFE, EFE and RUPS. And then PCA is performed among the two groups.

2.2 Design of a 2 Actuated DOFs Upper Limb Exoskeleton Rehabilitation Robot

Based on synergy characteristics analysis results above, which suggest that the five joint of the right arm can be clustered to 2 groups, we can design a 2 actuated DOFs upper limb exoskeleton rehabilitation robot, with one motor driving the SHR and SAA, the other driving the SFE, EFE and RUPS. Let $\theta = [\theta_1 \quad \theta_2 \quad \cdots \quad \theta_5] \in \Re^5$ denotes the joint angle vector of the exoskeleton robot. Then the affine map from the actuation space to the joint space can be denotes as (2)

$$\boldsymbol{\theta} = \bar{\boldsymbol{\theta}} + \boldsymbol{\eta}\mathbf{u} \tag{2}$$

where $\bar{\theta}$ is the biased joint angle vector from the zero position. $\boldsymbol{\eta} \in \Re^{5\times2}$ is the transmission matrix and the $\mathbf{u} \in \Re^{2\times1}$ is the actuator angle vector. Here we use tendon-driven pulley mechanism to realize the implementation of the 2 actuated DOFs upper limb exoskeleton rehabilitation robot. Pulleys of different radius are used to configure the transmission matrix. A CAD model of the underactuated rehabilitation robot for the right arm with is shown in Fig. 1.

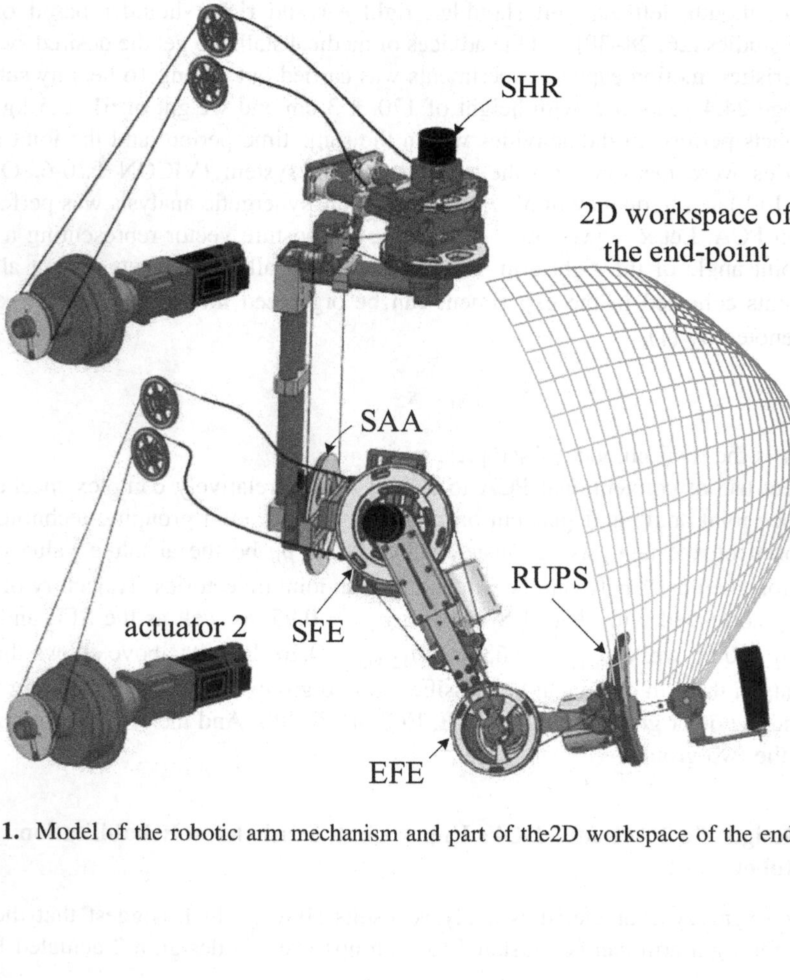

Fig. 1. Model of the robotic arm mechanism and part of the2D workspace of the end-point

3 Gaze-Based Interaction Method

With the development of an upper limb rehabilitation robot with 2 actuated DOFs, the intuitive rehabilitation training can be performed by the interaction between the robot and the 2D gaze information, namely that the end-point can be guided to move on its workspace of a 2D surface.

3.1 2D Gaze Tracking

Video-based Gaze tracking technology has gone far along the way of 2D point of regards (POR) on a monitor or the image of a scene camera on the gaze tracker device. Usually the methods of 2D POR estimation includes 3D model-based gaze estimation method and 2D regression-based gaze estimation method, and both of these two methods using the extracted features from the eye images. Three-dimensional model-based approaches model the common physical structure of human eye geometrically so as to calculate a 3D gaze direction vector and then the POR is computed as the intersection of the gaze direction vector with the nearest object of the scene, such as a monitor. However, most 3D-model based approaches rely on subject-specific information that may be difficult to measure directly such as the visual axis, refraction indexes, the distance between the cornea center and pupil center, and the angles of the visual axis and optical axis [31]. In addition, a sophisticated system needs to be configured which requires camera calibration and global geometric model of light source, camera and monitor position and orientation. All of these onerous preparation procedures may result in depression in the motivation of the patients to perform active rehabilitation robot. The 2D regression-based gaze estimation method use the extracted pupil center and glint to estimate the POR on a specific monitor via a linear or nonlinear mapping, which may usually need more than 16 calibration point for the user to look at. As a result, 2D regression-based gaze estimation method is not an elegant solution for patients deeply affected by stroke or disabilities due to heavy burden of 2D-mapping calibration.

For the real-time guiding of the rehabilitation robot and continual engagement in the reaching-task-specific rehabilitation training, we proposed a method that make use of the difference between the gaze direction and a reference direction to define an intention direction in the plane parallel to the plane of the face. Then the intention direction is used to interaction with 2-actuated-DOFs rehabilitation robot in the robot coordinate frame, namely guiding the end-point of the robot to move on its 2D workspace surface. The block diagram of the proposed method is demonstrated in Fig. 2.

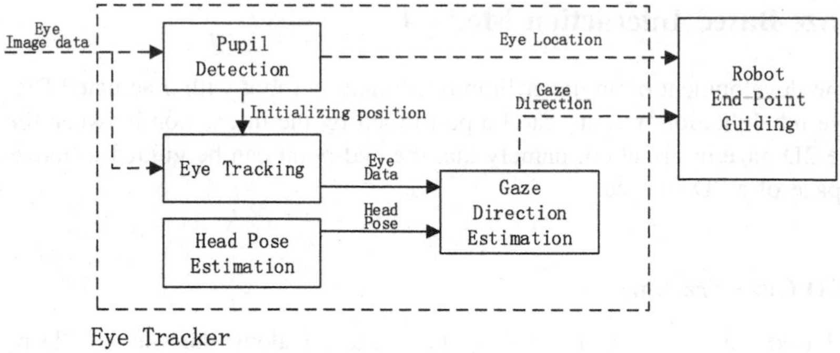

Fig. 2. Diagram of eye gaze-based interaction with the rehabilitation robot

3.2 Gaze Guiding of the Underactuated Rehabilitation Robot

In this eye gaze-based interaction between the patients and the rehabilitation robot, an iterative strategy is chosen to guide the end-point of the rehabilitation robot on its 2D surface workspace, with the gaze information as input, shown in Fig. 3.

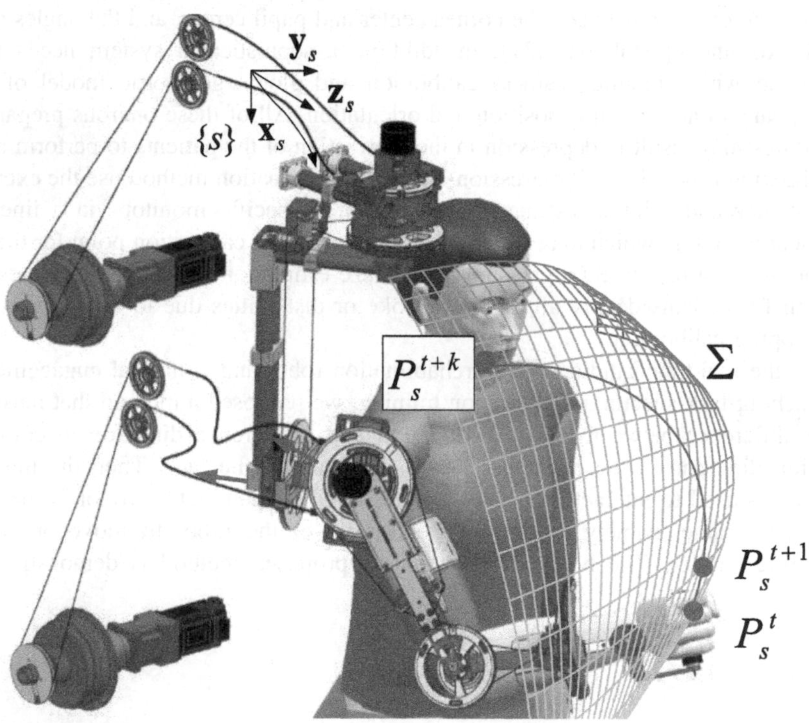

Fig. 3. Eye gaze-based interaction with the rehabilitation robot in iteratively

In details, the next position of the end-point on its workspace in computed iteratively by the formulation below (3):

$$P_S^{t+1} = \arg\min_p \left\| P_S^t + (k_x \Delta x^t, k_y \Delta y^t, 0) - p \right\|$$
$$s.t.$$
$$\begin{cases} p \in \Sigma \\ \left\| p - P_S^t \right\| < r \\ r > 0 \end{cases} \tag{3}$$

where the P_S^t and P_S^{t+1} denote the position of the end-point at discrete time ***t*** and ***t + 1*** respectively. And the Δx^t,Δy^t denote the x and y element of the intention direction vector, representing the gaze direction relative to the reference. And r is radius of the neighbor area to find a nearest point on the workspace to be the solution of the formulation (3).

Acknowledgement. This work was partially supported by the National Natural Science Foundation of China (Grant No. 91648203 and No. 51335004), the International Science & Technology Cooperation Program of China (Grant No. 2016YFE0113600), and the Science Foundation for Innovative Group of Hubei Province (Grant No. 2015CFA004).

References

1. Go, A.S., Mozaffarian, D., Roger, V.L., Benjamin, E.J., Berry, J.D., Blaha, M.J., et al.: Heart disease and stroke statistics-2014 update. Circulation **129**(3), e28–e292 (2014)
2. Wang, L.: Report on the Chinese stroke prevention. Peking Union Medical College Press, Beijing (2015). 53, 62
3. Brault, M.W.: Americans with disabilities: 2010. Current population reports (2012). 7, pp. 1–131
4. Broeks, J.G., Lankhorst, G., Rumping, K., Prevo, A.: The long-term outcome of arm function after stroke: results of a follow-up study. Disabil. Rehabil. **21**(8), 357–364 (1999)
5. Wing, K., Lynskey, J.V., Bosch, P.R.: Whole body intensive rehabilitation is feasible and effective in chronic stroke survivors: a retrospective data analysis. Top. Stroke Rehabil. **15** (3), 247–255 (2008)
6. Bayona, N.A., Bitensky, J., Salter, K., Teasell, R.: The role of task-specific training in rehabilitation therapies. Top. Stroke Rehabil. **12**(3), 58–65 (2005)
7. Kwakkel, G., Wagenaar, R.C., Koelman, T.W., Lankhorst, G.J., Koetsier, J.C.: Effects of intensity of rehabilitation after stroke. Stroke **28**(8), 1550–1556 (1997)
8. Bütefisch, C., Hummelsheim, H., Denzler, P., Mauritz, K.-H.: Repetitive training of isolated movements improves the outcome of motor rehabilitation of the centrally paretic hand. J. Neurol. Sci. **130**(1), 59–68 (1995)
9. Hogan, N., Krebs, H.I., Rohrer, B., Palazzolo, J.J.: Motions or muscles? Some behavioral factors underlying robotic assistance of motor recovery. J. Rehab. Res. Dev. **43**(5), 605 (2006)

10. Chen, S.H., Lien, W.M., Wang, W.W., Lee, G.D., Hsu, L.C., Lee, K.W., et al.: Assistive control system for upper limb rehabilitation robot. IEEE Trans. Neural Syst. Rehab. Eng. **24** (11), 1199–1209 (2016)
11. Ugurlu, B., Nishimura, M., Hyodo, K., Kawanishi, M., Narikiyo, T.: Proof of concept for robot-aided upper limb rehabilitation using disturbance observers. IEEE Trans. Hum. Mach. Syst. **45**(1), 110–118 (2015)
12. Kiguchi, K., Hayashi, Y.: An EMG-based control for an upper-limb power-assist exoskeleton robot. IEEE Trans. Syst. Man Cybern. Part B (Cybern.) **42**(4), 1064–1071 (2012)
13. Frisoli, A., Loconsole, C., Leonardis, D., Banno, F., Barsotti, M., Chisari, C., et al.: A new Gaze-BCI-Driven control of an upper limb exoskeleton for rehabilitation in real-world tasks. IEEE Trans. Syst. Man Cybern. Part C (Appl. Rev.) **42**(6), 1169–1179 (2012)
14. Kan, P., Huq, R., Hoey, J., Goetschalckx, R., Mihailidis, A.: The development of an adaptive upper-limb stroke rehabilitation robotic system. J. Neuroeng. Rehab. **8**(1), 1 (2011)
15. Lynch, D., Ferraro, M., Krol, J., Trudell, C.M., Christos, P., Volpe, B.T.: Continuous passive motion improves shoulder joint integrity following stroke. Clin. Rehab. **19**(6), 594–599 (2005)
16. Krebs, H.I., Volpe, B., Hogan, N.: A working model of stroke recovery from rehabilitation robotics practitioners. J. Neuroeng. Rehab. **6**(1), 6 (2009)
17. Lobo-Prat, J., Kooren, P.N., Janssen, M.M.H.P., Keemink, A.Q.L., Veltink, P.H., Stienen, A.H.A., et al.: Implementation of EMG- and force-based control interfaces in active elbow supports for men with duchenne muscular dystrophy: a feasibility study. IEEE Trans. Neural Syst. Rehab. Eng. **24**(11), 1179–1190 (2016)
18. Burger, B., Ferrané, I., Lerasle, F., Infantes, G.: Two-handed gesture recognition and fusion with speech to command a robot. Auton. Robot. **32**(2), 129–147 (2012)
19. Breazeal, C., Aryananda, L.: Recognition of affective communicative intent in robot-directed speech. Auton. Robot. **12**(1), 83–104 (2002)
20. Anderson, K., McOwan, P.W.: A real-time automated system for the recognition of human facial expressions. IEEE Trans. Syst. Man Cybern. Part B (Cybern.) **36**(1), 96–105 (2006)
21. Arkin, R.C., Fujita, M., Takagi, T., Hasegawa, R.: An ethological and emotional basis for human–robot interaction. Robot. Auton. Syst. **42**(3), 191–201 (2003)
22. Ji, Q., Zhu, Z. (eds.): Eye and gaze tracking for interactive graphic display. In: Proceedings of the 2nd International Symposium on Smart Graphics. ACM (2002)
23. Betke, M., Gips, J., Fleming, P.: The camera mouse: visual tracking of body features to provide computer access for people with severe disabilities. IEEE Trans. Neural Syst. Rehab. Eng. **10**(1), 1–10 (2002)
24. Li, S., Zhang, X., Webb, J.: 3D-gaze-based robotic grasping through mimicking human visuomotor function for people with motion impairments. IEEE Trans. Bio-Med. Eng. (2017)
25. Tostado, P.M., Abbott, W.W., Faisal, A.A. (eds.): 3D gaze cursor: continuous calibration and end-point grasp control of robotic actuators. In: 2016 IEEE International Conference on Robotics and Automation (ICRA), 16–21 May 2016
26. Magermans, D., Chadwick, E., Veeger, H., Van Der Helm, F.: Requirements for upper extremity motions during activities of daily living. Clin. Biomech. **20**(6), 591–599 (2005)
27. Koshland, G.F., Galloway, J.C., Nevoret-Bell, C.J.: Control of the wrist in three-joint arm movements to multiple directions in the horizontal plane. J. Neurophysiol. **83**(5), 3188–3195 (2000)

28. Coscia, M., Cheung, V.C., Tropea, P., Koenig, A., Monaco, V., Bennis, C., et al.: The effect of arm weight support on upper limb muscle synergies during reaching movements. J. Neuroeng. Rehab. **11**(1), 22 (2014)
29. Chen, W., Xiong, C., Huang, X., Sun, R., Xiong, Y.: Kinematic analysis and dexterity evaluation of upper extremity in activities of daily living. Gait Posture. **32**(4), 475–481 (2010)
30. van Andel, C.J., Wolterbeek, N., Doorenbosch, C.A., Veeger, D.H., Harlaar, J.: Complete 3D kinematics of upper extremity functional tasks. Gait Posture **27**(1), 120–127 (2008)
31. Hansen, D.W., Ji, Q.: In the eye of the beholder: a survey of models for eyes and gaze. IEEE Trans. Pattern Anal. Mach. Intell. **32**(3), 478–500 (2010)

Investigation of Phase Features of Movement Related Cortical Potentials for Upper-Limb Movement Intention Detection

Hong Zeng[1](✉), Baoguo Xu[1], Huijun Li[1], Aiguo Song[1], Pengcheng Wen[2], and Jia Liu[3]

[1] School of Instrument Science and Engineering, Southeast University, Nanjing 210096, China
hzeng@seu.edu.cn

[2] AVIC Aeronautics Computing Technique Research Institute, Xian 710068, China

[3] Jiangsu Collaborative Innovation Center of Atmospheric Environment and Equipment Technology (CICAEET), Nanjing University of Information Science and Technology, Nanjing 210044, China

Abstract. The movement related cortical potential (MRCP) is a well-known neural signature of humans self-paced movement intention, which can be exploited by future rehabilitation robots. Most existing studies have explored the amplitude representation for the detection. In this paper we have investigated the phase representation for such a task. On the data sets in which 15 healthy subjects executed a self-initiated upper limb center-out reaching task, we have evaluated the detection models with MRCP amplitude features, MRCP phase features and a concatenation of MRCP amplitude and phase features, respectively. The experimental results have demonstrated that the detector based on the concatenation of amplitude and phase features has not only attained the largest percentage of correct classified trials among the three models ($88.05\% \pm 8.80\%$ of trials), but also achieved the earliest detection of the upper-limb movement intention before the actual movement onset (634.58 ± 211.12 ms before the movement onset).

Keywords: Movement intention detection · Electroencephalogram signals · Movement related cortical potentials · Amplitude and phase features

1 Introduction

The development of Brain-Computer Interface (BCI) based on the electroencephalographic (EEG) activity for the functional rehabilitation of patients with motor disabilities has gained special interest over the last years [1]. In particular, it is crucial to detect the voluntary limb movement intention prior to its actual execution for ensuring a proprioceptive feedback timing, which will induce Hebbian associative neural plasticity for rehabilitation purposes [2]. One of the

Y. Huang et al. (Eds.): ICIRA 2017, Part I, LNAI 10462, pp. 350–358, 2017.
DOI: 10.1007/978-3-319-65289-4_34

most investigated neural correlates of movement intention, as imaged by EEG, is the low frequency narrow-band MRCP [3]. The relation between MRCP and movement intention has been recently studied with EEG-based BCIs in the context of self-paced movements or motor imagery [2,4–9].

Most existing studies that used MRCP information for the detection of movement intention explored the amplitude representation of the neural correlate [2,4–9]. However, the amplitude only expresses the intensity of the oscillatory activity of neuron populations in a brain area, the phase information of the neural oscillations has not been utilized for the MRCP detection. Until very recently, the researchers in neuroscience [10,11] have demonstrated on the grand average of trials level that phase-locked neural oscillations in the delta frequency band contralateral to the moving limb are prerequisites for the preparation and execution of motor actions, carring information on the temporal structure of the neural activity of the same neuron populations in the same condition. Such findings have inspired us that the phase information of the MRCPs is complementary to that in the amplitude domain, which may improve the movement intention detection by only the MRCP amplitude on the single trial level.

To this end, we have evaluated the movement intention detection models with MRCP amplitude features, MRCP phase features and a concatenation of MRCP amplitude and phase features, respectively, on the data sets in which 15 healthy subjects executed a self-initiated upper limb center-out reaching task. Experimental results have demonstrated that the MRCP phase information is indeed able to improve the movement intention detection accuracy as well as reducing the detection latency with only the MRCP amplitude.

2 Materials and Methods

2.1 Experimental Procedure

Fifteen volunteers (all right-handed males, mean age = 21.5 ± 2.6 years old) participated in the experiment. All subjects were healthy without any known neurological anomalies or musculo-skeletal disorders. The experimental protocol was approved by the ethical committee of the Southeast University, China and all subjects gave written informed consent before participating in the experiments. Subjects were seated in a comfortable chair in front of a computer screen holding on to a haptic manipulandum (PHANTOM Premium 1.5, Sensable Technologies) with their forearms resting comfortably on the table as shown in Fig. 1(a). The subjects were asked to perform the reaching task by moving the manipulandum that controls the position of a cursor (a red circle) on the computer screen (see Fig. 1(b)). The rest position is the condition when the red circle remains inside the red square box in the center of the screen. The task was to move the cursor to one of the 4 center-out target locations (up, down, left and right being projected as red-frame boxes) with the right upper limb. When the target location was cued, the subject was asked to wait at least 1.5 s before initiating the movement at their own pace in order to induce a self-paced movement. The role of the visual cue was to ensure equal distribution of the four target locations during

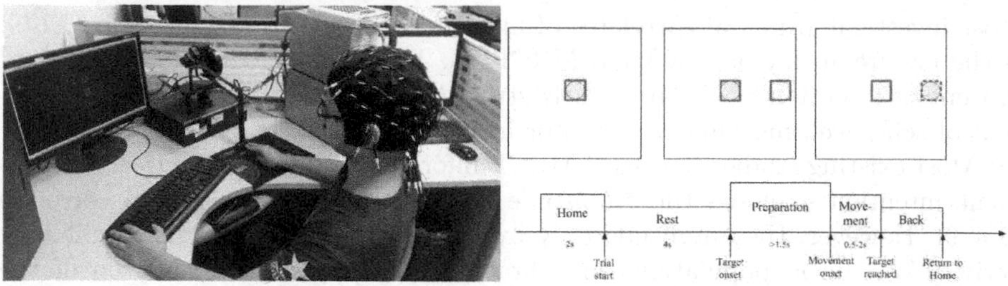

(a) (b)

Fig. 1. Experimental setup (a) and experimental protocol (b). (Color figure online)

the recording. If the subject moved before 1.5 s (i.e. an immediate reaction), the trial was stopped, discarded from the analysis and repeated until the subject successfully fulfilled the requirement of 1.5 s waiting period. The participants were asked to stay in a relaxed position during the idle period before initiating a reaching whenever they wish. For each subject, we recorded one session. A session is composed of 6 runs each consisting of 40 trials (there were break intervals with a duration adjusted to the need of the participants between each run), thus resulting of 240 trials for each subject. After discarding the early start and ocular artifacts, it remains an average of 230 trials in each session per subject.

2.2 Data Acquisition

The electroencephalogram (EEG) signals and electrooculography (EOG) signals were simultaneously recorded with a Neuroscan SynAmps II amplifier and a 64-channel EEG cap with 10/20 montage. There were 33 EEG channels[1] used for further processing and the electrodes impedance was kept below 5 kΩ during the recordings. EOG channels were placed above nasion and below the outer canthi of both eyes in order to capture horizontal and vertical EOG signals for assisting the rejection of EOG artifacts contaminated EEG trials. The EEG and EOG data were sampled at 1000 Hz, filtered at a lower cut-off frequency of 0.05 Hz and the signal bandwidth (highest frequency) is limited at 200 Hz by the acquisition system. In order to reduce the EOG artifact contaminated EEG trials, the subjects were also required to fixate on the central square box and avoid blinking in particular before stating the arm movement, they were encouraged to blink during the period of returning to the Home position. The movement onset is defined as the time when the cursor (i.e., the red circle) exits the center square box. Such events were then sent as triggers to the EEG acquisition system via parallel port.

[1] F3, F1, FZ, F2, F4, FC5, FC3, FC1, FCZ, FC2, FC4, FC6, C5, C3, C1, CZ, C2, C4, C6, CP5, CP3, CP1, CPZ, CP2, CP4, CP6, P5, P3, P1, PZ, P2, P4, P6.

2.3 Data Processing

MRCP was analyzed as an EEG neural signature of arm movement intention. For the analysis, EEG data was processed with a narrow band zero-phase second-order Butterworth filter at 0.1–1 Hz and then downsampled to 8 Hz. As shown in [2,4–9], this filter has reliable characteristics for the analysis of MRCP. The EEG signals during each trial was segmented into 4 s long epochs before the movement onset event (each trial lasted from −4 to 1 s relative to the movement onset), corresponding to the movement preparation and execution periods. Trials that containing EOG artifacts above 70 μV peak-peak in any channel were discarded from further analysis. The signals were further spatially filtered using the common average reference (CAR) procedure to remove the global background activity.

2.4 Phase Feature Extraction

In this paper, the phase information of the MRCP will be investigated for movement intention detection. The most commonly used method for the narrow-band signal phase estimation is to use the analytic signal representation. Accordingly, for the narrow-band signal $s(t)$, its analytical form is defined as follows:

$$z(t) = s(t) + jHT\{s(t)\}, \tag{1}$$

where HT{s(t)} is the Hilbert transform of the signal $s(t)$, defined as

$$HT\{s(t)\} = \frac{1}{\pi} P.V. \int_{-\infty}^{\infty} \frac{s(t)}{t-\tau} d\tau. \tag{2}$$

Then the analytic signal $z(t)$ can be rewritten as $z(t) = A(t)e^{j\phi(t)}$, where $A(t)$ and $\phi(t)$ represent the instantaneous amplitude and the instantaneous phase of the narrow-band signal $s(t)$, respectively. Ultimately, the phase features of MRCP will be represented by $[\cos(\phi(t)) \;\; \sin(\phi(t))]$, which is inspired by the Eulers formula $e^{j\phi(t)} = \cos(\phi(t)) + j\sin(\phi(t))$.

2.5 Neurophysiological Analysis and Intention Detection

Firstly, we conducted the neurophysiological analysis on the amplitude and phase dynamics in the segment of [−4 1] s and summarized the general trends that were common across trials. For computing the grand averages of MRCP amplitude, each epoch was baseline corrected with the average activity between [4 3.5] s before the movement onset. An important indicator of the phase dynamics between trials is the phase locking index (PLI), also denoted as inter-trial phase locking or inter-trial coherence. PLI is defined as

$$PLI(t) = |\frac{1}{N}\sum_{k=1}^{N} e^{j\phi_k(t)}|,$$

where N is the number of trials and $\phi_k(t)$ is the phase of trial k at time t. PLI is a measure of similarity of the phases of a signal over many repetitions. PLI ranges from 0 to 1. PLI = 1 means identical phase of the signal across trials. Low values of PLI suggest temporal heterogeneity of the phases between individual trials. Thus, PLI measures the degree of inter-trial variation in phase and thereby quantifies phase locking of the oscillatory activity irrespective of its amplitude [12].

Secondly, for the detection task of the movement intention, we studied the detector model with (1) Amplitude model, a single-view detection model based on low frequency MRCP amplitude features, (2) Phase model, a single-view detection model based on MRCP instantaneous phase features, and (3) Amplitude + Phase model, a multi-view detection model based on the concatenation of the amplitude and phase features. For each trial lasted from −4 to 0 s relative to the movement onset, features were extracted from the 33 electrodes with a 1 s-long sliding window in steps of 125 ms. Windows between −4 and −1.5 s were labeled as relaxation state and those from −1.5 to 0 s as pre-movement state. We analyzed a total of 25 windows per trial out of which the last 8 belonged to the pre-movement class. The linear discriminant analysis (LDA) classifier was used for the classification.

We evaluated the detection models in a 10-fold nested chronological cross-validation on each subject within a session. The classification probability threshold was automatically selected in the inner loop cross-validation as the one that maximizes the percentage of correctly classified trials (i.e., trials with a true positive and without false positives) on the validation set. Then the performance of detection models was evaluated for each of the ten test folds in the outer loop. We assessed the performance of the models as the percentage of trials detected at least a true positive window (TP), the number of false positive windows per minute (Fp/min), the percentage of correctly classified trials (CT), and the latency relative to the onset of motion (Latency). A correctly classified trial is the one contains at least a true positive window (i.e., pre-movement state detection) in the [−1.5, 0] s time interval, where 0 is the motion onset, is above the probability threshold selected in cross-validation and has no false positive windows in the [−3, −1.5] s time interval. This metric gives a conservative estimate of performance. Statistical analysis was done using Friedman's test with Bonferroni Correction ($p < 0.05$ was considered significant).

3 Results

3.1 Neurophysiological Analysis

Figure 2 shows the low-frequency EEG correlates of right upper-limb movement intention across the subjects. Figure 2(a) presents the MRCP amplitude features in grand average over subjects. One can observe the MRCP amplitude pattern showing a negative deflection in the activity recorded by electrodes located over contra-lateral brain regions (precentral, central, and postcentral) which are known to be involved in motor planning and motor execution, starting about 2 s prior to the movement onset. As shown in Fig. 2(a), the grand

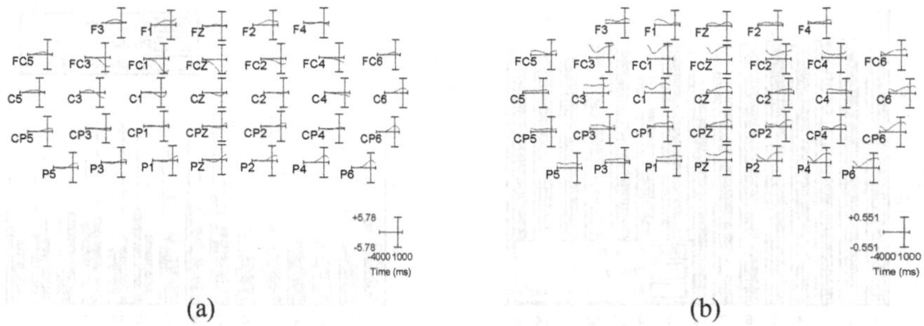

Fig. 2. The low-frequency EEG correlates of right upper-limb movement intention across the subjects and sessions. (a) the MRCP amplitude in grand average. (b) the MRCP phase locking index.

averaged potentials are in agreement with previous MRCP studies [2,4,5] based on the amplitude. Figure 2(b) shows the PLI or the phase synchronization. We can observe that there is an increase in PLI over the contra-lateral motor cortex electrodes, indicating a higher synchronization in the phases patterns starting about 2.5 s prior to the movement onset. Such a trend of PLI is also consistent with the latest work [11] in the neuroscience field.

3.2 Movement Intention Detection Performance

The performance indices of the three detectors are demonstrated in Fig. 3. Figure 3(a) demonstrates the performance of the three detection models at trial level across subjects (S1-S15), as percentage of trials detected at least a true positive window. The Amplitude + Phase model attains the highest TP (95.69% $\pm$ 6.28% in mean $\pm$ standard deviation), followed by the Amplitude model (93.03% $\pm$ 8.50%) and by the Phase model (80.11% $\pm$ 22.80%), and there is a significant main effect of the factor "the detection model" on performance ($p = 5.81 \times 10^{-61}$). Multiple comparisons with the Bonferroni critical value showed that all the the mean differences for the pairs of the detection models are significantly (Amplitude v.s. Amplitude + Phase: $p = 1.33 \times 10^{-6}$, Phase v.s. Amplitude + Phase: $p = 2.04 \times 10^{-25}$, Amplitude v.s. Phase: $p = 1.33 \times 10^{-7}$).

Figure 3(b) shows the comparison results as the number of false positive windows per minute. The FP/min's obtained by the Amplitude, Phase and Amplitude + Phase models are 0.54 ± 0.56, 0.33 ± 0.59 and 0.57 ± 0.68, respectively. There is significant main effect of the factor "the detection model" on performance ($p = 3.93 \times 10^{-9}$). After the Bonferroni correction, we found a statistically significant difference between Amplitude and Phase model ($p = 8.23 \times 10^{-8}$), as well as between Phase and Amplitude + Phase model ($p = 5.93 \times 10^{-7}$), but no statistically significant difference between the Amplitude and Amplitude + Phase model.

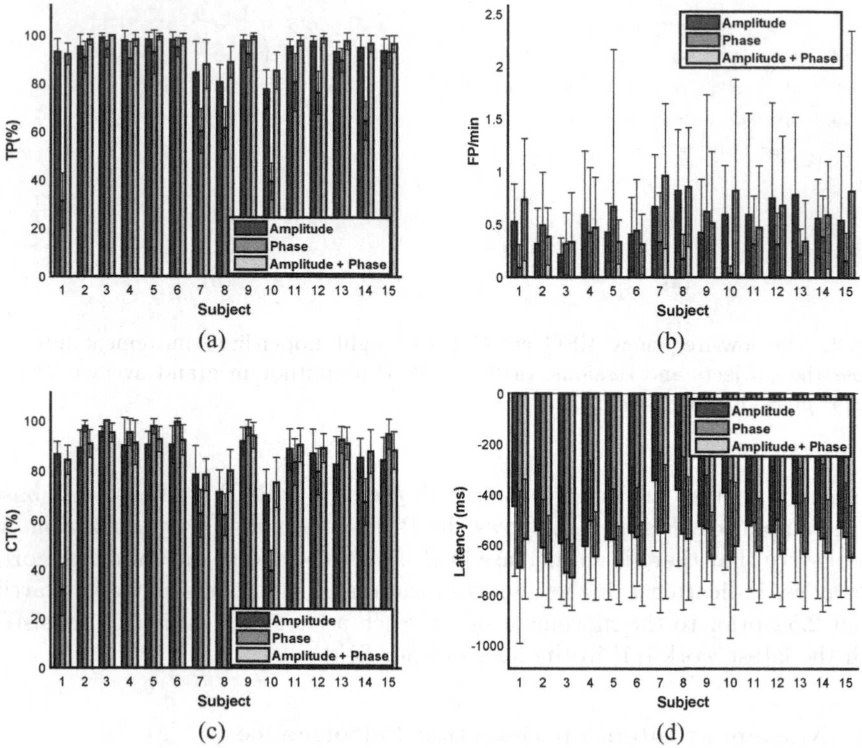

Fig. 3. Comparison of performance with the three detection models: (a) TP (b) FP (c) CT and (d) Latency.

Figure 3(c) depicts the detection results as the percentage of correctly classified trials. The highest averaged CT was attained for the Amplitude + Phase model (88.05% ± 8.80%), while the lowest CT was obtained using the Phase model (76.74% ± 21.66%). According to the Friedman's test, there is significant main effect of the factor "the detection model" on performance ($p = 1.44 \times 10^{-8}$). Furthermore, we found significant differences of CT for all pairs of the detection models (Amplitude v.s. Amplitude + Phase: $p = 3.10 \times 10^{-3}$, Phase v.s. Amplitude + Phase: $p = 5.92 \times 10^{-9}$, Amplitude v.s. Phase: $p = 0.02$) by multiple comparisons with the Bonferroni critical value.

Figure 3(d) summarizes the latency relative to the onset of motion across subjects. All the three models can detect the upper-limb movement intention earlier than the actual movement onset (Amplitude: -500.87 ± 280.70 ms, Phase: -576.74 ± 262.25 ms and Amplitude + Phase:-634.58 ± 211.12 ms), while the main effect of the factor "the detection model" on latency is significant ($p = 5.81 \times 10^{-61}$). Again, multiple comparisons with the Bonferroni critical value showed that all the the mean differences for all pairs of the detection models were significantly different (Amplitude v.s. Amplitude + Phase: $p = 3.98 \times 10^{-56}$, Phase v.s. Amplitude + Phase: $p = 1.50 \times 10^{-34}$, Amplitude v.s. Phase: $p = 1.37 \times 10^{-3}$).

As a summary of Fig. 3, the Amplitude + Phase detector has not only attained the largest percentage of correct classified trials, but also achieved the earliest detection of the upper-limb movement intention before the actual movement onset. These findings indicate that the Amplitude + Phase model is the most reliable one for the movement intention detection.

4 Conclusion

The study was focused on the single-trial asynchronous detection of the pre-movement state and we evaluated the results by envisioning its applicability to closed-loop BCI (e.g., rehabilitation scenarios). In particular, the phase representation of the MRCP and its usage for the detection of upper-limb movement intention was investigated. Firstly, we show that in addition to the amplitude pattern, the MRCP instantaneous phase pattern is also a correlate of upper-limb movement intention, which even emerges a little earlier than the MRCP amplitude pattern. Nextly, we built three detection models based on LDA by using either temporal pre-movement state correlates (Amplitude model), or phase pre-movement state correlates (Phase model), or a concatenation of temporal and phase features (Amplitude + Phase detection model). The experimental results have demonstrated that the Amplitude + Phase detector has not only attained the largest percentage of correct classified trials among the three models, but also achieved the earliest detection of the upper-limb movement intention before the actual movement onset.

References

1. Muralidharan, A., Chae, J., Taylor, D.: Extracting attempted hand movements from EEGS in people with complete hand paralysis following stroke. Front. Neurosci. **5**, 39 (2011)
2. Xu, R., Jiang, N., Lin, C., Mrachacz-Kersting, N., Dremstrup, K., Farina, D.: Enhanced low-latency detection of motor intention from EEGS for closed-loop brain-computer interface applications. IEEE Trans. Biomed. Eng. **61**(2), 288–296 (2014)
3. Shibasaki, H., Hallett, M.: What is the bereitschaftspotential? Clin. Neurophysiol. **117**(11), 2341–2356 (2006)
4. Bhagat, N.A., Anusha, V., Berdakh, A., Artz, E.J., Nuray, Y., Blank, A.A., et al.: Design and optimization of an EEG-based brain machine interface (BMI) to an upper-limb exoskeleton for stroke survivors. Front. Neurosci. **10**(564), 122 (2016)
5. Lew, E., Chavarriaga, R., Silvoni, S., Millán, J.D.R.: Detection of self-paced reaching movement intention from EEG signals. Front. Neuroeng. **5**(13) (2012)
6. Sburlea, A.I., Montesano, L., Minguez, J.: Continuous detection of the self-initiated walking pre-movement state from EEG correlates without session-to-session recalibration. J. Neural Eng. **12**(3), 036007 (2015)
7. Jochumsen, M., Niazi, I.K., Dremstrup, K., Kamavuako, E.N.: Detecting and classifying three different hand movement types through electroencephalography recordings for neurorehabilitation. Med. Biol. Eng. Comput. **54**(10), 1491–1501 (2016)

8. López-Larraz, E., Trincado-Alonso, F., Rajasekaran, V., Prez-Nombela, S., Del-Ama, A.J., Aranda, J., et al.: Control of an ambulatory exoskeleton with a brain Machine interface for spinal cord injury gait rehabilitation. Front. Neurosci. **10** (2016)
9. Norman, S.L., Dennison, M., Wolbrecht, E., Cramer, S.C., Srinivasan, R., Reinkensmeyer, D.J.: Movement anticipation and EEG: implications for BCI-contingent robot therapy. IEEE Trans. Neural Syst. Rehab. Eng. **24**(8), 911–919 (2016)
10. Igarashi, J., Isomura, Y., Arai, K., Harukuni, R., Fukai, T.: A θ-γ oscillation code for neuronal coordination during motor behavior. J. Neurosci. **33**(47), 18515–18530 (2013)
11. Popovych, S., Rosjat, N., Toth, T., Wang, B., Liu, L., Abdollahi, R., Viswanathan, S., Grefkes, C., Fink, G., Daun, S.: Movement-related phase locking in the delta-theta frequency band. NeuroImage **139**, 439–449 (2016)
12. Tass, P.A.: Phase Resetting in Medicine and Biology: Stochastic Modelling and Data Analysis. Springer, Heidelberg (2007)

Human-Machine Interaction

Mobile Terminals Haptic Interface: A Vibro-Tactile Finger Device for 3D Shape Rendering

Xingjian Zhong, Juan Wu(✉), Xiao Han, and Wei Liu

School of Instrument Science and Engineering, Southeast University, Sipailou 2#, Nanjing 210096, Jiangsu, China
{juanwuseu,220142658,lwseu}@seu.edu.cn

Abstract. As the increasing in popularity of mobile terminals, the research on the haptic rendering accessory becomes more and more essential. This paper presents the design and implement of a vibro-tactile device mounted on the finger that could interact with mobile terminals and provide vibro-tactile stimulus for 3D shape displaying. To generate distributed vibro-tactile stimulus, four piezoelectric actuators are settled around the finger. A force sensor is attached on the fingertip to gauge the active force during the interaction. The lateral force is calculated corresponding with the changes of gradient for 3D object and local surface orientation. When stroking across the virtual object, the lateral force in contact is simulated as the vibro-tactile stimulus. The distribution and intension of the vibro-tactile stimulus vary with the direction and amplitude of the lateral force. To evaluate the performance of the tactile display system of 3D shape, psychophysical experiment was carried out. Results showed that the device is effective to provide vivid tactile perception corresponding with different shape.

Keywords: Mobile terminals · Vibro-tactile device · Shape rendering

1 Introduction

In recent years, the development of mobile terminals with touch screens has changed interaction modalities [1], while tactile interaction on mobile devices is still insufficient compared with visual and auditory interaction. The function and the range of applications of mobile terminals are broadened when tactile interactive technology is applied to mobile terminals. Tactile feedback on mobile terminals can be used for various applications. For example, tactile feedback can offer a new option for the visually impaired to experience 3D art at museums [2]. Tactile technology can be implemented on mobile terminals for mobile e-commerce so that the user can feel some kind of feedback when a product is selected [3]. Moreover, social mobile pervasive games use sensors, tactile and networking technologies to provide immersive game experiences integrated with

Y. Huang et al. (Eds.): ICIRA 2017, Part I, LNAI 10462, pp. 361–372, 2017.
DOI: 10.1007/978-3-319-65289-4_35

the real world [4]. Therefore, it is of great promise to design tactile interfaces for mobile terminals.

Previous work on designing tactile interfaces for mobile terminals can be classified into two categories according to the principles of generating tactile stimulation. One kind is that films was adapted on the screen of mobile terminals to actuate various stimulus [5–8] such as electro vibration, electrostatic force, mechanical vibration. The other is the haptic accessory used to configure with mobile terminals [9,10], such as pen-like or wearable haptic feedback device. Compared with the former generating haptic feedback on the mobile terminals, the latter method uses independent module which is simple structure and good versatility.

Some literatures have discussed rendering shape using tactile interfaces. The easiest approach to display shape is simply pushing or pulling the user's finger by means of force feedback equipments [5,11–15]. The distinguishing characteristic of this shape display approach is that they need one motor for each component of the force to be independently rendered, and, for this reason, it is quite difficult to make them wearable and portable for interacting with mobile terminals. The second approach for shape rendering on mobile terminals maps film forces on the screen of mobile terminals, such as electrostatic force, electro-vibration force and friction force based on squeeze-film effect, to shape features [16,17]. This method allows for the creation of a feasible and lightweight tactile feedback apparatus suitable for touch screens. However it has not been used in practical application due to the power limit and safety. The third approach to rendering shape deals with dynamic pin arrays [18,19]. Although this approach is very flexible and effective, it usually employs a large number of actuators, which compromises the overall wearability and portability of the system.

Recent studies have shown vibro-tactile stimuli are fundamental in recognizing shape [20], curvature discrimination tasks [21], in surface orientation discrimination [22] and in improving the illusion of presence in virtual and remote environments. Thus, the vibro-tactile stimulus provides a new scheme for rendering shape on the mobile terminal. Moreover, the rapid development of materials and actuator design technology makes large-scale production of small actuators possible. The vibro-tactile stimulus generated by the small actuator can be applied to the mobile terminal, providing rich vibrating patterns for simulating rich contact interaction with virtual objects. In [23], Dapeng Chen presents a novel miniature multi-mode haptic pen for image interaction on mobile terminals. The haptic pen can provide two types of vibro-tactile feedback by linear resonant actuator and piezo-ceramic actuator. In [24], Ki-Uk Kyung presents the Ubi-Pen, a pen-like haptic interface providing texture and vibration stimuli with mobile terminals. Although, vibro-tactile stimulus is limited to possibility of simulating richer force patterns. However tactile interfaces with vibro-tactile stimulus are wearable and portable for interacting with mobile terminals, and can provide rich vibration pattern.

In this paper, a portable-wearable vibro-tactile finger device based on distributed vibro-tactile stimulus is designed for shape display on mobile terminals.

This is achieved by settling four small size piezoelectric actuators around finger and attaching a force sensor in proposed device. Furthermore, a virtual shape rendering algorithm is introduced based on the vibro-tactile finger device. The lateral force is chosen as shape feature. Its direction indicates direction of bump surface, and its amplitude is equal to gradient in touching position and the active force exerted by the operator. The lateral direction is mapped to distribution of vibro-tactile stimulus, and lateral force amplitude is modulated as a linear function of intension of vibro-tactile stimulus. The rest of the paper is organized as follows. The fingertip device is introduced in Sect. 2. Section 3 discusses driving piezoelectric actuators. Shape display algorithm is presented in Sect. 4. Experiment is conducted in Sect. 5. Finally, the conclusion and the future work are described in Sect. 6.

2 Design of the Fingertip Device

2.1 Mechanical Design

The completed portable-wearable fingertip device that interacts with the mobile terminals is shown in Fig. 1(a). The proposed device can be divided into two main modules: control module and interactive module. The control module belted onto the forearm takes charge of communication with mobile terminals, generating driver signals for actuators and supplying power for the entire device. The interactive module can be worn onto human finger for measuring active force as well as providing vibro-tactile feedback to the finger. The total mass of this device is under 100 g.

The interactive module should not only utilize the space rationally to place the four actuators and the force sensor, but also have good security and comfort for finger wearing. Figure 1(b) shows the schematic diagram of the interactive module. Capacitance nib was selected in order to interact with capacitive touchscreen.

The connector connects capacitance nib with a screw thread. The force sensor is mounted above the connector with a tight fit. A layer of 3 mm thick sponge is placed between the main shell and piezoelectric actuators to prevent the main

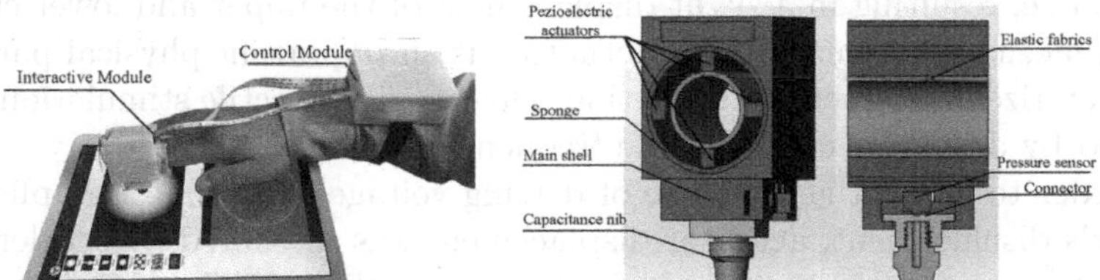

Fig. 1. (a) Complete wearable fingertip device. (b) Schematic diagram of the interactive module

shell vibrating with actuators. A layer of elastic fabric of 1 mm thick is mounted between the actuator and human finger. It is flexible to fit the size of human finger, as well as insulated to ensure safety. Actuators are distributed around the finger in a circle. The main shell and the connector are made up of resin printed by 3D printer.

2.2 Device Control Hardware

The primary electronics are mounted in the control module. It comprises five distinct physical units: WIFI communication unit, host MCU unit, active force processing unit, piezoelectric actuator driver unit and power supply unit. The device hardware diagram is showed in Fig. 2(a). WIFI communication unit transports shape features of the touch point from mobile terminals to host MCU unit. Active force processing unit handles gauged force data by means of a low-pass filter in order to remove noise from actuators' vibration. The host MCU unit integrated shape features and active force to generate vibration control signals. According to the vibration control signals, the driver unit generates driving voltage to piezoelectric actuators. The power supply unit converts battery voltage to various voltage (3.3 V, 5 V, 9 V) for each unit. Since power consumption of each piezoelectric actuator is less than 1 W, so the battery (3.7 V 2100 mAh) is capable of fully supplying the energy requirements for 2hrs. The battery is used lithium and is fully rechargeable in approximately 1 h.

3 Driving Piezoelectric Actuators

The device is comprised of four piezoelectric actuators for generating vibro-tactile stimulation. The four actuators are attached onto the four sides of finger in order to generate distributed vibro-tactile stimulus. In this section, first driving the piezoelectric actuators is discussed, then relationship between the amplitude, frequency of the driving voltage and piezoelectric actuators' displacement was studied to get rich vibro-tactile stimulation.

In this paper the piezoelectric actuators are stack multilayer actuators. Actuators' driving voltage is sine wave whose frequency, amplitude can be altered. When a sine wave voltage is applied to the piezoelectric actuator, the internal pressure rise, resulting in a slight displacement of the upper and lower changes. The displacement of piezoelectric actuators is an important physical parameter to characterize vibro-tactile stimulation. Diverse vibro-tactile stimulation can be produced by delicately changing the frequency and amplitude.

In order to invest the influence of driving voltage's frequency, amplitude on actuator's displacement, actuator displacement was measured by accelerometer through twice integral at different frequency, amplitude. The frequency of the driving signal was among 100 Hz, 200 Hz, 250 Hz, 300 Hz, and the amplitude was between 20 V and 100 V. There were 32 different driving voltages in total. Displacement at each driving voltage was measured three times. Figure 2(b) shows the actuators' displacement at different amplitude and frequencies of the driving voltage. To study how the amplitude and frequency would influence displacement of actuator, two-way repeated measure ANOVA was used to analyze the data. The results show that there are significant effects of amplitude on the displacement ([$F(4, 12) = 129.03$, $p < 0.001$]). Frequencies has no noticeable effect on the displacement ([$F(3, 12) = 2.01$, $p = 0.166$]). From the figure, large scale vibro-tactile stimulation can be obtained by changing amplitude of the driving voltage that has a basic frequency of 250 Hz. Moreover, frequency of 250 Hz is the optimal sensitivity of the Pacinian corpuscles which are the mechanoreceptors responsible for detection of higher frequency vibrational stimulation [25]. Consequently driving signal at a basic frequency of 250 Hz is more suitable for this device to generate vibro-tactile stimulation. As showed in Fig. 2(b), the relationship between displacement of piezoelectric actuator and the amplitude of the driving voltage at a basic frequency of 250 Hz can be fitted as a logarithmic function which can be described as:

$$D = 14.5 \ln(A) - 40.6 \tag{1}$$

where A represents amplitude of driving voltage, D is displacement. If the displacement of actuators is known to generate specific vibro-tactile stimulation, the amplitude of driving voltage could be calculated according to (1).

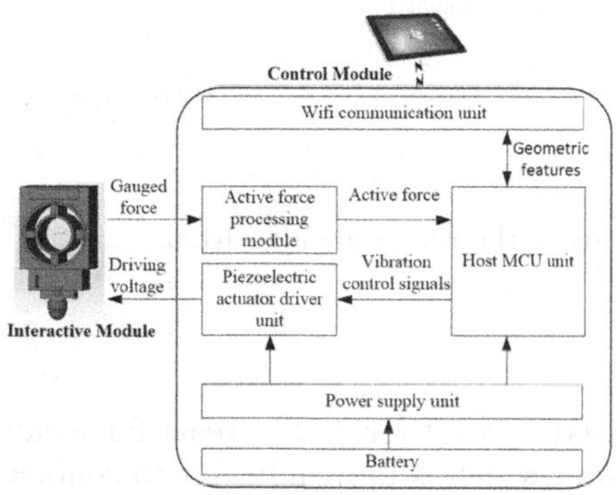

(a)

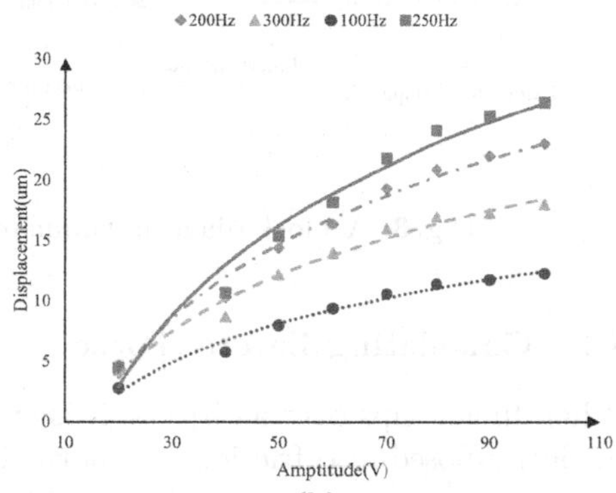

(b)

Fig. 2. (a) Device hardware diagram. (b) Actuators' displacement at different amplitude and frequencies

4 Shape Rendering Algorithm

In this section, a rendering algorithm is introduced for shape display based on lateral force. The lateral-force-based haptic illusion is well known in [26,27]. By presenting the lateral force distribution shown in the upper side of Fig. 4(a), the device presents the bump illusion shown in the lower side of Fig. 4(a) to the user. Figure 3 summarizes the rendering algorithm. As shown in Fig. 3, the proposed algorithm is made of two parts: calculating lateral force and generating vibro-tactile stimulus. The input to the algorithm is a gray level image of a surface to be rendered, the touch position and active force. The algorithm first extracts height distribution from images using SFS algorithm. It then detects touch position and active force. The lateral force is calculated with active force, height distribution and touch position. Its direction indicates direction of bump surface, and its amplitude equivalents with gradient in touching position and the active force exerted by the operator. In order to simulate lateral force, lateral force direction is mapped to vibro-tactile stimulus distribution, and lateral force amplitude is modulated as a linear function of intension of vibro-tactile stimulus (displacement of actuators). Finally the amplitude of the driving voltage is computed according to its displacement.

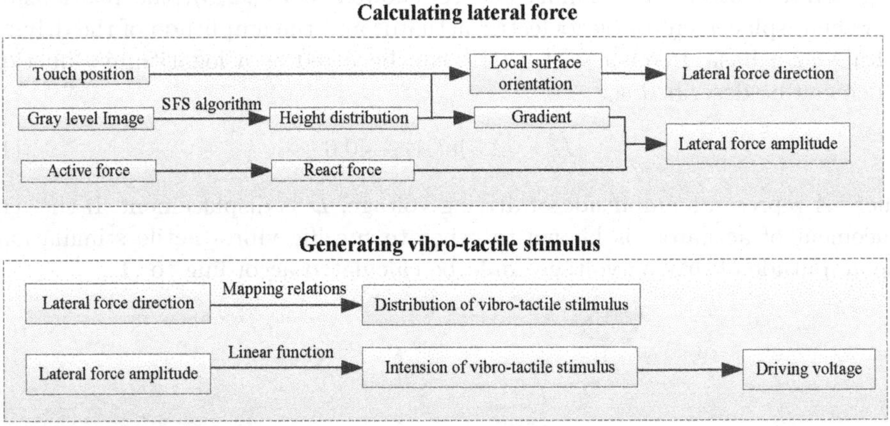

Fig. 3. A block diagram summarizing of the rendering algorithm.

4.1 Calculating Lateral Force

When finger slips over a objects, it is exposed to react force. The react force can be decomposed into friction and normal force which is perpendicular to contact point tangent plane. Similarly, react force can also be decomposed into lateral force within horizontal plane and vertical force along z axis. Force analysis of the touch point is in Fig. 4(b). From Fig. 4(b), we can see:

$$F_l = F\sin\theta_1 + f\cos\theta_1 \qquad with \quad F_r = F + f \tag{2}$$

$$f = \mu F \tag{3}$$

$$\theta_1 = \tan^{-1}\frac{\sqrt{(\partial H/\partial x)^2 + (\partial H/\partial y)^2}}{\partial H/\partial z} \tag{4}$$

where F_l and F_r is the lateral force and react force. F and f represent normal force and friction. μ is coefficient of dynamic friction. θ_1 is the angle between the normal force and the Z axis. $\partial H/\partial x$, $\partial H/\partial y$, $\partial H/\partial z$ are gradient components of 3D contour along x, y, z axis respectively. H denotes height distribution extracted from image. Since the SFS algorithm is effective and easy to implement

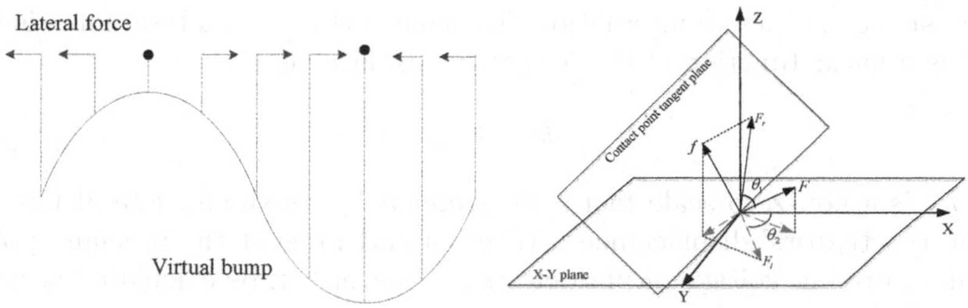

(a) (b)

Fig. 4. (a) Lateral-force-based haptic illusion. (b) Force analysis of contact point

in mobile terminal, so it is chosen to extract the surface height distribution of virtual objects in an image.

The contact process is assumed to be a quasi-static process, then the normal force is equal to active force. Here, active force is supposed to be the force applied to terminals' screen when the device slides on the screen which can be gauged by force sensor. But actuators' vibrations will add noise to the active force measurement, thus it is necessary to filter the gauged force data. The filter is designed to be Butterworth low - pass filter whose cutoff frequency is 1 Hz, because frequency of the active force is below in 1 Hz.

In most cases, the friction component is one order smaller than the normal force component. Hence influence of the friction component to lateral force can be ignored. Neglecting friction component, lateral force approximately is calculated as:

$$F_l = F \sin \theta_1. \tag{5}$$

From (5), lateral force is specified by active force and shape features. Thus it is reasonable to choose lateral force as physical parameters of objects' shape and active interaction.

As shown in Fig. 4(a), direction of bump surface and lateral force direction shifts with touching position in virtual environment. More specifically, when the surface of bump is convex or concave to the left and lateral force direction is along left direction. Similarly, when the surface of bump is convex or concave to the right and lateral force direction is along right direction. Thus it can be concluded that lateral force direction can indicates direction of bump surface. Lateral force direction is defined as the angle between the lateral force and the x-axis positive direction. Ignoring friction component, lateral force direction is calculated as:

$$\theta_2 = \tan^{-1}\left(\frac{\partial H/\partial y}{\partial H/\partial x}\right) \tag{6}$$

4.2 Generating Vibro-Tactile Stimulus

The main element in our rendering procedure is generating vibro-tactile feedback relative to the direction and amplitude of lateral force in order to render objects' shape and active interaction. To achieve this goal, lateral force is modulated as a linear function of the actuators displacement as:

$$D = k_1 F_l \tag{7}$$

where k_1 is a non-zero scale factor determined by maximum lateral force and maximum actuators' displacement. When lateral force at the touching point is calculated, driving voltage of piezoelectric actuator can be calculated according to (1) and (7):

$$A = 16.4e^{k_2 F_l} \tag{8}$$

where k_2 is a non-zero scale factor proportional to k_1.

In order to simulate the virtual bump surface direction information, the proposed rendering algorithm maps the lateral force direction to the distribution of vibro-tactile stimulus. The device has four actuators to generate distributed vibro-tactile stimulus, if one of the four actuators vibrates then there are four vibro-tactile stimulus distribution. So the direction of the lateral force can be divided into four equal parts and each mapped to the four vibro-tactile stimulus distribution. Table 1 shows the mapping of the lateral force direction and the distribution of vibration tactile stimulus. When the direction of the lateral force is close to x-axis positive direction, the right actuator vibrates, and the other cases are shown in Table 1.

Table 1. Lateral force direction mapping to distribution of vibro-tactile stimulus. The red actuator represents the vibrating actuator

Lateral force direction	$-45° \leq \theta_2 < 45°$	$45° \leq \theta_2 < 135°$
distribution of vibro-tactile stimulus	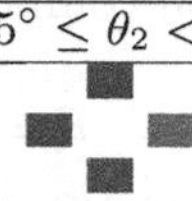	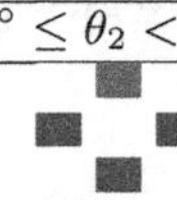
Lateral force direction	$135° \leq \theta_2 < 225°$	$-135° \leq \theta_2 < -45°$
distribution of vibro-tactile stimulus	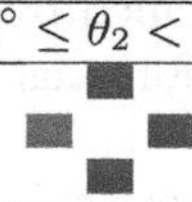	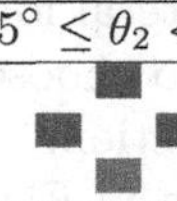

5 Evaluation of the Vibro-Tactile Device for 3D Shape Displaying

5.1 Methods and Procedures

To evaluate the performance of our developed fingertip device for 3D shape displaying, a psychophysical task was presented in a forced choice (6 choices) paradigm to 20 subjects, 8 females and 12 males. Six representative shapes: (a) the Gaussian shape bump, (b) Gaussian shape concave, (c) inclined shape bump, (d) inclined shape concave, (d) round flat surface (f) square flat surface were chosen for rendering. The six shapes represent the most common three types of shapes: quadric surface (a, b), inclined surface (c, d) and flat surface (d, e). Figure 5 (a) shows gray level images and its extracted height distributions. Because it would be difficult to cover any and all 3D shapes in such an experiment, those three types of shapes were presumed to be the 3D shape primitives. The experimental setup consisted in the finger device and a mobile terminals. When subjects wear the fingertip device sliding back and forward within screen of mobile terminals, the finger device generated corresponding vibro-tactile stimulus of the touch point to subjects.

Subjects performed the experiment in two phases: pre-experiment phase and formal experiment phase. The pre-experiment phase lasted about 5 min with view of the virtual shapes, no exploring strategy was suggested them. Subjects perceived two virtual shapes (not the six experimental shapes) in order to adapt to the finger device and familiar with distributed vibro-tactile stimulus. In formal experiment phase, every experimental shape repeated 5 times without view of the virtual shapes. At each trial, one virtual shape was selected randomly, and subjects match their perceived shapes with reference pictures (view of the virtual shapes) according to perceived vibro-tactile stimulus. The correct answer ration of perceiving virtual objects was calculated and recorded, as shown in Fig. 5 (b).

5.2 Results and Discussion

From Fig. 5(b), correct answer ration averaged over subjects of the six shapes (minimum value 0.81) are significantly higher than the chance level (0.167). Results of experiment demonstrated that subjects can identify vibro-tactile stimulus that represents the six virtual shapes. The performance of the experiments reported high variation of the six virtual shapes among subjects (maximum value 10.2). Such difference between subjects' performance might be addressed to different exploratory strategies, since no predefined strategy was suggested. The result of two-way repeated measure ANOVA shows that type of shape ([F (4, 16) = 92.7, $p < 0.001$]) has significant effects of correct ration. From Fig. 5(b), quadric surface has higher correct ration than inclined surface and flat surface.

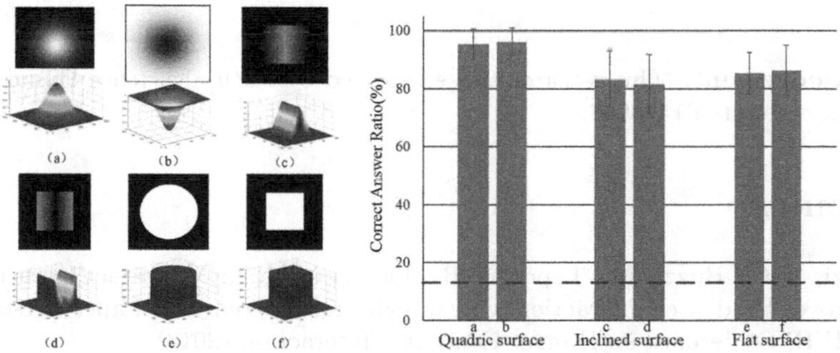

(a) (b)

Fig. 5. (a) Gray level images and its height distributions. (b) Correct answer ratio of perceiving virtual shapes. Dashed line represents the chance level.

All subjects said that when perceive quadratic surface, they can perceive the obvious changes of vibro-tactile stimulus. However, when they feel the inclined surface and flat surface, they can not perceive the change of vibro-tactile stimulus. This indicates that human are sensitive to changing vibro-tactile stimulus. The deeper reason is that our shape display algorithm is based on lateral force. The lateral force can provide sufficient information about height change rate rather than height. The experimental results show that the proposed 3D shape rendering algorithm has some validity, especially when rendering the quadratic surface, the effect is remarkable. However, height information of virtual shape is not display in this algorithm, weakening the sense of reality.

6 Conclusions

In this paper, a wearable and portable fingertip device is introduced for shape display. It comprises four small size piezoelectric actuators and force sensor which can measure active force during exploration. The four actuators are settled around finger to generate distributed vibro-tactile stimulus. In order to obtain appropriate actuators' driving voltage for this device, the relationship between the amplitude, frequency of the driving voltage and piezoelectric actuators' displacement was studied. Furthermore, a shape rendering algorithm is introduced. In contrast to previous shape display algorithm, the proposed algorithm not only provide sufficient information about shape information and active force, but also generating distributed vibro-tactile stimulus to simulate the direction of bump surface. It is achieved by mapping lateral force direction to vibro-tactile stimulus distribution, and modulating lateral force amplitude as a linear function of displacement of actuators. A perceptive experiment was carried out to verify feasibility of the device. Result show that the device is capable of generating different and real tactile sense regarding shape. However, lack of height information in this algorithm weaken the sense of reality. Therefore, in order to produce shape sensation as real as possible, the characteristics of piezoelectric actuators should further be studied so as to render more parameters of shape such as height.

Acknowledgement. This research was supported by Natural Science Foundation of China under grants 61473088.

References

1. Buzzi, M.C., Buzzi, M., Leporini, B., Paratore, M.T.: Vibro-tactile enrichment improves blind user interaction with mobile touchscreens. In: Paper Presented at the IFIP Conference on Human-Computer Interaction (2013)
2. Jansson, G., Bergamasco, M., Frisoli, A.: A new option for the visually impaired to experience 3d art at museums: manual exploration of virtual copies. Visual Impairment Res. **5**(1), 1–12 (2003)

3. Muniandy, M., Ee, W.K.: User's perception on the application of haptics in mobile e-commerce. In: 2013 International Conference on Paper Presented at the Research and Innovation in Information Systems (ICRIIS) (2013)
4. Sra, M., Schmandt, C.: Expanding social mobile games beyond the device screen. Pers. Ubiquitous Comput. **19**(3), 495–508 (2015). doi:10.1007/s00779-015-0845-0
5. Poupyrev, I., Rekimoto, J., Maruyama, S.: A tactile display for handheld devices. In: Paper Presented at the CHI 2002 Extended Abstracts on Human Factors in Computing Systems
6. Hoggan, E., Anwar, S., Brewster, S.A.: Mobile multi-actuator tactile displays. In: Paper Presented at the International Workshop on Haptic and Audio Interaction Design (2007)
7. Xu, C., Israr, A., Poupyrev, I., Bau, O., Harrison, C.: Tactile display for the visually impaired using TeslaTouch. In: Paper Presented at the CHI 2011 Extended Abstracts on Human Factors in Computing Systems, Vancouver, BC, Canada (2011)
8. Saga, S., Deguchi, K.: Lateral-force-based 2.5-dimensional tactile display for touch screen. In: 2012 IEEE Paper Presented at the Haptics Symposium (HAPTICS) (2012)
9. Kyung, K.-U., Park, J.-S.: Development of a compact tactile display module and its application to a haptic stylus
10. Frediani, G., Mazzei, D., De Rossi, D.E., Carpi, F.: Wearable wireless tactile display for virtual interactions with soft bodies. Front. Bioeng. Biotechnol. **2** (2014)
11. Salisbury, K., Brock, D., Massie, T., Swarup, N., Zilles, C.: Haptic rendering: programming touch interaction with virtual objects. In: Proceedings of the 1995 symposium on Interactive 3D graphics, pp. 123–130. ACM, April 1995
12. Weiss, M., Wacharamanotham, C., Voelker, S., Borchers, J.: FingerFlux: near-surface haptic feedback on tabletops. In: Proceedings of the 24th Annual ACM Symposium on User Interface Software And Technology, pp. 615–620. ACM, October 2011
13. Pacchierotti, C., Salvietti, G., Hussain, I., Meli, L., Prattichizzo, D.: The hRing: a wearable haptic device to avoid occlusions in hand tracking. In: 2016 IEEE Haptics Symposium (HAPTICS), pp. 134–139. IEEE, April 2016
14. Gleeson, B.T., Horschel, S.K., Provancher, W.R.: Design of a fingertip-mounted tactile display with tangential skin displacement feedback. IEEE Trans. Haptics **3**(4), 297–301 (2010)
15. Prattichizzo, D., Chinello, F., Pacchierotti, C., Malvezzi, M.: Towards wearability in fingertip haptics: a 3-dof wearable device for cutaneous force feedback. IEEE Trans. Haptics **6**(4), 506–516 (2013)
16. Wang, T., Sun, X.: Electrostatic tactile rendering of image based on shape from shading. In: 2014 International Conference on Audio, Language and Image Processing (ICALIP), pp. 775–779. IEEE, July 2014
17. Kim, S.C., Israr, A., Poupyrev, I.: Tactile rendering of 3D features on touch surfaces. In: Proceedings of the 26th Annual ACM Symposium on User Interface Software and Technology, pp. 531–538. ACM, October 2013
18. Yang, T.H., Kim, S.Y., Kim, C.H., Kwon, D.S., Book, W.J.: Development of a miniature pin-array tactile module using elastic and electromagnetic force for mobile devices. In: Third Joint EuroHaptics Conference, 2009 and Symposium on Haptic Interfaces for Virtual Environment and Teleoperator Systems, World Haptics 2009, pp. 13–17. IEEE, March 2009

19. Sarakoglou, I., Garcia-Hernandez, N., Tsagarakis, N.G., Caldwell, D.G.: A high performance tactile feedback display and its integration in teleoperation. IEEE Trans. Haptics **5**(3), 252–263 (2012)
20. Jansson, G., Monaci, L.: Identification of real objects under conditions similar to those in haptic displays: providing spatially distributed information at the contact areas is more important than increasing the number of areas. Virtual Reality **9**(4), 243–249 (2006)
21. Chinello, F., Malvezzi, M., Pacchierotti, C., Prattichizzo, D.: A three DoFs wearable tactile display for exploration and manipulation of virtual objects. In: 2012 IEEE Haptics Symposium (HAPTICS), pp. 71–76. IEEE, March 2012
22. Frisoli, A., Solazzi, M., Reiner, M., Bergamasco, M.: The contribution of cutaneous and kinesthetic sensory modalities in haptic perception of orientation. Brain Res. Bull. **85**(5), 260–266 (2011)
23. Chen, D., Song, A., Tian, L.: A novel miniature multi-mode haptic pen for image interaction on mobile terminal. In: 2015 IEEE International Symposium on Haptic, Audio and Visual Environments and Games (HAVE), pp. 1–6. IEEE, October 2015
24. Kyung, K.U., Park, J.S.: Ubi-Pen: development of a compact tactile display module and its application to a haptic stylus. In: Second Joint EuroHaptics Conference, 2007 and Symposium on Haptic Interfaces for Virtual Environment and Teleoperator Systems, World Haptics 2007, pp. 109–114. IEEE, March 2007
25. Chouvardas, V.G., Miliou, A.N., Hatalis, M.K.: Tactile displays: overview and recent advances. Displays **29**(3), 185–194 (2008)
26. Minsky, M., Ming, O.Y., Steele, O., Brooks Jr., F.P., Behensky, M.: Feeling and seeing: issues in force display. In: ACM SIGGRAPH Computer Graphics, vol. 24(2), pp. 235–241. ACM, February 1990
27. Robles-De-La-Torre, G., Hayward, V.: Force can overcome object geometry in the perception of shape through active touch. Nature **412**(6845), 445–448 (2001)

Towards Finger Gestures and Force Recognition Based on Wrist Electromyography and Accelerometers

Bo Lv, Xinjun Sheng(✉), Weichao Guo, Xiangyang Zhu, and Han Ding

State Key Laboratory of Mechanical System and Vibration,
Shanghai Jiao Tong University, Shanghai 200240, China
xjsheng@sjtu.edu.cn

Abstract. Surface electromyography (EMG) is widely used in hand gesture recognition for human-computer interface (HCI). This paper presents a finger gesture recognition scheme at two level of plane pressing force through the fusion of wrist EMG and accelerometers (ACC). The classification algorithm is evaluated on eight healthy subjects for identifying five finger gestures at two plane pressing force level. Experimental results show that frequency domain (improved discrete Fourier, iDFT) feature is better than time domain (TD) feature for wrist EMG classification. Moreover, it indicates that the fusion of EMG and ACC achieved improved recognition performance (85.77%) for finger gestures at two level of plane pressing force when compared to that obtained using EMG (80.65%) or ACC (56.86%) solely.

Keywords: Wrist EMG · Accelerometer · Finger gesture recognition · Plane pressing force

1 Introduction

Since mobile and wearable computing are of great importance in our daily life, the interaction between human and computer has been widely studied [1]. Hand gesture recognition is an intuitive and natural approach in this field and has wide-ranging applications [2]. Surface electromyography (EMG) is one of biological signals generated along with muscle contraction of the human body. Therefore, EMG provides intuitive biological information for hand gesture recognition. It is reported that the recognition accuracy has reached above 95% for nine kinds of gestures [3]. The use of EMG signals which require comparatively inexpensive sensors for the measurement of hand gesture force is practical and convenient [4]. Hoozemans et al. [5] used the EMG of up to six forearm muscles to predict handgrip force. However, the recognition of hand gesture from EMG signals is difficult due to the noisy nature of EMG signals, differences in EMG sensor placement and contact conditions, and inter-individual differences in performing a gesture [6]. Inertial measurement unit (IMU) is widely adopted to improve the performance of hand gesture recognition based on EMG. Chen et al. [7]

Y. Huang et al. (Eds.): ICIRA 2017, Part I, LNAI 10462, pp. 373–380, 2017.
DOI: 10.1007/978-3-319-65289-4_36

used EMG sensors and 2D-accelerometers (ACC) to improve the classification performance of multiple wrist and finger gestures. Xiong et al. [8] proposed a EMG-IMU based mouse controller. Haque et al. [9] demonstrated that consumer-level EMG and IMU sensing was practical for distant pointing and clicking on large displays. Georgi et al. [10] combined the signals of the EMG of muscles in the forearm and an IMU worn at the wrist to classify hand and finger movements. In general, EMG is usually measured from forearm muscles, which is inconvenient and limited for daily usage. Therefore, hand gesture recognition based on wrist EMG will extend the range of application furthermore [11]. Nagar et al. [12] achieved an accuracy of 88.3% in identifying three gestures. Oyama et al. [13] used wrist EMG to classify seven types of wrist motions (76.5%). However, the finger gestures and force recognition has rarely been evaluated using wrist EMG, which has promising application prospect to interact with mobile devices.

In this paper, we proposed a method using wrist EMG and ACC for improving the classification performance of finger gestures at two level of plane pressing force. The first purpose is to evaluate the performance of two kinds of typical time domain feature and frequency domain feature extracted from wrist EMG. The second purpose is to determine whether ACC provides features which are complementary to EMG for finger gesture recognition at two force level. The paper is organized as follows. Section 2 presents the experiment procedures and methods, Sect. 3 describes the results and discussion, and the last section is conclusion.

2 Methods

2.1 Data Acquisition

A total of eight subjects (20 to 30 years old) with no history of neuromuscular or joint diseases take part in the experiments. The experimental session of all the subjects is in accordance with the Declaration of Helsinki.

Five hand gestures with the finger pressing against a plane are considered in this study, which are distinguished between two level of plane pressing force. These gestures are intuitive to users, and convenient to interact with mobile device. The five surface gestures are: index finger (IF), middle finger (MF), thumb, index finger and middle finger pressing together (TIMF), index finger, middle finger and ring finger pressing together (IMRF), and all five fingers pressing together (AF), as shown in Fig. 1. The two level of plane pressing force are 10% and 50% of the maximal voluntary contraction (MVC), respectively. The two level of force depends on the subjects' intuition which is intuitive and can avoid muscle fatigue.

The commercial myoelectric signal system, TrignoTM Wireless system (Delsys Inc., 20–450 Hz band pass filter), is used to collect the EMG data from the subjects. The wrist has five primary muscles, which are the extensor carpi radialis longus (ECRL), extensor carpi radialis brevis (ECRB), extensor carpi ulnaris (ECU), flexor carpi ulnaris (FCU) and flexor carpi radialis (FCR). Two of the wrist muscles, ECRL and ECRB, have such similar lines of action and moment

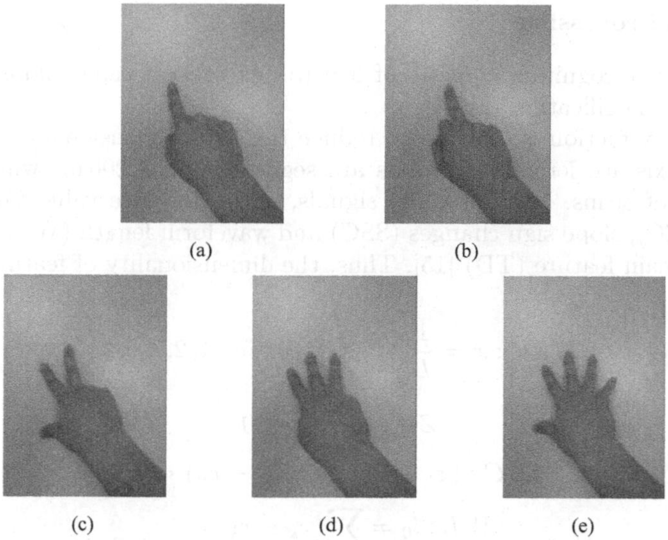

Fig. 1. Five finger gestures: (a) index finger (IF), (b) middle finger (MF), (c) thumb, index finger and middle finger pressing together (TIMF), (d) index finger, middle finger and ring finger pressing together (IMRF), and (e) all five fingers pressing together (AF).

arms that they are often modeled as one muscle [14]. Thus, four electrodes are placed around the wrist: two are on the anterior muscles, and the other two are on the posterior muscles (Fig. 2). Each electrode contains one EMG and one three-axis accelerometer (ACC) sensors. The sampling frequency is set as 2 kHz for the EMG signals, and 148 Hz for the acceleration signals.

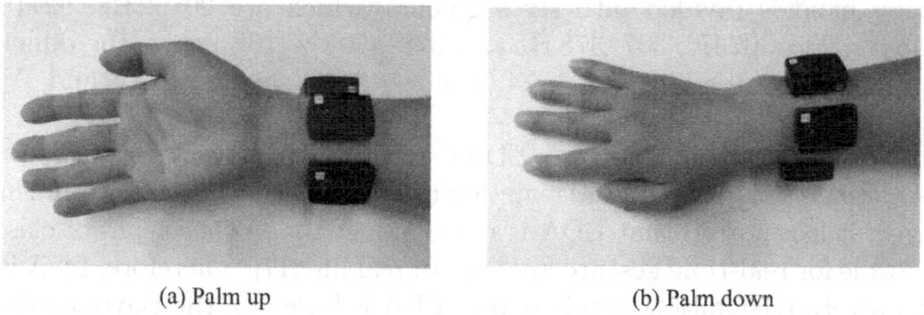

Fig. 2. Experiment setup: four electrodes are placed around the wrist: two are on the anterior muscles, and the other two are on the posterior muscles.

For each subject, twenty trials are performed, half of which are with 10% MVC and the remaining ten trials are with 50% MVC. In each trail, each finger gesture is sustained for 5 s and subjects have a 5 s rest between subsequent finger gestures to avoid fatigue.

2.2 Data Processing

Hand gesture recognition consists of features extraction using sliding windows and pattern classification.

Feature extraction is devoted to reduce the data dimensionality. The EMG and three-axis accelerometer signals are segmented into 200 ms windows with an overlap of 50 ms. For the EMG signals, mean absolute value (MAV), zero crossings (ZC), slope sign changes (SSC) and waveform length (WL) are chosen as time domain feature (TD) [15]. Thus, the dimensionality of feature vector is $4 \times 4 = 16$.

$$MAV : \bar{x} = \frac{1}{L}\sum_{k=I}^{L} |x_k|, for\ i = 1, 2, \ldots, I \tag{1}$$

$$ZC : x_k x_{k+1} < 0 \tag{2}$$

$$SCC : (x_k - x_{k-1})(x_{k+1} - x_k) < 0 \tag{3}$$

$$WL : l_0 = \sum (x_k - x_{k-1}) \tag{4}$$

where L is the length of samples, x_k is the kth sample in segment i, I is the number of the segment, and an improved discrete Fourier transform is extracted as frequency domain feature (iDFT). It is defined as

$$iDFT : iDFT_i = \log(\sum_{j=1}^{n_i} \frac{|X(f_{i,j})|}{n_i}), i = 1, 2, \ldots, L \tag{5}$$

where $X(\cdot)$ is the time domain signal after calculation of discrete Fourier transform (DFT), L is the segments of the spectrum, $f_{i,1}$ and f_{i,n_i} is the starting frequency and ending frequency of the ith segment.

Since the EMG signals are band pass filtered from 20 Hz to 450 Hz, the frequency band is divided into six segments, which are 20–92 Hz, 92–163 Hz, 163–235 Hz, 235–307 Hz, 307–378 Hz and 378–450 Hz [16]; hence, the dimensionality of feature vector is $6 \times 4 = 24$. For the ACC signals, MAV and WL are extracted($2 \times 4 = 16$).

Linear distriminant analysis (LDA) is commonly used for classification because of its low computational complexity and stable recognition performance. Moreover, it is reported that LDA is more robust than other types of classifiers and suitable for real-time gesture analysis in real life [17]. Therefore, LDA is used for the pattern recognition in this paper. LDA is based on the Bayesian decision rule and Gaussian assumption [18]. The discriminant function of LDA is:

$$g_i(x) = u_i^T \Sigma^{-1} x - \frac{1}{2} u_i^T \Sigma^{-1} u_i + \ln(p(\omega_i)) \tag{6}$$

where u_i is the mean vector of training samples of class i and $p(\omega_i)$ is the prior probability of class i, and Σ is the pooled sample covariance matrix.

The 2-fold cross validation method is used to evaluate the accuracy of recognition. The classification accuracy (CA) is used to compare the performance of different features of EMG and ACC, which is computed as:

$$CA = \frac{Number\ of\ correct\ testing\ samples}{Total\ number\ of\ testing\ samples} \times 100\% \quad (7)$$

3 Result and Discussion

Five classes of finger gestures at two level of MVC are analyzed to explore the roles of fusing wrist EMG and ACC in finger gesture recognition. The performance of the experiment is calculated with datasets consisting five different features: (1) TD; 2) iDFT; (3) ACC; (4) TD+ACC; (5) iDFT+ACC. The classification results (mean and standard deviation) for eight subjects are shown in Fig. 3. One-way analysis of variance (ANOVA) is used to compare the performance of different features, and the statistical significance level is set at $p < 0.05$. The p-values reflecting the differences between different features are listed in Table 1.

It can be seen that the average classification accuracies based on TD, iDFT, and ACC are (69.93% ± 7.02%), (80.65% ± 9.10%), and (56.86% ± 12.88%), respectively. The p-values between TD and iDFT is 0.020, TD versus ACC is 0.025, iDFT versus ACC is 0.000 (Table 1). These results demonstrate that iDFT feature performs significantly better than TD feature for the recognition of finger gesture based on wrist EMG signal, and EMG signal is significantly more

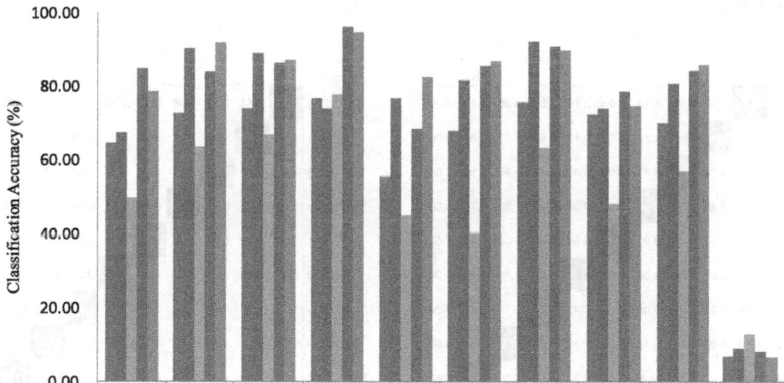

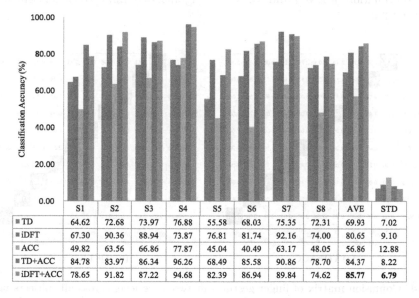

	S1	S2	S3	S4	S5	S6	S7	S8	AVE	STD
TD	64.62	72.68	73.97	76.88	55.58	68.03	75.35	72.31	69.93	7.02
iDFT	67.30	90.36	88.94	73.87	76.81	81.74	92.16	74.00	80.65	9.10
ACC	49.82	63.56	66.86	77.87	45.04	40.49	63.17	48.05	56.86	12.88
TD+ACC	84.78	83.97	86.34	96.26	68.49	85.58	90.86	78.70	84.37	8.22
iDFT+ACC	78.65	91.82	87.22	94.68	82.39	86.94	89.84	74.62	**85.77**	**6.79**

Fig. 3. Classification accuracy of eight subjects. For each subjects, classification accuracy is reported by using TD, iDFT, ACC, TD+ACC, iDFT+ACC features.

Table 1. The p-values of one-way ANOVA among five features.

	TD	iDFT	ACC	TD+ACC	iDFT+ACC
TD					
iDFT	0.020				
ACC	0.025	0.001			
TD+ACC	0.002	0.405	0.000		
iDFT+ACC	0.000	0.223	0.000	0.717	

accurate than ACC for finger gesture recognition at two force level. Furthermore, the results acquire from the fusion of EMG and ACC (TD+ACC, iDFT+ACC) are ($84.37\% \pm 8.22\%$), and ($85.77\% \pm 6.79\%$), respectively. The p-values between TD and TD+ACC is 0.002, iDFT versus iDFT+ACC is 0.223, TD+ACC versus iDFT+ACC is 0.717 (Table 1). These results indicate that the fusion of EMG and ACC sensors has a significant improving effect on the classification accuracies for TD feature, and improve slightly for iDFT feature. The feature of TD+ACC and iDFT+ACC are both suitable for the recognition of finger gesture at two level of MVC. The accelerometers are usually used to capture dynamic gesture. However, in this study, the finger gestures are static, and the fusion of EMG and ACC data is significantly better than either sensor type on its own. One possible reason is that the ACC signals contain the information of mechanomyography (MMG). MMG is the mechanical signal generated by contracting muscles in form of low-frequency vibration or sound, which could be measured by accelerometers [19]. The combination of EMG and MMG can significantly improve the classification accuracy [20].

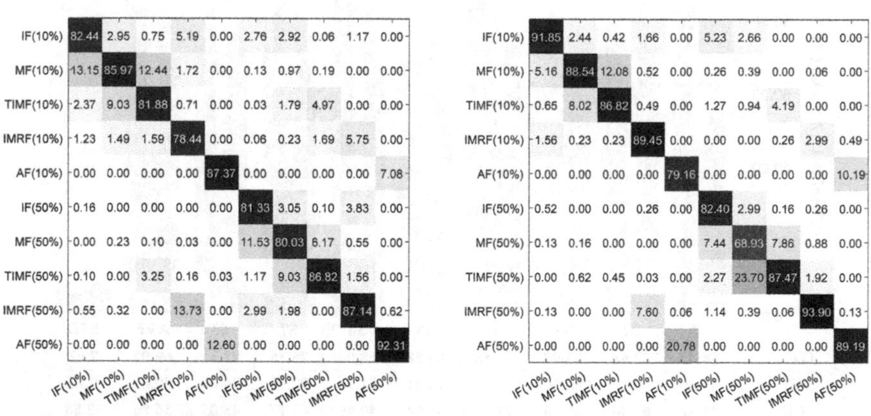

Fig. 4. Confusion matrix of finger gestures at two force level across all subjects using EMG and ACC, (a) TD+ACC; (b) iDFT+ACC.

The confusion matrix in Fig. 4 shows the performance of specific gesture classification with TD+ACC and iDFT+ACC features across all subjects. It can be seen that the adjacent finger gestures are easily confused (such as IF and MF). Moreover, for the same finger gesture at two level of MVC, IMRF (10%) and IMRF (50%), AF (10%) and AF (50%) are easily confused. This is partly due to the few muscle fibers in the wrist, and the EMG signal is weak. The standard deviation of TD+ACC (±3.30%) is smaller than that of iDFT+ACC (±7.41%). This indicates that TD+ACC is more stable than iDFT+ACC for specific gesture classification.

4 Conclusion

This paper presents the recognition of finger gestures at two force level based on wrist EMG and ACC. The experimental results show that EMG has advantage over ACC to classify static finger gestures and force. Furthermore, frequency domain feature (iDFT) is better than time domain feature (TD) for recognizing wrist EMG. It is likely that wrist ACC contains the information of MMG, thus the classification accuracy is significantly improved via fusing the TD feature and ACC feature. The performance of iDFT+ACC is slightly improved on the basis of iDFT. The effects of TD+ACC and iDFT+ACC are similar on the defined finger gesture recognition performance. In conclusion, classification accuracy fusing EMG and ACC is improved as compared to EMG or ACC alone for static finger gesture recognition.

Acknowledgments. This work was supported by the National Natural Science Foundation of China (Grant No. 51375296, 51620105002).

References

1. Kim, J., Kim, M., Kim, K.: Development of a wearable HCI controller through sEMG & IMU sensor fusion. In: 2016 13th International Conference on Ubiquitous Robots and Ambient Intelligence (URAI), pp. 83–87. IEEE (2016)
2. Mitra, S., Acharya, T.: Gesture recognition: a survey. IEEE Trans. Syst. Man Cybern. Part C (Appl. Rev.) **37**(3), 311–324 (2007)
3. Chu, J.U., Moon, I., Lee, Y.J., Kim, S.K., Mun, M.S.: A supervised feature-projection-based real-time EMG pattern recognition for multifunction myoelectric hand control. IEEE/ASME Trans. Mechatron. **12**(3), 282–290 (2007)
4. Mobasser, F., Eklund, J.M., Hashtrudi-Zaad, K.: Estimation of elbow-induced wrist force with EMG signals using fast orthogonal search. IEEE Trans. Biomed. Eng. **54**(4), 683–693 (2007)
5. Hoozemans, M.J., Van Dieen, J.H.: Prediction of handgrip forces using surface EMG of forearm muscles. J. Electromyogr. Kinesiol. **15**(4), 358–366 (2005)
6. Samadani, A.A., Kulic, D.: Hand gesture recognition based on surface electromyography. In: Engineering in Medicine and Biology Society (EMBC), 2014 36th Annual International Conference of the IEEE, pp. 4196–4199. IEEE (2014)
7. Chen, X., Zhang, X., Zhao, Z.Y., Yang, J.H., Lantz, V., Wang, K.Q.: Hand gesture recognition research based on surface EMG sensors and 2D-accelerometers. In: 2007 11th IEEE International Symposium on Wearable Computers, pp. 11–14. IEEE (2007)

8. Xiong, A., Chen, Y., Zhao, X., Han, J., Liu, G.: A novel HCI based on EMG and IMU. In: 2011 IEEE International Conference on Robotics and Biomimetics, pp. 2653–2657. IEEE (2011)
9. Haque, F., Nancel, M., Vogel, D.: Myopoint: pointing and clicking using forearm mounted electromyography and inertial motion sensors. In: Proceedings of the 33rd Annual ACM Conference on Human Factors in Computing Systems, pp. 3653–3656. ACM, Berlin (2015)
10. Georgi, M., Amma, C., Schultz, T.: Recognizing hand and finger gestures with IMU based motion and EMG based muscle activity sensing. In: BIOSIGNALS. pp. 99–108 (2015)
11. Oyama, T., Choge, H., Karungaru, S., Tsuge, S., Mitsukura, Y., Fukumi, M.: Identification of wrist EMG signals using dry type electrodes. In: ICCAS-SICE, pp. 4433–4436. IEEE (2009)
12. Nagar, A., Zhu, X.: Gesture control by wrist surface electromyography. In: 2015 IEEE International Conference on Pervasive Computing and Communication Workshops, pp. 556–561. IEEE (2015)
13. Oyama, T., Mitsukura, Y., Karungaru, S.G., Tsuge, S., Fukumi, M.: Wrist EMG signals identification using neural network. In: 35th Annual Conference of IEEE Industrial Electronics (IECON 2009), pp. 4286–4290. IEEE (2009)
14. Buchanan, T.S., Moniz, M.J., Dewald, J.P., Rymer, W.Z.: Estimation of muscle forces about the wrist joint during isometric tasks using an EMG coefficient method. J. Biomech. **26**(4–5), 547–560 (1993)
15. Englehart, K., Hudgins, B.: A robust, real-time control scheme for multifunction myoelectric control. IEEE Trans. Biomed. Eng. **50**(7), 848–854 (2003)
16. He, J., Zhang, D., Sheng, X., Meng, J., Zhu, X.: Improved discrete fourier transform based spectral feature for surface electromyogram signal classification. In: 2013 35th Annual International Conference of the IEEE Engineering in Medicine and Biology Society (EMBC), pp. 6897–6900. IEEE (2013)
17. Kaufmann, P., Englehart, K., Platzner, M.: Fluctuating EMG signals: investigating long-term effects of pattern matching algorithms. In: 2010 Annual International Conference of the IEEE Engineering in Medicine and Biology Society (EMBC), pp. 6357–6360. IEEE (2010)
18. Liu, H., Yuan, X., Tang, Q., Kustra, R.: An efficient method to estimate labelled sample size for transductive LDA (QDA/MDA) based on bayes risk. In: European Conference on Machine Learning. pp. 274–285. Springer, New York (2004)
19. Posatskiy, A., Chau, T.: The effects of motion artifact on mechanomyography: A comparative study of microphones and accelerometers. J. Electromyogr. Kinesiol. **22**(2), 320–324 (2012)
20. Guo, W., Sheng, X., Liu, H., et al.: Mechanomyography assisted myoeletric sensing for upper-extremity prostheses: a hybrid approach. IEEE Sens. J. **17**(10), 3001–3011 (2017)

Man-Machine Interaction for an Unmanned Tower Crane Using Wireless Multi-Controller

Songbo Ruan, Yeping Peng, Guangzhong Cao(✉), Sudan Huang, and Xiangyong Zhong

Shenzhen Key Laboratory of Electromagnetic Control, Shenzhen University, Shenzhen 518060, China
gzcao@szu.edu.cn, ruansongbo@gmail.com

Abstract. Operator cab is one of the important components of a tower crane, which is used to obtain sufficient quality feedback information from tower crane and ground observers. However, there are two main drawbacks of the tower crane system, low efficiency and labor costs, because the tower crane need to work with the assistance of the ground observers. In particular, the work environment of the cap operators is extremely bad and dangerous. In order to reduce the manual intervention and increase the efficiency of the tower crane, this paper presents a man-machine interaction for an unmanned tower crane using wireless multi-controller. Images of the crane cab are captured by cameras and then are coded and transmitted to wireless controllers and a server terminal through wireless local area network (WLAN). In this stage, the programmable logic controller (PLC) is used as a tower crane controller to collect sensor data. The data is also transmitted to the wireless controllers and the server terminal through WLAN. In addition, control parameters are transmitted to the server terminal and the tower crane controller via wireless controller transmission. After that, the tower crane controllers obtain the control commands and then perform them. Finally, testing results indicate that the unmanned tower crane is effective when it works with the man-machine interaction.

Keywords: Tower crane · Video transmission · WLAN · Multi-control

1 Introduction

In consideration of offering crane operator sufficient quality of the feedback from the tower crane and good visual control of the work arena, the operator sits in the operator cab atop the tower crane and operates the crane from this overview position. However, it's incommodious for the operator to climb up dozens of meters through the humble ladder inside of mast section to reach to operator cab at the beginning and end of the workday and often at break times. Long hours in the narrow operator cab during work time may cause the operator to lose focus and alertness. Furthermore, on the vast majority of construction sites, especially, those with abnormity or high-rise building, the line of sight of the operator was occasionally obstructed during the fine maneuvering of hook when approaching the pick-up/drop-off locations. The signalers are trained to

Y. Huang et al. (Eds.): ICIRA 2017, Part I, LNAI 10462, pp. 381–392, 2017.
DOI: 10.1007/978-3-319-65289-4_37

be the operator's eyes and ears on the ground and pass the information to the operator with alternate use of hand signals and wireless communications. The use of signalers reduces the construction efficiency enormously and increases construction cost.

Given this situation, Sany Heavy Industry proposed a wired remote system of tower crane [1]. The system including ground control unit, connecting cable and crane cab control box. The ground control unit is placed in a ground control room. The crane cab control box and ground control unit are connected with connecting cable. However, it's hard to implement the wired control due to the rotating structures and the dynamic nodes. For example, the top part of tower crane above jacking mechanism will rise with the number of mast section increased and jib will rotate horizontally with slewing mechanism. Comparing with the wired control, wireless control is more flexible and simple. Then radio remote control is implemented on crane [2, 3]. Radio remote control has a feature that the control unit (console with joysticks and buttons) is not physically connected to the crane which allowing operator unlimited mobility to get better visibility during cargo handling. However, for both wired and wireless control, lots of valuable view and feedback information about tower crane will disappear by losing the body-touch sensing system from tower crane. What's more, the operator moves around the construction site for seeking a better view of the transported load while concentrating on controlling the tower crane, which induce potential hazard easily. On the other hand, a comprehensive and meticulous research on tower crane cycle times between remote-control and cab-control operation is presented [4] that remote-control operation mode penetrates gradually with irreplaceable advantages.

This paper presents a wireless multi-control system of unmanned tower crane to provide an appropriate place for the tower crane operators using a wireless controller. The wireless controllers are implemented with tablet that was installed software with interactive interface. The bird's eye view on the operator cab is substituted by the video image on tablet, which is captured by the cameras mounted on tower crane. Additionally, the operator obtains feedback information about tower crane, in another way, "feel" the crane, with accurate sensor measurements which are displayed on the control tablet. Each tower crane is connected with up to three control tablets. Using those control tablets, operators distributed along the way to control the tower crane attentively without motion for seeking a better view for load during long distances transportation. Besides all that, a monitoring station is implemented to monitor and record the valuable data, such as video data, sensors data and control commands. All the devices are connected through the wireless local area network (WLAN). This system not only improves the operator's working environment but also operation efficiency and accuracy.

2 Proposed Man-Machine Interaction for an Unmanned Tower Crane

Man-machine interaction (MMI) is focused on the optimized interaction between people and machine, that is, between the tower crane and its operators in this study. The flow of information for tower crane operation is defined as the loop of interaction, and our main motivation is to optimize it [5, 6]. The information flow block diagrams are shown in Fig. 1. In terms of original tower crane, construction tasks are assigned to ground observers. It is important for the operators to communicate with the ground observers to act on signals from him in the operation of crane, especially, operation for high-precision movement of load. In addition, the operators need to "feel" the crane instinctively, such as determine whether the crane is overloaded by motor sound and vibration. However, using wireless controller, the construction tasks are assigned directly to the crane operator. With the wireless controller, the operator focuses more on controlling the tower crane with clear view of the load and sufficient quantity of accurate feedback of crane. The information flow block diagram of proposed system is more concise and efficient.

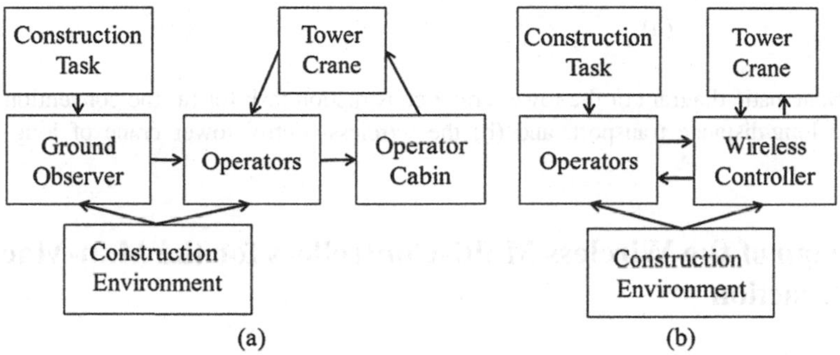

Fig. 1. Block diagrams of (a) the conventional tower crane, and (b) the wireless-control tower crane

In the proposed system, a tower crane is connected with three wireless controllers, which increases communication nodes between the operators and the tower crane. Tower crane in construction plays an indispensable role for hoisting and transporting building materials. The crane operation schematic diagrams of transporting building materials are shown in Fig. 2. Every crane operation is initially broken down into two segments: fast-travel part of the operation (the hook's fast motion throughout most of the travel path) and fine-maneuvering part of the operation (the hook's slow motion during the landing process, namely the slowing approach of the hook to the pick-up/drop-off point). To transport building materials in long-distance with original tower crane, as shown in Fig. 2(a), three crane observers are distributed in the way of transportation to assist the crane operator for fine maneuvering. However, three crane operators are able to finish the job with wireless controllers, like Fig. 2(b). The handover control of tower crane operation often occurs in the fast-travel part of the operation. Obviously, with the proposed system, construction costs are greatly reduced due to the cutting out of construction crew.

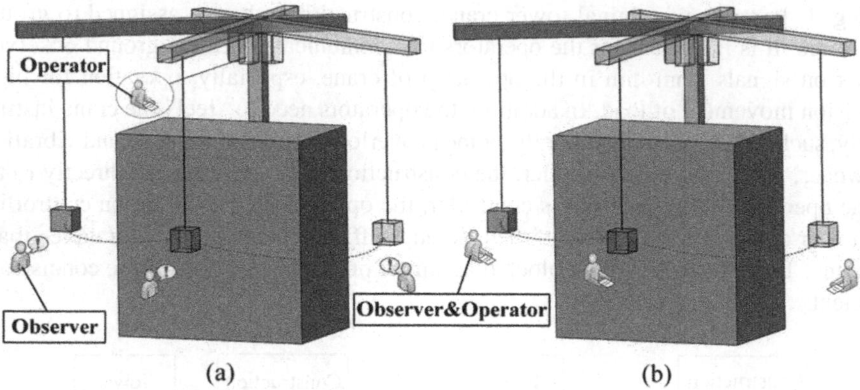

Fig. 2. Schematic diagram of the tower crane construction task for (a) the conventional tower crane for long-distance transport, and (b) the wireless-control tower crane of long-distance transport.

3 Design of the Wireless Multi-Controllers for the Man-Machine Interaction

This paper presents a wireless multi-control system of tower crane based on WLAN, including both the hardware and software. Regarding the hardware, the components utilized includes tower crane station, wireless bridge device, wireless controllers, monitoring station, and tower crane with sensors. The tower crane station consists of the wireless AP launcher, the tower crane controller, a router and several cameras. Wireless controllers are implemented with several control tablets. The monitoring stations include a wireless access device and a server. The whole system is shown in Fig. 3. The software comprises of two Windows software for wireless controllers and monitoring station.

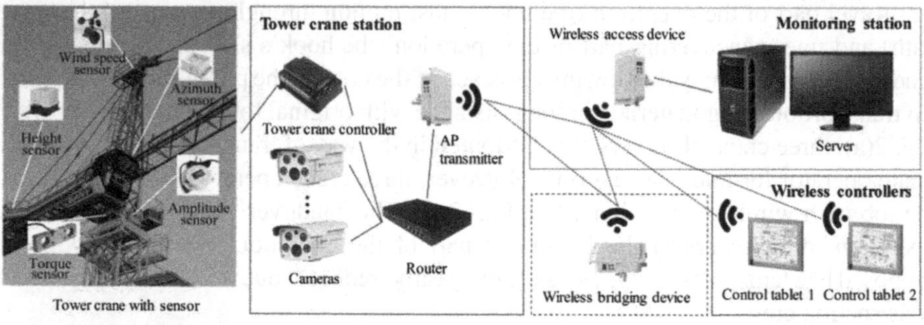

Fig. 3. Schematic diagram of the wireless multi-control system of tower crane

This section aims to illuminate the designing of wireless multi-control system which enhances the interaction between tower crane and operators. Firstly, the wireless local area network is examined for considering signal quality and network performance.

Secondly, we designed a high-performance streaming media server to implement video monitoring system. Then, the assembling of tower crane controller to tower crane is presented. Finally, all components of software system are explained in details.

3.1 Wireless Network Framework

In order to offer a long range, high performance WLAN over large areas for the system, the high power wireless access point (WAP) is deployed with Rocket M2 wireless bridge of Ubiquiti. The WAP provides wireless controllers and wireless access device of monitoring station with wireless connections. The WAP is connected to a router that provides wire connections using cables for tower crane controller and cameras, which in this case is connected to TL-R406 SOHO broadband router of TP-LINK [7]. However, the Wi-Fi signal always be blocked by the buildings and other obstacle, or the devices is out of range for connection. Then another Rocket M2 is used as repeater to wipe out the dead spots and extended the coverage area of the WAP. To ensure performance, IxChariot provides a solution for assessing and troubleshooting networks before and after deployment of WAP. The solution consists of an IxChariot Server and performance endpoints distributed throughout data centers, end-user PCs, and mobile devices. An important step for performance measurement is the choice of the right endpoint pair type, and we use video pair to test the WLAN. The video pair emulates a pair of endpoint computers that are streaming video data. Table 1 shows the maximum bandwidths between two mobile devices that were connected into WLAN of Tenda FH456 wireless router and Rocket M2 at different distance. It can be seen in Table 1, within the distance of 65 m, the bandwidth of Rocket M2 is in the range of [2.13, 3.04] Mbps, which is much higher than that of the compressed 720P video transmission (1–2 Mbps). Furthermore, the bandwidth at 65 m of Rocket M2 is approximate to that at 0 m of Tenda FH465.

Table 1. Bandwidth of Tenda FH456 and Rocket M2 at different distance.

Device	Distance	Bandwidth measurement (Mbps)				
		Pair 1	Pair 2	Pair 3	Pair 4	Pair 5
Rocket M2	0 m	3.04	3.04	3.03	3.04	3.03
	55 m	2.66	2.67	2.67	2.67	2.67
	65 m	2.14	2.15	2.15	2.3	2.14
Tenda	0 m	2.4	2.41	2.4	2.41	2.41

3.2 Video Monitoring System

Video image is one of the most important information carriers of human beings, and carries an intuitive and concrete form of information expression. The image is captured by cameras which are installed on the operator cab and trolleying mechanism of tower crane. Comparing with other types of data, the amount of raw video data is extremely large. The raw video data imposes a large transmission burden. However, due to the notable correlation between the pixels of each frame image, as well as, between adjacent frames, the raw video data is compressed to remove a large amount of redundant information. Human cannot perceive the distortion of the compressed video. The H.264 or MPEG-4 Part 10, Advanced Video Coding (MPEG-4 AVC) is the most widely used standard for video coding, the development of which was coordinated by the ITU-T Video Coding Experts Group (VCEG) and the ISO/IEC Moving Pictures Expert Group (MPEG). The H.265/HEVC video coding standard is the latest video coding standard developed by ITU-T following H.264. For H.265/HEVC-based encoding, HEVC encoder provides significant average bit-rate savings of 39.3% relative to x264 encoder which is one of the best representatives of publicly available H.264/AVC-based encoding implementations. On the other hand, the typical encoding times of the HEVC encoder are more than 800 times higher than those measured for the x264 encoder [8]. In considering reducing network transmission burden of video image for more than 20 cameras, the HEVC encoder is implemented to compressed video data in this system (Fig. 4).

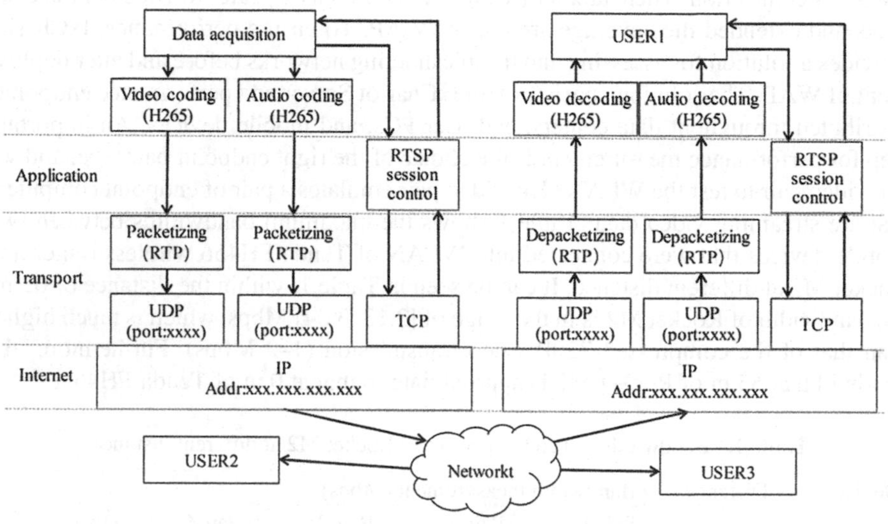

Fig. 4. Schematic diagram of the streaming media system

Although the amount of the raw video data in transmission is greatly reduced after compression, it is still important to choose the appropriate network transmission protocol to ensure the real-time performance and quality of video under the premise network. RTP (Real-time Transport Protocol) is used to carry data that has real-time properties, such as image data is recorded. Applications typically run RTP on top of the UDP protocol in transport layer, of course, replaced with TCP or ATM and other protocols according to the situation. RTP stream is monitored by RTP Control Protocol (RTCP) that provides end-to-end delivery services, including payload type identification, sequence numbering and time stamping. RTSP (Real Time Streaming Protocol) is an application layer protocol for controlling over the delivery of data with real time properties, used like handshake protocol between wireless controllers, server and the cameras [9].

3.3 Tower Crane Controller

Tower crane controller is realized by using PLC experimental platform. The PLC experimental platform is a highly integrated platform for PLC Fx2N-48MR which is the high-performance small programmable controller of Mitsubishi. The Fx series is the most widely used Mitsubishi PLC products. Moreover, the Fx2N series of PLC can be further expanded using Digital Expansion Blocks in various fields of application. The PLC experimental platform is extended with D-A module, A-D module and serial module. With this platform, control commands from wireless controller are captured via the extended serial module. Then the logic state of digital output pins or voltage of analog output pins are changed depending to the control commands. At the same times, the tower crane controller collects the sensor data via digital input pins and analog input pins, then the sensor data is encoded and sent out through the extended serial module. The serial server device bridges the TCP/IP layer to a serial port that transmits the data to the wireless controller and monitoring station by the WLAN.

The system is designed to control Sany Heavy Industries' SYT100 (T6515-6) tower cranes. The SYMC controller is applied to this type of tower crane electrical control system, which has good protection performance. And the tower crane is equipped with an exhaustive sensor system to detect the inclination, wind speed, torque, rotation angle and hook height.

In addition to the new way of control, also kept the old way of control from cab, a selector switch is added between operator cab, SYMC and tower crane controller. The hardware schematic is shown in Fig. 5.

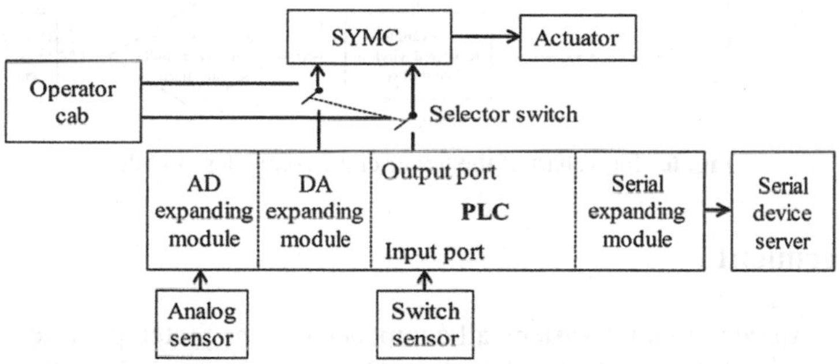

Fig. 5. Schematic diagram of the tower crane controller

3.4 Software Implementation for Multi-Controlling

The purpose of the system software is to control the data transmission between tower cranes controllers and wireless controllers through WLAN. There are two software programs. The controller software is installed on control tablet and the server software is installed on server. The control command is generated based on the operation with the controller software. The server software is used to assign the connection between tower cranes controllers and wireless controllers to reduce the computational burden of the tower crane controller and control tablets. The sensors data and commands of control is displayed and recorded through user interface. In the software, the image is available when users log in it using a valid user name and correct password.

All tower crane controllers establish TCP connection with server, then register an element in the tower crane control list, indexed by IP address. The control tablets also establish TCP connection with server and register an element in control tablet list, indexed by the number that assigned by the server. The server receives the sensors data and displays the sensors data in the software interface, and forwards the sensor data to the control tablets that are registered in its control tablet list. The server receives the control commands, decodes the data and audit authority according to its indexing number and control state. The control commands sent by the control tablets are forwarded to the selected tower crane controller and displayed in the software interface. When the connection of the element is faulty, the element will be deleted. The whole module flow is shown in Fig. 6.

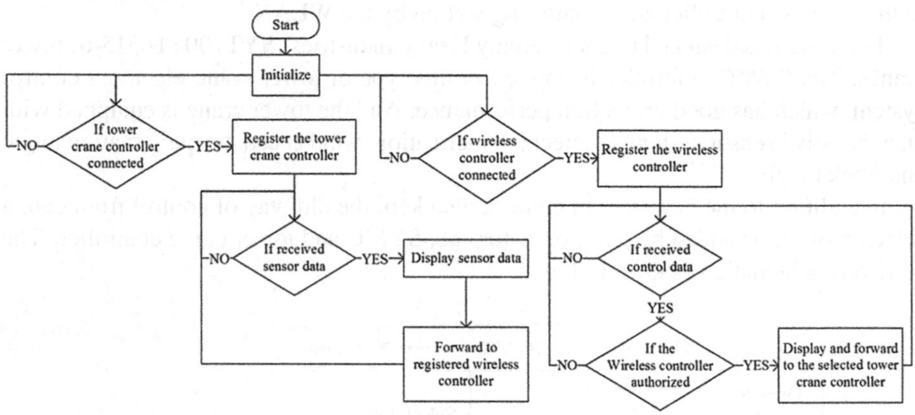

Fig. 6. Flowchart of the tower crane controller module

4 Experiment

During the experiment of the system, all components of the prototype were assembled together and tested. Due to the bandwidth at 65 m with Rocket M2 is approximate to that at 0 m with Tenda FH465, the WAP is deployed with Tenda FH465 that replace the composition of Rocket M2 and TL-R406. The control tablets and the wireless access device of monitoring station access the WLAN through wireless connection. The camera and the serial server device are connected with wireless router through cables. With several switches and LEDs simulating sensors and actuators, the PLC experimental platform connects to the serial server device via expanding serial port [10] (Fig. 7).

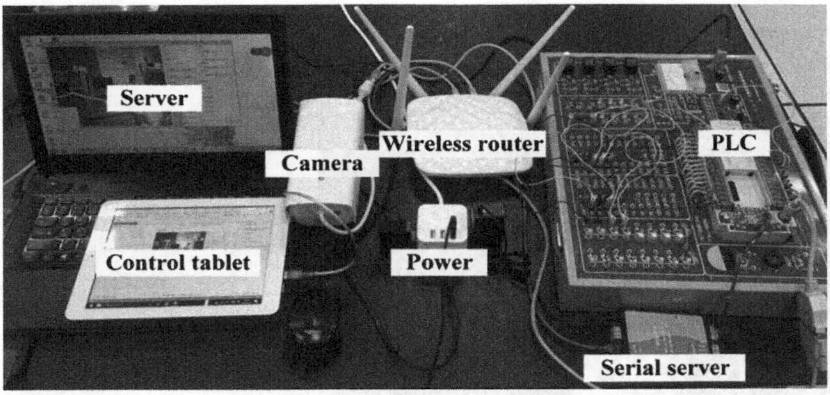

Fig. 7. The test system

When the entire system had been assembled, the server software and the controller software is executed and the real-time video image is transmitted and displayed by utilizing the interface of software respectively, like Figs. 8 and 9. When the wireless controllers establish connection with selected tower crane controller, the outputs of three switches are displayed in wireless controllers and server. The LED luminance is determined by the location of the slider on interface of controller software. When the entire system had been tested, it was observed that the system was operating correctly.

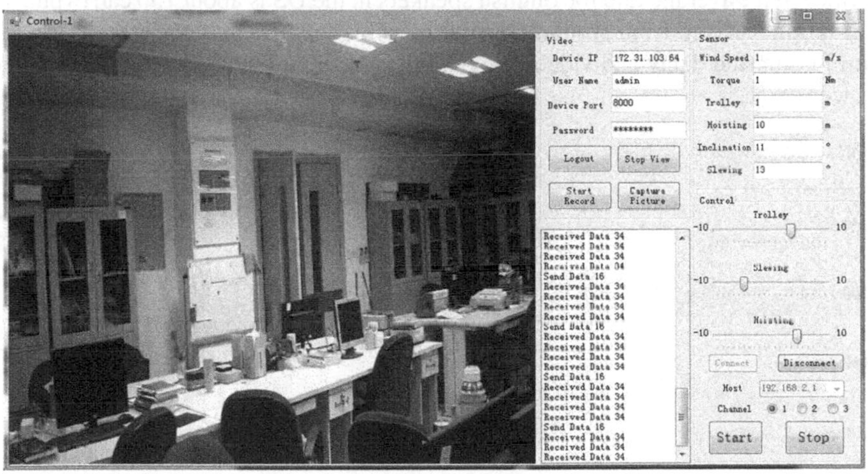

Fig. 8. The interface of server

Fig. 9. The interface of wireless controller

High-reliability and low-delay are two important properties to ensure the effectiveness of the proposed man-machine interaction. We perform experiments that test the transmission delay of control command delay and video delay and the measurements are shown in Fig. 10. The control command delay is about 90 ms and the video transmission delay is about 200 ms, on average. According to the National Center for Voice and Speech the average rate for English speakers in the US is about 150 carriwpm. The information received from ground crew with walkie-talkie is 500 ms later at least. The control command delay and video delay is low and slight fluctuation for the operation of tower crane.

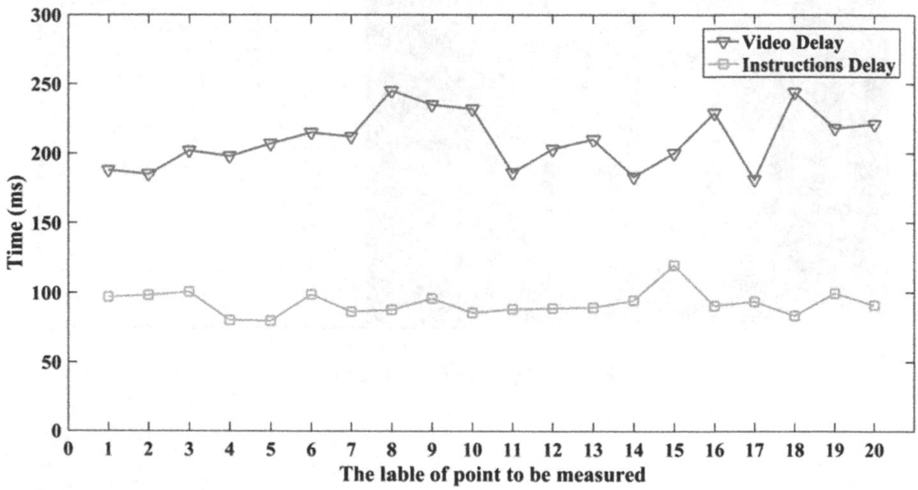

Fig. 10. Control command delay and video delay of proposed system

5 Conclusion and Future Work

This paper presents the man-machine interaction that implemented with proposed wireless multi-controller system for unmanned tower crane, which includes the specialty of remote controlling and video monitoring in real time. We perform experiments that test the transmission delay of control command delay and video delay to evaluate real-time performance which ensure the implication of the man-machine interaction saves cost of hiring crew and improves efficiency and work environment of operators.

In this project, the wireless multi-controller system has been tested indoor to control several LEDs. The condition of construction site is more complicated. We are planning to test the prototype in the outdoor environment of construction site. Another topic that appears to be of interest is the implement of Virtual-Environment modeling in this system. The geometric error of the virtual tower crane model could be corrected by fusing measurement of all the sensors mounted on tower crane. The controller interface will be a large graphic screen presenting corrected geometric model drawings of the tower crane, as well as video and sensor feedback from tower [6, 11].

Acknowledgment. This work was supported in part by the National Natural Science Foundation of China under Grant 51677120 and 51275312, and in part by the Shenzhen government fund under Grant KJYY20160428170944786.

References

1. JIANGLU machinery & electronics Group co., ltd.: Wired remote system of tower crane. China Patent No. CN202322061 U (2012)
2. Stepanov, P., Krosnjar, A., Pavlovic, G., Stepanova, A.: Implementation of wireless control on electrical bridge cranes and gantry crane. In: 23th Telecommunications Forum TELFOR, pp. 24–26. IEEE Press, Belgrade (2015)
3. Wei, P., Li, X., Liu, Z., Zhu, J.: Design of intelligent control system of eight-way wireless remote control crane based on RF technology. In: International Conference on Advanced Mechatronic System, pp. 18–21. IEEE Press, Tokyo (2012)
4. Shapira, A., Elbaz, A.: Tower crane cycle times: case study of remote-control versus cab-control operation. J. Constr. Eng. Manage. **140**(12), 307–318 (2014)
5. Xu, Y., Er, M.J.: Human-computer interaction in intelligent control of an unmanned aerial vehicle. In: International Conference on Intelligent Control Power and Instrumentation, pp. 42–46. IEEE Press, Kolkata (2016)
6. Hallset, J.O., Berre, G.: Modular integrated man-machine interaction and control. In: OCEANS '94'. Oceans Engineering for Today's Technology and Tomorrow's Preservation, pp. 108–112. IEEE Press, Brest (1994)
7. Ghittino, A., Maio, N.D., Tommaso, D.D.: WiFi network residual bandwidth estimation: a prototype implementation. In: 9th Annual Conference on Wireless On-Demand Network Systems and Services, pp. 43–46. IEEE Press, Italy (2012)
8. Grois, D., Marpe, D., Mulayoff, A., Itzhaky, B., Hadar, O.: Performance comparison of H. 265/MPEG-HEVC, VP9, and H.264/MPEG-AVC encoders. In: Picture Coding Symposium (PCS), pp. 394–397. IEEE Press, California (2013)

9. Chu, D., Jiang, H.J., Hao, Z.B., Jiang, W.: The design and implementation of video surveillance system based on H.264, SIP, RTP/RTCP and RTSP2013. In: 6th International Symposium on Computational Intelligence and Design, pp. 39–43. IEEE Press, Hangzhou (2013)
10. Howedi, A., Jwaid, A.: Design and implementation prototype of a smart house system at low cost and multi-functional. In: Future Technologies Conference (FTC), pp. 876–884. IEEE Press, San Francisco (2016)
11. Li, H., Song, A.: Virtual-environment modeling and correction for force-reflecting teleoperation with time delay. IEEE Trans. Ind. Electr. **54**(2), 1227–1233 (2007)

Web-Based Human Robot Interaction via Live Video Streaming and Voice

Jiahui Shi, Hongbin Ma(✉), Jialiang Zhao, and Yunxuan Liu

School of Automation, Beijing Institute of Technology,
Beijing 100081, People's Republic of China
mathmhb@bit.edu.cn

Abstract. Human-robot interaction (HRI) becomes one of the new technology coming into people's sight in recent years. A natural way to interact with robots is crucial in creating more intelligent robots. This paper shows a novel way to control and communicate with robots by monitoring live video captured from camera and doing voice interaction on web. Control commands from web are sent to control robots via Robot Operating System (ROS). The HRI process relies totally on web, users can interact with robots as long as they can connect to the wi-fi shared by robots.

Keywords: Human-robot interaction · Web · Live video streaming · Voice interaction · Robot operating system

1 Introduction

As the development of technologies, robots have been applied to many fields recently. Most of them are designed in advanced programming language, and they are utilized to execute preprogrammed tasks [1]. In this condition, they are usually isolated from humans. But now we want our robots to do various works; we need them to be intelligent enough to receive our commands in real time. That is why HRI is introduced and getting increasing attentions.

In recent studies, researchers focus on introducing vision and voice to tele-operate robots separately. Using voice to interact with robots is the easiest way to control robots [2] because we can use the language we understand to control the unfamiliar machine systems on robots [3] without employing any operation tools; it can also assist users in reducing their manual efforts. Based on vision, we can know better about the robots and make more accurate orders, which helps a lot in the whole HRI process [4].

In our system, we try to integrate them with web and ROS. Web browser is chosen as the interface because it is cross-platform and convenient. No applications are required for users to install, they can interact with robots simply by

This research was partially supported by the National Natural Science Foundation of China under Grant No. 61473038.

Y. Huang et al. (Eds.): ICIRA 2017, Part I, LNAI 10462, pp. 393–404, 2017.
DOI: 10.1007/978-3-319-65289-4_38

browser. ROS is useful for its ability to communicate between different tasks. So we can teleoperate robots by voice on web while monitoring the real-time vision passed from camera.

The objective of this research is to develop new user-friendly interface for robot control, which can be used by those have litter background in programming language.

2 Main Work

Based on the background, we decided to design a HRI system which contains vision and voice interface on web. Data communication at the back-end relies on ROS. The whole system contains three parts in general: streaming real-time video on web, voice interaction on web and teleoperation via ROS. The whole framework of the system is shown in Fig. 1.

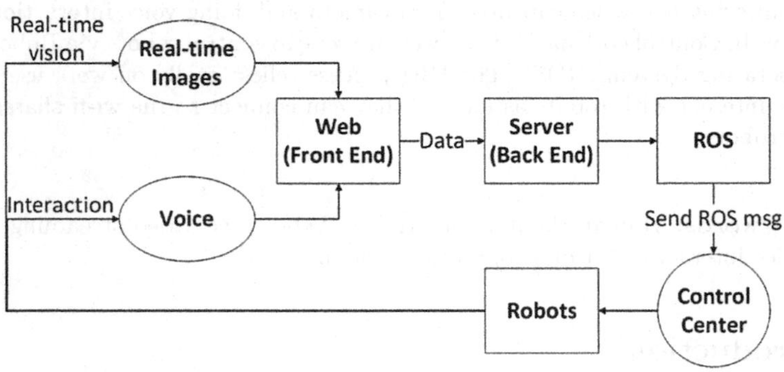

Fig. 1. An overview of the HRI system.

In Fig. 1, the users can see the real-time video from camera in browser. To send commands, there are two methods. One way is to press the buttons on web to control robots, which is manual teleoperation. Another way is to send commands via voice. Commands will be recorded on web in pcm formatted file and will be further transmitted to the back-end. This audio file will be recognized at the server, voice commands (e.g. "move forward", "turn left", etc.) will be captured and sent as ROS messages to control center, then the robots will move accordingly. The result for voice interaction will be sent back and be broadcast at the front-end.

3 Live Video Streaming

To display real-time video on web, we need to stream images captured from camera at server. A python micro web framework, namely flask, is chosen to build the web server. We choose Python language because it is supported by ROS and easy to be further developed.

3.1 Remote Monitoring

The remote monitoring system should be real-time and in low-latency, to implement that, we choose Socket.IO as the transmission media. Socket.IO is the cross-browser WebSocket for real-time applications, it enables real-time bidirectional event-based communication [5]. By using Socket.IO the system will be able to communicate between front-end and back-end in real-time. The whole process is shown in Fig. 2.

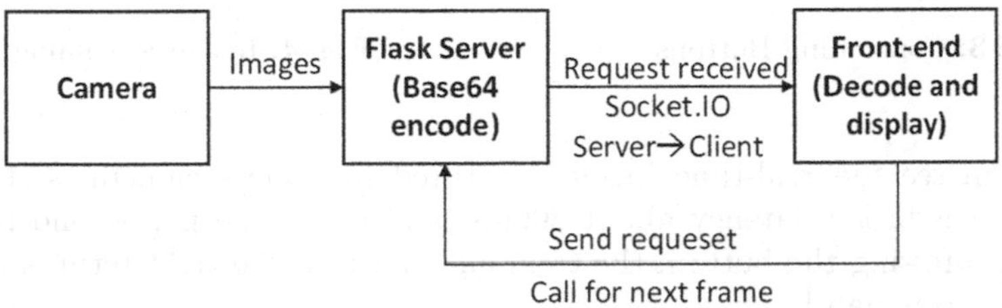

Fig. 2. Remote Monitoring System

Based on C/S framework, Socket.IO is driven by event. In our system, when users open the web the first time, a request for images will be sent from front-end to back-end. When receiving the request, the back-end server will collect images from the camera, after encoding them in Base64 format, image codes will be sent to front-end by Socket.IO. Then the Base64 codes will be decoded and displayed at the front-end, and a new request for next frame will be sent to server. The whole communication process will last until users close the web. The request-reply design makes full use of the characteristics of Socket.IO, which guarantee the low latency in transmission. Another advantages of this design is that the server sends images only if users visit the web. The transmission will be shut down when users close the web.

3.2 Remote Control

Based on the real-time images, users can teleoperate robots via buttons (shown in Fig. 3) on web. The left five buttons send moving commands(in four directions) and stop command. The right button is used for voice recording in voice interaction. When button is pressed, the corresponding color will be changed and a post request will be sent at front-end and detected at back-end. Commands will then be sent to ROS and robots will move accordingly.

3.3 Experiment

To test the image streaming system, we use a router to broadcast server, then we use mobile phone connected to the wi-fi shared by robots to visit the specific address. The result is shown in Fig. 4.

Fig. 3. Controlling Buttons

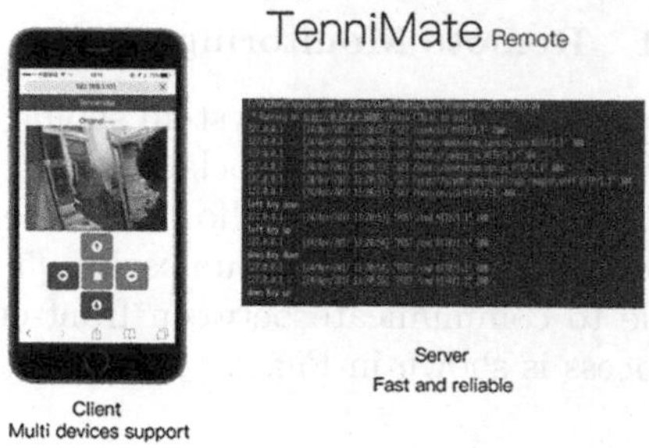

Fig. 4. Image Streaming

We can see the real-time images captured by robots camera is shown in browser, it is in low latency about 200ms, and the frame per second is about 30. When pressing the button, the terminal shown in the right returns the corresponding command.

4 Voice Interaction

4.1 Voice Recording

To record voice, two ways are adopted in the system. The first one is the typical method to set up microphones on robots to collect voice. This method is easy but the quality of the audio recorded is usually poor due to the effect of noise, which lower the accuracy of Automatic Speech Recognition (ASR) directly. To solve this, we adopt another way. We choose to record voice via web, the quality of audio will be much better because the voice will be gathered directly to the device connected to the web, rather than to a wild space, the effect of environment voice will be restrained to the minimum. Another advantage of this method is that users can complete voice interaction process with robots wherever they can connect to the web. Moreover, the voice response can be broadcast on web. Users do not need to speak near the robots anymore. The whole recording process is shown in Fig. 5.

We use WebRTC to record voice. WebRTC is a free, open-source project that provides browsers and mobile applications with Real-Time Communications (RTC) capabilities via simple Application Programming Interfaces (APIs)[6]. By using this APIs we can get the urls for the recorded audio file, through which the audio file can be uploaded to back-end server.

4.2 Voice Processing

To get better result in ASR, the recorded voice should be further processed. The whole voice processing has two sub procedures — speech enhancement and echo cancellation.

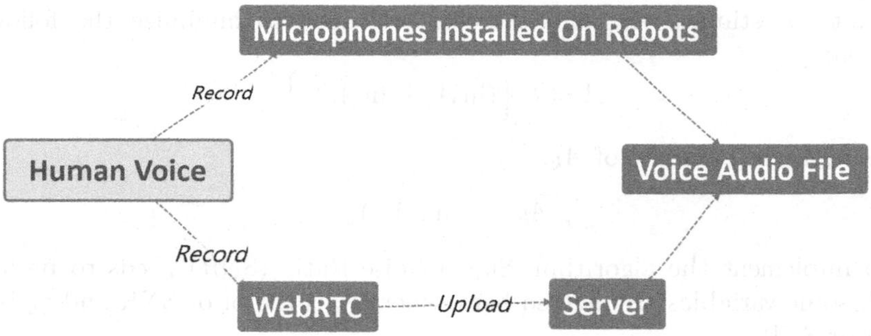

Fig. 5. Voice Recording

Speech Enhancement. The quality of voice that is recoded is usually poor due to the noise in environment, if not properly dealt with, it will decrease the accuracy of ASR and therefore influence the whole voice interaction, which is why noise reduction and speech enhancement are needed. We introduce an algorithm called Minimum Mean-Square Error Log-Spectral Amplitude (MMSE-LSA [7]). The whole process is shown in Fig. 6.

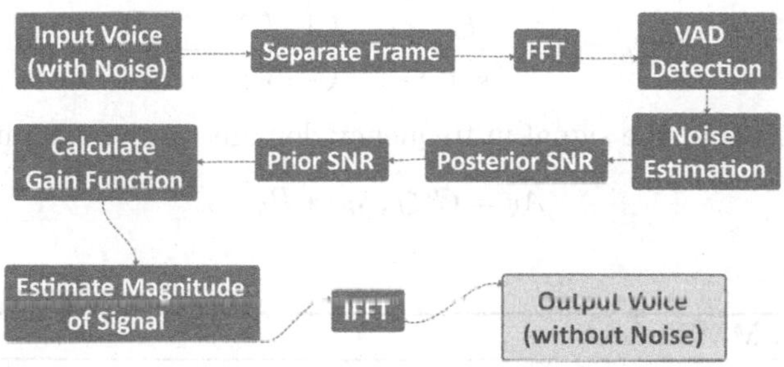

Fig. 6. Speech Enhancement

Suppose the voice input y(t) is consists of two parts: the pure signal x(t) a nd the noise signal d(t); x(t) and d(t) are independent with each other [8].

$$y(t) = x(t) + d(t) \tag{1}$$

Let X_k, Y_k, D_k denotes the k_{th} spectral component of the pure signal x(t), the noisy observation y(t) and the noise d(t), respectively.

$$X_k = A_k e^{j^{\alpha_k}} \tag{2}$$

$$Y_k = R_k e^{j^{\theta_k}} \tag{3}$$

Where k represents the frequency bin index after calculating the Fast Fourier Transformation (FFT). Our goal is to estimate x(t) from y(t), in another words, we want to estimate A_k, and therefore we need to minimize the following equation:

$$A = E\left\{(\ln A_k - \ln \widehat{A_k})^2\right\} \tag{4}$$

where $\widehat{A_k}$ is the estimate of A_k.

$$\widehat{A_k} = E\left[\ln A_k | Y_k\right] \tag{5}$$

To implement the algorithm, Signal-Noise Ratio (SNR) needs to be introduced, some variables are defined here, where ξ_k is the prior SNR and γ_k is the posterior SNR.

$$\xi_k = \frac{\lambda_x(k)}{\lambda_d(k)} \tag{6}$$

$$\gamma_k = \frac{R_k^2}{\lambda_d(k)} \tag{7}$$

$$v_k = \frac{\xi_k}{\xi_k + 1}\gamma_k \tag{8}$$

where $\lambda_x(k) = E\left[|X_k|^2\right]$ is the variance of the pure signal, $\lambda_d(k) = E\left[|D_k|^2\right]$ is the variance of the noise signal. After calculation we can get the gain function from ξ_k and γ_k:

$$G(\xi_k, \gamma_k) = \frac{\xi_k}{\xi_k + 1} \exp\left\{\frac{1}{2}\int_{v_k}^{\infty} \frac{e^{-t}}{t} dt\right\} \tag{9}$$

By gain function, the signal in frequency domain can be computed:

$$\widehat{A_k} = G(\xi_k, \gamma_k) * R_k \tag{10}$$

Algorithm 1. MMSE-LSA

Input: signal with noise y(t)
Output: pure signal after noise reduction x(t)
1: **for** $i = 1$ to n **do**
2: Compute prior SNR and posterior SNR from Eq. 6 and Eq. 7
3: Get Gain Function $G(\xi_k, \gamma_k)$ from Eq. 9
4: Compute $\widehat{A_k}$ and therefore the spectrum X_k
5: x(t) = IFFT(X_k)
6: Update k
7: **end for**
8: **return** x(t)

In summary, we first separate frame for the input signal, and then we calculate FFT for each frame to get Y_k, later we estimate prior SNR and posterior SNR to calculate the gain function $G(\xi_k, \gamma_k)$, and last we can get $X_k = A_k e^{j^{\alpha_k}}$, by using Inverse Fast Fourier Transformation (IFFT) we can get the pure signal x(t).

4.3 Echo Cancellation

Echo usually happens when we interrupt as robots speak, the microphone picks up not only the voice we speak but the voice robots speak. It will mess up the ASR process, so we must do echo cancellation to eliminate the influence. We choose to use Adaptive Echo Cancellation (AEC [9]), and the algorithm is called Normalized Least Mean Square (NLMS [10]). Some variables shown in Fig. 7 [11] is explained in Table 1:

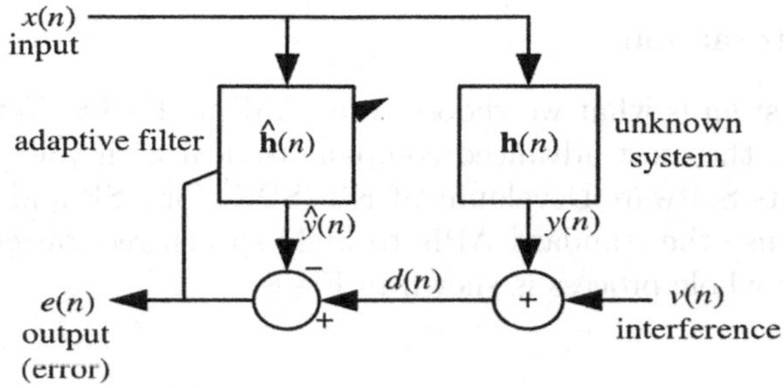

Fig. 7. NLMS Diagram

Table 1. Variables needed to be explained

Variable	Description
μ	Step size
$\boldsymbol{x}$	$[x(n), x(n-1)..., x(n-L+1)]^T$
$h(n)$	The filter coefficient vector
$e(n)$	The error signal
$y(n)$	Filter output

At first, the training signal is filtered with the estimated impulse response:

$$r(n) = \boldsymbol{x}(n)^T \widehat{h}(n-1) \tag{11}$$

Then we can calculate the error signal:

$$e(n) = d(n) - \widehat{h}(n)^T x(n) \tag{12}$$

Finally the coefficients should be updated:

$$\widehat{h}(n) = \widehat{h}(n-1) + \mu(n)\boldsymbol{x}(n)e(n) \tag{13}$$

Notice that the step size is dependent on the variance of the signal x; the varying step size speeds convergence while still maintaining stability [12].

Algorithm 2. NLMS

Input: $\boldsymbol{x}$, d(n), L, μ
Output: y(n), e(n)
1: **for** $i = 1$ to n **do**
2: e(i) = d(i) - $\widehat{h}(i)^T x(i)$
3: $\widehat{h}(n) = \widehat{h}(n-1) + \mu(n)\boldsymbol{x}(n)e(n)$
4: **end for**
5: $y(n) = \boldsymbol{x}(n)^T\widehat{h}(n)$

4.4 Voice Interaction

IFlytek voice system is what we choose to do ASR and TTS (Text To Speech). IFlytek is one of the most advanced company to deal with voice. It distributes free versions of its Software Development Kit (SDK) for ASR and TTS. It allows programmer to use the standard APIs to build speech recognizer commanding robots [13]. The whole process is shown in Fig. 8.

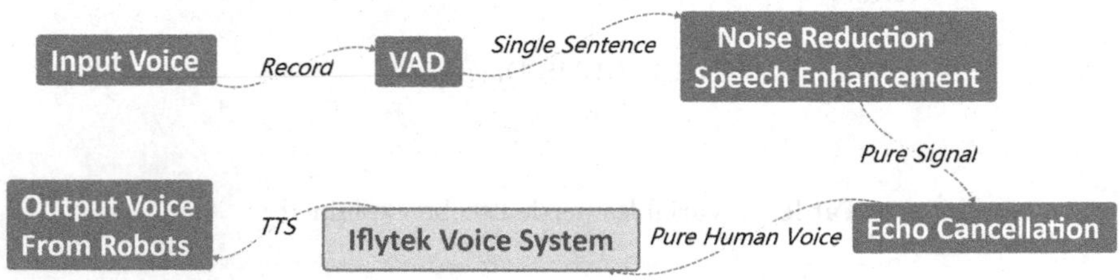

Fig. 8. Voice interaction

4.5 Experiment

To evaluate result of the voice processing algorithm, we use CoolEdit [14] to analyze the waveform of voice signal.

Speech Enhancement. In order to test the efficiency of the algorithm, we record an audio in a noisy environment, and there is an electric fan around the microphone to make noise. After we get the original audio, we run MMSE-LSA algorithm to reduce the noise, and the waveform of audios before and after being proceeded is shown in Figs. 9 and 11.

We can clearly see that sharp noises caused by the electric fan is reduced, the waveform after proceeding is much smoother.

Echo Cancellation. To test the efficiency of AEC, we record our voice by microphone when the computer plays a song, the song it plays causes an echo to the input voice, and we can get our original audio with the echo. Then we

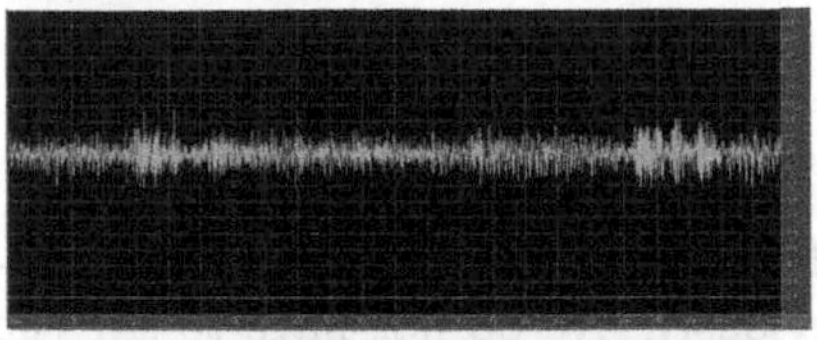

Fig. 9. Noisy audio before proceeding

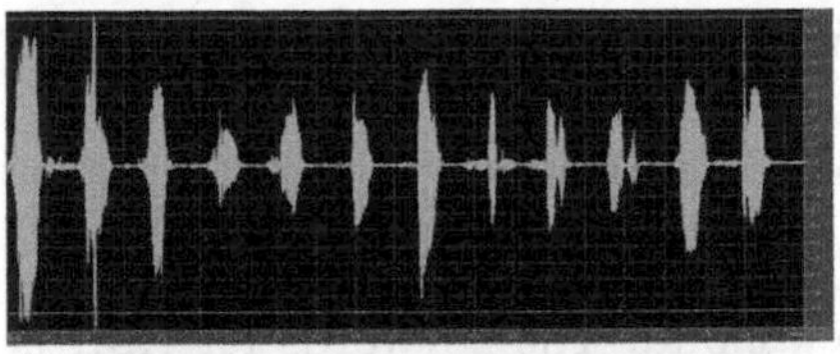

Fig. 10. Echo audio before proceeding

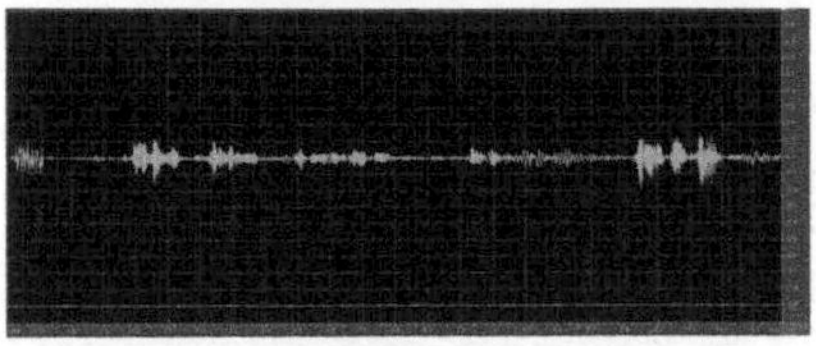

Fig. 11. Noisy audio after proceeding

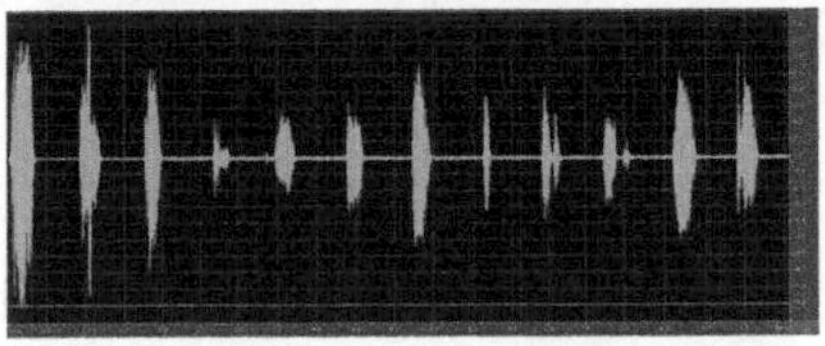

Fig. 12. Echo audio after proceeding

apply NLMS algorithm to the original audio to see how it works. Similarly, we record the waveform of two audio in Figs. 10 and 12. The oscillation between two waves is the background music, it sharpens the wave when we speak; however, in Fig. 12 the oscillation caused by the background music is almost eliminated, and the peak of waves when we speak become lower. In terms of listening, we can hardly hear the background music after proceeding. The effects of NLMS is pretty good.

Voice Interaction. At last we need to test the whole voice interaction system. We make many experiments in different environment; in each experiment, we give orders to the robot, if the robot gives correct responses, then this experiment is treated as successful. The final result is shown in Table 2.

Table 2. Test for voice interaction

Location	Test number	Successful cases (recorded by microphone/web)	Success rate (recorded by microphone/web)
Bedroom	50	47/49	94%/98%
Playground	50	42/48	84%/96%
Mall	50	38/47	76%/94%
Factory	50	32/45	64%/90%

From Table 2 we can find that the success rate is in average 80% when voice recorded by microphone set up on robots, which is satisfactory. Moreover, by recording voice on web, the success rate can be as high as 95% in average.

5 Teleoperation via ROS

5.1 Command

The response of voice interaction can be designed and uploaded to the iFlytek cloud. The keywords for ASR is set previously. We do not have to say the exact keywords; the voice interaction can be successful as long as keywords are contained in the sentences. The keywords [15] are shown in Table 3.

Table 3. Test for voice interaction

Keywords	Types	Description
Move/Go	Movement	Move linearly
Stop	Movement	Stop the robot
Turn	Movement	Make a turn
Velocity/Speed	State	Information about speed
Acceleration	State	Information about acceleration
Forward	Direction	Move forward
Backward	Direction	Move backward
Left	Direction	Turn left
Right	Direction	Turn right
Meters	Parameter	The distance robots move
Degrees	Parameter	The degrees robots turn

Four types of commands are provided to scale the commands, so we can control the robots precisely via these keywords.

6 Command Transmission by ROS

A ROS system is a computation graph consisting of a set of nodes communicating with one another over edges. The communication consists of messages that are organized by topics [16].

In our system, we use Arch Linux as the operating system; it is a lightweight and flexible Linux distribution. Arch Linux is the first pure rolling release distro and is targeted to advanced users [17]. For ROS distribution, we choose indigo for its stability.

The command received should be passed to the control center to drive robots. We use the topic and message mechanism in ROS to implement it.

As shown in Fig. 13, different processes in ROS are treated as different nodes. Nodes in ROS, like cells in human, are the smallest operating units. Nodes can subscribe topics; topics are like piping, they allows different nodes to communicate with each other by publishing messages. Messages in ROS are the key for

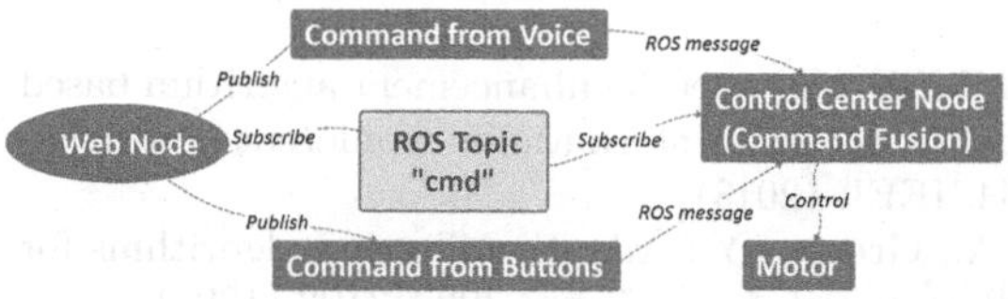

Fig. 13. Architecture of ROS

Fig. 14. Test in real robots

communication. For this HRI system, after command is received, the web node subscribes to a topic "cmd" and publishes a message from voice and buttons in "cmd". At the same time, control center node subscribes to topic "cmd" and it can therefore receive messages in this topic; by doing command fusion according to priority, the command can be then sent to control motor. The way ROS is used to pass commands is convenient and instantaneous. We do a test on real robots, the web interface is shown in Fig. 14, live video is shown in browser with the tennis ball detected, the buttons below can be used to control robots, or to do voice interaction by recording voice.

7 Conclusion

We have introduced a way how HRI through vision and voice can be implemented via web and ROS. It has the particularity that we can control and interact with robots at web end. In this way, we can control the robots whenever we have access to the Wi-Fi shared by robots. This user-friendly system brings much convenience to the those who have little programming experience, it can also be further developed because of the integration of ROS.

References

1. Kosuge, K., Hirata, Y.: Human-robot interaction. In: IEEE International Conference on Robotics and Biomimetics, ROBIO, pp. 8–11. IEEE (2004)
2. Gallardo-Estrella, L., Poncela, A.: Human/robot interface for voice teleoperation of a robotic platform. Lect. Notes Comput. Sci. **6691**, 240–247 (2011)
3. Ido, J., Takemura, K., Matsumoto, Y., Ogasawara, T.: Robotic receptionist aska: a research platform for human-robot interaction. In: 11th IEEE International Workshop on Robot and Human Interactive Communication, Proceedings, pp. 306–311 (2002)
4. Du, G.L., Zhang, P., Yang, L.Y., Su, Y.B.: Robot teleoperation using a vision-based manipulation method. In: International Conference on Audio Language and Image Processing, pp. 945–949 (2010)
5. Rai, R.: Socket. IO Real-Time Web Application Development. Packt Publishing Ltd., Mumbai (2013)
6. Johnston, A.B., Burnett, D.C.: WebRTC: APIs and RTCWEB protocols of the HTML5 real-time web. Digital Codex LLC (2012)

7. Ephraim, Y., Malah, D.: Speech enhancement using a minimum mean-square error log-spectral amplitude estimator. IEEE Trans. Acoust. Speech Signal Process. **33**(2), 443–445 (1985)
8. Zhang, Y., Liu, Y.: An improved MMSE-LSA speech enhancement algorithm based on human auditory masking property. In: 2013 International Conference on Asian Language Processing, pp. 151–154. IEEE (2013)
9. Benesty, J., Amand, F., Gilloire, A., Grenier, Y.: Adaptive filtering algorithms for stereophonic acoustic echo cancellation, vol. 5, no. 5, pp. 3099–3102 (1995)
10. Gilloire, A., Vetterli, M.: Adaptive filtering in subbands with critical sampling: analysis, experiments, and application to acoustic echo cancellation. IEEE Trans. Signal Process. **40**(8), 1862–1875 (1992)
11. Wikipedia, Least mean squares filter – wikipedia, the free encyclopedia (2017). https://en.wikipedia.org/w/index.php?title=Least_mean_squares_filter&oldid=782836501
12. Paleologu, C., Ciochină, S., Benesty, J., Grant, S.L.: An overview on optimized NLMS algorithms for acoustic echo cancellation. EURASIP J. Adv. Signal Process. **2015**(1), 1 (2015)
13. Xu, Y., Song, Y., Long, Y.-H., Zhong, H.-B., Dai, L.-R.: The description of iFlyTek speech lab system for NIST2009 language recognition evaluation. In: 7th International Symposium on Chinese Spoken Language Processing, pp. 157–161. IEEE (2010)
14. Lu, J., Miao, H.-T.: Applying cool edit Pro in EFL speaking and listening teaching research. Comput. Assist. Foreign Lang. Educ. Chin. **2**, 82–97 (2008)
15. Poncela, A., Gallardo-Estrella, L.: Command-based voice teleoperation of a mobile robot via a human-robot interface. Robotica **33**(1), 1–18 (2014)
16. Cousins, S.: Welcome to ROS topics [ROS topics]. IEEE Rob. Autom. Mag. **17**(1), 13–14 (2010)
17. Castro, J.D.: Arch linux. In: Introducing Linux Distros, pp. 235–252. Springer, New York (2016)

Object-Shape Recognition Based on Haptic Image

Yi Gong, Juan Wu(✉), Miao Wu, and Xiao Han

School of Instrument Science and Engineering, Southeast University, Sipailou 2#, Nanjing 210096, Jiangsu, China
{220142684,juanwuseu}@seu.edu.cn

Abstract. This paper introduces a concept of haptic image which contains sufficient haptic data acquired from haptic interaction. Deep mining and proper processing of haptic image may extend the applications to many fields. An approach of haptic shape recognition based on haptic image is presented. Firstly, a glove-like device mounted with pressure sensors and fiber sensors is utilized to acquire haptic image during the exploration of object shape. Secondly, pre-processing of haptic image is conducted including smoothing and standardization. Thirdly, haptic flow is extracted from haptic image as shape feature. Haptic flow proposed in this paper is the displacement of contact points between adjacent time intervals, which is inspired by optical flow. At last, a self-organizing map (SOM) is employed for the classification and recognition of the explored shapes. In the experiment, a recognition test of 4 different shapes, including cube, block, cylinder and sphere, is conducted and the mean recognition rate is approximately 90%.

Keywords: Haptic image · Shape recognition · Haptic flow · Self-organizing map

1 Introduction

The sense of touch is an important ability for human to explore the world [1]. When human interact with the environment, the information, including contact force, position, velocity, rotation and so on, is essential to haptic perception [2]. All these interactive information associated with haptic perception can be concluded as haptic data. Haptic data is an objective description of interactive process [3]. It contains abundant information of contacted objects and the operator. Effective mining and analysis of haptic data may extend applications to many fields, such as recognition of object properties, personal identity verification and evaluation of haptic tasks.

However, recent researches of haptic data processing and application mostly concentrated on data reduction and compression for high-efficiency remote transmission in real time [4–6]. Classical processing method [7] was down-sampling based on Weber's law and subsequently a data reconstruction model was proposed to recover the compressed data. These processing methods focused on

Y. Huang et al. (Eds.): ICIRA 2017, Part I, LNAI 10462, pp. 405–416, 2017.
DOI: 10.1007/978-3-319-65289-4_39

the accuracy and authenticity of acquired data. Several researchers have tried some different applications of haptic data [8,9]. For example, Namkee et al. [10] investigated strategies for recognizing user states such as frustration from users' patterns of pressure on a TouchPad. They made a comparison of frustration and non-frustration data including maximal force, minimal force and frequency to judge user state. Cemil et al. [11] designed an alphabet gesture recognition system by analyzing the hand shape and the trajectory. Haptic data, including angle and position, were collected to describe these features and then processed by an artificial neural network to generate the words and names. Besides, haptic shape recognition is also an important application, which can be utilized to enhance the accuracy of robot hand manipulation [12] and help the visual impaired individual explore the environment [13].

The present work concentrates on shape recognition from haptic data flow. In most robotic systems, the methods of shape recognition can be mainly categorized into vision-based method [14,15] and haptic-based method [16–18]. Vision-based method focuses on the processing of images obtained from operating scene. For example, Okada et al. [19] designed a vision-guided humanoid robot system. They extracted 2D straight edge and color histogram from visual images to recognize shape. Yabuta et al. [20] presented a binocular robot vision system. This system calculated the spatial coordinates of an object by using the correspondence of right and left images. Haptic-based method is conducted by analyzing haptic data acquired from robot hand. Bhattacharjee et al. [21] attached a tactile sensor on the forearm of a robot and selected features, including maximum force, contact area, and contact motion, for shape classification. Gorges et al. [22] combined kinesthetic information of robot joints and tactile information of contact area acquired by a palpation sequence to classify an object. Compared with vision-based method, recognizing object shape from the haptic sensor data is a direct way without building a 3D-model of the object. While vision-based method is an indirect method by reconstructing 3D shapes from 2D images which may lead to low accuracy and high computation complexity. In most haptic-based systems, the extracted feature of shape is mainly based on statistics. For example, Luzhnica et al. [23] calculated some statistical parameters, the maximum, minimum, mean value and standard deviation of original data, as features of time series. To some extent, statistical parameter method can describe the data variation of time series, but for complicated series, this method may be invalid as lack of sufficient features in time dimension.

In this paper, a haptic image concept similar to the visual image was proposed. Haptic image was organized as a matrix containing multidimensional spatio-temporal haptic data acquired from real haptic interaction. A brief introduction about the generation of haptic image and analysis of concealed haptic features in haptic image was presented. Based on haptic flow feature, an approach of shape recognition was conducted using a self-organizing map (SOM). The experiment has validated the effectiveness of haptic flow method, particularly in the recognition among similar shapes.

2 Haptic Image

Haptic image is defined as a description of interactive process which contains sufficient haptic information, such as force/pressure, position, angle and so on. The generation process of haptic image corresponds to the process of data acquisition by distributed and multiple sensors. For convenient processing, these data are arranged in a matrix changing with space and time according to certain methods. A probably general form of haptic image is shown in Formula (1).

$$\begin{cases} HI = \{HI_t\} \\ HI_t = [F_t^m \; A_t^n \; P_t^p \; ...] \end{cases} t = 1, 2, ..., T \tag{1}$$

where HI_t is the observation state of the haptic image at time t. F_t^m, A_t^n and P_t^p are common parameters measured in haptic interaction, where F_t^m is force information, A_t^n is angle information and P_t^p is position information. m,n and p are number of distributed sensors. The specific form of haptic image may differ according to different distribution of sensors and categories of acquired data.

Haptic features of contacted object or the operator can be obtained from the information in haptic image. For example, the softness of object is relative to the force and position information of contact area. The position and velocity information of exploring hand can be used to recognize object shape. The distribution and variation of contact forces may differ among different operators during the same interaction, which can be analyzed for identity recognition. The following section presents a processing method of haptic image for shape recognition.

3 Method of Shape Recognition

This section presents an approach of shape recognition based on haptic image. The overall process of recognition is described in the form of a flowchart in Fig. 1. The approach is to track the position of contact points while exploring different shapes. Haptic images containing distributed angel and force information are acquired and then haptic flow features are extracted as input of a SOM to recognize the shapes.

3.1 Haptic Image Acquisition

In this paper, a glove-like device shown in Fig. 2 is employed as the acquisition device which transforms hand and finger motions into real-time digital joint-angle data accurately with fiber sensors and measures the pressure on fingertips. Each finger is mounted with a type of thin-film flexible pressure sensor at fingertip and two fiber sensors at the Metacarpo Phalangel (MP) joint and Proximal Interphalangeal (PIP) joint. The sensors are also mounted between adjacent fingers for the measurement of adduction and abduction.

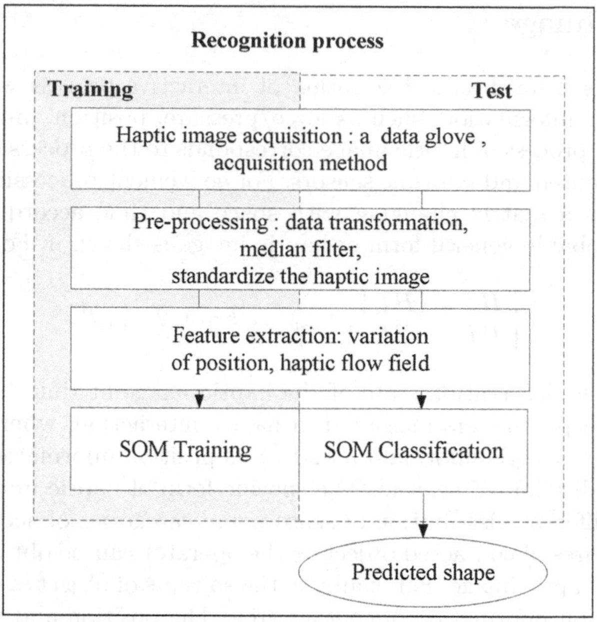

Fig. 1. The overall recognition process.

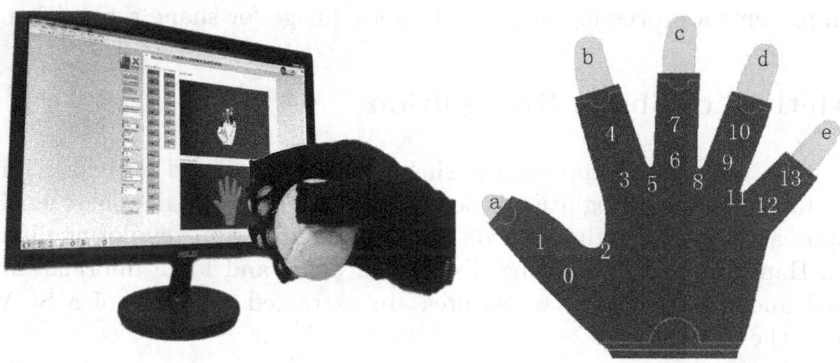

Fig. 2. Data glove device (left) and mounting locations of sensors (right): 0–13 represent fiber sensors, a-e represent pressure sensors.

For sufficient exploration of object shape, the acquisition process is divided into two periods: lateral exploration and vertical exploration shown in Fig. 3. Firstly, grasp the object on the desk from the lateral orientation and then move the hand around the object in a circle slowly. During the exploration, it is necessary to keep the fingers and palm contacted with object surface. Secondly, repeat the exploration from vertical orientation.

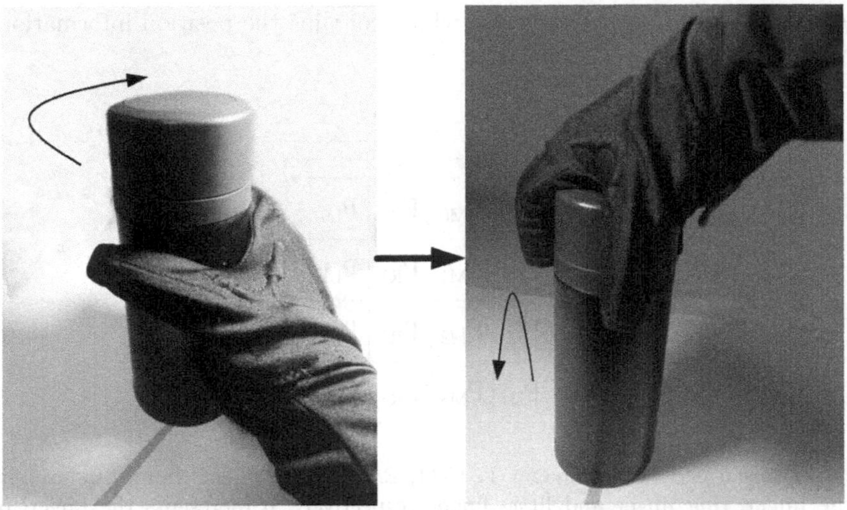

Fig. 3. Lateral exploration (left) and vertical exploration (right).

This acquisition process is the generation of haptic image and the acquired data is arranged in a matrix according to the distribution of sensors and hand structure. Haptic image in this application is described as:

$$HI_t = \begin{bmatrix} F_t^a & F_t^b & F_t^c & F_t^d & F_t^e \\ A_t^1 & A_t^4 & A_t^7 & A_t^{10} & A_t^{13} \\ A_t^0 & A_t^3 & A_t^6 & A_t^9 & A_t^{12} \\ A_t^2 & A_t^5 & A_t^8 & A_t^{11} & 0 \end{bmatrix} \tag{2}$$

The five columns of the matrix correspond to the five fingers of human hand, for example, the second column represents the index finger and it is the description of index finger motion. The first row describes the distribution of pressure. The second and third row describe the extension-flexion motion and the last row describes the adduction-abduction motion of human hand.

3.2 Pre-processing

The purpose of pre-processing is to ensure the authenticity of acquired data and form proper data format. This part includes three steps. Firstly, transform the acquired angle data into position data by Denavit-Hartenberg method [24]. The shape information can be expressed by the position of contact points only. While it is difficult to find correlation between shape and contact forces as the applied forces are greatly influenced by subjective factors. Therefore, in this application of shape recognition conducted by human hand, only position information is extracted for recognition. The transformed position information of each joint and fingertip can generate a new haptic image which expresses the movement of hand and positions of contact points, as shown in Fig. 4. Each pixel in the haptic image represents a joint or fingertip which contains the position information of current moment.

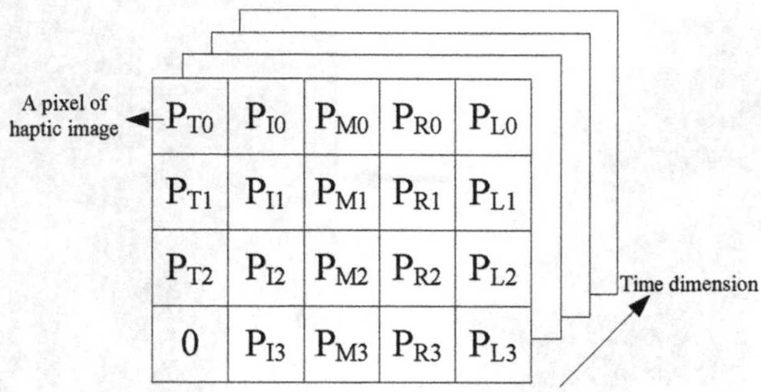

Fig. 4. Transformed haptic image: T, I, M, R and L represent thumb, index finger, middle finger, ring finger and little finger respectively. 0 represents the fingertip, 1 represents Distal Interphalangeal (DIP) joint, 2 represents MP joint and 3 represents PIP joint. Each pixel contains information of three orientations x, y, and z.

Secondly, smoothen the original data with median filter for more than one time. It can reserve the edge information of a sequence effectively. Each point P_t^i in haptic image can be replaced with mid-value MP_t^i when the window is set with the length of $2N+1$.

$$MP_t^i = Med\{P_{t-N}^i, \ldots, P_t^i, \ldots, P_{t+N}^i\}, i = 0, 1, \ldots, 13 \tag{3}$$

where i is the number of sensor and Med means calculating the mid-value of a sequence.

Thirdly, standardize the length of each sample. As the variation in exploring speed and object size, the length of original sample varies from each other which may lead to different dimensions of feature vectors. To solve this problem, the length of longest time series among all recorded dataset is marked as a standard and the others are extended to this length by the method of interpolation. If there are n interpolation points between MP_t^i and MP_{t+1}^i, the point m is calculated by Formula (4).

$$MP^i(m) = \frac{n+1-m}{n+1} MP_t^i + \frac{m}{n+1} MP_{t+1}^i \tag{4}$$

3.3 Feature Extraction

The shape feature is extracted by haptic flow which is inspired by optical flow [25]. Optical flow is the distribution of velocities of movement of brightness patterns in an image that arises from relative motion between object and viewer. The brightness of each pixel is described as $B(x, y, t)$, then the optic flow equation is written as:

$$\frac{dB}{dt} = \frac{\partial B}{\partial x}v_x + \frac{\partial B}{\partial y}v_y + \frac{\partial B}{\partial t} = 0 \tag{5}$$

Where the vector (v_x, v_y) is optical flow of this pixel. In simple terms, optical flow is the displacement of each pixel between adjacent frames. Vectors of all pixels in the image form an optical flow field of this moment. Continuous optical flow fields are effective description of object movement in time domain. In fact, the essential theory of optical flow is a frame difference method which can be also used in the processing of haptic images.

The main work of shape recognition is to distinguish hand motions from haptic images while exploring different shapes. In analogy with optical flow, a haptic flow method was proposed to extract hand movements. Each pixel in haptic image contains information of space and time, expressed as $P(x, y, z, t)$. The haptic flow is the displacement of each contact point in three-dimensional space between adjacent time intervals and can be calculated by:

$$\begin{cases} v_x = x(t+N) - x(t) \\ v_y = y(t+N) - y(t) \\ v_z = z(t+N) - z(t) \end{cases} \tag{6}$$

where (v_x, v_y, v_z) is haptic flow vector of a pixel and N is the time interval between frames.

The haptic flow vectors of all 19 pixels in haptic image represent the hand posture and variation trend of this moment. Continuous haptic flow fields contain the edge information of object which is mainly utilized by human to recognize the shape. Differing from visual images, haptic images are sampled with higher frequencies and the variation between adjacent frames is quite small. Therefore, calculating haptic flow of every two adjacent frames similar with optical flow will lead to redundancy and inefficiency. To select proper time interval N, analysis of the smallest time period T_m is conducted beforehand when finger joints move over a rigid edge. The maximum time interval must be smaller than T_m in order to ensure completeness of edge features. The extracted continuous haptic flow fields are input into a classifier as final feature vectors.

3.4 Classification and Recognition

A self-organizing map is used for the classification of extracted features. The SOM is a two-layer network. The first layer receives the input data, the second layer is entirely connected to the first layer and each neuron represents a currently unknown category. Since the input data is discrete points, the input layer of the SOM represents a matrix of the dimensions: number of input parameters multiplies by number of time steps. The parameter contains haptic flow element v_x, v_y and v_z of all 19 pixels.

During the training, the recorded samples are presented to the SOM in random order and after enough iterations only several neurons are activated corresponding to the classes of input samples. The recognition of a new sample is achieved by measuring the distance to existing classes. The new sample is considered as a member of the class with minimum distance.

4 Experiment

The proposed approach was validated by an experimental trial on four objects with four common shapes: cube, block, cylinder and sphere.

4.1 Data Collection

The first step of experiment is data collection. To eliminate the effect caused by different operation habits among operators, all the explorations were performed by the same subject. Before starting with the data recording, the process of shape exploration was explained. The participant was asked to explore the object shape with right hand at a constant speed. During the exploration, the recording of data began when the participant has grasped the object and been ready for the exploration. Recording ended while the two exploration periods were completed. The data were sampled at 150 Hz by data glove. Each object was explored 30 times to form a dataset with 120 samples.

4.2 Data Processing

The acquired 120 samples were processed for recognition. Firstly, the angel data were transformed into position data of each contact point. Then an 11-point median filter was employed to smoothen the raw data for 5 times. As exploring time of all 120 samples was between 2.5 s to 6.4 s, the samples were standardized with the length of 6.4 s by linear interpolation algorithm. The continuous haptic flow fields of each sample were calculated when the time interval was 0.4 s. Figure 5 is an example of data processing of cylinder and cube in X direction. At last, random 20 samples of each object were presented to the SOM for training and the remaining ones were text samples.

4.3 Results and Discussion

The result of recognition was depicted by the confusion matrix shown in Table 1. All samples of cylinder and sphere were classified correctly while the accuracy of cube and block was 80%. The wrong recognition only happened between cube samples and block samples. This phenomenon might be caused by the quantity of edges. Sphere had no rigid edges and cylinder only had four edges which could be detected during the vertical exploration. However, both of cube and block had eight rigid edges and they only differed in the spatial distance between edges: edges of the cube were equidistant while edges of the block were not. A rigid edge led to a great change of haptic flow, therefore, shapes with different quantity of edges could be easily distinguished. The distance feature between edges was described by the time interval between haptic flows with great changes. However,

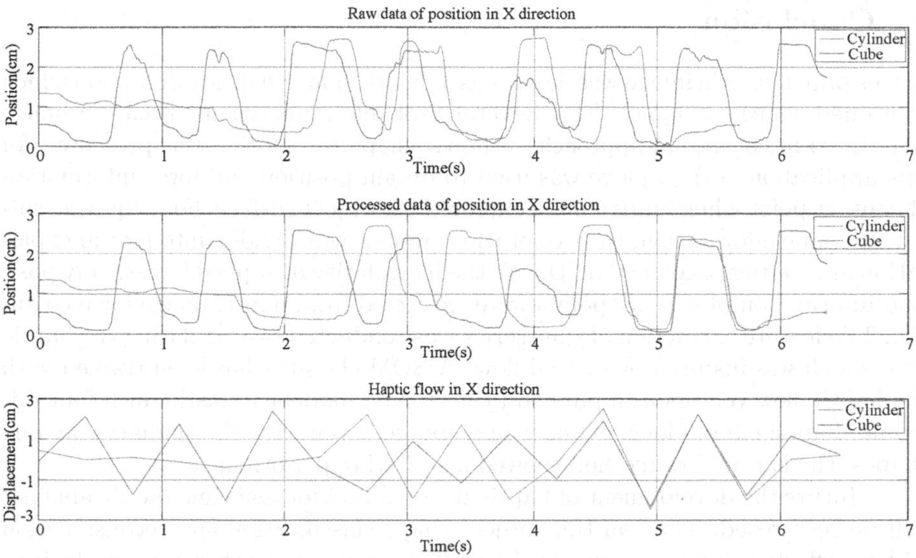

Fig. 5. An example of data processing: this figure shows the data processing of a cylinder sample and a cube sample in the X direction of MP joint of middle finger. The uppermost picture shows the raw data and middle picture shows processed data after smoothing and standardization. The last picture shows the extracted haptic flow component v_x.

Table 1. Confusion matrix of recognition result

Input	Output			
	Cube	Block	Cylinder	Sphere
Cube	8	2	0	0
Block	2	8	0	0
Cylinder	0	0	10	0
Sphere	0	0	0	10

the time interval was influenced by exploring speed. It was difficult to control the exploration at a constant speed for human which might lead to misclassification between cube and block. This problem could be solved in robotic systems due to the accuracy control of speed. Compared with the statistical method presented in [22], the haptic flow method showed a better performance for shape recognition. In [22], the recognition rate of cube and block was approximately 60% while only the statistical parameters of position were used. The statistical method ignored the correlation of data in time dimension while haptic flow contained sufficient features of time series.

5 Conclusion

In this paper, haptic image was defined as a description of haptic interaction which contained sufficient haptic data. Several probable applications of haptic image were introduced and an approach of haptic shape recognition was presented. In this application, a data glove was used to obtain position and force information of contact point when human hand explored the object surface. For acquiring sufficient shape information, the exploration process was divided into lateral exploration and vertical exploration. Due to the irregularity of applied forces, only position information of contact points were extracted from haptic image for recogniton. The feature extraction of time series was conducted by calculating the haptic flow which was inspired by optical flow. A SOM classifier has been trained with the haptic flow vectors as inputs and the trained model can distinguish four different shapes: cube, block, cylinder and sphere. Even with the similarity among shapes, the haptic flow method showed a satisfying performance.

In future the development of haptic flow method for real time tactile analysis will be researched. The real time object shape classification and reconstruction will be effective for incorporation in artificial arms and other robotic devices. Besides, the application of haptic flow method in other object property recognition, such as texture, roughness will be extended.

Acknowledgement. This research was supported by Natural Science Foundation of China under grants 61473088.

References

1. Schill, J., Laaksonen, J., Przybylski, M., Kyrki, V., Asfour, T., Dillmann, R.: Learning continuous grasp stability for a humanoid robot hand based on tactile sensing. In: 4th IEEE RAS & EMBS International Conference, pp. 1901–1906. IEEE (2012)
2. Shahabi, C., Kolahdouzan, M.R., Barish, G., Zimmermann, R., Yao, D., Fu, K., Zhang, L.: Alternative techniques for the efficient acquisition of haptic data. In: ACM SIGMETRICS Performance Evaluation Review, vol. 29, no. 1, pp. 334–335. ACM, Berlin (2001)
3. Gao, Y., Hendricks, L.A., Kuchenbecker, K.J., Darrell, T.: Deep learning for tactile understanding from visual and haptic data. In: 2016 IEEE International Conference on Robotics and Automation, pp. 536–543. IEEE (2016)
4. Nakano, T., Uozumi, S., Johansson, R., Ohnishi, K.: A quantization method for haptic data lossy compression. In: 2015 IEEE International Conference on Mechatronics, pp. 126–131. IEEE (2015)
5. Kaneko, T., Ito, S., Sakaino, S., Tsuji, T.: Haptic data compression for rehabilitation databases. In: 13th International Workshop on Advanced Motion Control, pp. 657–662. IEEE (2015)
6. Lee, J.-Y., Payandeh, S.: Haptic data compression. In: Haptic Teleoperation Systems, pp. 61–85. Springer, Cham (2015). doi:10.1007/978-3-319-19557-5_5
7. Nadjarbashi, O.F., Abdi, H., Nahavandi, S.: Applying inverse just-noticeable-differences of velocity to position data for haptic data reduction. In: 2015 IEEE International Conference on Systems, Man, and Cybernetics, pp. 440–445. IEEE (2015)

8. Burka, A., Hu, S., Helgeson, S., Krishnan, S., Gao, Y., Hendricks, L.A., Kuchenbecker, K.: Proton: a visuo-haptic data acquisition system for robotic learning of surface properties. In: 2016 IEEE International Conference on Multisensor Fusion and Intergration for Intelligent Systems, pp. 58–65. IEEE (2016)
9. Gemici, M.C., Saxena, A.: Learning haptic representation for manipulating deformable food objects. In: 2014 IEEE/RSJ International Conference on Intelligent Robots and Systems, pp. 638–646. IEEE (2014)
10. Park, N., Zhu, W., Jung, Y., McLaughlin, M., Jin, S.: Utility of haptic data in recognition of user state. In: Proceedings of HCI International, vol. 11. Las Vegas, Nevada (2005)
11. Oz, C., Leu, M.: Recognition of finger spelling of American sign language with artificial neural network using position/orientation sensors and data glove. In: Advances in Neural Networks ISNN 2005, pp. 812–812. Springer, New York (2005)
12. Ji, W., Zhao, D., Cheng, F., Xu, B., Zhang, Y., Wang, J.: Automatic recognition vision system guided for apple harvesting robot. Comput. Electr. Eng. **38**(5), 1186–1195 (2015)
13. Yatani, K., Banovic, N., Truong, K.: SpaceSense: representing geographical information to visually impaired people using spatial tactile feedback. In: Proceedings of the SIGCHI Conference on Human Factors in Computing Systems, pp. 415–424. ACM, New York (2012)
14. Hirano, Y., Kitahama, K.I., Yoshizawa, S.: Image-based object recognition and dexterous hand/arm motion planning using RRTs for grasping in cluttered scene. In: 2005 IEEE/RSJ International Conference on Intelligent Robots and Systems, pp. 2041–2046. IEEE (2005)
15. Choi, J., Seo, B.K., Lee, D., Park, H., Park, J.I.: RGB-D camera-based hand shape recognition for human-robot interaction. In: 2013 44th International Symposium on Robotics, pp. 1–2. IEEE (2013)
16. Johnsson, M., Balkenius, C.: Experiments with proprioception in a self-organizing system for haptic perception. In: Towards Autonomous Robotic Systems, pp. 239–245 (2007)
17. Dipietro, L., Sabatini, A.M., Dario, P.: A survey of glove-based systems and their applications. IEEE Trans. Syst. Man Cybern. Part C (Appl. Rev.) 38(4), 461–482 (2008)
18. Marion, G.C.: Wireless communication glove apparatus for motion tracking, gesture recognition, data transmission, and reception in extreme environments. In: Proceedings of the 2009 ACM symposium on Applied Computing, pp. 172–173. ACM, New York (2009)
19. Okada, K., Kojima, M., Tokutsu, S., Maki, T., Mori, Y., Inaba, M.: Multi-cue 3D object recognition in knowledge-based vision-guided humanoid robot system. In: 2007 IEEE/RSJ International Conference on Intelligent Robots and Systems, pp. 3217–3222. IEEE (2007)
20. Yabuta, Y., Mizumoto, H., Arii, S.: Binocular robot vision system with shape recognition. In: ICCAS2007 International Conference on Control, Automation and Systems, pp. 2299–2302. IEEE (2007)
21. Bhattacharjee, T., Rehg, J.M., Kemp, C.C.: Haptic classification and recognition of objects using a tactile sensing forearm. In: 2012 IEEE/RSJ International Conference on Intelligent Robots and Systems, pp. 4090–4097. IEEE (2012)
22. Gorges, N., Navarro, S.E., Goger, D., Worn, H.: Haptic object recognition using passive joints and haptic key features. In: 2010 IEEE International Conference on Robotics and Automation, pp. 2349–2355. IEEE (2010)

23. Luzhnica, G., Simon, J., Lex, E., Pammer, V.: A sliding window approach to natural hand gesture recognition using a custom data glove. In: 2016 IEEE Symposium on 3D User Interfaces, pp. 81–90. IEEE (2016)
24. Craig, J.J.: Introduction to robotics: mechanics and control, vol. 3, pp. 48–70. Pearson Prentice Hall, Upper Saddle River (2005)
25. Gibson, J.J.: The Perception of the Visual World. Houghton Mifflin, Oxford (1950)

Personal Desktop-Level Jet Fighter Simulator for Training or Entertainment

Xinye Zhao[1,2](✉), Yitao Wang[1], Wenming Zhang[3], and Xiaowei Zhang[4]

[1] PLA Naval Dalian Academy, Dalian 116010, People's Republic of China
zhaoxinye@nudt.edu.cn
[2] Naval Aeronautical and Astronautical University,
Yantai 264001, People's Republic of China
[3] PLA 63655, Urumqi 830001, People's Republic of China
[4] PLA 96401, Baoji 721000, People's Republic of China

Abstract. This paper describes the design of the software architecture for the jet fighter simulator on desktop computer, which provides a robust simulation framework that allows the simulator to fulfill training or entertainment goals. A human pilot math model was developed using LYRASim as the 6-DoF (6 Degree-of-Freedom) flight dynamic model (FDM). Additionally, we developed a Qt-based application comprising a number of HUD (Head Up Display) display systems that facilitated operations during flight, which sends control data to LYRASim and meanwhile reads realtime flight data from LYRASim over UDP sockets. LYRASim has also been extended to handle output of messages over a socket to a 2D and a 3D virtual application used at the same computer which provide a 2D and a 3D view similar to what the realtime control operator would see. The software architecture consists of above components, and the simulation results show that prototype system based on this architecture is flexible, valid and economic. Finally, suggestions are made for improvement of future desktop-based jet flight simulator.

Keywords: Flight simulation · 3D Visualization · 6-DoF flight dynamic model · Desktop-level simulator

1 Introduction

As technology has advanced and costs have fallen, the advantages of using simulators to train for safe, economical, and environmentally friendly driving have become more apparent [1,2]. Nowadays, the use of flight simulation in both civil aviation and military training is commonplace and widely accepted [3,4]. Jamson et al. evaluated the benefits of adaptive collision avoidance system compared to non-adaptive system utilizing the Leeds driving simulator [5]. AutoPW simulator was developed by the Warsaw University of Technology to test drivers abilities and reactions in standard and dangerous traffic situations [6]. Another kind of simulator, the HVDS, was primarily designed for training heavy-duty truck drivers for safe, economic, and environmentally friendly driving (Fig. 1)[7].

Y. Huang et al. (Eds.): ICIRA 2017, Part I, LNAI 10462, pp. 417–427, 2017.
DOI: 10.1007/978-3-319-65289-4_40

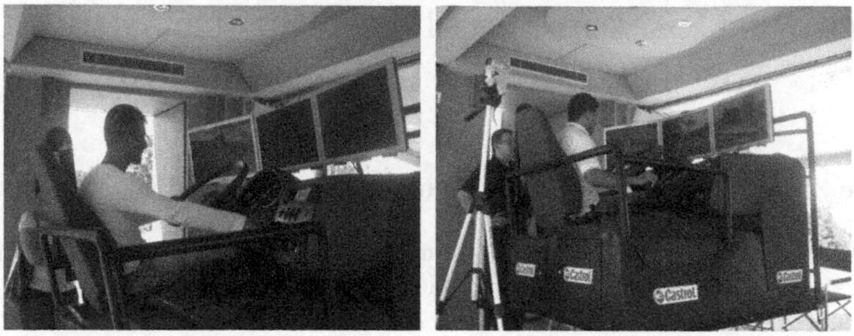

Fig. 1. The use of the HVDS by drivers [1]

Some simulators allow flight crews or even common people to practice potentially life-threatening manoeuvres or rather entertainment in the relative comfort of a training centre. All functional jet flight simulator is a complex and expensive test equipment, thus most of the jet flight enthusiasts can not afford to build a new simulator. We obtain an easy to use, low-cost desktop-level jet fighter simulator in a novel generic framework of orient object architecture; this enables us to gain a general training and entertainment experience [8].

There are numerous FDMs that are available for jet fighters, such as LaRCsim [9], JSBSim [10], YASim [11], UIUC [12], etc. There are also some proprietary softwares that are written without the benefit of a true operating system, which limits new adaption of different aircraft platforms [13]. We discuss the implementation of our overall architecture, and we provide insight into how the evolutionary FDM assists users in experiencing and analyzing proposed force and organizational structures in the jet fighter flight domain.

The remainder of this paper is organized as follows. Section 1 includes some preliminaries that will be used in the rest of the paper. Section 2 presents the main contribution of our work, showing how to design the system architecture of the simulator. More precisely, Sect. 2.1 is dedicated to the FDM (LYRASim), Sect. 2.2 to the HUD display system of the simulator, Sect. 2.3 to the 2D map GIS (Geographical Information System) viewer, and Sect. 2.4 to the 3D visualization viewer. Section 3 illustrates the methodology with a simulation application problem. Finally, Sect. 4 sets out some conclusions and ideas for future research. Due to a lack of space, some technical details have been left out of this paper.

2 System Architecture

System architecture is designed as in Fig. 2. The FDM (LYRASim) is designed to calculate the position of the aircraft in realtime and output the flight data to the fighter simulator over a UDP socket. The HUD display system takes in the flight data and displays the necessary information in the panel, in the meanwhile receives the control data from mouse, keyboard and joystick. In addition, the FDM sends the flight data of aircraft to the 2D map GIS viewer and the 3D visualization viewer. The 2D map GIS viewer is meant for a 2D geographic information display, while 3D visualization viewer is designed for six degrees of freedom of aircraft movement in 3D visualization.

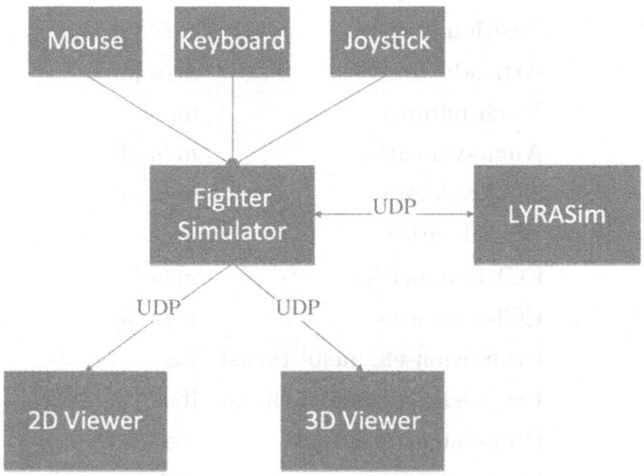

Fig. 2. Simulator system architecture

In order to drive fighter simulator through external commands, it is necessary to write the interface driver in accordance with the network communication protocol defined by LYRASim. It consists of 28 attribute variables, as shown in Table 1.

An important line of research within the flight dynamic research is to propagate and track the path of a flying craft over the surface of the Earth (or another planet), given the forces and moments that act on the vehicle. LYRASim refers to following frames of interest [4]. NED-velocities are the vehicle velocities in the NED (North, East, Down) frame, with the origin at the vehicle mass center, the X axis pointing North, the Y axis pointing east, and the Z axis positive downward by the right hand rule. BF-velocities are the vehicle velocities in the body-fixed frame, with the X axis positive forwards, with the X axis positive forwards out the nose of the aircraft, the Y axis positive out the right side of the aircraft, and the Z axis positive down. ECEF-velocities are the vehicle velocities in the Earth Centered, Earth Fixed (ECEF) frame, with the Z axis coincident with the spin axis and positive north, the X axis positive through 0 longitude and 0 latitude. This frame rotates with the Earth at a constant rate and does not translate. ECI-velocities are the vehicle velocities in the Earth Centered Inertial (ECI) frame, which is fixed in celestial space with the Z axis positive north and coincident with the spin axis, with the X and Y axes located in the equatorial plane.

Table 1. Variables of FDM

Variable	Unit
Simulation-time	s
Velocity	m/s
Position	m/s [3]
Attitude	m/s [3]
Mach number	mach
Angle-velocities	m/s [3]
NED-velocities	m/s [3]
BF-velocities	m/s [3]
ECEF-velocities	m/s [3]
ECI-velocities	m/s [3]
Propulsion-engine[0]-thrust	lbs
Propulsion-engine[1]-thrust	lbs
Propulsion-total-fuel	lbs
Gear-pos	[0,1]
Flap-pos	[0,1]
Speedbrake-pos	[0,1]
Right-aileron-pos	[0,1]
Elevator-pos	[0,1]
Rudder-pos	[0,1]
Inertia-weight	kg
Inertia-empty-weight	kg
Inertia/pointmass-weight	lbs

2.1 Flight Dynamic Model

LYRASim is a FDM of 6-DOF rigid body. The most important part of FDM is mainly for aircraft flight system mathematical modeling and calculation, to achieve the flight performance of the flight simulation, providing flight parameters for other components of the simulator. LYRASim calculates the forces and moments acting on the rigid body by initializing the input and updates the position and kinetic parameters of the rigid body with certain simulation step. Therefore, it is necessary to model the aerodynamic characteristics of the rigid body, and the aerodynamically dependent atmosphere (wind, temperature, barometric pressure, etc.) also needs to be modeled. For the aircraft itself, in addition to aerodynamic forces and moments, it is necessary to model propulsion, buoyancy, gravity, ground friction, and other external forces.

The general class library of dynamics simulation of jet fighter simulator includes the following main modules: data input module, flight control module, aerodynamic force and moment module, propulsion module, gravity module, atmospheric module, GNC (Guidance, Navigation, and Control) module, 6-DoF flight dynamics module, data output module. The relationship of core calculation modules is shown in Fig. 3.

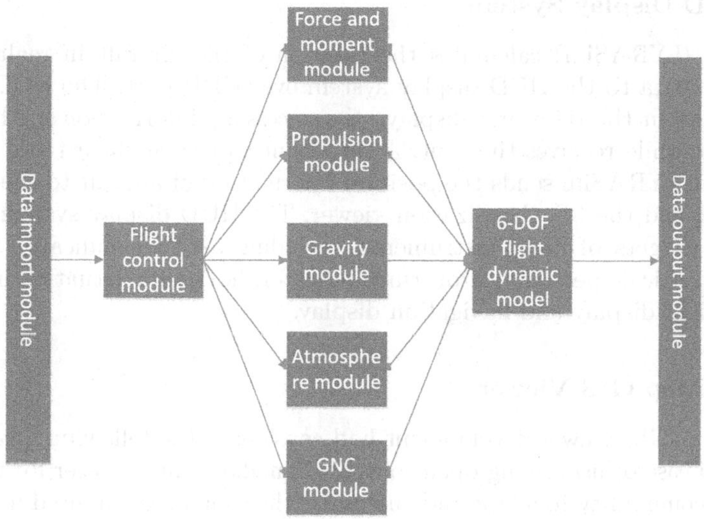

Fig. 3. Relationships of core calculation modules

The specific functions are as follows:

- Data input module mainly assigns all the initial variables of the aircraft to determine the intrinsic parameters of the aircraft (such as weight, span area, moment of inertia, position of center of gravity, aerodynamic data of wind tunnel test or actual data of the aircraft).
- Flight control module mainly controls the flight control law, the stability increment system, the automatic flight, the flight matching and so on.
- Aerodynamic force and moment module mainly calculates the real-time force and torque during the aircraft flight.
- Propulsion module mainly computes the forces converted by the engine and propeller.
- Gravity module mainly calculates the real-time gravity of the aircraft;
- Atmospheric environment module mainly calculate the atmospheric environment such as atmospheric disturbance, temperature, atmospheric pressure and so on.
- GNC module is the navigation control module, responsible for aircraft flight path control.
- 6-DoF dynamics module calculates the attitude, position, angle of attack, side slip angle, velocity and angular velocity of the aircraft based on the integrated force model of the aircraft. 6-DoF kinematics model is needed to

solve the computational and data storage requirements of several higher-order differential equations, which are solved step by step using standard fourth-order Runge-Kutta method Aircraft Flight Dynamics Equations.

2.2 HUD Display System

The FDM (LYRASim) calculates the position of the aircraft in realtime and output the data to the HUD display system over UDP port. The HUD display system takes in the data and displays the necessary information in the panel, in the meanwhile receives the control data from mouse, keyboard and joystick. In addition, LYRASim sends the position information of aircraft to the 2D map GIS viewer and the 3D visualization viewer. The HUD display system consists of several widgets of flight instruments, including attitude indicator, airspeed indicator, vertical speed indicator, turn indicator, horizontal situation indicator, primary flight display and navigation display.

2.3 2D Map GIS Viewer

The 2D map GIS viewer development had considered the following goals: open-source and ease of use. Being open source will make it much easier for the flight simulator community to adapt and customize the tool for the desired usage. The 2D map GIS viewer is developed based on an open source project - OpenMap [14,15], which is a Java Beans based toolkit for building applications and applets needing geographic information. Using OpenMap components, programmers can access data from legacy applications, in-place, in a distributed setting. To allow a wide range of users, specifically non computer savvy users, the tool has to be graphical and easy to use. The map viewers GUI is programmatically broken down by area of interest for the user.

Map interactive view C displaying multiple data layers on the map in addition to drawing objects, units and control measures. The MapHandler is the class in OpenMap responsible for managing the layers of geospatial data that need to be displayed on the MapPanel. The map viewer implements a RouteLayer class for drawing aircrafts geospatial data. The RoutLayers constructor accepts the array representing the file and the Lat-Lon coordinates are then extracted from the array based on the criteria that they will always be preceded by a GDC tag and transferred to a temporary array. MIL-STD-2525B symbol pallet is presented, as well as layers pallet that displays several different layers.

2.4 3D Visualization Viewer

The 3D visualization viewer is inspired by an open source project - myCesium-flight [16], which is a small Cesium engine based 3D webgl flight simulator written in html5, css and javascript. Cesium is an open-source JavaScript library for world-class 3D globes and maps. Our mission is to create the leading 3D globe and map for static and time-dynamic content, with the best possible performance, precision, visual quality, platform support, community, and ease of use [17]. Cesium is running on the basis of Node.js which is a JavaScript runtime built on Chrome's V8 JavaScript engine [18]. Node.js uses an event-driven, non-blocking I/O model that makes it lightweight and efficient. Node.js' package ecosystem, npm, is the largest ecosystem of open source libraries in the world.

Fly anywhere in the world with photorealistic high resolution terrain imagery and elevation heightmaps by accessing the google map information. 3D viewer uses cssscale to increase fps rate(50 fps fullscreen on my desktop computer) where standalone webgl and localhost modes available. Openstreetmap information is fetched to display airports on radar and generate airports radio messages. Also osm bridges and buildings save and reload on startup.

3 A Simulation Application

LYRASim sends longitude, latitude, altitude, pitch angle, yaw angle, roll angle and other flight data over UDP packets to the HUD system, the 2D map GIS viewer and the 3D visualization viewer, while receiving from real-time control data from the mouse, keyboard or joystick. An excerpt of the HUD system is shown in Fig. 4. The HUD system is collection of Qt5 widgets of flight instruments, including attitude indicator, airspeed indicator, vertical speed indicator, turn indicator, horizontal situation indicator, primary flight display and navigation display. The right panel in the simulator is designed to show flying information of the fighter in detail, such as altitude, speed, position and so on.

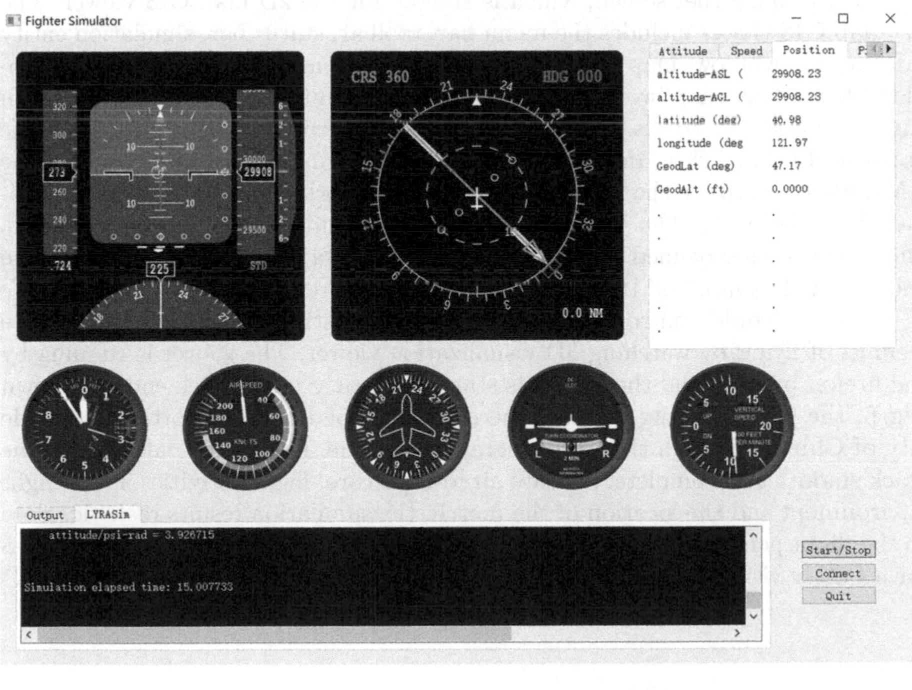

Fig. 4. HUD display systems

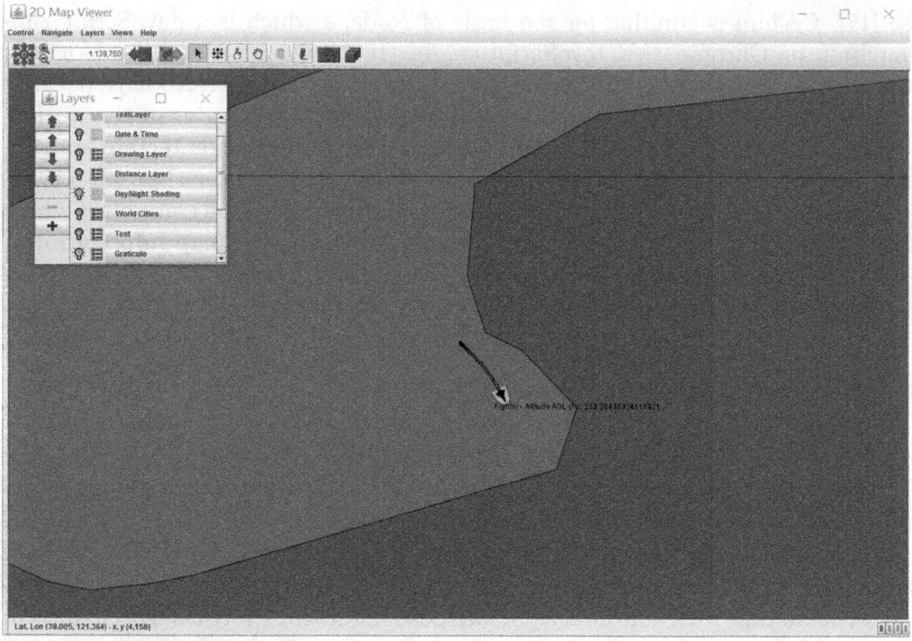

Fig. 5. 2D map GIS viewer

Turn into another screen, which is showed for the 2D map GIS viewer. The 2D map GIS viewer includes the menu bar, toolbar, status bar, simulation entity information window, map window. Toolbar and menu bar corresponding to provide file loading, map browsing, labeling, editing and other functions, as shown Fig. 5. The status bar is used to display the current mouse position and the horizontal field of view. In the entity information window, the user can load the XML file to complete the software configuration work, and set the display parameters of the entity. The Map window is used to load map files, display entities, annotations, user-defined areas, and so on. The trajectory of fighter could be showed in the map, and the altitude is updated in realtime.

The user could control the jet through the joystick or others to experience feelings of flying by watching 3D visualization viewer. The viewer is running by the firefox browser and the nodejs is started up early in the back end. As shown Fig. 6, the jet was flying through the downtown of Dalian, a northeast seaside city of China. Through the Cesium graphics engine, high-speed calculation, the track smooth and complete, delicate aircraft texture, flight activities in the flight environment and the location of the match, the simulation results realistic, close to the flight perspective to observe the real situation. Please refer to Figs. 7 and 8 for a better visualization of the trajectory of simulator's position and velocity.

Fig. 6. CesiumFlight3D visualization viewer

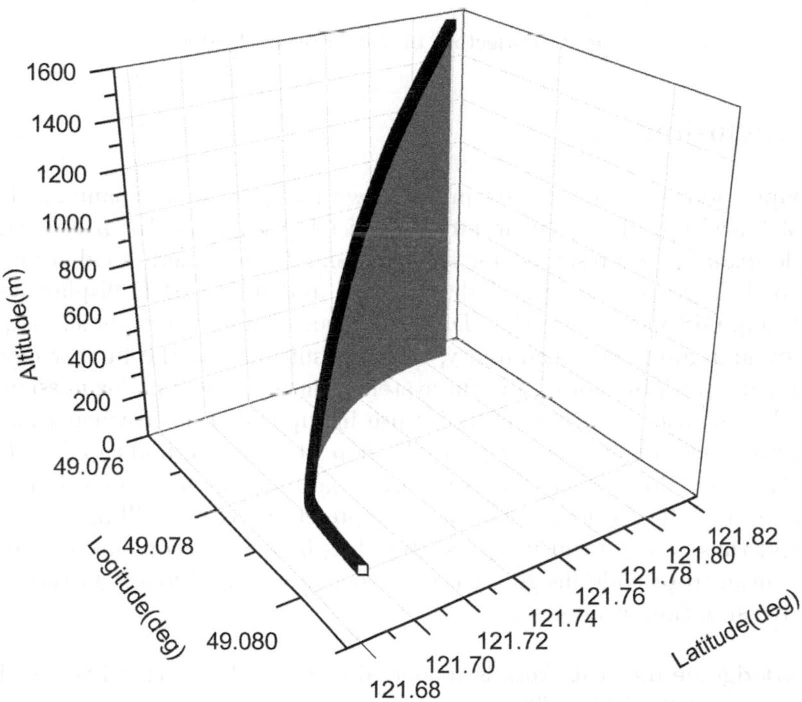

Fig. 7. Trajectory of simulator's position

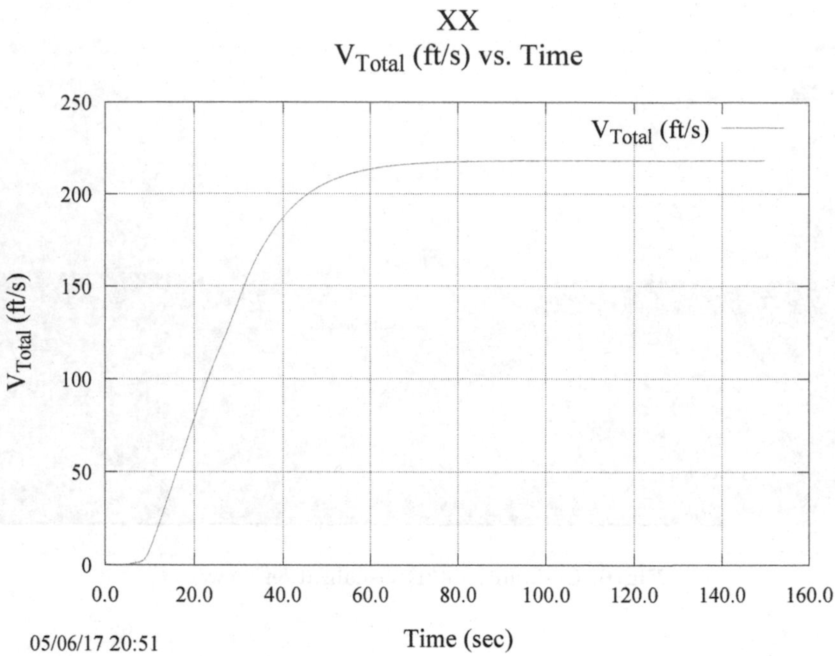

Fig. 8. Trajectory of simulator's velocity

4 Conclusion

This paper addresses a complex problem with requirements spanning diverse areas of knowledge. Personal desktop-level jet fighter simulator is studied and a simple example is presented for a fighter simulation where good results are achieved. It is proved that the flight dynamic model, the HUD display system, the 2D map GIS viewer and the 3D visualization viewer are realistic, fast, high efficiency and good code readability, which is suitable for the further development. Future work includes using the system to implement complex mission such as aerial acrobatics [13], perfecting its use in supersonic fixed-winged aircraft. Otherwise, a possible future work is to develop a testing method capable of comparing the physiological sensations felt in a real flight with the ones induced in the simulation [19]. Although the approach presented here is still in the research stage and thus is a work in progress, the ideas behind its implementation have the potential to provide insights to researchers and practitioners in the field of predictive situation awareness.

Acknowledgements. This work is supported by China Postdoctoral Science Foundation under Grant 2016M602962.

References

1. Eryilmaz, U., Tokmak, H.S., Cagiltay, K., Isler, V., Eryilmaz, N.O.: A novel classification method for driving simulators based on existing flight simulator classification standards. Transp. Res. Part C Emerg. Technol. **42**(42), 132–146 (2014)
2. Ippolito, C., Pritchett, A.: Software architecture for a reconfigurable flight simulator. In: Modeling and Simulation Technologies Conference (2013)
3. Allerton, D.J.: Flight simulation - past, present and future. Aeronaut. J. New Series **104**(1042), 651–663 (2000)
4. Allerton, D.: Principles of Flight Simulation (2009)
5. Jamson, A.H., Lai, F.C.H., Carsten, O.M.J.: Potential benefits of an adaptive forward collision warning system. Transp. Res. Part C Emerg. Technol. **16**(4), 471–484 (2008)
6. Zdanowicz, P., Jurecki, R.S., Stanczyk, T.L., Guzek, M., Lozia, Z.: Research on behaviour of drivers in accident situation conducted in driving simulator. J. Kones **16**(1), 173–183 (2009)
7. Barkenbus, J.N.: Eco-driving: an overlooked climate change initiative. Energy Policy **38**(2), 762–769 (2010)
8. Gerlach, T., Durak, U., Knppel, A., Rambau, T.: Running high level architecture in real-time for flight simulator integration. In: AIAA Modeling and Simulation Technologies Conference (2016)
9. Bruce, J.E.: Manual for a Workstation-Based Generic Flight Simulation Program (LaRCsim) Version 1.4. NASA Langley Technical Report Server (1998)
10. Berndt, J.S.: JSBSim: an open source flight dynamics model, pp. 2004–4923 (2004)
11. Gary “Buckaroo” Neely. Yasim. http://www.buckarooshangar.com/flightgear/yasimtut.html
12. Rozanov, E.V., Zubov, V.A., Schlesinger, M.E., Yang, F., Andronova, N.G.: The uiuc three-dimensional stratospheric chemical transport model: description and evaluation of the simulated source gases and ozone. J. Geophys. Res. Atmos. **104**(D9), 11755–11781 (1999)
13. Klenke, R., Mcbride, J., Nguyen, H.: A reconfigurable, linux based, flight control system for small UAVs. In: AIAA Infotech@Aerospace 2007 Conference and Exhibit (2013)
14. BBN technology. Openmap. https://ds.bbn.com/
15. Zhao, X.Y., Cai, Y., Yang, S.L., Huang, K.D.: Lessons learned from design and development of military scenario definition language scenario editing toolset. Adv. Mater. Res. **748**, 1041–1045 (2013)
16. Chung’s Blogspot. Mycesiumflight - a webgl cesium 3D flight simulator. https://sourceforge.net/projects/mycesiumflight/?source=directory
17. Smith, T.: Cesium - an open-source JavaScript library for world-class 3D globes and maps. http://cesiumjs.org/
18. Joyent. Node.js - an asynchronous event driven JavaScript runtime. https://nodejs.org/en/
19. Arjoni, D.H., Rocha, G., Pereira, R., Moreira, A.H., Nicola, R.M., Oliveira, W.R., Silva, A.V.S., Natal, G.S., Silveira, L., Silva, E.T.: Experimental evaluation of the human performance on a robotic flight simulator based on FOQA parameters. In: Aerospace Technology Congress (2016)

Full-Pose Magnetic Estimation Based on a Two-Stage Algorithm for Remote Hand Rehabilitation Training

Hui-Min Shen(✉)

University of Shanghai for Science and Technology, Shanghai 200093, China
shen_huimin@163.com

Abstract. This paper presents a two-stage algorithm for real-time full-pose magnetic localization in remote hand rehabilitation training. In the proposed two-stage algorithm, a database implying the location-orientation mapping relationship is established, which improves the non-linear optimization processing result by providing a more reliable initialization. A simulated experiment based on a link mechanism developed in SolidWorks is presented demonstrating the feasibility and effectiveness of the proposed algorithm. The results show that almost 90% improvement in full-pose estimation can be provided with a cost of 10% more processing time.

Keywords: Magnetic localization · Inverse problem · Mapping · Levenberg-Marquardt algorithm · Hand training

1 Introduction

As a result of the rapid growth in economy and science technology, people's average life expectancy is greatly extended, which leads to the aging of the world's population. One of the negative consequences that comes along is the increasing incidence of cardiac-cerebral vascular disease (CCVD), especially like stroke. Although most individuals suffered from CCVD may survive after surgery, kinds of sequelae, such as impaired motor function in one of the hands, often happen [1]. It is reported that in China more than 10% paralyzed individuals suffered from impaired hands motor function rely on assistive technology in daily life [2], which lead to heavy burden on their families and the society. The interest in rehabilitation services is increasing steadily with marked growth in the last 10 years with the graying of the population.

Hand motion analysis, including the flexion angles of lingers and space trajectory information, plays an important role in the proceeding of rehabilitation training of hand function recovery. The anatomy properties of hand include multi-joint, non-rigid motion, and seriously self-cover, which bring great challenge in real-time hand motion analysis. The researches in hand motion analysis focus in computer version technology [3] and wearable sensor technology [4, 5]. The former has entered into the stage of commercialization for human-computer interaction applications in virtual reality and computer games, such as Microsoft Kinect [6]. The later taking advantages of intuitive nature, free from light influence and self-cover, can provide accurate flexion angles and

Y. Huang et al. (Eds.): ICIRA 2017, Part I, LNAI 10462, pp. 428–437, 2017.
DOI: 10.1007/978-3-319-65289-4_41

motions [7]. Magnet marked tracking method is widely applied due to its contactless, compactness, low-cost, and no requirement for power source and connect wire [8, 9]. However, the nonlinear inverse model of magnetic tracking makes it difficult to derive an accurate solution in real-time.

Normally, efforts in magnetic localization are focused in two aspects: sensing system [10–12] and processing algorithm [13–15]. These methods relied on improvement of sensing system take advantages of the quantity and quality of sampled original magnetic signals, which requires dedicated equipment. According to algorithms employed, approaches relied on modification of processing algorithm can be divided into two categories: the direct approach and the statistic approach. The chief advantages the direct approach methods present are their realization simplicity and the fast processing time at the cost of estimation accuracy. Besides, these methods are limited as they are sensitive to environment noise. The latter approach can provide high accuracy results with prolonged response time, which is often used for post processing.

Constrained by the anatomical structure of the hand, the full-pose hand motion parameters, namely location and orientation, is relatively associated with each other, with fixed wrist location. This paper presents a novel magnetic estimation method for hand motion analysis by establishing a two-stage algorithm based on location-orientation mapping relationship. The reminder of this paper offers the following:

- A two-stage algorithm is introduced based on location-orientation mapping database (LOMD), where least-squares (LS) algorithm for solving the inverse magnetic localization problem with magnetic dipole model is presented.
- A simulated link mechanism with location-orientation mapping relationship is illustrated to validate the efficiency and accuracy of the proposed method.
- The proposed algorithm is solved numerically. As will be shown, improved magnetic estimation results can be derived with little sacrifice of the processing time.

2 Methodology

2.1 Inverse Magnetic Localization with Magnetic Dipole Model

Based on the fact that the size of the needle-like cylindrical magnet marker is very small as compared to its distance r from the sensor (it is normally accepted as a rule of thumb that if the marker diameter D satisfies $5D < r$, the modelling errors will below 1%), the marker can be modeled as a magnetic dipole which can be presented by five parameters of location and orientation in Eq. (1), as illustrated in Fig. 1. From a strict mathematical point of view, if we have five-independent equations then the unknown 5 dipole parameters (including location vector $\mathbf{r}_p(x_p, y_p, z_p)$ and unit magnetic moment $\mathbf{M}(\theta, \varphi)$) can be derived. These equations are provided by data measured by magnetic sensors.

$$\mathbf{B}(\mathbf{r}, \mathbf{M}) = \frac{\mu_0 m}{4\pi r^3}\left[\frac{3}{r^2}(\mathbf{M} \bullet \mathbf{r})\mathbf{r} - \mathbf{M}\right] \tag{1}$$

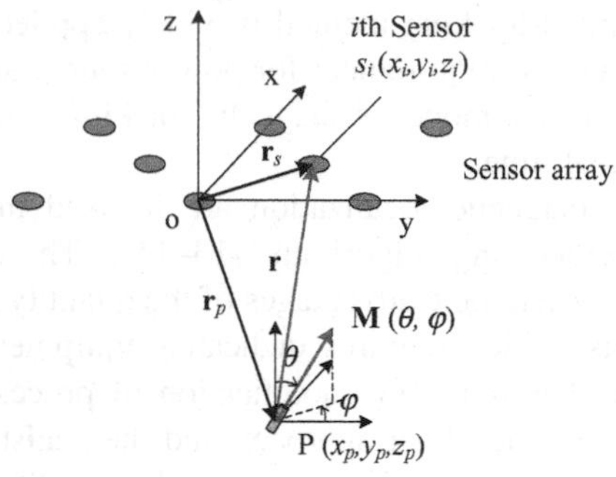

Fig. 1. Scheme illustrates localization system for PM maker modeled as magnetic dipole in spherical coordinate system

where μ_0 is the free space magnetic permeability ($=4\pi \times 10^{-7}$N/A^2); **M** is the unit orientation vector defined for the magnetic moment with the rotation θ () from the positive z-axis towards the xy-plane and the rotation φ () from the positive x-axis in counterclockwise direction towards the projection in the xy-plane; m (A·m^2) is the magnitude of the magnetic moment; **r** is the vector from the dipole $\mathbf{r}_p$ to the measured point $\mathbf{r}_s$ in Cartesian coordinates with the module r.

Nevertheless, the 5D high-order non-linear characteristic of the equations and the presence of noise, induce ambiguities that necessitate processing additional data. These data are obtained by adding sensors to the grid. This leads to a system of equations that become over determined. A nonlinear LS method is always employed by minimizing the objective function in Eq. (2) which relates to observations $\mathbf{B}_m$ measured by a sensor array. To derive high accuracy results, good initial guess of the parameters is always required in solving this problem.

$$\min f(\mathbf{r}_p, \mathbf{M}) = \min \left(\sum_{i=1}^{N} \|B_m(i) - B_c(\mathbf{r}_p, \mathbf{M}, i)\|^2\right) \tag{2}$$

where $B_m(i)$ is the magnetic flux density recorded by the ith sensor, $B_c(\mathbf{r}_p, \mathbf{M}, i)$ is the instantaneous magnetic field at the measured positions computed from the mathematic model of magnetic dipole presented in Eq. (1).

2.2 Two-Stage Algorithm Combined with LOMD

As mentioned in Subsect. 2.1, there is no closed form solution for the over determined high order nonlinear problem in Eq. (2). Taking advantages of good global convergence and iteration efficiency, Levenberg-Marquardt method (LMM) was chosen to estimate the position of the magnet marker based on the magnetic sensor output and dipole equation. LMM integrates the Newton's method and the steepest descent method. The algorithm in some way should take advantage of the safe, global convergence properties of the steepest descent method. On the other hand, the algorithm takes the quadratic convergence of Newton's method when the iterates get close enough to the solution. In every iteration step, the objective function computed by estimation of parameter $\mathbf{Q} = (\mathbf{r}_p, \mathbf{M})$ gets smaller, where Q is replaced by a new estimate $\mathbf{Q}+\mathbf{q}$. To determine $\mathbf{q}$, the function $\mathbf{B}(\mathbf{Q})$ is approximated by their linearization in Eq. (3):

$$\mathbf{B}(\mathbf{Q}+\mathbf{q}) \approx \mathbf{B}(\mathbf{Q}) + \mathbf{J}(\mathbf{q}) \tag{3}$$

where $\mathbf{J}$ is the Jacobian of $\mathbf{B}$ at $\mathbf{Q}$.

when the objective function in Eq. (2) finds a minimum where $\nabla_q f = 0$, we have

$$\left(\mathbf{J}^{\mathrm{T}}\mathbf{J} + \mu\mathbf{I}\right)\mathbf{q} = -\mathbf{J}^{\mathrm{T}}\mathbf{B} \tag{4}$$

from which $\mathbf{q}$ can be obtained. Thus, by proper adjustment of the damping parameter μ we have a method that combines the good qualities of the steepest descent method in the global part of the iteration process with the fast ultimate convergence of Newton's method.

It is known that the solution of this nonlinear LS optimization problem is closely related to the initializations of the unknown parameters. For different conditions, including static tracking and dynamic tracking, LMM initializations in the proposed algorithm are set separately. In static and discrete points tracking, a novel two-stage algorithm was devised based on the LOMD, as illustrated in Fig. 2. The LOMD is established according to the motion behaviors, which implies the mapping relationship between the location and orientation parameters of the marked magnet. To be specific:

(1) In the 1st stage, LMM with the *xy*-plane location of sensor providing the strongest measurement and zero *z* and moment as initialization and rough tolerance provides a pre-estimated result $\left(\bar{\mathbf{r}}_p, \bar{\mathbf{M}}\right)$.
(2) For a certain sensor, the measured magnetic field strength is more sensitive to magnet location compared to magnet orientation in Eq. (1). Thus, it is reasonable that the pre-determined location parameters are more reliable than pre-determined orientation. A mapped orientation $\mathbf{M}'$ according to the pre-estimated $\bar{\mathbf{r}}_p$ can be searched from LOMD, which is more reliable.
(3) In the 2nd stage, the solving process of objective function f speeds up by employing $\left(\bar{\mathbf{r}}_p, \mathbf{M}'\right)$ as initialization.
(4) Iteration of LMM stops when proper parameters are found.

The introduction of the database and two-stage algorithm in solving Eq. (2) provides more accurate results without sacrificing time spent on processing. In follow-up dynamic tracking, the estimation result of the former observation spot is taken as initialization with normal LMM process.

The efficiency and accuracy of solving results derived by the two-stage algorithm is closed related to the LOMD, whose construction needs particular attention. If the data density of LOMD is too large, the time spent on searching mapping orientation will be much. On the contrary, if the data interval becomes too large, the relationship between the mapping orientation and the pre-determined location will weak, which gives little

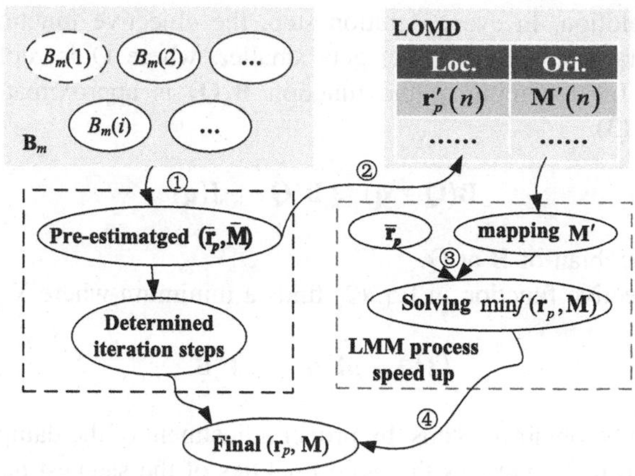

Fig. 2. Process of the proposed two-stage algorithm with LOMD

help in improving the result accuracy. Thus, construction of LOMD needs comprehensive consideration.

3 Simulated Experiment Validation

The validity and effectiveness of the proposed method were examined by simulations. A 2-DOF link mechanism was designed and developed in SolidWorks (SolidWorks, Version 2014, Dassault Systèmes SolidWorks Corporation, Massachusetts, the USA), whose output represents the trajectory of a marked magnet and was controlled with two linear inputs. Then, the outputs of the sensor array derived from the PM were computed according to the dipole model in Eq. (1). The estimated PM deduced from the proposed two-stage algorithm and traditional algorithm were compared.

3.1 Development of the Simulated Experimental Setup

Link mechanism model. The structure of the link mechanism, whose mechanism has a location-orientation mapping relationship, was developed by SolidWorks and used to simulate the trajectory of the marked PM. Figure 3(a) illustrates the mechanism schematics of experimental set, and Fig. 3(b) presents the SolidWorks model.

The real-time PM parameters, including locations $\mathbf{r}_p(x_p, y_p, z_p)$ and orientations $\mathbf{M}(\theta, \varphi)$, can be computed by the geometry structure relationship and the inputs (x_I, y_I):

$$\begin{cases} x_p = x_I - \frac{l_{IP}}{l_{IH}}(x_I - x_H) \\ y_p = y_I - \frac{l_{IP}}{l_{IH}}(y_I - y_H) \\ z_p = l_{IP}\cos\theta \end{cases} \tag{5a}$$

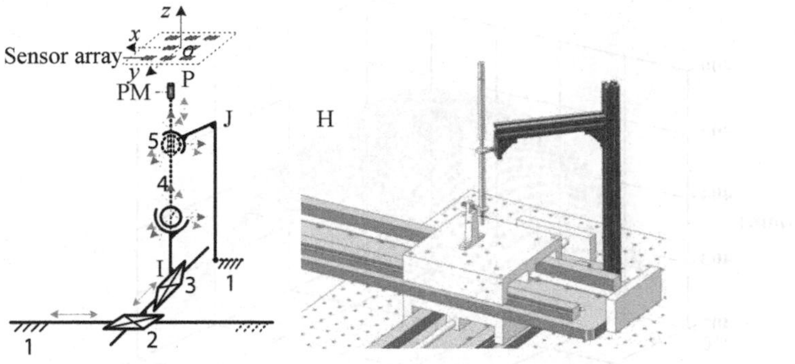

(a) Mechanism schematics of experimental setup

(b) SolidWorks model of experiment setup

Fig. 3. Illustration of kinematics simulation geometry model

$$\begin{cases} \theta = \arcsin\dfrac{\sqrt{(x_I - x_H)^2 + (y_I - y_H)^2}}{l_{HI}}; \\ \varphi = \begin{cases} \arctan\left(\frac{y_I - y_H}{x_H - x_I}\right); & \text{if } x_H > x_I \\ \pi - \arctan\left|\frac{y_I - y_H}{x_H - x_I}\right|; & \text{if } x_H < x_I \end{cases} \end{cases} \tag{5b}$$

where l_{IP}, l_{IH} are the length of corresponding bars; (x_H, y_H) is the location of point H in the xy-plane. Considering the pendulum angle of the selected ball bearings, $\theta \in (0, 10°)$.

Sensor array. As shown in Fig. 4, eight single-axis anisotropic magnetoresistive (AMR) sensors are employed, which provides the magnetic distribution along the z-axis.

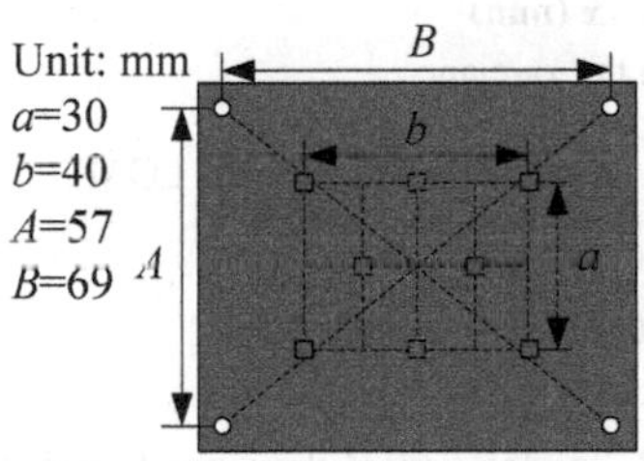

Fig. 4. Illustration of sensor array displacement

Establishment of LOMD. In this experiment, LOMD is established according to the movement range of the input linear guide, which provides input linear motion along x-axis (0–23 mm) and y-axis (0–17 mm). It is assumed that the input step interval is 0.5 mm. With the linear input motion, the link mechanism model in Eq. (5a) outputs the trajectory of PM and its orientation. As illustrated in Fig. 5, there are total 1645 sets of data in the established LOMD, and each set includes location and orientation parameters.

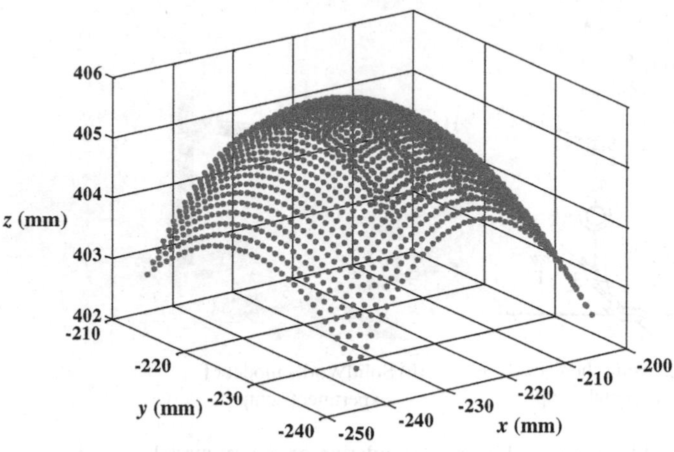

(a) Displacement of LOMD

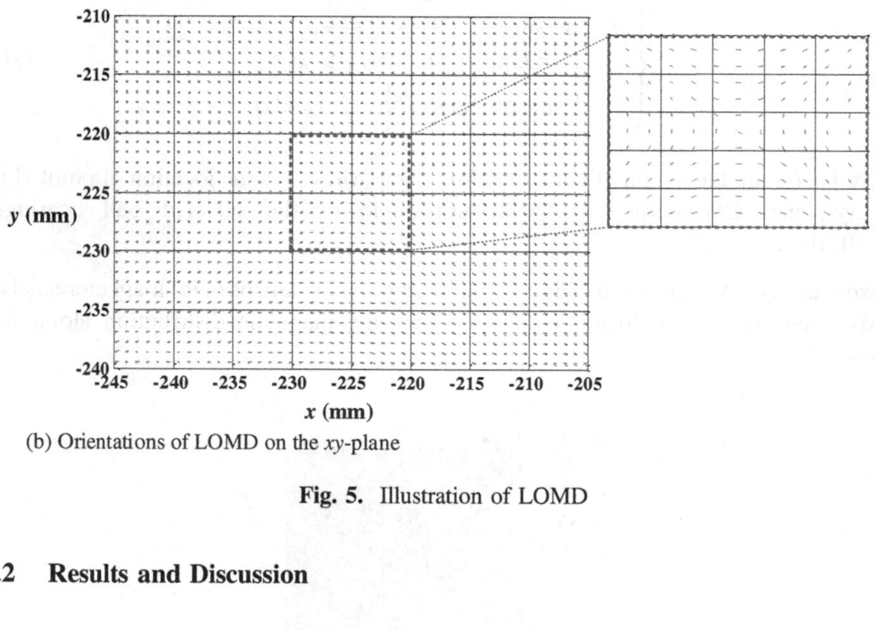

(b) Orientations of LOMD on the *xy*-plane

Fig. 5. Illustration of LOMD

3.2 Results and Discussion

Performance Evaluation. In order to validate the feasibility and effectiveness of the proposed two-stage algorithm in magnetic tracking, simulations were carried out, where the true localization parameters can be known. In order to focus on evaluation of the proposed algorithm, noise-free sensor measurements computed from Eq. (1) are employed for solving the inverse magnetic estimation problem. We generated 1000 uniformly distributed positions randomly in the positioning region.

Estimation results derived from three methods were compared:

- **Method 1-traditional method**: the estimation result was computed directly, with initializations in the 1st step of the proposed algorithm.

- **Method 2-the two-stage algorithm**: magnetic estimation was carried out based on the established LOMD.
- **Method 3-dual traditional method**: since the LS algorithm with different initializations runs twice in the proposed method (Method 2), the LS algorithm was processed again with the initialization computed from the traditional method.

The effectiveness of three methods were evaluated by orientation error $e_{\mathbf{M}}$ defined by the angle between estimated $\mathbf{M}'$ and true $\mathbf{M}$, and location error $e_{\mathbf{r}}$ defined by the norm of vector $\tilde{\mathbf{r}}$ points from estimated $\mathbf{r}'$ to true $\mathbf{r}$. Figure 6 presents the mean, standard deviation values of the positioning error. The mean time spent on algorithm processing of the three method is 80 ms, 91 ms and 80 ms, respectively.

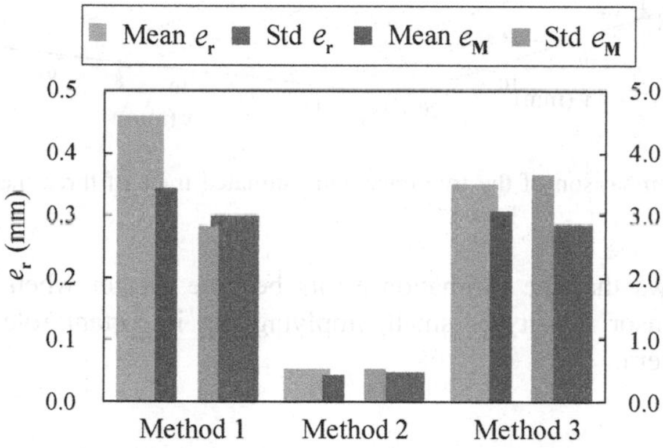

Fig. 6. Simulation results computed form three methods.

To demonstrate the performance of the proposed two-stage algorithm in continuous trace estimation, a section of arcs starts from (4.583, −7.63, −22.19) mm to (8.65, 14.98, −23.41) mm was constructed according to the link mechanism modeled in SolidWorks. The calculated trace estimated by three methods were compared with true trace as illustrated in Fig. 7.

Discussion of Results. Some observations can be made from the above results:

(1) In Fig. 6, the mean localization error and orientation error derived from the proposed two-stage algorithm are the smallest, which are about one ninth of the other two methods.
(2) The average processing time spent on the proposed two-stage algorithm is 10 ms more than the other methods, which is mainly spent on LOMD search.
(3) Comparison of results estimated by Method 3 and Method 1 in Figs. 6 and 7 shows little improvement, implying that multi-run of LS algorithm helps little and suggesting the improvement made by the employment of LOMD.
(4) The calculated trace in Fig. 7 shows that the estimated result of the proposed two-stage algorithm fits the true trace best, compared with the other two methods.

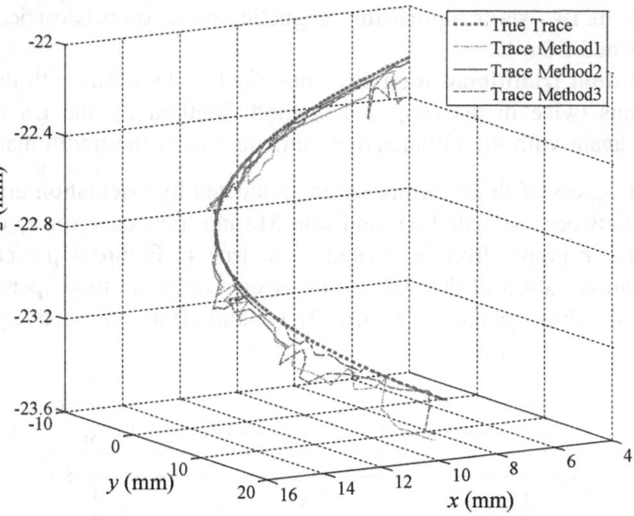

Fig. 7. Comparison of the true trace and estimated trace of three methods.

(5) Figure 7 shows that the estimation errors become greater when in fringe area where the sensor density is small, implying the important role of the signal sampling system.

4 Conclusion

A two-stage algorithm for real-time magnetic estimation, employing the location-orientation mapping relationship, has been presented. This method needs to establish a database implying the location-orientation mapping relationship. A simulated link mechanism is modeled for method validation. As demonstrated, the proposed algorithm can provide almost 90% improvement in both location and orientation estimation with a cost of 10% more processing time. The proposed two-stage algorithm can be used for hand motion analysis in remote hand function recovery.

Acknowledgments. This work was supported by the National Natural Science Foundation of China under Grant 51605291.

References

1. Krebs, H.I., Volpe, B.T.: Rehabilitation robotics. Handb. Clin. Neurol. **110**, 283 (2013)
2. Zhang, T.: Guideline for stroke rehabilitation in China. Chin. J. Rehabil. Theor. **18**(4), 301–318 (2012)
3. Thomas, M.C., Pradeepa, A.P.M.S.: A comprehensive review on vision based hand gesture recognition technology. Int. J. Res. Advent Technol. **2**(1), 303–310 (2014)

4. Ju, Z., Liu, H.: Human hand motion analysis with multisensory information. IEEE/ASME Trans. Mechatron. **19**(2), 456–466 (2014)
5. Schroder, M., Elbrechter, C., Maycock, J., et al.: Real-time hand tracking with a color glove for the actuation of anthropomorphic robot hands. In: Proceedings of the International Conference on Humanoid Robots, Humanoids, IEEE (2012)
6. Herrera-Acuña, R., Argyriou, V., Velastin, S.A.: A kinect-based 3d hand-gesture interface for 3D databases. J. Multimodal User Interf. **9**(2), 121–139 (2015)
7. Dipietro, L.S.A.M., Dario, P.: A survey of glove-based systems and their applications. IEEE Trans. Syst. Man Cybern. Part C Appl. Rev. **38**(4), 461–482 (2008)
8. Fahn, C.S., Sun, H.: Development of a fingertip glove equipped with magnetic tracking sensors. Sensors **10**(2), 1119–1140 (2010)
9. Kortier, H.G., Antonsson, J., Schepers, H.M., et al.: Hand pose estimation by fusion of inertial and magnetic sensing aided by a permanent magnet. IEEE Trans. Neural Syst. Rehabil. Eng. **23**(5), 796–806 (2015)
10. Hu, C., Li, M., Song, S., et al.: A cubic 3-axis magnetic sensor array for wirelessly tracking magnet position and orientation. IEEE Sens. J. **10**(5), 903–913 (2010)
11. Song, S., Qiu, X., Wang, J., et al.: Design and optimization strategy of sensor array layout for magnetic localization system. IEEE Sens. J. **17**(6), 1849–1857 (2017)
12. Kortier, H.G., Sluiter, V.I., Roetenberg, D., et al.: Assessment of hand kinematics using inertial and magnetic sensors. J. Neuroengineering Rehabil. **11**(1), 70 (2014)
13. Song, S., Li, B., Qiao, W., et al.: 6-D Magnetic localization and orientation method for an annular magnet based on a closed-form analytical model. IEEE Trans. Magn. **50**(9), 5000411 (2014)
14. Lee, K.M., Li, M.: Magnetic tensor sensor for gradient-based localization of ferrous object in geomagnetic field. IEEE Trans. Magn. **52**(8), 1–10 (2016)
15. Shen, H.M., Hu, L., Qin, L.H., et al.: Real-time orientation-invariant magnetic localization and sensor calibration based on closed-form models. IEEE Magn. Lett. **6**, 6500304 (2015)

Preprocessing and Transmission for 3D Point Cloud Data

Zunran Wang[1], Chenguang Yang[1,2(✉)], Zhaojie Ju[3], Zhijun Li[1], and Chun-Yi Su[4]

[1] Key Lab of Autonomous System and Network Control and College of Automation Science and Engineering, South China University of Technology, Guangzhou, China
cyang@ieee.org

[2] Zienkiewicz Centre for Computational Engineering, Swansea University, Swansea, UK

[3] School of Computing, University of Portsmouth, Portsmouth, UK

[4] School of Automation, Guangdong University of Technology, Guangzhou, China

Abstract. Robots play an increasingly important role in social life, especially the front desk robots. But the front desk robots seldom handle business in reality unless they have proper functionalities for Human-Robot Interaction (*HRI*). For enhancing the immersed sense in the interactive process, we consider the 3D image of the upper body of the operator in the remote control room as the upper body of front desk robots. However, it is a great challenge to transmit the 3D point cloud data in the way of remote interaction. The paper uses a simple method to deal with the problem of transmitting the 3D point cloud data and the idea of the method is to reduce the network data volume. In order to reduce the network data volume, we only consider that the 3D point cloud of interest will be transmitted. The filters are used to remove the noise and background of the 3D image, the segmentation algorithm will be used to acquire the 3D point cloud data of interest. The experiment result demonstrates that the method can reduce the network data volume and ensure high-quality image information. Thus, the method can reduce transmission time.

Keywords: 3D point cloud · Human-Robot Interaction (*HRI*) · Filter and segmentation of 3D Image · 3D image transmission

1 Introduction

In order to enhance the work performance of robots, people add HRI technology into robots, such as the authors use HRI technology to improve NAO robot's performance of understanding and interpretation [1]. With the rapid development of technologies about 3D point cloud data processing and HRI, people pay

Y. Huang et al. (Eds.): ICIRA 2017, Part I, LNAI 10462, pp. 438–449, 2017.
DOI: 10.1007/978-3-319-65289-4_42

more and more attention to the research of Virtual Reality (VR) technology, the VR technology are used in many products. For enhancing the immersed sense in the interactive process, the upper body of front desk robots is the 3D image which comes from the remote operator's upper body. The key problem of the remote interaction system is to process and transmit 3D point cloud data. So far many researches have been conducted on how to deal with the problem of processing and transmitting the 3D point cloud data.

There are many methods to acquire high-quality 3D point cloud data in the reality. The sensors such as Microsoft Kinect and ZED Stereo Camera can easily acquire 3D point cloud data, the processing procedure requires less external hardware and the work of dealing with the 3D image will be more simple. Moreover, a laser scanner can also be used to obtain 3D data, but there are many noises in the point cloud data. In this paper, the filters and the segmentation algorithm will be proposed to process 3D point cloud data.

Filtering algorithms are based on mathematical morphology, the triangulation, wavelet transform etc. Filtering method based on morphological operations have been used to remove outliers and filled missing data, the method can keep the terrain features unchanged [2]. A filtering algorithm based on Voxel Grid Statistical Outlier can quickly and effectively deal with the data of 3D point cloud [3]. In order to remove the noise, the authors in [4] firstly use the low-pass geometric filter.

The segmentation algorithms are mainly used to acquire the image of interest and perform the operation of image matching. In the recent years, many works have proposed point cloud segmentation methods, the methods mainly are based on edge, region, model, or graph [5]. Edge-based segmentation method will go wrong when a closed curve is produced. A region-based segmentation method is proposed for 2D and 3D image segmentation through using the information of edge, shape and intensity [6]. A method based on the smoothness constraint was designed by Rabbani [7]. Chen et al. [8] proposed a unseeded method for segmentation by using clustering to determine planar surfaces, which deal with the problems of excessive or inadequate segmentation. In order to obtain a more accurate performance and less noise sensitivity, a method based on graph considered FH (Felzenswalb-Huttenlocher) algorithm as segmentation algorithm [9] and FH algorithm was upgraded by Strom et al. [10] with an additional criterion. The authors in [11] proposed a method for 3D image segmentation through using the relation between the shape of an object and the gray level variation in an image. Based on the model, the methods that are fast and robust are widely used, Schnabel et al. [12] utilized RANSAC (RANdom SAmple Consensus) algorithm to segment 3D point cloud data of interest. In order to deal with problem of the brain MRI image segmentation, the authors in [13] presented segmentation method based on Gaussian mixture model.

In order to transmit 3D point cloud data to the front desk robot, the point cloud data obtained by Kinect in the remote control room must be sent through the network. However, the size of the 3D point cloud is very large, rendering it for real-time remote interaction is a challenge in poor network environments. Many methods are designed to solve this issue, the main method is through compressing the data of 3D point cloud. Birk et al. in [14] proposed an approach to transmitting the data of point cloud by compressing surface points to corresponding plane patches. Poppinga et al. [15] reduced the data volume up to 50% of the original size through extracting convex surface polygons. In order to enhance the compression ratio, Wiemann et al. [16] used a deeper polygonal search. However, the color information of the 3D image does not be took into account in many works about compressing the data of 3D point cloud. The transmission methods which data of interest is transmitted are more preferable than the methods that as much data as possible is passed [17].

This paper presents the procedure of preprocessing and transmission for 3D point cloud data, preprocessing methods of 3D point cloud data include filters and segmentation algorithms, the filters include the bandpass filter and the statistical filter, the RANdom SAmple Consensus (*RANSAC*) algorithm will be utilized to acquire 3D point cloud of interest. The next section introduces the theories of filters and segmentation algorithms. In Sect. 3 the theories will be realized through an experiment. In the Sect. 4, the conclusion will be introduced.

2 Theoretical Knowledge of Algorithm

The first, we use the bandpass filter to remove the background of the 3D image. Then, some outliers will be removed by the statistical filter. In the end, the segmentation algorithm will be used to process the 3D image so that some small block 3D images that are unnecessary for remote interaction are removed and the TCP protocol is used to transmit the data of point clouds through the internet, as shown in Fig. 1.

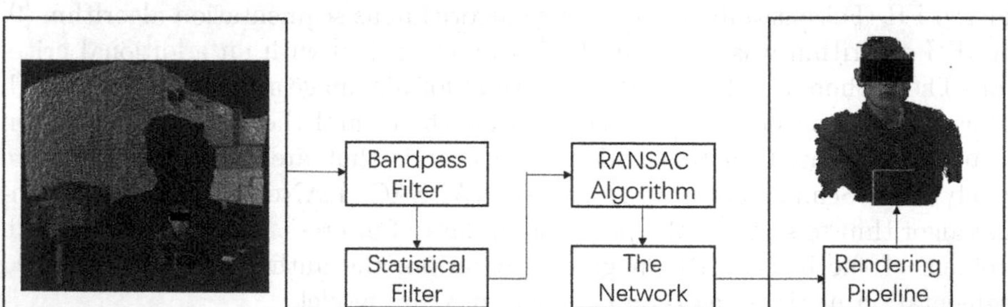

Fig. 1. The overall block diagram

In the section, the theories of filters and segmentation algorithms will be introduced, the filters include the bandpass filter and the statistical filter, the aim of the bandpass filter is to remove the background of the 3D image and the statistical filter is used to remove the outliers of the 3D point cloud data. The RANSAC algorithm is utilized to acquire the 3D point cloud of interest. The paper uses Point Cloud Library (*PCL*) to implement these algorithms.

2.1 Bandpass Filter and Statistical Filter

Bandpass Filter. Bandpass filter that allows a specific band of waves shields other bands at the same time, the paper will use the similar principle to remain point cloud data of interest based on constraints for one particular field of the point type. The particular field can be X axis, Y axis or Z axis. In the PCL, the filter is called PassThrough filter.

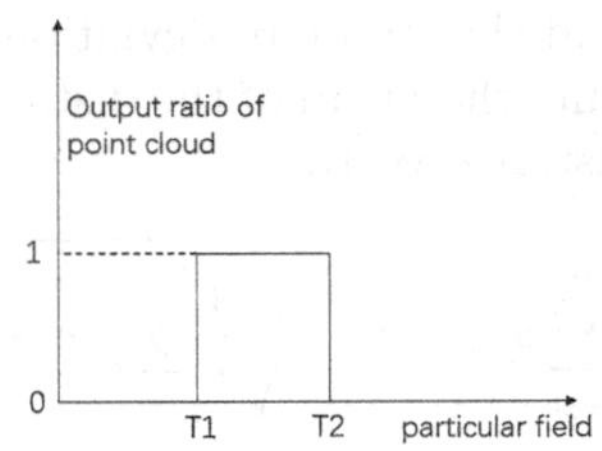

Fig. 2. The waveforms of bandpass filter outputting point cloud data

In Fig. 2, the particular field can is X axis, Y axis or Z axis. When the value of the particular field is greater than $T1$ and less than $T2$, the point cloud data will be all saved. When the value of the particular field is less than $T1$ or greater than $T2$, the point cloud data will be all removed. The mathematical expression of the bandpass filter can be represented by the following formula, x, y represents the particular field and output proportion of the point cloud, respectively.

$$y = \begin{cases} 0 & x < T1 \\ 1 & T1 \leq x \geq T2 \\ 0 & x > T2 \end{cases} \tag{1}$$

Statistical Filter. Statistical filter mainly is used to remove sparse outliers by computing the neighbour distances in the input dataset.

The RGB-D camera such as Microsoft Kinect typically will generate point cloud dataset whose point density is not uniform, this increases the difficulty of estimating local point cloud characteristic such as normal and curvature changes, which is not conducive to the following segmentation processing work. We can remove outliers so that point density of point cloud dataset is uniform through a statistical analysis on each point's neighbourhood [18].

Algorithm steps:
step 1: Compute the mean distances for all points with respect to their k nearest neighbours, assuming the mean distances of all points to x_1, x_2, x_3......x_n, n is the size of the point cloud data, d_j represents the distance between the point to the Jth of their k nearest neighbors.

$$x_i = (\sum_{j=1}^{k} d_j)/k, (i = 1, 2, 3......n) \tag{2}$$

step 2: Estimate the mean and the standard deviation of the distance vector (x_1, x_2, x_3......x_n). x_{mean} represents the mean of the distance vector, σ represents the standard deviation of the distance vector

$$x_{mean} = (\sum_{i=1}^{n} x_i)/n \sigma = \sqrt[2]{\frac{1}{n}\sum_{i=1}^{n}(x_i - x_{mean})^2} \tag{3}$$

step 3: Compute distance threshold, $d_{threshold}$ represents distance threshold, std_{mul} represents the standard deviation multiple and depends on the size of the analysed neighbourhood. The more number of neighbourhoods, the bigger std_{mul}.

$$d_{threshold} = x_{mean} + std_{mul} * \sigma \tag{4}$$

step 4: According to the distance threshold($d_{threshold}$), remove the outliers, Points having a too high average distance are outliers and will be removed. if

$$x_i \geq d_{threshold} \tag{5}$$

this point will be removed.

The Pseudocode of The Statistical Filter:

```
1.k ⟵ the number of nearest neighbors
2.i ⟵ 0
3.j ⟵ 0
4.n ⟵ the size of point cloud data
5.while i < n − 1
    1. do while j < k − 1
        1.do d_x ⟵ (point[i].x − point[j].x)^2
        2.d_y ⟵ (point[i].y − point[j].y)^2
        3.d_z ⟵ (point[i].z − point[j].z)^2
        4.d[j] ⟵ ²√(d_x + d_y + d_z)
        5.x[i] ⟵ d[j]/n + x[i]
        6.j ⟵ j + 1
    2. i ⟵ i + 1
6.i ⟵ 0;
7.while i < n − 1
    1. do x_mean ⟵ x[i]/n + x_mean
8.i ⟵ 0;
9.while i < n − 1
    1. do yy ⟵ (x[i] − x_mean)^2/n + yy
    2. i ⟵ i + 1
10.σ ⟵ ²√yy
11.d_threshold ⟵ x_mean + std_mul * σ
12.i ⟵ 0;
13.j ⟵ 0;
14.while i < n − 1
    if(x[i] <= d_threshold)
    then [
        index[j] ⟵ i
        j ⟵ j + 1
        i ⟵ 1 + i]
    else [i ⟵ 1 + i]
```

2.2 RANSAC Algorithm

The filters can remove outliers and unnecessary background in the 3D point cloud, but some unnecessary points that are on the same plane with points of interest can't be removed. Thus, the RANSAC Algorithm will be implemented to acquire the point cloud data of interest. In the recent years, many scholars use segmentation algorithms to acquire part information about the human body, such as the authors in the [19] extract human hand gesture features through using the earth movers distance and Lasso algorithms. In the paper, we use the RANSAC Algorithm to acquire the 3D image information of the upper body. The algorithm was first published by Fischler and Bolles in 1981. Unlike other algorithm building model, The RANSAC algorithm attempts to build a model that includes a minimum number of outliers. As shown in Fig. 3 RANSAC Algorithm

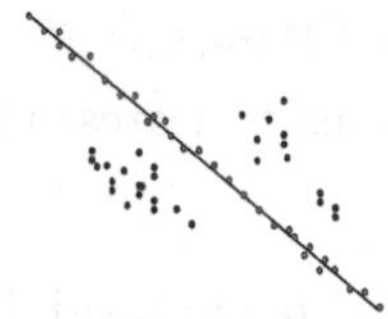

Fig. 3. The example of explaining RANSAC algorithm

can evaluate a model when the input data include the outliers and the inliers, the outliers will be removed but the inliers will be retained [20].

Algorithm steps:
step 1: For specific issue, a specific model is designed to determine whether a point is an outlier or not. A specific model can be a plane model, line model, circle model, sphere model and so on.
step 2: Compute the number of iterations (n). δ represents the probability that the RANSAC algorithm selects inliers from the input data set. k represents the number of points that at least can be used to estimate a model parameter. ε represents the probability of choosing an inlier when each time a single point is selected

$$n = \frac{log(1-\delta)}{log(1-\varepsilon^k)} \tag{6}$$

step 3: Randomly select n groups data from all point cloud data, each group include k points.
step 4: Use the n groups data to build n models.
step 5: According to a threshold (ξ), find out all outliers of the corresponding model, ξ will be designed by users. Reuse the inliers to build the model and re-find out the inliers.
step 6: Record the number of inliers, make the number to compare with the biggest number of inliers at present. If the number is larger than the biggest number of inliers at present, the parameters of the model will be saved and the number of inliers in this model will be the biggest number of inliers at present.
step 7: Repeat steps 4–6 until the number of Repeating reach the number of iterations.

3 Experimental Studies

The Kinect sensor can obtain 3D point cloud data and RGB image of scene, infrared transmitter transmits the infrared dot matrix and the traditional COMS image sensor with an infrared filter will detect the infrared information. The position and size of dots will change due to change of distance between the objects and Kinect sensor. The resolution of depth is 1280 × 1024 and the sensor can detect the range of distance between 1 m (below 1 m) and 3.5 m. The visual area of Kinect sensor is a rectangular cone whose horizontal and vertical directions are 58° and 45° respectively. The sensor also can obtain the RGB image through the RGB camera. The structure of the Kinect sensor will be shown in Fig. 4.

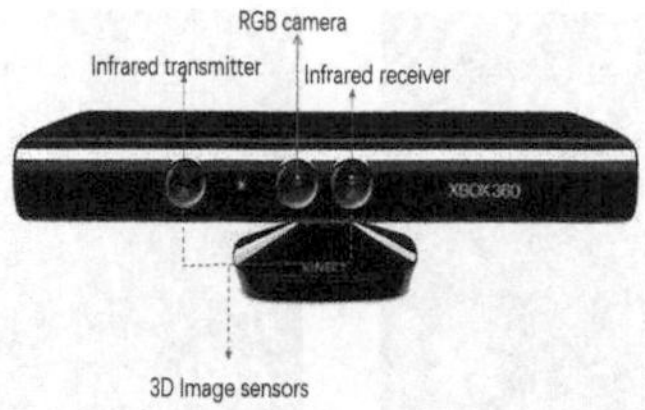

Fig. 4. The principle of kinect sensor

The Kinect sensor will be used to acquire the 3D point cloud data in our experiment. The results are produced in the paper through the machine with an Intel fourth-generation Core i5-4460 @3.20 GHz Processor with 8 GB RAM, the development environment is Microsoft Visual Studio 2010. We place the Kinect sensor on the desk, As shown in Fig. 5.

Fig. 5. Picture of the device of the experiment [taken at South China University of Technology]

In the bandpass filter, the particular filed will be Z axis, where $T1 = 0.0$ and $T2 = 1.0$, thus, we can retain the points whose depth are between 0.0 and 1.0. In the statistical filter, we make $k = 20$ and $std_{mul} = 1.0$. In the RANSAC algorithm, the specific model is a plane model and the threshold ξ is chosen to 0.1.

In the Fig. 6(a), the 3D point cloud data was not processed through any algorithm and was taken directly from the Kinect sensor. Obviously, the size of the 3D point cloud data is very large, there are many outliers in the 3D point cloud data. If the 3D point cloud data is transmitted over the network, it will cause serious network congestion. Moreover, a lot of information in the 3D point cloud data is unnecessary for the remote interaction. Thus, the 3D point cloud data must be processed through some algorithms.

The background of the 3D image is unnecessary for our remote interaction, then, the background must be removed through the Bandpass filter. As shown in Fig. 6(b), we already removed the background of the 3D image, it would be obvious that the number of point clouds data lessened. However, there are some

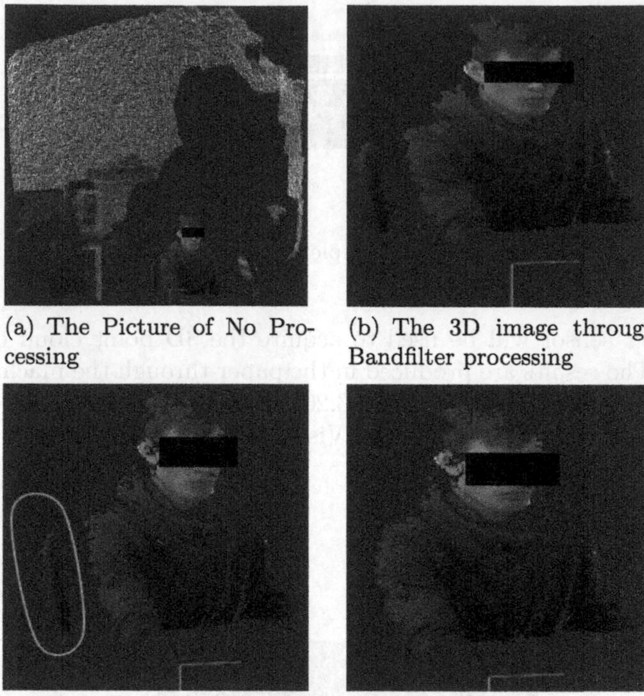

(a) The Picture of No Processing
(b) The 3D image through Bandfilter processing
(c) The 3D image through statistical filter processing
(d) The 3D image through RANSAC algorithm processing

Fig. 6. The procedure of the compression based on face detection algorithm (Color figure online)

outliers in the 3D point cloud data, which make the 3D image not smooth. The statistical filter can remove outliers and ensure the 3D image smooth. The result can be seen in Fig. 6(c).

To acquire the 3D point cloud data of the upper body of operator in the remote control room, the segmentation algorithm must be used to process the 3D point cloud data so that some small block 3D images that are unnecessary for remote interaction are removed. in the Fig. 6(c), the small block 3D image circled in a yellow circle was unnecessary for remote interaction. As shown in Fig. 6(d), we successfully removed some small block 3D images that were unnecessary for remote interaction through RANSAC segmentation Algorithm.

In the Table 1, the size of the original 3D point cloud data was 307200 (640×480). Through the bandpass filter processing, the size of 3D point cloud data drastically dropped to 60033. Then, through using the statistical filter, the size of 3D point cloud data was 52449. The last, the size of 3D point cloud data was 49667 through RANSAC segmentation Algorithm processing. As shown in Table 2, the proportion of decline of the point cloud data arrived 0.8045 when

Table 1. The change of the size of the point cloud data

The image processing	The size of points
Original picture	307200
Bandpass filter	60033
Statistical filter	52449
RANSAC algorithm	49667

Table 2. The proportion of decline of the size of the point cloud data

The image processing	The proportion of decline
Bandpass filter	0.8045
Statistical filter	0.8291
RANSAC algorithm	0.8383

we used the bandpass filter to process the 3D point cloud data, and proportion of decline in the whole process reached 0.8383. The experiment result demonstrates that our processing method can remove lots of unnecessary points, which ensures the smooth network and reduce transmission time. In the end, we used the TCP protocol to transmit the data of point clouds.

4 Conclusion

In the paper, in order to transmit the 3D point cloud data over the unstable network, we propose some methods for processing the 3D point cloud data. The purpose of the approaches of processing 3D point cloud data is to apply it in the front desk robots whose upper body is the 3D image from the upper body of operator in the remote control room. Because we only need the upper body information of the operator, we try to reduce the data volume of 3D image obtained by Kinect sensor. The first, we use the Bandpass filter to remove the background of the 3D image. Then, some outliers will be removed through the statistical filter. In the end, the segmentation algorithm will be used to process the 3D image so that some small block 3D images that are unnecessary for remote interaction are removed and the TCP protocol is used to transmit 3D point cloud data through the internet. The experiment result demonstrates proportion of decline in the whole process reaches probably 0.8383. Obviously, our processing methods can remove lots of unnecessary points, which ensures the smooth network.

Acknowledgments. This work was partially supported by National Nature Science Foundation (NSFC) under Grant 61473120, Guangdong Provincial Natural Science Foundation 2014A030313266 and International Science and Technology Collaboration Grant 2015A050502017, Science and Technology Planning Project of Guangzhou 201607010006, State Key Laboratory of Robotics and System (HIT) Grant SKLRS-2017-KF-13, and the Fundamental Research Funds for the Central Universities 2017ZD057.

References

1. Ju, Z., Ji, X., Li, J., Liu, H.: An integrative framework of human hand gesture segmentation for human-robot interaction. IEEE Syst. J. (2015). 10.1109/JSYST.2015.2468231
2. Chen, Q., Gong, P., Baldocchi, D., Xie, G.: Filtering airborne laser scanning data with morphological methods. Photogram. Eng. Remote Sens. **73**(2), 175–185 (2007)
3. Hu, F., Zhao, Y., Wang, W., Huang, X.: Discrete point cloud filtering and searching based on VGSO algorithm. In: ECMS, pp. 850–856 (2013)
4. Moreno, R., Garcia, M.A., Puig, D.: Graph-based perceptual segmentation of stereo vision 3D images at multiple abstraction levels. In: Escolano, F., Vento, M. (eds.) GbRPR 2007. LNCS, vol. 4538, pp. 148–157. Springer, Heidelberg (2007). doi:10.1007/978-3-540-72903-7_14
5. Bizjak, M.: The segmentation of a point cloud using locally fitted surfaces. In: 2016 18th Mediterranean Electrotechnical Conference (MELECON), pp. 1–6. IEEE (2016)
6. Wählby, C., Sintorn, I.M., Erlandsson, F., Borgefors, G., Bengtsson, E.: Combining intensity, edge and shape information for 2d and 3d segmentation of cell nuclei in tissue sections. J. Microsc. **215**(1), 67–76 (2004)
7. Rabbani, T., Van Den Heuvel, F., Vosselmann, G.: Segmentation of point clouds using smoothness constraint. Int. Arch. Photogram. Rem. Sens. Spat. Inf. Sci. **36**(5), 248–253 (2006)
8. Chen, J., Chen, B.: Architectural modeling from sparsely scanned range data. Int. J. Comput. Vis. **78**(2–3), 223–236 (2008)
9. Felzenszwalb, P.F., Huttenlocher, D.P.: Efficient graph-based image segmentation. Int. J. Comput. Vis. **59**(2), 167–181 (2004)
10. Strom, J., Richardson, A., Olson, E.: Graph-based segmentation for colored 3d laser point clouds. In: 2010 IEEE/RSJ International Conference on Intelligent Robots and Systems (IROS), pp. 2131–2136. IEEE (2010)
11. Yang, J., Duncan, J.S.: 3d image segmentation of deformable objects with joint shape-intensity prior models using level sets. Med. Image Anal. **8**(3), 285–294 (2004)
12. Schnabel, R., Wahl, R., Klein, R.: Efficient RANSAC for point-cloud shape detection. In: Computer Graphics Forum, vol. 26, pp. 214–226. Wiley Online Library (2007)
13. Balafar, M.: Gaussian mixture model based segmentation methods for brain MRI images. Artif. Intell. Rev. **41**(3), 429–439 (2014)
14. Birk, A., Schwertfeger, S., Pathak, K., Vaskevicius, N.: 3d data collection at disaster city at the 2008 NIST response robot evaluation exercise (RREE). In: 2009 IEEE International Workshop on Safety, Security & Rescue Robotics (SSRR), pp. 1–6. IEEE (2009)
15. Poppinga, J., Vaskevicius, N., Birk, A., Pathak, K.: Fast plane detection and polygonalization in noisy 3d range images. In: IEEE/RSJ International Conference on Intelligent Robots and Systems, IROS 2008, pp. 3378–3383. IEEE (2008)
16. Wiemann, T., Nüchter, A., Lingemann, K., Stiene, S., Hertzberg, J.: Automatic construction of polygonal maps from point cloud data. In: 2010 IEEE International Workshop on Safety Security and Rescue Robotics (SSRR), pp. 1–6. IEEE (2010)
17. Murphy, R.R.: Trial by fire [rescue robots]. IEEE Robot. Autom. Mag. **11**(3), 50–61 (2004)

18. Rusu, R.B., Marton, Z.C., Blodow, N., Dolha, M., Beetz, M.: Towards 3d point cloud based object maps for household environments. Robot. Auton. Syst. **56**(11), 927–941 (2008)
19. Ju, Z., Gao, D., Cao, J., Liu, H.: A novel approach to extract hand gesture feature in depth images. Multimedia Tools Appl. **75**(19), 11929–11943 (2016)
20. Chum, O., Matas, J.: Randomized RANSAC with Td, d test. In: Proceedings of British Machine Vision Conference, vol. 2, pp. 448–457 (2002)

Dexterous Hand Motion Classification and Recognition Based on Multimodal Sensing

Yaxu Xue[1,2,3], Zhaojie Ju[2(✉)], Kui Xiang[1], Chenguang Yang[4], and Honghai Liu[2]

[1] School of Automation, Wuhan University of Technology, Wuhan, China
[2] School of Computing, The University of Portsmouth, Portsmouth, UK
zhaojie.ju@port.ac.uk
[3] School of Electrical and Mechanical Engineering, Pingdingshan University, Pingdingshan, China
[4] College of Engineering, Swansea University, Swansea, UK

Abstract. Human hand motions analysis is an essential research topic in recent applications, especially for dexterous robot hand manipulation learning from human hand skills. It provides important information about gestures, moving, speed and the control force captured via multimodal sensing technologies. This paper presents a comprehensive discussion of the nature of human hand motions in terms of simple motions, such as grasps and gestures, and complex motions, *e.g.* in-hand manipulations and re-grasps. And then, a novel multimodal sensing based hand motion capture system is proposed to acquire the sensory information. By using an adaptive directed acyclic graph algorithm, the experimental results show the proposed system has a higher recognition rate compared with those with individual sensing technologies.

Keywords: Multimodal sensing · EMG · Contact force · Data glove · Support vector machine

1 Introduction

With the rapid development in the areas of artificial intelligence (AI) and robotics, and the related sensing technologies, various robots, such as humanoid robots, cleaning robots and porter robots, have been used to work in complex, heavy, and dangerous environments. Moreover, the increasing labor cost is another important factor for promoting the rapid progress of research in robotics. These new applications require dexterous robots to perform increasingly human-like manipulation tasks, such as re-grasps, rotation and the translation of objects with one hand (in-hand manipulation). However, the current level of development of a sophisticated multi-fingered robot hand is only in the beginning stages, because of its control complexity, and the immature synchronous cooperation between sensor-motor systems [1]. Hence, how to intensively investigate in-hand manipulation skills of human, and further transfer such skills into bionic multi-fingered dexterous robotic hand have been receiving considerable attention.

Y. Huang et al. (Eds.): ICIRA 2017, Part I, LNAI 10462, pp. 450–461, 2017.
DOI: 10.1007/978-3-319-65289-4_43

Realizing hand dexterity is a complex process, involving multimodal sensing and fine motor control. Only exacting skeleton postures from human hands is not enough for human hand motions (HHMs) analysis, more characteristics of HHMs, such as force, speed are also very important. As an essential research topic of autonomous dexterous robot hand manipulation, multimodal sensing based HHMs analysis is attracting more and more researchers in the areas of neuroscience, biomedical engineering, robotics and so on. Liu introduced a cycle of hand skill transfer from analysing and capturing the human hand motions to hand-centered applications in a computational context [2]. A large number of hand motions were reviewed and analysed to provide the natural human-computer interaction based on multiple algorithms [3,4]. Wheatland et al. provided an overview of the research and technologies for creating the hand and finger animation [5]. In addition to the literatures mentioned above, a large number of related scientific papers as well as technical demonstrations have already appeared in various robot journals and conferences.

Current hand motion capturing systems can be mainly categorised into: data glove based capturing, attached force based capturing, EMG based capturing, optical markers based capturing and vision based capturing. Because of the complex properties involved in the human hand motions, hand motion capture will be faced with some additional challenges, like large posture variations, different colored skin and severe occlusions of the fingers during movements. It is crucial to compensate the drawbacks of uni-modal sensors by using the combination of varied types of sensors. For instance, data gloves are not have the basic ability of the spatial location information collection, and thus vision based sensors like Kinect, Leap motion controller can be integrated into motion capture system to compensate the limitation of data glove. Some researchers have done a great work on the integration of the aforesaid multiple sensory information for analysing the hand motions. Ju et al. proposed a generalized framework integrating multiple sensors to study and analyze the hand motions that contained multimodal information [6]. Marin et al. proposed a novel hand gesture recognition scheme explicitly targeted to the hand motions analysis based on the vision sensors [7]. However, most references show that multiple sensory information is integrated through one or two types of sensors. There are few researchers to analyze the hand motions by using the integration of muscle signals with the finger trajectories and the contact forces.

This paper aims to provide a comprehensive classification of current hand motions taxonomy, as well as an analysis of motion recognition based on multimodal sensing. The rest of the paper is organized as follows. First, the natural classification of the human hand motion mechanism is reviewed and discussed, followed by a detailed description of human hand motions from two aspects in Sect. 2. Then, Sect. 3 presents a multimodal sensing based hand motion capture system, which includes EMG, data glove, and Finger Tactile Pressure Sensing (FingerTPS) for hand motion recognition. In addition, The capability of the proposed hand motion recognition algorithm and the comparative experimental results have been investigated in Sect. 4. Finally, Sect. 5 concludes this paper with further discussion.

2 Human Hand Motions

Human hand is one of the most complex and dexterous motor system of human body for communication and interaction [8]. Different hand movements include fingers of different numbers. Fingertips are used to maintain the grasp stability by using proper normal forces and tangential forces. That is why it is so important to find effective ways to design human-like robots and prosthetic hands for different tasks through learning and modeling of human hand skills. In order to design a dexterous human-like hands, the classification of hand motions is an initial and important research question. Elliott et al. first developed a detailed classification system to describe four broad classed of HHMs [9]. Exner proposed five types of HHMs based on the earlier work of Elliott and Connolly [10]. Pont et al. presented an adaptation of Exner's classification system and further described six types of HHMs [11]. More recently, Fougner et al. proposed a multimodal approach using surface electromyograph (sEMG) and accelerometers to realize eight classes of HHMs [12]. Bullock et al. designed a hand-centric and motion-centric classification scheme to create a descriptive framework [13], which was used to effectively depict HHMs during manipulation in a variety of contexts, and to be combined with existing object centric or other taxonomies to provide a complete description of a specific manipulation task. In this paper, hand motions are classified into two general categories based on the motion complexity: simple hand motions, including various grasps and gestures, and complex hand motions, such as the dynamic motions and in-hand manipulation.

2.1 Simple Hand Motions

Simple motions are very common in the real world, including grasp, lift, hold, put, rotation and various gestures. Most hands-on work can be done with simple motions: opening a cup to drink, operating a mouse, picking up a mobile phone to enter the password and so on. These motions are completed through one or several types of sub-actions and finger primitives. In order to analyse and recognize these simple motions for robot learning, researchers have processed variant degree research. Mitra et al. provided a survey on gesture recognition with particular emphasis on hand gestures [14]. They categorized five human gestures to describe the HHMs: gesticulation, sign languages, language-like gestures, pantomimes and emblems. Although this classification presents a summary of hand motions, there is no specific description of in-hand manipulation, such as in-hand object shifting.

A detailed taxonomy of hand manipulation was presented based on different behavior standard, such as object contact, grasps [15]. By using hand-centric and motion-centric taxonomy, a natural hand interaction with external objects was described, and fifteen subclasses of HHMs were defined. These subclasses were applied to "instantaneous" or "discrete" hand movements rather than longer time sequences, so HHMs were described as the composition of smaller subclasses. However, the detailed information of hand configuration or the object interaction is not captured, which could be solved by adding other subclasses, such as grasp. Five types of simple motions are proposed based on multi-fingered configuration as shown in Table 1. By analyzing the simple motions, it is easy to acquire some characteristics of hand and segment the motions into some sub-sections, but they are limited in complex or advanced tasks, and complex messages are not be transfered well. The complex motions will be discussed in the next session for purpose of solving these problems.

Table 1. Two proposed motion categories

Simple motions	
Classification	Examples
Static gestures	Victory sign
	Pointing a finger projection
	Thumb up
Touching	Pressing a button
	Pushing a closed door
	Sliding a pen on the table
	Flipping a light switch
Stable grasps	Holding a phone
	Grasping a coin on the palm
	Writing with a pencil
	Cutting a paper with a scissor
Simple shift	Lifting a water glass
	Pushing a key into a keyhole
	Taking a book from a shelf
	Putting a cover on
Rotating an object in hand	Screwing/unscrewing jar lids
	Rolling pingpang among fingertips
	Turning doorknob
	Spinning a small top
Complex motions	
Classification	Examples
Dynamic hand movements	Sign language movements
	Finger gymnastic
Complex shift	"Walking" fingers down the shaft of a pen to position it for write
Complex rotation	Turning over coins in-hand
	Spinning a pencil like a "helicopter" in the fingertips
Two-hand cooperation	Carrying a box with two-hand
	Telerobotic remote surgical service

2.2 Complex Hand Motions

Researchers have made a further comprehensive study based on the remarkable features, including multi-fingered movements, wrist movement cooperating closely with in-hand manipulation and changes in the hand's location and posture. Using the wavelet transform of multi-channel EMG signals, six complex finger movements were classified for dexterous hand prosthesis control in [16]. Ju et al. proposed and evaluated methods of nonlinear feature extraction and nonlinear classification to identify different hand manipulations [17]. By using the expectation maximization algorithm, Lu et al. identified the features of several in-hand manipulations, and recognized these hand manipulation signals based on BP neural network and support vector machine classifiers [18]. Complex HHMs show more flexible and dexterous human in-hand operations, so it is more difficult to describe the process for multi-fingered manipulation.

Dynamic gestures are one of the characteristics of complex HHMs. Based on temporal relationships, the dynamic gestures change continuously with respect to hand's location, and the related messages can be obtained in the temporal sequence through hand trajectories, orientations, the fingers' shape and flex angles. However, dynamic gestures just realize the flexible changes of hand. Human in-hand object manipulation consists of a sequence of sub-motions. During the process of manipulation, the correct sub-motion is selected based on the related parameters of object such as size, weight, shape and texture [19]. Complex shift combines shift with sequential patterns movements of an object where participating fingers move independently from each other to form a distinct pattern. In addition, discontinuous movements occur when repositioning some fingers on the object, while the others move together. It is evident that the thumb moves independently of other fingers, and all the involved fingers act as one single part to control the object for simple rotation. In this case, manipulation will create rotatory action which usually generates rotation around a single axis in the appropriately shaped objects. For the complex rotation, an object will be rotated around one or more of its axes. The participating fingers and the thumb are required to execute isolated and independent finger movements to complete a rotation. In addition to the above complex motions, two-hand cooperation is another important type of HHMs. This kind of interaction requires precise physical models so as to allow interaction among users who are manipulating objects at the same time. Advanced manipulations have been widely applied to telerobotics and surgical applications.

3 Hand Motion Capture System

The hand motion capture system consists a CyberGlove, a wireless tactile force measurement system from FingerTPS and a high frequency EMG capture system with Trigno Wireless Sensors, as shown in Fig. 1. It can capture the finger trajectories, the contact force signals, and the muscle signals simultaneously. Next the system architecture and the preprocessing module will be presented, followed by the data capturing in the end.

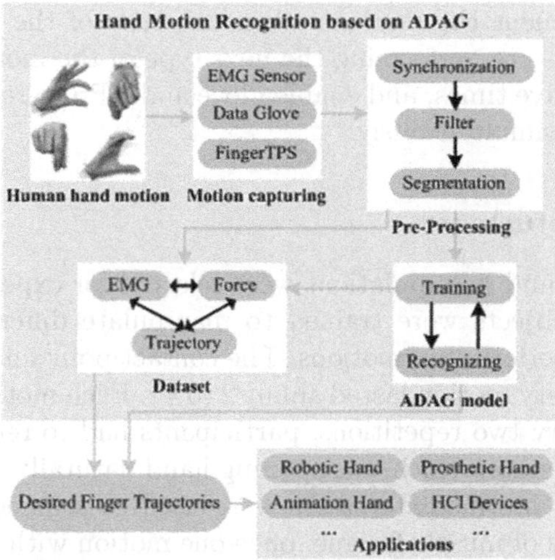

Fig. 1. Framework of multimodal sensing hand motion capture system, modified from [6].

3.1 System Configuration and Synchronization

The EMG capture system uses the Trigno wireless sensors, and has 16 EMG channels and 48 accelerometer analog channels for motion capture. The CyberGlove I is a fully instrumented glove, which provides up to 22 high-accuracy joint-angle measurements. The Finger Tactile Pressure Sensing (FingerTPS) system utilizes highly sensitive capacitive-based pressure sensors to reliably quantify forces applied by the human hand. A high-speed digital signal processor (DSP) has been used to acquire, process and send raw synchronized information digitally to a PC for analysis. The CPU speed of the DSP is far greater than 10 MHz for a fast and efficient data acquisition. We utilize USB whose maximum data transfer rate is 10 megabits per second to realize the interface connection between the DSP and the computer. The resolution is set to 16 bits, and the three devices are sampled simultaneously.

3.2 Motion Segmentation

Motion segmentation is the key issue to separate the current motion with the next motion in the same type. By extracting the features of the multimodal data collected from human demonstrations of manipulation tasks, segmentation of the action phases and trajectory classification was accomplished. Hand manipulation tasks typically take the intermediate state with short periods of time as a flag for the start and end of the action. Hence, we define the flat hand with no strength as the intermediate state. The motion begins when the finger angles change from the intermediate state, and ends when the finger angles change to the intermediate state. In this way, we utilize five-quick-grasp generated in the experiments to segment the motions when one type of the motions is finished and the participants are performing the next type of the motions. It will enable muscle to contract five times, and enable glove and TPS to record five maximum in the trajectories, simultaneously.

3.3 Motion Capturing

Ten commonly in-hand manipulations were selected for experimental data sampling, and eight subjects were trained to manipulate different objects. Every participant would perform ten motions. The contact points had been decided for every object and every motion lasted about 2 to 4 s. Each motion was repeated 10 times. Between every two repetitions, participants had to relax the hand for 2 s in the intermediate state, which was opening hand naturally without any muscle contraction. These intermediate states were used to segment the motions. To overcome the effects of muscle fatigue, once one motion with ten repetitions was finished, participants had to relax the hand for 2 min before the next motion started.

4 Hand Motion Recognition Method

Support Vector Machines (SVMs) are a novel large margin classifier used for classification and regression. SVMs map vectors onto a much higher dimensional space, and set up a maximum margin hyperplane that divides the clusters of vectors. New objects are then located into the same space and recognized to belong to a class based on which area they fall in. This method can minimize the upper bound of generalization error and provide excellent generalization ability, so it is effective in high dimensional space and compatible with different kernel functions specified for the decision function. More details about the theory of SVMs can be referred in [20–22].

Most of the problems encountered in reality require the discrimination for more than two categories, how to effectively extend SVM for multiclass classification is still an ongoing research issue. To solve these problems, a number of classification models were proposed, such as one-versus-rest approach, one-versus-one, direct multiclass SVM, and decision directed acyclic graph SVM (DAGSVM). In this paper, we use an novel adaptive directed acyclic graph (ADAG) [23]. This method improves the accuracy and reliability of DAGSVM, and avoids the error accumulation, especially in case of data sets with a large number of classes. A more detailed description of ADAG is summarized as follows.

DAGSVM has been shown to resolve the unclassifiable problem well, and have satisfactory results in [24]. However, error accumulation is the main problem, another drawback is that the final output is highly dependent on the ordering of the nodes. In this case, an adaptive DAG method is proposed to reduce the dependency on the sequence of nodes and lowers the depth of the DAG, and consequently the number of node evaluations for a correct class.

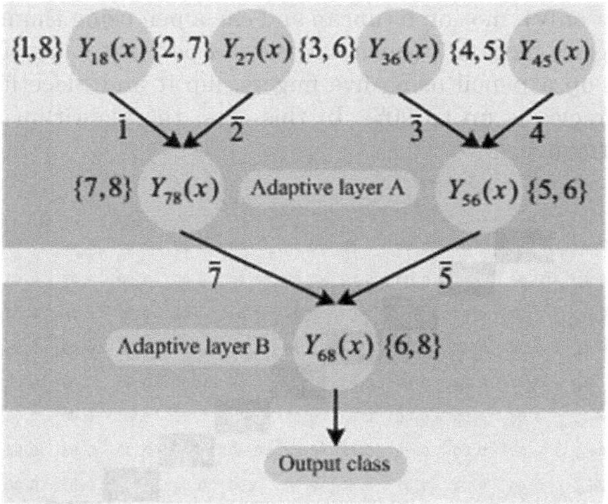

Fig. 2. Adaptive directed acyclic graph classification.

ADAG uses a decision tree with a reversed triangular structure in the testing stage. Training of a ADAG is the same as DAGSVM. In a k-class problem, it still contains $k * (k - 1) / 2$ binary classifiers, and has $k - 1$ internal nodes. An example is shown in Fig. 2, the nodes are arranged in a reversed triangle with $k/2$ nodes at the top, $k/2^2$ nodes in the second layer and so on until the lowest layer of a final node.

ADAG will start at the top level, the binary function at each node is evaluated. Like the DAGSVM, the node is then exited via the outgoing edge with a message of the preferred class. With the continued operation, the number of candidate classes will be reduced by half in each round, and then the ADAG process repeats until reaching the final node at the lowest level.

5 Experiment and Validation

In our research, we assess the recognition rates of ten hand motions gathered from the combined sensors, as shown in Fig. 3. The proposed method gives a high average recognition rate of 94.57%, indicating the capability of the proposed hand motion recognition algorithm. The recognition rates are different for each motion. Specifically, we find that the ADAG method presented a perfect performance when identifying motion 1 (grasp and lift a book using five fingers with the thumb abduction) and motion 3 (grasp and lift a can full of rice using five fingers with the thumb abduction). For motion 2 (grasp and lift a can full of rice using thumb, index finger, and middle finger only), motion 4 (grasp and lift a big ball using five fingers), motion 5 (grasp and lift a disc container using thumb and index finger only), motion 6 (uncap and cap a pen using thumb, index finger, and middle finger), motion 7 (open and close a pen box using five fingers) and motion 8 (pick up a pencil using five fingers, flip it and place it on the table), all of the accuracies is up to 90%. In this case, the algorithm also reveals an excellent performance.

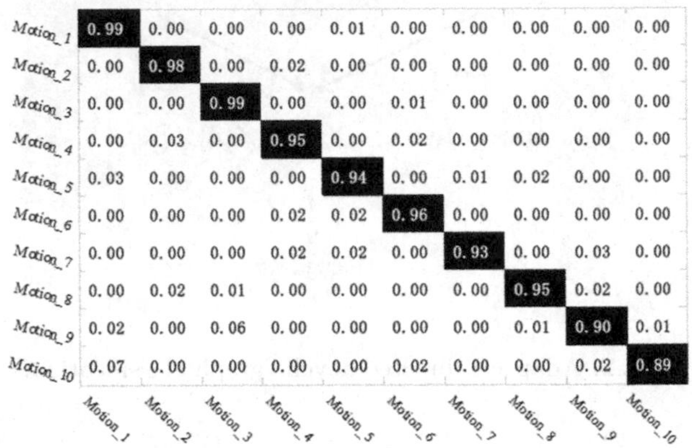

Motion_1 0.99 0.00 0.00 0.00 0.01 0.00 0.00 0.00 0.00 0.00
Motion_2 0.00 0.98 0.00 0.02 0.00 0.00 0.00 0.00 0.00 0.00
Motion_3 0.00 0.00 0.99 0.00 0.00 0.01 0.00 0.00 0.00 0.00
Motion_4 0.00 0.03 0.00 0.95 0.00 0.02 0.00 0.00 0.00 0.00
Motion_5 0.03 0.00 0.00 0.00 0.94 0.00 0.01 0.02 0.00 0.00
Motion_6 0.00 0.00 0.00 0.02 0.02 0.96 0.00 0.00 0.00 0.00
Motion_7 0.00 0.00 0.00 0.02 0.02 0.00 0.93 0.00 0.03 0.00
Motion_8 0.00 0.02 0.01 0.00 0.00 0.00 0.00 0.95 0.02 0.00
Motion_9 0.02 0.00 0.06 0.00 0.00 0.00 0.00 0.01 0.90 0.01
Motion_10 0.07 0.00 0.00 0.00 0.00 0.02 0.00 0.00 0.02 0.89
Motion_1 Motion_2 Motion_3 Motion_4 Motion_5 Motion_6 Motion_7 Motion_8 Motion_9 Motion_10

Fig. 3. Confusion Matrix for the ten motions using ADAG, where the total accuracy is 94.57%.

There are some reasons for good results. Firstly, more feature information is extracted to better capture hand movements for recognition. Secondly, five-quick-grasp and four-quick-grasp models identified using peak-detection algorithm, are utilized in the experiments to protect the integrity of raw data. Thirdly, the optimal kernels and parameters obtained will be conducive to gaining a better recognition rate, as well as the standard deviation of the accuracy.

Figure 4 shows the comparative experimental results and their variances of eight subjects using different sensor types. For different subjects performing all motions, the EMG based has the lowest average recognition rate of 85.55%, the data glove based average recognition rate is 88.30%, while the fingerTPS based average recognition rate is 89.20%. Compared with the results by using single sensor, multi-sensor based has the highest accuracy. However, the recognition time increased because the extracted keypoints increased from hand motions. On the other hand, the identification of motion 9 (hold and lift a dumbbell) and 10 (grasp and lift a cup using thumb, index finger, and middle finger) also performs well, but it indicates the relatively lower discrimination compared with other motions. The main reason may be caused by the noise when collecting the EMG and finger angle information during two-hand cooperation and complex in-hand manipulation.

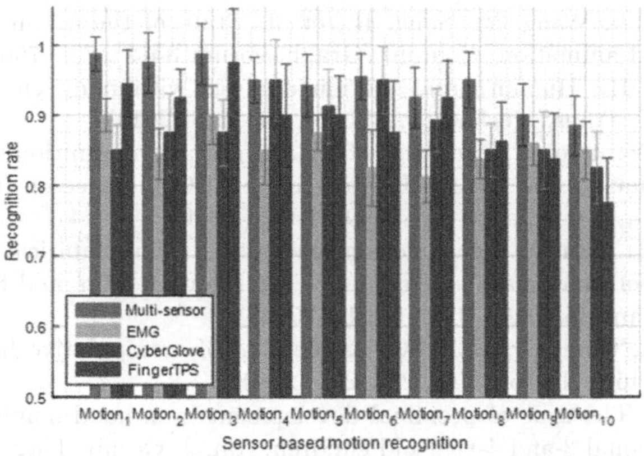

Fig. 4. Comparative experimental results.

6 Concluding Remarks

This paper provides a comprehensive human hand motion taxonomy, and proposes a novel hand motion capture system based on multimodal sensing techniques. As a novel and effective recognition method, AGAG has been used to recognize ten common hand motions from eight different subjects. From the experimental data, the recognition rate with multiple data fusion is up to 94.57%, which is much higher than individual sensor based. In future work, we will implement robust and real time hand motion recognition captured from vision based and contact based capturing systems, and further to be integrated with robots to serve as a friendly and natural human-machine interaction, and prosthetic hands control [25, 26].

Acknowledgment. This work is partially supported by Natural Science Foundation of China (Grant No. 51575412), the Fundamental Research Funds for the Central Universities (Grant No. 2016-JL-011). Also, the authors would like to acknowledge the reviewers for their valuable comments and suggestions that helped to improve the quality of the manuscript.

References

1. Saudabayev, A., Varol, H.A.: Sensors for robotic hands: a survey of state of the art. IEEE. Access **3**(9), 1765–1782 (2015)
2. Liu, H., Nguyen, K.C., Perdereau, V., et al.: Finger contact sensing and the application in dexterous hand manipulation. Auton. Robots **39**(1), 25–41 (2015)
3. Khan, R.Z., Ibraheem, N.A.: Survey on gesture recognition for hand image postures. Comput. Inf. Sci. **5**(3), 110–121 (2012)
4. Liu, H., Ju, Z., Ji, X., Chan, S., Khoury, M.: Human Motion Sensing and Recognition: A Fuzzy Qualitative Approach. Springer, Heidelberg (2017)

5. Wheatland, N., Wang, Y., Song, H., et al.: State of the art in hand and finger modeling and animation. Comput. Graph. Forum **34**(2), 735–760 (2015)
6. Ju, Z., Liu, H.: Human hand motion analysis with multisensory information. IEEE/ASME Trans. Mechatron. **675**(5), 456–466 (2014)
7. Marin, G., Dominio, F., Zanuttigh, P.: Hand gesture recognition with leap motion and kinect devices. In: 2014 IEEE International Conference on Image Processing (ICIP), Paris, France, pp. 1565–1569. IEEE (2014)
8. Ju, Z., Liu, H.: Human hand motion recognition using empirical copula. In: 2010 IEEE/RSJ International Conference on Intelligent Robots and Systems (IROS), Taipei, Taiwan, pp. 4625–4630. IEEE (2010)
9. Elliott, J.M., Connolly, K.J.: A classification of manipulative hand movements. Dev. Med. Child Neurol. **3**(6), 283–296 (1984)
10. Exner, C.E.: The zone of proximal development in in-hand manipulation skills of nondysfunctional 3-and 4-year-old children. Am. J. Occup. Ther. **44**(10), 884–891 (1990)
11. Pont, K., Wallen, M., Bundy, A.: Conceptualising a modified system for classification of inhand manipulation. Aust. Occup. Ther. J. **56**(2), 2–15 (2009)
12. Fougner, A., Scheme, E., Chan, A.D.C., et al.: A multi-modal approach for hand motion classification using surface EMG and accelerometers. In: 2011 Annual International Conference of the IEEE Engineering in Medicine and Biology Society, EMBC, Boston, USA, pp. 4247–4250. IEEE (2011)
13. Bullock, I.M., Dollar, A.M.: Classifying human manipulation behavior. In: IEEE International Conference on Rehabilitation Robotics (ICORR), Zurich, Switzerland, pp. 1–6. IEEE (2011)
14. Mitra, S., Acharya, T.: Gesture recognition: a survey. IEEE Trans. Syst. Man Cybern. Part C (Appli. Rev.) **33**(4), 311–324 (2007)
15. Bullock, I.M., Ma, R.R., Dollar, A.M.: A hand-centric classification of human and robot dexterous manipulation. IEEE Trans. Haptics **6**(9), 129–144 (2013)
16. Jiang, M.W., Wang, R.C., Wang, J.Z., et al.: A method of recognizing finger motion using wavelet transform of surface EMG signal. In: 27th Annual International Conference on Engineering in Medicine and Biology, Shanghai, China, pp. 2672–2674. IEEE (2006)
17. Ju, Z., Ouyang, G., Wilamowska-Korsak, M., et al.: Surface EMG based hand manipulation identification via nonlinear feature extraction and classification. IEEE Sens. J. **13**(4), 3302–3311 (2013)
18. Lu, Y., Lu, G., Bu, X., Yu, Y.: Classification of hand manipulation using bp neural network and support vector machine based on surface electromyography signal. IFAC-PapersOnLine **45**(12), 869–873 (2015)
19. Kappassov, Z., Corrales, J.A., Perdereau, V.: Tactile sensing in dexterous robot hands - Review. Robot. Autonom. Syst. **74**(7), 195–220 (2015)
20. Burges, C.J.C.: A tutorial on support vector machines for pattern recognition. Data Min. Knowl. Disc. **2**(6), 121–167 (1998)
21. Abe, S.: Support Vector Machines for Pattern Classification. Springer, London (2005)
22. Ben-Hur, A., Weston, J.: A users guide to support vector machines. Data Min. Tech. Life Sci. **609**(10), 223–239 (2010)
23. Kijsirikul, B., Ussivakul, N.: Multiclass support vector machines using adaptive directed acyclic graph. In: Proceedings of the 2002 International Joint Conference on Neural Networks, IJCNN 2002, Honolulu, HI, USA, pp. 980–985. IEEE (2002)

24. Platt, J.C., Cristianini, N., Shawe-Taylor, J.: Large margin DAGs for multiclass classification. In: Advances in Neural Information Processing Systems, pp. 547–553 (1999)
25. Ju, Z., Ji, X., Li, J., Liu, H.: An integrative framework of human hand gesture segmentation for human robot interaction. IEEE Syst. J. **99**(9), 1–11 (2015)
26. Ju, Z., Gao, D., Cao, J., Liu, H.: A novel approach to extract hand gesture feature in depth images. Multimedia Tools Appl. **75**(19), 11929–11943 (2017). Springer, Heidelberg

Static Hand Gesture Recognition with Parallel CNNs for Space Human-Robot Interaction

Qing Gao[1,2], Jinguo Liu[1(✉)], Zhaojie Ju[4], Yangmin Li[1,3(✉)], Tian Zhang[1], and Lu Zhang[5]

[1] The State Key Laboratory of Robotics, Shenyang Institute of Automation, Chinese Academy of Sciences, Shenyang 110016, China
liujinguo@sia.cn
[2] University of the Chinese Academy of Science, Beijing 100049, China
[3] The Department of Electromechanical Engineering, University of Macau, Taipa, Macao 999078, China
ymli@umac.mo
[4] School of Computing, University of Portsmouth, Portsmouth PO1 3HE, UK
[5] Key Laboratory of Space Utilization, Technology and Engineering Center for Space Utilization, Chinese Academy of Sciences, Beijing 100094, China

Abstract. As a new type of human-robot interaction (HRI), hand gesture has many advantages such as natural operation, rich expression and not subject to environ-mental constraints. So it is very suitable for space human-robot interaction tasks in special and harsh environment. Considering that static hand gesture is one of the main gesture expressions in human-computer interaction, so a parallel convolution neural networks (CNNs) is designed to improve the accuracy of static hand gesture recognition in the conditions of complex background and changing illumination. In addition, the method is applied to the operation of space human-robot system with hand gesture control. Various space HRI hand gestures from different subjects are evaluated and tested, and experimental results demonstrate that the proposed method outperforms the single-channel CNN methods and other popular methods with a higher accuracy.

Keywords: HRI · Hand gesture recognition · CNN · Space robot

1 Introduction

The traditional HRI methods have the shortcomings such as complicated and unnatural operation, and subject to environmental constraints. Especially for space HRI tasks, the characteristics such as microgravity environment, multi-degree-of-freedom space robots and multiple spatial ranges bring great challenges to the traditional HRI [1]. With the development of computer and AI, space HRI has gradually shifted from robot-centered methods to astronaut-centered ones. Among the new interactive ways, hand gesture interaction is a major interactive way [2, 3]. Its advantages, such as natural, intuitive and not subject to environmental constraints are very suitable for space HRI tasks. Hand gesture recognition is the core technology of hand gesture HRI. Therefore, many

Y. Huang et al. (Eds.): ICIRA 2017, Part I, LNAI 10462, pp. 462–473, 2017.
DOI: 10.1007/978-3-319-65289-4_44

scholars have studied the hand gesture recognition deeply. The current hand gesture recognition technology has been a significant development. But for the visual-based hand gesture recognition under the conditions of complex background and changed illumination. There are still a lot of difficulties and challenges.

The development of hand gesture recognition has gone from wearing-device-based hand gesture recognition to vision-based hand gesture recognition. Among them, data glove as a typical representative of the wearing device, has been developed very mature with a high degree of hand gesture recognition accuracy. Vision-based hand gesture recognition is still in the development stage. With the emergence of depth sensors such as Kinect, the method of fusing RGB images and depth images together has grown up to be a mainstream research direction. In the aspect of recognition algorithm, the traditional visual gesture recognition algorithms include Dynamic Time Warping (DTW), Artificial Neural Network (ANN) and Hidden Markov Model (HMM). But these methods have some problems to some extent. With the development and success of deep learning (DL) in the field of image recognition, more and more scholars have begun to apply the DL method to the study of visual-based hand gesture recognition and have made some research results [4, 5].

Takayoshi Yamashita proposed a hand gesture recognition method based on bottom-up structured deep CNN with curriculum learning. This method identified hand gestures under clutter background with illumination changes. The network contains a binarization layer to obtain the intermediate information and outputs binary images. This made the complex hand gesture recognition issue into binarization and classification. And this method made the average recognition accuracy of the 6 gestures reach 88.78% [6]. Pavlo Molchanov aimed at driver gestures, proposed a dynamic hand gesture recognition method with 3D CNN. The network was designed as a multi-channel architecture, and the recognition accuracy reached 77.5% on the VIVA challenge dataset [7]. Jawad Nagi aimed at the gesture interaction with a mobile robot, proposed a deep neural network combining convolution and max-pooling (MPCNN), and the recognition accuracy reached 96% of 6 hand gestures with colored gloves [8]. Hsien-I Lin proposed a CNN method to detect the hand gestures. The hand color model and hand pose were obtained as the training data. And the recognition accuracy reached 95.96% for 7 hand gestures [9]. Vijay John designed a recurrent convolution network (LRCN) for the control of a smart car. The recognition accuracy reached 91% and the average recognition time for each video sequence reached 110 ms [10].

In this paper, in order to improve the static hand gesture recognition accuracy under complex background and changing illumination, the color image and the depth image of the hand gesture are fused together. A two-column parallel network is developed. In this network, the deep convolution neural networks are developed for RGB images and depth images, respectively. And then the classification result of the RGB image is fused with the classification result of the corresponding depth image to obtain the final predicted results. In the experiments, the designed method and the single-channel CNN methods are verified on the American Sign Language (ASL) database, respectively. Then, the experiment result of the parallel CNNs is compared with the results of single-channel CNN methods and other scholars' methods.

The rest of this paper is structured as follows: in Sect. 2, the parallel CNNs proposed in this paper is introduced in detail. In Sect. 3, for the hand gesture recognition of the space HRI, the hand gesture recognition experiments are carried out using the above method. In Sect. 4, present the conclusions.

2 Proposed Method

Hands are non-rigid objects. The sizes, shapes and colors of different people's hand are not the same. Coupled with the complex background and changed illumination, make the hand gesture recognition as a multifarious task. Hand gesture has many information such as color, shape and depth. Using a variety of information fusion approach can make a higher recognition accuracy [11, 12].

Therefore, for the static hand gesture recognition, this paper designs a parallel CNNs from the perspective of information fusion. One of the channels has a CNN for RGB hand gesture images, the color of the hand as the main information to identify the hand gestures. The other one has a CNN for depth hand gesture images, the depth of the hand as the main information to identify the hand gestures. Finally, the outputs of the two channels are fused to achieve the final predictions of gesture semantics. The ultimate recognition accuracy of hand gesture will be higher than that of the single-channel CNN because the recognition results of two different characteristics are merged. The structure of the parallel CNNs is shown in Fig. 1.

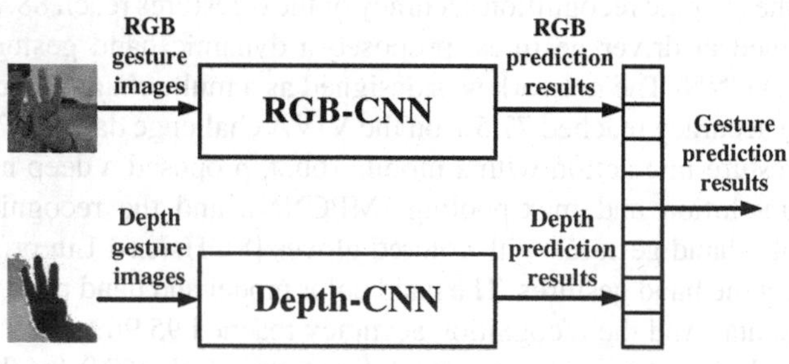

Fig. 1. The structure of the parallel CNNs

The dataset, network structure and network training method of the parallel CNNs are described as follows.

2.1 Dataset

The ASL database is selected as the hand gesture recognition database in this paper. The ASL database is presented in Fig. 2. The database contains a total of 24 hand gestures, representing the 24 letters (except J and Z, because J and Z are dynamic hand gestures). Each gesture contains 5000 sample images, which are 2500 RGB images and 2500 depth images, complete by 5 people in different backgrounds and different illumination. So there are a total of 120000 pictures, which are 60000 RGB images and 60000 depth images [13].

Fig. 2. The ASL database

2.2 Network Structure

The network structure of the proposed method is illustrated in Fig. 3. The network includes two sub-networks, which are RGB-CNN sub-network and Depth-CNN sub-network. The two subnetworks run in parallel. The semantics of RGB images and the corresponding depth images are predicted respectively. Then, the prediction results are merged to obtain the final prediction results [14]. The two sub-networks and the fusion method of forecast results are shown as follows.

2.2.1 RGB-CNN Sub-network

As Fig. 3 shows, the RGB-CNN sub-network has a total of 7 layers, the first four layers are convolution layers, the latter two layers are fully-connected layers, the last layer is a softmax layer. Specifically, the input is a RGB image with a size of 100 × 100. It translates into 9 feature maps with a size of 48 × 48 by passing through a convolution layer with 9 convolution kernels with a size of 5 × 5, a rectified linear unit (ReLU) layer and a max-pooling layer with a size of 2 × 2. Then, they translate into 18 feature maps with a size of 22 × 22 by passing through a convolution layer with 18 convolution kernels with a size of 5 × 5, a ReLU layer and a max-pooling layer with a size of 2 × 2. Next, they translate into 36 feature maps with a size of 9 × 9 by passing through a convolution layer with 36 convolution kernels with a size of 5 × 5, a ReLU layer and a max-pooling layer with a size of 2 × 2. And then, they translate into 72 feature maps with a size of 5 × 5 by passing through a convolution layer with 72 convolution kernels with a size of 5 × 5 and a ReLU layer. After that, they are followed by a fully-connected layer with 144 neurons and a fully-connected layer with 72 neurons. Finally, the softmax method is used for classification to obtain prediction probabilities for semantic hand gestures [15].

All the layers in the sub-network, except for the softmax layers, have ReLU layers. And its activation function is:

$$f(z) = \max(0, z) \tag{1}$$

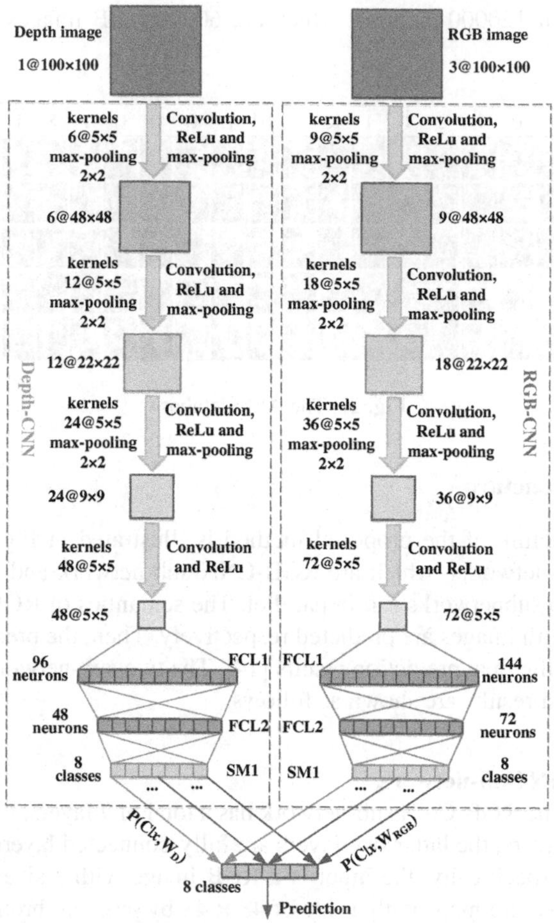

Fig. 3. The structure of the parallel CNNs

Where z is the value of output neural.

The classification layer uses the softmax method, its expression is:

$$P(C|x, \mathrm{W}) = \frac{\exp(z_C)}{\sum_q \exp(z_q)} \tag{2}$$

Where x is the observation of hand gesture. C is the neuron representing the actual semantic of the hand gesture. W is the weight parameters of the network. And z_q is the output of the neuron q.

2.2.2 Depth-CNN Sub-network

As Fig. 3 shows, the structure of Depth-CNN sub-network is similar to the structure of RGB-CNN sub-network. And finally it also gets the prediction probabilities of hand gesture semantics.

2.2.3 Results Fusion Method

In the RGB-CNN sub-network, W_{RGB} is the network parameter, x is the observation of hand gesture, thus $P(C|x, W_{RGB})$ is the prediction probability of gesture semantics C. In the Depth-CNN sub-network, W_D is the network parameter, x is the observation of hand gesture, thus $P(C|x, W_D)$ is the prediction probability of gesture semantics C. So the prediction probability of the final hand gesture classifier is:

$$P(C|x) = P(C|x, W_{RGB}) \times P(C|x, W_D) \tag{3}$$

2.3 Training

In the course of training, the data is divided into several batches for training. This can improve the efficiency of training. Set the batch size as N, so the loss function is:

$$L(W) = \frac{1}{N}\sum_{i}^{N} f_W(X^{(i)}) + \lambda r(W) \tag{4}$$

Where $f_W(x^{(i)})$ is the loss of $x^{(i)}$. Firstly, calculate the value of each individual sample x, and then sum them, and finally find the mean. $\lambda r(W)$ is the weight decay term. With the loss function, the loss and gradient can be solved iteratively to optimize the problem. In the neural network, forward pass is used to solve the loss, and backward pass is used to solve the gradient.

In this paper, the stochastic gradient descent (SGD) is used to solve the gradients. SGD updates W through a linear combination of the negative gradient $\nabla L(W)$ and the last weight update value V_t. The iteration formula is shown as follows [16].

$$V_{t+1} = \mu V_t - \alpha \nabla L(W_t) \tag{5}$$

$$W_{t+1} = W_t + V_{t+1} \tag{6}$$

Where α is the learning rate of the negative gradient. μ is the momentum of the last gradient value, and it used to weight influence that the previous gradient direction gives rise to the current gradient descent direction. These two parameters need to be tuned to get the best results, usually based on experience. t represents the current number of iterations.

The step method is chosen to adjust the learning rate. It can make the network preliminaries fast convergence and reduce the shock and become stable in the later period. The formula is displayed as follows.

$$\alpha = \alpha_0 \times \gamma^{(t/s)} \tag{7}$$

Where α_0 is the initial learning rate. γ is adjustable parameter, and it is generally set to 0.1. s represents the iterative length of the adjustment learning rate. That is to say, when the current iteration number t reaches an integer multiple of s, the learning rate will be adjusted.

3 Experiments

3.1 Hand Gesture Recognition Experiments

The experimental hardware device selected Intel Core i5-6400 CPU, NVIDIA GeForce GTX 1060 6 GB GDDR5, 16 GB RAM. The experimental development environment was selected Caffe in the Ubuntu 14.04 64 bit OS system.

Static hand gesture recognition experiments were carried out on the ASL dataset using single channel RGB-CNN, Single channel Depth-CNN and parallel CNNs respectively for the 24 hand gestures. Actual experimental hand gesture images are a total of 240000 through mirror processing. Select one of the 200000 images as training samples, the remaining 40000 images as test samples. The images input to the network after size conversion, data range changes and mean operation. The train loss, test loss, and test accuracy for the three networks were obtained. Comparisons of the three networks' experimental data are shown in Figs. 4, 5 and 6. The specific data for the test accuracy are given in Table 1.

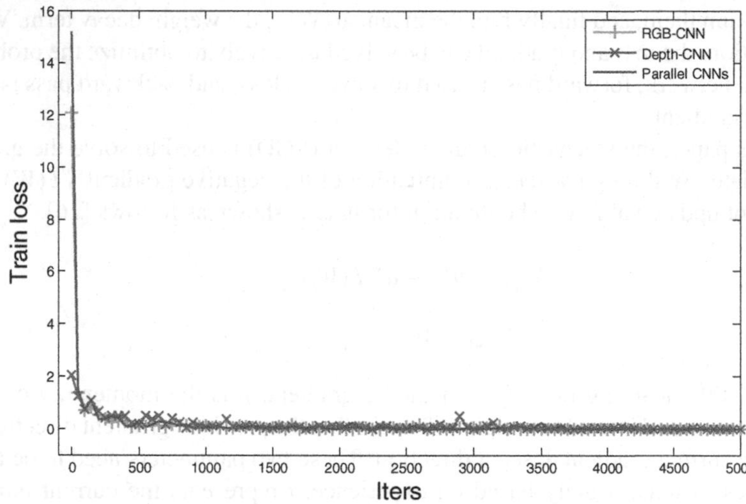

Fig. 4. The comparison of train losses

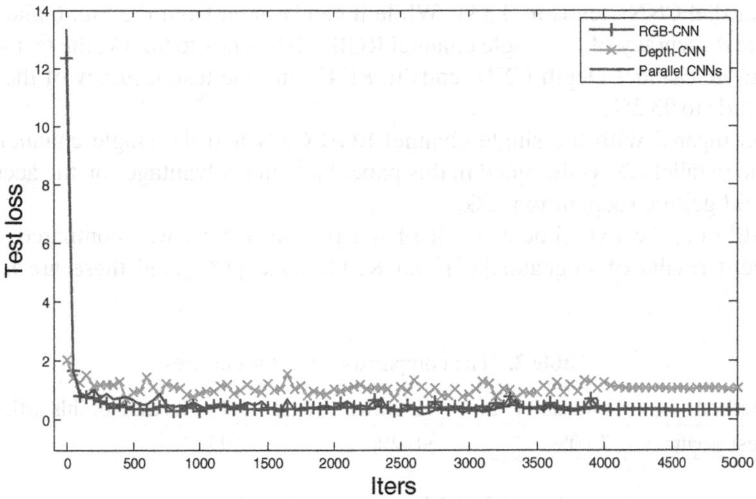

Fig. 5. The comparison of test losses

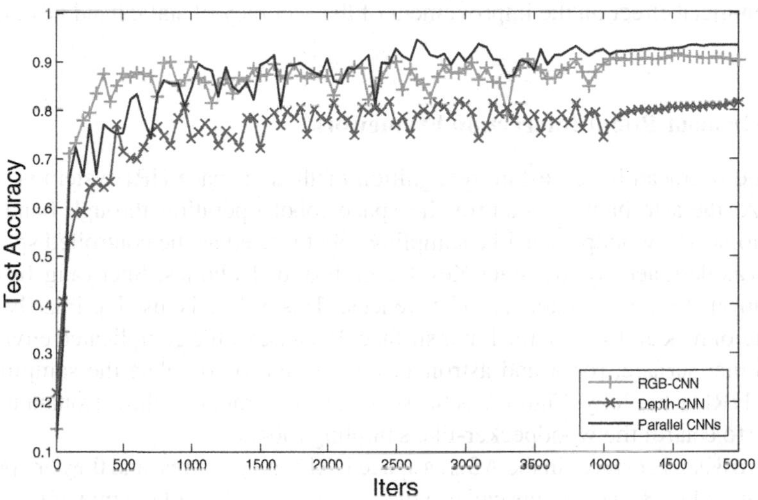

Fig. 6. The comparison of test accuracies

Table 1. The comparison of test accuracies

Method	RGB-CNN	Depth-CNN	Parallel CNNs
Test accuracy	90.3%	81.4%	93.3%

It can be seen from the Fig. 4 that the train losses of the three networks all tends to 0. It can be seen from the Fig. 5 that the test loss of the single channel RGB-CNN tends to 0.356, the test loss of the single channel Depth-CNN tends to 1.071 and the test loss of the parallel CNNs tends to 0.331. While it can be seen from the Fig. 6 and Table 1 that the test accuracy of the single channel RGB-CNN tends to 90.3%, the test accuracy of the single channel Depth-CNN tends to 81.4% and the test accuracy of the parallel CNNs tends to 93.3%.

So, compared with the single channel RGB-CNN and the single channel Depth-CNN, the parallel CNNs designed in this paper has some advantages on the accuracy of static hand gesture recognition task.

In addition, the experiment result of the parallel CNNs was compared with the experiment results of Pugeault [17] and K. Otiniano [18]. And these are shown in Table 2.

Table 2. The comparison of test accuracies

Method	Pugeault et al.	K. Otiniano et al.	The method of this article
Test accuracy	75.0%	90.2%	93.3%

These three method all used RGB images and deep image fusion methods on ASL dataset. It can be seen from Table 2 that the accuracy of the parallel CNNs proposed in this paper is higher than that of Pugeault and K. Otiniano. Therefore, the parallel CNNs has a beneficial effect on the improvement of the accuracy of static hand gesture recognition.

3.2 Astronaut-Robot Interaction Framework

Using the overhead hand gesture recognition method, a space HRI system is designed to realize the astronauts to control the space robot operation through hand gesture instructions. The woodpecker-like sampling robot is used as the controlled space robot which was designed by the State Key Laboratory of Robotics, Shenyang Institute of Automation, Chinese Academy of Sciences. This robot is used primarily for the sampling of rock and soil on the lunar surface. Because of the complicated environment on the lunar surface, robot and astronaut are needed to complete the sampling work through HRI technology. Figure 7 is the schematic diagram that the astronaut uses hand gestures to control the woodpecker-like sampling robot.

Chose 8 hand gestures in the ASL as space HRI hand gestures, and they are presented in the box of Fig. 8. Astronauts can use these gestures to control the movement or operation of the woodpecker-like sampling robot. Specific space HRI gesture semantics are shown in Table 3.

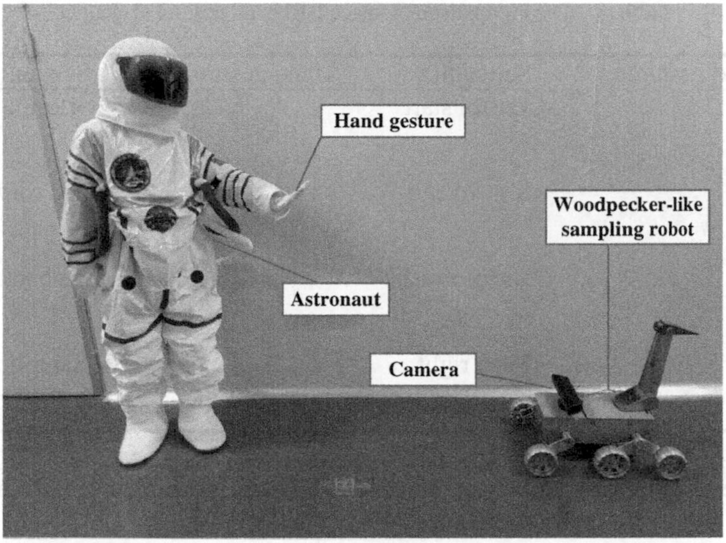

Fig. 7. The schematic diagram that the astronaut uses hand gestures to control the robot

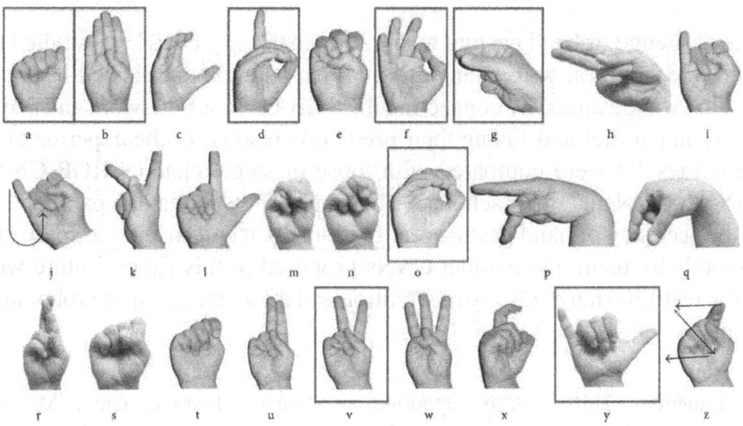

Fig. 8. ASL hand gestures

Table 3. The comparison of space HRI hand gesture semantics

Hand gestures	Semantics	Hand gestures	Semantics
	Control start		Turn left
	Control end		Move forward
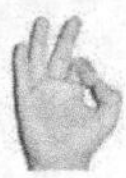	Control finish		Move backward
	Turn right		Sampling

At present, the astronaut-space robot interaction experiments have not been verified. This part will be the main work of the future.

4 Conclusions

Facing the challenge of hand gesture recognition in the space HRI, we studied the static hand gesture recognition with changing illumination and complicated background. A parallel CNNs was designed by connecting the RGB-CNN sub-network and Depth-CNN sub-network in parallel and fusing their prediction results. In the experiment part, the experimental results were compared with those of single channel RGB-CNN, single channel Depth-CNN and other scholars. The results showed that the parallel CNNs can improve the accuracy of hand gesture recognition. So, it can ensure the space HRI tasks more smoothly by using the parallel CNNs proposed in this paper. Future works will focus on the real-time hand gesture recognition and the astronaut-space robot interaction experiments.

Acknowledgments. This work is supported by Research Fund of China Manned Space Engineering (050102), the Key Research Program of the Chinese Academy of Sciences (Y4A3210301), the National Science Foundation of China (51175494, 61128008, and 51575412), and the State Key Laboratory of Robotics Foundation.

References

1. Fong, T., Nourbakhsh, I.: Interaction challenges in human-robot space exploration. J. Interact. **12**(2), 42–45 (2005)
2. Liu, J.G., Luo, Y.F., Ju, Z.J.: An interactive astronaut-robot system with gesture control. J. Comput. Intell. Neurosci. **2016** (2016)
3. Wolf, M.T., Assad, C., Stoica, A.: Decoding static and dynamic arm and hand gestures from the JPL BioSleeve. In: Proceedings of the IEEE Aerospace Conference (2013)

4. Pisharady, P.K., Saerbeck, M.: Recent methods and databases in vision-based hand gesture recognition: a review. J. Comput. Vis. Image Underst. **141**, 152–165 (2015)
5. Hasan, H., Abdul-Kareem, S.: Human-computer interaction using vision-based hand gesture recognition systems: a survey. J. Neural Comput. Appl. **25**(2), 251–261 (2014)
6. Yamashita, T., Watasue, T.: Hand posture recognition based on bottom-up structured deep CNN with curriculum learning. In: Proceedings of the 2014 IEEE International Conference on Image Processing (ICIP) (2014)
7. Molchanov, P., Mello, S.D., Kim, K., Kautz, J.: Hand gesture recognition with 3D convolutional neural networks. In: Proceedings of the IEEE Computer Vision and Pattern Recognition, CVPR 2015 (2015)
8. Nagi, J., Ducatelle, F., Caro, G.D., et al.: Max-pooling convolutional neural networks for vision-based hand gesture recognition. In: Proceedings of the 2011 IEEE International Conference on Signal and Image Processing Applications (ICSIPA) (2011)
9. Lin, H.I., Hsu, M.H., Chen, W.K.: Human hand gesture recognition using a convolution neural network. In: Proceedings of the 2014 IEEE International Conference on Automation Science and Engineering (CASE) (2014)
10. John, V., Boyali, A., Mita, S, et al.: Deep learning-based fast hand gesture recognition using representative frame. In: Proceedings of the 2016 International Conference on Digital Image Computing: Techniques and Applications (DICTA) (2016)
11. Ju, Z.J., Gao, D.X., Cao, J.T., Liu, H.H.: A novel approach to extract hand gesture feature in depth images. J. Multimedia Tools Appl. **75**(19), 11929–11943 (2015)
12. Ju, Z.J., Wang, Y.H., Zeng, W, Cai, H.B., Liu, H.H.: An integrative framework for human hand gesture segmentation in RGB-D data. IEEE Syst. J. (in Press). doi:10.1109/JSYST.2015.2468231
13. Gattupalli, S, Ghaderi, A., Athitsos, V.: Evaluation of deep learning based pose estimation for sign language recognition. In: Proceedings of the 9th ACM International Conference on PErvasive Technologies Related to Assistive Environments (2016)
14. Ciresan, D., Meier, U., Schmidhuber, J.: Multi-column deep neural networks for image classification. In: Proceedings of the CVPR, pp. 3642–3649 (2012)
15. Krizhevsky, A., Sutskever I., Hinton, G.: ImageNet classification with deep convolutional neural networks. J. Adv. Neural Inf. Process. Syst. **25** (2012)
16. LeCun, Y., Bengio, Y., Hinton, G.: Deep learning. J. Nat. **521**, 436–444 (2015)
17. Pugeault, N., Bowden, R.: Spelling it out: real-time ASL fingerspelling recognition. In: Proceedings of the IEEE International Conference on Computer Vision Workshops (ICCV Workshops), pp. 1114–1119 (2011)
18. Odriguez, K.O., Chavez, G.C.: Finger spelling recognition from RGB-D information using kernel descriptor. In: Proceedings of the 26th SIB-GRAPI-Conference on Graphics, Patterns and Images (SIBGRAPI), pp. 1–7 (2013)

Robust Human Action Recognition Using Dynamic Movement Features

Huiwen Zhang[1,2(✉)], Mingliang Fu[1,2], Haitao Luo[1], and Weijia Zhou[1]

[1] State Key Laboratory of Robotics, Shenyang Institute of Automation, Shenyang 110016, China
zhanghuiwen@sia.cn

[2] University of Chinese Academy of Science, Beijing, China

Abstract. Action recognition has been widely researched in video surveillance, auxiliary medical care and robotics. In the context of robotics, in order to program robots by demonstration (PbD), we not only need our algorithms to be capable of identifying different actions, but also to be able to encode and reproduce them. Dynamic movement primitives (DMPs), as a trajectory encoding method, are widely used in motion synthesize and generation. But at the same time it can also be applied to action recognition. With this idea, this paper extracts a kind of dynamic features from the original trajectory within DMP framework. The feature is temporal-spatial invariant. Based on the feature, FastDTW-KNN algorithm is proposed to solve the recognition task. Experiments tested on HAR dataset and handwritten letters dataset achieved an excellent recognition performance under a large data noise, which has verified the effectiveness of our method. In addition, comparative recognition experiments based on the original feature and our extracted dynamic feature are conducted. Results show that the dynamic feature is robust under temporal and spatial noise. As for classifiers, we compared our method with KNN, SVM and DTW-KNN followed with a detailed analysis of their advantages and disadvantages.

Keywords: Action recognition · DMP · DTW

1 Introduction

With the development of robot technology, we are increasingly expecting robots to help us in our daily lives. To achieve this goal, we need a simple and intuitive way to program robots. Learning from demonstration (LfD) [1] provides such a mechanism. But we met a lot of problems and challenges in LfD. First, in order to imitate human actions, the robot needs to "understand" them, which involves in action recognition. This problem has been widely researched in many different fields, such as image processing, computer vision, pattern recognition and machine learning.

If we solve action recognition using image inputs captured by vision sensor, which happens a lot in computer vision domain, we call them vision-based

Y. Huang et al. (Eds.): ICIRA 2017, Part I, LNAI 10462, pp. 474–484, 2017.
DOI: 10.1007/978-3-319-65289-4_45

methods. Generally, this kind of methods include the following three steps: (1) detecting the motion information from the image frame and extracting the underlying features; (2) modeling the behavior pattern or the action; (3) mapping low level visual features into high level semantic information like action categories. Usually quality of feature extraction is particularly important for visual-based methods. We can divide visual features into two categories: static and dynamic. Static features such as size, shape, color depth, etc., due to complex background, occlusion, lighting changes and other factors, it is difficult to ensure recognition performance. Dynamic features such as speed, trajectory, optical flow, etc., is essential for action recognition. But obtaining these features from video or image sequence is both non-trivial and difficult. Meanwhile, the quality of these features will suffer from the limitations mentioned before. Therefore, this article directly uses inertial measurement unit (IMU) or wearable sensors to obtain dynamic information, then recognition task is conducted based upon them, which we think is more reliable and suitable in robot behaviour learning domain.

Then, the action recognition problem becomes a pattern recognition problem of time series signal. For time series data analysis, HMM has obtained a large number of successful applications, especially in the field of speech recognition [2,3]. DTW [4], act as a signal preprocessing method, can also be used to action recognition when combined with other distance-based classification method. But all of these methods suffer from a high computational complexity. So in order to use a simple classifier while still maintains a high recognition performance, we need a kind of robust features. This paper constructs the robust features based on DMP, then uses FastDTW-KNN as our classifiers, which obtained a high and reliable recognition accuracy. The whole framework of our method is shown in Fig. 1.

The main contributions of this paper are as follows: (1) Constructing a kind of robust dynamic feature with DMP. The feature is easy to calculate and very suitable for recognition task; (2) The FastDTW-KNN method is proposed to classify the feature, which can improve the recognition accuracy, but also keep a low computation complexity; (3) A large number of comparative experiments are conducted to compare different classification algorithms and features.

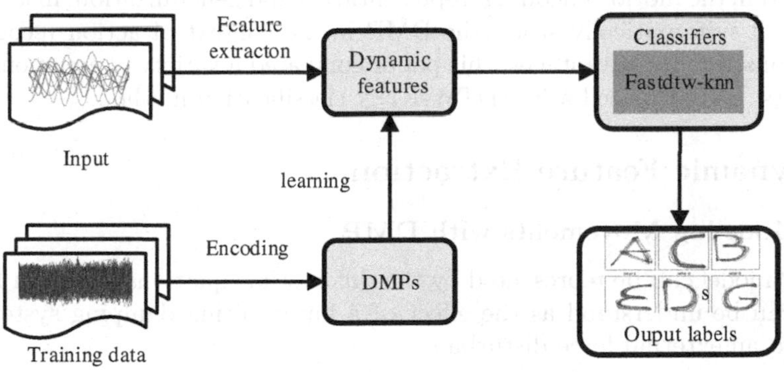

Fig. 1. The proposed recognition framework.

The subsequent chapters of this paper are organized as follows: The Sect. 2 introduces related work and the background of DMP. The Sect. 3 elaborates the method of using DMP to obtain dynamic features. Section 4 introduces our classifier followed with a lot of contrast experiments in Sect. 5. Finally, we give a brief conclusion and outlook of the future work.

2 Related Work

Most researches on action recognition focus on using computer vision methods [5,6]. Generally, they include two steps. First, extracting features from image or video data, and then a classifier is used to recognize actions in feature space. Static features is one kind of important features which we can obtain from the source data. They are mainly used for describing the size, color, edge, contour, shape and depth information of the human body. With these information, a good classifier can be constructed. Carsson et al. [7] implemented an action recognition algorithm by shape matching. Another kind of features are dynamic features, and they are considered to be a very important information in computer vision, such as optical flow, space-time features. Danafar et al. [8] used optical flow and SVM to recognize actions in the context of video surveillance. However, these dynamic feature extraction methods are complex in their own right. Besides, they are prone to be affected by environment. More importantly, in the context of robot behavior learning, we are not only seeking a mechanism for action recognition, but also hoping that it can be used for motion encoding and reproduction. DMP provides such a mechanism.

DMP was first proposed by Ijspeert, Schaal et al. [9]. Its main idea is to use our well-known second-order differential equations to represent movements, and then adjust it through a nonlinear term to achieve our desired attractor landscapes. Given the good characteristics of DMP, it is used in various fields such as control, planning, and learning [10,11]. Many improvements and extensions of DMP are also presented within a few decades, such as rhythmic DMP for gait research [12], DMP with obstacle avoidance [13], coupling multiple DMPs in Movement coordination [14] and probabilistic DMP [15]. As a powerful tool, DMP has also been used in imitation learning [16]. But all of these work are concentrated in the motion encoding, representation and generalization, instead, this paper will systematically study the DMP in the context of action recognition. To demonstrate its advantages, this paper compared a variety of commonly used classifiers, and proposed a FastDTW-KNN classification method.

3 Dynamic Feature Extraction

3.1 Modeling Movements with DMP

A DMP model can be represented by the differential equations shown in Eq. (1), which can be understood as the effect of a linear spring damping system subjected to an external force disturbance.

$$\begin{aligned} \tau\dot{v} &= \alpha_v\left[\beta_v\left(g-x\right)-v\right]+f \\ \tau\dot{x} &= v \end{aligned} \tag{1}$$

Where x and v represent position and velocity respectively; x_0 and g stand for the start and target position; τ is time scaling factor; α_v and β_v are constants used for control the damping characteristics of the system. f is a non-linear function which can generate complex movements by modifying the weights parameters in f. It is defined as follows:

$$f(s) = \frac{\sum_i w_i \Psi_i(s)}{\sum_i \Psi_i(s)} s(g - x_0) \tag{2}$$

Where, $\Psi_i(s) = exp\left(-h_i(s - c_i)^2\right)$. h_i, c_i represent the precision and center of Gaussian function.f doesn't depend on time directly, but rather depend on a phase variable s , which changes from 1 to 0 as defined in Eq. (3).

$$\tau \dot{s} = -\alpha_s s \tag{3}$$

$$f_{target}(s) = \tau^2 \ddot{x} - \alpha_v \left[\beta_v (g - x) - \tau \dot{x}\right] \tag{4}$$

The formula in Eq. (3) is called the canonical system, which is equivalent to an internal clock. The system defined in Eq. (1) is called a transformation system. If we know f, we will be able to get our desired output via the transformation system. From the Eq. (2) we can see, f is decided by weights w. So in order to encode the movement, we need to find w. The interesting thing is that we found w is invariant in temporal-spatial space, which means whether you extend or shorten a given path or scale it by multiplying a random factor, as long as its contour is similar, then w basically keeps the same. In other words, w encodes a class of trajectories with similar topologies. This is also the main reason that feature w can be used for classification. The next section we will introduce the method used to calculate w.

3.2 Feature Calculation

The process of feature learning is the process of solving weights, which can be summarized as follows: (1) Calculate from demonstration trajectories; (2) Integrate formula (3) to get the phase variable; (3) Use Eq. (1) to get Eq. (4). Comparing formulas (2) and (4), we can find it is a function approximation problem. To achieve a better performance, we need to estimate a set of weights which can minimize $J = \sum_i \left(f_{target}(s) - f(s)\right)^2$. This problem is equivalent to minimizing the weighted minimum quadratic error of the following equation:

$$J_i = \sum_{t=1}^{T} \Psi_i(t) \left(f_{target}(t) - w_i \xi(t)\right)^2 \tag{5}$$

Where $\xi(t) = s(t)(g - x_0)$, this is a weighted linear regression problem. Its solution is:

$$w_i = \frac{\xi^T \mathbf{\Gamma}_i \mathbf{f}_{t\arg et}}{\xi^T \mathbf{\Gamma}_i \xi} \tag{6}$$

Where,

$$\xi = \begin{bmatrix} \xi(1) \\ \xi(2) \\ \cdots \\ \xi(T) \end{bmatrix}, \mathbf{\Gamma}_i = \begin{bmatrix} \Psi_i(1) & & & \\ & \Psi_i(2) & & \\ & & \cdots & \\ & & & \Psi_i(T) \end{bmatrix}, \mathbf{f}_{t\arg et} = \begin{bmatrix} f_{t\arg et}(1) \\ f_{t\arg et}(2) \\ \cdots \\ f_{t\arg et}(T) \end{bmatrix}$$

According to Eq. (6), we can get the weight of each base function. The feature vector is constructed by concatenating all weights together and then fed to our classifier.

4 Classifier

In the previous section, we introduced how to extract our dynamic features. Once the feature is acquired, a classifier is required to complete the classification task. In principle, as long as the feature is good enough, we can achieve a good classification performance even that a simple classifier is used. So in this paper we use the most simple KNN classifier, and compared with the SVM. In addition, we find that the extracted dynamic features have a certain drift in the time dimension, which may reduce the classification effect. We know that DTW is a classic way of dealing with time drift, but it is computationally expensive. In order to maintain a good classification effect, as well as minimize the computational complexity, we use a method called Fast DTW [17] to align our dynamic features, and then fed into the KNN classifier. DTW is used to measure the similarity of two sequential signals with different length, and is widely used in time series signal analysis. The basic principle is calculating distance metric with dynamic programming method. Distance between two sequences with length is defined as:

$$D(i,j) = Dist(i,j) + \mathtt{min}\left[D(i-1,j), D(i,j-1), D(i-1,j-1)\right] \tag{7}$$

DTW is used to measure the similarity of two sequential signals with different length, and is widely used in time series signal analysis. The basic principle is calculating distance metric with dynamic programming method. Distance between two sequences with length i, j is defined as:

For traditional DTW, in order to obtain $D(i,j)$, we must traverse the entire D matrix and the time complexity of the algorithm is $o(N^2)$. To speed up the calculation, there are usually two ways: (1) reduce the search space in D matrix; (2) data abstraction, reduce the length of the data. FastDTW combines both of them. It consists of three steps: coarsening, projection and refinement. Coarsening devotes to shrinking the original signal length without a big loss of accuracy. Projection finds a minimum-distance warp path with the coarse curve. Refinement refines the warp path projected from a lower resolution through local adjustments of the warp path. Through all of these operations, we can reduce the time complexity to $o(N)$. Using FastDTW we can get the distance metric accurately and quickly, then KNN gives us the estimated labels. This constitutes the FastDTW-KNN algorithm.

5 Experiments

This section will conduct a series of comparative experiments and compare our methods from multiple perspectives. We have two purposes. The first one is to verify the time advantage of FastDTW-KNN. The second is to show the good recognition performance when using the proposed dynamic features introduced in Sect. 3. In order to verify the temporal-spatial invariance, the test data will be randomly extended or cut off 10% length to introduce the temporal noise. Mean while, the test data is scaled by multiply a random factor to introduce spatial noise. Therefore, our test data is divided into four classes: the original non-polluting test data, data with temporal noise, data with spatial noise and data with temporal-spatial noise. As shown in Fig. 2, which takes letter "C" as an example. For simplicity, in later sections, graphs and tables, we use words"clean, temporal, spatial, T-S" represent these four different noises, respectively.

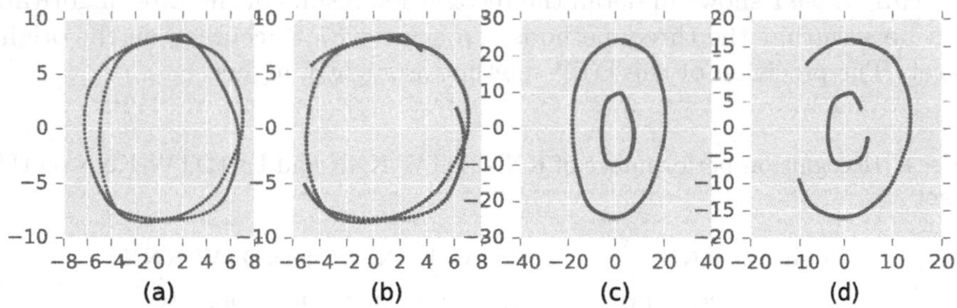

Fig. 2. Different kinds of noise added to sample character "C". (a) Stands for the original data, two samples; (b) Temporal noise is introduced by deleting some points randomly; (c) Add spatial noise by scaling the original data; (d) Both temporal noise and spatial noise are introduced

Experimental tasks can be summarized into three aspects. First, based on the HAR dataset, the recognition performance of KNN, DTW-KNN and FastDTW-KNN is investigated under the four test noise. Second, based on the hand written letters dataset, we investigate the recognition accuracy without any feature extraction. Thirdly, on the HW dataset, the recognition accuracy of different algorithms on different test noises is calculated based on the dynamic features introduced in our paper.

5.1 Datasets

This article uses two datasets. The first one is the UCI Human Activity Recognition (HAR) dataset [18]. The data is collected from 30 volunteers within an age bracket of 19–48 years. Each person performed six activities (walking, walking upstairs, walking downstairs, sitting, standing, laying). Using the embedded accelerometer and gyroscope in smartphone carried by volunteers, we captured 3-axial linear acceleration and 3-axial angular velocity signals. The data has been video recorded to label the data manually. In our experiments, we randomly choose 70% of data as training set and the left 30% is used for test.

Another dataset is a handwritten letter dataset provided by Calinon [19]. This dataset is located in a package that they implemented for LfD. The dataset includes 26 Latin characters contributing to a total of 340 sample trajectories in two-dimensional coordinates. Velocity and acceleration information are also recorded. As before, 70% of the data is used for training and the rest is used for testing.

5.2 Experiments Results

Using the HAR data set, we compare the KNN, DTW-KNN and FastDTW-KNN algorithms respectively. The metrics of performance include accuracy, precision and recall. Table 1 shows in detail the recognition results of the three algorithms. It can be seen that the three methods can achieve 87% accuracy on the original dataset. The precision of the DTW method is slightly higher.

Table 1. Recognition performance of KNN, DTW-KNN and FastDTW-KNN on HAR dataset

Noise	KNN			DTW-KNN			FastDTW-KNN		
	acc	pre	rec	acc	pre	rec	acc	pre	rec
clean	0.87	0.88	0.87	0.87	0.89	0.87	0.87	0.89	0.87
temporal	0.70	0.73	0.71	0.87	0.89	0.87	0.87	0.89	0.87
spatial	0.80	0.84	0.80	0.68	0.77	0.67	0.68	0.77	0.67
T-S	0.61	0.68	0.61	0.71	0.81	0.71	0.70	0.80	0.71

In order to observe the performance of the three algorithms on the noise data, we drew the bar graph, as shown in Fig. 3, from which we can see the advantages of various methods clearly. KNN performs well in the original data. However, once the noise is introduced, whether it is temporal noise or spatial noise, the recognition accuracy drops rapidly. For DTW and FastDTW-KNN, the recognition accuracy keeps the same when adding temporal noise, which shows that the two methods have good temporal robustness. This is also consistent with the facts. We know that the DTW method can find the best match points between two sequences automatically. Therefore, even if the sequence is noisy in time, it does not affect the calculation of distance, and thus the recognition accuracy is also maintained. However, the DTW method and the Fast-DTW method are poor when faced with spatial noisy data. The recognition accuracy has dropped to 68% under this condition.

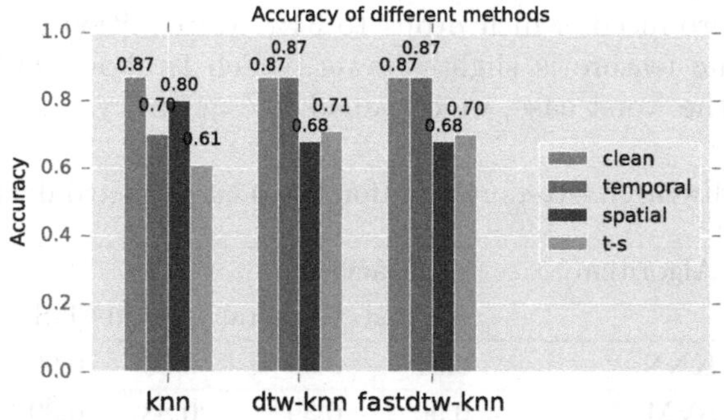

Fig. 3. Comparison of recognition accuracy on HAR dataset. "clean, temporal, spatial, T-S" stand for recognition results on different test datasets which is introduces in Fig. 1

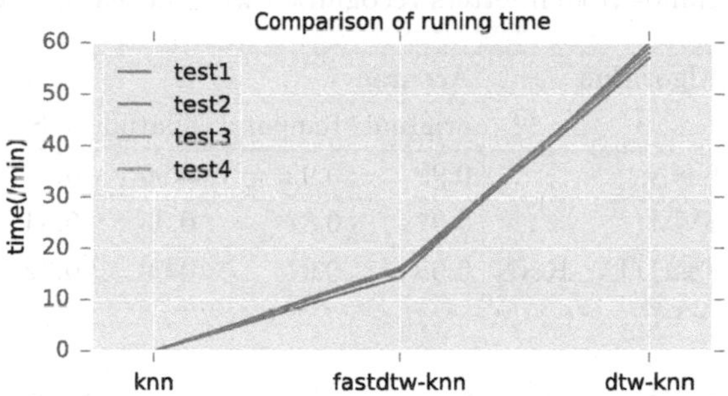

Fig. 4. Comparison of running time with four tests

In order to verify the time advantage of the Fast-DTW method compared to the DTW method, we conducted four-round tests on a computer with i5-4210 CPU, 1.7 GHz, 4 core. The test results are shown in Fig. 4. It can be seen that KNN is the fastest and FastDTW-KNN is between KNN and DTW-KNN. This shows that FastDTW-KNN not only has good temporal robustness. It's computational complexity is much lower than the traditional DTW algorithm. The larger the dataset is, the more obvious the advantage shows. But all three methods do not have a good spatial robustness. Therefore, the dynamic features are proposed to cure this drawback. The proposed feature is temporal-spatial invariant, so it can still maintain a high recognition accuracy for the signal with time and space noise. The following experimental results will prove this fact.

We compared the recognition effect with and without feature extraction. In order to prove the versatility of the extracted features, we compared different recognition methods, namely KNN, SVM and FastDTW-KNN. The experimental results are shown in the table. Table 2 shows the recognition accuracy of the three methods with extracted features. Table 3 shows the recognition accuracy of the three methods without feature extraction. Generally speaking, the accuracy is up to 98% when tested with clean data. But the volatility is very large when noise is introduced, which drops to 34%. Recognition accuracy with the proposed dynamic feature is slightly lower, which is about 90%. But it keeps steady. Even in the worst case, we obtained 88% accuracy.

Table 2. Hand-written letters recognition based on extracted dynamic features

Algorithm	Accuracy			
	original	temporal	spatial	T-S
KNN	0.91	0.90	0.90	0.90
SVM	0.90	0.89	0.88	0.89
FastDTW-KNN	0.91	0.91	0.91	0.91

Table 3. Hand-written letters recognition without feature extraction

Algorithm	Accuracy			
	original	temporal	spatial	T-S
KNN	0.98	0.92	0.90	0.77
SVM	0.97	0.87	0.35	0.34
FastDTW-KNN	0.98	0.97	0.60	0.62

Figure 5 clearly shows the advantages of our approach. The polyline with circle markers shows the performance of the three methods in the absence of noise, which acted as a standard reference. The polyline with polygon markers and square markers represents the recognition result with and without feature extraction under the condition of temporal and spatial noise respectively. It can be seen that the recognition accuracy of our method is slightly lower than that of methods without any noise. However, the performance of our method is significantly better than others in case of noisy data. Besides, our method

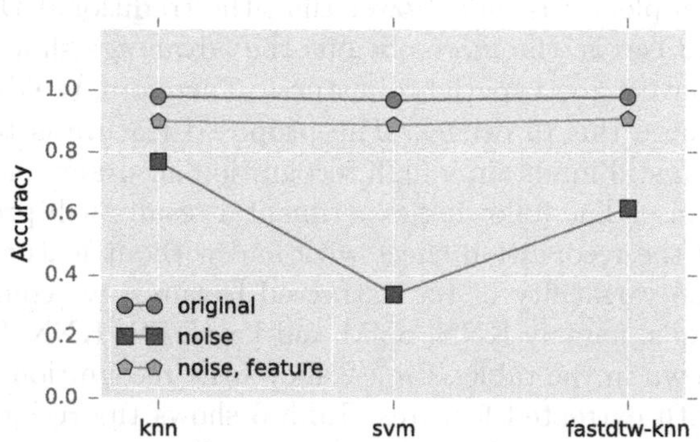

Fig. 5. Recognition performance with and without feature extraction

maintains a good consistency for the three methods. FastDTW-KNN performs slightly better, which can be attributed to the alignment of the extracted weights.

6 Conclusion and Future Work

This paper deals with the action recognition problem in the context of robot learning. We compared KNN method and DTW method. Experiments showed that the DTW method is robust to temporal noise, but computational expensive. To overcome this drawback, this paper proposed a FastDTW method to reduce the time complexity while still maintain the temporal robustness. Considering the recognition problem under spatial noise, this paper proposed a dynamic feature obtained in DMP framework. This feature is temporal-spatial invariant, so that we can maintain a high recognition accuracy even with noisy data. Besides, we found the extracted dynamic feature is prone to drift in time dimension. To fix this problem, we proposed the FastDTW-KNN algorithm, which contributes to a further improvement on the recognition accuracy. Finally, experiments were conducted on HAR dataset and handwritten letters dataset. Recognition results verified the proposed method.

Considering HMM is so popular in time sequential data analysis. Subsequent work plans to use HMM or its variants as a classifier, and the results will be compared to this paper. In addition, with the development of deep learning in recent years, especially the recursive neural network, which is very successful in dealing with sequential problems, it is also very meaningful and challenging to explore the application of RNN on action recognition.

Acknowledgments. This work is supported by National Natural Science Foundation of China (Grant Nos. 51505470).

References

1. Argall, B.D., Chernova, S., Veloso, M., Browning, B.: A survey of robot learning from demonstration. Robot. Auton. Syst. **57**(5), 469–483 (2009)
2. Huang, X.D., Ariki, Y., Jack, M.A.: Hidden Markov Models for Speech Recognition, vol. 2004. Edinburgh University Press, Edinburgh (1990)
3. Rabiner, L.R.: A tutorial on hidden Markov models and selected applications in speech recognition. Proc. IEEE **77**(2), 257–286 (1989)
4. Müller, M.: Information Retrieval for Music and Motion, vol. 2. Springer, Heidelberg (2007)
5. Poppe, R.: A survey on vision-based human action recognition. Image Vis. Comput. **28**(6), 976–990 (2010)
6. Weinland, D., Ronfard, R., Boyer, E.: A survey of vision-based methods for action representation, segmentation and recognition. Comput. Vis. Image Underst. **115**(2), 224–241 (2011)
7. Carlsson, S., Sullivan, J.: Action recognition by shape matching to key frames. In: Workshop on Models Versus Exemplars in Computer Vision, vol. 1 (2001)

8. Danafar, S., Gheissari, N.: Action recognition for surveillance applications using optic flow and SVM. In: Yagi, Y., Kang, S.B., Kweon, I.S., Zha, H. (eds.) ACCV 2007. LNCS, vol. 4844, pp. 457–466. Springer, Heidelberg (2007). doi:10.1007/978-3-540-76390-1_45
9. Ijspeert, A.J., Nakanishi, J., Schaal, S.: Learning attractor landscapes for learning motor primitives. In: Advances in Neural Information Processing Systems, pp. 1547–1554 (2003)
10. Schaal, S., Peters, J., Nakanishi, J., Ijspeert, A.: Learning movement primitives. In: Robotics Research, pp. 561–572 (2005)
11. Ijspeert, A.J., Nakanishi, J., Hoffmann, H., Pastor, P., Schaal, S.: Dynamical movement primitives: learning attractor models for motor behaviors. Neural Comput. **25**(2), 328–373 (2013)
12. Nakanishi, J., Morimoto, J., Endo, G., Cheng, G., Schaal, S., Kawato, M.: Learning from demonstration and adaptation of biped locomotion. Robot. Auton. Syst. **47**(2), 79–91 (2004)
13. Park, D.H., Hoffmann, H., Pastor, P., Schaal, S.: Movement reproduction and obstacle avoidance with dynamic movement primitives and potential fields. In: 8th IEEE-RAS International Conference on Humanoid Robots, Humanoids 2008, pp. 91–98. IEEE (2008)
14. Gams, A., Nemec, B., Ijspeert, A.J., Ude, A.: Coupling movement primitives: interaction with the environment and bimanual tasks. IEEE Trans. Rob. **30**(4), 816–830 (2014)
15. Paraschos, A., Daniel, C., Peters, J.R., Neumann, G.: Probabilistic movement primitives. In: Advances in Neural Information Processing Systems, pp. 2616–2624 (2013)
16. Schaal, S., Ijspeert, A., Billard, A.: Computational approaches to motor learning by imitation. Philos. Trans. Roy. Soc. Lond. B Biol. Sci. **358**(1431), 537–547 (2003)
17. Salvador, S., Chan, P.: Toward accurate dynamic time warping in linear time and space. Intell. Data Anal. **11**(5), 561–580 (2007)
18. Reyes-Ortiz, J.L., Anguita, D., Ghio, A., Parra, X.: Human activity recognition using smartphones data set. UCI Machine Learning Repository (2013)
19. Calinon, S.: A tutorial on task-parameterized movement learning and retrieval. Intel. Serv. Robot. **9**(1), 1–29 (2016)

Static Ankle Joint Stiffness Estimation with Relaxed Muscles Through Customized Device

Renjie Xiong[1], Cheng Sun[1(✉)], Muye Pang[1], Kui Xiang[1], and Zhaojie Ju[2]

[1] School of Automation, Intelligent System Research Institute, Wuhan University of Technology, Wuhan, Hubei, China
751004381@qq.com, pangmuye@163.com

[2] School of Computing, University of Portsmouth, Portsmouth, UK
Zhaojie.ju@port.ac.uk

Abstract. Human Ankle joints play significant role in physical activity. Studying on biomechanical characteristic of ankle joint helps to know the movement of human body. This paper reports a procedure to measure static ankle mechanical impedance with relaxed muscles. An experimental protocol using a self-made mechanical device makes it reliable to obtain torque and angle data. The machinery mainly consists of a multifunction controller, an encoder, key components to get torque and angle data, a supporting framework and a drive motor. A surface electromyographic (sEMG) system is used to determine whether ankle joint is static or not. In order to overcome the inertia and friction of the device, a calibration method based on error correcting was applied. Experiment with nine subjects keeping muscles of ankle as relaxed as possible shows that the static torque-angle relation presents a quasi-linear property and estimated the elastic parameter of ankle impedance form it. The static characterization of human ankle joint is important to understand human moving function during interaction with the environment and helps study more complex features of dynamic ankle mechanical impedance.

Keywords: Ankle joint · Static properties · Mechanical impedance · Surface electromyography

1 Introduction

While all joints in the kinematic chain between foot and trunk(ankle, knee, and hip) participate in lower-extremity functions, the contribution of the ankle is significant [1]. Previous works have shown that under stationary conditions the relation between angular displacement of the ankle and the resulting ankle torque may be described by a quasi-linear relation [2, 3]. Sinkjaer measured intrinsic and reflex contributions to the stiffness of the ankle dorsiflexors [4]. Lamontagneet studied the viscoelastic behavior of the plantarflexorsat rest [5]. H. Lee pointed out that the ankle joint is predominantly spring-like in the fully relaxed condition [6].

There are some different meanings between the words "stiffness" and "impedance". The stiffness reflects the static feature of ankle joint, while the mechanic impedance of

Y. Huang et al. (Eds.): ICIRA 2017, Part I, LNAI 10462, pp. 485–493, 2017.
DOI: 10.1007/978-3-319-65289-4_46

the ankle joint is a general term for the mechanical properties, including damping and inertia of the ankle joint. Therefore, the impedance of the ankle joint consists of the static stiffness component and dynamic damping component. The stiffness can also be called the static ankle mechanical impedance. The impedance can be used to evaluate the state of the ankle joint, and can also be used as a performance index for the design of the ankle joint. The mechanical impedance characteristics of the ankle joint is divided into passive and active properties, passive mechanical impedance is in a relaxed relationship between ankle show angle and resistance moment condition; active mechanical impedance is related to human calf muscle activation shown.

The measure instrument and method are important to get the parameter of ankle joint. Hunter and Kearney used the methods of Stochastic system identification to track the dynamic stiffness of the human ankle joint [3]. A wearable therapeutic robot was used to measure torque and angle data in multiple degrees of freedom simultaneously by H. Lee [6]. Zinder using a medial/lateral swaying cradle to obtain inversion/eversion ankle stiffness [7].

In this study we identify the biomechanical property of the ankle joint and estimated the elastic parameter of it. An experimental protocol is introduced to obtain the torque of ankle joint in direction of dorsiflexio-planrarflexion (DP) by a self-made mechanical device when volunteer's foot in a static and muscle-relaxed state.

2 Methods

2.1 Experimental Setup

The experimental setup includes two parts: a surface electromyographic (sEMG) system (ELONXI EMG 100-Ch-Y-RA, Elonxi Ltd.) and a mechanical device that is designed for measurement of ankle stiffness.

A. EMG and Choose the location of selected electrode site

EMG is the study of muscle function by the electrical signal of it. It can indicate the extent of muscle activation, showing direct perspectives of the muscle. A sEMG system, ELONXI EMG 100-Ch-Y-RA (Elonxi Limited, Hangzhou, China), is used to acquire EMG signal.

According to medical knowledge, surface muscles associated with ankle joint movement are mainly tibialis anterior and gastrocemius (Fig. 1). EMG signals of the two muscles demonstrate whether the volunteers' muscles are tense or relaxed.

To detect tibialis anterior and gastrocemius, two channels of sEMG system have been used. Channel 1 is for Tibialis anterior, and channel 2 is for Gastrocemius. Every channel has two electrodes attached to the same piece of muscle fiber. In addition, there is a Ref Electrode that is attached to the bone such as lateral malleolus or medial malleolus, acting as ground signal in a circuit.

In the experiment we followed the standard procedure that two electrodes on the same channel should be attached closely to the same piece of muscle fiber, and the direction of the electrode should be along the direction of the muscle fibers. These tips are helpful to reduce sEMG artifacts and get high quality signals.

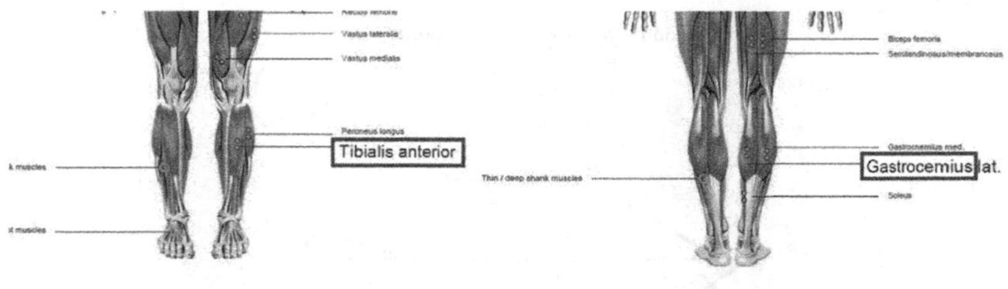

Fig. 1. Anatomical positions of selected electrode sites.

B. The Introduction of Mechanical Device

The machinery mainly consists of a multifunction controller and an encoder, key components to get torque and angle data, a supporting framework and a drive motor. The multifunction controller is the controller of the motor and it has additional function of getting output torque of motor. The encoder is coupled to the shafts of loaded pedal. A drive motor is arranged at the bottom of the frame, and the rotating shaft in the cantilever is connected with the rotating wheel through a belt pulley to control the rotation of the pedal (Fig. 2).

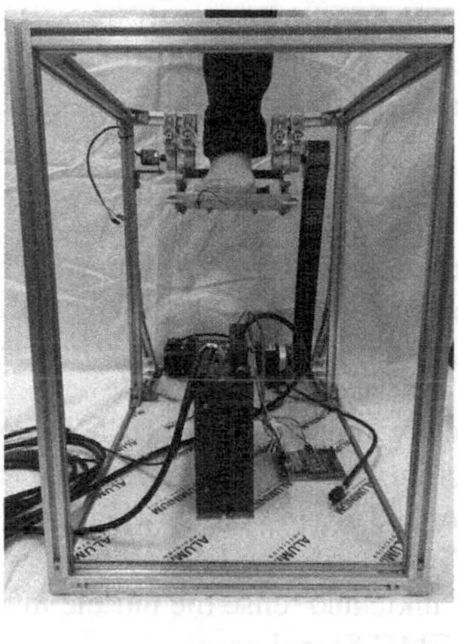

Fig. 2. General view of mechanical device.

The ankle joint is the region where the foot and the leg intersect. The ankle joint is a synovial hinge joint permitting two movements: plantarflexion and dorsiflexion. Plantarflexion's normal range is 20–50° and dorsiflexion's is 10–30° (Fig. 3). The angle range varies from person to person. In this experiment, ankle joint was measured from −45° to 35°. Therefore, the mechanical device is designed to rotate ankle joint in the sagittal plane from −45° to 35° relative to horizontal plane.

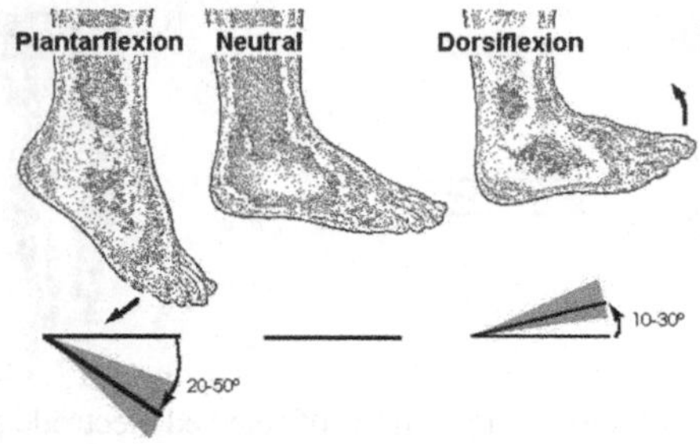

Fig. 3. Movements Angle of the Ankle Joint.

2.2 Experimental Protocol

In this experiment, static ankle joint stiffness is estimated by measuring several sets of ankle joint data from nine voluntaries. The self-made mechanical device made it reliable to obtain torque and angle data cooperated with the sEMG system. When voluntaries keep the ankle at a certain angle, we need to make sure that the muscles are in a relaxed state and measure the torque. If the sEMG signal has a wide range of fluctuations, exceeding a threshold, volunteers' ankles are regarded as activated. In this case, this set of data is invalid. Details of the experimental procedure are as follows.

Step1 Skin preparation
Skin preparation is important in the EMG measurement. Hair, skin oil, dead skin cells, dirt, sweat increase the impedance of skin which affects the recordings. The skin was shaved and cleaned with alcohol in order to remove this factor.

Step2 Paste surface electrodes and observe the EMG signal
As described in previous section, the part of EMG signal performing most intensely are the muscle of Tibialis anterior and Gastrocemius when ankle joint is in a tense condition or in voluntary movement. To identify the EMG system working properly, volunteers were asked to activate their ankle and tense the muscle in the shins. Muscle activation status can be detected from EMG Signal graph.

Step3 Adjustment according to different Subjects
Before the experiment, the high of load-board and the back-support supposed to be adjusted according to different subject and rotation center of foot and load-board must be on the same axis. The foot was fixed on the load pedal by the bandage extending form hole of the pedal. About the sitting posture, the volunteer should keep the lower-extremity in neural position. After the posture was fixed subjects cannot move again.

Step4 Obtain the experiment data through machinery device
Subjects were seated on a chair with proper height and their foot held by the load pedal in a correct posture we had discussed. They were instructed to relax while the pedal

moved to certain angle. The mechanical device has been designed to rotate ankle joint limited in a range of −45° to 35° relative to horizontal plane. The complete protocol divided the range of ankle movement into equally spaced angle of 5°, in other words, data of angle and torque would be read once every 5°. In the course of the experiment, the subjects remained their muscle relaxed all the time. And judging from the EMG Signal graph it can be checked whether volunteers' state meet the test requirements. It is necessary to repeat the experimental protocol for more credible data.

2.3 Analysis Methods

The ankle joint torque was acquired by measuring the output torque of motor. But the relationship between them exist multistage transmission. It is inevitable for any torque components to overcome the inertial and friction. Consequently, unless the influence of device itself was eliminated, the data of torque reading form instrument was incredible. In order to overcome the inertia and friction of the device, a calibrations method based on error correcting was applied.

The principle of the method is to build a map relationship between the torque of instrument and the actuarial torque of ankle joint. The specific means are as follows.

According to the experimental protocol explained previous chapter, there are not many differences but with human subject replaced by standard weight blocks. And the actual torque can be calculate by the derivation of formula

$$\tau = nGL\sin(\phi - \theta) \tag{1}$$

referring to Fig. (4).

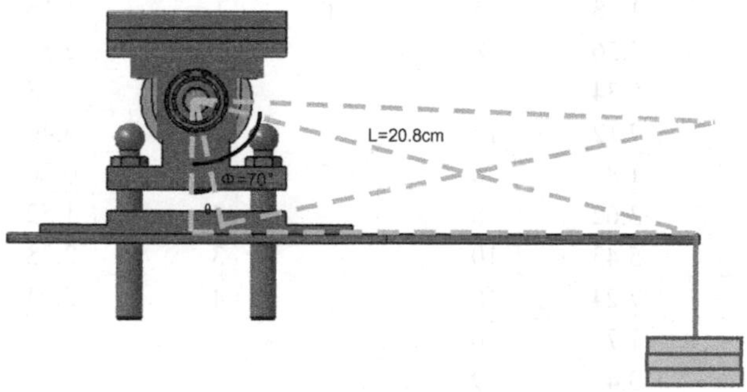

Fig. 4. Force diagram of pedal

There were seven positions chosen as calibration points, ranged from −45° to 45° with a increment of 15°. At each position, weight blocks were hung on end point of the pedal. The number of hung weight blocks increased from 1 to 4, one by one. The actual torque was converted form gravity of blocks based on Eq. (1). At the same time recording the data read from the motor controller, a correspondence table about the map relationship was established. Then the table can be described in a form of fitting curve which observed more directly the relationship between controller data and real torque and had value to estimate the blank area of map relationship.

3 Result

3.1 Calibrations

As discussed previously, what the facility actually reached is the motor output torque but not the torque of ankle joint. In order to construct the input-output connection, a test based on error correcting was applied to solve the problem. The Table 1 was the result of calibrations. The coordinate system was set up with the read torque of X and the actual torque of Y. A fitting curve laying the foundation of a least square method indicated that there is a linear relationship between them. The intersection of curve and x axis is supposed to be the inherent error of equipment. The curve equation, mapping relation of the displayed data to actual torque, is (Fig. 5)

$$y = 0.45x - 0.9 \tag{2}$$

Table 1. Result of calibrations

Degree	N	Actual	Sensor	Degree	N	Actual	Sensor
45°	1	1.65	3	−15°	1	1.49	8
	2	3.28	4		2	2.98	10
	3	4.92	6		3	4.47	14
	4	6.56	7		4	5.96	18
30°	1	1.78	5	−30°	1	1.16	8
	2	3.56	8		2	2.32	11
	3	5.34	9		3	3.48	15
	4	7.12	12		4	4.46	18
15°	1	1.81	6	−45°	1	0.76	8
	2	3.62	8		2	1.52	11
	3	5.43	10		3	2.28	13
	4	7.24	15		4	3.04	16
0°	1	1.7	6				
	2	3.4	9				
	3	5.1	13				
	4	6.8	16				

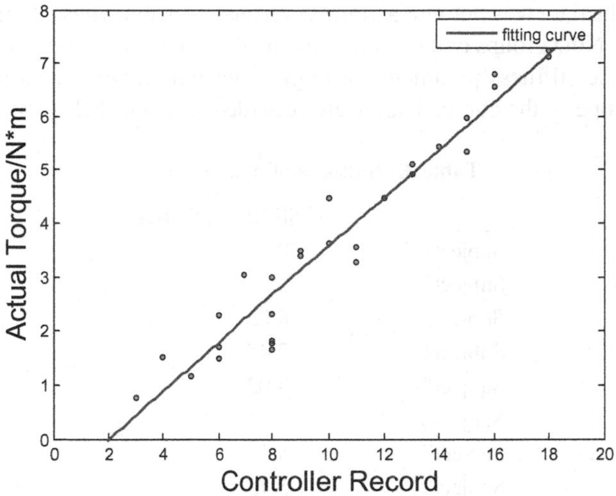

Fig. 5. Fitting curve of calibrations

3.2 The Stiffness of Ankle Joint

The machinery is designed for simultaneous measurement of angle and torque. And the main purpose of the experiment is to obtain static ankle joint stiffness at muscle relaxed statue. The raw data were measured coming form nine subjects. According to the result of calibrations, more valuable data demonstrating the link of torque-angle of ankle join was written in Table 2.

Table 2. Relationship of angle and actual torque for nine subjects

Subjects angle	1	2	3	4	5	6	7	8	9
−45°	−4.51	−1.81	−2.26	−1.36	−0.91	−1.81	−0.91	−1.81	−1.36
−40°	−1.81	−0.01	−0.46	−0.01	−0.46	0.44	−0.01	0.44	−0.91
−35°	−0.01	−0.01	−0.46	−0.01	−0.01	−0.46	−0.01	0.89	−0.01
−30°	0.44	1.34	0.89	−0.01	−0.01	−0.01	0.44	0.44	0.44
−25°	1.34	1.79	0.89	0.44	1.34	1.34	0.89	0.44	0.44
−20°	1.79	2.69	0.89	0.44	1.34	1.34	1.34	0.89	0.89
−15°	2.24	2.24	1.34	1.79	2.24	0.44	0.89	2.24	1.79
−10°	3.14	3.59	1.79	1.34	2.69	0.44	1.79	1.79	1.79
−5°	4.49	3.59	2.24	2.69	4.04	0.89	1.34	2.69	2.24
0°	5.39	4.04	3.59	3.14	3.59	1.34	2.24	2.69	2.69
5°	4.49	4.94	3.59	3.59	4.94	1.79	2.69	3.59	3.59
10°	5.84	5.84	4.04	4.04	5.39	2.24	3.14	3.59	4.04
15°	6.74	6.29	4.49	5.39	5.84	2.24	4.04	4.94	4.94
20°	8.54	6.74	6.74	6.29	7.19	4.04	4.49	4.49	4.49
25°	9.89	8.09	8.09	8.09	8.99	4.49	5.84	6.29	6.29
30°	11.24	9.89	9.44	6.74	9.44	6.29	8.09	6.74	6.29
35°	12.14	11.69	11.69	10.34	10.34	11.69	7.64	8.54	7.19

Similarly, the data was put into a plane coordinate system whose angle is x axis and torque is y axis. Nine groups of data belonging to all subjects were fitted to a linear curve respectively. The stiffness parameter of angle joint which we concerned most about equals to the slope of the curve. They were recorded to Table 3 (Fig. 6).

Table 3. Stiffness of nine subjects

	Stiffness (Nm/rad)
Subject1	10.32
Subject2	8.02
Subject3	8.02
Subject4	7.45
Subject5	8.02
Subject6	5.73
Subject7	5.73
Subject8	5.73
Subject9	5.73

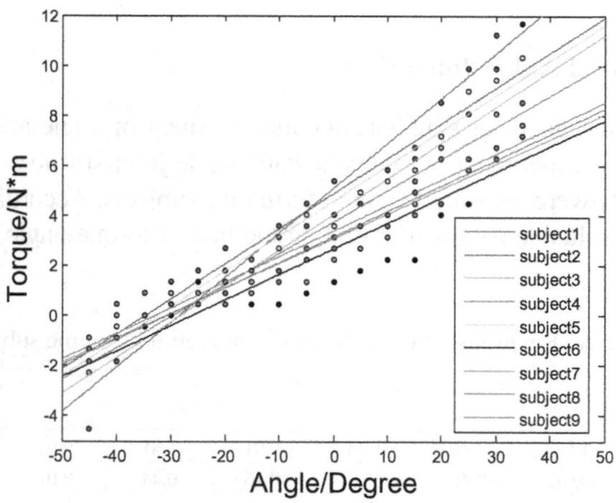

Fig. 6. Fitting curve of nine subjects

4 Summary

In this experiment, a customized machine device was used and the ankle joint parameters of the subjects were successfully measured by it. In the course of test, volunteers must relax their muscle and the EMG was used to detect the muscle activation level. In order to eliminate the disturbance, data wouldn't be read until it became stable. And this ensures the ankle joint in a state of static. A calibration method was applied to build a mapping relationship of reading torque and actual torque. In the foundation of the relationship, the most valuable data of ankle join was obtained.

The main goal of the experiment is to measure the stiffness of human ankle joint. The stiffness reflects the static feature of ankle joint and it is also indispensable part of the mechanical properties. Owing to its importance during the movement of human body, a quantitative analysis of property of the ankle joins would be valuable to understand the law of coordinated motion of human body. In the final result, we accurately measured the ankle stiffness of several subjects. Though Most of them had little difference in static ankle impedance but one subject's stiffness relatively high, at least we knew the average size of stiffness of ankle joint. The data are crucial to the further measurement of impedance of ankle joint and have instructive significance in the design of mechanical measuring device.

Acknowledgments. This work was partly supported by National Natural Science Foundation of China (61603284 & 51575412).

References

1. DeVita, P., Helseth, J., Hortobagyi, T.: Muscles do more positive than negative work in human locomotion. J. Exp. Biol. **210**(19), 3361–3373 (2007)
2. Agarwal, G.C.A.: GL Compliance of the human ankle. Trans. Am. Sot. Mech. Eng **99**, 166–170 (1977)
3. Hunter, I.W., Kearney, R.E.: Dynamics of human ankle stiffness: variation with mean ankle torque. J. Biomech. **15**(10), 747–752 (1982)
4. Sinkjaer, T., Toft, E., Andreassen, S.: Muscle stiffness in human ankle dorsiflexors: intrinsic and reflex components. J. Neurophysiol. **60**(3), 1110–1121 (1988)
5. Lamontagne, A., Malouin, F., Richards, C.L.: Viscoelastic behavior of plantar flexor muscle-tendon unit at rest. J. Orthop. Sports Phys. Ther. **26**(5), 244–252 (1997)
6. Lee, H., Ho, P., Rastgaar, M.A.: Multivariable static ankle mechanical impedance with relaxed muscles. J. Biomech. **44**(10), 1901–1908 (2011)
7. Gatt, A., Chockalingam, N.: Validity and reliability of a new ankle dorsiflexion measurement device. Prosthet. Orthot. Int. **37**(4), 289–297 (2013)

IMU Performance Analysis for a Pedestrian Tracker

Jianwei Zheng(✉), Minhui Qi, Kui Xiang, and Muye Pang

Wuhan University of Technology, Wuhan, Hubei, China
zjwwhut@163.com, pangmuye@163.com

Abstract. The performance of inertial measurement unit (IMU) is very important for a pedestrian tracker. It is possible for an object to keep track of changes in its own position using an IMU. It is well known that the inertial element acts as the main hardware of IMU, but the accuracy of the inertial sensors degrades with time. The errors are mainly caused by the imperfect structure of the inertial element itself, the change of the internal physical factors and the change of the operating environment, etc. The performances of three IMUs which includes MPU9250, LSM9DS0 and BMX055 are analyzed in this paper. The parameters of the three sensors are compared. The typical error items for inertial elements and the principle and algorithm of the Allan Variance are introduced. Allan Variance technique is used to analyze errors of the inertial sensors. Finally, the design of IMU and the three IMU experimental results are introduced. Final results show that the performance of MPU9250 is the best and BMX055 is the worst.

Keywords: IMU · Allan variance · Inertial element · Error analysis

1 Introduction

A pedestrian navigation system that tracks the location of a person is useful for finding and guiding emergency first responders, blind persons or for location-aware computing [1–3]. In the study of Fischer et al. [1], a pedestrian tracker using foot-mounted inertial sensors was introduced. Position tracking of human movement usually requires an unrestricted line-of-sight to an installed infrastructure consisting of one or more transmitters and/or receivers or optical-based systems, such as video tracking. Such systems require extensive setup and calibration of the tracking volume that may be of limited size and may suffer from occlusion [2]. Many researchers have focused on using IMU to estimate a pedestrian's location. Hassen Fourati et al. [2] provide a heterogeneous data fusion algorithm for pedestrian navigation via foot-mounted IMU and complementary filter.

The accuracy of IMU is important in an inertial navigation system. An inertial sensor typically outputs the vehicle's acceleration and angular rate, which are then integrated to obtain the vehicle's position, velocity, and attitude [4]. But the accuracy of the inertial element is not very high. In order to improve the accuracy of the element, we need to analyze the error terms which affect the precision of IMU [10].

Y. Huang et al. (Eds.): ICIRA 2017, Part I, LNAI 10462, pp. 494–504, 2017.
DOI: 10.1007/978-3-319-65289-4_47

Traditional tools for the analysis of random processes are the auto correlation function and power spectral density (PSD) [4], which are closely coupled. However, to analyze a long-term characteristic of the sensor, the processing of huge amount of data is necessary. In this case, determination of the PSD is very computationally intensive. Several time-domain methods have been devised. They are basically very similar and primarily differ in that various signal processing, by way of weighting functions, window functions, etc., The simplest is the Allan Variance.

The Allan Variance is a time-domain-analysis technique. It can provide information on the types and magnitude of various noise terms. Allan Variance is used as a measure of frequency stability in a variety of precision oscillators, such as crystal oscillators, atomic clocks and frequency stabilized lasers over a period of a second or more [4, 5].

2 Methodology

The output signal of the inertial element contains a variety of random errors. Allan Variance considers that the liner system output is caused by one or more normally distributed white noise.

Assuming that the inputs of the MEMS gyroscope and the accelerometer are zero and there are N consecutive data points, each have a sample time of t_0. N sampling data is devided into K = N/M independent data clusters. The length of each cluster is M. Associated with each cluster time $T = Mt_0$ and then, we can get the Allan Variance results from the mean of two adjacent data clusters [4, 6].

The average of each cluster is:

$$\bar{\omega}_k(\mathrm{M}) = \frac{1}{T}\sum_{i=1}^{n} \omega_{(k-1)\mathrm{M}+i}, (\mathrm{k} = 1, 2, \ldots \mathrm{K}) \tag{1}$$

where $\bar{\omega}_k(M)$ is the average of each cluster, M is the length of each cluster.

Thus, the Allan Variance is:

$$\sigma^2(\mathrm{T}) = \frac{1}{2}[\omega_{k+1}(\mathrm{T}) - \bar{\omega}_k(\mathrm{T})]^2 \tag{2}$$

Several common random errors are analyzed, including quantization noise, angle random walk, markov noise, sinusoidal noise, bias instability, rate random walk, and drift rate ramp. A typical Allan Variance analysis result is plotted in Fig. 1.

The following subsections will show the definition of random error terms. The markov noise and sinusoidal noise is not explained due to the difficulty in observing [4, 7].

Quantization noise. The quantization noise represents the lowest resolution level of the inertial element which is generated during the A/D conversion of the signal. The effect of the quantization noise is large at the beginning of sampling. The slope in the Allan standard deviation curve is −1. The Allan Variance of quantization noise is:

$$\sigma_Q^2(T) = \frac{3Q^2}{T^2} \tag{3}$$

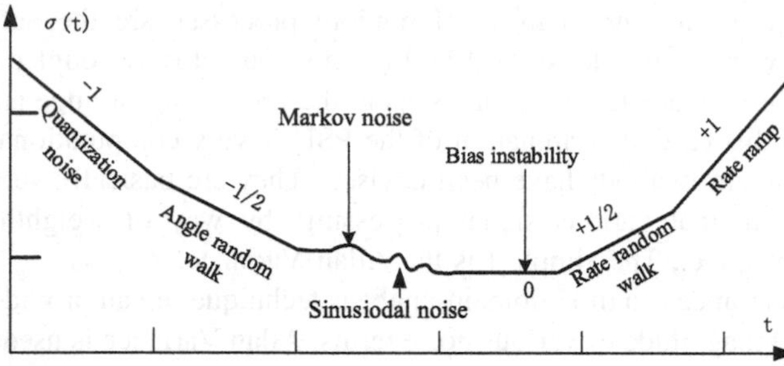

Fig. 1. Typical Allan deviation plot for a system

Angle random walk. Angle random walk is the main part of the MEMS gyro random error. The slope of the Allan standard deviation curve is −1/2. The Allan Variance of Angle random walk is:

$$\sigma_N^2(T) = \frac{N^2}{T^2} \tag{4}$$

Bias instability. Bias instability comes from circuit noise, ambient noise, and other internal components that may produce flicker noise. The bias instability error occupies the dominant position of the corresponding period with the extension of the measurement time. The Allan Variance of bias instability is:

$$\sigma_B^2(\mathrm{T}) = \left(\frac{B}{0.6648}\right)^2, \ \mathrm{T} \geq \mathrm{t_0} \tag{5}$$

Rate random walk. The source of the rate random walk error is uncertain, which may be caused by the limit of the exponential correlation noise with long correlation time or the aging effect of the internal crystal of the MEMS gyroscope. The slope in the Allan standard deviation curve is 1/2. The Allan Variance of rate random walk is:

$$\sigma_K^2(\mathrm{T}) = \frac{K^2 T}{3} \tag{6}$$

Rate ramp. The rate ramp error may be due to the random character. The slope in the Allan standard deviation curve is 1. The Allan Variance of rate ramp is:

$$\sigma_R^2(\mathrm{T}) = \frac{R^2 T^2}{2} \tag{7}$$

It is impossible to give the system integrated error model exactly due to the existence of cross coupling error. If the noise statistics are independent then Allan Variance can be expressed as the square sum of the various types of errors:

$$\sigma^2_{total}(T) = \sigma^2_Q(T) + \sigma^2_N(T) + \sigma^2_B(T) + \sigma^2_K(T) + \sigma^2_R(T) \tag{8}$$

The corresponding Allan Variance is obtained and the least square fitting is carried out at different relevant time T. The random error coefficients will be calculated. The variance is generally small. The fitting standard deviation can improve the accuracy. Thus, the model of Allan standard deviation is:

$$\sigma_{total}(T) = \sqrt{\sigma^2_{total}(T)} \tag{9}$$

The parameters of the various types of error estimation can be calculated according to Eqs. (8) and (9).

3 Experimental Preparation

3.1 The Parameter Comparison of Three Chips

Three different IMUs were involved in modeling inertial-sensor noise. The IMUs include MPU9250, LSM9DS0 and BMX055. The parameter comparison of three chips has been done, as shown in Table 1.

Table 1. The parameter comparison of three chips

		MPU9250	LSM9DS0	BMX055
Gyroscope	Range (deg/s)	2000	2000	2000
	Bias stability (deg/s)	5		
	Zero-rate offset (deg/s)	±5	±25	±1
	Noise density (deg/s/√Hz)	0.01		
	Resolution [bits]	16	16	16
Accelerometer	Range (g)	16	16	16
	Bias stability (ug)			
	Zero-rate offset (mg)	±60 (x, y) ±80 (z)	±60	±80
	Noise density (ug/√Hz)	300		150
	Resolution [bits]	16	16	12
Magnetometer	Range (gauss)	48	13	12
	Resolution [bits]	14	16	12

According to the parameters provided in the official data sheet, the gyroscope range and accelerometer range of three sensors are the same. The magnetometer range of MPU9250 is the largest and BMX055 is the smallest. By comparing the zero-rate offset of gyroscope, BMX055 performs better than other sensors. LSM9DS0 has the best performance of accelerometer zero-rate offset.

3.2 Allan Variance Analysis of Inertial Sensors

MPU9250 Allan Variance analysis. The 10 h of gyroscope and accelerometer data were collected at room temperature. The sensor are placed static and the data output rate is set to 100 Hz. The static sampling data of MPU9250 is shown in Fig. 2. By applying the Allan Variance to the 8 h of relative stable data, a log-log plot of the Allan standard deviation versus the cluster time is shown in Fig. 3 for gyroscope data and Fig. 4 for accelerometer data.

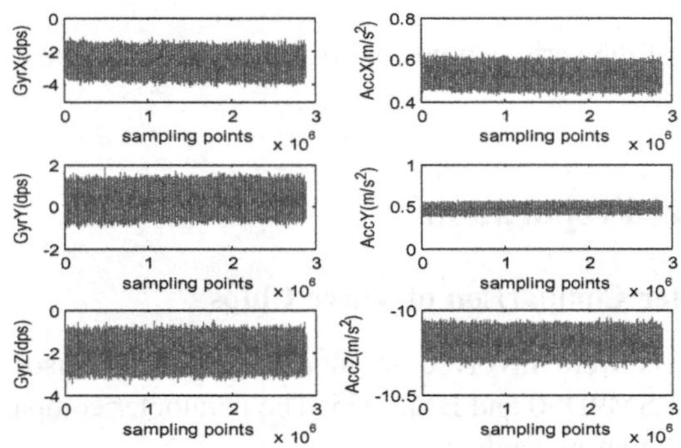

Fig. 2. MPU9250 sampling data

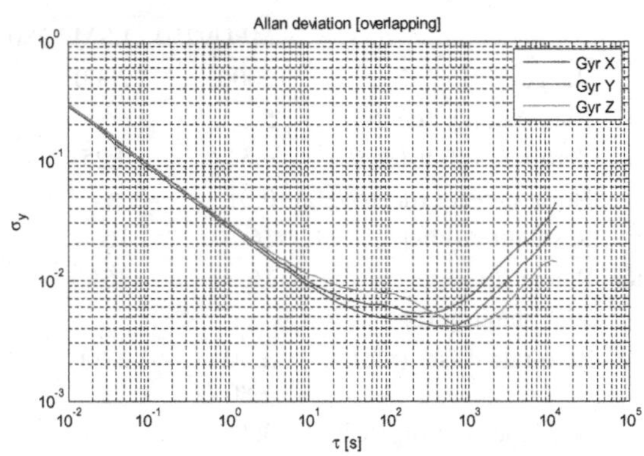

Fig. 3. MPU9250 gyroscope Allan Variance results

The Allan Variance results of MPU9250 gyroscope that are shown in Fig. 3 clearly show that the quantization noise and angle random walk is the prominent error term in the short cluster times(0 s–20 s), whereas it is the bias instability during long cluster times(20 s–800 s). Rate random walk and rate ramp are the main error terms after 800 s. For the accelerometer, The Allan Variance results of MPU9250 that are shown in Fig. 4 clearly show that quantization noise and angle random walk are the prominent error term in the short cluster times. Whereas it is the rate random walk in the long cluster times.

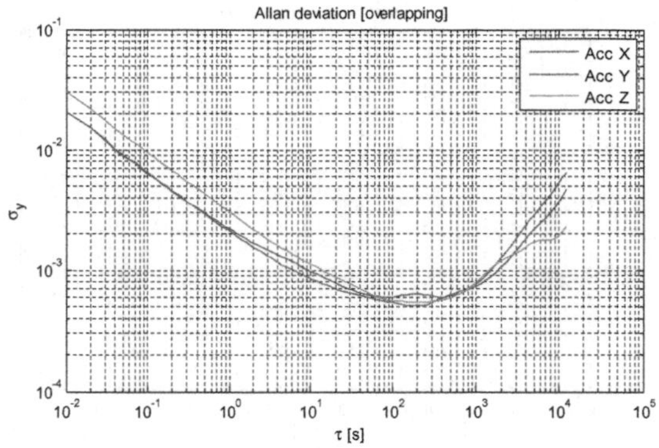

Fig. 4. MPU9250 accelerometer Allan Variance results

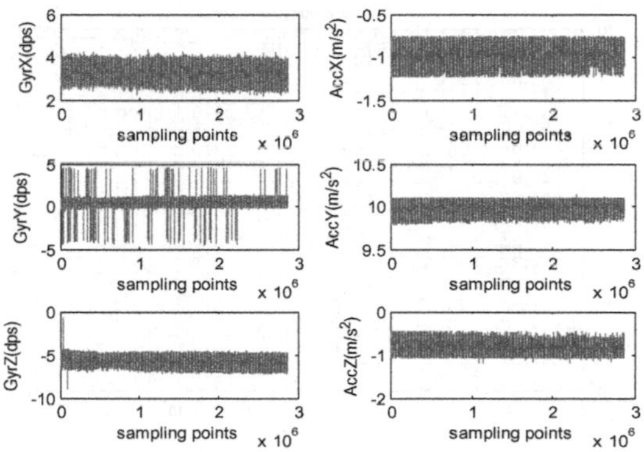

Fig. 5. LSM9DSO sampling data

LSM9DSO Allan Variance analysis. The 10 h of gyroscope and accelerometer data were collected at room temperature. The sensor are placed static and the data output rate is set to 100 Hz. The static sampling data of LSM9DSO is shown in Fig. 5. By applying the Allan Variance to the 8 h of relative stable data, a log-log plot of the Allan standard deviation versus the cluster time is shown in Fig. 6 for gyroscope data and Fig. 7 for accelerometer data.

The results of gyroscope and accelerometer in LSM9DSO clearly indicate that the quantization noise and random walk is the dominant errors for the short cluster times, whereas it is the bias instability during long cluster times. Rate random walk and rate ramp are the main error terms after 100 s.

BMX055 Allan Variance analysis. The 10 h of gyroscope and accelerometer data were collected at room temperature. The sensor are placed static and the data output rate is set to 100 Hz. The static sampling data of BMX055 is shown in Fig. 8. By applying the Allan Variance to the 8 h of relative stable data, a log-log plot of the Allan standard deviation versus the cluster time is shown in Fig. 9 for gyroscope data and Fig. 10 for accelerometer data.

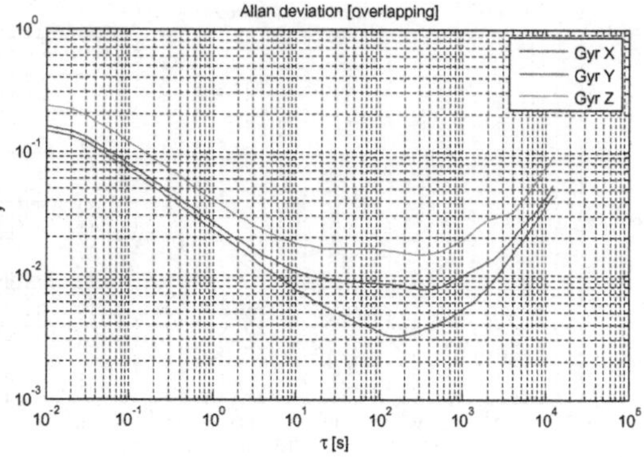

Fig. 6. LSM9DS0 gyroscope Allan Variance results

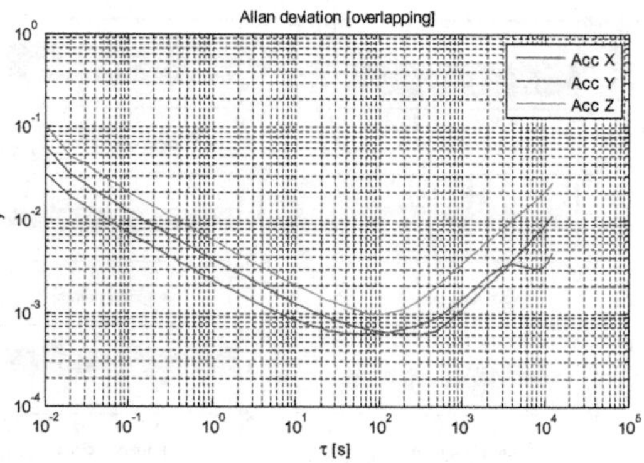

Fig. 7. LSM9DS0 accelerometer Allan Variance results

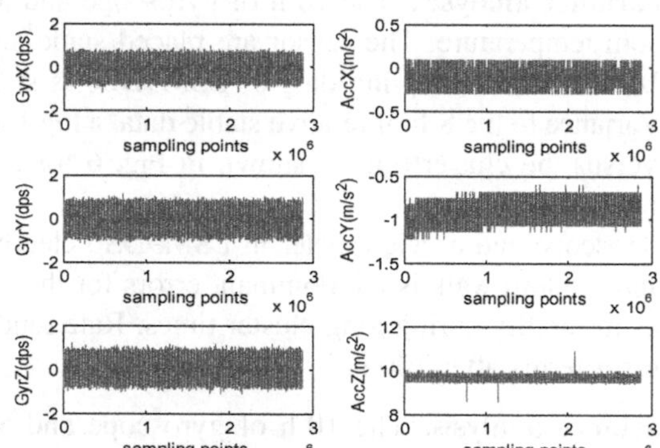

Fig. 8. BMX055 sampling data

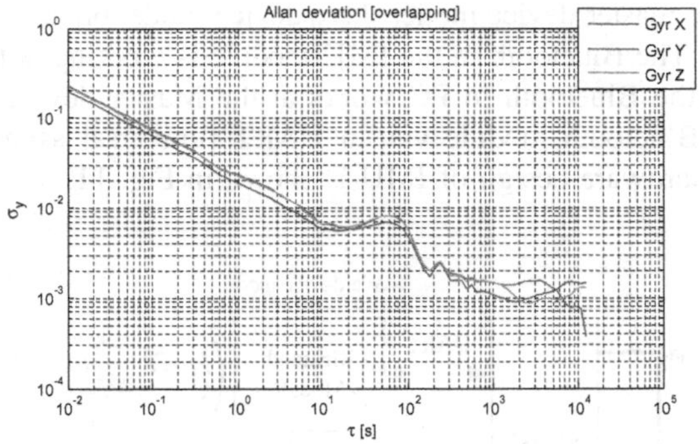

Fig. 9. BMX055 gyroscope Allan Variance results

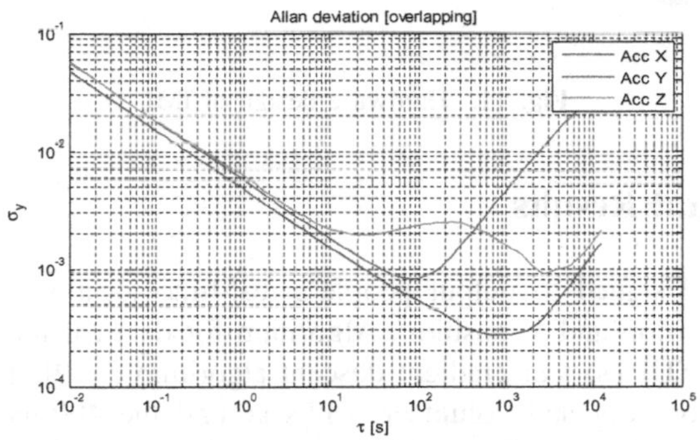

Fig. 10. BMX055 accelerometer Allan Variance results

It can be seen from the Allan variance analysis results of BMX055 that quantization noise and angle random walk are the main error terms of gyroscope and accelerometer in the short cluster time. The Allan Variance results of BMX055 gyroscope show that markov noise and sinusoidal noise is the dominant error source in the long cluster time. Starting from 100 s, the curve tends to be irregular. The Allan Variance results of BMX055 accelerometer indicate that rate ramp is the dominant error in the long cluster time.

Comparing the gyroscope error term coefficients among MPU9250, LSM9DS0 and BMX055, BMX055 is the best among three chips. Comparing the results of the accelerometer error term coefficients among the chips, it is obvious that the noise of MPU9250 accelerometer is smaller.

3.3 Design of IMU

An IMU employs a STM32, an inertial sensor and a Bluetooth module. Each inertial sensor contains 3 gyroscopes, 3 accelerometers and 3 magnetometers. Outputs of 3 gyroscopes, 3 accelerometers and 3 magnetometers are received by STM32 via I2C and the data is send out by the Bluetooth module which has four working modes. The four working modes are master device mode, slave device mode, broadcast mode and mesh networking mode. The Bluetooth in the IMU module is working at the mode of Slave device mode and the Bluetooth in PC works at the Master device mode. STM32 is connected to the Bluetooth module via an UART (universal asynchronous receiver transmitter). The hardware design of IMU is shown in Fig. 11.

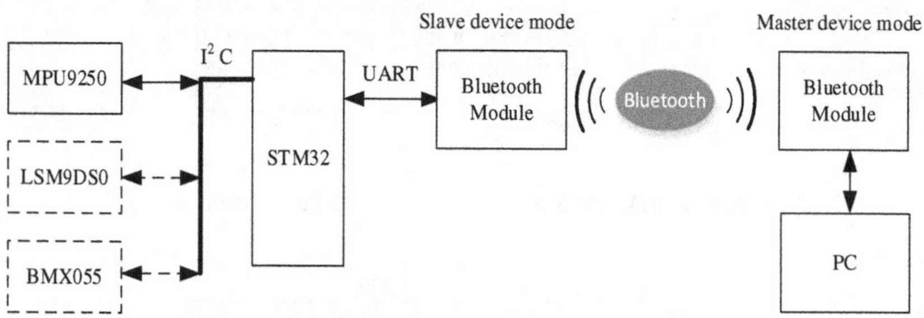

Fig. 11. Hardware design of IMU

4 Experimental Results

Three IMUs are designed to output people's real-time location information which includes the X position and Y position. Algorithm for data calculating used the zero velocity updates (ZUPTs) and gradient descent algorithm [8, 9]. Testing is done by putting the IMU on foot and volunteer walks around the 400-meters runway. The trajectory is drawn by Matlab. Figure 12 shows the experimental results of MPU9250. Figure 13 shows the results of LSM9DS0 and Fig. 14 shows the results of BMX055.

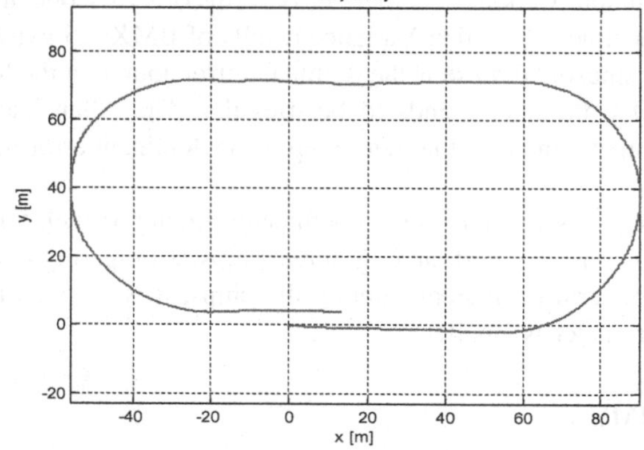

Fig. 12. Experimental results of MPU9250

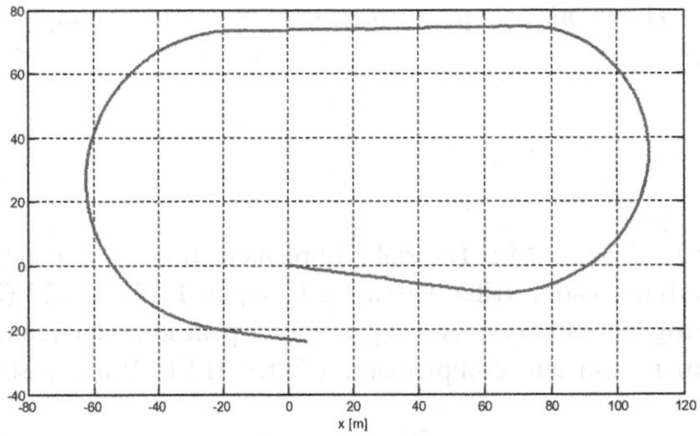

Fig. 13. Experimental results of LSM9DS0

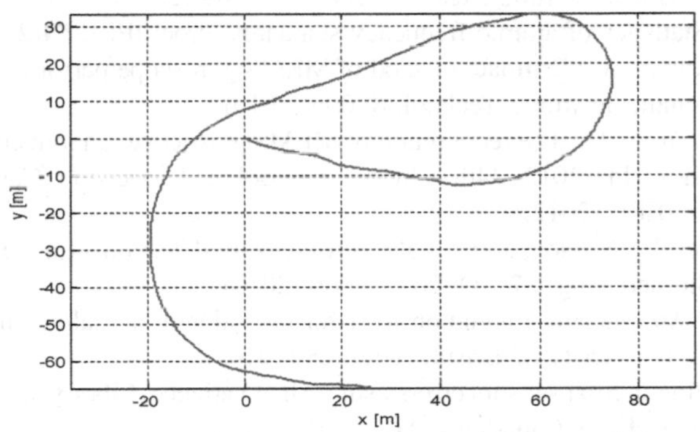

Fig. 14. Experimental results of BMX055

It is obvious that the trajectory of MPU9250 and LSM9DS0 are better than BMX055. From the experimental result of MPU9250 the distance between start point and end point is less than 10 m. As can be seen from Fig. 13, the distance between start point and end point is more than 20 m. It is clear from Fig. 14 that BMX055 cannot output the position information accurate. By comparing three trajectories, the performance of MPU9250 is the best and BMX055 is the worst.

5 Conclusions

In this paper, the typical error terms of the inertial elements, the principle and algorithm of Allan Variance were introduced. The parameters of the three sensors are compared. The random error terms of MPU9250, LSM9DS0 and BMX055 were analyzed via the method of Allan Variance. The Allan Variance analysis results of gyroscope indicates that the performance of BMX055 is the best. The performance of MPU9250 accelerometer is better than other sensors. Finally, a test is carried out to evaluate the capability of the three IMUs. Experimental results show that the performance of MPU9250 is the best and BMX055 is the worst.

Acknowledgements. This work was partly supported by National Natural Science Foundation of China (61603284).

References

1. Sukumar, C.F.P.T., Hazas, M.: Tutorial.: implementation of a pedestrian tracker using foot-mounted inertial sensors. IEEE Pervasive Comput. **12**(2), 17–27 (2012)
2. Fourati, H.: Heterogeneous data fusion algorithm for pedestrian navigation via foot-mounted inertial measurement unit and complementary filter. IEEE Trans. Instrum. Measur. **64**(1), 221–229 (2015)
3. Foxlin, E.: Pedestrian tracking with shoe-mounted inertial sensors. IEEE Comput. Graph. Appl. **25**(6), 38–46 (2005)
4. El-Sheimy, N., Hou, H., Niu, X.: Analysis and modeling of inertial sensors using Allan variance. IEEE Trans. Instrum. Measur. **57**(1), 140–149 (2008)
5. Allan, D.W.: Statistics of atomic frequency standards. Proc. IEEE **54**(2), 221–230 (1966)
6. Xuefeng, Z., Xinyan, L.: Estimate method of MEMS gyroscope performance based on allan variance. Micronanoelectronic Technol. **8**, 005 (2010)
7. Wen-hui, P., Wen-jie, C.: The recognition of MEMS gyroscope random error terms based on ALLAN variance. In: 2012 24th Chinese Control and Decision Conference (CCDC), pp. 1603–1606. IEEE (2012)
8. Skog, I., Handel, P., Nilsson, J.O., et al.: Zero-velocity detection—an algorithm evaluation. IEEE Trans. Biomed. Eng. **57**(11), 2657–2666 (2010)
9. Madgwick, S.: An efficient orientation filter for inertial and inertial/magnetic sensor arrays. Report x-io and University of Bristol (UK), 25 (2010)
10. Hou H.: Modeling inertial sensors errors using Allan variance. Library and Archives Canada, Bibliothèque et Archives Canada (2005)

A Remote Online Condition Monitoring and Intelligent Diagnostic System for Wind Turbine

Detong Kong[1], Wei Liu[2(✉)], Zhanli Liu[1], and Hongwei Wang[1,2]

[1] Huadian Electric Power Research Institute, Hangzhou 310030, China
[2] University of Portsmouth, Portsmouth PO1 3DJ, UK
Wei.Liu@myport.ac.uk

Abstract. Wind power has become one of the most popular renewable resources around the world. As the capacity of installed wind power increases, there is a constant need to minimise downtime and maximise productivity. Therefore, condition monitoring and fault diagnosis are significant for wind turbine operation and maintenance. This paper analyses typical faults in wind turbines and proposes a diagnostic process, and details the design and development of a remote online condition monitoring and intelligent diagnostic system based on the process. The system framework and the main functionality of the system are described in a real-world application of the system in some Chinese wind farms. Evaluation of the system shows that this system has achieved better reliability of wind turbines and hence saved maintenance cost to a certain extent.

Keywords: Wind turbine · Condition monitoring · Fault diagnosis · Intelligent system

1 Introduction

As a kind of green energy, wind power has replaced fossil fuels gradually due to the issues of global warming and environmental pollution. At the same time, as a major source of renewables around the world and it is going to have a remarkable share in the energy market [1]. The Global Wind Energy Council (GWEC) launched the Global Wind Report: Annual Market Update on 19 April 2016. This report revealed, at the end of 2015, the new global capacity reached 432.9 GW. Installations in Asia leapt to first place in the world, with Europe in the second spot, and North America ranked in the third [2]. Moreover, the installed wind power capacity in the world has achieved to 63,467 MW by the end of 2015, in the highest proposed growth scenario, it estimated that by 2020, wind power could supply 2.600 TWh, about 11.5–12.3% of global electricity supply, rising to 21.8% by 2030 [3]. The rapid development of wind power has brought great challenges to the wind equipment manufacturing industry and wind farms. Because most wind turbines are installed in remote and offshore areas with a height of above ground 70 m and terrible circumstance (e.g. unstable load condition), the failure rate of the wind turbine is inevitably high. In the meanwhile, operation

Y. Huang et al. (Eds.): ICIRA 2017, Part I, LNAI 10462, pp. 505–516, 2017.
DOI: 10.1007/978-3-319-65289-4_48

failure affects the safety and economy of wind power generation directly. For a turbine with a 20-year working life, the operation and maintenance (O&M) costs of 750 kW turbines might account for about 25%–30% of the overall energy generation cost or 75–90% of the investment costs [4, 5]. Therefore, condition monitoring and fault diagnosis have an important significance for a large-scale wind turbine, whether it is to reduce the cost of operation and maintenance or to improve the operating efficiency and reliability of wind turbines. Nowadays, more and more researchers are involved in developing advanced diagnostic approaches, whereas the limited success has been unable to keep pace with the development rate of the wind turbine. Although many experts have proposed a variety of diagnostic methods, most of them are still in the simulation test or off-line faults diagnosis stage. Hence these systems cannot provide timely and accurate fault diagnosis results and thus cannot address main problems that the wind power industry are facing. The remaining parts of the paper will put forward an intelligent diagnosis platform of the wind turbine to realise remote control, on-line monitoring and fault diagnosis, remote technical support and other functions. This diagnosis platform has demonstrated good functionality and has shown good potential in solving fault diagnosis problems in wind farms.

2 Typical Faults in Wind Turbines and Methodology

Compared with the traditional power plants, most wind farms are constructed in remote areas with bad conditions. Wind turbines work in a variable environment such as strong wind, snowstorm, lightning, dust corrosion, icing, etc. The most common failures have root causes in subsystems that contain blades, shaft bearings, electrical control, gearbox and generator [6]. Several previous studies considered the distribution of wind turbine failures in the main components [7–9]. Haln [7] reported a survey of 1500 wind turbines over 15 years and found that five component groups, i.e. electrical system, control system, hydraulic system, sensors and rotor blades, are responsible for 67% of failures in wind turbine [10].

2.1 Gearbox

Gearbox is an indispensable part of a non-direct drive wind turbine. The function is transmitting wind power into the generator from low speed to high rotor speed. It is one of the most rotational components occurs fault in wind turbines [6]. Failures of gearbox are generally due to gear tooth damage backlash, shaft misalignment, shaft imbalance, bearing damage, leaking oil, high oil temperature and poor lubrication [11].

2.2 Blades

With the continuous improvement of wind turbine technology, the diameter of blades are also increasing to capture more wind power. Also, due to blades rigidity is poor it is possible to cause blades failures when the wind speed is too high or irregular operation in the course of the rotation. Hence the reasons come from external and internal stress, fatigue, crack, damage, etc. These causes lead to the performance of wind generation deterioration [12]. The common faults on blades containing rotor imbalance, blades and hub corrosion and serious aeroelastic deflections in large wind turbines [13].

2.3 Electronic Control

The conventional failures are various including short-circuit, damages in generator winding and transformer wirings as well as over voltage of the subsystems. The root causes are lightning, poor electrical installation and technical defects [14]. Sensors are used for some monitoring of technologies, which may measure speed, vibrations, temperature, etc. These sensors are together coupled with algorithms and architectures, which allow for efficient monitoring running state of wind turbines [15].

2.4 Generator

Failures in the wind turbine generator will cause several issues such as abnormal noises and excessive vibration [10]. The function of a generator is to convert mechanical energy into electricity energy. There are various generators including cage induction generator, brushless doubly-fed generator, Alternating Current (AC) excited generator, synchronous generator, etc. [13]. However, double-fed asynchronous motors are widely used by most generator manufacturers. Generators operate in the electromagnetic environment for a long time, the common faults will be occurred due to either internal or external failures [16–18]. The short circuits in AC grid as a kind of external failure leads to generator overheating. Internal failures usually ascribed to generator excessive vibration, abnormal noises, insulation damage, etc. [19].

2.5 Yaw System

Yaw system is an essential part of horizontal axis wind turbine. The main rotary components of yaw consist drive motor, reducer and gear ring. Rotating nacelle to make sure wind rotors are in the windward state, the purpose is to make wind turbine achieving the maximum efficiency under the condition of uncertain wind direction. Yaw system is divided into active and passive yaw system. Passive yaw system is relying on wind to complete action on the wind rotor by the relevant agencies [13]. Active yaw adopts electric or hydraulic drag to achieve the windward action, the general structure of the types are gear-driven and sliding. At the same time, the gear driving of passive yaw system is generally applied in the large grid wind turbine. Typical failures on yaw system have motor vibration, yaw gear tooth wear, abnormal noise, the yaw limit switch fault, lubricant leakage and bearing damage [20].

2.6 Methodology

Health monitoring and fault diagnosis are playing a critical role in improving reliability and reducing maintenance cost. The general process of diagnosis is divided into four stages (Fig. 1), namely data collection, health status monitoring, analysis and diagnosis. In the first step, data source comes from existing integrated control platform called Supervisory Control and Data Acquisition (SCADA) such as wind speed, nacelle position, generation capacity, gearbox oil temperature, etc. Moreover, to obtain more accurate monitoring information, installing sensors on drive train of gearbox and generator collection vibration data for further analysis work. The following steps utilise the data are measured and processed by the integrated platform to monitor health status of wind turbines. Based on indexes analysis, standard energy utilisation, actual generation power, real-time load and other parameters to achieve preliminary screening for hidden fault in wind turbines from the massive data. The third stage is extracting the sensitive feature of indicators, usage data mining and data processing in-depth analysis. Finally, identifying the specific fault components by a series of analysis diagnostic methods.

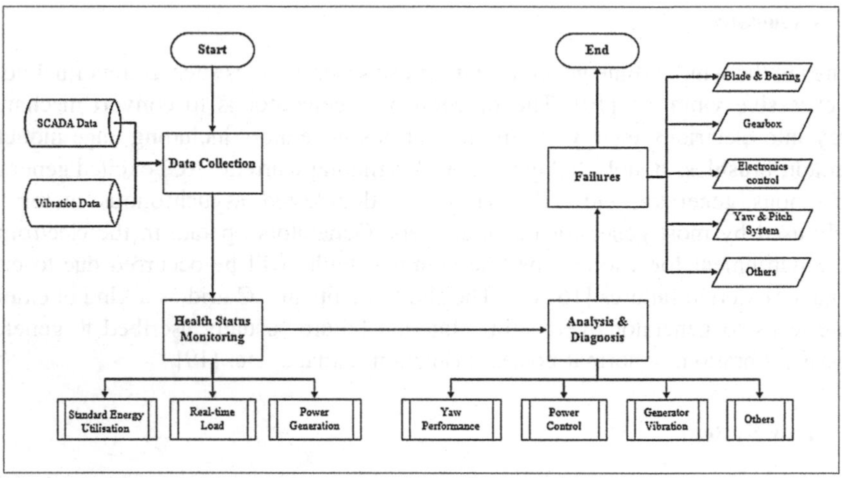

Fig. 1. General process of wind turbine fault diagnosis

3 Monitoring and Diagnosis Platform

In recent years, the wind power industry increasingly needs a system of wind power monitoring and fault diagnosis to increase power generation and improve economic efficiency. This section will introduce a remote diagnosis system for providing remote central control of wind farms, on-line monitoring, fault diagnosis, technical support and other functions.

3.1 System Architecture

The architecture of diagnosis platform provides services from bottom to top layer by layer. It is divided into three layers including data acquisition, application layer and presentation layer. Figure 2 shows the structure of the monitoring platform.

Metadata. Data collection from Condition Monitoring Systems (CMS), SCADA, wind power forecasting, boost station and other data by Kraftwerk Kennzeichen

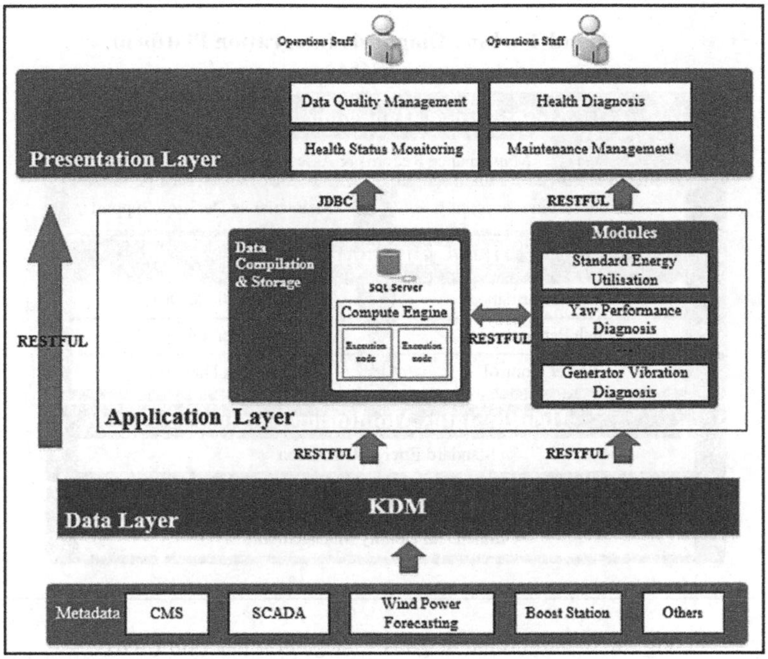

Fig. 2. System structure of the monitoring platform

System (KKS) Data Management (KDM) also can be called via the Representational State Transfer (RESTful) interface.

Data Layer. KDM data centre construction, integrated data acquisition interface, database and forwarding interfaces. KDM acquisition server is connected with metadata by the interface to achieve uniform data coding, as well as standardised data acquisition and transmission.

Application Layer. Application of data analysis is a pivotal segment of the platform, depending on the actual needs of the various layers. It enables embedding several services with no coupling between each other. This part is made up of a Service-Oriented Architecture (SOA) as shown in Fig. 3, there are four main services in this layer, namely data quality management, health status monitoring, health diagnosis and maintenance management service. Furthermore, every service covers some independent modules with the loose coupling of each other, and also the functional modules adopting unified database and interface can be conveniently reused in other systems. The computing engine obtains real-time data from the Application Programming Interface (API) of KDM RESTful, and the wind turbine account generates calculation index by performing data processing. Finally, the result will be exhibited in the presentation layer after the analysis of functional modules.

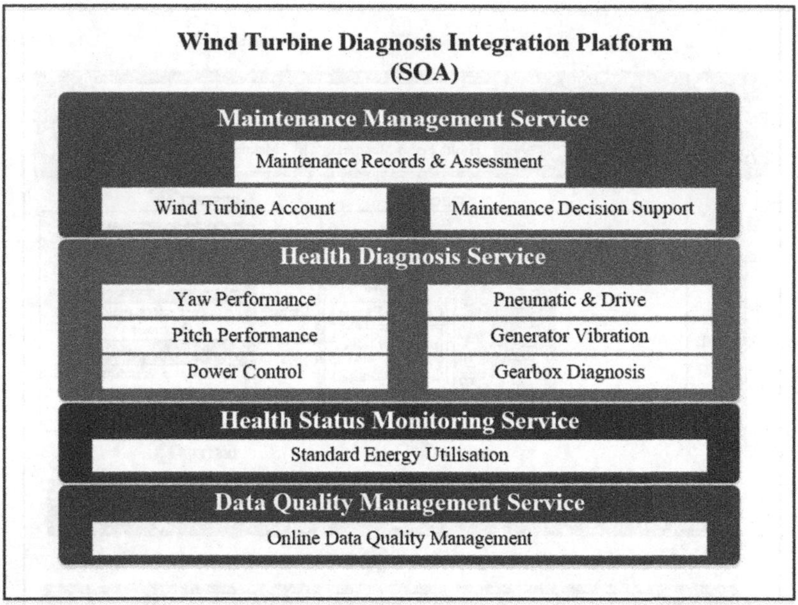

Fig. 3. Wind turbine diagnosis integration platform (SOA)

Presentation Layer. Displaying results of four core functions, namely data quality management, health status monitoring, health diagnosis and maintenance management by intuitive way or a certain logical calculation.

3.2 Data Quality Management Service

Data quality holds the key to a diagnosis platform. This service is used for displaying and alarming abnormal data. Data anomaly affects the diagnostic accuracy largely and in particular data interruption and data delay account for a major proportion. This service adopts online data quality management module to monitor health status. Also, this platform collects 10 SCADA data points (wind speed, power, power generation, power factor, generator speed, gearbox oil temperature, wind turbine state, nacelle, wind direction and nacelle position). In addition, it also collects 8 vibration data (main bearing vertical vibration, radial vibration of gearbox input, radial vibration of gearbox inner ring gear, low-speed and high-speed shaft radial vibration of the gearbox, the high-speed shaft of gearbox axial vibration, radial vibration of generator drive end and radial vibration of generator non - drive end.) Finally, the abnormal data can be screened out from the massive data and displayed on the platform in an intuitive way.

3.3 Health Status Monitoring Service

This function interface displays the wind turbines' real-time operation status. Through calculation and comparison of standard energy utilisation to identify problems. Based on the standard energy utilisation module, the wind turbine condition monitoring could be realised. The platform is developed in cooperation with China Huadian Corporation. Figure 4 displays different regional indexes, the bottom right corner shows installed capacity and power generation overview. According to the standard energy utilisation module, reveals energy indexes of various wind farms. At the same time, compared with the loss of power to find problems (such as power brownouts, failure, maintenance, limited load) and to achieve query of problematic turbines.

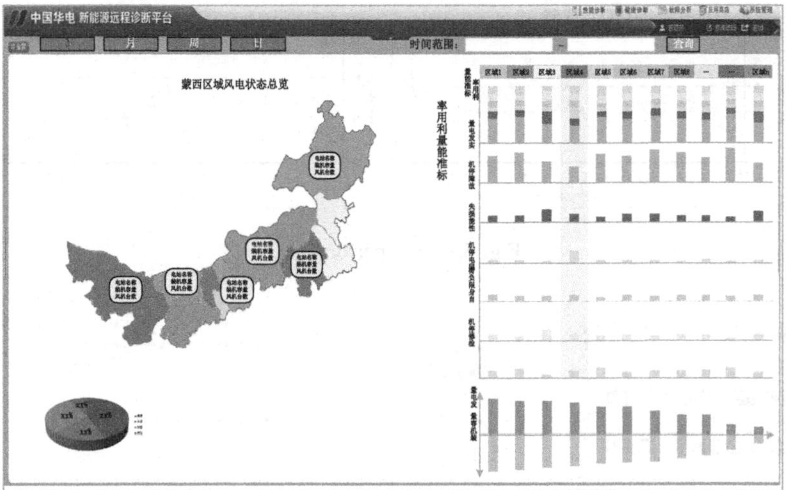

Fig. 4. Regional index

3.4 Health Diagnosis Service

The service of wind turbine health diagnosis is based on condition monitoring, through diagnostic modules to diagnose specific defects and define the causes of problems. It focuses on the wind turbines which have hidden failures. The diagnostic modules involve yaw and pitch performance, generator vibration and gearbox diagnosis. These modules carry out the comprehensive diagnosis and analysis, as well as providing optimisation and rectification opinions to the problematic turbines, which against on uncertainty of the wind direction, yaw angle error, pitch in advance and other issues. In order to avoid power loss caused by low wind energy, utilisation of yaw system diagnostic module and pitch performance module is needed to detect failures. Figure 5 shows the diagnosis module of yaw system performance, the interface displays yaw related parameters, mean and standard deviation of wind directions which wind indicator reveals are under different speed (two line charts, one on lower left and the other on lower right). Figure 6 shows pitch performance module, the top left of the interface describing the equipment, on the opposite, is the related parameters putting forward corresponding suggestions from the same parameters in the performance statistics. The scatter plot of power - pitch angle and speed - pitch angle are distributed on both sides of the bottom interface respectively. Moreover, the further relevant issue can be identified by analysing the graphics. For example, the turbine cannot reach the rated situation indication from the scatter plot power-pitch angle.

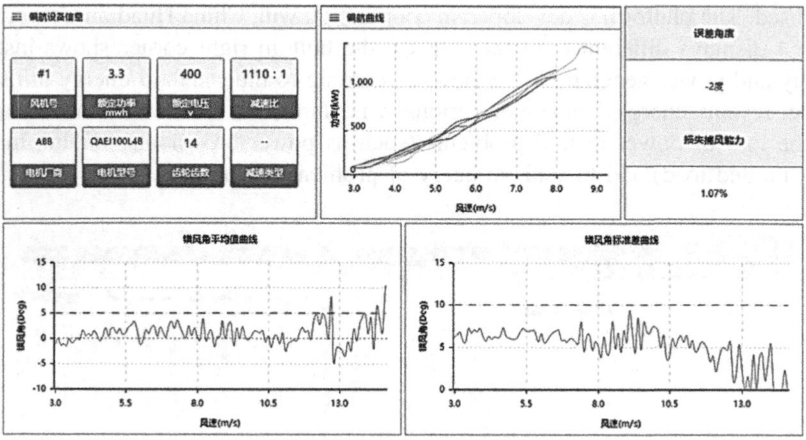

Fig. 5. Yaw performance

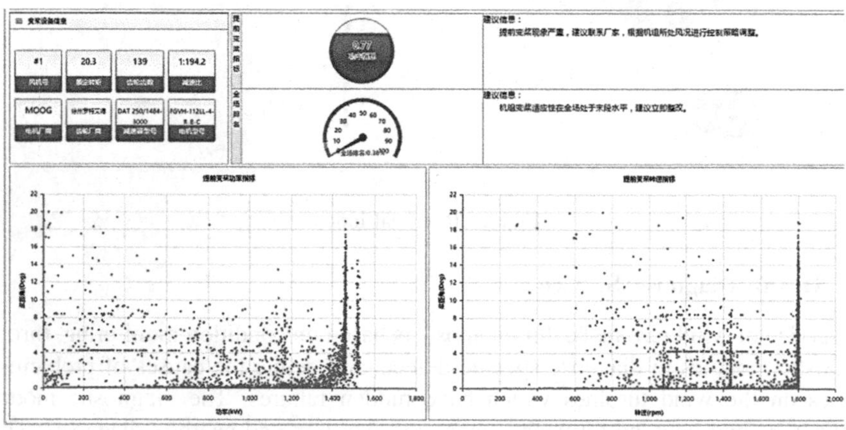

Fig. 6. Pitch performance

Pneumatic and Drive Performance Module. Similarly, the module of pneumatic and drive displays related parameters about pneumatic equipment as shown in Fig. 7. The line chart shows trends of the smallest pitch angle of the wind turbine, wind capture ability of pneumatic drive and the median index value of wind catching ability in whole site. From the tendency in line chart of Fig. 7, can recognise the performance fluctuations which facilitate further analysis. The line chart indicates running minimum pitch angle has an increasing tendency and the continuous indicator decline. Therefore, this unit may appear blade angle on zero which affects the unit aerodynamic performance. Furthermore, in Fig. 7 also presents the chart of unit wind speed - pitch angle exhibits on the lower left of the figure, as well as the unit power - pitch angle scatters plot on the lower right side. These two graphs were generated by filtered data points which assist the analyst to determine any further relevant questions by interpreting the figure.

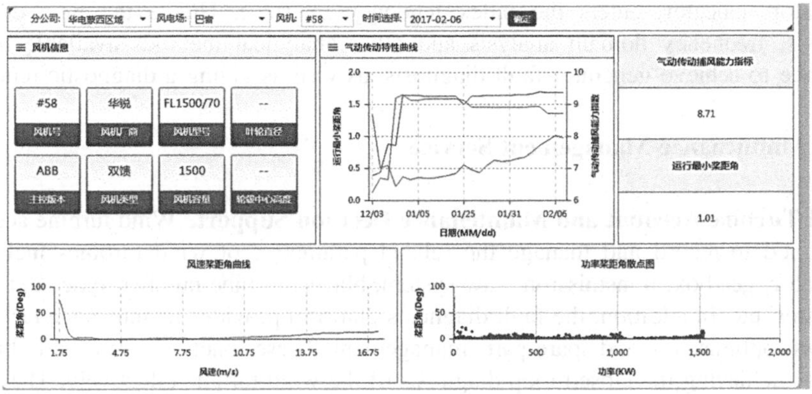

Fig. 7. Pneumatic and drive interface

Generator Vibration Diagnosis Module. In the interface of generator vibration diagnosis (Fig. 8) consists of the basic information of the generator, status of the indicator and its' related diagnostic recommendations and other changes records of the indicators as time goes by. Generators operate in the electromagnetic environment, the common failures including excessive generator vibration, generator overheating, bearing overheating, abnormal noises and insulation damage, etc. By clicking on the

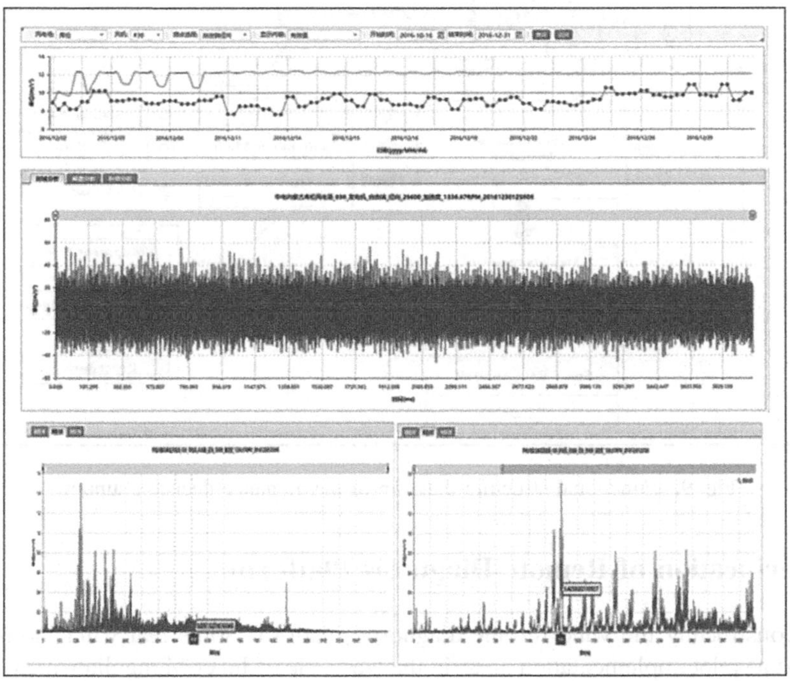

Fig. 8. Generator analysis-specific diagnosis

generator indicator, enters the index tendency interface. Utilise the time domain analysis, frequency domain analysis and other analytical methods available in the interface to achieve generator fault diagnosis, as well as giving a diagnostic report.

3.5 Maintenance Management Service

Wind Turbine Account and Maintenance Decision Support. Wind turbine account is applied to record and manage the related parameters of wind turbines including generator, gearbox, transmission, yaw system, blades, turbine number, type, and manufacturer, etc. In addition, the fault diagnosis platform provides maintenance technical support, optimisation and spare parts management. These functions are used to handle failures achieving the closed-loop diagnosis of the wind turbine effectively. The functions of platform structure are shown in the following Fig. 9. First of all, determine the fault mode based on the results of the diagnostic analysis, generating standardised maintenance work package according to the fault mode then moving to maintenance mode. The final decision requires professionals to confirm through the actual site and eventually to the maintenance and implementation stage. Other that, the platform can make statistical analysis for equipment defects, spare parts and other information forming a closed loop with monitoring, evaluation, and diagnostic modules.

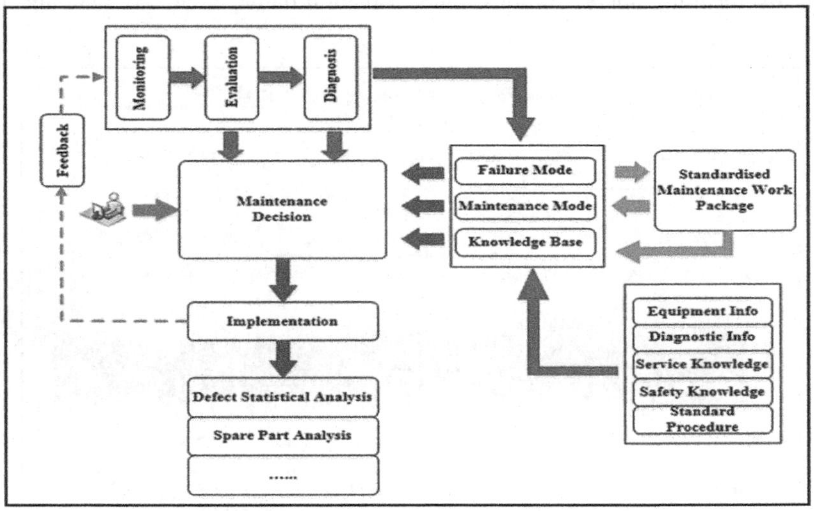

Fig. 9. Functional structure diagram of maintenance decision support

4 Verification of Remote Diagnosis Platform

After constructing the wind turbine diagnostic platform, some real wind farms were selected as pilot implementation areas in the west side of Inner Mongolia which belong to Huadian Corporation and the system begin running trials from July 2016.

The installed capacity of these parts reached to 2.3 MW, total generating capacity of 4.6 TWh. Taking a day as an example, the platform collects 9792 sets of data through the analysis of several diagnosis modules in the platform identifying that yaw, pitch, main driving system, and other fault modes total of 18 units faults in wind turbines. The platform is applied to practical production not only to avoid extra financial loss, as well as prompt technical support provision for maintenance staff. Through the preliminary evaluation gained that the remote intelligent diagnostic platform is effective for the condition monitoring and fault diagnosis of wind turbines. The interface is user-friendly, and this system can collaborate with different technical personnel utilisation. Therefore, the remote diagnostic platform has produced social and economic benefit during the trial operation.

5 Conclusion

In recent years, the maintenance of wind turbines has become the focus of the wind power industry. This paper has discussed the typical faults occurring in wind turbines. According to the analysis of the general process of wind turbine fault diagnosis, development of a remote online status monitoring and an intelligent diagnostic system is implemented for wind turbines. The novelty of this solution is two-fold. First, the conditional-monitoring solution combines a range of analysis methods to provide a means of evaluating the exploitation of energy as the turbines are running. Second, the system provides an integrated solution for both online monitoring and fault diagnosis, and thus offers better support for members of staff with different roles (e.g. management, technician and experts). Several independent and loosely-coupled modules are integrated into this system to implement operation monitoring of wind turbine, fault diagnosis and maintenance management. Additionally, this platform allows adding professional function modules for enhancing system accuracy and stability. The diagnostic platform has been applied to real-world wind farms, confirming the feasibility and viability of various functions the data collection, analysis and diagnosis and so on. As shown in the application, this platform provides effective and efficient technical support for operation staff of wind farm. In our future work, further verification work will be done to evaluate the diagnosis capability and data analysis capability of the intelligent diagnostic system.

References

1. Farahani, E., Hosseinzadeh, N., Ektesabi, M.: Comparison of fault-ride-through capability of dual and single-rotor wind turbines. Renew. Energy **48**, 473–481 (2012)
2. Global Wind Energy Council: GLOBAL WIND REPORT ANNUAL MARKET UPDATE, 13–14 (2015)
3. Soua, S., Van Lieshout, P., Perera, A., Gan, T., Bridge, B.: Determination of the combined vibrational and acoustic emission signature of a wind turbine gearbox and generator shaft in service as a pre-requisite for effective condition monitoring. Renew. Energy **51**, 175–181 (2013)

4. Márquez, F.G., Tobias, A., Pérez, J.P., Papaelias, M.: Condition monitoring of wind turbines: techniques and methods. Renew. Energy **46**, 169–178 (2012)
5. Milborrow, D.: Operation and maintenance costs compared and revealed. Windstats Newsl. **19**, 1–3 (2006)
6. Lau, B.C.P., Ma, E.W.M., Pecht, M.: Review of offshore wind turbine failures and fault prognostic methods. In: 2012 IEEE Conference on Prognostics and System Health Management (PHM), pp. 1–5. IEEE (2012)
7. Hahn, B., Durstewitz, M., Rohrig, K.: Reliability of wind turbines. In: 2007 Conference on Wind Energy, pp. 329–332. Springer, Heidelberg (2007)
8. Spinato, F., Tavner, P., van Bussel, G., Koutoulakos, E.: Reliability of wind turbine subassemblies. IET Renew. Power Gener. **3**, 387 (2009)
9. Amirat, Y., Benbouzid, M.E.H., Member, S., Bensaker, B., Wamkeue, R.: Condition monitoring and fault diagnosis in wind energy conversion systems: a review. In: Proceedings of the IEEE International Electric Machines and Drives Conference, IEMDC 2007, Antalya, Turkey, vol. 2, pp. 1434–1439. IEEE (2007)
10. Takoutsing, P., Wamkeue, R., Ouhrouche, M., Slaoui-Hasnaoui, F., Tameghe, T., Ekemb, G.: Wind turbine condition monitoring: state-of-the-art review, new trends, and future challenges. Energies **7**, 2595–2630 (2014)
11. Liu, W., Tang, B., Han, J., Lu, X., Hu, N., He, Z.: The structure healthy condition monitoring and fault diagnosis methods in wind turbines: a review. Renew. Sustain. Energy Rev. **44**, 466–472 (2015)
12. Amirat, Y., Benbouzid, M., Al-Ahmar, E., Bensaker, B., Turri, S.: A brief status on condition monitoring and fault diagnosis in wind energy conversion systems. Renew. Sustain. Energy Rev. **13**, 2629–2636 (2009)
13. Yang, M., Chengbing, H., Xinxin, F.: Institutions function and failure statistic and analysis of wind turbine. Phys. Procedia **24**, 25–30 (2012)
14. The Confederation of Fire Protection Association CFP A Europe: Wind turbines fire protection guideline (2010)
15. Benbouzid, M.: Bibliography on induction motors faults detection and diagnosis. IEEE Trans. Energy Convers. **14**, 1065–1074 (1999)
16. Yang, W., Tavner, P., Wilkinson, M.: Condition monitoring and fault diagnosis of a wind turbine synchronous generator drive train. IET Renew. Power Gener. **3**, 1 (2009)
17. Bennouna, H., Camblong, R.: Diagnosis of the doubly-fed induction generator of a wind turbine. Wind Eng. **29**, 431–447 (2005)
18. Rothenhagen, K., Fuchs, F.: Doubly fed induction generator model-based sensor fault detection and control loop reconfiguration. IEEE Trans. Ind. Electron. **56**, 4229–4238 (2009)
19. Attya, A., Hartkopf, T.: Penetration impact of wind farms equipped with frequency variations ride through algorithm on power system frequency response. Int. J. Electr. Power Energy Syst. **40**, 94–103 (2012)
20. Ghaedi, A., Abbaspour, A., Fotuhi-Firuzabad, M., Moeini-Aghtaie, M.: Toward a comprehensive model of large-scale DFIG-based wind farms in adequacy assessment of power systems. IEEE Trans. Sustain. Energy **5**, 55–63 (2014)

Rehabilitation Training for Leg Based on EEG-EMG Fusion

Heng Tang[1(✉)], Gongfa Li[1,2], Ying Sun[1,2], Guozhang Jiang[1,2], Jianyi Kong[1,2], Zhaojie Ju[3], and Du Jiang[1]

[1] Key Laboratory of Metallurgical Equipment and Control Technology, Ministry of Education, Wuhan University of Science and Technology, Wuhan 430081, China
464792431@qq.com

[2] Hubei Key Laboratory of Mechanical Transmission and Manufacturing Engineering, Wuhan University of Science and Technology, Wuhan 430081, China

[3] The Laboratory of Intelligent System and Biomedical Robotics, University of Portsmouth, Portsmouth PO1 3HE, UK

Abstract. Stroke is a kind of cerebral vascular disease with high death rate and high disability rate, most stroke patients lose a lot of physiological function. For example, motor function, language function, etc. Two data acquisition methods of lower limb rehabilitation system for patients with stroke were introduced in this paper that is EEG signal extraction based on BCI and lower limb muscle electrical stimulation system based on EMG model. Through the wavelet packet transform (WPT) to analyze the EEG signal and collect the effective EEG signal. The wavelet transform is used to analyze the time and frequency domain, which provides a good feature vector for the dynamic analysis and motion recognition of EMG signals.

Keywords: EEG signal · EMG signal · Feature extraction · Pattern recognition · Electrical stimulation system

1 Introduction

About tens of millions of cerebrovascular patients around the world today, who died of a stroke among them about ten million one year [1]. The medical definition of stroke is a kind of cerebrovascular disease due to cerebral vascular blockage or cerebral vascular rupture, causes the brain blood supply shortage and then affects the local brain area function damage. Medical theory and clinical Medicine shows that early surgical treatment and necessary drug treatment is not enough to these patients, the correct and scientific rehabilitation training has played a very important role in recovery and improvement of limb motor function [2]. The best time to recover limb function for stroke patients is within 11 weeks after stroke. Therefore, it is of great significance to take exercise rehabilitation training as early as possible to regain the ability of autonomous walking [3].

Y. Huang et al. (Eds.): ICIRA 2017, Part I, LNAI 10462, pp. 517–527, 2017.
DOI: 10.1007/978-3-319-65289-4_49

The traditional rehabilitative training relies on the physical therapist to carry on the real-time movement guidance to the patients. It can improve the patient's movement function better, its long-term training effect is more remarkable. Taking common weight loss walking training as an example, in the traditional way, three physiotherapists at least are required to complete a gait training. As a result, there are serious phenomenons of the lack of a physical therapist. At the same time, traditional artificial rehabilitation training needs to consume a lot of time and physical strength of physical therapist. The efficacy of treatment depends on the experience and level of physical therapist. For patients, because of the brain dysfunction, the training process is more passive treatment, as active therapy effect is obvious [4].

The lower limb rehabilitation training system introduced in this paper is a physical adjuvant therapy to help stroke patients recover after surgery. The system establishes an external skeletal machine that assists the patient's lower limbs movement, and stimulates passive physical training of human physical therapists, to alleviate the atrophy of muscle due to prolonged relaxation. At the same time, based on BCI technology [5] to extract brain electrical signals from patients and reading the consciousness of the human movement center, then to repair or replace the patient's nerve output. According to EMG modeling [6], an effective electrical stimulation model for lower limb muscles can be established. We can use functional electrical stimulation to stimulate the lower limb muscles, so that the patient's brain and lower limb muscles actively participate in active physical exercise [7].

2 EEG Signal Acquisition Based on BCI

BCI is a direct communication and control channel between the human brain and computer or other electronic devices. BCI technology is an interdisciplinary techniques which is involved in neurophysiology, signal processing, pattern recognition, control theory, computer science and rehabilitation medicine [8]. In the 1990s, research on BCI has rose gradually, and made some substantial progress. In 1995, there were no more than 6 groups and organizations studying BCI technology, and now it has grown to hundreds.

2.1 Signal Acquisition

At present, two common methods of signal acquisition in BCI system are invasive and non-invasive tapes. Invasive type requires specialist do surgery and place an electrode in the brain, detect EEG signals such as cortical images. There is a certain risk in detection of Electrocorticogram (ECoG) [9], also exists psychological and ethical issues. Non-invasive type is to put electrode cap on the head to detect EEG (EEG) signals. This method is simple, but the electrode is far away from the neuron, the signal-to-noise ratio of the measured is low, which brings trouble to the subsequent treatment. At present, non-invasive systems are more applied. That the signal collected will be amplified by the amplifier, and then preprocessed, finally converted to digital signal stored in the computer. At present, the signal acquired by BCI system has Visual Evoked Potentials(VEP) [10], Event – Related Potentials (ERP) [11] P300, Slow Cortical Potentials (SCP) [12], Alpha-wave of spontaneous EEG, μ-rhythm and β-rhythm signals [13], etc.

2.2 Feature Extraction

The signal characteristics of the users intentions are extracted from the EEG signals after preprocessing and digitized processing. At present, the common feature extraction methods of BCI system have autoregressive AR model [14], Wavelet Transform [15] and Wavelet Packet Energy [16], etc. Feature extraction can choose the corresponding time-domain or frequency-domain feature extraction method according to the characteristics of the signal. Such as, due to the P300 signal in the vicinity of 300 ms mixing seriously, CWT is needed to extract time domain characteristics as the main extraction [17]. Steady–State Visual Evoked Potential has good frequency domain characteristics, Fast Fourier Transform (FFT) can be used to extraction [18]. Feature extraction of the Event–Related Desynchronizatlon (ERD) of the μ rhythm and β rhythm needs lateral effects of brain movement to extract its airspace characteristics [19]. However, due to the low SNR of EEG signal, and some interference components have similar time-frequency characteristics to the signal. Simple time-frequency feature extraction can affect the classification effect by insufficient differentiation. Therefore, it is necessary to turn to a more complex time-domain-spatial analysis method for the synthesis of signal processing.

2.3 Signal Classification

Signal classification is accomplished by classifier, including linear and nonlinear types. Linear classifier including linear Discriminant Analysis (LDA) [20] and Fisher Discriminant Analysis (FDA) [21]. Nonlinear classifier includes Two Discriminant Analysis [22], Support Vector Machine (SVM) [23], Artificial Neural Nets (ANN) [24], Genetic Algorithm (GA) [25] and Fuzzy Algorithm (FA) [26], etc. The linear classifier is easy to learn and has high classification accuracy in case of low characteristic dimension, is widely applied now. But for nonlinear data with high dimensionality, the linear classifier is no longer applicable, the nonlinear classifier should be used.

2.4 Rehabilitation Technology Based on BCI Technology

There are three kinds of rehabilitation technology based on BCI technology. First one, using BCI systems to communicate directly with the outside world, as controlling nerve prostheses, smart wheelchairs, cursors on computer screens, etc. Second one, for those with nerve blocking but the limb still in the handicapped, BCI system can be used to directly control the limb muscles, make the limbs complete the basic action in daily life. Third one, due to the development of neuroscience, scientists have found that the function of the central nervous system in the whole life process can be reconstructed under the reasonable physiological potentials, so Neural recovery can be performed with BCI system, such as stimulating the recovery of nerve by observing the movement and generating the corresponding EEG [27]. Xu bao-guo et al. [28] has proposed a system of upper limb rehabilitation which is based on motion imagine EEG. Training in passive training, active training and active training with damped. And finally their online recognition rate as Table 1.

Table 1. Online recognition rate (passive training).

The subjects	The first group/(%)	The second group/(%)	The third group/(%)	On average/(%)
A	90.0	92.5	95.0	92.5
B	90.0	92.5	100.0	94.1
C	92.5	90.0	95.0	92.5
D	92.5	90.0	97.5	93.3
E	95.0	92.5	97.5	95.0
F	92.5	87.5	95.0	91.6

3 Muscle Electric Stimulation System of Lower Limb Based on EMG Model

3.1 Characteristics of EMG Signal

Surface EMG signal is the superposition of action potentials which many motion unit in muscles in time and space, reflecting the functioning state of nerves and muscles. There are differences in different individuals. However, there are common features, and surface EMG signals have typically characterized by the following [29].

1. EMG signal is a faint signal. The amplitude range of EMG signals is generally 0–5 mV, about 60–300 uV when muscles shrink, about 20–30 uV when relaxation and generally not exceeding the noise level. For healthy adults, peak peaks of EMG can reach 1–3 mV. For the remnant of the limb, the EMG amplitude is generally less than 350 uV, some even less than 1 uV which is less a few times to dozens of times than average person.
2. EMG signal is a kind of AC voltage, cross-degeneration exists in it. It is roughly proportional to the strength of the muscles in the amplitude. There is a good linear relationship between muscle relaxation and tension and the amplitude of the surface EMG voltage.
3. The characteristic of EMG can be reflected more in frequency domain parameter than time domain parameter. Because when the force size changes slightly, the time domain waveform is rather changeable, and the frequency domain characteristic is no big change.
4. When the surface electrode is used, the energy of EMG signal is mainly concentrated under 1000 Hz. Spectrum distribution between 20–500 Hz, mainly concentrated under 500 Hz, more than 300 Hz significantly weakened. Among them, most of the spectrum is concentrated between 50–150 Hz. The maximum frequency of power spectra varies with muscle, usually between 30–300 Hz.
5. Different muscle EMG signal exists difference [30]. In frequency domain, the amplitude of EMG signal of different muscles is not only the difference in numerical range, but also the distribution of power spectrum.
6. The amplitude spectral frequency characteristic curves of the same muscles are still similar in different motions. It shows that there is a certain regularity of EMG distribution in different muscles [31].

7. These typical characteristics provide theoretical basis for the application of EMG signal and the design of collecting system.

3.2 Feature Extraction Method

There are a lot of electromyographic signal feature extraction methods, mainly including time domain analysis, frequency domain analysis method, time - frequency domain analysis method, based on the parameter model and the nonlinear characteristic analysis method. Reddy [32] extracted the eigenvalues of sEMG by the time domain method, which was used to study the relationship between sEMG and motion displacement, thus realizing the control of finger and wrist joint model. Hou Wen-sheng et al. [33] studied the correlation between grip strength and the pattern of muscle activity on the forearm muscles by using mean square root as the characteristic parameter. Studies have shown that grip strength is related to muscle activity patterns, and suggested that sEMG can not only predict grip strength, but also can be used for motor function test and rehabilitation evaluation. Luo Zhi-zeng [34] proposed a new method for feature extraction of sEMG signals called power spectral ratio method. In this method, the power spectrum of surface EMG signal near the maximum amplitude at the frequency of a certain width of the power spectrum and the power spectrum integral integral ratio, as the feature of surface EMG signal and hand motion recognition, and it achieved ideal effect. Wavelet packet transforming can precisely describes signals and provides rich pattern information at different frequency bands. Cheng Bo [35] proposed the wavelet packet entropy method to analyze sEMG signals in different emotion states. Luo Zhizeng [36] used the order AR model coefficients as the input of the neural network, the output corresponding to the wrist rest, wrist cupped, wrist valgus and other movements, to achieve effective control of the prosthetic hand. Nonlinear research methods are gradually applied in the field of EMG signal processing, which mainly includes two aspects, nonlinear PCA and nonlinear dynamics.

3.3 Current Development of Electric Stimulation System

After years of development, the method has been invented by various electro-stimulating therapies. The frame of the electrical stimulation system is shown in Fig. 1. In foreign countries, the TENS-type nerve muscle power-promoting apparatus developed in Japan adopts single channel low-frequency current to shrink the muscle of paralyzed patients [37]. American Respond Select electric stimulation therapeutic apparatus have achieved dual channel low frequency electric stimulation [38]. In the domestic, some companies have produced electric stimulation therapeutic apparatus not only have multi-channel electric stimulation output also can realize the adjustment of parameter as stimulation frequency and stimulation intensity and so on [39], meet the treatment needs of a large number of patients in China and improve therapeutic efficiency. Bobath technology as a classical nerve development therapy, is the most commonly used treatment methods and techniques. But there are some problems within the method, for example, training efficiency and training intensity are difficult to guarantee, the therapeutic effect of the therapist level, is lacking objective data to evaluate the relationship between training parameters and rehabilitation effects, is

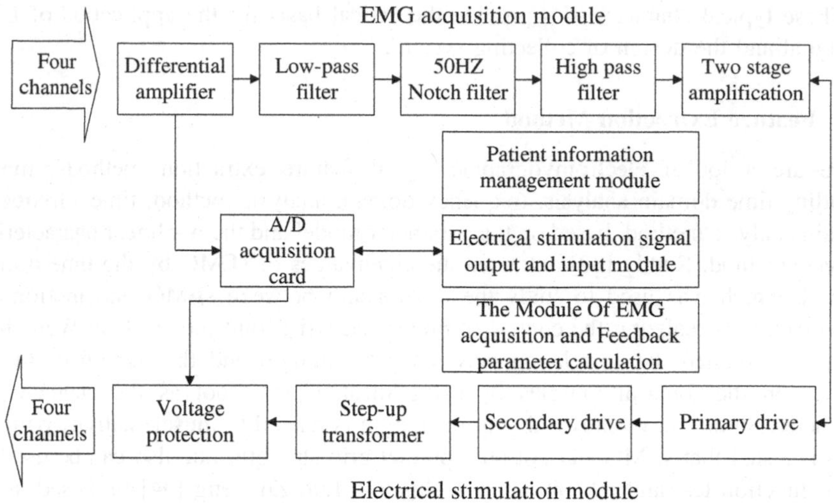

Fig. 1. Overall structure of the system

difficult to optimize the training parameters to obtain the best treatment scheme, poor in pertinence of training, etc. Surface electromyography (sEMG) is a set of voltage-time series signals that are obtained by the electro-electrical changes of the neuromuscular system when conducting random and non-random activities by the surface electrode to guide, enlarge, display and record. The basic research has shown that the changes of SEMG under the condition of good control can quantitatively reflect the local fatigue degree, muscle strength level, muscle activation mode, motor unit excited conduction velocity, multi muscle group coordination and other muscle activity and central control characteristics.

4 EEG-sEMG Fusion

Interaction control between robots and patients is a very important aspect in the research of lower limb rehabilitation robots. Because the rehabilitation robot of lower limbs is interacting with the injured limb, and the patient is the object of autonomic movement consciousness, the interaction control between the robot and the patient is indispensable. First, interactive control creates a safe, comfortable, natural and proactive training environment for patients to avoid the disease, such as spasm, trembling, abnormal muscle activity and the robot to resist, protect it from two times damage. Secondly, the interactive controlling will obtain the active intention of the patient from the sensor signal, encourage the patient to participate actively in the movement, realize the so-called active training, thus improve the healing effect.

Functional evaluation of motor nervous system is a key problem in the field of sports functional rehabilitation. At present, the sports function evaluation in the rehabilitation process of sports dysfunction mainly includes physical morphology evaluation, muscle strength and muscle tone grade evaluation, and evaluation of daily activity ability. And most of the judgment by the medical experience, subjective and inefficient. How to effectively identify the movement characteristics of neuromuscular system, and to construct an objective quantification evaluation system of rehabilitation function is especially important [40]. EEG and EMG signal includes the body movement control information and the muscle's function response information to the brain control intent, which can react directly to the functional state of the nervous system. Therefore, the study on the evaluation method of motion function based on EEG and EMG signals should be extracted directly from the physiological parameters of motor nervous system as evaluation index. This method becomes the research trend of sports function rehabilitation evaluation and is concerned with the researchers gradually.

Mima [41] found that there was no significant correlation between the brain and EMG in the lateral upper limb of the subcortical stroke. Fang [42] found that the brain and EMG coherence in the hand movement of the stroke patients were less than those in the gamma band.

Because the nervous system transmits motion control information and sensory feedback information through nerve oscillation, nerve damage hinders the conduction of nerve oscillation and causes motor dysfunction [43]. The correlation between the brain and EMG showed that the coherence between the brain and EMG can reflect the oscillation of the central nervous system and muscle, and then evaluate the function of the motor nerve. The study shows that in process of movement, the coherence of brain and EMG is mainly manifested in the EEG beta frequency and gamma frequency. The coherence of different frequencies reflects different functional relations. The EEG beta frequency oscillation is correlated with the muscle movement of the human body, and the coherence of the brain and EMG is correlated with the static force output of muscle. Gamma-band oscillation in EEG reflects the process of cerebral cortex information integration which is related to cognitive function [44]. Therefore, the analysis of the brain and EMG coherence in beta and gamma two bands contributes to evaluating the functional state of the nervous system. At the same time, the existing coherence analysis method is mainly based on Fourier transform to realize spectral estimation of EEG and EMG signals [45].

Due to the complexity of frequency components and many influencing factors in EEG signals, coherent analysis based on Fourier transform can bring spectral interference and influence the accuracy of coherent spectral estimation in beta and gamma two functional bands. In order to avoid interference between different frequency components of EEG signals, we need to study the coherent analysis method of brain and EMG based on effective time-frequency decomposition are needed to study, and the coherence between the specific EEG and EMG signals related to the control of motor muscle are needed to analyze [46].

Ma Pei-pei [47] has proposed a method of analyzing the coherence of brain and EMG based on wavelet decomposition. That is, making wavelet decomposition of EEG signal to select the corresponding wavelet coefficients than reconstruct beta and gamma band signal. Next step, to do the coherence analysis with EMG signal separately to avoid interference of irrelevant EEG components. To study the neural oscillation relation of the neural motion control system in the EEG characteristic frequency bands, further define the significant coherence area indicator quantization to describe the similarity and phase locking activity between EEG and EMG signals in specific frequency bands. Based on this, the EEG and EMG signals are collected in patients with stroke and sports dysfunction or health people to do synchronous signal acquisition experiment. Comparing the differences between the brain and EMG coherence between the patients and the healthy people and the patient side and the health side have provided a basis for exploring the mechanism of motor dysfunction and evaluating the sports function in the rehabilitation process.

5 Results

How to help patients with stroke rehabilitation is an important research topic in today's society. Compared with the passive rehabilitation, the rehabilitation effect of active rehabilitation is more obvious. The establishment of EMG model in healthy subjects, and an electrical stimulation system was built according to the EMG model. The electrical stimulation system is used as an executive terminal, using the coherence of EEG and EMG signals to extract the active signals as the control side of rehabilitation. Initial results have been achieved, but there are still more work to be carried out. The EEG of classification processing need high accuracy and real-time performance, compared with the wavelet package relative energy as the feature, based on wavelet packet coefficient variance has more advantages, and wavelet packet coefficient variance meaning is clear and its calculation is simple, the noise is not sensitive. But the wavelet packet variance directly applied to EEG and BCI research is relatively small. Rehabilitation training for stroke patients may involve a number of training methods, the accuracy of it exceeds the scope of application will be reduced, so the applicability of the EMG model will be more widely used in subsequent studies. The electrical stimulation system that based on the establishment of the EMG model of the subjects can be very good to meet the human electrical stimulation intensity and reduce the degree of muscle fatigue, but it still cannot meet the comfort requirement. So the research should be to improve the comfort the comfort of electrical stimulation system and increase feedback system in the future. Electrical stimulation system can be dynamically regulated for the recovery of the patients, let the patient can carry on the self-rehabilitation training in the daily life. CMC signal is used to extract the independent signal of stroke patients, and can better extract the signal of self-awareness of stroke patients. Accuracy is higher than that of the EEG and EMG signals, but requires more accurate extraction of the autonomic consciousness of the stroke patients. But the real time of data processing is not its advantage, the research should focus on improving the real-time signal in the future, so that it can be better applied to the rehabilitation system.

Acknowledgments. This work was supported by grants of National Natural Science Foundation of China (Grant Nos. 51575407, 51575338, 51575412) and the UK Engineering and Physical Science Research Council (Grant No. EP/G041377/1).

References

1. Naghavi, M., Wang, H., et al.: Global, regional, and national age-sex specific all-cause and cause-specific mortality for 240 causes of death, 1990–2013: a systematic analysis for the global burden of disease study 2013. Lancet **385**(9963), 71–117 (2015)
2. Luo, Y., Xu, C.-Y.: Effect of rehabilitation training and rehabilitation after discharge on the rehabilitation of cerebral infarction. Chin. J. Gen. Pract. **9**(10), 1646–1647 (2011)
3. Tesio, L.: Outcome measurement in behavioural sciences: a view on how to shift attention from means to individuals and why. Int. J. Rehabil. Res. **35**(1), 1–12 (2012)
4. Beldalois, J.M., Horno, S.M., Bermejobosch, I., et al.: Rehabilitation of gait after stroke: a review towards a top-down approach. J. Neuroengineering Rehabil. **8**(1), 1–20 (2011)
5. Chu, Y., Zhao, X., Han, J., et al.: SSVEP based brain-computer interface controlled functional electrical stimulation system for upper extremity rehabilitation. In: IEEE International Conference on Robotics and Biomimetics, pp. 2244–2249 (2014)
6. Do, A.H., Wang, P.T., King, C.E., et al.: Brain-computer interface controlled functional electrical stimulation device for foot drop due to stroke. In: International Conference of the IEEE Engineering in Medicine and Biology Society, pp. 6414–6417 (2012)
7. Takahashi, M., Takeda, K., Otaka, Y., et al.: Event related desynchronization-modulated functional electrical stimulation system for stroke rehabilitation: a feasibility study. J. Neuroengineering Rehabil. **9**(1), 56–61 (2012)
8. Rajan, B., Lokesh, J., Kiron, V., et al.: Differentially expressed proteins in the skin mucus of Atlantic cod (Gadus morhua) upon natural infection with Vibrio anguillarum. BMC Vet. Res. **9**(1), 1–11 (2013)
9. Xu, F., Zhou, W., Zhen, Y., et al.: Classification of motor imagery tasks for electrocorticogram based brain-computer interface. Biomed. Eng. Lett. **4**(2), 149–157 (2014)
10. Nawrocka, A., Holewa, K.: Brain - computer interface based on steady - state visual evoked potentials. In: Carpathian Control Conference, pp. 251–254. IEEE (2013)
11. Aloise, F., Schettini, F., Aricò, P., et al.: Asynchronous P300-based brain-computer interface to control a virtual environment: initial tests on end users. Clin. EEG Neurosci. **42**(4), 219–224 (2011)
12. Northoff, G.: Slow cortical potentials and "inner time consciousness" A neuro-phenomenal hypothesis about the "width of present". Int. J. Psychophysiol. **103**, 174–184 (2015)
13. Jamal, W., Das, S., Maharatna, K., et al.: Brain connectivity analysis from EEG signals using stable phase-synchronized states during face perception tasks. Physica A **434**, 273–295 (2015)
14. Atyabi, A., Shic, F., Naples, A.: Mixture of autoregressive modeling orders and its implication on single trial EEG classification. Expert Syst. Appl. **65**, 164–180 (2016)
15. Lekshmi, S.S., Selvam, V., Pallikonda Rajasekaran, M.: EEG signal classification using principal component analysis and wavelet transform with neural network. In: International Conference on Communications and Signal Processing, pp. 687–690 (2014)
16. Wang, D., Miao, D., Xie, C.: Best basis-based wavelet packet entropy feature extraction and hierarchical EEG classification for epileptic detection. Expert Syst. Appl. **38**(11), 14314–14320 (2011)
17. Duvinage, M., Castermans, T., Petieau, M., et al.: A subjective assessment of a P300 BCI system for lower-limb rehabilitation purposes. In: Engineering in Medicine and Biology Society, pp. 3845–3849 (2012)
18. Diez, P.F., Müller, S.M.T., Mut, V.A., et al.: Commanding a robotic wheelchair with a high-frequency steady-state visual evoked potential based brain–computer interface. Med. Eng. Phys. **35**(8), 1155–1164 (2013)

19. Maclean, M.H., Arnell, K.M.: Greater attentional blink magnitude is associated with higher levels of anticipatory attention as measured by alpha event-related desynchronization. Brain Res. **1387**(2), 99–107 (2011)
20. Kang, S.K., Choi, H.H., Chang, S.M., et al.: Comparison of k-nearest neighbor, quadratic discriminant and linear discriminant analysis in classification of electromyogram signals based on the wrist-motion directions. Curr. Appl. Phys. **11**(3), 740–745 (2011)
21. Sutarman, M.M.A., Zain, J.M.: A review on the development of Indonesian sign language recognition system. J. Comput. Sci. **9**(11), 1496–1505 (2013)
22. Jeong, E.C., Kim, S.J., Song, Y.R., et al.: Comparison of wrist motion classification methods using surface electromyogram. J. Cent. S. Univ. **20**(4), 960–968 (2013)
23. Zhang, T., Liu, T., Li, F., et al.: Structural and functional correlates of motor imagery BCI performance: insights from the patterns of fronto-parietal attention network. Neuroimage **134**, 475–485 (2016)
24. Yu, J.H., Sim, K.B.: Classification of color imagination using Emotiv EPOC and event-related potential in electroencephalogram. Opt. Int. J. Light Electron. Opt. **127**(20), 9711–9718 (2016)
25. Roy, R., Mahadevappa, M., Kumar, C.S.: Trajectory path planning of EEG controlled robotic arm using GA. Procedia Comput. Sci. **84**, 147–151 (2016)
26. Das, A.K., Suresh, S., Sundararajan, N.: A discriminative subject-specific spatio-spectral filter selection approach for EEG based motor-imagery task classification. Expert Syst. Appl. **64**, 375–384 (2016)
27. Ang, C.S., Sakel, M., Pepper, M., et al.: Use of brain computer interfaces in neurological rehabilitation. Br. J. Nurs. **7**(3), 523–528 (2011)
28. Xu, B., Song, A., Wang, A.: EEG feature extraction method based on wavelet packet energy. J. SE Univ. (Nat. Sci. Ed.) **40**(6), 1203–1206 (2011)
29. Wang, J., Zhang, C., Bai, B., et al.: Upper limb rehabilitation robot experimental platform for sEMG acquisition system's design. Electron. World **11**, 28–30 (2012)
30. Zhang, Q., Xi, X., Luo, Z.: A pattern recognition method for surface electromyography based on nonlinear features. J. Electron. Inf. Technol. **35**(9), 2054–2058 (2013)
31. Zhang, T., Wang, X.Q., Jiang, L., et al.: Biomechatronic design and control of an anthropomorphic artificial hand for prosthetic applications. Robotica **1**(10), 1–18 (2015)
32. Reddy, N.P., Gupta, V.: Toward direct biocontrol using surface EMG signals: control of finger and wrist joint models. Med. Eng. Phys. **29**(3), 398–403 (2007)
33. Hou, W., Xu, R., Zheng, X., et al.: Relationship between handgrip forces and surface electromyogram activities of forearm muscle. Space Med. Med. Eng. **20**(4), 264–268 (2007)
34. Luo, Z., Wang, R.: Study of myoelectric bionic artificial hand with tactile sense. Chin. J. Sens. Actuat. **18**(1), 23–27 (2005)
35. Cheng, B., Liu, G.: Emotion recognition based on wavelet packet entropy of surface EMG signal. Comput. Eng. Appl. **44**(26), 214–216 (2008)
36. Luo, Z., Yang, G.: Surface electromyography analytical method based on the parameter of AR model. Chin. J. Sens. Actuat. **16**(4), 384–387 (2003)
37. Li, Y., Mao, L.: Neuromuscular electrical promoting role in the treatment of general instrument in median nerve injury. Med. J. Present Clin. **29**(5), 2537 (2016)
38. Yang, B., Tian, R., Lianguo, C., et al.: Meta-analysis of transcranial magnetic stimulation to treat post-stroke dysfunction. Neural Regeneration Res. **6**(22), 1736–1741 (2011)
39. Liu, D., Liu, D., Hong, S., et al.: The effects of low frequency electrical stimulation on connectivity changes in the brain and motor function after ischemic stroke. Chin. J. Phys. Med. Rehabil. **34**(11), 821–824 (2012)
40. Veer, K., Sharma, T.: A novel feature extraction for robust EMG pattern recognition. J. Med. Eng. Technol. **40**(4), 149–154 (2016)

41. Meng, F., Tong, K.Y., Chan, S.T., et al.: Cerebral plasticity after subcortical stroke as revealed by cortico-muscular coherence. IEEE Trans. Neural Syst. Rehabil. Eng. **17**(3), 234–243 (2009)
42. Bayram, M.B., Siemionow, V., Yue, G.H.: Weakening of corticomuscular signal coupling during voluntary motor action in aging. J. Gerontol. **70**(8), 1037–1043 (2015)
43. Pavlidou, A., Schnitzler, A., Lange, J.: Beta oscillations and their functional role in movement perception. Transl. Neurosci. **5**(4), 286–292 (2014)
44. Seeber, M., Scherer, R., Wagner, J., et al.: EEG beta suppression and low gamma modulation are different elements of human upright walking. Front. Hum. Neurosci. **8**(485), 485–486 (2014)
45. Kamp, D., Krause, V., Butz, M., et al.: Changes of cortico-muscular coherence: an early marker of healthy aging? AGE **35**(1), 49–58 (2013)
46. Scafetta, N., Mazzarella, A.: Spectral coherence between climate oscillations and the $M \geq 7$ earthquake historical worldwide record. Nat. Hazards **76**(3), 1807–1829 (2015)
47. Ma, P., Chen, Y., Yihao, D., et al.: Analysis of corticomuscual coherence during rehabilitation exercises after stroke. J. Biomed. Eng. **5**, 971–977 (2014)

A Review of Gesture Recognition Based on Computer Vision

Bei Li[1(✉)], Gongfa Li[1,2], Ying Sun[1,2], Guozhang Jiang[1,2], Jianyi Kong[1,2], Zhaojie Ju[3], and Du Jiang[2]

[1] Key Laboratory of Metallurgical Equipment and Control Technology, Ministry of Education, Wuhan University of Science and Technology, Wuhan 430081, China
1198413210@qq.com

[2] Hubei Key Laboratory of Mechanical Transmission and Manufacturing Engineering, Wuhan University of Science and Technology, Wuhan 430081, China

[3] School of Computing, University of Portsmouth, Portsmouth PO1 3HE, UK

Abstract. With the improvement of computer performance and the development of image processing technology, Gesture recognition based on computer vision has become a hotspot. This paper introduces the main ways of gesture recognition including to data glove, EMG signal and computer vision. The basic principle and working process are focused on computer vision, and describe the technology of gesture segmentation, tracking and positioning, feature extraction and classification recognition, then the main problems existing in recognition method of the computer vision are analyzed. Finally, the future research area of gesture recognition technology in computer vision is prospected.

Keywords: Computer vision · Gesture recognition · Gesture segmentation · Tracking and positioning · Feature extraction

1 Introduction

At present, the main ways of human-computer interaction are keyboard interaction based on text mode and mouse interaction based on graph, and these interactions focuses on machine which makes people adapted to an input and output device of machine. Therefore, there are many inconveniences in the interaction process. With the rapid development of computer science, human computer interaction has been transferred from machine to human, and the interaction of human habits become hot. Gesture is a combination of hand or arm to produce a variety of postures or movements, and it can also include a broad sense of human expression, gait, and even any part of the body movement. Gestures can be divided into static gestures and dynamic gestures. Static gesture recognition takes into account the shape features of gestures at a certain time point, and the dynamic gestures focus on a series of actions for a group of people in a certain period of time, which increases the time information and motion characteristics.

Y. Huang et al. (Eds.): ICIRA 2017, Part I, LNAI 10462, pp. 528–538, 2017.
DOI: 10.1007/978-3-319-65289-4_50

It is the most commonly used to communication media, which is vivid, intuitive and easy to understand, and contains a wealth of information, and is an efficient way of communication. Human-computer interaction techniques based gesture recognition will offer simple and efficient Man-machine interface, which is convenient for people's life. Therefore, human-computer interaction technology based on gesture recognition has important research value.

2 The Main Methods of Gesture Recognition

According to the different input modes of gesture image, hand gesture recognition can be divided into three categories: gesture recognition based on data glove, gesture recognition based on EMG signal and gesture recognition based on vision.

2.1 Gesture Recognition Based on Data Glove

The method measures the bending degree and the joint angle of the finger by using a sensor, and combines the position tracking device to measure the spatial trajectory in hand. Grimes use data glove to make gesture recognition in Bell Laboratory in 1983, which set off a boom in gesture recognition research [1,2]. At home, Gaowen [3] in Harbin Institute of Technology uses data glove of 18 sensors and 3 position tracker as an input device, and had designed a sign language recognition system based on multi feature and multi classifier. The gesture recognition system based on data glove has high recognition rate and fast speed [4]. However, this method requires the user to wear the complex data glove and position tracker, which does not conform to the requirements of human-computer interaction. Besides, the data glove is expensive and not suitable for a large number of promotions. The typical applications include the dexterous hand grasping system of Zhejiang University, the space teleoperation simulation system of Beihang University, and the desktop virtual reality system of South China University of Technology [5].

2.2 Gesture Recognition Based on EMG Signals

The method mainly uses the sensor to collect the multi epidermis EMG signal, amplify it and processes the data, then extracts the characteristic parameters of each gesture, and finally realizes the gesture recognition. The advantage of this method is that the EMG signal is not affected by the external environment. There-fore, many experts and scholars at home and abroad have carried out the study of EMG control [6–8]. Fang Yinfeng et al. [9] not only designed a new electrode arrangement and developed a 16 channel surface EMG signal acquisition system but also designed the two-dimensional EMG to establish the relationship between the multi-channel EMG signal and the movement in hand. Liu Nangeng [10] by using the surface EMG signal acquisition instrument to extract internal and external rotation, fist, exhibition boxing, cut and undercut to make recognition from surface EMG signals. But, the influence of individual differences and the EMG electrode position will increase the difficulty of classification. Besides, it is inconvenient to wear equipment EMG acquisition. And, it is currently used to prosthetic hand training [11,12].

2.3 Gesture Recognition Based on Vision

Gesture recognition based on vision uses one or more cameras to collect gesture images or videos, and Image processing and machine vision are used to analyze the gesture information. The advantage of this method is that the input device is cheap, and the camera is more and more popular in all kinds of electronics products, and it does not require additional requirements for the human hand. The interaction between computer and human is more natural. Therefore, the research on vision based hand gesture recognition is more and more, which has a great improvement in recognition rate and real-time performance. Because computer vision for assistive technologies is very hot-topic, there has been a tremendous increase in demand for assistive technologies usefully to overcome functional limitations of individuals and to improve people's quality of life. And the potential impact on the assessment of assistive technologies considering users, medical, economic and social perspective is also addressed [13]. At present, the visual based human-computer interaction is also widely used, such as the field of smart home appliances, gaming and entertainment, virtual reality, robot control. What' more, a human robot interaction system is able to recognize gestures usually employed in human non-verbal communication is introduced [14–16].

3 Basic Process of Gesture Recognition Based on Vision

Gesture recognition based on vision can be divided into the following steps: gesture segmentation, tracking and positioning, feature extraction and classification recognition [17,18]. Figure 1 shows the basic process of gesture recognition.

3.1 Gesture Segmentation

Gesture segmentation is the process that the gesture region is separated from the complex background and other parts of the human body. The effect of gesture segmentation has a direct impact on the accuracy of gesture recognition [19,20]. At present, there are three main methods of hand gesture segmentation: Background subtraction is a commonly used method for hand detection and segmentation. In the assumptions of limiting the background completely stationary, it can separate the hand from the image to use the most simple static background subtraction method. The corresponding point of the background and gesture image is located in the same space as the target point. Otherwise it will cause great error. When the assumption is not established, some adaptive background modeling methods are used to segment the gesture, the reference [21]

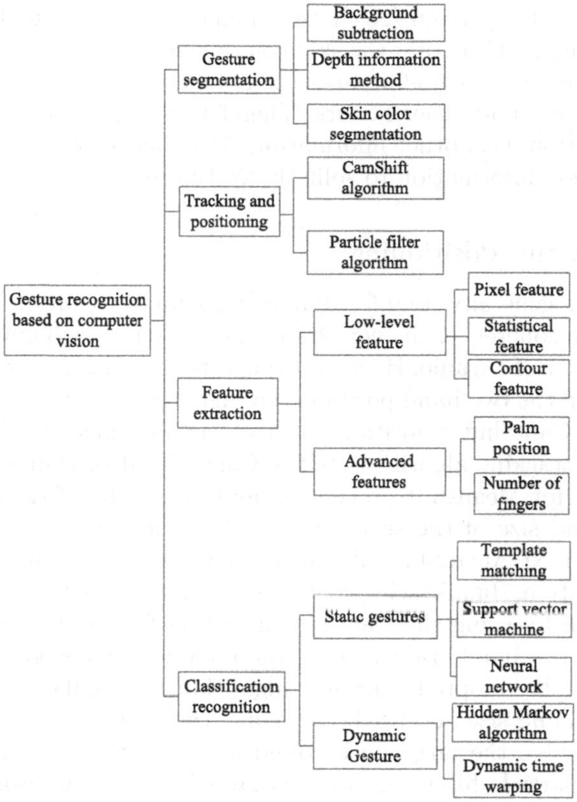

Fig. 1. Basic process of gesture recognition based on vision

summarized all kinds of the background modeling techniques and compared with each other from speed, storage space requirements and accuracy. Gesture segmentation based on depth information in depth images is a new method which had been developed in recent years. According to the distance from the hand to the camera, the method can divide the gesture region. Jagdish [22] directly set the threshold to split the gesture area. Yi Li [23] used the pixel clustering which is located in a certain interval to split gesture region. Cao Chuqing et al. [24] calculated the optimal segmentation threshold of the gesture region and background by the gray histogram of the depth image. The depth information method is simple to compute, but it usually needs to limit the distance between the hand and the camera. Skin color segmentation is the most common method of gesture detection and segmentation. Experiments show that the distribution of human skin color in the color space is concentrated in a certain area. The skin color detection is mainly based on the distribution of skin color in the color space to detect the skin area. Vladimir Vezhnevets et al. [25] Compared the accuracy of various skin color detection methods, and pointed out that the detection accuracy of the I channel threshold setting method in YIQ color space was up to 94.7%. Kakumanu [26] have summarized the methods of skin color detection, and pointed out that the skin color model is easy to be affected by the changes of illumination, camera parameters, human race, human age and gender, skin color area and other factors. Therefore, the skin color model is often used in conjunction with other information. Jonathan Alon [27] used skin color model and motion information to split the gesture region.

3.2 Tracking and Positioning

Hand tracking is generally used for dynamic gesture recognition system. As the dynamic gesture coherence, the next frame hand position will appear hand position in the vicinity of a frame. Hand tracking is to establish a corresponding relationship between the two hand positions, so that the subsequent target detection process is easy. CamShift algorithm and particle filter algorithm are commonly used as gesture tracking algorithm. In the CamShift algorithm, the video image of each frame is for Meanshift processing, and the results of the previous frame (the centroid and size of the search window) as the initial value of the search window of the next MeanShift algorithm. The target tracking is achieved in continuous iteration. JuanWachs et al. [28] developed the Gestix man-machine exchange system by using the Camshift algorithm for hand tracking and using the tracking of the hand trajectory as the direction of navigation commands. Peng Juanchun [29] adopted Camshift algorithm to calculate gesture tracking window position and size and predicted palm position with Kalman filter, which can effectively solve the large background color interference and gesture part sheltered. The particle filter algorithm is probability distribution with a set of weighted particle approximation system. It can achieve target tracking to update the state of particle and weight by iteration and predict the position of the target object from the average particle state. Yale Song [30] used to particle filter to predict the position of the gesture region in the next frame image. Calfeng Shan [31] combined with the advantages of particle filter and Meanshift algorithm to propose a Mean Shift Particle Filter method to achieve real-time tracking of gestures, which has good robustness. Gianni et al. [32] used three particle filters to track the hands and head simultaneously, improving the robustness to the rapid changes of limb and occlusion by introducing re-sampling and degradation update steps.

3.3 Feature Extraction

Feature extraction is the process of analyzing the original data of the image to get the advanced data with key information. The common features of gesture recognition can be divided into low-level features and high-level features. Low level features such as pixel features, statistical features, contour features are the basic image features, and the extraction process is relatively easy, but the classifier need to higher requires. Deng Rui [33] counted the interval distribution of the depth image pixels which is used as the feature to complete the recognition of digital gesture 1 5. Statistical moment features such as Hu moments,

Zernike moments have proportional invariance, translation invariance and rotation invariance. Luo Yuan [34] had added 3 expressions on the basis of Hu moments as gesture recognition features. Dong Lifeng et al. [35] had selected the static hand gesture features and mixed together with concave area and Perimeter area ratio of the gesture contour as well as the first four Hu moments. The Fourier descriptors has a good ability to describe the contour, Yu Ren [36] used 12 Fourier descriptors as the feature vectors of the 10 hand types. The gradient direction histogram calculates the statistic of the gradient direction in the local image as the feature, which is first used in pedestrian detection [37]. In recent years, researchers have begun to use it for gesture recognition [38] and had achieved good results. Advanced features need to extract the palm position, the number of fingers and other information, and sometimes need to establish a specific gesture model, the extraction process is more difficult, but these features are easy to classify and identify. The Reference [39] detected the number and relative position of the finger by the method of color histogram mapping, the gesture of l-9 can be distinguished. The average recognition rate is more than 90%. Weng Hanliang et al. [40] used Staffing profile and convex defect detection fingertip, then use the number and position of the finger to indicate a gesture. It combined with the geometric characteristics of the contour length and area to complete the gesture recognition.

3.4 Classification Recognition

After extracting the features of hand gestures, the meaning of gestures need to be classified to recognize. Static gesture classification methods mainly include template matching, support vector machine (SVM), neural network. Dynamic gesture related to changes of time and space, its trajectory can be regarded as a time series in space. Hidden Markov model and dynamic time warping is two models commonly. The core idea of template matching is to calculate the similarity between the gesture and template. Similarity metrics have Euclidean distance, Mahalanobis distance, Manhattan distance and so on [41,42] However, this method depends on the choice of similarity measure, and the same gesture has some differences, so the recognition rate is relatively low. Support vector machine is widely used in small sample, nonlinear and high dimensional pattern recognition. The key step is to find the optimal hyperplane in a high dimensional space, and the high dimensional space can be obtained to a nonlinear transformation. The reference [36] proposed an improved MEB-SVM algorithm based on SVM which reduces the computational complexity. The neural network has the ability of automatic learning and reasoning by simulating the function of biological neural system, and its biggest advantage is that it has strong adaptability and fault tolerance. Hasan et al. [43] used NN classifier to predict 10 kinds of static gestures for 5 people. The Reference [44]proposed a training algorithm based on quantum behaved particle swarm optimization BP neural network to improve the learning efficiency of BP neural network. Neural network method also has some inherent shortcomings such as the theory being imperfect, high computational complexity, poor real-time. HMM have two states- Hidden and observable states correspond to the standard states and observations of dynamic gestures respectively [45,46]. The advantage of HMM is that it has the invariance of time scale, and it can automatically segment the time series. DTW uses the idea of dynamic programming to find the optimal matching of the two models. And it does not need to have the same length of the two modes, which can automatically calculate an optimal path from the start and end of the model [47]. DTW needs a little of training samples, and can save a lot of training time.

4 The Main Problems of Gesture Recognition Based on Vision

Although there are many methods of gesture recognition, the gesture recognition based on vision still faces many serious problems in practice, such as the real-time and robustness of gesture recognition system especially in gesture segmentation and classification. It is always difficult to extract target gesture from complex background. What's more traditional classifiers are complex for training process and are sensitive to noise. Some are vulnerable to external interference, which leads to a low recognition rate.

5 Development Trend of Future Gesture Recognition Technology

More and more methods will be applied to gesture recognition such as sparse representation and compressed sensing as well as deep learning techniques. These methods are widely used in computer vision, and provide a new perspective to solve the problem of classification and recognition. Moreover, it is necessary to further study the gesture segmentation in different complex environments by the combination of various methods.

5.1 Sparse Representation Theories as the Power of Gesture Recognition

Sparse representation is used in the field of signal processing. The basic idea is to use the least possible elements to represent the same signal. And, it has been applied in the field of image in-painting. Moreover, the face recognition can be achieved in the complex environment successfully. Gestures can also be identified by sparse representation. This method has a simple training processing, and the selection of feature extraction has no specific requirements. It provides a theoretical basis for gesture recognition and a new perspective for gesture recognition technology, which will become a power of gesture recognition technology.

5.2 Depth Learning Technology Used in the Direction of Gesture Recognition

Deep learning is a new field in the research of machine learning, and its goal is to build and simulate the neural network of human brain. It can achieve automatic learning features through unsupervised learning. In depth learning, a machine learning model with many hidden layers is constructed, and the original sample space is transformed into a new feature space by a large amount of data to learn more useful features. This is similar to the human from the original signal as low-level abstraction gradually to the advanced Abstract iteration. The final classification or prediction is in the high-level abstraction layer to improve accuracy. It is better to describe the intrinsic information of the data by using the features of large data by comparing with the artificial rule construction method. Depth learning technology has been successful in the field of image processing. It will be very effective to introduce it into gesture recognition by using the characteristic of deep learning. And, it is undoubtedly the development direction of gesture recognition technology to Learn from the deep learning technology.

6 Conclusion

In this paper, the principle and main process of gesture recognition technology based on computer vision are discussed. The method of gesture segmentation, tracking and positioning, feature extraction and gesture classification are discussed. The application of deep learning technology in gesture recognition will improve the accuracy of learning gesture samples, reduce the recognition time and improve the recognition rate. Sparse representation theory provides a new theoretical basis for gesture recognition and develops a new way of gesture recognition. Gesture recognition technology used in automatic control, smart home, intelligent transportation and other fields will facilitate human life by controlling the device. Therefore, the research of gesture recognition technology based on computer vision has a wider social significance and practical application in human-computer interaction system.

Acknowledgments. This work is supported by National Natural Science Foundation under Grant 51575407, 51575338, 51575412 and the UK Engineering and Physical Science Research Council under Grant EP/G041377/1. This support is greatly acknowledged.

References

1. Nirmal, K.R., Mishra, N.: 3D graphical user interface on personal computer using P5 data glove. Int. J. Comput. Sci. Issues **8**(5), 155–160 (2011)
2. Camastra, F., Felice, D.D.: LVQ-based hand gesture recognition using a data glove. Neural Nets Surround. **19**, 159–168 (2013)

3. Gao, W., Chen, Y.Q., Fang, G.L., Yang, C.S., Jiang, D.L., Ge, C.B., Wang, C.L.: HandTalker II: a Chinese sign language recognition and synthesis system control. In: 8th International Conference on Control, Automation, Robotics and Vision, vol. 1, pp. 759–764 (2004)
4. Li, D.J., Li, J.X., Zhang, Y.: Gesture recognition of data glove based on PSO-improved BP neural network. Electr. Mach. Control **18**(8), 87–93 (2014)
5. He, J., Zhang, G.F., Dai, S.L.: Simulation system of space teleoperation based on dexterous robot hands. J. Syst. Simul. **21**(21), 6915–6919 (2009)
6. Khezri, M., Jahed, M.: A neuro-fuzzy inference system for sEMG-based identification of hand motion commands. IEEE Trans. Industr. Electron. **58**(5), 1952–1960 (2011)
7. Khan, M.S.: sEMG based human computer interface for robotic wheel. In: 2014 International Conference on Advances in Engineering and Technology Research, pp. 1–5 (2014)
8. Yang, D.P., Zhao, J.D., Li, N.: Recognition of hand grasp preshaping patterns applied to prosthetic hand electromyography control. J. Mech. Eng. **48**(15), 1–8 (2012)
9. Fang, Y.F., Liu, H.H., Li, G.F., Zhu, X.Y.: A multichannel surface EMG system for hand motion recognition. Int. J. Humanoid Rob. **12**(2), 1–13 (2015)
10. Liu, N.G., Lei, M.: Characterization of surface electromyography signal based on wavelet analysis and non-linear exponent. J. Clin. Rehabil. Tissue Eng. Res. **12**(17), 3285–3288 (2008)
11. Burck, J., Zeher, M.J., Armiger, R.: Developing the world's most advanced prosthetic arm using model-based design. In: The MathWorks News and Notes, pp. 1–4 (2009)
12. Lambrecht, J.M., Pulliam, C.L., Kirsch, R.F.: Virtual reality environment for simulating tasks with a myoelectric prosthesis: an assessment and training tool. J. Prosthet. Orthot. **23**(2), 89–94 (2011)
13. Leo, M., Medioni, G., Trivedi, M.: Computer vision for assistive technologies. Comput. Vis. Image Underst. **154**, 1–15 (2017)
14. Canal, G., Escalera, S., Angulo, C.: A real-time human-robot interaction system based on gestures for assistive scenarios. Comput. Vis. Image Underst. **149**, 65–77 (2016)
15. Tan, Q.S., Yuan, Z.P., Fan, X.C.: Visualized simulation system of a dexterous mechanical gripper. J. Shanghai Univ. **9**, 38–42 (2003)
16. Zhang, G.L., Wang, Z.N., Wang, T.: Survey on dynamic hand gesture recognition with computer vision. J. Huaqiao Univ. **35**(6), 653–658 (2014)
17. Wu, X., Zhang, Q., Xu, Y.X.: An overview of hand gestures recognition. Electron. Sci. Technol. **26**(6), 171–174 (2013)
18. Chen, D.S., Li, G.F., Sun, Y.: An interactive image segmentation method in hand gesture recognition. Sensors **17**(2), 1–16 (2017)
19. Chen, D.S., Li, G.F., Jiang, G.Z.: Intelligent computational control of multi-fingered dexterous robotic hand. J. Comput. Theor. Nanosci. **12**(12), 6126–6132 (2015)
20. Chen, Q., Georganas, N.D., Petriu, E.M.: Hand gesture recognition using haar-like features and a stoehastic context-free grammar. IEEE Trans. Instrum. Meas. **57**(8), 1562–1571 (2008)
21. Piceardi, M.: Background subtraction techniques: a review. IEEE Int. Conf. Syst. Man. Cybern. **4**, 3099–3104 (2004)

22. Raheja, J.L., Chaudhary, A., Sinal, K.: Tracking of fingertips and centres of palm using kinect. In: Proceedings of the 3rd IEEE International Conference on Computational Intelligence, Modelling and Simulation, pp. 248–252 (2011)
23. Yi, L.: Hand gesture recognition using kinect. In: Proceedings of 2012 IEEE 3rd International Conference on Software Engineering and Service Science, pp. 196–199 (2012)
24. Cao, C.Q., Li, R.F., Zhao, L.J.: Hand posture recognition method based on depth image technology. Comput. Eng. **38**(8), 16–19 (2012)
25. Vezhnevets, V., Sazonov, V., Andreeva, A.: A survey on pixel-based skin color detection techniques. In: Proceedings of Graphicon, vol. 3, pp. 85–92 (2003)
26. Kakumanu, P., Makrogiannis, S., Bourbakis, N.: A survey of skin-color modeling and detection methods. Pattern Recogn. **40**(3), 1106–1122 (2007)
27. Alon, J., Athitsos, V., Yuan, Q., Sclaroff, S.: A unified framework for gesture recognition and spatiotemporal gesture segmentation. IEEE Trans. Pattern Anal. Mach. Intell. **31**(9), 1685–1699 (2009)
28. Wachs, J., Stern, H., Edan, Y., Gillam, M., Feied, C., Smith, M., Handler, J.: Gestix: a doctor-computer sterile gesture interface for dynamnic environments. Soft Comput. Ind. Appl. **3**, 30–39 (2007)
29. Peng, J.C., Gu, L.Z., Su, J.B.: The hand tracking for humanoid robot using Camshift algorithm and Kalman filter. J. Shanghai Jiaotong Univ. **40**(7), 2161–2165 (2006)
30. Song, Y., Demirdjian, D., Davis, R.: Multi-signal gesture recognition using temporal smoothing hidden conditional random fields. In: Proceedings of the IEEE International Conference on Automatic Face and Gesture Recognition, pp. 1685–1699 (2011)
31. Shan, C., Tan, T., Wei, Y.: Real-time hand tracking using a mean shift embedded particle filter. Pattern Recogn. **40**(7), 1958–1970 (2007)
32. Gianni, F., Collet, C., Dalle, P.: Robust tracking for processing of videos of communication's gestures. In: Sales Dias, M., Gibet, S., Wanderley, M.M., Bastos, R. (eds.) GW 2007. LNCS, vol. 5085, pp. 93–101. Springer, Heidelberg (2009). doi:10.1007/978-3-540-92865-2_9
33. Deng, R., Zhou, L.L., Ying, R.D.: Gesture extraction and recognition research based on kinect depth data. Appl. Res. Comput. **30**(4), 1263–1265 (2013)
34. Luo, Y., Xie, Y., Zhang, Y.: Design and implementation of a gesture-driven system for intelligent wheelchairs based on the kinect sensor. Robot **34**(1), 110–114 (2012)
35. Dong, L.F., Ruan, J., Ma, Q.S.: The gesture identification based on invariant moments and SVM. Microcomput. Appl. **31**(6), 32–35 (2012)
36. Ren, Y., Zhang, F.M.: Hand gesture recognition based on MEB-SVM. In: International Conferences on Embedded Software and Systems, pp. 344–349 (2009)
37. Dalal, N., Triggs, B.: Histograms of oriented gradients for human detection. In: Proceedings of the IEEE Computer Society Conference on Computer Vision and Pattern Recognition, vol. 1, pp. 886–893 (2011)
38. Ren, Y., Gu, C.C.: Hand gesture recognition based on HOG characters and SVM. Bull. Sci. Technol. **27**(2), 211–213 (2011)
39. Ahmed, B.J., Walid, M., Yousra, B.J., Abdelmajid, B.H.: A new approach for digit recognition based on hand gesture analysis. Int. J. Comput. Sci. Inf. Secur. **2**(1), 1–8 (2009)
40. Weng, H.L., Zhan, Y.W.: Vision-based hand gesture recognition with multiple cues. Comput. Eng. Sci. **34**(2), 123–127 (2012)
41. Hu, R.X., Jia, W., Zhang, D., Gui, J., Song, L.T.: Hand shape recognition based on coherent distance shape contexts. Pattern Recogn. **45**(9), 3348–3359 (2012)

42. Belsare, S., Sujatha, K.: Robust part-based hand gesture recognition using kinect sensor. Int. J. Eng. Sci. Res. Technol. **4**(8), 75–80 (2015)
43. Hasan, H., Kareem, S.A.: Static hand gesture recognition using neural networks. Artif. Intell. Rev. **41**(2), 147–181 (2014)
44. Yang, Z.Q., Sun, G.: Gesture recognition based on quantum-behaved particle swarm optimization of back propagation neural network. J. Comput. Appl. **34**(S1), 137–140 (2014)
45. Kim, D., Song, J.Y., Kim, D.: Simultaneous gesture segmentation and recognition based on forward spotting accumulative HMMs. Pattern Recogn. **40**(11), 3012–3026 (2007)
46. Zheng, W., Shen, X.K.: Algorithm based on continuous data stream for dynamic gesture recognition. J. Beijing Univ. Aeronaut. Astronaut. **38**(2), 274–280 (2012)
47. Abdelkader, M.F., Abd-Almageed, W., Srivastava, A., Chellappa, R.: Silhouette-based gesture and action recognition via modeling trajectories on Riemannian shape manifolds. Comput. Vis. Image Underst. **115**(3), 439–455 (2011)

Hand Gesture Recognition Using Interactive Image Segmentation Method

Disi Chen[1(✉)], Gongfa Li[1,2], Jianyi Kong[1,2], Guozhang Jiang[1,2], Ying Sun[1,2], Du Jiang[2], and Zhaojie Ju[3]

[1] Key Laboratory of Metallurgical Equipment and Control Technology, Ministry of Education, Wuhan University of Science and Technology, Wuhan 430081, China
chendisi@foxmail.com

[2] Hubei Key Laboratory of Mechanical Transmission and Manufacturing Engineering, Wuhan University of Science and Technology, Wuhan 430081, China

[3] School of Computing, University of Portsmouth, Portsmouth PO1 3HE, UK

Abstract. In this paper, a novel hand gesture recognition method using interactive image segmentation algorithm is proposed. We applied Gaussian mixture model to build the model of color image and the iteration of expectation maximum algorithm learnt the parameters. Then the graph model of color image is built. Finally, the segmentation is achieved by minimizing the energy of graph model according to min-cut/max-flow algorithm. Segmentation results were quantitatively tested and compared, by evaluate the region accuracy and boundary accuracy of segmentation results. To apply interactive image segmentation method into a fully automatic recognition framework, we applied human skin feature and depth information to generate the initial seeds. We also built a hand gesture database which contains ten kind of hand gestures for recognition test, proving that the segmentation of hand gesture images improved the recognition accuracy.

Keywords: Image segmentation · Min-cut · Depth image · Segmentation seeds generation

1 Introduction

Hand gesture recognition, utilized as visual input of controlling computers, is one of the most important aspects in human-computer interaction. Compared with using mice, keyboards and data gloves [1], the use of hand gestures to control computers will optimize the user's learning curve and further expand the application scenario. To achieve hand gesture control [2], many research achievements have been conducted before. The hand gesture recognition process based on computer vision is illustrated in Fig. 1.

However, the vision-based hand gesture recognition method is highly dependent on the sensibility of image sensors. At the same time, the image processing algorithms are not robust enough, especially image segmentation algorithms.

Y. Huang et al. (Eds.): ICIRA 2017, Part I, LNAI 10462, pp. 539–550, 2017.
DOI: 10.1007/978-3-319-65289-4_51

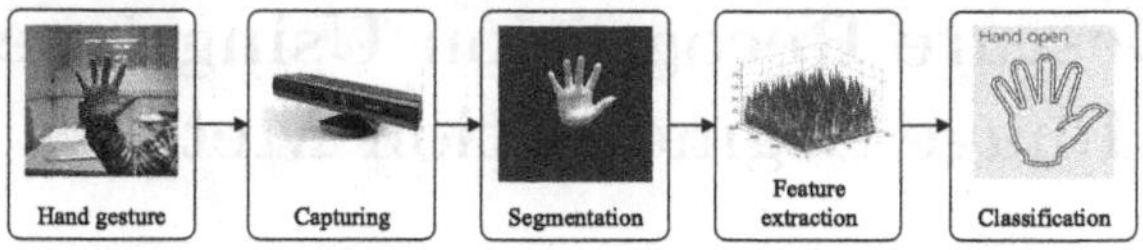

Fig. 1. Process of vision based hand gesture recognition.

To simplify the recognition process, on the one hand, advanced image sensors were used, such as Microsoft Kinect. On the other hand, developments of image processing algorithms contribute to improving segmentation accuracy, which also promoting the accuracy of classifiers to recognize different gestures. Boykov and Jolly [3] proposed an interactive image segmentation method, named graph cut. In this method gray histogram was established to describe the gray distribution of images, and then max-flow/min-cut theory was applied to turn segmentation problem into energy function minimization problem. Random walker [4] is another image segmentation method, where the image is viewed as an electric circuit. The edges in graph model were replaced by passive linear resistors, and the weight of each edge equals the electrical conductance. It proved to perform better then other method. Gulshan et al. [5] proposed a new image segmentation method, which regarded shape as a powerful cue for object recognition. The use of geodesic-star convexity made it achieved a much lower error rate compared with other interactive image segmentation methods. In the process of hand gesture recognition, feature extraction is a very important part. The features such as, Haar-like, Hu invariant and HOG [6] are widely used. For the sake of fairness, support vector machine will be applied as the only classifier to evaluate the recognition rates and prove the interactive segmentation of hand gesture images helps improving the recognition rate.

2 Modelling of Color Hand Gesture Image

2.1 Gaussian Mixture Model of RGB Image

In reality, the color distributions of a gesture image can be represented by three independent histograms [7]. With red, green and blue distributions. Gaussian mixture model was introduced to approximate the continuous probability distribution by increasing the number of single Gaussian models. The probability density function of GMM with k mixed Gaussian models was defined as:

$$p(\boldsymbol{x}) = \sum_{i=1}^{k} \pi_i p_i(\boldsymbol{x}; \theta_i) = \sum_{i=1}^{k} \pi_i N_i(\boldsymbol{x}_i, \boldsymbol{\mu}_i, \Sigma_i) \quad (1)$$

where $i \in \{1, 2, \cdots, k\}$ shows the component belongs to which single Gaussian models. And π_i is the mixing coefficients of k mixed component [8] or the prior probability of $\boldsymbol{x}$ belonging to the i-th single Gaussian model, and $\sum_{i=1}^{k} \pi_i = 1$. $p_i = (\boldsymbol{x}; \theta_i)$ is the probability density function of the i-th single Gaussian model, parameterized by $\boldsymbol{\mu}_i$ and Σ_i in $N_i(\boldsymbol{x}_i, \boldsymbol{\mu}_i, \Sigma_i)$. In order to be simple for remember, we introduce $\boldsymbol{\Theta}$ a parameters [9] set $\{\pi_1, \pi_2, \cdots, \pi_k, \theta_1, \theta_2, \cdots, \theta_k\}$, to represent the π_i and θ_i.

As mentioned above, one RGB hand gesture image could be described in the dataset $\boldsymbol{X} = \{\boldsymbol{x}_1, \boldsymbol{x}_2, \cdots, \boldsymbol{x}_n\}$, and if we regard $\boldsymbol{X}$ as a sample, its probability density is:

$$p(\boldsymbol{X}; \boldsymbol{\Theta}) = \prod_{j=1}^{n} p(\boldsymbol{x}_j; \boldsymbol{\Theta}) = L(\boldsymbol{X}; \boldsymbol{\Theta}), \boldsymbol{x}_j \in \boldsymbol{X} \tag{2}$$

where $L(\boldsymbol{X}; \boldsymbol{\Theta})$ is called likelihood function of parameters given the sample $\boldsymbol{X}$. Then we hope to find a set of parameter $\boldsymbol{\Theta}$ to finish modelling. According to Maximum likelihood method [10], our next task is to find $\widehat{\boldsymbol{\Theta}}$ where:

$$\widehat{\boldsymbol{\Theta}} = arg \max_{\Theta} L(\boldsymbol{\Theta}; \boldsymbol{X}) \tag{3}$$

The function $L(\boldsymbol{\Theta}; \boldsymbol{X})$ and $L(\boldsymbol{X}; \boldsymbol{\Theta})$ have the same equation form, but considering now we are going to use $\boldsymbol{X}$ to estimate $\boldsymbol{\Theta}$, the $\boldsymbol{\Theta}$ becomes variable and $\boldsymbol{X}$ are the fixed parameters, so it is denoted in the second form. The value of $p(\boldsymbol{X}; \boldsymbol{\Theta})$ is usually too small to be calculated by computer, so we need to replace it with the log-likelihood function:

$$\ln(L(\boldsymbol{\Theta}; \boldsymbol{X})) = \ln\left[\prod_{j=1}^{n} p(\boldsymbol{x}_j; \boldsymbol{\Theta})\right] = \sum_{j=1}^{n} \ln\left[\sum_{i=1}^{k} \pi_i p_i(\boldsymbol{x}_j; \theta_i)\right] \tag{4}$$

2.2 Expectation Maximum Algorithm

The Expectation Maximum (EM) Algorithm [11] is introduced in further calculation. EM algorithm is a method of acquiring the parameters set $\boldsymbol{\Theta}$ in maximum likelihood method. There are two steps in this algorithm, **E-step** and **M-step** respectively, to start the **E-step** we will introduce another probability $Q_i(\boldsymbol{x}_j)$. It is a posterior probability of π_i, in another word, the posterior probability of each $\boldsymbol{x}_j$ belonging to i-th single Gaussian model, as we already know the dataset $\boldsymbol{X}$.

$$Q_i(\boldsymbol{x}_j) = \frac{\pi_i p_i(\boldsymbol{x}_j; \theta_i)}{\sum_{t=1}^{k} \pi_i p_t(\boldsymbol{x}_j; \theta_t)} \tag{5}$$

where the definition of $Q_i(\boldsymbol{x}_j)$ is given according to Bayes' theorem, and $\sum_{i=1}^{k} Q_i(\boldsymbol{x}_j) = 1$. Then we use formula (5) to modify the log-likelihood function in (4).

$$\sum_{j=1}^{n} \ln\left[\sum_{i=1}^{k} Q_i(\boldsymbol{x}_j) \frac{\pi_i p_i(\boldsymbol{x}_j; \theta_i)}{Q_i(\boldsymbol{x}_j)}\right] \geq \sum_{j=1}^{n} \sum_{i=1}^{k} Q_i(\boldsymbol{x}_j) \ln\left[\frac{\pi_i p_i(\boldsymbol{x}_j; \theta_i)}{Q_i(\boldsymbol{x}_j)}\right] \tag{6}$$

In formula (6), the Jensen's inequality have been applied, since $\ln''(x) = -\frac{1}{x^2} < 0$, it is concave on its domain. Then:

$$\ln\left[\sum_{i=1}^{k} Q_i(\boldsymbol{x}_j) \frac{\pi_i p_i(\boldsymbol{x}_j; \theta_i)}{Q_i(\boldsymbol{x}_j)}\right] \geq \sum_{i=1}^{k} Q_i(x_j) \ln\left[\frac{\pi_i p_i(x_j; \theta_i)}{Q_i(x_j)}\right] \tag{7}$$

Maximizing the right side of formula (7), guaranteed that $\ln(L(\boldsymbol{\Theta};\boldsymbol{X}))$ is maximized. The iteration of EM algorithm estimating the new parameters in terms of the old parameters is given as follows:

Initialization: $\boldsymbol{\mu}_{i0}$ started with random numbers, and the unit matrices are used as covariance matrices Σ_{i0} to start the first iteration. The mixed coefficients or prior probability is assumed as $\pi_{i0} = \frac{1}{k}$.

E-step: Compute the posterior probability of π_i using current parameters,

$$Q_i(\boldsymbol{x}_j) := \frac{\pi_i p_i(\boldsymbol{x}_j;\theta_i)}{\sum_{t=1}^{k} \pi_t p_t(\boldsymbol{x}_j;\theta_t)} = \frac{\pi_i N(\boldsymbol{x}_j;\boldsymbol{\mu}_i;\Sigma_i)}{\sum_{t=1}^{k} \pi_t N(\boldsymbol{x}_j;\boldsymbol{\mu}_t;\Sigma_t)}. \tag{8}$$

M-step: Renew the parameters,

$$\pi_i := \frac{1}{n}\sum_{j=1}^{n} Q_i(\boldsymbol{x}_j), \tag{9}$$

$$\boldsymbol{\mu}_i := \frac{\sum_{j=1}^{n} Q_i(\boldsymbol{x}_j)\boldsymbol{x}_t}{\sum_{j=1}^{n} Q_i(\boldsymbol{x}_j)} \tag{10}$$

$$\Sigma_i := \frac{\sum_{j=1}^{n} Q_i(\boldsymbol{x}_j)(\boldsymbol{x}_j - \boldsymbol{\mu}_i)(\boldsymbol{x}_j - \boldsymbol{\mu}_i)^T}{\sum_{j=1}^{n} Q_i(\boldsymbol{x}_j)}. \tag{11}$$

For most hand gesture images, the number of iterations is usually defined as a certain number. In order to improve the segmentation quality and to take account of the efficiency, the number of iterations should be 8 [12].

3 Hnad Gesture Image Segmentation

3.1 Graph Model

To segment an image into one foreground and one background, we introduced an extra variable $\boldsymbol{\alpha}$, which acted as labels. As $\alpha_j \in \{1, 0\}$ is assigned to each pixel, the value 0 is taken for labelling background pixels and 1 for foreground pixels. Two independent k-component GMMs were also introduced, one for the foreground pixel set and one for the background pixel set. The parameters of each component now become $\theta_i = \{\pi_i(\alpha_j), \boldsymbol{\mu}_i(\alpha_j), \Sigma_i(\alpha_j = 0, 1; i = 1, 2, \cdots, k)\}$.

Graph model was a concept from graph theory, which was widely used in image processing, probability theory and statistics. To build the graph model, 2 extra nodes, S and T, which mean "source" and "sink" respectively, as shown in Fig. 2(b). In graph model, each node represented a pixel, if pixel $\boldsymbol{x}_j$ belonged to foreground or object (denoted by $\boldsymbol{x}_j \in \boldsymbol{O}$), was linked with S. Otherwise, background pixels, denoted by $\boldsymbol{x}_j \in \boldsymbol{B}$, were linked with T as shown in Fig. 2(c).

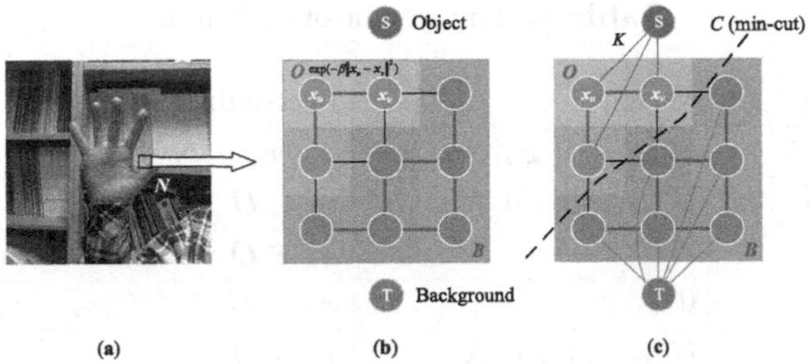

Fig. 2. Graph model and segmentation. **(a)** original image. **(b)** graph model. **(c)** segmentation.

We introduced the energy function $E(\boldsymbol{\alpha})$, $\boldsymbol{\alpha}$ was used to denote one kind of segmentation.

$$E(\boldsymbol{\alpha}) = E(\alpha, \boldsymbol{\Theta}, \boldsymbol{X}) = E(\alpha, i, \theta, \boldsymbol{X}) \tag{12}$$

$$U(\alpha, i, \theta, \boldsymbol{X}) + V(\alpha, \boldsymbol{X}) \tag{13}$$

The term $U(\alpha, i, \theta, \boldsymbol{X})$, called regional term, is defined taking account of GMM. It indicated the penalty of $\boldsymbol{x}_j$ classified to the background or foreground,

$$U(\alpha, i, \theta, \boldsymbol{X}) = \sum_{j=1}^{n} -ln[p_i(\boldsymbol{x}_j) \times \pi_i(\alpha_j)] \tag{14}$$

$$= \sum_{j=1}^{n} \{-ln[\pi_i(\alpha_j)] - ln[\frac{1}{2}log|\sum_i(\alpha_j)|] + \frac{1}{2}[\boldsymbol{x}_j - \boldsymbol{\mu}_i(\alpha_j)]^T \sum_i(\alpha_j)^{-1}[\boldsymbol{x}_j - \boldsymbol{\mu}_i(\alpha_j)]\} \tag{15}$$

$V(\alpha, \boldsymbol{X})$, the boundary term, was defined to describe the smoothness between pixel $\boldsymbol{x}_u$ and its neighbour pixels $\boldsymbol{x}_v$ in the pixel set $\boldsymbol{N}$,

$$V(\alpha, \boldsymbol{X}) = \gamma \sum_{\boldsymbol{x}_u, \boldsymbol{x}_v \in \boldsymbol{N}} [\alpha_u \neq \alpha_v] exp(-\beta \parallel \boldsymbol{x}_u - \boldsymbol{x}_v \parallel^2) \tag{16}$$

where the constant γ was obtained as 50 by optimizing the efficiency over training. $[\alpha_u \neq \alpha_v]$ is an indicator function taking values 0 or 1, by judging the formula inside. is a constant, which represents the contrast of the pixel set $\boldsymbol{N}$, to adjust the exponential term. $E(\boldsymbol{x})$ in the equation below is the expectation,

$$\beta = \frac{1}{2E[(\boldsymbol{x}_u - \boldsymbol{x}_v)^T(\boldsymbol{x}_u - \boldsymbol{x}_v)]}, \boldsymbol{x}_u, \boldsymbol{x}_v \in \boldsymbol{N}. \tag{17}$$

To achieve the segmentation we desired, we applied min-cut/max-flow algorithm [13]. There were 3 kinds of link in Fig. 2(c), considering one pixel and its neighbourhood pixel set $\boldsymbol{N}$, the links between them were denoted by $\overline{\boldsymbol{x}_u \boldsymbol{x}_v}$. Those links connected S with pixels and T with pixels were denoted by $\overline{\boldsymbol{x}_u S}$

Table 1. The weight of each link

Link type	Weight	Precondition
$\overline{\boldsymbol{x}_u \boldsymbol{x}_v}$	$exp(-\beta \parallel \boldsymbol{x}_u - \boldsymbol{x}_v \parallel^2)$	$\boldsymbol{x}_u, \boldsymbol{x}_v \in \boldsymbol{N}$
$\overline{\boldsymbol{x}_u S}$	$U(\alpha = 0, i, \theta, \boldsymbol{X})$	$\boldsymbol{x}_u \in \boldsymbol{U}$
	K	$\boldsymbol{x}_u \in \boldsymbol{O}$
	0	$\boldsymbol{x}_u \in \boldsymbol{B}$
$\overline{\boldsymbol{x}_u T}$	$U(\alpha = 1, i, \theta, \boldsymbol{X})$	$\boldsymbol{x}_u \in \boldsymbol{U}$
	0	$\boldsymbol{x}_u \in \boldsymbol{O}$
	K	$\boldsymbol{x}_u \in \boldsymbol{B}$
where: $K = 1 + \max_{\boldsymbol{x}_u \in \boldsymbol{X}} \sum_{\boldsymbol{x}_u, \boldsymbol{x}_v \in \boldsymbol{N}} exp\left(-\beta \parallel \boldsymbol{x}_u - \boldsymbol{x}_v \parallel^2\right)$		

and $\overline{\boldsymbol{x}_u T}$ respectively. Each link was assumed with a certain weight or cost for it being cut down, which detailed in Table 1.

From Table 1, $\boldsymbol{x}_u \in \boldsymbol{U}$ mean the pixels belongs to set $\boldsymbol{U}$, which were not yet classified into set $\boldsymbol{O}$ or $\boldsymbol{B}$. According to min-cut/max-flow algorithm, to achieve the best segmentation is to find the min-cut C or one $\boldsymbol{\alpha}$ to minimize the energy function: $\min_{\{\alpha_j; i \in U\}} [\min_i E(\alpha, i, \theta, \boldsymbol{X})]$.

Then the Gibbs energy could be minimized by using the min-cut defined above. The whole process of this segmentation is as follows: firstly, assign the GMM components i to each $\boldsymbol{x}_j \in \boldsymbol{U}$ according to the human select of the $\boldsymbol{U}$ region. Secondly, the parameters set $\boldsymbol{\Theta}$ is learned from the whole pixel set $\boldsymbol{X}$. Thirdly, use the min-cut to minimize the energy function of the whole image. Then jump to the first step to start another round, until convergence, the optimal segmentation will be achieved.

3.2 Automatic Seed Selection

To begin segmentation, we defined that the all the pixels in the picture were automatically marked as undefined, labeled $\boldsymbol{U}$. $\boldsymbol{B}$ was the background seed pixel set and $\boldsymbol{O}$ was the foreground seed set. After the training over training set $\boldsymbol{X}$, the set $\boldsymbol{O}$ was obtained as the segmentation result and $\boldsymbol{O} \subset \boldsymbol{U}$. Three pixel sets are shown in Fig. 3.

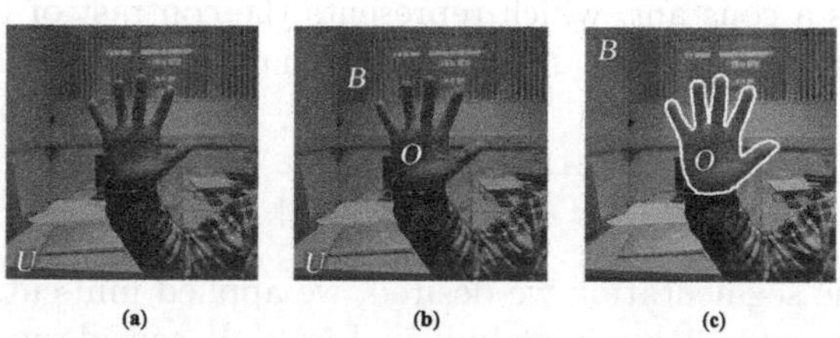

Fig. 3. The relationships between three pixel sets. **(a)** original image. **(b)** image with seeds. **(c)** segmentation result.

To achieve the segmentation automatically, we proposed an initial seeds selection method in hand gesture images. Considering that the human skin color was an elliptical distribution in $YCbCr$ color space, the image was transformed from RGB color space to $YCbCr$, All the pixels in skin color distribution area will be labelled as the foreground seeds, which belong to set $\boldsymbol{O}$. As for background seeds, which belong to set $\boldsymbol{B}$, we applied the depth image to mark all the image region out of hand gesture depth range as background seeds.

4 Experimental Comparison

To evaluate interactive segmentation quantitatively, an image dataset proposed by Gulshan [5], which contains 151 images. Those images covered all kinds of shapes, textures and backgrounds. The corresponding ground true images together with the initial seeds were also included in this dataset. The initial seed maps were made up of 4 manually generated brush-strokes all in 8 pixels wide, and 1 for foreground and 3 for background.

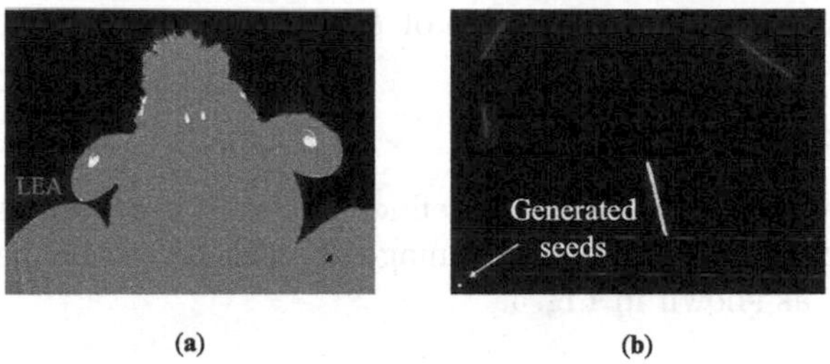

Fig. 4. Evaluation on the dataset. **(a)** the largest connected segmentation error area. **(b)** one generated seed on seed map.

To simulate the human interactions, after the first segmentation with initial seed map, one more seed would be generated in the largest connected segmentation error area (LEA) automatically. As shown in Fig. 4(a), the blue area is the segmentation result of the algorithm, while the white one is the ground true segmentation and the LEA is marked in yellow. From Fig. 4(b), the seed is a round dot (8 pixels in diameter), generated according to the LEA. Then we update the segmentation with all the seeds. After that, this step is repeated 20 times, and a sequence of segmentations will be obtained.

We introduced two different methods in evaluating the region accuracy (RA) and boundary accuracy (BA). Each evaluation will be conducted to a single segmentation, and all the images in Gushan's dataset will be tested to verify that our proposed method is suitable for interactive image segmentation.

4.1 Region Accuracy

The region accuracy (RA) of segmentation results is evaluated by a weighted $F_\beta - measure$ [14]. Compared with normal $F_\beta - measure$, the two terms $Precision$ and $Recall$ become,

$$Precision^w = \frac{TP^w}{TP^w + FP^w}, \qquad Recall^w = \frac{TP^w}{TP^w + NP^w} \tag{18}$$

where, TP denotes the overlap of ground truth and segmented foreground pixels. FP is the wrongly segmented pixels compared with ground true images and NP represent the wrongly segmented background pixels.

The $F_\beta^w - measure$ is defined as follows:

$$RA = F_\beta^w = (1+\beta)\frac{Precision^w \cdot Recall^w}{\beta^2 \cdot Precision^w + Recall^w}, \tag{19}$$

where, β signifies the effectiveness of detection with respect to a user who attaches β times as much importance to $Recall^w$ as to $Precision^w$, normally $\beta = 1$. Then, we applied $F_1^w - measure$ to calculate the region accuracy of different segmentation results. The value of RA showed the segmentation quality.

4.2 Boundary Accuracy

The boundary accuracy (BA) [15] is defined referring to the Hausdorff distance. The boundary pixels of ground truth image and segmented image are defined as B_{GT} and B_{SEG} as shown in Fig. 5.

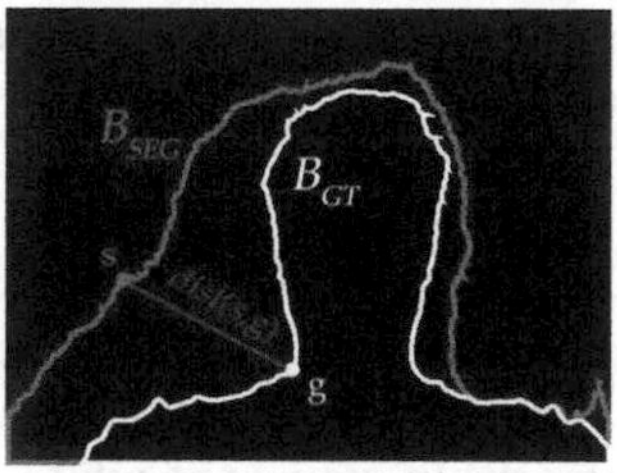

Fig. 5. Boundary extraction.

The definition of BA was given by:

$$BA = \frac{N(B_{SEG}) + N(B_{GT})}{\sum_s \min_g (dist(s,g)) + \sum_g \min_s (dist(g,s))}, \tag{20}$$

where, $g \in B_{GT}$ and $g \in B_{SEG}$, $dist(\bullet)$ denoted the Euclidean distance, $N(\bullet)$ calculated the pixel quantities in one set. The value of BA shows the segmentation accuracy of boundaries.

4.3 Results Analysis

We evaluated segmentations on the whole Gulshan's dataset and used the human interaction simulator to perform the interactions, which generated the seeds 20 times to further refine the segmentation results. The result of each simulation step has been tested on the experiment platform. The *RA* and *BA* scores are the mean values, shown in Fig. 6.

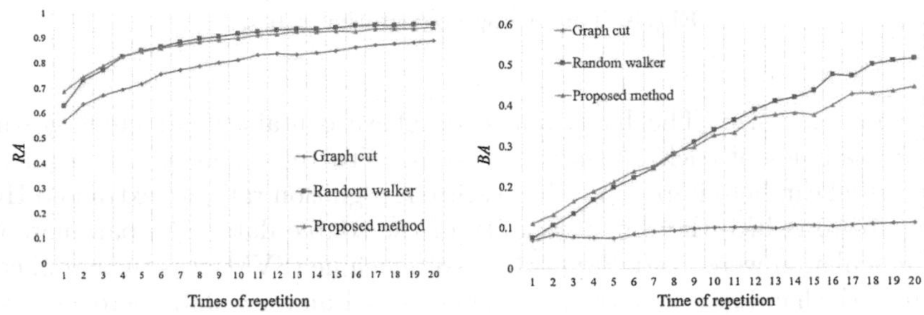

(a) (b)

Fig. 6. Segmentation results. **(a)** *RA* score. **(b)** *BA* score.

From Fig. 6, the segmentation quality shows an increase with simulated human interactions. When the seed number becomes high, a satisfactory segmentation will be achieved. Our method obtains the best segmentation quality with few human interactions. Since the seeds are generated once automatically in human hand image segmentation, our method is suitable for human image segmentation.

5 Hand Gesture Recognition

In order to further test the proposed hand gesture recognition method, we designed an automatic hand tracking and image acquisition system. We defined 10 different hand gestures as shown in Fig. 7.

Fig. 7. 10 kind of hand gestures.

Kinect was utilised as color and depth sensor, and the hand tracking algorithm was applied to track the hand and save the square region of recorded

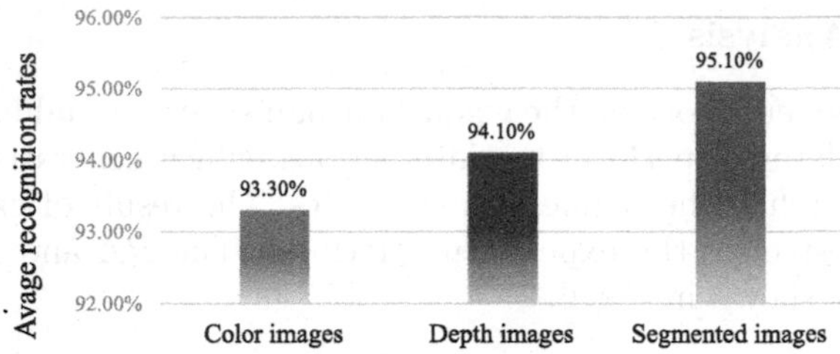

Fig. 8. The average recognition rates.

images in same size. The hand image database contain 10 different gestures and 100 samples of each gesture were stored. To test using image segmentation method can help improving the gesture recognition rate, we extracted HOG [6] feature to reduce the dimensionality of the source data. The parameters of HOG calculation were fixed. Support vector machine (SVM) was then applied to recognize the hand gestures, the parameters, $c = 1$ and $\gamma = 0.07$, were also fixed. We compared the recognition rate using different image samples, color images, depth images and segmented hand gesture images. The overall recognition result was represented in Fig. 8.

By segmenting the images, the recognition rates on our hand gestures dataset are increased compared with using color images and depth images as samples. The confusion matrices of 3 different recognition results were shown in Fig. 9.

	G00	G01	G02	G03	G04	G05	G06	G07	G08	G09
G00	0.99									
G01		1.00								
G02			1.00							
G03				1.00						
G04					1.00					
G05						1.00				
G06	0.01						0.98	0.03	0.08	
G07								0.88	0.17	0.15
G08								0.03	0.72	0.05
G09							0.02	0.06	0.03	0.80

(a)

	G00	G01	G02	G03	G04	G05	G06	G07	G08	G09
G00	0.98									
G01		1.00								
G02			1.00							
G03				1.00						
G04					1.00					
G05						1.00				
G06							0.93			
G07							0.06	0.96		
G08	0.01							0.04	0.99	0.38
G09	0.01						0.01		0.01	0.62

(b)

	G00	G01	G02	G03	G04	G05	G06	G07	G08	G09
G00	0.98						0.04		0.02	
G01		0.99					0.01			
G02		0.01	1.00						0.01	
G03	0.01			1.00						
G04					1.00					
G05						0.91	0.01	0.03	0.02	0.01
G06	0.01						0.94	0.01		0.01
G07						0.03		0.93	0.02	0.01
G08						0.04		0.01	0.85	0.06
G09						0.03		0.02	0.07	0.91

(c)

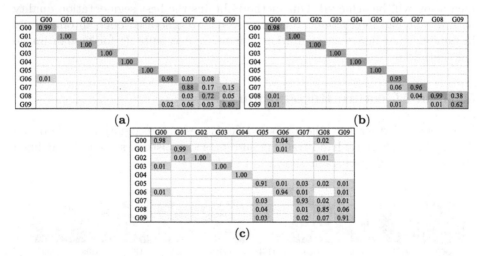

Fig. 9. Confusion matrices. **(a)** color image samples. **(b)** depth image samples. **(c)** segmentation samples

6 Conclusions and Future Work

In conclusion, the interactive hand gesture image segmentation method can perfectly meet the demands of segmenting hand gesture images without human interactions. The mechanism behind this method is carefully explored and deduced. Comparing the segmentation results of Gulshan's public image database with other popular image segmentation methods, our method can obtain a better segmentation accuracy and a higher quality, when there are limited seeds. Automatic seeds generation helps to reduce human interactions. The segmentation of hand images further improved the recognition rate. In future work, we could adapt this method to higher resolution pictures, which requires simplifying the calculation process. In seed selection, the automatic selection method could be improved to overcome various interferes, such as highlights, shadows and image distortion. Other future work will focus on improving the recognition rate by integrating the segmentation algorithm with more advanced recognition methods.

Acknowledgments. This work is supported by National Natural Science Foundation under Grant 51575407, 51575338, 51575412 and the UK Engineering and Physical Science Research Council under Grant EP/G041377/1.This support is greatly acknowledged.

References

1. Chen, D.C., Li, G.F., Jiang, G.Z., Fang, Y.F., Ju, Z.J., Liu, H.H.: Intelligent computational control of multi-fingered dexterous robotic hand. J. Comput. Theor. Nanosci. **12**, 6126–6132 (2015)
2. Ju, Z.J., Zhu, X.Y., Liu, H.H.: Empirical copula-based templates to recognize surface EMG signals of hand motions. Int. J. Humanoid Robot. **8**, 725–741 (2011)
3. Sinop, A.K., Grady, L.: A seeded image segmentation framework unifying graph cuts and random walker which yields a new algorithm. In: IEEE 11th International Conference on Computer Vision (ICCV), Rio de Janeiro, Brazil, pp. 1–8 (2007)
4. Grady, L.: Multilabel random walker image segmentation using prior models. In: IEEE Computer Society Conference on Computer Vision and Pattern Recognition, CVPR 2005, San Diego, USA, pp. 763–770 (2005)
5. Varun, G., Carsten, R., Antonio, C., Andrew, B., Andrew, Z.: Geodesic star convexity for interactive image segmentation. In: IEEE Computer Vision and Pattern Recognition, San Francisco, USA, CVPR 2010, pp. 3129–3136 (2010)
6. Xu, Y., Yu, G., Wang, Y., Wu, X., Ma, Y.: A hybrid vehicle detection method based on Viola-Jones and HOG+SVM from UAV images. Sensors **16**(8), 1325 (2016)
7. Bian, X., Zhang, X., Liu, R., Ma, L., Fu, X.: Adaptive classification of hyperspectral images using local consistency. J. Electron. Imaging. **23**, 063014–063014 (2014)
8. Song, H., Wang, Y.: A spectral-spatial classification of hyperspectral images based on the algebraic multigrid method and hierarchical segmentation algorithm. Remote Sens. **8**(4), 1–23 (2016)
9. Hatwar, S., Anil, W.: GMM based image segmentation and analysis of image restoration tecniques. Int. J. Comput. Appl. **109**, 45–50 (2015)

10. Banbura, M., Modugno, M.: Maximum likelihood estimation of factor models on datasets with arbitrary pattern of missing data. J. Appl. Econ. **29**, 133–160 (2014)
11. Simonetto, A., Leus, G.: Distributed maximum likelihood sensor network localization. IEEE Trans. Sig. Process. **62**(6), 1424–1437 (2013)
12. Zhang, Y., Brady, M., Smith, S.: Segmentation of brain MR images through a hidden Markov random field model and the expectation-maximization algorithm. IEEE Trans. Med. Imaging. **20**, 45–57 (2001)
13. Lee, G., Lee, S., Kim, G., Park, J., Park, Y.: A modified GrabCut using a clustering technique to reduce image noise. Symmetry **8**, 64 (2016)
14. Margolin, R., Zelnik-Manor, L., Tal, A.: How to evaluate foreground maps?. In: IEEE Conference on Computer Vision and Pattern Recognition, Columbus, USA, pp. 248–255 (2014)
15. Zhao, Y., Nie, X., Duan, Y., Huang, Y., Luo, S.: A benchmark for interactive image segmentation algorithms. In: IEEE Person-Oriented Vision, Kona, USA, pp. 3–38 (2011)

Simulation of 2-DOF Articulated Robot Control Based on Adaptive Fuzzy Sliding Mode Control

Feng Du[1(✉)], Gongfa Li[1,2], Zhe Li[2], Ying Sun[1,2], Jianyi Kong[1,2], Guozhang Jiang[1,2], and Du Jiang[1]

[1] Key Laboratory of Metallurgical Equipment and Control Technology, Ministry of Education, Wuhan University of Science and Technology, Wuhan 430081, China
305048751@qq.com
[2] Hubei Key Laboratory of Mechanical Transmission and Manufacturing Engineering, Wuhan University of Science and Technology, Wuhan 430081, China

Abstract. About the control of articulated robot, researchers put out an adaptive fuzzy sliding mode control algorithm by analyzing the classical sliding mode algorithm. There is an adaptive single input and output fuzzy to calculate the control gain. Meanwhile, researchers designed a adaptive law which is based on Lyapunov theory, and simulated the adaptive fuzzy sliding mode control in Simulink. Simulink results show that when the chattering exiting becomes weaker, the function of system is stronger. Since the adaptive algorithm joined, fuzzy sliding mode control can be different with the change of system's state and adjusted automatically. Steady-state convergence is constant, adaptive fuzzy sliding mode control algorithm still with a good robustness under the condition that articulated robot's parameters uncertain and external interference.

Keywords: Sliding mode control's theory · Fuzzy control · Adaptive control · 2-DOF articulated robot

1 Introduction

In the robot's control system, Control in joint drive moment is to make state variable can track the expected given trajectory [1,2]. By informing robots' instructions and relative parameters, robot's mathematical model can described its' dynamic characteristics. However, it's hard to get a specific mathematical model in practice, so some uncertain factors would be ignored in process [3–5].

Sliding mode controls [6] traits show that systems structure changes with systems condition. This kind of structure leads system to move along with trajectory around sliding surface [7]. Sliding mode control and system's parameters are none of external interference's business. So, sliding mode control with a well robustness can get widely focus by scholars at home and abroad [8–10].

In recent researches, especially in nonlinear and undefined robot system, sliding mode control gets widely used and developed [11–14]. At the same time,

Y. Huang et al. (Eds.): ICIRA 2017, Part I, LNAI 10462, pp. 551–559, 2017.
DOI: 10.1007/978-3-319-65289-4_52

serious defects displayed that condition trajectory moves again and again on both sliding surfaces. It's hard to move to balance point and cause chattering exiting. Because of chattering exiting, system's control performance is deteriorated. At the worst, which would causes a shaky state system, and it's hard to use in practice. In recent years, different kinds of control methods' advantages and disadvantages, the function of system is improved [15].

About articulation robot control, this article connected adaptive control, fuzzy control and sliding mode control to make the sliding made control system's gain adjusted and system's high-frequency buffeting weaken obviously. At the same time, Adaptive fuzzy sliding mode control algorithm not only can keep a good robustness but can weaken in influence from chattering exiting.

2 Sliding Mode Controls Basic Theory

2.1 Sliding Mode Controls Basic Problem

Suggesting the control system

$$\dot{x} = f(x, u, t) \quad x \in R^n, u \in R^m, t \in R \tag{1}$$

Switching functions need to be confirmed

$$s(x), d \in R^m \tag{2}$$

We can draw the conclusion:

$$u = \begin{cases} u^+(x) & s(x) > 0 \\ u^-(x) & s(x) < 0 \end{cases} \tag{3}$$

Among this $u^+(x) = u^-(x)$

(1) If the sliding mode is real, formula (3) is come into existence.
(2) Accessibility conditions are satisfied. Sports point that lies out of switching surface $s(x) = 0$ needs to move into it in limited time.
(3) Make sure the stability of sliding mode movements.
(4) Reach the control system's dynamic quality's demand.

Such a name of sliding mode can be called until three tips above are reached.

One prerequisites of sliding mode application is to match system's sliding mode. Under the circumstances of this way, there is one more called switching functions also needed. When system's initial point $x(0)$ stays in optional position, system's movement must goes to sliding mode. Switching surface $s = 0$, otherwise, the movements of sliding mode can't be activated. Because of thousand kinds of sliding mode control stratagem, system's accessibility condition's realization form is different. Sliding mode's existence conditions can be expressed as formula (4).

$$\lim_{s \to 0^+} \dot{s} < 0, \lim_{s \to 0^-} \dot{s} > 0 \tag{4}$$

Formula (4) means the path of particle can get to the switching surface in sliding mode surface and limited time. Thus, these also are called sliding mode structure control's partial condition.

$$s\dot{s} < 0 \tag{5}$$

It equivalent of the form, switching function $s\left(x\right)$ should fulfill the following:

(1) Differentiable
(2) Over origin, $s\left(x\right) = 0$

Because the distance between x and switching surface is not sure, formula (5) called sliding mode variable structure's global arrival condition. In order to arrive to the destination in time, researchers change the formula (5) as:

$$s\dot{s} < -\delta \tag{6}$$

Among this, $\delta > 0$, δ can be arbitrarily small.

Normally, formula (5) can be expressed as the Lyapunov function's arrival condition:

$$\dot{V}\left(x\right) = \frac{1}{2}s^2 < 0 \tag{7}$$

Among this, $V\left(x\right)$ belongs to a defined Lyapunov function.

2.2 Sliding Mode Controls Invariance

Sliding mode control system's movement is made up of sliding mode movement and approach motion. After reaching the sliding mode surface, system is in sliding situation. At the same time, system dynamics behavior is decided by formula $s\left(x\right) = 0$, and there is no relationship with control law. System dynamic behavior is not influenced by system's inner parameters and chargeable environment, that is to say the robustness.

Take the following uncertain system into consideration:

$$\dot{x} = Ax + Bu + d \tag{8}$$

Among this $A = A^* + \varDelta A$, $B = B^* + \varDelta B$ and $d = d^* + \varDelta d$,and A^*, B^*, d^* means system's nominal value, $\varDelta A$, $\varDelta B$, $\varDelta d$ means system's uncertainty.

Design sliding mode surface function is $s\left(x, t\right)$, so:

$$\dot{s}\left(x, t\right) = \frac{\partial s}{\partial t} + \frac{\partial s}{\partial x}\left(\left(A^* + \varDelta A\right)x + \left(B^* + \varDelta B\right)u + \left(d^* + \varDelta d\right)\right) \tag{9}$$

Suppose there is parameter matrix u_1, u_2 and u_3, system's uncertain term should be satisfied.

$$\varDelta A = B^* u_1, \varDelta B = B^* u_2, \varDelta d = B^* u_3 \tag{10}$$

It would be well if $\frac{\partial s}{\partial x}(B^* + \Delta B)$ is reversible, make $\dot{s} = 0$, so that we would get equivalent control on sliding mode surface:

$$u_{eq} = -\left[\frac{\partial s}{\partial x}(B^* + \Delta B)\right]^{-1}\left[\frac{\partial s}{\partial t} + \frac{\partial s}{\partial x}((A^* + \Delta A)x + (d^* + \Delta d))\right] \quad (11)$$

Formulas (10) and (11) are substituted in formula (8), which comes out:

$$\begin{aligned}\dot{x} = -B^*\left[\tfrac{\partial s}{\partial x}B^*\right]^{-1}\tfrac{\partial s}{\partial t} + \left(I - B^*\left[\tfrac{\partial s}{\partial x}B^*\right]^{-1}\tfrac{\partial s}{\partial x}\right)A^*x \\ + \left(I - B^*\left[\tfrac{\partial s}{\partial x}B\right]^{-1}\tfrac{\partial s}{\partial x}\right)d^*\end{aligned} \quad (12)$$

From formula (12), we can see that sliding mode surface system is none relationship with parameter perturbation, this kind of sliding mode anti-perturbation is normally called perturbation invariance.

3 Modeling for Two Degree of Articulated Robot

Figure 1 shows articulation robot's diagram.

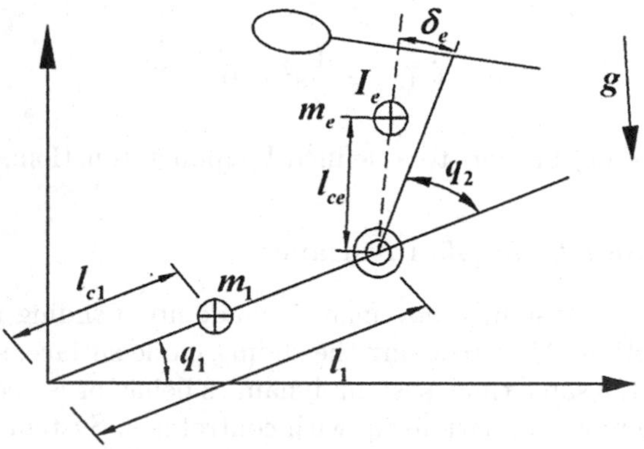

Fig. 1. Articulation robot's diagram

Link rod 1's length is l_1, mass regarded as m_1, the length between barycenter and articulation is l_{c1}, the moment of inertia is I_1, link rod 2 and the load can be regarded as a whole, its' mass is m_e, the moment of inertia is I_e, the length between barycenter and articulation is l_{ce}, the intersection angel between barycenter and link rod 2 is δ_e.

Multi joint robot's system dynamics model expression defined as:

$$H(q)\ddot{q} + C(q,\dot{q})\dot{q} + G(q) + \tau_d = \tau \quad (13)$$

Chosen an articulation robot system, as following:

$$\begin{aligned}&\begin{bmatrix}\alpha + 2\varepsilon\cos(q_2) + 2\eta\sin(q_2) & \beta + \varepsilon\cos(q_2) + \eta\sin(q_2)\\ \beta + \varepsilon\cos(q_2) + \eta\sin(q_2) & \beta\end{bmatrix}\begin{bmatrix}\ddot{q}_1\\ \ddot{q}_2\end{bmatrix}\\ &+\begin{bmatrix}\varepsilon Y_1 + \eta Y_2 + (\alpha - \beta + e_1)e_2\cos(q_1)\\ \varepsilon Y_3 + \eta Y_4\end{bmatrix} + \begin{bmatrix}\tau_{d1}\\ \tau_{d2}\end{bmatrix} = \begin{bmatrix}\tau_1\\ \tau_2\end{bmatrix}\end{aligned} \quad (14)$$

Among them:

$$Y_1 = -2\sin(q_2)\dot{q}_1\dot{q}_2 - \sin(q_2)\dot{q}_2^2 + e_2\cos(q_1+q_2)$$
$$Y_2 = 2\cos(q_2)\dot{q}_1\dot{q}_2 + \cos(q_2)\dot{q}_2^2 + e_2\sin(q_1+q_2)$$
$$Y_3 = \sin(q_2)\dot{q}_1^2 + e_2\cos(q_1+q_2)$$
$$Y_4 = -\cos(q_2)\dot{q}_1^2 + e_2\sin(q_1+q_2)$$
$$e_1 = m_1 l_1 l_{c1} - I_1 - m_1 l_1^2, e_2 = g/l_1$$
$$\tau_{d1} = \begin{cases} 0 & 0 \le t < 4 \\ 500 & 4 \le t \le 4.1 \\ 0 & 4.1 < t \le 6 \end{cases}, \tau_{d2} = \begin{cases} 0 & 0 \le t < 4 \\ 500 & 4 \le t \le 4.1 \\ 0 & 4.1 < t \le 6 \end{cases}$$

Among them, represents acceleration of gravity, $\alpha = I_1 + m_1 l_{c1}^2 + I_e + m_e l_{ce}^2 + m_e l_1^2$, $\beta = I_e + m_e l_{ce}^2$, $\varepsilon = m_e l_1 l_{ce}\cos\delta_e$, $\eta = m_e l_1 l_{ce}\sin\delta_e$

Confirm the parameters: $m_1 = 1.0, l_1 = 1.0, l_{c1} = 0.5, I_1 = 0.0833, m_e = 3.0, l_{ce} = 1.0, I_e = 0.4, \delta_e = 0.0$

Simulation result shows in Fig. 2.

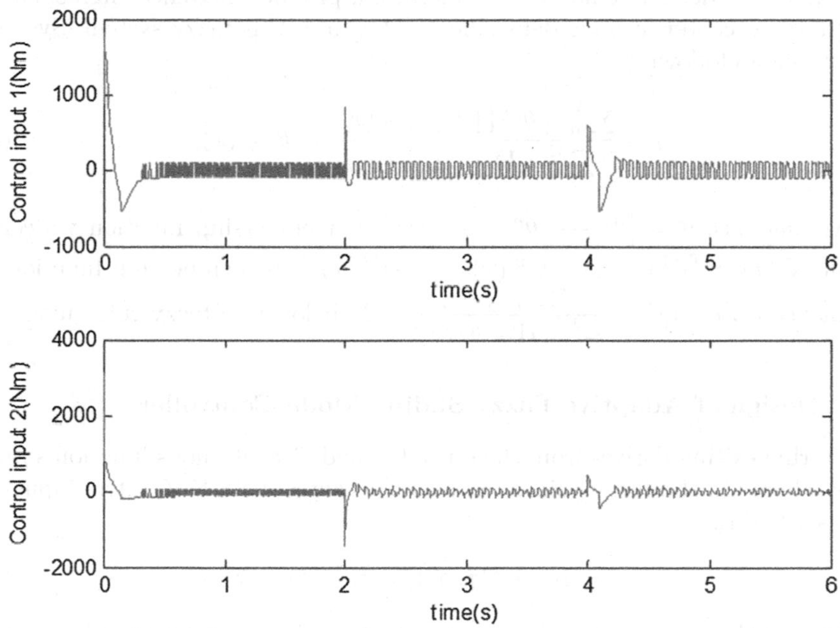

Fig. 2. Joint 1 and Joint 2's control output moment

In above pictures, chattering exiting is reflected apparently. Buffeting phenomenon will use up more system's energy. Situations would be worse that system's stability destroyed, equipment are destroyed as well.

4 Simulations Based on Adaptive Fuzzy Sliding Made Control Algorithm

4.1 Basic Theory of Fuzzy Control

Figure 3 is fuzzy controller's basic structure.

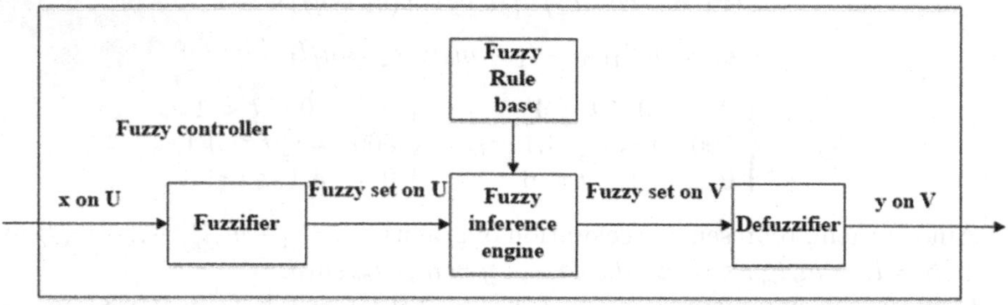

Fig. 3. Fuzzy controller's structure

Due to the design demands' consideration, product inference engine, singleton fuzzifier, center average defuzzifier is chosen in this fuzzy system. System's output shows following:

$$y = \frac{\sum_{m=1}^{M} \theta^m \prod_{i=1}^{n} \mu_{A_i^m}(x_i^*)}{\sum_{m=1}^{M} \prod_{i=1}^{n} \mu_{A_i^m}(x_i^*)} = \theta^T \psi(x) \tag{15}$$

Among this, $\theta = \left[\theta^1, \cdots, \theta^m, \cdots, \theta^M\right]^T$ is membership function y's center vector, $\Psi(x) = \left[\psi^1(x), \cdots, \psi^m(x), \cdots, \psi^M(x)\right]^T$ is membership function y's height vector. $\psi^m(x) = \frac{\prod_{i=1}^{n} \mu_{A_i^m}(x_i^*)}{\sum_{m=1}^{M} \prod_{i=1}^{n} \mu_{A_i^m}(x_i^*)}$, M belongs to fuzzy rule number.

4.2 Design of Adaptive Fuzzy Sliding Mode Controller

Chattering exiting derives from the set date$k$ and discontinuous function $sgn(s)$, Researchers use the fuzzy gain K replace the control gain K. Control input can be described is:

$$\tau = \hat{H}\ddot{q}_r + \hat{C}\ddot{q}_r + \hat{G} + \tau_d - As - k \tag{16}$$

Among this $k = [k_1, \cdots, k_i, \cdots, k_n]$, each of them is estimated by a single fuzzy system.

(1) Extraction of the rule base

Fuzzy gain k is makes up for the system's uncertain and the decrease of s' energy design the rule base like:$k_i$increases, with the s_i increased. When s_i is small, k_i can be small if $|k_i| > |\Delta f_i|$, when s_i is zero, k_i follows. Rule base's choice shows in Table 1.

Table 1. Fuzzy rule base table

Fuzzy system output/input	Fuzzy set						
s_i	NB	NM	NS	ZE	PS	PM	PB
k_i	NB	NM	NS	ZE	PS	PM	PB

In above rules, s_i and k_i equipped with the same kind of membership function. In NB, NM, NS, ZE, PS, PM, PB, N represents negative, P represents positive, B represents big, M represents middle, S represents small, at last, ZE represents zero. s_i and k_i' same point due to the same name of their membership function, while the different point is due to the membership function's corresponding core and its height. What's more, the controller's adaption can be seen easily. That is s_i membership function's parameters is future fixed, k_i' membership function's parameters is updated in line.

(2) Design of control law

Formula (16) is substituted in formula (13), which comes out:

$$H\dot{s} = -(C + A)\,s + \Delta f - k \tag{17}$$

Define $\theta_{k_{id}}$, so that $k_i = \theta_{k_{id}}^T \psi_{ki}(s_i)$ is the Δf_i best supplement. Adaptive selection law is:

$$\tilde{\theta}_{k_i} = s_i \psi_{k_i}(s_i) \tag{18}$$

Choose Lyapunov candidate function is:

$$V = \frac{1}{2} s^T H s + \frac{1}{2} \sum_{i=1}^{n} \left(\tilde{\theta}_{k_i}^T \tilde{\theta}_{k_i} \right) \tag{19}$$

After analyzing, the whole system's adaptive law is relative stable to S. So $\lim_{t\to\infty} s = \lim_{t\to\infty} (\dot{e} + \lambda e) = 0$, $\lim_{t\to\infty} q = q_d$ and $\lim_{t\to\infty} \dot{q} = \dot{q}_d$, which improves that by using adaptive fuzzy sliding mode control's real joints' angle can get the expected value.

4.3 Simulation Researches

Emulation condition is equal to traditional sliding mode control's emulation condition. Simulation results showed in Fig. 4.

Something showed in control moment diagram, chattering exiting become less powerful and the function of the system can get improved. Even though the articulation robot has the uncertain parameters and external interference, adaptive fuzzy sliding mode control algorithm still with a good robustness and a high tracing accuracy, it's very useful in real control situation. Due to the simple control structure, high computational efficiency and non-real time sports parameters of robots.

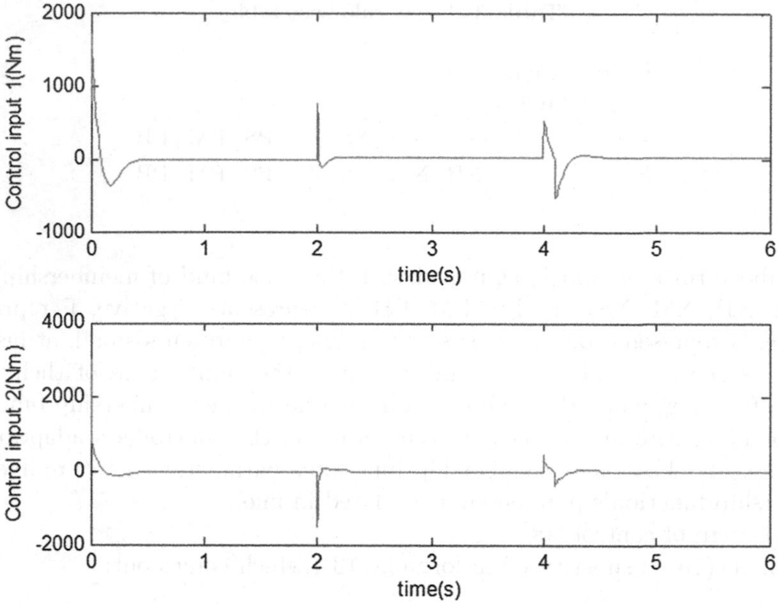

Fig. 4. Joint 1 and Joint 2's controlling output moment

5 Conclusion

By establishing the articulation robot's dynamic model, researchers have done many simulation researches on traditional sliding mode control and adaptive fuzzy sliding mode control. Simulation results show that there comes out a chattering exiting situation in traditional sliding mode control. While in adaptive fuzzy sliding mode control, system's uncertain parameters are decided by control gain and membership function's updates. There is no need of system's uncertain parameter's future knowledge, and system's stability can be sure with the chattering exiting become weaker, the function of the system is promoted.

Acknowledgement. This work is supported by National Natural Science Foundation under Grant 51575407, 51575338, 51575412 and the UK Engineering and Physical Science Research Council under Grant EP/G041377/1. This support is greatly acknowledged.

References

1. Ning, X.T., Chen, J., Shi, Q., Pan, Y.T.: Simulation of trajectory planning for robot workspace. Comput. Simul. **33**(2), 367–372 (2016)
2. Yang, L.J., Shen, L.Y., Ding, H.: The trajectory planning simulation of 4-DOF upper limb rehabilitation robot. Comput. Simul. **33**(8), 332–337 (2016)
3. Gao, C.: Trajectory planning of welding robot based on terminal priority planning. Sens. Transducers **169**(4), 111–116 (2014)

4. Li, H.K., Shi, A.J., Dai, Z.D.: A trajectory planning method for sprawling robot inspired by a trotting animal. J. Mech. Sci. Technol. **31**(1), 327–334 (2017)
5. Li, Z.X., Shi, S.Z., Wang, H., Zhao, J.Q.: Simulation of optimal control model for robot motion. Comput. Simul. **33**(5), 280–284 (2016)
6. Liu, J.K., Sun, F.H.: Research and development on theory and algorithms of sliding mode control. Control Theor. Appl. **24**(3), 407–418 (2007)
7. Hu, S.B., Lu, M.: Adaptive double fuzzy sliding mode control for three-links spatial robot. J. Tongji Univ. (Nat. Sci.) **40**(4), 622–628 (2012)
8. Wu, J.F., Li, Y.: The reseach on the application of two-wheeled self-balancing robot based on the method of variable structure control. J. Harbin Univ. Sci. Technol. **18**(2), 95–100 (2013)
9. Rossomando, F.G., Soria, C.M.: Adaptive neural sliding mode control in discrete time for a SCARA robot arm. IEEE Lat. Am. Trans. **14**(6), 2556–2564 (2016)
10. Hashem, Z.S.M., Khorashadizadeh, S., Fateh, M.M., Hadadzarif, M.: Optimal sliding mode control of a robot manipulator under uncertainty using PSO. Nonlinear Dyn. **84**(4), 2227–2239 (2016)
11. Zhu, S.Q., Jin, X.L., Yao, B., Chen, Q.C., Pei, X., Pan, Z.Q.: Non-linear sliding mode control of the lower extremity exoskeleton based on human-robot cooperation. Int. J. Adv. Rob. Syst. **13**(5), 1–10 (2016)
12. Zhu, S.Q., Chen, Q.C., Wang, X.Y., Liu, S.G.: Dynamic modelling using screw theory and nonlinear sliding mode control of serial robot. Int. J. Robot. Autom. **31**(1), 63–75 (2016)
13. Ayman, A.K., Najib, E., Abdelaziz, H., Frdric, N., Janan, Z.: Type-2 fuzzy sliding mode control without reaching phase for nonlinear system. Eng. Appl. Artif. Intell. **24**(1), 23–38 (2011)
14. Alouia, S., Pages, O., El Hajjaji, A., Chaari, K.Y.: Improved fuzzy sliding mode control for a class of MIMO nonlinear uncertain and perturbed systems. Appl. Soft Comput. J. **11**(1), 820–826 (2011)
15. He, J., Luo, M.Z., Zhang, X.L., Ceccarelli, M., Fang, J., Zhao, J.H.: Adaptive fuzzy sliding mode control for redundant manipulators with varying payload. Ind. Robot **43**(6), 665–676 (2016)

Dynamical System Algorithm Specification Analysis and Stabilization

Charles C. Phiri[1,2,3], János Botzheim[2,3,4], Cristina Valle[2,3], Zhaojie Ju[1,5(✉)], and Honghai Liu[1]

[1] School of Computing, University of Portsmouth, Portsmouth, UK
zhaojie.ju@port.ac.uk
[2] CC Initiative Ltd., Newbury, UK
[3] CC Initiative Ltd., Tokyo, Japan
[4] Department of Automation, Széchenyi University, Győr, Hungary
[5] School of Automation, Wuhan University of Technology, Wuhan, China

Abstract. This paper investigates approaches to deliberately designing systems whose controllability can be quantified. Preliminary findings of ongoing research are presented on complex dynamical system control algorithms. The specification analysis and quality of the pressure control algorithm applied to a Topical Negative Pressure Wound Therapy device are conducted, with further discussion on self-regulation mechanism and characterization of both the partially observable and partially controllable workspace represented by the negative pressure chamber. Statistical methods are employed to understand the device physics and fuzzy logic and bacterial memetic algorithm are utilised to explore and optimize the existing algorithms and also extract the rule base.

Keywords: Specification analysis · Fuzzy inference · Bacterial memetic algorithm

1 Introduction

In highly regulated environments, such as the medical device industry, designing complex dynamical systems is usually prescribed as an exact art. This, in reality, is not the case and many medium sized companies settle for the minimum acceptable solution in order to minimize Research and Development costs or the cost of hiring domain expert help. On the other hand, the complexity of the system may not be fully understood by the development team such that decisions based on a dogmatic view of the world usually pass as the pragmatic solution. In this paper, we examine approaches to deliberately designing systems whose controllability can be quantified. We present the preliminary findings of ongoing research on complex dynamical system control algorithms. We further discuss the specification analysis and quality of the pressure control algorithm applied to a Topical Negative Pressure Wound Therapy device, NPWT [1]; we also look at self-regulation mechanism and characterization of both the partially observable and partially controllable workspaces represented by the negative pressure chamber.

Y. Huang et al. (Eds.): ICIRA 2017, Part I, LNAI 10462, pp. 560–569, 2017.
DOI: 10.1007/978-3-319-65289-4_53

Dynamical systems theory and chaos theory deal with the long-term qualitative behavior of dynamical systems [6]. The task in this paper is not to try and solve the intractable problem of finding a precise solution to the set of equations describing the dynamical system, but rather we are seeking to provide answers to the questions of whether the system settles down to a steady state in the long term. In the event that the system does indeed settle down, what are the possible steady states and whether the long-term behavior of the system depend on its initial condition. We use statistical methods to understand the device physics; we apply fuzzy logic [11] and Bacterial Memetic Algorithm (BMA) [3,4] to explore and optimize the existing pressure control algorithms. BMA is used to infer the fuzzy rule base to identify the optimal control parameters and strategy.

2 Topical Negative Pressure Wound Therapy Pump as a Dynamical Systems

The concept of a dynamical system has its origins in Newtonian mechanics [6]. The evolution rule of dynamical systems is an implicit relation that gives the state of the system for only a short time into the future. The rule describes the time dependence of the position of a point in its ambient space represented as a vector mapped to the coordinates of a manifold (geometrical space). To determine the state for all future times requires iterating the relation many times for each advancing small time step. The subsequent step from the current step are described by a fixed rule known as the evolution rule. The rule may be deterministic such that for a given time interval only one future state follows from the current state; or it may be stochastic such that the evolution of the state is subject to random shocks. Solving the dynamical system (or integrating the system) to determine the states at all future time frames involves iterating the map many times for each advancing small time step. If the system can be solved, given an initial point, it is possible to determine all its future positions, a collection of points known as a trajectory or orbit. Sharkovsky's theorem [2,10] is an interesting statement about the number of periodic points of a one-dimensional discrete dynamical system. A dynamical system can be viewed as a study of iterations of a continuous function from an interval into itself.

Figure 1 shows the lumped parameters model of the system. This is a logical model of the system capturing both the controllable and observable components of the pneumatic circuit. The pressure present at the wound site ideally represents the delta between the pressure measurements in pressure sensor alpha and pressure sensor beta. In the control loop, these can either be used as individual input parameters or combined by observing the delta; this describes a completely different characteristic to the individual inputs. Due to the positioning of the pressure sensors, the presence of pressure release valves, the fact that the wound contact dressing does not form a perfect seal, the effects of gravity on fluids in the lumen and changes in its diameter; the varying ratios of exudate to air, viscosity of the fluids and their biochemical composition and reactions; the system under observation is not fully observable and exhibits turbulent flow.

The observed pressure includes noise introduced from both intrinsic variances and nonlinearities in the sensors and ADC's in the loop which contribute directly to the accuracy of the measurement method and the statistical characteristics of the observable ambient space. The top left of the figure shows the model of the effects fluid flow in a lumen.

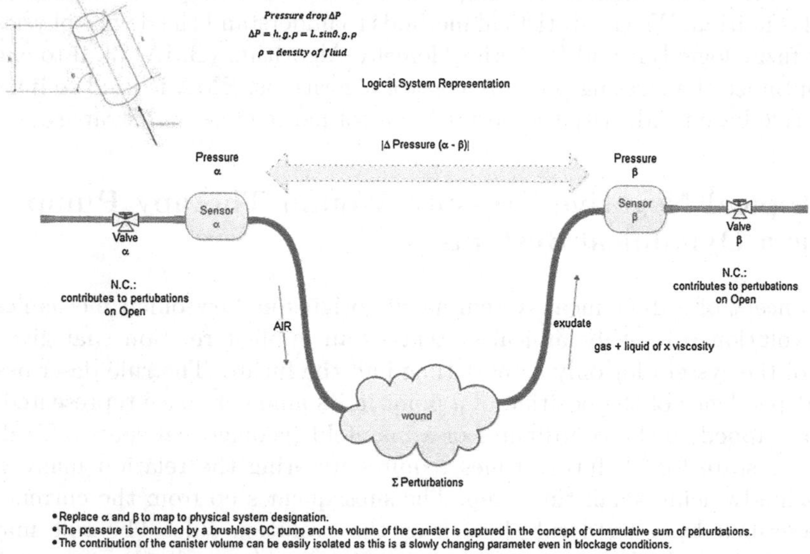

Fig. 1. Lumped parameters model

In NPWT a double lumen (or multi lumen) system has the possibility to (1) introduce a defined air leak into the system to create the pressure delta required to move fluid and (2) independently monitor the pump pressure and wound pressure to allow the system to determine to adjust the pump pressure accordingly. The air leak is not always necessary, on the contrary a large constant air leak increases the load on the vacuum pump resulting in more noise and drawing more power from the battery. Monitoring both the pump pressure and wound pressure helps determine when the introduction of an air leak would be beneficial. The balancing exercise is to let air in when it is needed, meaning when the risk exists for exudate standing still in the tubing. To accommodate this proposed self-regulation, the system is equipped with 2 pressure sensors and 2 valves in the pneumatic circuit.

Figure 2 shows the system decomposed further to a conceptual model that looks at the mapping between the physical world to control theory and dynamic programming. This allows the solutions to be developed without too much noise from the details. The two worlds are coupled and adjusted in tandem to make sure that the model evolves in step with the availability of expert knowledge of the physical system. This also provides a mapping of the specific terms and environments owing to the pedigree of the theories themselves. The system is modeled using function approximators that treat the concept as a discrete system. This demands that a mathematical model of the system exists beforehand. While dynamic programing model requires that a model is present in the first place, reinforcement learning model does not require that the model exist beforehand. The features of the system are learnt during the evolution cycles of the system. The state space is summarized in the bottom left of the figure; this shows how the transitions occur during the iterations. The bottom right captures the essence of the entire problem space in terms of an artificial intelligent agent. In this view the system is collapsed into a collection of sensors perceiving the configuration space and a set of control actions influencing the configuration space. The models are all contingent on the existence of the feedback loop.

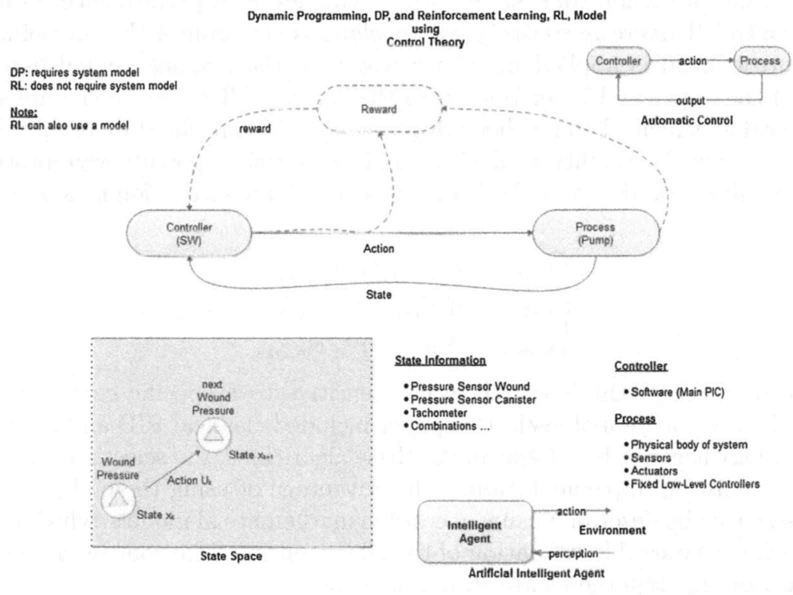

Fig. 2. State space model

The concept starts by exploring whether a method exists to consistently and efficiently determine if a model fits the specification and if so, by what metric to make that determination. In order to appreciate the extent of the challenge, we start by restating the problem space and bounding it to the following terms:

- Obtain a quantifiable description of the pump and the control algorithm
- Identifying dominant features in the pump performance and characteristics
- Describe the relationships between the features (duty cycle of the PWM control signal for the pump, pump speed, wound and canister pressure readings)
- Describe characteristic performance at the steady state
- Describe characteristic performance in the transient state.

The NPWT system describes a leaky system, the control trajectory includes an error term. The flow of fluids in the pneumatic circuit of the NPWT may exhibit chaotic behavior where small initial errors in a trajectory may diverge exponentially. The control algorithm focuses on moving the system towards a stable operating point. On the other hand, the effects of the control mechanisms are not immediately applied (temporal term). The system is unstable without control. The dynamics of the system are unstable. The effect of the control commands have a temporal delay. The effects are observable.

We state here without proof that the NPWT is a nonlinear dynamical system. The regulation algorithm for the NPWT tries to achieve the asymptotic stability of the nonlinear systems. Plenty of literature exists describing the control theory in terms of the Lyapunov Stability [8]. The range of control mechanisms in the industry range from Bang-Bang control through evolved versions of PID to discrete finite state control system with varying levels of performance. In Bang-Bang control (hysteresis control), an envelope is set around the set-point and the control signal is applied in such a way that the response signal oscillates between the upper and lower bounds of the envelope. The thresholds are set as of crisp values which when reached trigger an alteration in the state of the control signal. Lyapunov's stability analysis is applied to design globally asymptotically stable, nonlinear feedback controllers. The control law saturation is described as follows:

$$u(t) = \begin{cases} U_{max} & \text{if } u_{des}(t) \geq U_{max} \\ u_{des} & \text{if } U_{min} < u_{des}(t) < U_{max} \\ U_{min} & \text{if } u_{des}(t) \leq U_{min} \end{cases}$$

The outcome of this research is not sensitive to a specific control mechanism. The control algorithms in this paper included classical PID and FSM. The methodology however is not specific to these algorithms, the selection was determined by existing implementations. The advantage of using these algorithms is they can easily be described using accessible mathematical models, which in turn allowed easily accessible validation of the effectiveness of the control mechanism.

The working hypotheses are stated as below:

- Given a set point (the nominal therapy condition at a suction of −120 mmHg delivered at the wound site):
  - it is possible derive the function mapping the input parameters to the output parameters
  - it is possible to map out an optimal control strategy
- The control signal can be made adaptive based on the deviation of the output compared to the steady-state error at the wound site
- The regulation algorithm can be modified to maintain the range of the delta of the pressure sensor outputs within bounded limits.

3 Specification Analysis Using Fuzzy Rule Base

The performance criteria of the NPWT and the majority of medical devices is strictly regulated by the relevant authorities especially in situations where there is a risk of injury or death to the patient. The control software development has to comply with the IEC/ISO/IEEE 62304 Medical Device Software Life Cycle process, at the very least. The risk is mitigated via the application of clinical domain expert knowledge as well as the resolution of the ISO 14971 Medical Devices Application of Risk Management to Medical Devices. Institutions such as the Food and Drug Authority, FDA, provide guidance and enforcement for products entering the US markets. These standards and risk control measures and the subsequent test methods and methodologies care about whether the system works for the specified range of safe parameters or if it is broken. In this research case, however, focus on metric with which to measure the if the model fits to the specification. Combining the critical components in the standards and system specification, in our research, the linguistic variables (computable information granules) could be extracted and modeled. These variables were then manually translated into a Fuzzy Rule Base. Figure 3 shows the surface graph. From this specification it was established that pump was over-specified by a factor of approximately 60%. This directly contributed to tighter control measures being applied to remove the risk of a runaway system that could result in extreme injuries to the patient. Hardware controls feature have been implemented to minimize this risk.

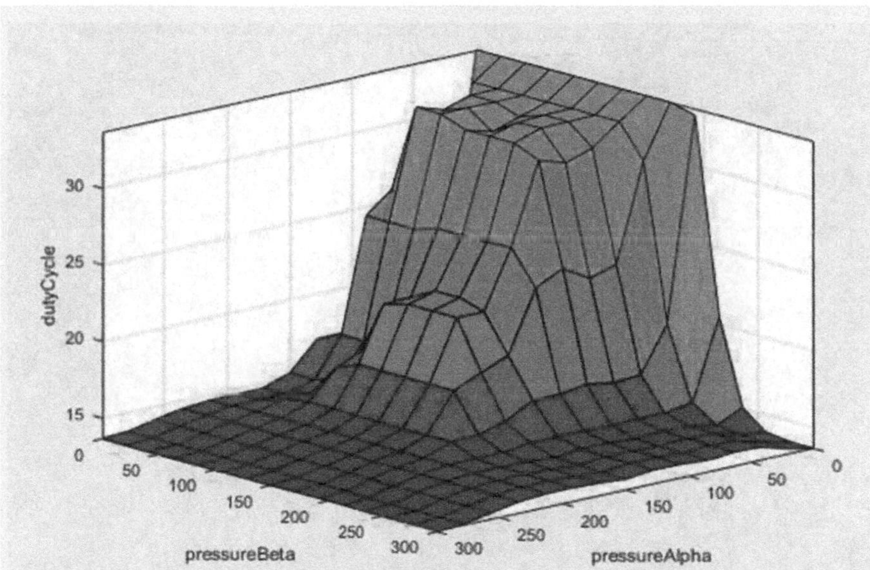

Fig. 3. Specification analysis

Also from Fig. 3, it can be shown that the system performance will always fall within the surface graph. The boundary of discrimination is defined by the aforementioned linguistic variables and coincide with the boundary for safely operating the pump in this configuration to deliver adequate therapy. Further, as there exists a direct relationship between the control voltage applied to the pump and the duty cycle; the pump being one of the largest contributors to the power budget of the unit, the description of the duty cycle in this figure closely approximates the power performance response graph. Further, this also defines the surface area over which the system could be safely customized.

3.1 Checking Assumptions

Given that no mathematical model exists, the control system has to be validated by using experimental data. Figure 4 shows the relationship between duty cycle and pump speed (RPM). It is therefore not surprising that there is a very strong correlation between the pump duty cycle and the pump speed. Figure 5 looks at the relationship between the pressure sensors alpha and beta. The lumina (lumens), represented by the two pressure sensor reading are controlled by a suction generated from a single pump. It should be noted that owing to the physics of the device, the lumina are out of phase. Pressure Sensor Alpha response has the effect of applying a low pass filter on the Pressure Sensor Beta response graph. In other words, the perturbations in beta lumen are attenuated (and inverted) in alpha lumen response graph.

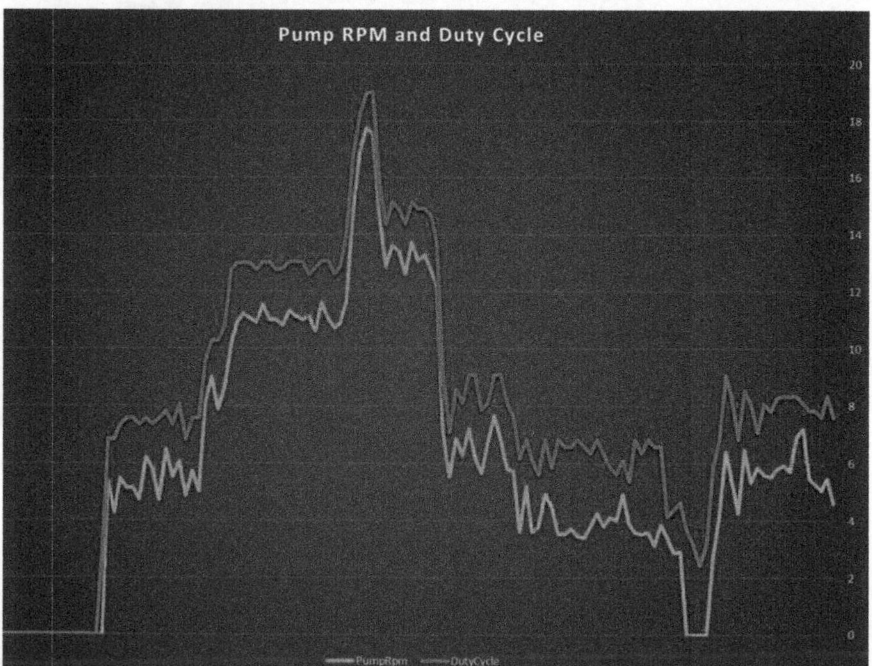

Fig. 4. Pump RPM and duty cycle

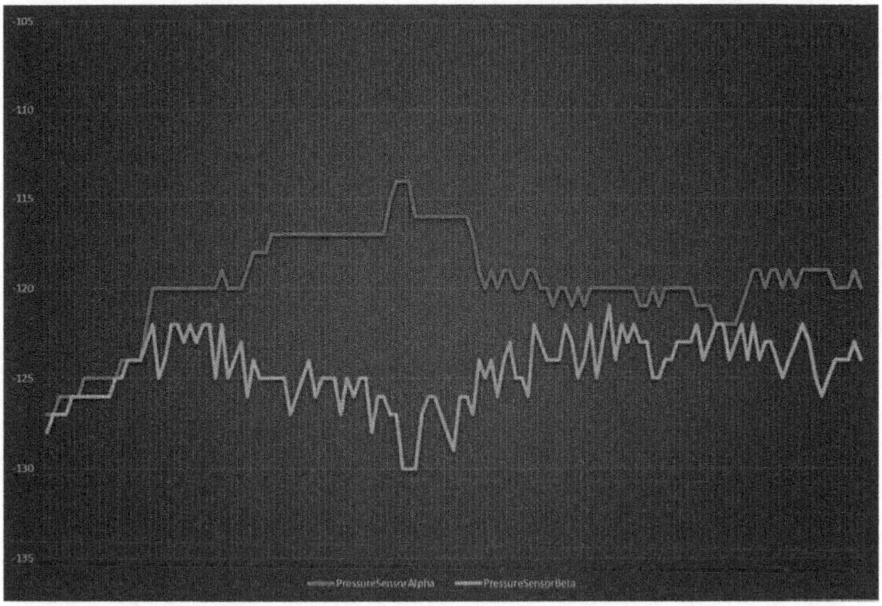

Fig. 5. Pressure alpha and pressure beta relationship

3.2 Computational Intelligence Based Model

Based on above assumptions, a computational intelligence based model is proposed as illustrated in Fig. 6. In this model, a fuzzy rule base is extracted and optimized. The inputs of the fuzzy rule base are the pressure sensors alpha and beta. The pump RPM value can be omitted from the inputs because its relationship to the duty cycle as shown in Fig. 4. The output of the fuzzy rule base is the percentage duty cycle. Mamdani type fuzzy rules and inference method are applied in the fuzzy model [5].

The optimization of the fuzzy rule base is performed by Bacterial Memetic Algorithm which is a population based stochastic nature-inspired optimization algorithm [3,4]. BMA effectively combines global and local search in order to find a quasi-optimal solution for the given problem. The bacterial operators (bacterial mutation and gene transfer) are applied in the global search. The role of the bacterial mutation is the optimization of the bacteria's chromosome. The gene transfer allows the transfer of information in the population among the different bacteria. As a local search technique, the Levenberg-Marquardt method is applied by a certain probability for each individual. One rule base corresponds to one bacterium in the BMA when applied for fuzzy rule base optimization. The fuzzy rule bases (bacteria) are evaluated by the mean square error criterion. BMA updates the fuzzy rule base parameters (membership functions) by the bacterial mutation, Levenberg-Marquardt, and gene transfer operations [3].

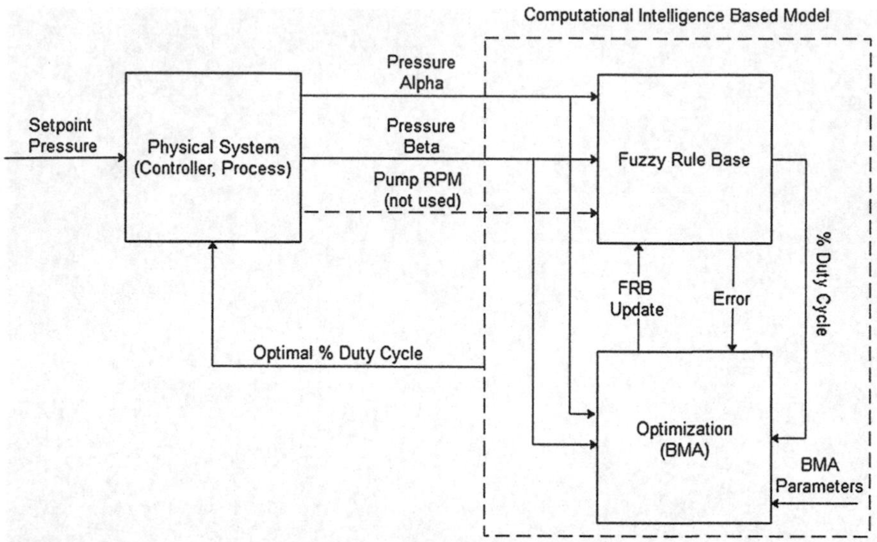

Fig. 6. Computational intelligence based model

The output of the computational intelligence based model is the optimal percentage duty cycle which is fed back into the physical system.

4 Conclusion

In this work, we have shown approaches to deliberately designing systems which are partially controllable and partially observable. We have presented the preliminary findings of ongoing research on complex dynamical system control algorithms. This research uses data from the medical industry. We have discussed the specification analysis and quality of the control algorithm applied to a Topical Negative Pressure Wound Therapy device, and we have also looked at self-regulation mechanism and characterization of both the partially observable and partially controllable workspace represented by the negative pressure chamber. We have proposed a computational intelligence based model to determine the optimal percentage duty cycle value for the physical system. In the computational intelligence based model a fuzzy rule base approach has been used together with a nature-inspired optimization algorithm. Although the proposed approach was not able to fully solve the optimal control of the pump, it encourages us for further research. As a future work our aim is to deeply investigate the relationship between the duty cycle and the pressure of the pump by computational intelligence based models [7,9].

Acknowledgments. The authors would like to acknowledge support from CC Initiative Ltd., project of NSFC (51575412), DREAM EU FP7-ICT (611391), State Key Laboratory of Digital Manufacturing Equipment & Technonlogy (DMETKF2017003), and the exchange program from School of System Design, Kubota Laboratories, Japan.

References

1. Berrevoet, F., Vanlander, A., Sainz-Barriga, M., Rogiers, X., Troisi, R.: Infected large pore meshes may be salvaged by topical negative pressure therapy. Hernia **17**(1), 67–73 (2013)
2. Borwein, J., Bailey, D.: Mathematics by Experiment: Plausible Reasoning in the 21st Century. A K Peters, Wellesley (2003)
3. Botzheim, J., Cabrita, C., Kóczy, L.T., Ruano, A.E.: Fuzzy rule extraction by bacterial memetic algorithms. Int. J. Intell. Syst. **24**(3), 312–339 (2009)
4. Botzheim, J., Toda, Y., Kubota, N.: Bacterial memetic algorithm for offline path planning of mobile robots. Memet. Comput. **4**(1), 73–86 (2012)
5. Mamdani, E.H.: S.A.: An experiment in linguistic synthesis with a fuzzy logic controller. Int. J. Man Mach. Stud. **7**(1), 1–13 (1975)
6. Gotoda, H., Shinoda, Y., Kobayashi, M., Okuno, Y., Tachibana, S.: Detection and control of combustion instability based on the concept of dynamical system theory. Phys. Rev. E **89**(2), 022910 (2014)
7. Ju, Z., Liu, H., Xiong, Y.: Fuzzy empirical copula for estimating data dependence structure. Int. J. Fuzzy Syst. **16**(2), 160–172 (2014)
8. Jung, J.W., Leu, V.Q., Do, T.D., Kim, E.K., Choi, H.H.: Adaptive PID speed control design for permanent magnet synchronous motor drives. IEEE Trans. Power Electron. **30**(2), 900–908 (2015)
9. Phiri, C.C., Ju, Z., Liu, H.: Accelerating humanoid robot learning from human action skills using context-aware middleware. In: Kubota, N., Kiguchi, K., Liu, H., Obo, T. (eds.) ICIRA 2016. LNCS, vol. 9834, pp. 563–574. Springer, Cham (2016). doi:10.1007/978-3-319-43506-0_49
10. Weisstein, E.: Sharkovsky's theorem. Mathworld-a Wolfram web resource. http://mathworld.wolfram.com/SharkovskysTheorem.html
11. Zadeh, L.: Fuzzy sets. Inf. Control **8**, 338–353 (1965)

A Review of Upper and Lower Limb Rehabilitation Training Robot

Wenlong Hu[1(✉)], Gongfa Li[1,2], Ying Sun[1,2], Guozhang Jiang[1,2], Jianyi Kong[1,2], Zhaojie Ju[3], and Du Jiang[1]

[1] Key Laboratory of Metallurgical Equipment and Control Technology, Ministry of Education, Wuhan University of Science and Technology, Wuhan 430081, China
942599328@qq.com, ligongfa@wust.edu.cn

[2] Hubei Key Laboratory of Mechanical Transmission and Manufacturing Engineering, Wuhan University of Science and Technology, Wuhan 430081, China

[3] The Laboratory of Intelligent System and Biomedical Robotics, University of Portsmouth, Portsmouth, PO1 3HE, UK

Abstract. With the aging of society, the number of patients with limb disorders caused by stroke has increased year by year, it is necessary to introduce more advanced technology into the field of rehabilitation treatment. Rehabilitation training based on the brain plasticity has been proved by clinical medical practice as an effective treatment method, and because of the serious lack of professional rehabilitation therapists, a large number of rehabilitation training robot have been designed so far. This article analyzed and described the research status on upper and lower limbs rehabilitation training robot, and at last the paper forecasts the future development trend of rehabilitation robot.

Keywords: Rehabilitation training · Robot · Stroke · Brain plasticity

1 Introduction

Stroke, also known as "apoplexy", is a kind of acute cerebral vascular disease, and it is caused by cerebrovascular sudden rupture or vascular obstruction which will make the blood can't get into the brain. The stroke can cause persistent brain tissue damage and dysfunction, and there are 85% of patients with hemiplegia symptoms, especially in the elderly. At the same time, stroke is the world's second largest, China's largest fatal disease, it not only has high mortality, but also has high incidence rate, high recurrence rate and high disability rate, all of which have serious impact on the patients' health and quality of life [1].

Hemiplegia caused by stroke is a kind of central neurological movement disorder. After the onset of stroke, only a few people can recover after treatment, most patients will be accompanied by behavior disorder and severe cases with hemiplegia. Now rehabilitation training based on the brain plasticity after treatment has been proved by clinical medical practice as an effective treatment method [2, 3]. Traditional medical rehabilitation training is mainly dominated by professional rehabilitation master hand operation, this method for patients, their family members and medical staff is a time-consuming,

Y. Huang et al. (Eds.): ICIRA 2017, Part I, LNAI 10462, pp. 570–580, 2017.
DOI: 10.1007/978-3-319-65289-4_54

laborious and tedious work. And because of the too large number of patients in our country and a serious lack of professional rehabilitation therapists (about 1%), so that the intensity of rehabilitation training in many patients is insufficient and their rehabilitation cannot reach the ideal state. Meanwhile, rehabilitation robots as a kind of artificial machine are able to perform tasks automatically, which make it can be used to replace or assist the body with certain function and ensure the effect of the patient rehabilitation training, there is no doubt that them can play an important role in the process of medical rehabilitation. Therefore, it is necessary to introduce the robotics into rehabilitation field, let the robot replace physical therapist to do some heavy work.

2 Research Status of Rehabilitation Robots

Rehabilitation robot is the product of robot technology applied in rehabilitation medicine, involving the multi-discipline knowledge, its main principle is to promote the reorganization of the cerebral cortex by the right amount and specific repetitive training, and make patients through the profound experience to learn and store the right movement patterns. Today, rehabilitation robots have a rather comprehensive development.

2.1 Upper Limb Rehabilitation Robots

In 1991, MIT have developed a rehabilitation training robot MIT-MANUS, which can help patients with hemiplegia to carry out rehabilitation training. The subject of the device is five link mechanism, with two degrees of freedom movement, it can help patients to complete the planar motion of the arm and can get the arm movement synchronization parameters such as speed, Angle, it also can feedback the information to the patient through the microcomputer interface [4, 5]. Patients were divided into two groups for traditional treatment methods and the comparative analysis of auxiliary rehabilitation training robot test, the experiment result shows that: the rehabilitation robot can effectively help patients for treatment, treatment effect have significantly improved comparing to the traditional method, and it do not have any damage and other adverse effects on the patients.

At Stanford university, upper limbs rehabilitation robot system MIME (mirror image motion enabler) based on PUMA500, 560 industrial robots was developed [6, 7], this system can gather the trajectory of the healthy side upper limb mirrored to the other side, and through the industrial robot assisted with lateral movement, and which has been applied in clinical rehabilitation.

In 1999, the University of California and the rehabilitation institute of Chicago developed a robot named ARM – Guide [8], the robot has an active DOF, and it can be used to measure range of patients with ARM, at the same time, it also can help upper limb with rehabilitation exercise, such as arm flexion and extension.

In 2002, the European international research team led by the University of Reading shoulder elbow developed a GENTLE upper limb rehabilitation robots for shoulder and elbow joint [9]. Patients' arm are connected to the Haptic Master robot with sling, and its driving force and gravity are provided by industrial robot, it also can provide visual feedback combined with a virtual technology, so which has a remarkable curative effect on the rehabilitation training of patients.

At the University of Leeds, Raymond Holt et al. developed the iPAM arms rehabilitation robots, there are three joints on each of the mechanical arm, and it uses pneumatic drive device to provide torque traction in patients with forearm and upper arm to complete the shoulder, elbow composite movement [10].

Britain's university of Southampton developed the famous five - DOF SAIL upper limbs rehabilitation robot, and it has no driver, but there is a torsional spring elasticity auxiliary support system in the shoulder, elbow, revolute joint, this robot by using virtual reality (VR) technology combined with electrical stimulation arm muscle technology, and can perform rehabilitation training of shoulder, elbow and wrist, which have achieved good effect of rehabilitation [11].

CADEN - 7 is an exoskeletons of a 7 DOF robot devised by the University of Washington's Joel Perry and others. It can help patients with shoulder, elbow, wrist joint rehabilitation exercise, but also can be used for the healthy body of the experimental study.

At present in China, while strengthening the people's emphasis on rehabilitation medicine, but compared with foreign, rehabilitation robot research level is still relatively backward. Most of the rehabilitation robot research and development is carried out by the school, and at the shallow level stage, the research results of industrialization are mostly based on the simple single joint rehabilitation training device.

Tsinghua university is one of China's earlier research units engaged in rehabilitation robot, it has developed by the end of the 2 - DOF traction type upper limb rehabilitation robot system, the system has a variety of functions such as a passive training, auxiliary active training and constrained impedance training, etc. [12]. Southeast university Song Aisong and others based on the remote control of compound 3 - DOF end traction type rehabilitation robot has carried on the system research [13], it can be in two modes for upper limb rehabilitation training, and lower limb rehabilitation training can also be carried out. Harbin industrial university, Harbin engineering university, Beijing University of technology, Taiwan University carried out a similar study [14, 15].

2.2 Lower Limb Rehabilitation Robot

Lower limb rehabilitation training robot as a rehabilitation training robot, is mainly used to help patients with lower limb motor function obstacle for rehabilitation training. Research has shown that lower limb movement training assisted by robots, their sensory and motor information synchronization is better, which for the nervous system remodeling, the formation of a correct sense and motion loop is very helpful [16].

The structure of the lower limb rehabilitation training robot configuration and rehabilitation motion control strategy is the key two technologies of the research [17]. In recent years, the lower limb rehabilitation training robot has a great development in the structural configuration. According to the different postures of the patients while training, the lower limb rehabilitation training robot is divided into two types: standing and sitting.

2.2.1 Standing Lower Limb Rehabilitation Training Robot

The design of the lower limb rehabilitation training robot must comply with the basic rule of kinesiology. Because the lower limb function of most of the patients is poor, which means they cannot load the normal weight, so generally standing lower limb rehabilitation training robot is equipped with weight loss. One of the typical structures of the standing leg rehabilitation training robot is to use lower limb exoskeleton and medical treadmill to coordinate with the patients' gait training. In 1999 the Swiss HOCOMA Medical Equipment Co., Ltd. cooperate with Swiss Zurich Balgrist medical rehabilitation center developed LOKOMAT walking rehabilitation training robot, which is the first set of products that can assist patients with lower limb movement disorders in weight loss walking training on a medical treadmill [18]. The training evaluation system is composed of the exoskeletal lower limb gait rectification drive, the intelligent weight system and medical treadmill. The exoskeletal lower limb gait rectification drive is connected to a quadrilateral structure which is supported by a spring, and it is the core part of the training system. The exoskeleton are left and right symmetrical mechanical legs, and the motor is respectively installed on the waist rack of the mechanical leg and the leg pole of the thigh, and each motor drives a screw nut mechanism, so as to promote mechanical legs finish walking motion.

At present, there are many studies show that LOKOMAT has a significant effect on the rehabilitation of patients with gait disturbance. The Swiss federal institute of technology in Zurich and the University of Zurich have used the LOKOMAT rehabilitation robot and its cooperation training mode to carry out an observation contrast training, which is combined with 2 cases of chronic patients with incomplete spinal cord injury and 2 cases of chronic stroke patients, and the duration of observation is four weeks [19]. Finally, under different conditions on foot 10 m tests to assess patients with fruit found that participants can easily adapt to the training mode of cooperation, and also found that this kind of collaborative training mode is better in improving the training enthusiasm and their muscle activity, there is one case have Increase in pace after treatment, the other 3 cases found no significant change. This shows that LOKOMAT's cooperation training mode is feasible in clinical practice, and the training mode can make the training more positive, variety and nature. However, due to the limited number of subjects, it cannot be concluded that the cooperative training model has an effective conclusion on the walking ability of the patients. This problem needs to be solved by more clinical trials.

The overseas clinical applications of LOKOMA are widespread [20–22], there are also many domestic [23, 24]. North China University of Science and Technology Institute of Rehabilitation designed a comparative experiment with 80 stroke patients with lower limb motor dysfunction, which was divided into the intervention group and the control group, and intervention group uses the LOKOMAT rehabilitation training robot to train, the control group using conventional rehabilitation treatment; after 10 weeks of observation, the exercise parameters and physiological parameters of the two groups were statistically analyzed, and the results were found that the rehabilitation effect of the intervention group was good [25]. In addition, the product currently has 3 types of configurations, respectively, based type, professional type and child-shape [26]. Although this product is fully functional, its dynamic model is relatively complex, and it is difficult to control, by the way the price is also too expensive. There are many similar products, such as the United States MOTORICA [27], the LOKOHELP [28], Germany's GAIT TRAINER [29], this kind of robot aided by the auxiliary suspension weight reduction device drive with the robot arm, the movement, the foot board and other ways to drive the patient's lower limb, and make it swing, so as to complete the whole or part of the gait track training [30].

The second typical structure of the standing leg rehabilitation training robot is by controlling the foot pedal to drive the whole lower limb movement. The mechanical gait rehabilitation trainer (MGT) developed by Free University Berlin in 1996 is the first prototype of the foot pedal powered lower limb rehabilitation training robot in the world. In 2003, according to the results of the study, Hesse put forward the idea of Haptic Walker [31], and it was developed the German Fraunhofer Institute, combining the latest robot technology, computer technology and virtual reality technology [32, 33]. Haptic Walker also adopts the suspension method to reduce weight to keep the patient body balance. Patients use the programmable control pedal to carry out rehabilitation training, and it can provide a variety of possible foot trajectory; the programmable control pedal is composed of two fully symmetrical but mutually independent mechanical arms, and it is arranged at the bottom of the mechanical arm. The robot arm is installed on the linear guide rail, and the horizontal direction of the pedal is controlled by the linear motor, and the position of the vertical direction is controlled by the motor which is installed at the 3 joints of the mechanical arm. The system can provide the trainer with the ground, up and down the stairs and so on a variety of training scenarios, which is helpful to improve the training interest; and this robot has passed the Rhine in Germany TUV medical product safety testing and is approved by Charlie Berlin university hospital clinical ethics committee. All the functions of the machine were successfully tested by 20 healthy volunteers and 40 patients with stroke, the results show that the rehabilitation robot can provide an infinite variety of training path for patients with stroke, and the performance of the machine can be well reflected. At present, the combined motion of the machine is the emphasis and difficulty of the research, and the experiment requires a lot of algorithms [34]. In addition during the test of a spinal cord injury patient, the patient can independently accomplish the stair climbing training, and after a period of training patients found himself spasm phenomenon reduced; training 30 min a day, six weeks later, the results showed that the maximum walking capacity increased by 1 times, and the patient can climb up a flight of stairs with a slight help. It is obvious that the device has a good development prospect [35]. But the device is too large, and the price is too expensive [36].

In our country, there are a lot of research on the lower limb rehabilitative robot which adopts the pedal mechanism as the main body, for example, Harbin Engineering University, [37]. At present most of the researches of gait motion mechanism focus on the main movement performance of different structure forms [38, 39], rarely involved in foot posture adjustment [40]; but in fact, the ankle joint is a fine tuning hinge of human stability, its dorsiflexion has extremely important significance for lower limb motor function, balance and gait [41]. The adjustment of the foot posture can lead to the ankle joint movement, so the adjustment of the foot posture is very important in the gait movement. Liu Junkai et al. design a gait motion mechanism based on pedal, the emphasis of which is the design of the foot posture adjusting mechanism. The mechanism is composed of two sets of single side structure with 180° phase difference, which are connected through a main shaft. The main moving parts is offset slider-crank mechanism and the driving link is crank driven by the main motor. Foot posture adjusting component is a swing decelerating mechanism, which is composed of screw nut pair, transmission parts, foot shaft, foot board, etc., the screw as a prime mover driven by the two auxiliary motors on both sides. Its Structural parameters of the main moving parts of the gait motion mechanism ensure the trajectory, velocity and acceleration characteristics of the main motion, the equivalent speed of the main motor ensures the period of the main motion, and the equivalent speed of the auxiliary motor ensures the coordination relation between the foot posture adjustment movement and the main movement. The simulation results show that the mechanism is suitable for patients with lower limb rehabilitation training. But the gait parameters to achieve the best rehabilitation effect should be studied further.

The third kind of typical standing lower limb rehabilitation training robot is pneumatic exoskeleton lower limb rehabilitation robot proportional controlled by lower limb EMG signal. The university of Michigan human neural mechanics laboratory has developed a pneumatic exoskeleton lower limb rehabilitation robot proportional controlled by lower limb EMG signal [42]. The robot is mainly composed of the exoskeletons of ankles, knees and buttocks, the Shell attached to exoskeletons and drive. By analyzing the motion data and physiological data of the walking training of two patients with incomplete spinal cord injury using the device, the results show that all the data have been changed after using the device, the activity of the muscle of the patient was improved after using this device for 24 min, which means the patient could get greater thrust. This improvement is still valid when the support force of the exoskeleton is closed, so the rehabilitation effect is quite obvious. However, this robot is only suitable for patients with certain athletic ability.

Standing lower limb rehabilitation training robots often use the method of suspension to reduce weight, but there are some exceptions. In 2014, based on the M300 series of anti-gravity treadmill launched by AlterG [43], a lower limb rehabilitation training robot which uses airbags to reduce weight was developed by University of Shanghai for Science and Technology [44]. The device is mainly composed of walking device and weight support mechanism. And the weight support mechanism is composed of an upper airbag, a lower airbag, the brackets of the airbags and tight shorts which is connected to the upper and lower airbags. The inflated air bags can provide the training person with an upward lift force, so that the patient is in a weight loss environment. This design uses the airbags to reduce weight, which is a very good idea, but its volume is too large and it is too massive, so it needs further improvement in the future, and further clinical studies are also needed to test the practicality and safety of the machine.

2.2.2 Horizontal Lower Limb Rehabilitation Training Robot

The horizontal lower limb rehabilitation training robots take the seat or the bed as the carrier of the patient, so the weight support mechanism is simplified. And currently the most representative of the lower limb rehabilitation robot is MotionMaker, which is developed by SWORTEC [45]. The MotionMaker is mainly composed of two two-degree-of-freedom manipulators, seat, base, closed loop control system and functional electrical stimulation module; the manipulators are symmetrically arranged on both sides of the base, and the movement of the hip joint, the knee joint and the ankle joint is respectively driven by the screw nut mechanism driven by the DC motor; angle sensor is installed in each joint and the pressure sensors are installed in the connecting position of the screw rod and the crank, and according to the feedback data of the sensor, the closed-loop control of the manipulator is formed; the length of manipulators can be adjusted to accommodate patients with different height. In 2008, the first generation of MotionMaker was introduced into clinical application in Switzerland. There are relevant research on the MotionMaker motion monitoring system for testing. For instance, one incomplete paraplegic patient was assessed with the MotionMaker in week 10 and week 18 after his injury. He performed three repetitive leg press movements in extension in three different conditions: with his residual strength, with electrical stimulation of the glutei, quadriceps and triceps surae muscles and finally with a combination of residual strength and electrostimulation. For each condition, horizontal forces developed at the foot and hip, knee and ankle moments were recorded and averaged across the three trials. The results show that the mean forces and moments were the highest when the movement was performed with electrostimulation and residual strength, this is very meaningful for clinical analysis [46]. But the robot cannot achieve the hip abduction and adduction movements and lower extremity movement patterns is not comprehensive and lack of mechanical limit. These are yet to be improved in the later development. In 2014, Yanshan University and Shanghai Jiaotong University jointly developed an exoskeleton lower limb rehabilitation robot which is similar to MotionMaker [47]. This robot is more intelligent, and there are triple protection function of software, hardware, and mechanical limit. At present, the rehabilitation robot has started trials in China Rehabilitation Research Center, its feasibility and safety has been verified, but the rehabilitation effect still needs to be further verified.

In 2011, Marmara University jointly developed the lower limb rehabilitation training robot Physiotherabot [48], which is mainly used for the rehabilitation training of the patients who lost lower extremity motor function because of spinal cord injury, stroke, muscle disorders, or surgical procedures. The robot manipulator can perform all active and passive exercises as well as learn specific exercise motions and perform them through the Human–Machine interface, and the patient's state of motion can be fed back to the patient. All the motors are installed on the base, so that the manipulators can get better performance. And all movements are controlled by the Human–Machine interface, which is very convenient. But the robot can only be trained on one leg at a time, and it cannot carry out the movement of the ankle joint, there is no clinical trials or device for monitoring physiological parameters of patients.

In 2014, North China University of Science and Technology proposed a lower limb rehabilitation robot that is based on 2-URS&UPS parallel mechanism [49]. The mechanism has six DOF and can do three-dimensional motion. During the training, the patient sits on the chair, and the foot and foot pedal fixed together. The rationality of the design of the rehabilitation robot has been verified by simulation. Currently, this robot is in the theory stage, and needs more research in the future.

The other structure of the horizontal rehabilitation training robot is to use the bed body to reduce weight. For instance, University of Shanghai for Science and Technology designed a body weight loss multi-function system in patients with hemiplegia rehabilitation training and assessment, which is mainly used for functional training and assessment of patients with lower limb dysfunction [50]. The system can help patients in different stages of rehabilitation with their rehabilitation training and the system also has real-time detection function on body kinematics and biomechanical parameters, so it is convenient to evaluate the patient's recovery and develop a more reasonable treatment. The system had passed the ethical review of Huashan Hospital Affiliated to Fudan University, and has more than 100 cases of clinical trials, the doctors and patients reflect the effect is good. However, the exterior design of the system is not so reasonable, the patient up and down the training bed is not convenient,so it must be improved further.

3 Summarize and Prospect

This article analyzed and described the research status on upper and lower limbs rehabilitation training robot. There are some simple descriptions about the typical robot in the process of the development of upper and lower limbs, and some of their advantages and disadvantages are also analyzed. We can know that rehabilitation robots perfectly combine intelligent control with limb movement by robotic system, and it can help patients to complete the rehabilitation exercise. And it is in-comparably superior to the traditional rehabilitation method: it can reduce the labor intensity of rehabilitation therapists and make up for the present situation of insufficient professional therapists, and the most important thing is that each patient has an objective and balanced treatment. The control system can exert more accurate driving torque, and the therapist can make an objective evaluation, improve the treatment plan in a timely manner based on the detailed records of treatment data and exercise parameters and times of training. Researchers can also link parameters and the curative effect, so as to find more rules about robot assisted rehabilitation therapy.

In the future, it is possible to carry out the research work of rehabilitation robots in the following directions:

(1) The mechanical structure of the robot should be compact, easy to control and multiple functional;
(2) It is necessary to consider differences in patients' pathologies and physical conditions, in order to improve the applicability of the rehabilitation robot;
(3) From the hardware structure and software system design, the robot must set up a reasonable protection device to ensure the safety of patients.
(4) In addition, the training mode of intelligent, interesting is also the trend of development of rehabilitation robots in the future.

Acknowledgments. This work was supported by grants of National Natural Science Foundation of China (Grant Nos. 51575407, 51575338, 51575412) and the UK Engineering and Physical Science Research Council (Grant No. EP/G041377/1).

References

1. Wang, L.: Report on the Chinese Stroke Prevention. Peking Union Medical College Press, Beijing (2015)
2. Lou, J., Zeng, Z.: Stroke and brain plasticity. J. Henan Univ. (Med. Sci.) **29**(1), 1–4 (2010)
3. Yang, D., Zeng, X.: Research progress of brain plasticity for stroke. Chin. J. Cerebrovasc. Dis. **8**(4), 221–224 (2011)
4. Charles, S.K., Krebs, H.I., Volpe, B.T., et al.: Wrist rehabilitation following stroke: Initial clinical results. In: IEEE International Conference on Rehabilitation Robotics, pp. 13–16 (2005)
5. Krebs, H.I., Dipietro, L., Levy-Tzedek, S., et al.: A paradigm shift for rehabilitation robotics. IEEE Eng. Med. Biol. Mag. **27**(4), 61–70 (2008)
6. Lum, P.S., Burga, C.G., Shor, P.C.: Use of the MIME robotic system to retrain multi-joint reaching in post-stroke hemiparesis: why some movement patterns work better than others. in: Annual International Conference of the IEEE Engineering in Medicine and Biology, pp. 1475–1478 (2003)
7. Burgar, C.G., Lum, P.S., Shor, P.C., et al.: Development of robots for rehabilitation therapy: The Palo Alto VA/Stanford experience. J. Rehabil. Res. Dev. **37**(6), 663–673 (2000)
8. Reinkensmeyer, D.J., Kahn, L.E., Averbuch, M., et al.: Understanding and treating arm movement impairment after chronic brain injury: progress with the ARM guide. J. Rehabil. Res. Dev. **37**(6), 653–662 (2000)
9. Amirabdollahian, F., Gradwell, E., Loureiro, R., et al.: Effects of the GENTLE/S robot mediated therapy on the outcome of upper limb rehabilitation post-stroke: analysis of the battle hospital data. In: 8th International Conference on Rehabilitation Robotics, pp. 55–58 (2003)
10. Kemna, S., Culmer, P.R., Jackson, A.E, et al.: Developing a user interface for the iPAM stroke rehabilitation system. In: IEEE International Conference on Rehabilitation Robotics, pp. 879–884 (2009)
11. Cai, Z., Tong, D., Meadmore, K.L., et al.: Design & control of a 3D stroke rehabilitation platform. In: IEEE International Conference on Rehabilitation Robotics, p. 5975412 (2011)
12. Zhang, Y.B., Wang, Z.X., Ji, L.H., et al.: The clinical application of the upper extremity compound movements rehabilitation training robot. In: IEEE International Conference on Rehabilitation Robotics, pp. 91–94 (2005)
13. Xu, B.G., Peng, S., Song, A.G., et al.: Robot-aided upper-limb rehabilitation based on motor imagery EEG. Int. J. Adv. Rob. Syst. **8**(4), 88–97 (2011)
14. Yang, Y., Wang, L., Tong, J., et al.: Arm rehabilitation robot impedance control and experimentation. In: IEEE International Conference on Robotics and Biomimetics, pp. 914–918 (2006)
15. Mao, K., Li, X.Y., Jiang, N.F., et al.: The design and implementation of the passive upper limb rehabilitation robot based on magnetic powder brakes. In: International Symposium on IT in Medicine and Education, pp. 341–345 (2011)
16. Hu, X.L., Tong, K.Y., Song, R., et al.: A comparison between electromyography-driven robot and passive motion device on wrist rehabilitation for chronic stroke. Neurorehabil. Neural Repair **23**(8), 837–846 (2009)
17. Li, J., Wang, J., Zhao, H., et al.: Critical technologies of lower limb rehabilitation training robot. Mach. Des. Manuf. **9**, 220–223 (2013)
18. Hocoma: Lokomat-Enhanced Functional Locomotion Therapy with Augmented Performance Feedback. http://www.hocoma.com/en/products/lokomat. Accessed 5 May 2015

19. Schück, A., Labruyère, R., Vallery, H., et al.: Feasibility and effects of patient- cooperative robot - aided gait training applied in a 4-week pilot trial. J. Neuroeng. Rehabil. **9**(1), 1–14 (2012)
20. Mirbagheri, M.M., Niu, X., Kindig, M., et al.;The effects of locomotor training with a robotic-gait orthosis (Lokomat) on neuromuscular properties in persons with chronic SCI. In: Engineering in Medicine and Biology Society (EMBC), 2012 Annual International Conference of the IEEE, pp. 3854–3857 (2012)
21. Krewer, C., Rieß, K., Bergmann, J., et al.: Immediate effectiveness of single - session therapeutic interventions in pusher behaviour. Gait Posture **37**(2), 246–250 (2013)
22. Westlake, K.P., Patten, C.: Pilot study of Lokomat versus manual-assisted treadmill train-ing for locomotor recovery post-stroke. J. Neuroeng. Rehabil. **6**(1), 18 (2009)
23. Guo, S.M., Li, J.M., Wu, Q.W., et al.: Effect of gait training and assessment system of Lokomat automatic robot on walking ability of patients with incomplete spinal cord injury. Chin. J. Tissue Eng. Res. **16**(13), 324–327 (2012)
24. Guo, S.M., Li, J.M., Wu, Q.W., et al.: Clinical application of Lokomat automatic robot gait training and assessment system. China Med. Devices **26**(3), 94–96 (2011)
25. Ma, S.H., Liu, D., Hao, Z.W., et al.: Effect of Lokomat on lower limb motor function recovery in patients with stroke. Shandong Med. J. **52**(28), 52–54 (2012)
26. Hocoma. Lokomat- full automatic robot gait evaluation training system. http://www.soreha.net/Product/content/id/58.html. Accessed 5 May 2015
27. Wang, J., Yang, Z.H., Liu, H.B., et al.: Application and research progress of lower limb rehabilitation robot in patients with stroke. Chin. J. Rehabil. Med. **29**(8), 784–788 (2014)
28. Freivogel, S., Mehrholz, J., Husak-Sotomayor, T., et al.: Gait training with the newly developed 'LokoHelp' - system is feasible for non-ambulatory patients after stroke, spinal cord and brain injury. A feasibility study. Brain Inj. **22**(7–8), 625–632 (2008)
29. Hesse, S., Uhlenbrock, D., Werner, C., et al.: A mechanized gait trainer for restoring gait in nonambulatory subjects. Arch. Phys. Med. Rehabil. **81**(9), 1158–1161 (2000)
30. Mantone, J.: Getting a leg up? Rehab patients get an assist from devices such as HeathSouth's AutoAmbulator, but the robots' clinical benefits are still in doubt. Mod. Healthc. **36**(7), 58–60 (2006)
31. Schmidt, H., Hesse, S., Bernhardt, R., et al.: HapticWalker-a novel haptic foot device. ACM Trans. Appl. Percept. **2**(2), 166–180 (2005)
32. Schmidt, H., Sorowka, D., Hesse, S., et al.: Development of a robotic walking simulator for gait rehabilitation. Biomedizinische Technik Biomed. Eng. **48**(10), 281–286 (2003)
33. Medical, P.: Robots help to heal: G-EO-System. http://www.medicalpark.de/en/main/g-eo-system.htm. Accessed 5 Jun 2015
34. Schmidt, H., Krger, J., Hesse, S.: HapticWalker-haptic foot device for gait rehabilitation. In: Grunwald, M. (ed.) Human Haptic Perception: Basics and Applications. Birkhäuser Basel, Basel (2008)
35. Hesse, S., Werner, C.: Connecting research to the needs of patients and clinicians. Brain Res. Bull. **78**(1), 26–34 (2008)
36. Zhang, J.J., Hu, X.F., Xu, X.L.: Research progress of lower limb rehabilitation training robot. Chin. J. Rehabil. Theory Pract. **18**(8), 728–730 (2012)
37. Zhang, X.C.: Research on Key Technologies of Lower Limb Rehabilitation Training Robot. Harbin Engineering University, Harbin (2009)
38. Hesse, S., Uhlenbrock, D.: A mechanized gait trainer for restoration of gait. J. Rehabil. Res. Dev. **37**(6), 701–708 (2000)

39. Hesse, S., Uhlenbrock, D.: An electromechanical gait trainer for restoration of gait in hemiparetic stroke patients: preliminary results. Neurorehabil. Neural Repair **15**(1), 39–50 (2001)
40. Liu, J.K., Sun, N., Huang, M.F.: Design of a gait mechanism of a lower limbs rehabilitative robot. Mach. Des. Res. **22**(5), 59–62 (2006)
41. Luo, H.M., Sun, L.B.: Clinical observation of rehabilitation training with low frequency electric stimulation on patients with stroke. Chin. J. Gerontol. **9**(29), 2390–2391 (2009)
42. Ferris, D.P., Lewis, C.L.: Robotic lower limb exoskeletons using proportional myoelectric control. In: Engineering in Medicine and Biology Society, Annual International Conference of the IEEE, pp. 2119–2124 (2009)
43. Alter, G.: A Revolution in Sports Rehabilitation and Athlete Training [EB/OL]. http://www.alterg.com/products/antigravitytreadmills/m320-320/athletic-train-er#content. Accessed 5 Dec 2010
44. Zhao, J., Zou, R.L., Xu, X., et al.: Design and analysis of body weight support based treadmill for lower limb rehabilitation training. Shanghai Biomed. Eng. **35**(4), 187–190 (2014)
45. Metrailler, P., Blanchard, V., Perrin, I., et al.: Improvement of rehabilitation possibilities with the MotionMaker TM. In: The First IEEE/RAS - EMBS International Conference on Biomedical Robotics and Biomechatronics, pp. 359–364 (2006)
46. Reynard, F., Gerber, F., Favre, C., et al.: Movement analysis with a new robotic device-the MotionMaker: a case report. Gait Posture **30**(2), S149–S150 (2009)
47. Shi, X.H., Wang, H.B., Sun, L., et al.: Design and dynamic analysis of an exo-skeletal lower limbs rehabilitation robot. J. Mech. Eng. **50**(3), 41–48 (2014)
48. Akdoğan, E., Adli, M.A.: The design and control of a therapeutic exercise robot for lower limb rehabilitation: Physiotherabot. Mechatronics **21**(3), 509–522 (2011)
49. Cui, B.Y., Liang, X., Li, Z.X., et al.: Design and simulation of 2-URS&UPS lower limb rehabilitation robot. J. Mech. Transm. **38**(8), 96–99 (2014)
50. Xu, X.L., Zou, R.L., Lu, R.R., et al.: A design of body weight loss multi-function system in patients with hemiplegia rehabilitation training and assessment. Chin. J. Biomed. Eng. **29**(6), 882–888 (2010)

External Force Detection for Physical Human-Robot Interaction Using Dynamic Model Identification

Dewen Wu[1,2(✉)], Quan Liu[1,3], Wenjun Xu[1,2], Aiming Liu[1,3], Zude Zhou[1,3], and Duc Truong Pham[4]

[1] School of Information Engineering, Wuhan University of Technology, Wuhan 430070, China
{wdw201509,quanliu,xuwenjun,zudezhou}@whut.edu.cn

[2] Hubei Key Laboratory of Broadband Wireless Communication and Sensor Networks, Wuhan University of Technology, Wuhan 430070, China

[3] Key Laboratory of Fiber Optic Sensing Technology and Information Processing, Wuhan University of Technology, Ministry of Education, Wuhan 430070, China
liuaiming@cbmi.com.cn

[4] Department of Mechanical Engineering, School of Engineering, University of Birmingham, Birmingham, B15 2TT, UK
d.t.pham@bham.ac.uk

Abstract. Nowadays as more and more tasks require humans to collaborate with robots in modern industry, and the focus of many robotic researchers worldwide has turned towards human-robot collaboration. In human-robot interaction, ensuring the safety issues has the absolute priority for all other research work. In this context, sensorless collision detection and fast response researches in robotics contribute significantly to solve the safety issues. However, existing approaches for collision detection involve in the usage of external sensors, not fit for closed industrial robots or the offline observer based on robot's the generalized momentum, poor in the real time response. In this study, a different method of external forces detection for sensor-less industrial robots using dynamics model identification is proposed. The main idea of our method is to identify the external torques by the comparison of the actual motor torques with the predicted joint torques based on dynamics model. Without using any extra sensors, a strict dynamics model including the parameterized friction torques has been formulated only by utilizing the measurements of the joint angles and joint torques. In addition, the essential response strategies in the post-contact stage are the main directions for our following research. Finally, the model accuracy and performance of the proposed method were evaluated in a 6-DOF manipulator. The experimental results demonstrated the reliability of our detection method basically.

Keywords: Physical human-robot interaction · External force detection · Dynamics model identification · Safety-aware

1 Introduction

In the custom manufacturing industry such as the auto-mobile production lines, the industrial manipulators are widely used to execute most simple and repeated tasks

Y. Huang et al. (Eds.): ICIRA 2017, Part I, LNAI 10462, pp. 581–592, 2017.
DOI: 10.1007/978-3-319-65289-4_55

independently of humans in order to ensure safety. However, when it comes to some elaborate and skilled operations, tasks are beginning to involve humans collaborating with robots more frequently [1], such as tasks in some human-robot assembly or disassembly workshops. In this case, the safety issues of operators as well as facilities have to be considered for these technical workers and researchers. In case of malfunctioning of the robot itself or misuse of the manipulator, both preventive and reactive effective measures are supposed to be taken. Among preventive approaches, primordial safety concerns contributed to the development of lightweight mechanical designs and compliant joint for robots [2, 3]. Besides, it is suggested to avoid the collision in the first place by using a vision sensor system [4]. There have been some another suggestions regarding this issue including installing a passive mechanism for the absorption of damage [5], and the design and control scheme of manipulators based on collision analysis and safety evaluations [6].

The above schemes may show their effectiveness in certain aspect but none of a single approach is the best solution to deal with the collision safety issue. Meanwhile, considering the closed control architecture of the manipulator, it is more "cost-efficient" to improve the safety performance just through a detection and reaction scheme. Therefore, most researchers tend to focus on collision detection for robots without any additional sensors. In literature [7], an end-user approach to distinguish human-robot collision from intentional contacts was proposed through the suitable online processing of the motor currents. Also in [8], a collision detection algorithm using the disturbance observer was proposed.

In fact, the problem about human-robot collision detection without external sensors can be regarded as an application extension in the field of fault detection and isolation [9]. The detection strategies can be mainly split into two steps, the residual generation and evaluation. Various methods for residual generation exist in the control theory. A classical method for residual generation consists in comparing the actual applied joint torques with the estimated ones. Recently, some researchers proposed a way by constructing an external F/T observer for robots based on the estimation of the generalized momentum for the robot [10, 11]. This method doesn't need to calculate the joint acceleration of manipulators and thus reduces the influence of the measure noise. In addition, another approaches of impact detection based on torque filtering or joint current filtering which also doesn't require measurements or estimates of joint acceleration are described in [12]. The most simple residual evaluation method is bound to compare the generated residual with a constant threshold, which must be a little greater than the highest disturbance error affecting the residual.

In general, the effectiveness and accuracy of all strategies for model-based residual generation may rely on modeling uncertainties and measurement noise. Besides, another important challenge is how the identified signal and control commands can feedback to the robot controller in real time. In our present research, it is assumed that an industrial manipulator equipped with a fixed tool is used in a specific task scenario, thus the model parameters may basically hold steady. Although no a priori information about all the inertial parameters is given, a relatively accurate dynamics model can be established through an integrated experiment. As long as the estimated dynamics parameters are accurate enough, a collision detection approach based on dynamics model identification can be realized to address the issue of discriminating actual collision accident at a less threshold. Besides, we construct the real-time communication between the robot controller and the PC to transmit the collision information and reaction commands.

The remainder of this paper is organized as follows. Firstly, the overall system architecture of the collision detection scheme is described in Sect. 2. Then, Sect. 3 presents the identification process of our robot inertial parameters, in which experiment design, data acquisition and signal processing are elaborated briefly. Later on, our experimental platform, identification results and collision validation are shown in Sect. 4. Finally, the main conclusions are given in Sect. 5.

2 Integrated Detection Scheme

The proposed detection scheme is implemented on an industrial manipulator of KUKA KR-6-R700 SIXX available in our lab. This is a small size 6R manipulator, weighting 50 kg itself, having a 6 kg of max payload and a 0.7067 m of maximum motion radius. Since this robot is equipped with the KUKA Robot Sensor Interface (RSI), the transient data of joint positions and torques derived from the embedded joint sensors can be collected by communicating with the controller on an external PC every 12 ms.

As described in [13], a dynamic model relates the manipulator's motion to its joint torques. That is, the state information of a robot controller is closely relevant to its kinematics parameters. Therefore, an accurate dynamics model suggests the commanded joint torques during the robot's normal motion. Once there exist external torques applied to any part of the robot body, the instant difference between commanded torques and actual motor torques is able to indicate the collided links and moment for the manipulator.

In general, a standard parameter identification procedure includes these steps: robot modeling, excitation trajectories design, data acquisition and processing, parameter estimation and model validation. The kinematic and geometric information of the robot are the inputs to the identification procedure. Meanwhile, the detection threshold is selected as a constant signal just derived from the mean value of the residual signal in the absence of collision in many repeated experiments. At last, the moment when any external forces are applied to the manipulator can be easily found by comparison the residual with a threshold signal. In a word, a brief schematic block diagram of the external force detection is shown in Fig. 1(a). Since the industrial manipulator is one kind of serial robot, the motion of these joints close to its base will have an influence on kinematic parameters of their subsequent link. Meanwhile, the Denavit-Hartenberg (D-H) representation was initially proposed to describe the position of each link relative to its neighbors [14]. Therefore, the kinematic problem of the end-effector can be defined as transformation from the joint space to the Cartesian space using D-H representation, which contributes to construct linear dynamics model and make limitations to the manipulator's work space in searching optimal trajectory. In this case, the frame attached to each link firstly needs to be defined as shown in Fig. 1(b), and a modified D-H parameter table is shown in Table 1, where L1 = 0.4 m, L2 = 0.315 m, L3 = 0.035 m, L4 = 0.025 m, L5 = 0.365 m, L6 = 0.08 m.

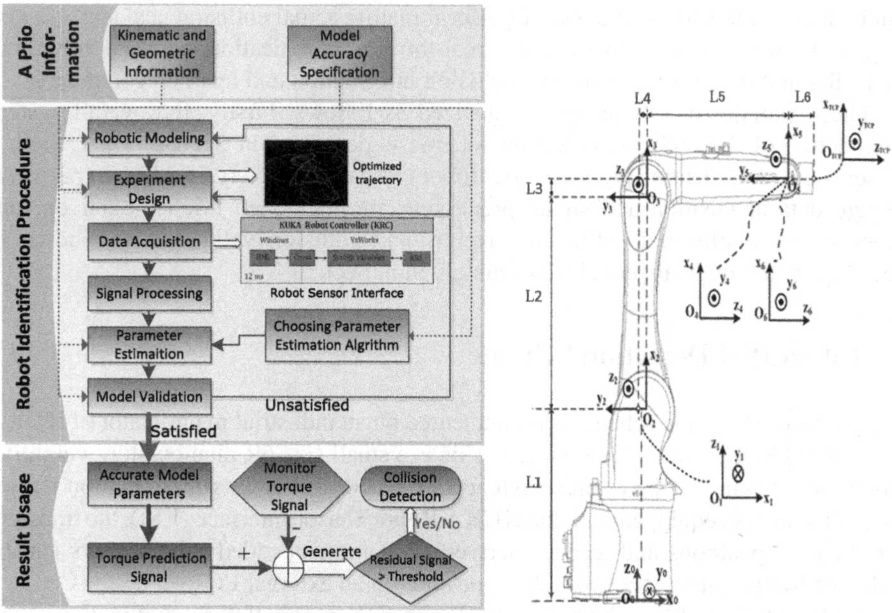

Fig. 1. (a) A block diagram for external force detection (b) Modified D-H coordinate frames

Table 1. Modified D-H parameters table

j	a_i	α_i	d_i	θ_i
1	0	π	–L1	q1
2	L4	$\pi/2$	0	q2
3	L2	0	0	$-\pi/2 + q3$
4	L3	$\pi/2$	–L5	q4
5	0	$-\pi/2$	0	q5
6	0	$\pi/2$	0	q6
TCP	0	0	–L6	0

3 Dynamics Model Identification

3.1 Dynamics Model for Sensorless Collision Detection

Since the robot dynamics model lays the basis for the following procedure, a unified and applied dynamics formulation is bound to be established. As in [13], for a n-DOF industrial robot, an inverse dynamics model derived by the Newton-Euler or Lagrangian method can be written as follows:

$$\boldsymbol{\tau}_{dyn} = \boldsymbol{M}(q, \vartheta)\ddot{q} + \boldsymbol{C}(q, \dot{q}, \vartheta) + \boldsymbol{G}(q, \vartheta) \quad (1)$$

In order to simplify the process of parameter identification, Eq. (1) can be rewritten as a linear function of the barycentric vectors according to the modified Newton-Euler parameters [15], as follows:

$$\tau_{dyn} = W(q, \dot{q}, \ddot{q})\theta_{dyn}, \tag{2}$$

where **W** is the observation matrix, which only depends on the motion data. This property considerably simplifies the procedure of the parameters identification.

However, apart from the effects of barycentric parameters, the torques caused by joint friction, dynamic coupling of actuator rotors, etc. contribute notably to the dynamic behaviors of the manipulator. In fact, the effect of joint friction is a complex nonlinear model, but a simplified and acceptable friction model only consisting of Coulomb and viscous friction can be expressed as:

$$\tau_{fric} = f_c \mathrm{sign}(\dot{q}) + f_v \dot{q}, \tag{3}$$

Therefore, the integrated dynamics model of robots can be rewritten as follows:

$$\tau_s = W_s(q, \dot{q}, \ddot{q})\theta_s, \tag{4}$$

θ_s is a 12n-vector of unknown dynamic parameters. For instance, the dynamics parameters of link i are given in the form:

$$\theta_s^i = [I_{xxi}, I_{xyi}, I_{xzi}, I_{yyi}, I_{yzi}, I_{zzi}, m_i r_{xi}, m_i r_{yi}, m_i r_{zi}, m_i, f_{ci}, f_{vi}]^T, \tag{5}$$

where $I_{\zeta\zeta i}(\zeta = x, y, z)$ is the inertial tensor of link i. Similarly, $m_i r_{\zeta i}$ denotes the first-order mass moment and m_i is the mass of link i.

3.2 Optimized Excitation Trajectory Design

In the procedure of dynamics parameters identification, the excitation trajectories are often used to actuate the robot sufficiently to provide accurate and fast parameter estimation. Although there exist couples of optimized methods for robot excitation trajectories such as five-order polynomials [16], etc., periodic and bandlimited joint position sequences results in bandlimited measurements with the same period easy to obtain more accurate velocities and accelerations with less noises [17]. In general, a finite periodic Fourier series is selected as excitation trajectories as follows:

$$q_i(t) = q_{i,0} + \sum_{k=1}^{N} \left(a_{i,k} \sin\left(kw_f t\right) + b_{i,k} \cos(kw_f t)\right), \tag{6}$$

in which t represents time, and $w_f = 2\pi / T_f$ is the same for all joints. Each Fourier series has $2N + 1$ unknown coefficients.

Firstly, the choice of fundamental frequency w_f and the harmonic order N have to account that good coverage of the robot workspace and including high frequencies to provide high accelerations should be both satisfied. Then, the selection of the trajectory coefficients can be attributed to solve a nonlinear optimization problem with motion constraints on robot. The optimization criterion should just correspond to a less condition number for the matrix $\boldsymbol{\Phi}_s'$. The motion constraints imposed on the robot in case of collisions during the excitation motion. The constraints can be expressed in mathematic form as follows:

Minimize:

$$f_0(x) = cond(\boldsymbol{\Phi}_s'), \tag{7}$$

Subject to:

$$\boldsymbol{q}_{min} \leq \boldsymbol{q}(x) \leq \boldsymbol{q}_{max}, \tag{8}$$

$$|\dot{\boldsymbol{q}}(x)| \leq \dot{\boldsymbol{q}}_{max}, \tag{9}$$

$$|\ddot{\boldsymbol{q}}(x)| \leq \ddot{\boldsymbol{q}}_{max}, \tag{10}$$

$$\boldsymbol{W}(\boldsymbol{q}(\boldsymbol{x})) \subset \boldsymbol{W}_0, \tag{11}$$

in which $\boldsymbol{f}_0(\mathrm{x})$ is the cost function for optimization, $\boldsymbol{q}_{min}$ and $\boldsymbol{q}_{max}$ are the upper and lower limits for joint positions, $\dot{\boldsymbol{q}}_{max}$ and $\ddot{\boldsymbol{q}}_{max}$ are the peak value for joint velocities and accelerations respectively, x denotes all the feasible solution, $\boldsymbol{W}_0$ is the available workspace for the TCP (Tool Point Center).

3.3 Data Acquisition and Processing

The basic inputs to the identification model include the measured joint positions, joint velocities, joint accelerations and motor torques. The actual joint angles are measured by the encoders mounted on the actuator shafts and joint torques are calculated by the embedded sensors of each joint. And, the joint velocities and accelerations are offline obtained using DFT and IDFT. Since the control architecture is closed to user [7], the discretized joint positions and velocities are assigned to the controller by programming the robot using the KRL language on the PC. The required data of joint positions and torques are collected on the external PC communicating with the controller every 12 ms through the KUKA RSI.

Firstly, it is noted that a KRL program can only run on the KRC and have no support for advanced mathematical tools, while lots of sampled datum for the use of parameter estimation have to be stored in a file. So the communication between a user program and a KRL program can be implemented using a Java open-source cross-platform communication interface [18] that makes it possible to read and write all of the controlled manipulator variables. The block diagram of the communication interface is shown in Fig. 2(a). More importantly, it is allowed to implement the real-time communication as well as transmitting control commands to the robot using the Kuka CrossComm.

After obtaining the measured data, it is necessary to clean up the measurements due to the presence of noise. Then we have to calculate the joint velocity and acceleration estimates. And since the data are periodic, it is efficient to remove outliers by data averaging firstly. Especially for measured joint torques, a classical method of Wavelet denoising can be utilized properly to smooth the raw data. Since velocities and accelerations can't be measured directly from the KRC. However, a traditional calculation method of numerical differentiation for joint angles is bound to amplify the measurement noise. There, another simple and efficient method based on DFT can be adopted to improve the accuracy of input data. A brief block diagram of the method can be shown in Fig. 2(b).

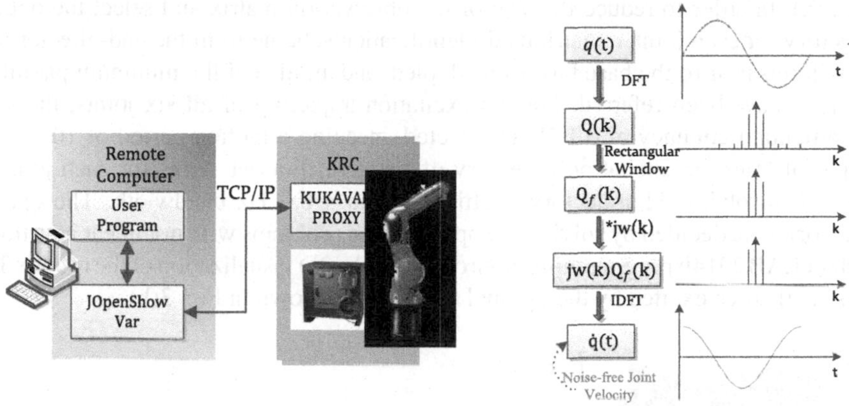

Fig. 2. (a) A block diagram for communication between PC and KRC (b) The flow chart for obtaining noise-free velocities and accelerations

3.4 Parameter Estimation

When an acceptable dynamics model and all the input data have been ready, a method near to unbiased estimates tends to be selected. In general, two popular methods are usually adopted to estimate the unknown parameters, that is, linear least squares parameter estimation (LLSE) and the maximum likelihood estimation (MLE). The LLSE method is a non-iterative method that finds the estimated parameter in a single step of singular value decomposition and may result in biased estimates when existing inaccurate input data. In contrast, the MLE method, with which a cumbersome problem of guessing the initial value of the unknown parameters has to be faced, provides the unbiased estimates with minimal uncertainty regardless of the measurement noise.

After the signal processing procedures, the noise in the raw data can be greatly suppressed. At the same time, in order to simplify the estimation work, an acceptable accuracy has been proved to obtain by utilizing the LLSE method. According to the linear dynamics formulation for joint i, parameters for link i can be calculated as follows:

$$\hat{\theta}_i = (W_{s,i}^T * W_{s,i})^{-1} * W_{s,i}^T * \tau_i, \tag{12}$$

in which $\hat{\theta}_i$ is the estimated value of the unknown parameters for link i.

4 Model Validation and Experiment Results

4.1 Parameters Identification Results

As mentioned above, the KUKA KR-6-R700 SIXX, a 6-DOF industrial manipulator available in our lab without payload, is used as the experiment object, as shown in Fig. 3(a). In order to reduce the size of the observation matrix and select the optimal trajectory for each joint, a distributed identification scheme from the end-effector joint to the joints near to the base has been adopted, and an idea of the minimum parameter sets [19] have been referred. For the excitation trajectory of all six joints, the same fundamental frequency of 0.1 Hz is selected, meaning a motion period of 10 s. After couple of trials, an excitation trajectory of five-term Fourier series for each joint are selected, involving 11 trajectory coefficients and a 0.5 Hz bandwidth. The optimal trajectories are decided by solving the optimization problems with nonlinear constraints in MATLAB 2014b programming environment. The 3D visualization of the robot's TCP in its workspace excited by the optimal trajectory is shown in Fig. 3(b).

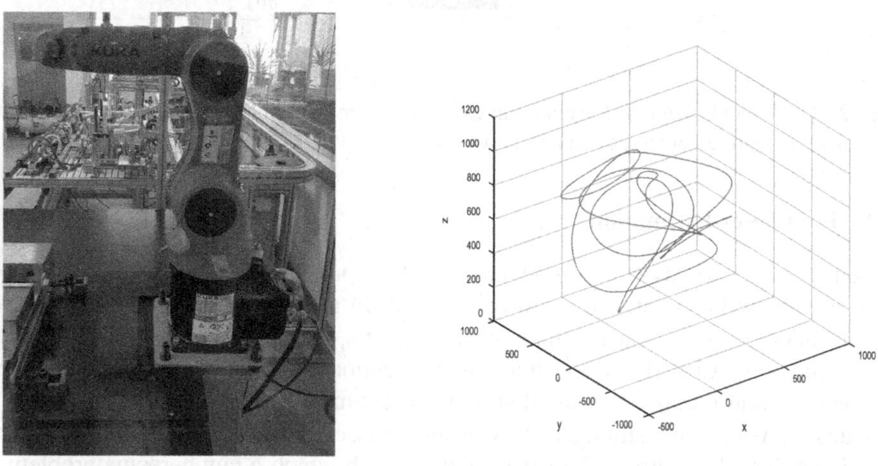

Fig. 3. (a) The KUKA robot available in our lab (b) The 3D visualization of optimal excitation trajectory for our robot

Since it is unable to set the acceleration of each joint for the KUKA robot, the excitation trajectory can just be fitted by dividing each Fourier series into huge amounts of linear segments. Then, each segment is assign with different velocities, and these discrete motion points with different velocities are programmed to the KUKA controller. After the transients die out, the data of joint angles and torques are collected and the total measurement time is 60 s, corresponding to 6 periods of excitation trajectory. When all the 6 joints are moving together, the model parameters can be calculated using the LLSE method. The identification results show that the first 3 joints contain 21 model parameters, able to identify, 15 base inertial parameters and 6 friction parameters. The accuracy of the identified parameters is to be validated by comparing the predicted joint torques with the corresponding measurements of torque.

4.2 Model Validation

The actuator torque-prediction of a robot model is important for offline simulation and advanced robot control. In our studies, the more accurate the model is, the more accurate the predicted torque residual is. Thus, the required detection threshold can be lower to discriminate the collision accidents from the noise disturbance. A good validation trajectory should be different from the excitation trajectory, while being representative of the intended application. We compare the measured torques with the model-based torques by inputting a random trajectory composed of a set of joint angles, velocities to the robot controller. The result about the measured and predicted torques for the last three joints is shown in Fig. 4.

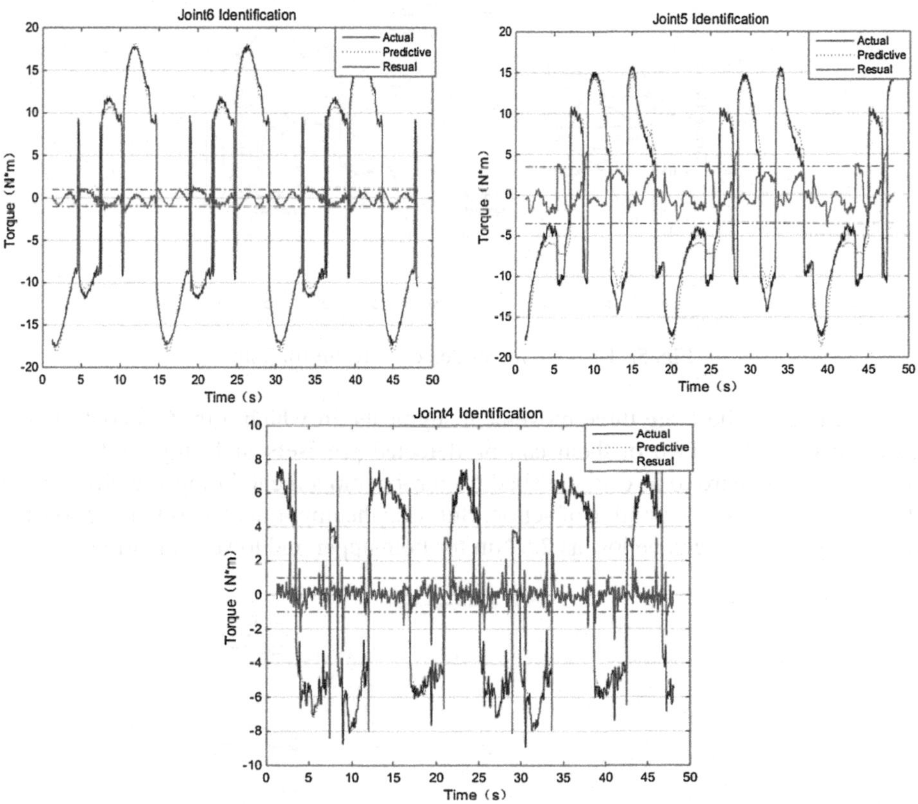

Fig. 4. The validation about the measured and predicted torques for the last three joints

The above results show that the predicted torques of all three joints have the similar trend basically with the measured torques to a great extent although the predicted error is slightly big during velocity reversal. Considering that our study focuses on the detection of robot's collision accident, and, the "false peak" can be distinguished from the collision residual signal by comparing the duration of action.

4.3 Collision Experiments and Results

In this section, two given crash cases that combine with the task execution of our robot are designed to validate the efficiency of the proposed scheme for collision detection. Firstly, since tool point center of the robot is exposed to the outermost end portion, and, thus it is more likely to collide with other parts. Therefore, the first collision case that joint 6 is applied by external forces is considered, the experiment result is shown as in Fig. 5.

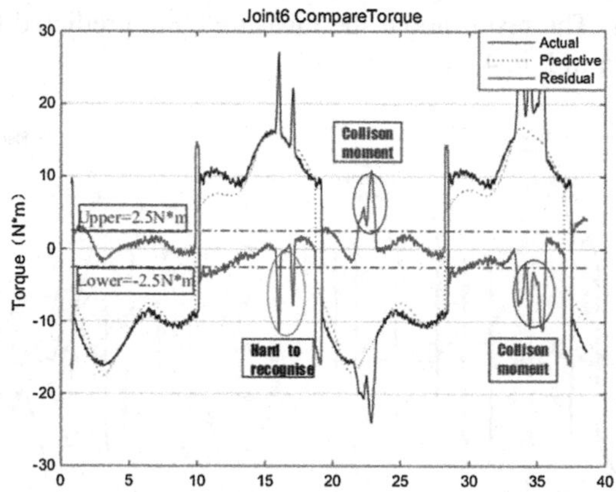

Fig. 5. External forces detection result for joint 6

In this case, there are three movement segments, in which joint 6 is contacted by external obstacles. Two of them can be detected precisely and duly, and the other moment is very hard to be distinguished from robot movement during velocity reversal due to just a very short duration of action. But what the important fact is that the constant detection threshold can be low as 2.5 Nm for both upper and lower limitation.

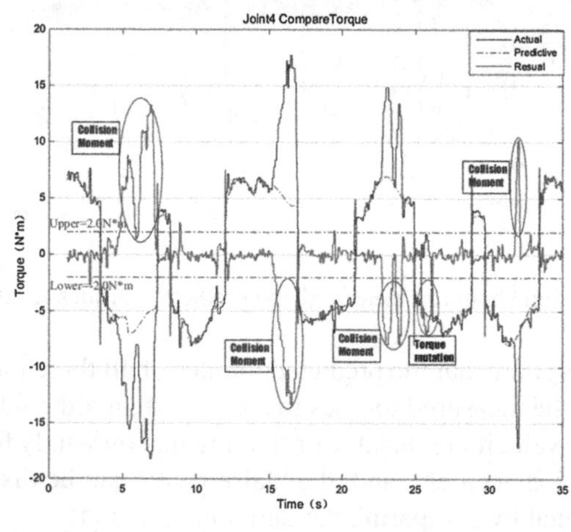

Fig. 6. External forces detection for joint 4

In another case, an obstacle is intermittently placed on the way that the robot performs a random motion task with joint 4 being contacted a few times. The experimental performance about our detection method in this case is shown in Fig. 6. As we have seen, almost all the collision moments have been detected except for an un-normal saltus of measured toques as long as the contact forces are obvious enough. Besides, the constant detection threshold can be low as 2.0 Nm. In summary, the collision experiments not only show an accurate dynamics model we built but also basically demonstrate that our detection method for basic collision is reliable enough.

5 Conclusions

In this paper, a different and efficient method for sensorless collision detection based on dynamics model identification for a 6-DOF industrial robot has been proposed. We design an integrated experiment including optimal excitation trajectory design, data acquisition and processing, parameter estimation and model validation. All the model parameters have been identified by our excitation trajectory experiment. And, when comparing the predicted torques with the measured joint torques, it can be found that the dynamics model built by us can accurately estimate the motor torques based on the kinematic data. Besides, the following collision tests based on our model furtherly demonstrate the efficiency and reliability of our collision detection method. In our future studies, we will focus on making corresponding reaction strategies after detecting solid collision through the communication channel between the PC and the robot controller. What is more, the safety issues and intelligent interaction in human-robot collaboration will be the main directions for our study.

Acknowledgments. This research is supported by National Natural Science Foundation of China (Grant No. 51675389), the International Science & Technology Cooperation Program, Hubei Technological Innovation Special Fund (Grant No. 2016AHB005), the Fundamental Research Funds for the Central Universities (Grant No. 2017III5XZ), and Engineering and Physical Sciences Research Council (EPSRC), UK (Grant No. EP/N018524/1).

References

1. Lee, S.D., Kim, M.C., Song, J.B.: Sensorless collision detection for safe human-robot collaboration. In: IEEE/RSJ International Conference on Intelligent Robots and Systems, pp. 2392–2397. IEEE, Hamburg (2015)
2. Makarov, M., Grossard, M., Rodriguez-Ayerbe, P., Dumur, D.: Generalized predictive control of an anthropomorphic robot arm for trajectory tracking. In: 2011 IEEE/ASME International Conference on Advanced Intelligent Mechatronics (AIM), pp. 948–953. IEEE, Budapest (2011)
3. Albu-Schäffer, A., Haddadin, S., Ott, C., Stemmer, A., Wimböck, T., Hirzinger, G.: The DLR lightweight robot: design and control concepts for robots in human environments. Ind. Robot **34**, 376–385 (2007)

4. Flacco, F., Kröger, T., Luca, A.D., Khatib, O.: A depth space approach to human-robot collision avoidance. In: IEEE International Conference on Robotics and Automation, pp. 338–345. IEEE, Saint Paul (2012)
5. Park, J.J., Kim, H.S., Song, J.B.: Safe robot arm with safe joint mechanism using nonlinear spring system for collision safety. In: IEEE International Conference on Robotics and Automation, pp. 571–576. IEEE, Kobe (2009)
6. Haddadin, S., Haddadin, S., Khoury, A., Rokahr, T.: A truly safely moving robot has to know what injury it may cause. In: IEEE/RSJ International Conference on Intelligent Robots and Systems, pp. 5406–5413. IEEE, Vilamoura (2012)
7. Geravand, M., Flacco, F., De Luca, A.: Human-robot physical interaction and collaboration using an industrial robot with a closed control architecture. In: IEEE International Conference on Robotics and Automation, pp. 4000–4007. IEEE, Karlsruhe (2013)
8. Ho, C.-N., Song, J.-B.: Collision detection algorithm robust to model uncertainty. Int. J. Control Autom. Syst. **11**, 776–781 (2013)
9. Dixon, W.E., Walker, I.D., Dawson, D.M., Hartranft, J.P.: Fault detection for robot manipulators with parametric uncertainty: a prediction-error-based approach. IEEE Trans. Robot. Autom. **16**, 689–699 (2000)
10. De Luca, A., Albu-Schaffer, A., Haddadin, S., Hirzinger, G.: Collision detection and safe reaction with the DLR-III lightweight manipulator arm. In: IEEE/RSJ International Conference on Intelligent Robots and Systems, 2006, pp. 1623–1630. IEEE, Beijing (2006)
11. Haddadin, S., Albu-SchäFfer, A., De Luca, A., Hirzinger, G.: Collision detection and reaction: a contribution to safe physical human-robot interaction. In: IEEE/RSJ International Conference on Intelligent Robots and Systems, pp. 3356–3363. IEEE, Nice (2008)
12. Lewis, F.L., Dawson, D.M., Abdallah, C.T.: Robot Manipulator Control: Theory and Practice, 2nd edn. Marcel Dekker, Basel (2003)
13. Swevers, J., Verdonck, W., De Schutter, J.: Dynamic model identification for industrial robots. IEEE Control Syst. **27**, 58–71 (2007)
14. Ayob, M.A., Zakaria, W.N.W., Jalani, J.: Forward kinematics analysis of a 5-axis RV-2AJ robot manipulator. In: Electrical Power, Electronics, Communications, Controls and Informatics Seminar (EECCIS), 2014, pp. 87–92. IEEE, Malang (2014)
15. Calanca, A., Capisani, L.M., Ferrara, A., Magnani, L.: MIMO closed loop identification of an industrial robot. IEEE Trans. Control Syst. Technol. **19**, 1214–1224 (2011)
16. Armstrong, B.: On finding exciting trajectories for identification experiments involving systems with nonlinear dynamics. Int. J. Robot. Res. **8**, 28–48 (1987)
17. Pintelon, R., Schoukens, J.: System Identification: A Frequency Domain Approach, 2nd edn. Wiley, Hoboken (2012)
18. Sanfilippo, F., Hatledal, L.I., Zhang, H., Fago, M., Pettersen, K.Y.: JOpenShowVar: an open-source cross-platform communication interface to Kuka robots. In: IEEE International Conference on Information and Automation, pp. 1154–1159. IEEE, Hailar (2014)
19. Gautier, M., Khalil, W.: Exciting trajectories for the identification of base inertial parameters of robots. Int. J. Robot. Res. **11**(4), 494–499 (1992). United States

Mechanical Design and Human-Machine Coupling Dynamic Analysis of a Lower Extremity Exoskeleton

Bo Li[1(✉)], Bo Yuan[2], Jun Chen[1], Yonggang Zuo[1], and Yifu Yang[1]

[1] Department of Petroleum Supply Engineering, Logistical Engineering University, Chongqing, China
doctor_libo@163.com

[2] Department of Machinery and Electrical Engineering, Logistical Engineering University, Chongqing, China

Abstract. Wearable exoskeleton enhances human abilities by integrating human body with machines. In order to design an anthropomorphic exoskeleton, the anatomical structure and biomechanical characteristics of lower limb are studied, the degrees of freedom of movement on each lower limb is 7, and the movement range is given out. The kinematic and dynamic characteristic of walking gait analyzed by OpenSim indicates that the ankle joint demands asymmetric unidirectional actuators, the knee and hip joint demands bidirectional actuators. The mechanical structure was designed according to the anatomy structure and gait analysis of the lower limb. The human-machine interaction forces at different interface position were analyzed. It demonstrates that, in order to improve wearability, the passive exoskeleton's choices of human-machine interaction positions on the shank and thigh limb are the down and middle position on each part respectively, and the active exoskeleton's choices of human-machine interaction positions on the shank and thigh limb are the middle and down position on each part respectively.

Keywords: Exoskeleton · Mechanical design · Biomechanics · Interaction force

1 Introduction

The exoskeleton is a species of wearable robot that integrate human body with machines, which enhances human function and capability by empowering the human limb where it is worn. Exoskeleton can be categories into three broads according to its main function [1]. The first one is human performance augmentation exoskeleton, which is also called extender that extends the strength of the human limb beyond its natural ability. The second broad is assistive device for individuals with disabilities, and the third broad is therapeutic exoskeleton for rehabilitation. Both of the second and third category is bi-function exoskeletons which can be helpful for therapy as well as increasing the physical capabilities of the wearer.

Y. Huang et al. (Eds.): ICIRA 2017, Part I, LNAI 10462, pp. 593–604, 2017.
DOI: 10.1007/978-3-319-65289-4_56

The studies on exoskeletons began in the early 1960s [2]. The first active exoskeleton named Hardiman was developed at General Electric Co., it was a man-amplifiers-manipulators controlled by an electrical actuated master-slave follower system in order to enhance the strength of a human manipulator. However, Bulky and complex structure caused the Hardiman project to be failed.

In 2000, the US Department of Defense Advanced Research Projects Agency (DARPA) launched the “exoskeleton for human performance augmentation (HPA)” program [3], which had funded a number of projects successively, and greatly promoted the theory and application research of exoskeleton. Under the guidance of the United States, other developed countries had also increased investment for exoskeleton research, hence a large number of outstanding exoskeletons had been developed, and their function covered performance augmentation, rehabilitation, industrial manufacturing and so on. The most successful human performance augmentation exoskeleton was the Berkeley Lower Extremity Exoskeleton (BLEEX) [4], which had actuating joints at the hip, knee, and ankle joint. Lockheed Martin acquired the HULC exoskeleton technology from Berkeley Bionics. HULC was developed based on the BLEEX, and it could allow soldiers to carry heavy loads over long distances. Like most of the exoskeletons in development, there was a lack of quantitative assessments of the device available to the user.

Until now, a large number of exoskeletons had been developed to assist disable individuals. The HAL exoskeleton developed by Cyberdyne in Japan was one of the few commercial devices for patient with deficits walking function [5]. HAL had both unilateral and bilateral version that actuate the knee and hip joints, which were controlled by using the eEMG signal. The limitations of HAL was lack of performance data to assess. There are a number of literatures for the design and control of the exoskeleton [6], but little literature talks about quantitative performance assessments of the exoskeleton, which is very importance for user as well as the designer. The human-machine interaction is a key issue, and the kinematic compliance plays an important role to achieve ergonomic human-machine interfaces.

The rest of this paper is organized as follows. Section 2 studies the anatomy and biomechanical characteristics of human lower limb. Section 3 develops an exoskeleton based on the kinematic and dynamics of human in detail. Section 4 analyses the human-machine coupling dynamics with different interaction positions. Finally, Sect. 5 concludes.

2 Anatomical and Biomechanical Study of Lower Limb

The wearable lower extremity exoskeleton is worn on the body surface of lower limbs, along with the human body to complete various locomotion, reproducing human functions or copying human actions, there is a one-to-one correspondence between human anatomical joints and the exoskeleton joints. Therefore, design and modeling of the exoskeleton should consider the factor of human firstly. The mechanism should be designed according to the static lower limb structure characteristics and dynamic lower limb movement characteristics.

2.1 Locomotive Anatomy of Lower Limb

The study of human kinematics mainly uses the three planes and reference axis shown in Fig. 1, the sagittal plane, coronal plane and horizontal plane, the sagittal axis, coronal axis and vertical axis. The lower limb locomotion system is comprised of fixed bones of lower limb (hip bone, etc.), free bones of lower limb (femur, tibia, fibula, patella, and foot bones), the hip joint, knee joint, ankle joint and corresponding skeletal muscle connecting with them. Figure 2 shows the major bones and joints associated with lower limb locomotion. In the reference frame of Fig. 1, the joint movement of human is divided into flexion/extension, adduction/abduction, and internal rotation/external rotation.

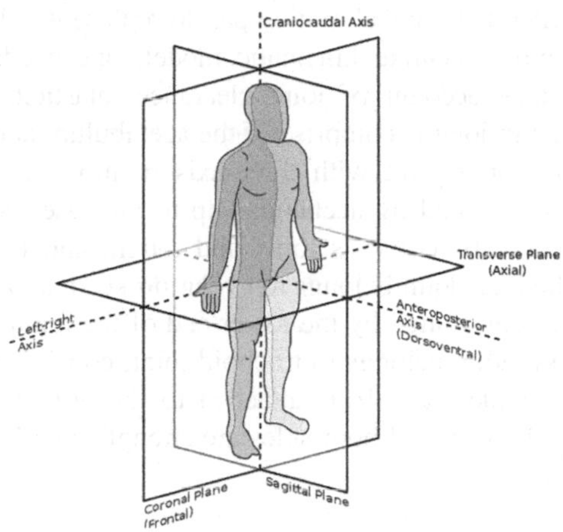

Fig. 1. The reference anatomy axis and planes

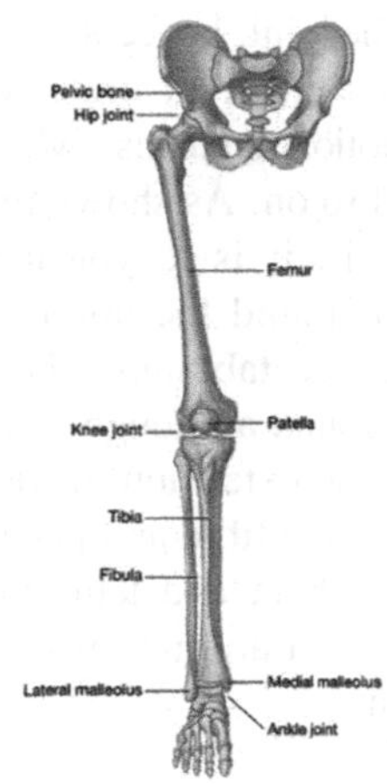

Fig. 2. Bones and joints of lower limb

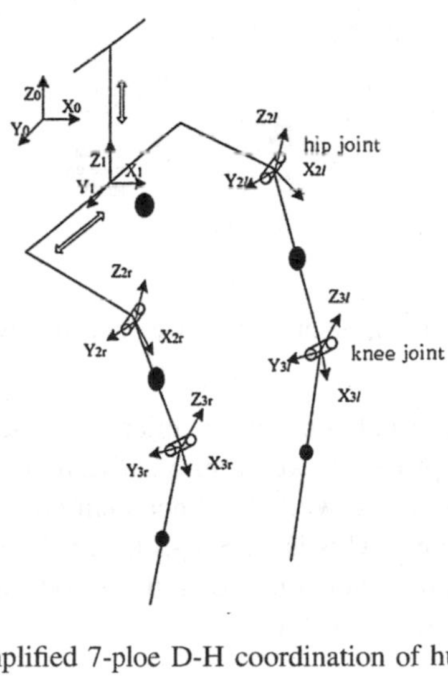

Fig. 3. The simplified 7-ploe D-H coordination of human lower limb

In order to facilitate the mechanical design of the exoskeleton, some researchers simplified the lower limbs to be a 7-pole model, shown in Fig. 3, which only taken into account the degree of freedom (DOF) of the hip, knee and ankle joint in the sagittal plane around the joint axis, namely the R-R-R kinematic chain. This can reduce the complexity of the design of exoskeleton mechanism, but the designed exoskeleton can only complete the normal sagittal linear walking movement, when there are movements on other reference surface, such as turning on the slope ground, the DOF of the exoskeleton is not enough, which will result in man-machine interference.

In fact, the joint of human is mainly comprise of the articular surface, articular capsule and articular cavity, it has a complex structure. Joint is separated with the bone surface and the synovial fluid cavity, it is connected through the surrounding connective tissue, which ensures its flexibility. The complex structure makes complex motion in the joint, hence the use of simplified mechanical motion pair to replace it will get a greater error. In order to achieve more accurate kinematic model, one needs detailed motion analysis, which should take account of joint clearance, elasticity, friction and so on. As shown in Fig. 4, the hip joint is comprise of the acetabulum and caput femoris, it is a typical pestle and mortar joint with three-axis rotation. The acetabulum around the fibrous cartilage is surround by acetabular lip to increase the depth of the acetabulum, The capsula articularis coxae is tonic and tough, and the ligaments around are tough and tensile. The knee joint belongs to cyclarthrosis, mainly for flexion and extension movement, it was constituted by the lower end of the femur, upper end of the tibia and patella. The ankle joint belongs to trochoid joint, consist of distal tibia, fibula and talus blocks, its articular capsule is attached to the articular surface, its front and rear walls are thin and loose, and both sides are strengthened by the ligament.

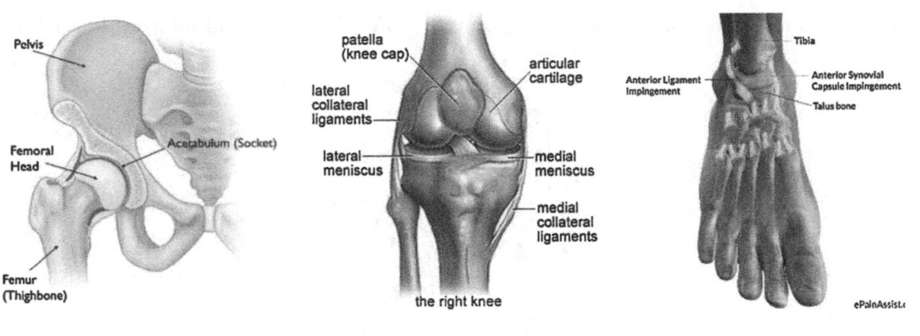

Fig. 4. Joint of lower limb: hip joint (left), knee joint (middle) and ankle joint (right)

In order to make the exoskeleton have similar movement characteristics with the human lower limbs, to improve its wearability, researchers usually consider the hip joint 3 DOFs, ankle joint 3 DOFs, while the knee joint only consider flexion/extension movement in the sagittal plane. This is an S-R-S kinematic chain. The joint DOFs and joint movement range of the lower limbs of human body are shown in Table 1, and their D-H coordinates are shown in Fig. 5.

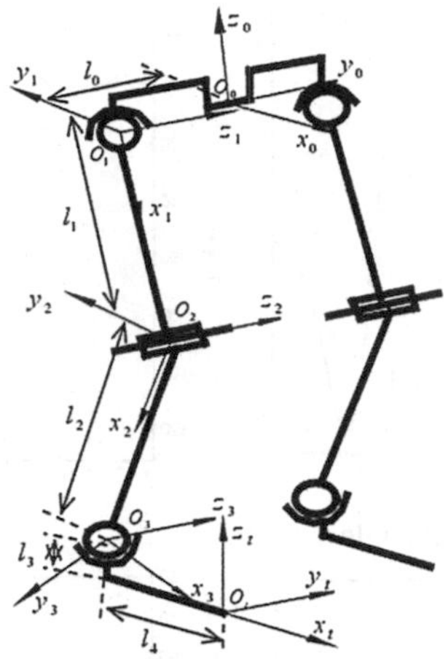

Fig. 5. The 15 DOFs and D-H coordination of human lower limb

Table 1. DOF and movement range of lower limb

Joint	DOF	Axis of rotation	Movement range (°)
Hip joint	adduction/abduction	Sagittal axis	(−30 ~ −35) ~ 40
	flexion/extension	frontal axis	−120 ~ 65
	internal/external rotation	Vertical axis	(−15 ~ −30) ~ 60
Knee joint	flexion/extension	Frontal axis	(−120 ~ −160) ~ 0
Ankle joint	eversion/inversion	Sagittal axis	(−30 ~ −35) ~ (15 ~ 20)
	dorsiflexion/plantar flexion	frontal axis	−20 ~ (40 ~ 50)
	Internal/external rotation	Vertical axis	−15 ~ (30 ~ 50)

2.2 Walking Gait Analysis Based on OpenSim

A gait cycle is defined as the period between two consecutive heel strikes, which is composed of a stance phase where the foot is on the ground and a swing phase where the foot is off the ground. As shown in Fig. 6, the stance phase covers about 60% while swing phase covers 40% of one gait cycle. The stance phase can be subdivided into

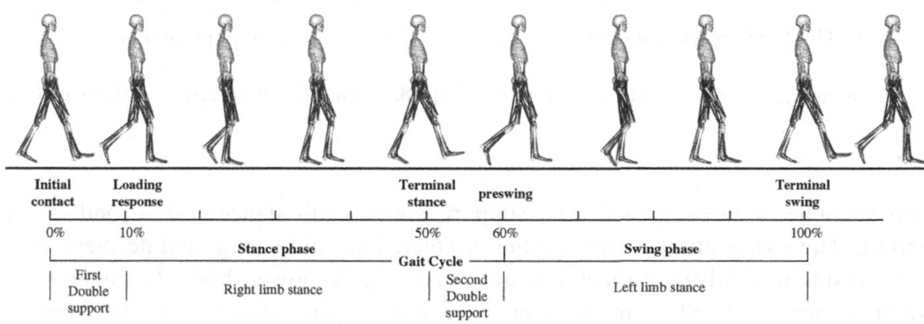

Fig. 6. Gait analysis based on OpenSim

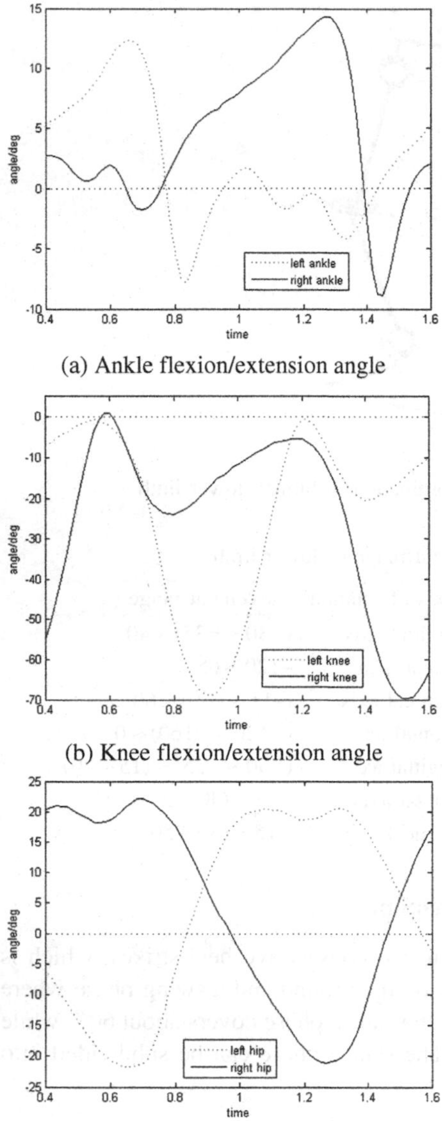

(a) Ankle flexion/extension angle

(b) Knee flexion/extension angle

(c) Hip flexion/extension angle

Fig. 7. Kinematic characteristics of lower limb

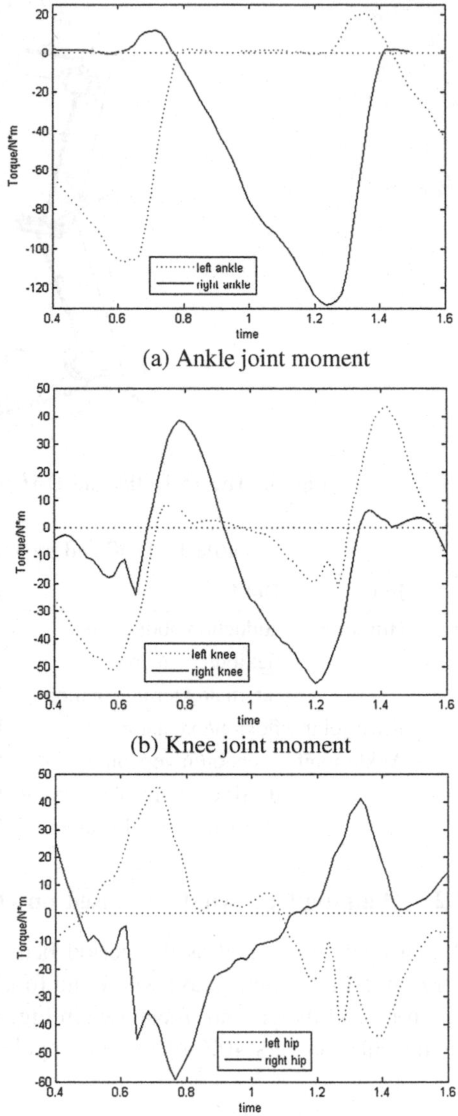

(a) Ankle joint moment

(b) Knee joint moment

(c) Hip joint moment

Fig. 8. Dynamic characteristics of lower limb

three separated phases: first double support, single limb stance and second double support. The swing phase events contain acceleration, mid-swing, and deceleration.

In order to analyze normal human walking gait, lower body kinematics and dynamics are simulated by the software of OpenSim [7], the subject in the OpenSim software platform is chosen to be a 180 cm tall, 75 kg weighted male, walking on flat ground at 1.2 m/s in the Sagittal plane. The simulation results of hip, knee, and ankle joint's angles and moments are shown as follow. As shown in Fig. 7, the angles of ankle joint is ranged from 15° to 10°, the heel strike occurs at the beginning of the gait cycle and toe-off occurs at 60% of the gait cycle. The minimum angle of knee joint is about 70°, occurring in early-mid swing phase to enable the foot to clear the ground when swing forward. The joint angle of hip has an approximately sinusoidal pattern that the thigh flexed upward at heel strike to allow foot-ground contact in front, then followed by an extension through most of the stance phase and a flexion through swing. The kinematic characteristics of lower limb is useful for exoskeleton mechanical design in Sect. 3.

Figure 8 details the joint moment of lower limb, the ankle joint torque is almost entirely negative, this asymmetry implies a preferred orientation for asymmetric uni-directional actuator. The knee joint torque has both negative and positive values, and the highest peak is extension in early stance, indicating the need for asymmetric bidirectional actuator. The hip joint torque has both negative and positive values, and is relatively symmetric, demands bidirectional actuator.

3 Mechanical Design

In order to move smoothly and have a good assistive performance, the exoskeleton structure should map on to the human limb anatomy, there is a one-to-one correspondence between human anatomical joints and the exoskeleton joints [8]. This provides a kinematic compliance to achieve ergonomic human-exoskeleton interfaces, promises an effective transfer of power between the human and machine.

According to the anatomy structure of human lower limb analyzed in Sect. 2, the main structure of the exoskeleton is shown in Fig. 9(left), which have 3 DOFs on the hip joint, one DOF on the knee joint, and 2 DOFs on the ankle joint, and all the joint are designed to be rotation joint. The whole structure is almost anthropomorphic, it is kinematically similar to the human lower limb. The human-exoskeleton interface is designed on the heel, shank, thigh, and waist part, shown in Fig. 9(right).

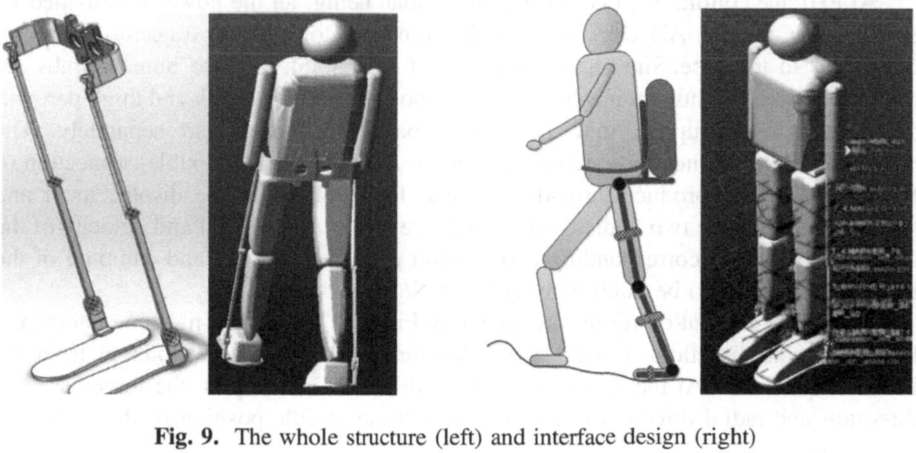

Fig. 9. The whole structure (left) and interface design (right)

In Sect. 2, we learn that the hip joint is a pestle and mortar joints with three-axis rotation pass through the ball and socket. This limits the structure design of exoskeleton's hip joint, such that the rotation joint has to be moved, because it does not align with the human's hip joint. The Internal rotation/external rotation and flexion/extension joints are both placed at the lateral side, and the adduction/abduction joint are placed at the backside, shown in Fig. 10(a). The knee joint's rotation range is restricted by the angle of circular arc shown in Fig. 10(b) according to Table 1. The ankle joint is designed to be two DOFs in the sagittal plane and the coronal plane, shown in Fig. 10(c).

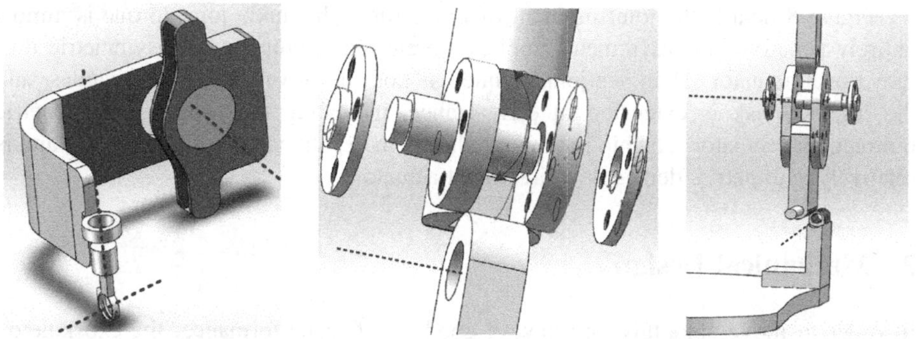

(a)hip joint (b)Knee joint (c)ankle joint

Fig. 10. Joint design details

4 Human-Machine Coupling Dynamic Analysis

4.1 Active Exoskeleton

Active exoskeleton transmit forces to the human musculoskeletal system through the physic interaction of bandage designed on the exoskeleton. The active exoskeletons are always used for rehabilitation and human performance augmentation.

At extreme conditions, like a paralysis human being, all the power is provided by exoskeleton. In the ADAMS, the exoskeleton actuator's angle trajectories are set according to the OpenSim angle trajectory of each joint, and the human limbs are totally passive. The human-machine interface positions on the shank and thigh part will be located at the upper, mid, and lower position of each part separately. The human-machine connections are equated in the ADAMS by the flexible connection of BUSHING which produce three-dimensional force related to the displacement and velocity between the two bodies, here it is the relative displacement and velocity of the lower limb with its corresponding exoskeleton part. The stiffness and damping of the BUSHING are set to be 5000 N/m and 100 N/(m/s).

Part of the simulation results are shown in Fig. 11. The human-machine interaction forces at axial direction (Y axis) and radial direction (Z axis) are different from the interface positions. At the up position of the thigh and shank part, the forces at axial direction and radial direction is lower, while at the middle position of the thigh and

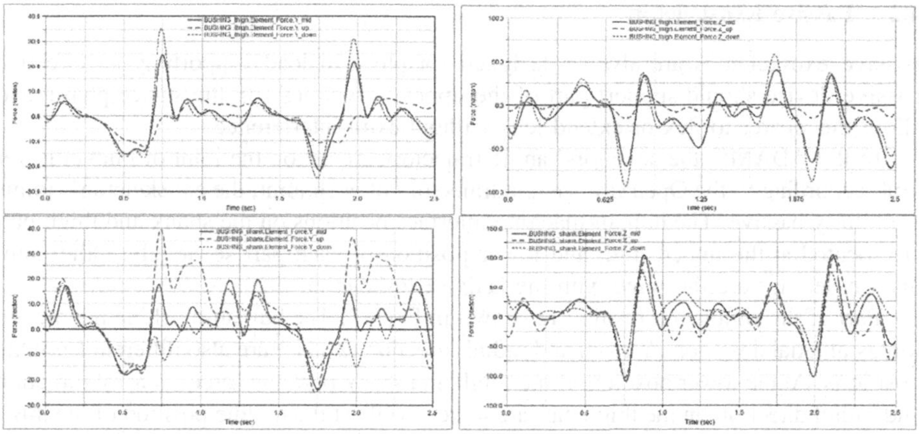

Fig. 11. Human-machine interaction forces at different connection position by active exoskeleton

Table 2. The RMS of the interaction force on thigh

RMS of the force at axis direction (N)				RMS of the force at radial direction (N)			
Thigh / Shank	Up	Mid	Down	Thigh / Shank	Up	Mid	Down
Up	5.8100	6.7584	6.9404	Up	13.7894	21.0754	21.8147
Mid	9.0377	9.2358	8.4628	Mid	26.1203	30.7241	27.6846
Down	14.9062	14.1273	10.0871	Down	35.8528	40.1673	31.1072

Table 3. The RMS of the interaction force on shank

RMS of the force at axis direction (N)				RMS of the force at radial direction (N)			
Thigh / Shank	Up	Mid	Down	Thigh / Shank	Up	Mid	Down
Up	13.6196	12.4512	11.6675	Up	43.1935	33.3483	24.2856
Mid	13.3240	9.6057	8.8053	Mid	59.8045	36.1679	25.8887
Down	20.7206	11.7018	8.8845	Down	79.6372	41.7892	26.6740

shank part, the forces at axial direction and radial direction is lower. Changing the interface position of the thigh and shank, more results will be achieved, Tables 2 and 3 shown the RMS of the interaction force on the thigh and shank part respectively. In order to prevent high human-machine interaction force, especially, on the axial direction (Y axis) which causes uncomfortable, the interface position on the thigh and shank part are suggested to be at the down and middle part respectively.

4.2 Passive Exoskeleton

Passive exoskeletons are always designed for physical load supporting, an effective passive device should support part of the upper body weight in the stance phase, and allow the wearer to flex or extend legs without extra interference.

In the ADAMS, the actuators' angle trajectory are set on the joint of human lower limb according to the OpenSim's angle trajectory of each joint, the exoskeleton's joints are totally passive. The human-machine interface positions on the shank and thigh part are located at the upper, mid, and lower position of each part separately. Other simulation settings are the same with the active exoskeleton.

Part of the simulation results are shown in Fig. 12. The human-machine interaction forces at axial direction (Y axis) and radial direction (Z axis) are also different from the positions. At the lower position of the thigh and shank part, the forces at axial direction and radial direction on the thigh part are lower, while at the middle position of the thigh and shank part, the forces at axial direction and radial direction on the shank part are lower. Changing the interface position of the thigh and shank, more results will be achieved, Tables 4 and 5 shows the RMS of the interaction forces on the thigh and shank part respectively. In order to prevent high human-machine interaction force, especially, on the axial direction (Y axis) which causes uncomfortable, the interface position on the thigh and shank part are suggested to be at the middle and down part respectively.

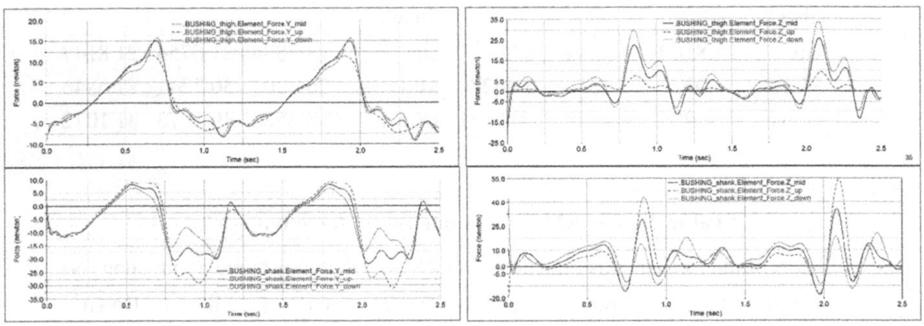

Fig. 12. Human-machine interaction forces at different connection position by passive exoskeleton

Table 4. The RMS of the interaction force on thigh

RMS of the force at axis direction (N)				RMS of the force at radial direction (N)			
Thigh Shank	Up	Mid	Down	Thigh Shank	Up	Mid	Down
Up	6.1839	6.1787	6.1085	Up	4.2561	7.2953	8.5110
Mid	6.6555	6.6802	6.4626	Mid	5.5197	8.5122	9.3347
Down	7.0132	6.6320	6.2461	Down	8.9983	9.9892	10.2713

Table 5. The RMS of the interaction force on shank

RMS of the force at axis direction (N)				RMS of the force at radial direction (N)			
Thigh / Shank	Up	Mid	Down	Thigh / Shank	Up	Mid	Down
Up	13.6915	13.7224	13.8092	Up	12.0842	9.3846	7.0530
Mid	11.0954	11.2127	11.4394	Mid	12.6145	10.0685	8.4923
Down	9.1182	9.0130	9.5462	Down	15.4660	11.1956	9.7306

5 Conclusion and Future Work

The wearable exoskeleton enhancing human abilities, such as perception, manipulation, walking and so on. In order to design an anthropomorphic exoskeleton, the anatomical structure and biomechanical characteristics of lower limb were studied, there were 7 degrees of freedoms on each lower limb, and the movement range was given out in Table 1. The kinematic and dynamic characteristic of walking gait were analyzed by OpenSim. It indicates that the ankle joint demands asymmetric unidirectional actuator, the knee joint demands asymmetric bidirectional actuator, and the hip joint demands bidirectional actuator.

The mechanical structure were designed according to the anatomy structure and gait analysis of the lower limb. Human-machine interaction forces at different interface positions were analyzed. The study demonstrates that the passive exoskeleton's choices of human-machine interaction position on the shank limb is the down position on its part, and thigh limb is middle position on its part. The active exoskeleton's choices of human-machine interaction position on the shank limb is the middle position on its part and thigh limb is down position on its part.

In this paper, the analysis of human-machine coupling dynamics are only simulated without extra loads and plantar forces on the exoskeleton which could cause more complicated results. Only mechanical structure are designed, the design of actuators, and control system for active exoskeleton is our future work.

Acknowledgement. Funding for this research were provided by National Natural Science Foundations of China (Grant No. 51505494 and 11504427).

References

1. Young, A., Ferris, D.: State-of-the-art and future directions for robotic lower limb exoskeletons. IEEE Trans. Neural Syst. Rehabil. Eng. **25**(2), 171–182 (2016)
2. Dollar, A.M., Herr, H.: Lower extremity exoskeletons and active orthoses: challenges and state-of-the-art. IEEE Trans. Robot. **24**(1), 144–158 (2008)
3. Garcia, E., Sater, J.M., Main, J.: exoskeleton for human performance augmentation (EHPA): a program summary. J. Robot. Soc. Japan **20**(8), 44–48 (2002)

4. Kazerooni, H., Steger, R.: The Berkeley lower extremity exoskeleton. J. Dyn. Syst. Measur. Control **128**(1), 14 (2006)
5. Kawamoto, H., Sankai, Y.: Power assist method based on phase sequence and muscle force condition for HAL. Adv. Robot. **19**(7), 717–734 (2005)
6. Chen, B., Ma, H., Qin, L.-Y., et al.: Recent developments and challenges of lower extremity exoskeletons. J. Orthop. Transl. **5**, 26–37 (2016)
7. Delp, S.L., Anderson, F.C., Arnold, A.S., et al.: OpenSim: open-source software to create and analyze dynamic simulations of movement. IEEE Trans. Biomed. Eng. **54**(11), 1940–1950 (2007)
8. Zoss, A.B., Kazerooni, H., Chu, A.: Biomechanical design of the Berkeley lower extremity exoskeleton (BLEEX). IEEE/ASME Trans. Mechatron. **11**(2), 128–138 (2006)

Intention-Based Human Robot Collaboration

Guoqiang Liang, Xuguang Lan(✉), Hanbo Zhang, Xingyu Chen, and Nanning Zheng

Institute of Artificial Intelligence and Robotics, Xi'an Jiaotong University, Xi'an 710049, China
gqliang@stu.xjtu.edu.cn, xglan@mail.xjtu.edu.cn

Abstract. In most of the current human-robot collaboration systems, the motion of robots is based on some predefined instructions, which can just deal with several specific situations. It is difficult to predefine a complete set of instructions for situations in the area of assistance and rehabilitation due to their high complexity and large variety. As a result, the instruction-based human-robot collaboration cannot be applied in these areas directly. In this paper, we propose a new human-robot collaboration framework based on the understanding of human intention. In this framework, the behavior of robots is based on the actively recognition and understanding of the dynamic scene and human intention, which is one of the core character of the coexisting robot. Based on this framework, we have accomplished some initial experiments, whose results show the effectiveness of our proposed system.

Keywords: Human intention · Scene understanding · Human robot collaboration

1 Introduction

Today, robots have been applied in many areas, such as manufacture and remote reconnaissance. Most robots are deployed in complete isolation from humans and just execute a series of predefined instructions. However, the lack of human–robot collaboration restricts applied areas of robots [1]. In assistance and rehabilitation, robots are designed to collaborate with people to accomplish some tasks and their roles should be changed from purely automated machines to autonomous companions. In these human robot collaboration systems, human is generally in the center and their behavior determines the action of robots. Compared with the instruction-based robots, new autonomous robots should collaborative with people actively.

In human robot collaboration system, one of the most important problems is to make the robot understand the intention of human so that the robot can actively collaborate with human. In the force control [2] and impedance control [3], robots are trained to follow the motion of human passively. However, these control methods are likely to over fitting on the training data. As a result, it cannot be applied in situation where the motion of robots is highly complex and variable. In actual human-human cooperation, the action of one people highly depends on the intention of his or her partner. To imitate this, many literatures have investigated the role of human intention in human robot collaboration. In most works, human intention is defined as motion intention, which is the future position of human joints. This kind of intention

Y. Huang et al. (Eds.): ICIRA 2017, Part I, LNAI 10462, pp. 605–613, 2017.
DOI: 10.1007/978-3-319-65289-4_57

is estimated by Markov process [4], Kalman filter [5], and neural network [6]. However, human intention is much beyond the motion intention. Complete inference of human intention needs the entire understanding of motion intention, scene understanding, even the work or living custom. The ability of dynamic scene understanding and human intention inference is the core character of next generation of robots.

For this, we propose a new human robot collaboration frame, whose core is the understanding of human intention. Firstly, the robot needs to understand the scene where human and robots locate. The objects in the scene may play an important role in the human-robot collaboration. As a result, scene understanding aims at detecting and recognizing objects. Secondly, the human behavior should be parsed accurately and completely, which compress human detection, pose estimation and motion tracking. Then the human intention is inferred from the knowledge of scene and human behavior according to the cooperation pattern of human. Finally, the response of robots is generated based on the human intention. All response generation rules are directly learnt from the real world instead of hand- crafted instructions.

Our main contribution includes two aspects: (1) we propose an intention-based human robot collaboration frame (2) Based on this frame, we have completed some initial experiments and achieve good performance.

2 Related Work

Since research on human robot collaboration includes lots of contents, a complete review is beyond the scope of this paper. Please refer to [7, 8] for a detailed review. In this paper, we will just review the most related works on scene understanding and human intention inference for human robot collaboration.

2.1 Scene Understanding

In almost every daily event, human needs to interact with the surrounding environment. In this process, human needs to understand the scene, which includes object detection and recognition. These problems are very hot research topic in Computer Vision and Robot community. With the development of Convolutional Neural Network (CNN), the accuracy of image classification [9], scene labeling [10] and object recognition [11] has improved a lot. After simple classification of objects, the robot needs to know their function and operating area, like grasp position of a cup. This can be inferred from the appearance [12] or learnt from real scene [13]. Sun et al. [12] proposed to identify color, shape, material, and name attributes of objects from the RGB-D data. Ref. [13] formulated the problem of localizing and recognition of functional areas from an arbitrary indoor scene as a two-stage deep learning based detection pipeline. Ref. [14] cascaded two auto encoders to detect the grasp position from coarse to fine. Aksoy et al. [15] construct a dynamic graph sequence of tracked image segments from human demonstrations and this representation is used by the robot for manipulating objects.

2.2 Human Intention Inference

In many previous works, human intention inference is defined as human motion prediction. For example, Ravichandar [16] defined human intention as goal locations in 3-dimensional (3D) space. The dynamic of human motion is learned by a neural network. To represent the dependence between motion trajectory and task, ref. [17] modeled the conditional likelihood of a trajectory given a motion class and task. Ref. [18] employed a conditional variational auto encoder to predict a window of future human motion given a window of past frames. This method can predict the motion up to 1660 ms.

Besides the motion intention prediction, there are some works on human intention inference from the whole environment. Caire P. et al. [19] declared the importance of the ability to identify causes and explanations for changes to the environment, which could represent the human intention. To anticipate a belief about possible future human actions in contextually-rich environments, Koppula [20] construct a graph to represent the human motion and interaction with objects. The Markov decision process (MDP) is used to model the human's and robot's behavior. To deal with variability in the real environments, Liu [21] developed an Object Functional Role Perspective method to endow a robot with comprehensive behavior understanding.

Compared with these works, our intention-based frame aims at active human robot collaboration. To achieve this, it integrates scene understanding and human behavior analysis. Based on these, the robot can infer the human intention and react accurately.

3 Intention-Based Human Robot Collaboration

As analyzed in Sect. 1, our intention-based human-robot collaboration system consists of four modules: scene understanding, human behavior understanding, human intention inference, and robot's response generation. Figure 1 shows the proposed frame and gives the relationship of these four modules. In the following, we will give a detailed description of each module.

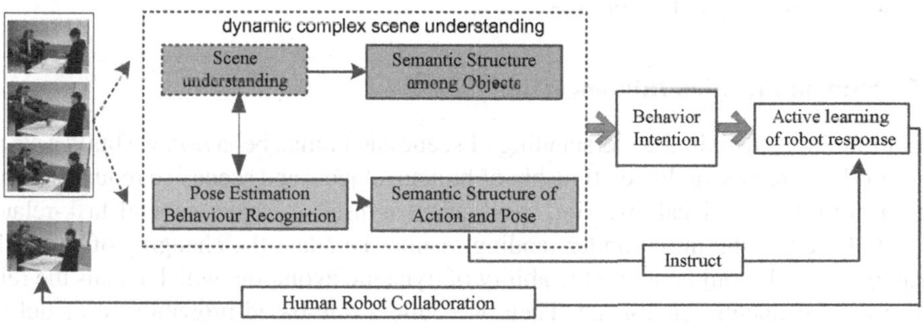

Fig. 1. The proposed intention-based human robot collaboration frame

3.1 Scene Understanding

The scene is the environment, where human and robots collaboration occurs. It plays an important role in human-robot collaboration. Generally, the scene contains lots of objects, which will participate or impact the collaboration in different degree. In scene understanding, there are two key points: obtain information of objects and build rough relationship among objects. The object's information includes the location, category and shape and so on. The relationship can contain spatial, temporal and semantic relation. Different from doing detection, classification and segmentation individually, we integrate all these tasks in a unified frame [22]. Due to time limitation, it is impossible to get all information of all objects, especially for complex scene. As a result, attention mechanism is used to seek several key objects, which are highly related to human and robots. Relationship among objects helps to build a structure model for human robot collaboration.

3.2 Human Behavior Understanding

Human intention is mainly expressed via the behavior of human and the operating objects. As a result, human behavior understanding is a prerequisite condition for human intention inference. Most of current researches on human behavior analysis just focus on action classification, which just classify the human action into several rough categories, like running, walking [23]. This kind of action classification is too coarse for accurate inference of human intention in human robot collaboration.

In order to understand the human behavior completely, we analyze human behavior from two aspects: fine-grained action recognition [24] and human pose estimation [25], both of which can be replaced with new models. These two tasks can recognize the category of human and estimate the location of human joints, which will give a sufficient representation of human behavior. To utilize the close relation between human pose estimation and human action recognition, we integrate these two works in a single framework. Combined with the information of operation objects, human behavior can be understood more thoroughly [24].

3.3 Human Intention Inference

After obtaining the initial understanding of scene and human behavior, we build a structure model to represent the relationship of human, objects and scene. To reduce computation complexity, visual attention mechanism is used to select several task-related objects based on the initial understanding results. To adapt the changing of tasks, the structure model should possess the ability of dynamic evolution, which means the relationship of elements can change. Then, we build a rule-based probabilistic model for human intention inference, whose input includes structure model and information of task, human behavior, and objects. Using Bayes' theorem, the human intention can be inferred. Compared with [26], our model includes more complete information of scene, human behavior. Therefore, our model is more robust.

3.4 Robot's Response Generation

After getting the human intention, the next step is to generate the robot's response. This includes two parts: what to do and how to do. The first part is to generate suitable response according to the human intention. It is difficult to learn the response pattern from the human activity datasets. Our current response pattern is defined as several rules and is learnt from the tasks. The second part focuses on the execution of specified action. To accomplish the action, a robot needs to integrate the information of scene and produce the motion route. Through building suitable reward functions, we can use deep reinforcement learning to generate the robot's motion route [27]. Using online learning, the model can be refined according to the executed result and feedback of human.

In summary, each module in the proposed frame accomplishes an individual task. With this division, new algorithms for each module can be quickly adopted, which can lead to quick performance increase. To demonstrate the effect of this frame, we design a simple task, which will give in the next section.

4 Experiments

Based on the proposed framework, we have implemented a simple experiment, where a robot pours water into a cup and gives the cup to a person. In this process, the robot needs to recognize and locate a person, two cups firstly, which belongs to scene understanding. Then, the robot understands human behavior and infers his intention. Finally, the robot generates and accomplishes the response: pour water and give the mug to the person.

The whole experiment is completed on a Baxter robot, which is equipped with two mechanical arms with 7 degree freedom. The gripper of our Baxter consists of two parallel clips, whose distance can vary from 0 cm to 4 cm. We use a Kinect to capture the RGB image. The Kinect is fixed on the top of the head of Baxter, which means the Kinect is at the height of 1.75 m on the ground. The cups are places in a desk before the Baxter. Detailed environment is shown in Fig. 5.

4.1 Cup Detection

The aim of cup detection is to get the rough position of cup from an input image. For simplicity and speed, the discriminatively trained part-based model (DPM) [28] is used to detect the cup currently. The input to the DPM is a RGB image captured by a Kinect. Since there is no available model for cup detection, a new model is trained. For training the parameter of DPM, we collect a small cup detection dataset, which includes 120 images with different cups, illumination, and occlusion. Figure 2 shows some examples of the positive images. 100 images without cups are selected from PASCAL VOC 2007 dataset and are regarded as negative images. For positive images, the position of cups is annotated with a rectangle, which encircles the cups. We achieve 85% detection accuracy on our dataset. The detection speed is 12.6 s per frame when the size of images is 640×480. In Fig. 3, we show some samples of the cup detection results.

Fig. 2. Some images of the cup dataset

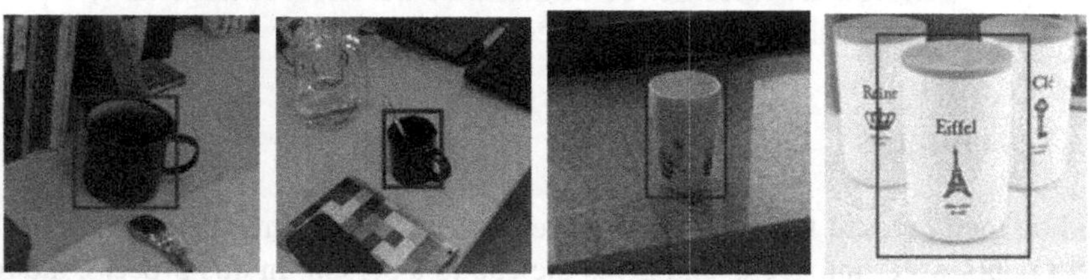

Fig. 3. The sample results of cup detection

4.2 Grasp Position Detection

Grasp position detection is to locate the specific position for grasping. We employ the Lenz's method [14], which cascades two auto encoders to detect the grasp position. The first one aims at rough grasp position detection. The second one is used to choose the best grasp position from the rough candidates. For detailed architecture, refer to [14]. The parameters of auto encoder are trained on the Cornell grasping dataset [29], which contains 1035 images with 280 grasp objects. In each image, there are multiple grasp position annotations. The annotations which can be grasped by robots are regarded as positive examples and the remaining are negative examples. On the Cornell grasping dataset, the accuracy of binary classification is 94.03% while the position accuracy is 73.06% based on the rectangle metric [14]. Some grasps detection results on our dataset are shown in Fig. 4.

4.3 Final Results

For human behavior analysis, we employ the pose estimation method from [25], whose accuracy is 93% on FLIC dataset. Human pose estimation can give the specific location of human joints. Combining them with cup detection, we can know a person holds a cup and puts it on a desk. Our current human intention inference is defined as several rules, which will be trained in future. Based on these rules, human intention can be inferred.

The action of robot is generated via MDP. The final results are shown in Fig. 5, where the result of cup detection and human pose estimation is shown.

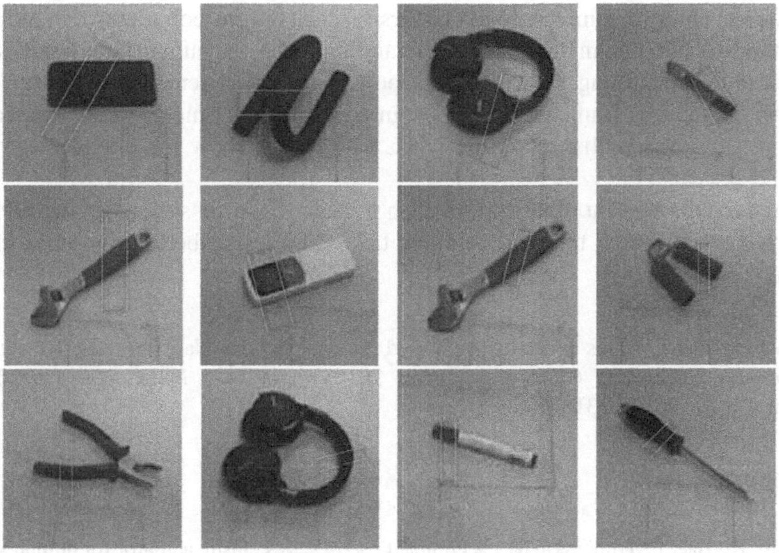

Fig. 4. The grasps detection result

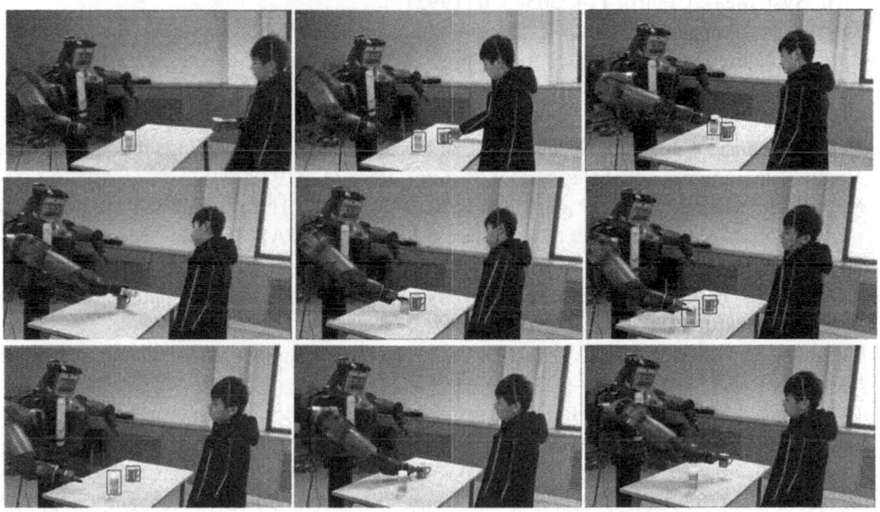

Fig. 5. The experimental result of our cup grasp and delivery

5 Conclusion

In summary, we propose a new human robot collaboration system, which is based on the understanding of human intention. Firstly, this system completes dynamic scene understanding and human behavior understanding. Then, the human intention is inferred from these understanding. Finally, the response of robot is generated based on human intention. Based this framework, we design a simple experiment, where human robot collaborates to accomplish a simple work. The results show the effectiveness of the proposed system.

However, the current algorithm in each module is a little simple. In the future, we will extend this system to more complex tasks and design better algorithm for each module.

Acknowledgement. This work is supported in part by the National Key Research and Development Program of China under grant 2016YFB1000903, NSFC No. 61573268 and Program 973 No. 2012CB316400.

References

1. Kong, K., Bae, J., Tomizuka, M.: A compact rotary series elastic actuator for human assistive systems. IEEE/ASME Trans. Mechatron. **17**(2), 288–297 (2012)
2. Newman, W.S.: Stability and performance limits of interaction controllers. Trans. ASME J. Dyn. Syst. Meas. Control **114**, 563–570 (1992)
3. Hogan, N.: Impedance control: an approach to manipulation—part I: theory; part II: implementation; part III: applications. Trans. ASME J. Dyn. Syst. Meas. Control **107**(1), 1–24 (1985)
4. Wang, Z., Peer, A., Buss, M.: An HMM approach to realistic haptic human-robot interaction. In: Proceedings of 3rd Joint Eurohaptics Conference and Symposium on Haptic Interfaces for Virtual Environment and Teleoperator Systems, pp. 374–379 (2009)
5. Wakita, K., Huang, J., Di, P., Sekiyama, K., Fukuda, T.: Human-walking intention-based motion control of an omnidirectional-type cane robot. IEEE/ASME Trans. Mechatron. **18**(1), 285–296 (2013)
6. Li, Y., Ge, S.: Human-robot collaboration based on motion intention estimation. IEEE/ASME Trans. Mechatron. **19**(3), 1007–1014 (2014)
7. Thomaz, A., Hoffman, G., Cakmak, M.: Computational Human-Robot Interaction. Found. Trends R Robot. **4**(2–3), 105–223 (2013)
8. Royakkers, L., van Est, R.: A literature review on new robotics: automation from love to war. Int. J. Soc. Robot. **7**(5), 549–570 (2015)
9. He, K., Zhang, X., Ren, S., Sun, J.: Deep residual learning for image recognition. In: Proceedings of the IEEE Conference on Computer Vision and Pattern Recognition, pp. 770–778 (2016)
10. Farabet, C., Couprie, C., Najman, L., LeCun, Y.: Learning hierarchical features for scene labeling. IEEE Trans. Pattern Anal. Mach. Intell. **35**(8), 1915–1929 (2013)
11. Girshick, R., Donahue, J., Darrell, T., Malik, J.: Rich feature hierarchies for accurate object detection and semantic segmentation. In: Proceedings of IEEE Conference on Computer Vision and Pattern Recognition, pp. 580–587 (2014)

12. Sun, Y., Bo, L., Fox, D.: Attribute based object identification. In: Proceedings of IEEE International Conference on Robotics and Automation, pp. 2096–2103 (2013)
13. Ye, C., Yang, Y., Fermuller, C., Aloimonos, Y.: What can I do around here? Deep functional scene understanding for cognitive robots. arXiv preprint, arXiv:1602.00032
14. Lenz, I., Lee, H., Saxena, A.: Deep learning for detecting robotic grasps. Int. J. Robot. Res. (IJRR) **34**(4–5), 705–724 (2013)
15. Aksoy, E., Abramov, A., Dorr, J., Ning, K., Dellen, B., Worgotter, F.: Learning the semantics of object-action relations by observation. J. Robot. Res. **30**(10), 1229–1249 (2011)
16. Ravichandar, H., Dani, A.: Human intention inference and motion modeling using approximate em with online learning. In: Proceedings of IEEE/RSJ International Conference on IEEE Intelligent Robots and Systems (IROS), pp. 1819–1824 (2015)
17. Pérez-D'Arpino, C., Shah, J.: Fast target prediction of human reaching motion for cooperative human-robot manipulation tasks using time series classification. In: Proceedings of 2015 IEEE International Conference on Robotics and Automation, pp. 6175–6182 (2015)
18. Bütepage, J., Kjellström, H., Kragic, D.: Anticipating many futures: online human motion prediction and synthesis for human-robot collaboration. arXiv preprint, arXiv:1702.08212
19. Caire, P., Cornelius, G., Voos, H.: Collaborative explanation and response in assisted living environments enhanced with humanoid robots. In: Collaborative Explanation and Response in Assisted Living Environments Enhanced with Humanoid Robots (2016)
20. Koppula, H., Jain, A., Saxena, A.: Anticipatory planning for human-robot teams. In: Experimental Robotics, pp. 453–470 (2016)
21. Liu, R., Zhang, X.: Understanding human behaviors with an object functional role perspective for robotics. IEEE Trans. Cogn. Dev. Syst. **8**(2), 115–127 (2016)
22. Wu, Y., Zheng, N., Liu, Y., Yuan, Z.: Fine-grained and layered object recognition. Int. J. Pattern Recogn. Artif. Intell. **26**(2), 1255006 (2012)
23. Feichtenhofer, C., Pinz, A., Zisserman, A.: Convolutional two-stream network fusion for video action recognition. In: Proceedings of the IEEE Conference on Computer Vision and Pattern Recognition, pp. 1933–1941 (2016)
24. Wei, P., Zhao, Y., Zheng, N., Zhu, S.: Modeling 4D human-object interactions for joint event segmentation, recognition, and object localization. IEEE Trans. Pattern Anal. Mach. Intell. **39**(6), 1165–1179 (2016)
25. Liang, G., Lan, X., Wang, J., Wang, J., Zheng, N.: A limb-based graphical model for human pose estimation. IEEE Trans. Syst. Man Cybern. Syst. **PP**(39), 1–13 (2017)
26. Wang, Z., Mülling, K., Deisenroth, M.P., Ben Amor, H., Vogt, D., Schölkopf, B., Peters, J.: Probabilistic movement modeling for intention inference in human–robot interaction. Int. J. Robot. Res. **32**(7), 841–858 (2013)
27. Gu, S., Holly, E., Lillicrap, T., et al.: Deep reinforcement learning for robotic manipulation with asynchronous off-policy updates. arXiv preprint arXiv:1610.00633 (2016)
28. Felzenszwalb, P.F., Girshick, R.B., McAllester, D., Ramanan, D.: Object detection with discriminatively trained part-based; models. IEEE Trans. Pattern Anal. Mach. Intell. **32**(9), 1627–1645 (2010)
29. Jiang, Y., Moseson, S., Saxena, A.: Efficient grasping from RGBD images: learning using a new rectangle representation. In: Proceedings of IEEE International Conference on Robotics and Automation (ICRA), Shanghai, China, pp. 3304–3311 (2011)

Swarm Robotics

A Knowledge-Based Intelligent System for Distributed and Collaborative Choreography

Xinle Du[1], Haoqin Ma[2], and Hongwei Wang[2](✉)

[1] Qufu Normal University, Rizhao 276826, China
[2] School of Engineering, University of Portsmouth, Portsmouth, PO1 3DJ, UK
1608119@qq.com, Hongwei.Wang@port.ac.uk

Abstract. As a knowledge-intensive process, choreography is increasingly done in a collaborative and distributed environment. However, current research on digital choreography is mainly focused on developing digital tools to facilitate the design of movements, spaces, effects, etc. while little work has been done on capturing and supplying knowledge to support decision-making in particular when multiple experts are involved in the choreography process. In this research, a knowledge-based intelligent system is proposed to effectively capture useful knowledge from a complex choreography process and to provide intelligent support to various experts in a distributed environment. Specifically, the system framework of a distributed and collaborative choreography scheme enabled by advanced computing technologies is proposed. A knowledge representation model is developed based on design rationale to represent both procedural and causal knowledge. A prototype system is also designed in this work and its key technologies are developed. A literature search has shown that little work has been found on knowledge-based intelligent support for choreographers. A preliminary evaluation of the methods in a real-world example shows that they have a range of advantages and more importantly this works opens up opportunities for this new area of work.

Keywords: Intelligent system · Distributed choreography · Knowledge representation · Design rationale

1 Introduction

Choreography can be referred to as design itself as it involves deliberation and embodiment of ideas and transformation of these ideas into motions, forms and spaces. Modern choreography generally involves the use of novel technologies to generate special effects and thus entails an interdisciplinary team to complete various tasks. In this sense, choreographing can be viewed as a complex collaborative process whereby professional choreographers, dancers, musicians, special effects experts need to work together to solve problems. The convergence of artistic practices with multimedia software, computer-mediated communication, distance education, multiple site performance, and collaboration, is bringing about dramatic change in many fields including dancing [1]. In this context, digital choreography has become an interdisciplinary research field with

Y. Huang et al. (Eds.): ICIRA 2017, Part I, LNAI 10462, pp. 617–627, 2017.
DOI: 10.1007/978-3-319-65289-4_58

an emphasis on utilizing information technologies in choreographing. The collaboration between experts from multiple disciplines has raised the need of supporting distributed development of plans in particular for large-scale choreographing projects. Nevertheless, this distribution feature has not been well addressed by the digital choreographing tools currently available. Existing research on distributed choreography has been mainly focused on video conferencing and immersed environments [1, 2].

Collaborative work of multiple experts in a distributed environment also means that choreography is a knowledge-intensive process. Knowledge from various sources is generated in this process, which needs to be shared within a choreographing team and more important needs to be reused in future projects. Intelligent support enabled by advanced computational intelligence and knowledge management has been applied to various fields. For example, a vast amount of research has been done for capturing and reusing engineering design knowledge [3, 4]. To facilitate effective capture of knowledge, a structure is needed to represent the knowledge space, which is often called a knowledge representation model [5]. Research has also been done to study how design knowledge can be captured in a collaborative working environment [6]. However, a literature search has found that little work has been done on management of choreographing knowledge despite the importance of knowledge reuse in this collaborative problem-solving process.

This work is motivated by this gap and aims to introduce knowledge management and intelligent technologies to this traditional field. It has great potential for applying the findings and results from other fields such as engineering design to supporting decision making in distributed choreography. For example, design rationale [3] is actually quite general-purpose and its structure can also naturally describe the solution generation process in choreographing. Therefore, the very first step is to extend and adapt the results from other disciplines to make them useful for choreographing and the ultimate goal is to develop knowledge-based intelligent support for effective and efficient choreographing in a distributed environment.

The rest of this paper is organized as follows. Section 2 reviews published work related to this research. Section 3 introduces the framework of a collaborative and distributed choreographing system so as to identify the key research areas required for achieving the ultimate goal. Section 4 details a knowledge representation model based on design rationale. Section 5 describes the key technologies for a prototype system together with its preliminary evaluation. Section 6 discusses the main conclusions of this paper as well as potential future work.

2 Literature Review

2.1 Choreography and Computers

Dancing and choreography is not commonly linked to computers at least not as much as other disciplines such as the IT and manufacturing industries. Nevertheless, early research on the convergence between computers and art can be dated back to 1960s [7]. The capability of computers has been exploited to facilitate decision making in choreography. This exploitation has never stopped since then [1] and in more recent decades the dance field has continued to manifest itself as an experimental playground in which dance practitioners, designers and dance researchers explore the opportunities of emergent digital technologies for the communication of dance knowledge [8]. In [9], it is indicated that the pervasiveness of information and communications technologies has produced new levels of thought, new concepts, and new types of human interaction. The application of these technologies to this traditional discipline has become a common practice, resulting in the research field of digital choreography. For example, digital literature has been adopted by the dancing and choreography curriculum [10].

The importance of digital technology in dancing practice has become more prominent in recent years and it seems that a more integrated use of digital technology in both professional dance practice and within the research community via the use of multimedia research networks with intuitive could have the potential to inform and help shape both the perception of experimental digital tools by new generations of dance artists [11]. There has been argument about the role of computers in dancing and choreography and it has been widely accepted that technology can only assist designers in the choreography process rather than replacing human's intelligence [7]. Then the question has become what sort of things computer can support dancing practitioners and designers. Most of the tools currently available are mainly focused on facilitating movement analysis and synthesis, analyze interactions and enable creation of virtual experiments. Little work has been done to address the nature of collaborative and distributed development of choreographing.

2.2 Collaborative and Distributed Choreography

The distributed and collaborative development nature of choreography has also been emphasized in recent research. The term distributed choreography has been used in recent literature. For example, a framework to support the design of computer-based artefacts for choreographers with special reference to Brazil was proposed [12]. The use of a virtual environment to provide a new context for practice-led research in dance [13]. However, research in this area is still not much and most of the published work has been focused using video-conferencing technology to facilitate communication and collaboration [1, 2]. These studies are mainly focused on providing a way for collaboration in a distributed environment while little has been done in terms of how the content of collaboration can be mediated by computer as well. This content is highly important as it involves the creation and sharing and understanding of useful design knowledge which is critical in the modern context of doing large-scale choreography projects involving a lot of artefacts.

There is an interesting idea of using dance notes which facilitates knowledge sharing. The preference for a movement-oriented approach to dance knowledge and the aim of looking for ways to transmit the 'experiential knowledge' of dance is also apparent in the reappraisal of dance notes [8]. However, how computers can improve capturing, sharing and reusing design knowledge remains an area requiring much more work. In particular, knowledge management and its computer tools have been widely studied in a wide range of disciplines such as engineering design, management, etc. [3, 4]. How these research methods and tools can be used in choreography remain an open challenge and thus open up the opportunities for interdisciplinary research in this area.

3 A Knowledge-Based System for Collaborative and Distributed Choreographing

3.1 System Framework for Knowledge Management in Collaborative and Distributed Artefact Creation

In an attempt to exploit advanced technologies such as knowledge management and computational intelligence to achieve innovation application in a relatively traditional discipline, i.e. choreography, this research moves a first step towards developing a knowledge-based collaborative and distributed choreographing system. Such a system can support the artefact (i.e. a dancing performance) deliberation, argumentation and creation process in which multiple developers work together to capture and reuse knowledge about problem solving and decision making in a collaborative and distributed environment. As discussed earlier, this process is in essence similar to any other design processes that involve problem analysis, task identification, solution generation and decision making. To achieve this aim, a range of technologies are needed to address the requirements of different developers and from different stages of the process. Therefore, it would be useful t firstly analyze this process as well as its developers with various roles. Figure 1 shows the framework for such a system and gives an overview of the process, the key users involved and their interactions with the system.

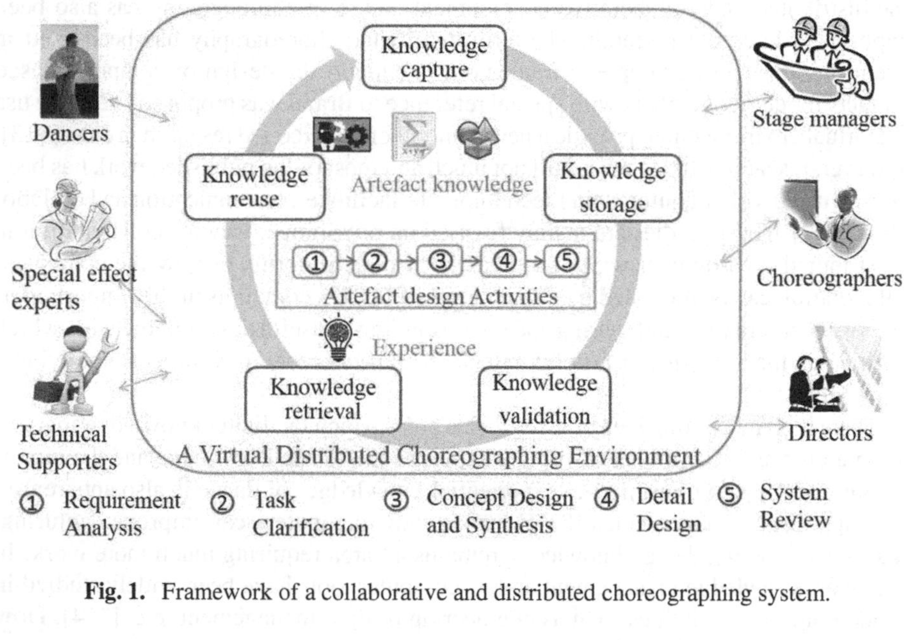

Fig. 1. Framework of a collaborative and distributed choreographing system.

As shown in the figure, the collaborative and distributed choreographing paradigm involves a process consisting of a few key stages, namely requirement analysis, task clarification, concept design, synthesis, detail design and system review. Specifically, requirement analysis involves an analysis of the artefact to be choreographed and identification of the key requirements to address in terms of functions (e.g. aesthetics, movements and spaces), constraints (e.g. stages, length of performance and budget) and considerations. Task clarification is undertaken to identify the key activities for the whole project and to specify the key inputs (e.g. requirements), outputs (e.g. concepts and detailed plans), time constrains and resources needed for their undertaking. Concept design involves the generation and evaluation of a number of concepts to realize the functions identified and synthesis means the combination of the generated concepts to form a complete artefact (or an important part of an artefact). When some concepts are finally chosen, the detail design stage involves creating specific forms (e.g. specification of movements, spaces, musical components and special effects) for implementing the concepts. Eventually, the system review stage refers to a formal review of the whole process to evaluate the artefact created.

The knowledge management circle involves a few stages, namely from knowledge capture, storage, validation, retrieval and reuse. These stages come across with all the stages of the artefact creation process. In this case, the system can capture knowledge across the whole choreographing process. All these knowledge management stages are carried out through collaboration of various developers with different roles. Each developer can have a range of assigned tasks to complete while they can also work together on a specific task. The developers of a collaborative and distributed choreographing project include professional choreographer, musician, special effect experts, dancer, lighting expert, stage manager and director. Specifically, a director deals with the whole project and coordinates key decisions. Choreographers are responsible for developing concepts and forms for which dancers, special effect experts, lighting experts and stage management can give some useful inputs.

3.2 Requirements of the Proposed System

The proposed system is very complicated as it spans across a few disciplines, namely information technology, choreography and knowledge management. In this sense, it entails a range of research to be undertaken. This research moves a first step through identifying the system framework and the key components for such a system. To support effective collaborative and distributed choreographing, a number of technological requirements needs to be addressed. Firstly, a collaboration engine is needed to coordinate the collaborative work of multiple users in a distributed environment. This engine will provide a collaborative working space for each user, control the undertaking of activities, and facilitate the interactions between different users. Secondly, a virtual working environment is needed, which will be provided as a Web-based tool accessible to any user distributed on the Internet. Thirdly, this tool should support creation of dancing artefacts and enable the combination of a range of multimedia resources. Last but not least, the system should support knowledge capture and reuse throughout the whole artefact creation process. In this sense, it should provide a knowledge operation engine which facilitates creation and description of knowledge records as well as storage of these records. More importantly, it should do another two important things: (1) to develop effective knowledge retrieval methods which are critical when a huge amount of knowledge records accumulate; (2) to support effective knowledge reuse using recommendation methods enabled by computational intelligence.

4 A Knowledge Representation Model Based on Design Rationale

To enable effective knowledge capture and reuse, a knowledge representation model is needed to describe knowledge elements in a structured way. Such a model not only provides a way for human users to naturally understand knowledge elements but also supports effective processing of these elements by computers. This function is particularly useful in the context of big data when data in various formats such as pictures, videos, audio recording, sketches and drawings are accumulated. Design rationale is proposed in the 1970s as a novel means of constructing issue based information systems, which has the power of describing problem solving processes for various disciplines. It has been widely used in a variety of fields such as software engineering and engineering design [3]. The structure of design rationale fits naturally with the need of describing knowledge for choreography as it is in essence a design process involving a lot of issues, options and decision making.

Therefore, a design rationale based knowledge representation model is proposed in this research to capture the process of choreographing with a particular focus on the issues, solutions, considerations and arguments involved in this process. Figure 2 shows the structure of such a model. The whole dance artefact creation process can be divided into several areas such as movement, space, musical, etc. This provides a way of categorizing knowledge elements generated throughout a choreographing process. For the

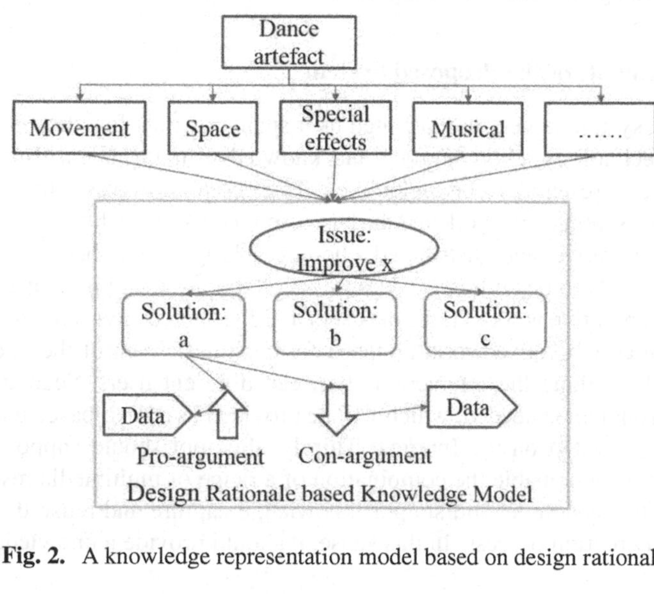

Fig. 2. A knowledge representation model based on design rationale

knowledge elements to be captured using the model, a detailed structure is shown in the figure which involves a diagram-based description using a number of simple elements.

The meanings of these elements are summarized in Table 1 along with some examples to explain their usage. Specifically, the Issue element means an issue to be addressed. This issue can be further divided and thus can have a variety of information granularity. The Solution element refers to a potential solution for the issue and such an element can be created for anyone who has got an idea. The Pro-argument element is used to express a pro opinion for a particular solution while the Con-argument is used to provide evidence in support of the solution. The Data element can be embedded to the diagram in the form of a picture, an equation, a few words or a recording of voice, which is used to provide evidence for a particular argument. There is a freedom for the linking between different elements, which enables a user to freely capture the deliberation and argumentation process.

Table 1. Elements in the design rationale based model

Element	Meaning	Example
Issue	An issue to be addressed	Improve the balance of a gesture
Solution	A potential solution for an issue	Introduce two consecutive movements
Pro-argument	A pro-argument for a solution	Can make a natural transition without losing balance
Con-argument	A con-argument for a solution	Can cause a difficulty for two dancers to coordinate movement
Data	Data as evidence for an argument	A picture showing a previous example

5 A Prototype System and Key Technologies

To evaluate the proposed method and model, a prototype system has been developed. While this research aims to move a first step towards a collaborative and distributed choreographing system, a partial implementation of the system has been completed at this stage. This section introduces the prototype system, describes the key enabling technologies for a full system implementation, and discusses the preliminary evaluation of the system using a real-world example.

5.1 A Prototype System

The prototype system has been designed as a Web-based system to enable multiple users to access the system without the need of installing any client software. It is developed using PHP and HTML5 technologies and the whole system is deployed in an Apache server. Figure 3 shows the function design of the system while Fig. 4 gives a screenshot of the system. As shown in Fig. 3, the system consists of three layers, namely user interface layer, technology layer and resources layer. Specifically, the user interface layer provides interactive operation for various users. The technology layer involves the development of various enabling technologies and the resources layer deals with the coding and storage of information for the system. The graphic user interface is supported by HTML5 which enables rich contents such as graphic information and 3D models. The current version of the system mainly uses this technology for developing the graphic modeler. The system enables multiple users to work on a project. First, multiple users can register on the system and log in to the system to see an individual working space. Second, the system can support the key stages of an artefact creation project. Third, the structure of an artefact creation process knowledge is shown on the interface as a tree. Fourth, a user can create a knowledge element for a specific part of the tree structure and edit its content using a modeling toolkit which implements the design rationale based knowledge representation model. Additionally, more functions are proposed in the design, which will be implemented in future work. These include: (1) modeling toolkit with integration of multimedia contents such as audio clips and videos; (2) a cooperative working engine that supports simultaneous operation of multiple users on the same knowledge element; (3) advanced retrieval methods; and (4) computational methods to provide intelligent support to users.

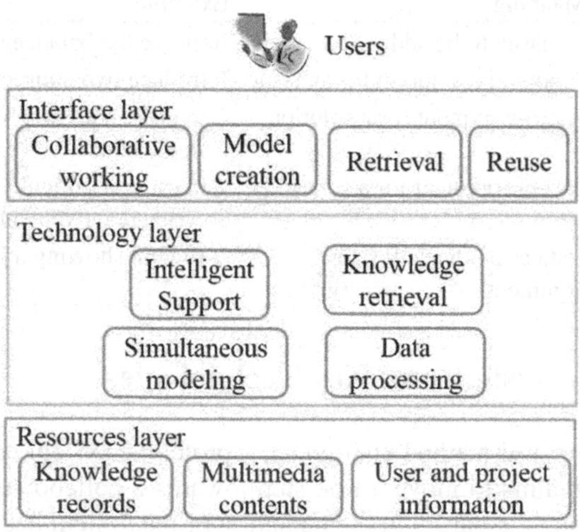

Fig. 3. Function design of the system.

5.2 Key Enabling Technologies

To implement the functions of the system, a number of technologies are needed. First, HTML5 and advanced Web programming technologies are needed to develop rich user interfaces. Development of Web technologies has enabled the development of better graphical user interfaces as well as the display of a variety of information such as 3D model, pictures and videos on these interfaces. Second, a cooperative modeling engine requires management of ownerships of model elements, updating of changes in model, and rolling back of modeling operation events. Ownership of model elements means that at one time only one user can do operations on a model element. Updating of changes should be done as soon as a model element is changed so that other users can see them.

Third, traditional retrieval methods rely on keyword matching. While this method can find matched records, it suffers bad performance in terms of efficiency and precision in particular for the cases with a large number of records. The representation model proposed involved a graphical structure with a variety of information and thus advanced retrieval methods that can exploit the structure and the content of the information will be more useful. To match users' needs of information, computational intelligence methods such as neural network is needed to quickly match the needs to records. The process of data to extract knowledge also requires a lot of data mining techniques.

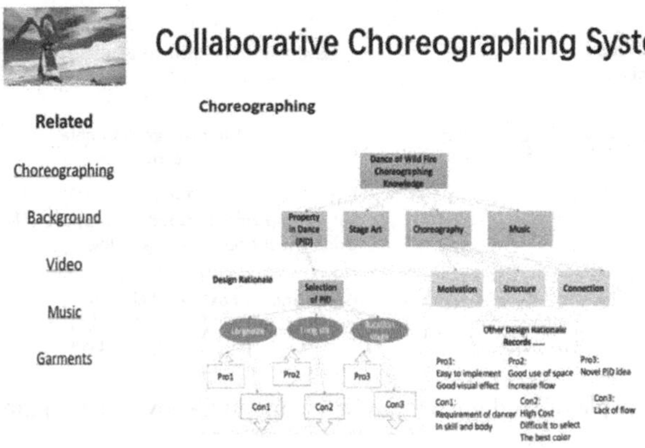

Fig. 4. Screenshots of the prototype system

5.3 Preliminary Evaluation

To evaluate the prototype system, it is applied to a simple example of wild fire dance choreographing. As this work is still at an early stage, this preliminary evaluation aims to find out: (1) whether the proposed system can well support users in artefact creation and (2) whether the proposed knowledge model can effective capture choreographing knowledge. This evaluation involves a number of users, namely a choreographer, a music expert, a stage art expert and a Property in Dance (PiD) expert. The system user interface is shown in Fig. 4. These users can log in to the system with different roles each of which involves different levels of authority. Once a project is created, its key tasks can be defined and linked to different users. These users can create knowledge records using the design rationale based model propose in this work such as the example in Fig. 5.

As shown in Fig. 5, the knowledge record can be constructed by multiple users through creating and editing the elements. The system can also quickly make records of the operations made by these users. Although only a small part of the whole choreographing project is completed in the example, it is shown that the knowledge model provides a clear structure for organizing the tasks of a project and can effectively describe the considerations and communications of the users. The users also confirm that the structure is easy to understand and naturally fit with the purpose of recording the choreographing knowledge. This preliminary evaluation provides evidence that the proposed model can effectively capture knowledge and the prototype system can be accepted by the users with little knowledge about information technology.

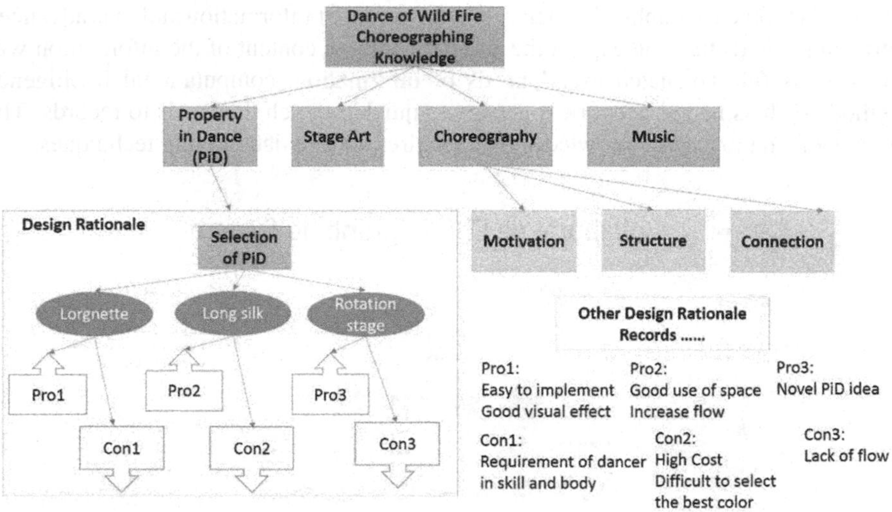

Fig. 5. An example of wild fire dance choreographing knowledge capture

6 Conclusions and Future Work

In this paper, a method for distributed and collaborative choreographing is proposed. A system framework enabled by advanced distributed computing and computational intelligence is developed to identify the process along with interactions between the various users and the system. A knowledge representation model for capturing choreographing knowledge is developed based on design rationale. The methods and model are partially implemented in a prototype system as a first step towards developing a distributed and collaborative choreographing system. The implementation shows that the proposed method is feasible in the sense that a range of enabling technologies will support its full implementation. The proposed knowledge model is useful in describing choreographing knowledge in a distributed and collaborative development process. This process is similar to other design processes in the sense it also involves problem-solving by an integrated team and the elements in design rationale fit with the purpose of choreographing development. Preliminary evaluation of the system in a choreographing example shows that the system is useful for supporting the development process and as well as for capturing knowledge as this process proceeds. A literature search shows that this is a new idea in choreographing and more development will be carried out in our future work.

References

1. Naugle, L.M.: Distributed choreography: a video-conferencing environment. PAJ J. Perform. Art **24**(2), 56–62 (2002)
2. Sheppard, R., et al.: New digital options in geographically distributed dance collaborations with TEEVE: tele-immersive environments for everybody. In: Proceedings of the 15th ACM International Conference on Multimedia, pp. 1085–1086. ACM, New York (2007)
3. Bracewell, R., Wallace, K., Moss, M., Knott, D.: Capturing design rationale. Comput. Aided Des. **41**(3), 173–186 (2009)
4. Wang, H., Johnson, A.L., Bracewell, R.H.: The retrieval of structured design rationale for the re-use of design knowledge with an integrated representation. Adv. Eng. Inf. **26**(2), 251–266 (2012)
5. Qin, H., Wang, H., Johnson, A.L.: A RFBSE model for capturing engineers' useful knowledge and experience during the design process. Robot. Comput. Integr. Manuf. **44**, 30–43 (2017)
6. Peng, G., Wang, H., Zhang, H., Zhao, Y., Johnson, A.L.: A collaborative system for capturing and reusing in-context design knowledge with an integrated representation model. Adv. Eng. Inform. (2017, in press)
7. DeLahunta, S.: Software for dancers: coding forms. Performance Research **7**(2), 97–102 (2002)
8. Karreman, L.: The dance without the dancer. Perform. Res. **18**(5), 120–128 (2013)
9. Naugle, L.: Digital dancing. IEEE Multimedia **5**(4), 8–12 (1998)
10. Risner, D., Anderson, J.: Digital dance literacy: an integrated dance technology curriculum pilot project. Res. Dance Educ. **9**(2), 113–128 (2008)
11. Whatley, S., Varney, R.: Born digital; dance in the digital age. Int. J. Perform. Arts Digital Media **5**(1), 51–63 (2009)
12. Schulze, G.B.: Distributed choreography: a framework to support the design of computer-based artefacts for choreographers with special reference to Brazil. University of Surrey (2005)
13. Bailey, H.: Ersatz dancing: negotiating the live and mediated in digital performance practice. Int. J. Perform. Arts Digital Media **3**(2–3), 151–165 (2007)

Distributed Consensus Control of Multi-USV Systems

Bin Liu, Hai-Tao Zhang(✉), Yue Wu, and Binbin Hu

School of Automation and the State Key Lab of Digital Manufacturing Equipment and Technology, Huazhong University of Science and Technology, Wuhan 430074, People's Republic of China

zht@mail.hust.edu.cn

Abstract. This paper presents a distributed consensus control scheme of unmanned surface vehicle (USV) to achieve speed and orientation consensus with the connected communication topology. To deal with the features including nonlinear, coupling and underactuation, the feedback linearization method is used when design the consensus control. Furthermore, the Lyapunov theory is used to proof the stability of the multi-USV systems. Finally, numerical simulation demonstrates the effectiveness of the distributed consensus control.

Keywords: Unmanned surface vehicle · Multi-agent system · Consensus · Feedback linearization

1 Introduction

In recent years, many researchers devoted to the investigation of swarm intelligence in biological group dynamics [1] such as fish schools, insect colonies, bird flocks, self-driven particles, as well as industrial multi-agent systems (MAS). The swarm intelligence has great potential in unmanned area. As the robustness and adaptability are more strong when more individuals execute a task collaboratively. From the engineering application point of view, collective motion control can greatly improve the performances of multi-robot systems cooperation, smart grid load balancing, unmanned air vehicles (UAV), wireless sensor networks, communication congestion alleviation, etc. USV as a robot can achieve efficient stable performance with the collective motion control in the vast waters.

Multi-agent system theory has been developed several decades. In the literature, several important contributions are taken into account for this area. In 1995, Vicsek et al. [1] proposed a dynamics model for self-propelled particles, where each agent aligns its velocity to their neighbors. By injecting inter-agent repulsive and attractive forces, a more comprehensive three-sphere model was constructed by Couzin et al. in [2] to yield three typical collective behaviors, i.e., flocking, swarming and rotation. To theoretically support the Vicsek model [1], Jadbabaie et al. [3] proposed a joint connectivity condition for proximity

Y. Huang et al. (Eds.): ICIRA 2017, Part I, LNAI 10462, pp. 628–635, 2017.
DOI: 10.1007/978-3-319-65289-4_59

networks of such alignment-based MAS. This condition guarantees the convergence of all agents to consensus. Then a general frame work of the consensus problem for networks of integrator agents with fixed and switching topologies is addressed in [4]. Afterwards, the joint connectivity condition was alleviated by Ren et al. [5] to the existence of a rooted directed spanning tree over consecutive time intervals. Then, Lu and Chen [6] and Yu et al. [7] showed that the results of [8] hold for coupling configuration and cooperative agents with nonlinear dynamics. MAS theory has been a rapid development, and have been achieved in many applications of USV. In 2011, Maclaurin et al. [9] presented a consensus control framework for configuration of two underactuated vehicles. In 2012, K.D. Do [10] presented a desired formation control of underactuated ships with elliptical shape approximation. In 2013, Zhouhua Peng et al. [11] presented a leader-follower formation of autonomous surface vehicles. In 2017, Wenjing Xie et al. [12] presented a new formation control of multiple underactuated surface vessels. Those works greatly promote the multi-USV systems research.

This paper presents a distributed consensus control scheme of unmanned surface vehicle (USV) to achieve speed and orientation consensus with the connected communication topology. Each USV just obtain information on the neighbors, and the distribution characteristics can be achieved. The feedback linearization method and consensus algorithm is used to design the controller to achieve consensus. To confirm the effectiveness, numerical simulation with appropriate parameters is finished. The rest of the paper is organized as follows. In the next section, the dynamic of USV and problem formulation are presented. In Sect. 3, consensus algorithm and feedback algorithm is provided. And the distributed consensus control is designed. In Sect. 4, the stability of presented algorithm is proofed. In Sect. 5, numerical simulation is provided.

Throughout the paper, the following notations will be used: $\mathbb{Z}$, $\mathbb{Z}^+$, $\mathbb{R}$ and $\mathbb{R}^+$ denote the integer, positive integer, real number and positive real number sets, respectively. $\mathbf{0}$ and $\mathbf{I}$ denote zero and unit matrices with compatible dimensions, respectively. $\mathbf{1}_n = [1, 1, \cdots, 1]^{\mathsf{T}}_{n\times 1}$ and $\mathbf{0}_n = [0, 0, \cdots, 0]^{\mathsf{T}}_{n\times 1}$.

2 Problem Formulation

2.1 Multi-agent System

Let $\mathcal{G} = (\mathcal{V}, \mathcal{E}, \mathcal{A})$ be a underacted graph of order n with the set of nodes $\mathcal{V} = \{v_1, \cdots, v_n\}$, set of edges $\mathcal{E} \subseteq \mathcal{V} \times \mathcal{V}$, and a weighted adjacency matrix $\mathcal{A} = [a_{ij}]$ with nonnegative adjacency elements a_{ij}. The node indexes belong to a finite index set $\mathcal{I} = \{1, 2, \cdots, n\}$. An edge of G is denoted by $e_{ij} = (v_i, v_j)$. The adjacency elements associated with the edges of the graph are positive, i.e., $e_{ij} \in \mathcal{E} \Longleftrightarrow a_{ij} > 0$. Moreover, we assume $a_{ii} = 0$ for all $i \in \mathcal{I}$. The set of neighbors of node v_i is denoted by $N_i = \{v_j \in \mathcal{V} : (v_i, v_j) \in \mathcal{E}\}$.

Let $x_i \in \mathbb{R}$ denote the value of node v_i. We refer to $G_x = (G, x)$ with $x = (x_1, \cdots, x_n)^T$ as a network with value $x \in (R)^n$ and topology G. We say nodes of a network have reached a consensus if and only if $x_i = x_j$ for all $i, j \in \mathcal{I}, i \neq j$. Whenever the nodes of a network are all in agreement, the common value of all nodes is called the group decision value. The Laplacian matrix $L \in \mathbb{R}^{n\times n}$ of $\mathcal{G}$ is defined as

$$l_{ij} = \begin{cases} \sum_{k=1,k\neq i} a_{ik}, & j = i \\ -a_{ij}, & j \neq i \end{cases} \tag{1}$$

2.2 USV Dynamic

Since we seek to control the ship motion in the horizontal plane, we neglect the dynamics associated with the motion in heave, roll, and pitch when modeling the ship. Moreover, as a first step toward finding a solution to the achieving consensus problem, we do not include the environmental forces due to wind, currents and waves in the model. Furthermore, we assume that the inertia, added mass and damping matrices are diagonal. In this case, the USV's dynamics described by the differential equation (see [13]).

$$\begin{aligned} \dot{u}_i &= k_1 v_i r_i + k_2 u_i + k_3 \tau_{ui} \\ \dot{v}_i &= k_4 u_i r_i + k_5 v_i \\ \dot{r}_i &= k_6 u_i v_i + k_7 r_i + k_8 \tau_{ri} \\ \dot{x}_i &= u_i \cos\psi_i - v_i \sin\psi_i \\ \dot{y}_i &= u_i \sin\psi_i + v_i \cos\psi_i \\ \dot{\psi}_i &= r_i \end{aligned} \tag{2}$$

where $k_1 = \frac{m_{22}}{m_{11}}$, $k_2 = -\frac{d_{11}}{m_{11}}$, $k_3 = \frac{1}{m_{11}}$, $k_4 = -\frac{m_{11}}{m_{22}}$, $k_5 = -\frac{d_{22}}{m_{22}}$, $k_6 = \frac{m_{11}-m_{22}}{m_{33}}$, $k_7 = -\frac{d_{33}}{m_{33}}$, $k_8 = \frac{1}{m_{33}}$, and u_i, v_i and r_i are the velocities in surge, sway, and yaw, respectively. And x_i, y_i, ψ_i denote the position and orientation of the ship in the earth-fixed frame. The parameters $m_{ii} > 0$ are given by the ship inertia and added mass effects, The parameters $d_{ii} > 0$ are given by the hydrodynamic damping. The available controls are the surge force τ_{ui}, and the yaw moment τ_{ri}.

It's the problem to control all of the ships achieving consensus, indicated as

$$v_{xi} = v_{xj}, \quad v_{yi} = v_{vj}, \quad \psi_i = \psi_j, \tag{3}$$

where $v_{xi} = \dot{x}_i, v_{yi} = \dot{y}_i$ and $i, j \in \mathbb{N}$, $i \neq j$. This means that all of the ship drive to the same direction with the same speed.

In practice, lots USVs should move to a specified completing a task. An important step is that all USVs must run with the same speed and orientation to avoid crash and gall behind. It's significative in military, transportation and detection.

3 Distributed Consensus Scheme

3.1 Feedback Linearization

For system (2), the differential equation

$$\dot{u}_i = k_1 v_i r_i + k_2 u_i + k_3 \tau_{ui},$$

and

$$\dot{r}_i = k_6 u_i v_i + k_7 r_i + k_8 \tau_{ri}$$

satisfy the condition

$$\dot{x} = f(x) + G(x)u \tag{4}$$

where $f(x) = k_1 v_i r_i, G(x) = k_3$, and $f(x) = k_6 u_i v_i, G(x) = k_8$. So using feedback linearization controller, τ_{ui} and τ_{ri} can be designed as

$$\begin{aligned} \tau_{ui} &= \tfrac{1}{k_3}(\tau_{ui}^* - k_1 v_i r_i), \\ \tau_{ri} &= \tfrac{1}{k_8}(\tau_{ri}^* - k_6 u_i v_i). \end{aligned} \tag{5}$$

The system (2), can be transformed as

$$\begin{aligned} \dot{u}_i &= k_2 u_i + \tau_{ui}^* \\ \dot{v}_i &= k_4 u_i r_i + k_5 v_i \\ \dot{r}_i &= k_7 r_i + \tau_{ri}^* \\ \dot{x}_i &= u_i \cos\psi_i - v_i \sin\psi_i \\ \dot{y}_i &= u_i \sin\psi_i + v_i \cos\psi_i \\ \dot{\psi}_i &= r_i \end{aligned} \tag{6}$$

Remark 1. The feedback linearlization method is effective to eliminate the nonlinear term and improve the control performance when the dynamic is precise. This method is useful with the robustness of the system when the dynamic exists a little error.

3.2 Coordinate Transformation

The coordinate diagram is shown as Fig. 1. At first, we need some lemmas.

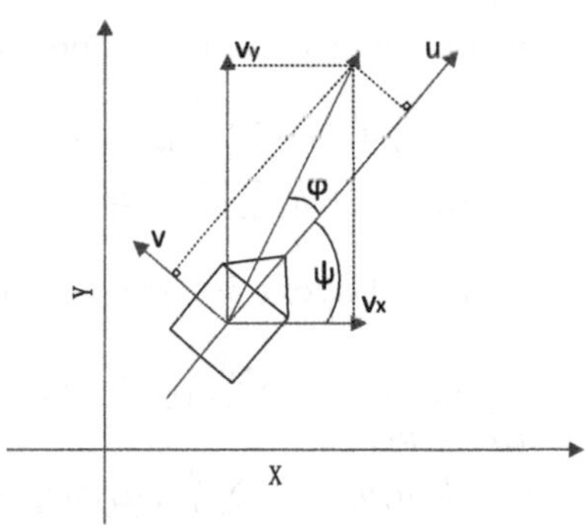

Fig. 1. Boat direction.

Lemma 1. *For system Eq. (2), when v_x and v_y is constant, v and r will convergence to* 0. ∎

Proof. As $v_x = v_{x0}$ and $v_y = v_{y0}$, u, v and ψ will convergence constant u_0, v_0 and ψ_0. It's obvious that $r = \dot{\psi} \to 0$. So

$$\dot{v} = k_5 v,$$

where $k_5 < 0$ and $v \to 0$. This completes the proof. ∎

From Lemma (1), when the system is stable, Eq. 6 gets

$$\begin{aligned} v_{xi} &= u_i \cos\psi_i - v_i \sin\psi_i = u_i \cos\psi_i \\ v_{yi} &= u_i \sin\psi_i + v_i \cos\psi_i = u_i \sin\psi_i. \end{aligned} \tag{7}$$

The consensus condition $v_{xi} = v_{xj}$ and $v_{yi} = v_{yj}$ converts to

$$u_i \cos\psi_i = u_i \cos\psi_j, \quad u_i \sin\psi_i = u_i \sin\psi_j.$$

So the condition Eq. 3 is equivalent to

$$u_i = u_j, \quad \psi_i = \psi_j, \quad v_i = v_j = 0, \quad r_i = r_j = 0, \tag{8}$$

where $i, j \in \mathbb{N}$, $i \neq j$.

3.3 Controller Designed

For the single-integrator modeled agent $\dot{x}_i(t) = u_i(t)$ with fixed or switching topology and zero communication time-delay, the following linear consensus protocol is used (see [4])

$$u_i = \sum_{v_j \in N_i} a_{ij}(x_j - x_i). \tag{9}$$

For multi-USV systems, the controller can be designed as

$$\begin{aligned} \tau_{ui}^* &= \quad k_a u_i + k_b \textstyle\sum_{j=1}^{N_i}(u_j - u_i), \\ \tau_{ri}^* &= -k_4 u_i v_i + k_c \textstyle\sum_{j=1}^{N_i}(\psi_j - \psi_i). \end{aligned} \tag{10}$$

where N_i is the neighbor number of agent i. So the system (6) can indicated as

$$\begin{aligned} \dot{u}_i &= (k_2 + k_a)u_i + k_b \textstyle\sum_{j=1}^{N_i}(u_j - u_i) \\ \dot{v}_i &= k_4 u_i r_i + k_5 v_i \\ \dot{r}_i &= k_7 r_i - k_4 u_i v_i + k_c \textstyle\sum_{j=1}^{N} a_{ij}(\psi_j - \psi_i) \\ \dot{\psi} &= r_i, \end{aligned} \tag{11}$$

where if and only if $u_i = u_j$, $\psi_i = \psi_j$, $v_i = v_j = 0$, $r_i = r_j = 0, \dot{u}_i = 0, \dot{v}_i = 0, \dot{r}_i = 0, \dot{\psi}_i = 0$. This mean that the equilibrium of system (11) is equivalent to the consensus.

Let $U = [u_1 \quad u_2 \cdots u_N]^T$, $\Psi = [\psi_1 \quad \psi_2 \cdots \psi_N]^T, V = [v_1 \quad v_2 \cdots v_N]^T$, $R = [r_1 \quad r_2 \cdots r_N]^T, W = [u_1 r_1 \quad u_2 r_2 \cdots u_N r_N]^T$, $H = [u_1 v_1 \quad u_2 v_2 \cdots u_N v_N]$.

The system can be indicated as

$$\begin{aligned} \dot{U} &= (k_2 + k_a)U - k_b LU \\ \dot{V} &= k_4 W + k_5 V \\ \dot{R} &= k_7 R - k_4 H - k_c L\Psi \\ \dot{\Psi} &= R \end{aligned} \tag{12}$$

4 Theorem Proof

The stability of the system is the mast important features. To proof the stability, we need some lemmas at first.

Lemma 2. *Let $G = (\mathcal{V}, \mathcal{E}, \mathcal{A})$ be a weighted digraph with Laplacian L. If G is strongly connected, then rank$(L) = n - 1$, all the eigenvalues $\lambda_i \geq 0$, and* $\mathbf{1}^T \mathbf{L} = \mathbf{0}$.

Proof. See [4].

Theorem 1. *When $\mathbb{G}(q)$ is connected, system (12) will stable and achieve consensus with the controller (10) and (5) when $k_a \leq -k_2, k_b > 0, k_c > 0$.* ■

Proof. Designed the Lyapunov function as

$$V = \frac{1}{2}(U^T LU + k_c \Psi^T L \Psi + V^T V + R^T R) \tag{13}$$

where L is Laplacian matrix and $k_c > 0$. $V \geq 0$ and if and only if the system has reached consensus $V = 0$. Basing on Lyapunov theory, to proof the stability of system, we just need obtain $\dot{v} < 0$ except in equilibrium point.

$$\begin{aligned} \dot{V} &= U^T L\dot{U} + k_c \Psi^T L \dot{\Psi} + V^T \dot{V} + R^T \dot{R} \\ &= U^T L((k_2 + k_a)U - k_b LU) + k_c \Psi^T LR + V^T (k_4 W + k_5 V) \\ &\quad + R^T (k_7 R - k_4 H - k_c L\Psi) \\ &= (k_2 + k_a)U^T LU - k_b U^T LLU + k_c \Psi^T LR + k_4 V^T W + k_5 V^T V \\ &\quad + k_7 R^T R - k_4 R^T H - k_c R^T L \Psi \end{aligned} \tag{14}$$

where $\Psi^T LR = R^T L \Psi$ and $k_4 V^T W = k_4 R^T H$. So

$$\dot{V} = (k_2 + k_a)U^T LU - k_b U^T LLU + k_5 V^T V + k_7 R^T R.$$

According to Lemma 2, $\lambda_i \geq 0$ and $\mathbf{1}^T \mathbf{L} = \mathbf{0}$. So if and only if in the equilibrium point, $V = 0$ and $\dot{V} = 0$. So system (2) is stable. ■

5 A Numerical Example

In this section, we finish some computer simulations to verify the effectiveness of the scheme (10). For system (2), consider the parameters as in $m_{11} = 1.956 \pm 0.019, m_{22} = 2.405 \pm 0.117, m_{33} = 0.043 \pm 0.0068, d_{11} = 2.436 \pm 0.023, d_{22} = 12.992 \pm 0.297, d_{33} = 0.0564 \pm 0.00085$ (see [14]). Choosing the parameters as $k_a = 1.2454, k_b = 10$ and $k_c = 30$, and the Laplacian matrix as $\mathcal{A} = [0,1,0,1,0;1,0,0,0,0;0,0,0,1,1;1,0,1,0,0;0,0,1,0,0]$, the system (12) will be stable and achieve consensus.

Remark 2. To guarantee the surge speed not being convergent to zero or divergent to infinite, we set $k_a = -k_2 = 1.2454$. The effect of k_b and k_c is to control the rate of convergence of surge speed and orientation. Here, we choose appropriate convergence rate, and we let $k_b = 10, k_c = 30$.

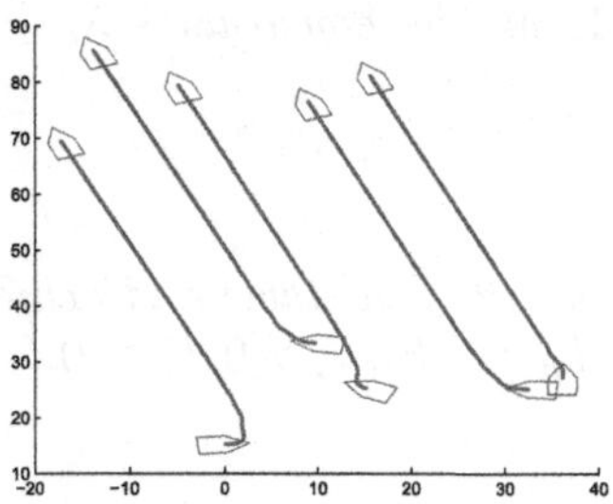

Fig. 2. The process of consensus

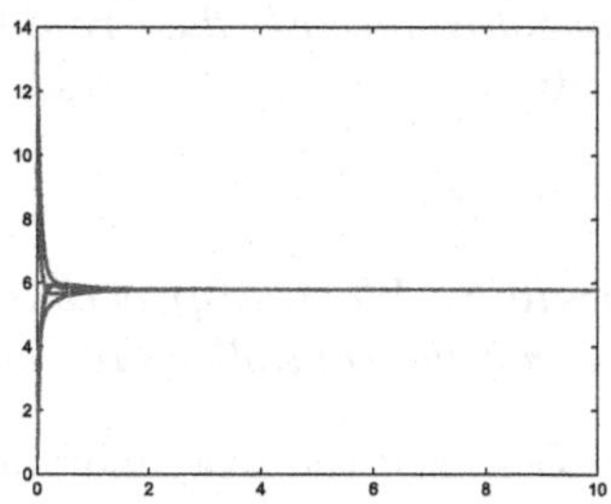

Fig. 3. The surge speed

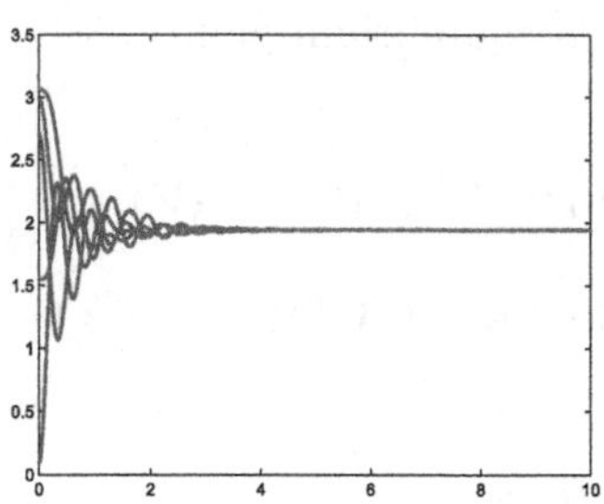

Fig. 4. The orientation

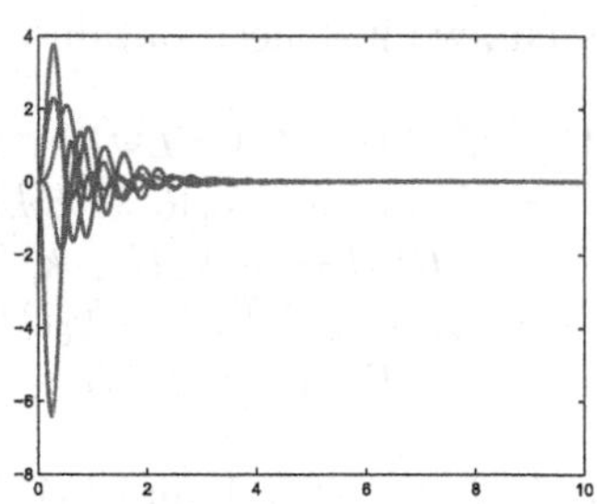

Fig. 5. The sway speed

The process of consensus is shown as Fig. 2 and surge speed, orientation, sway speed are shown as Figs. 3, 4 and 5 respectively. From Figs. 2, 3, 4 and 5, the orientation and speed of the USV are random at first. After a period of time, the orientation and speed tended to consensus. And in Fig. 2, it's obvious that all boat run with the same orientation and speed at last.

6 Conclusion

This paper presents a distributed consensus control scheme of multi-USV systems with the connected communication topology. Here, we use feedback linearization method to solve the nonlinear, coupling and underactuation features. And the simulations verify the effectiveness of the presented scheme. The further work is to considering the uncertain part of the system such as the disturbance from wind, wave and flow and so on.

References

1. Vicsek, T., Czirók, A., Ben-Jacob, E., Cohen, I., Shochet, O.: Novel type of phase transition in a system of self-driven particles. Phys. Rev. Lett. **75**(6), 1226–1229 (1995)
2. Couzin, I.D., Krause, J., James, R., Ruxton, G.D., Franks, N.R.: Collective memory and spatial sorting in animal groups. J. Theor. Biol. **218**, 1–11 (2002)
3. Jadbabaie, A., Lin, J., Morse, A.S.: Coordination of groups of mobile agents using nearest neighbor rules. IEEE Trans. Autom. Control **48**(6), 988–1001 (2003)
4. Olfati-Saber, R., Murray, R.: Consensus problems in networks of agents with switching topology and time-delays. IEEE Trans. Autom. Control **49**(9), 1520–1533 (2004)
5. Ren, W., Beard, R.W., Arkins, E.M.: Information consensus in multivehicle cooperative control. IEEE Control Syst. Mag. **71**(2), 71–82 (2007)
6. Lu, J., Chen, G.: A time-varying complex dynamical network model and its controlled synchronization criteria. IEEE Trans. Autom. Control **50**(6), 841–846 (2005)
7. Yu, W., Chen, G., Cao, M.: Consensus in directed networks of agents with nonlinear dynamics. IEEE Trans. Autom. Control **56**(6), 1436–1441 (2011)
8. Ren, W., Beard, R.W.: Consensus seeking in multiagents systems under dynamically changing interaction topologies. IEEE Trans. Autom. Control **50**(5), 655–661 (2005)
9. Hutagalung, M., Hayakawa, T.: Consensus seeking in multiagents systems under dynamically changing interaction topologies. Preprints of the 18th IFAC World Congress Milano (Italy), 28 August–2 September 2011
10. Do, K.D.: Formation control of underactuated ships with elliptical shape approximation and limited communication ranges. Automatica **48**, 1380–1388 (2012)
11. Peng, Z., Wang, D., Chen, Z., Hu, X., Lan, W.: Adaptive dynamic surface control for formations of autonomous surface vehicles with uncertain dynamics. IEEE Trans. Control Syst. Technol. **21**(2), 513–520 (2013)
12. Xie, W., Ma, B., Fernando, T., Iu, H.-H.C.: Adaptive dynamic surface control for formations of autonomous surface vehicles with uncertain dynamics. Int. J. Control (2017)
13. Fossen, T.I.: Guidance and Control of Ocean Vehicles. Wiley, New York (1994)
14. Ashrafiuon, H., Muske, K.: Sliding mode tracking control of surface vessels. In: Proceedings American Control Conference, Seattle, Washington, USA, pp. 556–561 (2008)

Distributed Event-Triggered Consensus Control of Neutrally Stable Linear Multi-agent Systems

Bin Cheng and Zhongkui Li(✉)

State Key Laboratory for Turbulence and Complex Systems,
Department of Mechanics and Aerospace Engineering,
College of Engineering, Peking University, Beijing 100871, China
{bincheng,zhongkli}@pku.edu.cn

Abstract. This paper considers the distributed event-triggered consensus control problem of neutrally stable linear multi-agent systems. We design distributed event-triggered protocols, consisting of consensus control laws and triggering functions, under which consensus is achieved and the Zeno behavior can be excluded. Compared to the previous related works, our main contribution is that the proposed protocols are fully distributed and scalable, whose design does not rely on any global information of the network graph. Meanwhile, continuous communications are not required for either control laws updating or triggering functions monitoring so as to reduce the communication frequency.

Keywords: Neutrally stable agents · Consensus · Distributed control · Event-based control

1 Introduction

The consensus problem of multi-agent systems has attracted increasing attention in the recent years. Many researchers have devoted their efforts to studying the consensus problem, due to its broad applications in various areas, including flocking, formation control, and distributed sensor networks; see the works [1–7]. To achieve consensus, communications among the agents in these works are required to be carried out at each time for continuous-time algorithms or at all iterations for discrete-time algorithms. However, when the signal does not change significantly, sampling the signal and transmitting the samples over the communication channels are evidently wasting energy. In practice, the bandwidth of the communication network and the power source of the agents are inevitably constrained, because many practical systems have becoming increasingly networked, wireless, and spatially distributed. Thus, one recent hot research topic is how to reduce the communication frequency, with a tradeoff kept between control performance and resource utilization.

In order to reduce the load of the networks, event-based control is currently being developed as an important means for avoiding continuous communications. Significant research efforts were initiated by a few important research

Y. Huang et al. (Eds.): ICIRA 2017, Part I, LNAI 10462, pp. 636–647, 2017.
DOI: 10.1007/978-3-319-65289-4_60

works [9,10]. The development of event-based control claims the reformulation of answers as to when to sample, when to update controller, or when to transmit information. Based on the event-triggered control, communication is not required unless it is really necessary in system operation. The event-triggered control protocol is composed of two components: the consensus control laws and the triggering functions, which are used for, respectively, regulating the performance of system and reducing the communication frequency. For single-integrator agents, periodic event-triggered consensus algorithm was studied in [14] and a self-triggered control algorithm was proposed in [10]. Decentralized event-based consensus protocols were designed for single- and double-integrator multi-agent systems in [12]. Event-based consensus problem of general linear multi-agent systems was studied in [8,13,15–19]. However, continuous information still has to be used in [10,19] to judge whether or not the event-triggering functions are violated or not. Nonzero eigenvalues of the Laplacian matrix associated with the communication graph, which are actually global information of the network, are required to determine some parameters of the protocols in [8,10,12,14,15,18]. How to design fully event-triggered consensus protocols for multi-agent systems is still a challenging problem to be solved.

Motivated by the aforementioned discussions, in this paper we consider the distributed event-triggered consensus problem for neutrally stable linear multi-agent systems. We propose event-based consensus protocols, under which consensus is achieved and the Zeno behavior can be excluded. Here are our main contributions. (i) The proposed protocols are fully distributed, whose design does not require any global information, such as the nonzero eigenvalue information of the Laplacian matrix of the communication graph. (ii) Based on the event-triggered strategy, communications are not required for either control laws updating or triggering functions monitoring. When use these protocols, we can attain the objective of saving the limited resources of communication capacity and energy. (iii) We utilize a Lyapunov function approach to solve the distributed event-triggered consensus problem. The advantage of this approach is that we can design the triggering functions according to the proof of the consensus, as long as the Zeno behavior can be excluded. (iv) In our theoretical framework, the distributed event-triggered consensus problem of single integrators or the harmonic oscillators can be easily solved just as special cases.

The rest of this paper is organized as follows. The problem statement is given in Sect. 2. In Sect. 3, we introduce our main results. Finally, Sect. 4 concludes this paper.

Notations: Let $\mathbf{R}^{m\times n}$ be the set of $m \times n$ real matrices, $\mathbf{1}_n$ and $\mathbf{0}_n$ denote a $n \times 1$ column vector with the all entries equal to 1 and 0, respectively. I_p represents the identity matrix of dimension p and $\text{diag}\{a_1, \cdots, a_n\}$ represents a diagonal matrix with elements $a_i, i = 1, \cdots, n$, on its diagonal. For a vector $x \in \mathbf{R}^n$, let $\|x\|$ denotes its 2-norm. For a symmetric matrix A, $\lambda_{\min}(A)$ and $\lambda_{\max}(A)$ denote, respectively, the minimum and maximum eigenvalues of A. The Kronecker product of matrices A and B is denoted by $A \otimes B$.

2 Problem Statement

Consider a team of N identical agents with continuous-time linear dynamics. The dynamics of the i-th agent are described by

$$\dot{x}_i = Ax_i + Bu_i, i = 1, \cdots, N, \tag{1}$$

where $x_i \in \mathbf{R}^n$ and $u_i \in \mathbf{R}^p$ denote the state and the control input of the i-th agent, respectively, A and B are constant matrices with compatible dimensions.

To solve the distributed consensus problem, the agents need to exchange information. Let the information flow among the N agents be represented by an undirected graph $\mathcal{G} = (\mathcal{V}, \mathcal{E})$, where $\mathcal{V} = \{v_1, \cdots, v_N\}$ is the node set and $\mathcal{E} \subseteq \mathcal{V} \times \mathcal{V}$ is the edge set, in which an edge is represented by an unordered pair of distinct nodes. Because $\mathcal{G}$ is undirected, $(v_i, v_j) \in \mathcal{E} \Longleftrightarrow (v_j, v_i) \in \mathcal{E}$. If $(v_i, v_j) \in \mathcal{E}$, node v_i is called a neighbor of node v_j. A path from node v_{i_1} to node v_{i_l} is a sequence of adjacent edges of the form $(v_{i_k}, v_{i_{k+1}})$, $k = 1, \cdots, l-1$. An undirected graph is connected if there exists a path between every pair of distinct nodes, otherwise is disconnected. The adjacency matrix associated with $\mathcal{G}$ is represented by $\mathcal{A} = [a_{ij}] \in \mathbf{R}^{N \times N}$, where $a_{ii} = 0$, $a_{ij} = 1$, if $(v_j, v_i) \in \mathcal{E}$, and $a_{ij} = 0$ otherwise. The Laplacian matrix $\mathcal{L} = [l_{ij}] \in \mathbf{R}^{N \times N}$ is defined as $l_{ii} = \sum_{j=1, j \neq i}^{N} a_{ij}$ and $l_{ij} = -a_{ij}, i \neq j$. The degree of agent i is defined as $d_i = l_{ii}$.

Assumption 1. The graph $\mathcal{G}$ associated with the N agent is undirected and connected.

Lemma 1 [7]. *For an undirected graph $\mathcal{G}$, 0 is a simple eigenvalue of $\mathcal{L}$ if and only if $\mathcal{G}$ is connected. The smallest nonzero eigenvalue $\lambda_2(\mathcal{L})$ of $\mathcal{L}$ satisfies*

$$\lambda_2(\mathcal{L}) = \min_{x \neq 0, \mathbf{1}^T x = 0} \frac{x^T \mathcal{L} x}{x^T x}.$$

The objective of this paper is to solve the distributed event-triggered consensus problem for the agents in (1), i.e., to design distributed event-based protocols to ensure that the states of the N agents reach agreement in the sense of $\lim_{t \to \infty} \|x_i(t) - x_j(t)\| = 0$, $\forall i, j = 1, \cdots, N$, and the Zeno behavior can be excluded, i.e., there is no trajectory of the system with an infinite number of events with a finite period of time [8].

3 Main Results

In this paper, we will consider the distributed event-triggered consensus of multi-agent systems, satisfying the following assumption:

Assumption 2. The matrix A is neutrally stable[1].

[1] A matrix $A \in \mathbf{C}^{n \times n}$ is neutrally stable in the continuous-time sense if it has no eigenvalue with positive real part and the Jordan block corresponding to any eigenvalue on the imaginary axis is of size one.

Let $\tilde{x}_i(t) \triangleq e^{A(t-t_k^i)}x_i(t_k^i)$, $\forall t \in [t_k^i, t_{k+1}^i)$, $i = 1, \cdots, N$. For each agent, we define a measurement error $e_i(t)$ using the state error between the current time t and the latest event time t_k^i, i.e.,

$$e_i(t) \triangleq \tilde{x}_i(t) - x_i(t), i = 1, \cdots, N. \tag{2}$$

Based on the relative states of neighboring agents, the following distributed event-triggered control law is proposed for each agent:

$$u_i(t) = cK \sum_{j=1}^{N} a_{ij}(\tilde{x}_i(t) - \tilde{x}_j(t)), i = 1, \cdots, N, \tag{3}$$

where $c \in \mathbf{R}$ is a positive constant coupling gain, $K \in \mathbf{R}^{p\times n}$ is the feedback gain matrix, and a_{ij} denotes the (i, j)-th entry of adjacency matrix associated with the graph $\mathcal{G}$.

The triggering function for each agent i is given by

$$f_i(t) = 4d_i\|Ke_i\|^2 - \delta \sum_{j=1}^{N} \|K(\tilde{x}_i - \tilde{x}_j)\|^2 - \mu e^{-\nu t} \leq 0, \tag{4}$$

where d_i is the degree of agent i, δ, μ, and ν are positive constants. As long as the triggering condition $f_i(t) \geq 0$ is fulfilled, an event is triggered for agent i. At this time instant, agent i updates its consensus control law (3) using its current state to its neighbors. Meanwhile, the measurement error $e_i(t)$ is reset to zero. When agent i receives new information from its neighbors, the control law for agent i will also be updated immediately. If $f_i(t) < 0$ for $\forall i = 1, \cdots, N$, there is no communication occurring until the next event is triggered.

Define $\xi = \begin{bmatrix} \xi_1 \\ \vdots \\ \xi_N \end{bmatrix} \triangleq \begin{bmatrix} x_1 - \frac{1}{N}\sum_{j=1}^N x_j \\ \vdots \\ x_N - \frac{1}{N}\sum_{j=1}^N x_j \end{bmatrix}$ and $\tilde{\xi} = \begin{bmatrix} \tilde{\xi}_1 \\ \vdots \\ \tilde{\xi}_N \end{bmatrix} \triangleq \begin{bmatrix} \tilde{x}_1 - \frac{1}{N}\sum_{j=1}^N \tilde{x}_j \\ \vdots \\ \tilde{x}_N - \frac{1}{N}\sum_{j=1}^N \tilde{x}_j \end{bmatrix}$.

Then, we can get that

$$\xi = (M \otimes I_n)x, \tag{5-1}$$

$$\tilde{\xi} = (M \otimes I_n)\tilde{x}, \tag{5-2}$$

where $M = I_N - \frac{1}{N}\mathbf{1}\mathbf{1}^T$. It is clear that 0 is a simple eigenvalue of M with $\mathbf{1}$ as the corresponding right eigenvector, and 1 is the other eigenvalue with multiplicity $N-1$. Then, it follows that $\xi = 0$ if and only if $x_1 = \cdots = x_N$. Thus, we refer to ξ as the consensus error. Using (3) for (1), it follows that ξ satisfies the following closed-loop network dynamics:

$$\dot{\xi} = (I_N \otimes A)\xi + (c\mathcal{L} \otimes BK)\tilde{\xi}. \tag{6}$$

Before moving forward, we introduce the following lemmas.

Lemma 2 [7]. *A complex matrix $A \in \mathbf{C}^{n\times n}$ is Hurwitz if and only if there exist a positive-definite matrix $Q = Q^H$ and a matrix $B \in \mathbf{C}^{n\times m}$ such that (A, B) is controllable and $A^HQ + QA = -BB^H$.*

Lemma 3 [7]. *For matrices $S \in \mathbf{R}^{n\times n}$ and $H \in \mathbf{R}^{n\times m}$, where S is skew-symmetric and (S, H) is controllable, the matrix $S - (x + \iota y)HH^T$ is Hurwitz for all $x > 0$ and $y \in \mathbf{R}$.*

From Lemmas 2 and 3, we can choose $U \in \mathbf{R}^{n_1\times n}$ and $W \in \mathbf{R}^{(n-n_1)\times n}$ such that

$$\begin{bmatrix} U \\ W \end{bmatrix} A \begin{bmatrix} U \\ W \end{bmatrix}^{-1} = \begin{bmatrix} S & 0 \\ 0 & X \end{bmatrix}, \tag{7}$$

where $S \in \mathbf{R}^{n_1\times n_1}$ is skew-symmetric and $X \in \mathbf{R}^{(n-n_1)\times(n-n_1)}$ is Hurwitz.

Let $z = \left(I_N \otimes \begin{bmatrix} U \\ W \end{bmatrix}\right)\xi$ and $\tilde{z} = \left(I_N \otimes \begin{bmatrix} U \\ W \end{bmatrix}\right)\tilde{\xi}$. Then the dynamics of z can be obtained as

$$\dot{z} = \left(I_N \otimes \begin{bmatrix} S & 0 \\ 0 & X \end{bmatrix}\right) z + \left(c\mathcal{L} \otimes \begin{bmatrix} U \\ W \end{bmatrix} BK \begin{bmatrix} U \\ W \end{bmatrix}^{-1}\right)\tilde{z}, \tag{8}$$

Let $H = UB$. Because (A, B) is stabilizable, it is easy to see that (S, H) is controllable by using Lemma 1. Let $U^+ \in \mathbf{R}^{n\times n_1}$ and $W^+ \in \mathbf{R}^{n\times(n-n_1)}$ be such that $\begin{bmatrix} U^+ & W^+ \end{bmatrix} = \begin{bmatrix} U \\ W \end{bmatrix}^{-1}$, where $UU^+ = I$, $WW^+ = I$, $WU^+ = 0$, and $UW^+ = 0$. By choosing $K = -B^TU^TU$, then we can get

$$\dot{z} = \left(I_N \otimes \begin{bmatrix} S & 0 \\ 0 & X \end{bmatrix}\right) z - \left(c\mathcal{L} \otimes \begin{bmatrix} UBB^TU^T & 0 \\ WBB^TU^T & 0 \end{bmatrix}\right)\tilde{z}, \tag{9}$$

Define $z_I = (I_N \otimes U)\xi$, $\tilde{z}_I = (I_N \otimes U)\tilde{\xi}$, $z_{II} = (I_N \otimes W)\xi$ and $\tilde{z}_{II} = (I_N \otimes W)\tilde{\xi}$. Then, (9) can be rewritten as

$$\dot{z}_I = (I_N \otimes S)z_I - (c\mathcal{L} \otimes HH^T)\tilde{z}_I, \tag{10-1}$$

$$\dot{z}_{II} = (I_N \otimes X)z_{II} - (c\mathcal{L} \otimes WBB^TU^T)\tilde{z}_I. \tag{10-2}$$

Here, we introduce two lemmas to be used later.

Lemma 4 [20]. *If x and $\dot{x}$ are bounded, and $\int_0^{+\infty} x^T(\tau)x(\tau)d\tau < +\infty$, then $x(t) \to 0$ as $t \to +\infty$.*

Lemma 5 (Young's Inequality, [21]). *If a and b are nonnegative real numbers and p and q are positive real numbers such that $\frac{1}{p} + \frac{1}{q} = 1$, then $ab \le \frac{a^p}{p} + \frac{b^q}{q}$.*

The following theorem designs the distributed event-triggered consensus protocol.

Theorem 1. *Suppose that Assumptions 1 and 2 hold. The distributed consensus problem of the agents described by (1) is solved under the controllers (3) and the triggering conditions (4) with $c > 0$, $0 < \delta < 1$, and $K = -B^TU^TU$. Furthermore, the Zeno behavior can be excluded.*

Proof. Let

$$V_1 = \frac{1}{2} z_I^T z_I. \tag{11}$$

The time derivative of V_1 along the trajectory of (10-1) can be obtained as

$$\dot{V}_1 = \frac{1}{2} z_I^T [I_N \otimes (S + S^T)] z_I - z_I^T (c\mathcal{L} \otimes HH^T) \tilde{z}_I. \tag{12}$$

Because S is skew-symmetric, it is easy to get that $z_I^T [I_N \otimes (S + S^T)] z_I = 0$. From the definition of z_I, we can get that $(\mathbf{1}^T \otimes I) z_I = 0$. Because $\mathcal{G}$ is connected, it then follows from Lemma 1 that $z_I^T (c\mathcal{L} \otimes HH^T) z_I \geq c\lambda_2(\mathcal{L}) z_I^T (I_N \otimes RR^T) z_I$. Then, we have

$$\begin{aligned}
\dot{V}_1 &= -\frac{1}{2} z_I^T (c\mathcal{L} \otimes HH^T) z_I - \frac{1}{2} \tilde{z}_I^T (c\mathcal{L} \otimes HH^T) \tilde{z}_I + \frac{1}{2} e^T (c\mathcal{L} \otimes K^T K) e \\
&\leq -\frac{c\lambda_2(\mathcal{L})}{2} z_I^T (I_N \otimes HH^T) z_I + \frac{c}{2} [e^T (\mathcal{L} \otimes K^T K) e - \tilde{x}^T (\mathcal{L} \otimes K^T K) \tilde{x}].
\end{aligned} \tag{13}$$

Let

$$V_2 = \frac{1}{2} z_{II}^T (I_N \otimes P) z_{II}, \tag{14}$$

where P is a solution of the following ARE:

$$PX + X^T P + 2Q = 0, \tag{15}$$

where Q is a positive-definite matrix. The time derivative of V_2 along the trajectory of (10-2) can be obtained as

$$\dot{V}_2 = \frac{1}{2} z_{II}^T [I_N \otimes (PX + X^T P)] z_{II} - z_{II}^T (c\mathcal{L} \otimes PWBB^T U^T) \tilde{z}_I. \tag{16}$$

Using Lemma 5 gives

$$\begin{aligned}
&- z_{II}^T (c\mathcal{L} \otimes PWBB^T U^T) \tilde{z}_I \\
&\leq \frac{1}{2} z_{II}^T (I_N \otimes Q) z_{II} + \frac{c^2 \lambda_N(\mathcal{L})}{2\lambda_{\min}(Q)} \tilde{z}_I^T (\mathcal{L} \otimes UBB^T W^T PPWBB^T U^T) \tilde{z}_I \\
&\leq \frac{1}{2} z_{II}^T (I_N \otimes Q) z_{II} + \frac{c\alpha_1}{2} \tilde{x}^T (\mathcal{L} \otimes K^T K) \tilde{x},
\end{aligned} \tag{17}$$

where $\alpha_1 = \frac{c\lambda_N(\mathcal{L}) \|PWB\|^2}{\lambda_{\min}(Q)}$, with $\lambda_N(\mathcal{L})$ being the largest eigenvalue of $\mathcal{L}$.

Consider the following Lyapunov function candidate:

$$V = \frac{\alpha_1}{1 - \delta} V_1 + V_2. \tag{18}$$

Evidently, V is positive definite.

$$\begin{aligned}\dot{V} &\leq \frac{c\alpha_1}{2(1-\delta)}[-\lambda_2(\mathcal{L})z_I^T(I_N\otimes HH^T)z_I + e^T(\mathcal{L}\otimes K^TK)e - \tilde{x}^T(\mathcal{L}\otimes K^TK)\tilde{x}]\\ &\quad + \frac{1}{2}z_{II}^T[I_N\otimes(PX+X^TP+Q)]z_{II} + \frac{c\alpha_1}{2}\tilde{x}^T(\mathcal{L}\otimes K^TK)\tilde{x}\\ &\leq -\alpha_2 z_I^T(I_N\otimes HH^T)z_I - \frac{1}{2}z_{II}^T(I_N\otimes Q)z_{II}\\ &\quad + \alpha_3[e^T(\mathcal{L}\otimes K^TK)e - \delta\tilde{x}^T(\mathcal{L}\otimes K^TK)\tilde{x}]\end{aligned} \tag{19}$$

where $\alpha_2 = \frac{c\alpha_1\lambda_2(\mathcal{L})}{2(1-\delta)}$ and $\alpha_3 = \frac{c\alpha_1}{2(1-\delta)}$.

Because $a_{ij} = a_{ji}$, it is not difficult to verify that

$$\begin{aligned}e^T(\mathcal{L}\otimes K^TK)e &= \sum_{i=1}^N\sum_{j=1}^N a_{ij}e_i^TK^TK(e_i-e_j)\\ &\leq 2\sum_{i=1}^N\sum_{j=1}^N a_{ij}e_i^TK^TKe_i\\ &= 2\sum_{i=1}^N d_ie_i^TK^TKe_i\\ &= 2\sum_{i=1}^N d_i\|Ke_i\|^2,\end{aligned} \tag{20}$$

and

$$\begin{aligned}\tilde{x}^T(\mathcal{L}\otimes K^TK)\tilde{x} &= \sum_{i=1}^N\sum_{j=1}^N a_{ij}\tilde{x}_i^TK^TK(\tilde{x}_i-\tilde{x}_j)\\ &= \frac{1}{2}\sum_{i=1}^N\sum_{j=1}^N a_{ij}(\tilde{x}_i-\tilde{x}_j)^TK^TK(\tilde{x}_i-\tilde{x}_j)\\ &= \frac{1}{2}\sum_{i=1}^N\sum_{j=1}^N a_{ij}\|K(\tilde{x}_i-\tilde{x}_j)\|^2.\end{aligned} \tag{21}$$

By substituting (20) and (21) into (19), we have

$$\begin{aligned}\dot{V} &\leq -\alpha_2 z_I^T(I_N\otimes HH^T)z_I - \frac{1}{2}z_{II}^T(I_N\otimes Q)z_{II}\\ &\quad + \frac{\alpha_3}{2}\sum_{i=1}^N\{4d_i\|Ke_i\|^2 - \delta\sum_{j=1}^N a_{ij}\|K(\tilde{x}_i-\tilde{x}_j)\|^2\}\\ &\leq -\alpha_2 z_I^T(I_N\otimes HH^T)z_I - \frac{1}{2}z_{II}^T(I_N\otimes Q)z_{II} + \frac{\alpha_3}{2}N\mu e^{-\nu t},\end{aligned} \tag{22}$$

where we have substituted the event-triggered functions (4) into (22) to get the last inequality. From (22), we can get that

$$0 \leq V(t) \leq \frac{\alpha_3}{2} N\mu \int_0^t e^{-\nu\tau} d\tau. \tag{23}$$

Thus, V is bounded. Invoking (18), z_I and z_{II} are all bounded, which further implies that $\dot{z}_I$ and $\dot{z}_{II}$ are bounded according to (10). Besides, (22) enforces that

$$\begin{aligned} V(\infty) - V(0) \leq & -\alpha_2 \int_0^\infty z_I^T(\tau)(I_N \otimes HH^T) z_I(\tau) d\tau \\ & - \frac{1}{2} \int_0^\infty z_{II}^T(\tau)(I_N \otimes Q) z_{II}(\tau) d\tau + \frac{\alpha_3}{2} N\mu \int_0^\infty e^{-\nu\tau} d\tau. \end{aligned} \tag{24}$$

Note that (24) can be rewritten as

$$\int_0^\infty z_I^T(\tau)(I_N \otimes HH^T) z_I(\tau) d\tau \leq \frac{1}{\alpha_2}[V(0) - V(\infty) + \frac{\alpha_3}{2\nu} N\mu], \tag{25-1}$$

$$\int_0^\infty z_{II}^T(\tau)(I_N \otimes Q) z_{II}(\tau) d\tau \leq 2[V(0) - V(\infty) + \frac{\alpha_3}{2\nu} N\mu]. \tag{25-2}$$

Thus, using Lemma 4 yields that $z_I(t) \to 0$ and $z_{II}(t) \to 0$ as $t \to \infty$. That is, the consensus can be achieved.

Next, we will show that the closed system (6) does not exhibit the Zeno behavior. Since $e_i(t_k^i) = 0$, i.e., $\|Ke_i(t_k^i)\| = 0$, when agent i is triggered. From the triggering function (4), the $(k+1)$-th triggering time instant for agent i is defined as the first time when $\|Ke_i(t)\|$ exceeds $\sqrt{\frac{1}{4d_i}\left(\sum_{j=1}^N a_{ij}\|K(\tilde{x}_i(t) - \tilde{x}_j(t))\|^2 + \mu e^{-\nu t}\right)}$. In other words, the time interval between the k-th and $(k+1)$-th triggering time instants is not less than the time for $\|Ke_i(t)\|$ to increase from 0 to $\sqrt{\frac{1}{4d_i}\left(\sum_{j=1}^N a_{ij}\|K(\tilde{x}_i(t) - \tilde{x}_j(t))\|^2 + \mu e^{-\nu t}\right)}$.

The following two cases are considered.

(i) In the first case, there are no neighbors of agent i being triggered during the interval between the k-th and the $(k+1)$-th events of agent i. It follows from (2) and (3) that the right-hand Dini derivative of e_i can be written as

$$D^+ e_i(t) = Ae_i(t) - cBK \sum_{j=1}^N a_{ij}(\tilde{x}_i(t) - \tilde{x}_j(t)). \tag{26}$$

By noting $e_i(t_k^i) = 0$, the solution of Eq. (26) can be obtained as

$$e_i(t) = -\int_{t_k^i}^t e^{A(t-\tau)} cBK \sum_{j=1}^N a_{ij}[e^{A(\tau - t_k^i)} x_i(t_k^i) - e^{A(\tau - t_m^j)} x_j(t_m^j)] d\tau. \tag{27}$$

Then, we then have

$$\begin{aligned} \|Ke_i(t)\| &\leq c\|K\|\|BK\| \left\| \int_{t_k^i}^{t} \sum_{j=1}^{N} a_{ij}[\tilde{x}_i(t) - \tilde{x}_j(t)] d\tau \right\| \\ &\leq (t - t_k^i) c\|K\|\|BK\| \sum_{j=1}^{N} a_{ij} \|\tilde{x}_i(t) - \tilde{x}_j(t)\|. \end{aligned} \tag{28}$$

By the triggering condition (4), the agent i will not be triggered until the following equation holds

$$\begin{aligned} &(t - t_k^i) c\|K\|\|BK\| \sum_{j \in \mathcal{N}_i} \|\tilde{x}_i(t) - \tilde{x}_j(t)\| \\ &= \sqrt{\frac{1}{4d_i} \left(\sum_{j=1}^{N} a_{ij} \|K(\tilde{x}_i(t) - \tilde{x}_j(t))\|^2 + \mu e^{-\nu t} \right)}. \end{aligned} \tag{29}$$

Note that the right-hand side of (29) is positive, whenever consensus is not yet achieved. In this case, $t - t_k^i$ must be positive such that (29) holds. This implies that the event intervals are strictly positive, before the agents reach consensus.

(ii) In the second case, there is at least one neighbor of agent i being triggered during the interval of agent i. Without loss of generality, assume that agent j, one neighbor of agent i, is triggered after t_k^i, say $t_{k^*}^j$. In other words, $t_{k^*}^j - t_k^i > 0$. Then, the interval between the k-th and the $(k+1)$-th events of agent i must be more than $t_{k^*}^j - t_k^i$, which is strictly positive.

Therefore, by analyzing the above two cases, we can get that the closed-loop system does not exhibit the Zeno behavior. □

Remark 1. To save the limited resources of communication capacity and energy supply, we design the fully distributed protocol based on the event-triggered strategy. It is worth mentioning that continuous communications are not required for either the control laws updating or the triggering functions monitoring. In other word, we just use the local state information in the event time instants.

Remark 2. Compared to the works [8,12], we consider the distributed event-triggered consensus problem by using a Lyapunov function approach. The advantage of this approach is that we can design the triggering functions according to the proof of the consensus, as long as the Zeno behavior can be excluded, which is similar to the adaptive control idea in certain sense.

In the following, we consider two special cases.

Case 1. We consider the consensus problem for the single-integrator multi-agent systems, whose dynamics are described as:

$$\dot{x}_i = u_i, i = 1, \cdots, N, \tag{30}$$

where $x_i \in \mathbf{R}$ and $u_i \in \mathbf{R}$ denote, respectively, the state and the control input of the i-th agent.

The distributed consensus control law for each agent is designed as

$$u_i(t) = c\sum_{j=1}^{N} a_{ij}(\tilde{x}_i(t) - \tilde{x}_j(t)), i = 1, \cdots, N, \tag{31}$$

where $c \in \mathbf{R}$ is a positive constant coupling gain and $\tilde{x}_i(t) \triangleq x_i(t_k^i)$. The triggering function for each agent i is given by

$$f_i(t) = 4d_i\|e_i\|^2 - \sum_{j=1}^{N} a_{ij}\|\tilde{x}_i - \tilde{x}_j\|^2 - \mu e^{-\nu t} \leq 0, \tag{32}$$

where μ and ν are positive constants.

Corollary 1. *Suppose that Assumption 1 holds. The distributed consensus problem of the agents described by (30) is solved under the control law (31) and the triggering condition (32) with $c > 0$. Furthermore, the Zeno behavior can be excluded.*

Remark 3. Contrary to [11], we need not to implement the adaptive control strategy as in (31). Our algorithm is much simpler, compared to those nonlinear adaptive protocols in [11].

Case 2. Here, we consider the consensus problem for the harmonic oscillators, whose dynamics are described by:

$$\dot{r}_i = v_i, \tag{33-1}$$

$$\dot{v}_i = -\alpha v_i + u_i, \tag{33-2}$$

where $r_i \in \mathbf{R}$ and $v_i \in \mathbf{R}$ denote the position and velocity of the i-th oscillator respectively, $\sqrt{\alpha}$ is the frequency of the oscillators, and $u_i \in \mathbf{R}$ is the control input of the i-th agent to be designed. Define the state $x = \begin{bmatrix} x_1 \\ \vdots \\ x_N \end{bmatrix}$, where $x_i \triangleq \begin{bmatrix} r_i \\ v_i \end{bmatrix}$, $i = 1, \cdots, N$. Then, the system (33) can be written as

$$\dot{x}_i = Ax_i + Bu_i, i = 1, \cdots, N, \tag{34}$$

where $A = \begin{bmatrix} 0 & 1 \\ -\alpha & 0 \end{bmatrix}$ and $B = \begin{bmatrix} 0 \\ 1 \end{bmatrix}$.

The distributed consensus control law for each agent is designed as

$$u_i(t) = -cB^T\sum_{j=1}^{N} a_{ij}(\tilde{x}_i(t) - \tilde{x}_j(t)), i = 1, \cdots, N, \tag{35}$$

where $c \in \mathbf{R}$ is a positive constant coupling gain and $\tilde{x}_i(t) \triangleq e^{A(t-t_k^i)}x_i(t_k^i)$. The triggering function for each agent i is given by

$$f_i(t) = 4d_i\|B^T e_i\|^2 - \sum_{j=1}^{N} a_{ij}\|B^T(\tilde{x}_i - \tilde{x}_j)\|^2 \leq 0. \quad (36)$$

Corollary 2. *Suppose that Assumption 1 holds. The distributed consensus problem of the agents described by (33) is solved under the control law (35) and the triggering condition (36) with $c > 0$. Furthermore, the Zeno behavior can be excluded.*

4 Conclusion

In this paper, we have considered the distributed event-triggered consensus problem of neutrally stable linear multi-agent systems. Compared to the previous related works, our main contribution is that the proposed protocols are fully distributed and scalable, independent of the network's scale and relying on none global information. In our protocols, continuous communications are not required among neighboring agents and the Zeno behavior can be excluded. It should be admitted that only the special case of neutrally stable linear agents is considered in this paper. Designing fully distributed event-triggered consensus protocols for general linear multi-agent systems is an interesting topic for future study.

Acknowledgement. This work was supported by the National Natural Science Foundation of China under grants 61473005 and 11332001.

References

1. Olfati-Saber, R., Fax, J., Murray, R.: Consensus and cooperation in networked multi-agent systems. Proc. IEEE **95**(1), 215–233 (2007)
2. Olfati-Saber, R., Murray, R.: Consensus problems in networks of agents with switching topology and time-delays. IEEE Trans. Autom. Control **49**(9), 1520–1533 (2004)
3. Ren, W., Beard, R.W., Atkins, E.M.: Information consensus in multivehicle cooperative control. IEEE Control Syst. Mag. **27**(2), 71–82 (2007)
4. Li, Z., Duan, Z., Chen, G., Huang, L.: Consensus of multiagent systems and synchronization of complex networks: a unified viewpoint. IEEE Trans. Circ. Syst. **57**(1), 213–224 (2010)
5. Li, Z., Ren, W., Liu, X., Xie, L.: Distributed consensus of linear multi-agent systems with adaptive dynamic protocols. Automatica **49**(7), 1986–1995 (2013)
6. Li, Z., Liu, X., Ren, W., Xie, L.: Distributed tracking control for linear multiagent systems with a leader of bounded unknown input. IEEE Trans. Autom. Control **58**(2), 518–523 (2013)
7. Li, Z., Duan, Z.: Cooperative Control of Multi-agent Systems: A Consensus Region Approach. CRC Press, Boca Raton (2014)

8. Yang, D., Ren, W., Liu, X., Chen, W.: Decentralized event-triggered consensus for linear multi-agent systems under general directed graphs. Automatica **69**, 242–249 (2016)
9. Tabuada, P.: Event-triggered real-time scheduling of stabilizing control tasks. IEEE Trans. Autom. Control **52**(9), 1680–1685 (2007)
10. Dimarogonas, D.V., Frazzoli, E., Johansson, K.H.: Distributed event-triggered control for multi-agent systems. IEEE Trans. Autom. Control **57**(5), 1291–1297 (2012)
11. Xie, D., Xu, S., Zhang, B., Li, Y., Chu, Y.: Cosensus for multi-agent systems with distributed adaptive control and an event-triggered communication strategy. IET Control Theor. Appl. **10**(13), 1547–1555 (2016)
12. Seyboth, G.S., Dimarogonas, D.V., Johanasson, K.H.: Event-based broadcasting for multi-agent average consensus. Automatica **49**(1), 245–252 (2013)
13. Garcia, E., Cao, Y., Yu, H., Giua, A., Antsaklis, P., Casbeer, D.: Decentralised event-triggered cooperative control with limited communication. Int. J. Control **86**(9), 1479–1488 (2013)
14. Meng, X., Chen, T.: Event based agreement protocols for multi-agent networks. Automatica **49**(7), 2125–2132 (2013)
15. Zhu, W., Jiang, Z., Feng, G.: Event-based consensus of multi-agent systems with general linear models. Automatica **50**, 552–558 (2014)
16. Guo, G., Ding, L., Han, Q.: A distributed event-triggered transmission strategy for sampled-data consensus of multi-agent systems. Automatica **50**, 1489–1496 (2014)
17. Liuzza, D., Dimarogonas, D.V., Bernardo, M.D., Johansson, K.H.: Distributed model based event-triggered control for synchronization of multi-agent systems. Automatica **73**, 1–7 (2016)
18. Zhang, Z., Hao, F., Zhang, L., Wang, L.: Consensus of linear multi-agent systems via event-triggered control. Int. J. Control **87**(6), 1243–1251 (2014)
19. Zhang, H., Feng, G., Yan, H., Chen, Q.: Observer-based output feedback event-triggered control for consensus of multi-agent systems. IEEE Trans. Industr. Electron. **61**(9), 4885–4894 (2014)
20. Ioannou, P., Sun, J.: Robust Adaptive Controls. Prentice-Hall, New York (1996)
21. Bernstein, D.: Matrix Mathematics: Theory, Facts, and Formulas. Princeton University Press, Princeton (2009)

Approximate Dynamic Programming for Relay Deployment in Multi-robot System

Song Yao, Yunlong Wu, and Bo Zhang(✉)

State Key Laboratory of High Performance Computing, College of Computer, National University of Defense Technology, Changsha 410073, Hunan, China
zhangbo10@nudt.edu.cn

Abstract. We consider the scenario where a robot squadron navigates along pre-set routes, collects monitored data and sends the collected data to a remote base-station. When the robot squadron moves distant from the base station, a communication relay may be applied in order to provide satisfactory communication quality. Meanwhile, the energy consumption by the relay should be minimized in order to prolong the operation of the system. Hence, this paper aims at optimally deploying the communication relay so that the total energy consumption is minimized under the premise of ensuring the quality of communication. The initial idea is to employ dynamic programming (DP), but DP incurs a high computational complexity and may not be scalable. Therefore, we proposed two approximate optimization techniques, namely the relay-range limited DP (RRL-DP) and multi-phase optimization scheme (MPOS). It is shown that both RRL-DP and MPOS may greatly reduce the computation cost while maintaining the optimality of DP. Finally, we propose an amalgamated MPOS-RRL-DP method that exploits the benefits of both the RRL-DP and MPOS, and it is shown that MPOS-RRL-DP may strike a beneficial tradeoff between computational complexity and total energy optimization in order to meet various practical requirements. Specifically, for the considered scenario, MPOS-RRL-DP achieves a $\sim 50\times$ speed up at a cost of marginal $\sim 0.04\%$ energy loss in comparison to conventional DP.

Keywords: Robot squadron · Energy minimization · Dynamic programming · Multi-phase optimization scheme

1 Introduction

Robotic technology has developed rapidly and has achieved a great breakthrough in the industry and academic research in recent years. A great number of research contributions have been devoted to robotic sensing, cognition, motion/path planning and control [1–3]. In recent years, taking multi-robot systems to finish some specific works attracts more and more attentions. Compared with single robot, a multi-robot system has the advantages of achieving challenging tasks, and further enhances the system's reliability and robustness. There are many challenging and complex problems which are suitable for multi-robot systems, such as mapping, exploration, rescue and remote data collection [4]. Inter-robot communication is one of the key technology of multi-robot systems and attracts a lot of contributions. For example, a communication

Y. Huang et al. (Eds.): ICIRA 2017, Part I, LNAI 10462, pp. 648–658, 2017.
DOI: 10.1007/978-3-319-65289-4_61

system based on the ZigBee network has been designed for solving the problem of the cooperation between multiple robots in [5]. Researchers discussed the impacts of different standards in the IEEE 802.11 wireless local area network (WLAN) family on the communication performance in multi-robot systems [6].

In this paper, we focus on a multi-robot squadron team, which detects an area and transmits the collected data to a remote base station through a mobile relay. In detection tasks, the robots of the squadron team may need to continuously change its position to fulfill the task requirement. Therefore, in order to keep a high communication quality, the relay should adjust its positions accordingly. Considering that unmanned platforms are energy-constrained, how to manage energy resources to improve endurance ability becomes a key problem. In this paper, the squadron team is assumed moving along pre-set routes, so under the premise of ensuring the quality of communication, how to adjust the positions of the relay for reducing the energy consumption is a problem.

We start by using the dynamic programming (DP) to solve the problem, but the computational complexity of DP is not scalable and may not meet the requirements of practical applications. Therefore, we then propose two approximate DP schemes, the relay-range limited DP (RRL-DP) and multi-phase optimization scheme (MPOS), which are shown to greatly reduce the computational complexity at the cost of marginal energy loss. Finally, we propose an amalgamated MPOS-RRL-DP method that exploits the benefits of both the RRL-DP and MPOS.

2 Problem Formulation

2.1 System Model

The detection scenario that considered by this paper is shown in Fig. 1. There are multiple robots executing the detection tasks, and the communication between the squadron team and the base station is guaranteed by a mobile relay. The robots are responsible for collecting information with the fixed sampling rate, and the relative positions between each robot are fixed. After a period of time T, the robots stop and send the collected information to the base station via the relay before moving on.

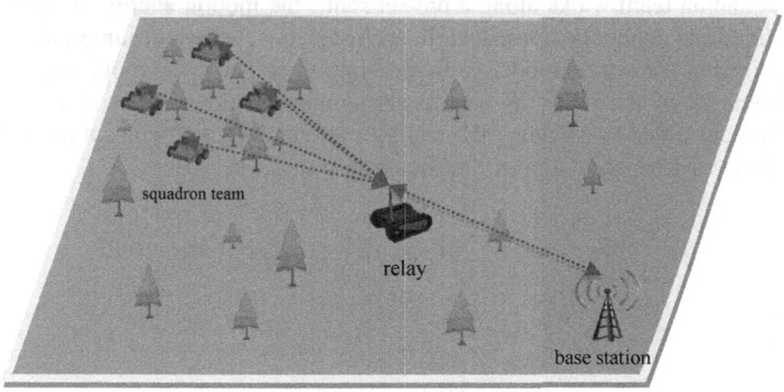

Fig. 1. A squadron team detection scenario

During the time period of T, the squadron team performs the probing tasks. When sending information, the squadron team transmits information to the relay, then the relay follows the decode-forward strategy by decoding the data, re-encoding and forwarding to the base station using same modulation and coding.

For the sake of convenience, we define a step, which includes the time to collect information and transmit information, and a whole task may contain multiple steps. The purpose of this paper is to minimize the total energy consumption of the whole task. Since the squadron team moves along the pre-set route, the squadron team's motion energy is fixed. Therefore, the main concern is to reduce the communication energy and relay motion energy.

2.2 Communication Energy Analysis

The communication energy is mainly determined by the packet error rate (PER) and the wireless channel. Here, we quantify the relationship between the communication energy and PER. According to [7, 8], at high SNR regions, the end-to-end PER can be closely approximated by

$$PRE_{E_i,R,BS}^{end-end'} = \alpha_n \left(\frac{1}{G_{E_i,R} P_{E_i,TX}} + \frac{1}{G_{R,BS} P_{R,TX}} \right), \tag{1}$$

where $G_{R,BS} = \|s_R - s_{BS}\|^{-\beta} / N_0 B$, $G_{E_i,R} = \|s_{E_i} - s_R\|^{-\beta} / N_0 B$, s_R, s_{E_i}, s_{BS} are the positions of the mobile relay, the i-th robot and the base station respectively. N_0 is the noise power spectral density (PSD). β is the path loss index. B is the channel bandwidth. $\alpha_n = [\gamma_{pn} + (a_n/g_n)\exp(-g_n \gamma_{pn})]$. n is the model index and a_n, g_n are relative to $n.\gamma$ is the received signal-to-noise ratio (SNR). γ_{pn} is the SNR threshold. $P_{E_i,TX}$ and $P_{R,TX}$ are the transmit powers of the i-th robot and of the mobile relay.

2.3 Energy Optimization Problem Formulation

As the squadron team walks along a pre-set route, the motion energy of the squadron team is fixed and cannot be optimized. In each step, the energy consumption E that can be optimized in the whole model is divided into three parts: the motion and communication energy of the relay, as well as the communication energy of the squadron team. Here, in order to calculate the energy conveniently, we will optimize the total energy generated by each step of the squadron team as in

$$\begin{aligned} \min_{\substack{s_R, P_{R,TX},\\ P_{E_i,TX}, \forall i}} \quad & E = \sum_{i=1}^{N} \left(P_{E_i,TX} t_{E_i} + P_{R,TX} t_{R,i} \right) + k_M l_R(s_R) \\ \text{s.t.} \quad & l_R(s_R) \le V_{\max} T \\ & PER_{E_i,R,BS}^{end-end'}\left(P_{E_i,TX}, P_{R,TX}\right) \le PER_{\lim} \quad \forall i \end{aligned} \tag{2}$$

In Eq. (2), S_R is the position of the relay. $t_{E_i} = T\eta_i/BR_{E_i}$, $t_{R,i} = T\eta_i/BR_R$ are the time at which the i-th robot and the relay transmit the $T\eta_i$ bits of the data in BR_{Ei} and BR_R (bit/s) respectively. B is the bandwidth and $R(bit/s/Hz)$ is the spectral efficiency. $k_M l_R(S_R)$ is the motion energy consumption of the relay, where $l_R(S_R)$ is the distance covered by the relay from the single step movement. Because the relay has the maximum movement speed $V_{\max}$, $l_R(S_R) \le V_{\max} T \cdot k_M$ is the motion energy parameter. In order to ensure the quality of communication, the end-to-end PER of the two-hop link from the robot via the relay to the base station should be less than the predetermined threshold $PER_{\lim}$.

If a feasible position of relay s_R is given, the motion energy may be derived directly by $k_M l_R(s_R)$ and we may transform the problem in Eq. (2) into a conventional communication planning problem, which is formulated as

$$\begin{aligned} \min_{P_{R,TX}, P_{E_i,TX}, \forall i} \quad & E' = \sum_{i=1}^{N} (P_{E_i,TX} t_{E_i} + P_{R,TX} t_{R,i}) \\ \text{s. t.} \quad & PER_{E_i,R,BS}^{end-end'} \left(P_{E_i,TX}, P_{R,TX}\right) \le PER_{\lim} \quad \forall i \end{aligned} \tag{3}$$

By analyzing Eq. (3), we can simply use the Hessian matrix to prove that Eq. (3) is a convex optimization problem, so we can get the optimal transmit power $P'_{E_i,TX}$ and $P'_{R,TX}$ in closed-form expressions using KKT (Karush-Kuhn-Tucker) conditions. Therefore, we may substitute the optimal power in Eq. (2) and reduce the problem as a motion-planning problem, which is formulated as

$$\begin{aligned} & \min_{s_R} E = \sum_{i=1}^{N} \left[P'_{E_i,TX} t_{E_i}(s_R) + P'_{R,TX} t_{R,i}(s_R) \right] + k_M l_R(s_R) \\ & \text{s. t. } l_R(s_R) \le V_{\max} T \end{aligned} \tag{4}$$

Where

$$P'_{E_i,TX}(s_R) = \left(\frac{\alpha_n}{G_{E_i,R}} \left[\sqrt{v (\sum_{j=1}^{N} t_{E_j} \frac{\alpha_n}{G_{Ej,R}}) (\sum_{j=1}^{N} t_{R,j})} + (\sum_{j=1}^{N} t_{E_j} \frac{\alpha_n}{G_{Ej,R}}) \right] \right) \Big/ PER_{\lim} (\sum_{j=1}^{N} t_{E_j} \frac{\alpha_n}{G_{Ej,R}}),$$

$$P'_{R,TX}(s_R) = \left(v \sum_{j=1}^{N} t_{R,j} + \sqrt{v (\sum_{j=1}^{N} t_{E_j} \frac{\alpha_n}{G_{Ej,R}}) (\sum_{j=1}^{N} t_{R,j})} \right) \Big/ PER_{\lim} (\sum_{j=1}^{N} t_{R,j}), (i = 1, \ldots, N),$$

$v = \alpha_n / G_{R,BS}$.

The next section would be devoted to solving the motion-planning problem in Eq. (4), we would start with the conventional DP solutions and identify the problems of DP, which are dealt with by proposing novel optimization techniques.

3 Motion-Planning Solutions

3.1 DP Solutions

In the whole model, the route of the squadron team is pre-set series of time-ordered positions, which involves making decisions in each step and the decision of a step may affect the following steps. Therefore, the problem may be modeled as a multi-step optimization problem, which can be solved by DP.

The state variable of the system is defined as $s, s \in S$, where S is the set of all possible states. The displacement of the mobile relay in the single step is defined as the action (control) variable a. As the movement time and movement speed of the relay in a single step are limited, $|a| \leq a_{\max}$, where $a_{\max}$ is the displacement threshold. A is defined as a collection of all candidate action variables. The definition of the strategy π is a series of mappings from the state space to the action space: $S \to A, \pi(s_R) = a_R$ means that if the state is S_R and the strategy π is chosen, then action a_R should be taken. The whole process is assumed to be persistent and the robot continues to collect data, the single-step energy consumption for each state is calculated using Eq. (5), so minimizing the energy consumption problem can be modeled as an optimization problem for an infinite problem.

$$\min V^{\pi}(s) = E_{\pi,0} + \sum_{t=1}^{\infty} \gamma^t E_{\pi,t} \quad s \in S \tag{5}$$

$V^{\pi}(s)$ means the total energy consumption at the initial state $S \cdot \gamma$ is the discount factor that balances the weight of the current and future single-step energy consumption. $E_{\pi,t}$ is the single-step energy consumption in the t-th step generated by the strategy π. The core idea of using DP to solve the problem is value iteration. Our idea is to use equation $V^{\pi}(s) = \min(E_{\pi,0} + \gamma V^{\pi}(s'))$ to iteratively optimize $V^{\pi}(s)$ for each state until the difference between two iterations is less than the threshold $\eta \cdot S'$ is the follow-up state of the state s under the action $a = \pi(s)$. The details of the iteration can be found in [9].

3.2 Approximate Optimization Solutions

Relay-Range Limited DP (RRL-DP)

In DP, traversing all possible actions for each state and storing $V^{\pi}(s)$ and $\pi(s)$ result in time complexity and spatial complexity, which is impractical in large-scale problems. Also, in the actual problem, the region of the next state which each state can reach is limited. In addition, according to the knowledge of decode-forward based relaying communications, the relay is generally located near the midpoint of the squadron team and the base station in order to achieve good communication quality.

Therefore, we may constrain the relay movement range between the base station and the squadron team for the sake of reducing the time complexity and spatial complexity. Specifically, the squadron's trajectory has multiple positions for transmitting information, and we may calculate the average of the midpoints spanning from each position and the base station as the center of the relay movement range, which is denoted as p_{mid}. For the sake of convenience, we use rectangle as the shape of the relay movement range and gradually narrow the relay movement range. As shown in Fig. 2, it should be noted that the relay has an initial position s_{init}, and the maximum length $l_{\max}$ of the relay in each step is limited. Therefore, the range of the rectangle with p_{mid} as the center must intersect the range that the first step of the relay can reach, which means that the side length of the rectangle has a minimum value $2l_{\min}$, satisfying $l_{\max} + l_{\min} \geq |p_{mid} - s_{init}|$.

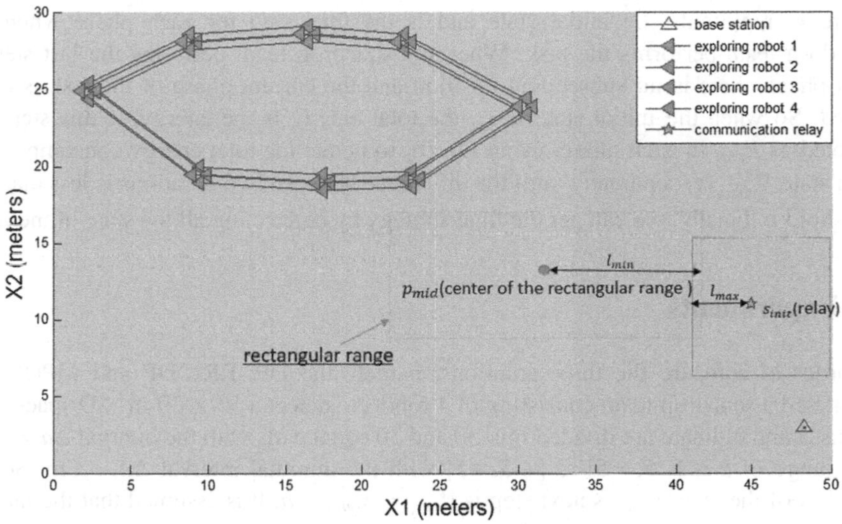

Fig. 2. Limited range of the relay

There is another issue of determining the aspect ratio of the rectangular range. We propose two different methods: (1) different side of the square. (2) As mentioned above, in order to achieve good communication quality, the relay is generally located in the midpoint of the squadron team and the base station according to the communication knowledge. With the continuous movement of the squadron team, the position of the relay also changes. Therefore, we may assume that the movement range of the relay is of a similar shape to that of the squadron's trajectory, so we determine the aspect ratio of the relay movement range based on the aspect ratio of the rectangle, in which the squadron's trajectory is located.

Multi-phase Optimization Scheme (MPOS)

In RRL-DP, how to control the range of the relay is a difficult problem. If the range is too small, it may lead to an increase in total energy consumption, which violates the original intention of the problem. Therefore, we propose another approach, namely the multi-phase optimization scheme (MPOS) to reduce the computational complexity of

DP. The overall idea is to divide the entire lasting detection process into multiple phases. The number of steps in each phase is assumed equal and limited. In this way, the infinite detection task is divided into multiple phases. Reducing the total energy consumption can be achieved by finding the optimal solution of the energy in each phase, while greatly reducing the overall computational complexity. The phase-oriented iteration equation is as follows:

$$V^{\pi}_{\text{phase}}(s) = \begin{cases} \min\left[E_{\pi,0} + \gamma V^{\pi}_{\text{phase}}(s')\right] & \text{if } s \neq s_x \\ \min(E_{\pi,0}), & \text{if } s = s_x \end{cases} \tag{6}$$

where s_x represents the initial state and is the final step for each phase when the squadron team performs the task. When the squadron team performs the last step of each phase, there is no subsequent position and the current phase of the task is completed. So when the initial state is s_x, the total energy is the energy of this step and denoted as $E_{\pi,0}$. In each phase, using Eq. (6) to iterate the total energy consumption at each state $V^{\pi}_{\text{phase}}(s)$ separately until the difference between two iterations is less than the threshold η. Finally, we can get the final strategy by connecting all the state of the $\pi(s)$.

4 Experiments

In order to compare the three solutions, namely the DP, RRL-DP and MPOS, we simulated a squadron team consisting of 4 robots to detect a $50 \times 30\ \text{m}^2$ 2D space. The abscissa and ordinate are divided into 50 and 30 equal parts with the interval $\Delta d = 1$ m. The range of a is [−5, −5] to [+ 5, +5] with the minimal interval $\Delta a = 1$ m. So the position of the relay s_{relay}'s next step is $s'_{relay} = s_{relay} + a$. It is assumed that the motion trajectory of the squadron team is fixed and is annular, including 8 way-points for transmitting data. The data sampling rate for each robot in the squadron $\eta_1 = \eta_2 = \eta_3 = \eta_4 = 10$ Mbps is identical. The physical layer specification follows the 802.11 g protocol with the bandwidth $B = 20$ MHz Channel path loss index $\beta = 3.68$, N_0 is set to -100 dBm/Hz. The motion parameter $k_M = 7.69$ applies to the Pioneer 3DX robot measurement [10].

4.1 Performance Analysis of RRL-DP

According to the movement trajectory of the squadron team and the position of the base station, we calculate the midpoint coordinates of the relay motion range, which is [32.3, 13.5]. The initial position of the relay is s_{init} [45, 11], so we have $l_{\min} = 8.7$ m, it means that the minimum horizontal side length of the rectangle equals $2l_{\min} = 17.6$ m. In this paper, the aspect ratio of the squadron's trajectory range is 1:3. Therefore, when taking rectangular range as the relay motion range, the aspect ratio is also 1:3. We experiment with square ranges and rectangular ranges of different side lengths, respectively. The time consumption and total energy consumption are shown below.

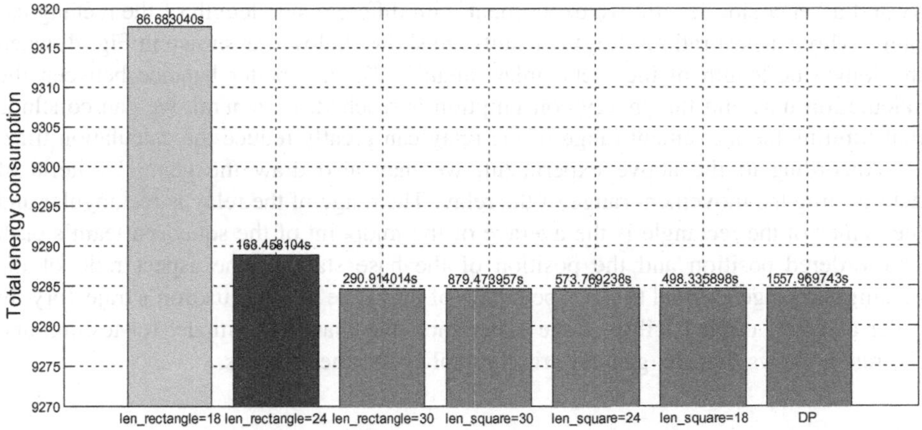

Fig. 3. Total energy consumption and time consumption comparison of RRL-DP for square and rectangular shapes. (Color figure online)

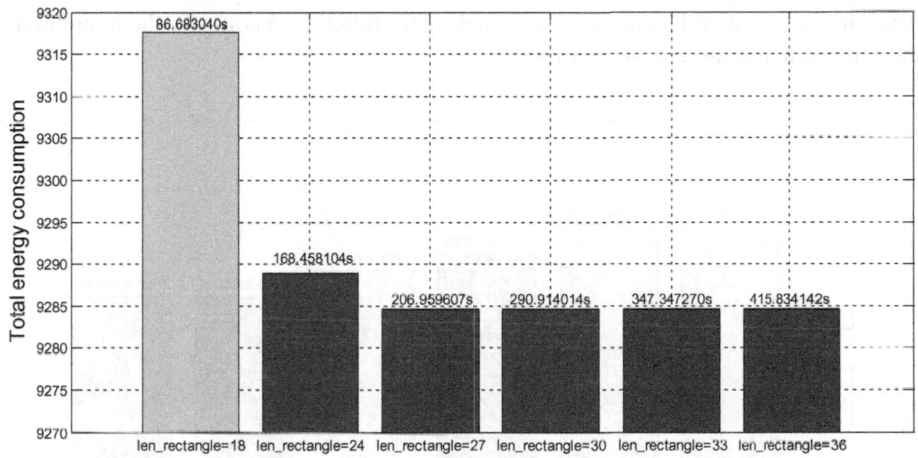

Fig. 4. Total energy consumption and time consumption comparison of different side length for rectangular relay ranges.

The column chart of Fig. 3 shows the energy consumption and the red number above each histogram represents the calculation time. We can conclude as long as the range can be controlled well, the energy consumption may approach that of the DP, regardless of chosen shape of the relay movement range, i.e. square or rectangular. However, if the movement range of the relay is too small, the total energy consumption increases noticeably, as seen in the left column that uses a rectangular with a short side-length of 18 m.

From Fig. 3, it may be observed that with the reduction of the relay range, the calculation time is greatly reduced, and the calculation time of the rectangular range with the aspect ratio of 1:3 is less than the calculation time of the square range. In order to find a better side length, we experiment with different side length of the rectangular range whose aspect ratio is 1:3, the results are shown below. As shown in Fig. 4, when the long side length of the rectangular range is 27 m, a better balance between the calculation time and the energy consumption is reached. In general, we can conclude that limiting the movement range of the relay can greatly reduce the calculation time.

According to the above experiment, we may also draw the general criteria of determining the movement range of the relay. The range of the relay is rectangular and the center of the rectangle is the average of the midpoint of the squadron team's each time-ordered position and the position of the base station. The aspect ratio of the rectangular range is equal to the aspect ratio of the range of the squadron's trajectory. In order to verify the feasibility of the experience, we change the model for testing, and the result shows that the general criteria applies to other models.

4.2 Performance Analysis of MPOS

In MPOS, we will compare MPOS by dividing it into four different phases. The model is tested according to the above three different solutions. Among them, the rectangular range has a long side length of 27 m in RRL-DP. Below, we compare the total energy consumption and the calculation time.

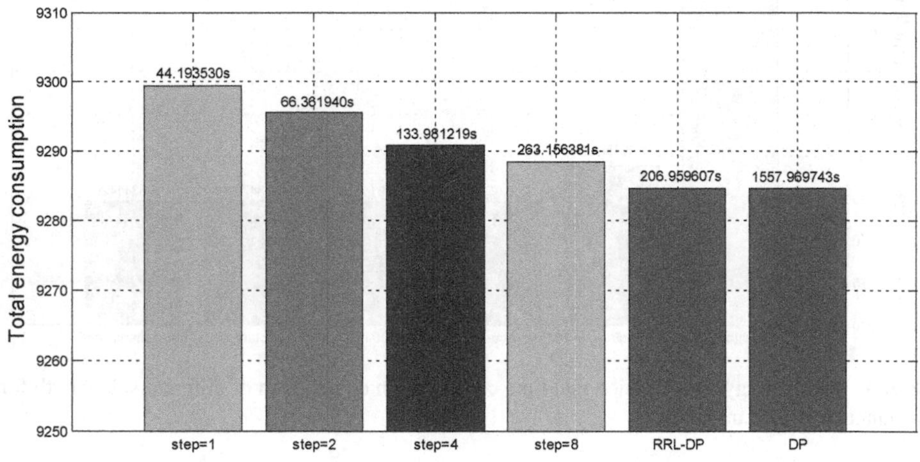

Fig. 5. Total energy consumption and time consumption comparison

The total energy consumption and calculation time is shown in Fig. 5. Because the trajectory of the squadron team is pre-set, the motion energy consumption is fixed. So the total energy consumption here is the sum of the motion energy of the mobile relay and the communication energy. From the results, the total energy consumption of DP and RRL-DP are the same, but the calculation time of RRL-DP is much smaller than that of the DP. When the number of steps is very small, the total energy consumption of MPOS is higher than the total energy consumption of DP, but the calculation time is much less than DP and RRL-DP. On the other hand, as the step increases, the total energy consumption of MPOS decreases and approaches the total energy consumption of DP but the calculation time also increases. Specifically, the calculation time increases exponentially with the number of steps. When the steps for each phase are equal to the number of way-points, the calculation time of MPOS even exceeds the calculation time of RRL-DP, but the total energy consumption of MPOS is also closer to that of the DP.

4.3 Performance Analysis of MPOS-RRL-DP

From the above results we can see that MPOS can effectively save calculation time, but may increase energy consumption. RRL-DP can save only parts of calculation time but energy consumption is minimized. Therefore, a natural question arises on how to achieve a beneficial trade-off between effective calculation efficiency and energy consumption minimization.

In order to address the above question, we propose an amalgamated MPOS-RRL-DP method that exploits the benefits of both RRL-DP and MPOS. In MPOS-RRL-DP, We inherit the benefits of both RRL-DP and MPOS simultaneously. The number of steps for each phase is equal to the number of way-points, which ensures that the total energy consumption of MPOS-RRL-DP is close to that of the DP. In addition, the range of the relay is limited to a most suitable interval. So the calculation time may also be greatly reduced without increasing the energy consumption.

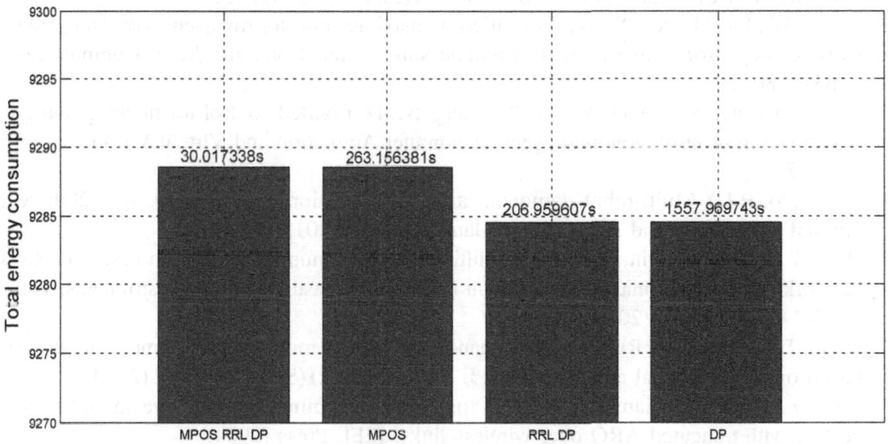

Fig. 6. Total energy consumption and time consumption for different solutions

Figure 6 compares the calculation time and energy consumption of MPOS-RRL-DP with that of the DP, RRL-DP and MPOS. We still use the same model and experimental set-ups. In MPOS, the number of steps for each phase *Step* = 8, and the rectangular range having a long side length of 27 m in RRL-DP is used as the relay motion range.

Figure 6 shows that the energy consumption of these four solutions is similar. The energy consumption of MPOS-RRL-DP and MPOS is marginally higher than that of the DP by ~0.04%, while the calculation time of MPOS-RRL-DP is much smaller than that of the other three solutions by a speed up of ~52, ~6.9 and ~8.8 times in comparison to DP, RRL-DP and MPOS. Therefore, we may conclude that MPOS-RRL-DP can effectively reduce the calculation time than other three solutions and effectively save the energy consumption.

5 Conclusion

This paper considers a scenario that a robot squadron detects an area along pre-set routes and sends collected data to a remote base station through a relay. We propose DP, RRL-DP, MPOS and MPOS-RRL-DP solutions to minimize the total energy consumption. Simulation results show that compared with other three solutions, MPOS-RRL-DP can greatly reduce the computation cost while maintaining the optimality of DP, which is more suitable in solving practical problems.

References

1. Kehoe, B., Patil, S., Abbeel, P., Goldberg, K.: A survey of research on cloud robotics and automation. IEEE Trans. Autom. Sci. Eng. **12**(2), 398–409 (2015)
2. Liang, W., Luo, J., Xu, X.: Network lifetime maximization for time-sensitive data gathering in wireless sensor networks with a mobile sink. Wirel. Commun. Mob. Comput. **13**(14), 1263–1280 (2013)
3. Cai, Z., Chang, X., Wang, Y., Yi, X., Yang, X.: Distributed control for flocking and group maneuvering of nonholonomic agents. Computer Animation and Virtual Worlds. Accepted in 2017
4. Avanzato, R.L.: Multi-robot communication for education and research. In: 120th ASEE Annual Conference and Exposition, Atlanta, Georgia (2013)
5. Wan, J., Wang, Y., Qin, Q., Li, Y.: Multi-robots' communication system based on ZigBee network. In: International Conference on Electronic Measurement and Instruments, vol. 35, pp. 3-14–3-19. IEEE (2009)
6. Yuan, L.I., Yuan, K., Rui, Z.: Performance analysis of multi-robot communication system based on wireless local area networks. J. Syst. Simul. **21**(8), 2218–2219 (2009)
7. Liu, Q., Zhou, S., Giannakis, G.B.: Cross-Layer combining of adaptive modulation and coding with truncated ARQ over wireless links. IEEE Press (2004)
8. Wu, Y., Zhang, B., Yi, X., Tang, Y.: Communication-motion planning for wireless relay-assisted multi-robot system. IEEE Wirel. Commun. Lett. **5**(6), 568–571 (2016)
9. Geramifard, A., Walsh, T.J., Tellex, S.: A tutorial on linear function approximators for dynamic programming and reinforcement learning. Foundations and Trends®. Mach. Learn. **6**(4), 375–454 (2013)
10. Yan, Y., Mostofi, Y.: To go or not to go: on energy-aware and communication-aware robotic operation. IEEE Trans. Control Netw. Syst. **1**(3), 218–231 (2014)

GSDF: A Generic Development Framework for Swarm Robotics

Xuefeng Chang, Zhongxuan Cai, Yanzhen Wang(✉), Xiaodong Yi, and Nong Xiao

HPCL, School of Computer, National University of Defense Technology, Changsha 410073, Hunan, China
yzwang@nudt.edu.cn

Abstract. Programming swarm robots is considered to be more challenging than developing single robot applications, mostly due to the unpredictable behaviors and complex coordination of the swarm. In this paper, we present *GSDF*, a generic development framework for swarm robotics, to promote both programmability and flexibility. In its core lies a lightweight runtime, which maintains all the swarm-related data structures in a decentralized manner. The runtime is based on the design of a *two-level communication graph*, which decouples intra- and inter-robot communications. On the intra-robot level, existing popular messaging mechanisms, such as that of the Robot Operating System (ROS), can be adopted to reuse programming resources in the ROS eco-system. On the inter-robot level, the framework presents an abstract communication layer. It provides a unified communication interface to the framework runtime, and hence makes the framework agnostic to underlying communication mechanisms. *GSDF* also provides a clear interface to facilitate programming robot swarms, including prescribed tools for swarm management and information sharing. Finally, we demonstrate the feasibility and efficiency of *GSDF* using a couple of experiments in robotic simulators.

Keywords: Swarm robotics · Programming framework · Robot Operating System

1 Introduction

In recent years, swarm robotics has received increasing attention, in terms of both theoretical research and practical applications. Swarm robotic systems are inspired by social animals, such as ants and bees, which are excellent examples of how a large number of simple individuals create complex collective behaviors merely by local interactions. It is a novel approach to coordinate large numbers of robots and the coordination capability is beyond the reach of current multi-robot system [1]. In addition, the collective behaviors of swarm robotics appear to be robust, scalable, and flexible [2].

On the other hand, programming distributed robot systems is complicated for the large numbers of robots and the difficulties of network programming [3].

Y. Huang et al. (Eds.): ICIRA 2017, Part I, LNAI 10462, pp. 659–670, 2017.
DOI: 10.1007/978-3-319-65289-4_62

For swarm robotic systems, there is no centralized or hierarchical control and communication systems to coordinate robots' behavior [18], which makes programming swarm robotic systems more difficulty than multi-robot systems. Tools such as the *buzz* programming language [4] are proposed in recent years, but it is typically difficult for them to reuse the existing programming resources in popular robotics communities, such as the Robot Operating System (ROS) eco-system.

In this paper, We propose a generic development framework for swarm robotics, namely *GSDF*. It provides a decentralized runtime and swarm programming interfaces, following a component-based development scheme. Currently, *GSDF* is completely compatible with ROS, enabling developers to use resources in ROS eco-system. However, we base *GSDF* on an abstracted communication layer, which means that users are not confined within any specific robotic middlewares. Other messaging mechanisms, such as the Data Distribution Services (DDS), can be easily incorporated into the framework, making the framework flexible and extensible. *GSDF* is released as open-source software under the BSD license which allows the development of both non-commercial and commercial projects. The full source code of *GSDF* is available at http://wiki.ros.org/micros_swarm_framework/.

2 Related Work

ROS is an flexible framework consisting of libraries, tools and conventions that simplifies the development of robotic applications [5]. It encourages collaborative robotics software development and is convenient for code reuse. However, ROS was designed mainly for single-robot applications, and it is not suitable for large-scale robot swarms. ROS 2.0, the next generation of ROS, has been initiated recently, with new design considerations including teams of multiple robots, real-time system, and non-ideal networks. DDS is used as the default communication middleware. These design choices makes ROS 2.0 more suitable for swarm robotics, but it is still under heavy development.

COROS is a multi-agent software architecture for cooperative and service robots. The objective of COROS is to make the design and development of multi-robot applications easier [6]. In addition, COROS is implemented on top of ROS and has the advantage of reusing ROS packages. However, COROS is highly coupled with ROS and can not be easily extended to swarm robotic systems.

ALLIANCE [8] is a behavior-based model which is fully distributed and fault-tolerant. There is research effort combining the ALLIANCE model with ROS to provide a multi-robot development framework [7] for ROS users. However, this framework tends to create a large number of CPU processes and relies on reliable underlying communication, which hamper its application to large-scale robot swarms.

SWARMORPH system is proposed in [20]. It was designed to form arbitrary morphologies of sizes and shapes with self-propelled robots for solving specific tasks, such as navigation-based challenges. In particular, it presented a behavior-based script language called SWARMORPH-Script which can generate multiple morphologies. Recently, a new programming language called *buzz* for heterogeneous swarm robotics is proposed in [4]. It provides primitives to define collective behaviors both from the perspective of individual robot and the swarm. SWARMORPH-Script and *buzz* share the same problem that existing software resources such as ROS packages cannot be easily incorporated.

3 Framework Overview

The structure of *GSDF* is presented in Fig. 1. In its core lies a decentralized and lightweight runtime, which maintains all the swarm-related data as well as processes messages. On top of the runtime, *GSDF* provides multiple programming interfaces as well as a swarm library which enable developers to define swarm behaviors more easily. At the bottom, *GSDF* presents an abstracted communication layer, which is intended to make the framework flexible and extensible. Besides, *GSDF* adopts component-based application development approach.

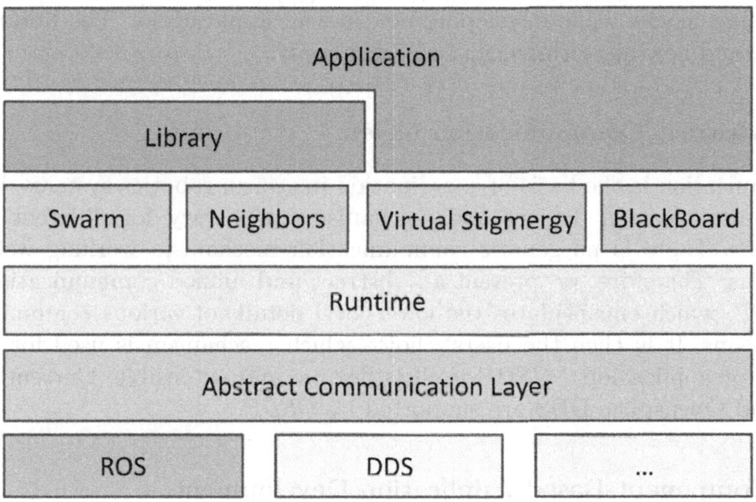

Fig. 1. Framework of *GSDF*

3.1 Decentralized Runtime

The runtime is used to maintain swarm-related data, including information about individuals, neighbors, and the entire swarm. It also process all kinds of messages being passed in the system. Unlike single robot or multi-robot system, swarm robotics system is generally large-scale and there is no centralized or hierarchical control and communication systems to coordinate the robots' behaviors. Decentralized runtime are more robust and extensible, and hence more suitable for swarm robotics system. As shown in Fig. 3, runtime is based on a *two-level communication graph.* On the intra-robot level, existing popular messaging mechanisms, such as that of ROS, can be adopted to reuse resources in the community. Finally, the runtime also should be lightweight enough, due to the fact that individual robots in typical swarms possesses very limited computational resource.

3.2 Programming Interfaces and Library

To facilitate the design and implementation of swarm behaviors, *GSDF* provides several generic programming interfaces enlightened by the *buzz* programming language, such as *Neighbor*, *Swarm*, and *Virtual Stigmergy.* Besides, we introduce a new efficient near-range information sharing method called *BlackBoard.* Using these interfaces, users can define swarm behaviors more conveniently, without paying attention to the low-level details.

On top of the programming interfaces, *GSDF* introduces a swarm library that contains commonly-adopted algorithms for typical swarm behaviors, such as flocking [9], Particle Swarm Optimization [10]. Users could directly use them as building blocks while developing new swarm applications. The library itself is open, and new algorithms can be added easily.

3.3 Abstract Communication Layer

Communication is the basis of coordination in swarm robotics systems. In practice, communication devices and mechanisms may vary for different robotic platforms. There is no generic communication mechanism working well in all situations. Therefore, we present an abstract and unified communication layer in *GSDF*, which encapsulates the lower-level details of various communication mechanisms. It is then the users' choice which mechanism is used for specific swarms or applications. *GSDF* itself is flexible and extensible. Currently, both ROS and Opensplice DDS are supported by *GSDF.*

3.4 Component-Based Application Development

Component-based application development, which is recommended in ROS 2.0, is more conducive to modularity and software reuse. It facilitates adding common concepts to existing code, runs multiple applications in a single process with low overhead, and presents more efficient intra-process communication [11]. Moreover, it is more lightweight than the approach in which each application possesses a new process. In *GSDF*, an application is called a component and it can be loaded or unloaded at runtime.

4 Implementation Details

GSDF follows a modular development scheme to ensure its extensibility and scalability. In this section, we introduce the implementation details of each module in the framework, including the abstract communication layer, runtime, programming interfaces, and application development.

4.1 Abstract Communication Layer

We assume that each robot is equipped with an omni-directional communication device that broadcasts messages at a certain frequency and receives messages within a limited range. This can be satisfactorily realized through WIFI, Bluetooth, or ZigBee. According to this communication model, we define a unified underlying communication interface, which contains three functions: *init*(), *broadcast*() and *receive*(). We could utilize different communication mechanisms to implement this interface, and integrate them with *GSDF* easily. *GSDF* itself currently supports both ROS message protocols (TCPROS and UDPROS) and OpenSplice DDS, which supports high-performance, scalable, and QoS-assuring message delivery. For applications which have a high demand for real-time performance, users could choose DDS as the underlying message protocol to achieve better real-time performance.

4.2 Runtime

Each robot has an independent process which is called a *daemon node* in *GSDF*. It has an integer ID to identify the corresponding robot uniquely. The *daemon node* publishes kinds of messages at different frequencies and processes received messages automatically. Different kinds of output messages are put in specific output queues which were scheduled using Weight-Round-Robin (WRR) strategy. All the input messages entered the only First-In-First-Out (FIFO) input queue which was scheduled sequentially. In addition, the *daemon node* also has the duty of maintaining swarm-related data, including information about individuals, neighbors, as well as the entire swarm. The specific structure of the *daemon node* is shown in Fig. 2.

ROS relies on the concept of computational graph, which represents the network of ROS processes. In ROS, each process is called a node and they can communicate with each other through ROS-provided message passing mechanism. For the purpose that *GSDF* has better scalability and be compatible with ROS completely, the *daemon node* is based on a *two-level communication graph*, as shown in Fig. 3. On the intra-robot level, the *daemon node* communicates with other ROS nodes which are in the same ROS computational graph through TCPROS or UDPROS message passing mechanism. This intra-robot level enables us to reuse software resources in the ROS eco-system. On the inter-robot level, *GSDF* presents an abstract communication layer and the *daemon node* of each robot communicates with each other through the unified communication interface, hence makes the framework agnostic to underlying communication mechanism.

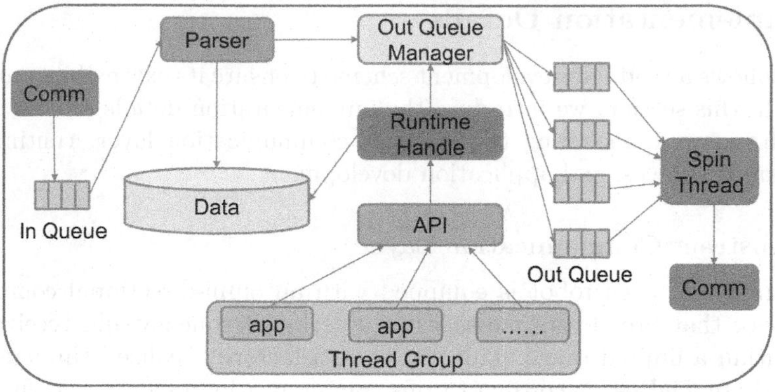

Fig. 2. The structure of the Daemon node.

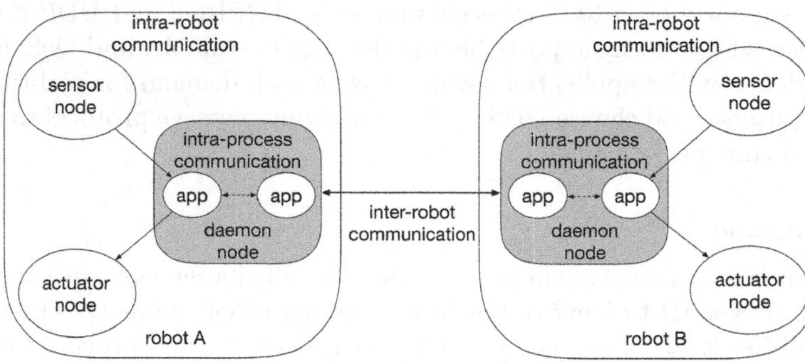

Fig. 3. The two-level communication graph of *GSDF* runtime.

4.3 Programming Interfaces

GSDF provides several programming interfaces for developers to define swarm behaviors both from the perspective of single robot and the entire swarm, which are enlightened by *buzz*. We defined similar programming interfaces to *buzz*, e.g., *Neighbors*, *Swarm*, *Virtual Stigmergy*, but implemented them using different method. In *buzz*, they are *primitives*, in *GSDF*, they are *classes*. We use C++ language and object-oriented programming to implement them. Each of them contains multiple methods. We can use *Neighbors* to manipulate neighborhood data including position and velocity, use *Swarm* to manage robot teams dynamically, e.g. creating a swarm, joining in a swarm or leaving a swarm, and use *Virtual Stigmergy* which essentially works as a distributed tuple space to share information globally across the swarm.

Virtual Stigmergy works well when we want to agree on the values of a set of variables across the swarm globally, but for near-range robot teams, it is costly. When reading or writing tuples in *Virtual Stigmergy*, *daemon node* needs to broadcast messages which contain the tuples. Nearby robots once receive the messages, check whether the received tuples is newer than local. If this is the case, the robots also needs to propagate the messages. This is quite costly for near-range robot teams. We propose a simpler but more efficient method called *BlackBoard* for near-range robot teams to agree on the values of sets of variables. In *BlackBoard*, the data is stored centrally on a single robot. So it it centralized. Similar to *Virtual Stigmergy*, *BlackBoard* also provides three method: *create*(), *put*(), *get*(). Note that when we create a new *BlackBoard* structure, we need to specify its unique id and on which robot the data will be stored.

4.4 Application Development

GSDF defined a generic interface named *Application* for users to develop behaviors for swarm robotics. What we need to do is creating a new class which inherited *Application* and implemented the *start*() function defined in the *Application*. The *start*() function is the unified entry for application and the main logic of the application should be written in it. Note that, there should be no main function in source files. The application development is based on pluginlib and further details on the usage of plugin can be found in [12]. As plugins, applications can be loaded into the *daemon node* at runtime, without creating a new process, which is more lightweight. Moreover, different applications could communicate with each other through more efficient intra-process communication in *daemon node*, as shown in Fig. 3.

In order to load or unload applications at runtime conveniently, we provide an user-friendly tool which is called *app_loader* in *GSDF*. It is an independent

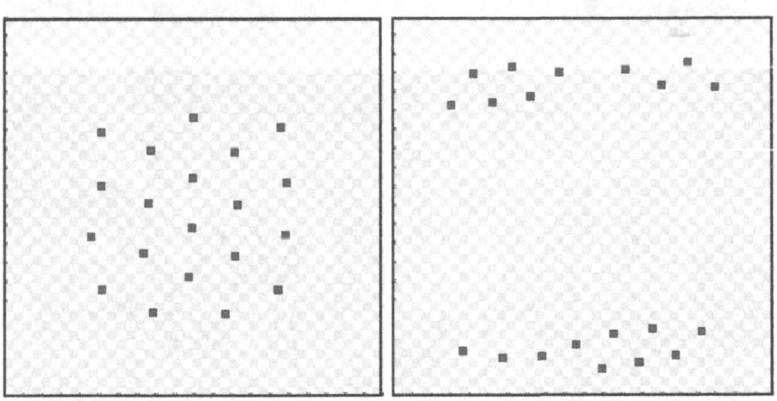

(a) (b)

Fig. 4. Snapshots of the swarm behaviors simulated in stage: (a) patter formation, (b) segregation. The swarm size is 20.

executable program which needs two string arguments, one is application name, the other one is application type. When the loader process was created, it would load the application which was specified uniquely according to the name and the type arguments. On the contrary, once the loader was destroyed, it would unload the application which was loaded by itself automatically. In addition, if the *daemon node* was destroyed first of all, all the applications would be unloaded by the strong hand and all the loaders would be destroyed automatically. In *GSDF*, *app_loader* is implemented through ROS services (Fig. 4).

5 Experiments

In this section, we design and implement some common swarm behaviors based on *GSDF*, such as pattern formation, segregation, and collective flocking, to test and verify the proposed framework. Moreover, we also carry out the performance statistics and analysis to analyze the performance and scalability of *GSDF* (Fig. 5).

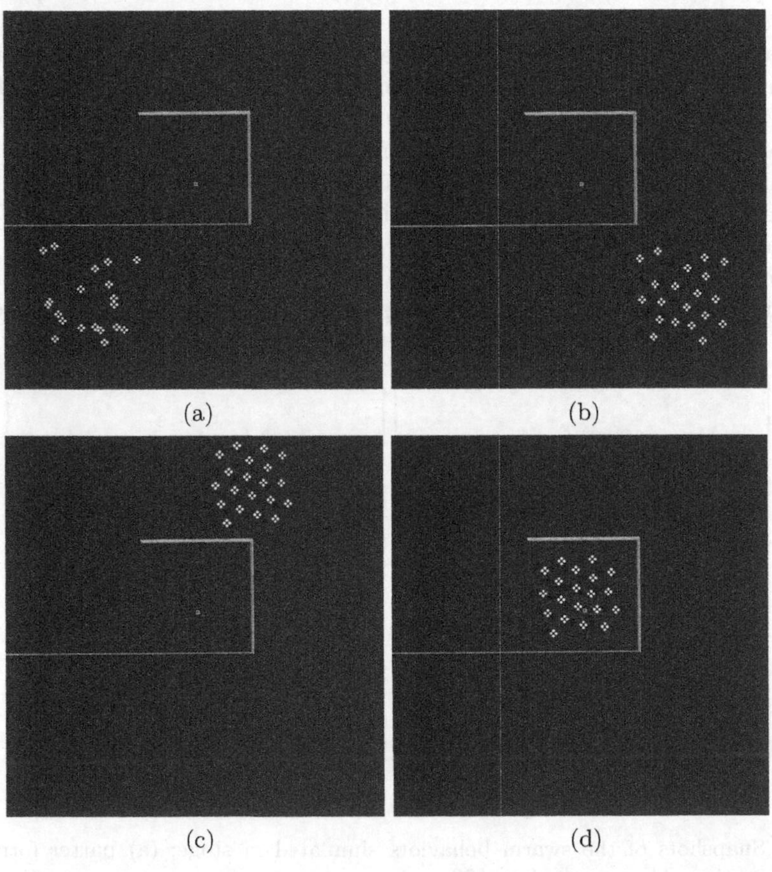

Fig. 5. Collective flocking behavior simulated in Gazebo. The swarm size is 20.

5.1 Swarm Behaviors

Pattern Formation and Segregation. Pattern formation is a common and widely used swarm behavior. In this section, we choose to use virtual physics [15] which is the most common approach to implement the pattern formation application. Each neighbor of a robot will produce a virtual force against it due to the local interaction, and the sum of the virtual forces is its direction vector. We use *Neighbors* class to implement this behavior.

Dividing a swarm of robots into multiple sub-groups is called segregation, which is also a basic swarm motion behavior. In order to implement this behavior, we use the method proposed by [16]. What we need to do is defining two swarms, and use pattern formation to impose a short distance between robots which belong to the same swarm, and a long distance between robots not in the same swarm. We use *Neighbors* and *Swarm* class to implement this behavior.

Collective Flocking. Flocking is a collective behavior of a robot swarms. There are many flocking phenomena in nature, e.g., swarming of fish, birds, bacteria and so on. In many engineering applications including design of mobile sensor networks, cooperative tasks of robot swarms, flocking is also very important. It attracts researchers from various disciplines and has been the focus of various fields and research [17], e.g., in [22], Tang and Mao studied the disease transmission by the flocking behavior of birds, and in [21] they did research on information entropy-based metrics for measuring emergences in artificial societies. In this section, we use Olfati-Saber's algorithm [17] to implement flocking behavior, in which an robot's acceleration is calculated based on the position and velocity of its neighbors. *Neighbors* and *Swarm* is suitable to implement Olfati-Saber's algorithm.

5.2 Performance Analysis

In this sub-section, we analyze the performance and scalability of *GSDF*. We conduct different experiments under different conditions, e.g., different size of robot swarms, component-based or node-based application development. These experiments are conducted using the Stage [3] simulator and we mainly focus on the CPU utilization and occupied network bandwidth. The hardware platform is a mainstream desktop with a quadcore Intel Core i7-4790 CPU, GTX-1050Ti GPU and 16 GB main memory. In addition, ROS version is Indigo and Opensplice DDS version is the community edition 6.4.140407.

Scalability. In order to analyze the scalability of *GSDF*, we simulate pattern formation and segregation behaviors under different size of robot swarms, from 10 to 70. *GSDF* uses UDPROS as the underlying message passing mechanism in this experiment. As shown in Fig. 6, as the size of the swarm grows, CPU as well as Network usage grows linearly approximatively and we can simulate a swarm containing as many as 70 robots using a mainstream desktop computer, which indicates that *GSDF* is extensible and lightweight.

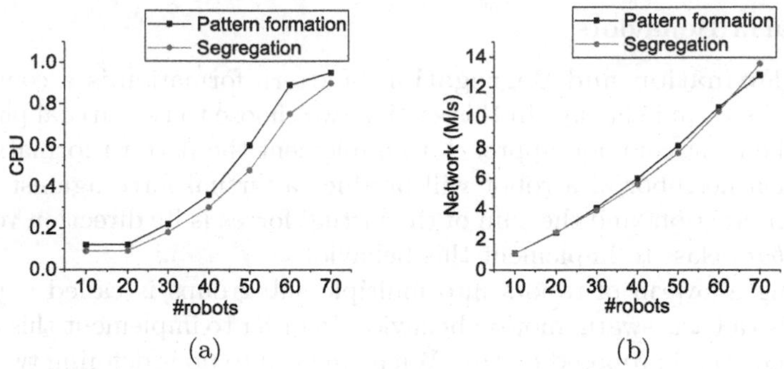

Fig. 6. Performance statistics of pattern formation and segregation applications (a) CPU utilization, (b) network bandwidth occupation. The underlying message passing mechanism is UDPROS.

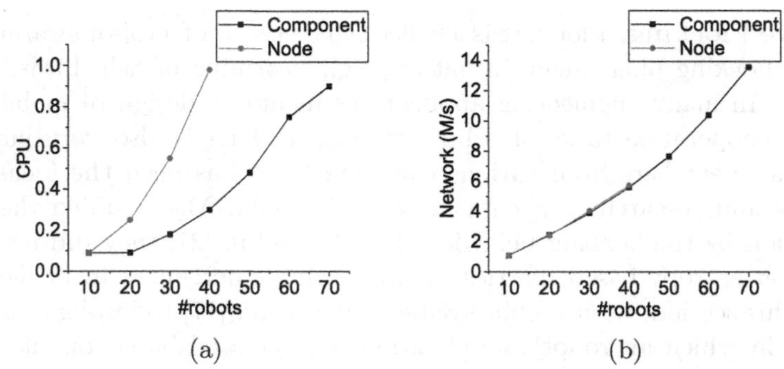

Fig. 7. Performance statistics of component-based and node-based development schemes. (a) CPU utilization, (b) network bandwidth occupation. The underlying message passing mechanism is UDPROS and the test case is segregation application.

Component vs. Node. To analyze the performance of component-based application development, we compare it with the approach in which each application possesses a new process. In the former approach application is called *Component* and in the latter one it is called *Node*. In this experiment we use segregation as the test case. As shown in Fig. 7, component-based development has great advantages both of the CPU and the Network. What we need to pay attention to is that due to the lack of resources, we can only simulate up to 40 robots if each application creates a new process.

BlackBoard vs. Virtual Stigmergy. In this experiment we assume that there are some near-range robots which can communicate with each other. We test three scenarios, i.e., writing, reading, simultaneous reading and writing. As shown in Table 1, when writing tuples BlackBoard is far better than Virtual Stigmergy but it is slightly worse when reading. On average, Blackboard has performance advantages for near-range robot teams.

Table 1. Performance statistics of BlackBoard and Virtual Stigmergy. (M/s)

#robots	Virtual Stigmergy			BlackBoard		
	Write	Read	Write & Read	Write	Read	Write & Read
10	0.34	0.23	0.48	0.21	0.30	0.43
20	1.96	0.94	2.50	0.93	1.30	1.83
30	5.77	2.15	7.01	2.09	3.02	4.25
40	12.48	3.85	14.85	3.78	5.49	7.70
50	22.89	6.10	26.25	5.96	8.62	12.17
60	35.78	8.75	38.76	8.67	12.54	17.54

6 Conclusions and Future Work

This paper presents *GSDF*, a generic development framework for swarm robotics, which is built upon a two-level communication graph. It is compatible with multiple communication mechanism and provides various programming interfaces for developing swarm robotic applications. On top of *GSDF*, developers can use the abundant resources in the ROS eco-system and define swarm behaviors easily. In addition, other message passing mechanism, such as DDS, can be easily incorporated into the framework, which makes it flexible and extensible. Various experiments in robotic simulators demonstrate the effectiveness and efficiency of the framework. In the future, we would like to provide more programming interfaces and extend the swarm library in *GSDF*. Visualization and logging tools will also be taken into consideration in our future work.

Acknowledgment. This work was supported by National Science Foundation of China under Grant No. 91648204 and No. 61601480, and Research on Foundations of Major Applications, Research Programs of NUDT under Grant No. ZDYYJ-CYJ20140601.

References

1. Şahin, E.: Swarm robotics: from sources of inspiration to domains of application. In: Şahin, E., Spears, W.M. (eds.) SR 2004. LNCS, vol. 3342, pp. 10–20. Springer, Heidelberg (2005). doi:10.1007/978-3-540-30552-1_2
2. Brambilla, M., Ferrante, E., Birattari, M., et al.: Swarm robotics: a review from the swarm engineering perspective. Swarm Intell. **7**(1), 1–41 (2013)
3. Gerkey, B., Vaughan, R.T., Howard, A.: The player/stage project: tools for multi-robot and distributed sensor systems. In: Proceedings of the 11th International Conference on Advanced Robotics, vol. 1, pp. 317–323 (2003)
4. Pinciroli, C., Buzz, B.G.: An extensible programming language for heterogeneous swarm robotics. In: 2016 IEEE/RSJ International Conference on Intelligent Robots and Systems (IROS), pp. 3794–3800. IEEE (2016)
5. Robot Operating System Wiki: ROS Documentation-Introduction, May 2014. http://wiki.ros.org/ROS/Introduction

6. Kouba, A., Sriti, M.F., Bennaceur, H., et al.: COROS: a multi-agent software architecture for cooperative and autonomous service robots. In: Koubâa, A., Martínez-de Dios, J. (eds.) Cooperative Robots and Sensor Networks 2015, pp. 3–30. Springer, Cham (2015)
7. Li, M., Cai, Z., Yi, X., Wang, Z., Wang, Y., Zhang, Y., Yang, X.: ALLIANCE-ROS: a software architecture on ROS for fault-tolerant cooperative multi-robot systems. In: Booth, R., Zhang, M.-L. (eds.) PRICAI 2016. LNCS, vol. 9810, pp. 233–242. Springer, Cham (2016). doi:10.1007/978-3-319-42911-3_19
8. Parker, L.E.: ALLIANCE: an architecture for fault tolerant multirobot cooperation. IEEE Trans. Robot. Autom. **14**(2), 220–240 (1998)
9. Ferrante, E., Turgut, A.E., Stranieri, A., et al.: A self-adaptive communication strategy for flocking in stationary and non-stationary environments. Nat. Comput. **13**(2), 225–245 (2014)
10. Shi, Y.: Particle swarm optimization: developments, applications and resources. In: Proceedings of the 2001 Congress on Evolutionary Computation, vol. 1, pp. 81–86. IEEE (2001)
11. ROS2 Wiki: Composing multiple nodes in a single process, December 2016. https://github.com/ros2/ros2/wiki/Composition
12. Robot Operating System Wiki: ROS Documentation-pluginlib, May 2015. http://wiki.ros.org/pluginlib
13. Marder-Eppstein, E., Berger, E., Foote, T., et al.: The office marathon: robust navigation in an indoor office environment. In: 2010 IEEE International Conference on Robotics and Automation (ICRA), pp. 300–307. IEEE (2010)
14. Foote, T.: tf: The transform library. In: 2013 IEEE International Conference on Technologies for Practical Robot Applications (TePRA), pp. 1–6. IEEE (2013)
15. Spears, W.M., Spears, D.F., Hamann, J.C., et al.: Distributed, physics-based control of swarms of vehicles. Autonom. Robots **17**(2), 137–162 (2004)
16. Pinciroli, C., Lee-Brown, A., Buzz, B.G.: An extensible programming language for self-organizing heterogeneous robot swarms. arXiv preprint 2015. arXiv:1507.05946
17. Olfati-Saber, R.: Flocking for multi-agent dynamic systems: algorithms and theory. IEEE Trans. Autom. Control **51**(3), 401–420 (2006)
18. Şahin, E., Winfield, A.: Special issue on swarm robotics. Swarm Intell. **2**(2), 69–72 (2008)
19. Wu, Y., Zhang, B., Yi, X., et al.: Communication-motion planning for wireless relay-assisted multi-robot system. IEEE Wirel. Commun. Lett. **5**(6), 568–571 (2016)
20. O'Grady, R., Christensen, A.L., Dorigo, M.: Swarmorph: morphogenesis with self-assembling robots. In: Doursat, R., Sayama, H., Michel, O. (eds.) Morphogenetic Engineering, pp. 27–60. Springer, Berlin (2012). doi:10.1007/978-3-642-33902-8_2
21. Tang, M., Mao, X.: Information entropy-based metrics for measuring emergences in artificial societies. Entropy **16**(8), 4583–4602 (2014)
22. Tang, M., Mao, X., Guessoum, Z.: Research on an infectious disease transmission by flocking birds. Sci. World J. (2013)

A Case Study on the Performance of Gazebo with Multi-core CPUs

Hai Yang(✉) and Xuefei Wang

State Key Laboratory of High Performance Computing, School of Computer, National University of Defense Technology, Changsha 410073, Hunan, China
yanghai11@nudt.edu.cn

Abstract. Gazebo has become a popular platform for the simulation of robots and their working environment. However, it is inefficient as the number of robots increases or when the environment becomes more complex. In this paper, we investigate the program hotspot (the most time consuming program units) of two typical cases using the Intel VTune Amplifier. The results show that the simulation time increases quadratically with respect to the number of robots in the swarm robots scenario, and increases linearly as the complexity of environment becomes higher in the large-scale campus scenario.

Keywords: Hotspots analyze · Case study · Gazebo

1 Introduction

Simulators play a critical role in robotic research. They are used for quickly and efficiently testing of new concepts, strategies and algorithms [1]. One of the most popular platform is Gazebo, a 3D multi-robot simulator with dynamics. Gazebo offers the ability to simulate multi-robot in complex indoor and outdoor environment with very high fidelity. Gazebo has been integrated into ROS (Robotic Operating System), which is widely used by robotic developers.

Recently, micros group from the State Key Laboratory of High Performance Computing used Gazebo to demonstrate the correctness of algorithms, including the flocking algorithm for nonholonomic agents [2] and autonomous navigation algorithm for robots in large scale environment [3]. The real time factor is the ratio of simulation time to real-time. In the former test, the real time factor of Gazebo dropped down drastically to 0.2 when the swarm size is over 50. In the latter, the real time factor declined to 0.1 when adding models to the simulation environment for the construction of large-scale environment. Thus the optimization of Gazebo becomes necessary for these cases with multi-robot in complex environment.

The physical engine is the key function component of Gazebo, while ODE (Open Dynamics Engine) is the default and commonly used physics engine. Parallelization of the physics engine is an option to improve Gazebo's performance. Michelle Goodstein et al. [4] from CMU identified the hotspots of the ODE

Y. Huang et al. (Eds.): ICIRA 2017, Part I, LNAI 10462, pp. 671–682, 2017.
DOI: 10.1007/978-3-319-65289-4_63

physics engine using GNU gprof. They further parallelized the collision detection process of ODE. Ashley-Rollman et al. [5] parallelized the Hash Table construction process in ODE using OpenMP in the DPRSim2 simulator. In addition, the Intel TBB tutorial [6] used the VTune Performance Analyzer to identify the hotspots in ODE library and multi-threading processIslandFast() of ODE using Threading Building Blocks. These optimization strategies for ODE physics engine mentioned above have not yet been applied to Gazebo.

The Open Source Robotics Foundation [7] implemented a parallelize version of the ODE physics engine, and incorporated it into Gazebo, which was named paraGazebo. They used an *island thread* strategy to parallelize simulation of non-interacting entities. Simulated entities are interacting if they are connected by an articulated joint (such as a revolute or universal joint) or are connected via contact. They clustered the interacting entities into islands that are mentally decoupled from each other and simulated it in parallel. After each step, the clustering of island is recalculated. This parallel strategy can allocated computation into different islands when simulated models do not react very often. A performance test report of this parallel strategy is given. But all the test scenario used in the performance test didn't considering simulation with multi-robot interact through communication or large-scale environment.

Our aim is to locate the most time-consuming functions in Gazebo when simulating these two scenarios and identify significant factors that affect the performance. Performance analysis with paraGazebo is also conduct. We expect such an analysis could provide us directions for further optimization of the simulation platform.

2 Methodology

Our work focus on profiling the code using the Intel VTune Amplifier and identifying hotspots for two typical test cases: the swarm robots scenario and large-scale campus scenario. The Intel VTune Amplifier 2017 is a low overhead performance profiler which can identify and analyze hotspots in serial or parallel application by performing a series of steps in a workflow. It could also provide Elapsed Time, i.e. wall time from the beginning to the end of collection, in the basic hotspots analysis result.

Simulations is performed on a high-performance computing node with 12 CPU cores, 2.1 GHz CPU frequency, and 16 GB RAM. The operating system is Ubuntu 16.04.1 LTS.

In a typical simulation, Gazebo launches two different kinds of processes: gzserver and gzclient. However, we only analyzed hotspots for gzserver since gzclient is only responsible for visualization, which is beyond the aim of this paper. Parameters provided by VTune for the performance analysis and their definitions are shown in Table 1.

Table 1. Definitions of the hotspots analysis parameters

Elapsed time	Wall clock time from the beginning to the end of collection
CPU time	The amount of time a thread spends executing on a logical processor. For multiple threads, the CPU Time of the threads is summed. The application CPU Time is the sum of the CPU Time of all the threads that run the application
Wait time	Time when software threads are waiting due to APIs that block or cause synchronization. Wait Time is per-thread, therefore the total Wait Time can exceed the application Elapsed Time
Spin time	Wait Time during which the CPU is busy

3 Hotspots Analysis of Gazebo

3.1 The Swarm Robots Scenario

The swarm robots case, as shown in Fig. 1, consists of several *quadrotors* provided by the Gazebo platform. This case is used for testing flocking algorithms for multi-agent systems with nonholonomic kinematics. Flocking algorithm is used in many occasions. For example, Tang studied the H7N9 virus transmission by flocking birds [8] and proposed metrics to measure the degree of emergence (outbreak of H7N9 influenza) [9]. During the simulation, each robot is controlled by a Gazebo node and they communicate through boost::asio library. The simulation time increases drastically as the size of the swarm grows. The real time update factor drop down to 0.2 when swarm size reaches 50.

We run the Basic Hotspots Analysis of Vtune Amplifer on the swarm robots scenario with swarm size of 10, 20, 30, 40 and 50 respectively to study the impact of swarm size on hotspots. Each simulation is carried out with 40000 time-steps. The top hotspots are sched_yield(), syscall(), operator

Fig. 1. The swarm robots scenario

new(), sdf::readFile(), tbb::internal::allocate_root_proxy::allocate(), sendmsg(), _GI_pthread_mutex_lock(), _GI_() and _GI_pthread_mutex_unlock().

The Elapsed Time and CPU Time of the whole simulation process are quadratic functions of swarm size as shown in Fig. 2. The CPU Time of sched_yiled(), syscall() also increase quadratically (Fig. 3). The CPU Time of operator new() and sdf::readFile() increase almost linearly with respect to the swarm size (Fig. 4(a)). The CPU Time of _GI_pthread_mutex_lock(), _GI_pthread_mutex_unlock() and _GI_() are rather close and increase almost linearly as the swarm size increases (Fig. 4(b)). The CPU Time of the tbb::internal::allocate_root_proxy::allocate() and sendmsg() show an upward trend but show no obvious relation with swarm size (Fig. 4(c)).

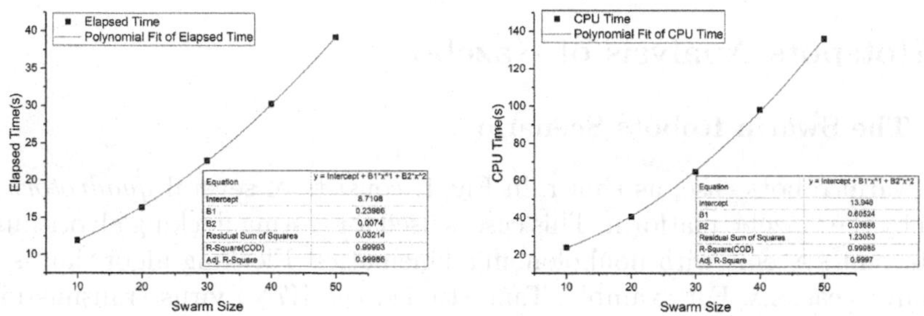

Fig. 2. Influence of swarm size on Elapsed Time and CPU Time of gzserver

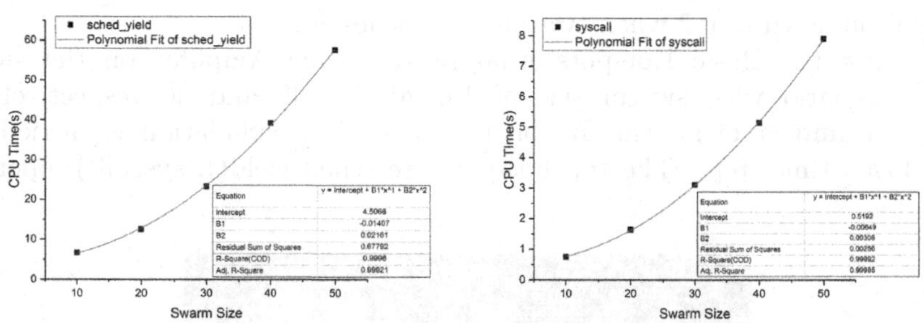

Fig. 3. Influence of swarm size on CPU Time of sched_yield() and syscall()

Assuming the CPU Time and Elapsed Time of gzserver to swarm size $C(x)$ and $E(x)$ respectively where x represents the swarm size. Then $C(x)$ and $E(x)$ can expressed as follows:

$$C(x) = a \times x^2 + b \times x + c$$
$$E(x) = d \times x^2 + e \times x + f$$

The quadratic term of these equations are mainly contributed by syscall() and sched_yield().

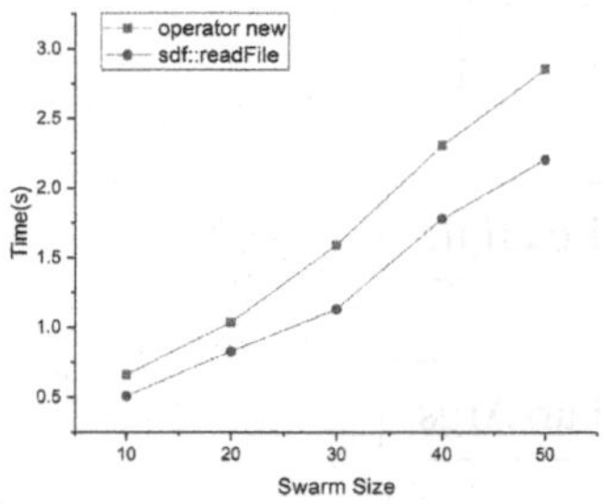

(a) operator new() and sdf::readFile()

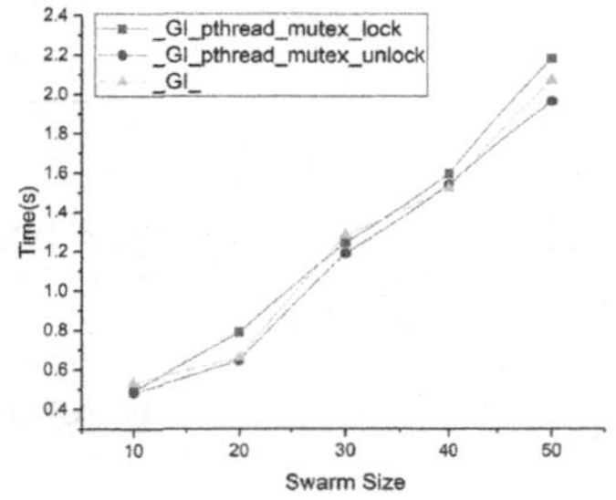

(b) _GI_(), lock() and unlock()

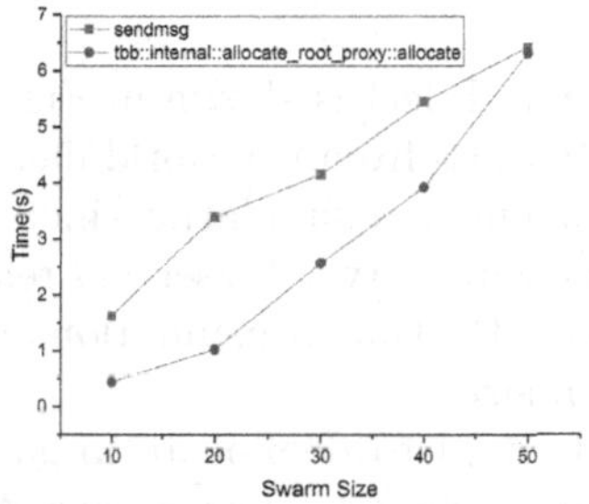

(c) allocate() and sendmsg()

Fig. 4. Influence of swarm size on CPU Time of different functions

syscall() is a small library function that invokes the system call whose assembly language interface has the specified number with the specified arguments [10]. System calls provide an essential interface between a process and the operating system. A process can request a service from the operating system through this interface. This may include hardware-related services, creation and execution of new processes, and communication with integral kernel services such as process scheduling.

sched_yield() is system call which can be invoked by syscall(). It is worth noting that the CPU Time of sched_yield() is spend on spinning. When sched_yield() is called in a process, the process will then be moved to the end of the process queue with the same static priority and a new process gets to run [11]. Strategic calls to sched_yield() can improve performance by giving other threads or processes a chance to run when contended resources (e.g. mutexes) have been released by the caller. However, sched_yield() causes great Spin Time rather than improve performance in this test scenario.

operator new() is used to allocate memory for objects from a pool called the free store [12]. It was called by different module of Gazebo like: gazebo:: transport::Connection, micros_flocking_gazebo_msgs, gazebo::transport::Publisher and so on to allocate memory for them.

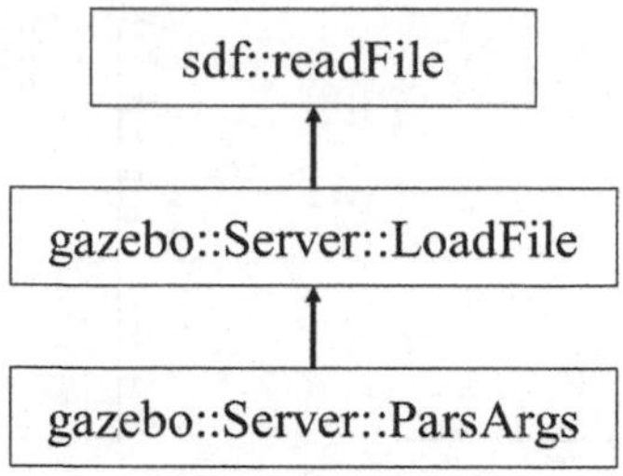

Fig. 5. Call stack of sdf::readFile(). This function is used for reading values for the SDF class from the world file and is called by function gazebo::Server::LoadFile().

The call stack of sdf::readFile() is shown in Fig. 5. This function is used for reading values for the SDF class from the world file. The world file describes the parameters for all the element in a simulation, including robots, lights, sensors, and static objects. The Gazebo server (gzserver) reads this file to generate and populate a world. Therefore, the time consumption of this function grows linearly with respect to the swarm size.

The function _GI_() is used to deallocate msgs of Gazebo. The sendmsg() is used to transmit a message to another socket and is called by boost:: asio library in Gazebo. The tbb::internal::allocate_root_proxy::allocate() is called by operator new(). This function is used to allocate memory for gazebo:: transport::Connection::OnReadData() especially. All these three functions mentioned in this paragraph are related with Gazebo communication among threads. The larger the swarm size the more communications between UAVs. So the CPU Time of these three functions show an upward trend as the swarm size grows.

3.2 The Large-Scale Campus Scenario

The large-scale campus scenario as shown in Fig. 6 is constructed for testing the autonomous navigation algorithm of robot in large-scale environment. Its dimension is 400 m × 300 m. The campus has houses, gas station, cafe environment, etc. Simulation for this scenario with Gazebo encounters problems of long startup time and sharp decline of real time factor when adding models to the simulation case at run-time. In many cases, crash could occur during the simulation. Thus adding models to the existing environment becomes difficult because Gazebo only saves the initial configuration when crashes happen.

We collect the hotspot data of this static scenario within 100000 time iterations. This test scenario contains 201 models. It take 19.427 s from the beginning to the end of collection. 75 threads are spawned during the simulation. The top hotspots are listed in Table 2. sdf::readFile() and sched_yield() are still hotspots in this scenario. Other hotspots functions in the swarm robots case are also used for this simulation. However, they only take 12.7% of total CPU Time.

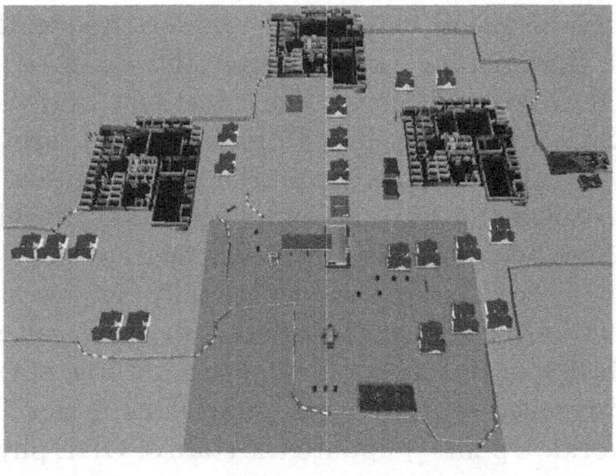

Fig. 6. Part of the large-scale campus scenario

Table 2. Top hotspots of the large campus scenario

Function	Module	CPU time
dxHashSpace::collide()	libgazebo_ode.so.7	4.93 s
sched_yield()	libc.so.6	1.968 s
sdf::readFile()	libsdformat.so.4	1.332 s
ComputeRows()	libgazebo_ode.so.7	1.123 s

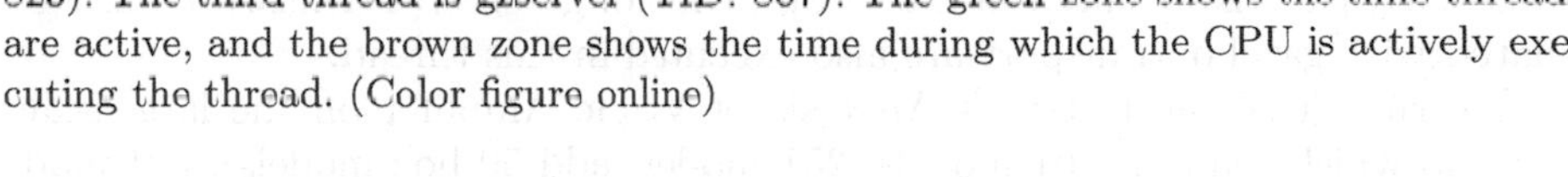

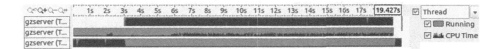

Fig. 7. Distribution of CPU Time utilization of three most computation intensive threads. The first thread is gzserver (TID: 897). The second thread is gzserver (TID: 823). The third thread is gzserver (TID: 807). The green zone shows the time threads are active, and the brown zone shows the time during which the CPU is actively executing the thread. (Color figure online)

There are three threads in 75 threads that account for 87.8% of the CPU workload and the CPU usage of these three threads are given in Fig. 7.

It is shown that the gzserver thread (TID: 807) is frequently executed during the beginning and end of the running process. Check the functions of this thread we find that the sdf::readFile() is executed in this thread. As described in Sect. 3.1, this function is responsible for populating the values of the SDF class from the world file. When the world file in this case contains 201 static objects, it take 1.332 s for sdf::readFile() to read through the world file and populate the SDF class. This process results in long startup time.

The first gzserver thread with TID 897 is frequently executed during the middle of the running process after the gzserver thread (TID:807) populated the SDF class. Checking the functions of this thread we find that the dxHashSpace::collide() and ComputeRows() are executed in this thread and they are all functions from the ODE physics engine. ComputeRows() is used to solve the Linear Complementarity Problem(LCP) in iterative projected Gauss-Seidel Solver. dxHashSpace::collide() determines which pairs of *geoms* in a *hashspace* may potentially intersect, and calls dCollide() callback function with each candidate pair. The call stack of ComputeRows() and dxHashSpace::collide() are shown in Fig. 8(a) and (b) respectively. They are both called by gazebo::physics::ODEPhysics::UpdateCollision(). gazebo::physics::World::Update() update the state of ODE physics engine during every iteration.

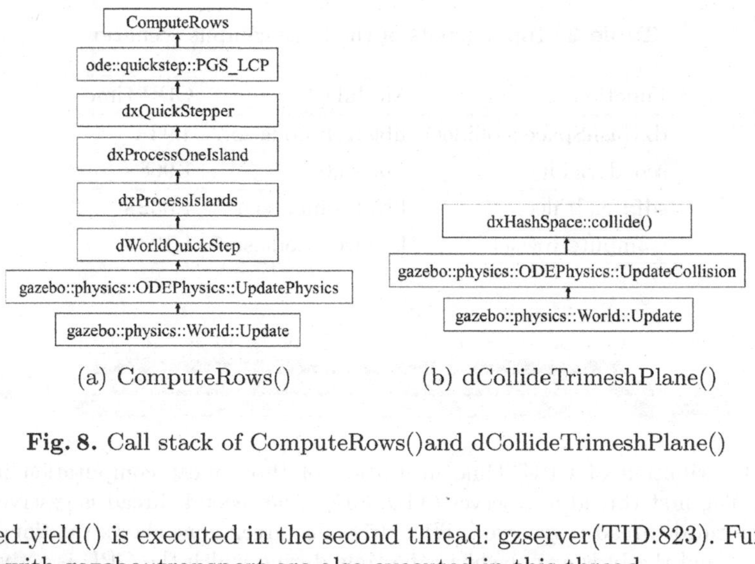

(a) ComputeRows() (b) dCollideTrimeshPlane()

Fig. 8. Call stack of ComputeRows()and dCollideTrimeshPlane()

sched_yield() is executed in the second thread: gzserver(TID:823). Functions related with gazebo::transport are also executed in this thread.

We run the Basic Hotspots Analysis of Vtune Amplifer on the large-scale scenario which contains 201 models, 251 models(add 50 box models), 301 models(add 100 box models), 351 models(add 150 box models), 401 models(add 200 box models) respectively to study the performance of Gazebo as the number of models in the environment increases. Each simulation run 400000 iterations and the top hotspots are: dxHashSpace::collide()sched_yield(), sdf::readFile() and ComputeRows().

The Elapsed Time and CPU Time of the whole simulation process and the CPU Time of sdf::readFile() increase linearly as the models contained in the simulation environment increases as shown in Figs. 9 and 11. The CPU Time of dxHashSpace::collide() is quadratical function of the number of models (Fig. 10). The CPU Time of ComputeRows() shows an upward trend but doesn't show an obvious relation with the number of models as shown in Fig. 11. The CPU Time

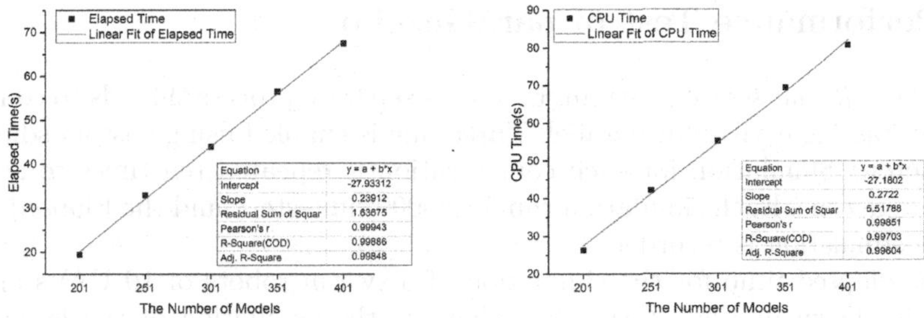

Fig. 9. Influence of the number of models on Elapsed Time CPU Time of gzserver

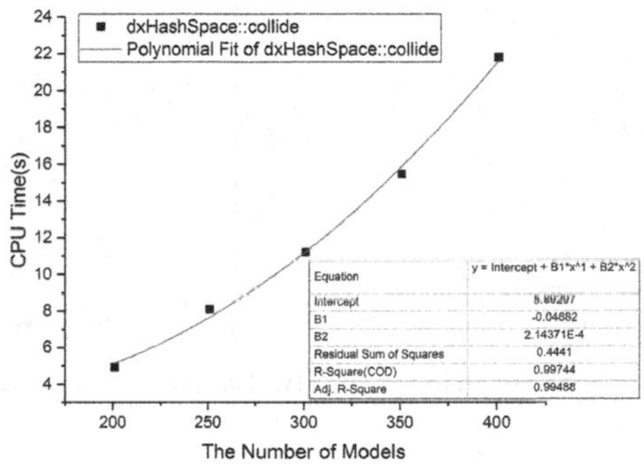

Fig. 10. Influence of the number of models on CPU Time of dxHashSpace::collide()

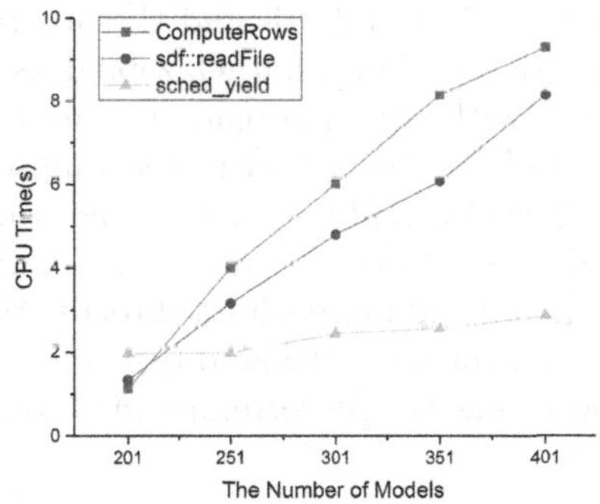

Fig. 11. Influence of the number of models on CPU Time of different functions

results of sched_yield() doesn't change a lot as models contained in the simulation environment increases as shown in Fig. 11.

Considering the increasing influence of quadratical growth on the whole simulation process, the CPU Time and Elapsed Time of gzserver applicant will gradually increase in the form of quadratic function as the number of models increases.

4 Performance Test of paraGazebo

The *island thread* strategy attempts to uses separate process threads to compute non-interacting entities in parallel. Threading is enabled using custom sdformat parameters. Simulation for each configuration is repeated ten times to achieve an average time. Each simulation run for 40000 time-steps and the Elapsed Time for the simulation is recorded.

The elapsed time for the simulation of a swarm robots of 10 UAVs and 30 UAVs is shown in Fig. 12(a), while that for the simulation of the large-scale campus scenario with 201 models and 301 models is shown in Fig. 12(b).

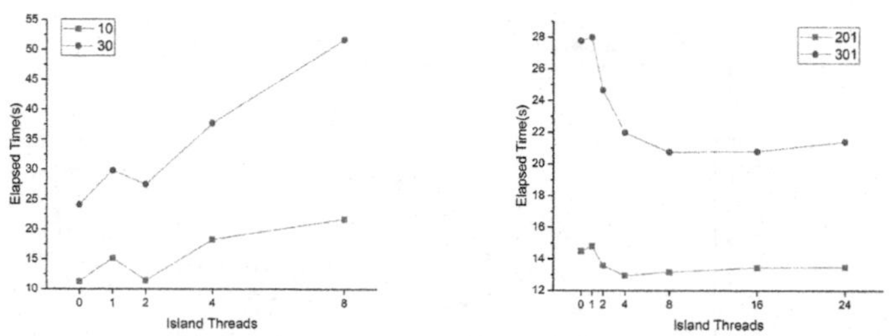

(a) Swarm robot scenario (b) Large-scale campus scenario

Fig. 12. Elapse Time of swarm robot scenario and large-scale campus scenario

The swarm robots scenario has 105, 106, 107, 109, 113 threads when it contains 30 UAVs and uses 0, 1, 2, 4 and 8 island threads respectively. 30 threads among them are control nodes of UAVs. paraGazebo doesn't work well on swarm robot scenario. Simulation with island threads is found more time-consuming. Specifically, when one island thread is spawn, the Elapsed Time rise 23.6% for 30 UAVs and 34.2% for 10 UAVs. This is due to the overhead of data transfer between the single island thread and the main thread. However, better performance can be achieved when using more island threads. We can see in Fig. 12(a), there is an obvious improvement when 2 island threads are used. When the number of island threads is over four, the performance of Gazebo become worse again.

Table 3. CPU Time for hotspot functions when simulating 30 UAVs with none island thread and 8 island threads.

Function	None island thread	8 island threads	Difference
sched_yield()	26.458 s	76.862 s	50.404 s
_GI_pthread_mutex_unlock()	1.318 s	4.88 s	3.562 s
_GI_pthread_mutex_lock()	1.413 s	1.99 s	0.577 s
pthread_cond_wait()	0 s	3.056 s	3.056 s

Table 3 shows the CPU Time for hotspot functions when simulating 30 UAVs. It is worth noticing that most of the CPU Time of sched_field() and pthread_cond_wait() is spent on spinning. We can see there is a significant increment of overhead with multi-threading. _GI_pthread_mutex_unlock() is used to release the mutex object referenced by mutex. _GI_pthread_mutex_lock() is used to lock the mutex object referenced by mutex. The pthread_cond_wait() function is used to block on a condition variable. This function atomically releases mutex and causes the calling thread to block on the condition variable.

The large-scale campus scenario has 75, 76, 77, 79, 83, 91, 99 threads when it uses 0, 1, 2, 4, 8, 16 and 24 island threads respectively. Figure 12(b) shows an improvement of performance with the number of island thread up to 4 for 201 models and 8 for 301 models. The time saved by thread-level parallelization is greater than the overhead from thread operations. When the number of island threads is further increased, the elapsed time increases again due to the competition of hardware resources. The simulation time with 201 models and 301 models shows a similar trend. The speedup with 301 models is higher, reaching a maximum of 1.21 when 8 island threads are used. The island thread mechanism works better when the simulation environment is more complex.

5 Conclusion

The hotspots analysis of Gazebo is carried out with case studies of two typical scenarios using the VTune Amplifier. Our analysis shows that the hotspot functions are quite dependent on the simulation scenarios. We also investigate the effect of the number of models to the simulation time. The Elapsed Time and the CPU Time increase quadratically in the swarm robot case while they increase linearly in the large-scale campus case. The performance of paraGazebo for the two cases is tested. With the *island thread* mechanism provided by Gazebo, a maximum speedup of 1.21 is achieved in the large-scale campus scenario and the simulation of the swarm robot case even slows down due to the thread spinning problem. As future work, we will investigate the causes of the trend of simulation time. Further study on the optimization of the ODE update loop and the threading strategy should also be considered for improving the performance of Gazebo.

References

1. Koenig, N., Howard, A.: Design and use paradigms for Gazebo, an open-source multi-robot simulator. In: Proceedings of 2004 IEEE/RSJ International Conference on Intelligent Robots and Systems, IROS 2004, vol. 3, pp. 2149–2154. IEEE (2004)
2. Cai, Z., Chang, X., Wang, Y., et al.: Distributed control for flocking and group maneuvering of nonholonomic agents. Computer Animation and Virtual Worlds (2017)
3. Manhui, S., Shaowu, Y., Xiaodong, Y., Hengzhu, L.: Autonomous navigation of robot in large-scale environments based on GIS and SLAM. Chin. J. Sci. Instrum. **3**, 586–592 (2017)

4. Michelle, G., Michael, A., Paul, Z.: Parallelizing the open dynamics engine [EB/OL]. http://www.cs.cmu.edu/mpa/ode/final_report.html
5. Ashley-Rollman, M.P., Pillai, P., Goodstein, M.L.: Simulating multi-million-robot ensembles. In: 2011 IEEE International Conference on Robotics and Automation (ICRA), pp. 1006–1013. IEEE (2011)
6. Reinders, J.: Intel threading building blocks: outfitting C++ for multi-core processor parallelism. O'Reilly Media, Inc. (2007)
7. Open Source Robotics Foundation: DARPA robotics challenge OSRF Gazebo parallel physics report (2015)
8. Tang, M., Mao, X., Guessoum, Z.: Research on an infectious disease transmission by flocking birds. Sci. World J. **2013**(2013) (2013). Article ID 196823
9. Tang, M., Mao, X.: Information entropy-based metrics for measuring emergences in artificial societies. Entropy **16**(8), 4583–4602 (2014)
10. man7.org: Linux/UNIX system programming training [EB/OL]. http://man7.org/linux/man-pages/man2/syscall.2.html
11. man7.org: Linux/UNIX system programming training [EB/OL]. http://man7.org/linux/man-pages/man2/sched_yield.2.html
12. Microsoft: Developer network/new and delete operators [EB/OL]. https://msdn.microsoft.com/en-us/library/kftdy56f.aspx

CNP Based Satellite Constellation Online Coordination Under Communication Constraints

Guoliang Li(✉), Lining Xing, and Yingwu Chen

College of Information System and Management, National University of Defense Technology, No.109, Deya Rd., Kaifu District, Changsha, Hunan, China
worldchinali@126.com

Abstract. This paper focuses on online managing Earth observation satellite constellation under dynamic environment, like detecting, observing, and tracking forest fires or volcanic eruptions without ground interval. In reality, the inter-satellite communication is limited by practical reasons. The objective is to maximize the total profit of the satellite constellation by increasing the efficiency of onboard resources and coordinating the different satellite to provide a greater level of responsiveness and adaptability, subject to communication time window constraints and observation time window constraints. Firstly, the online scheduling algorithm for a single satellite is proposed on basis of revision techniques and progressive techniques. Then, we propose the novel online coordination mechanism and the core algorithm based on Contract Net Protocol.

Computational experiments indicate that the profit of the proposed online coordination scheduling mechanism is much higher than online individually scheduling mode and offline scheduling mode. Especially, to the stochastic arrival of urgent task, the proposed online coordination mechanism outperforms the other two modes both on the number and the percentage of successfully scheduled urgent tasks under different spatial and temporal distribution.

Keywords: Earth observation satellite constellation · Online scheduling · Dynamic environment · Coordination · Contract Net Protocol

1 Introduction

Space mission planning & scheduling system is historically implemented on the ground segment for the operation complexity and the computation power required [1]. Two attracting advantages, reducing operational costs and increasing the efficiency, are pushing the realization of planning and scheduling on board [2].

In the last decade, the interest for space missions involving multiple spacecrafts is rapidly growing. Distributed systems of small satellites offer interesting capabilities to complement traditional Earth observation satellites with respect to increasing temporal and spatial resolution. Some remarkable examples are Disaster Monitoring Constellation, Rapid Eye and SAR-Lupe. A number of studies

Y. Huang et al. (Eds.): ICIRA 2017, Part I, LNAI 10462, pp. 683–695, 2017.
DOI: 10.1007/978-3-319-65289-4_64

have recently shown interest for the disaster management, focusing on sensor web [3] or just on Earth Observation constellations [4–6]. In these works, the plan is generated on ground and uploaded to the constellation, so the need of autonomy is shifted from onboard to the ground segment. Furthermore, these works neglected the dynamics of the problem itself. As the low earth orbit (LEO) satellite first detects the happening of the disaster on-board, many images are needed in a very short time frame. The coordination on constellation level is therefore foreseen to run on-board to respond in a shorter time. Additionally, each satellite is within direct communication of a ground station only 2%–5% of the total time. The time between two successive visibility windows between a given LEO satellite and a given ground station depends on the station latitude, but is very irregular along time: from 100 min (one revolution) to more than 15 h. The communication link among these Earth observation satellites is a critical resource. In this case, the real big challenge is online coordination and optimization in face of disaster happening.

A limited amount of literature exists that deals explicitly with online coordination in a distributed satellite system. Grégory Bonnet [7] realized on-board planning for a satellite swarm via Inter-Satellite Links inspired from the Fuego mission [8,9], through a hybrid agent approach. At the individual level, agents are deliberative in order to create a local plan but at the collective level, they use normative decision rules in order to coordinate with one another. The satellites referred are placed in different sun-synchronous orbits and only communicate with each other in the polar areas. J. van der Horst [10] extended a market-based mechanism to manage task allocation in the distributed satellite system under Keplerian mobility model, but didn't consider the scheduling problem at all.

The contract net protocol (CNP) has been developed to achieve task assignment by a negotiation process in multi-agent systems. So far, most methods for Multi-robot task allocation (MRTA) are based on the CNP model. The protocol is composed of a sequence of three main steps:

Step 1: a task is published to the agents by some entity, usually called manager.

Step 2: agents suited to the tasks reply to the manager with a bid indicating the capabilities to execute the announced task, these agents are called bidders.

Step 3: the manager may receive several bids, chooses the best bid and awards the contract to the corresponding bidder. At the same time, the manager rejects the other bids.

The rest of the paper is organized as follows. Section 2 describes the space mission in need of online coordination, aiming at disaster management. Section 3 describes the online scheduling problem in the constellation, presents the single satellite online scheduling algorithm and proposes the online coordination mechanism. Section 4 sets up a scenario including a LEO satellite constellation, conducts the experiments and presents the results. Finally, Sect. 5 provides conclusions.

2 Case Study

Inspired from the Bird (http://spacesensors.dlr.de/-SE/bird/) and Fuego [8,9] projects, the space mission in this paper aims to detect, to observe, and to track forest fires or volcanic eruptions. In case of these urgent events happening, we focus on increasing the system reactivity and improving the global return by use of online coordination.

The online arrival of observation requests are from three main sources: task self-generation of the satellites in the constellation, task generation and broadcasting from the satellites out of the constellation, and the emergency upload by the ground station.

The Earth observation satellites are in the low earth orbits. The inter-satellite links (ISL) are not always available. For example, RapidEye is a commercial multispectral Earth observation mission including a constellation of five small satellites. The satellites are placed equally spaced in a single 630 Km sun-synchronous orbit to ensure consistent imaging conditions and a short revisit time (about 19 min intervals). We can notice that there is no inter-satellite link among the five satellites, because the number and the altitude of satellites cannot permit the existence of ISL. So we can only operate and coordinate these satellites offline in the ground station. Inspired by [8], we make full use of the geostationary orbit (GEO) satellites and relax the communication shortage to a certain extent. The first big change is that the communication time window between GEO and LEO is much broader than that between the ground station and the LEO satellite, and another change is that the GEO can keep the continuous request upload and data download with the ground station located in its coverage area. So in this way, the constellation online coordination becomes realistic, supporting a greater level of disaster management.

Figure 1 illustrates the communication links between space and ground components.

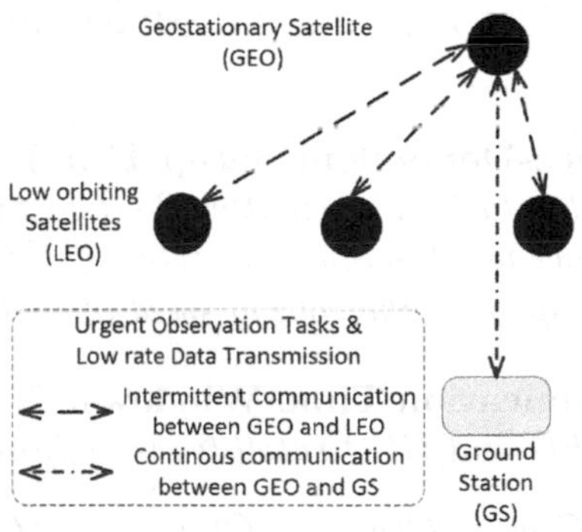

Fig. 1. Possible communications between space and ground components.

3 Proposed Approach

3.1 Problem Representation

In the Online Scheduling Problem for the satellite constellation, there are two types of tasks, normal tasks and urgent tasks. The ground control center receives observation requests from users and schedules them, then uploads the generated observation plan through the space-ground link. In this paper, these observation requests are called as normal tasks, and those stochastically arrived on-board from the satellite inside or outside the constellation itself as urgent tasks. The unforeseen urgent tasks need to be scheduled online without the ground interval. The profit of urgent task is usually higher than normal task for the need of rapid responsiveness and high value.

Definition 1 (Urgent Task). *An urgent task has two forms, an original one and a specific one. When generated on-board or uploaded from ground, the original form of urgent task j in batch o is a tuple* $<id_{oj}, longtitude_{oj}, latitude_{oj}, p_{oj}, e_{oj}>$*:*

- id_{oj} *is an identifier;*
- $longtitude_{oj}$ *is the longitude value of urgent observation target on the earth;*
- $latitude_{oj}$ *is the latitude value of urgent observation target on the earth;*
- p_{oj} *is the imaging time needed;*
- e_{oj} *is the profit if the task is completed.*

When referring to a specific LEO satellite, the urgent task transforms its form into a specific one. This transforming process is essential for further scheduling, including Orbit Prediction, Ephemeris Calculation and Observation Time Window Calculation. The specific form is a tuple $<id_{oj}, lsat_i, oes_{ioj}, olf_{ioj}, p_{oj}, e_{oj}, \theta_{ioj}>$*, corresponding to identifier of this urgent task, identifier of the specific LEO satellite, earliest start date, latest finish date, imaging time, profit and observation angle. The observation time window is the time interval between earliest start date and latest finish date, as* $otw_{ioj} = [oes_{ioj}, olf_{ioj}]$.

Definition 2 (Sequence-Dependent Setup Time). *Assuming that task b is executed successively after task a, no matter the two tasks are normal type or urgent type, sequence-dependent setup time between the two tasks is calculated as:* $s_{ab} = \frac{|\theta_a - \theta_b|}{sp_{roll}}$*, where* sp_{roll} *is the rolling speed of LEO satellite.*

Definition 3 (Communication Time Window). *The set of Communication time windows among LEO and GEO satellites is defined as:*

$$CTW = \{Ctw_{11}, ..., Ctw_{1n_2}; Ctw_{21}, ..., Ctw_{2n_2}; ......; Ctw_{n_1 1}, ..., Ctw_{n_1 n_2}\}$$

And the set of Communication Time Windows for LEO satellite i and GEO satellite g is defined as:

$$Ctw_{ig} = \{ctw_{ig1}, ctw_{ig2}, ..., ctw_{igq}, ..., ctw_{igm_{ig}}\}$$

where the communication time window ctw_{igq} *is defined as:*

$ctw_{igq} = [ces_{igq}, clf_{igq}]$
where ces_{igq} *is the communication earliest start date,* clf_{igq} *is the communication latest finish date.*

Definition 4 (Batch Availability and Urgent task Time-Validity). *The urgent task batch o can reach to LEO satellite i, if and only if there exists one or more communication time window(s) between LEO satellite i and GEO satellite(s). We define the indicator of batch availability as* w_{io}*:*

$$w_{io} = \begin{cases} 1, & if \exists g, \exists q, \exists tr_{io}, (ces_{igq} < tr_{io} < clf_{igq}) \cap (tr_{io} < r_o) = 1 \\ 0, & otherwise. \end{cases}$$

where tr_{io} *is the arrival time of urgent task batch o at LEO satellite i.*

Urgent task j in batch o is time-valid for scheduling and execution for LEO satellite i, if and only if the urgent task batch o arrives at LEO satellite i before the observation latest start time ols_{ioj}*. So we define the indicator of urgent task time-validity* w_{ioj} *as:*

$$w_{ioj} = \begin{cases} 1, & if (ols_{ioj} < tr_{io}) \cap (w_{io} == 1) = 1 \\ 0, & otherwise. \end{cases}$$

where $ols_{ioj} = max\{olf_{ioj} - p_{oj} - (2 \cdot max\theta)/sp_{roll}, oes_{ioj}\}$.

The objective is as follows:

$$Max \sum_{i=1}^{n_1} \left(\sum_{o=1}^{l} \sum_{j=1}^{u} w_{ioj} \cdot x_{ioj} \cdot e_j + \sum_{k=1}^{v_i} z_{ik} \cdot e_{ik}\right)$$

Decision variables

$$x_{ioj} = \begin{cases} 1, & if\ urgent\ task\ j\ from\ batch\ o\ is\ scheduled\ into\ LEO\ satellite\ i \\ 0, & otherwise. \end{cases}$$

$$z_{ik} = \begin{cases} 1, & if\ normal\ task\ k\ is\ maintained\ in\ LEO\ satellite\ i \\ 0, & otherwise. \end{cases}$$

3.2 The Online Scheduling Algorithm for Single Satellite

As described in [11], there are three main families of techniques for scheduling under uncertainty: proactive techniques, revision techniques, and progressive techniques. In our paper, the online scheduling is executed on basis of the complete schedule generated and uploaded by the ground operator. This case is fit for the application of revision techniques. Additionally, progressive techniques can be applied to limit the rescheduling horizon regarding to the arrival of urgent tasks. Given the set of normal tasks, the plan can be generated progressively with the period T. So we define the rescheduling time point as the arrival date of urgent task batch, and the rescheduling horizon is constrained by predefined timestamps.

The online rescheduling algorithm can be described as follows (Fig. 2).

Input:

$sc_i^{\sim t}$- the schedule of LEO satellite i preserved until the online rescheduling time point t, covering the rescheduling horizon between time point t and the next predefined timestamp;

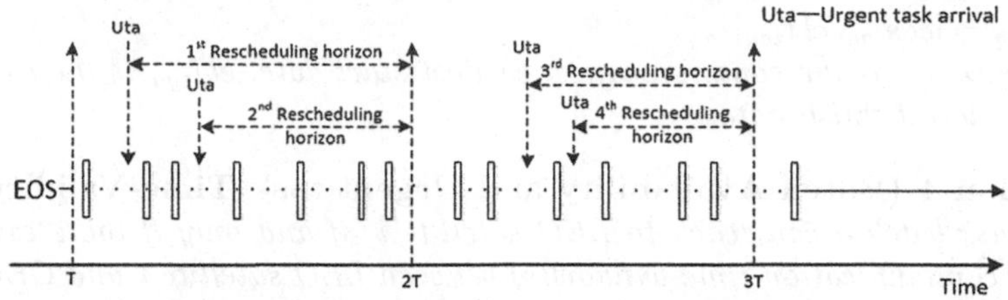

Fig. 2. The illustration of online scheduling mechanism for single satellite.

$Ub^{\sim t}$- the batch of urgent tasks received by the satellite and being unscheduled until the time t;

Output:

$\overline{sc}_i^{t\sim}$- the revised schedule at time point t.

$\Delta E^{t\sim}$- the value change on the total profit of revised schedule $sc_i^{t\sim}$ compared with $sc_i^{\sim t}$.

Begin Procedure

Step 1 Sort all the received urgent tasks $Ub^{\sim t}$ by the modified Weighted Shortest Imaging Time First (m-WSITF) heuristic, selecting and scheduling the task that has the highest $e_{ioj}/(p_{ioj} + \overline{s}_{previousTask,ioj})$ ratio among the available tasks;

Step 2 Revise the schedule $sc_i^{\sim t}$ by selecting the urgent task one by one following the above new order until no more urgent tasks can be successfully scheduled. The operators are insertion and replacement. The scheduling criteria is improving the total profit.

Step 3 Output the revised schedule $sc_i^{t\sim}$ and the value change $\Delta E^{t\sim}$ on total profit.

End Procedure

3.3 Online Coordination Mechanism and Algorithm

In our paper, GEO satellite plays the role of manager, responsible for announcing the urgent tasks, evaluating the bids and responding the result to all bidders. If there is more than one GEO satellite, the role of manager can be transferable among these GEO satellite, and the reliability improves. Seen as resource agent, LEO satellite cannot communicate with other LEO satellites directly, and the connection link between manager and resource agent is intermittent.

The pseudo-code of proposed online coordination mechanism can be described as follows.

Require: LEO satellite set $Lsat = \{lsat_1, lsat_2, ..., lsat_{n_1}\}$;
GEO satellite set $Gsat = \{gsat_1, gsat_2, ..., gsat_{n_2}\}$;
Urgent task set $Ut = \{ut^1, ut^2, ..., ut^{(u \cdot l)}\}$;
Original schedule set $Sc = \{sc_1, sc_2, ..., sc_{n_1}\}$;
Communication time window set CTW;
Size of an urgent task batch u.

Ensure: Revised schedule set $\overline{Sc} = \{\overline{sc_1}, \overline{sc_2}, ..., \overline{sc_{n_1}}\}$

Begin Procedure
$t \leftarrow 0$; $j \leftarrow 0$; $o \leftarrow 1$; $t \leftarrow (t+1)$;
if $0 < t < tl$ **then**
  if new urgent task $ut \in Ut$ arrives at $Gsat$ **then**
    $j \leftarrow (j+1)$
    $ut \leftarrow < id_{oj}, longtitude_{oj}, latitude_{oj}, p_{oj}, e_{oj} >$
    if $j = u$ **then**
      $r_o \leftarrow t$;
      $Ub \leftarrow \{ut^{o \cdot u-u+1}, ut^{o \cdot u-u+2}, ..., ut^{o \cdot u}\}$;
      $\overline{Ub} \leftarrow Sort(Ub)$;
      $\overline{Sc}^{t\sim} \leftarrow ContractNetProtocol(Lsat, Gsat, \overline{Ub}, Sc^{\sim t}, r_o, u, o)$;
      $j \leftarrow 0$;
      $o \leftarrow (o+1)$;
    end if
    $t \leftarrow t+1$;
  end if
end if
End Procedure

Algorithm 1. CNP-based Online Coordination Scheduling Algorithm

ContractNetProtocol()
Begin Procedure
for $i \leftarrow 1$ *to* n_1 **do**
  $(w_{io}, tr_{io}) \leftarrow BatchAvailability(Gsat, ut_{io}, ctw_{ig}, r_o)$;
end for
for $j \leftarrow 1$ *to* u **do**
  $temp\Delta E_{oj} \leftarrow 0$; $Contractor_{oj} \leftarrow 0$;
  for $i \leftarrow 1$ *to* n_1 **do**
    $ut_{ioj} \leftarrow < id_{oj}, lsat_i, oes_{ioj}, olf_{ioj}, p_{oj}, e_{oj}, \theta_{ioj} >$;
    $w_{ioj} \leftarrow TimeValidity(ut_{ioj}, tr_{io})$;
    if $w_{ioj} = 1$ **then**
      $(\overline{sc_j}, \Delta E_{ioj}) \leftarrow OnlineScheduling(sc_i, ut_{ioj}, tr_{io})$;
      if $\Delta E_{ioj} > temp\Delta E_{oj}$ **then**
        if $Contractor_{oj} \neq 0$ **then**
          $tempID \leftarrow Contractor_{oj}$
          $\overline{sc_{tempID}} \leftarrow sc_{tempID}$;
        end if
        $temp\Delta E_{oj} \leftarrow \Delta E_{ioj}$;
        $Contractor_{oj} \leftarrow i$;
      end if
    end if
  end for
end forreturn $\overline{Sc}$
End Procedure

The bubble sorting for the urgent tasks in one batch needs $O(u^2)$, and the time validity check for a specific batch to a specific LEO satellite by using of Binary-Search method takes $O(log2(\overline{m}) \cdot n_2)$, $\overline{m} = max_{g=1}^{n_2}(m_{ig})$. Here the online scheduling algorithm for one urgent task on a specific satellite requires $O((\frac{v_i \cdot T}{tl})^2)$. So the overall computational complexity of online coordination scheduling mechanism is $O(l \cdot u^2 + l \cdot n_1 \cdot (log2(\overline{m}) \cdot n_2 + u \cdot (\frac{v_i \cdot T}{tl})^2))$.

4 Computational Studies

4.1 Scenario Setting

The simulation scenario implements 3 LEO satellites on the same orbit, and 3 GEO satellites above the equator. The time length of simulation is set as 6 h (Table 1).

Table 1. The setting of orbit parameters for 3 LEO and 3 GEO satellites.

Orbit parameter	LEO_A	LEO_B	LEO_C	GEO_3	GEO_2	GEO_1
Semimajor axis (/km)	6878.14			42164.20		
Eccentricity	0.001			0		
Inclination (/degree)	45			0		
Argument of perigee (/degree)	0			0		
RAAN (/degree)	0			119.32	239.32	359.32
Mean anomaly (/degree)	134.92	89.88	44.92	0		

The other related characteristics of this scenario are as follows:

1. To the three serial LEO satellites on the same orbit, when referring to imaging the same target on the earth, the time interval between LEO_A and LEO_B is about 12 min, so as to the time interval between LEO_B and LEO_C.
2. The communication coverage area of GEO satellite is the zone in its circular cone with cone angle being 7.5°.
3. The observation coverage area of LEO satellite is rectangular. The maximum rolling angle to one side apart from the center point is 45°. The Satellite Rolling speed is 5°/s. The change value of imaging angle between LEO_A and LEO_B is around 20°, so as to the change on imaging angle between LEO_B and LEO_C.

4.2 Experimental Results

We compare three modes of managing the constellation:

1. The offline scheduling mode, where the observation plan for the constellation is completely generated on ground, and each sub-plan is uploaded to the corresponding LEO satellite at the beginning of the whole scheduling horizon.

2. The online individually scheduling mode. Once GEO satellites have collected enough urgent tasks (exceeded a prescribed threshold), they will find the earliest communication link to transmit these urgent tasks to the available LEO satellite in the constellation. The LEO satellite receives the batch of urgent tasks and schedules them immediately.
3. The online coordination scheduling mode. Based on Contract Net Protocol, GEO satellite takes the role of manager, and LEO satellites in the constellation are potential contractors due to limited communication between GEO and LEO satellites.

The computational complexity of online scheduling mechanism without coordination is $O(l \cdot u^2 + l \cdot (n_1 \cdot log2(\overline{m} \cdot n_2 + (\frac{v_i \cdot T}{tl})^2))$. Compared with online coordination scheduling mechanism, the different point is in the third item.

To evaluate the performance of the three modes on various types of problems, we focus on the responsiveness of satellite constellation to the arrival of urgent tasks, so the two task-relevant factors, Arrival rate and Pointing angle distribution, are important for generating test instances. These two factors represent the temporal distribution and the spatial distribution of urgent tasks respectively. Moreover, all other factors with less of an impact were generated from a given uniform distribution. The details are in Table 2.

Table 2. The factor setting for the tasks and mechanism

(a) Normal task-relevant factor setting	
Number of normal tasks for generating offline plan	720
Imaging time of normal task	Uniform(5,10)
Observation time window of normal task	Uniform(15,20)
Profit of normal task	Randi(1,4)
Rolling angle distribution of normal task	Uniform(−45,45)
(b) Mechanism-relevant factor setting	
The number of urgent tasks in one batch u	adaptive threshold
The interval between two neighbor timestamps T	6 min
(c) Urgent task-relevant factor setting	
Arrival rate λ	{0.003,0.008,0.016,0.025,0.033}
Imaging time	Uniform(3,8)
Observation time window	Uniform(10,20)
Profit	Randi(10,15)
Pointing angle distribution	Uniform{(−45,45), (−60,60), (−80,80)}

Figures 3, 4 and 5 shows that for the satellite constellation, both the proposed online coordination mode and online individually scheduling mode outperform

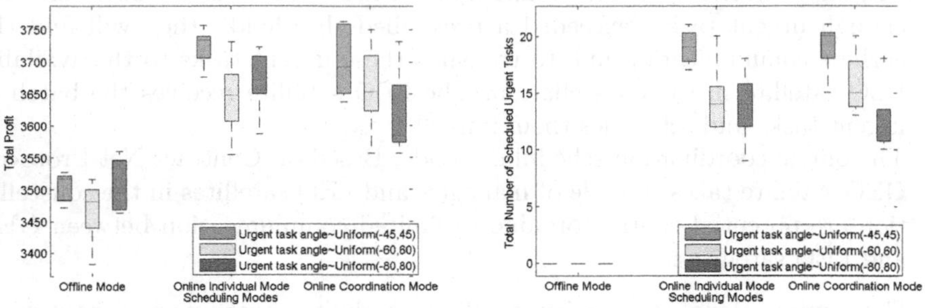

(a) Profit (b) Scheduled Urgent Tasks

Fig. 3. The results by different modes for urgent task arrival rate being 0.003.

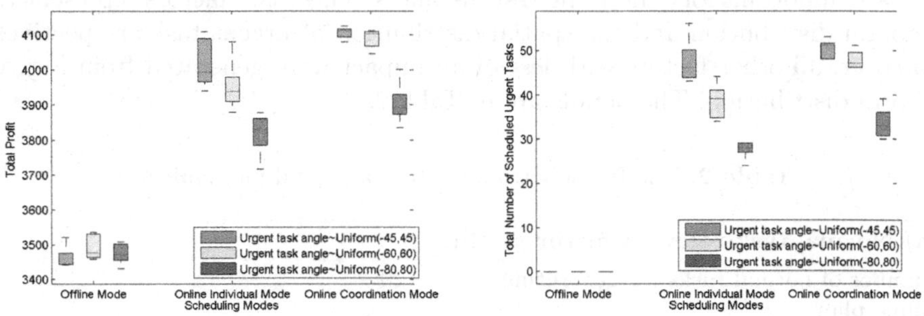

(a) Profit (b) Scheduled Urgent Tasks

Fig. 4. The results by different modes for urgent task arrival rate being 0.008.

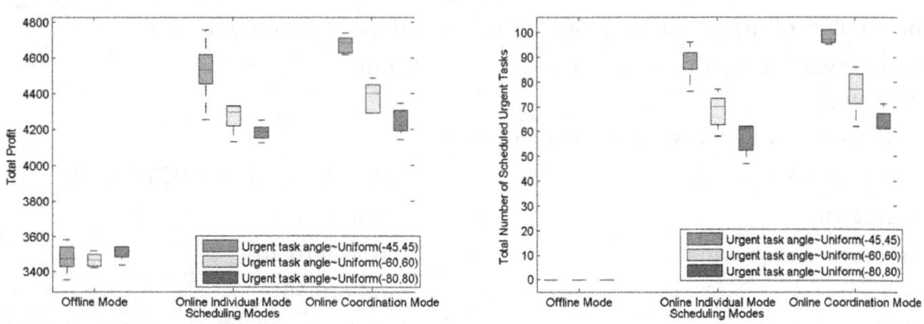

(a) Profit (b) Scheduled Urgent Tasks

Fig. 5. The results by different modes for urgent task arrival rate being 0.016.

Table 3. Profit improvement (%) by online individual/coordination mode compared with offline mode

Urgent task arrival rate	Pointing angle	Offline mode	Online individual mode	Online coordination mode
0.003	Uniform(−45,45)	0	6.22	5.6
	Uniform(−60,60)	0	5.09	5.57
	Uniform(−80,80)	0	4.69	3.21
0.008	Uniform(−45,45)	0	16.04	18.51
	Uniform(−60,60)	0	13.09	16.98
	Uniform(−80,80)	0	9.8	12.29
0.016	Uniform(−45,45)	0	30.04	34.24
	Uniform(−60,60)	0	22.96	26.2
	Uniform(−80,80)	0	19.39	21.02
0.033	Uniform(−45,45)	0	45.86	53.89
	Uniform(−60,60)	0	38.58	51.34
	Uniform(−80,80)	0	33.37	43.73

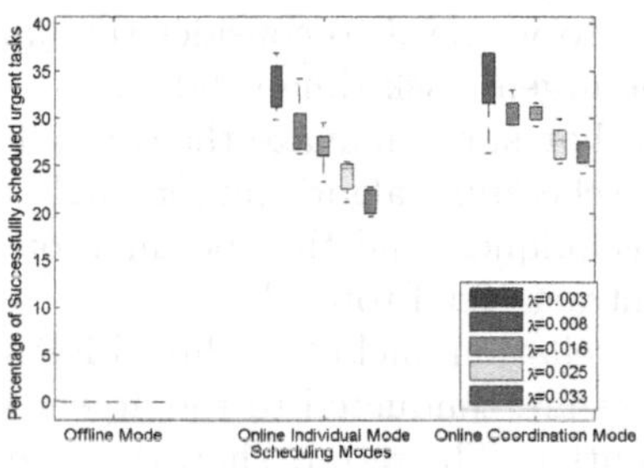

(a) Uniform(-45,45)

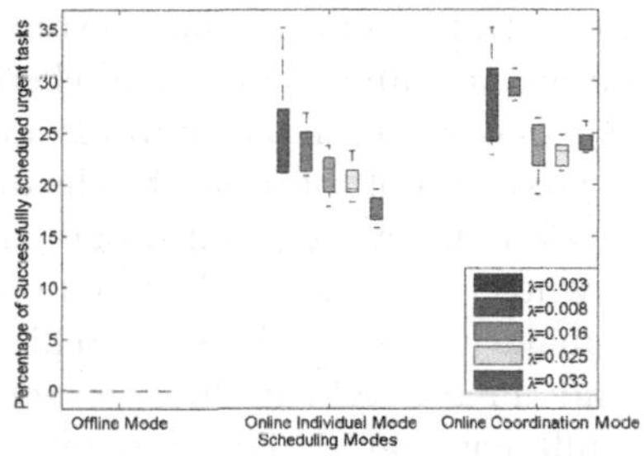

(b) Uniform(-60,60)

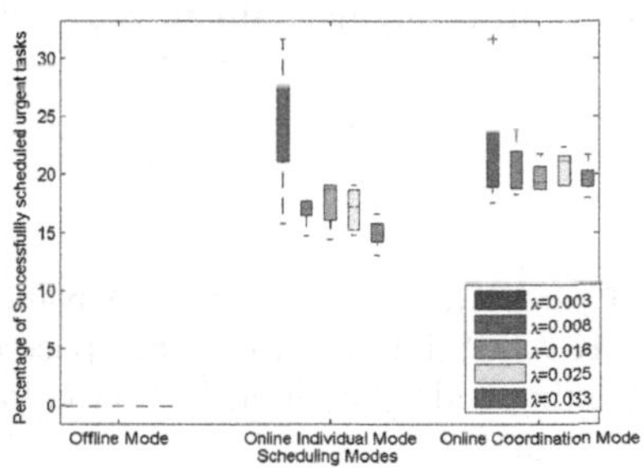

(c) Uniform(-80,80)

Fig. 6. The results by different modes for urgent tasks under different spatial and temporal distribution.

offline scheduling mode on total profit and total number of successfully scheduled urgent tasks, and the lead becomes wider as the arrival rate of urgent task rate increases. Additionally, the proposed online coordination mode performs better than online individually scheduling mode.

Table 3 describes the details of profit improvements by two online scheduling modes. We can notice that for different pointing angle distributions, the proposed online coordination mode performs best among these three modes, especially in the distribution between −60 and 60. It shows that the proposed method bitterly improves the utilization of the resources in the constellation.

Figure 6 presents the percentage of successfully scheduled urgent tasks under different spatial and temporal distribution by the three modes. Obviously, the offline mode can't respond to any urgent task arriving at the constellation throughout the whole scheduling horizon. The online coordination scheduling mode outperforms the online individually scheduling mode under any spatial and temporal distribution.

5 Conclusions

This paper described the online scheduling problem for earth observation satellite constellation under the stochastic arrival of urgent tasks. There exist communication constraints among the constellation, so we need to consider the limit of communication time window and check the urgent task time validity.

We proposed an online coordination mechanism to manage the constellation with a maximum of total profit. The online scheduling algorithm for single satellite is based on revision and progressive techniques, and the coordination part for the whole constellation is based on Contract Net Protocol.

We make up a scenario of a satellite constellation, including three LEO satellites, and three GEO satellites. Experiments are conducted to the urgent tasks under different spatial and temporal distribution. The results show that the proposed online coordination mechanism outperforms the other two modes both on the number and the percentage of successfully scheduled urgent tasks.

Future we will investigate into the mechanism-relevant factors to improve the efficiency of mechanism.

References

1. Steel, R., Niezette, M., Cesta, A., Fratini, S., et al.: Advanced planning and scheduling initiative: MrSPOCK AIMS for XMAS in the space domain. In: 6th International Workshop on Planning and Scheduling for Space, Pasadena, California, USA (2009)
2. Montenbruck, O., Markgraf, M., Naudet, J., et al.: Autonomous and precise navigation of the PROBA-2 spacecraft. In: AIAA/AAS Astrodynamics Specialist Conference, Honolulu, Hawaii (2008)
3. Chien, S., Doubleday, J., McLaren, D., et al.: Combining space-based and in-situ measurements to track flooding in Thailand. Geosci. Remote Sens. Symp. **58**(11), 3935–3938 (2011)

4. Pralet, C., Verfaillie, G., Olive, X.: Planning for an ocean global surveillance mission. In: 7th International Workshop on Planning and Scheduling for Space, ESOC, Darmstadt (2011)
5. Grasset-Bourdel, R., Verfaillie, G., Flipo, A.: Building a really executable plan for a constellation of agile earth observation satellites. In: 7th International Workshop on Planning and Scheduling for Space, ESOC, Darmstadt (2011)
6. Raghava-Murthy, D.A., Kesava-Raju, V., Srikanth, M., Ramanujappa, T.: Small satellite constellation planning for disaster management. In: International Astronautical Federation, Prague, Czech Republic (2010)
7. Bonnet, G., Tessier, C.: Collaboration among a satellite swarm. In: 6th International Joint Conference on Autonomous Agents and Multiagent Systems, Honolulu, Hawaii (2007)
8. Damiani, S., Verfaillie, G., Charmeau, M.C.: An Earth watching satellite constellation: how to manage a team of watching agents with limited communications. In: 4th International Joint Conference on Autonomous Agents and Multiagent Systems, Utrecht, The Netherlands (2005)
9. Escorial, D., Tourne, I.F., Reina, F.J.: FUEGO: a dedicated constellation of small satellites to detect and monitor forest fires. Acta Astronaut. **52**(9–12), 765–775 (2003)
10. Van Der Horst, J., Noble, J.: Task allocation in dynamic networks of satellites. In: 22nd International Joint Conference on Artificial Intelligence, Barcelona, Spain (2011)
11. Bidot, J., Vidal, T., Laborie, P., Beck, J.C.: A theoretic and practical framework for scheduling in a stochastic environment. J. Sched. **12**, 315–344 (2009)

Underwater Robotics

Research on Fault-Tolerant Control Method of UUV Sensor Using Walcott-Zak Observer

Zheping Yan, Yingming Bi(✉), and Tao Chen

Harbin Engineering University, Harbin 150001, Heilongjiang, China
sdubiyingming@126.com

Abstract. When UUV performs a task in a complex marine environment, it is very dangerous to have a sensor failure. So it is essential to study a method of sensor fault tolerant control in a UUV. In order to study the fault-tolerant control problem of UUV sensor, the 3-DOF motion mathematic model of UUV is founded. In this paper, the expert S controller is used as the nominal controller of UUV, it has better adaptability than the intelligent integral controller. Based on the virtual sensor reconfiguration strategy, the Walcott-Zak observer is added to the UUV model to solve the uncertain interference. An expert S surface controller has been used to control the speed and the heading angle of UUV. Through a series of simulations, It is effective of the sensor fault tolerant control based on the Walcott-Zak observer. UUV can continue to maintain a stable voyage even when some sensor is faulty.

Keywords: Walcott-Zak observer · Expert S surface controller · Fault tolerant control

1 Introduction

With the continuous development of science and technology, UUV has been used widely in many fields, like ocean exploration. UUV is playing a more and more important role in the fast development of ocean exploration. The reliability of UUV which can perform more complicated missions is very important. The fault tolerant control ability is the indispensable constituent which ensure the UUV can safely and steadily navigate in the event of fault.

The concept of active fault tolerant control comes from the idea of active handling of the fault and the key point of active fault handling is the acquisition of fault information and the change of the system state. Active fault tolerant control needs to adjust the parameters of the controller and the structure of the controller [1, 2]. The original controller is debugged and tested many times, which contains a large amount of equipment information and experience. The method of virtual sensor recombination for sensor fault can be used to maintain the feature of normal sensor. This method realizes the reconstruction on the basis of original controller. A fault tolerant control strategy using a virtual sensor for linear parameter varying system is proposed, this method is to reconfigure the control loop such that the nominal controller could still be used without need of changing it [3]. Zhu Daqi used finite impulse response filter instead of fault sensor output [4]. Zhang Mingguang combined the nonlinear sliding

Y. Huang et al. (Eds.): ICIRA 2017, Part I, LNAI 10462, pp. 699–710, 2017.
DOI: 10.1007/978-3-319-65289-4_65

mode observer and virtual sensor, and the improved system can be equivalent to the original nonlinear uncertain systems [5]. Lan, JL and Patton, RJ propose an integrated fault estimation and fault-tolerant control design for Lipschitz non-linear systems subject to uncertainty, disturbance, and sensor faults [6]. Ben Zina H, Allouche M present an active fuzzy fault-tolerant control strategy for induction motor that ensures the performances of the field-oriented control [7]. Sun Hui proposed a method for addressing the problem of sensor fault-tolerant control for anti-skid braking systems [8]. Zhu JW investigates the observer-based fault-tolerant control problem for linear system with multiple faults and external disturbances [9].

2 Mathematic Model of the UUV

The UUV is a nonlinear and coupling system. It is under the influence of both external forces and internal forces. During the sensor fault tolerant research, a series of simulation experiments were conducted by the mathematic model of UUV. So establishing a proper mathematic motion model was the first step of the research [10].

2.1 Establish Coordinate System

First of all, a special reference frame must be establish for analyzing the spatial motion. In this paper two coordinate systems were established [8]. Using the ground coordinate systems $E - \xi\eta\zeta$ as fixed reference coordinate system, its movement was relatively static with the earth. $O - xyz$ is the motion coordinate system. The origin of motion system was set at the barycenter of UUV. The coordinate systems are shown in Fig. 1.

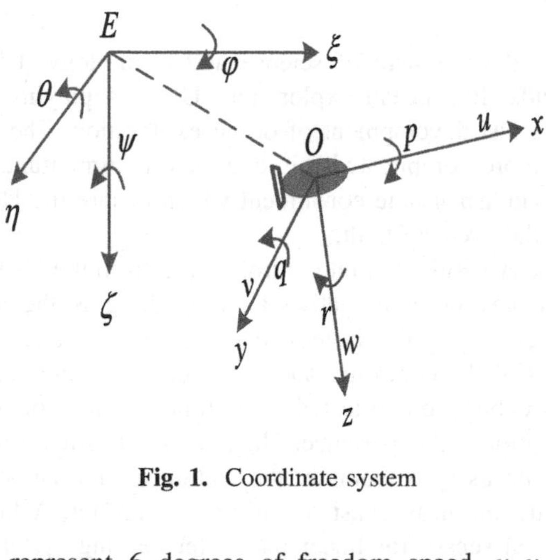

Fig. 1. Coordinate system

u, v, w, p, q, r represent 6 degrees of freedom speed, u, v, w represent linear velocity and p, q, r represent the angular velocity.

2.2 Establish Coordinate System

The equation which describes the dynamic movement of the UUV can be found in various literature, based on [11] we have:

$$\begin{aligned} M\dot{v} + C(v)v + D(v)v &= \tau \\ \dot{\eta} &= J(\psi)v \end{aligned} \tag{1}$$

Where M is the matrix of inertia of the UUV, including the rigidity matrix and addition mass. $C(v)$ is the vector with the terms relating to the centrifugal effects and Coriolis forces. $D(v)$ is the damping matrix. τ is the vector of generalized forces and moments of UUV. M, $D(v)$ and transformation matrix of angular velocity and velocity is given by:

$$M = \begin{bmatrix} m - X_{\dot{u}} & 0 & 0 \\ 0 & m - Y_{\dot{v}} & -Y_{\dot{r}} \\ 0 & -Y_{\dot{r}} & I_z - N_{\dot{r}} \end{bmatrix} \tag{2}$$

$$D = \begin{bmatrix} -X_u & 0 & 0 \\ 0 & -Y_v & -Y_r \\ 0 & -Y_r & -N_r \end{bmatrix} \tag{3}$$

$$J(\psi) = \begin{bmatrix} \cos\psi & -\sin\psi & 0 \\ \sin\psi & \cos\psi & 0 \\ 0 & 0 & 1 \end{bmatrix} \tag{4}$$

Where m represent the mass of the UUV; I_z represent the moment of inertia in relation to the Oz axis; $X_{\dot{u}}, Y_{\dot{v}}, N_{\dot{r}}$ are the added mass caused by acceleration in relation to Ox and Oy axis and angular acceleration in relation to Ox axis. $Y_{\dot{r}}$ is the added mass by the coupling of acceleration in relation to Oy axis and angular acceleration in relation to Ox axis. X_u, Y_v, Y_r, N_v, N_r are the hydrodynamic linear damping coefficient of UUV in each direction.

3 Expert S Surface Controller

3.1 S Surface Controller Characteristics

This paper adopts a kind of simple practical control method, the S surface controller. S surface controller using nonlinear function fitting the control object which has strong nonlinearity. This control method has simple structure, less input and suitable for nonlinear system. S surface control is similar to fuzzy control, it uses the Sigmod surface to replace the fuzzy rule. As shown in Fig. 2, the abscissa and ordinate represents the deviation and the deviation change rate, the Z axis represents the control force. The control force is between +1and −1. When the deviation and deviation change rate are small, the control force is small too. When they tend to zero, the control

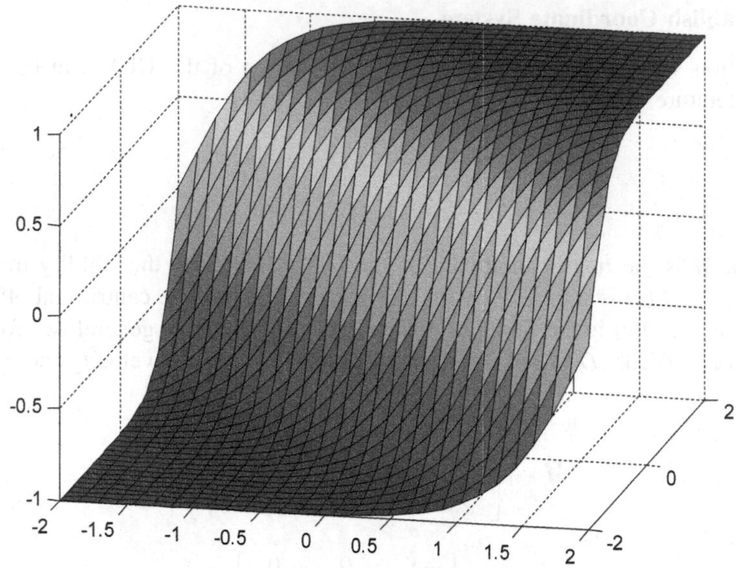

Fig. 2. Sigmod surface

force is close to zero. When the deviation and deviation change rate are big, the control force is close to +1. The control force output is very smooth.

3.2 S Surface Controller Characteristics

When UUV navigating in the ocean, there will be steady-state error due to changes in buoyancy and the impact of the ocean flow field. The steady-state error is eliminated by adding intelligent integral term to the S surface controller. The intelligent integral S surface controller can be described as:

$$\begin{aligned} & u_i = 2.0/(1.0 + e^{(-k_{i1}e - k_{i2}\dot{e})}) - 1.0 + k_{i3}\int edt \\ & e * \dot{e} > 0 \quad or \quad \dot{e} = 0 \, and \, e \neq 0 \\ & u_i = 2.0/(1.0 + e^{(-k_{i1}e - k_{i2}\dot{e})}) - 1.0 \\ & e * \dot{e} < 0 \quad or \quad e = 0 \\ & f_i = C_i * u_i \end{aligned} \tag{5}$$

Where e and $\dot{e}$ represent the deviation and deviation change rate of i degree of freedom. k_{i1} and k_{i2} are the control parameters of i degree of freedom. We design expert S surface controller to solve this problem by the adaptive adjustment of parameters.

The expert S controller has two levels of real-time coordination controller which is composed of the basic intelligent integral S control level and the expert intelligent integral S control level and expert intelligent coordination level. The basic control level and the controlled object UUV form a closed-loop control system. The expert intelligent coordination level is mainly composed of three parts: database; knowledge base; intelligent coordinator. The database mainly stores the adaptive adjustment parameters, thresholds and the range of controller parameters. The knowledge base stores the presupposed adaptive rules. The intelligent coordinator is the influence engine. According to the detected system data information, The inference engine completes the on-line adjustment of the S surface controller parameters. The following real-time adjustment algorithm as shown in Table 1 is obtained by summing up the experience of S surface controller.

Table 1. Adaptive rule of S surface controller

Rules	K_1	K_2	k_{i3}
$e > 0, \dot{e} < 0, e > \delta$	Δk_{i1}	0	0
$e > 0, \dot{e} < 0, e < \delta$	$-\Delta k_{i1}$	$-\Delta k_{i2}$	0
$e < 0, \dot{e} < 0$	$-\Delta k_{i1}$	Δk_{i2}	k_{i3}
$e < 0, \dot{e} > 0$	$-\Delta k_{i1}$	$-\Delta k_{i2}$	0
$e > 0, \dot{e} > 0$	Δk_{i1}	0	k_{i3}
$e * \dot{e} > 0, or\ \dot{e} = 0$	0	0	k_{i3}
$e * \dot{e} < 0$ or $e = 0$	0	0	0

In the above rule, Δk_{i1} and Δk_{i2} represent a very small increment of controller parameters. K_1 and K_2 are the dynamic correction factor of k_{i1}, k_{i2}. The correction algorithms is given by:

$$\begin{aligned} k_{i1}(t+1) &= k_{i1}(t) + K_1 \\ k_{i2}(t+1) &= k_{i2}(t) + K_2 \end{aligned} \tag{6}$$

$k_{i1}(t)$, $k_{i2}(t)$ are the controller parameters for the current time. $k_{i1}(t + 1)$, $k_{i2}(t + 1)$ are the next time control parameters. The initial values $k_{i1}(1)$, $k_{i2}(1)$ are selected according to the accumulated control experience and the characteristics of control system.

We designed the expert S surface controller and intelligent integral S surface controller for the control of longitudinal velocity combined the established UUV model. The longitudinal velocity curve is shown in Fig. 3.

It can be seen from the figure that the response speed of intelligent integral S surface controller is slower than the expert S surface controller. The expert S surface controller achieves speed stable with a faster response. It shows that the expert S surface has a better adaptive adjustment capability.

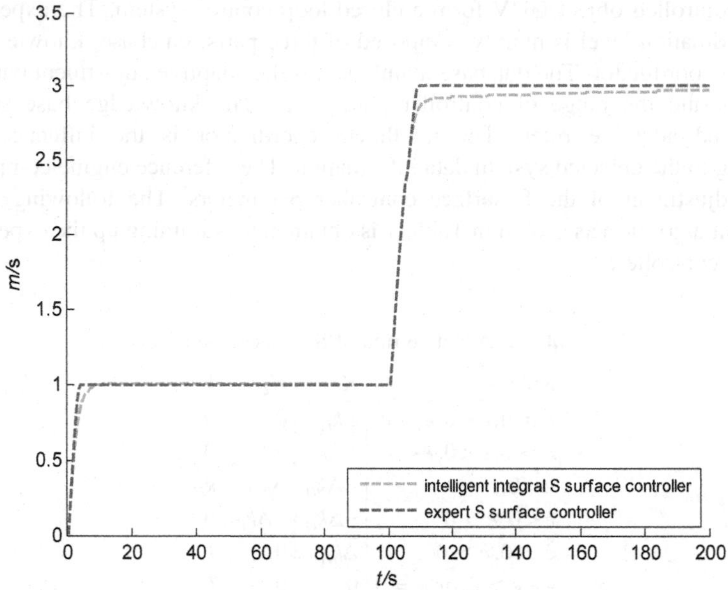

Fig. 3. Control curve of longitudinal velocity

4 Design of Walcott-Zak Observer

The main description of the three degrees of freedom of UUV horizontal movement can be described by vector $\eta = [x \quad y \quad \psi]^T$ and $v = [u \quad v \quad r]^T$. Where x, y is the position of UUV in fixed reference coordinate system. ψ represents the heading angle, and the parameters in v represent the longitudinal velocity, lateral velocity, and the angular velocity. As shown in the second section of this paper. The UUV horizontal plane system is given by:

$$\begin{cases} M\dot{v} + (C(V) + D(V))v = \tau + f(t, x, u) \\ \dot{\eta} = J(\psi)v \end{cases} \tag{7}$$

The equation of state for the corresponding nominal system can be described as:

$$\begin{cases} \dot{x}(t) = Ax(t) + Bu(t) + f(t, u, x) \\ y(t) = Cx(t) \end{cases} \tag{8}$$

The corresponding matrix is:

$$
\begin{aligned}
& A = -(C(V) + D(V)) \\
& B = M^{-1} \\
& u(t) = \tau \\
& C = E_{3\times 3} \\
& f(t,x,u) = B\xi(t,x,u)
\end{aligned}
\tag{9}
$$

Where *f*(*t*, *x*, *u*) is the uncertainty of the system. The fault system model can be described as:

$$
\begin{cases} \dot{x}_f(t) = Ax_f(t) + Bu_f(t) + f(t,u,x) \\ y_f(t) = C_f x_f(t) \end{cases}
\tag{10}
$$

The system state model of virtual sensor can be described as:

$$
\begin{aligned}
& \dot{x}_v = A_v x_v(t) + B_v u_c(t) + L y_f(t) \\
& u_f(t) = u_c(t) \\
& y(t) = C_v x_v(t) + P y_f(t) \\
& x_v(0) = x_{v0} \\
& A_v = A - LC_f \\
& C_v = C - PC_f
\end{aligned}
\tag{11}
$$

Where *L*, *P* are the parameter matrices, which can be arbitrarily chosen. When the matrix is zero matrix, the faulted sensor dose not participate in the reconstruction. We introduced the Walcott-Zak nonlinear robust observer to solve the problem of uncertain interference. The virtual sensor can be designed as follows:

$$
\begin{cases}
\dot{x}_v = A_v x_v(t) + B_v u_c(t) + L y_f(t) + Bv \\
v = \begin{cases} -\rho \dfrac{FCe}{\|FCe\|} & \|FCe\| \neq 0 \\ 0 & \|FCe\| = 0 \end{cases} \\
y(t) = C_v x_v(t) + P y_f(t) \\
A_v = A - LC_f \\
C_v = C - PC_f \\
e = C_f x_v(t) - y_f(t)
\end{cases}
\tag{12}
$$

Where the parameter ρ is designed as follow formula shown:

$$\rho \geq r_1 \|u(t)\| + \alpha(t, y) + \eta \tag{13}$$

Assuming the parameter matrix L is existing, we select parameter matrix L to ensure the matrix $A - LC$ is Hurwitz matrix. There are positive definite symmetric matrix satisfy the equation as follow:

$$\begin{aligned} C^T F^T &= PB \\ A_0^T P_F + P_F A_0 &= -Q(P_F) \end{aligned} \tag{14}$$

The method of finding the parameter matrices is described as follows:

(1) Select the spectrum of matrix A_v and calculate the parameter matrix L.
(2) According to the expression $C^T F^T = PB$, use each element of matrix F to express the matrix P and get the matrix P_F. Ensure the matrix P_F is a positive definite symmetric matrix.
(3) According to the formula $A_v^T P_F + P_F A_v = -Q(P_F)$, get the matrix $Q(P_F)$ calculated by P_F and A_v.
(4) Select each component of the matrix F and P to ensure the matrix $Q(P_F)$ is a positive definite symmetric matrix.

The specific structure is shown in Fig. 4. Increasing the integral term and the Gaussian white noise to the structure, then we obtained the heading angle virtual sensor, as shown in Fig. 5.

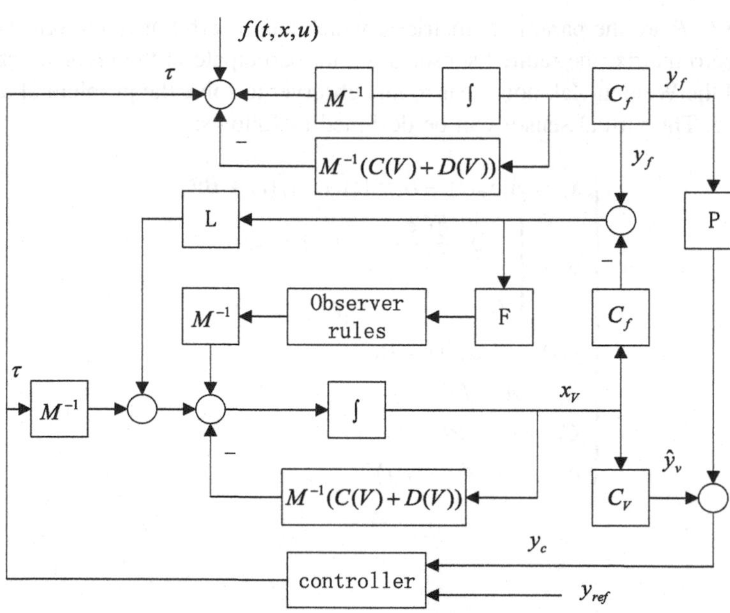

Fig. 4. Walcott-Zak robust virtual velocity sensor

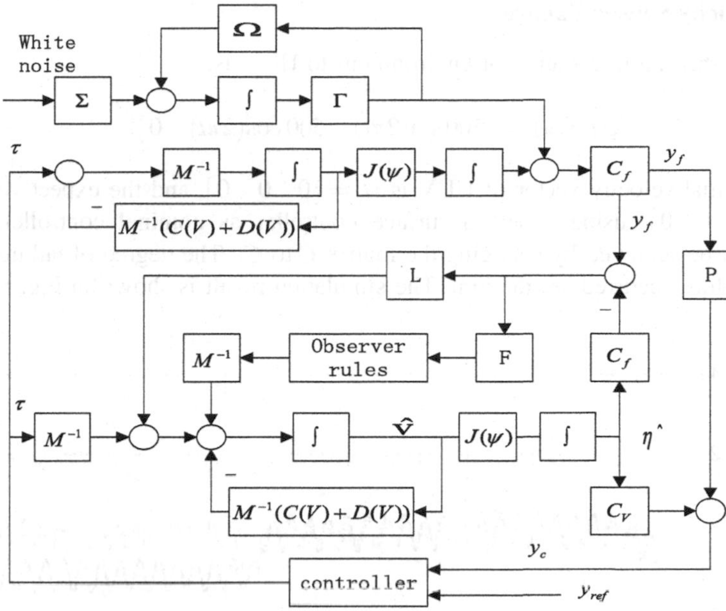

Fig. 5. Walcott-Zak robust virtual heading angle sensor

5 Fault Tolerant Control Simulation of UUV Sensor

We use UUV for simulation which is developed by Harbin Engineering University. We approximates the inertial matrix of UUV in the simulation process, as follows.

$$B = M^{-1} = \begin{bmatrix} 6 & 0 & 0 \\ 0 & 8 & 1 \\ 0 & 1 & 1 \end{bmatrix} \quad (15)$$

We get matrix F according the rules is given by:

$$F = \begin{bmatrix} 12 & 6 & -6 \\ 9 & 15 & -5 \\ 2 & 1 & 2 \end{bmatrix} \quad (16)$$

We take the velocity sensor failure and the heading angle sensor failure as examples to verify the method proposed. Assuming that the fault diagnosis system can accurately detect the fault and fault degree.

5.1 Velocity Sensor Failure

Assuming that the uncertainty of environment to UUV is:

$$\xi(t, x, u) = [500\sin(2\pi t) \quad 500\cos(2\pi t) \quad 0]^T \tag{17}$$

The initial velocity vector of UUV is $v_0 = [0 \quad 0 \quad 0]$, and the expect velocity is $v_{ref} = [1 \quad 0 \quad 0]$, using expert S surface controller as nominal controller. Sensor failure can be achieved by changing the matrix C to C_f. The degree of failure and the time of failure occurred are random. The simulation result is shown in Fig. 6.

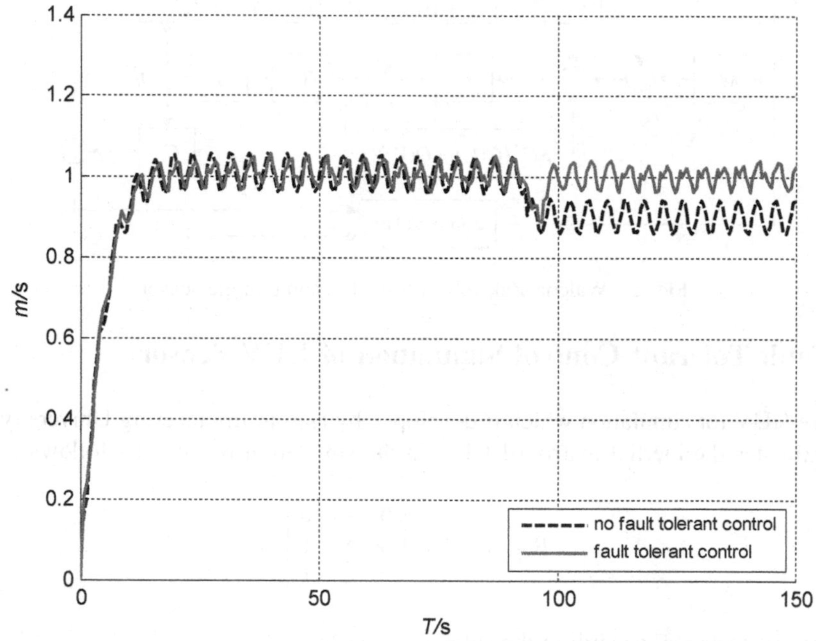

Fig. 6. Fault tolerant control curve of longitudinal velocity sensor

It can be seen from the figure that the observer has a good tracking effect on the sensor before the fault occurs. When UUV velocity sensor fails, after three simulation steps, the fault tolerant control system works. The value of observer is closed to the true value.

5.2 Heading Angle Sensor Failure

The initial position vector of UUV is $\eta_0 = [0 \quad 0 \quad 0]$, and the expect position vector to be $\eta_{ref} = [0 \quad 0 \quad \pi/4]$. The longitudinal velocity of UUV is set to 1 m/s. The degree of failure and the time of failure occurred are random.. When UUV heading

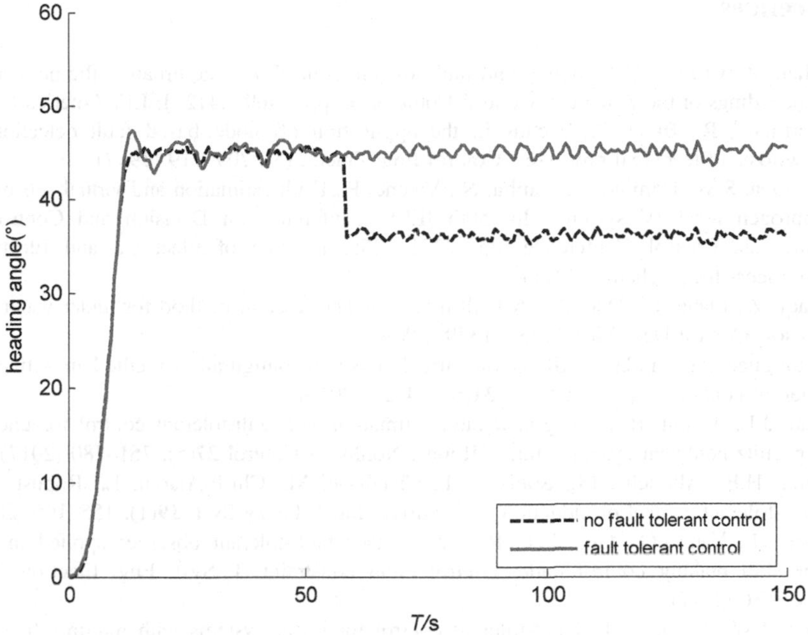

Fig. 7. Fault tolerant control curve of heading angle sensor (Color figure online)

angle sensor fails, and after one simulation steps, the fault tolerant control system works. The simulation results is shown in Fig. 7.

When the UUV heading angle sensor fails and no fault tolerant control, the UUV's heading angle deviates greatly. This situation is very dangerous. The black dotted line indicates a non fault tolerant measurement, and the red solid line indicates the output of virtual sensor. When the fault occurs, UUV discards the fault sensors value and use the value of reconstructed virtual sensor. It can be seen from the figure that the reconstructed virtual sensor signal stabilizes near the expected value, that indicates the designed virtual sensor is valid.

6 Conclusions

Based on the dynamic model of UUV, the Walcott-Zak virtual sensor is designed for the uncertain disturbance. The robust Walcott-Zak observer can eliminates the influence of the system's nonlinearity and uncertainty. We designed the expert S surface controller to control the UUV. Taking the UUV velocity sensor fault and the heading angle sensor fault as examples to verify the validity of virtual sensor. The simulation results show that the UUV can use the virtual sensor to complete the task after the sensor fails. It has great significance for the safety of UUV navigation.

References

1. Chen, Z., Chang, T.: Modeling and fault-tolerant control of large urban traffic network. In: Proceedings of the American Control Conference, pp. 2469–2472. IEEE, America (1997)
2. Isermann, R., Balle, P.: Trends in the application of model-based fault detection and diagnosis of technical processes. Control Eng. Pract. **5**(5), 709–719 (1997)
3. De Oca, S.M., Damino, R., Fatiha, N., Vicenc, P.: Fault estimation and virtual sensor FTC approach for LPV systems. In: 50th IEEE Conference on Decision and Control and European Control Conference, pp. 2251–2256. Institute of Electrical and Electronics Engineers Inc., Orlando (2011)
4. Daqi, Z., Liang, C., Qian, L.: A fault diagnosis and tolerant method for under water robot sensor. Control Des. **24**(9), 1335–1339 (2009)
5. Mingguang, Z., Yida, J.: Study on virtual sensor reconfiguration method in active fault tolerant control. Autom. Instrum. **31**(6), 24–26 (2010)
6. Lan, J.L., Patton, R.J.: Integrated fault estimation and fault-tolerant control for uncertain Lipschitz nonlinear systems. Int. J. Robust Nonlinear Control **27**(5), 761–780 (2017)
7. Zina, H.B., Allouche, M., Souissi, M., Chaabane, M., Chrifi-Alaoui, L.: Robust sensor fault-tolerant control of induction motor drive. Int. J. Fuzzy Syst. **19**(1), 155–166 (2017)
8. Sun, H., Yan, J.G., Qu, Y.H., Ren, J.: Sensor fault-tolerant observer applied in UAV anti-skid braking control under control input constraint. J. Syst. Eng. Electron. **28**(1), 126–136 (2017)
9. Zhu, J.W., Yang, G.H.: Fault-tolerant control for linear systems with multiple faults and disturbances based on augmented intermediate estimator. IET Control Theor. Appl. **11**(2), 164–172 (2017)
10. Feijun, S., Edgar, P.A., Folleco, A.: Modeling an simulation of autonomous underwater vehicle: design and implementation. IEEE J. Oceanic Eng. **28**(2), 283–296 (2003)
11. Chang, W.J., Liu, J.C., Yu, H.N.: Mathematic model of AUV motion control and simulation. Ship Eng. **3**, 58–60 (2002)

The Tracking Control of Unmanned Underwater Vehicles Based on QPSO-Model Predictive Control

Wenyang Gan[1], Daqi Zhu[1(✉)], Bing Sun[1], and Chaomin Luo[2]

[1] The Laboratory of Underwater Vehicles and Intelligent Systems, Shanghai Maritime University, Shanghai 201306, China
zdq367@aliyun.com
[2] The Department of Electrical and Computer Engineering, University of Detroit Mercy, Detroit, USA
luoch@udmercy.edu

Abstract. For the trajectory tracking control of Unmanned Underwater Vehicles (UUV), an improved Model Predictive Control (MPC) method based on Quantum-behaved Particle Swarm Optimization (QPSO) is proposed. The concept of trajectory tracking is given firstly in this paper. Then QPSO-MPC is employed to realize the tracking control. The QPSO problem is suggested to optimization problem of minimizing the objective function with the conditions of satisfying the control constraints. The simulation results which is under the two-dimensional situation show that QPSO-MPC can effectively solve the speed jump problem. More effective and feasible for trajectory tracking problem compared with backstepping control method.

Keywords: Unmanned Underwater Vehicle · Trajectory tracking · Quantum-behaved particle swarm optimization · Model predictive control · Backstepping control

1 Introduction

The target of trajectory tracking control is making UUV track the expected path by controlling velocity and angular velocity. The error between the desired state and the actual state for tracker is finally converged to zero. The related research in this challenging filed is not too much [1]. From the point of view of the control algorithm in this filed, the various control methods reported in former literatures could be categorized into five groups: PID control [2, 3], sliding mode control [4, 5], neural network [6, 7], backstepping control [8, 9], model predictive control [10, 11] etc.

Backstepping control method is proverbially used to control tracking for mobile robots. It has been applied to the control system of UUV. It can handle the large error of initial state. But its shortcoming is also clear. The large initial state error will generate large velocity change, which will cause speed jump problem. This means that the required force may beyond the control constraints at the turning points, which is the problem to be solved in backstepping control.

Y. Huang et al. (Eds.): ICIRA 2017, Part I, LNAI 10462, pp. 711–720, 2017.
DOI: 10.1007/978-3-319-65289-4_66

For the speed jump problem in backstepping control method, model predictive control method is proposed to solve the problem with the constraints of the speed. Model predictive control method is easy to model and itself can improve the robustness of UUV system. Moreover, the rolling optimization strategy can achieve better dynamic control performance. QPSO [12, 13] is combined with model predictive control method owning to its global search capability and the constraints of particle's position. QPSO-MPC can guarantee the speed within a specified range, more flexibly to track the desired trajectory.

The rest parts of this paper are organized as follows: In Sect. 2, the description of trajectory tracking of UUV is introduced and QPSO-MPC is employed to realize the tracking control based on the optimization process with QPSO. In Sect. 3, the simulation results tracked by QPSO-MPC compared with backstepping control method are shown with the relevant comparisons and analyses. In Sect. 4, concluding remarks are summarized, about the trajectory tracking research of UUV in the future is discussed.

2 The Trajectory Tracking Control Based on QPSO-MPC

Trajectory tracking is one part of UUV's control, which is important and meaningful for detecting the ocean resource. Model predictive control based on QPSO is applied to implement UUV's trajectory tracking control as follows. The diagrammatic sketch of trajectory tracking control of UUV is shown in Fig. 1.

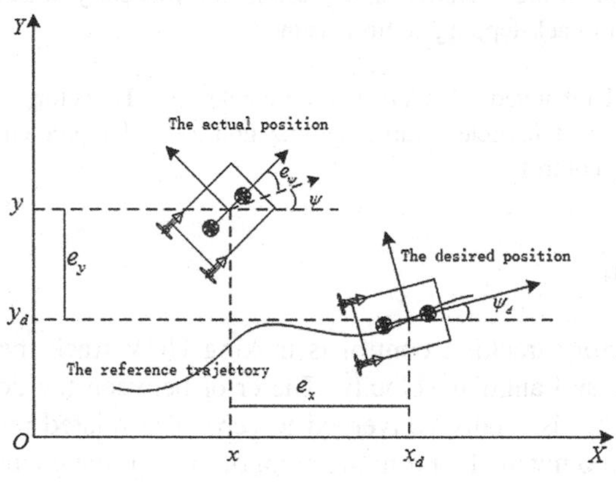

Fig. 1. The diagrammatic sketch of trajectory tracking control of UUV

The reference position of UUV is $\boldsymbol{\eta}_d(t) = [x_d(t) \quad y_d(t) \quad \psi_d(t)]^T$ in two-dimensional space, and its actual position is $\boldsymbol{\eta}(t) = [x(t) \quad y(t) \quad \psi(t)]^T$. The goal of trajectory tracking is making the error $\boldsymbol{e}(t)$ between actual trajectory and desired trajectory converge to zero.

$$\boldsymbol{e}(t) = \boldsymbol{\eta}_d(t) - \boldsymbol{\eta}(t) \tag{1}$$

$$\lim_{t \to \infty} \boldsymbol{e}(t) = 0 \tag{2}$$

Backstepping control method may encounter the speed jump problem when achieved to tracking control. In this section, model predictive control is proposed to solve the speed jump problem with the speed constraints provided by quantum-behaved particle swarm optimization. The scheme of the proposed control system of UUV is illustrated in Fig. 2.

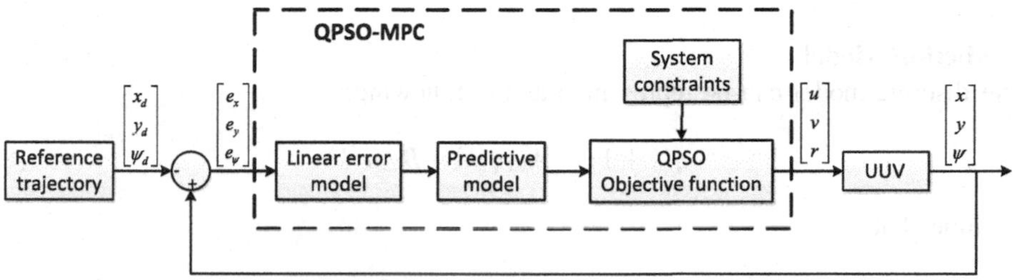

Fig. 2. The scheme of the QPSO-MPC tracking control

2.1 The Linear Error Model

The system can be seen as a control system with input $\boldsymbol{u} = [u \quad v \quad r]^T$ and the state $\eta = [x \quad y \quad \psi]^T$. The general form is:

$$\dot{\boldsymbol{\eta}} = f(\boldsymbol{\eta}, \boldsymbol{u}) \tag{3}$$

And the general form of reference trajectory is:

$$\dot{\boldsymbol{\eta}}_d = f(\boldsymbol{\eta}_d, \boldsymbol{u}_d) \tag{4}$$

where $\boldsymbol{\eta}_d = [x_d \quad y_d \quad \psi_d]^T$, $\boldsymbol{u}_d = [u_d \quad v_d \quad r_d]^T$.

Using Taylor series expansion and ignoring the higher order terms in formula (3) at the reference trajectory point, then it can be obtained as:

$$\dot{\boldsymbol{\eta}} = f(\boldsymbol{\eta}_d, \boldsymbol{u}_d) + \left.\frac{\partial f(\boldsymbol{\eta}, \boldsymbol{u})}{\partial \boldsymbol{\eta}}\right|_{\substack{\eta=\eta_d \\ u=u_d}} (\boldsymbol{\eta} - \boldsymbol{\eta}_d) + \left.\frac{\partial f(\boldsymbol{\eta}, \boldsymbol{u})}{\partial \boldsymbol{u}}\right|_{\substack{\eta=\eta_d \\ u=u_d}} (\boldsymbol{u} - \boldsymbol{u}_d) \tag{5}$$

The error model of UUV can be obtained when formula (5) subtract formula (4):

$$\dot{\tilde{\boldsymbol{\eta}}} = \begin{bmatrix} \dot{x} - \dot{x}_d \\ \dot{y} - \dot{y}_d \\ \dot{\psi} - \dot{\psi}_d \end{bmatrix} = \begin{bmatrix} 0 & 0 & -u\sin\psi - v\cos\psi \\ 0 & 0 & u\cos\psi - v\sin\psi \\ 0 & 0 & 0 \end{bmatrix} \begin{bmatrix} x - x_d \\ y - y_d \\ \psi - \psi_d \end{bmatrix} + \begin{bmatrix} \cos\psi & -\sin\psi & 0 \\ \sin\psi & \cos\psi & 0 \\ 0 & 0 & 1 \end{bmatrix} \begin{bmatrix} u - u_d \\ v - v_d \\ r - r_d \end{bmatrix} \tag{6}$$

Formula (6) should be discretized based on the linear error model, which can be applied in MPC controller design. So it can be derived as:

$$\tilde{\boldsymbol{\eta}}(k+1) = \boldsymbol{A}_{k,t}\tilde{\boldsymbol{\eta}}(k) + \boldsymbol{B}_{k,t}\tilde{\boldsymbol{u}}(k) \tag{7}$$

where $\boldsymbol{A}_{k,t} = \begin{bmatrix} 1 & 0 & (-u\sin\psi - v\sin\psi)T \\ 0 & 1 & (u\cos\psi - v\cos\psi)T \\ 0 & 0 & 1 \end{bmatrix}$, $\boldsymbol{B}_{k,t} = \begin{bmatrix} T\cos\psi & -T\sin\psi & 0 \\ T\sin\psi & T\cos\psi & 0 \\ 0 & 0 & T \end{bmatrix}$, T is sampling time.

2.2 Optimization Problems Based on QPSO

Prediction Model
The discrete model can be represented as the following:

$$\tilde{\boldsymbol{\eta}}(k+1) = \boldsymbol{A}_{k,t}\tilde{\boldsymbol{\eta}}(k) + \boldsymbol{B}_{k,t}\tilde{\boldsymbol{u}}(k) \tag{8}$$

Assume that:

$$\xi(k|t) = \begin{bmatrix} \tilde{\boldsymbol{\eta}}(k|t) \\ \tilde{\boldsymbol{u}}(k-1|t) \end{bmatrix} \tag{9}$$

Then a new state space expression can be summarized as:

$$\xi(k+1|t) = \tilde{\boldsymbol{A}}_{k,t}\xi(k|t) + \tilde{\boldsymbol{B}}_{k,t}\Delta\boldsymbol{u}(k|t) \tag{10}$$

$$\boldsymbol{\eta}(k|t) = \tilde{\boldsymbol{C}}_{k,t}\xi(k|t) \tag{11}$$

where $\tilde{\boldsymbol{A}}_{k,t} = \begin{bmatrix} \boldsymbol{A}_{k,t} & \boldsymbol{B}_{k,t} \\ \boldsymbol{0}_{m\times n} & \boldsymbol{I}_m \end{bmatrix}$, $\tilde{\boldsymbol{B}}_{k,t} = \begin{bmatrix} \boldsymbol{B}_{k,t} \\ \boldsymbol{I}_m \end{bmatrix}$, $\tilde{\boldsymbol{C}}_{k,t} = [\,C_{k,t} \quad 0\,]$, n, is state dimension, m is control dimension.

To simplify the calculations, the following assumptions are done:

$$\boldsymbol{A}_{k,t} = \boldsymbol{A}_t,\ k = 1, \cdots t + \mathrm{N} - 1 \tag{12}$$

$$\boldsymbol{B}_{k,t} = \boldsymbol{B}_t,\ k = 1, \cdots t + \mathrm{N} - 1 \tag{13}$$

If the prediction domain and control domain of system are Np and Nc, then the state and output of system can be obtained by computing with the following formulas:

$$\xi(t+N_p|t) = \tilde{\boldsymbol{A}}_t^{N_p}\xi(t|t) + \tilde{\boldsymbol{A}}_t^{N_p-1}\tilde{\boldsymbol{B}}_t\Delta\boldsymbol{u}(t|t) + \ldots + \tilde{\boldsymbol{A}}_t^{N_p-N_c-1}\tilde{\boldsymbol{B}}_t\Delta\boldsymbol{u}(t+N_c|t) \tag{14}$$

$$\eta(t+N_p|t) = \tilde{\boldsymbol{C}}_t\tilde{\boldsymbol{A}}_t^{N_p}\xi(t|t) + \tilde{\boldsymbol{C}}_t\tilde{\boldsymbol{A}}_t^{N_p-1}\tilde{\boldsymbol{B}}_t\Delta\boldsymbol{u}(t|t) + \ldots + \tilde{\boldsymbol{C}}_t\tilde{\boldsymbol{A}}_t^{N_p-N_c-1}\tilde{\boldsymbol{B}}_t\Delta\boldsymbol{u}(t+N_c|t) \tag{15}$$

In order to make the whole relationship more explicit, the output of system in the future is expressed in the matrix form:

$$Y(t) = \Psi_t \xi(t|t) + \Theta_t \Delta U(t) \tag{16}$$

Rolling Optimization Based on QPSO
Here, take the form of the objective function as follows when designing tracking controller:

$$J(k) = \sum_{i=1}^{N_p} \left\| \eta(k+i|t) - \eta_{ref}(k+i|t) \right\|_Q^2 + \sum_{i=1}^{N_c-1} \left\| \Delta u(k+i|t) \right\|_R^2 \tag{17}$$

where Q and R are weight matrixes, N_p is predictive domain, N_c is control domain, the difference of the speed of UUV between the previous time and the current time is Δu. Meanwhile, in the actual control system, the constraint for speed shows as follows:

Control constraint:

$$u_{\min}(t+k) \le u(t+k) \le u_{\max}(t+k),\ k = 0, 1, \cdots N_c - 1 \tag{18}$$

Plugging formula (16) into (17), the full form of the objective function can be deduced as:

$$J(\xi(t), u(t-1), \Delta U(t)) = \Delta U(t) H_t \Delta U(t)^T + G_t \Delta U(t)^T \tag{19}$$

where $H_t = \begin{bmatrix} \Theta_t^T Q \Theta_t + R & 0 \end{bmatrix}$, $G_t = \begin{bmatrix} 2E(t)^T Q \Theta_t & 0 \end{bmatrix}$

Therefore, solving optimization problems with constrains of model predictive control in every step are equivalent to solving QPSO problems as follows:

$$\min_{\Delta V(t)} \Delta U(t) H_t \Delta U(t)^T + G_t \Delta U(t)^T \tag{20}$$

The process using QPSO to solve the formula (19) is as follows:

(1) Setting $t = 0$, the current position $X_i(0)$ of every particle is initialized in the problem space. And the individual best position as $P_i(0) = X_i(0)$. To find the minimum objective function and the corresponding particle g, that means the best global position $G(0) = P_g(0)$.
(2) Computing the average best position of the particle swarm according to formula (21).

$$C_j(t) = \frac{1}{M} \sum_{i=1}^{M} P_{i,j}(t) \tag{21}$$

(3) Steps (4)–(7) are performed for each particle $i\ (1 \le i \le M)$ in particle swarm.
(4) The objective function value of the current position $X_i(t)$ is computed. The best individual position of particle is updated according to formula (22).

$$P_i(t) = \begin{cases} X_i(t), & \text{if } J[X_i(t)] < J[P_i(t-1)] \\ P_i(t-1), & \text{if } J[X_i(t)] \geq J[P_i(t-1)] \end{cases} \quad (22)$$

(5) For particle i, the objective function value of $P_i(t)$ is compared with the objective function value of the best global position $G(t-1)$. If $J(P_i(t)) > J(G(t-1))$, then $G(t) = P_i(t)$, otherwise $G(t) = G(t-1)$.

(6) For each dimension of particle i, the position of random point is achieved by calculating the formula (23) based on weighing the best individual position and global position.

$$p_{i,j}(t) = \varphi_j(t) \bullet P_{i,j}(t) + [1 - \varphi_j(t)] \bullet G_j(t), \varphi_j(t) \sim U(0,1) \quad (23)$$

(7) The new positions of particles are calculated according to the formula (24).

$$X_{i,j}(t+1) = p_{i,j}(t) \pm \alpha \bullet |C_j(t) - X_{i,j}(t)| \bullet \ln[\frac{1}{u_{i,j}(t)}], u_{i,j}(t) \sim U(0,1) \quad (24)$$

(8) Setting $t = t+1$, return step 2. The whole optimization process by the end of $t = MAXITER$. Finally, the best global position is returned to the system as the control input increment.

Feedback Mechanism

A series of input increments in control domain after optimization can be got as:

$$\Delta U_t^* = [\Delta u_t^* \quad \Delta u_{t+1}^* \quad \cdots \quad \Delta u_{t+N_c-1}^*]^T \quad (25)$$

The first element of control sequence is the input increment of actual control to act on UUV, which is:

$$u(t) = u(t-1) + \Delta u_t^* \quad (26)$$

System carries out this controlling value until the next time. In the new time, system predicts the output next period of time domain again according to the state information, and gets a new sequence controlling increment by optimizing the process. So on ad infinitum, until the system completes the control process.

3 Experimental Results

In this section, backstepping control method and QPSO-MPC are used to track triangle in two dimensional space with MATLAB environment. As a result of length reason, here one simulation case provided to analysis the tracking performance. Triangle tracking is carried out because it can not only reflected the tracking performance for the continuous trajectory (the straight line portion), but also can reflect the tracking performance for the discrete trajectory (the trajectory is continuous but not differentiable). So then the effectiveness of QPSO-MPC is verified by comparing the two algorithms' simulation results in various environments.

3.1 Triangle Tracking

The actual initial position of UUV is $\boldsymbol{\eta}_0 = [x_0 \quad y_0 \quad \psi_0]^T = [-1 \quad -1 \quad 0]^T$ while the desired initial position is $\boldsymbol{\eta}_{d0} = [x_{d0} \quad y_{d0} \quad \psi_{d0}]^T = [0 \quad 0 \quad 0]^T$, so the initial position error is $[1 \quad 1 \quad 0]^T$. Setting the simulation time is T = 30 s and the sampling time is 0.1 s. The reference trajectory $\boldsymbol{\eta}_d = [x_d \quad y_d \quad \psi_d]^T$ and speed $\boldsymbol{u}_d = [u_d \quad v_d \quad r_d]^T$ are set as follows:

$$0 \le t \le 10\,\mathrm{s}: \begin{cases} x_d(t) = 0.6 * t \\ y_d(t) = 0 \\ \psi_d(t) = 0 \end{cases} \begin{cases} u_d(t) = 0.6 \\ v_d(t) = 0 \\ r_d(t) = 0 \end{cases}$$

$$10\,\mathrm{s} \le t \le 20\,\mathrm{s}: \begin{cases} x_d(t) = -0.3 * (t-10) + 6 \\ y_d(t) = -0.3\sqrt{3} * (t-10) \\ \psi_d(t) = 2 * \pi/3 \end{cases} \begin{cases} u_d(t) = 0.3 \\ v_d(t) = 0 \\ r_d(t) = 0 \end{cases}$$

$$20\,\mathrm{s} \le t \le 30\,\mathrm{s}: \begin{cases} x_d(t) = -0.3 * (t-20) + 3 \\ y_d(t) = -0.3\sqrt{3} * (t-20) + 3\sqrt{3} \\ \psi_d(t) = \pi/3 \end{cases} \begin{cases} u_d(t) = -0.3 \\ v_d(t) = 0 \\ r_d(t) = 0 \end{cases}$$

The simulation results tracked by QPSO-MPC and backstepping control are shown in Fig. 3. Figure 4 shows the error of tracking results. The tracking results of both backstepping control and QPSO-MPC are ideal after tracking the reference trajectory and the error curve is close to zero when the tracking is stable.

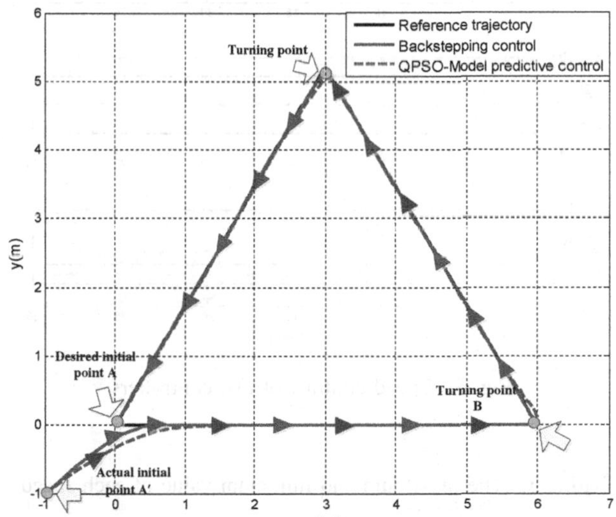

Fig. 3. Polyline tracking

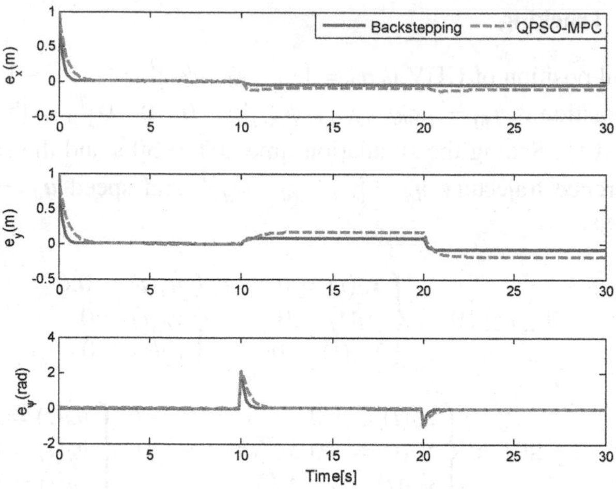

Fig. 4. Error contrast of two controllers

Figure 5 shows the comparison of two algorithms' surge speed $u(t)$, sway speed $v(t)$ and yaw speed $r(t)$, the maximum and minimum of each speed are shown in Table 1.

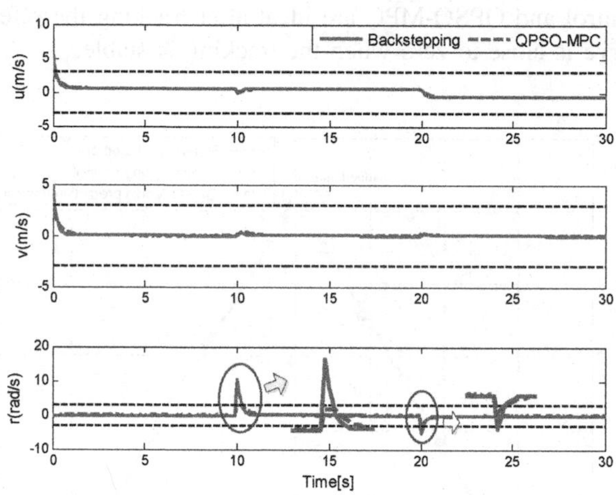

Fig. 5. Speed contrast of two controllers

Table 1. The maximum and minimum value of each speed

	$u(t)$ (m/s)	$v(t)$ (m/s)	$r(t)$ (rad/s)
Maximum	3	3	3
Minimum	−3	−3	−3

For the piecewise line like triangle, the surge speed, sway speed and yaw speed of UUV are mostly smooth. But the backstepping control has the speed jump problem when tracking the initial point A (0,0) and the inflection points B (6,0), C $(3, 3\sqrt{3})$. Because the state change of triangle means the change of yaw speed value, so the analysis of speed jump will focus on yaw speed value for the turning points B and C.

When tracking the desired initial point A, $u(t)$ and $v(t)$ of backstepping control have the speed jump problem and also out of bounds in Fig. 5. The speed jump values at the actual initial point A' are compared in Table 2. For example, the velocity u of backstepping algorithm is 5.6 m/s at the first reference trajectory point 1 and becomes 3.62 m/s at next point 2, so the speed jump value is 1.98 m/s which is far beyond permissible range. But the velocity u of QPSO-MPC are suppressed within the range of permissible speed.

Table 2. The speed jump values at actual initial point A'

Controller	A'					
	$u(t)$ (m/s)		$v(t)$ (m/s)		$r(t)$ (rad/s)	
	1	2	1	2	1	2
Backstepping	5.6	3.62	5	3.02	0	0
QPSO-MPC	2.96	2.45	2.36	1.87	−0.32	−0.2

Table 3. The speed jump values at turning points B and C

Controller	B		C	
	$r(t)$ (rad/s)		$r(t)$ (rad/s)	
	200	201	400	401
Backstepping	0	10.472	0	−5.236
QPSO-MPC	0.001	3	0	−2.644

For the turning points B and C, it can be seen from the black ellipse marking part of yaw speed r in Fig. 5, the speed jump of QPSO-MPC controller is less than backstepping controller. The yaw speed jump values at the turning points B and C are compared in Table 3. For example, the velocity r of backstepping algorithm is 0 rad/s at the reference trajectory point 200 and becomes 10.472 rad/s at next point 201, hence the speed jump value is 10.472 rad/s. However, the velocity r of QPSO-MPC is −0.001 rad/s at the reference trajectory point 200 and becomes 3 rad/s at next point 201 with the speed jump value 2.999 rad/s only, the speed jump which is smaller than backstepping control.

The comparisons and analyses from the triangle simulation results between backstepping control method and QPSO-MPC, the backstepping control method has the speed jump problem with large initial posture error. It should be noted that in the simulation study, the velocity control signal is generated in a theoretical way without any physical limit. However, for speed jump problem with backstepping control method, UUV cannot offer such big speeds in a very short time in actual situation. On the contrary, QPSO-MPC can solve the speed jump problem because of the constraints in the rolling optimization process. That is, under the circumstance of large initial error, QPSO-MPC can eliminate the speed jump well and avoid excessive speed which is faster than actual situation.

4 Conclusion

In this paper, kinematics trajectory tracking control problem consider in two-dimensional space for UUV is studied. QPSO is used to the rolling optimization part of MPC, combined with the speed constraints of UUV. The results comparison between QPSO-MPC and backstepping control method can indicate that QPSO-MPC can solve well of the speed jump problem and make the speed within a specified range.

References

1. Santhakumar, M., Asokan, T.: Investigations on the hybrid tracking control of an underactuated autonomous underwater robot. Adv. Robot. **24**(11), 1529–1556 (2012)
2. Wu, X.P., Feng, Z.P., Zhu, J.M.: Application on AUV control based on fuzzy PID strategy. Ship Sci. Technol. **29**(1), 95–98 (2007)
3. Guo, X.X., Li, G., Yan, W.J., et al.: PID controller based on GA for AUV depth control. J. Changchun Univ. Sci. Technol. **33**(3), 37–39 (2010)
4. Bagheri, A., Moghaddam, J.J.: Simulation and tracking control based on neural-network strategy and sliding-mode control for underwater remotely operated vehicle. Neurocomputing **72**(7–9), 1934–1950 (2009)
5. Zhu, D.Q., Sun, B.: The bio-inspired model based hybrid sliding-mode tracking control for unmanned underwater vehicles. Eng. Appl. Artif. Intell. **26**(10), 2260–2269 (2013)
6. Bagheri, A., Karimi, T., Amanifard, N.: Tracking performance control of a cable communicated underwater vehicle using adaptive neural network controllers. Appl. Soft Comput. **10**(3), 908–918 (2010)
7. Pan, C.Z., Lai, X.Z., Yang, S.X., et al.: An efficient neural network approach to tracking control of an autonomous surface vehicle with unknown dynamics. Expert Syst. Appl. **40**(5), 1629–1635 (2012)
8. Dong, W.: Flocking of multiple mobile robots based on backstepping. IEEE Trans. Syst. Man Cybern. Part B Cybern. **41**(2), 414–424 (2011)
9. Xu, J., Wang, M., Qiao, L.: Backstepping-based controller for three-dimensional trajectory tracking of underactuated unmanned underwater vehicle. Control Theor. Appl. **11**, 1589–1596 (2014)
10. Wang, L.L., Wang, H.L., Pan, L.X.: Autonomous underwater vehicle motion planning via sampling based model predictive contorl. Appl. Mech. Mater. **670–671**, 1370–1377 (2014)
11. Hirose, N., Tajima, R., Koyama, N., Sukigara, K., Tanaka, M.: Following control approach based on model predictive control for wheeled inverted pendulum robot. Adv. Robot. **30**(6), 1–12 (2016)
12. Li, R.F., Dokgo, M.C., Hu, L., Han, C.: Mobile robot trajectory planning based on QPSO algorithm and experiment. Control Decis. **29**(12), 2151–2157 (2014)
13. Li, J.: A review of quantum-behaved particle swarm optimization. Open J. Appl. Sci. **5**(6), 240–250 (2015)

Motion Analysis of Wave Glider Based on Multibody Dynamic Theory

Xiao-tao Li(✉), Fang Liu, Li Wang, and Hu-qing She

Yichang Research Institute of Testing Technology, Yichang 443003, China
lixiaotao710@163.com

Abstract. A wave glider has been developed and field tested. Differ from the conventional Unmanned Surface Vehicles (USVs) or Unmanned Underwater Vehicle (UUVs), the wave glider consists of the surface ship, the connecting cable and the underwater gliding body. It can roughly be regarded as an interrelated multibody system which. Therefore, the traditional dynamic methods and analyzing process of the USVs and the UUVs are inapplicable to the wave glider. In this paper, a dynamic model of wave glider is derived using multibody dynamic theory. The mathematic models of the surface ship, the underwater gliding body and the connection cable are respectively established based on the mechanical relationship. Motion analysis according to the dynamic model is also conducted by using numerical simulation software ADAMS and MATLAB. The navigation motions on water surface are simulated. These simulations are important to the performance analysis and parameter optimization for wave glider. The simulation results show a good agreement with sea trials.

Keywords: Wave glider · Multibody dynamic · Motion analysis · Numerical simulation

The wave glider is a new kind of wave-propelled autonomous ocean-going platform for deploying a series of sensor systems [1]. It makes use of ocean wave energy for platform propulsion. It uses solar panels to harvest energy from the sun to supply the electric devices and sensors, allowing for the Wave Glider to travel long distance without needing to refuel. The wave gliders have an autonomous navigation system based on GPS. It also can be remotely operated from shore-based control center via by satellite. The navigation methods can make sure the wave glider work for a long time [2]. Due to the difference in forward propulsion and navigation method, wave glider can work continuously for several months at the horizontal speeds of about 1 knot, while traditional propeller driven AUVs achieve an endurance ranging from hours to days at a speed of 2 knots [3]. The wave glider can deploy many kinds of scientific sensors to collect the ocean environment data from near free-surface to superficial water. The collected data and system state parameters can transmit to the shore-based control center in real time or nearly real time. The wave gliders having the characteristics of long fight range, high carrying capacity and real time. They are therefore proposed for oceanic observations with multiple mission and long endurance.

The wave gliders are attracting scientists' attention and a great deal of related scientific research has been carried out. N. Kraus and B. Bingham [4] understand the

Y. Huang et al. (Eds.): ICIRA 2017, Part I, LNAI 10462, pp. 721–734, 2017.
DOI: 10.1007/978-3-319-65289-4_67

capabilities and limitations of the wave glider's novel propulsion mechanism, consisting of a surface float and submerged glider to harvest wave energy for forward motion, them propose a simplified dynamic model appropriate for real-time implementation. Zhan Feng Qi et al. [5] analyze wave glider's operating mechanism and builds the dynamic model of wave glider. By simplifying the model into 3 DOS in longitudinal plane and selecting three generalized velocity, the kinematic equations and the generalized force can be confirmed. Kraus and Nicholas David [6] present a six degree of freedom set of nonlinear dynamic equations of motion for the wave glider. This work also identifies the key hydrodynamic parameters using analytical and experimental methods, and this simulation output is compared against a set of field trials to verify the ability of the model to emulate observed wave glider motion.

1 Overall Structure

As shown in Fig. 1, the wave glider system consists of three parts: surface float, underwater glider and umbilical cable [7]. The surface float is composed of solar panels, sealed compartment with floating material surrounded, charging system, scientific sensors, antenna mast and rechargeable battery. The underwater glider part contains rotatable gliding wings, framework, rudder module, magnetic compass and steering gear. The umbilical cable is a flexible cable that connects the surface float to the submerged glider part. The umbilical cable also supply power and control signal to magnetic compass and rudder module of the glider.

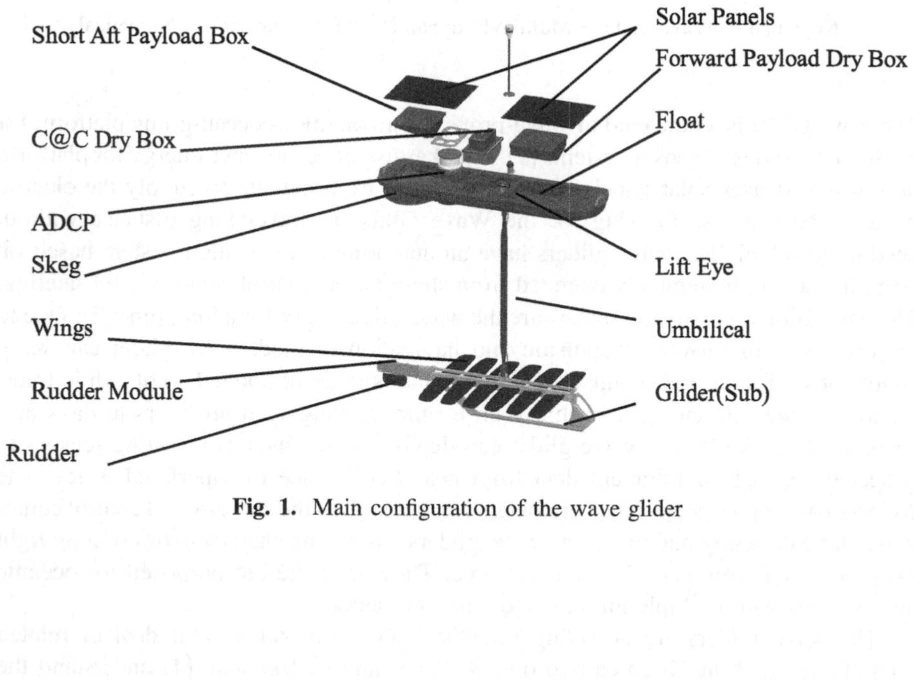

Fig. 1. Main configuration of the wave glider

2 Motion Mechanism

According to Airy's wave theory [8], water particle makes a circle motion in a certain depth. The vibrating amplitude of the water particles gets larger as underwater depth increases. The vibrating amplitude and of the water particles can be expressed as:

$$r = \mathrm{a}e^{kz} \tag{1}$$

where a and z represent the vibrating amplitude of the water particles at surface and the depth of water, k is the wave number.

And the wave energy of water particles in unit width can be expressed as

$$E_\mathrm{r} = 1/2\rho g r^2 L \tag{2}$$

Where L represents the wavelength of the water particles, ρ is density of water.

Hence, the wave energy difference between surface and the certain depth in unit width [9] is

$$\Delta E_C = E_a - E_r = 1/2\rho g L a^2\left(1 - \mathrm{e}^{2kz}\right) \tag{3}$$

According to the equations, the wave energy difference between surface and the certain depth in unit width is proportional to the depth and the square of surface wave's amplitude, then inversely proportional to wave numbers. The wave energy gap is generated when the amplitude of underwater particles larger than the surface particle. The wave glider can convert the wave energy gap to sustained forward force of system.

Figures 2 and 3 show the typical operation process of underwater wings. During the surface floatage moving downwards, wave energy gap drives the glider's wings make a clockwise motion. The resultant force of gravity and hydrodynamic force on the glider's wings point to horizontal direction. It drives the wave glider moves forward [10]. When the floatage moves upwards, the wings of underwater glider circle anticlockwise under the energy gap function. The underwater glider is moves forward under the resultant force of cable's pulling force and hydrodynamic force.

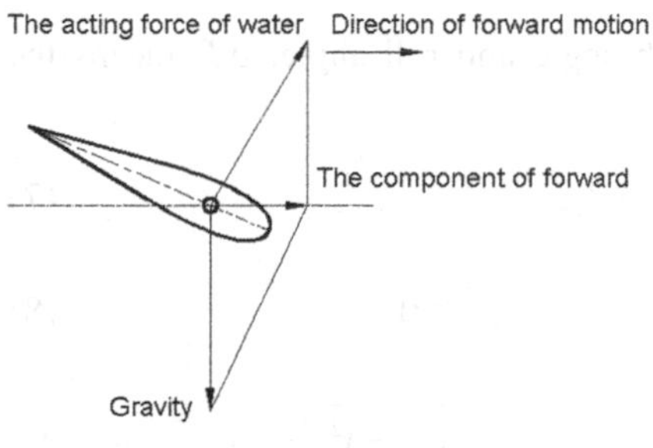

Fig. 2. Force analysis during moving downwards

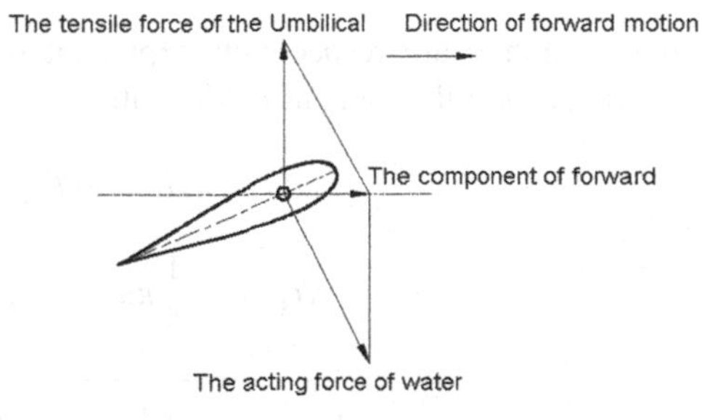

Fig. 3. Force analysis during moving upwards

According to the before analyze, the wave glider can efficiently convert the wave energy gap to sustained power.

3 Dynamic Modeling of Wave Glider

According to the system structure and motion principle, the dynamic model of wave glider divide into three subsystems models (the surface floatage, connecting cable, underwater glider) based on multibody dynamic theory. During the numerical modeling, the movement relationship and interaction forces between the subsystems are considered carefully. According to the dynamic model, motion analysis is also conducted by using numerical simulation software ADAMS and MATLAB [11] in this paper.

3.1 Modeling of the Connecting Cable

In this paper, the connecting cable is described as a set of connected branch segments. Then, the force condition of i-th segment is analyze using the finite element method. The i-th segment is acted upon by gravity, buoyancy, fluid resistance, tension of the cable in the motion. According to Newton's second law of motion, the balanced motion equation of i-th segment can be described [12] as follows:

$$M_i\ddot{x} = \Delta T_i + B_i + G_i + f_i \tag{4}$$

where M_i means mass matrix of the k-th segment.

$$M_i = m_i I + M_{oi} = \frac{1}{4}[C] \bullet k_a \rho g \pi d^2 \bullet [C]^{-1} \tag{5}$$

C is transport matrix from the local coordinate system to the inertial coordinate system of the cable.

$$C = \begin{bmatrix} \cos\theta\cos\phi & -\cos\theta\sin\phi & -\sin\phi \\ \sin\phi & \cos\phi & 0 \\ \sin\theta\cos\phi & -\sin\theta\sin\phi & \cos\theta \end{bmatrix} \tag{6}$$

where θ and ϕ are respectively represent pitch angle and roll angle. ΔT_i means the tension of the i-th segment of the cable.

$$\Delta T_i = \Delta T_{i+1} - \Delta T_{i-1} \tag{7}$$

$$\Delta T_{i+1} = \frac{1}{4}\pi d^2 E \varepsilon_{i+1} \bullet \tau_{i+1} \quad \varepsilon_{i+1} \geq 0 \tag{8}$$

$$\varepsilon_{i+1} = \frac{\sqrt{(x_{i+1} - x_i)^2 + (y_{i+1} - y_i)^2 + (z_{i+1} - z_i)^2}}{l_{i+1}} - 1 \tag{9}$$

$$B_i + G_i = -\frac{1}{8}\pi d^2 \rho g l_i + m_i g I \; ; \tag{10}$$

B_i, G_i,f_i means the buoyancy, gravity and fluid resistance of the i-th segment of the cable.

According to Bedendender's of Wilson resistance model [13], f_i can be expressed as

$$f_i = \begin{bmatrix} 0.5\rho d C_n v^2 \sin\alpha \mid \sin\alpha \mid \frac{v_x}{\sqrt{v_x^2 + v_y^2}} \\ 0.5\rho d C_n v^2 \sin\alpha \mid \sin\alpha \mid \frac{v_y}{\sqrt{v_x^2 + v_y^2}} \\ 0.5\rho\pi d C_t v^2 \cos\alpha \mid \cos\alpha| \end{bmatrix} \tag{11}$$

3.2 Modeling of the Surface-Water Floatage

The movement of surface-water floatage involves two kinds of medium, air and water. And, it is easily affected by wave motion and cable tension. In order to facilitate the analysis, we assume the side of the surface-water floatage is always parallel to the horizontal plane. According to the second law of Newton, the dynamic motion equation of floatage [14] can be expressed as,

$$M_t \ddot{x} = \Delta B \; + \; R_T \; + \; T_N \tag{12}$$

where $M_t \Delta B$, R_T means the mass matrix, the net buoyancy and fluid resistance of the surface-water floatage, T_N is the pulling force of k-th segment of the cable act on the surface-water floatage;

When the drought of the surface-water floatage increase Δy, ΔB is

$$\Delta B = \rho g V(\Delta y) = \rho g S_y \Delta y \tag{13}$$

Where S_y means the area of the water plane.

When there is no wind, the surface-water floatage drifts along with the waves will not move forward. According to Airy's micro amplitude wave theory, the motion equation of the surface-water floatage in longitudinal plane can be expressed as,

$$\Delta y = a[\cos(\sigma(t + \Delta t)) - \cos(\sigma t)] \tag{14}$$

The formulation of the surface-water floatage's fluid resistance [15] is

$$R_T = \left(C_F + \Delta C_F + C_{pr}\right)\rho V^2 S/2 \tag{15}$$

Where C_F means frictional resistance coefficient. According to the formulation of Prandtl-Schlichting, C_F is

$$C_F = \frac{0.455}{(\lg Re)^{2.58}} \tag{16}$$

ΔC_F means roughness allowance coefficient; C_{pr} is viscosity coefficient and can be expressed as

$$C_{\mathrm{pr}} = 0.09\frac{A_{\mathrm{m}}}{S}\sqrt{A_{\mathrm{m}}/2L_r} \tag{17}$$

where A_{m} is area of middle cross section, S, L_r means wetted areas and run length.

3.3 Modeling of the Underwater Glider

To ensure the motion stability, the connection between underwater glider with the cable adopt two points fasten method. The force analysis of the glider is shown in Fig. 4. $\xi 0$, $\xi 1$ is respectively the angle between section xoz of the glider and the l_0-th, l_1-th segment of the cable. $\gamma 0$, $\gamma 1$ is respectively the angle between the projection of l_0-th and l_1-th in xoz and the axis x. l_0, l_1 is the distance between connecting point and the buoyant centre of the glider, and h is the vertical distance between them.

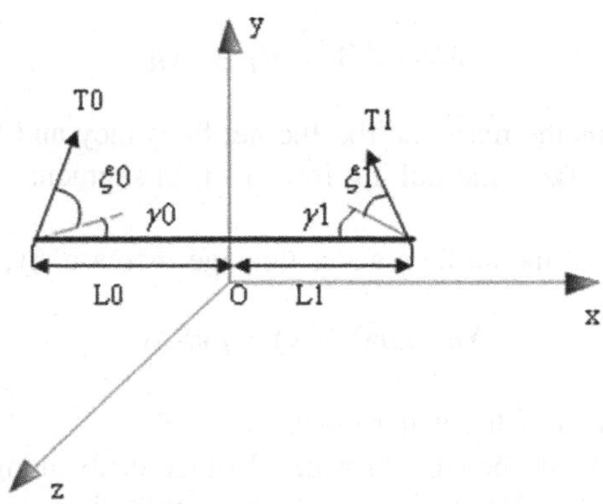

Fig. 4. Force act on the glider

Thus, the force equations of the glider can be expressed as

$$\begin{cases} T_x = T_0 \cos\zeta_0 \cos\gamma_0 - T_1 \cos\zeta_1 \cos\gamma_1 \\ T_y = T_0 \sin\zeta_0 + T_1 \sin\zeta_1 \\ T_z = -T_0 \sin\zeta_0 \sin\gamma_0 - T_1 \sin\zeta_1 \sin\gamma_1 \\ M_x = h(T_0 \sin\zeta_0 \sin\gamma_0 + T_1 \sin\zeta_1 \sin\gamma_1) \\ M_y = l_1 T_1 \sin\zeta_1 \sin\gamma_1 - l_0 T_0 \sin\zeta_0 \sin\gamma_0 \\ M_z = -h(T_0 \cos\zeta_0 \cos\gamma_0 + T_1 \cos\zeta_1 \cos\gamma_1) \\ \qquad -l_0 T_0 \sin\zeta_0 + l_1 T_1 \sin\zeta_1 \end{cases} \tag{18}$$

According to momentum theorem, the dynamic equation of the underwater glider can be expressed as [16]:

$$[\mathbf{A}_m + \mathbf{A}_\lambda]\dot{\mathbf{V}} + \frac{d[\mathbf{A}_m + \mathbf{A}_\lambda]}{dt}\mathbf{V} + \mathbf{A}_B\mathbf{V} = \mathbf{f} \tag{19}$$

$$\mathbf{A}_\lambda = \begin{bmatrix} \lambda_{11} & \lambda_{12} & 0 & 0 & 0 & \lambda_{16} \\ \lambda_{21} & \lambda_{22} & 0 & 0 & 0 & \lambda_{26} \\ 0 & 0 & \lambda_{33} & \lambda_{34} & \lambda_{35} & 0 \\ 0 & 0 & \lambda_{43} & \lambda_{44} & \lambda_{45} & 0 \\ 0 & 0 & \lambda_{53} & \lambda_{54} & \lambda_{55} & 0 \\ \lambda_{61} & \lambda_{62} & 0 & 0 & 0 & \lambda_{66} \end{bmatrix}$$

$$\mathbf{A}_\mathrm{m} = \begin{bmatrix} m & 0 & 0 & 0 & mz_c & -my_c \\ 0 & m & 0 & -mz_c & 0 & mx_c \\ 0 & 0 & m & my_c & -mx_c & 0 \\ 0 & -mz_c & my_c & J_{xx} & -J_{xy} & -J_{xz} \\ mz_c & 0 & -mx_c & -J_{yx} & J_{yy} & -J_{yz} \\ -my_c & mx_c & 0 & -J_{zx} & -J_{zy} & J_{zz} \end{bmatrix}$$

$$\mathbf{A}_\mathrm{B} = \left[\begin{matrix} 0 & -m\omega_z & m\omega_y \\ m\omega_z & 0 & -m\omega_x \\ -m\omega_y & m\omega_x & 0 \\ -m(y_c\omega_y + z_c\omega_z) & my_c\omega_x & mz_c\omega_x \\ mx_c\omega_y & -m(z_c\omega_z + x_c\omega_x) & mz_c\omega_y \\ mx_c\omega_z & my_c\omega_z & -m(x_c\omega_x + y_c\omega_y) \end{matrix}\right.$$
$$\left.\begin{matrix} m(y_c\omega_y + z_c\omega_z) & -mx_c\omega_y & -mx_c\omega_z \\ -my_c\omega_x & m(z_c\omega_z + x_c\omega_x) & -my_c\omega_z \\ -mz_c\omega_x & -mz_c\omega_y & m(x_c\omega_x + y_c\omega_y) \\ 0 & -J_{yz}\omega_y - J_{xz}\omega_x + J_{zz}\omega_z & J_{yz}\omega_z + J_{xy}\omega_z - J_{yy}\omega_y \\ J_{yz}\omega_y + J_{xz}\omega_x - J_{zz}\omega_z & 0 & -J_{zx}\omega_z - J_{xy}\omega_y + J_{xx}\omega_x \\ -J_{yz}\omega_z - J_{xy}\omega_x + J_{yy}\omega_y & J_{xz}\omega_z + J_{xy}\omega_y - J_{xx}\omega_x & 0 \end{matrix}\right]$$

$$\mathbf{f} = \begin{bmatrix} X_{\alpha\mu} + X_B + X_G + T_x \\ Y_{\alpha\mu} + Y_{\omega\mu} + Y_B + Y_G + T_y \\ Z_{\alpha\mu} + Z_{\alpha\mu} + Z_B + Z_G + T_z \\ M_{\alpha\mu x} + M_{\omega\mu x} + M_{Gx} + M_x \\ M_{\alpha\mu y} + M_{\omega\mu y} + M_{Gy} + M_y \\ M_{\alpha\mu z} + M_{\omega\mu z} + M_{Gz} + M_z \end{bmatrix}$$

Where **f** means forces and moments of forces as mentioned above, including fluid inertial force, net buoyancy, fluid viscous drag, fluid viscous force and so on. T, M means the forces and moments between the glider and the cable.

4 Co-simulation Model

As the system involves multiple disciplines, it can make full use of the research application results of single subject by applying the joint simulation method to studying the movement of wave gliders in water. Moreover, the appropriate data exchange enables the realization of multidisciplinary coupling calculation, in which the model and calculation results can better manifest the state of motion of physical system. The modeling core of wave gliders system can be divided into two parts, namely the complex computation part and multi-body dynamics part, the former being solved by MATLAB and the latter by ADAMS (Fig. 5).

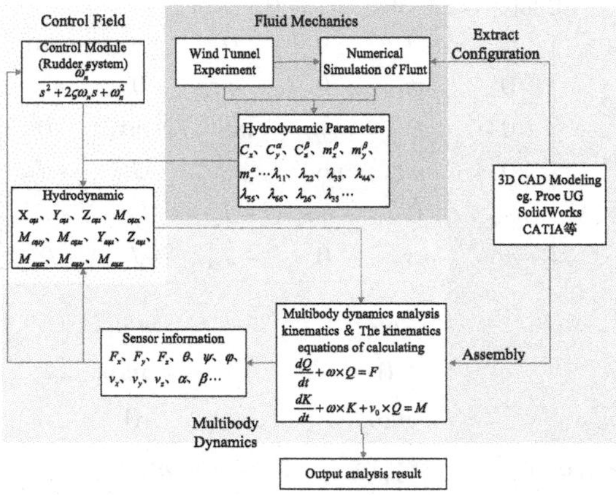

Fig. 5. The data flow diagram of multi-disciplinary co-simulation platform

With the aid of ADAMS, the model of multi-body dynamics is built up (see Fig. 6), and the control model and other required auxiliary modules are established (see Fig. 7). Based on the docking between software, the quantitative change of input and output

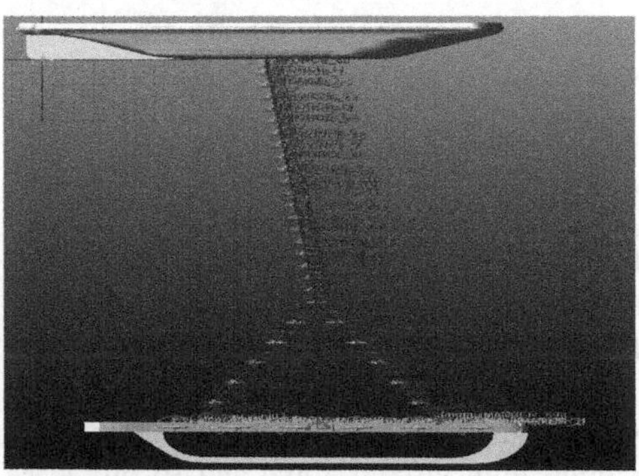

Fig. 6. Kinetics model of multi-body

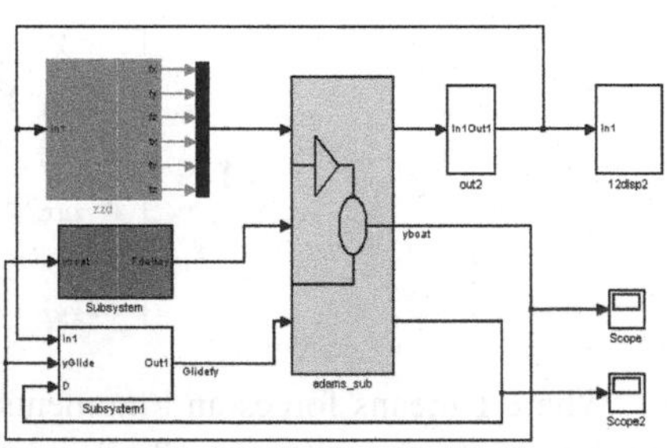

Fig. 7. Computing model by matlab

state is interchanged between the software. The paper takes the numerical iterative algorithm for estimation of the later system kinematics, kinematics characteristic and control characteristic at a given timing.

5 Motion Simulations and Analysis

According to the derived dynamic equations, we set cable length is 4 m, the number of wings is 6 pairs and the sea conditions is 3 levels (which wave height of 1 m, period of 4 s). The typical motions of wave glider are simulated in this paper to guide the design.

As illustrated in Fig. 8, the wave glider achieves approximate straight line motion. The average velocity of navigation in sea condition of levels 3 is nearly 1.125 m/s.

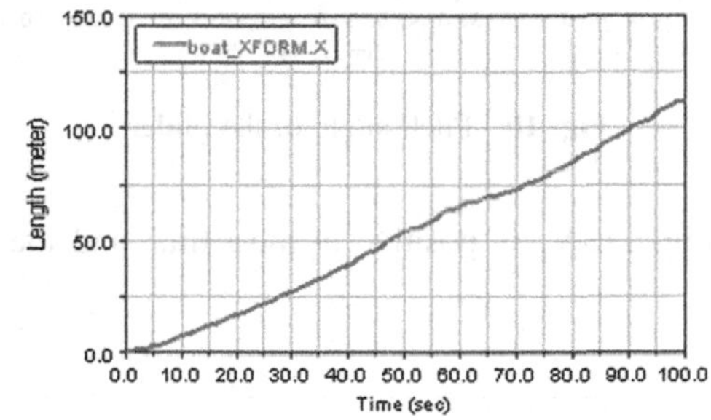

Fig. 8. Navigation curve of the wave glider

The underwater glider can switch the wave energy to propulsion power. It drives the floatage forward by the connected cable. As illustrated in Fig. 9, the floatage's maximum velocity can achieve 2.5 m/s and its average velocity is more than 1 m/s (the influence of wind isn't considered). It keeps a positive horizontal velocity (in global coordinate system) at most time. It is indicated that the floatage can continuous move forward in sea condition of level 3.

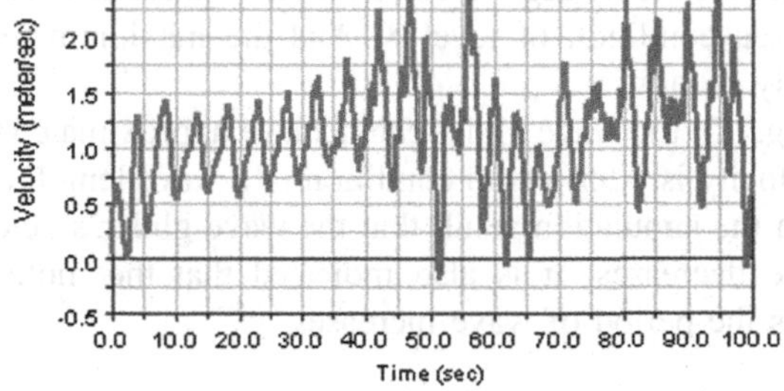

Fig. 9. Velocity along axis X of the floatage (v_x)

As illustrated in Fig. 10, the pitch angle of the glider varies greatly during moving forward. The maximum pitch angle is nearly 0.55 rad. This angle is benefit to the forward motion of the wave glider.

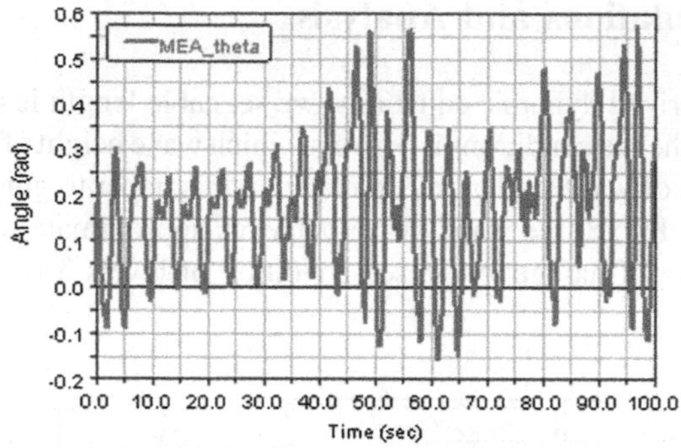

Fig. 10. Pitch angle of the glider

As illustrated in Fig. 11, V_x is positive at most time and the average velocity is 1.25 m/s.

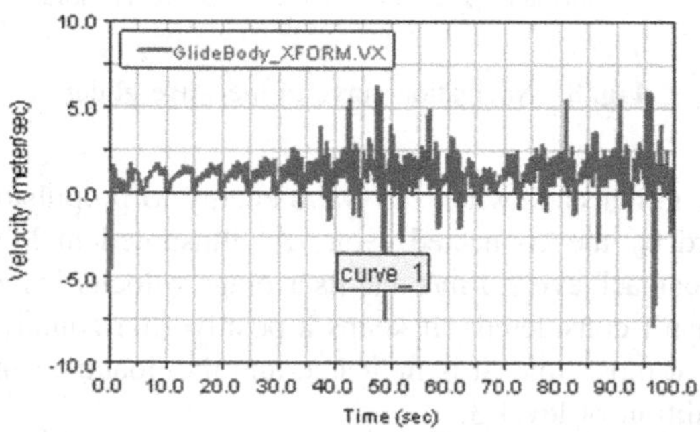

Fig. 11. Velocity along axis X of the glider (V_x)

As shown in Fig. 12, the average force of the cable is about 1000 N when the wave glider working at sea condition of level 3. And the maximum tension is less than 4000 N when steady motion.

As shown in Fig. 13, the wave glider can achieve approximate straight line motion and the average velocity is 0.55 m/s in condition of a wave length of 1 m, a period of 10 s. It is shown in the simulation result that the wave glider's velocity gets faster as the period of wave decreases. It is also indicated that the motion of wave glider becomes steadier as the period of wave increases.

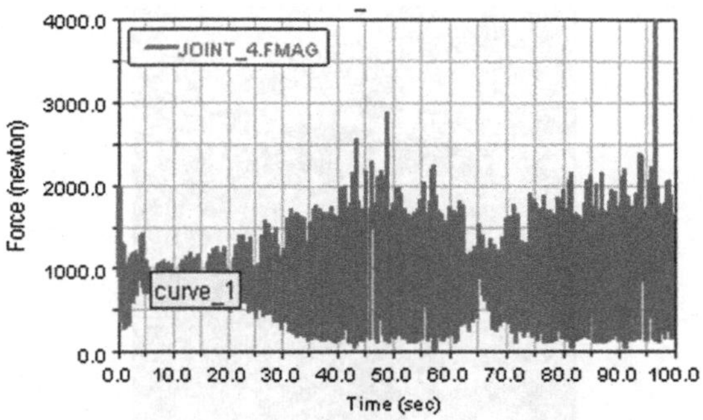

Fig. 12. The tension of the cable

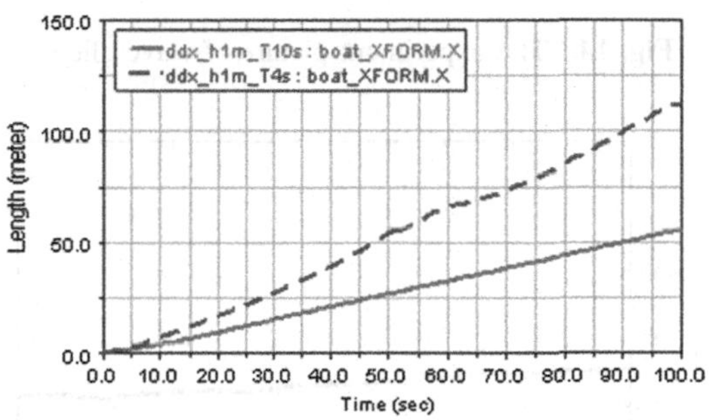

Fig. 13. Navigation curves of the wave glider in condition of a wave length of 1 m, a period of 10 s

6 Test Result

Test result in the ocean August 2015, the wave glider was tested in the offshore waters in Sanya. Test is complied with conditions of third-level sea. Figure 14 is a photograph of test sample of the wave glider in the sea.

Figure 15 is sea trail of the wave glider in certain course. It is to set the sailing line as the two points at the same latitude and the maximum angle of the steering wheel as 10°, which shows that the wave glider can continue to effectively navigate forward. It also can be seen from Fig. 16, the average navigational speed of the wave glider is 0.88 m/s on average and its variation range 0.12 m/s.

Figure 17 is the floating body of the underwater glider and its pitching angle, which can be seen that the pitching angle of the floating body is 3.6° on average and its variation range 2.2°, the pitching angle of the underwater glider is −11.7° on average and its variation range 7.4°. In the head-bent state, it is more conducive for underwater glider to dive under the gravity access to faster speed and greater sailing.

Fig. 14. The experiment picture of wave glider

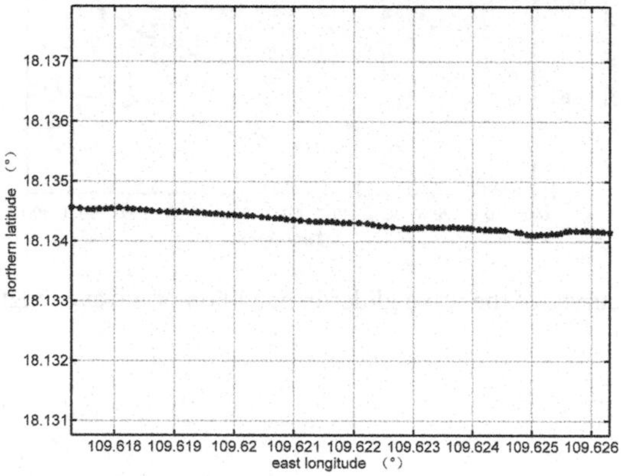

Fig. 15. The trajectory of wave glider

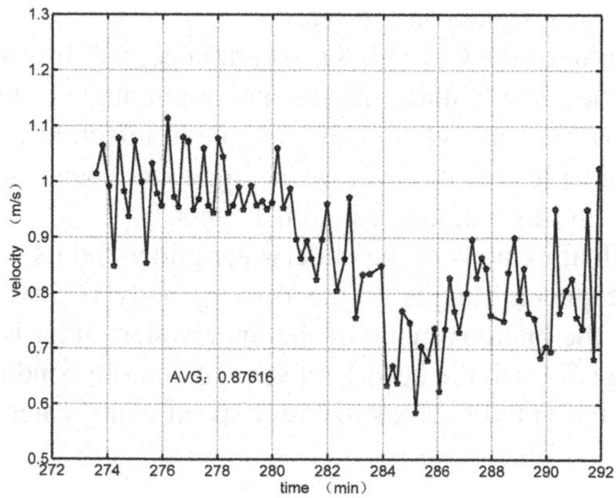

Fig. 16. The velocity of wave glider

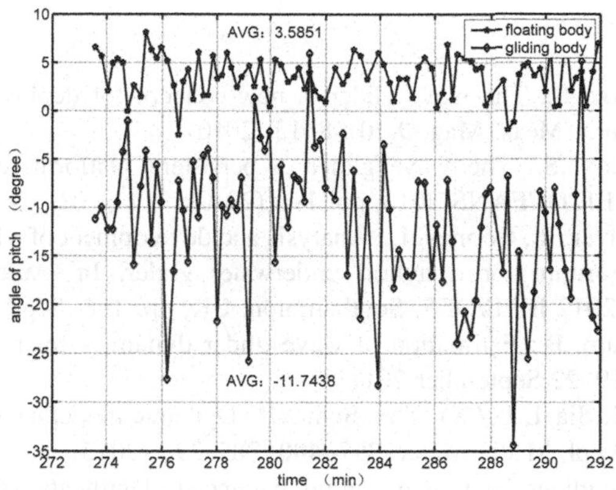

Fig. 17. The pitch angle of float and gliding body

The result of sea trail shows that the wave glider is able to continue to navigate forward in the third-lever sea with 0.88 m/s in speed which is a little different from the real speed 1.125 m/s though, and the simulation result is acceptable with the current and wind taken into account.

7 Conclusions

A multi-mission wave glider is developed as a low-drag, lightweight, large range rating unmanned vehicle for oceanic oceanography missions. In this paper, a dynamic model was established for wave glider utilizing multibody dynamic theory. The dynamic model of wave glider divide into three subsystems models – the surface floatage, connecting cable, underwater glider. Based on the model, the traditional motion patterns of wave glider have been simulated by using ADMAS and MATLAB software. The simulation results agree well with the experimental results, verifying feasibility and effectiveness of the dynamic model. Both the simulation and the experimental results show that the vehicle achieves stable status in sea condition of level 3. And its average velocity is more than 1.25 m/s. The average and maximum tense of the cable is nearly 1000 N and 4000 N. This work provides guidance for dynamical behavior prediction and system design improvements for the wave glider.

Acknowledgements. The research was supported by the National Hi-tech Research and Development Program of China (Grant No. 2014AA09A508) and the National Key Research and Development Program of China (Grant No. 2016YFC0301101). The authors gratefully acknowledge the contributions of the members in the R&D team for wave glider.

References

1. Manley, J., Willcox, S.: The wave glider: a new concept for deploying ocean instrumentation. IEEE Instrum. Meas. Mag. **9**(10), 8–13 (2010)
2. Manley, J., Willcox, S.: The wave glider: a persistent platform for ocean science. In: Proceedings of IEEE OCEANS 2010, pp. 1–6 (2010)
3. Claus, B., Bachmayer, R., Cooney, L.: Analysis and development of a buoyancy-pitch based depth control algorithm for a hybrid underwater glider. In: Autonomous Underwater Vehicles (AUV), 2012 IEEE/OES, Southampton, UK, pp. 1–6, September 2012
4. Kraus, N., Bingham, B.: Estimation of wave glider dynamics for precise positioning. In: OCEANS 2011, 19–22 September 2011
5. Qi, Z.F., Liu, W.X., Jia, L.J., Qin, Y.F., Sun, X.J.: Dynamic modeling and motion simulation for wave glider. Appl. Mech. Mater. **397–400**, 285–290 (2013)
6. Kraus, N.D.: Wave glider dynamic modeling, parameter identification and simulation. M.S. - Mechanical Engineering, May 2012
7. Frolov, S., Bellingham,J., Anderson, W., Hine, G.: Wave glider—a platform for persistent monitoring of algal blooms. In: Proceedings of MTS/IEEE Oceans 2011 Conference (2011)
8. Willcox, S., Meinig, C., Sabine, C., Lawrence-Slavas, N., Richardson, T., Hine, R., Manley, J.: An autonomous mobile platform for underway surface carbon measurements in open-ocean and coastal waters. In: Proceedings MTS/IEEE OCEANS 2009 Conference, Biloxi, MS, 26–29 October (2009)
9. Huang, X., Xinshen, L.: Ocean engineering fluid mechanics and structural response analysis. Shanghai Jiao Tong University Press, Shanghai (1992). (in Chinese)
10. Kubo, Y., Shimoyama, I., Kaneda, T., Miura, H.: Study on wings of flying microrobots. In: Proceedings of the IEEE International Conference on Robotics and Automation, vol. 2, pp. 834–839 (1994)
11. Liu, G., Wang, M., He, B.: Cooperative simulation based on Adams and Matlab/Simulink for autonomous underwater vehicle. J. Mech. Eng. **45**(10), 22–29 (2009). (in Chinese)
12. Shyy, W., Udaykumar, H.S., Ran, M.M., Smith, R.W.: Computational Fluid Dynamics with Moving Boundaries. Taylor & Francis, Washington DC (1996)
13. Li, X.: Multibody system dynamics modeling and application in the underwater cable. Doctoral Dissertation of Tianjin University (2004). (in Chinese)
14. Shuxin, W., Xiaoping, L., Yanhui, W., et al.: Dynamic modeling and analysis of underwater gliders. Ocean Technol. **24**(1), 5–9 (2005)
15. Shabana, A.A.: Dynamics of Multibody Systems. Cambridge University Press, Cambridge (1998)
16. Yan, W.: The Torpedo Navigation Mechanics. Northwestern Polytechnical University Press, Xian (2005). (in Chinese)

Distributed Formation Control of Autonomous Underwater Vehicles Based on Flocking and Consensus Algorithms

Wuwei Pan, Dapeng Jiang(✉), Yongjie Pang, Yuda Qi, and Daichao Luo

Science and Technology on Underwater Vehicle Laboratory, Harbin Engineering University, Harbin 150001, China
jdpl03@hrbeu.edu.cn

Abstract. In this paper, a distributed formation controller of autonomous underwater vehicles is presented. An artificial potential field is proposed based on the Lennard-Jones potential. A weight function is designed to make the potential smoothly die off at the boundaries. A control algorithm is proposed according to the three rules of Reynolds: flock centering, collision avoidance and velocity matching. A single-step target tracking method is proposed to translate the desired acceleration, produced by the potential force, to the desired velocity and heading for the AUV. To make the algorithm distributed, each AUV holds a copy of the virtual leader and calculates the guidance information individually. A consensus algorithm is proposed to make the mismatches convert to zeros. A fold line and a curve line path following of the fleet are simulated, with random initial positions of the AUVs. Under the formation control algorithms proposed, the AUVs are uniformly distributed and form a lattice-like formation. We increase the number of AUVs, and the algorithm serves well, which shows good availability and flexibility.

Keywords: Autonomous underwater vehicle · Distributed control · Formation control · Flocking algorithm · Consensus algorithm

1 Introduction

Multiple autonomous underwater vehicles (AUVs) system is getting increasingly attention by researchers, which has amount of potential applications, such as environment monitoring, cooperative location and mapping.

However, formation control of multi-AUVs is still a challenging research area. The model of the AUV is underactuated with highly nonlinear hydrodynamic parameters, and the formation system is much more complex. Algorithms such as leader and followers [1–4], virtual structure [5, 6], artificial potential [7] and behavioral rules [8, 9] have been proposed. The formation control problem is often decoupled into formation keeping problem and single vehicle control problem. Leader and followers algorithm is often applied. The leader tracks a desired path, while the followers maintain predefined offsets of the leader's position. Information, such as position and velocity of the leader

Y. Huang et al. (Eds.): ICIRA 2017, Part I, LNAI 10462, pp. 735–744, 2017.
DOI: 10.1007/978-3-319-65289-4_68

or generalized path length, is broadcasted from the leader to the followers to achieve cooperation.

Disadvantages of the methods above are less expansible and flexible when the number of vehicles increases, as the distance and relationship between AUVs is configured by the designers, and the controllers are different. Furthermore, problems before the completion of the formation are not studied, such as collision avoidance between AUVs. In order to solve these problems, we propose a distributed collision free formation control algorithm, which is flexible and self-organized.

The organization of this paper is as follows: In Sect. 2, the model of AUV is presented. In Sect. 3, we design a potential field to organize the formation. In Sect. 4, the flocking algorithm of formation control is presented. In Sect. 5, we propose a consensus algorithm to make the method distributed and give the system diagram. Then, in Sect. 6 we apply our method to a simulation example and draw conclusions in Sect. 7.

2 Model of AUV

Consider a system consists of N AUVs, moving on the horizontal plain with constant depth. The i-th AUV can be described as [10]

$$\begin{cases} \dot{\boldsymbol{\eta}}_i = \boldsymbol{J}(\boldsymbol{\eta}_i)\boldsymbol{v}_i \\ \boldsymbol{M}_i\dot{\boldsymbol{v}}_i + \boldsymbol{C}_i(\boldsymbol{v}_i)\boldsymbol{v}_i + \boldsymbol{D}_i\boldsymbol{v}_i = \boldsymbol{\tau}_i \end{cases}, \tag{1}$$

where $\boldsymbol{\eta}_i = \left[\boldsymbol{q}_i^T, \psi_i\right]^T$ and $\boldsymbol{v}_i = \left[\boldsymbol{p}_i^T, \omega_i\right]^T$ are the generalized position and velocity of the AUV, $\boldsymbol{M}_i$, $\boldsymbol{C}_i(\boldsymbol{v}_i)$ and $\boldsymbol{D}_i$ are the inertia matrix, the Coriolis and centripetal matrix and damping matrix, $\boldsymbol{\tau}_i = \left[\tau_x, 0, \tau_\psi\right]^T$ is the control input. The position and velocity are $\boldsymbol{q}_i = [x_i, y_i]^T$, $\boldsymbol{p}_i = [u_i, v_i]^T$. Then, the matrix can be expressed as

$$\boldsymbol{J}(\boldsymbol{\eta}_i) = \begin{bmatrix} \cos\psi_i & -\sin\psi_i & 0 \\ \sin\psi_i & \cos\psi_i & 0 \\ 0 & 0 & 1 \end{bmatrix}, \quad \boldsymbol{M}_i = \begin{bmatrix} m_{11} & 0 & 0 \\ 0 & m_{22} & m_{23} \\ 0 & m_{32} & m_{33} \end{bmatrix}, \quad \boldsymbol{C}_i(\boldsymbol{v}_i) = \begin{bmatrix} 0 & 0 & C_{13} \\ 0 & 0 & C_{23} \\ C_{31} & C_{32} & 0 \end{bmatrix},$$

$$\boldsymbol{D} = \begin{bmatrix} d_{11} & 0 & 0 \\ 0 & d_{22} & d_{23} \\ 0 & d_{32} & d_{33} \end{bmatrix},$$

where $m_{23} = m_{32}$, $C_{13} = -C_{31} = -m_{22}v_i - (m_{23} + m_{32})\omega_i/2$, $C_{23} = -C_{32} = m_{11}u_i$.

3 Lennard-Jones Potential

We use the artificial potential field to organize the formation of AUVs for two purposes. One is to make the formation has the ability of self organization and the other is to make the AUVs form a uniform distribution. Here, we choose a potential function, called Lennard-Jones bi-reciprocal potential [11]. Let

$$\phi^{LJ}(r) = \frac{-e_0}{m-n}\left(m\left(\frac{r_0}{r}\right)^n - n\left(\frac{r_0}{r}\right)^m\right), \tag{2}$$

where e_0 is called the dissociation energy and r_0 is called the equilibrium separation distance, m and n are design parameters. Then, the potential energy between the i-th AUV and the j-th AUV is

$$\phi_{ij}^{LJ} = \phi^{LJ}(\|\boldsymbol{q}_j - \boldsymbol{q}_i\|). \tag{3}$$

The force acting on the i-th AUV, produced by the j-th AUV, is the gradient of ϕ_{ij}^{LJ} at the position of the i-th AUV. It can be expressed as

$$\boldsymbol{f}_{ij}^{LJ}(\|\boldsymbol{q}_j - \boldsymbol{q}_i\|) = \nabla_{\boldsymbol{q}_i}\phi_{ij}^{LJ}(\|\boldsymbol{q}_j - \boldsymbol{q}_i\|), \tag{4}$$

where $f^{LJ}(r) = \frac{e_0 mn}{m-n}\left(\frac{r_0^n}{r^{n+1}} - \frac{r_0^m}{r^{m+1}}\right)$.

The Lennard-Jones potential was initially used to model forces between molecules. The potential increases infinitely when two molecules are very close, and dies off at an infinite distance. As we want to design a distributed control algorithm, an AUV should not be influenced by other AUVs outside a boundary. The neighborhood of the i-th AUV is defined as $N_i = \{j \mid \|\boldsymbol{q}_j - \boldsymbol{q}_i\| \le r_{boundary}\}$. A two-times smooth step function is designed to die off the potential at the neighborhood boundary, which is expressed as

$$s(r) = \begin{cases} 1 & , r < r_a \\ \rho(z) \quad , z = \frac{r-r_a}{r_b-r_a} & , r_a \le r \le r_b \\ 0 & , r > r_b \end{cases}, \tag{5}$$

where r_a and r_b are parameters, and $\rho(z)$ satisfies

$$\begin{cases} \rho(0) = 1 \\ \rho(1) = 0 \end{cases}, \begin{cases} \dot{\rho}(0) = 0 \\ \dot{\rho}(1) = 0 \end{cases}, \begin{cases} \ddot{\rho}(0) = 0 \\ \ddot{\rho}(1) = 0 \end{cases}.$$

We use the cosine functions as bases, and get $\rho(z) = 0.5 + 0.5625\cos \pi z - 0.0625\cos 3\pi z$.

The potential increases infinitely when two AUVs are very close, while the actuators have physical limitations. We map the distance r to a new form as

$$r_\sigma = \sqrt{r^2 + \sigma^2}, \tag{6}$$

where σ is a designed parameter. When the distance r decreases to zero, r_σ will be σ. When r is greater than some value, the map is almost linear and $r_\sigma \approx r$.

Now, we can get the potential function and force function between the i-th AUV and the j-th AUV.

$$\begin{cases} \phi_{ij}(\|\boldsymbol{q}_j - \boldsymbol{q}_i\|) = s(r_\sigma)\phi_{ij}^{LJ}(r_\sigma) \\ \boldsymbol{f}_{ij}(\|\boldsymbol{q}_j - \boldsymbol{q}_i\|) = s(r_\sigma)\boldsymbol{f}_{ij}^{LJ}(r_\sigma) + \dot{s}(r_\sigma)\phi_{ij}^{LJ}(r_\sigma) \\ r_\sigma(\|\boldsymbol{q}_j - \boldsymbol{q}_i\|) = \sqrt{\|\boldsymbol{q}_j - \boldsymbol{q}_i\|^2 + \sigma^2} \end{cases}. \tag{7}$$

4 Flocking Algorithm

In 1986, Reynolds [12] introduced three rules that led to flocking, and conducted a computer animation. The three rules are: flock centering, collision avoidance and velocity matching. In Olfati-Saber [13], a flocking algorithm was proposed

$$\boldsymbol{u}_i = \boldsymbol{f}_i^g + \boldsymbol{f}_i^d + \boldsymbol{f}_i^\gamma. \tag{8}$$

It consists of three terms, where $\boldsymbol{f}_i^g$ is a gradient-based term, $\boldsymbol{f}_i^d$ is a velocity consensus term that acts as a damping force, and $\boldsymbol{f}_i^\gamma$ is a navigation feedback of the fleet objective. With this control input, the system becomes a dissipative particle system, which consists of potential energy and kinetic energy. Finally, the system converts to a stable state, which is a local minimum of system energy. The particles form a lattice-like formation, with equal distance between each other.

Inspired by this work, we suppose the i-th AUV is motivated by three kinds of forces

$$\boldsymbol{f}_i = \boldsymbol{f}_i^g + \boldsymbol{f}_i^d + \boldsymbol{f}_i^\gamma, \tag{9}$$

where $\boldsymbol{f}_i^g$ is a force produced by the potential energy, $\boldsymbol{f}_i^d$ is a damping force produced by the energy of velocity mismatch, $\boldsymbol{f}_i^\gamma$ is an attractive force produced by the virtual leader of the fleet. The three terms can be expressed as

$$\begin{cases} \boldsymbol{f}_i^g = c_g \sum\limits_{j \in N_i} \boldsymbol{f}_{ij}(\|\boldsymbol{q}_j - \boldsymbol{q}_i\|) \\ \boldsymbol{f}_i^d = \sum\limits_{j \in N_i} c_{ij}^d(\boldsymbol{p}_j - \boldsymbol{p}_i) \\ \boldsymbol{f}_i^\gamma = c_1(\boldsymbol{q}_r - \boldsymbol{q}_i) + c_2(\boldsymbol{p}_r - \boldsymbol{p}_i) \end{cases}, \tag{10}$$

where $\boldsymbol{q}_r \in R^2$ and $\boldsymbol{p}_r \in R^2$ are position and velocity of the virtual leader, c_g, c_{ij}^d, c_1 and c_2 are parameters.

c_{ij}^d is designed as a smooth step function of the distance between the i-th AUV and the j-th AUV. This is because that we want the force to change smoothly, when an AUV across the neighborhood boundary. It is chosen as the same with Eq. (5). c_1 is designed as a saturation function of the distance between the i-th AUV and the virtual leader. When the fleet follows the virtual leader, the AUVs will distribute around. If c_1 is a constant value, the attractive forces of the virtual leader will be too large for the AUVs on the boundary of the groups. Then, the AUVs in the center will be compressed by the AUVs outside, and the group will not distribute uniformly. We design another step function

$$s'(r) = \begin{cases} 0 & , r < r_a \\ \rho'(z) \quad , z = \frac{r - r_a}{r_b - r_a} & , r_a \le r \le r_b \, , \\ 1 & , r > r_b \end{cases} \tag{11}$$

where $\rho'(z) = 0.5 - 0.5625\cos\pi z + 0.0625\cos 3\pi z$. Then we can get c_1 as

$$c_1 = c_1(r) = k_1[s(r)r + (r_a + r_b)s'(r)/2], \tag{12}$$

where k_1 is the slop of the linear section.

In the sections above, the total virtual forces are calculated. As AUVs are rigid bodies and the control inputs are rudder angles and thruster speeds, a method should be designed to translate the virtual force to the control inputs.

A single-step target tracking method is proposed. At each time step, we suppose there is a virtual particle having the same position and velocity with the i-th AUV. The force acting on the AUV also acts on the particle. Then, at this time step, the velocity and position can be calculated with integration

$$\begin{cases} \dot{\boldsymbol{q}}_i^{particle} = \boldsymbol{p}_i^{particle} \\ \dot{\boldsymbol{p}}_i^{particle} = \boldsymbol{f}_i \end{cases}, \tag{13}$$

where $\boldsymbol{q}_i^{particle} \in R^2$ and $\boldsymbol{p}_i^{particle} \in R^2$ are position and velocity of the i-th particle.

Then, the particle will be a little distance ahead of the AUV, which predicts the desired states of the i-th AUV for the next time step. We set the particle as the tracking target for the AUV, and solve the problem with a constant bearing guidance algorithm, shown as Fig. 1.

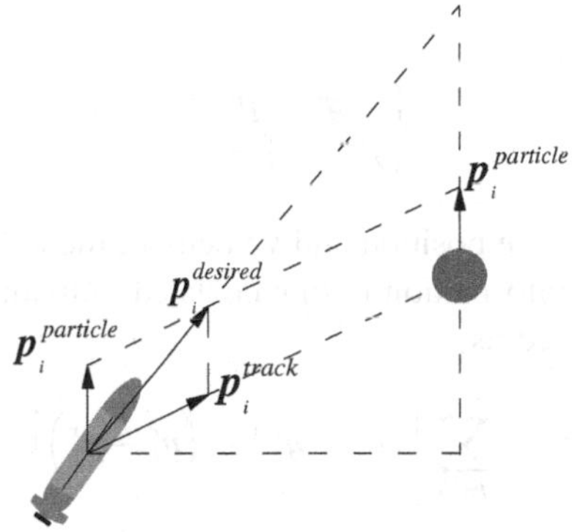

Fig. 1. Constant bearing target tracking.

where the red circle is the tracking target with a velocity $p_i^{particle}$, $p_i^{desired}$ is velocity desired of the AUV, and p_i^{track} is relative velocity between AUV and the target. $p_i^{particle}$, p_i^{track} and $p_i^{desired}$ follow the parallelogram rule of vectors.

The position mismatch between the i-th AUV and the particle is

$$\boldsymbol{q}_i^{mismatch} = \boldsymbol{q}_i^{particle} - \boldsymbol{q}_i^{AUV}. \tag{14}$$

Then, the desired tracking velocity, which is the relative velocity between the i-th AUV and the particle, can be designed as

$$\boldsymbol{p}_i^{track} = \dot{\boldsymbol{q}}_i^{mismatch} = -k_2 \boldsymbol{q}_i^{mismatch}, \tag{15}$$

where $k_2 = \frac{U_{\max}}{\sqrt{\left\|\boldsymbol{q}_i^{mismatch}\right\|^2 + \delta^2}}$. $U_{\max}$ is the maximum tracking speed of the i-th AUV, and $\delta > 0$. The desired velocity of i-th AUV can be expressed as

$$\boldsymbol{p}_i^{desired} = \boldsymbol{p}_i^{track} + \boldsymbol{p}_i^{particle}. \tag{16}$$

In this way, the desired speed and heading of the i-th AUV can be calculated. We separate the AUV system into a guidance system and a control system, which is called the backseat-driver paradigm. The desired speed and heading serve as inputs of the control system, and the rudder angle and thruster speed can be calculated, which is not presented for short.

5 Consensus Algorithm and System Diagram

The algorithm proposed still is centralized, as there is a virtual leader of the group. To make the algorithm distributed, a local copy of the virtual leader is maintained by each AUV

$$\begin{cases} \dot{\boldsymbol{q}}_i^r = \boldsymbol{p}_i^r \\ \dot{\boldsymbol{p}}_i^r = \boldsymbol{\tau}_i^r + \boldsymbol{\tau}_i^c \end{cases}, \tag{17}$$

where $\boldsymbol{q}_i^r \in R^2$ and $\boldsymbol{p}_i^r \in R^2$ are position and velocity of the i-th local copy of the virtual leader, $\boldsymbol{\tau}_i^r$ is the guidance information term calculated individually, $\boldsymbol{\tau}_i^c$ is the consensus term,which can be expressed as

$$\boldsymbol{\tau}_i^c = \sum_{j \in N_i} \left[\left(\boldsymbol{q}_j^r - \boldsymbol{q}_i^r \right) + \left(\boldsymbol{p}_j^r - \boldsymbol{p}_i^r \right) \right]. \tag{18}$$

It is shown in [14] and [15] that consensus is achieved if the underlying directed graph, of the AUVs' time-varying graph topologies, has a directed spanning tree.

Now, we propose the diagram of the AUV formation system, shown as Fig. 2.

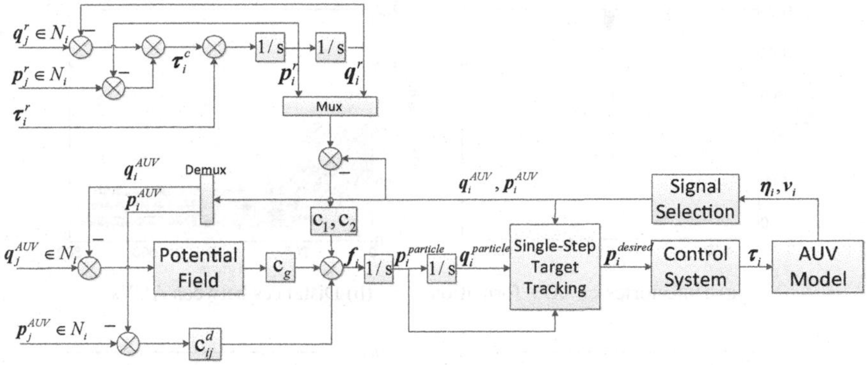

Fig. 2. Diagram of AUV formation system

The information communicated within an AUV and its neighbors is that the states of its local copy of the virtual leader and the velocity and position of the AUV itself.

We should keep in mind that the position and velocity of the virtual particle, $\boldsymbol{q}_i^{particle}$ and $\boldsymbol{p}_i^{particle}$, is reset the same with the i-th AUV at every time step. The idea behind this is that the virtual particle only serves to predict the desired states for the AUV at next time step. It is a method to translate the virtual force to control inputs for the AUVs.

6 Simulations

We choose the hydrodynamic parameters of the AUVs as: $m_{11} = 25.8$ kg, $m_{22} = 33.8$ kg, $m_{33} = 2.76\ \text{kg} \cdot \text{m}^2$, $m_{23} = m_{32} = 6.2\ \text{kg} \cdot \text{m}$, $d_{11} = 27$ kg/s, $d_{22} = 17$ kg/s, $d_{33} = 0.5\ \text{kg} \cdot \text{m}^2/\text{s}$, $d_{23} = 0.2\ \text{kg} \cdot \text{m/s}$, $d_{32} = 0.5\ \text{kg} \cdot \text{m/s}$. The control parameters is set as: $e_0 = 0.002\ \text{kg} \cdot \text{m}^2/\text{s}^2$, $r_0 = 50$ m, $m = 6$, $n = 9$, $\sigma = 35$ m, $k_1 = 0.2$, $c_2 = 9$, $U_{\max} = 1$ m/s, $\delta = 0.05$ m, and the amplitude of c_{ij}^d is 0.2. $r_a = 50$ m, $r_b = 55$ m for the potential function (7), while $r_a = 20$ m, $r_b = 30$ m for the saturation function (12). The initial positions of the group are set randomly in an area of 200 m × 200 m, and the initial headings of the AUVs are set randomly in a interval $[-\pi/2, \pi/2]$. The initial position of the virtual leader is (200 m, 0 m), and each copy of the virtual leader has a random initial mismatch from −15 m to 15 m.

First, we simulate a 12-AUV formation, tracking a straight line. The simulation results are shown as Fig. 3(a)–(d). The trajectories of the formation are shown in Fig. 3(a), where the blue lines are trajectories of AUVs and red lines are trajectories of local copies of the virtual leaders in each AUV. The diamond-like symbols are the final locations of the AUVs, and the vectors ahead are the velocities. The legend holds the same in Fig. 3(g) and (h).

The AUVs track their local copies of the virtual leader, form a steady formation from random initial positions and the headings convert to zeros gradually. The distances between an AUV and its neighbors are almost the same, which make a lattice-like formation. The minimum distance is about 10 meters, and collisions are avoided.

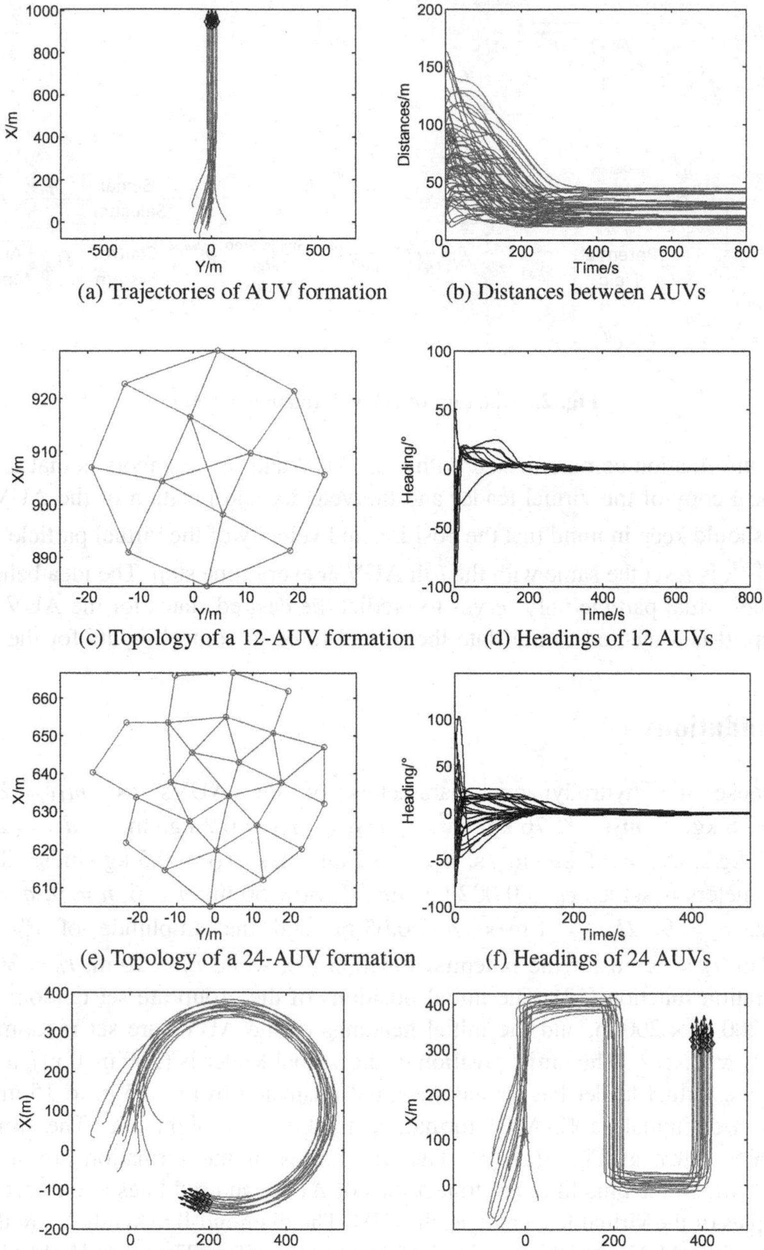

(a) Trajectories of AUV formation (b) Distances between AUVs

(c) Topology of a 12-AUV formation (d) Headings of 12 AUVs

(e) Topology of a 24-AUV formation (f) Headings of 24 AUVs

(g)Path following of a curve line (h) Path following of a fold line

Fig. 3. Simulation results of formation path following

We increase the number of AUVs to 24 to verify the scalability of the algorithm, with parameters remain the same. The results are shown as Fig. 3(e)–(f). The difference is that the topology of more AUVs is a little compressed, as AUVs on the formation boundary cause virtual forces on the AUVs in the center.

A curve line and a fold line formation path following of 12 AUVs are simulated, and the results are shown as Fig. 3(g)–(h). In Fig. 3(g), the virtual leader tracks a circle path with a radius of 250 m. In Fig. 3(h), the path points are $(400\ \mathrm{m}, 0\ \mathrm{m})$, $(400\ \mathrm{m}, 200\ \mathrm{m})$, $(0\ \mathrm{m}, 200\ \mathrm{m})$, $(0\ \mathrm{m}, 400\ \mathrm{m})$ and $(400\ \mathrm{m}, 400\ \mathrm{m})$, and the guidance algorithm is LOS(Line Of Sight), which is often used in a ship guidance system. In Fig. 3(g) and (h), the formation of the group remains steady while tracking the path. Even when the group passes the path points with right angle, there are little disturbs in the formation.

7 Conclusions

In this paper, we propose a distributed collision free formation control algorithm for AUVs. From the simulation results, we can draw the conclusions that, the artificial potential field is properly designed to guarantee the distances between AUVs, the flocking algorithm serves well which makes the algorithm insensitive with the number and initial states of AUVs, the single-step target tracking method can map the field forces to desired heading and speed to guidance the AUV, and with the help of the consensus algorithm, the formation control algorithm becomes distributed. The AUVs can form a lattice-like formation and remain steady while the group follows the path. With proper guidance of the virtual leader, the team can follow both curve lines and fold lines, which can be used in applications such as sea floor surveys.

References

1. Park, B.S.: Adaptive formation control of underactuated autonomous underwater vehicles. Ocean Eng. **96**, 1–7 (2015)
2. Xianbo, X., Bruno, J., Olivier, P.: Coordinated formation control of multiple autonomous underwater vehicles for pipeline inspection. Int. J. Adv. Rob. Syst. **7**(1), 75–84 (2010)
3. Rongxin, C.U.I., Shuzhi, G.E., Voon, H.B., et al.: Leader-follower formation control of underactuated autonomous underwater vehicles. Ocean Eng. **37**(17–18), 1491–1502 (2010)
4. Yang, E., Dongbing, G.: Nonlinear formation-keeping and mooring control of multiple autonomous underwater vehicles. IEEE/ASME Trans. Mechatron. **12**(2), 164–178 (2007)
5. Wang, Y., Yan, W., Li, J.: Passivity-based formation control of autonomous underwater vehicles. IET Control Theor. Technol. **6**(4), 518–525 (2012)
6. Do, K.D.: Practical formation control of multiple underactuated ships with limited sensing ranges. Robot. Auton. Syst. **59**, 457–471 (2011)
7. Hou, S.P., Cheah, C.C.: Can a simple control scheme work for a formation control of multiple autonomous underwater vehicles. IEEE Trans. Control Syst. Technol. **17**, 1090–1100 (2011)
8. Balch, T., Arkin, R.C.: Behavior-based formation control for multirobot teams. IEEE Trans. Robot. Autom. **14**(6), 926–939 (1988)

9. Djapic, V.: Unifying behavior based control design and hybrid stability theory for AUV application. Ph.D. thesis, University of California, Riverside (2009)
10. Fossen, T.I.: Handbook of Marine Craft Hydrodynamics and Motion Control, 1st edn. Wiley, United Kingdom (2011)
11. Tonnesen, D.L.: Dynamically coupled particle systems for Geometric modelling, reconstruction, and animation. PhD thesis, University of Toronto (1998)
12. Reynolds, C.W.: Flocks, herds, and schools: a distributed behavioral model. Comput. Graph. **21**(4), 25–34 (1987)
13. Olfati-Saber, R.: Flocking for multi-agent dynamic systems: algorithms and theory. IEEE Trans. Autom. Control. **51**(3), 401–420 (2006)
14. Ren, W., Beard, R.W.: Consensus seeking in multiagent systems under dynamically changing interaction topologies. IEEE Trans. Autom. Control **50**(5), 655–661 (2005)
15. Ren, W., Beard, R.W., Kingston, D.B.: Multi-agent Kalman consensus with relative uncertainty. In: Proceedings of the American Control Conference, Portland, pp. 1865–1870 (2005)

Pitch Angle Active Disturbance Rejection Control with Model Compensation for Underwater Glider

Dalei Song[1,2], Tingting Guo[1], Hongdu Wang[1], Zhijian Cui[1], and Liqin Zhou[1](✉)

[1] Ocean University of China, Qingdao 266100, Shandong, China
liqin72@126.com

[2] Qingdao Collaborative Innovation Center of Marine Science and Technology, Qingdao 266100, Shandong, China

Abstract. Underwater glider is a strong coupling and nonlinear system. Most current methods neglect the influences from the buoyancy adjustment system to pitch angle so that there always a large overshoot in the pitch angle control loop. In order to improve the control accuracy for pitch angle, a model compensation (MC) based on the Active disturbance rejection control (ADRC) was proposed in this paper. The Extended State Observer (ESO) estimated system comprehensive disturbances and avoided the influences from the perturbation by giving disturbance compensation. ADRC obtained segmental models through system modeling, estimation and physical sensors measurement. The estimation pressure of ESO was greatly reduced and the estimation precision was improved significantly. Simulations in the MATLAB indicated that MC-ADRC have a good control precision and low overshoot with settling time for glider systems. It reduced 4.5% overshoot and dropped the settling time to 90 s for pitch angle control than the traditional ADRC.

Keywords: Underwater glider · Pitch angle control · Model compensation · Active disturbance rejection control

1 Introduction

Underwater glider has broad applications in marine scientific research environmental monitoring, which is significant for coastal environmental observation worldwide [1, 2]. Applying slight change to its buoyancy and converting vertical motion to horizontal motion through matching the wings, its forward motion can be driven with low electric consumption as well as super-long voyage and high endurance. The most mature gliders including: Slocum [3], Seaglider [4], Spray [5] and Petrel [6].

The most effective control method for pitch angle is that a moveable weight such as battery packs called pitch angle adjustment mechanism. Battery packs positions determine the pitch angle for glider. In addition, by changing its buoyancy through a variable volume oil sac called buoyancy adjustment mechanism, an underwater glider can move up and down in the horizontal plane, Fig. 1. The oil sac is located in the central axis of the head of the glider and the battery packs is seated in the latter half the cabin.

Y. Huang et al. (Eds.): ICIRA 2017, Part I, LNAI 10462, pp. 745–756, 2017.
DOI: 10.1007/978-3-319-65289-4_69

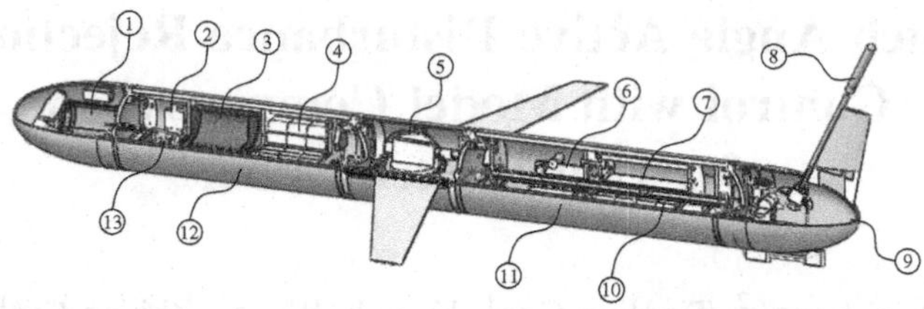

Fig. 1. Structure of glider. 1. External oil capsule. 2. Plunger pump. 3. Internal oil capsule. 4. Battery pack. 5. Mid-bin. 6. Pitching adjustment mechanism. 7. Rolling adjustment mechanism. 8. Antenna. 9. Dome 10. Control module 11. Rear-bin. 12. Fore-bin. 13. High-pressure valve.

Zigzag motion trajectory decides the pitch angle control is very important for the glider [7]. However, there always exists a strong coupling in pitch angle control loop. The pitch angle is not only controlled by the positions of the battery packs, but also it is susceptible towards to multiple parameters such as the oil sac volume and external disturbances from ocean currents. The coupled relationship makes actuators act frequently which consumes added energy during the pitch angle control process.

Some algorithms have been proposed in the past to improve the control effect, such as the linear quadratic regulator (LQR) [8] and the classical proportional, integral, and derivative (PID) [9, 10]. However, these algorithms are based on linear equations. The controller performances would be weak when system state far from the equilibrium point, let alone there exist strong parameter perturbations and disturbances.

Active Disturbance Rejection Control (ADRC), which was proposed by Chinese scholar, Han Jingqing [11]. It voids large overshoot by arranging the transition process through Tracking Differentiator (TD) and optimizes the control instruction. It observes the system uncertainty and disturbance total sum through the Extended State Observer (ESO). By designing the Nonlinear State Error Feedback (NLSEF) and combing the observation of the ESO for the system status and unknown disturbance, the linearization of the dynamic compensation is sequentially realized. The best characteristic of ADRC is that it does not need any mathematical models of the controlled object. Despite this, it would be difficult to paly advantages if ADRC doesn't make full use of the controlled object model in some complicated systems. For the glider, there are many coupled factors such as the battery packs position, buoyancy, velocity, etc. All of these factors are taken for perturbation terms and they are variable during glide motion. It is a heavy burden for ESO to estimate these perturbation terms.

For this reason, the model compensation ADRC (MC-ADRC) was proposed in this paper. By estimating parameters α and m_b for compensation model, the ESO needn't to estimate the total disturbances but only need to estimate the parameters that are not compensated. The ESO estimation pressure was reduced largely and the ADRC would have a higher control precision and stronger anti-interference ability.

This paper was organized as follows. Section 2 established the kinetic equations for the glider that considered pitch angle disturbance from the buoyancy adjustment mechanism. Section 3 introduced the ADRC based on the model compensation. Section 4 described the results of simulation in the MATLAB platform. In the Sect. 5, the experimental results were discussed and Sect. 6 concluded the paper.

2 Modelling of Underwater Glider

The general popular glider kinetic model was derived by the Gaver JG [12]. However, it did not consider that the variable buoyancy m_b has an impact on pitch angle when the positions of m_b is inconsistent with buoyancy centre. This paper deduced kinetic model based on the influence from the variable buoyancy for the pitch angle control.

2.1 Coordinate Frame Definition

Inertial coordinate $E - \xi\eta\zeta$, body coordinate $O\text{–}xyz$ was showed in Fig. 2. Other parameters were descripted in literature [12].

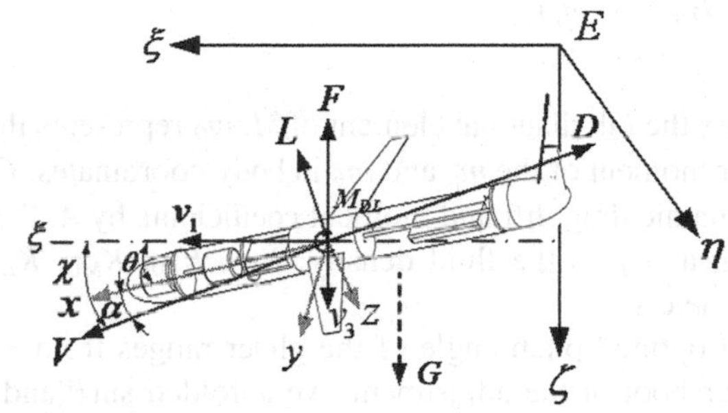

Fig. 2. Glider coordinate and force definition.

2.2 Glider Kinetic Equations

Glider is balanced with gravity G, buoyancy F, water resistance D, lift force L, and viscous moment M_{DL}. In the vertical plane, the glider moves with translational velocity $(v_1, 0, v_3)$ relative to the inertial coordinate $E - \xi\eta\zeta$. The variables were described as follows:

$$\mathbf{B} = \begin{bmatrix} x \\ 0 \\ z \end{bmatrix}, V = \begin{bmatrix} v_1 \\ 0 \\ v_3 \end{bmatrix}, \mathbf{\Omega} = \begin{bmatrix} 0 \\ \mathbf{\Omega}_2 \\ 0 \end{bmatrix}, r_p = \begin{bmatrix} r_{p1} \\ 0 \\ const \end{bmatrix}, U_p = \begin{bmatrix} u_{p1} \\ 0 \\ 0 \end{bmatrix},$$

$$r_b = \begin{bmatrix} const \\ 0 \\ 0 \end{bmatrix}, P_b = \begin{bmatrix} P_{p1} \\ 0 \\ P_{b3} \end{bmatrix}.$$

V is the velocity and $\mathbf{\Omega}$ denotes the pitch angle. $\mathbf{B}$ denotes the glider position in inertial coordinates and u_{p1} is the input variable. The battery packs position vector $r_{p3} = 0$. For buoyancy adjustment mechanism: $r_{b1} = const$ and $r_{b3} = 0$. The motion equations can be simplified as follows based on the constraint conditions above:

$$\dot{x} = v_1 \cos\theta + v_3 \sin\theta \tag{1}$$

$$\dot{z} = v_1 \sin\theta + v_3 \cos\theta \tag{2}$$

$$\dot{v}_1 = \frac{1}{m_1}\left[-m_3 v_3 \boldsymbol{\Omega}_2 - P_{p3}\boldsymbol{\Omega}_2 - m_0 g \sin\theta + L\sin\alpha - D\cos\alpha - u_{p1}\right] \tag{3}$$

$$\dot{v}_3 = \frac{1}{m_1}\left[m_1 v_1 \boldsymbol{\Omega}_2 + P_{p3}\boldsymbol{\Omega}_2 + P_{b1}\boldsymbol{\Omega}_2 + m_0 g\cos\theta - L\cos\alpha - D\sin\alpha\right] \tag{4}$$

α represents the angle of attack(AOA). χ is the path angle and $\chi = \theta + \alpha$.

According to the $\ddot{\theta} = \dot{\boldsymbol{\Omega}}_2$, it is easy to obtain the pitch angle control equation:

$$\dot{\boldsymbol{\Omega}}_2 = \frac{1}{J_2}[(m_3 - m_1)v_1 v_3 - (r_{p1}P_{p1} + r_{p3}P_{p3})\dot{\theta} - r_{b1}P_{b1}\dot{\theta} - m_p g(r_{p1}\cos\theta + r_{p3}\sin\theta) \\ - m_b g r_{b1}\cos\theta + M_{DL} - r_{p3}u_{p1}] \tag{5}$$

Where m_i denotes the i th diagonal element of M, m_0 represents the excess mass. P_p and P_b are the linear moment of the m_p and m_b in body coordinates. C_D, C_L and C_M are the standard aerodynamic drag, lift and moment coefficients by A, the maximum glider cross sectional area, and ρ is the fluid density. K_{D0}, K_{L0}, K_{M0}, K_D, K_L and K_M are hydrodynamic parameters.

The general and optimal pitch angle of the glider ranges from −30° to +30° [13]. Considering the overshoot of the adjustment, we unfolded $\sin\theta$ and $\cos\theta$ range from −40° to 0° (descend phase) and 0° to +40° (ascend phase) according to the Taylor formula $\sin\theta \approx a_1 + b_1\theta + c_1\theta^2$ and $\cos\theta \approx a_2 + b_2\theta + c_2\theta^2$. Where a, b and c are the coefficients of the fitting curves, Table 1.

Table 1. Coefficients of the fitting curves for $\sin\theta$ and $\cos\theta$.

Angle range	$\sin\theta$			$\cos\theta$		
	a_1	b_1	c_1	a_2	b_2	c_2
(−40°, 0°)	−0.011	0.0162	0	1.0008	2.205E-4	−1.418E-4
(0°, +40°)	0.011	0.0162	0	1.0008	2.205E-4	−1.418E-4

Eliminated the high order terms and transformed Eq. (5) as a standard equation $\ddot{\theta} = f(\theta, \dot{\theta}, w) + bu$, where w is the disturbances and u is the inputs variable, parameter b is the coefficient.

$$\ddot{\theta} = \frac{1}{J_2}\left[\begin{array}{l} -(r_{p1}P_{p1} + r_{p3}P_{p3} + r_{b1}P_{b1})\dot{\theta} - (b_2 m_p g + \mathrm{b}_1 m_p g r_{p3} + b_2 m_b g r_{b1})\theta - (a_2 m_p g r_{p1} \\ + a_1 m_p g r_{p3} + a_2 m_b g r_{b1}) + (m_3 - m_1)v_1 v_3 + M_{DL} - r_{p3}u_{p1} \end{array}\right] \tag{6}$$

It is obvious that the pitch angle control system is nonlinear and influenced by both of the battery packs position r_p, the oil sac volume m_b and the glider velocity v.

3 ADRC for Pitch Angle Control

3.1 ADRC Principle

ADRC absorbs the kernel from classic PID controller that the adjustment is based on the feedback error, and it references a state observation theory as well as constructs a new controller by nonlinear combinations.

3.2 ADRC Design

The additive perturbation and the external disturbances of the system are attributed to the general disturbances in pitch angle control. The ESO estimates the general perturbations and compensates in form of the feedforward. Considering two order nonlinear object as follows:

$$\begin{cases} \ddot{x} = f(x, \dot{x}) + w(t) + bu(t) \\ y = x(t) \end{cases}$$

Tracking Differentiator. TD has the function of extracting the continuous signal and its differential signal. The input signal is $v_0(t)$ and output signals are $v_1(t)$ and $v_2(t)$. Signal $v_1(t)$ tracks the signal $v_0(t)$ and $v_2 = \dot{v}_1$.

$$\begin{cases} e = v_1 - v_0 \\ \dot{v}_1 = v_2 \\ \dot{v}_2 = f_{han}(e, v_2, r_0, h_0) \end{cases}$$

Parameter r_0 denotes the speed factor and h_0 denotes the filter factor; function f_{han} represents the optimal control which is defined as follows:

$$f_{han}(e, v, r, h) = \begin{cases} -r sign(a), & |a| > d \\ -ra/d, & |a| \le d \end{cases}$$

Where $d = rh$, $d_0 = hd$, $y = e + hv$, $a_0 = \sqrt{d^2 + 8r|y|}$,

$$a = \begin{cases} v + sign(y)(a_0 - d)/2, & |y| > d_0 \\ v + y/h, & |y| \le d_0 \end{cases}$$

Extended State Observer

$$\begin{cases} e = z_1 - x \\ \dot{z}_1 = z_2 - \beta_{01} e \\ \dot{z}_2 = z_3 - \beta_{02} f_{al}(e, 0.5, h) + b_0 u \\ \dot{z}_3 = -\beta_{03} f_{al}(e, 0.25, h) \end{cases}$$

Parameter u is the input variable. z_1 is the estimation for the output state variable and z_2 is the speed estimation for output. Parameter h represents the sampling step and β_{01}, β_{02}, β_{03} are controller parameters. Let nonlinear function:

$$f_{al}(e,\alpha,\delta) = \begin{cases} e/\delta^{1-\alpha}, & |e| \leq \delta \\ |e|\alpha sign(e), & |e| > \delta \end{cases}$$

Each state tracks their status variable that was extended, $z_1(t) \rightarrow x(t)$, $z_2(t) \rightarrow \dot{x}(t)$, $z_3(t) \rightarrow \ddot{x}(t) \rightarrow f(x,\dot{x},w)$.

Non-Linear State Error Feedback. NLSEF makes good use of the past, present and future information for errors. The state estimation output z_i from ESO was assumed as the state feedback variable in the ADRC. It is compared with the output v_i from TD. The error amount $e_i = v_i - z_i$ constitutes the nonlinear combination for system error feedback.

$$\begin{cases} e_1 = v_1 - z_1 \\ e_2 = v_2 - z_2 \\ u_0 = \beta_{11} f_{al}(e_1, 0.5, h_1) + \beta_{12} f_{al}(e_2, 1.5, h_1) \end{cases}$$

Parameter β_{11} and β_{12} are gain coefficients.

Disturbances Compensation

$$u = u_0 - z_3/b_0$$

Parameter b_0 represents the compensation coefficient.

The control law brings a disturbance compensation control in time for the system. The disturbances cannot make the system go bad in that case.

3.3 MC-ADRC for Glider

ADRC is a control method that doesn't depend on the model of the controlled object. However, the ADRC may decrease the estimators for uncertainties by importing the model compensation. The model compensation could reduce the estimation pressures for ESO, and it improves the dynamic control performances.

From the structure of ADRC, the state estimation of the three orders ESO is used for estimating both of the parameter changes from the model and the external disturbances. When the object model changes acutely or the external disturbances are oversized. The ESO should have a good tracking performance, which means that the parameter z_3 have a very high estimation precision. The best way to improve tracking performance for ESO is that to decrease the estimators for z_3 as much as possible. That means the observed pressure of ESO from uncertainties would be reduced substantially by means of substituting known models into the observation function.

Suppose that $f(x,\dot{x}) = f_1(x,\dot{x}) + f_2(x,\dot{x})$, where $f_1(\cdot)$ and $f_2(\cdot)$ denote the known and unknown model respectively. On the basis of the ESO, let $x_3 = f_2(x,\dot{x},w)$, so the

$z_3 \rightarrow f_2(x, \dot{x}, \cdots, x^{(n-1)}, t) + w(t)$. The ESO observation equations were transformed as following forms:

$$\begin{cases} \dot{z}_1 = z_2 - \beta_{01} e \\ \dot{z}_2 = z_3 - \beta_{02} f_{al}(e, 0.5, h) + f_1(x, \dot{x}) + b_0 u \\ \dot{z}_3 = -\beta_{03} f_{al}(e, 0.25, h) \end{cases}$$

The system was approximated as the "Integrator Series Mode". The feedback compensation is $u(t) = u_0(t) - [z_3 + f_1(x, \dot{x})]/b$.

In the ESO, z_3 is the estimation value for generalized disturbances and it has many parameters estimation missions including θ, $\dot{\theta}$, r_p, m_b, v_1 and v_3. It can be seen that the system has proposed exorbitant requirements for ESO when the parameters m_b, θ and v changes tempestuously. In another words, z_3 must possesses a strong tracking and observation abilities.

Rewrite the pitch angle control equation: $\ddot{\theta} = f_1(\cdot) + f_2(\cdot) + bu$.

Function $f_1(\cdot)$ is the given model that can be estimated. Function $f_2(\cdot)$ is the unknown model in the pitch angle control loop.

The ESO estimation equation can be written as: $z_3 = f_2(\cdot) + (b - b_0)u$.

According to the structural characteristics of the glider, parameters θ and $\dot{\theta}$ can be measured by the Micro Inertial Measurement Unit (MIMU) named MTi sensor. Battery packs position can be measured by the film potentiometer. Due to the α, the horizontal component v_1 and vertical component v_3 of the velocity V cannot be acquired accurately according to the pitch angle θ.

Toward to the control equation, the input is: $bu = -r_{p3} u_{p1}$.

Where $f_1(\cdot) = -(r_{p1} P_{p1} + r_{p3} P_{p3})\dot{\theta} - (b_2 m_p g + b_1 m_p g r_{p3})\theta - (a_2 m_p g r_{p1} + a_1 m_p g r_{p3})$ and $f_2(\cdot) = -r_{b1} P_{b1} \dot{\theta} - b_2 m_b g r_{b1} \theta - a_2 m_b g r_{b1} + (m_3 - m_1) v_1 v_3 + M_{DL}$.

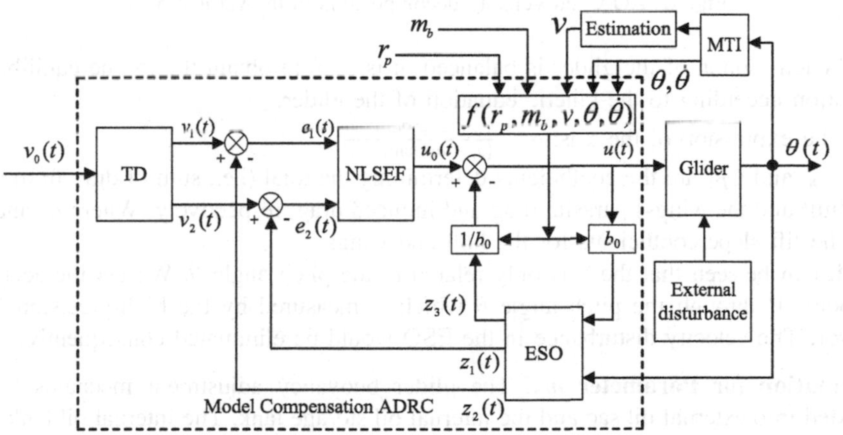

Fig. 3. Pitch angle control based on MC-ADRC for glider diagram.

On the basis of the ESO, the estimator is equal to zero if these four conditions are established: (1). Parameter z_1 is able to track pitch angle θ commendably. (2) Parameter m_b can be measured accurately. (3). Velocity v_1 and v_3 could be estimated precisely based on the pitch angle θ. (4). $b = b_0$. Off course, the conditions above are strict. In fact, even if the $b \neq b_0$ or there are some estimate errors in conditions (1), (2) and (3). What need estimated of the z_3 is only $(b - b_0)u$ left. It is much smaller than before. The observation burden and the disturbances of the ESO have been reduced greatly which would have a great improvement for estimation precision. The diagram of pitch angle control based on MC-ADRC shows in Fig. 3.

3.4 Parameters Estimation for MC-ADRC

Parameter Estimation for α and v. The external surface of the glider is asymmetric in the vertical direction. The α always exists as well as it is difficult to acquire.

Traditional methods for calculating velocity V as follows: obtaining the vertical velocity component v_3 via depth pressure sensor. $\chi \approx \theta$ is established by supposing that α is small enough. In that case, the horizontal velocity component $v_1 \approx v_3$ and we believe $V \approx V'$ in some way, Fig. 4. In fact, this assumption is not exactly right.

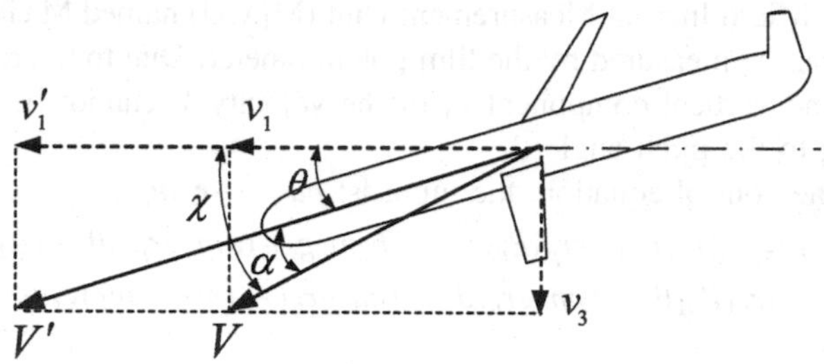

Fig. 4. AOA and velocity decomposition in the vertical plane.

When motion of the glider is balanced, it is easy to obtain the torque equilibrium equation according to the kinetic equation of the glider.

The expression of the α is: $\alpha = \frac{C_{D0} + C_{D1}\alpha^2}{(a_w + a_h)\tan(\theta + \alpha)}$

C_{D0} and C_{D1} are the coefficients determining the total (i.e., sum of drag from both the hull and the wings) parasite drag and induced drag, respectively. Where a_h and a_w are the lift-slope coefficients for the hull and wings.

It can be seen that the α is only related to the pitch angle θ. We get the accurate velocity V through the pitch angle θ which is measured by the high-precision MTi sensor. The velocity disturbance in the ESO would be eliminated consequently.

Estimation for Parameter m_b. The glider buoyancy adjustment mechanism was divided into external oil sac and the internal oil storage tank. The internal oil tank was designed as the inerratic bellows structure and it was installed a cable displacement sensor VXY30. The cable displacement sensor captured a variable quantity when the net buoyancy m_b changed.

Measurement for Parameter θ, $\dot{\theta}$ and r_p. Pith angle θ can be measured by the MTi-10 sensor. Its internal low-power signal processor provides drift-free 3D orientation as well as calibrated 3D acceleration, 3D rate of turn (rate gyro) and 3D earth-magnetic field data. Depended by the MTi, we get the precise pitch angle θ in an easy way. The pitch angular velocity is the differential signal of pitch angle θ which is defined as $\dot{\theta} = \frac{d\theta}{dt}$.

Battery packs position r_{p1} can be measured by the ThinPot film potentiometer. The battery packs were fixed at the sliding rail with servo motor. The motor drives battery packs and film potentiometer shuttling back and forth, the position signal r_{p1} of the battery packs could be obtained precisely.

4 Simulations Test in MATLAB

For testing the effects of the MC-ADRC algorithm, four groups of comparison experiments were simulated in the MATLAB platform. This control method was contrasted with the traditional ADRC. Control conditions for these four comparison tests as follows: Stabilizing the glider at a certain velocity V and changing the battery packs position r_{p1} independently, observing the results and regulation processes of the pitch angle θ between ADRC and MC-ADRC. The initial pitch angle is $-26°$ and desired pitch angle is $-34°$.

I. (1) ADRC algorithm: with no model compensation, α (as well as the v) or m_b. (2) MC-ADRC: parameter α was estimated accurately in compensation model but parameter m_b wasn't compensated.
II. (1) ADRC algorithm: with no model compensation, any α or m_b. (2) MC-ADRC: parameter m_b was estimated accurately but parameter α wasn't compensated.
III. (1) ADRC algorithm: with no model compensation, α or m_b. (2) MC-ADRC: both of the parameter m_b and α were estimated precisely.
IV. (1) ADRC algorithm: with no model compensation for α or m_b. (2) MC-ADRC: both of the parameter m_b and α were estimated precisely. In this comparison test, we added a disturbance in the adjustment for pitch angle. More specifically, we changed the buoyancy from 400 ml to 300 ml when the pitch changed from $-26°$ to $-34°$. In that case, the control process was encountered a strong disturbance.

4.1 ADRC Parameters Tuning [14]

Parameters r_0, h_0, β_{01}, β_{02}, β_{03} and h_1 are related to the sampling step length in the ADRC. They can be determined as follows: $r_0 = 0.001/h^2$, $h_0 = 5h$, $\beta_{01} = 1/h$, $\beta_{02} = 1/(3h^2)$, $\beta_{03} = 1/(32h^3)$, $h_1 = 10h$. In addition, gain coefficient β_{11} and β_{12} are equal to the proportional differential gains of PID controller; compensation coefficient b_0 corresponds to the integral gain. Combining to the simulation experiences, let $\beta_{11} = 2$, $\beta_{12} = 0.9$ and $b_0 = 0.4$.

4.2 Simulation Results

The comparison results shows in Fig. 5. The ADRC and MC-ADRC intervened at 180 s in the four control processes above. The disturbance of the variable buoyancy intervened between the 180 s and 200 s.

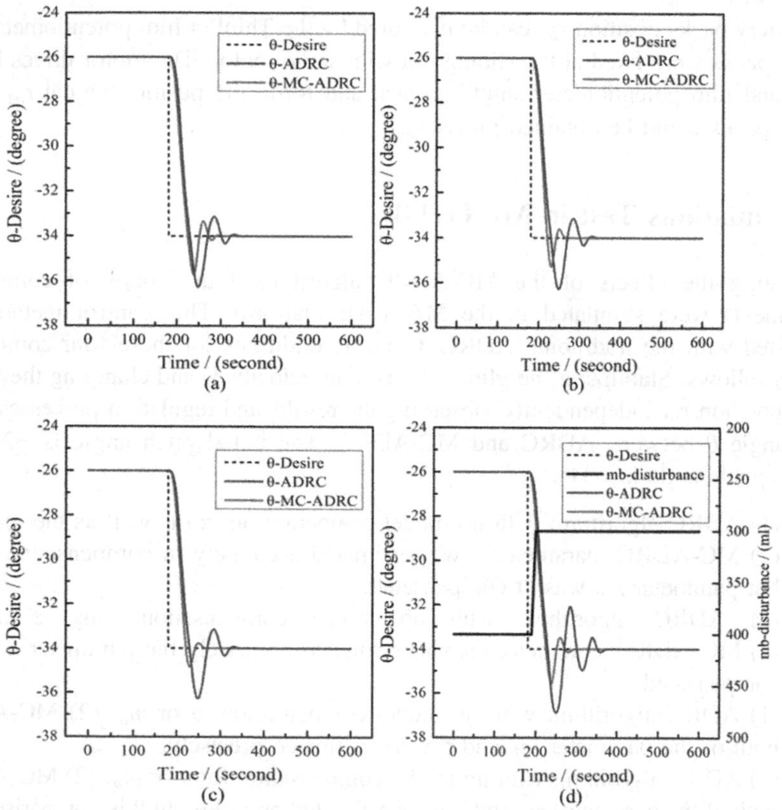

Fig. 5. Adjustment processes comparisons between the ADRC and the MC-ADRC algorithms. (a). AOA α estimated but buoyancy m_b was unknown; (b). Buoyancy m_b was estimated but AOA α was unknown; (c). Both of the buoyancy m_b and AOA α were estimated; (d). Buoyancy m_b and AOA α were estimated and there existed a buoyancy disturbance.

Notably, the MC-ADRC controller has good performances. In the Fig. 5(a) with test I, the MC-ADRC has a 110 s settling time and 5.3% overshoot; yet, the ADRC controller's performances were 160 s and 6.5%. Regarding to the Fig. 5(b) with test II, the performances of MC-ADRC and ADRC were 80 s vs. 160 s and 3.5% vs. 6.5%. For Fig. 5(c), the settling time and overshoot were reduced to 70 s and 2.0% in the test III. For comparison test with disturbance in Fig. 5(d) for test IV, the performances contrast was very obvious. MC-ADRC has a shorter settling time 105 s and a small overshoot 4.1%. However, the ADRC have a large overshoot process, the settling time was up to 190 s and the overshoot was reached 8.6%.

5 Discussions

Table 2 shows control performances comparison between ADRC and MC-ADRC algorithms. Where M_P denotes the overshoot and t_s represents the settling time in the pitch angle control process.

Table 2. Performances comparison between the ADRC and MC-ADRC algorithms.

Desired control	Performance index	ADCR	MC-ADRC
I	M_P	6.5%	5.3%
	t_s	160 s	110 s
II	M_P	6.5%	3.5%
	t_s	160 s	80 s
III	M_P	6.5%	2.0%
	t_s	160 s	70 s
IV	M_P	8.6%	4.1%
	t_s	190 s	105 s

It is very clear that the MC-ADRC has good control effects than the ADRC, no matter for the settling time or the overshoot. The former three groups of comparison tests (I, II and III) show that the model compensation method have improved the performances for ADRC significantly. Above all, the more models we know about the system, the better control performances we will obtain. Experiment IV shows that the MC-ADRC has excellent control performances for external disturbances. It reduced overshoot and settling time distinctly in pitch angle control.

6 Conclusions

A new approach to control pitch angle of an underwater glider using model compensation based on the ADRC has been demonstrated. The parameter estimation and parameter measurement methods were putted forward to acquire the compensation model in this paper. For the MC-ADRC method, the more parameters we know in the control model, the better control performances we would obtain. Simulation tests in the MATLAB proved its feasibility and high efficiency: the MC-ADRC algorithm has good control precision and low overshoot with settling time in pitch angle control for glider system. In the future, the MC-ADRC algorithm will be applied to our actual glider system.

References

1. Bachmayer, R., Leonard, N.E., Graver, J., Fiorelli, E.: Underwater gliders: recent developments and future applications. In: 4th International Symposium on Underwater Technology, pp. 195–200. International Symposium on Underwater Technology, Taipei, Taiwan (2004)
2. Rudnick, D.L., Davis, R.E., Eriksen, C.C., Perry, M.J.: Underwater gliders for ocean research. Mar. Technol. Soc. J. **38**(2), 73–84 (2004)
3. Glenn, S., Schofield, O., Kohut, J., Mcdonnell, J., Ludescher, R., Seidel, D.: The trans-atlantic slocum glider expeditions: a catalyst for undergraduate participation in ocean science and technology. Mar. Technol. Soc. J. **45**(1), 52–67 (2011)
4. Eriksen, C.C., Osse, T.J., Light, R.D., Wen, T., Lehman, T.W., Sabin, P.L.: Seaglider: a long-range autonomous underwater vehicle for oceanographic research. IEEE J. Oceanic Eng. **26**(4), 424–436 (2001)
5. Sherman, J., Davis, R., Owens, W.B., Valdes, J.: The autonomous underwater glider "Spray". IEEE J. Oceanic Eng. **26**(4), 437–446 (2001)
6. Wang, S.X., Sun, X.J., Wu, J.G., Wang, X.M.: Motion Characteristic Analysis of a Hybrid-driven Underwater Glider, pp. 1–9. Oceans IEEE (2010)
7. Mahmoudian, N., Woolsey, C.: Underwater glider motion control. In: 47th IEEE Conference on Decision and Control, pp. 552–557. IEEE Conference on Decision and Control, Cancun, Mexico (2008)
8. Noh, M.M., Arshad, M.R., Mokhtar, R.M.: Depth and pitch control of USM underwater glider: performance comparison PID vs. LQR. Indian J. Geo-Mar. Sci. **40**(2), 200–206 (2011)
9. Min, J.K., Joo, M.G.: Depth control of an underwater glider by using PID controller. J. Korean Inst. Inf. Technol. **13**(4), 1–7 (2015)
10. Muniraj, M., Arulmozhiyal, R.: Modeling and simulation of control actuation system with fuzzy-PID logic controlled brushless motor drives for missiles glider applications. Sci. World J. **2015**(5), 1–11 (2015)
11. Han, J.: From PID to active disturbance rejection control. IEEE Trans. Industr. Electron. **56** (3), 900–906 (2009)
12. Graver, J.G., Leonard, N.E.: Underwater glider dynamics and control. In: 12th International Symposium on Unmanned Untethered Submersible Technology, pp. 1–14 (2001)
13. Hussain, N.A.A., Arshad, M.R., Mohd-Mokhtar, R.: Underwater glider modelling and analysis for net buoyancy, depth and pitch angle control. Ocean Eng. **38**(16), 1782–1791 (2011)
14. Li, S.Q., Zhang, S.X., Liu, Y.N., Zhou, S.W.: Parameter-tuning in active disturbance rejection controller. Control Theor. Appl. **29**(1), 125–129 (2012)

Numerical Simulation Research in Flow Fields Recognition Method Based on the Autonomous Underwater Vehicle

Xinghua Lin, Jianguo Wu(✉), Dong Liu, and Lili Wang

School of Mechanical Engineering,
Hebei University of Technology, Tianjin 300130, China
wjg_6518@163.com

Abstract. The lateral line system (LLS) is an important organ of fish to sense the surrounding environment, and studying the mechanism of LLS is useful for the application of the artificial LLS on the mini autonomous underwater vehicle (mini-AUV). Computational fluid dynamics (CFD) is employed to study the AUV hydrodynamics at different speeds, and the spatial distribution of the near-body pressure of the AUV is studied over the whole computational domain. Furthermore, the structure of pressure field is studied quantificationally, and the relationship of two essential quantities: R_0/R (the radius coefficient) and E_p (the pressure coefficient) is studied emphatically to describe the spatial distribution pattern of pressure field. The simulation results demonstrate that the proposed computational scheme and corresponding algorithm are both effective to predict the flow field by using the speed of AUV, and these conclusions are useful to enhance the environmental adaptation of mini-AUV by using the artificial lateral line system.

Keywords: AUV · The flow field · CFD · LLS · Pressure sensing

1 Introduction

Intelligence and miniaturization is the main development direction of AUV, and when considering the environmental adaptability of mini-AUV, manoeuvrability is often considered as an important design criterion, which requires acute sensing capabilities. The traditional techniques such as the visual and acoustic sensing, have proved to be effective in the most commonly encountered configurations, but become inoperative in highly confined, turbid and murky environments. The lateral line system (LLS) is a complex, multi-branched, mechanoreceptive organ came from Long-term evolution of fishes, which can help fish possess a good environmental adaptability. According to the difference of the distribution position and the performance of nerves, the LLS includes the superficial neuromasts and the canal neuromasts, as shown in Fig. 1, the symbol "→" indicates the superficial neuromasts and the symbol "*" indicates the canal neuromasts [1]. The researches show that the superficial neuromasts respond to changes in external flow velocity, while the canal neuromasts respond to changes in external flow acceleration [2]. The stimulating factors of both the pressure and velocity of the

Y. Huang et al. (Eds.): ICIRA 2017, Part I, LNAI 10462, pp. 757–765, 2017.
DOI: 10.1007/978-3-319-65289-4_70

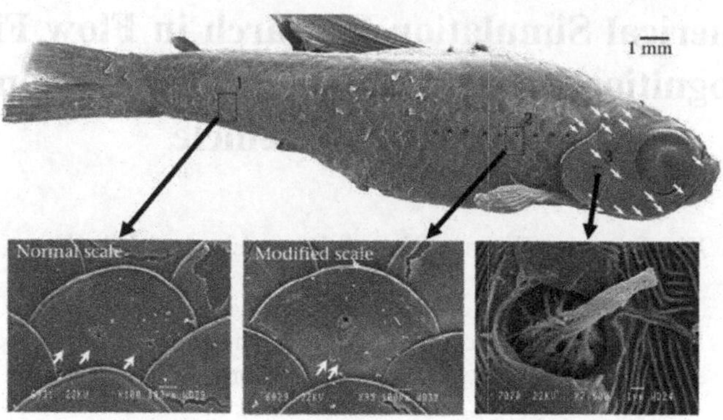

Fig. 1. The distribution diagram of LLS at the surface of fish and the structure of neuromast [1].

flow field are detected by the LLS and transduced into nerve impulses which are transmitted to the central nervous system to guide the fish to avoid obstacle, predation and courtship [3–5].

Many biologists studied the LLS in detail in the last years [1, 6, 7]. In recent years, the researchers have tried to put the LLS into the engineering application, for example, as an alternative way of sensing the fluid environment for AUV when visual and acoustic sensing are limited. Vicente et al. [8] made the first attempt at studying the mechanism of the LLS by the pressure sensor arrays, which was made by the micro-electromechanical systems (MEMS), and the results indicated that it is in general possible to extract specific shape information from measurements on a linear pressure sensor array for shape identification and vortex tracking. Ajay et al. [9] designed the polymer micro-electromechanical systems (MEMS) pressure sensors which was fabricated using liquid crystal polymer (LCP) as the sensing membrane material, and the experimental results demonstrate the ability of pressure sensor arrays to detect the velocity of underwater objects towed past by with high accuracy, and an average error of only 2.5%. Adrian et al. [10] designed an artificial lateral line canals equipped with optical flow sensors, which can be used to detect the water motions generated by a stationary vibrating sphere, the vortices caused by an upstream cylinder or the water (air) movements caused by a passing object, and the hydrodynamic information can be used to calculate bulk flow velocity, the size and the position of the cylinder. Despite the pervasive character of hydrodynamic sensing in nature and the growing body of cutting-edge biomimetic applications, the hydrodynamic fundamentals underpinning this sensory mode are still not clearly established. However, with such complex flows one has to resort to numerical simulations in order to access the solution of the problem unlike in the case considered here.

Since then, CFD models have been extensively exploited to understand the specific mechanism of the LLS. Charlotte B. and Humphrey [11] used a Navier–Stokes solver to calculate the unsteady flow past an elongated rectangular prism and a fish downstream of it, then investigate the motion-sensing characteristics of the LLTC and how it may be used by fishes to detect wakes. Rapo M.A. et al. [12] used the CFD to calculate the stimulus to the lateral line of a fish in still water, and the result shows good agreement with the experiments results. Han Zhou et al. [13] used the CFD to study the bio-inspired fish swimming hydrodynamics, then the spatial distribution and temporal variation of the near-body pressure of fish were studied over the whole computational domain, and a filtering algorithm is designed to fuse near-body pressure of one or multiple points for the estimation on the external flow. Windsor S.P. et al. [14, 15] used the blind Mexican cave fish (BMCF) as model to study two flow fields: Open water and heading towards a wall and Gliding parallel to a wall by the CFD method, and inspiringly, it was validated that pressure sensing values matched approximately between those by respective CFD simulations and experiment.

In this paper, CFD is employed to study the hydrodynamics of AUV at different speeds, and the spatial distribution of the near-body pressure of the AUV is studied over the whole computational domain. Furthermore, the structure of pressure field is studied quantificationally. The simulation results demonstrate that the proposed computational scheme and its corresponding algorithm are both effective to predict the flow field by using near-body pressure at distributed spatial points. These conclusions are useful to enhance the environmental adaptation of mini-AUV.

2 Model Building

The three-dimensional flow generated by an AUV at different speed is calculated by the Fluent software, which has been extensively tested and applied to a number of complex flows. The two-dimensional outline profiles for the AUV model is shown in Fig. 2. The one for the AUV, corresponding to the spindle apparatus, whose long axis is instead by $a = 900\,\text{mm}$ and the short axis is instead by $b = 150\,\text{mm}$, that is are in this paper. The geometry of computational domain is one cylinder, whose diameter $D = 60b$ and the length $L = 24a$, where the AUV is in the central location of domain.

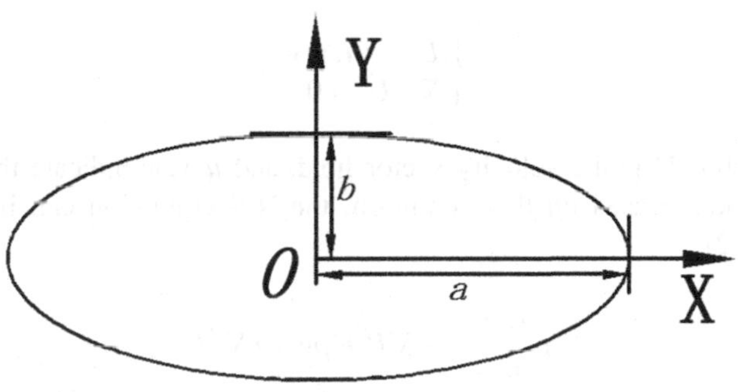

Fig. 2. The two-dimensional outline profiles for the AUV model.

A regular-tetrahedron grid is used to divide the domain, especially refined near the surfaces defining the AUV body. And a size function was created which the size of grid increases from the surface of AUV at a constant growth rate, the minimum of grid size is 1.0, and the growth rate is 1.2. This method can control the number of grids in a reasonable range to increase the calculated speed, while can guarantee computational precision at the same time. The intersecting surface of domain is shown in Fig. 3. The boundary condition of the inlet used for the calculation domain is velocity-inlet, and the exit plane is outflow. An impermeable and no-ship velocity condition is used for the flank of domain and the surface of AUV.

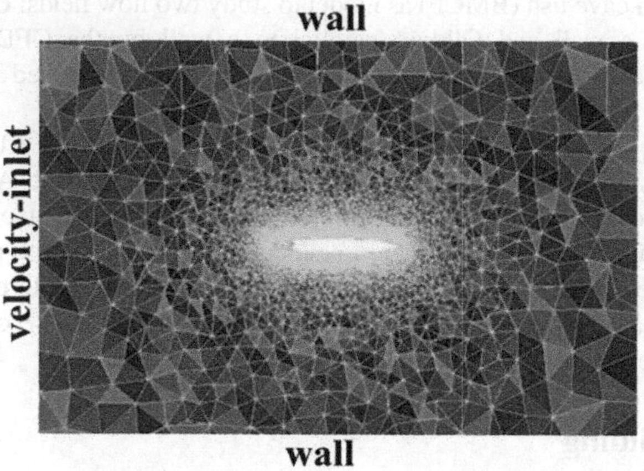

Fig. 3. The distribution of grids at the intersecting surface of domain.

A three-dimensional, steady and pressure based code solver is used to calculate the domain, and a SIMPLEC algorithm is used to describe the way of Pressure-Velocity Coupling. The use of second order upwind discredited forms and the RNG $k - \varepsilon$ turbulence model ensures the calculation accuracy. The differential forms of the mass and momentum conservation equations in vector notation for an incompressible and viscous newtonian fluid are shown in Eq. (1).

$$\begin{cases} \vec{U} = (u, v, w) \\ \nabla \cdot \vec{U} = 0 \end{cases} \tag{1}$$

where the symbol $\vec{U}$ is the velocity vector field, and u, v, w indicate the speed at the different direction. Based on this condition, the N-S equitation can be described as shown in Eq. (2).

$$\rho \frac{D\vec{U}}{Dt} = -\nabla P + \rho g + \mu \nabla^2 \vec{U} \tag{2}$$

where ρ and μ are the density and kinematic viscosity of water respectively; $\vec{U}$ is the vector velocity; P is the modified pressure; t and g are respectively the time and the gravity coefficient.

From the Eq. (2), we can fine that if the pressure $-\nabla P$ decrease, the momentum $\nabla^2 \vec{U}$ will increase when the gravity is a constant. Based on this relationship between the pressure and velocity, we can use the pressure signal mainly to help AUV to identify the underwater environment.

3 Results and Discussion

Calculations of the three-dimensional flow were performed until it converged to a steady–periodic state. Contours of the pressure in the XOY-cross-section when the inlet speed is 1 m/s, is provided in Fig. 4, illustrating the spatial distribution of pressure field have three obvious domain: the head high-pressure area (HHA), the middle low-pressure area (MLA) and the tail high-pressure area (THA). In the head and tail high-pressure area, there are obvious pressure gradient in the spatial environment. In this paper, we chose the THA as the object of study, whose isobaric line is shown in Fig. 5 when the inlet speed is 1 m/s.

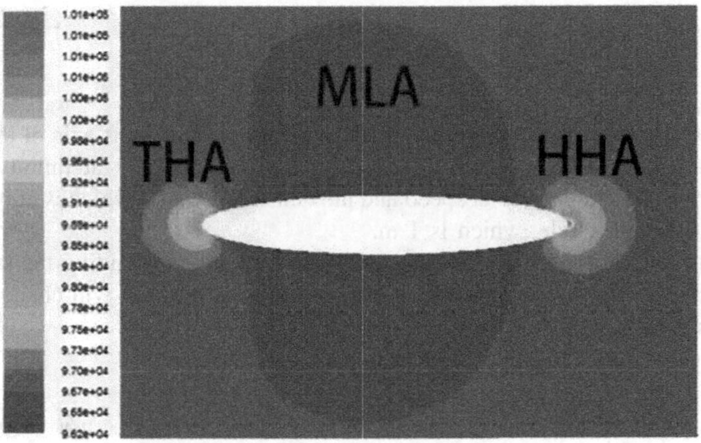

Fig. 4. The contours of the pressure in the XOY-cross-section when the inlet speed is 1 m/s.

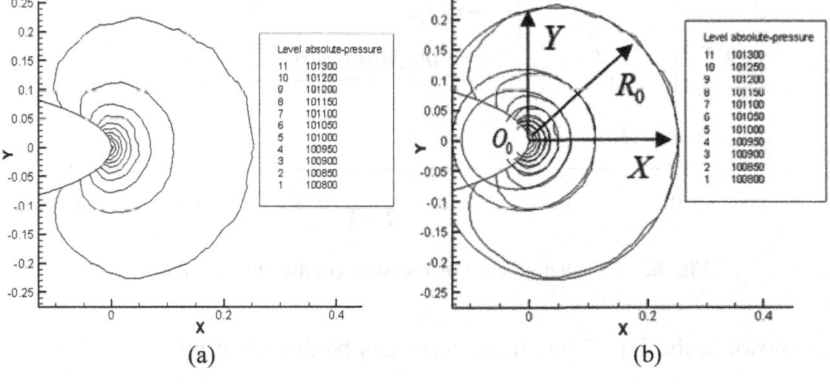

Fig. 5. The distribution of isobaric line in THA(a) and the schematic of direct mappings for the pressure field (b) when the inlet speed is 1 m/s.

In our sensing problem, the domain corresponds to the fluid domain surrounding the AUV delimited by mapping cure C, which has to be detected and whose shape has to be discriminated. Finding a simple way of representing and classifying its elements is key to discriminate the pressure field. As shown in the Fig. 6(b), the mapping cure is oval, so we chose the mapping radio R_0 and the position of mapping centre O_0 as the essential geometric elements. In order to the convenience of study, we see the mapping cure C as a circle approximatively, whose radio coefficient R_0/R is described as Eq. (3). And the pressure at different cure C is described as the pressure coefficient E_p, which is shown as the Eq. (4).

$$R_0/R = \frac{a' + b'}{2R} \tag{3}$$

$$E_p = \frac{2(p_\infty - p)}{\rho \cdot v^2} \tag{4}$$

where a' and b' are, respectively the length of long-axis and short-axis of the mapping cure; p and p_∞ are the pressure at the cure and the pressure at infinity, which is 101325 Pa; v and ρ are the inlet speed and the density of water, which is 1000 kg/m^3; R is the radio of unit circle, which is 1 m.

We can find that there are obvious pressure gradient surrounding the tail of AUV from the Fig. 6. The pressure coefficient E_p at different position will change by R_0/R, which is shown in Fig. 7.

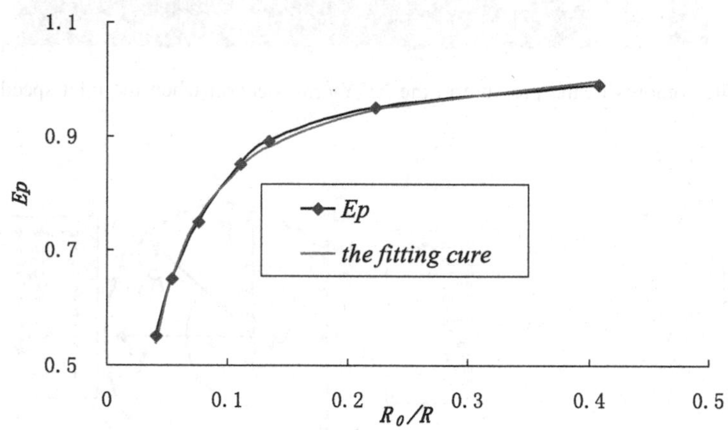

Fig. 6. The change cure of pressure coefficient E_p by R_0/R.

As shown in the Fig. 7, the fitted curve can be described by

$$E_p = A \cdot (R_0/R)^B + C \tag{5}$$

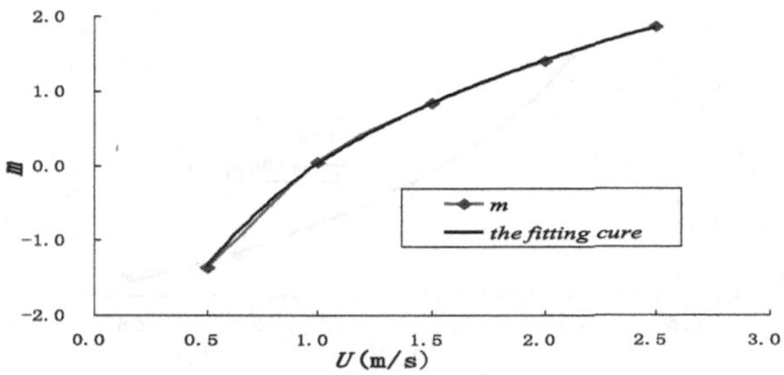

Fig. 7. The change cures of m with inlet speed U.

When the inlet speed is 1 m/s, the parameters A, B and C are −0.034, –0.86 and 1.07, whose fitting factor is 0.9981. We can find the spatial distribution of pressure showed exponent relation, and near the AUV, the pressure gradient is obvious; when the distance reaches a critical value, the pressure gradient is unconspicuous. We call the critical value as effective detection range (EDR), which is 0.25 m in this condition. We conclude that flow field sensing is simply not effective based on static single-point pressure sensing. One has to be so close to the object, whose distance should be little than EDR, that can catch the effective pressure signal. To overcome these serious constraints and extend the range of EDR, it is necessary to resort to a dynamically moving multi-point sensory device such as the LLS.

Taking double logarithmic at both sides of Eq. (5), it can be described as

$$\ln(E_p) = \ln(A + C) + B \ln C \cdot \ln(R_0/R) \tag{6}$$

We define $m = \ln(A + C)$ and $n = B \ln C$, so the Eq. (6) can be shown by Eq. (7).

$$\ln(E_p) = n \cdot \ln(R_0/R) + m \tag{7}$$

When the inlet speed changes from 0.5 m/s to 2.5 m/s, the n and m are shown in Table 1. And the change cures of m and n with inlet speed U are shown in Figs. 7 and 8.

Table 1. The numbers of m and n at different inlet speeds.

U(m/s)	0.5	1.0	1.5	2.0	2.5
m	−1.365	0.035	0.832	1.403	1.858
n	1.412	−0.058	−0.820	−1.451	−1.701

As shown in Figs. 7 and 8, the fitting cures can be described as Eq. (8). The coefficients m and n are linear relation with logarithmic of inlet speed U.

$$\begin{cases} m = 2 \ln U + 0.025 \\ n = -2 \ln U - 0.0029 \end{cases} \tag{8}$$

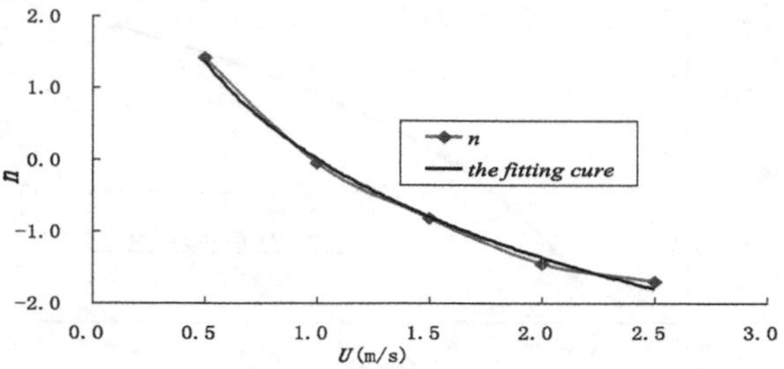

Fig. 8. The change cures of with inlet speed *U*.

So we can calculate the numbers of *m* and *n* at different inlet speeds by Eq. (8). Furthermore, we can know the pressure at different positions or the distance from object by Eq. (7). These findings seem to explain the documented ‘accelerate and glide past’ behaviour of BCF when placed in the presence of a new obstacle and suggest that man-made vehicles using pressure sensing for hydrodynamic mapping should probably adopt similar strategies.

4 Conclusions

(1) The spatial distribution of pressure showed exponent relation, and near the AUV, the pressure gradient is obvious; when the distance reaches a critical value, the pressure gradient is unconspicuous;

(2) Flow field sensing is simply not effective based on static single-point pressure sensing. To overcome these serious constraints and extend the range of EDR, it is necessary to resort to a dynamically moving multi-point sensory device such as the LLS.

(3) The coefficients *m* and *n* are linear relation with logarithmic of inlet speed *U*. And if we know the speed of AUV, we can map the pressure field.

References

1. Faucher, K., Parmentier, E., Becco, C.: Fish lateral system is required for accurate control of shoaling behavior. Anim. Behav. **79**, 679–687 (2010)
2. Sheryl, C., Paul, P.: Lateral line stimulation patterns and prey orienting behavior in the Lake Michigan mottled sculpin (Cottus bairdi). J. Comp. Physiol. **195**(3), 279–297 (2009)
3. Van Netten, S.M.: Hydrodynamic detection by cupulae in a lateral line canal: functional relations between physics and physiology. Biol. Cybern. **94**, 67–85 (2006)
4. Chagnaud, B.P., Bleckmann, H., Hofmann, M.H.: Lateral line nerve fibers do not code bulk water flow direction in turbulent flow. Zoology **111**, 204–217 (2008)
5. Bleckmann, H.: Role of the lateral line in fish behavior. In: Pitcher, T.J. (ed.) Behavior of Teleost Fishes, pp. 20–246. Chapman & Hall, London (1993)

6. Zhu, Y.D.: The Research of Lateral Line Pipe System and Laurene-urn & Lawrence System Based on Chinese Cartilaginous Fish. Shanghai Science and Technology Press, Shanghai (1980)
7. Chen, H.: The lateral-line sense organs of andrias davidianus. Acta Zool. Sin. **3**, 235–242 (1995)
8. Vicente, I.F.: Performance Analysis for Lateral-Line-Inspired Sensor Arrays. Massachusetts Institute of Technology (2011)
9. Kottapalli, A.G.P., Asadnia, M., Miao, J.M.: A flexible liquid crystal polymer MEMS pressure sensor array for fish-like underwater sensing. Smart Mater. Struct. **21**(11), 454–462 (2012)
10. Klein, A., Bleckmann, H.: Determination of object position, vortex shedding frequency and flow velocity using artificial lateral line canals. Beilstein J. Nanotechnol. **2**, 276–283 (2011)
11. Barbier, C., Humphrey, J.A.: Drag force acting on a neuromast in the fish lateral line trunk canal. I. numerical modelling of external-internal flow coupling. J. Royal Soc. Interface **6**, 627–640 (2009)
12. Rapo, M.A., Jiang, H.S., Grosenbaugh, M.A., Coombs, S.: Using computational fluid dynamics to calculate the stimulus to the lateral line of a fish in still water. J. Exp. Biol. **212**, 1494–1505 (2009)
13. Han, Z., Hu, T.J., Kin, H.L., et al.: Bio-inspired flow sensing and prediction for fish-like undulating locomotion: a CFD-aided approach. J. Bionic Eng. **12**, 406–417 (2015)
14. Windsor, S.P., Norris, S.E., Cameron, S.M., et al.: The flow fields involved in hydrodynamic imaging by blind Mexican cave fish (Astyanax fasciatus). Part I: open water and heading towards a wall. J. Exp. Biol. **213**, 3819–3831 (2010)
15. Windsor, S.P., Norris, S.E., Cameron, S.M., et al.: The flow fields involved in hydrodynamic imaging by blind Mexican cave fish (Astyanax fasciatus) part II: gliding parallel to a wall. J. Exp. Biol. **213**, 3832–3842 (2010)

Trajectory-Keeping Control of AUV Based on RNM-ADRC Method Under Current Disturbances for Terrain Survey Mission

Tao Chen[1,2], Hang Gao[1], Da Xu[1(✉)], Chuang Wan[1], Yuzhu Wang[1], and Zheping Yan[1]

[1] College of Automation, Harbin Engineering University, Harbin 150001, China
chentao_7777@163.com, xudakl@163.com
[2] Key Laboratory of Underwater Robot Technology, Harbin 150001, China

Abstract. Trajectory-keeping is the main motion mode of AUV (Autonomous Underwater Vehicle) for terrain survey mission. However, it is not easy to realize an accurate trajectory-keeping control especially under current disturbances. Auto disturbance rejection control (ADRC) is an effective control method with good robustness, fast response and precise compensation of disturbance, but is difficult to select the parameters. So, the recurrent networks model (RNM) is used to optimize the parameters of ADRC in this paper. By RNM, the parameters of ADRC can realize real-time optimizing to reject the disturbances much better. Then, the RNM-ADRC controller is designed for heading control of AUV with cross tracking error (CTE) heading guidance method. Simulations show that the controller proposed in this paper has strong adaptability, fast response and good disturbance rejection. The trajectory-keeping effect using RNM-ADRC is better than ADRC under strong current disturbance.

Keywords: AUV · Trajectory-keeping · RNM-ADRC · Current disturbances

1 Introduction

AUV is a very useful ocean tool and is conformable for many ocean survey missions [1]. For this paper, the terrain survey mission using an AUV to obtain seafloor topography is focused [2]. For this mission, the main motion of AUV is trajectory-keeping along a 2D predefined survey. How accurate the AUV can keep tracking the desired trajectory decides the terrain survey effect, especially under current disturbances. Many researches reported on trajectory keeping or tracking control of AUV in the presence of current. Ocean current observer was presented in [3, 4] to estimate the unknown constant current. Then, integrator back-stepping and Lyapunov based techniques are used to extend

This paper is supported by National Natural Science Foundation (NNSF) of China under Grant (51609046, 51609048), Harbin Science and Technology Bureau of China under Grant (2016RQQXJ089, 2016RAQXJ080), Fundamental Research Funds for the Central Universities (HEUCF XXX).

Y. Huang et al. (Eds.): ICIRA 2017, Part I, LNAI 10462, pp. 766–778, 2017.
DOI: 10.1007/978-3-319-65289-4_71

the kinematic controller dealing with current uncertainty. In [5], a modified line of sight (LOS) guidance law with integral action and three adaptive feedback control is researched to allow underactuated underwater vehicles to perform horizontal path following tasks for vertical irrotational ocean currents. In [6], a guidance system for 2-D straight path tracking is developed. Two adaptive nonlinear observers are designed to estimate current components w.r.t. the path-fixed frame and LOS guidance is designed for minimization of the cross-track error. Further, 3D straight line path following control strategy in the presence of ocean currents is developed using a modified three-dimensional LOS guidance law with two integrators to counteract the drifting caused by the unknown current [7]. For Curved Path following in horizontal plane, literature [8] proposed a closed-loop system consists of a guidance law and an adaptive feedback linearizing controller under unknown currents disturbances.

In this paper a integrated control system with current disturbances-rejection heading controller and conventional CTE heading guidance method is researched of AUV for trajectory-keeping. For disturbances-rejection heading control, ADRC technology is applied and RNM is used for real-time self-turning and optimizing the ADRC coefficients in the trajectory-keeping control of AUV.

The remainder of this paper is organized as follows. In Sect. 2, the problem statement of AUV trajectory-keeping control for terrain survey mission is described. In Sect. 3, kinematics and dynamics in horizontal plane of UUV are modeled. RNM-ADRC heading controller design of AUV is present in Sect. 4. Trajectory-keeping control system and heading guidance method is introduced in Sect. 5. Section 6 presents the simulation results. The end is the conclusion.

2 Problem Statements

Terrain survey mission is that AUV equipped with side scan sonar or multi-beam sonar performs an ocean area survey to obtain seafloor topography image or terrains bathymetry mapping. For the terrain survey mission, there are two motion modes of AUV: way-point tracking and trajectory-keeping which can be described as Fig. 1.

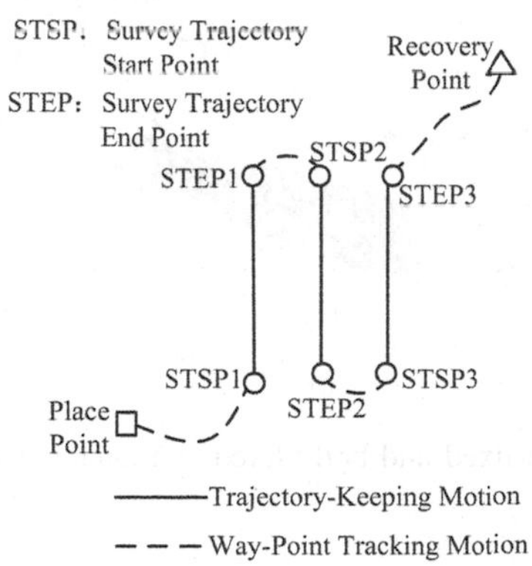

Fig. 1. Trajectory-keeping motion of AUV for terrain survey mission

The way-point tracking motion is used to let AUV navigate from place point to a survey trajectory start point, or from a survey trajectory end point to another survey trajectory start point, or from a survey trajectory end point to recovery point. The trajectory-keeping motion is used to let AUV navigate along the predefined survey trajectory to perform the terrain mapping.

Let AUV accurately keep on a predefined trajectory, the trajectory-keeping motion can be described as a 5-tuple [9]: $TTask = <P, S, C, \Omega, E>$. P is the way-points to denote 2D position of predefined survey trajectory S. C is the set of control behaviors, such as heading and velocity control. But in this paper, heading control is the mainly considered and the velocity of AUV is supposed fixed. Ω is the constraint set when AUV is trajectory-keeping along S, including constraints:

(1) Position constraint: for $\forall t$, there is $\varepsilon(t) = 0$. Where, $\varepsilon(t)$ is the distance from current position of AUV to trajectory S. Usually, $\varepsilon(t)$ is called cross track error.
(2) Heading constraint: for $\forall t$, there is $|\psi(t) - \psi_{trk}| = 0$. Where, $\psi(t)$ is the current heading of AUV; ψ_{trk} is the direction of trajectory S.
E is the terminate condition of trajectory-keeping control of AUV along S. Usually, the distance between AUV and the end point of S is used to judge the finish of trajectory-keeping control.

3 Vehicle Kinematics and Dynamics Models in Horizontal Plane

3.1 Kinematics Model

To describe the kinematics, two reference frames are employed, earth reference frame $\{E\}$ and body-fixed frame $\{B\}$, showed in Fig. 2. The body-fixed frame is chosen so as to coincide with the center of buoyancy (CB) and the center of gravity (CG).

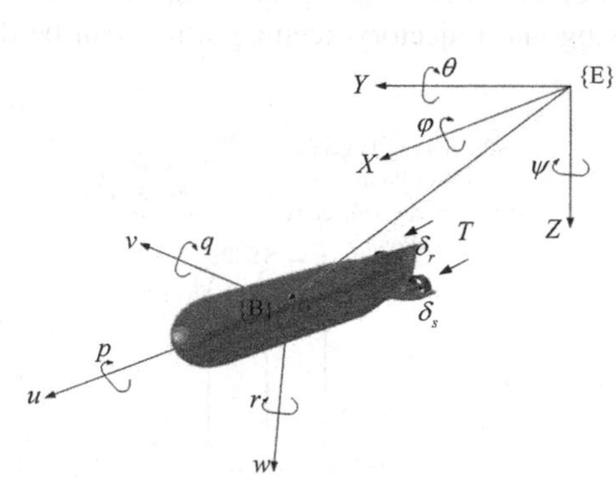

Fig. 2. Earth-fixed and body-fixed coordinate systems of AUV

Then, the general motion of vehicle in 6 DOF can be described as:

$$\dot{\boldsymbol{\eta}} = \boldsymbol{J}(\boldsymbol{\eta})\boldsymbol{v} \tag{1}$$

Where, $\boldsymbol{\eta} = [\boldsymbol{\eta}_1, \boldsymbol{\eta}_2]^T$; $\boldsymbol{\eta}_1 = [x, y, z]^T$ denotes inertial position; $\boldsymbol{\eta}_2 = [\varphi, \theta, \psi]^T$ denotes the orientation; $\boldsymbol{v} = [\boldsymbol{v}_1, \boldsymbol{v}_2]^T$; $\boldsymbol{v}_1 = [u, v, w]^T$ denotes linear velocity; $\boldsymbol{v}_2 = [p, q, r]^T$ denotes angular velocity. $\boldsymbol{J}(\boldsymbol{\eta}) = diag\{\boldsymbol{J}_1(\boldsymbol{\eta}_2), \boldsymbol{J}_2(\boldsymbol{\eta}_2)\}$ is the transformation between the $\{B\}$ and the $\{E\}$ frames is expressed as:

$$\boldsymbol{J}_1(\boldsymbol{\eta}_2) = \begin{bmatrix} c\psi c\theta & c\psi s\theta s\varphi - s\psi c\varphi & c\psi s\theta c\varphi + s\psi s\varphi \\ s\psi c\theta & s\psi s\theta s\varphi + c\psi c\varphi & s\psi s\theta c\varphi - c\psi s\varphi \\ -s\theta & c\theta s\varphi & c\theta c\varphi \end{bmatrix} \tag{2}$$

$$\boldsymbol{J}_2(\boldsymbol{\eta}_2) = \begin{bmatrix} 1 & s\varphi t\theta & c\varphi t\theta \\ 0 & c\varphi & -s\varphi \\ 0 & s\varphi/c\theta & c\varphi/c\theta \end{bmatrix} \tag{3}$$

Where, $s\bullet \triangleq \sin(\bullet)$, $c\bullet \triangleq \cos(\bullet)$, $t\bullet \triangleq \tan(\bullet)$.

Only consider the horizontal plane, and assume that the vehicle is always restoring itself to be upright, i.e. $\varphi = 0$, substituted into (1) using (2) and (3), the finally kinematics equations of the vehicle in horizontal plane are:

$$\begin{aligned} \dot{x} &= u\cos\psi\cos\theta - v\sin\psi + w\cos\psi\sin\theta \\ \dot{y} &= u\sin\psi\cos\theta + v\cos\psi + w\sin\psi\sin\theta \\ \dot{\psi} &= r \end{aligned} \tag{4}$$

3.2 Dynamics Model

The dynamic model of AUV presented in [10] is employed here. And, only horizontal motion is considered and the speed of AUV is assumed fixed at a certain working point. So, there only lateral and yaw equation for heading control is adopted:

$$\begin{aligned} m[\dot{v} - wp + ur] &= \tfrac{1}{2}\rho L^4[Y'_{\dot{r}}\dot{r} + Y'_{pq}pq + Y'_{qr}qr + Y'_{r|r|}r|r|] \\ &+ \tfrac{1}{2}\rho L^3[Y'_{\dot{v}}\dot{v} + Y'_{ur}ur + Y'_{wp}wp + Y'_{vq}vq + Y'_{v|r|}v|r|] \\ &+ \tfrac{1}{2}\rho L^2[Y'_{uv}uv + Y'_{vw}vw + Y'_{v|v|}v|v|] + (W - B)\cos\theta\sin\varphi + \tfrac{1}{2}\rho L^2 Y'_{\delta_r}u^2\delta_r \end{aligned} \tag{5}$$

$$\begin{aligned} I_z\dot{r} + (I_y - I_x)pq &= \tfrac{1}{2}\rho L^5[N'_{\dot{r}}\dot{r} + N'_{r|r|}r|r| + N'_{pq}pq + N'_{qr}qr] \\ &+ \tfrac{1}{2}\rho L^4[N'_{\dot{v}}\dot{v} + N'_{ur}ur + N'_{wp}wp + N'_{vq}vq + N'_{|v|r}|v|r] \\ &+ \tfrac{1}{2}\rho L^3[N'_{uv}uv + N'_{vw}vw + N'_{v|v|}v|v|] + \tfrac{1}{2}\rho L^3 N'_{\delta_r}u^2\delta_r \end{aligned} \tag{6}$$

Where, L, m, I_x, I_y, I_z denote length, mass and moment of inertia. u, v, w, p, q, r denote linear velocities and angular velocities round the three coordinate axes. φ, θ, ψ are the attitude of AUV describing roll, pitch and yaw angles respectively. $Y'_{(\bullet)}, N'_{(\bullet)}$ are the hydrodynamic coefficients. X_{prop} is the propulsive force of thruster. δ_r is the rudder angle. ρ is the water density, and W, B denote gravity and buoyancy respectively.

Suppose that the gravity and buoyancy of AUV is equal. Neglect the influences of vertical and rolling motion. Then, simplified and linear models of AUV are obtained:

$$\begin{aligned}(m - \tfrac{1}{2}\rho L^3 Y'_{\dot{v}})\dot{v} + (-\tfrac{1}{2}\rho L^4 Y'_{\dot{r}})\dot{r} = (\tfrac{1}{2}\rho L^2 Y'_{uv} u)v \\ + (\tfrac{1}{2}\rho L^3 Y'_{ur} u - mu)r + (\tfrac{1}{2}\rho L^2 Y'_{\delta_r} u^2)\delta_r\end{aligned} \tag{7}$$

$$\begin{aligned}(-\tfrac{1}{2}\rho L^4 N'_{\dot{v}})\dot{v} + (I_z - \tfrac{1}{2}\rho L^5 N'_{\dot{r}})\dot{r} = (\tfrac{1}{2}\rho L^3 N'_{uv} u)v \\ + (\tfrac{1}{2}\rho L^4 N'_{ur} u)r + (\tfrac{1}{2}\rho L^3 N'_{\delta_r} u_2)\delta_r\end{aligned} \tag{8}$$

4 RNM-ADRC Heading Controller Design

4.1 ADRC Control Method

ADRC is a modern control method improving the shortage of PID control by the combination of observations and compensated, with which disturbance can be compensated by observations. ADRC does not depend on the accurate mathematical model of object. Especially in the harsh environment, with require of high control precision, the advantage of ADRC is obvious. The general structure of a standard ADRC controller can be shown as Fig. 3, including: extended state observer (ESO), tracking differentiator (TD) and nonlinear combination of control law (NLSEF). The core is ESO which is the premise of rejection of disturbance. Since the output controlled is observable, it must contain the signal of disturbance. So, the information of disturbance can be extracted from the whole system.

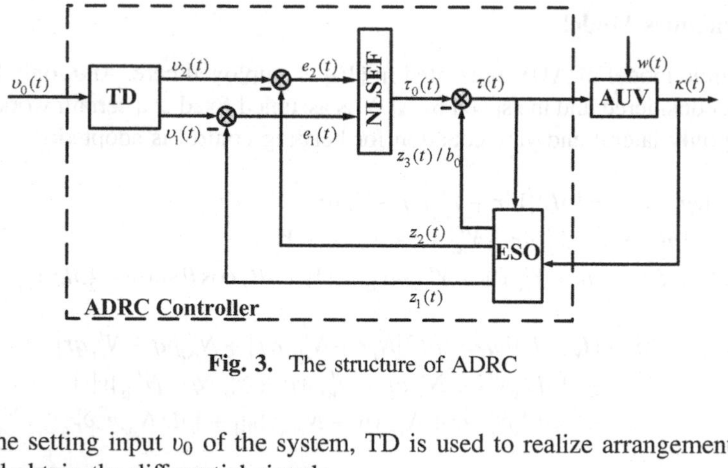

Fig. 3. The structure of ADRC

For the setting input υ_0 of the system, TD is used to realize arrangement of transition and obtain the differential signal:

$$\begin{cases} e = \upsilon_1 - \upsilon_0 \\ fh = fhan(e, \upsilon_2, r_0, h) \\ \upsilon_1 = \upsilon_1 + h\upsilon_2 \\ \upsilon_2 = \upsilon_2 + hfh \end{cases} \tag{9}$$

Where, υ_1 tracks υ_0, and υ_2 is the generalized derivative of υ_1.

ESO is to estimate state and disturbance of system output κ and control input τ.

$$\begin{cases} \varepsilon = z_1 - \kappa \\ z_1 = z_1 + h(z_2 - b_1 e) \\ z_2 = z_2 + h(z_3 - b_2 fal(\varepsilon, r_1, d) + b_0 \tau) \\ z_3 = z_3 + h(-b_3 fal(\varepsilon, r_2, d)) \end{cases} \tag{10}$$

Where, b_1, b_2, b_3 are the parameters of ESO. z_1, z_2 denote estimated system state; z_3 denotes estimated total disturbance of the system.

Nonlinear combination of control law is:

$$\begin{cases} e_1 = \upsilon_1 - z_1 \\ e_2 = \upsilon_2 - z_2 \\ \tau_0 = \beta_1 fal(e_1, a_{1,} d) + \beta_2 fal(e_2, a_{2,} d) \end{cases} \tag{11}$$

Where, β_1, β_2 is control parameters.

The compensation process of disturbance is:

$$\tau = \tau_0 - z_3 / b_0 \tag{12}$$

$fal(e, a, \delta)$ is select as:

$$fal(e, a, \delta) = \begin{cases} |e|^a sign(e), & |e| > \delta \\ e/\delta^{1-a}, & |e| \leq \delta \end{cases} \tag{13}$$

$fhan(x_1, x_2, r, h)$ is select as:

$$\begin{cases} a = \begin{cases} x_2 + (a_0 - rh)sign(\kappa)/2, & |\kappa| > d_0 \\ x_2 + \kappa/h, & |\kappa| \leq d_0 \end{cases} \\ fhan = \begin{cases} rsign(a), & |a| > d \\ ra/d, & |a| \leq d_0 \end{cases} \end{cases} \tag{14}$$

Where, $d = rh$, $d_0 = hd\,\kappa = x_1 + hx_2$, $a_0 = \sqrt{d^2 + 8r|\kappa|}$.

4.2 ADRC Parameters Self-turning Based on RNM

ADRC has many parameters which are difficult to make sure their values. Especially, β_1 and β_2 are greatly influential on the effect of control. In this paper RNM is used to optimize β_1 and β_2 at every sampling time to improve the adaptability of ADRC.

The objective function and errors are shown below:

$$E(k) = 0.5(r(k) - \kappa(k))^2 \tag{15}$$

$$error(k) = r(k) - \kappa(k) \tag{16}$$

Then, the gradient descent method is adopted as the learning algorithm of RNM:

$$\begin{cases} \beta_1(k) = \beta_1(k-1) + \eta_1\, error(k) \dfrac{\partial \kappa}{\partial u} fal(e_1, a_1, \delta) \\ \beta_2(k) = \beta_2(k-1) + \eta_2\, error(k) \dfrac{\partial \kappa}{\partial u} fal(e_2, a_2, \delta) \end{cases} \tag{17}$$

The Structure of RNM used to identify $\frac{\partial \kappa}{\partial \tau}$ is showed in Fig. 4:

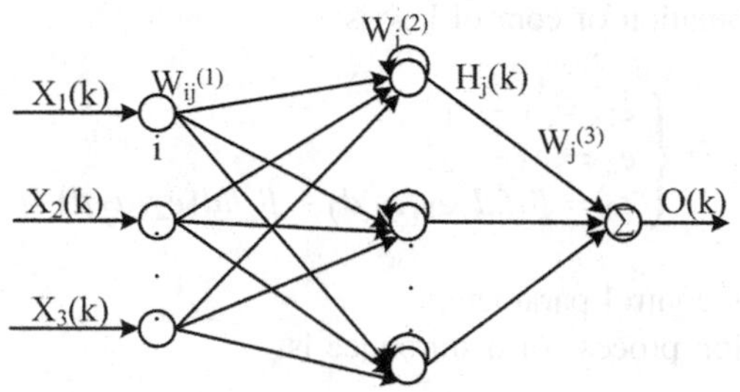

Fig. 4. The structure of recurrent networks model

Input layer:

$$I_i(k) = X_i(k) \tag{18}$$

Hidden layer:

$$H_j(k) = F_j(net_j(k)) \tag{19}$$

$$net_j(k) = W_j^{(2)} H_j(k-1) + \sum_i W_{ij}^{(1)} X_i(k) \tag{20}$$

Output layer:

$$O(k) = \sum_j W_j^{(3)} H_j(k) \tag{21}$$

Where, $X_i(k)$ is the net input node, $net_j(k)$ is the output of the j th node in hidden layer, $O(k)$ is the output node, $F(x) = 1/(1 + e^{-x})$ is the activation function, $W_{ij}^{(1)}$, $W_j^{(2)}$, $W_j^{(3)}$

are weight coefficients of input layer to hidden layer, recurrent layer, hidden layer to output layer respectively. $u(k)$, $y(k)$ are used as the input of RNM to identify.

The adjustment index of RNM weight coefficients is select as:

$$E_M = 0.5(y(k) - \hat{y}(k))^2 \tag{22}$$

Where, $\hat{y}(k)$ is the output of RNM identifying system, $e_m(k) = y(k) - \hat{y}(k)$. The process to adjust the weight coefficients of RNM using gradient descent method is:

$$\frac{\partial E_M}{\partial W_j^{(3)}} = -\, e_m(k)\frac{\partial \hat{y}(k)}{\partial W_j^{(3)}} = -\, e_m(k)H_j(k) \tag{23}$$

$$\frac{\partial E_M}{\partial W_j^{(2)}} = -\, e_m(k)\frac{\partial O(k)}{\partial W_j^{(2)}} = -\, e_m(k)W_j^{(3)}\delta_j^M(k) \tag{24}$$

$$\frac{\partial E_M}{\partial W_{ij}^{(1)}} = -\sum_j e_m(k)\frac{\partial O(k)}{\partial W_{ij}^{(1)}} = -\sum_j e_m W_j^{(3)}\beta_{ij}^M(k) \tag{25}$$

$$\delta_j^M(k) = F'\left(net_j(k)\right)\left(H_j(k-1) + W_j^{(2)}\delta_j^M(k-1)\right) \tag{26}$$

$$b_{ij}^M(k) = F'\left(net_j(k)\right)\left(X_j(k) + W_j^{(2)}b_{ij}^M(k-1)\right) \tag{27}$$

Where, $F'\left(net_j(k)\right) = \left(H_j(k)\left(1 - H_j k\right)\right)$.

Then, the weight coefficients can be expressed as:

$$W(k+1) = W(k) + \eta(k)\left(-\frac{\partial E_M}{\partial W}\right) + \alpha\Delta W(k) \tag{28}$$

Where, $\eta(k)$ is adaptive learning rate; α is momentum factor.

At last, $\frac{\partial y}{\partial \tau}$ can be obtained after weight coefficients adjustment:

$$\frac{\partial y(k)}{\partial \tau(k)} = \sum_j W_j^{(3)}F'\left(net_j(k)\right)W_{ij}^{(1)} \tag{29}$$

5 Trajectory-Keeping Control System and Heading Guidance

In this section, the trajectory-keeping control system using indirect control mode is adopted. Then, the heading guidance using cross tracking error method to obtain the AUV heading command is introduced.

5.1 Trajectory-Keeping Control System

There are two control modes of trajectory-keeping: direct control and indirect control. In this paper, the indirect control mode is select. The indirect control mode has advantage of relative separating of trajectory keeping function and heading control function. With this advantage, the trajectory-keeping control system is easily to design and realize in practice. The trajectory-keeping control system is consisted of three closed loops nested with each other including: heading guidance, heading control and rudder control (see Fig. 5). In addition, the speed of AUV for trajectory-keeping is assumed fixed and not need to control in this paper.

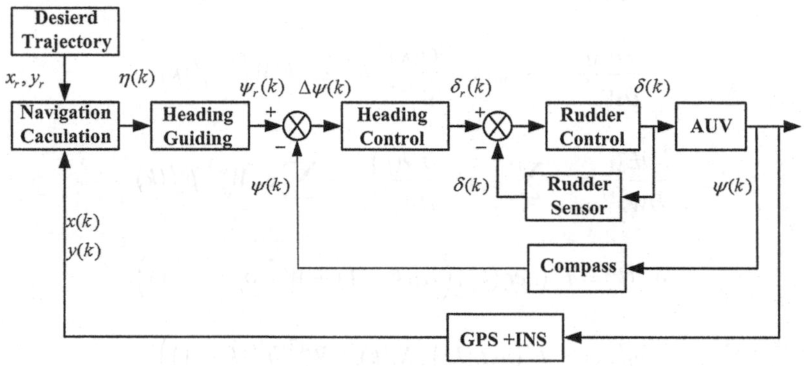

Fig. 5. The trajectory-keeping control system structure of AUV

The heading guidance loop is to obtain the heading command $\psi_r(k)$ by calculating trajectory keeping error between AUV and trajectory. In this loop, the heading guidance method is important which will guide AUV to eliminate keeping error continuously. The position of AUV can be obtained by GPS on surface and by INS underwater. The function of heading control loop is to control AUV heading using RNM-ADRC method and give rudder command $\delta_r(k)$. The heading error is $\Delta\psi(k) = \psi_r(k) - \psi(k)$. The real heading of AUV $\psi(k)$ is obtained by compass. The rudder control loop is a servo control to drive rudder $\delta(k)$ following rudder command $\delta_r(k)$.

5.2 Heading Guidance Method

Usually, there are two heading guidance method: cross tracking error (CTE) and LOS. In this paper, the CTE heading guidance method is used, by which the AUV adjusts position to reduce the keeping error to the desired trajectory. Cross tracking error $\varepsilon(t)$ is defined as the vertical distance between the AUV position $(X(t), Y(t))$ to the desired trajectory. Then, the purpose of CTE guidance method is to reduce the error $\varepsilon(t)$, which is introduced below and showed in Fig. 6.

See Fig. 6, $\overrightarrow{W_{i-1}W_i}$ is desired trajectory for AUV to track. W_i with coordinates $(X_{wt(i)}, Y_{wt(i)})$ is current waypoint AUV need to reach, W_{i-1} with coordinates

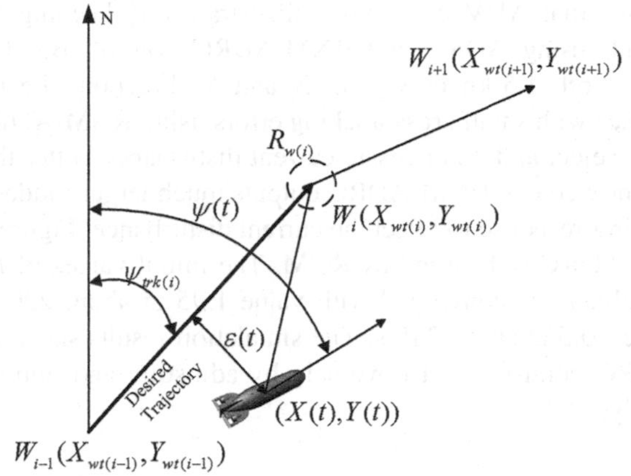

Fig. 6. Heading guidance method of AUV

$(X_{wt(i-1)}, Y_{wt(i-1)})$ is previous waypoint, and W_{i+1} with coordinates $(X_{wt(i+1)}, Y_{wt(i+1)})$ is next waypoint, $(X(t), Y(t))$ is position of AUV at time t. define: $\Delta x = X_{wt(i)} - X_{wt(i-1)}$, $\Delta y = Y_{wt(i)} - Y_{wt(i-1)}$, $\tilde{x} = X_{wt(i)} - X(t)$, $\tilde{y} = Y_{wt(i)} - Y(t)$. The trajectory length of $W_{i-1}W_i$ is $Li = \sqrt{\Delta x^2 + \Delta y^2}$ and the cross-track error is $\varepsilon(t) = (\tilde{x}\Delta y - \tilde{y}\Delta x)/Li$. The guiding law to obtain trajectory error revised angle is $\psi_m(t) = k_{pp} \cdot \varepsilon(t)$.

Then the desired heading command is:

$$\psi_r(t) = \psi_m(t) + \psi_{trk(i)} \tag{30}$$

Where, $\psi_{trk(i)}$ is the trajectory angle of $W_{i-1}W_i$ is:

$$\psi_{trk(i)} = a\tan 2(Y_{wt(i)} - Y_{wt(i-1)}, X_{wt(i)} - X_{wt(i-1)}) \tag{31}$$

6 Simulations

To illustrate the control performance of RNM-ADRC, simulations of trajectory-keeping control are carried out under current disturbances and the performances of ADRC and RNM-ADRC are compared. To avoid being a complex control system, it is assumed that the rudders of AUV can response accurately. So the trajectory-keeping control system showed in Fig. 5 is reduced to two closed loops. In simulation, the speed of AUV is set fixed as 5 kn, the initial position is (450, 150) and the initial heading angle is 90°. The current velocity in X and Y direction are both set 1.5 kn. Figure 7 shows the trajectory-keeping results of AUV based on RNM-ADRC and ADRC as contrasts. Figure 8 shows heading and rudder control responses of AUV using the two control method. The β_1 and β_2 parameters self-turning results based on RNM is showed in Fig. 9.

Figure 7 shows that AUV can both realize trajectory-keeping of a cropper-type terrain survey path using ADRC and RNM-ADRC control. But for relative strong disturbances with both 1.5 kn current in X and Y direction, the trajectory-keeping effect is much better with small cross tracking errors using RNM-ADRC. That is to say, RNM-ADRC can reject and compensate current disturbance better than ADRC. From Fig. 8, it can be showed that RNM-ADRC outputs much larger rudder range and brings much larger heading response to reject the current disturbance. Figure 9 presents the β_1 and β_2 values of ADRC self-turned by RNM. The initial values of β_1 and β_2 is set 1 and 30 respectively. β_1 is converged with value 1.95 at about 220 s, and β_1 is converged with value 30.5 at about 245 s. The simulation results show that the RNM can improve the ADRC control effect obviously by adjusting and optimizing the ADRC parameters quickly.

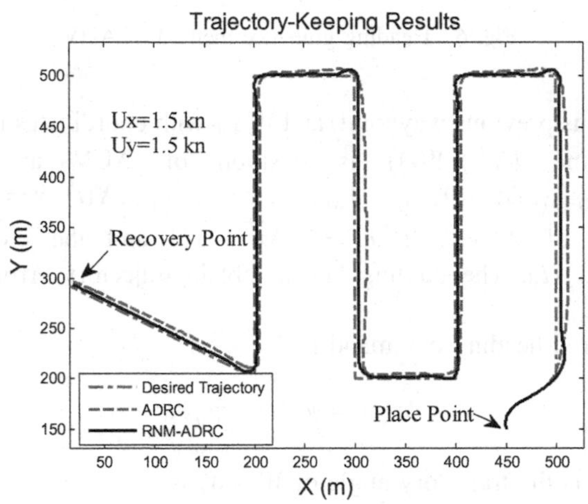

Fig. 7. Trajectory-keeping results based on ADRC and RNM-ADRC

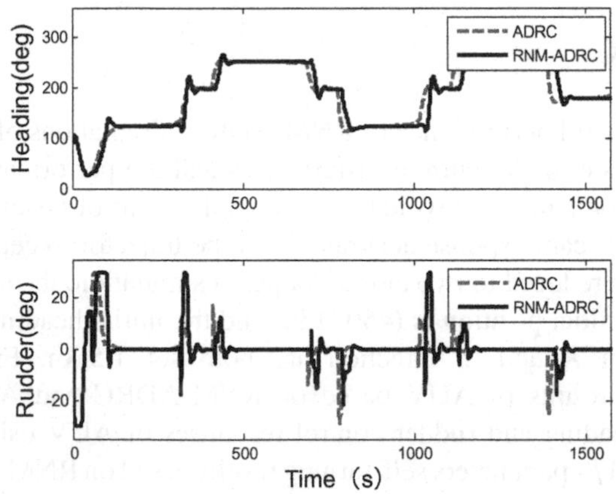

Fig. 8. Heading and rudder control responses of AUV

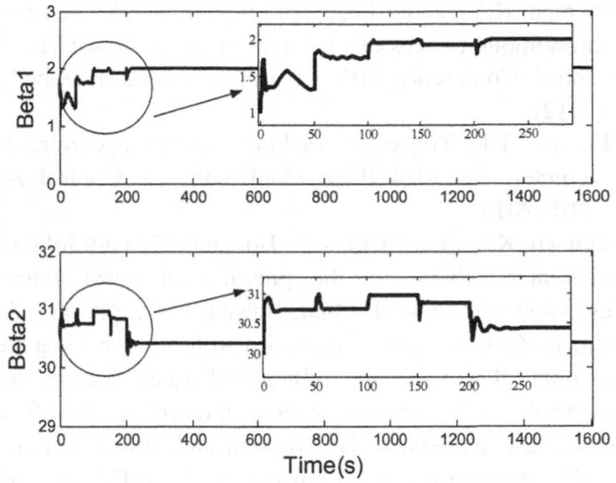

Fig. 9. β_1 and β_2 parameters self-turning results based on RNM

7 Conclusions

This paper addresses the problem of trajectory-keeping control of AUV under current disturbances. The kinematics and dynamics models in horizontal plane of AUV are established, simplified and linearized. The ADRC control method and parameters self-turning method using RNM are adopted to realize the heading control. Trajectory-keeping control system using indirect control mode and CTE heading guidance of AUV for terrain survey mission are introduced. At last, simulation of trajectory-keeping under strong current disturbances both using ADRC and RNM-ADRC are performed and compared. The simulation results show that RNM can enhance the adaptability of ADRC for optimizing the heading control output and increasing trajectory-keeping accuracy by self-turning the parameters quickly.

References

1. Bovio, E., Cecchi, D.: Autonomous underwater vehicles for scientific and naval operations. Ann. Rev. Control **30**, 117–130 (2006)
2. Caress, D.W., Thomas, H., Kirkwood, W.J.: High-resolution multibeam, sidescan, and subbottom surveys using the MBARI AUV D. Allan B. Mar. Habitat Mapp. Technol. Alsk. **4**, 47–69 (2008)
3. Bi, F.Y., Wei, Y.J., Zhang, J.Z., Cao, W.: Position-tracking control of under actuated autonomous underwater vehicles in the presence of unknown ocean currents. IET Control Theor. Appl. **4**(11), 2369–2380 (2010)
4. Tao, C., Da, X., Jiajia, Z., Anzuo, J.: Observer based 3-D path following control of under actuated underwater vehicle in the presence of ocean currents. In: Proceedings of the 28th Chinese Control and Decision Conference, Yinchuan, China, pp. 5268–5273 (2016)

5. Caharija, W., Pettersen, K.Y., et al.: Integral LOS guidance for horizontal path following of under actuated autonomous underwater vehicles in the presence of vertical ocean currents. In: American Control Conference 2012, Fairmont Queen Elizabeth, Montréal, Canada, pp. 5427–5434 (2012)
6. Lekkas, A.M., Fossen, T.I.: Trajectory tracking and ocean current estimation for marine under actuated vehicles. In: 2014 IEEE Conference on Control Applications, Antibes, France, pp. 905–910 (2014)
7. Caharija, W., Pettersen, K.Y., Gravdahl, J.T., Borhaug, E.: Path following of under actuated autonomous underwater vehicles in the presence of ocean currents. In: 51th IEEE Conference on Decision and Control, Maui, Hawaii, USA, pp. 528–535 (2012)
8. Moe, S., Pettersen, K.Y.: Line-of-sight curved path following for underactuated USVs and AUVs in the horizontal plane under the influence of ocean currents. In: 24th Mediterranean Conference on Control and Automation, Athens, Greece, pp. 38–45 (2016)
9. Tao, C., Yuzhu, W., Da, X., Zhang, H.: Mission and motion control of AUV for terrain survey mission using discrete event system theory. In: IEEE International Conference on Mechatronics and Automation 2016, Harbin, China, pp. 1012–1017 (2016)
10. Fossen, T.I.: Guidance and Control of Ocean Vehicles. Wiley, New York (1994)

UUV-Six Degrees of Freedom Positioning Method Based on Optical Vision

Wei Zhang, Ximeng Wang(✉), Lifeng Gao, and Shilin Wei

Harbin Engineering University, Harbin 150000, China
wangximeng409@126.com

Abstract. In the process of underwater dynamic recovery of UUV (unmanned underwater vehicle), UUV instantly needs to know the position information of UUV. The position information of UUV is relative to the target light source while accelerating the tracking of the target light source. The four degrees of freedom does not take the UUV's own information into account completely. It is easy to lead to UUV tracking error of target light source, when UUV rotates itself. Therefore, 6-DOF (six degrees of freedom) positioning method for light vision based on triangle target light source array is proposed. There are two cases, they are respectively: the light source array plane is parallel to the camera image plane and the light source array plane is not parallel to the camera image plane.

Keywords: Unmanned underwater vehicle · 6-DOF (six degrees of freedom) · Target light source

1 Introduction

UUV has high intelligence, autonomy, strong endurance and other characteristics [1]. So it can detect hypsography and the object search in the deep sea, it can also detect the target sea area closely for a long time. It is particularly important to recycle UUV when it completes the underwater operations or uploads the operation data. Especially the process of UUV and recovery mother ship from accurate identification to fast and secure docking recovery is more important [2–4].

Considering that in the process of accelerating into the docking station, UUV needs to know the 6-DOF relative to the dock. The paper uses the method of placing the triangle target light source array in the dock to guide the UUV [5]. Underwater camera in the middle of the front of UUV tracks light source array in recovery dock. UUV accelerates tracking light source. During the tracking, the camera generates the image sequences from the light source array [6]. After obtaining the three triangular coordinate in the target light source array, 6-DOF information between the UUV and the target light source array is obtained by coordinate transformation. Then, the 6-DOF information data is sent to the control computer, and the control computer is used to control the UUV movement. In this paper, 6-DOF calculation algorithm is proposed for the two cases: (1) the light source array plane is parallel to the camera image plane. (2) The light source array plane is not parallel to the camera image plane.

Y. Huang et al. (Eds.): ICIRA 2017, Part I, LNAI 10462, pp. 779–789, 2017.
DOI: 10.1007/978-3-319-65289-4_72

2 Triangle Target Light Source Array

The triangle light source array has the following characteristics: (1) the three light sources are on the vertex of an inverted isosceles triangle. It is shown as in Fig. 1. The distance between the two upper triangle light sources is 0.16 m, and the distance between the light source of the lower vertex and the center of the connection of the upper two light sources is 0.18 m. The triangle light source on the lower vertex of the dotted triangle is the center position of the recovery dock. After recovery, it is also placed at the opposite position of the camera. The camera is at the center of the UUV's head. (2) The light intensity and shape size of three light sources are the same.

The installation location between the three target light sources does not affect the target tracking of the target tracking algorithm [7, 8]. Because the target light source image is extracted as a template before tracking. It will return the coordinates of the three target light sources after template matching. In the tracking process, the moving target tracking algorithm predicts the current location based on the location of the target light source in the previous frame [9–11]. So it is not affected by the installation location between the light sources [12]. Of course, the distance between the three target light sources cannot be too small. When the UUV is far away from the target light source, it will cause that the three target light sources are misjudgment into a light source, which leads to the tracking failure. The light source array is set up as an isosceles triangle. It can calculate the UUV 6-DOF relative to the target light source array conveniently. So the calculation speed will be improved [13].

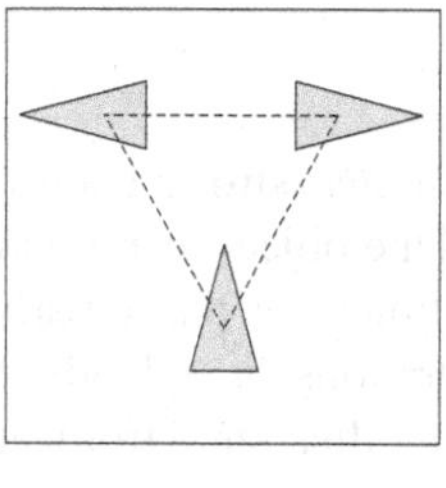

Fig. 1. Triangle target light source array

3 6-DOF Calculation

In this paper, three triangle light source arrays are used as the target light source array for UUV recovery. The array is fixed on the recovery dock. And the light intensity of the three triangle light sources can be adjusted. The single-eye camera is installed on the center of the front of the UUV, so the straight line of the camera optical center and the center axis of the circular UUV are on the same line. This paper mainly considers 6-DOF between the target light source array and the UUV when recovering the UUV. The 6-DOF is respectively relative distance, heading angle, pitch angle, sway, heave and spin angle [14]. In the target light source array, image coordinates of the upper left corner, the upper right corner and the lower side are $I_1(u_1, v_1)$, $I_2(u_2, v_2)$, $I_3(u_3, v_3)$. The coordinates in the camera coordinate system are $D_1(x_1, y_1, z_1)$, $D_2(x_2, y_2, z_3)$, $D_3(x_3, y_3, z_3)$.

3.1 The Camera Plane and the Light Source Plane Are Parallel to Each Other

Calculate the sway position, heave position and the relative distance H between the target light source array and the camera lens. They are calculated by D_1 and D_2. The distance between the two light sources in the world coordinate system is L, the corresponding coordinates in the image coordinate system are $I_1(u_1, v_1)$ and $I_2(u_2, v_2)$. As shown in Fig. 2(a), (b), the distance between I_1 and I_2 is R. The formula is obtained by similar principle of triangle:

$$\frac{R}{L} = \frac{f}{H} \tag{1}$$

Where $R = \sqrt{(u_2 - u_1) + (v_2 - v_1)^2} = \sqrt{\Delta u^2 + \Delta v^2}$

The relative distance between the target light source array and UUV is calculated as follows:

$$H = \frac{L \cdot f}{R} = \frac{L \cdot f}{\sqrt{\Delta u^2 + \Delta v^2}} \tag{2}$$

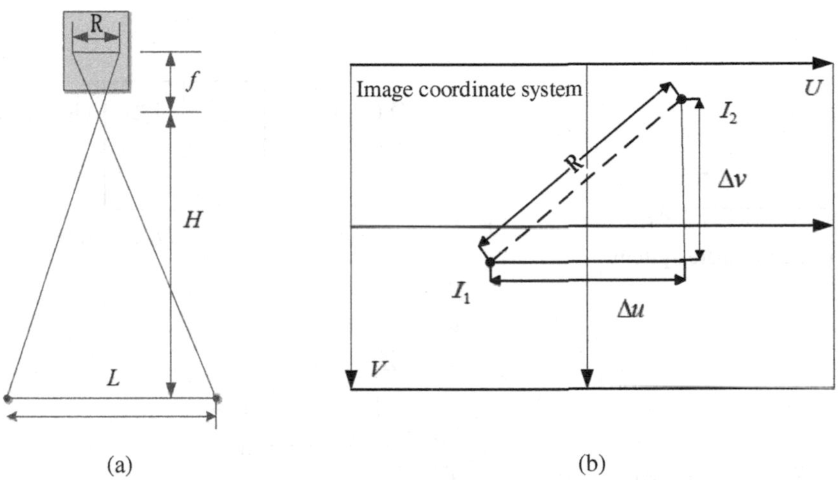

Fig. 2. (a) Triangle similarity principle (b) The position in the image coordinate system

Calculate sway and heave position. The camera plane is parallel to the source array plane. So $z_1 = z_2 = z_3 = H$, the coordinate of D_3 is calculated by formula (1).

$$\begin{pmatrix} x_3 \\ y_3 \end{pmatrix} = \frac{H}{f}\begin{pmatrix} u_0 \\ v_0 \end{pmatrix} = \frac{z_3}{f}\begin{pmatrix} u_0 \\ v_0 \end{pmatrix} \tag{3}$$

We can determine the coordinate $D_3(x_3, y_3, z_3)$. The sway distance x_3 and the surge distance y_3 between the target light source and the camera's optical center are obtained.

Assuming that the distance between the camera's optical center and the UUV center is Δm, so the matrix between the camera and the UUV is $H_u = (0, 0, \Delta \mathrm{m})^T$. The matrix between the camera's optical center and the light source array is $(x_3, y_3, z_3)^T$. So the distance $(x_3^u, y_3^u, z_c^u)^T$ between the light source array and the UUV center is

$$(x_3^u, y_3^u, z_c^u)^T = (x_3, y_3, z_3)^T + H_u \tag{4}$$

The relative distance between the target light source array and the UUV is z_c^u.

Calculate the heading angle α. After the successful recovery of UUV, the camera center and D_3 are on the same line. So D_3 is just projected onto $X_u O_u Z_u$ plane, we connect D_3 with the origin O_u in UUV coordinate system. Then we calculate the angle between D_3 and the axis Z_u. As shown in Fig. 3, the formula (5) is as follows:

$$\alpha = \arctan\left(\frac{x_3^u}{y_3^u}\right) \tag{5}$$

Where heading angle $\alpha \in (-\pi/2, \pi/2)$.

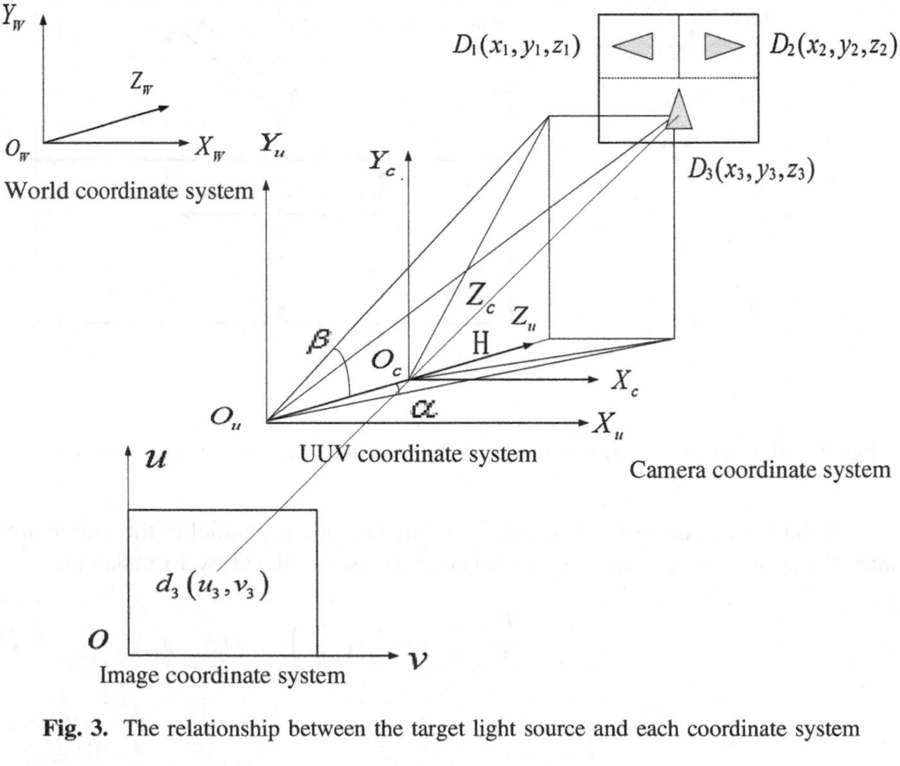

Fig. 3. The relationship between the target light source and each coordinate system

Calculate the Pitch Angle β [15]. Similarly, D_3 is projected onto $Y_uO_uZ_u$ plane, we connect D_3 with origin O_u in UUV coordinate system. Then we calculate the angle between D_3 and the axis Z_u.The formula (6) is as follows:

$$\beta = \arctan\left(\frac{y_3^u}{z_3^u}\right) \tag{6}$$

Where pitch angle $\beta \in (-\pi/2, \pi/2)$.

Calculate the Spin Angle γ. As shown in Fig. 3, we only need to calculate the angle between $\overrightarrow{D_1D_2}$ and the axis X_c. The formula (7) is as follows:

$$\gamma = \arctan\left(\frac{y_1^c - y_0^c}{x_1^c - x_0^c}\right) \tag{7}$$

Where spin angle $\gamma \in (-\pi/2, \pi/2)$.

3.2 The Camera Plane and the Light Source Plane Are not Parallel to Each Other

In the advance process of recovering the dock, curve movement such as turning, floating or sinking, causes the angle that is more than 0°. The angle is between the source array plane and the camera plane. As shown in Fig. 4, the distance between D_1 and D_2 in the actual target plane is R. The false plane, the camera plane and the image plane are parallel to each other. The intersection of the false planes the actual target plane is D_2. Two triangle light sources emit light into the camera, and the intersection of the light and the false plane are at two points. That is, the light emits from the two points coincides with the light from D_1 and D_2 in the actual target plane. In the image plane, the coordinate and the distance are equal. The distance is L. Let H represent the actual distance between the camera and the light source array. The distance between the false plane and camera plane is H' $(H' = H)$.

In this paper, we assume a heading angle between the camera plane and the light source array plane exits. After the false plane appears, the relative position and gesture between the UUV and the light source array will change. To solve this problem, an assuming camera rotation method is introduced. It is assumed that the camera rotates around the axis Y_c. The pitch angle is smaller and smaller, until the heading angle equal to 0. At this point, the camera plane is parallel to the light source plane. The process is discussed below.

When the UUV rotates around axis Y_c in the camera coordinate system, the heading rotation axis and the assumed rotation axis coincide. As shown in Fig. 5(a), the camera plane can be directly parallel to the light source plane using the method of camera rotation. When some parts of UUV are regarded as the rotation axis, there is no overlap between the heading rotation axis and the axis Y_c, as shown in Fig. 5(b). The camera plane can still be parallel to the light source plane using it. The distance between the heading rotation axis and the assumed rotation axis is Δz. The sway distance of assuming camera is about Δx, and the moving distance of the camera lens on the axis O_cZ_c is ΔH.

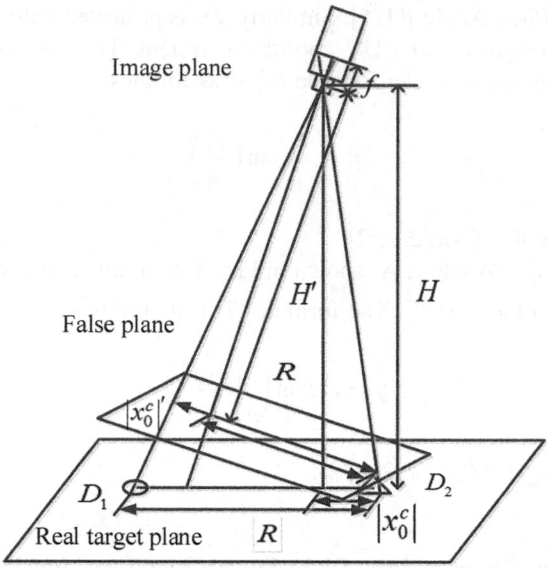

Fig. 4. Camera plane and the target light source array plane are not parallel.

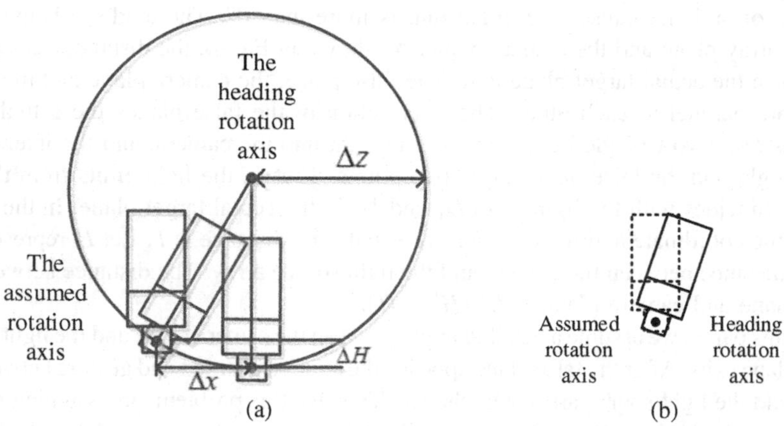

Fig. 5. (a) Two centers do not coincide. (b) Two centers coincide

When the UUV is rotating at the heading angle α_1, it can get the coordinates of D_1 and D_2, calculate the angle between D_1 and D_2, and the distance H' between the camera and the target light at this time. According to H', in the previously collected images (the camera plane is parallel to the source plane), UUV searches for the corresponding angle θ''. The angle θ'' is between D_1 and D_2. The camera offset angle θ in the horizontal direction is shown in the formula (8):

$$\theta = \theta'' - \theta' \tag{8}$$

That is the camera rotates θ in the horizontal direction.

The camera lens from the original position moves to the point of O'. As shown in Fig. 6(a), when the UUV rotates at the heading angle α_1, the corresponding image coordinate system after rotation is (x^1, y^1, z^1). Assuming that UUV rotated at θ, and the camera plane is parallel to the target light source. The assumed image coordinate system is (x^2, y^2, z^2). The coordinate of the light source D_3 in the image coordinate system (x^1, y^1, z^1) is (x', y'), and the corresponding coordinate in the image coordinate system (x^2, y^2, z^2) is (x'', y'').

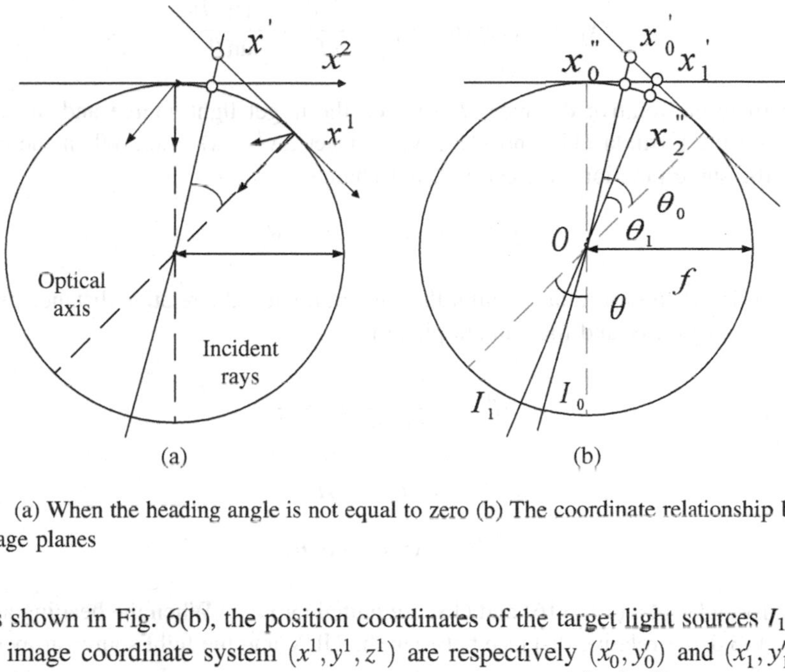

Fig. 6. (a) When the heading angle is not equal to zero (b) The coordinate relationship between the image planes

As shown in Fig. 6(b), the position coordinates of the target light sources I_1 and I_2 in the image coordinate system (x^1, y^1, z^1) are respectively (x'_0, y'_0) and (x'_1, y'_1). The corresponding coordinates in image coordinate system (x^2, y^2, z^2) are respectively (x''_0, y''_0) and (x''_1, y''_1). When the UUV rotates at the heading angle α_1, the angle between the two light sources I_0, I_2 and the axis Z_c in the camera coordinate system is respectively θ_0 and θ_1. Then the geometric relationship between the two coordinate systems can be obtained from Fig. 6(b).

$$\tan(\theta - \theta_0) = -\frac{x''_0}{f} \tag{9}$$

$$\tan(\theta - \theta_1) = -\frac{x''_1}{f} \tag{10}$$

Due to

$$\tan\theta_0 = \frac{x_0'}{f} \tag{11}$$

$$\tan\theta_1 = \frac{x_1'}{f} \tag{12}$$

Then

$$x_0'' = -f\tan(\theta - \theta_0) = \frac{f(x_0' - f\tan\theta)}{f + x_0'\tan\theta} \tag{13}$$

$$x_1'' = -f\tan(\theta - \theta_1) = \frac{f(x_1' - f\tan\theta)}{f + x_1'\tan\theta} \tag{14}$$

Calculate the relative distance H between the target light source and the camera lens. From the formula (11) and (12), we can get $\Delta x'' = x_1'' - x_0''$. When the camera rotates, the surge position of UUV will not change.

$$\Delta y'' = y_1'' - y_0'' = y_1' - y_0' \tag{15}$$

According to the formula (1), after the camera rotates, the relative distance between the light source array and the camera plane is

$$H'' = \frac{L \cdot f}{\sqrt{\Delta x''^2 + \Delta y''^2}} \tag{16}$$

$$H = H'' - \Delta H \tag{17}$$

$$\Delta H = \Delta z(1 - \cos\theta) \tag{18}$$

The formulas (4), (13), (16) and (18) are putted into (17).When the heading angle is not equal to 0, the relative distance between the UUV and the light source array can be calculated.

$$H' = H + H_u \tag{19}$$

Calculate the sway and heave position. First, assuming that the camera rotated, the position of D_3 can be calculated. The D_3 is relative to the origin O_3 in the image coordinate system.

Sway position is

$$x_3^{c'} = \frac{H}{f}x'' + \Delta z\sin\theta \tag{20}$$

Heave position is unchanged:

$$y_3^{c'} = \frac{H}{f} y_3'' \tag{21}$$

The heading angle α is

$$\alpha = \arctan(\frac{x_3^{c'}}{z_3^{c'}}) \tag{22}$$

Where $z_3^{c'} = H'$, $\alpha \in (-\pi/2, \pi/2)$.
The pitch angle β is unchanged:

$$\beta = \arctan\left(\frac{y_3^{c'}}{z_3^{c'}}\right) \tag{23}$$

Where $\beta \in (-\pi/2, \pi/2)$.
The spin angle γ is unchanged:

$$\gamma = \arctan\left(\frac{y_1^{c'} - y_0^{c'}}{x_1^{c'} - x_0^{c'}}\right) \tag{24}$$

Where $\gamma \in (-\pi/2, \pi/2)$.

When there is the pitch angle between the camera plane and the light source array plane, the image coordinates of the three light sources are obtained by using the assumed camera rotation method. Assuming that the UUV has a pitch angle α_2, the pitch angle of D_3 in the camera coordinate system is θ''. At this time the sway position is unchanged, the formula (25) is shown:

$$x_3^{c'} = \frac{H}{f} x_3'' \tag{25}$$

The surge position is changed, and the corresponding formula is

$$y_3^{c'} = \frac{H}{f} y'' + \Delta z \sin \theta'' \tag{26}$$

Calculating the remaining 4-DOF, we only need to assume that the camera rotated. The changed coordinate of D_3 is putted into the formula (19) and the formulas (22)–(24).

4 Algorithm Verification

As is shown in Fig. 7(a), (b) when the camera plane and the light source plane are parallel and their distance is 5 m, the camera collects the template pictures. The coordinates of the three underwater lights are returned by the template matching.

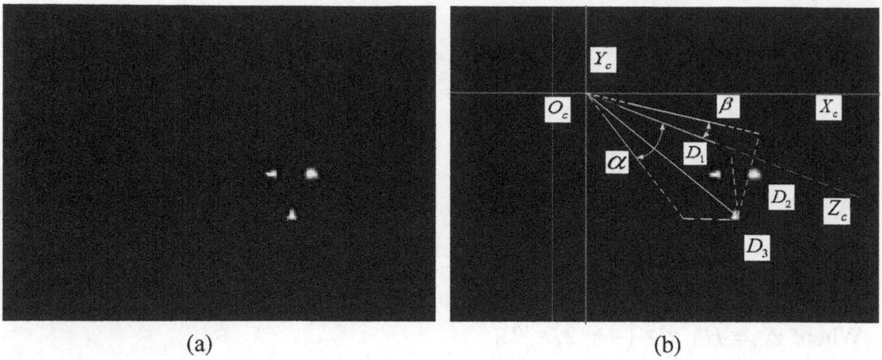

Fig. 7. (a) The camera plane and the light source plane are parallel (b) Their distance is 5 m (Color figure online)

The 6-DOF of the camera relative to the light source is obtained. There is no specific value of H_u in formula (2 and 3).

As is shown in Fig. 7(b), the intersection of cross-shaped red lines represents the optical center O_c. The coordinates relative to of the three light sources are respectively $D_1(x_1, y_1, z_1)$ $D_2(x_2, y_2, z_2)$ and $D_3(x_3, y_3, z_3)$. The distance of UUV and the light source in the direction O_cZ_c is H, then $z_1 = z_2 = z_3 = H$. We can calculate the heading angle α and the pitch angle β by connecting D_3 and O_c. The spin angle γ is the angle between $\overrightarrow{D_1D_2}$ and the axis X_c (Table 1).

Table 1. Shows the 6-DOF information when the UUV moves to the light source at the speed of 1 knot

x_1 and y_1 coordinate of D_1	x_2 and y_2 coordinate of D_2	x_3 and y_3 coordinate of D_3	Relative distance H	Heading angle α	Pitch angle β	Spin angle γ
(60, −22.5)	(70.9, −22.5)	(65.7, −35)	5.0	41.4	−44.7	0
(58.7, −22)	(71.2, −22)	(63, −36)	4.8	40.9	−44.7	0
(58.7, −22)	(71.2, −22)	(63, −36)	4.6	40.9	−44.74	0
(59, −22.4)	(72, −22)	(64, −36)	4.4	41.0	−44.78	1.45
(58, −21.5)	(70.5, −21.5)	(62.5, −35)	4.2	41.1	−44.79	0
(55, −19.4)	(68, −20)	(59.6, −34)	4.0	40.9	−44.8	−0.046
(51.4, −17.5)	(64.2, −17.5)	(55.8, −32.5)	3.8	40.7	−44.8	0
(46, −15)	(59, −16)	(51, −30)	3.6	4.7	−44.79	−4.3
(42.5, −15)	(55, −15.7)	(46, −30)	3.4	39.9	−44.8	−3.2
(37, −12)	(51.6, −13.4)	(40, −28)	3.2	39.3	−44.8	−5.4
(30, −10)	(49, −12.4)	(36.6, −27.4)	3.0	38.6	−44.8	−7.19
(30, −6)	(46, −8)	(34, −25)	2.8	38.8	−44.8	−7.15
(28, −5.6)	(45.4, −9)	(32.6, −27)	2.6	38.0	−44.85	−11
(27, −5)	(45, −8)	(32, −33)	2.4	37.3	−44.8	−9.4
(22, −8)	(42, −12)	(27, −38.45)	2.2	32.3	−44.9	−11.3
(14, −12)	(38, −17)	(19.5, −46)	2.0	24.33	−45	−11.7
(5, −16.3)	(32, −22)	(11, −50)	1.8	13.1	−44.5	−11.9
(−3.6, −15)	(27.5, −21.3)	(2.5, −56)	1.6	2.8	−45	14.7
(−14, −14)	(22, −22)	(−6, −68)	1.4	6.0	−45	12.5
(−26, −16)	(18, −26)	(−16, −72)	1.2	12.9	−44.9	12.8

5 Conclusion

In this paper, in the case of the dynamic motion of the dock and the low speed motion of UUV, the recovery of UUV decides the 6-DOF information. The 6-DOF information is relative to the light source array. According to the case of the camera image plane and the light source array plane (parallel or non-parallel), we can get the 6-DOF information formula. The UUV can know the 6-DOF information of itself relative to the light source.

References

1. Lan, Z.L., Zhou, J.B.: Development and application of unmanned underwater vehicles. Natl. Defense Sci. Technol. **29**(2), 11–15 (2008)
2. Sun, X.Y., Jiao, Z., Song, C.: Development and application of unmanned underwater vehicles. Mine Warfare Ship Self Defence **2**, 55–58 (2012)
3. Tan, K.M., Liddy, T., Anvar, A., et al.: The advancement of an autonomous underwater vehicle (AUV) technology. In: 3rd IEEE Conference on Industrial Electronics and Applications (ICIEA), pp. 336–341. IEEE Xplore, Singapore (2008)
4. Fletcher, B.: UUV master plan: a vision for navy UUV development. In: Oceans, pp. 65–71. IEEE Xplore, USA (2000)
5. Wang, X.J.: Study on Vision-Based AUV Underwater Recovery Guidance and Positioning Technology. Harbin Engineering University, Harbin (2011)
6. Zhang, M., Li, X., Wang, Y.: Underwater color image segmentation based on weight adjustment for color-to-gray. Harbin Eng. Univ. **36**(5), 707–713 (2015)
7. Li, P.H.: Moving Target Tracking Method in Sequence Image. Science Press, Beijing (2010)
8. Zhang, K., Zhang, L., Yang, M.H., et al.: Fast Tracking via Spatio-Temporal Context Learning. Computer Science, USA (2013)
9. Gong, P., Zhang, Q., Zhang, A.: Stereo vision based motion estimation for underwater vehicles. In: 2nd International Conference on Intelligent Computation Technology and Automation (ICICTA), pp. 745–749. IEEE, China (2009)
10. Shibata, M., Kobayashi, N.: Image-based visual tracking for moving targets with active stereo vision robot. In: International Joint Conference on SICE-ICASE, pp. 5329–5334. IEEE Xplore, Korea (2006)
11. Lots, J.F., Lane, D.M., Trucco, E., et al.: A 2D visual serving for underwater vehicle station keeping. In: IEEE International Conference on Robotics and Automation, pp 2767–2772. IEEE, Korea (2001)
12. Garcia, R., Cufi, X., Carreras, M., et al.: Correction of shading effects in vision-based UUV localization. In: International Conference on Robotics and Automation (ICRA), pp. 989–994. IEEE, Taiwan (2003)
13. Eustice, R.M., Pizarro, O., Singh, H.: Visually augmented navigation for autonomous underwater vehicles. Oceanic Eng. IEEE J. **33**(2), 103–122 (2008)
14. Jian, Y.J.: Modeling and Simulation of 6-DOF Motion of Underwater Platform. Northwestern Polytechnical University, Xian (2013)
15. Li, Y., Liu, Y., Huang, P.: Compensation method of binocular vision image with UUV roll and pitch. In: International Conference on Mechatronics and Automation (ICMA), pp 2352–2357. IEEE, China (2015)

Hydrographic and Meteorological Observation Demonstration with Wave Glider "Black Pearl"

Can Li[1], Hongqiang Sang[2], Xiujun Sun[2(✉)], and Zhanhui Qi[2]

[1] Tianjin Polytechnic University, Tianjin, China
[2] National Ocean Technology Center, Tianjin, China
sunxiujun@yahoo.com

Abstract. Wave Glider is one type of unmanned surface vehicles featuring long duration and large scale cruising capability by translating wave energy to forward motion and generating electricity from solar panels. This paper proposes a new developed wave glider named "Black Pearl", which equipped wave sensor, conductivity and temperature sensor and micro weather station. Firstly, we present its architectures and discuss the difference between the "Black Pearl" and the Wave Glider made in Liquid Robotics Ltd.; secondly, we propose the field trial plan and analyze the hydrological and meteorological data during the process of the sea experiment, in addition, we carry out a contrast trial to analyze the wave data measured by the "Black Pearl" and the data measured by wave-measuring buoy; thirdly, we talk about the vehicle's performance in a variety of ocean conditions. The sea trial data is beneficial to the environmental monitoring of the sea area, and provides reference for optimization of the structural design of "Black Pearl".

Keywords: Wave glider · "Black Pearl" · Island circumnavigating · Hydrology and meteorology

1 Introduction

Wave Glider is one type of unmanned surface vehicles featuring long duration and large scale cruising capability by translating wave energy to forward motion and generating electricity from solar panels. In 2005, Roger Hine and Joseph D. Rizzi created Wave Glider to record the songs or "singing" of Humpback Whales as they migrated along the coasts of the Big Island. In January 2007, Liquid Robotics was founded to develop wave glider technology and its applications. Until now, the Company has employed over 110 employees and produced a total of over 350 Wave Gliders. At any one time, there is approximately 1/4 of the fleet at sea [1]. Because of the wide applications in oceanographic surveys, many government agencies, research institutes or companies began to cooperate with Liquid Robotics Ltd. Monterey Bay Aquarium Research Institute operated the WG along three pre-determined synoptic sampling paths in northern and central Monterey Bay to monitor algal blooms [2]. Scripps Institution of Oceanography and Liquid Robotics combine High-frequency Acoustic Recording Package in wave glider to demonstrate a new approach to marine

Y. Huang et al. (Eds.): ICIRA 2017, Part I, LNAI 10462, pp. 790–800, 2017.
DOI: 10.1007/978-3-319-65289-4_73

mammal monitoring [3]. The NOAA Pacific Marine Environmental Laboratory and Liquid Robotics Ltd. are collaborating to address an urgent need for long-term in-situ observation of carbon parameters by integrating a suite of state-of-the-art pCO2, pH, and CTD sensors onto a Wave Glider [4]. The Wave Glider was equipped with CTD, CDOM, turbidity and refined fuels fluorescence, to capture the southern Tyrrhenian Sea major surface oceanographic features [5]. Depending on the characteristics of the stealth and silence of the wave glider, the U.S. Navy uses it to patrol the coast, spy on enemy ships, and track Somali pirates. A towed hydrophone array was successfully deployed behind the Wave Glider's submerged glider, demonstrating the very low noise levels of the vehicle configured in this manner for acoustic data collection [6].

Wave Gliders withstand the toughest conditions time and time again to complete customers' most critical missions, for example, it has experienced 17 hurricanes navigation. They're far cheaper than manned vessels and far more capable than alternative technologies. One of the longest voyages of the wave glider has reached 9380+ nm. Wave Gliders are the most experienced, autonomous surface vehicle in the ocean today. Time tested and ocean proven, all wave gliders running at sea have accumulated more than 30731+ days. From short to long duration missions, Wave Gliders have traveled the equivalent of 58 cycles around the Earth [1] (Fig. 1).

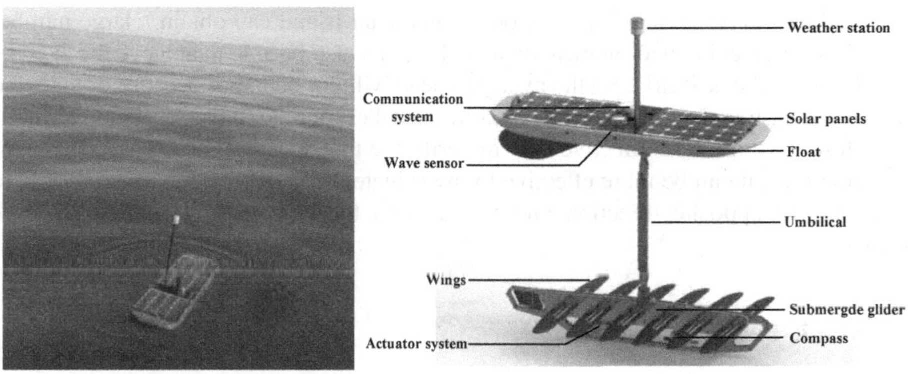

Fig. 1. Experimental field of "Black Pearl" and the composition of Black Pearl system

However, there are few domestic researches on the long period sea trials of wave gliders in China. Funded by the Project of National High Tech Development (863 Project) [7, 8], Team of Ocean Mobile Observation Technology (TOMOT) independently do the research and develop a Wave Glider named "Black Pearl". There are three models for the "Black Pearl", ranging from 1.6 to 2.4 m. The greater model takes greater speed and payload capacity, as well as more power generation capability for the on board sensors. Auxiliary electric propulsion is available for calm conditions and maneuvering. In the last 3 years, we have finished a lot of field trials about the performance of "Black Pearl". In 2014, the first prototype of the "Black Pearl" completed a 300 km trajectory test in the South China Sea. In 2015, "Black Pearl" completed the 1300 km path test within 62 days. In 2016, the acoustic communication relay test between "Black Pearl" and seafloor seated instrument was carried out for 3 days in Qingdao. In addition, China Shipbuilding Industry Corp 710 Research Institute, Shenyang Institute of Automation Chinese Academy of Sciences, Harbin Engineering University and other research units also have carried out lots of work on the development of wave glider [9] (Table 1).

Table 1. Differences between "Black Pearl" and Wave Glider

Items		"Black Pearl"	WG (SV2)	WG (SV3)
Size (L * W * H) [m * m * m]	Float	1.6 * 0.5 * 0.145	2.08 * 0.6 * 0.18	3.05 * 0.81 * 0.23
	Fin	0.513 * 0.004 * 0.15	0.6 * 0.0032 * 0.25	N/A
	Glider	1.6 * 0.046 * 0.27	1.91 * 0.025 * 0.40	2.13 * 0.040 * 0.21
	Wings	0.16 * 1.0 * 0.004	0.17 * 1.07 * 0.016	Width = 1.42
	Rudder	0.2 * 0.004 * 0.1	0.21 * 0.0040 * 0.13	N/A
	Umbilical	0.040 * 0.010 * 6	0.045 * 0.007 * 7	Length = 8
Mass(kg)	Float	22	60	Total 150
	Glider	26	41	
	Umbilical	2.5	2.5	

Field trial of circumnavigating around Qianliyan Island can obtain a large number of hydrological and meteorological data, and we can also be able to achieve the average speed of the "Black Pearl". So the average speed will not be larger because of the wind or the downstream, and will not become smaller due to adverse wind currents. Therefore, it is the most effective way to verify the true speed of the "Black Pearl". At the same time, it can be more effective to investigate the ability of the "Black Pearl" to move in the opposite direction and to maintain the course or its ability of track following.

2 Black Pearl

2.1 Structural Components

The "Black Pearl" is a two-part vehicle, consisting of a float in the surface connected to a submerged glider via a flexible tether. The surface float, made of elastic buoyancy materials, has been designed to provide buoyancy and protect electrical integration rooms in it. In addition, the electrical integration rooms, two or three solar panels, communication module and its antenna are also configured. In this paper, the standard configuration sensors for the "Black Pearl" are wave sensor, conductivity and temperature sensor and micro weather station. The specific parameters of the sensor are shown in the following Tables 2, 3 and 4.

Table 2. Main technical performance index of CT

Parameters	Range
Conductivity range	0–65 mS/cm
Conductivity accuracy	±0.005 mS/cm
Conductivity resolution	±0.001 mS/cm
Temperature range	−2–40 °C
Temperature accuracy	±0.003 °C
Temperature resolution	±0.001 °C

Table 3. Main technical performance index of Weather station

Parameters	Range	Accuracy	Resolution
Wind speed	0−40 m/s	0.5 m/s ± 10% (0−5 m/s); 1 m/s ± 5% (5−40 m/s)	0.1 m/s
Wind direction	0−350°	5°RMS (2−5 m/s); 2°RMS (>5 m/s);	0.1°
Air temperature	−40−+55°	±1.1 °C @ > 2 m/s	0.1 °C
Air pressure	300−1100 mbar	±1 mbar	0.1 mbar

Table 4. Main technical performance index of Wave sensor

Parameters	Range	Accuracy
Wave height	0.2 m−20 m	(0.3 ± 10%) m
Wave period	2.0 s−20 s	±0.5 s
Wave direction	0°−360°	±15°

Embedded central control unit and battery packs are hold in the electrical integration rooms. Solar panels continually replenish the batteries to power the vehicle's control electronics, communications systems, and payloads. Iridium 9602 is the core module for remote communication and ZigBee module is for Near-field debugging. Global Positioning System (GPS) and Compass combine together for navigation.

Submerged glider, the Black Pearl's wave-power propulsion system, is purely mechanical. The overall framework of glider is made of stainless steel and its wings are carbon fiber composite materials. In front of the glider, an electronic compass and the subsidiary control unit are arranged to capture the actual heading and provide navigation. At the stern of the mechanical system, a servo system is fixed up to provide a right steering for desired heading. Underwater gliders also have sufficient capacity to carry acoustic modem, conductivity, temperature, depth (CTD) sensor or other transducer.

2.2 Platform Performance

The "Black Pearl" can get enough dynamic energy from the waves. The main factors that affect the forward speed of the "Black Pearl" can be divided into two parts. Firstly, its own factors, the traveling wave resistance, buoyancy, and weight of "Black Pearl" and the length of the rope. Secondly, the impact of sea conditions, the wave height, wave period and current. In order to get superior performance of "Black Pearl" in various sea conditions, the parameters of the "Black Pearl" should be optimized to match the sea state parameters. The specific parameters of the "Black Pearl" are shown in the following Table 5:

Table 5. Specific parameters of the "Black Pearl"

Physical characteristics	Configuration
Endurance	up to 1 year, traveling 10000 km
Speed	SS1(sea state 1): ≥ 1.0 kts; SS4: ≥ 1.5 kts
Propellant	Wave energy
Battery	The peak power of 80 W, the average power of 8 W
Communication systems	Iridium satellite modem
Positioning accuracy	24 h virtual anchoring point error <200 m (CEP50), path tracking deviation <200 m (CEP80) (sea state 3)
payload capacity	Additional power load 15 W $\times$ 12 volts, additional structural load of 20 kg * 18 L

The "Black Pearl" absorbs the kinetic energy entirely from the wave kinetic energy and bears sever sea environmental conditions. Therefore, the life of a "Black Pearl" depends almost on the durability of its structure. High design requirements for "Black Pearl" promise that it should withstand extreme weather. It should work safely under the condition of wave height more than 10 m and Wind speed more than 40 knots. This requires that all accessories of our platform are very strong. It is ideal to ensure that the "Black Pearl" is still working properly even if some of the parts are damaged. Therefore, in the design process, we add some redundant design, multiple fault tolerance mechanisms and multiple beacons. Through these measures, the structural safety of the "Black Pearl" can be ensured.

Compared to bad weather, particularly calm sea conditions for the "Black Pearl" is a greater challenge. If there were no waves, the "Black Pearl" would not be able to move forward. We considered the problem in the design, the floating boat slightly response to any minor fluctuations and the submerged glider takes the platform forward. This requires that wave energy conversion equipment is very sensitive.

3 Field Trials

Qianliyan island, located in the south of the yellow sea in China (121° 23′09″E, 36° 15′57″N), is an uninhabited island with an ocean station of state oceanic administration and an international beacon light tower. And Qianliyan Sea is approved by the Ministry of Agriculture on November 25th in 2010 to establish National Aquatic Germplasm Resources Conservation Area. It is classified as the transitional maritime climate characteristics in summer but continental in winter. The annual average temperature is 25.4 °C in August but 0.1 °C in January. The range of the depth around the Qianliyan sea is 28–32 m, its water temperature changes between 5.2 and 29.1 °C. Salinity changes between 30.17 and 31.80. Qianliyan Island and its surrounding ocean circumstance is a typical warm representative system. Waters around the preservation of the Marine ecosystem integrity and naturalness on the island environment with good service function, and can provide Marine ecosystem background for similar parts [10].

3.1 Experiment Target

The main purpose of this experiment is to test the stability of the "Black Pearl" which are seen as the comprehensive platforms of marine environmental mobile observation. Via this experiment, we test the capability to survive in harsh environments and validate its precision of the sensors on them. The stability of the platform and the ability to fight against bad environment can be verified by using feedback parameters of "Black Pearl". The precision of the sensors, however, need to be corrected with other marine equipment. The contrast marine instruments in this experiment are the Wave Buoy developed by National Ocean Technology Center. In the process of experiment, we test the speed of the "Black Pearl" in different Sea State. The sailing speed performance targets for speed is greater than 0.25 m/s under the condition of Sea State 1, and 0.60 m/s under the Sea State 4.

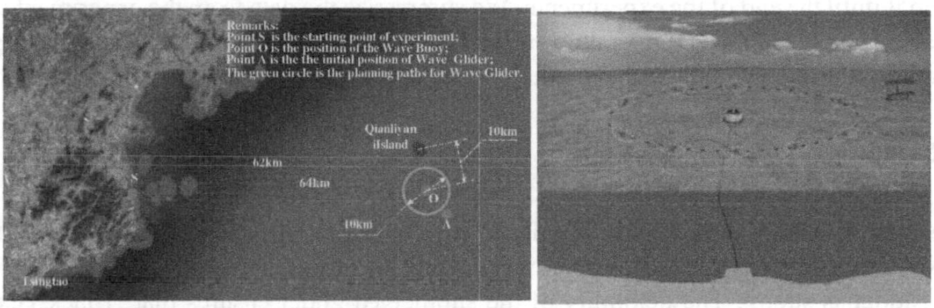

Fig. 2. Specific location and scheme and specific experimental plan

3.2 The Specific Plan

Experimental program requires "Black Pearl" to travel along a circular path for movement. Latitude and longitude coordinates is the circle's center, and it is also position of Wave Buoy. The radius of this circle is the length between the gliders and Wave Buoy. Specific experimental sea area is shown in Fig. 2. In order to compare the merits of the two different marine platforms with sensors, Wave Buoy carries on the same sensor with Black Pearl' s. Wave Buoy has passed certification from National Center of Ocean Standards and Metrology and its measuring accuracy can reach the international advanced level. The detailed experimental program is shown below in Fig. 2.

"Black Pearl" relies on iridium satellite system for communication, while Wave Buoy communicates through the Beidou satellite system. The two platforms will collect and transmit data about hydrologic and meteorological in this area at the same time. These data are gathered according to the different latitude and longitude coordinates. This paper analyzes the data mainly from "Black Pearl". The wave data of the Wave Buoy and the contrast results between the two platforms will be described in a future article.

At 6:30 am, in July 7, 2016, after leaving Laoshan wharf, we spend about 4 h by boat to reach the destination (121.503194 E, 36.287450 N) where Wave Buoy is deployed. Taking this location as a center, and five kilometers as radius to draw a circle, and this circle is our planning circular trajectory of "Black Pearl". On this path, we plan ten via way points for navigation of the "Black Pearl". So we put the "Black Pearl" in the water near the first target (121.503194 E, 36.287450 N). When we have confirmed that the operation of the "Black Pearl" is normal by the shore based operating system, the experiment began.

4 Results

After 24 days of experiment, the "Black Pearl" completes its circumnavigating mission. In the course of the experiment, the experimental data transmitting via the iridium satellite system, the current situation of the platform and the data of the sensor are uploaded to the headquarters. The communication period is about 15 min. When the experiment lasts a week, we turn the communication frequency into half an hour and keep it until the end of the experiments. We summarize the data from the experiment and the key parameters of the operation of the "Black Pearl" platform, including the target heading, desired heading, voltage, pitch angle, roll angle and the speed of the platform.

4.1 "Black Pearl" Parameters

Desired heading and actual heading. By comparing the actual heading with the desired heading, we find that the following capability of the submerged glider by turning its rudder is strong. The flexibility of the submerged glider ensures that "Black Pearl" approaches the preset courses immediately. the excellent approaching curve tell us that "Black Pearl" takes the intention to follow the preset course, but the actual course approaching effectiveness depends on the forward velocity of the "Black Pearl". The comparison between the desired heading and the actual heading is divided into two stages.

As shown in Fig. 3 below, the data for six consecutive days from July 8 to 13, 2016 was drawn. The red curve indicates the desired heading of the "Black Pearl", and the blue curve represents the actual heading of the "Black Pearl". The 0° measured by the electronic compass means that the "Black Pearl" points to the north direction. The heading angle of "Black Pearl" increases clock wisely. In this period of time, the frequency of communication is 4 times per hour. More than 700 data points can be seen from the graph, they show that the desired heading of the "Black Pearl" is well correlated with the actual heading. The results show that navigation performance of the platform is stable and Steering function is normal. There are several jumps for the desired heading and the actual heading. One kind of jumps means that the "Black Pearl" reaches the target point and switches to the next target. The other jumps belong to false ones, which only represents the transition between the 0 and 360°.

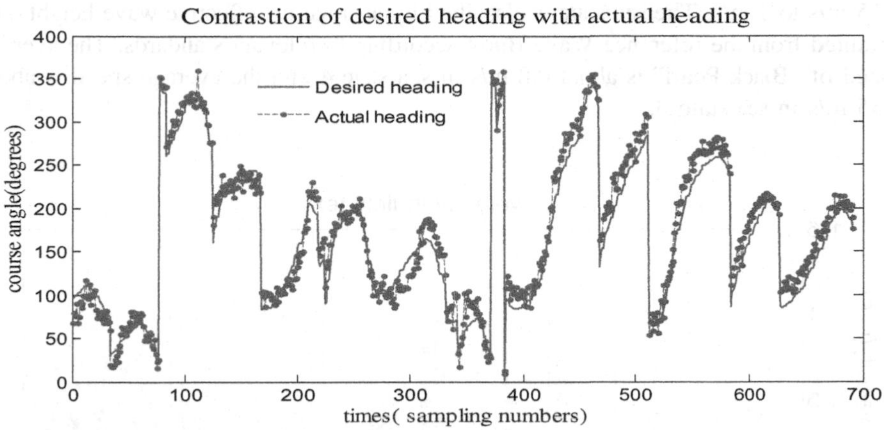

Fig. 3. Comparison curves between desired heading and actual heading (Color figure online)

Voltage conversion. It is not enough for the "Black Pearl" to realize the long life and long distance operation only by capturing wave energy, so the large size "Black Pearl" is equipped with 3 solar panels, supplying peak power up to 180 W. As is shown in Fig. 4 is the power supply voltage changes of "Black Pearl" with time during 24 days. There is a voltage cycle changing from high to low, which represents the power variation of one day. From 6:00 pm to 6:00 am the second day, the voltage gradually decreased to 16 V below, while, the voltage grows slowly again after 6am. When the battery pack voltage is lower than 16 V, the solar controller for charging and discharging will charge the battery pack. When the battery pack is fulfilled the solar controller will stop charging the lithium battery. If there is a plenty of solar energy for controller, navigation, communication, sensors and actuator, then the surplus electric can be used for driving an electrical propeller.

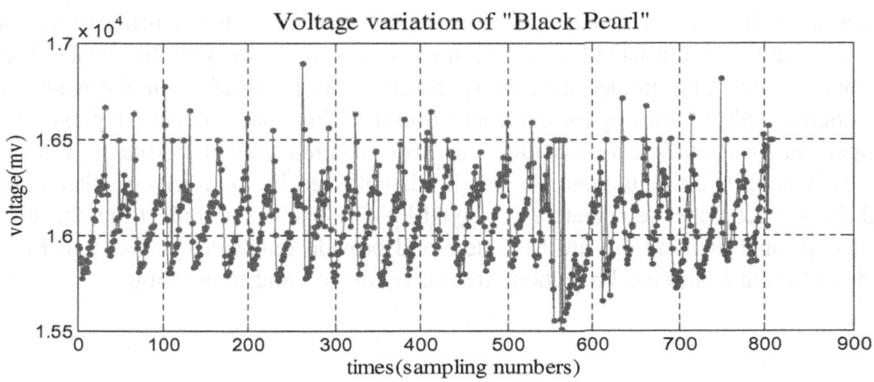

Fig. 4. 24 output voltage of circuit board

Platform speed. The average speed of the platform for six consecutive days is analyzed. In this paper, we calculate the distance between the two coordinate points by means of geodesic coordinate algorithm [9]. The running speed of the glider is roughly estimated by the distances and times in Fig. 5. The change of velocity is mainly from 0.15 m/s to 1 m/s. The sea state is classified by analyze the effective wave height data obtained from the reference Wave Buoy according to relevant standards. The average speed of "Black Pearl" is about 0.8 m/s in sea state 4 and the average speed is about 0.55 m/s in sea state 3.

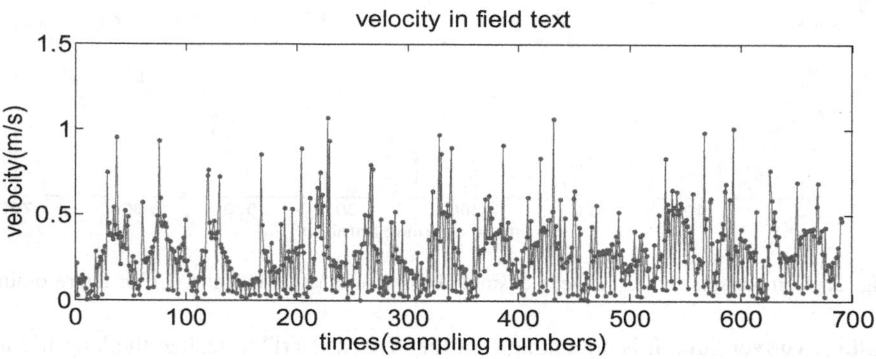

Fig. 5. Average speed curve of the platform from July 20, 102016 to 25

4.2 Hydrological and Meteorological Data

Hydrological and meteorological data. As is shown in Fig. 6, it can be seen that the "Black Pearl" has carried out an effective investigation on meteorology within 24 days of demonstration in the sea area, a region not more than 10 km wide. The air humidity ranged at 68.90–97.5 and most of the time maintained at about 90. This is directly related to the summer sea climate.

As is shown in Fig. 6 is the wind direction and wind speed collected by the "Black Pearl" "Black Pearl". The color of the arrow is used to represent the wind speed, the direction of the arrow to indicate the direction of wind. From the statistics of the data, we find that the strong wind has some influence on the trajectory of the "Black Pearl". In the early stage, the influence on the speed of the "Black Pearl" from the wind is not considered. Other hydrological and meteorological data, such as the air pressure, air temperature and surface temperature are also acquired and processed as drawn in Fig. 6. The air temperature and air pressure is collected by the micro weather station, and the surface water temperature is logged by CT. These data reflect the changes of hydrology and meteorology in the waters and provide guidance for National Marine Ranch (National Aquatic Protection Area) construction and monitoring.

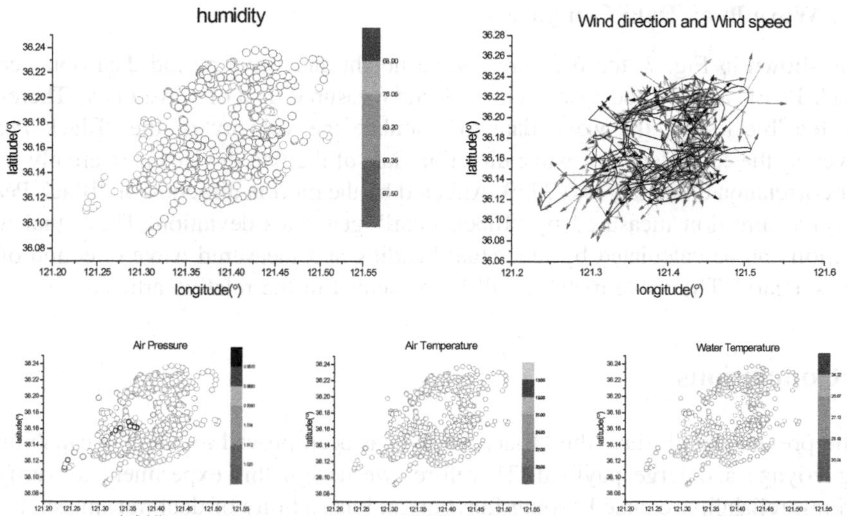

Fig. 6. Hydrological and meteorological data in experimental sea area

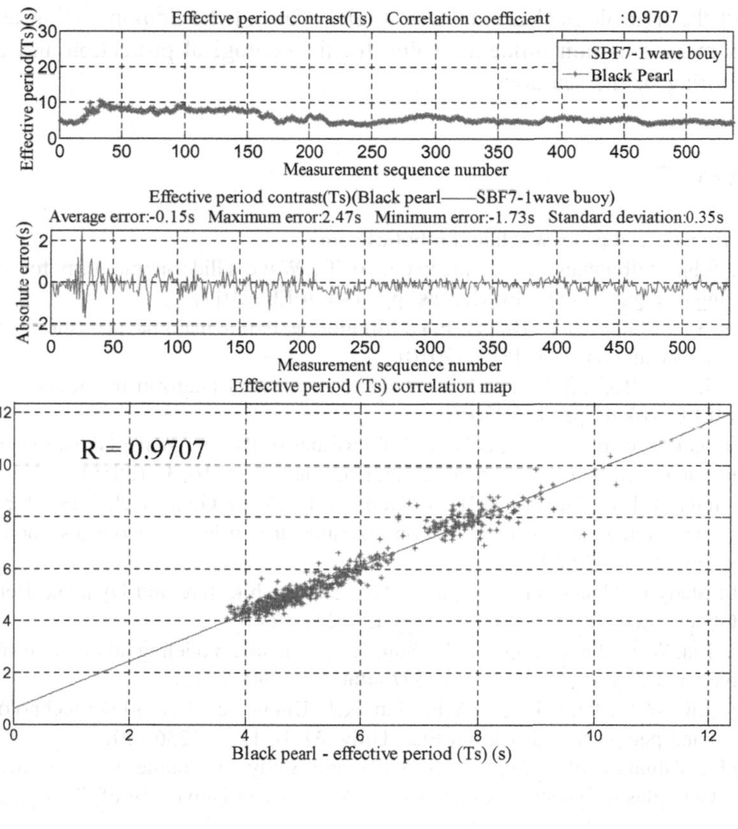

Fig. 7. Wave buoy compared with "Black Pearl"

4.3 Wave Buoy Data Comparison

As is shown in Fig. 7, the measured wave height and wave period data obtained by "Black Pearl" are in good agreement with the measured data of wave buoy. Therefore, it is feasible to use the wave data to calculate the velocity of the "Black Pearl". However, the consistency of wave direction data of the two devices is relatively weak, their correlation efficiency is 0.7359. Affected by the motion direction of "Black Pearl", the wave direction measured by "Black Pearl" generates deviation. The actual wave direction can be calculated by the actual heading and measured wave direction of the "Black Pearl". The solve method will be presented in the relative article.

5 Conclusions

In the previous field trials, the "Black Pearl" has been proved to take the capability of long voyage and large payload. Therefore, we design this experiment to verify its working reliability, course keeping function and its stability of data transmission. This experiment reflects the current technology status of the "Black Pearl" in China and provides a reference to the glider's attitude control improvement and its performance optimization. It can be seen that the information is complete and accurate, which can fully reflect the hydrological and meteorological data of the demonstration area. The paper does make a certain influence value for the ecological protection and environment monitoring of the sea area.

References

1. Liquid Robotics. https://www.liquid-robotics.com/
2. Frolov, S,F., Bellingham, J,S., Anderson, W,T.: Wave Glider-a platform for persistent monitoring of algal blooms. In: Oceans, pp. 1–5. IEEE (2011)
3. Wiggins, S,F., Manley, J,S., Brager, E,T.: Monitoring marine mammal acoustics using Wave Glider. In: Oceans, pp. 1–4. IEEE (2010)
4. Manley, J,F., Willcox, S,S.: The Wave Glider: a persistent platform for ocean science. In: Oceans IEEE Xplore, pp. 1–5 (2010)
5. Aulicino, G.F., Cotroneo, Y.S., Lacava, T.T.: Results of the first Wave Glider experiment in the southern Tyrrhenian Sea. Adv. Oceanogr. Limnol. **7**(1), 16–35 (2016)
6. Meyer-Gutbrod, E.L., Greene, C.H., McGarry, L.P.: Wave Glider technology for fisheries research new integrated instrumentation expands the fisheries acoustics toolbox. Sea Technol. **56**(12), 15–20 (2015)
7. Jia, L.J.: Study of Operation Principle of Two-Part Architecture and Dynamic Behavior of Wave Glider. National Ocean Technology Center (2014)
8. Qi, Z.F., Liu, W.X., Jia, L.J., Qin, Y.F., Sun, X.J.: Dynamic modeling and motion simulation for Wave Glider. Appl. Mech. Mater. **397–400**, 285–290 (2013)
9. Qi, Z.F., Liu, W.X., Jia, L.J., Qin, Y.F., Sun, X.J.: Unmanned wave glider technology: state of the art and perspective. J. Harbin Eng. Univ. **37**(9), 1227–1236 (2016)
10. Lin, Q.F.: Valuation of Ecosystem Services and Study of Countermeasures in National Aquatic Germplasm Resources Conservation Area. Ocean University of China (2014)

Simulation for Path Planning of OUC-II Glider with Intelligence Algorithm

Yuhai Liu[1], Xin Luan[1], Dalei Song[2], and Zhiqiang Su[2(✉)]

[1] College of Information Science and Engineering, Ocean University of China, no. 238 Songling Road, Laoshan District, Qingdao City, Shandong Province, China

[2] College of Engineering, Ocean University of China, no. 238 Songling Road, Laoshan District, Qingdao City, Shandong Province, China

szqouc@126.com

Abstract. Addressing the need for exploration of benthic zones utilising autonomous underwater gliders, this paper presents a simulation for an optimised path planning from the source node to the destination node of the autonomous underwater glider OUC-II Glider in near-bottom ocean environment. Near-bottom ocean current data from the College of Oceanic and Atmospheric Science, Ocean University of China have been used for this simulation. A cost function is formulated to describe the dynamics of the autonomous underwater glider in near-bottom ocean currents. This cost function is then optimised using various biologically-inspired algorithms such as genetic algorithm, Ant Colony optimisation algorithm and particle swarm optimisation algorithm. The simulation of path planning is also performed using Q-learning technique and the results are compared with the biologically-inspired algorithms. The results clearly show that the Q-learning algorithm is better in computational complexity than the biologically-inspired algorithms. The ease of simulating the environment is also more in the case of Q-learning techniques. Hence this paper presents an effective path planning technique, which has been tested for the OUC-II glider and it may be extended for use in any standard autonomous underwater glider.

Keywords: OUC-II glider · Path planning · Simulation · Intelligence algorithm

1 Introduction

With the recent development in technology, autonomous underwater gliders (AUGs) are increasingly being used to efficiently explore the various natural and artificial environments present on seabed [1]. The AUGs are efficient and effective for a variety of missions within the depths of waterbodies. One such AUG is the OUC-II Glider, manufactured by the Underwater Glider Laboratory of Ocean University of China. This AUG boasts the following features: dynamic buoyancy, seven hundred and twenty hours of endurance at nominal load, and a capacity to work at 1500 m of depth. Addressing the need of the study of benthic zones, this paper develops a simulation for the path planning of the OUC-II so that it may be used in near-bottom ocean currents. Defining the near-bottom ocean current

Y. Huang et al. (Eds.): ICIRA 2017, Part I, LNAI 10462, pp. 801–812, 2017.
DOI: 10.1007/978-3-319-65289-4_74

are the Gully marine protected area (MPA) under Canada's Oceans Act, the Gully becomes Canada's second Oceans Act MPA, and the first in the Atlantic region [2].

The Gully MPA regulations and accompanying regulatory impact analysis statement can be viewed. The data for the magnitude and direction of the velocity of the ocean currents at near-bottom ocean depth is taken from the College of Oceanic and Atmospheric Science, Ocean University of China. The simulation is done using MATLAB platform. Using the above data, a cost function was developed which accurately describes the dynamics of the OUC-II and its interaction with the near-bottom ocean environment. Cost function is optimised to obtain the shortest path between the source node and destination node [3]. This optimisation was achieved using various biologically-inspired algorithms such as genetic algorithm (GA), ant colony optimisation (ACO) algorithm and particle swarm optimisation (PSO) algorithm [4–6]. Optimisation was also done using Q-learning technique and the results obtained from both the biologically-inspired algorithms and Q-learning techniques were compared. The results clearly show that the Q-learning technique is computationally less expensive as compared to biologically-inspired algorithms, also using Q-learning the environment can be simulated much easily [7].

Since the advent of AUGs much work has been done to improve their path planning and obstacle-avoidance abilities. Some experts of underwater glider published a series of works pertaining to the path planning and obstacle-avoidance of the glider. They mainly used graph-based methods for the same. The gliders used in International scenario are gulper, wave, spray, and sea glider [8]. It utilised various evolutionary algorithms for path planning and collision avoidance of an underwater vehicle [9]. The nature of computing of ACO algorithm was introduced several years ago. Experts also gives much insight on the PSO algorithm [10]. The research works is a result of an inspiration from the above-mentioned works and has proposed a method for path planning of the OUC-II using tabular Q-learning technique and have compared it with other optimisation algorithms.

2 OUC-II Glider

A long-term underwater glider called OUC-II, is developed by our lab. The main specifications of OUC-II are shown in Table 1.

Table 1. Main specifications of OUC-II

Feature	Description
Length	2.7 m
Diameter	240 mm
Weight in air	90 kg
Operating depth	0–1500 m
Ballast system	Piston-cylinder
Communications	Wireless and Iridium
Sensors	Altimeter, MTI, CTD, Depth Sensor, Hydrophone, GPS
Battery	Lithium battery
Operating in glider mode	0.50–0.75 m/s

Figure 1 shows the details internal structure of OUC-II and the real glider. As shown in Fig. 1(a), OUC-II can be divided into four sections: Bow Section, Main Section, Electronic Section, and Thruster Section. Specifically, the Bow Section is in contact with water and changes float of underwater glider, which holds an altimeter and the outside oil bag. The Main Section contains the attitude control system and the sealed part of the ballast system. The attitude control system regulates the pitch angles of the vehicle by moving an internal mass. The ballast system changes the net weight of the vehicle by pumping oil inside oil bag and outside oil bag the vehicle. The Main Section also contains the control and signal processing boards, the navigation devices as well as the battery. The GPS and communication terminals together with the antennas are all fixed in the Electronic Section.

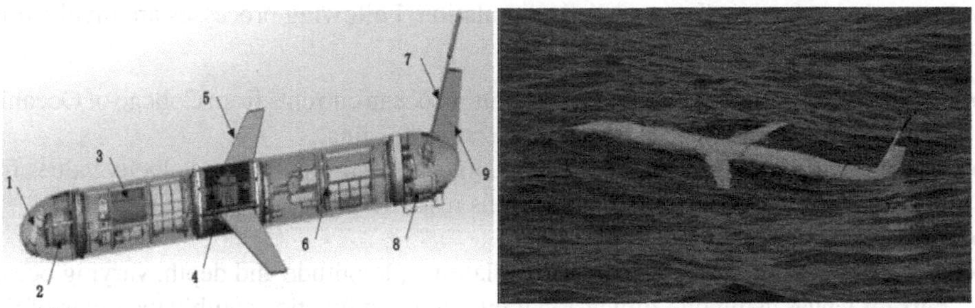

(a) (b)

Fig. 1. The details structure of the glider. (a) 1. Altimeter 2. Outside Oil-Bag 3. Inside Oil-Bag 4. Hydrophone 5. Horizontal wings 6. Posture adjustment mechanism 7. Communication antenna 8. CTD 9. Vertical tail. (b) Real glider

3 Cost Function

A cost function [11] was required to accurately describe the dynamics of the OUC-II and its interaction with the near-bottom ocean environment, especially with the near-bottom ocean currents. The work is inspired from the works of Joshua [11] and this research paper develops a cost function which takes into consideration the dynamics of Slocum, its principle of locomotion, and the spatially-varying velocity of ocean currents at near-bottom depths of the ocean. The cost function takes the input as the coordinates of the source node and the coordinates of the destination node and gives the output as time taken to travel from the source node to destination node.

The time taken from source node to destination node denoted by T is found using the equation of kinematics:

$$S = UT + \frac{1}{2}aT^2 \tag{1}$$

where S is the displacement, U is the initial velocity, a is the acceleration, and T is the time. Now as the OUC-II has low cruising speed (0.5 m/s–0.75 m/s), taking acceleration, a to be zero, this gives as

$$T = \frac{S}{U} \tag{2}$$

here T is the time taken from source node to destination node, S is the distance between the source node and destination node, and U is the average velocity with which the AUG travels from the source node to destination node.

3.1 Ocean Current Determination

The data for the near-bottom ocean current velocity is obtained. The data consist of the magnitude and direction of the ocean current velocity at various latitudes, longitudes and depths near the bottom surface of the ocean. These data are processed to obtain the values suitable for creating a realistic simulation. Following processes are involved in obtaining of the ocean current values:

- Obtaining the data regarding the near-bottom ocean currents from College of Oceanic and Atmospheric Science, Ocean University of China.
- Segregating the data to obtain separate databases of ocean current velocity values for different latitudes, longitudes, and depths respectively.
- Identifying the region of interest.
- Interpolating the separate databases of latitude, longitude and depth, varying ocean current values using one-dimensional interpolation functions within the range of the region of interest.
- The ocean current value at any particular coordinate is obtained by averaging the interpolated value of the ocean current at the required latitude, longitude, and depth.

As the data obtained has ocean current values at haphazard coordinates, it is necessary to interpolate within the range of region of interest to obtain ocean current values at regular intervals of the coordinates, thereby making the simulation more accurate.

3.2 Determination of Average Velocity

The velocity of the OUC-II at a particular coordinate, denoted by V_{ef} depends on its cruising speed, denoted by V_c, the magnitude and direction of the ocean current velocity, denoted by $V_{current}$ and the direction of the path from the previous node to the present node, denoted by V_{path} (for the source node the previous node is taken as the origin). Considering the shape of the OUC-II, the interaction of the ocean currents with it can be approximately modelled as an intersection between a line and a circle or a sphere. The point of intersection gives the V_{ef} as clarified by the following Equation:

$$Line: x(V_{ef}) = V_{ef} V_{path} \tag{3}$$

$$Circle/Sphere: V_c^2 = \|x - V_{current}\|^2 \tag{4}$$

$$D = (V_{path}^T \cdot V_{current})^2 + V_c^2 - V_{current}^T \cdot V_{current} \tag{5}$$

The discriminant given by Eq. (5) determines whether a particular path should be completely avoided or not, if the D becomes negative V_{ef} has no real value, hence, it would be Not a Number, *NaN*.

$$V_{ef} = V_{path}^{T} + \sqrt{D},\ for\, D > 0 \quad (6)$$

$$V_{ef} = NaN,\ otherwise \quad (7)$$

The average velocity $U(t)$ is determined by taking the average of the V_{ef1} and V_{ef2}, which are the effective velocities at source node and destination node, respectively.

$$U(t) = \frac{(V_{ef1} + V_{ef2})}{2} \quad (8)$$

3.3 Calculation of Distance

The distance between the source node and the destination node is calculated using the distance formula.

$$S = \sqrt{((x2 - x1)^2 + (y2 - y1)^2 + (z2 - z1)^2)} \quad (9)$$

where $x1, y1, z1$ are the coordinates of source node and $x2, y2, z2$ are the coordinates of destination node.

3.4 Dive Profile

The OUC-II changes its buoyancy to either ascend or descend within the waterbody. This defines its locomotion principle. Owing to its locomotion principle, the OUC-II has a characteristic dive profile which is sinusoidal in shape when observed in coarser approximation. As simulating the exact dive profile is computationally complex, we have approximated it and the result is a saw tooth shaped profile. For the sake of simulation we sample this dive profile and the discrete coordinates, thus obtained are given as input to the optimisation algorithms to establish shortest paths traversing all these points. By doing so we make sure that the characteristic dive profile of the OUC-II is accommodated in the path planning. The results presented in the paper also show that the optimised paths obtained from the algorithms are in sync with the dive profile of the OUC-II.

Before the cost function is optimised using the various heuristic algorithms, it is first optimised within itself. The OUC-II can divide its path from a source node to a destination node into various saw tooth shaped segments owing to its dive profile, a function within the cost function calculates the various time taken for the number of segments from one to the number of segments which is maximum. The minimum of the time taken is chosen as the final output of the cost function. As the division of a single path into n number of segments can be both advantageous and disadvantageous for the OUC-II depending on whether it avoids highly turbulent ocean currents or not, the pre-optimisation takes care that all the possible cases are considered, hence, making the simulation more realistic and the path planning more efficient.

4 Optimisation Algorithm

Once the cost function is developed, various optimization algorithms are used to minimise the cost function that is to minimise the time taken from source node to destination node and hence plan the shortest possible path from the source to the goal. Sampled discrete coordinates from the dive profile are chosen randomly and given to the optimisation algorithms. The results are observed to obey the dive profile of the OUC-II. These results are then compared. The inference from these comparisons is elucidated in the paper.

4.1 Genetic Algorithm

Genetic algorithm uses the techniques such as inheritance, mutation, selection and crossover recurrently till only the fittest individual, which is analogous to the best solution to the cost function remains and other less fit individuals or less optimised solutions wither away. For implementing the genetic algorithm firstly, a cost function must be defined that can evaluate the fitness of a solution. The potential solutions must be represented in a particular format (binary, real, gray coded, etc.). The Genetic Algorithm considers these potential solutions of the cost function as the genes of a chromosome and keeps modifying these randomly to get the fittest gene that is the optimised solution to the cost function. The pool of potential solutions is called a population [12]. This population generally consists of randomly generated solutions for the cost function. The crossover operator selects 'parents' from this population using Roulette Wheel method and randomly recombines these to produce 'offspring' which are also valid solutions of the cost function. If these 'offspring' solutions yield better results, these survive else these are discarded. The mutation operator also randomly selects a candidate from the population and makes random modifications to it, again this mutated solution is checked using the cost function, and if it yields optimised results, it survives, else it is discarded. Thus, imitating the principles of natural evolution, the most optimum result is selected by the genetic algorithm.

4.2 Ant Colony Optimisation Algorithm

Ant colony optimisation has been inspired by the inherent ability of the ants to find the shortest distance between their nest and their goal, which is mostly the food. Ants leave a trail of pheromone deposits wherever they travel. Assume a case where one ant leaves the nest in search of food, it takes random steps and finally reaches the food. Now, all along those random paths pheromone has been deposited. Now, as is obvious, once the ant has reached the goal, while returning, it takes the shortest path, therefore the amount of pheromone deposited in the shortest path is more as compared to other random paths.

Now when the subsequent ants travel to the goal, they sense the path with the maximum amount of pheromone. By doing so, not only do they find the shortest path but also deposit more pheromones on the shortest path compensating for any loss in pheromone deposits caused due to evaporation. To utilise this trait of an ant colony in computationally optimising cost functions, first a set of m random solutions to the cost function is chosen, these m random solutions represent m ants. Now each ant (solution) is evaluated on the basis of the output it begets from the cost function. According to this evaluation pheromone deposit concentration linked with the route taken by each ant is modified using the following equation:

$$\tau_{ij}(t) = \rho\tau_{ij}(t-1) + \Delta\tau_{ij}; t = 1, 2, 3 \ldots T \tag{10}$$

where T is the number of iterations, $\tau_{ij}(t)$ is the revised concentration of pheromone associated with path option I_{ij} at iteration t; $\tau_{ij}(t-1)$ is the concentration of pheromone at previous iteration $(t-1)$, $\Delta\tau_{ij}$ is the change in pheromone concentration and ρ is the pheromone evaporation rate having a value between zero and one. $\Delta\tau_{ij}$ is calculated using the following equation:

$$\begin{aligned} &\Delta\tau_{ij} = \textstyle\sum_{k}^{m} = 0\frac{R}{fitnessk} && \textit{if option } I_{ij} \textit{ is chosen by ant } k \\ &\Delta\tau_{ij} = 0 && \textit{otherwise} \end{aligned} \tag{11}$$

Here R is a constant called the pheromone reward factor and *fitnessk* is the value of inverse of the cost function for the kth ant (solution). After the pheromone deposit concentration is refreshed for one iteration, a different path is chosen for the next iteration, this choice of path is randomised using a roulette method of selecting the path.

4.3 Particle Swarm Optimisation Algorithm

Particle swarm optimisation algorithm owes its origin to the behavior of the birds in a flock while flying to reach a particular destination [13]. While flying in the search space, each bird of the flock looks in a particular direction and also each bird keeps on communicating with the other birds of the flock. This communication helps the birds to identify the one member of the flock which is at the best location with respect to the goal. Once this bird is identified, other birds fly towards the best location with a velocity that depends on their current position. After reaching the new positions, the birds again infer their positions to figure out the best position, this process is repeated until the flock reaches its destination. In this paper, authors have utilised the global optimising model proposed by Shi and Eberhart [13], given as:

$$\begin{aligned} V_{i+1} &= w * V_i + rand * C1 * (P_{best} - x_i) \\ &+ rand * C2 * (G_{best} - x_i) \end{aligned} \tag{12}$$

$$x_{i+1} = x_i + V_{i+1} \tag{13}$$

here V_i is the velocity of the ith member of the flock, x_i is its position, *C1* and *C2* are positive constant parameters called the acceleration coefficients, rand is uniformly distributed random number generating functions that generate random numbers in the range [0 1], P_{best} is the best position of the ith particle and G_{best} is best position among all particles of the flock and w is the inertial weight. Weight factor used is given by,

$$W_{i+1} = w_{\max} - \frac{w_{\max} - w_{\min}}{N_{iteration}} * i \tag{14}$$

where $N_{iteration}$ is the maximum number of iterations and w_{max} and w_{min} are maximum and minimum values of w, respectively.

4.4 Q-Learning Technique

Q-learning is a form of reinforcement learning which further belongs to the super class of machine learning. A computer system is said to learn from a data set denoted by *D* to perform the task denoted by *T* if after learning the system's performance on *T* improves as measured by a performance measurement index, denoted by *M*. Reinforcement learning utilises trial and error method to learn, it takes random actions to make a system achieve its goal and then it obtains feedback in the form of rewards or penalties that determine whether the action taken was correct or incorrect, therefore, after a number of iterations, the algorithm learns which steps are beneficial and which are not wrt to the goal to be achieved. This paper utilises Tabular Q-learning Technique to optimise the path between the source node and goal node. To implement the Q-learning, first the cost function is converted to a matrix form, termed as the reward matrix and denoted by R. The Q matrix was initialized as all zeroes. Now for a large number of iterations, the Q matrix is modified using the following equation:

$$Q(state, action) = R(state, action) + \gamma.Max(Q(next.state, all.action)) \tag{15}$$

The initial state is the coordinates of the source node and the next state is chosen randomly. Then for all possible paths, from the next state to the goal state, are checked and given rewards accordingly. After some iterations, the coordinates of the optimised path have the highest Q values, and hence, are chosen as the optimised path.

5 Results and Discussions

Random points sampled from the dive profile are given to the GA. These points include the starting node and goal node. The GA is programmed in MATLAB in such a way that it finds the shortest distance with reference to the cost function of the OUC-II, from the starting node to goal node, whilst traversing all the given random points and also finding its way back to the starting node. The results obtained show that not only does the GA find the shortest path, the path thus obtained also replicates the dive profile of the OUC-II. The time taken by the GA to compute the optimal function was 12.765700 CPU seconds. Figure 2 clearly shows the optimised path generated by the GA.

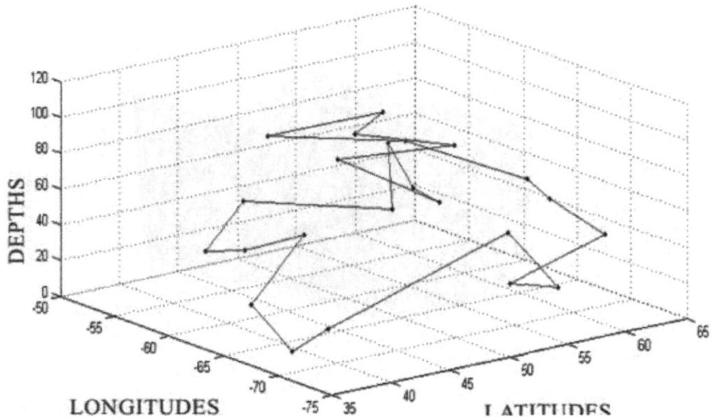

Fig. 2. The profile of glider optimised using genetic algorithm.

Similarly, random points sampled from the dive profile are given as input to the ACO algorithm. The starting node in this case is taken as origin. The ACO also gives a similar result with shortest path being evaluated incorporating all the random points given as input. The resulting path also was as expected from the dive profile of the OUC-II (evident from Fig. 3).

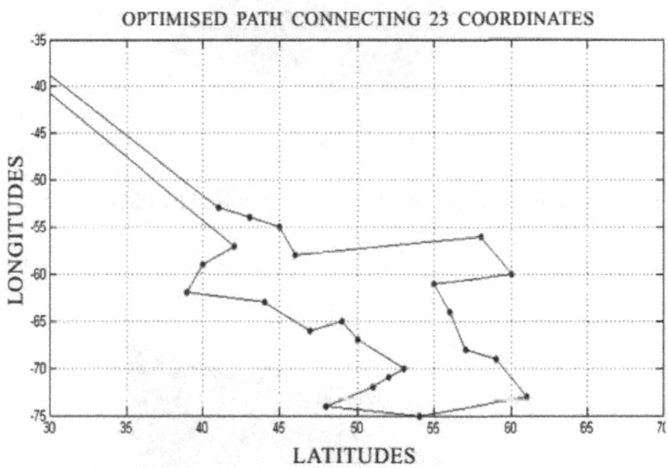

Fig. 3. Glider underwater flight profile optimised using ant colony optimisation.

The time taken by the ACO to compute the optimal function was 120.280000 CPU seconds. The Fig. 3 showcases the optimised path generated by the ACO algorithm. The value of ρ is taken as 0.15. The numbers of ants used in the simulation was 25 and the value of R used was 0.33. Figures 4, 5, and 6 show the impact of the PSO algorithm in optimizing Foxhole, Dejong F4 and f6 functions, respectively.

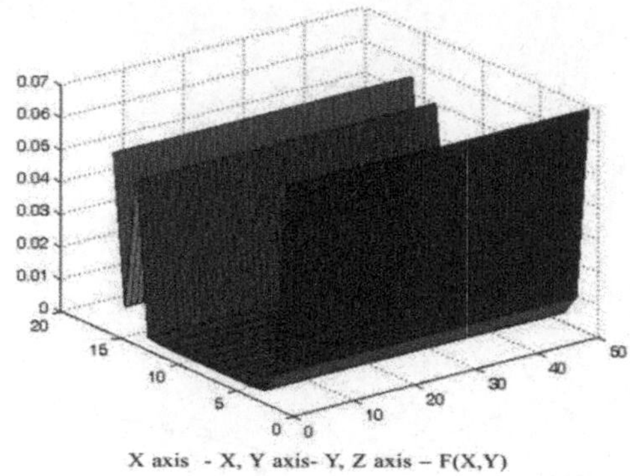

Fig. 4. Foxhole function optimised by PSO.

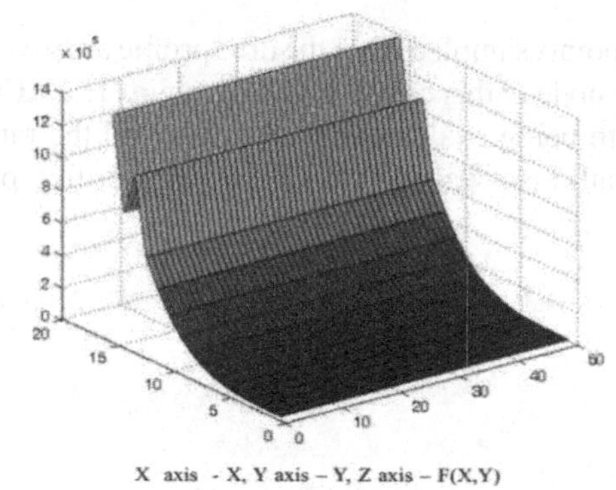

Fig. 5. Function Dejong F4 optimised by PSO.

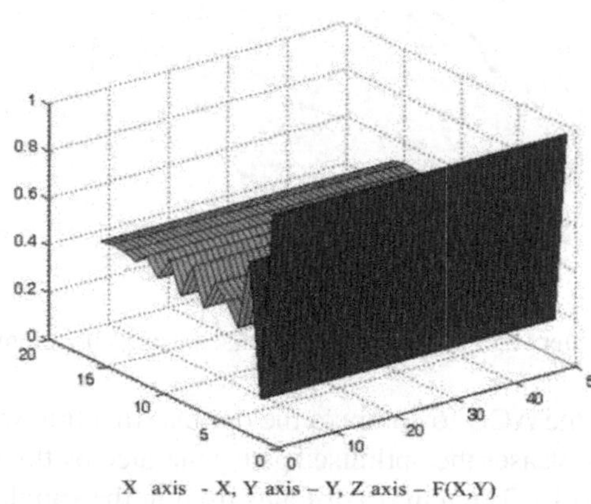

Fig. 6. Function f6 optimised by PSO.

It is evident from Fig. 7 that the cost function reaches saturation as the generation number increases. The number of particles of the swarm is taken as 40, the values of *C1* and *C2* are set as 2.5 each, the value of w_{max} is 0.1 and that of w_{min} is 0.9.

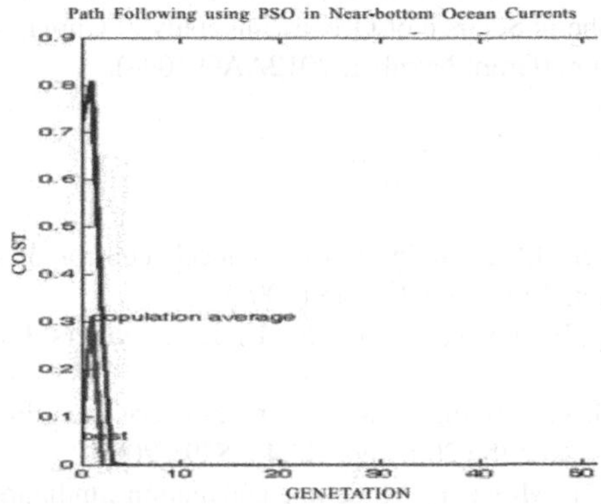

Fig. 7. Plot of cost function *vs* number of generations.

Figure 8 shows the maximum *Q* value of path obtained after standard one thousand iterations in red color and the other *Q* value of the tabular column shown in different colors. The coordinates corresponding to the maximum *Q* values (in the *Q* matrix) depict the shortest path with reference to the cost function as optimised by the Q-learning technique. The computational time taken by the Q-learning was 7.466598 CPU seconds. It is also seen from Fig. 8 that the *Q* values have saturated for only the optimised path after a standard number of iterations. The γ is used with a constant value of 0.8.

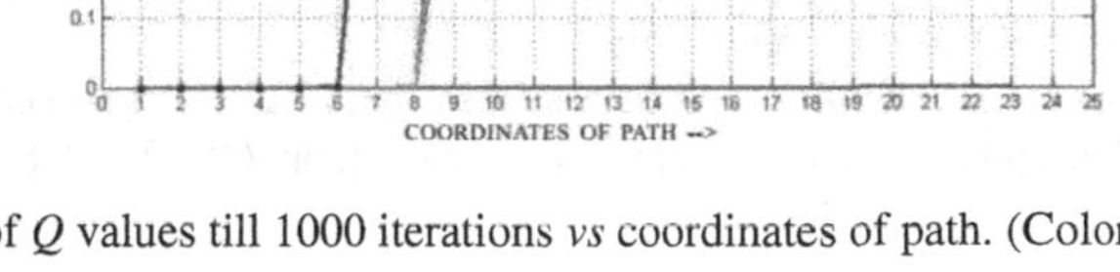

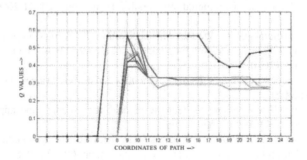

Fig. 8. Plot of *Q* values till 1000 iterations *vs* coordinates of path. (Color figure online)

6 Conclusion

It can be concluded from the paper that the Q-learning approach is better in path planning of the OUC-II to computational complexity and ease of environment simulation. The paper, therefore presents an effective method of path planning using tabular Q-learning technique. Concept of function approximation can be used instead of tabular Q-learning so as to improving computational complexity.

Acknowledgement. This work has been supported by the Underwater Glider Research Center and Oceanic and Atmospheric Science of Ocean University of China, 863 Plan Acoustic Glider System Development Team. (Grant Number: 2012AA091004).

References

1. Leonard, N.E., Graver, J.G.: Model-based feedback control of autonomous underwater gliders. IEEE J. Oceanic Eng. **26**, 633–645 (2001)
2. Griffiths, G., Jones, C., Ferguson, J., Bose, N.: Undersea gliders. J. Ocean Technol. **2**, 64–75 (2007)
3. Woolsey, C., Leonard, N.: Moving mass control for underwater gliders. In: American Control Conference, Proceedings of the 2002, pp. 2824–2829, (2002)
4. Arima, M., Ichihashi, N., Miwa, Y.: Modelling and motion simulation of an underwater glider with independently controllable main wings. In: OCEANS 2009-EUROPE, pp. 1–6 (2009)
5. Leighton, J.: System Design of an Unmanned Aerial Glider (UAV) for Marine Environmental Sensing. Massachusetts Institute of Technology, Cambridge (2013)
6. Aghababa, M.P.: 3-D path planning for underwater vehicles using five evolutionary optimisation algorithms avoiding static and energetic obstacles. Appl. Ocean Res. **38**, 48–62 (2012)
7. Dorigo, M., Maniezzo, V., Colorni, A.: Ant system: optimisation by a colony of cooperating agents. IEEE Trans. Sys. Man Cyber. Part B Cybern. **29**, 26–41 (1996)
8. Eichhorn, M.: A new concept for an obstacle-avoidance system for the AUV SLOCUM Glider operation under ice. In: Oceans 2009. IEEE, Bremen, Germany (2009)
9. Eichhorn, M.: Optimal path planning for AUVs in time varying ocean flows. In: 16th Symposium on Unmanned Untethered Submersible Technology, UUST 2009, Durham, NH, USA (2009)
10. Eichhorn, M., Williams, C.D., Bachmayer, R., deyoung, B.: A mission planning system for the AUV Slocum Glider for the Newfoundland and Labrador Shelf. In: Oceans 2010. IEEE, Sydney, Australia (2010)
11. Joshua, G.G.: Underwater gliders: Dynamics, control and design. Princeton University, (PhD, Dissertation) (2005)
12. Man, K.F., Tang, K.S., Kwong, S.: Genetic algorithms: concepts and applications. IEEE Trans. Ind. Electro. **43**(5), 519–534 (1996)
13. Shi, Y., Eberhart, A.R.: Modified particle swarm optimiser. In: Proceedings of the IEEE International Conference on Evolutionary Computation, pp. 69–73 (1998)

RETRACTED CHAPTER: The Summary of Underwater Gliders Control Strategies

Yuhai Liu[1], Xin Luan[1], Dalei Song[2], and Zhiqiang Su[1](✉)

[1] College of Information Science and Engineering, Ocean University of China, Qingdao, China
szqouc@126.com

[2] College of Engineering, Ocean University of China, no. 238, Songling Road, Laoshan, Qingdao, Shandong, China

Abstract. The underwater glider is a new type of system that combines buoy technology and underwater robotics technology. By changing the net buoyancy as a driving force and changing its own center of gravity position, the attitude angle can be changed. Many advantages, such as gliding with high efficiency, wide cruise range, less power consumption, low noise, and no pollution, make the underwater glider an important platform for marine environment observation and ocean resource exploration. In this paper, control strategies for existing underwater gliders are reviewed. A total of 3[illegible] papers indexed by Scopus with keywords control and underwater gliders were reviewed from 1989 to 2016. The majority of gliders use classical controllers, which cannot dynamically compensate for un-modeled hydrodynamic forces and unknown variations in water current and wind. With increasing operational depths and larger payloads, control strategies will become an increasingly important aspect for these gliders. Control strategies implemented in underwater gliders were reviewed and alternative control strategies are proposed.

Keywords: Underwater gliders · Control strategies · Summary

1 Introduction

An underwater glider is a type of autonomous underwater vehicle (AUV) that uses buoyancy control in conjunction with wings to convert vertical motion to horizontal motion in a saw-tooth pattern and thereby propelling itself in forward direction with very

---

The editors have retracted this chapter [1] because it shows significant overlap with a previously published chapter by Ullah et al, 2015 [2]. The contents of this chapter are therefore redundant. All of the authors agree to this retraction.

1. Liu, Y., Luan, X., Song, D., Su, Z.: The Summary of Underwater Gliders Control Strategies. In: Huang, Y. et al. (Eds.) ICIRA 2017, Part I, LNAI, vol. 10462, pp. 813–824. Springer International Publishing (2017). doi: 10.1007/978-3-319-65289-4_75
2. Ullah, B., Ovinis, M., Baharom, M.B., Javaid, M.Y., Izhar, S.S.: Underwater Gliders Control Strategies: A Review. In: 10th Asian Control Conference (ASCC). IEEE (2015). doi: 10.1109/ASCC.2015.7244859

The erratum to this chapter is available at https://doi.org/10.1007/978-3-319-65289-4_81

Y. Huang et al. (Eds.): ICIRA 2017, Part I, LNAI 10462, pp. 813–824, 2017.
DOI: 10.1007/978-3-319-65289-4_75

low power consumption [1]. They use buoyancy and wings to produce forward motion for long duration ranging from weeks to months and cover distances over 3000 km [2]. There are more than 50 different AUGs developed for research and commercial purposes. Much theoretical and experimental analysis has been carried out on glider dynamics [3, 4], glider design and development for payload and maneuverability [5, 6], performance and stability analysis [7, 8], trajectory control analysis [9, 10] and path planning [9, 11] based on different controller techniques [12]. The majority of these gliders use classical controllers such as Proportional Integral-Derivative (PID) and Linear Quadratic Regulator (LQR). However, These control methods cannot provide simultaneously easy control implementation and fast convergence speed for stabilization in the presence of continuously varying water current [13].

In this paper, existing underwater glider control strategies are reviewed with emphasis on their actuation system that includes wings and rudder control, followed by external effects of water current and variable loads which influence their motion.

No work to date achieved the control of fully under actuated vehicles via dynamic modelling with the presence of external varying forces such as water current and advanced controller design to compensate these problems and then implement them to underwater gliders. In order to overcome these issues, serious design and implementation considerations are required for robust control approach.

2 Underwater Glider Motion Control

The motion of an underwater gliders depends upon a buoyancy control system combined with a set of wings and rudders to move in vertical and horizontal direction through a water column [14]. The glider descends and ascends in water column due to control of buoyancy engine, which consists of oil filled bladder or piston evacuated cylinder. Buoyancy is controlled to be slightly positive or negative due to which a glider can move with almost zero power consumption with limited speed [15]. For example, the Spray glider as shown in Fig. 1, has a maximum speed of 0.35 m/s but can operate for up to six months.

Fig. 1. 1-Slocum [8], 2-Spray [16], 3-Seaglider [8]

Regardless of the design, gliders depend on vertical space to operate efficiently but cannot resist with strong currents, which ultimately creates a problem to control motion and trajectory. On the other hand, the average horizontal speed of underwater gliders is about 0.3 m/s. The low speed limitation has created challenges in controller performance such as control of vehicle orientation, position and velocity [14]. There is a direct effect of environmental forces including winds, wave and surface currents upon

Table 1. Laboratory scale gliders control controllers

Type of Glider	Institution	Micro controllers	Speed & depth	Remarks
ALBAC glider	University of Tokyo	Two NEC V50 CPUs controlled by math processor Intel 80 [8]	0.5 to 1.0 m/s, 300 m	Trajectory is controlled by changing pitch and roll angle
ZJU glider	Zhejiang University (ZJU) glider	TMS320F28335 controller for real time motion control [4]. AMESim software is used to investigate dynamic performance of buoyancy control system	0.6 m/s, 200 m	Dynamic analysis, stability analysis and root locus for varying ballast mass for pitch control are performed by ZJU glider
Sea-Wing underwater glider	Shenyang Institute of Automation	NXP LPC series ARM processor chip with ARM7TDMI-S based high-performance 32-bit RISC Microcontroller [17]	0.8 m/s,1200 m	The buoyancy-regulating mechanism alters the buoyancy of underwater glider by feeding hydraulic oil into or bleeding hydraulic oil out of the external oil bladder to produce the driving buoyancy
Seaglider	University of Washington	Onset Computer Corporation's TT8 controller. Motorola MC68332 microcontroller combined with a 12-bit A/D converter [18]	0.25 m/s,1500 m	Pitch and roll are controlled by using 16 mm Maxon neodymium magnet motors
Alex glider	Department of Marine System engineering Osaka Prefecture University	MAVC2 (Micro Aerial Vehicle Controller which has the ability to control 12 servo motors and a program to control the motion of glider is stored in this controller [7]	0.6 m/s, 200 m.	Pitch and roll motion is controlled by twisting of wings and steering of rudder
USM glider	USM	Arduino Mega (Primary controller) PC 104 (Secondary controller) LQR, MPC, and NNPC. [19]	0.4 m/s, 30 m	Internal sliding mass and ballast pump which are used as internal actuators control the pitch and buoyancy. Wings and rudder control maneuverability

(continued)

Table 1. (*continued*)

Type of Glider	Institution	Microcontrollers	Speed & depth	Remarks
Miniature underwater glider prototype (A robotic Fish (Grace))	Developed in the Smart Microsystems Laboratory at MSU	dsPIC[illegible]F6[illegible]4A [illegible]us program and provide st[illegible]rage memory. IMU including gyros(ST [illegible]Y503AL), accelerometer and [illegible]igital compass (ST LSM303DLMRT) are [illegible] equipped [13, 20]	0.36 m/s, 20 m	Movable battery mechanism for Pitch control. Pump is used with internal cylindrical tank system. For sealing O-rings and for wings Aluminum wings are used
FOLAGA	Combined research effort: IMEDEA, ISME and University of Genova, Italy [8]	PC104 card, GPS receiver, a depth sensor inclinometer [illegible]	1.01 m/s, 50 m	A robust PD control system followed by back-stepping methodology is implemented by FOLAGA
OUC glider	Ocean University of China	STM32F107 controller for underwater glider based high-performance 32-bit RISC Microcontroller [51]	[illegible] 1500 m	Fuzzy-PID controller was designed to control the pitch angle by OUC glider

the motion of underwater gliders and AUVs [21]. These effects may be compensated through an effective control strategy. Therefore the control strategy should be chosen to reduce the effect of these disturbances. However, the challenges associated with control of underwater gliders include a highly nonlinear dynamic system and complex hydrodynamic effects. Different control strategies for laboratory scale and commercial scale have been implemented to underwater vehicles which are discussed in Table 1.

The ALBAC glider used its payload to glide downward and upon releasing its payload, glided upward [2]. However most gliders such as Slocum [22], Spray [23] and Seaglider [18] use a buoyancy engine instead. These three gliders are fixed winged buoyancy propelled gliders which shift their internal ballast mass to control their attitude [1].

Leonard and Graver [1] developed a laboratory scale ROGUE glider to implement a model based feedback open loop approach. Another laboratory scale glider, the ALEX glider, developed by Arima et al. [7] designed an independently controllable wings and rudder (Fig. 2).

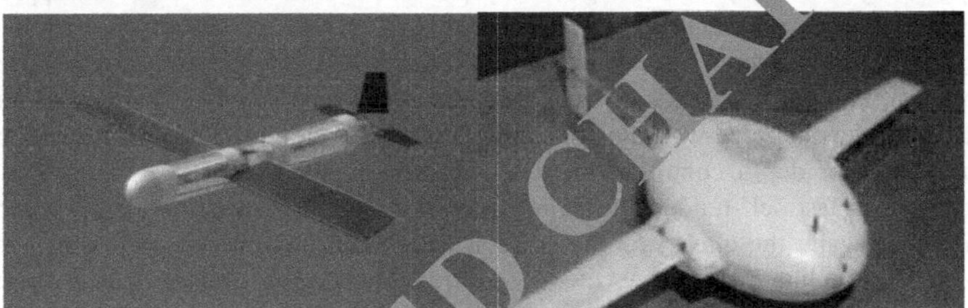

Fig. 2. Laboratory scale gliders, ALEX glider [7] and ROGUE glider [1]

3 Control Strategies

Several control strategies have been proposed to control the position, velocity and trajectory of UGs such as PID controllers by Bhatta and Noh [24, 25], Linear Quadratic Regulator (LQR) by Nina and Graver [9, 26], Sliding Mode Controllers (SMC) by Yang and Song [27, 28], Neural Network (NN) by Isa [19] and Dong [29], Fuzzy logic for self-tuning by Loc [30] and Model Predictive Controllers by Cooney [31] and Steenson [32]. PID control loops used for pitch and roll control by Panish [16] allows their UG to be guided in a variety of ways. In such control strategies, vertical and horizontal modes are decoupled to control the pitch, angle of attack and depth of the glider. The robustness of PID controller using nonlinear time invariant aspects of glider designs is investigated by Bender [33]. AUG dynamics are nonlinear but PID controllers are based-on linear control law. Therefore, PID controller cannot dynamically compensate for un-modeled vehicle hydrodynamic forces and unknown disturbances. As such, combinations of control strategies to form hybrid controllers are used to handle the complexity that is produced due to nonlinearity and hydrodynamics effect.

Fuzzy-PID controller for underwater glider pitch control has been used by Liu and Su [51]. Fuzzy-PID can adjust the control parameters to adapt to the change of environment. It can keep the stability of underwater glider. However, fuzzy rules, fuzzification and defuzzification are very complex.

Sliding mode control (SMC) for underwater trajectory tracking control has been used by Yang and Bessa [27, 34]; and Liu [21]. SMC is a robust system with low sensitivity to environmental disturbances. However, chattering is a well known downside of this control strategy, which can degrade the performance of gliders and may lead to instability.

A multivariable Multiple input Multiple Output (MIMO) control system have been proposed for AUVs and ROVs including self-tuning and position control by Goheen [35]. The dynamic model using LQR technique was implemented in a laboratory scale ROGUE glider in order to examine its motion. However, LQR has poor switching performance between upward and downward glide paths [26].

In a comparative study, Nag [36] discussed the use of an adaptive fuzzy control strategy for tracking trajectory system combined with PID controller. However, they considered the glider as a SISO system only and neglected nonlinearities. A recurrent NN based on MPC control strategy for motion control problem in longitudinal plane is proposed by Shan [37] for convergence to optimal solution. However, in this approach the effect of water current has not been considered.

Due to the inherent non-linearity in the dynamics, and hence difficulties in determining an accurate mathematical model of the vehicle, an adaptive fuzzy logic based control strategy coupled with dynamic compensators have been designed. The effect of fuzzy logic for depth controller has been implemented on KAUV-1 which has the advantage of fast self-tuning [30]. The simulation results showed satisfactory performance but again, the effect of external disturbance and nonlinearity were not investigated. The nonlinearity of the Odyssey IV for the purpose of disturbance rejection using Model predictive controller was discussed by Cooney [31] and a hybrid PIDMPC controller is suggested for improving and predicting future values. However, a communication problem between commanded input and actual input was observed during implementation of this strategy in Odyssey IV. In [27], Yang addressed the robustness issue produced due to nonlinearities and proposed the adaptive sliding mode controller. However, the resulting chattering effect lead to instability.

Neural network (NN) control strategy [19] have been proposed as a solution to the lack of robustness due to nonlinear effect in dynamic models. Due to the ability to handle nonlinearity and adapt to changing conditions as is the case of varying currents in glider dynamics, NN has been implemented in USM hybrid underwater glider [19]. In proposed model, all glider parameters e.g. pitch, speed, maneuverability, sliding mass, wing, rudder and propeller were controlled. A neural network self-tuning PID controller [29] for a spherical AUV was proposed for the purpose of controlling the vehicle velocity and tracking the desired target in the vehicle fixed coordinate system.

4 Hybrid Control Strategy

The performance of individual control strategies can be improved by combining it with another control strategy to form a hybrid control strategy. Different implementations of this concept have been applied successfully in AUVs such as Adaptive fuzzy sliding mode control strategy.

4.1 Combining Fuzzy Logic and Sliding Mode

Fuzzy logic has been used to design a sliding mode controller [28, 38]. However, the FLC was implemented in a SISO system only. Alternatively, PID and FLC have been combined to design a new controller. To eliminate the chartering effect of sliding mode, a boundary layer is introduced around the sliding surface by Song [28]. A general block diagram of adaptive fuzzy sliding mode controller is shown in Fig. 3.

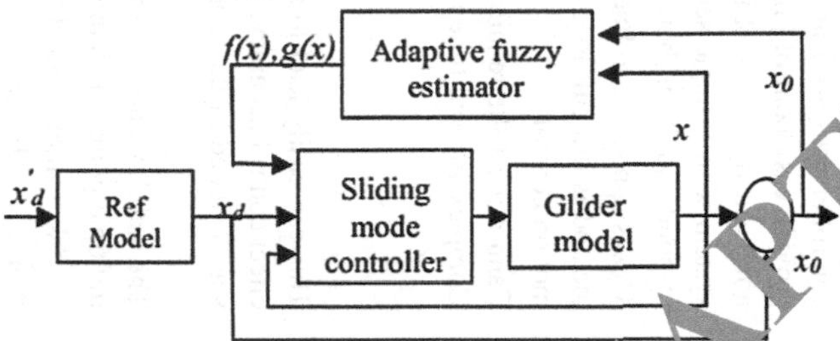

Fig. 3. Adaptive fuzzy sliding mode control [38]

4.2 Sliding Mode Fuzzy Controller

In this approach, sliding mode is used to design functions of the fuzzy logic controller and a switching function is formed which can eliminate the chattering effect. This type of approach was experimentally applied to [illegible]EX series AUVs for pitching and heading control. The inputs to SMFC heading controller were defined as heading error while the output for heading controller was to control the rudder [39].

The control strategy can be further improved using a fuzzy sliding mode controller has only been investigated for combination of nonlinear controllers for the motion of the Snorkel AUV [38]. Recent work concerning different autonomous underwater gliders based on added mass and control techniques for AUVs and AUGs is summarized in inertia effect of external varying disturbances and rudder Table 2 control inputs [40]. However the effectiveness of an adaptive.

5 Major Limitation and Challenges

Very few control schemes are proposed to measure the sensitivity analysis of environmental disturbances such as Isa [12] and Fan [50], but this is a new idea in undersea vehicles to consider both water current estimation and payload factor in dynamic modelling and control for UGs. The main attempt is to choose a controller that can guarantee the suppression or at least the limitation of the overshoot in the system response. An appropriate control strategy is required that will damp out the external disturbances and will follow trajectory with minimum error for varying payloads.

Table 2. Control strategies

Controller design	Implementation	Advantages	Limitation
PID control [41, 44]	Bluefin AUV [16] Seaglider [45], SLOCUM glider ALEX glider	The system adopts itself in a limited time and has the ability to decouple the vertical and horizontal modes. [16] PID controller regulator is used to follow the desired trajectory for Seaglider	The AUG dynamics are nonlinear and that a PID controller uses a linear control law and cannot provide easy control implementation due to nonlinear dynamics: It is difficult to tune the system with a single set of control gains to respond equally in both directions [16]
LQR [46] feed forward and feedback control	ROGUE Laboratory scale glider [47],	Using this method, comparatively less tuning is required. It has the ability to handle the MIMO system through state space representation	This is applicable to linear control laws only. This method is difficult to implement on unsteady motion controls
Neural network control [19, 29]	USM hybrid underwater glider, PETREL, hybrid underwater glider	It has the ability to handle the MIMO system with low rates of control update. Two planes, vertical and horizontal, can be controlled using a single controller [19]. Fast convergence and self-tuning effect	Cannot meet the requirement of rapid response. Complex for real time application
Sliding mode control [27, 39]	NPS ARIES AUV Grace(Gliding robotic fish)	Stability of SMC is in-built. Invariance property and ability to decouple high dimensional system into subsystems with lower dimensionality	This can easily lead to system chattering and effect control accuracy. Limited disturbance effect was included in the model [27]
Fuzzy logic control [36, 39]	KAUV-1 [36]	Needs expert knowledge to optimize. Easily implemented on several AUVs,	The stability and overshoot problem occurred at multiple points, including tuning fuzzy rules [48].
Combination of fuzzy logic and sliding mode AFSMC) [28]	OEX Series AUVs Snorkel AUV [38]	The errors found in FLC and SMC is easily eliminated by using combined effect of both of these. Chattering effect is minimized [34]	This was implemented on experimental AUV for limited depth. [28]

(*continued*)

Table 2. *(continued)*

Controller design	Implementation	Advantages	Limitation
Nonlinear model predictive control (MPC)	Odyssey IV (AUV) USM	MPC vs PID controller comparison and PID-MPC hybrid controller is designed for minimum cost. It has the ability to handle constrains and nonlinear systems	There was a disconnect problem between actual input and command input due to communication problem with the software during implementation of MPC
Fuzzy-PID control [51]	OUC glider	Fuzzy-PID controller can solve non-linear and linear control problem. It has the ability to finish the control of complex system	Fuzzy control rules, fuzzification and defuzzification are very complex
Nonlinear, passivity-based controller [13, 49]	Miniature UWG Grace(Gliding robotic fish)	Fast convergence, Easy implementation	

6 Conclusion

Different control strategies have been demonstrated for acceptable control performance in different AUVs and AUGs, but they still have limitations in compensating for the high nonlinearity of underwater gliders including external disturbances and carrying large payloads which are the main reasons that make it a challenging problem to control the underwater glider. Due to this it is highly desirable to design a controller that is robust and will give real time response. For this, combinations of two or more controllers have been developed by some researchers for AUVs to form a hybrid controller which can minimize the negative effect and will improve the performance.

Acknowledgement. This work has been supported by the Underwater Glider Research Center and Oceanic and Atmospheric Science of Ocean University of China, 863 Plan Acoustic Glider System Development Team (Grant Number: 2012AA091004).

References

1. Leonard, N.E., Graver, J.G.: Model-based feedback control of autonomous underwater gliders. IEEE J. Oceanic Eng. **26**, 633–645 (2001)
2. Griffiths, G., Jones, C., Ferguson, J., Bose, N.: Undersea gliders. J. Ocean Technol. **2**, 64–75 (2007)
3. Woolsey, C., Leonard, N.: Moving mass control for underwater vehicles. In: Proceedings of the 2002 American Control Conference, pp. 2824–2829 (2002)
4. Fan, S.-S., Yang, C.-J., Peng, S.-L., Li, K.-H., Xie, Y., Zhang, S.-Y.: Underwater glider design based on dynamic model analysis and prototype development. J. Zhejiang Univ. Sci. **2013**, 07–12 (2013)
5. Rudnick, D.L., Davis, R.E., Eriksen, C.C., Fratantoni, D.M., Perry, M.J.: Underwater gliders for ocean research. Marine Technol. Soc. J. **38**, 73–84 (2004)
6. Javaid, M.Y., Ovinis, M., Nagarajan, T., Hashim, F.B.: Underwater gliders: a review. In: MATEC Web of Conferences, p. 02020 (2014)
7. Arima, M., Ichihashi, N., Miwa, Y.: Modelling and motion simulation of an underwater glider with independently controllable main wings. In: OCEANS 2009-EUROPE, pp. 1–6 (2009)
8. Wood, S.: Autonomous underwater gliders. Underwater Vehicles, pp. 499–524 (2009)
9. Mahmoudian, N.: Efficient motion planning and control for underwater gliders. Virginia Polytechnic Institute and State University (2009)
10. Isa, K., Arshad, M.R.: Motion simulation for propeller-driven USM underwater glider with controllable wings and rudder. In: 2011 2nd International Conference on Instrumentation Control and Automation (ICA), pp. 316–321 (2011)
11. Yang, H., Ma, J.: Nonlinear feedforward and feedback control design for autonomous underwater glider. J. Shanghai Jiaotong Univ. (Sci.) **16**, 11–16 (2011)
12. Isa, K., Arshad, M.: Modeling and motion control of a hybriddriven underwater glider. IJMS **42**, 971–979 (2013)
13. Zhang, F., Tan, X., Khalil, H.K.: Passivity-based controller design for stablization of underwater gliders. In: American Control Conference (ACC), pp. 5408–5413 (2012)

14. Claus, B.R., Bachmayer, Cooney, L.: Analysis and development of a buoyancy-pitch based depth control algorithm for a hybrid underwater glider. In: Autonomous Underwater Vehicles (AUV), 2012 IEEE/OES, pp. 1–6 (2012)
15. Leighton, J.: System design of an unmanned aerial vehicle (UAV) for marine environmental sensing. Massachusetts Institute of Technology (2013)
16. Panish, R.: Dynamic control capabilities and developments of the Bluefin robotics AUV fleet. In: Proceedings of the International Symposium on Unmanned Untethered Submersible Technology (UUST), pp. 23–26, August 2009
17. Yu, J.-C., Zhang, A.-Q., Jin, W.-M., Chen, Q., Tian, Y., Liu, C.-J.: Development and experiments of the Sea-Wing underwater glider. China Ocean Eng. **25**, 721–736 (2011)
18. Eriksen, C.C., Osse, T.J., Light, R.D., Wen, T., Lehman, T.W., Sabin, P.L., et al.: Seaglider: a long-range autonomous underwater vehicle for oceanographic research. IEEE J. Oceanic Eng. **26**, 424–436 (2001)
19. Isa, K., Arshad, M.R.: Neural network control of buoyancy-driven autonomous underwater glider. In: Sen, G.G., Bailey, D., Demidenko, S., Carnegie, D. (eds.) Recent Advances in Robotics and Automation. Studies in Computational Intelligence, vol. 480, pp. 15–35. Springer, Berlin (2013). doi:10.1007/978-3-642-37387-9_2
20. Zhang, F., Tan, X.: Passivity-based Stabilization of Underwater Gliders with a Control Surface. J. Dyn. Syst. Meas. Control (2014)
21. Naeem, W.: Model predictive control of an autonomous underwater vehicle. In: Proceedings of UKACC 2002 Postgraduate Symposium, Sheffield, UK, September, pp. 19–23 (2002)
22. Webb, D.C., Simonetti, P.J., Jones, C.P.: SLOCUM: an underwater glider propelled by environmental energy. IEEE J. Oceanic Eng. **26**, 447–452 (2001)
23. Sherman, J., Davis, R., Owens, W., Valdes, J.: The autonomous underwater glider Spray. IEEE J. Oceanic Eng. **26**, 437–446 (2001)
24. Bhatta, P., Leonard, N.E.: Stabilization and coordination of underwater gliders. In: Proceedings of the 41st IEEE Conference on Decision and Control, pp. 2081–2086 (2002)
25. Noh, M.M., Arshad, M.R., Mokhtar, R.M.: Depth and pitch control of USM underwater glider: performance comparison PID vs LQR. Indian J. Geo-Mar. Sci. **40**, 200–206 (2011)
26. Graver, J.G.: Underwater gliders: Dynamics, control and design. Citeseer (2005)
27. Yang, H., Ma, J.: Sliding mode tracking control of an autonomous underwater glider. In: 2010 International Conference on Computer Application and System Modeling (ICCASM), pp. V4-555–V4-558 (2010)
28. Song, F., Smith, S.M.: Combine sliding mode control and fuzzy logic control for autonomous underwater vehicles. In: Bai, Y., Zhuang, H., Wang, D. (eds.) Advanced Fuzzy Logic Technologies in Industrial Applications. Advances in Industrial Control, pp. 191–205. Springer, London (2006). doi:10.1007/978-1-84628-469-4_13
29. Dong, E., Guo, S., Lin, X., Li, X., Wang, Y.: A neural networkbased self-tuning PID controller of an autonomous underwater vehicle. In: 2012 International Conference on Mechatronics and Automation (ICMA), pp. 898–903 (2012)
30. Loc, M.B., Choi, H.-S., Kim, J.-Y., Kim, Y.-H., Murakami, R.-I.: Design of Fuzzy PD Depth Controller for an AUV (2013)
31. Cooney, L.A.: Dynamic response and maneuvering strategies of a hybrid autonomous underwater vehicle in hovering. DTIC Document (2009)
32. Steenson, L.V., Turnock, S., Phillips, A., Furlong, M.E., Harris, C., Rogers, E., et al.: Model predictive control of a hybrid autonomous underwater vehicle with experimental verification. In: Proceedings of the Institution of Mechanical Engineers, Part M: Journal of Engineering for the Maritime Environment, p. 1475090213506185, (2014)

33. Bender, A., Steinberg, D.M., Friedman, A.L., Williams, S.B.: Analysis of an autonomous underwater glider. In: Proceedings of the Australasian Conference on Robotics and Automation, p. 110 (2008)
34. Bessa, W.M., Dutra, M.S., Kreuzer, E.: Depth control of remotely operated underwater vehicles using an adaptive fuzzy sliding mode controller. Robot. Auton. Syst. **56**, 670–677 (2008)
35. Goheen, K.R., Jefferys, E.R.: Multivariable self-tuning autopilots for autonomous and remotely operated underwater vehicles. IEEE J. Oceanic Eng. **15**, 144–151 (1990)
36. Nag, A., Patel, S.S., Akbar, S.: Fuzzy logic based depth control of an autonomous underwater vehicle. In: 2013 International Multi-Conference on Automation, Computing, Communication, Control and Compressed Sensing (iMac4 s), pp. 117–123 (2013)
37. Shan, Y., Yan, Z., Wang, J.: Model predictive control of underwater gliders based on a one-layer recurrent neural network. In: 2013 Sixth International Conference on Advanced Computational Intelligence (ICACI), pp. 328–333 (2013)
38. Sebastián, E.: Adaptive fuzzy sliding mode controller for the snorkel underwater vehicle. In: Nolfi, S., Baldassarre, G., Calabretta, R., Hallam, John C.T., Marocco, D., Meyer, J.-A., Miglino, O., Parisi, D. (eds.) SAB 2006. LNCS, vol. 4095, pp. 855–866. Springer, Heidelberg (2006). doi:10.1007/11840541_70
39. Song, F., Smith, S.M.: Design of sliding mode fuzzy controllers for an autonomous underwater vehicle without system model. In: Oceans 2000 MTS/IEEE Conference and Exhibition, pp. 835–840 (2000)
40. Sebastián, E., Sotelo, M.A.: Adaptive fuzzy sliding mode controller for the kinematic variables of an underwater vehicle. J. Intell. Rob. Syst. **49**, 189215 (2007)
41. Graver, J.G., Bachmayer, R., Leonard, N.E., Fratantoni, D.M.: Underwater glider model parameter identification. In: Proceedings of the 13th International Symposium on Unmanned Untethered Submersible Technology (2003)
42. Isa, K., Arshad, M.: Buoyancy-driven underwater glider modelling and analysis of motion control. Indian J. Geo-Mar. Sci. **41**, 516–526 (2012)
43. Gonzalez, L.A.: Design, modelling and control of an autonomous underwater vehicle. BE Thesis, The University of Western Australia, Australia (2004)
44. Mahmoudian, N., Woolsey, C.: Underwater glider motion control. In: 47th IEEE Conference on Decision and Control. CDC 2008, pp. 552–557 (2008)
45. Sliwka, J., Clement, B., Probst, I.: Sea glider guidance around a circle using distance measurements to a drifting acoustic source. In: 2012 IEEE/RSJ International Conference on Intelligent Robots and Systems (IROS), pp. 94–99 (2012)
46. Kan, L., Zhang, Y., Fan, H., Yang, W., Chen, Z.: MATLAB-based simulation of buoyancy-driven underwater glider motion. J. Ocean Univ. Chin. **7**, 113–118 (2008)
47. Graver, J.G., Leonard, N.E.: Underwater glider dynamics and control. In: 12th International Symposium on Unmanned Untethered Submersible Technology, pp. 1742–1710 (2001)
48. Azis, F., Aras, M., Rashid, M., Othman, M., Abdullah, S.: Problem identification for underwater Remotely Operated Vehicle (ROV): a case study. Procedia Eng. **41**, 554–560 (2012)
49. Zhang, F., Tan, X.: Nonlinear observer design for stabilization of gliding robotic fish. In: American Control Conference (ACC), pp. 4715–4720 (2014)
50. Fan, S., Woolsey, C.A.: Dynamics of underwater gliders in current. Ocean Eng. **84**, 249–258 (2014)
51. Liu, Y.-H., Su, Z.-Q., Luan, X., Song, D.-L., Han, L.: Motion analysis and fuzzy-PID control algorithm designing for the pitch angle of an underwater glider. J. Math. Comput. Sci. **17**, 133–147 (2017)

System Construction for Distributedly Controlling the Thrusters of X4-AUV

Xiongshi Xu[1(✉)], Keigo Watanabe[2], and Isaku Nagai[2]

[1] Division of Mechanical and Systems Engineering,
Graduate School of Natural Science and Technology, Okayama University,
3-1-1 Tsushima-naka, Kita-ku, Okayama 700-8530, Japan
xu@usmm.sys.okayama-u.ac.jp

[2] Department for Intelligent Mechanical Systems,
Graduate School of Natural Science and Technology, Okayama University,
3-1-1 Tsushima-naka, Kita-ku, Okayama 700-8530, Japan
http://usm.sys.okayama-u.ac.jp/index.html

Abstract. A small-sized X4-AUV is a kind of AUVs that do not use any rudders, and is driven by four thrusters arranged equally around the fuselage. In order to drive the X4-AUV more efficiently and precisely, a distributed control system of thrusters is proposed, in which the command of a main microcontroller is distributed to four sub-microcontrollers, one of which is mounted on each thruster. By utilizing the serial communication between the PC and the sub-microcontroller, the P, I and D gains of thruster can be easily adjusted on a Graphical User Interface (GUI). After the gains are adjusted, the performance of distributed control of the thrusters can be verified in real time through another designed GUI.

Keywords: Distributed control · Thruster · AUV · GUI

1 Introduction

Recently, the research and development of underwater robots have been prosperously conducted, especially in the field of replacing human beings to do the research of marine resources, ocean observations, or to do any work at the relatively shallow waters, and so on. Among such underwater robots, there is a kind of Remotely Operated Vehicle (ROV) equipped with power cables and communication cables. However, it has a limit of activity area according to those cables. Therefore, the research and development of Autonomous Underwater Vehicle (AUV), which does not use any cables and has no limit of activity area, is expected to become promising [1].

Multiple thrusters and rudders are commonly utilized by most of AUVs. However, when cruising at a low speed, the rudders arranged around the surface of the fuselage will not be efficient for the position and attitude control of such kinds of AUVs. From this fact, Watanabe et al. [2–4] proposed an X4-AUV that controls the position and attitude of the airframe only through the output of four

Y. Huang et al. (Eds.): ICIRA 2017, Part I, LNAI 10462, pp. 825–833, 2017.
DOI: 10.1007/978-3-319-65289-4_76

thrusters, without using any rudders. According to the control of four thrusters, the motions of total six degree-of-freedoms (DOFs) of X4-AUV, which has the position in three DOFs and the attitude in three DOFs, can be controlled by a noholonomic control method [5]. It has a merit that it can treat more number of DOFs than the number of inputs. Even in low speed cruising, the X4-AUV can maintain the excellent motion performance and can also be miniaturized.

In order to verify the proposal in an actual machine, Geri Letu et al. [6] designed and manufactured an X4-AUV to perform the control of the four thrusters with a 9-DOF sensor by a microcontroller. However, the PD control of thrusters and the amount of computations required for the calculation such as angle, angular velocity, and acceleration of the 9-DOF sensor are heavy, so that it cannot be said that it is the best method to deal with these calculations, if viewed from the efficiency and stability of the control of microcontroller.

In this research, as trying to drive the robot more efficiently, a microcontroller-mounted thruster is proposed, which distributes the PD control of the main microcontroller to the four sub-microcontrollers. It aims at improving the performance of speed control of thrusters by utilizing the PID control constructed by the sub-microcontrollers, each of which is mounted to each thruster. To adjust the P, I and D gains of each motor in real time, a GUI is used to communicate between the PC and the microcontrollers. After that, by sending the target speed values from another GUI via the main microcontroller to the four thrusters, it can verify the operation of each thruster. In this paper, it describes the construction of a control system for the thruster distributed control of X4-AUV.

2 Overview of X4-AUV

Figure 1 shows the designed fuselage of an X4-AUV. The X4-AUV has four thrusters arranged vertically and horizontally to control the position and attitude of the fuselage, and controls itself using the difference generated by the thrusters. Moreover, since a microcontroller, a battery, etc. are carried in the body, which are positioned at the center of the four thrusters, it in fact can seem to be a waterproofed container. Note here that a vinyl pipe is assumed to be used for the simplification of manufacturing an ideal spheroid, as the body to be proposed in this research.

As shown in Fig. 2, the three attitude angles of the X4-AUV are defined as (ϕ, θ, ψ). The body coordinate system {B} is composed of the orthogonal axes {X, Y, Z}, which is attached to the vehicle. The inertial coordinate system {E} is composed of the orthogonal axes {Ex, Ey, Ez}, in which Ex and Ey form a horizontal plane and the direction of Ez is set to the gravitational field, which is commonly placed at a fixed place on the Earth.

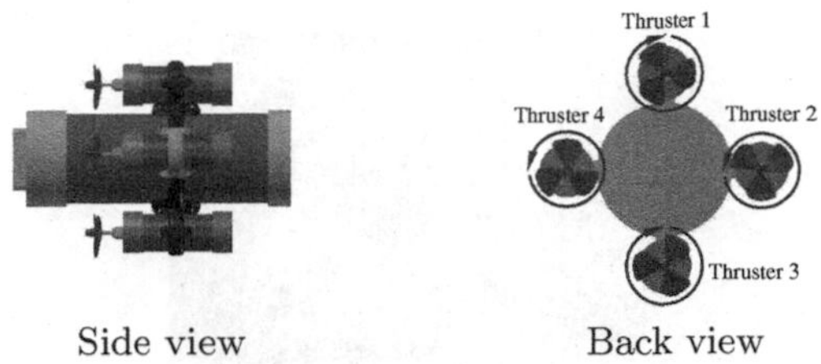

Fig. 1. Schematic of an X4-AUV

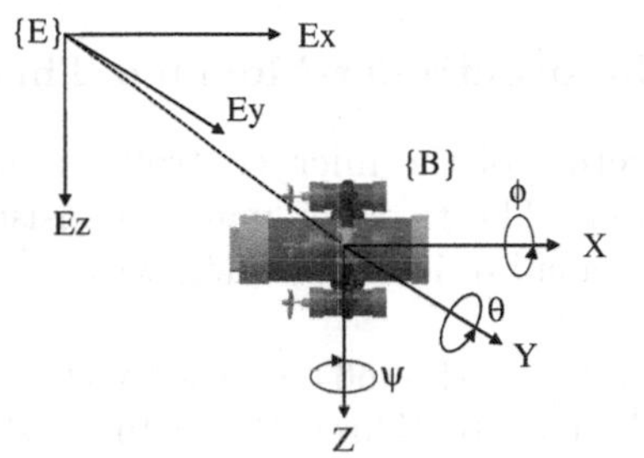

Fig. 2. Definition of the coordinate systems

3 Overview of Distributed Control of X4-AUV

In order to perform the distributed control of the thrusters, the servo control of each thruster is very necessary. Therefore, it constructs a distributed control system by using serial communication between the main microcontroller and sub-microcontrollers mounted in four thrusters. The outline of the main microcontroller and microcontroller-mounted thruster will be described as below.

3.1 Overview of Main Microcontroller

The main microcontroller of X4-AUV is RX63N of GR-SAKURA board produced by Renesas Electronics Corporation. The GR-SAKURA board is shown in the Fig. 3. The RX63N has digital IO ports, AD converters, PWMs, serial communication ports, USB connectors, an Ethernet controller, and a built-in floating-point arithmetic unit. When trying to see the present and target speeds of each thruster from a GUI on PC, five serial communication ports of the main microcontroller are used to realize full-duplex channel serial communication between the PC and the sub-microcontrollers. By utilizing their five serial communication ports, it is feasible to verify the performance of the proposed distributed control of the thrusters.

Fig. 3. Overview of GR-SAKURA board

3.2 Structure of Microcontroller-Mounted Thruster

Figure 4 shows the structure of the microcontroller-mounted thruster that was developed in this research. The thruster mainly consists of a motor driver, an encoder board, a motor, a sub-microcontroller, wires, waterproof parts, a screw propeller, and so on.

The sub-microcontroller is H8-3664 of the AKI-3664 board. The selected DC motor is TP-3644J made by Three Peace Inc., where the rated torque is 0.12 kg · cm. The motor driver used is an H-bridge integrated circuit L6203 manufactured by ST Microelectronics. The encoder is BTE030, which is the product of the Best Technology Co., Ltd., containing a transmission-type of photo-interrupter and a code wheel, where it is used in an incremental method. To fix the microcontroller into the thruster, a part made by 3D printer used to connect the motor and the microcontroller is built into the thruster.

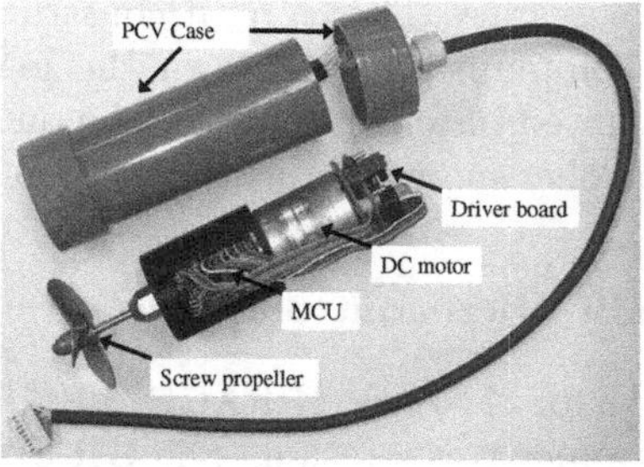

Fig. 4. Structure of thruster

3.3 Overview of Rotational Speed Control System

The parts related to the rotational speed control system are shown in Fig. 5. An encoder is fixed to the bottom of the driver board. A code wheel of the encoder is fixed to the shaft of a DC motor. The number of slits of the code wheel is 90. The rotation speed of the motor can be represented by using the slit numbers counted by the microcontroller per interrupt time.

Fig. 5. Motor driver and encoder board

3.4 Manufactured Thruster

The size of the microcontroller-mounted thruster manufactured is shown in Fig. 6. The weight of the thruster that contains wires is about 390 g. The length including the screw propeller is 200 mm. The diameter of both ends of the fuselage is 54 mm. Table 1 shows the specification of the thruster.

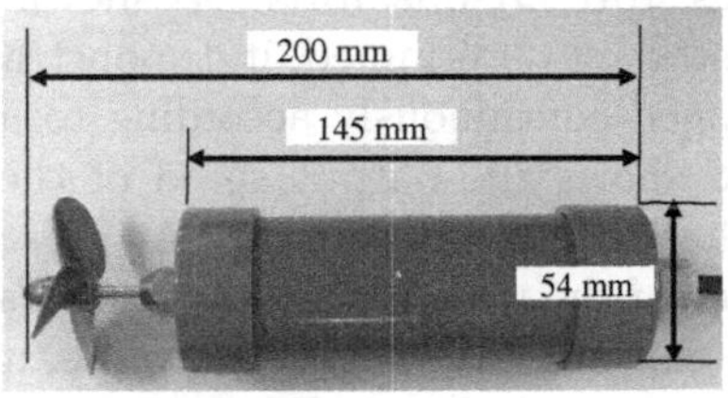

Fig. 6. Size of the manufactured thruster

Table 1. Specification of the manufactured thruster

Voltage	Maximum current	Output
12 V	4 A	4 W

4 Operational Verification of Distributed Control by GUI

Before testing the distributed control of the thrusters, it is necessary to find out the proper P, I and D gains. By adjusting the values of P, I and D gains through a GUI, the sub-microcontroller receives their values in an interrupt mode. At the same time, the sub-microcontroller calculates the PID speed for the output of the motor and returns it back to GUI through serial communication. Meanwhile, the dynamic changing graphs of the target speed and the current speed are shown on the GUI.

After the P, I and D gains are determined and written to the program of each sub-microcontroller, another GUI is used to verify the distributed control performance of each thruster. Since the main microcontroller can not receive the present speeds of each thruster at the same time by one serial port, three more serial ports are used. One more serial port is used for the main microcontroller to communicate with this GUI. The graphs of the target and current speeds at each thruster are drawn on this GUI.

4.1 Experiment of Gain Selection in the Air

Experimental Method. At first, use a serial communication cable to connect a microcontroller-mounted thruster with a GUI for adjusting the gains on PC. Then, set the target value of the speed through the control interface placed on the GUI appropriately. After that, adjust the value of P, I and D gains with the control interfaces. By checking the graph of the current and target speeds in the chart control, the appropriate gains can be determined quickly. After the P, I and D gains of the thruster are adjusted, the effect of the PID control can be confirmed, by checking whether the current output is converged, irrespective of the existence of a disturbance.

Experimental Results and Discussions. As shown in Fig. 7, after the P, I and D gains of the thruster are determined, it demonstrates a diagram of results of the target value changed continuously according to a double resolution way of the encoder. Figure 8 shows the output speed of the thruster when adding a disturbance to the shaft of the thruster. From Figs. 7 and 8, the values of adjusted P, I and D gains of the thruster are shown in Table 2. Since the current speed of the thruster is converged to the target value, it confirmed the validity of the P, I and D gains as shown in Fig. 7. Furthermore, referring to Fig. 8, it can be confirmed that the effect of output of the speed is small, even if a disturbance is added to the thruster.

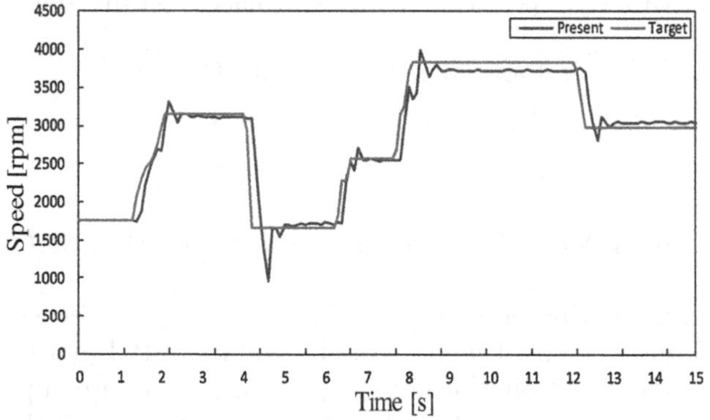

Fig. 7. Output of the thruster

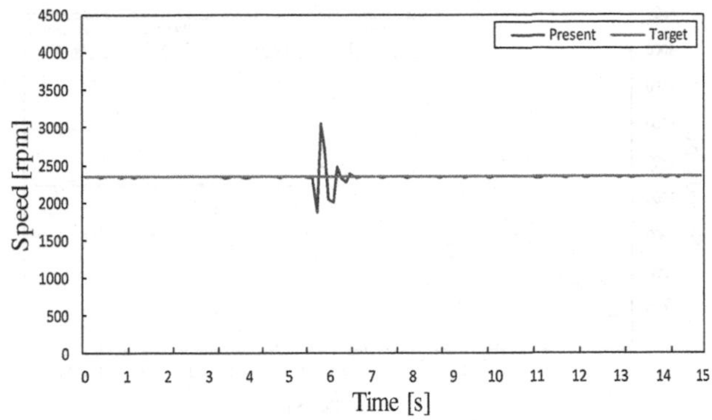

Fig. 8. For the case of an added disturbance

Table 2. PID gains

Gain P	Gain I	Gain D
1.0	0.1	1.3

4.2 Verification Experiment of Distributed Control in the Air

Experimental Method. Figure 9 shows the devices used in the experiments. It consists of a battery, a manufactured X4-AUV, a PC, a support frame, a circuit board, a serial communication cable, and so on. After the P, I and D gains of four thrusters are set at the same process discussed above, it can confirm the tracking ability of PID control of each thruster in real time, by using a designed distributed control verification GUI on PC.

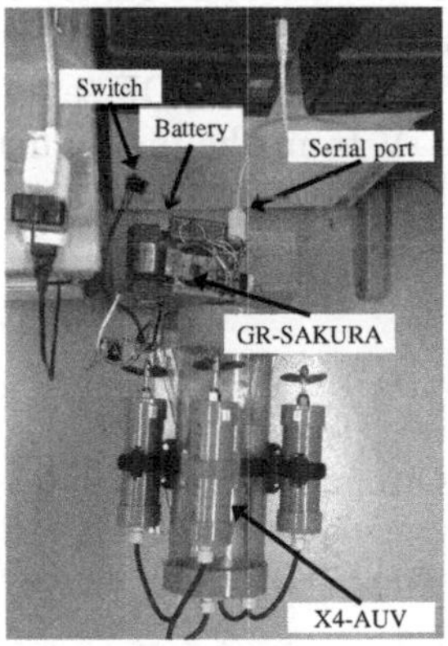

Fig. 9. Experimental setup

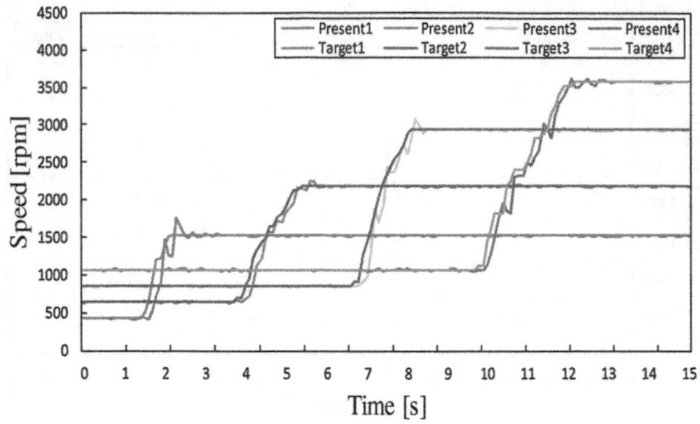

Fig. 10. Distributed control experiment

Experimental Results and Discussions. From the results shown in Fig. 10, by receiving the target speed values from the GUI via the main microcontroller, it can be seen that the current speeds of the four thrusters follow the target values, so that the stability at selected P, I and D gains is confirmed. It also confirmed the effectiveness of the operational verification in the distributed control between the main microcontroller and the sub-microcontrollers.

5 Conclusion

In this paper, it has described the design, manufacture and control of microcontroller-mounted thrusters to construct a distributed control system for the thrusters of X4-AUV. Experiments on the performance of four thrusters of X4-AUV have also been conducted to verify the effectiveness of the proposed distributed control system of the thrusters for X4-AUV. As to the future work, it is going to test the performance of four thrusters under a water environment. It is also going to analyze the data of 9-DOF sensor through GUI and perform the attitude control of X4-AUV from the main microcontroller.

References

1. Ura, T.: Robots for underwater world. J. Robot. Soc. Jpn. (JRSJ) **22**(6), 692–696 (2004)
2. Watanabe, K., Izumi, K., Okamura, K., Syam, R.: Discontinuous underactuated control for lateral X4 autonomous underwater vehicles. In: 2nd International Conference on Underwater System Technology, Theory and Applications 2008, Bali (2008)
3. Zain, Z.M., Watanabe, K., Nagai, I., Izumi, K.: The stabilization control of a position and all attitudes for an X4-AUV. In: 5th International Conference on Soft Computing and Intelligent Systems and 11th International Symposium on Advanced Intelligent Systems (SCIS and ISIS 2010), Okayama, pp. 1265–1270 (2010)

4. Zain, Z.M., Harun, N.F., Watanabe, K., Nagai, I.: Comparison of an X4-AUV performance using a direct Lyapunov - PD controller and backstepping approach. In: 10th Asian Control Conference (ASCC2015), Kota Kinabalu (2015)
5. Zain, Z.M., Watanabe, K., Izumi, K., Nagai, I.: A nonholonomic control method for stabilizing an X4-AUV. Artif. Life Robot. **16**(2), 202–207 (2011)
6. Letu, G., Tobita, S., Watanabe, K., Nagai, I.: The design and production of an X4-AUV. In: SICE Annual Conference 2015, Hangzhou, pp. 1383–1386 (2015)

A Localization Method Using a Dynamical Model and an Extended Kalman Filtering for X4-AUV

Keigo Watanabe[1(✉)], Takanori Yamaguchi[2], and Isaku Nagai[1]

[1] Department of Intelligent Mechanical Systems, Graduate School of Natural Science and Technology, Okayama University, 3-1-1 Tsushima-naka, Kita-ku, Okayama 700-8530, Japan
{watanabe,in}@sys.okayama-u.ac.jp

[2] Division of Mechanical and Systems Engineering, Graduate School of Natural Science and Technology, Okayama University, 3-1-1 Tsushima-naka, Kita-ku, Okayama 700-8530, Japan

Abstract. The self-position estimation problem of X4-AUV, which is an autonomous underwater vehicle (AUV) driven by four thrusters, is considered. Since a self-position cannot be underwater measured directly using GPS etc., we have to consider any method for realizing it by an indirect method. The AUV treated by this research has a mechanical structure that a self-position is controlled by changing the attitude from the feature of drive mechanism, and it can observe an attitude angle from an internal sensor, so that based on the dynamical model of the present AUV, a method for estimating the self-position is proposed by applying an extended Kalman filter. The usefulness of this technique is demonstrated by checking the feasibility in the simulation of the position control that used the position estimate.

Keywords: X4-AUV · Nonholonomic control · Localization · Extended kalman filter

1 Introduction

In addition to the development of marine resources and fishing resources, the demand of sea investigation, such as investigation of a marine structure, is increasing. However, the inside of the sea is the severe environment where people are not easily allowed to come near. Then, the activity of underwater robots is expected [1].

When an underwater robot is generally used in sea investigation, remotely operated vehicles (ROVs) are often applied [2]. However, there is a problem that a behavior range is restricted by the cable length, a big fluid resistance is applied to the cable of such a vehicle due to the stream flow, etc. Then, autonomous underwater vehicles (AUVs) [3] are being developed to solve such a problem appeared in ROVs.

Y. Huang et al. (Eds.): ICIRA 2017, Part I, LNAI 10462, pp. 834–845, 2017.
DOI: 10.1007/978-3-319-65289-4_77

Multiple thrusters and rudders are used for controlling most of AUVs [4]. Note however that in this technique, in order to obtain a high mobility performance for a low-speed cruise, it needs to enlarge the area of rudders of AUVs. Moreover, a large-sized AUV is unsuitable to activities in a narrow space. For this reason, an AUV which is small-sized and has a high mobility performance even for a low-speed cruise is requested.

Then, Zain [5] proposed an X4-AUV which controls a position and three attitudes only by using four thrusters, instead of using any rudders. The X4-AUV can realize a miniaturization, maintaining a high mobility performance for a low-speed cruise, because any rudders are not needed for controlling a position and three attitudes. Moreover, the positions in 3-DOFs, i.e., the X-, Y-, and Z-axis positions and the attitudes in 3-DOFs, i.e., the roll, pitch and yaw motions of the body, can be controlled for the X4-AUV by adjusting the thrusts of four thrusters, if any underactuated control approach would be adopted.

When using the X4-AUV for sea investigation, the self-position measurement and control are required, as well as the record of the picture of investigated results etc., but the position measurement under underwater is not easy. Since an electric wave is unreceivable underwater, it is impossible to use GPS [6]. Moreover, the sea environment needs the mechanical and electrical properties which are robust against the water pressure, the corrosion, the large change of humidity and temperature, etc. For this reason, any sensing by audio equipment, visual equipment, etc. is difficult [7].

In this research, the dynamical model of X4-AUV is used and it is aimed at carrying out the estimation and control of the self-position. Under the use of such a dynamical model, a position is computable based on the attitude angle of the body. However, when a noise is added to an attitude sensor, it is thought that it is difficult to estimate a self-position by any deterministic estimation technique. Therefore, a stochastic estimation technique is proposed here.

In this paper, the attitude angle acquired from the gyroscope sensor mounted on the X4-AUV is used as observations, and it is then possible to estimate the position when taking a translational motion in the X-axis direction by applying an extended Kalman filter. The usefulness of the proposed technique is checked through a simulation by verifying whether it can realize the position control by using the resultant estimate.

2 Overview of X4-AUV

2.1 X4-AUV

The appearance of X4-AUV is shown in Fig. 1. The X4-AUV mounts four thrusters. The position and attitude are controlled by adjusting appropriately the thrusts generated by these thrusters. Moreover, as shown in Fig. 1, they are defined as Thruster 1, Thruster 2, Thruster 3, and Thruster 4. The counter-torque generated by the revolutions of four thrusters is canceled by rotating the thrusters 1 and 3 clockwise, while rotating the thrusters 2 and 4 counterclockwise, and it prevents the body from an unexpected roll motion.

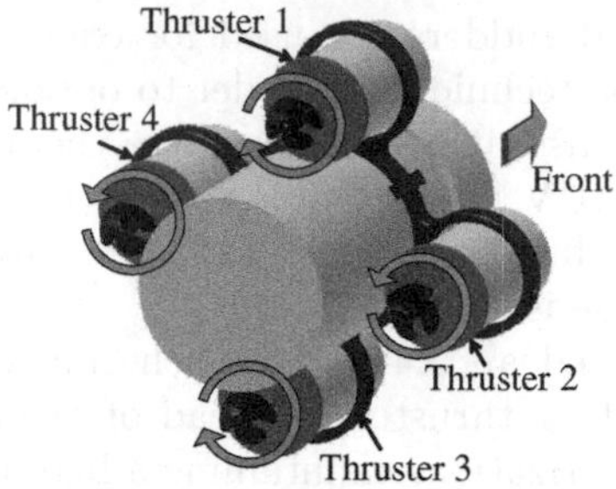

Fig. 1. The schematic diagram of an X4-AUV

2.2 Motion Mechanism

The coordinate system of X4-AUV which can be set horizontally is shown in Fig. 2. This coordinate system consists of an inertial coordinate system **E** and a body coordinate system **C**. The inertial coordinate system **E** is a right-hand coordinate system **E** (E_x, E_y and E_z) whose perpendicular facing down is set as the positive direction of the Z-axis. The body coordinate system **C** takes the center of gravity of the body at its origin, where such a system is a right-hand coordinate system **C** (X, Y, Z), but the body front is set to the positive direction of X-axis, the right of the body is set to the positive direction of Y-axis, and the perpendicular down for the body is set to the positive direction of Z-axis.

– Translational (or X-axis) Motion

The translational motion is achieved to the X-axis by making the thrust of all the thrusters increase or decrease simultaneously.

– Roll (or ϕ-axis) Motion

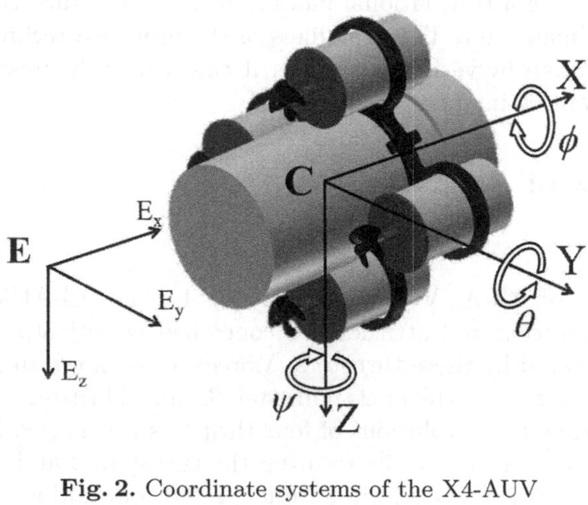

Fig. 2. Coordinate systems of the X4-AUV

Make an equal thrust generate in a pair of the thrusters 1 and 3, while the thrusts generated in a pair of the thrusters 2 and 4 are made equal. The roll motion is realized by giving a thrust difference between these two pairs of thrusters.

- Pitch (or θ-axis) Motion

The pitch motion is realized by making a thrust difference between the thrusts by the thrusters 1 and 3, while generating the thrusts in a pair of the thrusters 2 and 4 equally.

- Yaw (or ψ-axis) Motion

The yaw motion is realized by making a thrust difference between the thrusts by the thrusters 2 and 4, while generating the thrusts in a pair of the thrusters 1 and 3 equally.

3 Control of X4-AUV

3.1 Nonholonomic Constraints

An X4-AUV is a nonholonomic system consisting of four inputs and six states.

If the thrust of four thrusters is controlled appropriately, i.e., any under-actuated controller is used, X4-AUV can control the six states, i.e., the position (x, y, z) and the attitude angles (ϕ, θ, ψ). However, it has a demerit that a complicated control approach is needed. On the other hand, although only four states can be controlled at most by nonholonomic control, there is a merit that a relatively simple control approach can be used to stabilize the body.

Thus, the position and attitude of X4-AUV are controlled to aim at stabilizing the body.

3.2 State Space Representation for Dynamical Model

Dynamical Model of X4-AUV. Letting the gravitational acceleration be g, the distance from each thruster to the center of the body be l, the moment of inertia of the thruster be J_t, the control input for the translational position be u_1, and the control inputs for the roll, pitch, and yaw angles be u_2, u_3, and u_4, respectively, the dynamical model of X4-AUV can be described by the following formulas:

$$m_1\ddot{x} = \cos\theta\cos\psi u_1 \tag{1}$$

$$m_2\ddot{y} = \cos\theta\sin\psi u_1 \tag{2}$$

$$m_3\ddot{z} = -\sin\theta u_1 \tag{3}$$

$$I_x\ddot{\phi} = \dot{\theta}\dot{\psi}\left(I_y - I_z\right) + u_2 \tag{4}$$

$$I_y\ddot{\theta} = \dot{\phi}\dot{\psi}\left(I_z - I_x\right) - J_t\dot{\psi}\Omega + lu_3 \tag{5}$$

$$I_z\ddot{\psi} = \dot{\phi}\dot{\theta}\left(I_x - I_y\right) + J_t\dot{\theta}\Omega + lu_4 \tag{6}$$

Here, m_1, m_2, and m_3 are the mass of X-, Y-, and Z-axis directions, considering each added mass, and similarly I_x, I_y, and I_z are the moment of inertia of X-, Y-, and Z-axis directions, considering each added moment of inertia.

Moreover, introducing the thrust coefficient as b, the drag coefficient as d, the thrust for each thruster as f_i, the revolution velocity for each thruster as ω_i, the torque for each thruster as τ_{M_i}, the inputs u_1, u_2, u_3, and u_4 are reduced to

$$f_i = b\omega_i^2 \tag{7}$$

$$\tau_{M_i} = d\omega_i^2 \tag{8}$$

$$\Omega = (\omega_2 + \omega_4 - \omega_1 - \omega_3) \tag{9}$$

$$\begin{aligned} u_1 &= f_1 + f_2 + f_3 + f_4 \\ &= b\left(\omega_1^2 + \omega_2^2 + \omega_3^2 + \omega_4^2\right) \end{aligned} \tag{10}$$

$$u_2 = d\left(-\omega_2^2 - \omega_4^2 + \omega_1^2 + \omega_3^2\right) \tag{11}$$

$$u_3 = f_1 - f_3 = \left(\omega_1^2 - \omega_3^2\right) \tag{12}$$

$$u_4 = f_2 - f_4 = \left(\omega_2^2 - \omega_4^2\right) \tag{13}$$

State Space Representation. The equation rewritten from the dynamical model of X4-AUV to a state space representation is as follows:

$$\dot{\boldsymbol{X}} = f(\boldsymbol{X}, \boldsymbol{U}) \tag{14}$$

where the state vector $\boldsymbol{X} = (x_1, ..., x_{12})^{\mathrm{T}} \in \Re^{12}$ is given by

$$\begin{cases} x_1 = x & x_2 = \dot{x}_1 = \dot{x} \\ x_3 = y & x_4 = \dot{x}_3 = \dot{y} \\ x_5 = z & x_6 = \dot{x}_5 = \dot{z} \\ x_7 = \phi & x_8 = \dot{x}_7 = \dot{\phi} \\ x_9 = \theta & x_{10} = \dot{x}_9 = \dot{\theta} \\ x_{11} = \psi & x_{12} = \dot{x}_{11} = \dot{\psi} \end{cases} \tag{15}$$

Using this state vector given in Eq. (15) and the control input vector $\boldsymbol{U} = (u_1, u_2, u_3, u_4)^T$, the function $f(\boldsymbol{X}, \boldsymbol{U})$ is shown by

$$f(\boldsymbol{X}, \boldsymbol{U}) = \begin{pmatrix} x_2 \\ (\cos x_9 \cos x_{11})\frac{1}{m}u_1 \\ x_4 \\ (\cos x_9 \sin x_{11})\frac{1}{m}u_1 \\ x_6 \\ (-\sin x_9)\frac{1}{m}u_1 \\ x_8 \\ x_{10}x_{12}(\frac{I_y - I_z}{I_x}) + \frac{1}{I_x}u_2 \\ x_{10} \\ x_8 x_{12}(\frac{I_z - I_x}{I_y}) - \frac{J_t}{I_y}x_{12}\Omega + \frac{l}{I_y}u_3 \\ x_{12} \\ x_8 x_{10}(\frac{I_x - I_y}{I_z}) + \frac{J_t}{I_z}x_{10}\Omega + \frac{l}{I_z}u_4 \end{pmatrix} \tag{16}$$

Although the translational components of X4-AUV are independent of the attitude components from the above-mentioned formula, it is found that the latter components are dependent on the former components.

3.3 Controllers

The controller of X4-AUV is shown using the state vector given in Eq. (15). Defining the desired value of the translational position in the X-axis direction as x_1^d, its desired speed as x_2^d, and the positive gains as k_1 and k_2, the controller u_1 for the translational position in the X-axis direction can be described as follows:

$$u_1 = \frac{m\{k_1(x_1^d - x_1) + k_2(x_2^d - x_2)\}}{\cos x_9 \cos x_{11}} \tag{17}$$

When defining the desired values of attitude angles ϕ, θ, and ψ, as x_7^d, x_9^d, and x_{11}^d, and introducing the positive constant gains as k_3, k_4, and k_5, the controllers u_2, u_3, and u_4 related to the attitudes can be expressed as follows:

$$\begin{aligned} u_2 &= I_x(x_7^d - x_7) - k_3 x_8 \\ u_3 &= \frac{I_y}{l}(x_9^d - x_9) - k_4 x_{10} \\ u_4 &= \frac{I_z}{l}(x_{11}^d - x_{11}) - k_5 x_{12} \end{aligned} \tag{18}$$

4 Overview of Extended Kalman Filter

4.1 State Estimation by Extended Kalman Filter

When acquiring an attitude angle from the gyroscope sensor mounted on the X4-AUV, it is thought that a measurement noise occurs, so that a translational position is not estimated correctly. Then, the application of an extended Kalman filter is considered as the probable technique.

In this research, the attitude angles for the body of X4-AUV, (ϕ, θ, ψ), are observed through a gyroscope sensor, and the 12 states of $\hat{\boldsymbol{X}} = (\hat{x}_1, ..., \hat{x}_{12})^{\mathrm{T}} \in \Re^{12}$ are acquired as the estimates. It is aimed at controlling the position and attitude by using such estimates.

4.2 Derivation of Discreet-Time Model

A nonlinear state-space model for the X4-AUV is given in continuous-time t by

$$\dot{\boldsymbol{X}}(t) = f^*(\boldsymbol{X}(t), \boldsymbol{U}(t)) \tag{19}$$

Approximating this equation by an Euler method with the sample width Δt, it follows that

$$\frac{\boldsymbol{X}(t+\Delta t) - \boldsymbol{X}(t)}{\Delta t} = f^*(\boldsymbol{X}(t), \boldsymbol{U}(t)) \tag{20}$$

From this, it results in

$$\boldsymbol{X}_{k+1} = \boldsymbol{X}_k + f^*(\boldsymbol{X}_k, \boldsymbol{U}_k) \cdot \Delta t \tag{21}$$

where $\boldsymbol{X}_{k+1} \triangleq \boldsymbol{X}(t + \Delta t)$ and it is defined that

$$\begin{aligned} \boldsymbol{X}_k^T &\triangleq [x_1(t), ..., x_{12}(t)] \\ \boldsymbol{U}_k^T &\triangleq [u_1(t), ..., u_4(t)] \end{aligned} \tag{22}$$

The function $f^*(\boldsymbol{X}_k, \boldsymbol{U}_k)$ indicated here is provided by the dynamics given in Eq. (16)

Thus, the state-space model in the discrete time-step $k \geq 0$ is represented by the following formula:

$$\boldsymbol{X}_{k+1} = f_k(\boldsymbol{X}_k, \boldsymbol{U}_k) + \boldsymbol{G}_k \boldsymbol{w}_k \tag{23}$$

$$\boldsymbol{y}_k = \boldsymbol{H}_k \boldsymbol{X}_k + \boldsymbol{v}_k \tag{24}$$

where

$$f_k(\boldsymbol{X}_k, \boldsymbol{U}_k) \triangleq \boldsymbol{X}_k + f^*(\boldsymbol{X}_k, \boldsymbol{U}_k) \cdot \Delta t \tag{25}$$

The system noise $\boldsymbol{w}_k$ and the measurement noise $\boldsymbol{v}_k$ are mutually independent, which are subject to noises with the Gaussian distribution of a zero average, and with covariances $\boldsymbol{Q}_w$ and $\boldsymbol{R}_v$, respectively. Moreover, $\boldsymbol{G}_k$ is the distribution matrix of a noise vector and $\boldsymbol{H}_k$ is the observation matrix, each of which is assumed to be known in advance.

$$G_k \triangleq I_{12\times 12} \tag{26}$$

$$H_k = \begin{bmatrix} 0\,0\,0\,0\,0\,0\,1\,0\,0\,0\,0\,0 \\ 0\,0\,0\,0\,0\,0\,0\,0\,1\,0\,0\,0 \\ 0\,0\,0\,0\,0\,0\,0\,0\,0\,0\,1\,0 \end{bmatrix} \tag{27}$$

4.3 Extended Kalman Filter

The Kalman filter, which is applied in this research and has the gain $\boldsymbol{K}_k$, is shown below.
[Filtering Step]

$$\begin{aligned} \hat{\boldsymbol{X}}_{k|k} &= \hat{\boldsymbol{X}}_{k|k-1} + \boldsymbol{K}_k(\boldsymbol{y}_k - \boldsymbol{H}_k \hat{\boldsymbol{X}}_{k|k-1}) \\ \boldsymbol{K}_k &= \boldsymbol{P}_{k|k-1} \boldsymbol{H}_k^T (\boldsymbol{H}_k \boldsymbol{P}_{k|k-1} \boldsymbol{H}_k^T + \boldsymbol{R}_v)^{-1} \\ \boldsymbol{P}_{k|k} &= \boldsymbol{P}_{k|k-1} - \boldsymbol{K}_k \boldsymbol{H}_k \boldsymbol{P}_{k|k-1} \end{aligned} \tag{28}$$

[Prediction Step]

$$\begin{aligned} \hat{\boldsymbol{X}}_{k+1|k} &= f_k(\hat{\boldsymbol{X}}_{k|k}, \boldsymbol{U}_k) \\ \boldsymbol{P}_{k+1|k} &= \boldsymbol{F}_k \boldsymbol{P}_{k|k} \boldsymbol{F}_k^T + \boldsymbol{G}_k \boldsymbol{Q}_w \boldsymbol{G}_k^T \end{aligned} \tag{29}$$

where

$$\begin{aligned} \boldsymbol{F}_k &= \boldsymbol{I} + \boldsymbol{F}_k^* \\ \boldsymbol{F}_k^* &= \left.\frac{\partial f(\boldsymbol{x}_k, \boldsymbol{U}_k)}{\partial \boldsymbol{x}_k} \Delta t \right|_{\boldsymbol{x}_k = \hat{\boldsymbol{X}}_{k|k}, \boldsymbol{U}_k} \end{aligned} \tag{30}$$

5 Simulations

In this section, the extended Kalman filter described in the previous Sect. 4 is applied in the situation where there are a system noise and a measurement noise, to X4-AUV. It is proved through numerical simulations that the position of the body which carries out the translational motion in the X-axis direction can be estimated and the control is possible by using its estimates.

5.1 Setup Assumed in Simulations

Body of X4-AUV Used. The appearance and size of X4-AUV used in this simulation are shown in Fig. 3. The X4-AUV is a small underwater robot, and the weight in the air is 3.211 kgf, adjusted so that the underwater weight may become 0.0 kgf. Moreover, the physical parameters used in the simulation are shown in Table 1.

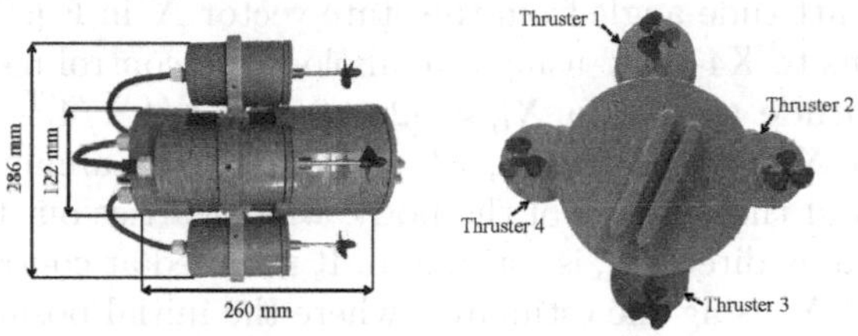

(a) Top view (b) Back view

Fig. 3. X4-AUV used in the simulation

Table 1. Model parameters of X4-AUV

Parameter	Description	Value	Unit
m_1	X-axis mass	3.334	kg
m_2	Y-axis mass	5.065	kg
m_3	Z-axis mass	5.065	kg
l_1	X-distance	0.13	m
l_2	Y-distance	0.061	m
l_3	Z-distance	0.061	m
I_x	Roll inertia	0.0182	$kg \cdot m^2$
I_y	Pitch inertia	0.0346	$kg \cdot m^2$
I_z	Yaw inertia	0.0346	$kg \cdot m^2$
J_t	Thruster inertia	1.194e−4	$kg \cdot m^2$
ρ	Fluid density	1023.0	$kg \cdot m^2$
b	Thrust factor	0.068	—
d	Drag factor	3.617e−4	—

MPU-9250. In order to estimate the current state, it needs to detect the attitude angle as the amount of observations. The MPU-9250 sensor module made from InvenSense carried in the body of X4-AUV is used for detecting the attitude. This sensor module is a 9-axis motion sensor equipped with 3-axis acceleration sensor, 3-axis gyroscope sensor, and 3-axis geomagnetism sensor. In the present X4-AUV, only 3-axis gyroscope sensor is used for detecting the attitude angle, and the attitude angle is assumed to be acquirable as the amount of observations using a gyroscope sensor in this simulation. Also, when acquiring the attitude angle, it is assumed that the measurement noises are additive, and such measurement noises $\boldsymbol{v}_k$ in Eq. (24) of Sect. 4 are determined by changing the RMS values.

5.2 Simulation Results

Define a new state vector as $\boldsymbol{X}^* = [x, y, z, \phi, \theta, \psi]^{\mathrm{T}}$, excluding their derivatives of the position and attitude angle from the state vector $\boldsymbol{X}$ in Eq. (15) of Sect. 4. In all the simulations to X4-AUV using a nonholonomic control method, the initial position and attitude are set to $\boldsymbol{X}_{\boldsymbol{0}}^* = [2, 0, 0, \pi/4, \pi/4, \pi/4]^{\mathrm{T}}$ and their target values are set to $\boldsymbol{X}_{\boldsymbol{d}}^* = [10, 0, 0, 0, 0, 0]^{\mathrm{T}}$. An extended Kalman filter is applied simultaneously and the position of the body, which carries out the translational motion in the X-axis direction, is estimated. It is aimed at controlling the target position of X4-AUV using the estimates, where the initial position and attitude in the estimates are set to $\hat{\boldsymbol{X}}_{\boldsymbol{0}}^* = [0, 0, 0, \pi/6, \pi/6, \pi/6]^{\mathrm{T}}$.

In the simulation, the Gaussian distribution noise with a standard deviation 0.1 deg is assumed to be added to the attitude angles ϕ, θ and ψ of the body, acquired as the amount of observations. The body state is estimated by applying a Kalman filter, and it is verified whether the control to the desired value

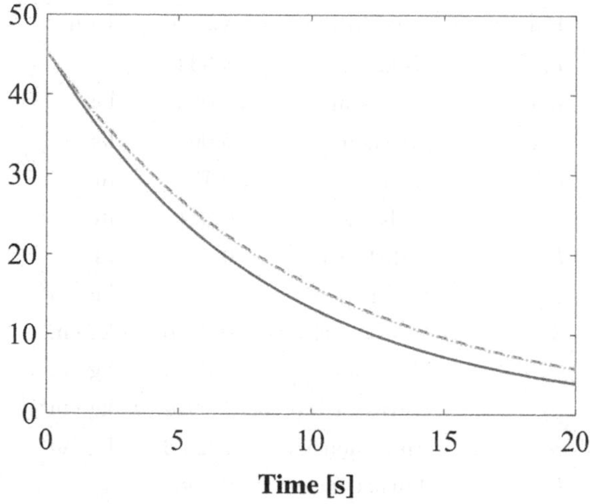

Fig. 4. True value of attitude angles with S.D. of 0.1 [deg]

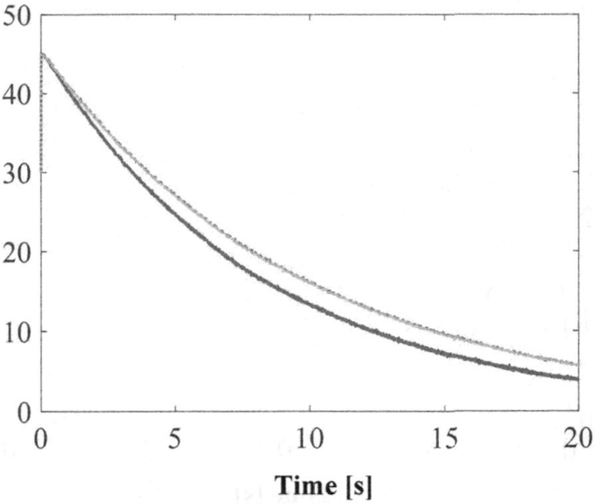

Fig. 5. Estimate of attitude angles with S.D. of 0.1 [deg]

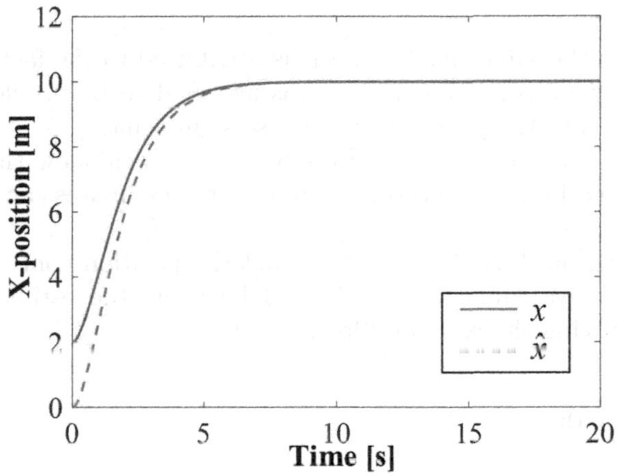

Fig. 6. X-positional state with S.D. of 0.1 [deg]

is possible using the estimates, comparing the true values in the translational position and attitude with their estimates.

Simulation Results and Considerations. The simulation result of one sample is shown in Figs. 4, 5, 6 and 7. Figure 4 shows that the true values of attitude angles of X4-AUV, i.e., ϕ, θ, and ψ converge from the initial attitude $\boldsymbol{X_0}$ to the target attitude $\boldsymbol{X_d}$.

It is turned out from Fig. 5 that the attitude state can be estimated, and that it can be controlled to the target attitude. The true value and the estimate

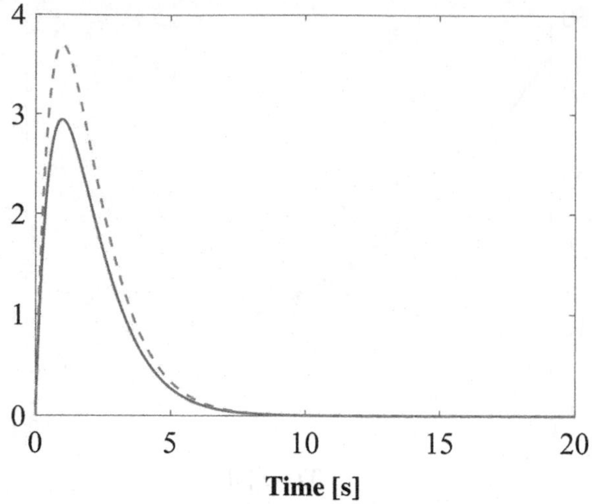

Fig. 7. X-positional speed state with S.D. of 0.1 [deg]

designate almost the same changes. This is attributed to the fact that the standard deviation of measurement noises was assumed to be 0.1 deg, so that the influence of noises to the estimate was not so significant.

Moreover, it is turned out from Figs. 6 and 7, which show the translational position and speed in the X-axis direction, that the estimates can follow the true values.

It is checked from Figs. 4, 5, 6 and 7 that the position control to the translational direction was able to be achieved by using the estimate, which was obtained by applying the Kalman filter.

6 Conclusions

This paper has described the dynamical model and a controller to the X4-AUV, which controls the position and attitude of the body using four thrusters. The extended Kalman filter was applied, assuming the system of an actual X4-AUV, and it was verified from simulations whether the body state was able to be estimated. Especially, the position of the body of X4-AUV which carries out the translational motion in the X-axis direction was estimated by using the attitude angle acquired as the amount of observations, and its position was controlled to a target position. It was checked from this result that the self-position was estimated from the dynamical model and its position control was able to be carried out with such a stochastic approach.

It needs to check whether the technique actually proposed here is effective using the system of X4-AUV. Moreover, it is thought that more accurate position control will be able to be performed by combining the present method with other localization methods.

References

1. Ura, T.: Robots for underwater world. J. Robot. Soc. Jpn. (JRSJ) **22–6**, 692–696 (2004)
2. Nakata, Y., Otsuka, M., Ozawa, H., Konosu, M.: Development of the hull inspection robot (RTV-SHIP). J. Soc. Naval Architects Jpn. **181**, 233–239 (1997)
3. Asakawa, K., Kojima, J.: Autonomous underwater vehicles exploring deep seas. Inf. Process. Soc. Jpn. (IPSJ) Mag. **42–5**, 491–497 (2001)
4. Aoki, T., Tsukioka, S., Yoshida, H., Hyakudome, T., Ishibashi, S., Sawa, T., Ishikawa, A., Tahara, J., Yamamoto, I., Ohkusu, M.: Advanced technologies for cruising AUV URASHIMA. Int. J. Offshore Polar Eng. **18–2**, 81–90 (2008)
5. Zain, Z.M., Watanabe, K., Izumi, K., Nagai, I.: A nonholonomic control method for stabilizing an X4-AUV. Artif. Life Robot. **16–2**, 202–207 (2011)
6. Takada, Y., Nakamura, T., Koyama, K., Wakisaka, T.: Self-position estimation of small fish robot based on visual information from camera. J. Jpn. Ins. Mar. Eng. (JIME) **47–3**, 437–443 (2012)
7. Iwamoto, Y.: The ocean and sensors - Use of the sound technology in the ocean. J. Inst. Electr. Eng. Jpn. **102–5**, 425–427 (1982)
8. Leonard, N.E.: Stability of a bottom-heavy underwater vehicle. Automatica **33–3**, 331–346 (1997)

Experiment Study of Propulsion Property of Marine Mobile Buoy Driven by Wave

Zongyu Chang[1,2](✉), Guiqiao Lu[1], Guangchao Du[1], Zhongqiang Zheng[1], Yuanguang Tang[1], Jiliang Wang[1], and Xin Lu[1]

[1] Engineering College, Ocean University of China, Qingdao 266100, People's Republic of China
zongyuchang@qq.com

[2] Key Ocean Engineering Lab of Shandong Province, 266100 Qingdao, People's Republic of China

Abstract. Mobile observation platform is an important tool to obtain the parameters of ocean environment in a large range for long time, but the energy supply for locomotion is the bottleneck to limit its long journey. The Marine Mobile Buoy (MMB) mentioned in this paper can harness the wave energy and propel itself forward. The experiment study MMB driven by wave is studied to get the propulsion property. The mechanism of propulsion driven by wave is introduction briefly. The physical prototype of the propeller of MMB is designed and built. Based on the flow flume an experiment rig to measure the thrust force of propeller is developed. The propeller is hoist by a cable to in a rotation reel. The step motor is used to control the rotation of reel and adjust the length of cable to simulate the heaving motion of buoy induced by the wave. The thrust force is measured by compensating the friction between sliding platform and the guidance bar. The results of experiment are also compared with simulation and agree fairly well. The work can be used to estimate the propulsion property and be helpful for the design of MMB.

Keywords: Marine mobile buoy · Underwater propeller · Wave driven propulsion

1 Introduction

Marine observations are widely used in the field of oceanographic research, ocean resource exploration and naval defense. Moored buoy is one of the most important equipment to implement marine environmental observation in special location for long time [1–4]. But it can only obtain the environment information of a fixed location. Mobile observation can measure the oceanographic parameters in a large range for long time and play an important role on marine observation [5, 6]. However, energy supply for propulsion of locomotion limits the function of marine observation platform [7]. Some researchers proposed the idea that renewable energy such as wave or wind can be used to propel the marine observation platform and it becomes a feasible solution for propulsion of mobile platform [8, 9].

Y. Huang et al. (Eds.): ICIRA 2017, Part I, LNAI 10462, pp. 846–856, 2017.
DOI: 10.1007/978-3-319-65289-4_78

The idea to directly propel marine vehicles by harvesting wave energy was proposed decades ago [10]. Some propulsion devices for marine vehicle have been invented by scientists of US, Norway USSR and Japan et al. [10, 13]. However, the technology in this area is far from mature in theory, and faced with many difficulties in application. Some of them have been abandoned for years due to issues in application [10]. However, in recent years great progress has been made. Japanese adventurer Kenichi Horie completed an adventure journey using wave power boat, the "Suntory Mermaid II" from Hawaii, USA to Tsuneishi in 2007 [11]. A wave glider, which is an unmanned vehicle propelled by wave energy, completed a journey across the Pacific in 2008, setting a Guinness World Record. It has become, since then, a powerful tool for oceanography research, oil spill detection and so on [12].

Research on wave powered vehicle has been implemented in recent years. Nicholas and Kraus [14] proposed a dynamic model of wave gliders; the same group carried out a estimation algorithm for the position of vehicle based on the Extended Kalman Filter [15]; In references [16, 17], models to predict the speed and other condition parameters of the floating platform were developed. Politis and Politis [18] presented a method of active pitch control for the foil to extract the wave energy and propel the boat forward and gave the simulation of hydrodynamics.

In the paper experiment of propulsion induced by wave of MMB is studied. By development of propeller of MMB and experiment rig, the thrust of propeller is measured and the relationship between thrust force and wave amplitude and frequency is studied.

2 The Propulsion Mechanism of Propulsion Driven by Wave in MMB

MMB is mainly composed of the surface floating body and the device of underwater propulsion and the flexible tether or cable connecting the two parts. When the floating body moves upward along with the wave, the fins in underwater propeller will rotate clockwise under the hydrodynamic force. The thrust force is then generated on the hydrofoil, pushing the platform forward to the left as shown in Fig. 1. While the floating body moves downward along with the wave, the fins will rotate anticlockwise, and the thrust force generated on the hydrofoil will push the platform forward to the left, too.

So whether the buoy moves upward or downward with the wave, the thrust force generated pushes the floating body forward. The floating body on the surface suffering from the oscillation of wave can absorb the wave energy to transfer to mechanical thrust force. This is the basic principle for the marine mobile buoy propelled by the wave motion [19].

Therefore, we can design the MMB to achieve the observation task by applying the wave energy. It has the advantage that free of energy supply, no carbon submission and quietness without noise [20]. From the principle of MMB, general speaking, the more severe wave there are, the more wave energy can be absorbed and maybe the larger thrust force and higher speed of MMB can be achieved.

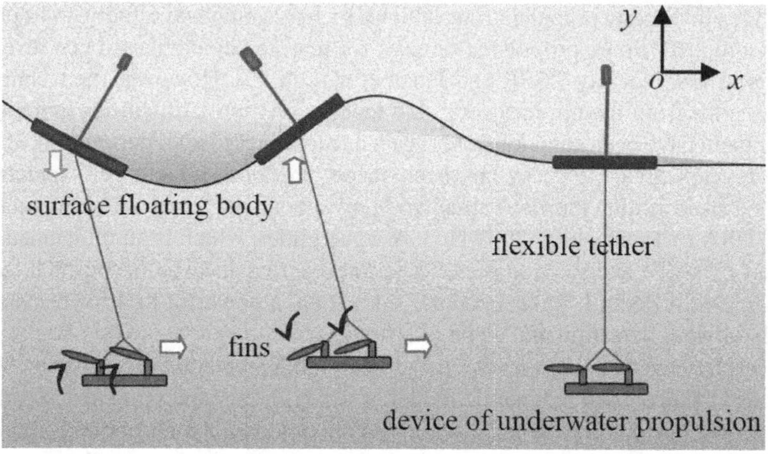

Fig. 1. The propulsive principle of the underwater propeller

3 Experiment of Propeller of MMB

3.1 Experiment Rig for Propulsion Force of Underwater Propeller

The physical prototype of underwater propeller is developed firstly and shown in Fig. 2. It is a frame with 4 pairs of fins which can be rotated about the corresponding individual axes. The propeller is connected with floating body by cable and thrust itself and floating body forward by transferring the wave motion.

Fig. 2. Underwater propeller

The experiment rig is installed on the flow flume. It includes guidance bar, sliding platform, tension force sensor, reels, cable, stepping motor, propeller and so on shown in Fig. 3. Force sensor and digital camera are individually used to measure the thrust force and recorder the motion condition of propeller and cable.

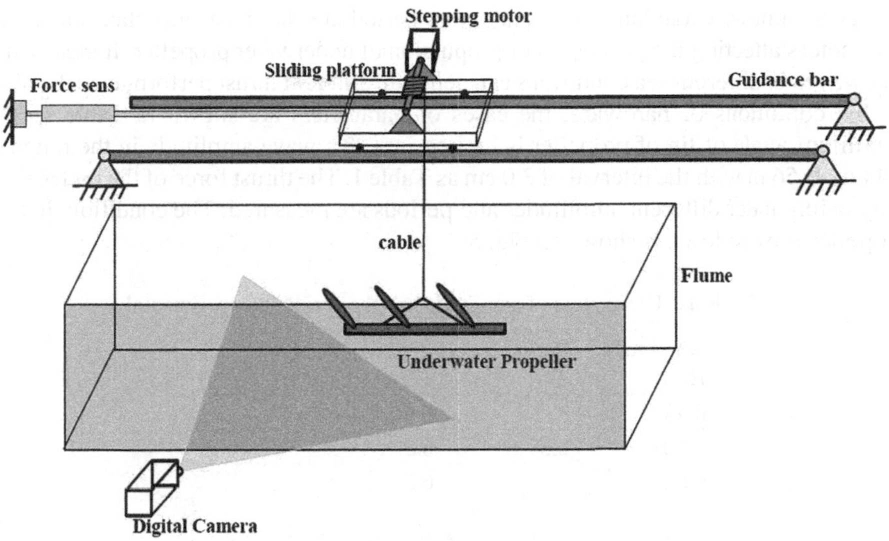

Fig. 3. Illustration of experiment rig

A control system is established to implement the heaving motion of propeller which is immersed in water. It is combined by industrial computer, control card for stepping motor and driver for stepping motor. By the control system, different heaving motion can be generated with different magnitudes and different periods. The reel with coil cable and stepping motor are installed on sliding plate which can slide along the guidance bar. The tension force sensor is connected with the sliding plate and fixed frame. Because the thrust force induced by heaving motion is fairly small, the friction force between sliding plate and guidance bar should be compensated. In order to compensate the friction force, the tilt angle on guidance is slight adjusted to friction angle of the interface between sliding plate and guidance bar (Fig. 4).

Fig. 4. Experiment rig on the flow flume

As we know, wave amplitude and wave period are the most important kinematic parameters affecting the properties of propulsion of underwater propeller. It means also that waves in different sea conditions can generate different thrust performance. Limited by the conditions of hardware, the cases of parameters are shown in Table 1. The maximum angle of fin of propeller is limited in ±28°; wave amplitude in the range of 0.44 m–0.56 m with the interval of 3.0 cm as Table 1. The thrust force of the underwater propulsion under different amplitudes and periods are measured. The condition that the propeller moves down is shown in Fig. 5.

Table 1. Parameters of amplitude and the period in experimental

Amplitude H/m	Period T/s	Period T/s	Period T/s
0.44		6.5	
0.50	5.3	6.5	7.7
0.56		6.5	

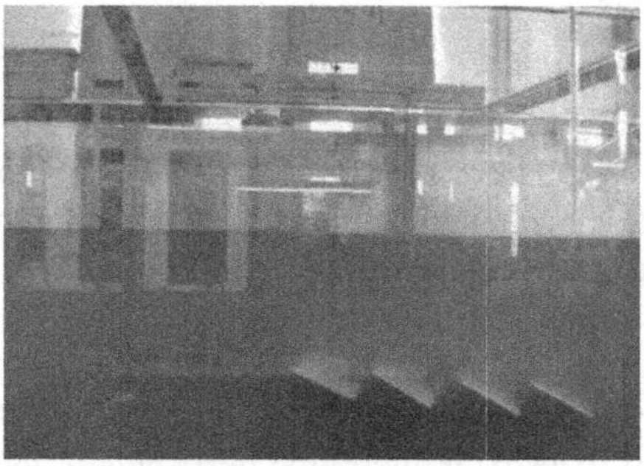

Fig. 5. The condition of propeller going down

The thrust force results are shown as Fig. 6(a)–(e).

From Fig. 6(a)–(c) we can found that under the condition of a certain wave period, the forward pull of underwater propulsion mechanism is positively correlated with the wave amplitude. Obviously, in the case of fixed wave period, the greater amplitude the wave has, the higher wave energy can be harnessed and larger propulsive force in the underwater propulsion mechanism can be generated. In the experimental cases, when the wave period is 6.5 s and the wave amplitude is 0.56 m, there is the largest propulsive force with 16.3 N. And the average pulling force is about 10.7 N. From Fig. 6(d)–(e), we can found that under the condition of a fixed wave amplitude, the forward pull of the underwater propulsion mechanism is negatively correlated with the wave period, this is because in the case of a certain wave amplitude, the greater the wave period, the smaller the wave energy, the lower the propulsion force of underwater propulsion mechanism; the underwater propulsion mechanism has the best effect when the wave amplitude is 0.50 m and the wave period is 5.3 s, at this point, the driving force peak is about 18 N, the average pulling force is about 11 N.

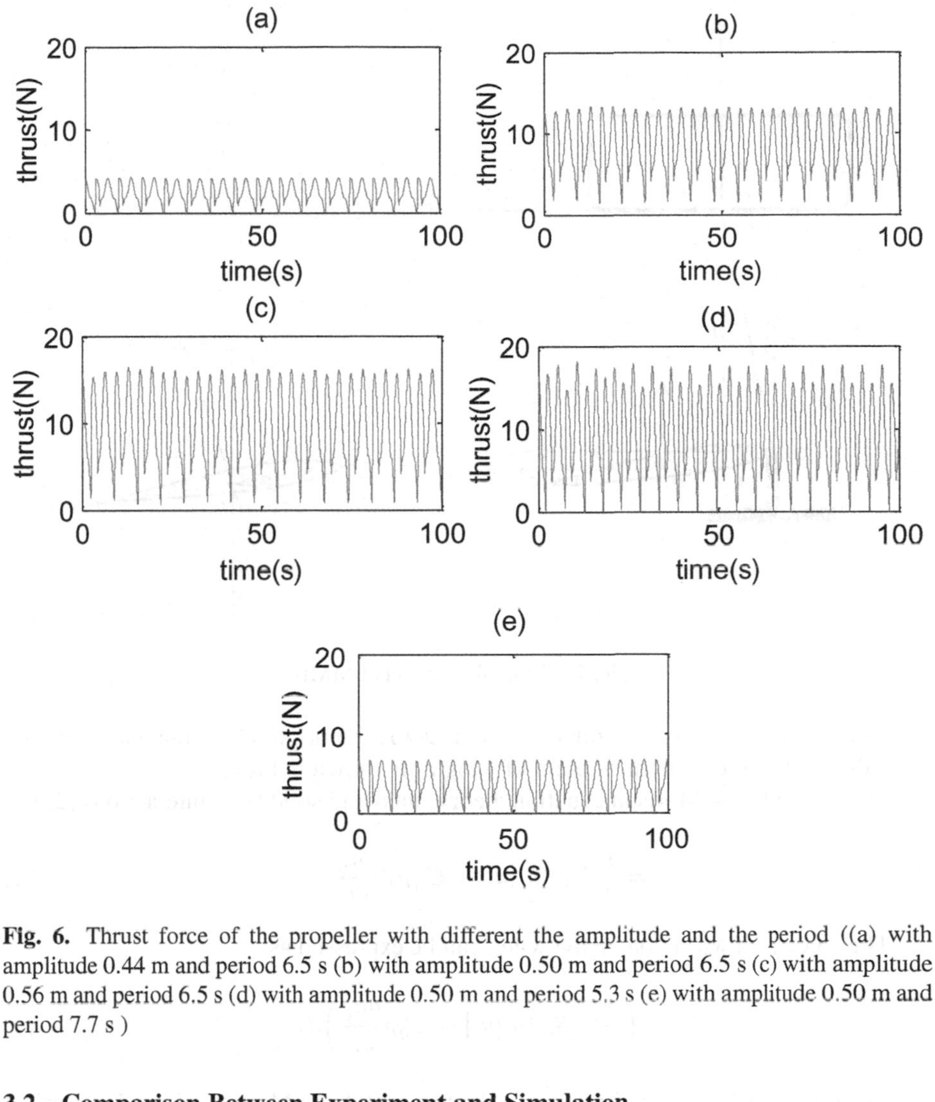

Fig. 6. Thrust force of the propeller with different the amplitude and the period ((a) with amplitude 0.44 m and period 6.5 s (b) with amplitude 0.50 m and period 6.5 s (c) with amplitude 0.56 m and period 6.5 s (d) with amplitude 0.50 m and period 5.3 s (e) with amplitude 0.50 m and period 7.7 s)

3.2 Comparison Between Experiment and Simulation

The MMB slickly use the energy of wave heaving to drive forward. And, the performance of underwater propulsion mechanism determines the speed of motor energy utilization and navigation buoys. So, this paper makes simulation and analysis to the key factors which influence the underwater propulsion. Meanwhile, verified the operating results by experiments in different sea conditions.

The rise of underwater propulsion is similar as the decline, this paper choose the rise process to make kinematics and dynamics analysis: Propulsion mechanism in the water with floating upward movement, tail fins have been forcing down swing as is shown in Fig. 7.

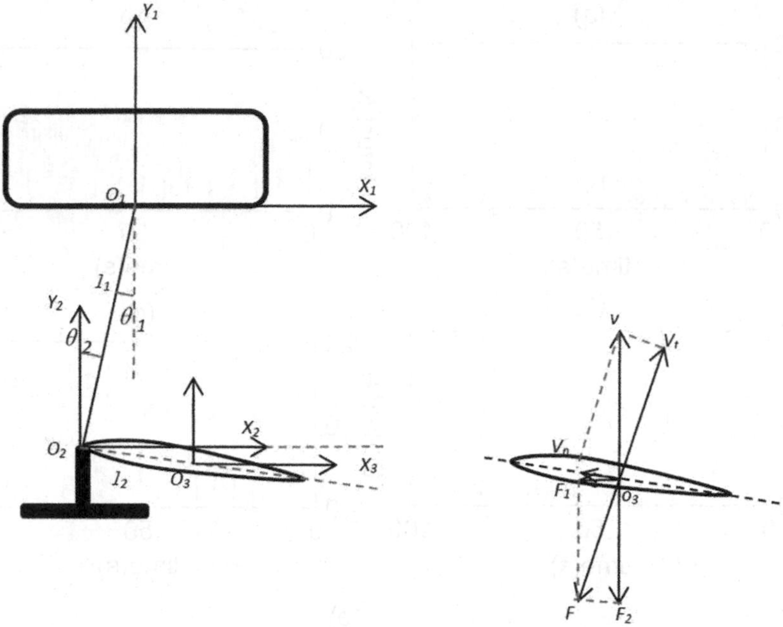

Fig. 7. Simplified model of MMB

Establishing the base coordinate system *O-XY*, the origin *O* is coincided with the axis. Point O_3 is the equipment point of wing as is shown in Fig. 7.

According to the Morison equation theory, fluid and solid have interactions [21],

$$f = \frac{1}{2} C_D \rho A u_x |u_x| + C_M \rho V \frac{\partial u_x}{\partial t} \tag{1}$$

The force *F* of Morrison on the wing, can be expressed as:

$$F = \int_0^1 \left(\frac{1}{2} C_D S \rho A u_x |u_x| + C_M \rho \frac{\partial u_x}{\partial t} \right) ds \tag{2}$$

The tension *T* is vertical direction on the horizontal, the Morrison force *F* is vertical right below the wing, Morrison force *F* can be broken down into vertical direction horizontal component forces F_1 and F_2. So, the fins can generate forward thrust F_1 during up exercise.

$$F_1 = F \sin \theta \tag{3}$$

The vertical component force F_2 is balance with the pulling force *T*, the equations of dynamics and dynamics of the underwater propulsion mechanism is shown as below:

$$m\ddot{x} + c\dot{x}|\dot{x}| - F \sin \theta = 0 \tag{4}$$

$$m\ddot{y} + F\cos\theta + G + F_{float} + F_w = 0 \tag{5}$$

$$I\ddot{\theta} = M_F \tag{6}$$

The two equations describe fin movement in *XY* in the absolute coordinate system. This section mainly analyzes the fin propulsion, the force of forward direction. Suppose the motion in the *Y* direction is known, so mainly considered the tangential component of wave forces on wing.

According to formula (4) and (6), under the condition of input sine wave function, the motion process is simulated. Take the two waves as the reference, select the following simulation scheme: the motion cycle is 6.5 s and the amplitude of the wave is 0.5 m. The navigation speed and speed of the MMB can be obtained, as shown in Figs. 8 and 9; at the same time, it can also get the forward pull force of the underwater propulsion mechanism to the surface floating body, as shown in Fig. 10.

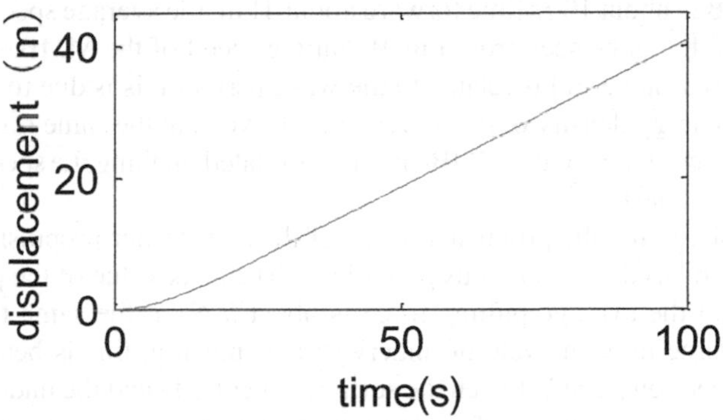

Fig. 8. The displacement of MMB

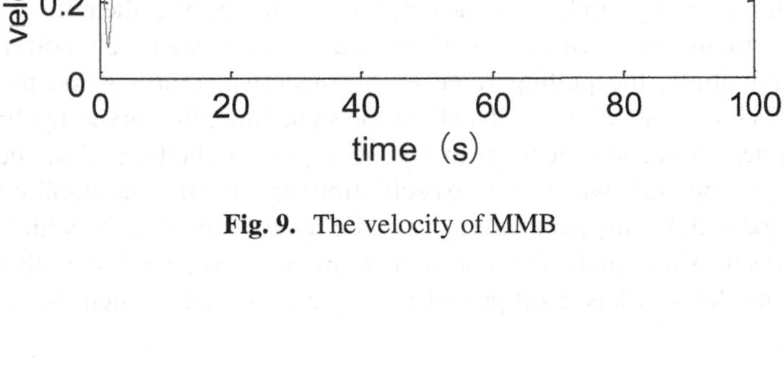

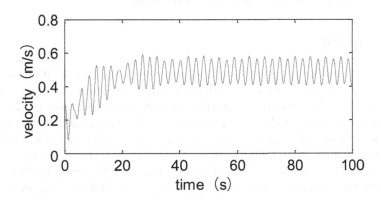

Fig. 9. The velocity of MMB

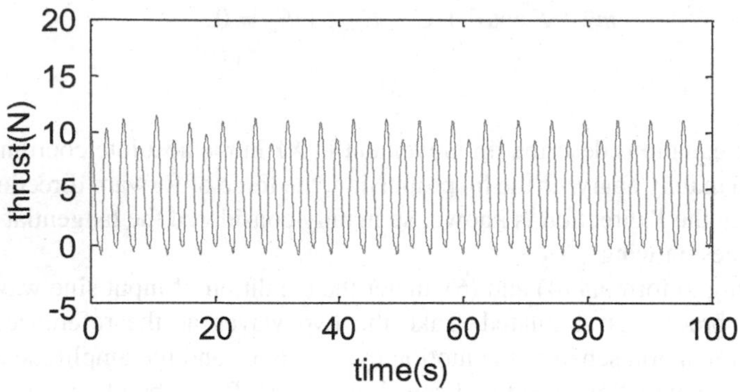

Fig. 10. The propulsion force

It can be seen from Fig. 8 that the wave amplitude is 0.5 m and the wave period is 6.5 s, The MMB in about 100 s time forward about 41 m, the average speed of navigation in the 0.41 m/s; It can be seen from Fig. 9 that the speed of the MMB is periodic, and the velocity variation period is related to the wave period, this is due to the continuous changes in the energy density of the waves in each cycle; at the same time, At the same time, due to the resistance of the MMB and speed-related, making the speed of the MMB is stabilized at about 0.41 m/s.

Figure 10 show that the propulsive force of the underwater propulsion mechanism of the MMB also has a more obvious periodicity. The peak value of the pulling force is about 12 N, and the average pulling force is about 5 N; at the same time, there is a negative pull force in each cycle of underwater propulsion, this is because the hinge joint is the connection point between the surface water body and the underwater propulsion mechanism, In the process of navigation, the water body is less affected by water flow. At the end of each cycle, the floating body is located in front of the underwater thruster, which leads to the negative force of the hinge.

4 Conclusion

In order to obtain the property of propulsion driven by wave of MMB, experimental study is carried out in the paper. Experimental rigs for propulsion force measurement are developed and the comparisons between simulation and experiment are given. In the condition of the same period, when the amplitude is increasing, the pulling force of the underwater propulsion mechanism is increased, which is positively correlated; under the same amplitude, the pulling force of the underwater propulsion mechanism is reduced, which is negatively correlated. At the same time, the forward pulling force of the underwater propulsion mechanism has obvious periodicity, and the period of the tension change coincides with the wave oscillation period; when the amplitude is 0.50 m, and the period is 5.3 s, the peak value of tensile force is about 18 N, which has a good effect. The results show, under the condition of different wave conditions, the underwater propulsion mechanism has good propulsive effect; especially when the wave has the appropriate amplitude and period, the high speed can be obtained for the MMB. The research is of great significance to guide the design and development of MMB.

References

1. Curcio, J.A., Mcgillivary, P.A., Fall, K., et al.: Self-positioning smart buoys, the "Un-Buoy" solution: logistic considerations using autonomous surface craft technology and improved communications infrastructure. IEEE, Boston (2006)
2. Wood, S., Rees, M., Pfeiffer, Z.: An autonomous self-mooring vehicle for littoral & coastal observations. In: OCEANS 2007-Europe, pp. 1–6. IEEE, Aberdeen (2007)
3. Orton, P.M., McGillis, W.R., Moisan, J.R., et al.: The mobile buoy: an autonomous surface vehicle for integrated ocean-atmosphere studies. AGU Spring Meeting Abstracts, San Francisco (2009)
4. Yoshie, M., Matsuzaki, Y., Fujita, I., et al.: At-sea trial test of an autonomous buoy which tracks drifting oil and observation of in-situ data tracking drifting markers on the sea for predicting location of the spilled heavy oil. In: The Nineteenth International Offshore and Polar Engineering Conference, Osaka (2009)
5. Crimmins, D.M., Patty, C.T., Beliard, M., et al.: Long-endurance test results of the solar-powered AUV system. In: OCEANS 2006. IEEE (2006)
6. Wang, Y.H., Zhang, H.W., Wu, J.G.: Design of a new type underwater glider propelled by temperature difference energy. Ship Eng. **31**(3), 51–54 (2009)
7. Wang, X., Shang, J., Luo, Z., et al.: Reviews of power systems and environmental energy conversion for unmanned underwater vehicles. Renew. Sustain. Energy Rev. **16**(4), 1958–1970 (2012)
8. Mikhail, D., Ageev, D., Richard, B.: Results of the evaluation and testing of the solar powered AUV and its subsystems. In: 11th International Symposium on Unmanned Un-tethered Submersible Technology, pp. 137–145. IEEE (1999)
9. Stommel, H.: The SLOCUM mission. Oceanography **2**(1), 22–25 (1989)
10. PESN Homepage. http://pesn.com/2005/09/21/9600170_Wave_Propulsion/. Accessed 06 Mar 2017
11. TSUNEISHI Homepage, http://www.tsuneishi.co.jp/english/horie/index.html. Accessed 06 Mar 2017
12. LIQUIDR Homepage. http://www.liquidr.com. Accessed 06 Mar 2017
13. Shang, J., Wang, X., Luo, Z., et al.: Design of a multi-propulsion ocean vehicle based on wave energy. In: Proceedings of the 8th International Conference on Manufacturing Research, pp. 76–80 (2010)
14. Kraus, N.: Wave Glider Dynamic Modeling, Parameter Identification and Simulation. Master thesis. University of Hawaii (2012)
15. Kraus, N., Bingham, B.: Estimation of Wave Glider dynamics for precise positioning. In: OCEANS-2011, pp. 1–9. IEEE, Santander (2011)
16. Wang, Y., Anvar, A., Hu, E.: A feasibility study on the design, development and operation of an automated oceanic wave surface glider robot. International Congress on Modelling and Simulation, Adelaide (2013)
17. Ngo, P., Al-Sabban, W., Thomas, J., et al.: An analysis of regression models for predicting the speed of a wave glider autonomous surface vehicle. In: Australasian Conference on Robotics and Automation, Sydney, pp. 1–9 (2013)
18. Politis, G., Politis, K.: Biomimetic propulsion under random heaving conditions, using active pitch control. J. Fluids Struct. **47**, 139–149 (2014)

19. Chang, Z.Y., Dai, Y., Chang, D.H., et al.: Design of propulsion mechanism of mobile buoy directly driven by wave. J. Ocean Univ. China **44**(3), 100–105 (2014). (in Chinese)
20. Yu, Z., Zheng, Z., Yang, X., Chang, Z., et al.: Dynamic analysis of propulsion mechanism directly driven by wave energy for marine mobile buoy. Chin. J. Mech. Eng. **29**(4), 710–716 (2016)
21. Berteaux, H.O.: Buoy Engineering. Wiley, New York (1976)

Design of Thermal Power Generation System Based on Underwater Vehicles

Rui Wang, Hongwei Zhang(✉), Guohui Wang, and Zhesong Ma

Tianjin University, Tianjin 300072, China
zhanghongwei@tju.edu.cn

Abstract. In this paper, a thermal power generation system is proposed to transform ocean thermal energy into electrical energy. Provide energy for marine exploration equipment, such as underwater glider, buoy, and can recycles. Marine exploration equipment driven by the novel system can extend the endurance, enhanced the detection capacity and duration, and reduce the impact of energy problems. The intermediate accumulator collects a slow volume change of Phase Change Material (PCM), completes a one-time release after the completion of phase change. This improves volume change rate of phase change material. An experimental method of hydraulic test system, temperature control system and phase change volume measurement system is proposed. The phase change material hexadecane was tested from the solidification state to the liquid. The liquid phase volume change parameter of hexadecane obtained by experiment was prepared for the next overall power generation experiment.

Keywords: Thermal engine · Hexadecane phase change · Power generation · Thermal energy

1 Introduction

Ocean thermal energy is the thermal energy produced by the difference in water temperature at different depths. Due to solar radiation, the temperature of the seawater decreases with the increase of water depth, and the thermal contains a large amount of energy [1]. The underwater glider driven by thermal energy have been employed in underwater survey plications, including deep sea inspections, oceanographic mapping and detection, location and neutralization of undersea mines [2]. The driving principle of the underwater glider to obtain energy from ocean thermal energy is that the large water temperature changes from the surface of the sea to the deep, through power system cycle principle, can be converted into mechanical energy and drive mechanical system work [3]. One of the biggest limitations of these underwater robots is the lack of power supply at seas away from the coast and service ships. All the underwater robots described above are powered by primary battery onboard [4]. Recently AUV development has been focused on improving the operation range and endurance for long term data collection [5, 6]. Thermal energy is a kind of renewable natural energy. It is an ideal energy source for underwater gliders in the deep sea and long working environment [7]. In this paper, a thermal energy acquisition system is proposed, which

Y. Huang et al. (Eds.): ICIRA 2017, Part I, LNAI 10462, pp. 857–866, 2017.
DOI: 10.1007/978-3-319-65289-4_79

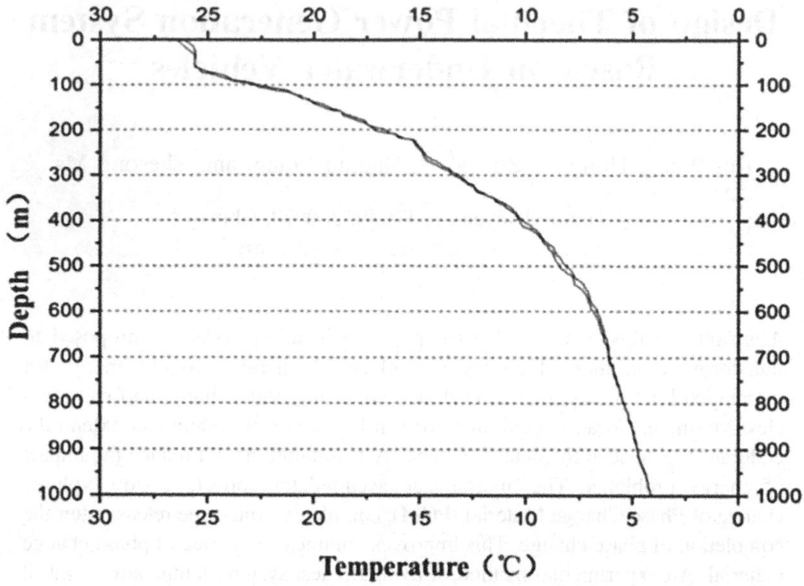

Fig. 1. Vertical temperature distribution. (1: Cooling-water machine(warm and cold); 2: Water tank; 3: Temperature recorder (can record the temperature of the liquid storage tank in real time); 4: Heat exchanger tube (contains a phase-change material and oil); 5: Lift pump; 6: One-way relief valve; 7: One-way valve; 8: Variable displacement hydraulic pump; 9: Oil tank; 10: One-way valve; 11: Pressure gauge; 12: Ball valve; 13: Bidirectional hydraulic cylinder; 14: One-way throttle valve; 15: One-way variable displacement hydraulic pump; 16: electromagnetic bidirectional relief valve.)

The temperature distribution is different in different thermocline [9]. It is assumed that the experimental sea area is the deep water thermocline. Figure 1 (between 118.671 ~ E, 21.311 ~ N China sea waters southwest of Taiwan and Dongsha Islands) is the curve of temperature with depth measured at the point of CTD.

It can be used to obtain the ocean thermal design for the thermal energy glider, with the positive hexadecane (see Table 1) as the phase change material. The phase transition temperature is 18.15 °C, corresponding to the depth of the ocean 170.2 m. From the water surface to 170.2 m is the melting heat absorption, while below 170.2 m is the solidification heat release. The change of sea water temperature is simulated to obtain the change of the liquid volume fraction of hexadecane. To provide experimental data support for the next phase transformation, loading pressure (5 MPa, 15 MPa and 22.1 MPa) obtained the curves of the hexadecane volume by measuring and processing the data, analyzed the change of liquid volume fraction of hexadecane under different pressures.

Table 1. Physical properties of hexadecane [8].

Melting point/°C	Latent heat/Kj.$(kg)^{-1}$	Specific heat capacity cps/cp1/Kj.$(kg.K)^{-1}$	Density ρs/ρ1/kg.m^{-3}	Thermal conductivity ks/keff/w. $(m.k)^{-1}$
18.15	236	1.64/2.09	864/773	0.14/1.48

2 Thermal Power Generation System Model

The thermal power generation system is mainly composed of the phase change material working unit, the mechanical transmission unit and the circuit unit. The accumulate stores the slow phase change process. After energy storage to complete, the pressure is released than the hydraulic cylinder piston promotes the ball screw and accelerator drivers general to produce electricity. As shown in Fig. 2, the phase change material starts to expand (the solid state melts into the liquid state), the hydraulic oil in the heat transfer tube 1 flows into the left end of the double acting hydraulic cylinder 2, drives the piston to move to the right. Open the solenoid valve 4 and close the solenoid valve 6, presses the low pressure liquid at the right end of the double acting cylinder into the accumulator3. The pressure of the accumulator 3 increases. After the phase change is completed, the accumulator 3 is finished and the solenoid valve 6 is opened to push the piston of the one-way hydraulic cylinder 7 to the right and push the screw nut mechanism 8 to drive the screw. The generator is rotated by the speed mechanism 9. When the phase change material begins to solidify (solidified by liquid solid), the rubber hose of the heat transfer tube 1 is in a vacuum state. The left side oil of the double acting cylinder 2 is sucked into the heat transfer pipe 1, while opening the solenoid valve 6 closing and the solenoid valve 4. The oil in the single acting hydraulic cylinder 7 is sucked into the right side of the double acting hydraulic cylinder 2, and the ratchet wheel 9 works as a one-way rotation to disconnect the lead screw 9 from the drive mechanism of the speed increasing mechanism 10. Above is completed a round of power generation.

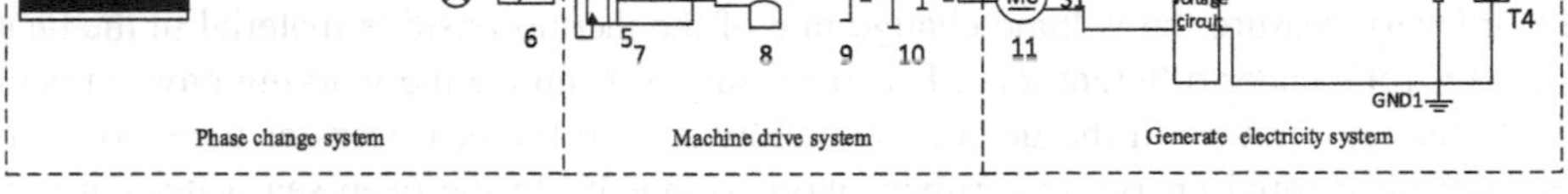

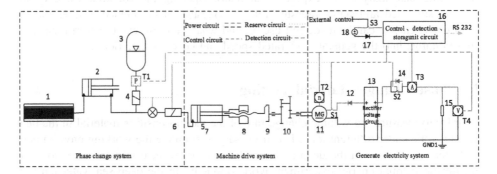

Fig. 2. Principle diagram system. (1: heat transfer tube; 2: double acting hydraulic cylinder; 3: accumulator; 4: solenoid valve; 5: pressure gauge; 6: solenoid valve; 7: single acting hydraulic cylinder; 8: ball screw; 10: growth rate institutions; 11: permanent magnet brushless DC motor; 12: diode; 13: rectified voltage circuit. 14: diode; 15: dissipation resistor. 16: control, detection, storage unit; 17 diode. 18 control, detection, storage chip power; T1: hydraulic valve high pressure detection sensor. T2: motor speed sensor; T3: load current sensor; T4: load voltage sensor; S2: load power control switch. S3: control, detection, memory chip power switch; RS 232: Communication serial port.)

2.1 Power Generation Control System

After the single chip microcomputer control system is initialized, the switch of parallel regulator is closed to achieve the circuit's turning on. Detecting the rotate speed of generator 11 with a motor speed sensor, when rotate speed of generator 11 is equal or greater than 5500 r/min, the generator switch S1 is closed and the battery charge switch S1 is closed as well, the batteries can be charged. When the battery power reaches the set maximum power of the battery, the switch S1 is turned off and the batteries stop charging. If the battery charge does not reach the maximum battery charge set, the battery charge switch S1 remains closed. When the generator speed is less than 5500 r/min, the battery charge switch S1 is disconnected, the battery stops charging, and the generator switch S1 is disconnected.

2.2 Accumulator System with Function of Intermediate Energy Storage

The accumulator is a hydraulic element which can store the pressure liquid, which can not only be used as an auxiliary hydraulic source, but also be used as a hydraulic device to absorb and store excess energy of the hydraulic system. What's more, it can ease pressure and shock of the system [10]. As the phase transition process is slow, the amount of oil produced by the phase change is not sufficient to reach the rated speed of the generator by using the ball screw and the speed increasing mechanism. Therefore the accumulator can play a role of intermediate storage. By opening the solenoid valve 4 and closing the solenoid valve 6, the accumulator 3 can store the oil from the phase change material at a time (Specific liquid phase change can be seen in chap 3.4 the processing and analysis for experimental data). Opening the solenoid valve 6 can achieve sustained and stable pressure release. Because of the device ensures the oil from phase change can be made full use of, and what's more it can transform the slow phase change into the generator 11 rated speed and generate electricity.

3 Phase Change Material Testing

In order to measure the volume change rate of the thermosensitive material in the heat transfer tube under different load. It is necessary to simulate the working environment of water temperature in the deep and shallow sea for the heat transfer tube, so as to ensure the thermal energy experiment work smoothly. In the deep sea water temperature of 4 °C. while the ocean surface temperature of about 30 °C. The thermal is large, which provide energy for expansion and contraction of the phase change material. The phase change volume and the liquid fraction were obtained under different load pressures by testing the Hexadecane phase change system. In this experiment, the test system is mainly composed of thermal control section, thermal phase change section, hydraulic test section and measurement record section (Fig. 3).

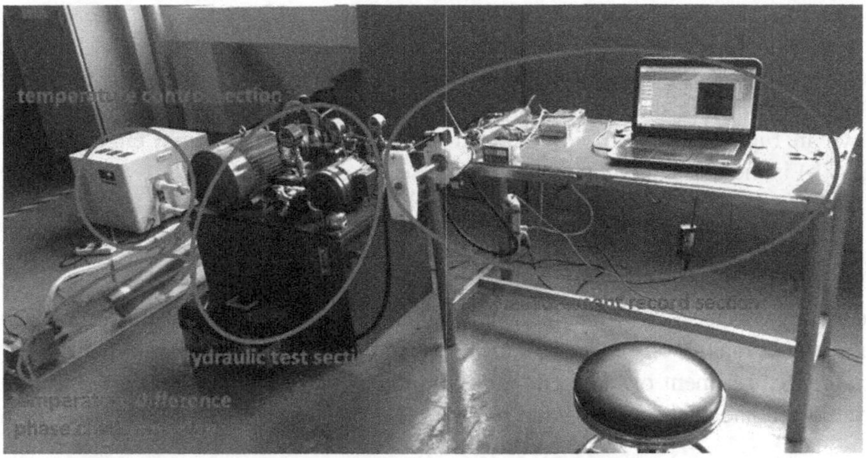

Fig. 3. Experimental test system platform.

3.1 Phase Change Section

Phase change is the mutual transformation of the different phases of the material system. In the mutual change, substance often accompanied by endothermic, exothermic and volume mutation so on. Techniques that are converted to the required energy using phase change volume changes are being continually improved [11]. The thermal engine with a double-tube structure is developed from the above analysis. The flexible tube running down the center while outside of the thermal engine is a metal shell, as shown in Fig. 4. Figure 4 shows that the thermal material fills the space between the flexible tube and the metal shell while the flexible tube is filled with oil. During the melting process,

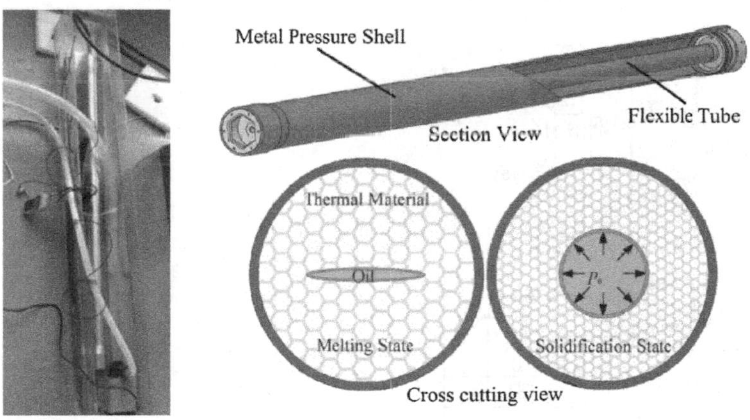

(a) (b)

Fig. 4. Thermal engine.

the thermal material expands to deflate the flexible tube and export the oil to the accumulator. During the contracting process, referencing the powder pressing theory, the isostatic pressing technology is adopted to improve the density of the thermal material by pressing the oil back to the flexible tube using a supercharging device, as shown in Fig. 4(a) section view and cross cutting view of thermal engine. And the Fig. 4(b) is the thermal engine which was used in experience.

3.2 Hydraulic Test Section

The heat exchange tube contains a phase-change material named hexadecane, which changes the volume of oil through the extrusion of the intermediate oil pipe. And the variable displacement of bidirectional hydraulic cylinder 13 could be converted to the variable volume of the hexadecane by measuring. The one-way variable displacement

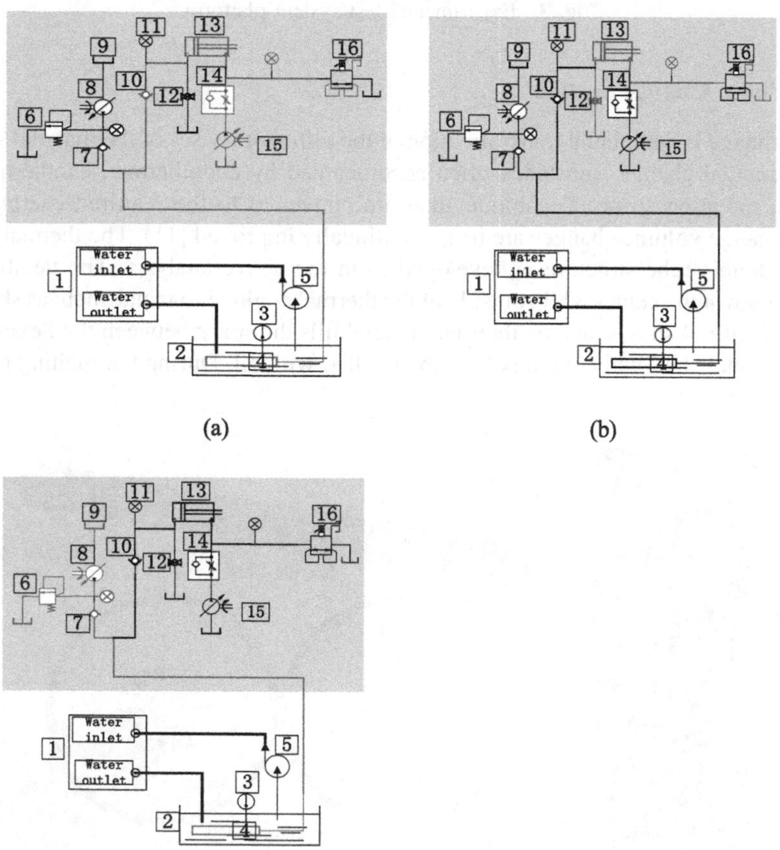

Fig. 5. Hydraulic test procedure.

hydraulic pump 15 provides the load pressure and the liquid volume fraction can be obtained by adjusting the load pressure.

Step 1: As shown in Fig. 5(a), the oil from the heat exchanger tube 4 flows through a one-way valve into the bidirectional hydraulic cylinder 13, because the phase-change material expands. At this point, the ball valve 12 closes and the hydraulic oil puts bidirectional hydraulic cylinder 13 right push. Then one-way variable displacement hydraulic pump brings the hydraulic oil as the load pressure from the oil tank into the right side of the bidirectional hydraulic cylinder 13. And the hydraulic oil flows from the right end of the hydraulic cylinder to the electromagnetic bidirectional relief valve 16 and then flows into the oil tank.

Step 2: As shown in Fig. 5(b), electromagnetic bidirectional relief valve 16 reverses connection and hydraulic oil flows into the bidirectional hydraulic cylinder 13 in opposite direction. At this point, the ball valve 12 opens and hydraulic oil flows into the oil tank.

Step 3: As shown in Fig. 5 (c), when the energy storage material is solidified from liquid, the heat exchanger tube 4 creates a state of partial vacuum. Then ball valve 12 closes and the hydraulic oil flows from the oil tank through the variable displacement hydraulic pump 8 and one-way valve 7 into the heat exchanger tube 4. In this circuit, variable displacement hydraulic pump 8 and one-way relief valve play a role in helping oil flow into the heat exchanger tube 4.

3.3 Measurement and Data Record Section

Experiment regards the cylinder as the carrier for a phase change volume measurement. The conversion of the carrier expansion of the piston rod shows the oil changes. The length of the piston rod is collected by the cable encoder. The data collected by the encoder is recorded by the data acquisition card, and the displacement with time curve of the piston rod are obtained. As shown in the Fig. 6.

3.4 Experimental Testing and Analysis

Experiment set the initial temperature of 5 °C. After 2 h of constant temperature coagulation, the temperature increases to 30 °C and heat to melt at constant temperature until the piston rod is no longer elongation. Through steps of the hydraulic test system 2.2 respectively, then compared the phase change of hexadecane under different pressures (5 MPa, 15 MPa and 22.1 MPa). The volume fraction of hexadecane liquid is changed with time in Fig. 7. As can be seen from Fig. 7, in the initial stage of the melting process, the volume expansion rate is small, that is, the slope of the curve is small. As the phase transition process goes, the rate of change increases rapidly and then decreases gradually until it finally approaches zero. With the progress of the phase transition process, solid hexadecane began to melt, and liquid hexadecane is

The Sudden inflection point 1, 2, and 3 of each curve are formed by unloading the load pressure.

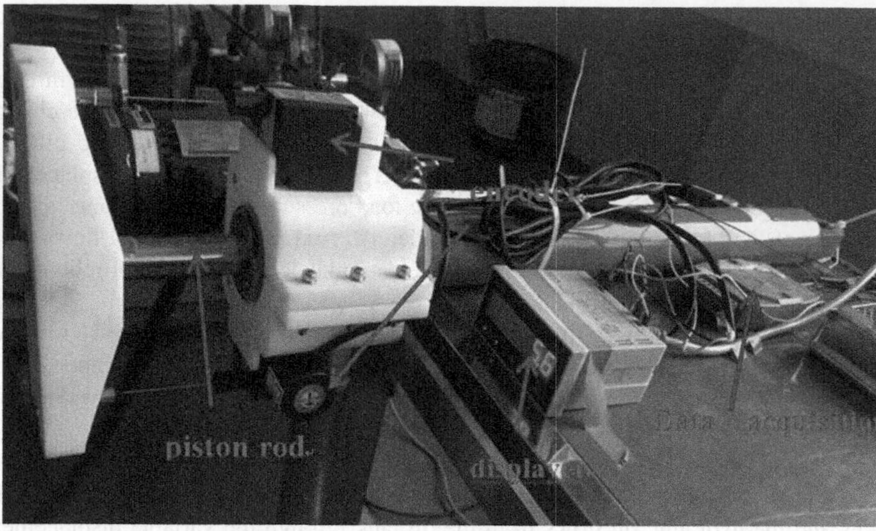

Fig. 6. Measurement and data record section.

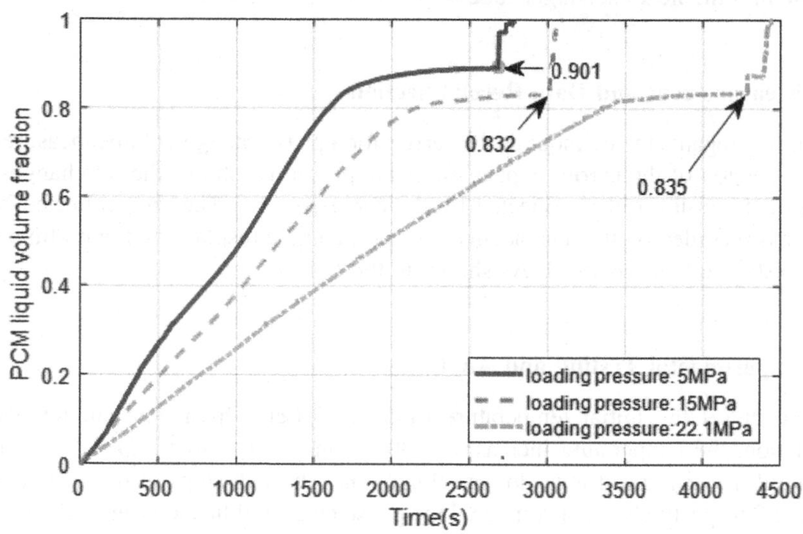

Fig. 7. Linear relationship between wall thickness and inner diameter in different internal design pressure condition.

formed on the surface. The thermal conductivity of solid hexadecane is greater than that of liquid hexadecane, in the initial stage of thermal resistance is relatively large and slow growth rate of volume change. With the gradual increase of liquid hexadecane content, due to the role of gravity, the temperature liquid hexadecane internal natural convection, and the settlement movement caused by solid-liquid density, the melting process is accelerated, the volume change rate is increased as well.

As the melting process goes, the composition of the solid hexadecane is less and less, and the amount of hexadecane melted at each time interval is less and less. Until the final hexadecane is completely melted, the volume change is reduced to zero.

The greater the pressure, the slower the phase transition, and the final phase change volume decreases as the pressure increases. The volume of the oil and the hexadecane liquid is only compressed due to the increase in the load pressure. At the load pressure of (5 MPa, 15 MPa, 22.1 MPa), the final liquid volume fraction is 0.901, 0.835 and 0.832.

4 Concentration

In this paper, an experimental platform is proposed to verify the phase ransition principle, and the feasibility of the phase change work of hexadecane under different load pressures is verified by using the experimental system of thermal power generation system. In the experiment, it also can be seen that the rate of transformation of hexadecane decreases as the pressure increases, and the phase change volume is affected by the compression amount of hexadecane and oil. The phase change volume slows as the pressure increases. The phase transition under this experimental model can be carried out at a load pressure of 22.1 MPa, which provides experimental feasibility for subsequent phase change power generation.

Acknowledgments. This work is funded by the National Major Scientific Instruments Development Program of China (No. 41527901). We thank other members of the project team to provide detailed sea trial data, and the discussions and pertinent criticisms of colleagues are helpful to us.

References

1. Xue, G., et al.: Discussion on ocean thermal energy and its sustainable utilization. J. Ocean Univ. China **02**, 15–19 (2008). (Social Science Edition)
2. Tossen, T.I.: Guidance and Control of Ocean Vehicle. Wiley Interscience, New York (1994)
3. Ni, Y.: Study on the Performance of Underwater Glider Driven by Temperature Difference. Shanghai Jiao Tong University, Shang Hai (2008)
4. Chao, Y.: Autonomous Underwater Vehicles and Sensors Powered by Ocean Thermal Energy. Institute of Electrical and Electronics Engineers, Shang Hai (2016)
5. Husaini, M., Samad, Z., Arshad, M.R.: Autonomous underwater vehicle propeller simulation using computational fluid dynamic, In: InTech 10, pp. 293–312 (2011)
6. Yan, X.: The world's largest solar collector – analysis of ocean thermal energy. Sol. Energy **02**, 8–9 (1998)
7. Yuh, J.: Design and control of autonomous underwater robots: a survey autonomous. Robots **8**(1), 7–24 (2000)
8. Kumano, H., Saito, A., Okawa, S., et al.: Study of direct contact melting with hydrocarbon mixtures as the PCM. Int. J. Heat Mass Transf. **48**(15), 3212–3220 (2005)
9. Wang, S., et al.: Design and experimental study of underwater glider driven by temperature difference energy. Ocean Technol. **25**(01), 1–5 (2006)

10. Sun, C., et al.: Selection of accumulator in hydraulic system retrofit. Fluid Power Transm. Control **06**, 48–51 (2005)
11. Kong, Q.: Study on the Phase Transition Process and Dynamic Performance of an Underwater Glider Using Ocean Thermal Energy. Shanghai Jiao Tong University, Shanghai (2010)
12. Yang, Y., et al.: A thermal engine or under water gilder driven by ocean thermal energy. Appl. Therm. Eng. **99**(25), 455–464 (2016)

Design and Simulation of a Self-adaptive Fuzzy-PID Controller for an Autonomous Underwater Vehicle

Jianhong Zhao, Wei Yi, Yuanxi Peng, and Xuefeng Peng(✉)

State Key Laboratory of High Performance Computing, College of Computer, National University of Defense Technology, Changsha 410073, China
zhaojianhong15@nudt.edu.cn, pengxf@163.com

Abstract. In recent years, AUV (autonomous underwater vehicle) has been applied to many fields, such as offshore oil exploitation, underwater target detection, military applications and so on, which raises a higher demand for accuracy control of AUV. In this paper, for the rim-drive dead zone and system nonlinearity problem of "Swordfish II" autonomous underwater vehicle in the actual control process, according to the shape and motion characteristics of the AUV, the six DOF (degrees of freedom) dynamic model is established under MATLAB/Simulink. The fuzzy control theory and the traditional PID control algorithm are combined to design a fuzzy-PID controller. Finally, we construct the simulation system. The simulation results show that the designed controller can control the AUV model and solve the dead zone problem well, and the fuzzy PID controller has better control effect than the traditional pid controller.

Keywords: AUV · Modeling · Simulation · Fuzzy-PID · Dead zone

1 Introduction

Ocean is the cradle of life, the ocean is rich in resources and energy. With the rapid development of science and technology, many countries have gradually increased the intensity of marine development, in which AUV is one of the important tools. Nowadays, AUV plays an irreplaceable role in the offshore oil exploitation, seabed detection and many other aspects. In order to ensure the task to be completed safely and smoothly, how to accurately control the AUV movement is the first problem to be solved. However, the AUV system is a complex, strong coupling, non-linear time-varying system, and the traditional linear control method is difficult to get a good control effect, so the study of appropriate control method is very necessary.

The research platform of this paper is "Swordfish II" unmanned submarine. The length of the submarine is about 5.085 m, the cylinder diameter of about 0.9 m, the cabin diameter of about 0.77 m, weighing about 2 tons. The submarine

Y. Huang et al. (Eds.): ICIRA 2017, Part I, LNAI 10462, pp. 867–878, 2017.
DOI: 10.1007/978-3-319-65289-4_80

relies on six rim-drives to provide the thrust required for movement, and the distribution of rim-drives is shown in the Fig. 1. During the actual component test, it was found that the rim-drives had a problem of dead zone, i.e., the rim-drives output was zero when the advancing pusher was less than 150 rpm(revolutions per minute) or the other pushers were less than 100 rpms. The nonlinearity of the system is exacerbated because of the dead zone and saturation limits of the thruster [7], and the control effect of the traditional PID controller can not meet the actual task requirements, mainly reflected in large overshoot, obvious oscillation, long stabilization time and so on.

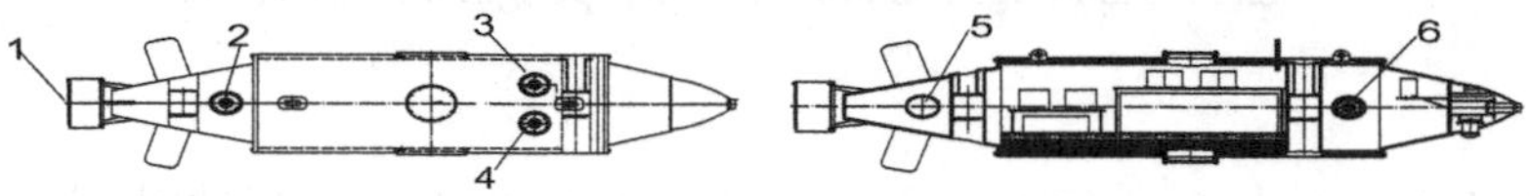

(a) Top view (b) Side view

Fig. 1. Distribution of the six rim-drives (numbered 1–6)

Through the study of literature, AUV control methods mainly include PID control [1,3,4,13,14,20], fuzzy control [5,12,18,19], neural network control [6,10,11] and a variety of hybrid control methods composed of two or more methods [8,9,15,17,21] and so on. PID control algorithm is widely used in various industrial control because of its simple structure, good stability, high reliability and easy realization. However, it has poor control effect on nonlinear, time-varying, complex and uncertain system. The fuzzy control method does not need the accurate system model, and has a satisfactory control effect on the parameter uncertain system. It is suitable for nonlinear, coupling and parameter uncertain systems. The neural network control method has the advantages of nonlinear approximation ability, good robustness and fault tolerance. It is suitable for uncertain system, but the method requires a lot of training data and long training time, moreover, in the early stage of entering the new environment, because there is not enough training data, the control effect will be reduced. The mixed control method will focus on the advantages of two or more control methods, learn from each other, to achieve better control effect.

In this paper, the self-adaptive fuzzy-PID control method combined with traditional PID algorithm and fuzzy control method is used to control the six degree-of-freedom AUV model and compared with the traditional PID controller. The simulation results show that the designed controller havs a good effect in reducing the influence of dead zone and non-linear, and overall, control effect is better than the traditional PID controller.

2 Modeling

2.1 Coordinate System and Parameters Definition

The analysis of the AUV motion is mainly based on two right-hand coordinate systems: fixed coordinate system (inertial coordinate system) E-$\xi\eta\zeta$ and body-fixed coordinate system (motion coordinate system) O-xyz. The origin of body-fixed coordinates can be set at the buoyancy center of the AUV, as shown in Fig. 2.

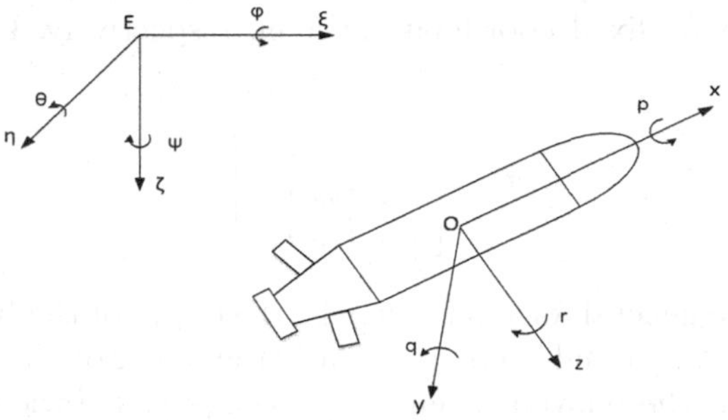

Fig. 2. Coordinate systems

In the fixed coordinate system, $[x, y, z]^T$ represents the position, $[\varphi, \vartheta, \psi]^T$ represents the rotation angle around x, y, z axis. In the body-fixed coordinate system, $V = [u, v, w]^T$ represents the velocity in three directions, $\Omega = [p, q, r]^T$ represents the angular velocity of rotation around the axis, $[X, Y, Z]^T$ represents the resultant force in three directions, $[K, M, N]^T$ represents the resultant moment.

The transfer matrix between the body-fixed angular velocity vector $[p, q, r]^T$ and the rate of change of the Euler angles $[\varphi, \vartheta, \psi]^T$ is as follows [9]:

$$\begin{bmatrix} \dot{\varphi} \\ \dot{\vartheta} \\ \dot{\psi} \end{bmatrix} = \begin{bmatrix} 1 & (\sin\varphi \tan\vartheta) & (\cos\varphi \tan\vartheta) \\ 0 & \cos\varphi & -\sin\varphi \\ 0 & (\sin\varphi \sec\varphi) & (\cos\varphi \sec\varphi) \end{bmatrix} \begin{bmatrix} p \\ q \\ r \end{bmatrix} \tag{1}$$

2.2 Kinematic and Dynamic Equations

In the assumption that the submarine is rigid and the water flow rate is zero, base on momentum theorem

$$\frac{dB}{dt} = F_\Sigma \tag{2}$$

and theorem of moment of momentum

$$\frac{dL}{dt} = M_\Sigma \tag{3}$$

the translational and rotational dynamic motion equations of AUV can be described as follows [16]:

$$m[\frac{dV}{dt} + \Omega \times V + \frac{d\Omega}{dt} \times R_G + \Omega \times (\Omega \times R_G)] = F_\Sigma \tag{4}$$

$$I\frac{d\Omega}{dt} + \Omega \times (I\Omega) + mR_G \times (\frac{dV}{dt} + \Omega \times V) = M_\Sigma \tag{5}$$

where B is momentum, L is moment of momentum, m is the mass of the AUV, R_G is the position of the center of gravity in the body-fixed coordinate system, F_Σ and M_Σ representing the resultant external force and resultant external moment in the body-fixed coordinate system, respectively. I is the rotational inertia matrix:

$$I = \begin{bmatrix} I_{xx} \; I_{xy} \; I_{xz} \\ I_{yx} \; I_{yy} \; I_{yz} \\ I_{zx} \; I_{zy} \; I_{zz} \end{bmatrix}$$

(4) and (5) are general formulas, that is, the origin of the body-fixed coordinate system is any point and is not necessarily the center of gravity. In this paper, the origin is set at the buoyancy center. F_Σ consists of static force(gravity and buoyancy), controlling force(the thrust provided by the rim-drive) and hydrodynamic force. With reference to the structure and hydrodynamic parameters of "Sparus II" [2], the hydrodynamic parameters and rotational inertia of "Swordfish II" are obtained as follows:

$$\begin{aligned} F &= (x_u + x_{uu} * |v|) * v \\ M &= (y_u + y_{uu} * |v|) * v \\ x_u &= [-110, -417, -417] \\ x_{uu} &= [-165, -417, -417] \\ y_u &= [-5.5, -56, -56] \\ y_{uu} &= [0, -70, -70] \\ I &= \begin{bmatrix} 269 & 0 & 0 \\ 0 & 2414 & 0 \\ 0 & 0 & 2414 \end{bmatrix} \end{aligned}$$

2.3 Modeling in Simulink

In this paper, the 6 DOF (Euler Angles) module is used to model the submersible under MATLAB/Simulink. The module (Euler Angles module) assumes that the origin of the body-fixed coordinate system is the center of gravity of the submersible:

$$m(\frac{dV}{dt} + \Omega \times V) = F_\Sigma \tag{6}$$

$$I\frac{d\Omega}{dt} + \Omega \times (I\Omega) = M_\Sigma \tag{7}$$

the origin of the body-fixed coordinate system is set to the buoyancy center of the AUV in this article, then $m(\frac{d\Omega}{dt} \times R_G + \Omega \times (\Omega \times R_G))$ and $mR_G \times (\frac{dV}{dt} + \Omega \times V)$ are the correction for the body-fixed coordinate origin is not the center of gravity. Finally, the dynamic model established is shown in Fig. 3.

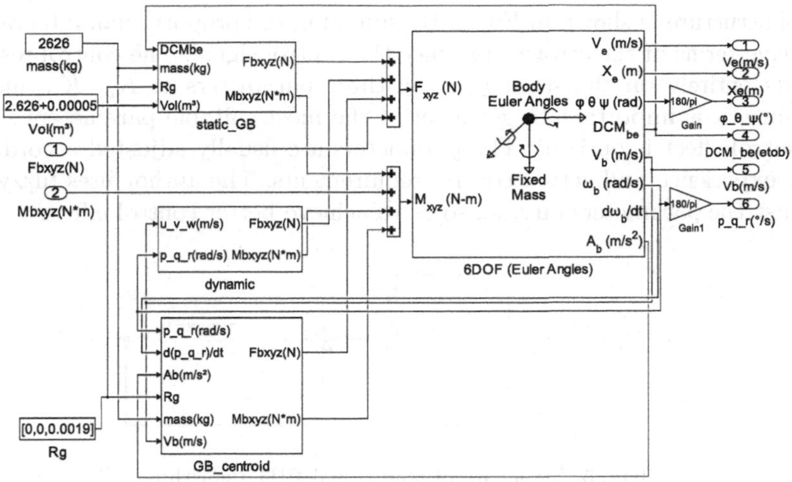

Fig. 3. Dynamic model under Simulink

Since the rim-drive has a problem of dead zone, we set the model output to 0 rpm when the number of revolutions is less than 150 (advance propeller)/100 (other propellers)rpm; the model outputs 1300 rpm when it is greater than 1300 rpm. As shown in Fig. 4.

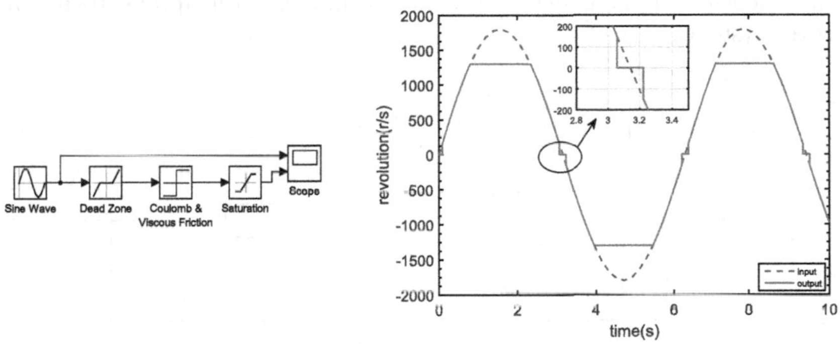

(a) (b)

Fig. 4. Non-linear module and simulation result

3 Control Method

3.1 Traditional PID Control Algorithm

PID control method is widely used in various industrial control processes because of its simple algorithm, high stability and good robustness. The typical PID control structure is shown in Fig. 5. By summing the proportional, integral, and derivative terms of the error to produce the control signal. The control response depends entirely on the setting of the three parameters of K_p, K_i, and K_d. Therefore, it is important to get a set of the most suitable parameters so that the control effect is optimal. The parameters are usually adjusted according to expert experience and actual control requirements. The author uses fuzzy logic to adjust the parameters on-line, so as to achieve better control effect.

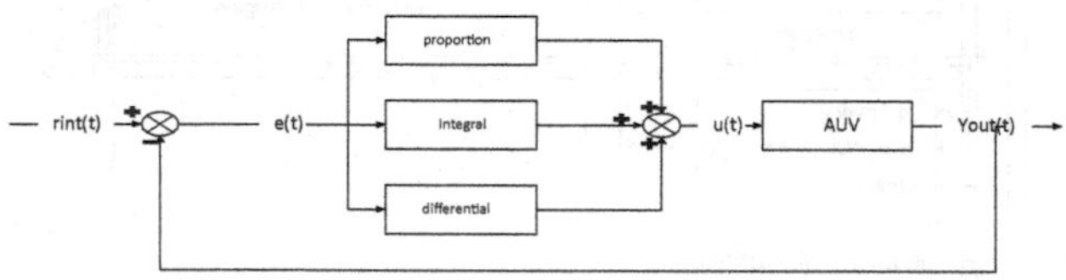

Fig. 5. Structure of traditional PID algorithm

3.2 Fuzzy-PID Control Method

The structure of the fuzzy-PID controller is shown in Fig. 6. The fuzzy reasoning part obtains the adjustment amount of K_p, K_i, and K_d according to the error(e) and the change of error(ec), so as to realize the on-line adjustment of the parameters. The fuzzy control theory relies only on the input e and ec, and does not depend on the exact mathematical model, so it applies to nonlinear, uncertain systems.

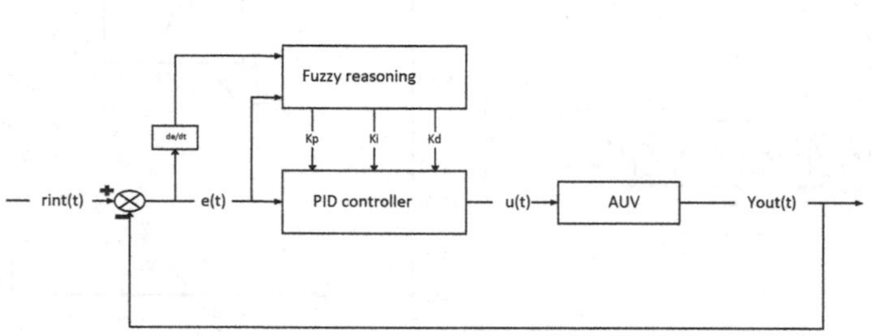

Fig. 6. Structure of fuzzy-PID algorithm

The structure of the fuzzy inference system (FIS) is shown in Fig. 7. Firstly, the input e and ec are fuzzified, and then the fuzzy value of the output is obtained according to the fuzzy reasoning rules in the fuzzy domain. Then, the exact output value is calculated by the defuzzification algorithm.

3.3 Design of the Controller

In this article, the toolbox in the MATLAB/Simulink is used to design the controller.

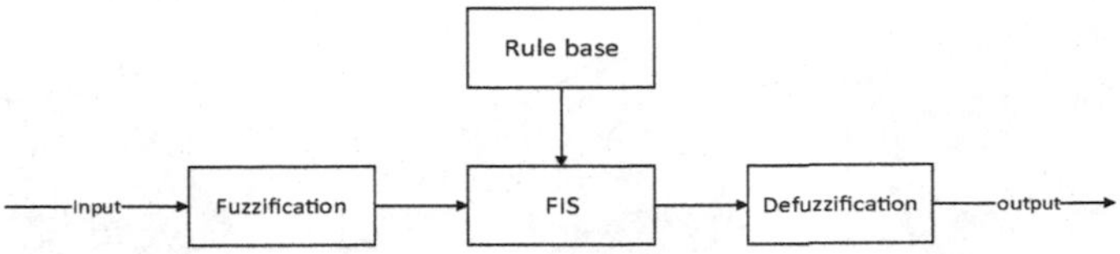

Fig. 7. Structure of fuzzy inference system

Fuzzification. The fuzzy set is e, ec = $\{nl, nm, z, pm, pl\}$, the domain is $[-1, 1]$; K_p, K_i, $K_d = \{s, m, l\}$, The range of output is $[0, 1]$. For the input and output, the combination of the s-shaped membership function and the triangular membership function is chosen, as shown in Fig. 8.

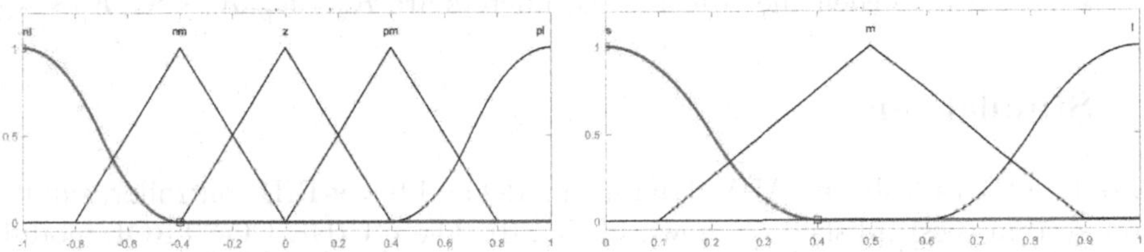

(a) Memberfunction of e and ec (b) Memberfunction of K_p, K_i, and K_d

Fig. 8. Memberfunctions

Fuzzy Rules. According to the analysis of PID parameters, the design of fuzzy rules need to consider the following points:
(a) if e is large, in order to improve the response speed, K_p should be large and K_i, K_d should be small;
(b) If e and ec are medium, in order to reduce the overshoot and maintain a good response speed, K_p should be medium;
(c) If ec is small, in order to stabilize the response, K_d should be large; but if ec is large, K_d should be small to reduce oscillation; usually keep medium.

With reference to the above, the fuzzy rule base is designed as follows: Table 1, Fig. 9.

Table 1. The rule table of K_p, K_i, K_d

(a) K_p

e \ ec	nl	nm	z	pm	pl
nl	l	l	l	l	l
nm	l	l	m	m	s
z	s	s	s	s	s
pm	s	m	m	l	l
pl	l	l	l	l	l

(b) K_i

e \ ec	nl	nm	z	pm	pl
nl	s	s	s	s	s
nm	m	m	m	m	m
z	l	l	l	l	l
pm	m	m	m	m	m
pl	s	s	s	s	s

(c) K_d

e \ ec	nl	nm	z	pm	pl
nl	s	s	s	s	s
nm	s	s	m	m	m
z	l	l	l	l	l
pm	m	m	m	s	s
pl	s	s	s	s	s

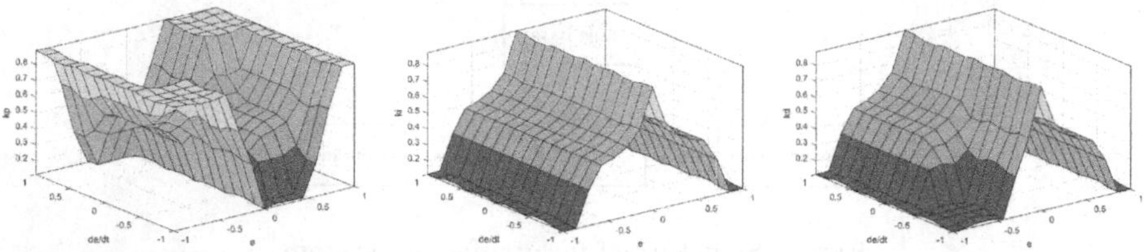

(a) K_p (b) K_i (c) K_d

Fig. 9. The rule surface of K_p, K_i, K_d

Defuzzification. Centroid method is used to calculate the exact $K_p^{'}$, $K_i^{'}$, $K_d^{'}$ parameters in the defuzzification process. Assuming that the original parameters are K_p, K_i, K_d, so then the adjusted parameters are $K_p^{'} * K_p, K_i^{'} * K_i, K_d^{'} * K_d$.

4 Simulation

Based on the established AUV dynamic model and fuzzy-PID controller, simulation system is established as shown in Fig. 10. The GUIDE_CONTROL module contains six independent control channels: $x, y, z, \varphi, \vartheta, \psi$. The x channel block diagram is shown in Fig. 11, and the other channels are similar to it.

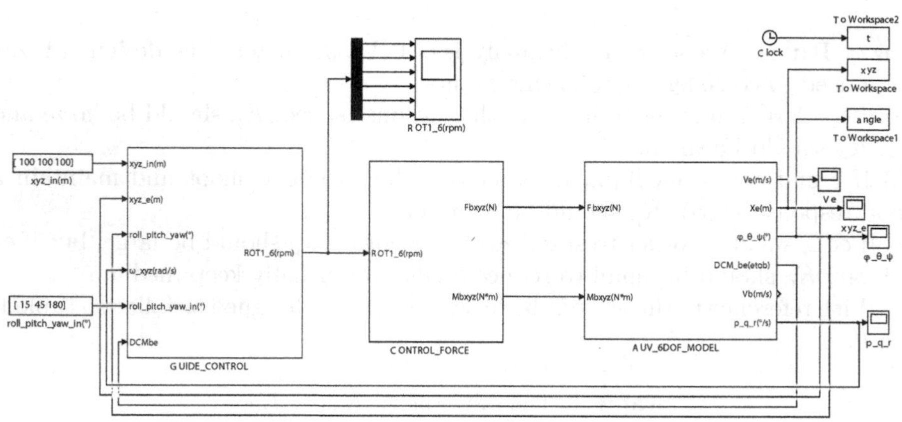

Fig. 10. Simulation system

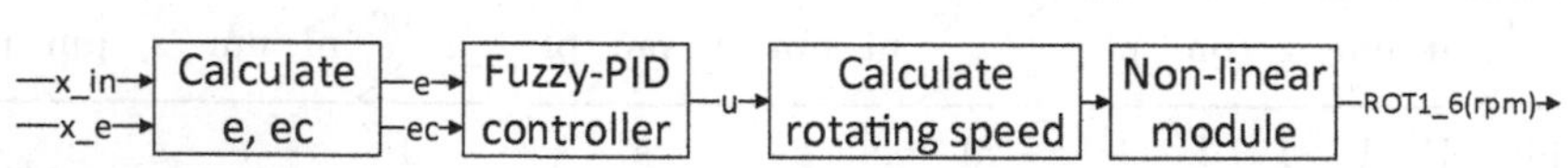

Fig. 11. x control channel in GUIDE_CONTROL module

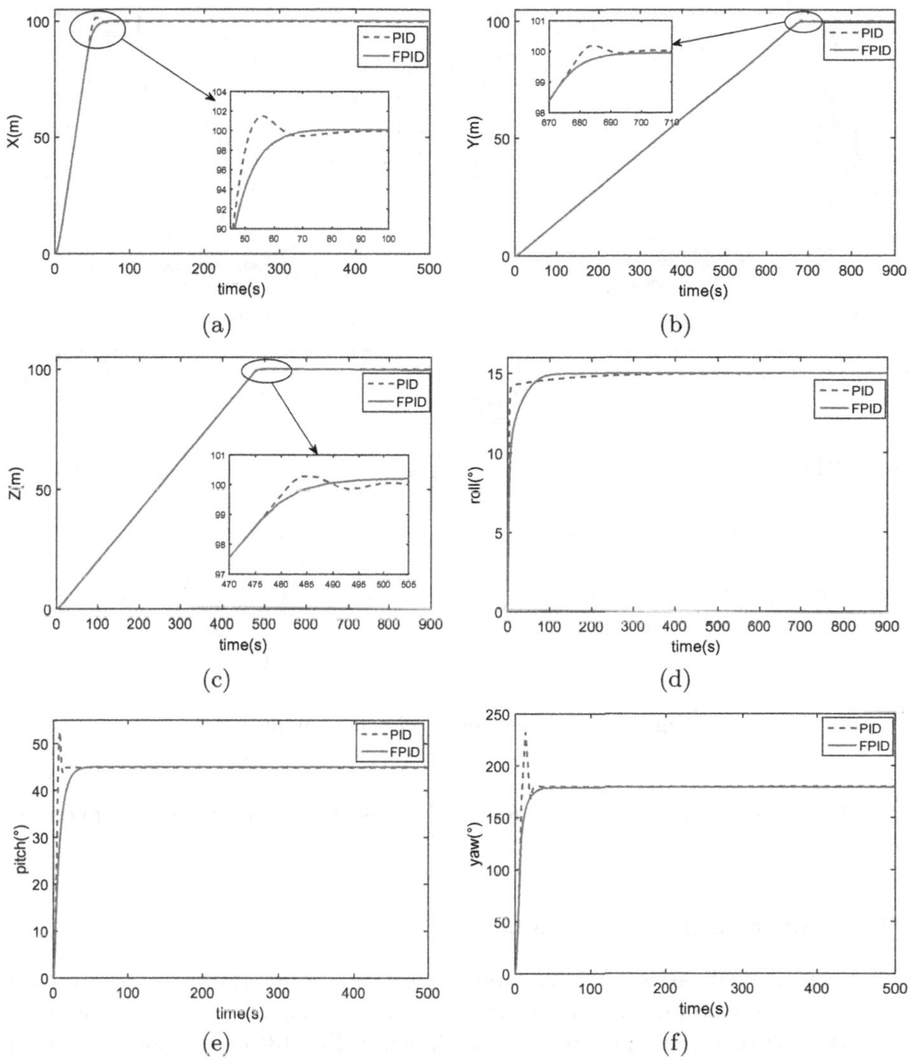

Fig. 12. Results of single channel simulation

4.1 Single Channel Simulation

The six degrees of freedom: $x, y, z, \varphi, \vartheta, \psi$ are simulated respectively. $[x, y, z, \varphi, \vartheta, \psi]$ is set to $[100, 0, 0, 0, 0, 0]$ for the simulation of x channel and $[0, 100, 0, 0, 0, 0]$ for y, $[0, 0, 100, 0, 0, 0]$ for z, $[0, 0, 0, 15, 0, 0]$ for φ, $[0, 0, 0, 0, 45, 0]$ for ϑ, $[0, 0, 0, 0, 0, 180]$ for ψ. The simulation results of traditional PID controller and fuzzy-PID controller are compared. The results are shown in Fig. 12.

As can be seen from the simulation results, compared with the traditional PID controller, the response curve obtained by the fuzzy-PID controller is

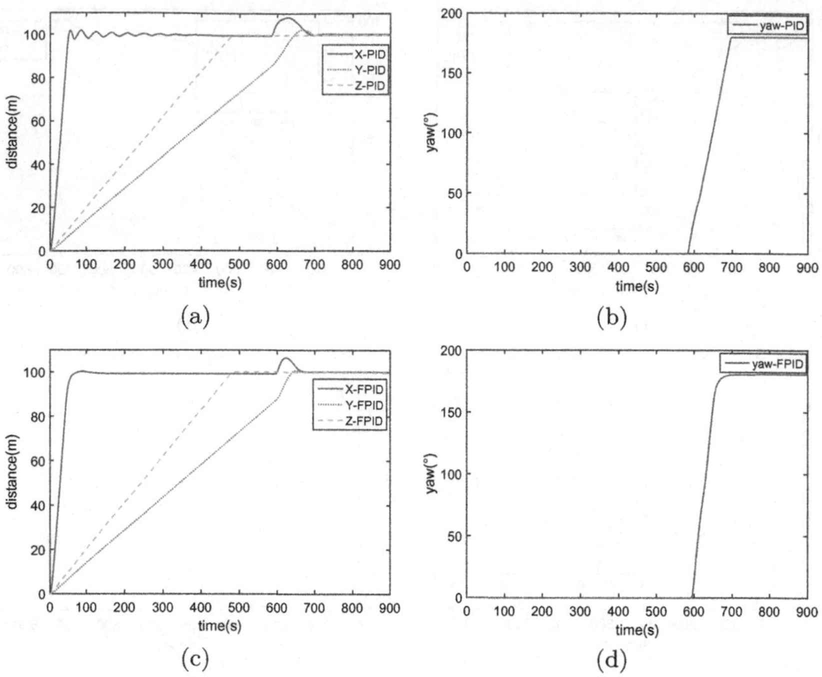

Fig. 13. Results of multi-channel simulation

smoother, the overshoot of x, y, z, θ, ψ is decreased, and the response speed of φ is improved.

4.2 Multi-channel Simulation

In practical applications, x, y, z and ψ are the most important four channels, so the four channels are simulated together and $[x, y, z, \varphi, \vartheta, \psi]$ is set to $[100, 100, 100, 0, 0, 180]$. The results are shown in Fig. 13(a), (b) and (c), (d) are the simulation results of traditional PID controller and fuzzy-PID controller respectively.

Simulation results show that, the designed controller can control the AUV model with dead zone nonlinearity effectively. Compared with the traditional PID controller, the response curve of yaw is smoother and the oscillation of X and overshoot of X and Y are obviously reduced.

5 Conclusion

In this paper, because the traditional PID algorithm has poor control effect on the "Swordfish II" AUV and the rim-drive has a problem of dead zone, a self-adaptive fuzzy-PID controller based on the fuzzy control theory and traditional

PID control algorithm is designed to control the established 6 DOF AUV model. The simulation results show that the designed fuzzy-PID controller can control the AUV effectively and compared with the traditional PID controller. It shows that control effect has improved significantly.

References

1. Ajmal, M., Labeeb, M., Dev, D.V.: Fractional order PID controller for depth control of autonomous underwater vehicle using frequency response shaping approach. In: 2014 Annual International Conference on Emerging Research Areas: Magnetics, Machines and Drives (AICERA/iCMMD), pp. 1–6. IEEE (2014)
2. Carreras, M., Candela, C., Ribas, D., Mallios, A., Magí, L., Vidal, E., Palomeras, N., Ridao, P.: Sparus ii, design of a lightweight hovering auv. In: Proceedings of the 5th International Workshop on Marine Technology (MARTECH), Girona, Spain, vol. 911, p. 163164 (2013)
3. Chemori, A., Kuusmik, K., Salumäe, T., Kruusmaa, M.: Depth control of the biomimetic U-CAT turtle-like auv with experiments in real operating conditions. In: 2016 IEEE International Conference on Robotics and Automation (ICRA), pp. 4750–4755. IEEE (2016)
4. Chen, Q., Chen, T., Zhang, Y.: Research of GA-based PID for AUV motion control. In: International Conference on Mechatronics and Automation, ICMA 2009, pp. 4446–4451. IEEE (2009)
5. Dong, Z., Wan, L., Liu, T., Zhuang, J.: Heading control of an AUV based on mamdani fuzzy inference (2015)
6. Guo, J., Chiu, F., Wang, C.C.: Adaptive control of an autonomous underwater vehicle testbed using neural networks. In: OCEANS 1995. MTS/IEEE. Challenges of Our Changing Global Environment, Conference Proceedings, vol. 2, pp. 1033–1039. IEEE (1995)
7. Hanai, A., Choi, H.T., Choi, S.K., Yuh, J.: Experimental study on fine motion control of underwater robots. Adv. Robot. **18**(10), 963–978 (2004)
8. Hu, B., Tian, H., Qian, J., Xie, G., Mo, L., Zhang, S.: A fuzzy PID method to improve the depth control of AUV. In: 2013 IEEE International Conference on Mechatronics and Automation (ICMA), pp. 1528–1533. IEEE (2013)
9. Khodayari, M.H., Balochian, S.: Modeling and control of autonomous underwater vehicle (AUV) in heading and depth attitude via self-adaptive fuzzy PID controller. J. Mar. Sci. Technol. **20**(3), 559–578 (2015)
10. Li, J.H., Lee, P.M.: A neural network adaptive controller design for free-pitch-angle diving behavior of an autonomous underwater vehicle. Robot. Auton. Syst. **52**(2), 132–147 (2005)
11. Lorentz, J., Yuh, J.: A survey and experimental study of neural network AUV control. In: Proceedings of the 1996 Symposium on Autonomous Underwater Vehicle Technology, AUV 1996, pp. 109–116. IEEE (1996)
12. Nag, A., Patel, S.S., Akbar, S.: Fuzzy logic based depth control of an autonomous underwater vehicle. In: 2013 International Multi-Conference on Automation, Computing, Communication, Control and Compressed Sensing (iMac4s), pp. 117–123. IEEE (2013)
13. Radmehr, N., Kharrati, H., Bayati, N.: Optimized design of fractional-order PID controllers for autonomous underwater vehicle using genetic algorithm. In: 2015 9th International Conference on Electrical and Electronics Engineering (ELECO), pp. 729–733. IEEE (2015)

14. Rout, R., Subudhi, B.: Inverse optimal self-tuning PID control design for an autonomous underwater vehicle. Int. J. Syst. Sci. **48**(2), 367–375 (2017)
15. Song, F., Smith, S.M.: Design of sliding mode fuzzy controllers for an autonomous underwater vehicle without system model. In: Oceans 2000 MTS/IEEE Conference and Exhibition, vol. 2, pp. 835–840. IEEE (2000)
16. Wang, B., Wan, L., Xu, Y.R., Qin, Z.B.: Modeling and simulation of a mini AUV in spatial motion. J. Mar. Sci. Appl. **8**(1), 7–12 (2009)
17. Wang, J.S., Lee, C.G.: Self-adaptive recurrent neuro-fuzzy control of an autonomous underwater vehicle. IEEE Trans. Robot. Autom. **19**(2), 283–295 (2003)
18. Wang, Y., Shen, Y., Wang, K., Sha, Q., He, B., Yan, T.: Fuzzy controller used smoothing function for depth control of autonomous underwater vehicle. In: OCEANS 2016-Shanghai, pp. 1–5. IEEE (2016)
19. Wu, J., Han, J., Yin, Y., Chen, G.: Variable universe based fuzzy control system design for AUV. In: OCEANS 2016-Shanghai, pp. 1–5. IEEE (2016)
20. Zhang, G., Du, C., Huang, H., Wu, H.: Nonlinear depth control in under-actuated AUV. In: 2016 IEEE International Conference on Mechatronics and Automation (ICMA), pp. 2482–2486. IEEE (2016)
21. Zhang, W., Wang, H., Bian, X., Yan, Z., Xia, G.: The application of self-tuning fuzzy PID control method to recovering AUV. In: Oceans, 2012, pp. 1–5. IEEE (2012)

Retraction Note to: The Summary of Underwater Gliders Control Strategies

Yuhai Liu, Xin Luan, Dalei Song, and Zhiqiang Su

Retraction Note to: Chapter "The Summary of Underwater Gliders Control Strategies" in: Y. Huang et al. (Eds.): *Intelligent Robotics and Applications,* Part I, LNAI 10462, https://doi.org/10.1007/978-3-319-65289-4_75

The editors have retracted this chapter [1] because it shows significant overlap with a previously published chapter by Ullah et al, 2015 [2]. The contents of this chapter are therefore redundant. All of the authors agree to this retraction.

1. Liu, Y., Luan, X., Song, D., Su, Z.: The Summary of Underwater Gliders Control Strategies. In: Huang, Y. et al. (Eds.) ICIRA 2017, Part I, LNAI, vol. 10462, pp. 813–824. Springer International Publishing (2017). doi:10.1007/978-3-319-65289-4_75
2. Ullah, B., Ovinis, M., Baharom, M.B., Javaid, M.Y., Izhar, S.S.: Underwater Gliders Control Strategies: A Review. In: 10th Asian Control Conference (ASCC). IEEE (2015). doi:10.1109/ASCC.2015.7244859

The retracted online version of this chapter can be found at
https://doi.org/10.1007/978-3-319-65289-4_75

Y. Huang et al. (Eds.): ICIRA 2017, Part I, LNAI 10462, p. E1, 2017.
https://doi.org/10.1007/978-3-319-65289-4_81

Author Index

Abad, Vanessa II-671, II-683
Abaroa, Thomas II-695
Aguilar, Wilbert G. II-671, II-683, II-695, III-287, III-298
Ahmad, Rafiq II-3
Ai, Qingsong I-301, II-812
Ai, Shangyou III-377
Albrecht, Stefan Tobias II-405
Álvarez, Leandro III-298
Andaluz, Víctor H. II-889, III-354, III-740
Angulo, Cecilio III-287
Arteaga, Oscar III-354

Bai, Dianchun I-206
Bai, Luoyu III-203
Bai, Wenzhi II-101
Bi, Qingzhen II-157
Bi, Weiyao II-856
Bi, Yingming I-699
Bian, Changsheng I-120
Botzheim, János I-560
Boyd, Peter III-98

Cai, Zhongxuan I-659
Cao, Baoshi II-286
Cao, Guangzhong I-381
Cao, Rongyun I-173
Cao, Yu II-879
Cao, Zixiang III-753
Carvajal, Christian P. III-354, III-740
Chan, Sixian III-449
Chang, Xuefeng I-659
Chang, Yuanyuan III-686
Chang, Zongyu I-846
Chao, Yongsheng III-193
Chen, Bing II-79
Chen, Chongyu II-634
Chen, Chuangrong II-634
Chen, Disi I-539
Chen, Gang III-551
Chen, Guangda III-267
Chen, Haojie III-203
Chen, Hualing I-120
Chen, Jun I-593
Chen, Lianrui III-538
Chen, Pinggen III-109
Chen, Qiaohong II-199
Chen, Shengyong III-449
Chen, Tao I-699, I-766, II-359
Chen, Wanmi III-335
Chen, Weidong II-588
Chen, Wenbo III-320
Chen, Xiaoping I-173, III-267, III-275
Chen, Xingyu I-605
Chen, Xuedong II-846
Chen, Yanyan II-800
Chen, Ying I-273
Chen, Yingfeng III-267, III-275
Chen, Yingwu I-683
Chen, Yong II-273
Chen, Yongfu III-166
Chen, Yunan II-869
Chen, Zhihao III-320
Chen, Zhong II-543
Cheng, Bin I-636
Cheng, Hui II-634
Cheng, Shili II-441
Cheng, Xiangliang III-650
Cheng, Yuanchao III-686
Cheng, Zhigang III-25
Cong, Qian I-63
Costa-Castello, Ramón III-287
Cui, Guowei III-267
Cui, Hongwei III-717
Cui, Shipeng II-238
Cui, Zhijian I-745

Dai, Jian S. I-71
Dai, Zhengchen II-298, II-311
Dang, Linan II-764
Deng, Jiaming II-371
Deng, Songbo III-686
Ding, Guangzheng II-879
Ding, Han I-373, II-14, II-24, II-59, II-157, III-309, III-377, III-765
Ding, Li II-789
Ding, Wan II-856
Ding, Ye II-184

Dong, Huimin II-135, II-147
Dong, Wei III-551, III-563
Dong, Wentao III-77
Dong, Xiaomin III-109
Dong, Zhurong III-717
Du, Baosen III-637
Du, Feng I-551
Du, Guangchao I-846
Du, Li II-453
Du, Wenliao II-429
Du, Xinle I-617
Du, Xu II-171
Du, Zhijiang I-195, II-417

Erazo, Yaritza P. II-889
Estrella, Jorge I. II-683

Fan, Shaowei I-217
Fan, Wei II-383
Fang, Fang III-141
Fang, Lijin II-225
Fang, Yinfeng III-98
Feng, Jingkai III-613
Feng, Kai II-576
Feng, Rui II-727
Feng, Yi Min III-563
Fu, Mingliang I-474
Fu, Xuelei III-808, III-831
Fu, Zhiyu II-111
Fuertes, Ronnie II-695

Gan, Haitao I-313
Gan, Wenyang I-711
Gao, Bo II-717
Gao, Farong I-313
Gao, Hang I-766
Gao, Lifeng I-779
Gao, Peng II-124
Gao, Qing I-462
Gao, Yunfeng II-659
Ge, Yong III-130
Ge, Yunhao I-14, I-26
Gong, Changyang II-776
Gong, Hu III-819
Gong, Jixiang II-486
Gong, Yi I-405
Gong, Zeyu III-309
Gu, Guo-Ying I-140
Gu, Yikun II-286, III-343, III-484, III-729
Guan, Enguang I-26, II-124, III-527
Guo, Bingjing III-495, III-506
Guo, Tingting I-745
Guo, Weichao I-373
Guo, Weizhong II-261
Guo, Zhao III-44

Han, Jianhai III-495, III-506
Han, Jong Ho III-563
Han, Liang III-650
Han, Liya II-101
Han, Xiao I-361, I-405
Hang, Lubin II-111
Hao, Yuan I-206
Hao, Yufei I-151
Hasegawa, Yasuhisa I-242
He, Fengjing II-359
He, Hong III-788, III-808, III-831
He, Jingfeng II-334
He, Leiying II-199
He, Mingchang I-162
He, Ping III-717
He, Qingchao III-707
He, Shanshan III-819, III-877
He, Wei II-869
He, Xiuyu II-869
He, Yu III-673
Heo, Jae III-69
Hou, Zhenghua II-659
Hu, Binbin I-628
Hu, Jian II-739
Hu, Jianhua II-48
Hu, Jiwei I-301
Hu, Pengcheng III-855
Hu, Songhua III-717
Hu, Wenlong I-570
Hu, Yongheng II-634
Huang, Bo I-330
Huang, Dongliang I-94
Huang, Hailin II-405
Huang, Haoguang II-634
Huang, Jian II-879
Huang, Jie II-171
Huang, Ju II-273
Huang, Panfeng III-662, III-698
Huang, Qi I-217
Huang, Qiang III-637
Huang, Sudan I-381
Huang, Yanjiang II-576, II-647

Huang, Yao I-273
Huang, YongAn III-77
Huang, Yuancan II-823

Ibarra, Alexander II-695

Ji, Feixiang III-843
Ji, Jingjing III-120
Jia, Lei II-36, III-377
Jian, Wang III-401
Jiang, Dapeng I-735
Jiang, Du I-517, I-528, I-539, I-551, I-570
Jiang, Guozhang I-517, I-528, I-539, I-551, I-570
Jiang, Haijun II-508
Jiang, Hongzhou II-334, II-359
Jiang, Jiaoying III-120
Jiang, Li I-217, III-343, III-413, III-729
Jiang, Min II-273
Jiang, Qingyuan II-752
Jiang, Tian II-520
Jiang, Xiaoqiang II-846
Jiang, Yinlai I-206
Jiao, Niandong I-110
Jin, Minghe II-286, III-707
Jin, Zhiyong III-166
Jing, Xiaolong II-776
Ju, Zhaojie I-438, I-450, I-462, I-485, I-517, I-528, I-539, I-560, I-570, III-449

Ke, Da II-812
Ke, Yin II-823
Kim, Dong-Eon III-215
Kong, Detong I-505
Kong, Jianyi I-517, I-528, I-539, I-551, I-570
Kuang, Yongcong II-611
Kubota, Naoyuki II-703
Kusano, Takamasa II-70
Kwon, Kyuchang III-237
Kwon, Younggoo III-69, III-237, III-246

Lan, Xuguang I-605
Lee, Chen-Han III-819, III-855, III-877
Lee, Jang-Myung III-215
Lee, Kok-Meng III-120
Li, Bei I-528
Li, Bin I-14
Li, Bing II-405
Li, Bo I-120, I-593, II-273
Li, Can I-790
Li, Chao II-486
Li, Chaodong II-346
Li, Chongqing II-334
Li, Chunxue III-517
Li, Congji III-788
Li, Fangxing II-823
Li, Gen II-101
Li, Gongfa I-517, I-528, I-539, I-551, I-570
Li, Guangyun III-254
Li, Guoliang I-683
Li, Hai II-576
Li, Huijun I-350
Li, Jiayu II-371, II-611
Li, Junfeng I-162
Li, Junwei III-449
Li, Kai II-599
Li, Kaichao I-39
Li, Ke II-739, III-686
Li, Miao III-472
Li, Ming I-63
Li, Minglei III-254
Li, Mingyu III-130
Li, Pan I-39
Li, Peixing III-527
Li, Qian I-162
Li, Qinchuan II-199
Li, Quanlin I-340
Li, Shan II-79
Li, Shuai II-543, II-823
Li, Tao III-388
Li, Wentao II-111
Li, Wenxiao III-506
Li, Xiangfei II-24
Li, Xiangpan III-495, III-506
Li, Xiaoqin III-788
Li, Xiao-tao I-721
Li, Xin III-179
Li, Xing I-51
Li, Xiyan III-855
Li, Xudong II-776
Li, Xun II-835
Li, Yang I-3
Li, Yangmin I-462
Li, Yihao II-429
Li, Ying I-313
Li, Zhe I-551
Li, Zhijun I-438
Li, Zhiqi III-460
Li, Zhiqiang I-120

Li, Zhongkui I-636
Lian, Shujun III-495
Liang, Bo II-311
Liang, Guoqiang I-605
Liang, Jinglun II-611
Liang, Le II-800
Liang, Shaoran III-576, III-602
Liang, Songfeng III-717
Liao, Bin II-250
Liao, Zhiwei I-3
Limaico, Alex III-298
Lin, Rongfu II-261
Lin, Xinghua I-757
Lin, Xueyan III-87
Lin, Yunhan III-425
Lin, Zhihao III-154
Ling, Luxiang II-543
Liu, Aiming I-581, III-3
Liu, Bin I-628
Liu, Dong I-757
Liu, Dongyu III-673
Liu, Fang I-721
Liu, Fei I-185, III-517
Liu, Hong I-217, II-238, II-286, III-413, III-460, III-484, III-673
Liu, Honghai I-450, I-560, III-98
Liu, Huan II-311
Liu, Jia I-350
Liu, Jiayu III-637
Liu, Jihao I-14, I-26, III-527
Liu, Jinguo I-462, III-613
Liu, Kai I-340, III-831
LIU, Kun I-242
Liu, Lei I-120
Liu, Lianqing I-110
Liu, Peijun I-290
Liu, Quan I-301, I-581, II-812
Liu, Wei I-361, I-505, III-437
Liu, Weiqi III-109
Liu, Xin-Jun II-3, II-212, II-856
Liu, Xiuhua I-262
Liu, Yang I-84, III-484
Liu, Yanjie II-800
Liu, Yanyang I-290
Liu, Yiming III-472
Liu, Yinhua III-843
Liu, Yiwei II-238, III-673
Liu, Yu II-486, II-623
Liu, Yufei II-91
Liu, Yuhai I-801, I-813
Liu, Yunxuan I-393
Liu, Zhanli I-505, III-437
López, William II-683
Lou, Yunjiang II-250
Loza, David II-695
Lu, Guiqiao I-846
Lu, Haojian I-84
Lu, Jiuru II-111
Lu, Qinghua II-394, III-154
Lu, Shaotian III-343, III-729
Lu, Xin I-846
Lu, Zongxin I-3
Luan, Xin I-801, I-813
Luna, Marco A. II-671
Luna, Marco P. II-671
Luo, Can III-650
Luo, Chao III-673
Luo, Chaomin I-711
Luo, Daichao I-735
Luo, Haitao I-474
Luo, Ming III-388
Luo, Qingsheng II-717
Luo, Xin II-846
Luo, Yifan III-335
Luo, Zirong I-51
Lv, Bo I-373

Ma, Bin II-273
Ma, Haifeng III-15
Ma, Haoqin I-617
Ma, Hongbin I-393
Ma, Letong III-765
Ma, Xudong III-141
Ma, Zhao II-776
Ma, Zhesong I-857
Mai, Jingeng I-262, I-280
Mao, Yangyang II-14
Mao, Yongfei III-506
Mei, Tao III-130
Meng, Fanjing I-290
Meng, Fanwei II-323
Meng, Qinmei II-371
Meng, Qizhi II-212
Meng, Wei I-301, II-812
Meng, Zhongjie III-698
Miao, Huihua I-14
Miao, Yu I-195
Min, Huasong III-425

Mo, Xixian III-765
Molina, María F. II-889
Moya, Julio F. II-671

Nagai, Isaku I-825, I-834
Nagata, Fusaomi II-70
Nian, Peng III-576, III-589, III-602
Ning, Yu II-429
Niu, Jie I-273

Okada, Yudai II-70
Ortiz, Jessica S. II-889
Ouyang, Gaofei II-611

Pan, Erzhen III-538
Pan, Gen II-124
Pan, Wuwei I-735
Pang, Jiawei III-753
Pang, Muye I-485, I-494
Pang, Yongjie I-735
Parra, Humberto II-671
Peng, Chen I-3
Peng, Fei III-563
Peng, Lei II-334, II-359
Peng, Xuefeng I-867, II-520
Peng, Yeping I-381
Peng, Yuanxi I-867, II-520
Pérez, José A. III-740
Pham, Duc Truong I-581, III-3
Phillips, Preetha III-367
Phiri, Charles C. I-560
Proaño, Luis E. III-740

Qi, Jingchen II-764
Qi, Junde II-79
Qi, Minhui I-494
Qi, Yuda I-735
Qi, Zhanhui I-790
Qian, Kun III-141
Qiao, Hong I-251
Qin, Tao I-290
Qingsheng, Luo III-624
Qiu, Yao I-195
Qiu, Zhicheng III-777
Qu, Chunlei III-166
Quan, Wei II-703
Quisaguano, Fernando III-298

Ren, Anye II-124
Ren, Guang II-346
Ren, Lei I-71
Rodic, Aleksandar II-739
Rodríguez, Guillermo A. III-298
Ruan, Songbo I-381
Ruiz, Hugo II-671

Sainkai, Yosiyuki I-242
Sandoval, Sebastián III-298
Sang, Hongqiang I-790
Saotome, Kousaku I-242
Segura, Luis II-695
Shan, Wentao II-789
Shang, Jianzhong I-51
Shao, Jia-Xin III-215
Shao, Zhufeng II-453
She, Hu-qing I-721
Shen, Hui-Min I-428
Shen, Huiping II-371
Shen, Jiaqi III-166
Shen, Mengzhu III-166
Shen, Xu II-3
Shen, Yajing I-84
Sheng, Xin Jun III-563
Sheng, Xinjun I-373, III-57, III-551
Shi, Chunyuan III-413
Shi, Jiahui I-393
Shi, Shicai III-707
Shi, Xiaodong II-564
Song, Aiguo I-350
Song, Bifeng III-576, III-589, III-602
Song, Dalei I-745, I-801, I-813
Song, Rong I-273
Song, Shuang II-752
Su, Chun-Yi I-438
Su, Guihua II-441
Su, Zhiqiang I-801, I-813
Sun, Bing I-711
Sun, Changyin II-869
Sun, Cheng I-485
Sun, Jingyuan II-417
Sun, Kui II-286
Sun, Lining I-94
Sun, Wenlei III-193
Sun, Xiangyi II-508
Sun, Xiujun I-790
Sun, Ying I-517, I-528, I-539, I-551, I-570

Sun, Yinghao II-311
Sun, Yongjun II-238

Tan, Bilian II-273
Tang, Chao I-120
Tang, Heng I-517
Tang, Li I-185, III-517
Tang, Shuai II-273
Tang, Yuanguang I-846
Tao, Bo III-309
Tao, Yi-Dan I-140
Tian, Weijun I-63
Tong, Xiaoqin III-35
Tong, Zhizhong II-334
Tong, ZhiZhong II-359
Tung, Steve I-110

Valle, Cristina I-560

Wan, Chuang I-766
Wan, Wenfeng I-84
Wan, Zhonghua I-323
Wang, Bin III-460
Wang, Bingheng III-698
Wang, Bo II-383
Wang, Boya I-217
Wang, Caidong II-429
Wang, Chenggang II-147
Wang, Danlin III-87
Wang, Daoming III-753
Wang, Delun II-135
Wang, Guohui I-857
Wang, Hesheng II-588
Wang, Hongdu I-745
Wang, Hongwei I-505, I-617, III-437
Wang, Jiliang I-846
Wang, Jinguo III-203
Wang, Jingyi I-110
Wang, Jiyue I-63
Wang, Li I-721, III-254
Wang, Liangwen II-429
Wang, Lili I-757
Wang, Liping II-453
Wang, Liqing II-835
Wang, Mingliang II-250
Wang, Naichen III-320
Wang, Nianfeng II-554, II-564
Wang, Qifu III-166
Wang, Qining I-251, I-262, I-280
Wang, Rong III-551
Wang, Rui I-230, I-857
Wang, Ruizhou II-464
Wang, Shaoning II-383
Wang, Shoujun III-226
Wang, Shuihua III-367
Wang, Shuo II-417
Wang, Tao II-383
Wang, Tianmiao I-151
Wang, Wenchang III-808
Wang, Ximeng I-779
Wang, Xu III-203
Wang, Xuefei I-671
Wang, Xuhong III-109
Wang, Yan II-225
Wang, Yanbin III-517
Wang, Yanbo III-686
Wang, Yang III-335
Wang, Yanzhen I-659
Wang, Yili III-320
Wang, Yitao I-417
Wang, Youhua III-77
Wang, Yue II-800
Wang, Yuzhu I-766
Wang, Zegang III-788
Wang, Zhe II-36
Wang, Zhenyu II-776
Wang, Zhichao III-460, III-484
Wang, Zhijie II-273
Wang, Zunran I-438
Watanabe, Keigo I-825, I-834, II-70
Wei, Fanan I-103
Wei, Guowu I-71
Wei, Linghua II-475
Wei, Shilin I-779
Wei, Yi II-273
Wei, Yifan I-39, II-727
Wei, Yimin III-517
Wen, Li I-151
Wen, Pengcheng I-350
Wu, Baohai III-388
Wu, Dewen I-581
Wu, Feng I-173, III-267, III-275
Wu, Guanglei II-135
Wu, Guohen I-51
Wu, Hao II-496
Wu, Hongtao II-441
Wu, Jianguo I-757
Wu, Jianhua II-599, III-15
Wu, Juan I-361, I-405

Wu, Miao I-405
Wu, Mofei III-25
Wu, Yu II-147
Wu, Yue I-628
Wu, Yunlong I-648

Xi, Wanqiang II-789
Xi, Zhigang III-154
Xiang, Kui I-450, I-485, I-494
Xiang, Li III-624
Xiao, Liang II-311
Xiao, Lin III-77
Xiao, Nong I-659
Xiao, Su II-475
Xiao, Xiaohui III-44, III-472
Xie, Fugui II-3, II-212, II-856
Xie, Jie II-647
Xie, Lingbo III-777
Xie, Shaobo II-79
Xie, Tao III-57
Xie, Zongwu III-484
Xing, Lining I-683
Xiong, Caihua I-323, I-330, I-340, II-532
Xiong, Gang II-184
Xiong, Renjie I-485
Xiong, Xin II-441
Xiong, Zhenhua II-599, III-15
Xu, Baoguo I-350
Xu, Benyan III-44
Xu, Binbin I-127
Xu, Changyu II-147
Xu, Da I-766
Xu, Dezhang II-91
Xu, Duo II-659
Xu, Kai II-298, II-311
Xu, Lamei III-25
Xu, Lingmin II-199
Xu, Peng III-788
Xu, Shangkun II-147
Xu, Wenfu III-538, III-650
Xu, Wenjun I-581, III-3
Xu, Xiangrong II-739
Xu, Xiongshi I-825
Xu, Yundou II-48, III-87
Xue, Yaxu I-450

Yamaguchi, Takanori I-834
Yan, Changya III-819, III-877
Yan, Guangwei III-796
Yan, Jihong I-127
Yan, Weixin I-14, I-26, III-527
Yan, Yao II-800
Yan, Zheping I-699, I-766
Yan, Zhiyuan I-195
Yang, Chenguang I-438, I-450
Yang, Dapeng I-217, III-413
Yang, Dongjun III-246
Yang, Fan II-124, III-686
Yang, Fangzhao III-855
Yang, Guobin II-111
Yang, Hai I-671
Yang, Haitao III-484
Yang, Hong-an III-203
Yang, Hua III-309
Yang, Huang III-401
Yang, Junyou I-206, I-230
Yang, Lin III-25
Yang, Lu III-226
Yang, Meng-Meng III-367
Yang, Wenqing III-576, III-589, III-602
Yang, Yanyi III-855
Yang, Yifu I-593
Yang, Yongliang I-110
Yang, Zhan I-94
Yang, Zhen I-63
Yao, Jiantao II-48, III-87
Yao, Lan III-753
Yao, Ligang I-3
Yao, Lin III-57
Yao, Song I-648
Ye, Congcong II-59
Ye, Dong III-77
Ye, Huanpeng III-57
Ye, Hunian III-796
Ye, Kehan III-831
Ye, Mei III-796
Yi, Wangmin II-323
Yi, Wei I-867
Yi, Xiaodong I-659
Yin, Haibin I-162
Yin, Zhouping III-309
Yokoi, Hiroshi I-206
You, Aimin III-495
Yu, Changle I-103
Yu, Chunwei I-230
Yu, Hengbin III-226
Yu, Hongjian II-417
Yu, Jianqiang III-109

Yu, Jing II-464
Yu, Juan III-877
Yu, Liang II-111
Yu, Xi II-24
Yu, Yao II-869
Yuan, Bo I-593, II-273
Yuan, Lei III-867
Yuan, Qingdan III-154
Yuan, Weitao II-453
Yuan, Yuan II-157

Zambrano, Víctor D. III-354
Zan, Wenpei III-203
Zeng, Hong I-350
Zeng, Shasha III-867
Zeng, Xiongzhi III-819, III-877
Zhang, Bainan III-673
Zhang, Bing III-538
Zhang, Bo I-648, III-765
Zhang, Chang-Hua II-623
Zhang, Chi I-301
Zhang, Congsheng II-812
Zhang, Dinghua III-388
Zhang, Dongsheng II-48
Zhang, Fan III-662
Zhang, Feiyu III-613
Zhang, Hai-Tao I-628
Zhang, Hanbo I-605
Zhang, Hao I-290
Zhang, Haochong I-173
Zhang, Hong III-87
Zhang, Hongwei I-857
Zhang, Huajie III-413
Zhang, Huichao II-298
Zhang, Huiwen I-474
Zhang, Jun I-3
Zhang, Lei II-36
Zhang, Lu I-462
Zhang, Qi I-63, II-532, III-3
Zhang, Qiang III-44
Zhang, Qinghua II-394
Zhang, Quan II-346
Zhang, Shaohua III-472
Zhang, Shengnan III-425
Zhang, Shiming III-843
Zhang, Songsong II-520
Zhang, Tian I-462
Zhang, Wei I-779
Zhang, Wenming I-417
Zhang, Wenzeng II-739, II-752, II-764
Zhang, Xi II-91
Zhang, Xianmin II-464, II-543, II-554, II-564, II-576, II-647, III-777
Zhang, Xiaomei III-3
Zhang, Xiaowei I-417
Zhang, Xiaoxing II-846
Zhang, Xinbin I-127
Zhang, Xu II-346, II-496
Zhang, Xuman III-686
Zhang, Yanduo II-835
Zhang, Yonghong II-59
Zhang, Yudong III-367
Zhang, Zhaokun II-453
Zhang, Zhendong I-280
Zhang, Zhifei II-554
Zhao, Bin II-311
Zhao, Donghui I-230
Zhao, Huan II-14, II-24, II-59
Zhao, Jialiang I-393
Zhao, Jianghong III-179
Zhao, Jiangran II-298, II-311
Zhao, Jianhong I-867
Zhao, Jianwen III-130
Zhao, Jie I-127
Zhao, Jingdong III-343, III-729
Zhao, Rui II-717
Zhao, Shaoan II-623
Zhao, Xin II-14
Zhao, Xinye I-417
Zhao, Yakun III-662
Zhao, Yanzheng I-14, I-26, III-527
Zhao, Yongsheng II-48, III-87
Zhao, Yu-Dong III-215
Zhao, Zhanfeng III-130
Zheng, Enhao I-251
Zheng, Jianghong I-103
Zheng, Jianwei I-494
Zheng, Kuisong III-267, III-275
Zheng, Liupo II-588
Zheng, Nanning I-605
Zheng, Xiao I-313
Zheng, Yanglong II-647
Zheng, Yu II-323
Zheng, Zhongqiang I-846
Zhong, Jingyang III-589
Zhong, Xiangyong I-381
Zhong, Xingjian I-361
Zhou, Bo III-141
Zhou, Chuan II-789
Zhou, Haibo III-226

Zhou, Haopeng II-647
Zhou, Haotian III-425
Zhou, Liqin I-745
Zhou, Weijia I-474
Zhou, Xiaolong III-449
Zhou, Xu III-226
Zhou, Yanglin III-254
Zhou, Ying II-475
Zhou, Zhihao I-262
Zhou, Zude I-581, III-3
Zhu, Chen III-77
Zhu, Daqi I-711
Zhu, Limin II-157, II-496
Zhu, Li-Min II-171
Zhu, LiMin II-184
Zhu, Xiangyang I-373, III-57
Zhuang, Chungang II-36, II-599, III-377
Zhuo, Zhenyang II-250
Zi, Bin II-91
Zilberstein, Shlomo I-173
ZiYang, Wang III-401
Zong, Wenpeng III-254
Zuo, Lin II-623
Zuo, Yonggang I-593

Printed in the United States
By Bookmasters